CLYMER
SUZUKI
OUTBOARD SHOP MANUAL
2-225 HP • 1985-1991 (Includes Jet Drives)

The World's Finest Publisher of Mechanical How-to Manuals

P.O. Box 12901, Overland Park, KS 66282-2901

Copyright ©1992 PRIMEDIA Business Magazines and Media Inc.

FIRST EDITION
First Printing February, 1992
Second Printing September, 1994
Third Printing August, 1996
Fourth Printing, April, 1998
Fifth Printing March, 2002

Printed in U.S.A.

CLYMER and colophon are registered trademarks of PRIMEDIA Business Magazines and Media Inc.

ISBN: 0-89287-556-9

Library of Congress: 91-77841

Tools shown in Chapter Two courtesy of Thorsen Tool, Dallas, Texas. Test equipment shown in Chapter Two courtesy of Dixson, Inc., Grand Junction, Colorado.

Technical illustrations courtesy of American Suzuki Motor Corporation with additional illustrations by Mitzi McCarthy.

COVER: Photo courtesy of American Suzuki Motor Corporation.

All rights reserved. Reproduction or use, without express permission, of editorial or pictorial content, in any manner, is prohibited. No patent liability is assumed with respect to the use of the information contained herein. While every precaution has been taken in the preparation of this book, the publisher assumes no responsibility for errors or omissions. Neither is any liability assumed for damages resulting from use of the information contained herein. Publication of the servicing information in this manual does not imply approval of the manufacturers of the products covered.

All instructions and diagrams have been checked for accuracy and ease of application; however, success and safety in working with tools depend to a great extent upon individual accuracy, skill and caution. For this reason, the publishers are not able to guarantee the result of any procedure contained herein. Nor can they assume responsibility for any damage to property or injury to persons occasioned from the procedures. Persons engaging in the procedure do so entirely at their own risk.

General Information	1
Tools and Techniques	2
Troubleshooting	3
Lubrication, Maintenance and Tune-up	4
Timing, Synchronizing and Adjustment	5
Fuel System	6
Electrical Systems	7
Power Head	8
Gearcase	9
Automatic Rewind Starter	10
Power Trim and Tilt System	11
Oil Injection System	12
Jet Drives	13
Index	14
Wiring Diagrams	15

CLYMER PUBLICATIONS
PRIMEDIA Business Magazines & Media

Chief Executive Officer Timothy M. Andrews
President Ron Wall
Vice President, PRIMEDIA Business Directories & Books Rich Hathaway

EDITORIAL

Editors
Mike Hall
James Grooms

Associate Editor
Dustin Uthe

Technical Writers
Ron Wright
Ed Scott
George Parise
Mark Rolling
Michael Morlan
Jay Bogart
Rodney J. Rom

Production Supervisor
Dylan Goodwin

Lead Editorial Production Coordinator
Shirley Renicker

Editorial Production Coordinators
Greg Araujo
Shara Pierceall

Editorial Production Assistants
Susan Hartington
Holly Messinger
Darin Watson

Technical Illustrators
Steve Amos
Robert Caldwell
Mitzi McCarthy
Bob Meyer
Michael St. Clair
Mike Rose

MARKETING/SALES AND ADMINISTRATION

Publisher
Randy Stephens

Marketing Manager
Elda Starke

Advertising & Promotions Coordinator
Melissa Abbott

Associate Art Directors
Chris Paxton
Tony Barmann

Sales Manager/Marine
Dutch Sadler

Sales Manager/Manuals
Ted Metzger

Sales Manager/Motorcycles
Matt Tusken

Operations Manager
Patricia Kowalczewski

Customer Service Manager
Terri Cannon

Customer Service Supervisor
Ed McCarty

Customer Service Representatives
Susan Kohlmeyer
April LeBlond
Courtney Hollars
Emily Osipik
Jennifer Lassiter
Ernesto Suarez

Warehouse & Inventory Manager
Leah Hicks

The following product lines are published by PRIMEDIA Business Directories & Books.

More information available at *primediabooks.com*

Contents

QUICK REFERENCE DATA IX

CHAPTER ONE
GENERAL INFORMATION 1

Manual organization 1
Notes, cautions and warnings 1
Torque specifications 2
Engine operation 2
Fasteners 2
Lubricants 8
Gasket sealant 10
Galvanic corrosion 11
Protection from galvanic corrosion 12
Propellers 14

CHAPTER TWO
TOOLS AND TECHNIQUES 21

Safety first 21
Basic hand tools 21
Test equipment 26
Service hints 28
Special tips 30
Mechanic's techniques 31

CHAPTER THREE
TROUBLESHOOTING 33

Operating requirements 33
Starting system 34
Lighting system 37
Charging system 37
Ignition system 39
Breaker-point ignition component testing
  (1985-1989 DT 2) 41
Transistorized ignition system (1990-on DT 2) 42
Pointless electronic ignition (PEI)
  component testing 43
Integrated circuit (IC) and
  Micro Link ignition systems 53
Fuel system (carburetted models) 74
Engine temperature and overheating 76
Engine 76

CHAPTER FOUR
LUBRICATION, MAINTENANCE AND TUNE-UP 88

Lubrication 88	Galvanic protection and bonding wires 99
Storage 96	Engine flushing 100
Complete submersion 97	Tune-up 100
Anti-corrosion maintenance 98	

CHAPTER FIVE
TIMING, SYNCHRONIZATION AND ADJUSTMENT 116

Engine timing and synchronization 116	Throttle valve sensor adjustment 136
Timing adjustment 118	DT 225 EFI 139
Throttle adjustment 132	

CHAPTER SIX
FUEL SYSTEM 140

Loss of high speed rpm (all DT 15) 140	Integral fuel pump carburetor (1988-on DT 8; 1985-on DT 9.9, DT 15) 157
Off idle lean condition (1988 DT 30 and 1989 DT 65) 140	Centerbowl single-throat carburetor (DT 20 through DT 85; DT 115, DT 140) 161
Fuel pump and filter 141	Centerbowl dual-throat carburetor (V4 and V6) ... 166
Carburetor removal/installation 148	Carburetor cleaning and inspection 171
Carburetor overhaul/adjustment 150	Reed valve assembly 173
Throttle plunger carburetor (DT 2) 151	Fuel tank 176
Centerbowl carburetor (DT 4, DT 6; 1985-1987 DT 8) 154	Fuel line, connector and primer bulb 179
	Fuel injection system (DT 225 V6) 179

CHAPTER SEVEN
ELECTRICAL SYSTEMS 195

Electrical connectors 195	Magneto breaker point ignition (1985-1989 DT 2) .. 220
Battery 195	Breakerless transistorized ignition (1990-on DT 2) .. 222
Lighting system (1985-1986 DT 9.9, DT 15, DT 40) 203	PEI ignition 223
Battery charging system 205	Integrated circuit (I.C.) ignition system 229
Electric starting systems 206	Micro Link ignition system (V4 and V6) 234
Starter motor 207	Ignition coil (all models) 238
	CDI unit (all models) 240

CHAPTER EIGHT
POWER HEAD 249

Engine serial number 250	Flywheel 251
Fasteners and torque 250	Power head 261

CHAPTER NINE
GEARCASE 373

Propeller 373	Pinion gear adjustment (1986-on DT 115, DT 140; V4 and V6) 451
Water pump 376	Propeller shaft end play (DT 8 through DT 65 models only) 453
Gearcase 393	Clutch pushrod dimension 454
Propeller shaft clutch 437	
Pinion gear depth and forward/reverse gear backlash 444	

CHAPTER TEN
AUTOMATIC REWIND STARTERS **458**

Bendix starter 458
Overhead starter 462
Adjustment 481

CHAPTER ELEVEN
POWER TRIM AND TILT SYSTEM **484**

Hydraulic pump 491
Troubleshooting 494
Pump oil pressure test 496
Component replacement 502

CHAPTER TWELVE
OIL INJECTION SYSTEM **506**

System components 507
Operation 508
Low oil level warning light and buzzer 508
Oil flow sensor 509
Air/oil mixing valve 509
Oil pump service 510
Component replacement 515

CHAPTER THIRTEEN
JET DRIVES **526**

Maintenance 526
Water pump 530
Jet drive 536
Bearing housing 538
Intake housing liner 541

INDEX **543**

WIRING DIAGRAMS **546**

Quick Reference Data

MODEL HISTORY

	1985	1986	1987	1988	1989	1990	1991
DT 2	X	X	X	X	X	X	X
DT 4	X	X	X	X	X	X	X
DT 6	X	X	X	X	X	X	X
DT 8	X	X	X	X	X	X	X
DT 8 SAIL					X	X	X
DT 9.9	X	X	X	X	X	X	X
DT 9.9 SAIL				X	X	X	X
DT 15	X	X	X	X	X	X	X
DT 20		X	X				
DT 25	X	X	X	X	X	X	X
DT 30	X	X	X	X	X		
DT 35			X	X	X		
DT 40	X	X	X	X	X	X	X
DT 55	X	X	X	X	X	X	X
DT 65	X	X	X	X	X	X	X
DT 75	X	X	X	X	X	X	X
DT 85	X	X	X	X	X	X	X
DT 90					X	X	X
DT 100					X	X	X
DT 100 SUPER FOUR						X	X
DT 115	X	X	X	X	X	X	X
DT 140	X	X	X	X	X	X	X
DT 150			X	X	X	X	X
DT 150 SUPER SIX		X	X	X	X	X	X
DT 175			X	X	X	X	X
DT 200		X	X	X	X	X	X
DT 200 EXANTE				X	X	X	X
DT 225						X	X

MAINTENANCE SCHEDULE*

At first 10 hours	Change gearcase lubricant.
Every 10 hours	Retighten bolts and nuts.
	Check wire harness connections.
	Check idle speed.
	Check and adjust carburetors (DT 20-DT 200).
	Check propeller for damage.
	Lubricate propeller shaft splines.
	Lubricate jet drive bearings.
	Check fuel lines for leakage.
	Check intake manifold hose for deterioration.
	Lubricate steering handle.
	Check neutral start interlock switch operation.
	Check emergency switch operation (if so equipped).
	Check and adjust remote control linkage (if so equipped).
	Check engine key and choke operation (if so equipped).
	Check starter button and choke operation (if so equipped).

(continued)

MAINTENANCE SCHEDULE* (continued)

Every 50 hours	Clean and regap spark plugs. Decarbonize the piston(s), cylinder and cylinder head. Change gearcase lubricant. Check fuel strainer or filter. Check steering handle preload. Check starter rope condition. Check tilt mechanism preload. Lubricate jet drive bearings.
Every 100 hours	Check and adjust ignition timing. Check water pump impeller.
Every week	Check oil injection lines (if so equipped).
Every month	Lubricate carburetor and choke linkage. Lubricate clamp screws.
Every 3 months	Lubricate swivel bracket. Lubricate support tube. Lubricate shift lever.
Once each season	Change starter rope. Check fuel tank condition. Replace water pump impeller.

* Not all items apply to all engines. Perform only those pertaining to your engine. Lubricate jet drive bearings after each operation.

RECOMMENDED LUBRICANTS, SEALANTS AND ADHESIVES

Type	Part No.
Water-resistant grease	99000-25170
Outboard Motor Gear Oil	99000-22540
Suzuki CCI 50:1 Outboard Oil	99105-00153
Super Grease "A"	99000-25010
Silicone Seal	99000-31120
Bond No. 4	99000-31030
Cemedine 366E	99000-31090
Thread Lock 1342	99000-32050
Thread Lock Super 1333B	99000-32020
DEXRON automatic transmission fluid	—
Jet Grease	99954-53885

RECOMMENDED SPARK PLUGS

	NGK part No.	Gap (in.)
DT 2		
1985-1990	BR4H	0.020-0.024
1991	BR5HS	0.024-0.028

(continued)

RECOMMENDED SPARK PLUGS (continued)

	NGK part No.	Gap (in.)
DT 4		
1985-1989	BP6HS	0.024-0.028
1990-on	BP5HS	0.024-0.028
DT 6		
1985-1986	BP6HS, BPR6HS	0.031-0.035
1987-on	BR6HS-10	0.035-0.039
DT 8		
1985-1987	BPR6HS	0.035-0.039
1988-on	B6HS-10	0.035-0.040
DT 9.9		
1985-1987	BR7HS-10	0.035-0.039
1988-on	B6HS-10	0.035-0.040
DT 15	BR7HS-10	0.035-0.039
DT 20, DT 25, DT 30		
(2-cylinder)	BR7HS	0.035-0.039
DT 25, DT 30		
(3-cylinder)	B7HS-10	0.035-0.039
DT 35, DT 40	B8HS	0.031-0.035
DT 55, DT 65	B8HS	0.031-0.035
DT 75, DT 85	B8HS	0.031-0.035
DT 115, DT 140		
1985-1990	BR8HS	0.031-0.035
1991	BR8HCS	0.031-0.035
V4	BR8HS-10	0.035-0.039
V6		
1986		
DT 150, DT 200	BR8HS-10	0.035-0.039
DT 150 SS	B8HS	0.024-0.028
1987-1988		
DT 150, DT 175, DT 200	B8HS-10	0.035-0.039
DT 150 SS, DT 200 AT	B8HS	0.024-0.028
1989-on		
DT 150-DT 200	BR8HS-10	0.035-0.039
1990-on DT 225	BR8HS-10	0.035-0.039

SPARK PLUG CROSS-REFERENCE CHART*

NGK	Champion	AC
B4HS, BR5HS	L81, L88A	44F, 44FF
B6HS, BR6HS	L9J, QL7J, RL7J	42F, 42FF
BP6HS, BPR6HS	RL12Y, RL87Y, L66Y	42FS, 43FS, R43FS
B7HS, BR7HS	L5, L7J	M42FF, S42FR
BR7HS-10	—	—
B8HS, BR8HS	L4J, RL4J, L78, RL78	S41FR, S40FR, M41FF
BR8HS-10	—	—

* The cross-referenced spark plugs are not exact replacements for original plug heat range and should be used only as a temporary replacement.

IDLE AIR SCREW ADJUSTMENT

Model	Turns out from lightly seated position
DT 2	
1985-1989	NA
1990-1991	1 1/4-1 3/4
DT 4	1-1 1/2
DT 6	
1985-1986	1 3/4-2 1/4
1987-on	
S-type	7/8-1 3/8
L and UL type	1-1 1/2
DT 8	
1985-1986	3/4- 1 1/4
1987	1/2-1
1988-on	1 3/4-2 1/4
DT 9.9	
1985-1987	1 1/4-1 3/4
1988-on	1 3/4-2 1/4
DT 15	
1985-1987	1 1/4-1 3/4
1988-on	1 1/2-2
DT 20	1 3/4-2 1/4
DT 25 (2-cylinder)	
1985	3/4-1 1/4
1986-1988	1 1/4-1 3/4
DT 30 (2-cylinder)	
1985	1 1/4-1 3/4
1986-1987	1 1/2-2
DT 25, DT 30 (3-cylinder)	
1988 (DT 30)	1 1/2-2
1989 (DT 25 and DT 30)	
MC	1 1/4-1 3/4
LE	1 1/2-2
1990	1 1/2-2
1991	1-1 1/2
DT 35	1 1/2-2
DT 40	
1985-1986	1 1/8-1 5/8
1987-on	1 1/2-2
DT 55	
1985	1 1/8-1 5/8
1986-1989	1 1/4-1 3/4
1990-on	1-1 1/2
DT 65	
1985	1 1/2-2
1986-1989	1 1/4-1 3/4
1990-on	1-1 1/2
DT 75	
1985-1986	1-1 1/2
1987	1 3/4-2 1/4
1988-on	1 3/4-2 1/4
DT 75, DT 85	
1985-1986	1-1 1/2
1987	1 3/4-2 1/4
1988-on	1 5/8-2 1/8

(continued)

IDLE AIR SCREW ADJUSTMENT (continued)

Model	Turns out from lightly seated position
DT 90	1 1/8-1 5/8
DT 100	1 1/8-1 5/8
DT 115	
1985	1 1/2
1986-1988	1 1/4-1 3/4
1989	7/8
1990-on	1-1 1/2
DT 140	
1985	1 3/8
1986-1988	1-1 1/2
1989-on	1 1/8-1 5/8
DT 150	
1986	1 1/4-1 3/4
1987-1988	1-1 1/2
1989-on	1 1/2-2
DT 150 SS	1 1/4-1 3/4
DT 175	
1987-1988	1 1/2-2
1989-on	1 1/4-1 3/4
DT 200	
1986	1 1/2-2
1987-on	1 1/4-1 3/4
DT 200 AE	1 1/4-1 3/4

INTRODUCTION

This detailed, comprehensive manual covers the Suzuki 2-225 hp outboard engines from 1985-on.

The expert text gives complete information on maintenance, tune-up, repair and overhaul. Hundreds of illustrations guide you through every step. The book includes all you will need to know to keep your Suzuki running right.

A shop manual is a reference. You want to be able to find information fast. As in all Clymer books, this one is designed with you in mind. All chapters are thumb tabbed. Important items are extensively indexed at the rear of the book. All procedures, tables, photos, etc., in this manual are for the reader who may be working on the outboard engine for the first time or using this manual for the first time. All the most frequently used specifications and capacities are summarized in the *Quick Reference Data* pages at the front of the book.

Having a well-maintained engine will increase your enjoyment of your boat as well as assure your safety when offshore. Keep the book handy in your tool box. It will help you better understand how your engine runs, lower repair costs and make yours a reliable, top-performance boat.

Chapter One

General Information

This detailed, comprehensive manual contains complete information on maintenance, tune-up, repair and overhaul. Hundreds of photos and drawings guide you through every step-by-step procedure.

Troubleshooting, tune-up, maintenance and repair are not difficult if you know what tools and equipment to use and what to do. Anyone not afraid to get their hands dirty, of average intelligence and with some mechanical ability, can perform most of the procedures in this book. See Chapter Two for more information on tools and techniques.

A shop manual is a reference. You want to be able to find information fast. Clymer books are designed with you in mind. All chapters are thumb tabbed and important items are indexed at the end of the book. All procedures, tables, photos, etc., in this manual assume that the reader may be working on the machine or using this manual for the first time.

Keep this book handy in your tool box. It will help you to better understand how your machine runs, lower repair and maintenance costs and generally increase your enjoyment of your marine equipment.

MANUAL ORGANIZATION

This chapter provides general information useful to marine owners and mechanics.

Chapter Two discusses the tools and techniques for preventive maintenance, troubleshooting and repair.

Chapter Three describes typical equipment problems and provides logical troubleshooting procedures.

Following chapters describe specific systems, providing disassembly, repair, assembly and adjustment procedures in simple step-by-step form. Specifications concerning a specific system are included at the end of the appropriate chapter.

NOTES, CAUTIONS AND WARNINGS

The terms NOTE, CAUTION and WARNING have specific meanings in this manual. A NOTE provides additional information to make a step or procedure easier or clearer. Disregarding a NOTE could cause inconvenience, but would not cause damage or personal injury.

A CAUTION emphasizes areas where equipment damage could result. Disregarding a CAUTION could cause permanent mechanical damage; however, personal injury is unlikely.

A WARNING emphasizes areas where personal injury or even death could result from negligence. Mechanical damage may also occur. WARNINGS *are to be taken seriously.* In some cases, serious injury or death has resulted from disregarding similar warnings.

TORQUE SPECIFICATIONS

Torque specifications throughout this manual are given in foot-pounds (ft.-lb.) and either Newton meters (N•m) or meter-kilograms (mkg). Newton meters are being adopted in place of meter-kilograms in accordance with the International Modernized Metric System. Existing torque wrenches calibrated in meter-kilograms can be used by performing a simple conversion: move the decimal point one place to the right. For example, 4.7 mkg = 47 N•m. This conversion is accurate enough for mechanics' use even though the exact mathematical conversion is 3.5 mkg = 34.3 N•m.

ENGINE OPERATION

All marine engines, whether 2- or 4-stroke, gasoline or diesel, operate on the Otto cycle of intake, compression, power and exhaust phases.

4-stroke Cycle

A 4-stroke engine requires two crankshaft revolutions (4 strokes of the piston) to complete the Otto cycle. **Figure 1** shows gasoline 4-stroke engine operation. **Figure 2** shows diesel 4-stroke engine operation.

2-stroke Cycle

A 2-stroke engine requires only 1 crankshaft revolution (2 strokes of the piston) to complete the Otto cycle. **Figure 3** shows gasoline 2-stroke engine operation. Although diesel 2-strokes exist, they are not commonly used in light marine applications.

FASTENERS

The material and design of the various fasteners used on marine equipment are not arrived at by chance or accident. Fastener design determines the type of tool required to work with the fastener. Fastener material is carefully selected to decrease the possibility of physical failure or corrosion. See *Galvanic Corrosion* in this chapter for more information on marine materials.

Threads

Nuts, bolts and screws are manufactured in a wide range of thread patterns. To join a nut and bolt, the diameter of the bolt and the diameter of the hole in the nut must be the same. It is just as important that the threads on both be properly matched.

The best way to determine if the threads on two fasteners are matched is to turn the nut on the bolt (or the bolt into the threaded hole in a piece of equipment) with fingers only. Be sure both pieces are clean. If much force is required, check the thread condition on each fastener. If the thread condition is good but the fasteners jam, the threads are not compatible.

Four important specifications describe every thread:
 a. Diameter.
 b. Threads per inch.
 c. Thread pattern.
 d. Thread direction.

Figure 4 shows the first two specifications. Thread pattern is more subtle. Italian and British

GENERAL INFORMATION

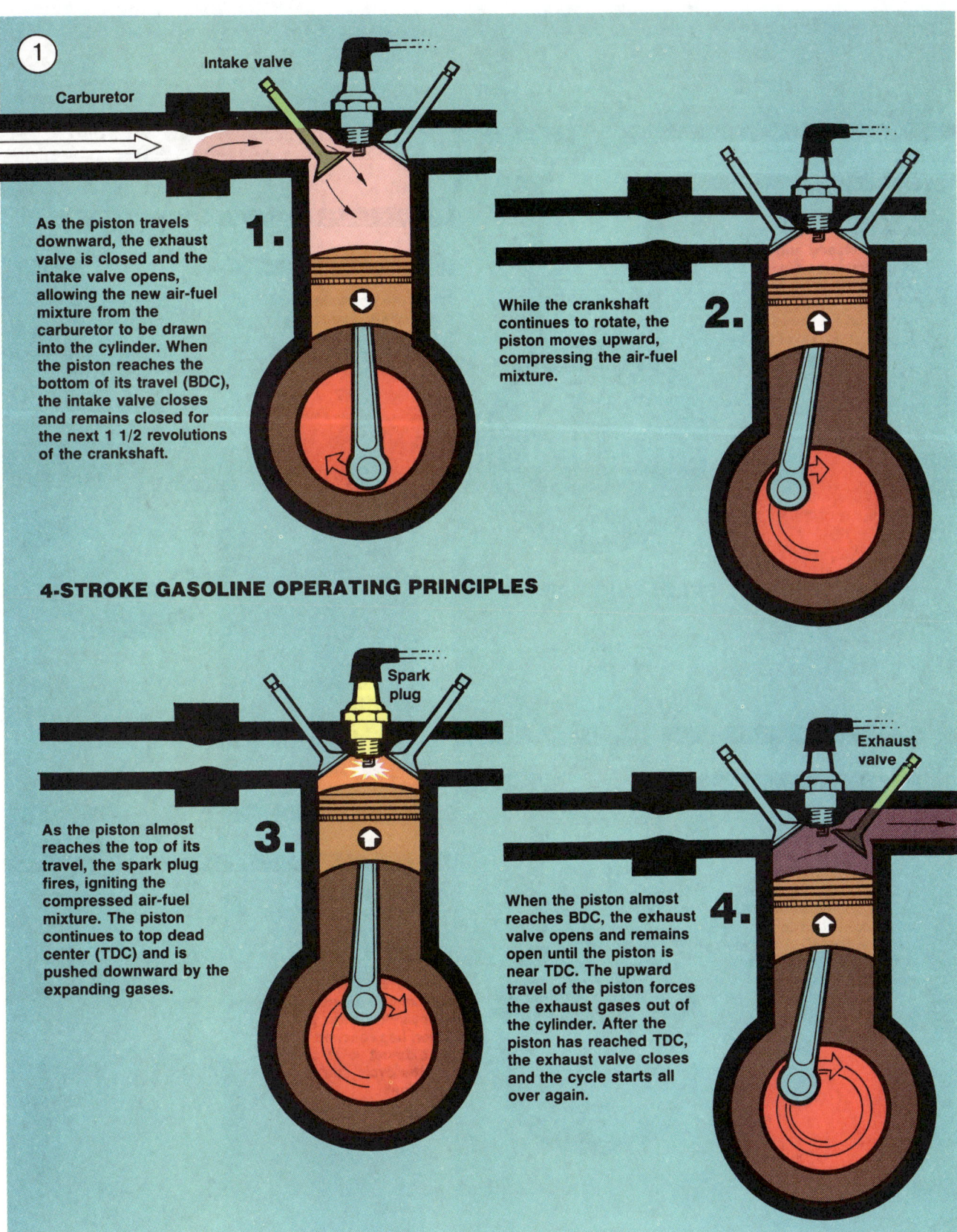

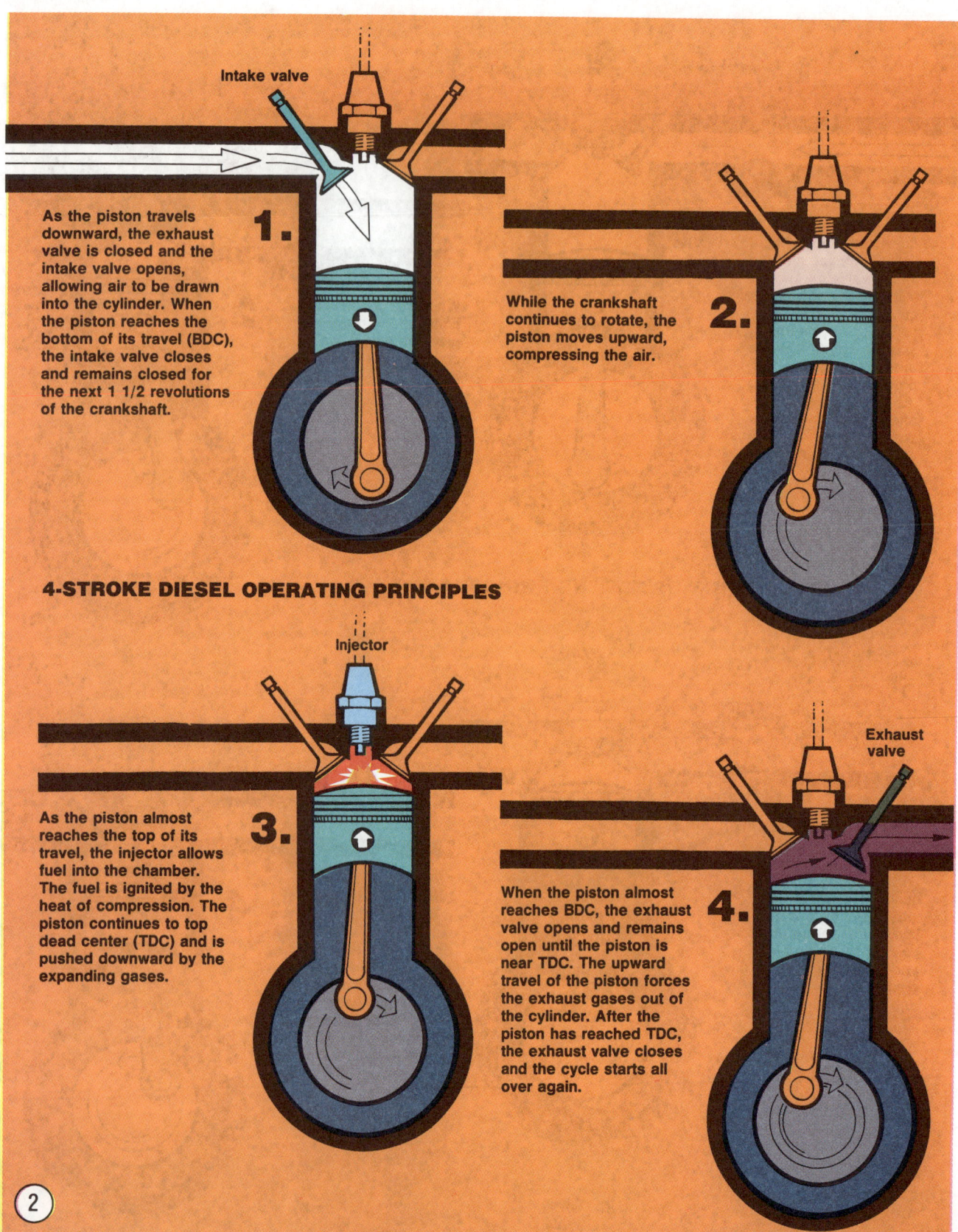

GENERAL INFORMATION

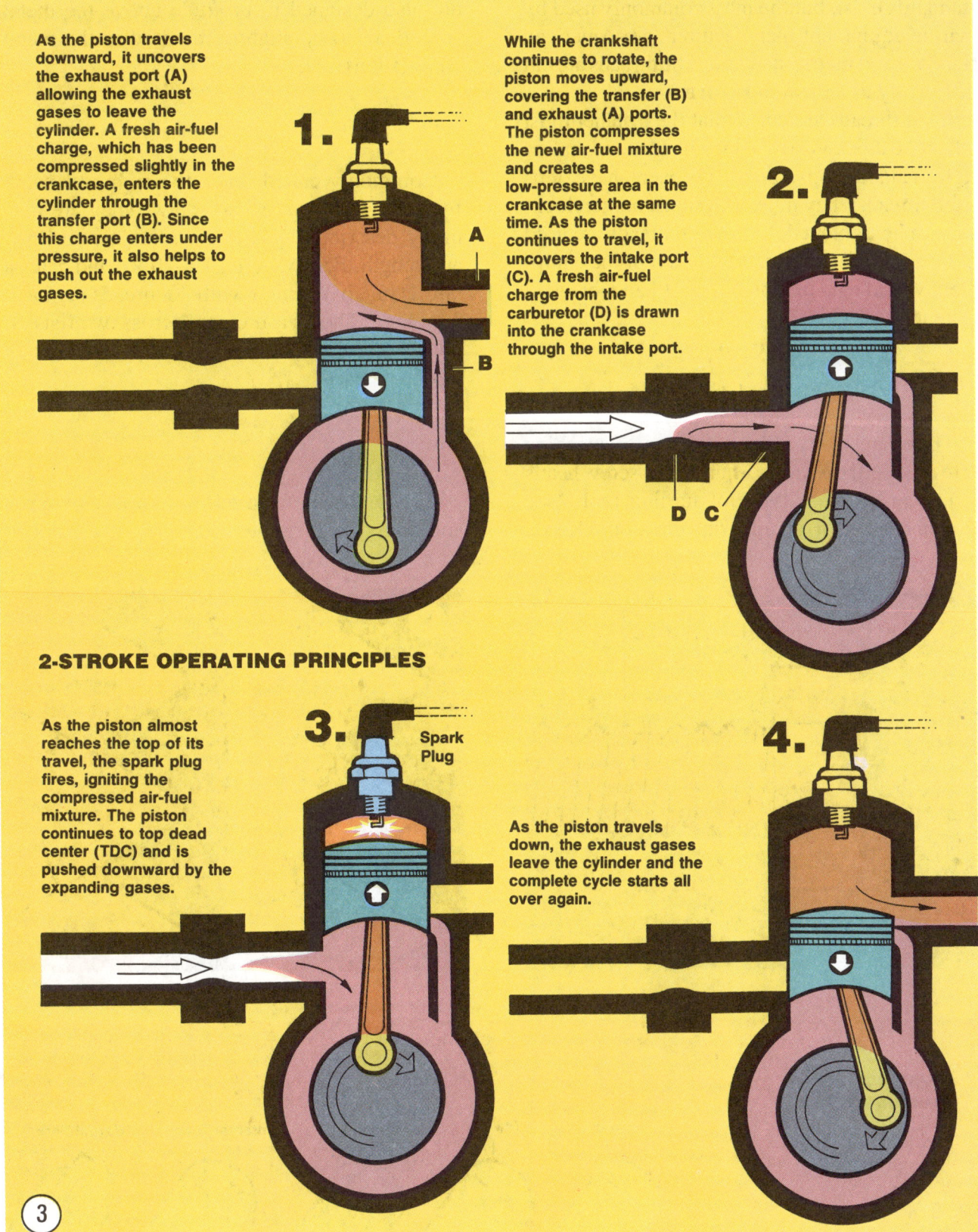

1. As the piston travels downward, it uncovers the exhaust port (A) allowing the exhaust gases to leave the cylinder. A fresh air-fuel charge, which has been compressed slightly in the crankcase, enters the cylinder through the transfer port (B). Since this charge enters under pressure, it also helps to push out the exhaust gases.

2. While the crankshaft continues to rotate, the piston moves upward, covering the transfer (B) and exhaust (A) ports. The piston compresses the new air-fuel mixture and creates a low-pressure area in the crankcase at the same time. As the piston continues to travel, it uncovers the intake port (C). A fresh air-fuel charge from the carburetor (D) is drawn into the crankcase through the intake port.

2-STROKE OPERATING PRINCIPLES

3. As the piston almost reaches the top of its travel, the spark plug fires, igniting the compressed air-fuel mixture. The piston continues to top dead center (TDC) and is pushed downward by the expanding gases.

4. As the piston travels down, the exhaust gases leave the cylinder and the complete cycle starts all over again.

CHAPTER ONE

standards exist, but the most commonly used by marine equipment manufacturers are American standard and metric standard. The threads are cut differently as shown in **Figure 5**.

Most threads are cut so that the fastener must be turned clockwise to tighten it. These are called right-hand threads. Some fasteners have left-hand threads; they must be turned counterclockwise to be tightened. Left-hand threads are used in locations where normal rotation of the equipment would tend to loosen a right-hand threaded fastener.

Machine Screws

There are many different types of machine screws. **Figure 6** shows a number of screw heads requiring different types of turning tools (see Chapter Two for detailed information). Heads are also designed to protrude above the metal (round) or to be slightly recessed in the metal (flat) (**Figure 7**).

Bolts

Commonly called bolts, the technical name for these fasteners is cap screw. They are normally described by diameter, threads per inch and length. For example, 1/4-20 × 1 indicates a bolt 1/4 in. in diameter with 20 threads per inch, 1 in. long. The measurement across two flats on the head of the bolt indicates the proper wrench size to be used.

Nuts

Nuts are manufactured in a variety of types and sizes. Most are hexagonal (6-sided) and fit

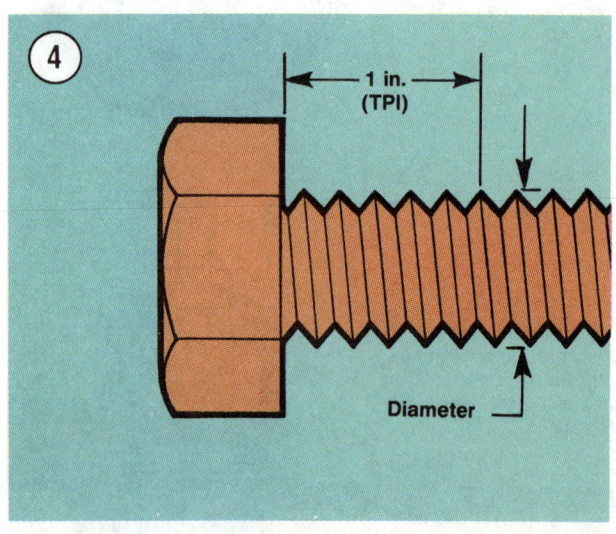
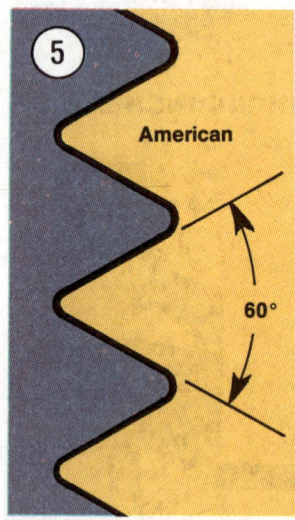
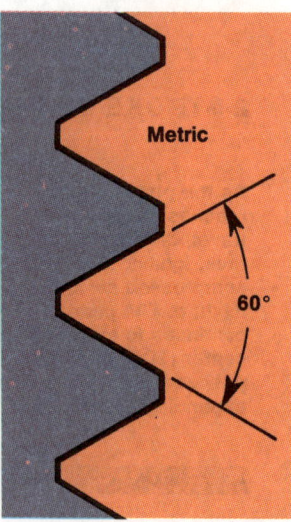
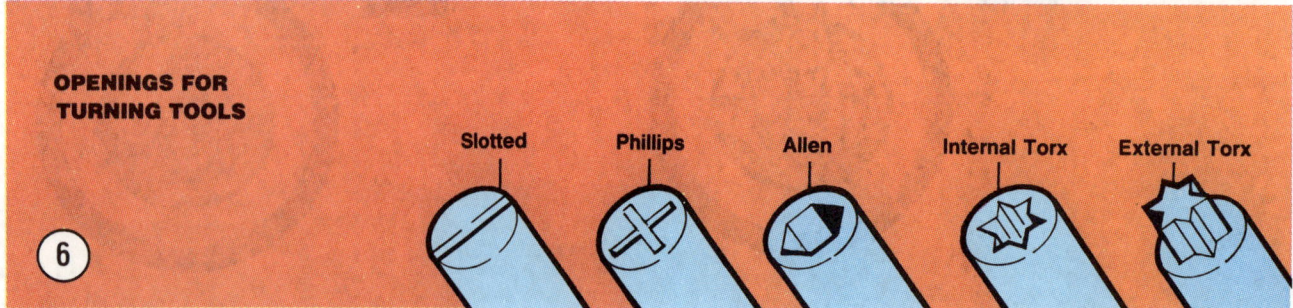

GENERAL INFORMATION

on bolts, screws and studs with the same diameter and threads per inch.

Figure 8 shows several types of nuts. The common nut is usually used with a lockwasher. Self-locking nuts have a nylon insert that prevents the nut from loosening; no lockwasher is required. Wing nuts are designed for fast removal by hand. Wing nuts are used for convenience in non-critical locations.

To indicate the size of a nut, manufacturers specify the diameter of the opening and the threads per inch. This is similar to bolt specification, but without the length dimension. The measurement across two flats on the nut indicates the proper wrench size to be used.

Washers

There are two basic types of washers: flat washers and lockwashers. Flat washers are simple discs with a hole to fit a screw or bolt. Lockwashers are designed to prevent a fastener from working loose due to vibration, expansion and contraction. **Figure 9** shows several types of lockwashers. Note that flat washers are often used between a lockwasher and a fastener to provide a smooth bearing surface. This allows the fastener to be turned easily with a tool.

Cotter Pins

Cotter pins (**Figure 10**) are used to secure special kinds of fasteners. The threaded stud

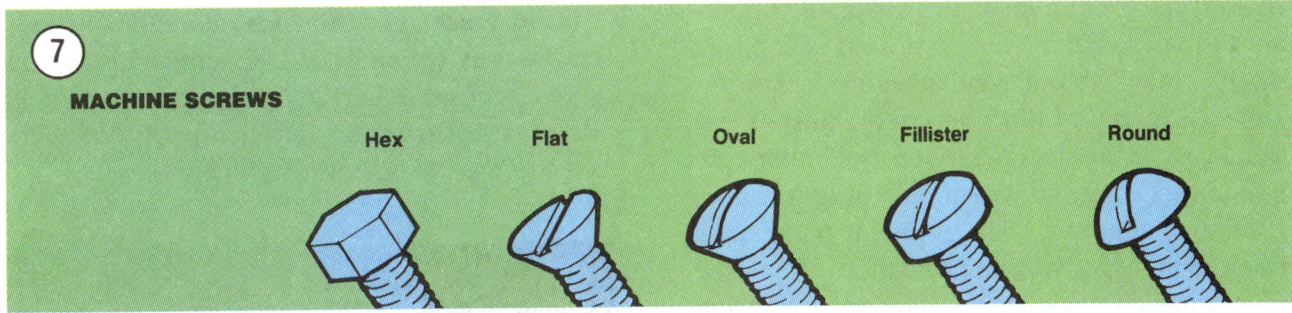

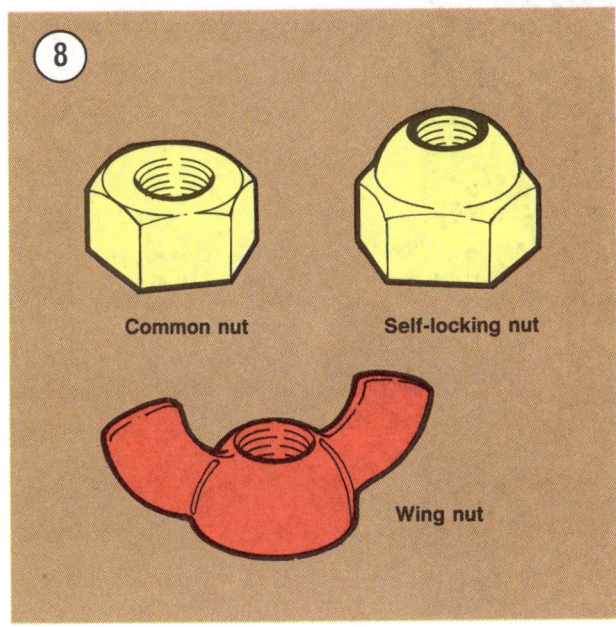

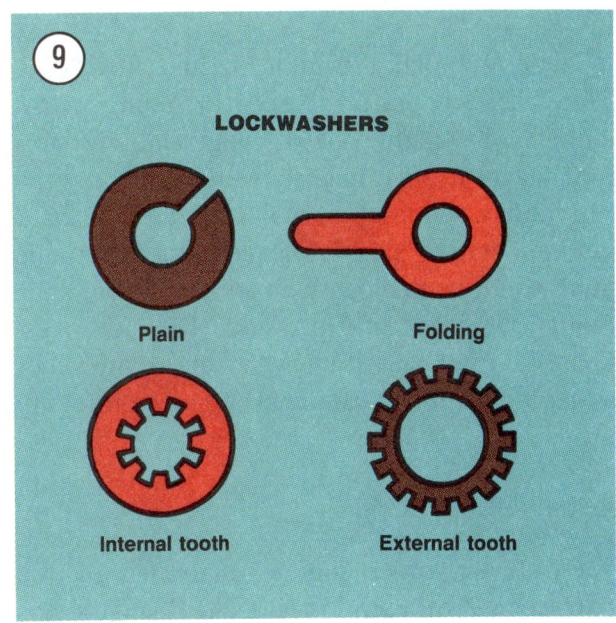

must have a hole in it; the nut or nut lock piece has projections that the cotter pin fits between. This type of nut is called a "Castellated nut." Cotter pins should not be reused after removal.

Snap Rings

Snap rings can be of an internal or external design. They are used to retain items on shafts (external type) or within tubes (internal type). Snap rings can be reused if they are not distorted during removal. In some applications, snap rings of varying thickness can be selected to control the end play of parts assemblies.

LUBRICANTS

Periodic lubrication ensures long service life for any type of equipment. It is especially important to marine equipment because it is exposed to salt or brackish water and other harsh environments. The *type* of lubricant used is just as important as the lubrication service itself; although, in an emergency, the wrong type of lubricant is better than none at all. The following paragraphs describe the types of lubricants most often used on marine equipment. Be sure to follow the equipment manufacturer's recommendations for lubricant types.

Generally, all liquid lubricants are called "oil." They may be mineral-based (including petroleum bases), natural-based (vegetable and animal bases), synthetic-based or emulsions (mixtures). "Grease" is an oil which is thickened with a metallic "soap." The resulting material is then usually enhanced with anticorrosion, antioxidant and extreme pressure (EP) additives. Grease is often classified by the type of thickener added; lithium and calcium soap are commonly used.

4-stroke Engine Oil

Oil for 4-stroke engines is graded by the American Petroleum Institute (API) and the Society of Automotive Engineers (SAE) in several categories. Oil containers display these ratings on the top or label (**Figure 11**).

API oil grade is indicated by letters, oils for gasoline engines are identified by an "S" and oils for diesel engines are identified by a "C." Most modern gasoline engines require SF or SG graded oil. Automotive and marine diesel engines use CC or CD graded oil.

Viscosity is an indication of the oil's thickness, or resistance to flow. The SAE uses numbers to indicate viscosity; thin oils have low numbers and thick oils have high numbers. A "W" after the number indicates that the viscosity testing was done at low temperature to simulate cold weather operation. Engine oils fall into the 5W-20W and 20-50 range.

Multi-grade oils (for example, 10W-40) are less viscous (thinner) at low temperatures and more viscous (thicker) at high temperatures. This allows the oil to perform efficiently across a wide range of engine operating temperatures.

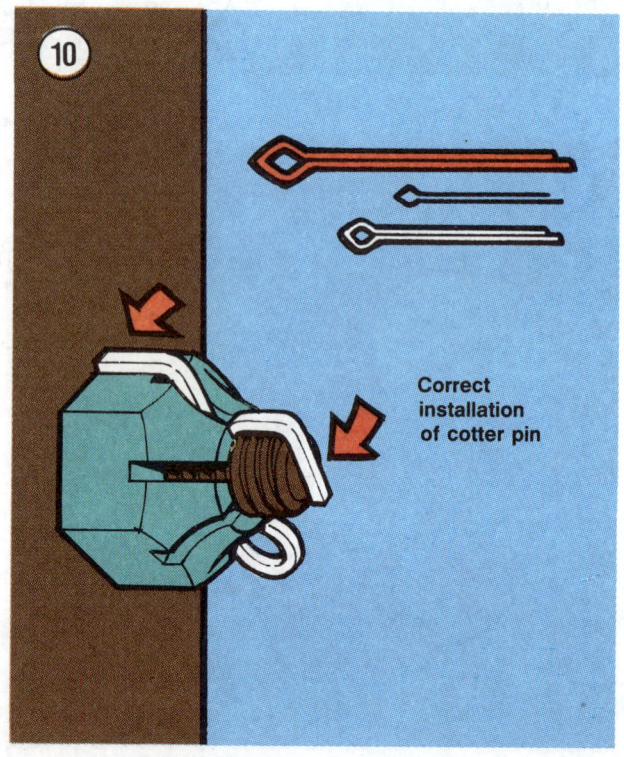

Correct installation of cotter pin

GENERAL INFORMATION

2-stroke Engine Oil

Lubrication for a 2-stroke engine is provided by oil mixed with the incoming fuel-air mixture. Some of the oil mist settles out in the crankcase, lubricating the crankshaft and lower end of the connecting rods. The rest of the oil enters the combustion chamber to lubricate the piston, rings and cylinder wall. This oil is then burned along with the fuel-air mixture during the combustion process.

Engine oil must have several special qualities to work well in a 2-stroke engine. It must mix easily and stay in suspension in gasoline. When burned, it can't leave behind excessive deposits. It must also be able to withstand the high temperatures associated with 2-stroke engines.

The National Marine Manufacturer's Association (NMMA) has set standards for oil used in 2-stroke, water-cooled engines. This is the NMMA TC-W (two-cycle, water-cooled) grade (**Figure 12**). The oil's performance in the following areas is evaluated:

 a. Lubrication (prevention of wear and scuffing).
 b. Spark plug fouling.
 c. Preignition.
 d. Piston ring sticking.
 e. Piston varnish.
 f. General engine condition (including deposits).
 g. Exhaust port blockage.
 h. Rust prevention.
 i. Mixing ability with gasoline.

In addition to oil grade, manufacturers specify the ratio of gasoline to oil required during break-in and normal engine operation.

Gear Oil

Gear lubricants are assigned SAE viscosity numbers under the same system as 4-stroke engine oil. Gear lubricant falls into the SAE 72-250 range (**Figure 13**). Some gear lubricants are multi-grade; for example, SAE 85W-90.

Three types of marine gear lubricant are generally available: SAE 90 hypoid gear lubricant is designed for older manual-shift units; Type C gear lubricant contains additives designed for electric shift mechanisms; High viscosity gear lubricant is a heavier oil designed to withstand the shock loading of high-performance engines or units subjected to severe duty use. Always use a gear lubricant of the type specified by the unit's manufacturer.

Grease

Greases are graded by the National Lubricating Grease Institute (NLGI). Greases are graded by number according to the consistency of the grease; these ratings range from No. 000 to No. 6, with No. 6 being the most solid. A typical multipurpose grease is NLGI No. 2 (**Figure 14**). For specific applications, equipment manufacturers may require grease with an additive such as molybdenum disulfide (MOS^2).

GASKET SEALANT

Gasket sealant is used instead of pre-formed gaskets on some applications, or as a gasket dressing on others. Two types of gasket sealant are commonly used: room temperature vulcanizing (RTV) and anaerobic. Because these two materials have different sealing properties, they cannot be used interchangeably.

RTV Sealant

This is a silicone gel supplied in tubes (**Figure 15**). Moisture in the air causes RTV to cure. Always place the cap on the tube as soon as possible when using RTV. RTV has a shelf life of one year and will not cure properly when the shelf life has expired. Check the expiration date

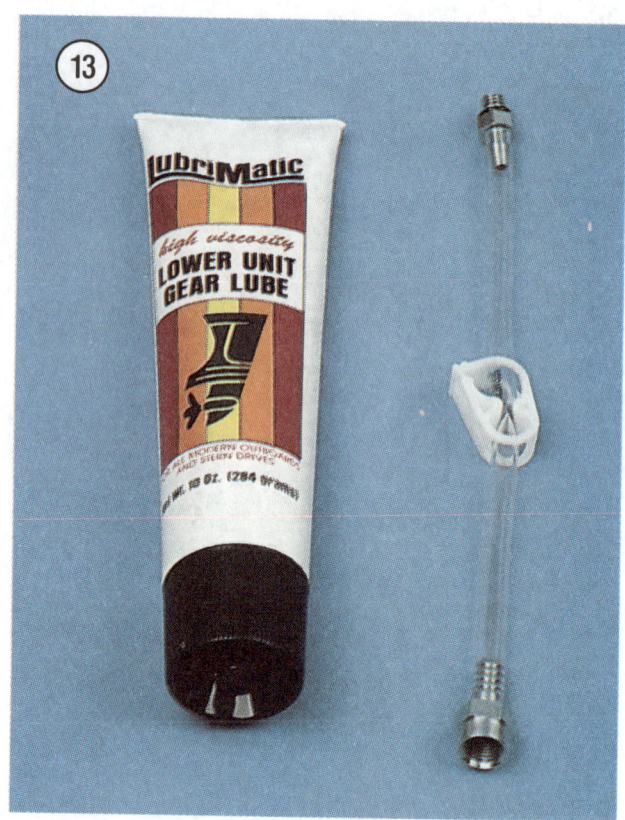

GENERAL INFORMATION

on RTV tubes before using and keep partially used tubes tightly sealed. RTV sealant can generally fill gaps up to 1/4 in. (6.3 mm) and works well on slightly flexible surfaces.

Applying RTV Sealant

Clean all gasket residue from mating surfaces. Surfaces should be clean and free of oil and dirt. Remove all RTV gasket material from blind attaching holes because it can create a "hydraulic" effect and affect bolt torque.

Apply RTV sealant in a continuous bead 2-3 mm (0.08-0.12 in.) thick. Circle all mounting holes unless otherwise specified. Torque mating parts within 10 minutes after application.

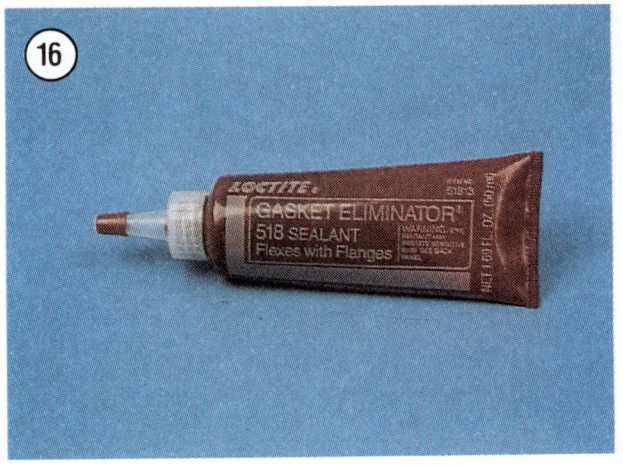

Anaerobic Sealant

This is a gel supplied in tubes (**Figure 16**). It cures only in the absence of air, as when squeezed tightly between two machined mating surfaces. For this reason, it will not spoil if the cap is left off the tube. It should not be used if one mating surface is flexible. Anaerobic sealant is able to fill gaps up to 0.030 in. (0.8 mm) and generally works best on rigid, machined flanges or surfaces.

Applying Anaerobic Sealant

Clean all gasket residue from mating surfaces. Surfaces must be clean and free of oil and dirt. Remove all gasket material from blind attaching holes, as it can cause a "hydraulic" effect and affect bolt torque.

Apply anaerobic sealant in a 1 mm or less (0.04 in.) bead to one sealing surface. Circle all mounting holes. Torque mating parts within 15 minutes after application.

GALVANIC CORROSION

A chemical reaction occurs whenever two different types of metal are joined by an electrical conductor and immersed in an electrolyte. Electrons transfer from one metal to the other through the electrolyte and return through the conductor.

The hardware on a boat is made of many different types of metal. The boat hull acts as a conductor between the metals. Even if the hull is wooden or fiberglass, the slightest film of water (electrolyte) within the hull provides conductivity. This combination creates a good environment for electron flow (**Figure 17**). Unfortunately, this electron flow results in galvanic corrosion of the metal involved, causing one of the metals to be corroded or eaten away by the process. The amount of electron flow (and, therefore, the amount of corrosion) depends on several factors:

a. The types of metal involved.
b. The efficiency of the conductor.
c. The strength of the electrolyte.

Metals

The chemical composition of the metals used in marine equipment has a significant effect on the amount and speed of galvanic corrosion. Certain metals are more resistant to corrosion than others. These electrically negative metals are commonly called "noble;" they act as the cathode in any reaction. Metals that are more subject to corrosion are electrically positive; they act as the anode in a reaction. The more noble metals include titanium, 18-8 stainless steel and nickel. Less noble metals include zinc, aluminum and magnesium. Galvanic corrosion becomes more severe as the difference in electrical potential between the two metals increases.

In some cases, galvanic corrosion can occur within a single piece of metal. Common brass is a mixture of zinc and copper, and, when immersed in an electrolyte, the zinc portion of the mixture will corrode away as reaction occurs between the zinc and the copper particles.

Conductors

The hull of the boat often acts as the conductor between different types of metal. Marine equipment, such as an outboard motor or stern drive unit, can also act as the conductor. Large masses of metal, firmly connected together, are more efficient conductors than water. Rubber mountings and vinyl-based paint can act as insulators between pieces of metal.

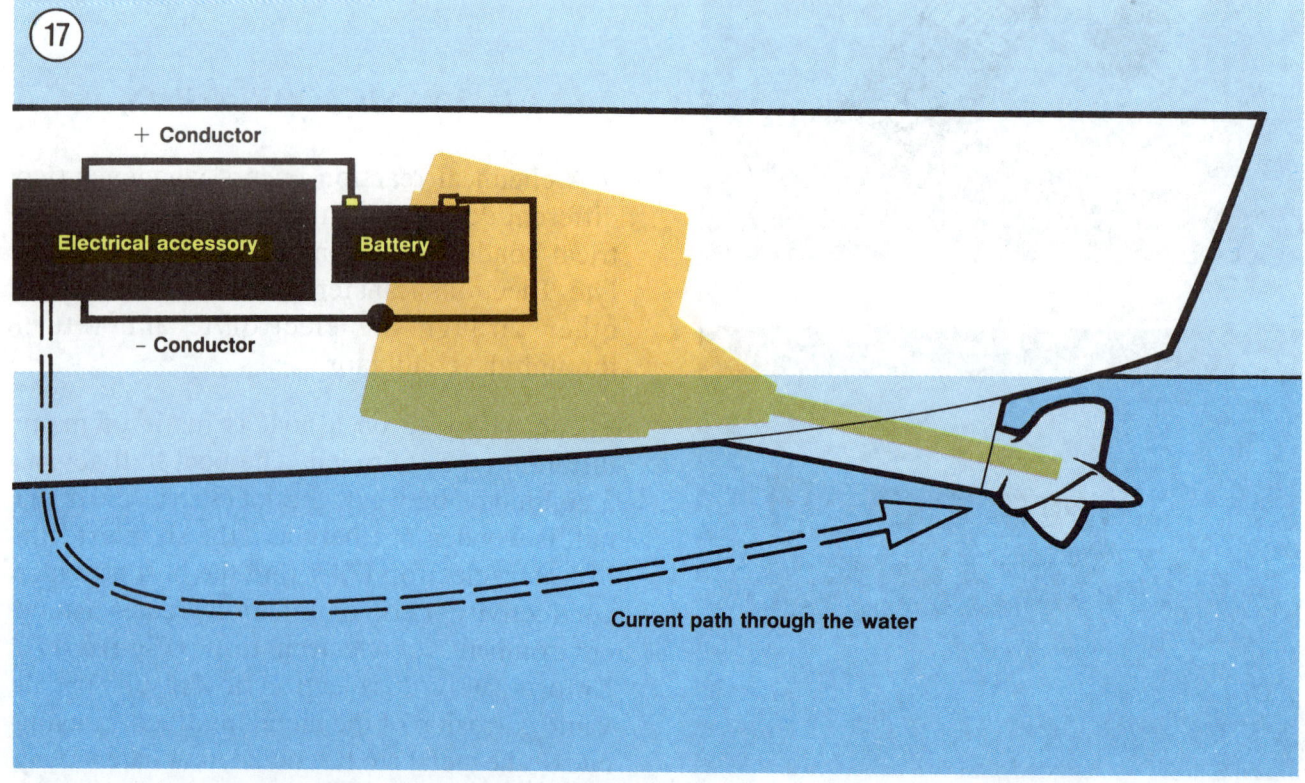

GENERAL INFORMATION

Electrolyte

The water in which a boat operates acts as the electrolyte for the galvanic corrosion process. The better a conductor the electrolyte is, the more severe and rapid the corrosion.

Cold, clean freshwater is the poorest electrolyte. As water temperature increases, its conductivity increases. Pollutants will increase conductivity; brackish or saltwater is also an efficient electrolyte. This is one of the reasons that most manufacturers recommend a freshwater flush for marine equipment after operation in saltwater, polluted or brackish water.

PROTECTION FROM GALVANIC CORROSION

Because of the environment in which marine equipment must operate, it is practically impossible to totally prevent galvanic corrosion. There are several ways by which the process can be slowed. After taking these precautions, the next step is to "fool" the process into occurring only where *you* want it to occur. This is the role of sacrificial anodes and impressed current systems.

Slowing Corrosion

Some simple precautions can help reduce the amount of corrosion taking place outside the hull. These are *not* a substitute for the corrosion protection methods discussed under *Sacrificial Anodes* and *Impressed Current Systems* in this chapter, but they can help these protection methods do their job.

Use fasteners of a metal more noble than the part they are fastening. If corrosion occurs, the larger equipment will suffer but the fastener will be protected. Because fasteners are usually very small in comparison to the equipment being fastened, the equipment can survive the loss of material. If the fastener were to corrode instead of the equipment, major problems could arise.

Keep all painted surfaces in good condition. If paint is scraped off and bare metal exposed, corrosion will rapidly increase. Use a vinyl- or plastic-based paint, which acts as an electrical insulator.

Be careful when using metal-based antifouling paints. These should not be applied to metal parts of the boat, outboard motor or stern drive unit or they will actually react with the equipment, causing corrosion between the equipment and the layer of paint. Organic-based paints are available for use on metal surfaces.

Where a corrosion protection device is used, remember that it must be immersed in the electrolyte along with the rest of the boat to have any effect. If you raise the power unit out of the water when the boat is docked, any anodes on the power unit will be removed from the corrosion cycle and will not protect the rest of the equipment that is still immersed. Also, such corrosion protection devices must not be painted because this would insulate them from the corrosion process.

Any change in the boat's equipment, such as the installation of a new stainless steel propeller, will change the electrical potential and could cause increased corrosion. Keep in mind that when you add new equipment or change materials, you should review your corrosion protection system to be sure it is up to the job.

Sacrificial Anodes

Anodes are usually made of zinc, a far from noble metal. Sacrificial anodes are specially designed to do nothing but corrode. Properly fastening such pieces to the boat will cause them to act as the anode in *any* galvanic reaction that occurs; any other metal present will act as the cathode and will not be damaged.

Anodes must be used properly to be effective. Simply fastening pieces of zinc to your boat in random locations won't do the job.

You must determine how much anode surface area is required to adequately protect the equipment's surface area. A good starting point is provided by Military Specification MIL-A-818001, which states that one square inch of new anode will protect either:

a. 800 square inches of freshly painted steel.
b. 250 square inches of bare steel or bare aluminum alloy.
c. 100 square inches of copper or copper alloy.

This rule is for a boat at rest. When underway, more anode area is required to protect the same equipment surface area.

The anode must be fastened so that it has good electrical contact with the metal to be protected. If possible, the anode can be attached directly to the other metal. If that is not possible, the entire network of metal parts in the boat should be electrically bonded together so that all pieces are protected.

Good quality anodes have inserts of some other metal around the fastener holes. Otherwise, the anode could erode away around the fastener. The anode can then become loose or even fall off, removing all protection.

Another Military Specification (MIL-A-18001) defines the type of alloy preferred that will corrode at a uniform rate without forming a crust that could reduce its efficiency after a time.

Impressed Current Systems

An impressed current system can be installed on any boat that has a battery. The system consists of an anode, a control box and a sensor. The anode in this system is coated with a very noble metal, such as platinum, so that it is almost corrosion-free and will last indefinitely. The sensor, under the boat's waterline, monitors the potential for corrosion. When it senses that corrosion could be occurring, it transmits this information to the control box.

The control box connects the boat's battery to the anode. When the sensor signals the need, the control box applies positive battery voltage to the anode. Current from the battery flows from the anode to all other metal parts of the boat, no matter how noble or non-noble these parts may be. This battery current takes the place of any galvanic current flow.

Only a very small amount of battery current is needed to counteract galvanic corrosion. Manufacturers estimate that it would take two or three months of constant use to drain a typical marine battery, assuming the battery is never recharged.

An impressed current system is more expensive to install than simple anodes but, considering its low maintenance requirements and the excellent protection it provides, the long-term cost may actually be lower.

PROPELLERS

The propeller is the final link between the boat's drive system and the water. A perfectly

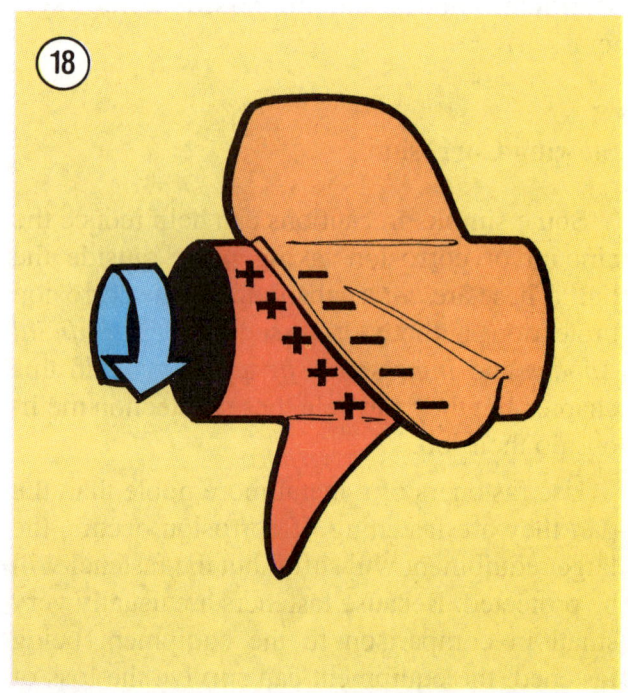

GENERAL INFORMATION

maintained engine and hull are useless if the propeller is the wrong type or has been allowed to deteriorate. Although propeller selection for a specific situation is beyond the scope of this book, the following information on propeller construction and design will allow you to discuss the subject intelligently with your marine dealer.

How a Propeller Works

As the curved blades of a propeller rotate through the water, a high-pressure area is created on one side of the blade and a low-pressure area exists on the other side of the blade (**Figure 18**). The propeller moves toward the low-pressure area, carrying the boat with it.

Propeller Parts

Although a propeller may be a one-piece unit, it is made up of several different parts (**Figure 19**). Variations in the design of these parts make different propellers suitable for different jobs.

The blade tip is the point on the blade farthest from the center of the propeller hub. The blade tip separates the leading edge from the trailing edge.

The leading edge is the edge of the blade nearest to the boat. During normal rotation, this is the area of the blade that first cuts through the water.

The trailing edge is the edge of the blade farthest from the boat.

The blade face is the surface of the blade that faces away from the boat. During normal rotation, high pressure exists on this side of the blade.

The blade back is the surface of the blade that faces toward the boat. During normal rotation, low pressure exists on this side of the blade.

The cup is a small curve or lip on the trailing edge of the blade.

The hub is the central portion of the propeller. It connects the blades to the propeller shaft (part of the boat's drive system). On some drive systems, engine exhaust is routed through the hub; in this case, the hub is made up of an outer and an inner portion, connected by ribs.

The diffuser ring is used on through-hub exhaust models to prevent exhaust gases from entering the blade area.

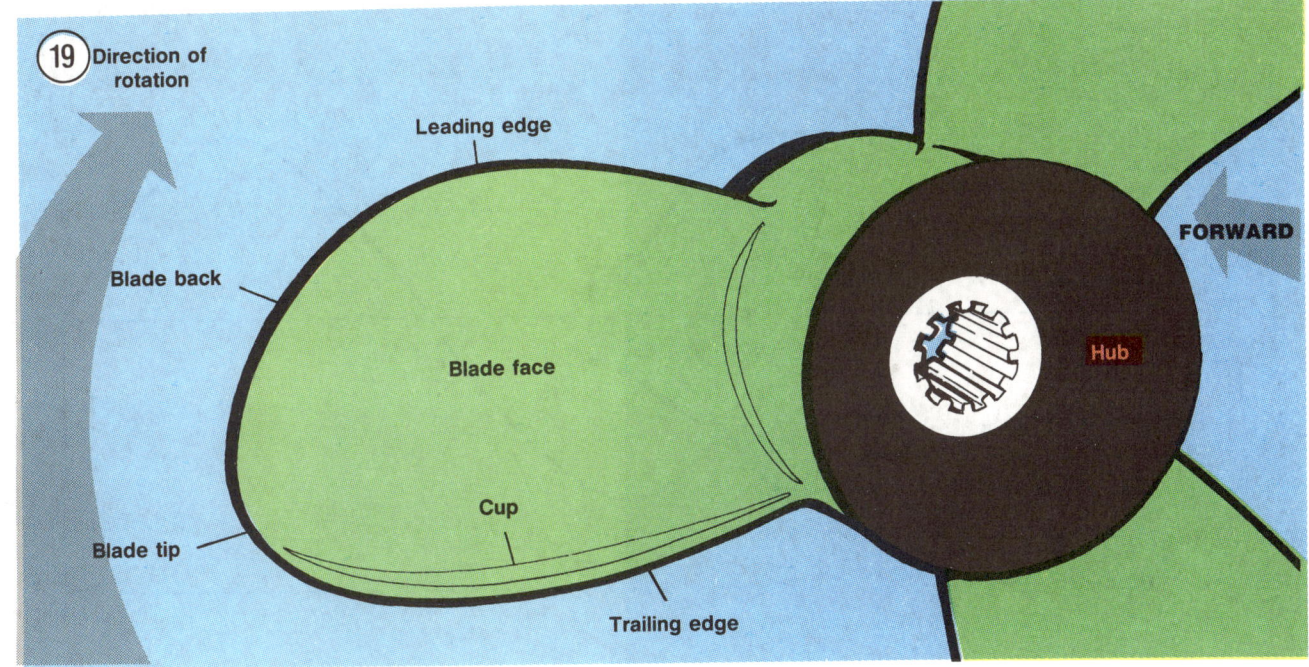

CHAPTER ONE

Propeller Design

Changes in length, angle, thickness and material of propeller parts make different propellers suitable for different situations.

Diameter

Propeller diameter is the distance from the center of the hub to the blade tip, multiplied by 2. That is, it is the diameter of the circle formed by the blade tips during propeller rotation (**Figure 20**).

Pitch and rake

Propeller pitch and rake describe the placement of the blade in relation to the hub (**Figure 21**).

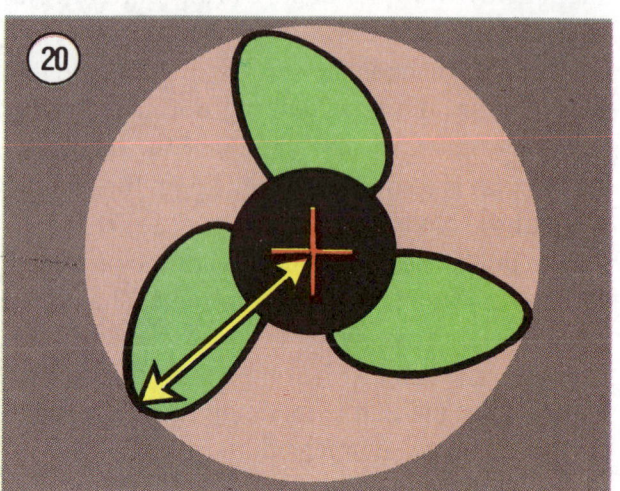

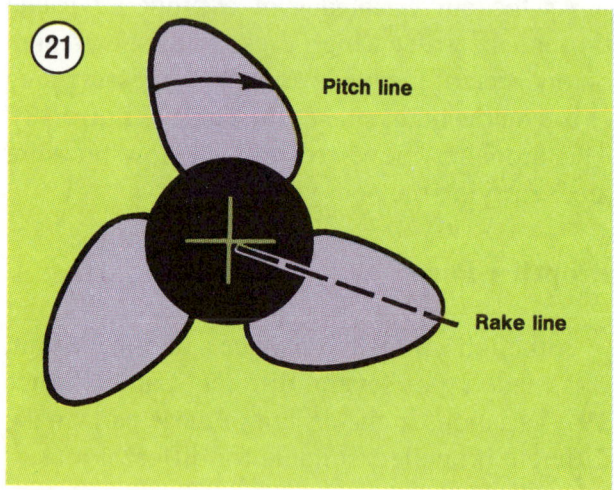

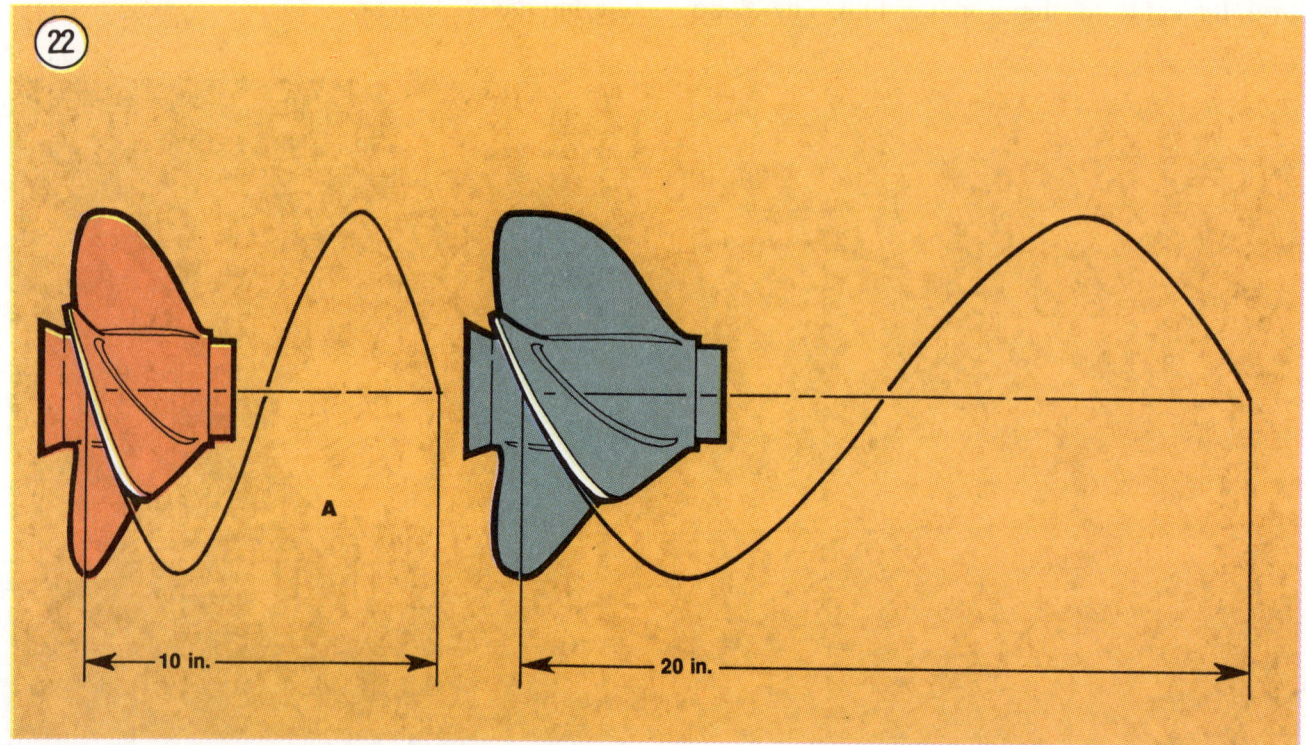

GENERAL INFORMATION

Pitch is expressed by the theoretical distance that the propeller would travel in one revolution. In A, **Figure 22**, the propeller would travel 10 inches in one revolution. In B, **Figure 22**, the propeller would travel 20 inches in one revolution. This distance is only theoretical; during actual operation, the propeller achieves about 80% of its rated travel.

Propeller blades can be constructed with constant pitch (**Figure 23**) or progressive pitch (**Figure 24**). Progressive pitch starts low at the leading edge and increases toward to trailing edge. The propeller pitch specification is the average of the pitch across the entire blade.

Blade rake is specified in degrees and is measured along a line from the center of the hub to the blade tip. A blade that is perpendicular to the hub (A, **Figure 25**) has 0° of rake. A blade that is angled from perpendicular (B, **Figure 25**) has a rake expressed by its difference from perpen-

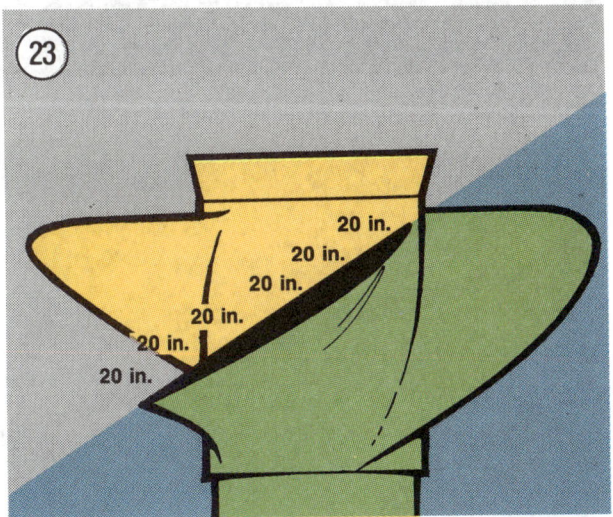

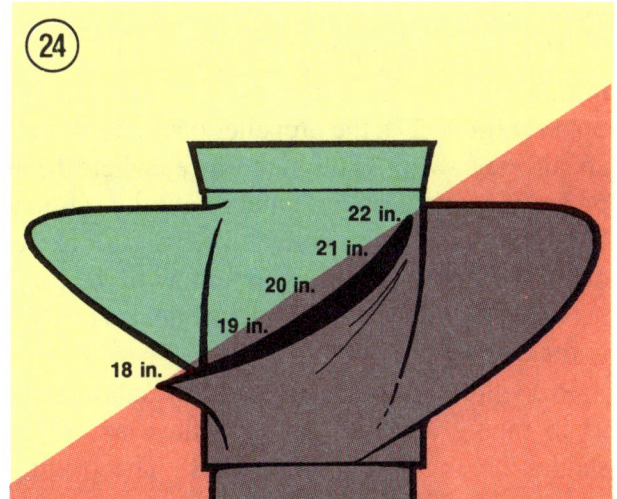

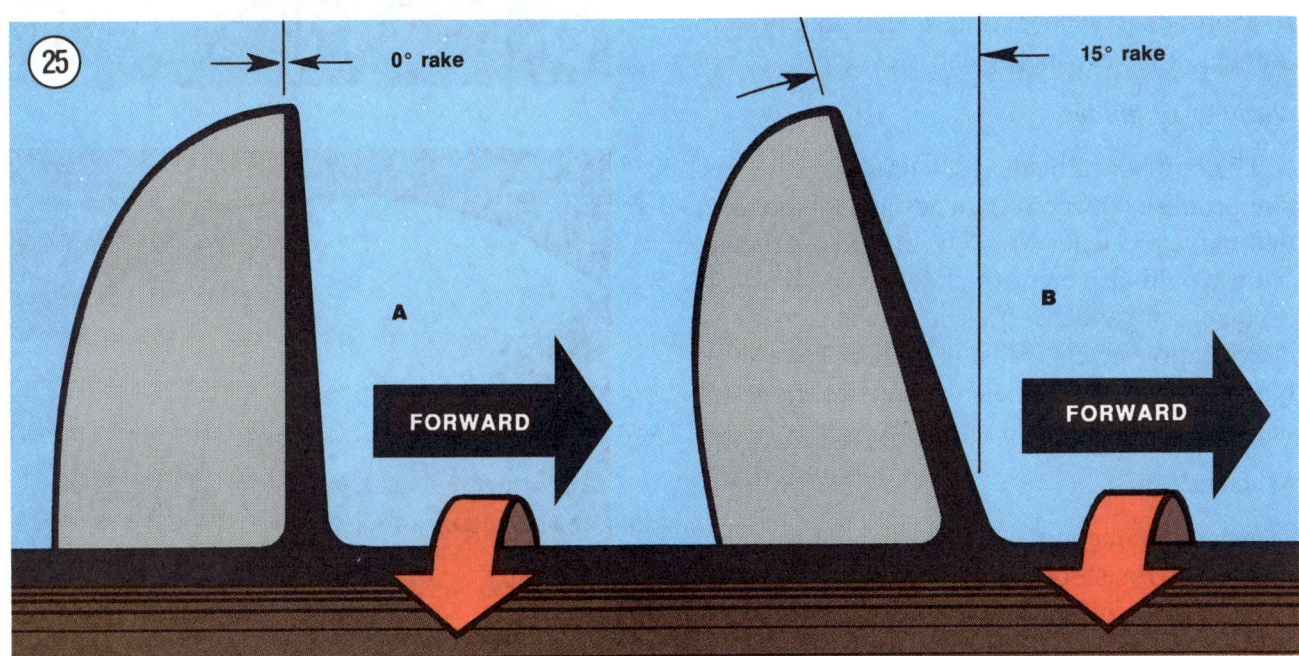

dicular. Most propellers have rakes ranging from 0-20°.

Blade thickness

Blade thickness is not uniform at all points along the blade. For efficiency, blades should be as thin as possible at all points while retaining enough strength to move the boat. Blades tend to be thicker where they meet the hub and thinner at the blade tip (**Figure 26**). This is to support the heavier loads at the hub section of the blade. This thickness is dependent on the strength of the material used.

When cut along a line from the leading edge to the trailing edge in the central portion of the blade (**Figure 27**), the propeller blade resembles an airplane wing. The blade face, where high pressure exists during normal rotation, is almost flat. The blade back, where low pressure exists during normal rotation, is curved, with the thinnest portions at the edges and the thickest portion at the center.

Propellers that run only partially submerged, as in racing applications, may have a wedge-shaped cross-section (**Figure 28**). The leading edge is very thin; the blade thickness increases toward the trailing edge, where it is the thickest. If a propeller such as this is run totally submerged, it is very inefficient.

Number of blades

The number of blades used on a propeller is a compromise between efficiency and vibration. A one-blade propeller would be the most efficient, but it would also create high levels of vibration. As blades are added, efficiency decreases, but so do vibration levels. Most propellers have three blades, representing the most practical trade-off between efficiency and vibration.

Material

Propeller materials are chosen for strength, corrosion resistance and economy. Stainless steel, aluminum and bronze are the most commonly used materials. Bronze is quite strong but

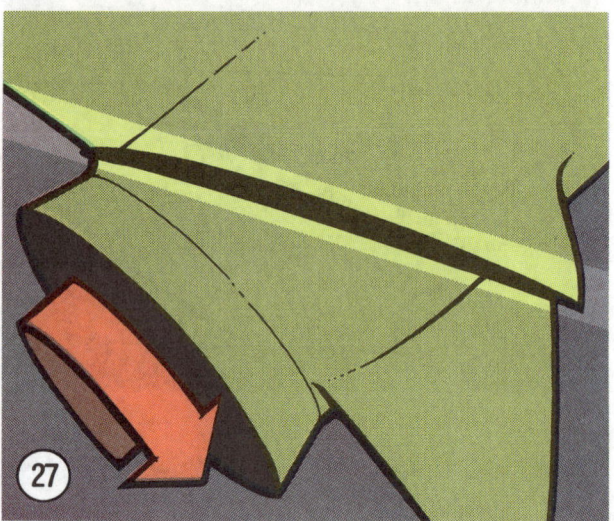

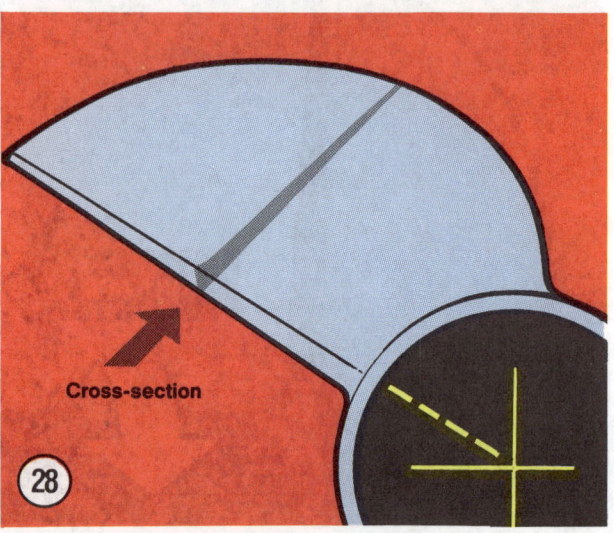

GENERAL INFORMATION

rather expensive. Stainless steel is more common than bronze because of its combination of strength and lower cost. Aluminum alloys are the least expensive but usually lack the strength of steel. Plastic propellers may be used in some low horsepower applications.

Direction of rotation

Propellers are made for both right-hand and left-hand rotation although right-hand is the most commonly used. When seen from behind the boat in forward motion, a right-hand propeller turns clockwise and a left-hand propeller turns counterclockwise. Off the boat, you can tell the difference by observing the angle of the blades (**Figure 29**). A right-hand propeller's blades slant from the upper left to the lower right; a left-hand propeller's blades are the opposite.

Cavitation and Ventilation

Cavitation and ventilation are *not* interchangeable terms; they refer to two distinct problems encountered during propeller operation.

To understand cavitation, you must first understand the relationship between pressure and the boiling point of water. At sea level, water will boil at 212° F. As pressure increases, such as within an engine's closed cooling system, the boiling point of water increases—it will boil at some temperature higher than 212° F. The opposite is also true. As pressure decreases, water will boil at a temperature lower than 212° F. If pressure drops low enough, water will boil at typical ambient temperatures of 50-60° F.

We have said that, during normal propeller operation, low-pressure exists on the blade back. Normally, the pressure does not drop low enough for boiling to occur. However, poor blade design

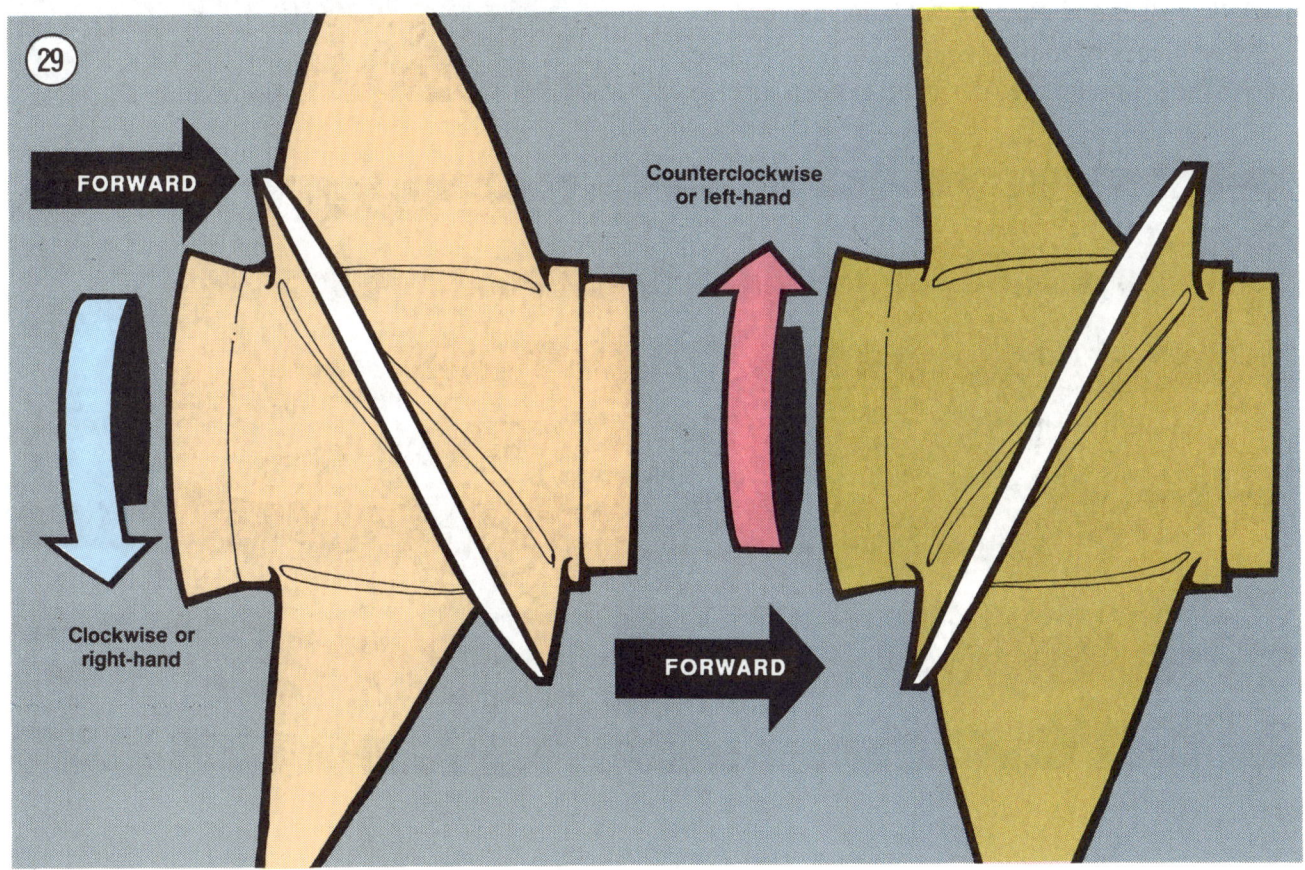

or selection, or blade damage can cause an unusual pressure drop on a small area of the blade (**Figure 30**). Boiling can occur in this small area. As the water boils, air bubbles form. As the boiling water passes to a higher pressure area of the blade, the boiling stops and the bubbles collapse. The collapsing bubbles release enough energy to erode the surface of the blade.

This entire process of pressure drop, boiling and bubble collapse is called "cavitation." The damage caused by the collapsing bubbles is called a "cavitation burn." It is important to remember that cavitation is caused by a decrease in pressure, *not* an increase in temperature.

Ventilation is not as complex a process as cavitation. Ventilation refers to air entering the blade area, either from above the surface of the water or from a through-hub exhaust system. As the blades meet the air, the propeller momentarily over-revs, losing most of its thrust. An added complication is that as the propeller over-revs, pressure on the blade back decreases and massive cavitation can occur.

Most pieces of marine equipment have a plate above the propeller area designed to keep surface air from entering the blade area (**Figure 31**). This plate is correctly called an "antiventilation plate," although you will often *see* it called an "anticavitation plate." Through hub exhaust systems also have specially designed hubs to keep exhaust gases from entering the blade area.

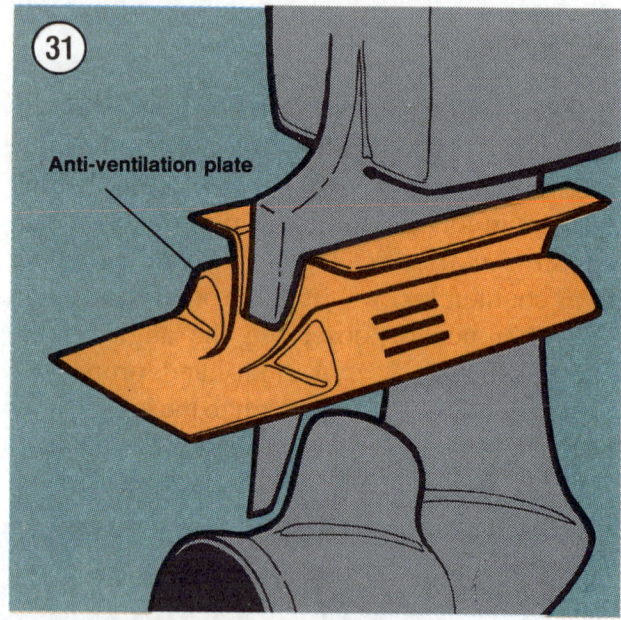

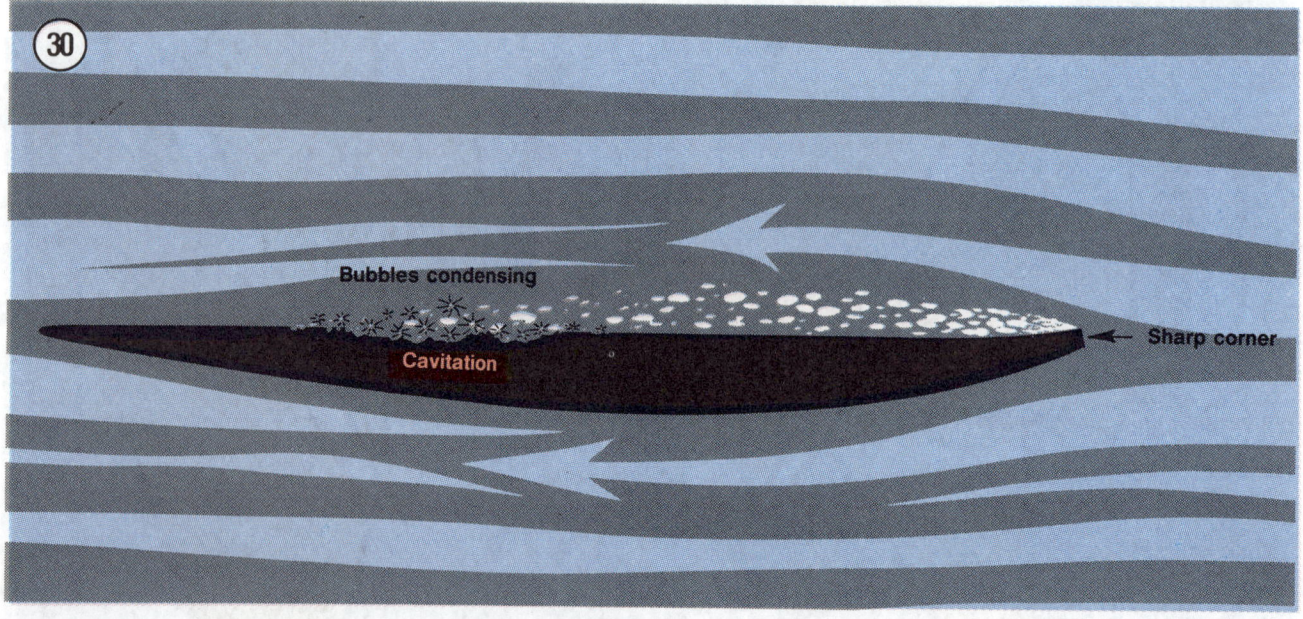

Chapter Two

Tools and Techniques

This chapter describes the common tools required for marine equipment repairs and troubleshooting. Techniques that will make your work easier and more effective are also described. Some of the procedures in this book require special skills or expertise; in some cases, you are better off entrusting the job to a dealer or qualified specialist.

SAFETY FIRST

Professional mechanics can work for years and never suffer a serious injury. If you follow a few rules of common sense and safety, you too can enjoy many safe hours servicing your marine equipment. If you ignore these rules, you can hurt yourself or damage the equipment.
1. Never use gasoline as a cleaning solvent.
2. Never smoke or use a torch near flammable liquids, such as cleaning solvent. If you are working in your home garage, remember that your home gas appliances have pilot lights.
3. Never smoke or use a torch in an area where batteries are being charged. Highly explosive hydrogen gas is formed during the charging process.
4. Use the proper size wrenches to avoid damage to fasteners and injury to yourself.
5. When loosening a tight or stuck fastener, think of what would happen if the wrench should slip. Protect yourself accordingly.
6. Keep your work area clean, uncluttered and well lighted.
7. Wear safety goggles during all operations involving drilling, grinding or the use of a cold chisel.
8. Never use worn tools.
9. Keep a Coast Guard approved fire extinguisher handy. Be sure it is rated for gasoline (Class B) and electrical (Class C) fires.

BASIC HAND TOOLS

A number of tools are required to maintain marine equipment. You may already have some of these tools for home or car repairs. There are also tools made especially for marine equipment repairs; these you will have to purchase. In any case, a wide variety of quality tools will make repairs easier and more effective.

Keep your tools clean and in a tool box. Keep them organized with the sockets and related drives together, the open end and box wrenches together, etc. After using a tool, wipe off dirt and grease with a clean cloth and place the tool in its correct place.

The following tools are required to perform virtually any repair job. Each tool is described and the recommended size given for starting a tool collection. Additional tools and some duplications may be added as you become more familiar with the equipment. You may need all standard U.S. size tools, all metric size tools or a mixture of both.

Screwdrivers

The screwdriver is a very basic tool, but if used improperly, it will do more damage than good. The slot on a screw has a definite dimension and shape. A screwdriver must be selected to conform with that shape. Use a small screwdriver for small screws and a large one for large screws or the screw head will be damaged.

Two types of screwdriver are commonly required: a common (flat-blade) screwdriver (**Figure 1**) and Phillips screwdrivers (**Figure 2**).

Screwdrivers are available in sets, which often include an assortment of common and Phillips blades. If you buy them individually, buy at least the following:

 a. Common screwdriver—5/16 × 6 in. blade.
 b. Common screwdriver—3/8 × 12 in. blade.
 c. Phillips screwdriver—size 2 tip, 6 in. blade.

Use screwdrivers only for driving screws. Never use a screwdriver for prying or chiseling. Do not try to remove a Phillips or Allen head screw with a common screwdriver; you can damage the head so that the proper tool will be unable to remove it.

Keep screwdrivers in the proper condition and they will last longer and perform better. Always keep the tip of a common screwdriver in good condition. **Figure 3** shows how to grind the tip to the proper shape if it becomes damaged. Note the parallel sides of the tip.

Pliers

Pliers come in a wide range of types and sizes. Pliers are useful for cutting, bending and crimping. They should never be used to cut hardened objects or to turn bolts or nuts. **Figure 4** shows several types of pliers.

Each type of pliers has a specialized function. General purpose pliers are used mainly for holding things and for bending. Locking pliers are used as pliers or to hold objects very tightly, like a vise. Needlenose pliers are used to hold or bend small objects. Adjustable or slip-joint pliers can

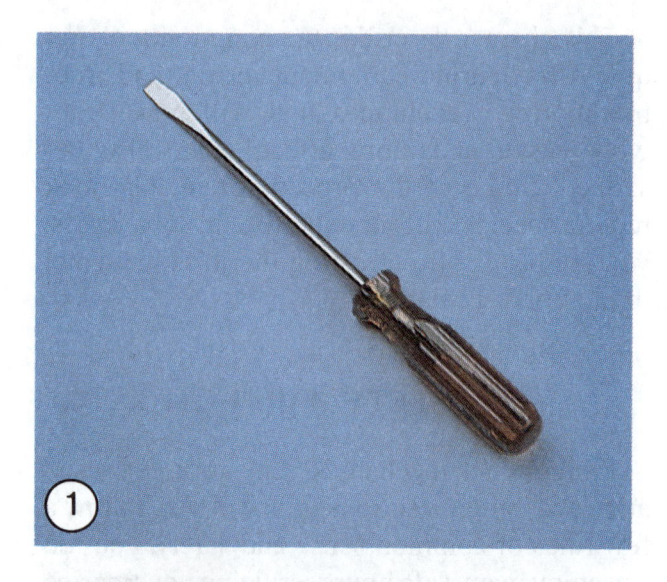

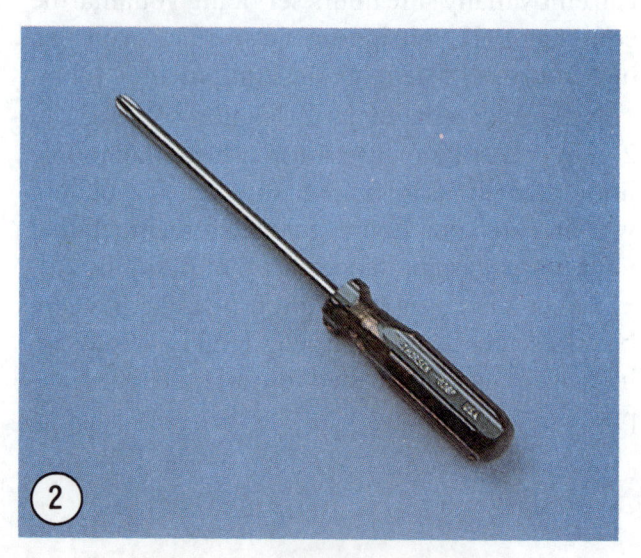

TOOLS AND TECHNIQUES   23

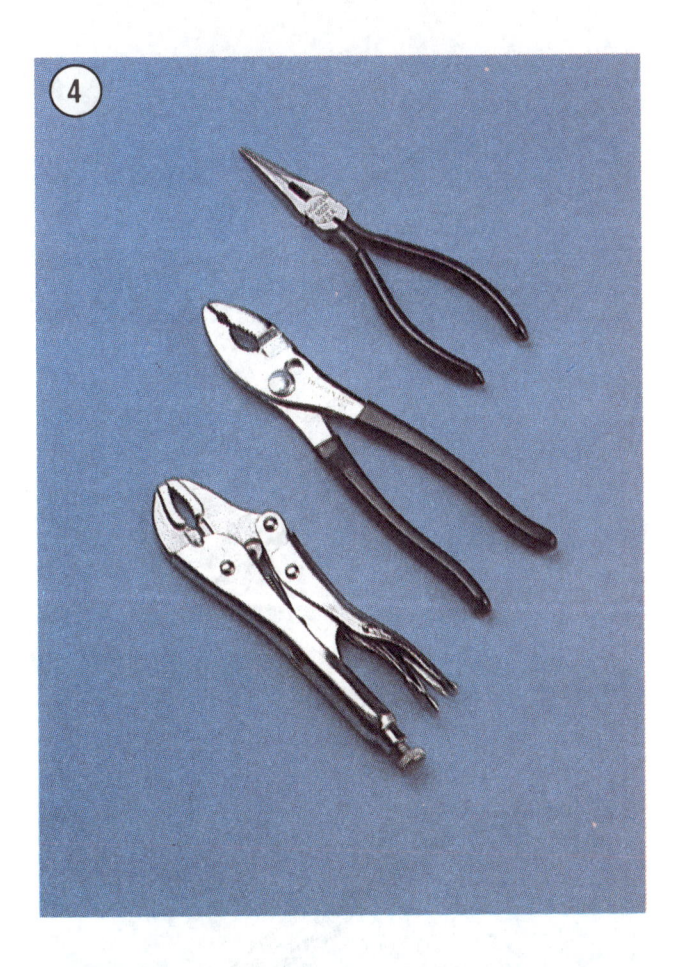

be adjusted to hold various sizes of objects; the jaws remain parallel to grip around objects such as pipe or tubing. There are many more types of pliers. The ones described here are the most commonly used.

Box and Open-end Wrenches

Box and open-end wrenches are available in sets or separately in a variety of sizes. See **Figure 5** and **Figure 6**. The number stamped near the end refers to the distance between two parallel flats on the hex head bolt or nut.

Box wrenches are usually superior to open-end wrenches. An open-end wrench grips the nut on only two flats. Unless it fits well, it may slip and round off the points on the nut. The box wrench grips all 6 flats. Both 6-point and 12-point openings on box wrenches are available. The 6-point gives superior holding power; the 12-point allows a shorter swing.

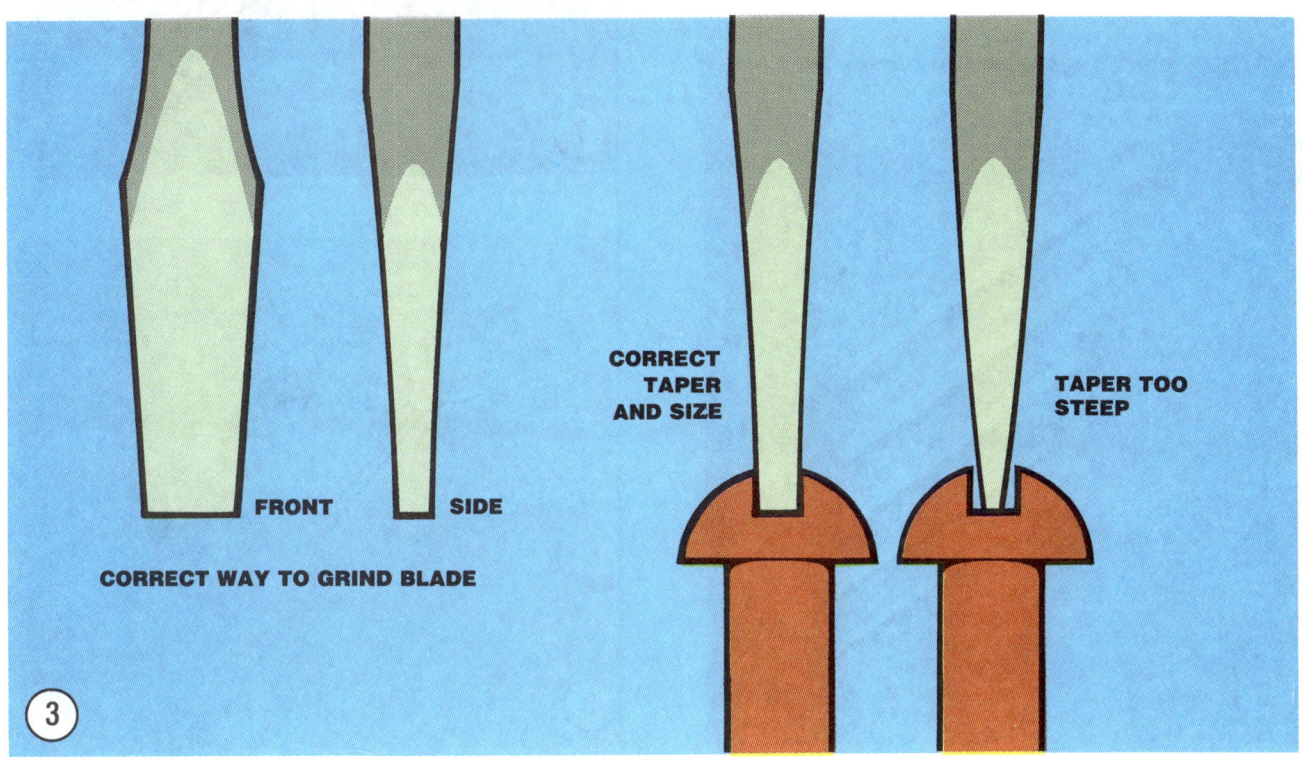

CHAPTER TWO

Combination wrenches, which are open on one side and boxed on the other, are also available. Both ends are the same size.

Adjustable Wrenches

An adjustable wrench can be adjusted to fit nearly any nut or bolt head. See **Figure 7**. However, it can loosen and slip, causing damage to the nut and maybe to your knuckles. Use an adjustable wrench only when other wrenches are not available.

Adjustable wrenches come in sizes ranging from 4-18 in. overall. A 6 or 8 in. wrench is recommended as an all-purpose wrench.

Socket Wrenches

This type is undoubtedly the fastest, safest and most convenient to use. See **Figure 8**. Sockets, which attach to a suitable handle, are available with 6-point or 12-point openings and use 1/4, 3/8 and 3/4 inch drives. The drive size indicates

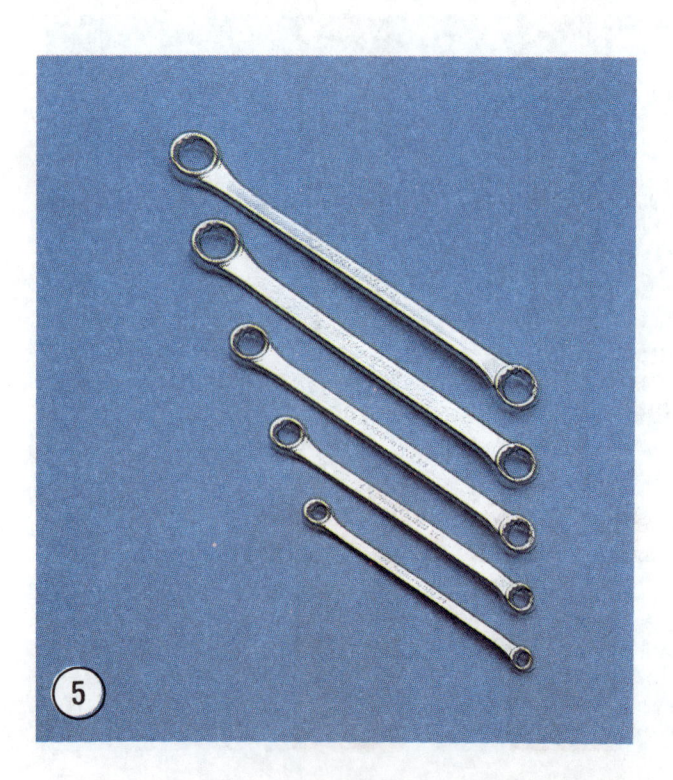

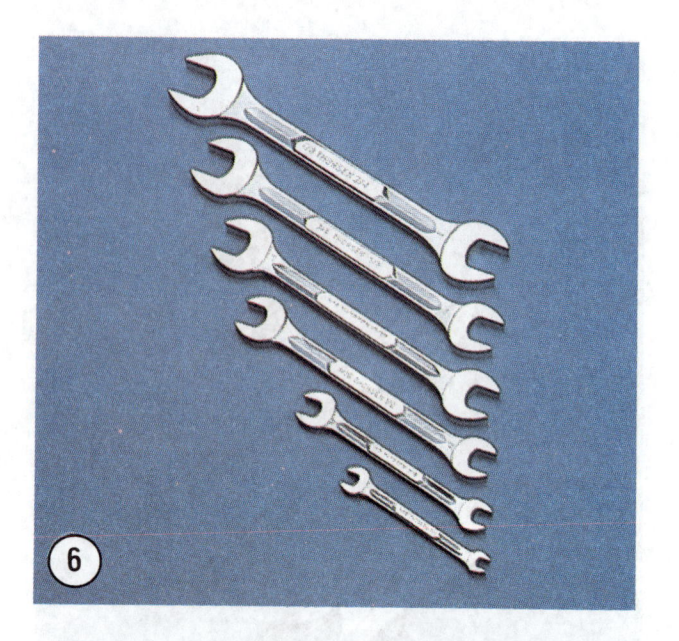

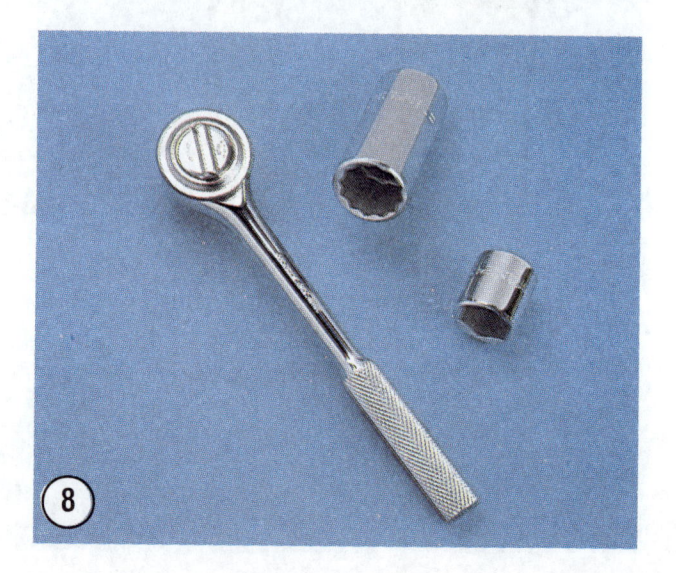

TOOLS AND TECHNIQUES

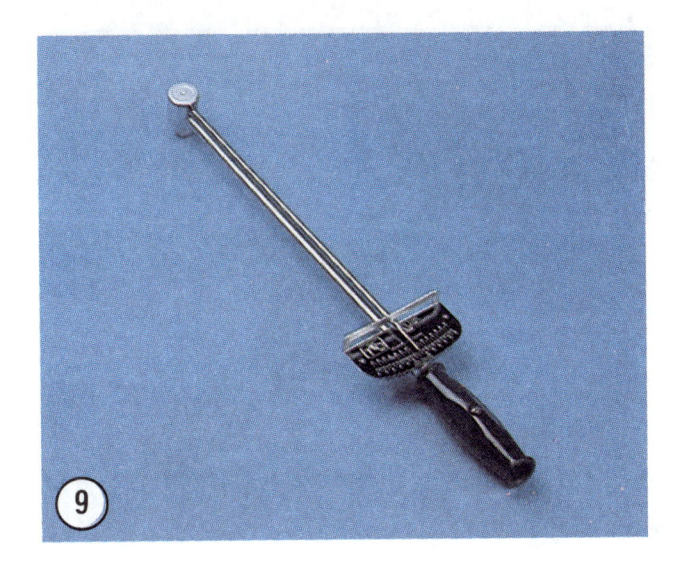

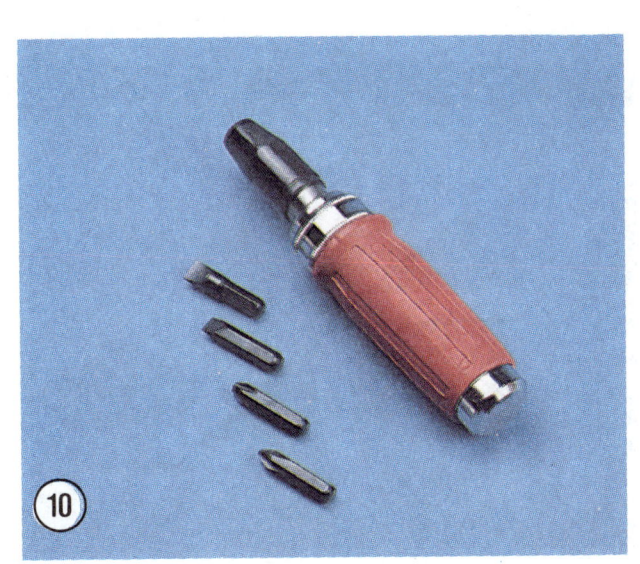

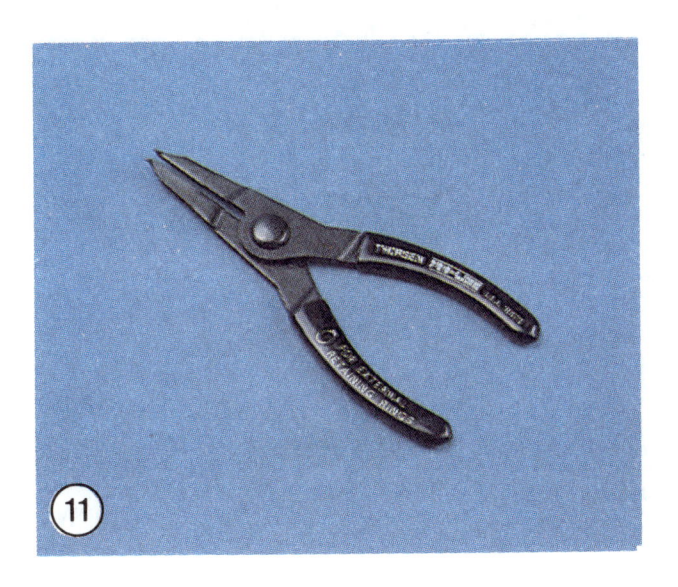

the size of the square hole that mates with the ratchet or flex handle.

Torque Wrench

A torque wrench (**Figure 9**) is used with a socket to measure how tight a nut or bolt is installed. They come in a wide price range and with either 3/8 or 1/2 in. square drive. The drive size indicates the size of the square drive that mates with the socket. Purchase one that measures up to 150 ft.-lb. (203 N•m).

Impact Driver

This tool (**Figure 10**) makes removal of tight fasteners easy and eliminates damage to bolts and screw slots. Impact drivers and interchangeable bits are available at most large hardware and auto parts stores.

Circlip Pliers

Circlip pliers (sometimes referred to as snap-ring pliers) are necessary to remove circlips. See **Figure 11**. Circlip pliers usually come with several different size tips; many designs can be switched from internal type to external type.

Hammers

The correct hammer is necessary for repairs. Use only a hammer with a face (or head) of rubber or plastic or the soft-faced type that is filled with buckshot (**Figure 12**). These are sometimes necessary in engine tear-downs. *Never* use a metal-faced hammer as severe damage will result in most cases. You can always produce the same amount of force with a soft-faced hammer.

Feeler Gauge

This tool has either flat or wire measuring gauges (**Figure 13**). Wire gauges are used to measure spark plug gap; flat gauges are used for all other measurements. A non-magnetic (brass) gauge may be specified when working around magnetized parts.

Other Special Tools

Some procedures require special tools; these are identified in the appropriate chapter. Unless otherwise specified, the part number used in this book to identify a special tool is the marine equipment manufacturer's part number.

Special tools can usually be purchased through your marine equipment dealer. Some can be made locally by a machinist, often at a much lower price. You may find certain special tools at tool rental dealers. Don't use makeshift tools if you can't locate the correct special tool; you will probably cause more damage than good.

TEST EQUIPMENT

Multimeter

This instrument (**Figure 14**) is invaluable for electrical system troubleshooting and service. It combines a voltmeter, an ohmmeter and an ammeter into one unit, so it is often called a VOM.

Two types of multimeter are available, analog and digital. Analog meters have a moving needle with marked bands indicating the volt, ohm and amperage scales. The digital meter (DVOM) is ideally suited for troubleshooting because it is easy to read, more accurate than analog, contains internal overload protection, is auto-ranging (analog meters must be recalibrated each time the scale is changed) and has automatic polarity compensation.

TOOLS AND TECHNIQUES

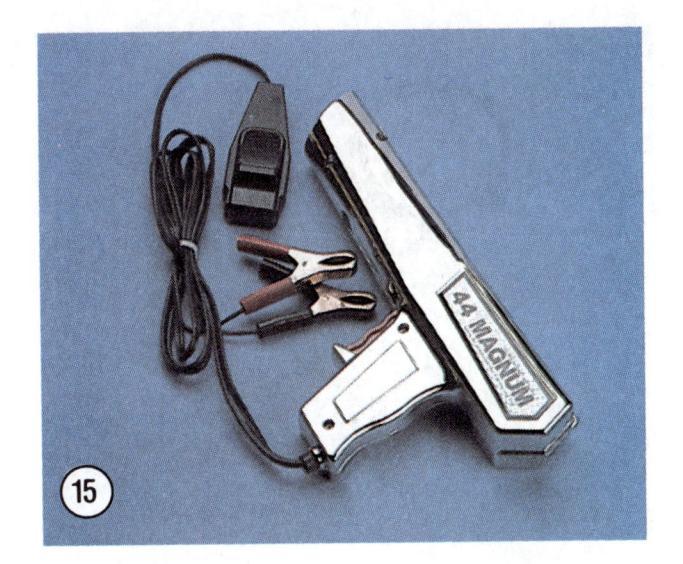

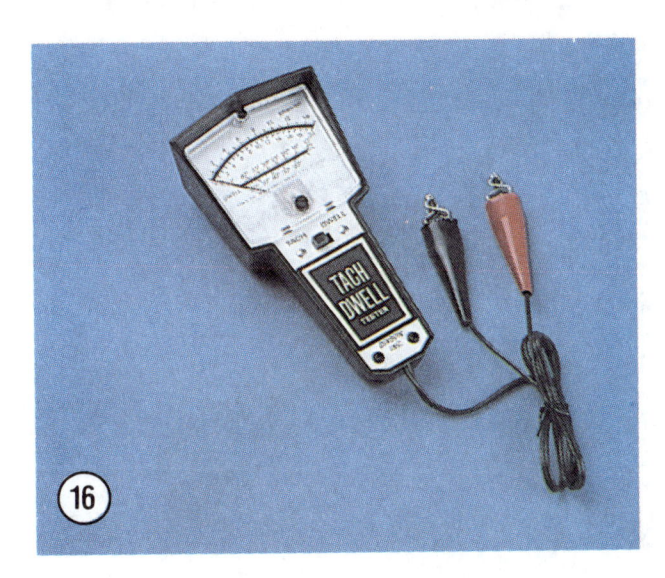

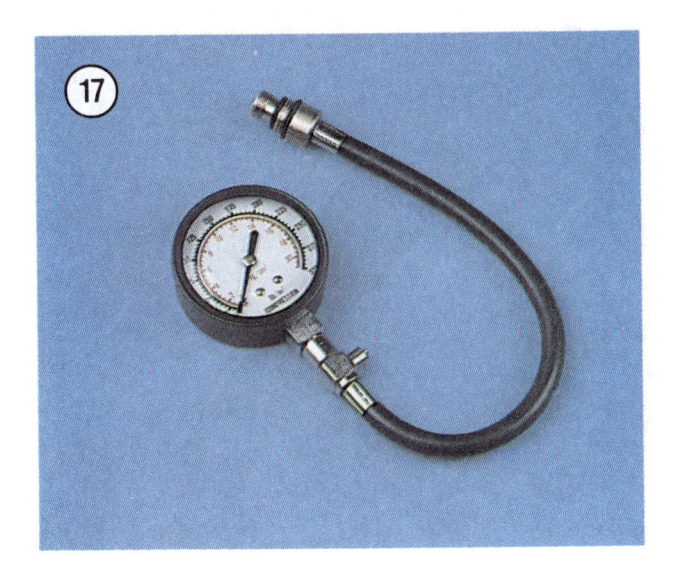

Strobe Timing Light

This instrument is necessary for dynamic tuning (setting ignition timing while the engine is running). By flashing a light at the precise instant the spark plug fires, the position of the timing mark can be seen. The flashing light makes a moving mark appear to stand still opposite a stationary mark.

Suitable lights range from inexpensive neon bulb types to powerful xenon strobe lights. See **Figure 15**. A light with an inductive pickup is best because it eliminates any possible damage to ignition wiring.

Tachometer/Dwell Meter

A portable tachometer is necessary for tuning. See **Figure 16**. Ignition timing and carburetor adjustments must be performed at the specified idle speed. The best instrument for this purpose is one with a low range of 0-1000 or 0-2000 rpm and a high range of 0-6000 rpm. Extended range (0-6000 or 0-8000 rpm) instruments lack accuracy at lower speeds. The instrument should be capable of detecting changes of 25 rpm on the low range.

A dwell meter is often combined with a tachometer. Dwell meters are used with breaker point ignition systems to measure the amount of time the points remain closed during engine operation.

Compression Gauge

This tool (**Figure 17**) measures the amount of pressure present in the engine's combustion chamber during the compression stroke. This indicates general engine condition. Compression readings can be interpreted along with vacuum gauge readings to pinpoint specific engine mechanical problems.

The easiest type to use has screw-in adapters that fit into the spark plug holes. Press-in rubber-tipped types are also available.

Vacuum Gauge

The vacuum gauge (**Figure 18**) measures the intake manifold vacuum created by the engine's intake stroke. Manifold and valve problems (on 4-stroke engines) can be identified by interpreting the readings. When combined with compression gauge readings, other engine problems can be diagnosed.

Some vacuum gauges can also be used as fuel pressure gauges to trace fuel system problems.

Hydrometer

Battery electrolyte specific gravity is measured with a hydrometer (**Figure 19**). The specific gravity of the electrolyte indicates the battery's state of charge. The best type has automatic temperature compensation; otherwise, you must calculate the compensation yourself.

Precision Measuring Tools

Various tools are needed to make precision measurements. A dial indicator (**Figure 20**), for example, is used to determine run-out of rotating parts and end play of parts assemblies. A dial indicator can also be used to precisely measure piston position in relation to top dead center; some engines require this measurement for ignition timing adjustment.

Vernier calipers (**Figure 21**) and micrometers (**Figure 22**) are other precision measuring tools used to determine the size of parts (such as piston diameter).

Precision measuring equipment must be stored, handled and used carefully or it will not remain accurate.

SERVICE HINTS

Most of the service procedures covered in this manual are straightforward and can be performed by anyone reasonably handy with tools.

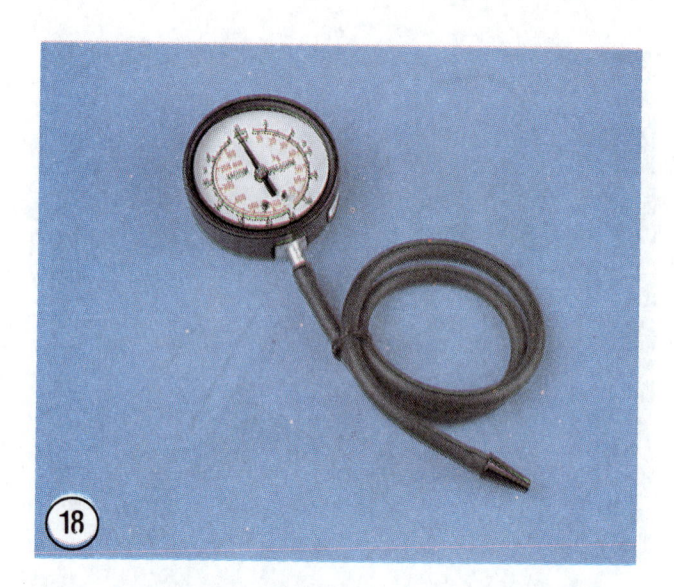

TOOLS AND TECHNIQUES

It is suggested, however, that you consider your own skills and toolbox carefully before attempting any operation involving major disassembly of the engine or gearcase.

Some operations, for example, require the use of a press. It would be wiser to have these performed by a shop equipped for such work, rather than trying to do the job yourself with makeshift equipment. Other procedures require precise measurements. Unless you have the skills and equipment required, it would be better to have a qualified repair shop make the measurements for you.

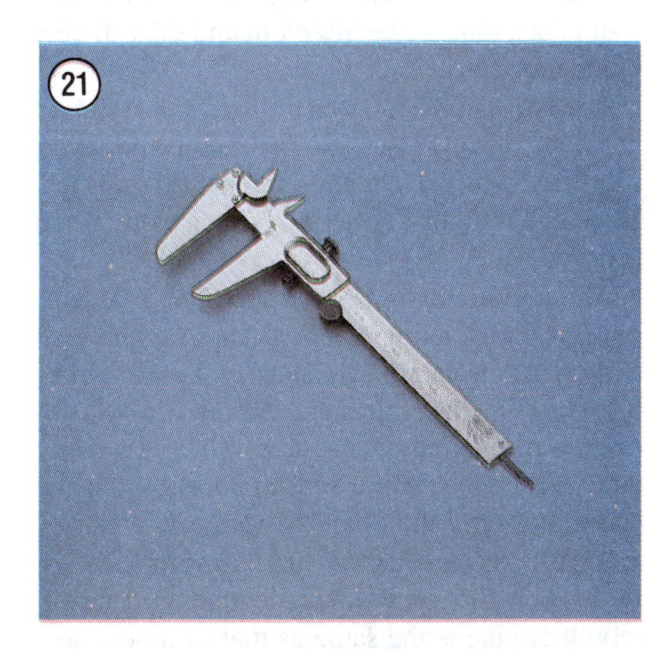

Preparation for Disassembly

Repairs go much faster and easier if the equipment is clean before you begin work. There are special cleaners, such as Gunk or Bel-Ray Degreaser, for washing the engine and related parts. Just spray or brush on the cleaning solution, let it stand, then rinse away with a garden hose. Clean all oily or greasy parts with cleaning solvent as you remove them.

> *WARNING*
> *Never use gasoline as a cleaning agent. It presents an extreme fire hazard. Be sure to work in a well-ventilated area when using cleaning solvent. Keep a Coast Guard approved fire extinguisher, rated for gasoline fires, handy in any case.*

Much of the labor charged for repairs made by dealers is for the removal and disassembly of other parts to reach the defective unit. It is frequently possible to perform the preliminary operations yourself and then take the defective unit in to the dealer for repair.

If you decide to tackle the job yourself, read the entire section in this manual that pertains to it, making sure you have identified the proper one. Study the illustrations and text until you have a good idea of what is involved in completing the job satisfactorily. If special tools or replacement parts are required, make arrangements to get them before you start. It is frustrating and time-consuming to get partly into a job and then be unable to complete it.

Disassembly Precautions

During disassembly of parts, keep a few general precautions in mind. Force is rarely needed to get things apart. If parts are a tight fit, such as a bearing in a case, there is usually a tool designed to separate them. Never use a screwdriver to pry apart parts with machined surfaces (such as cylinder heads and crankcases). You will mar the surfaces and end up with leaks.

Make diagrams (or take an instant picture) wherever similar-appearing parts are found. For example, head and crankcase bolts are often not the same length. You may think you can remember where everything came from, but mistakes are costly. There is also the possibility you may be sidetracked and not return to work for days or even weeks. In the interval, carefully laid out parts may have been disturbed.

Cover all openings after removing parts to keep small parts, dirt or other contamination from entering.

Tag all similar internal parts for location and direction. All internal components should be reinstalled in the same location and direction from which removed. Record the number and thickness of any shims as they are removed. Small parts, such as bolts, can be identified by placing them in plastic sandwich bags. Seal and label them with masking tape.

Wiring should be tagged with masking tape and marked as each wire is removed. Again, do not rely on memory alone.

Protect finished surfaces from physical damage or corrosion. Keep gasoline off painted surfaces.

Assembly Precautions

No parts, except those assembled with a press fit, require unusual force during assembly. If a part is hard to remove or install, find out why before proceeding.

When assembling two parts, start all fasteners, then tighten evenly in an alternating or crossing pattern if no specific tightening sequence is given.

When assembling parts, be sure all shims and washers are installed exactly as they came out.

Whenever a rotating part butts against a stationary part, look for a shim or washer. Use new gaskets if there is any doubt about the condition of the old ones. Unless otherwise specified, a thin coat of oil on gaskets may help them seal effectively.

Heavy grease can be used to hold small parts in place if they tend to fall out during assembly. However, keep grease and oil away from electrical components.

High spots may be sanded off a piston with sandpaper, but fine emery cloth and oil will do a much more professional job.

Carbon can be removed from the cylinder head, the piston crown and the exhaust port with a dull screwdriver. *Do not* scratch either surface. Wipe off the surface with a clean cloth when finished.

The carburetor is best cleaned by disassembling it and soaking the parts in a commercial carburetor cleaner. Never soak gaskets and rubber parts in these cleaners. Never use wire to clean out jets and air passages; they are easily damaged. Use compressed air to blow out the carburetor *after* the float has been removed.

Take your time and do the job right. Do not forget that the break-in procedure on a newly rebuilt engine is the same as that of a new one. Use the break-in oil recommendations and follow other instructions given in your owner's manual.

SPECIAL TIPS

Because of the extreme demands placed on marine equipment, several points should be kept in mind when performing service and repair. The following items are general suggestions that may improve the overall life of the machine and help avoid costly failures.

1. Unless otherwise specified, use a locking compound, such as Loctite Threadlocker, on all bolts and nuts, even if they are secured with lockwashers. Be sure to use the specified grade

TOOLS AND TECHNIQUES

of thread locking compound. A screw or bolt lost from an engine cover or bearing retainer could easily cause serious and expensive damage before its loss is noticed.

When applying thread locking compound, use a small amount. If too much is used, it can work its way down the threads and stick parts together that were not meant to be stuck together.

Keep a tube of thread locking compound in your tool box; when used properly, it is cheap insurance.

2. Use a hammer-driven impact tool to remove and install screws and bolts. These tools help prevent the rounding off of bolt heads and screw slots and ensure a tight installation.

3. When straightening the fold-over type lockwasher, use a wide-blade chisel, such as an old and dull wood chisel. Such a tool provides a better purchase on the folded tab, making straightening easier.

4. When installing the fold-over type lockwasher, always use a new washer if possible. If a new washer is not available, always fold over a part of the washer that has not been previously folded. Reusing the same fold may cause the washer to break, resulting in the loss of its locking ability and a loose piece of metal adrift in the engine.

When folding the washer, start the fold with a screwdriver and finish it with a pair of pliers. If a punch is used to make the fold, the fold may be too sharp, thereby increasing the chances of the washer breaking under stress.

These washers are relatively inexpensive and it is suggested that you keep several of each size in your tool box for repairs.

5. When replacing missing or broken fasteners (bolts, nuts and screws), always use authorized replacement parts. They are specially hardened for each application. The wrong 50-cent bolt could easily cause serious and expensive damage.

6. When installing gaskets, always use authorized replacement gaskets *without* sealer, unless designated. Many gaskets are designed to swell when they come in contact with oil. Gasket sealer will prevent the gaskets from swelling as intended and can result in oil leaks. Authorized replacement gaskets are cut from material of the precise thickness needed. Installation of a too thick or too thin gasket in a critical area could cause equipment damage.

MECHANIC'S TECHNIQUES

Removing Frozen Fasteners

When a fastener rusts and cannot be removed, several methods may be used to loosen it. First, apply penetrating oil, such as Liquid Wrench or WD-40 (available at any hardware or auto supply store). Apply it liberally and allow it penetrate for 10-15 minutes. Tap the fastener several times with a small hammer; do not hit it hard enough to cause damage. Reapply the penetrating oil if necessary.

For frozen screws, apply penetrating oil as described, then insert a screwdriver in the slot and tap the top of the screwdriver with a hammer. This loosens the rust so the screw can be removed in the normal way. If the screw head is too chewed up to use a screwdriver, grip the head with locking pliers and twist the screw out.

Avoid applying heat unless specifically instructed because it may melt, warp or remove the temper from parts.

Remedying Stripped Threads

Occasionally, threads are stripped through carelessness or impact damage. Often the threads can be cleaned up by running a tap (for internal threads on nuts) or die (for external threads on bolts) through threads. See **Figure 23**.

CHAPTER TWO

Removing Broken Screws or Bolts

When the head breaks off a screw or bolt, several methods are available for removing the remaining portion.

If a large portion of the remainder projects out, try gripping it with vise-grip pliers. If the projecting portion is too small, file it to fit a wrench or cut a slot in it to fit a screwdriver. See **Figure 24**.

If the head breaks off flush, use a screw extractor. To do this, centerpunch the remaining portion of the screw or bolt. Drill a small hole in the screw and tap the extractor into the hole. Back the screw out with a wrench on the extractor. See **Figure 25**.

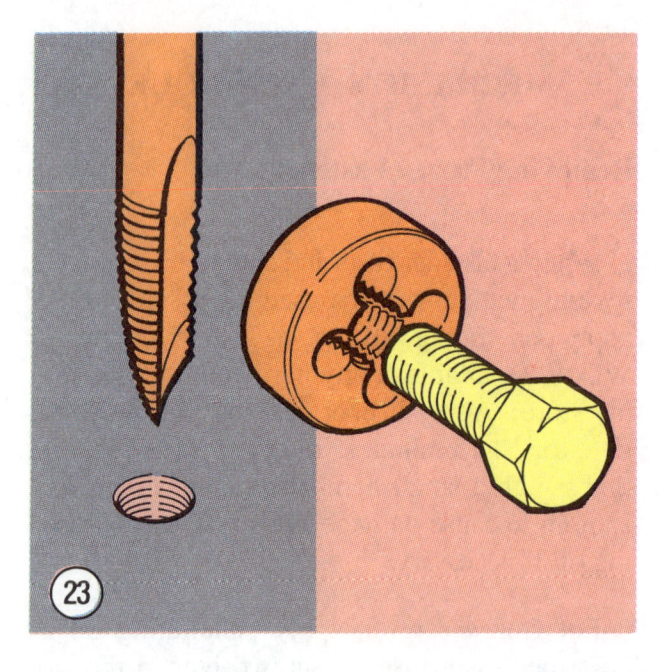

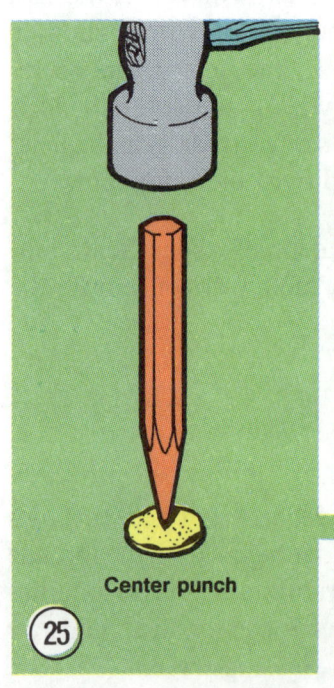

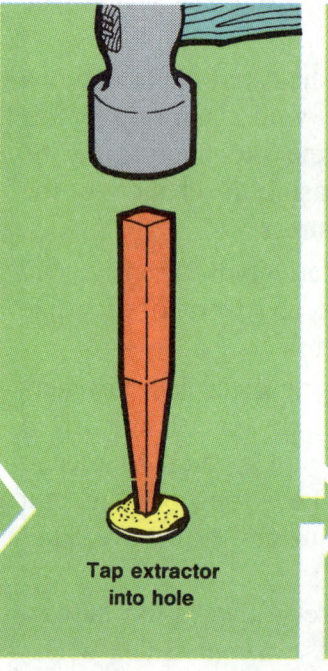

(23)

(24) Filed — Slotted

(25) Center punch → Drill hole → Tap extractor into hole → Remove screw

Chapter Three

Troubleshooting

Troubleshooting is a relatively simple matter when it is done logically. The first step in any troubleshooting procedure is to define the symptoms as closely as possible and then localize the problem. Subsequent steps involve testing and analyzing those areas which could cause the symptoms. A haphazard approach may eventually solve the problem, but it can be very costly in terms of wasted time and unnecessary parts replacement.

Proper lubrication, maintenance and periodic tune-ups as described in Chapter Four will reduce the necessity for troubleshooting. Even with the best of care, however, an outboard motor is prone to problems which will require troubleshooting.

This chapter contains brief descriptions of each operating system and troubleshooting procedures to be used. **Tables 1-3** present typical starting, ignition and carburetted fuel system problems with their probable causes and solutions. **Tables 1-7** are at the end of the chapter.

OPERATING REQUIREMENTS

Every outboard motor requires 3 basic things to run properly: an uninterrupted supply of fuel and air in the correct proportions, proper ignition at the right time and adequate compression. If any of these are lacking, the motor will not run.

The electrical system is the weakest link in the chain. More problems result from electrical malfunctions than from any other source. Keep this in mind before you blame the fuel system and start making unnecessary carburetor adjustments.

If a motor has been sitting for any length of time and refuses to start, check the condition of the battery first to make sure it has an adequate charge, then look to the fuel delivery system. This includes the gas tank, fuel pump, fuel lines and carburetor(s). Rust may have formed in the tank, obstructing fuel flow. Gasoline deposits may have gummed up carburetor jets and air passages or fuel injectors. Gasoline tends to lose its potency after standing for long periods. Condensation may contaminate it with water. Drain the old gas and try starting with a fresh tankful.

STARTING SYSTEM

Description

An electric starter motor (**Figure 1**) is optional on Suzuki DT 8 through DT 40 outboards and standard on DT 55 through DT 225 models. The motor is mounted vertically on the engine. When battery current is supplied to the starter motor, its pinion gear is thrust upward to engage the teeth on the engine flywheel. Once the engine starts, the pinion gear disengages from the flywheel. This is similar to the method used in cranking an automotive engine.

The electric starting system requires a fully charged battery to provide the large amount of current required to operate the starter motor. The battery may be charged externally or by a lighting coil on the magneto stator base which keeps the battery charged while the engine is running.

The starting circuit consists of the battery, a key ignition or push-button starter switch, an interlock switch, a stop switch, the starter motor, starter relay and the connecting electrical wiring.

The starter relay (**Figure 2**) carries the heavy electrical current to the motor. Depressing the starter switch or turning the key ignition switch to the START position allows current to flow through the relay coil. The relay contacts close and allow current to flow from the battery through the relay to the starter motor.

An interlock switch in all electric starting circuits prevents current flow to the starter motor if the shift mechanism is not in NEUTRAL. Most Suzuki models sold in the United States have an NSI circuit (Neutral Start Interlock) which prevents ignition unless the shift mechanism is in NEUTRAL. Models without the NSI circuit have a mechanical interlock in the rewind starter. This interlock device is connected to the interlock switch by a cable.

The stop switch shorts out the stator charge coil on some models. The models not equipped with this feature incorporate a low voltage stop circuit which prevents a high voltage leak when the voltage is diverted to ground. **Figure 3** is a schematic of a typical Suzuki electrical system showing the starting and stop circuits.

> *CAUTION*
> *Do not operate an electric starter motor continuously for more than 10 seconds. Allow the motor to cool for at least 2 minutes between attempts to start the engine.*

TROUBLESHOOTING

Troubleshooting

Refer to **Table 1** at the end of the chapter. Before troubleshooting the starting circuit, make sure that:

a. The battery is fully charged.
b. The battery cables are the proper size and length. Replace cables that are undersize or relocate the battery to shorten the distance between battery and the starter relay.
c. The shift mechanism is in NEUTRAL and the emergency switch cap is properly installed on remote control models.
d. All electrical connections are clean, free of corrosion and are tight.
e. The wiring harness is in good condition, with no worn or frayed insulation or loose harness electrical connectors.
f. The electrical circuit fuse is in good condition.
g. The fuel system is filled with an adequate supply of fresh gasoline that has been properly mixed with Suzuki CCI 50:1 Outboard Oil. See Chapter Four.

Troubleshooting is intended only to isolate a malfunction to a certain component. If further bench testing is necessary, remove the suspected component and have it tested by an authorized service center. Refer to Chapter Seven for component removal and installation procedures.

Starter Relay Continuity Test

1. Disconnect the yellow/green and black starter relay leads.

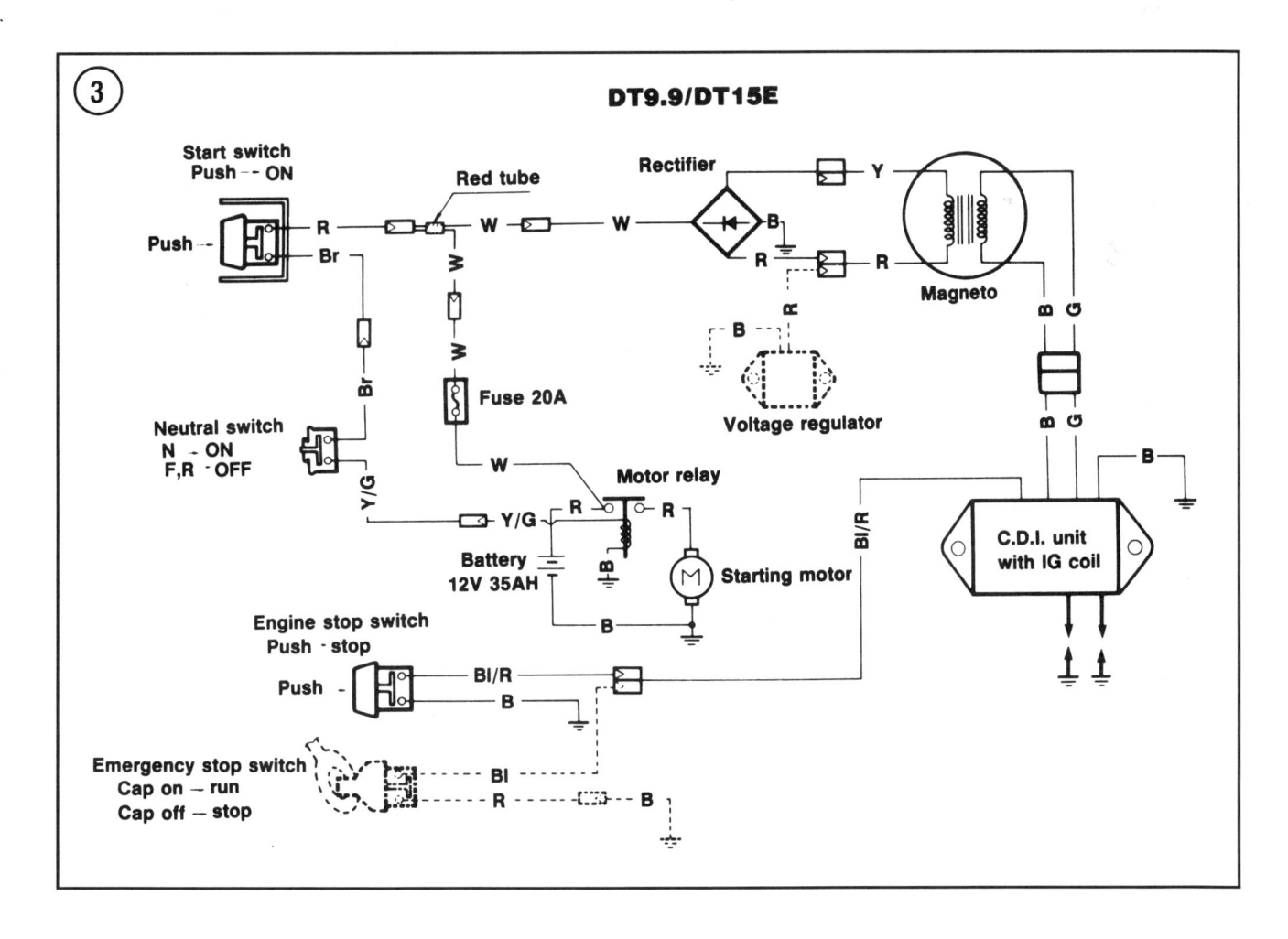

③ **DT9.9/DT15E**

[Wiring diagram showing: Start switch Push–ON, Red tube, Rectifier, Magneto, Neutral switch N–ON F,R–OFF, Fuse 20A, Voltage regulator, Motor relay, Battery 12V 35AH, Starting motor, C.D.I. unit with IG coil, Engine stop switch Push–stop, Emergency stop switch Cap on–run Cap off–stop]

2. Connect an ohmmeter between the 2 leads. See **Figure 4**, typical. If the meter does not read 3.2-3.8 ohms, replace the starter relay. See Chapter Seven.

Neutral Start Interlock Switch Continuity Test

1. Disconnect the yellow/green and brown interlock switch leads.
2. Connect an ohmmeter between the 2 leads (**Figure 5**, typical). There should be continuity (low resistance) with the shift lever in NEUTRAL and no continuity when the shift lever is moved to FORWARD or REVERSE.
3. If there is no continuity with the shift lever in NEUTRAL, manually depress the interlock switch plunger. If there is still no continuity (infinite resistance), replace the switch.
4. If continuity is shown in Step 3, turn the switch adjusting bolt until the switch performs as described in Step 2.

Push Button Start Switch Continuity Test

1. Disconnect the red and black start button leads.
2. Connect an ohmmeter between the 2 leads (**Figure 6**, typical). There should be no continuity (infinite resistance).
3. Depress the start button. The meter should show continuity (low resistance).
4. If the switch does not perform as described in Step 2 and Step 3, replace it.

Key Ignition Switch Continuity Test

Test the key switch with an ohmmeter or self-powered test lamp. If there is no continuity (infinite resistance) in each position as shown in **Figure 7**, replace the switch.

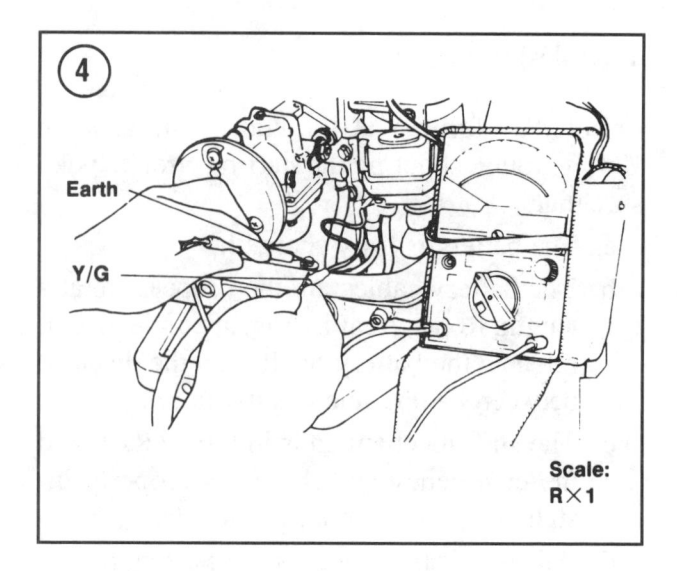

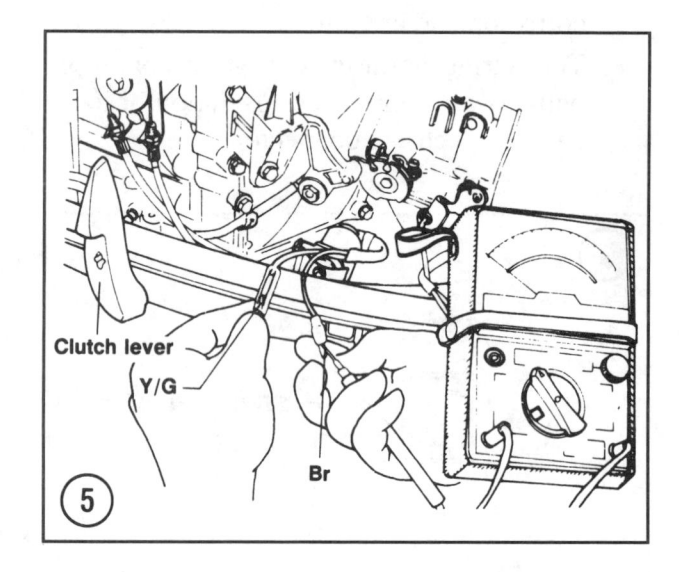

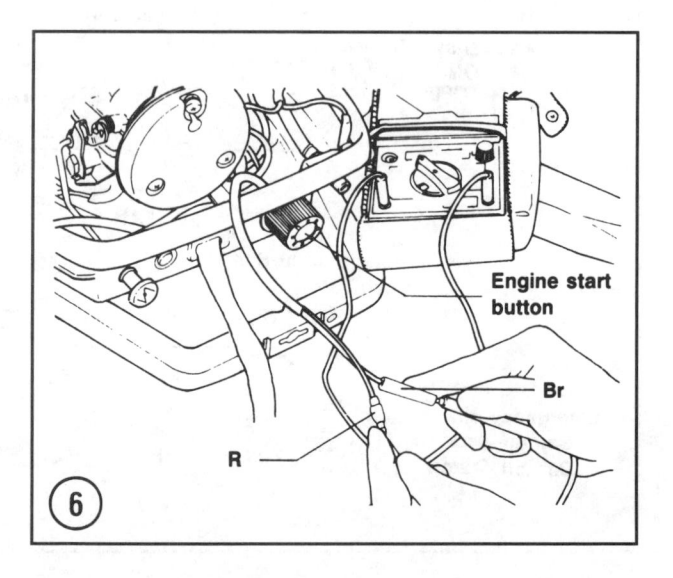

TROUBLESHOOTING

Stop Switch Continuity Test

1. Disconnect the black and blue/red stop switch leads.

2. Connect an ohmmeter between the 2 leads (**Figure 8**, typical). There should be no continuity (infinite resistance).

3. Depress the stop switch. The meter should show continuity (low resistance).

4. If the switch does not perform as described in Step 2 and Step 3, replace it.

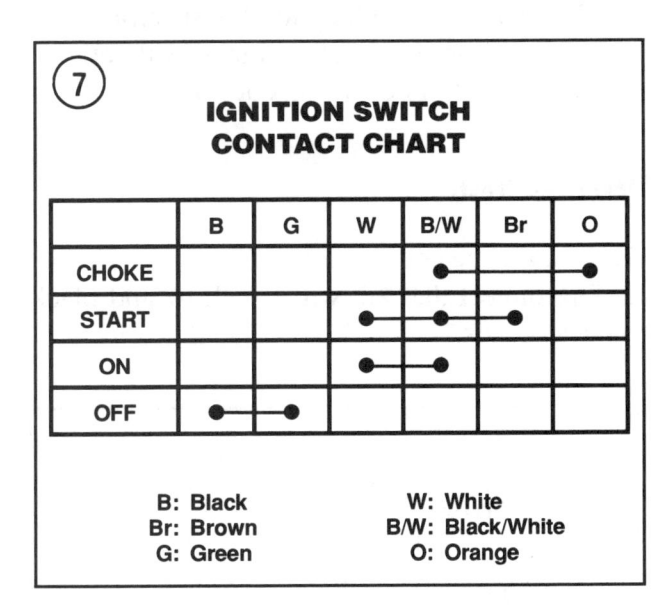

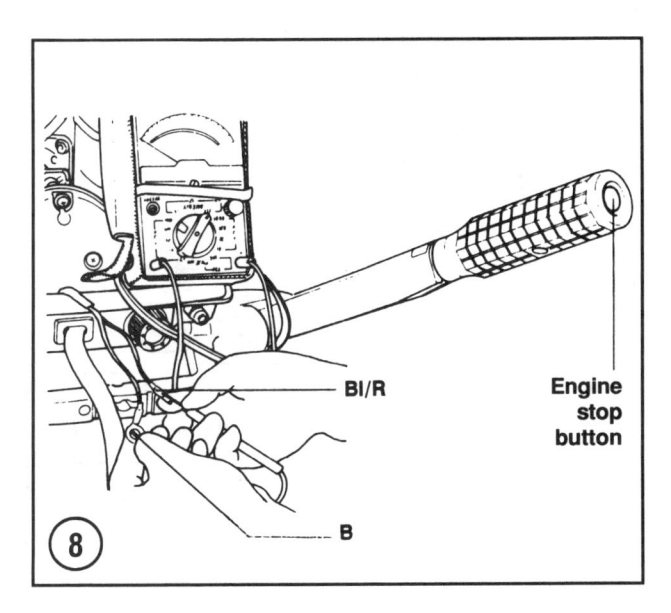

⑦ IGNITION SWITCH CONTACT CHART

	B	G	W	B/W	Br	O
CHOKE				●	●──●	
START			●─┬─●	●		
ON			●─●			
OFF	●─●					

B: Black W: White
Br: Brown B/W: Black/White
G: Green O: Orange

⑧ Bl/R — Engine stop button — B

LIGHTING SYSTEM

The AC lighting system consists of a lighting coil mounted on the magneto base assembly and permanent magnets located within the flywheel rim. The system provides alternating current to operate accessories such as boat lights only when the engine is running. Poor or corroded electrical connections, defective or worn through wiring insulation or the use of too many accessories are major causes of AC lighting system problems.

The AC lighting system can be converted to a battery charging system by installing a rectifier (with fuse) to convert the AC lighting voltage to DC battery charging voltage.

To check the lighting coil resistance, see *Battery Charging Coil Resistance Test* in this chapter.

CHARGING SYSTEM

Description

The standard charging system consists of a lighting or battery charging coil mounted on the magneto base assembly, permanent magnets located within the flywheel rim, a rectifier to change alternating current (AC) to direct current (DC), the battery, a 20 or 25 amp fuse and connecting wiring. **Figure 9** is a schematic of a *typical* Suzuki charging system. The components within the different systems vary along with the wire colors on the different models and years.

A malfunction in the charging system generally causes the battery to remain undercharged. Since the battery charging coil is protected by its location underneath the flywheel, it is more likely that the battery, rectifier, fuse or connecting wiring will cause problems. The following conditions will cause rectifier damage:

 a. The battery leads reversed.
 b. Running the engine with the battery leads disconnected.

c. A broken wire or loose connection resulting in an open circuit.

Troubleshooting

Before performing a battery charging coil or rectifier test, visually check the following.

1. Make sure the battery cables are properly connected. If polarity is reversed, check for a damaged rectifier.

NOTE
A damaged rectifier will generally have a discolored or a burned appearance.

2. Carefully inspect all wiring between the magneto base and battery for worn or cracked insulation. Makes sure all electrical connectors are free of corrosion and are pushed together tightly. Replace any defective wiring and/or clean and tighten connections as required.
3. Check the battery condition. Clean the terminals and recharge as required as described in Chapter Seven.

Battery Charging Coil Output Test

CAUTION
The engine must be provided with an adequate supply of water while performing this procedure. Install a flushing device, place the engine in a test tank or perform the test with the boat in the water.

1. Make sure the battery is fully charged.
2. Remove the engine cover.
3. Connect a tachometer according to manufacturer's instructions.
4. Start the engine and warm it to normal operating temperature.
5. Connect an ammeter between the battery and the fuse (**Figure 10**, typical).
6. Gradually increase engine speed to approximately 5,000 rpm and note the ammeter reading.

If it is not at least 6 amps (standard system) or 14 amps (optional system), replace the battery charging coil.

Battery Charging Coil Resistance Test

1. Remove the engine cover.
2. Disconnect the battery charging coil leads at their bullet connectors.
3. Connect an ohmmeter between the 2 disconnected leads.
4. Compare the reading to the specifications given in **Table 4**. If not within specifications, replace the battery charging coil(s).

Rectifier Test

1. Remove the engine cover.
2. Disconnect the red, white, yellow and black rectifier leads.

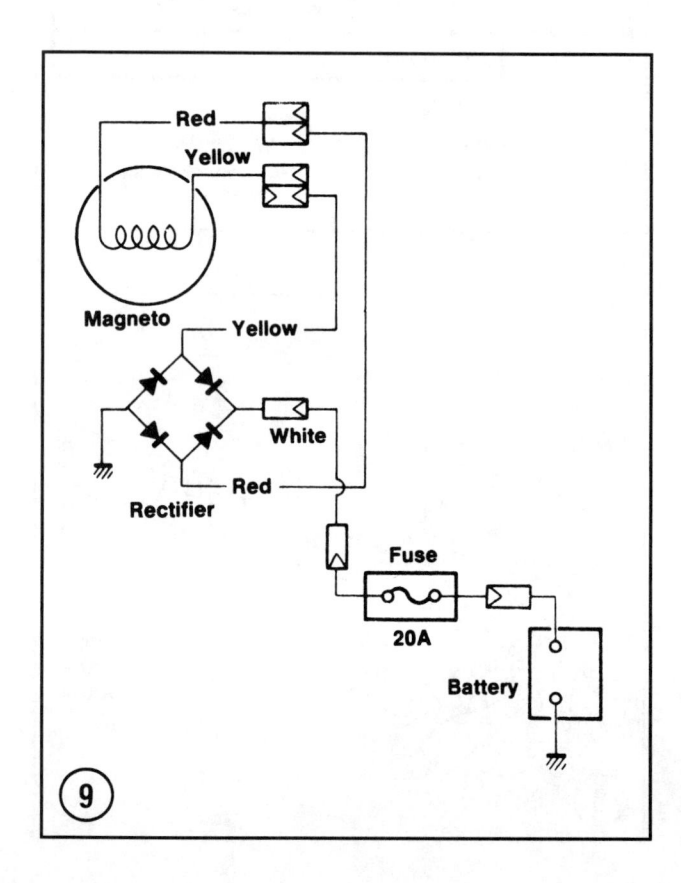

TROUBLESHOOTING

3. Connect a self-powered test lamp between the yellow and white leads, then reverse the test lamp connections. The lamp should light in one direction but not in the other.

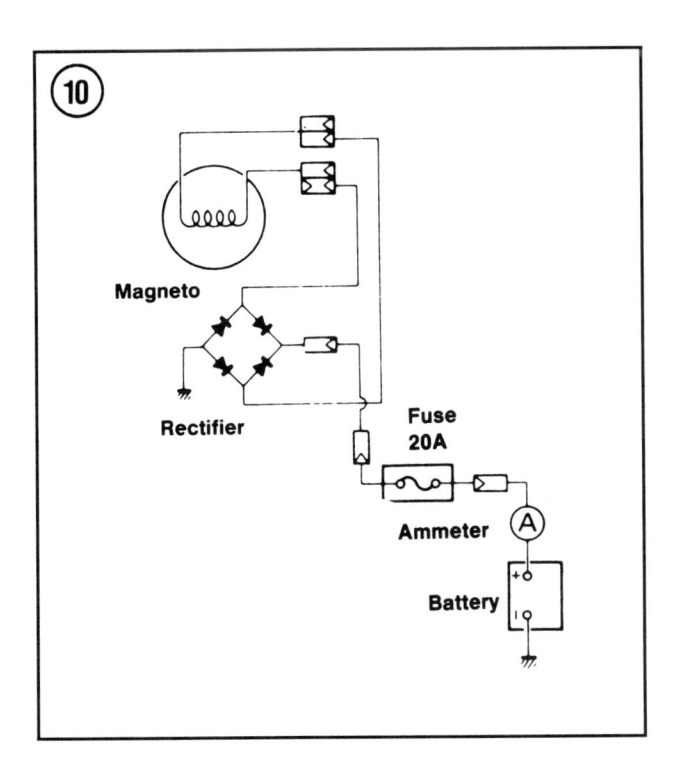

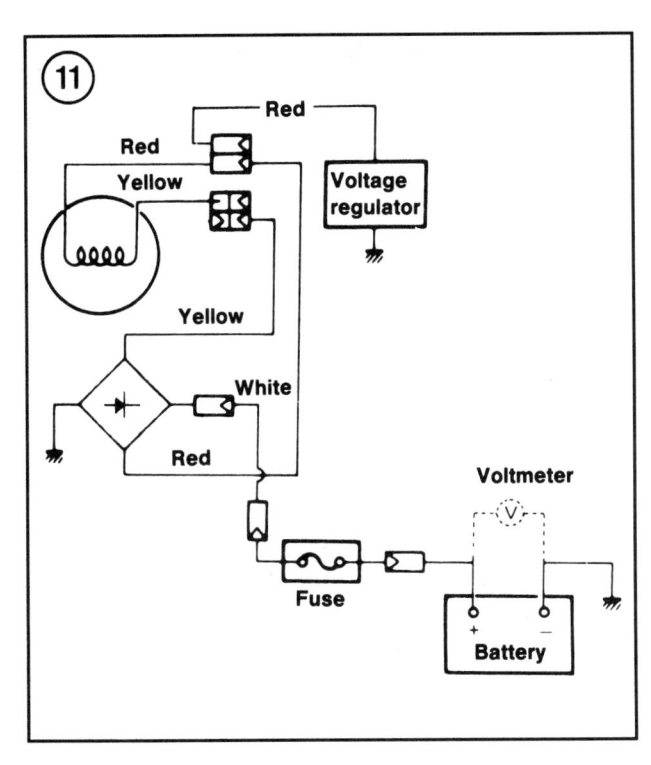

4. Repeat Step 3 to test the red and white, black and red and yellow and black leads.

5. If the lamp does not perform as described at any one of the diode connections, replace the rectifier.

Voltage Regulator Test

CAUTION
The engine must be provided with an adequate supply of water while performing this procedure. Install a flushing device, or place the engine in a test tank or perform the test with the boat in the water.

1. Make sure the battery is fully charged. Refer to Chapter Seven.
2. Remove the engine cover.
3. Connect a tachometer according to manufacturer's instructions.
4. Start the engine and warm it to normal operating temperature.
5. Connect a voltmeter across the battery terminals (**Figure 11**).
6. Gradually increase engine speed to approximately 5,000 rpm and note the voltmeter reading. If it is not 14-15 volts, replace the voltage regulator.

IGNITION SYSTEM

The wiring harness used between the ignition switch and engine is adequate to handle the electrical needs of the outboard motor. It will *not* handle the electrical needs of accessories. Whenever an accessory is added, install *new wiring* between the battery and the new accessory. Install a separate fuse panel to the instrument panel.

If the ignition switch requires replacement, *never* install an automotive-type switch. A marine-type switch must always be used.

Description

Variations of 5 different ignition systems have been used on Suzuki outboards since 1985. A full description of each system is covered in Chapter Seven. The ignition systems used on 1985-on Suzuki outboard motors are as follows:

a. A flywheel magneto breaker-point ignition.
b. Transistorized Ignition.
c. Pointless Electronic Ignition (PEI).
d. Integrated Circuit (IC) Ignition.
e. Micro Link Ignition.

General troubleshooting procedures are provided in **Table 2**.

Precautions

Several precautions should be strictly observed to avoid damage to the ignition system.

1. Do not reverse the battery connections. This reverses polarity and can damage the rectifier or CDI unit on the electronic ignition systems.
2. Do not "spark" the battery terminals with the battery cable connections to check polarity.
3. Do not disconnect the battery cables with the engine running.
4. Do not crank engine if the CDI unit is not grounded to engine.
5. Do not touch or disconnect any ignition components when the engine is running, while the ignition switch is ON or while the battery cables are connected.
6. Do not run an engine that has an electronic ignition system without the battery connected to the electrical harness.

Troubleshooting Preparation (All Ignition Systems)

NOTE
To test the wiring harness for poor solder connections in Step 1, bend the molded rubber connector while checking each wire for resistance.

1. Check the wiring harness and all plug-in electrical connectors to make sure that all terminals are free of corrosion, all connectors are tight and the wiring insulation is in good condition.
2. Check all electrical components that are grounded to the engine for a good ground. Disconnect, thoroughly clean and reconnect ground leads.
3. Make sure that all ground wires are properly connected and that the connections are clean and tight. Disconnect ground wires, thoroughly clean then reconnect.
4. Check all remaining ignition circuitry for disconnected wires and short or open circuits.
5. Make sure there is an adequate supply of fresh and properly mixed fuel available to the engine.
6. Check the battery condition (if so equipped). The battery must be fully charged. If necessary, clean the terminals and recharge battery, as described in Chapter Seven.
7. Check the spark plug cable routing. Make sure the cables are properly connected to their respective spark plugs.
8. Remove all spark plugs, keeping them in order. Check the condition of each plug as described in Chapter Four.

WARNING
If it is necessary to hold the spark plug cable during the next step, do so with an insulated pair of pliers. The high voltage generated by the ignition system can produce serious or fatal shocks.

9. Reconnect the proper plug cable to one plug. Lay the plug against the cylinder head so its base makes a good connection, then crank the engine. If weak or no spark is noted at the spark plug, check for loose connections at the ignition coil and battery. Repeat the check with each remaining plug. If all external wiring connections are good, the problem is most likely in the ignition system.

TROUBLESHOOTING

BREAKER-POINT IGNITION COMPONENT TESTING (1985-1989 DT 2)

The manufacturer recommends using Suzuki Pocket Tester (part 09900-25002) or a suitable equivalent ohmmeter to test breaker-point ignition components.

Ohmmeter testing should be performed with the engine cold. Resistance increases approximately 10 ohms per each degree of temperature increase. Ohmmeter tests performed on hot components may result in unnecessary parts replacement without solving the basic problem.

To perform ignition coil power and leakage tests, Suzuki Electro-tester (part 09900-28106) or a Merc-O-Tronic ignition analyzer should be used. Each analyzer includes detailed instructions for component testing as well as complete component specifications according to engine model and year of manufacture. Refer to the instructions provided with the particular analyzer to be used for the proper testing procedure.

Condenser Test

1. Remove the flywheel. See Chapter Eight.
2. Disconnect the condenser lead from the breaker points.
3. Connect one analyzer test lead to the condenser lead. Connect the remaining test lead to the stator base.

WARNING
High voltage is involved in a condenser leakage test. Handle the analyzer leads carefully and turn the analyzer switch to DISCHARGE before disconnecting the test leads from the condenser.

4. Set the analyzer controls according to the manufacturer's instructions. Check the condenser for leakage, resistance and capacity.
5. Compare the results in Step 4 with the specifications provided with the ignition analyzer. Replace the condenser if it fails any of the 3 tests.

Breaker Point Test

1. Remove the flywheel. See Chapter Eight.
2. Disconnect the breaker point leads from the stator base.
3. Connect one analyzer test lead to the breaker arm. Connect the other test lead to the breaker point screw terminal.
4. Set the analyzer controls according to the manufacturer's instructions.
5. If the breaker points are in acceptable condition, the analyzer needle will rest in the OK scale.
6. If the analyzer needle does not fall within the OK scale, clean the contacts with alcohol and recheck the analyzer leads to make sure that the connections are tight before discarding the breaker points. The low current present in this test makes clean points and proper connections very important.

Ignition Coil Resistance Test (1985-1989 DT 2)

This test checks the primary and secondary ignition coils for circuit continuity.

1. Remove the flywheel. See Chapter Eight.
2. For coil primary resistance, proceed as follows:
   a. Disconnect the black primary ignition coil lead at the bullet connector.
   b. Disconnect the secondary coil lead (spark plug wire) at the spark plug.
   c. Calibrate the ohmmeter on the low-ohm scale. Connect the meter between the primary coil lead and a good engine ground.
   d. Compare the reading to specifications (**Table 5**).
3. For coil secondary resistance, proceed as follows:
   a. Calibrate the ohmmeter on the high-ohms scale.
   b. Connect the meter between the secondary coil lead and a good engine ground.

c. Compare the reading to specifications (**Table 5**).

4. Replace the ignition coil if primary or secondary resistance is not as specified (**Table 5**). See Chapter Seven.

Ignition Coil Power and Leakage Tests (1985-1989 DT 2)

WARNING
Ignition coil power and leakage tests should be performed on a wooden or insulated bench top to prevent shock hazards.

The ignition coil must be removed from the power head before testing. See Chapter Seven.

1. Connect the ignition analyzer according to the manufacturer's instructions.
2. Check the coil for power and leakage according to the analyzer manufacturer's instructions. Compare the results to the specifications provided with the analyzer. Replace the ignition coil if it fails either of the tests. See Chapter Seven.

TRANSISTORIZED IGNITION SYSTEM (1990-ON DT 2)

DT 2 models (1990-on) are equipped with a transistorized ignition system. A full description is covered in Chapter Seven.

The manufacturer recommends using Suzuki Pocket Tester (part 09900-25002) or a suitable equivalent ohmmeter to test the ignition system.

Resistance tests should be performed with the engine cold. Resistance increases approximately 10 ohms per each degree of temperature increase. Resistance readings taken on hot components will indicate increased resistance and may result in unnecessary parts replacement without solving the basic problem.

Ignition Coil/Igniter Assembly Resistance Test (1990-on DT 2)

The ignition coil/igniter assembly must be disconnected before testing. See Chapter Seven.

1. Remove the flywheel. See Chapter Eight.
2. For primary winding resistance, proceed as follows:
   a. Calibrate the ohmmeter on the low-ohm scale.
   b. Connect the meter between the blue/red coil terminal and ground as shown in **Figure 12**.
   c. Compare the reading to specifications (**Table 5**).

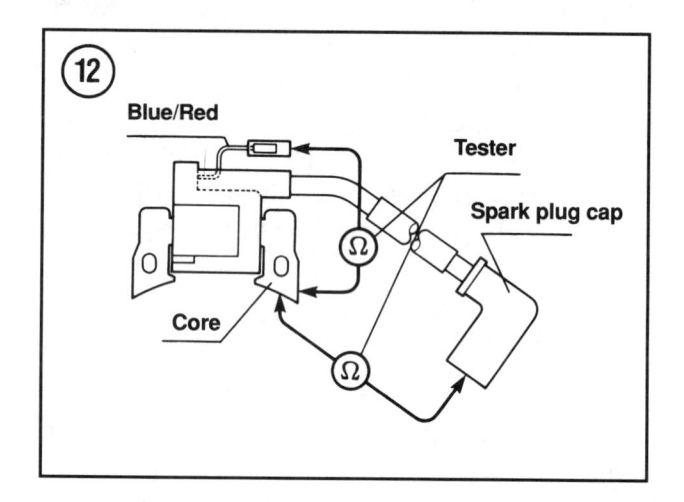

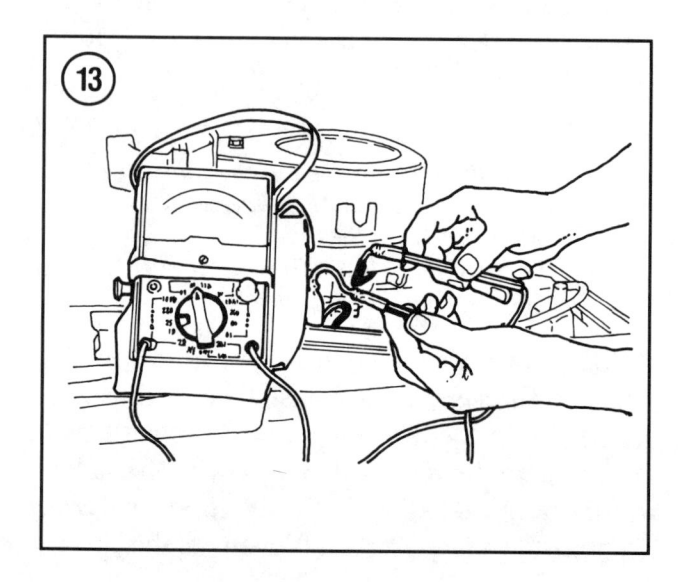

TROUBLESHOOTING

3. For secondary winding resistance, proceed as follows:
   a. Calibrate the ohmmeter on the high-ohms scale.
   b. Connect the meter between the spark plug terminal and ground as shown in **Figure 12**.
   c. Compare the reading to specifications (**Table 5**).
4. Replace the coil/igniter assembly if primary or secondary resistance is not as specified.

POINTLESS ELECTRONIC IGNITION (PEI) COMPONENT TESTING

Resistance Tests

An ohmmeter, although useful, is not always a good indicator of ignition system condition. This is primarily because resistance tests do not simulate actual operating conditions. For example, the power source in most ohmmeters is only 6-9 volts. A CDI charge coil, however, commonly produces 100-300 volts during normal operation. Such high voltage can cause coil insulation leakage that can not be detected with an ohmmeter.

Resistance tests should be performed with the engine cold. Resistance increases approximately 10 ohms per each degree of temperature increase. Ohmmeter tests on hot components will indicate increased resistance and may result in unnecessary parts replacement without solving the basic problem.

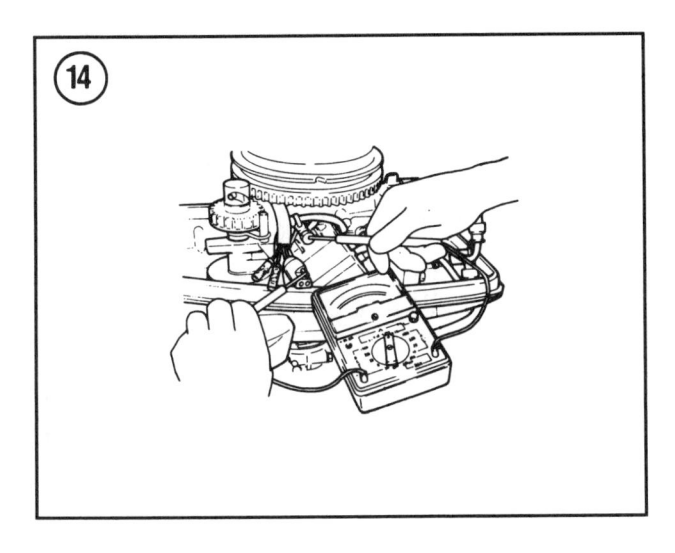

The manufacturer recommends using Suzuki Pocket Tester (part No. 09900-25002) or an equivalent ohmmeter to perform resistance tests.

Charge Coil Resistance Test

DT 4; 1987-on DT 6, DT 8

The condenser charge coil is referred to as the primary coil on some early models.
1. Remove the engine cover.
2. Disconnect the black/red and black charge coil wires at the bullet connectors located between the stator plate and the CDI unit.
3. Connect the tester between the black/red and black wires (**Figure 13**).
4. Compare charge coil resistance to the specification in **Table 6**. Replace charge coil if resistance is not within the specification. See Chapter Seven.

1985-1986 DT 6, DT 8

1. Remove the engine cover.
2. Disconnect the 3-pin connector located between the stator assembly and the CDI unit.
3. Connect the tester between the black/red terminal and a good engine ground (**Figure 14**).
4. Compare the charge coil resistance to the specification in **Table 6**. Replace the charge coil if the resistance is not as specified. See Chapter Seven.

1985-1987 DT 9.9; 1985-1988 DT 15

1. Remove the engine cover.
2. Disconnect the black wire and green wire 2-pin connector between the stator plate and CDI unit.

3. Connect the tester between the black and green wire terminals of the 2-pin connector (**Figure 15**). Compare charge coil resistance with the specification in **Table 6**. Replace the charge coil if resistance is not as specified. See Chapter Seven.

1989-on DT 15

1. Remove the engine cover.
2. Disconnect the negative battery cable on models so equipped.
3. Disconnect the 4-pin connector between the stator plate and the CDI unit.
4. Connect the tester between the green and black/red wire terminals (**Figure 16**). Compare the charge coil resistance with the specification in **Table 6**. Replace the charge coil if the resistance is not as specified. See Chapter Seven.

1988-on DT 8, DT 9.9

1. Remove the engine cover.
2. Disconnect the negative battery cable on models so equipped.
3. Disconnect the black/red and green charge coil wires at the bullet connectors between the stator plate and the CDI unit.
4. Connect the tester between the black/red and green wires (**Figure 17**). Compare charge coil resistance with the specification in **Table 6**. Replace the charge if resistance is not as specified. See Chapter Seven.

DT 20, DT 25, DT 30 (2-cylinder)

1. Remove the engine cover.
2. Disconnect the negative battery cable on models so equipped.

NOTE
The charge coil wires on 2-cylinder models DT 20, DT 25 and DT 30 are

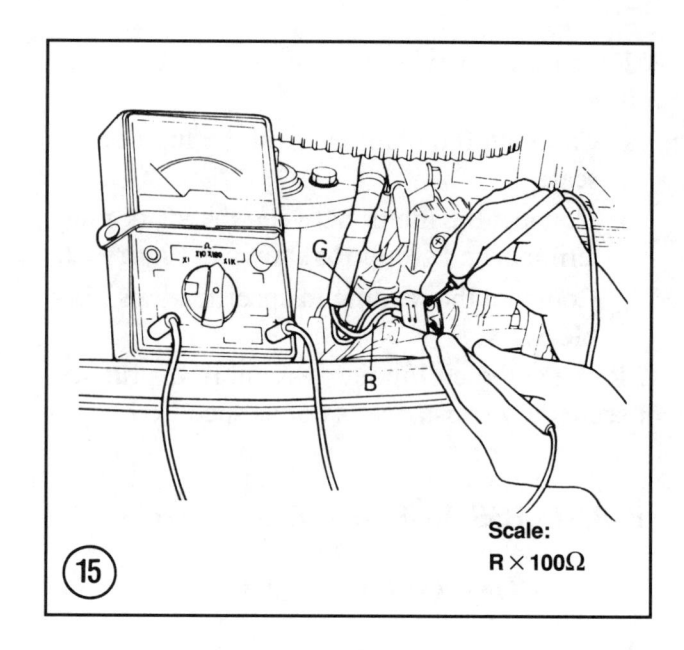

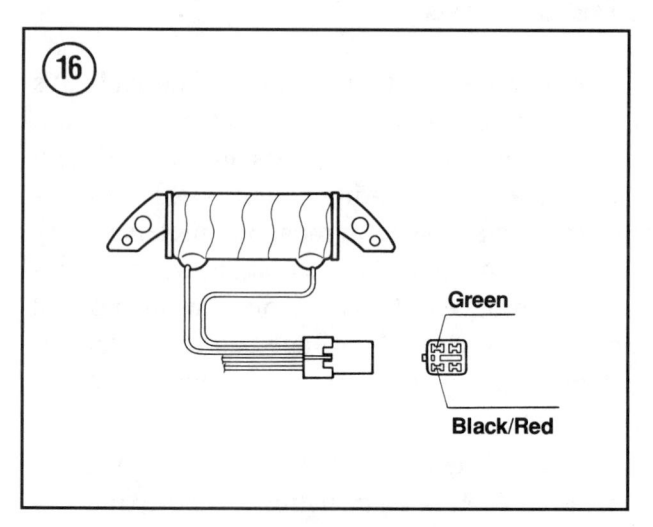

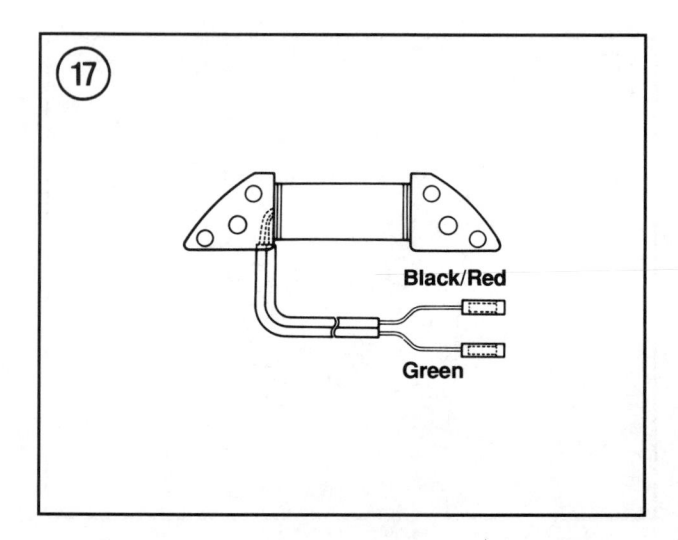

TROUBLESHOOTING

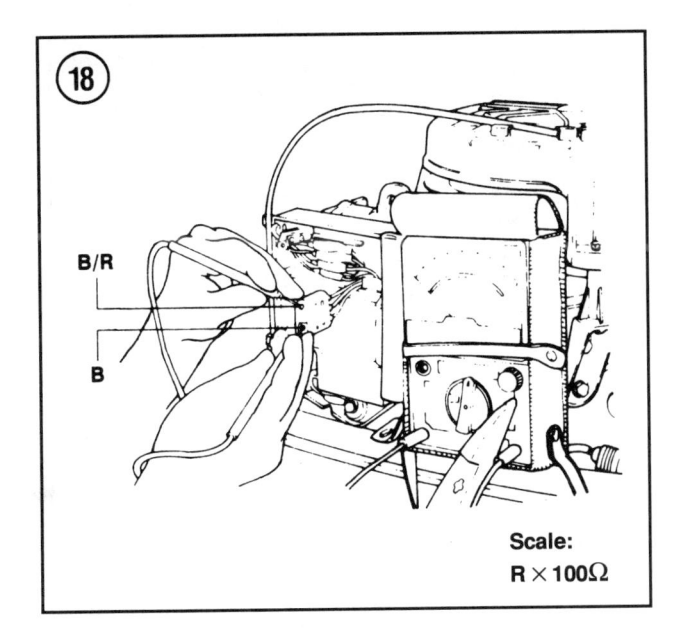

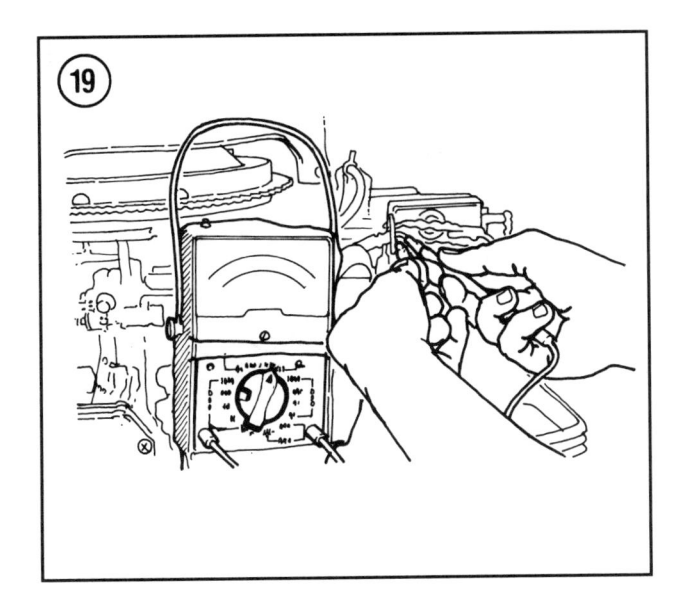

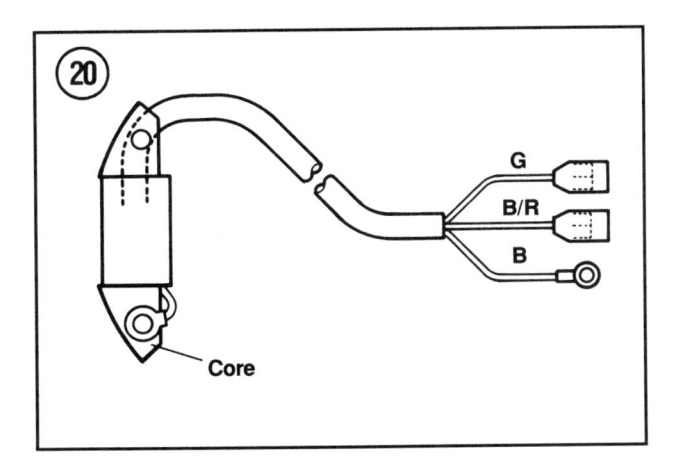

located inside the junction box. See **Figure 18**.

3A. 1985-1986—Disconnect the 3-pin connector between the stator plate and the CDI unit.

3B. 1987—Disconnect the black/red and black wires at the bullet connectors between the stator plate and CDI unit.

4. Connect the tester between the black/red and black wire terminals (**Figure 18**). Compare the resistance reading to the specification in **Table 6**. Replace the charge coil if the resistance is not as specified. See Chapter Seven.

DT 35, DT 40

1. Remove the engine cover.
2. Disconnect the negative battery cable on models so equipped.
3. Disconnect the green and black charge coil wires at the bullet connectors located inside the CDI unit mounting bracket.
4. Connect the tester between the green and black wires (**Figure 19**). Compare the resistance reading to the specifications in **Table 6**. Replace the charge coil if resistance is not as specified. See Chapter Seven.

1985 DT 55, DT 65

1. Remove the engine cover.
2. Disconnect the negative battery cable.
3. Disconnect the green and black/red charge coil wires between the stator plate and the CDI unit.
4. Connect the tester between the green and the black/red charge coil wires (**Figure 20**). Compare the resistance reading to the specifications in **Table 6**. Replace the charge coil if resistance is not as specified. See Chapter Seven.
5. Connect the tester between the green charge coil wire and the charge coil core (**Figure 20**). No continuity should be indicated. If continuity is noted between the green wire and the core, the

charge coil is shorted to ground and must be replaced.

1985-1987 DT 75, DT 85

1. Remove the engine cover.
2. Disconnect the negative battery cable.
3. Disconnect the 3-pin charge coil connector between the stator plate and the CDI unit.
4. To test low-speed charge coil resistance, connect the tester between the black/red and red/white wire terminals in the 3-pin connector (**Figure 21**).
5. To test high-speed charge coil resistance, connect the tester between the red/white and black wire terminals in the 3-pin connector (**Figure 21**).
6. Compare the resistance readings with the specifications in **Table 6**. Replace the charge coil if low-speed or high-speed resistance is not as specified. See Chapter Seven.

1985 DT 115, DT 140

1. Remove the engine cover.
2. Disconnect the negative battery cable.

NOTE
The charge coil wires are located inside the connector junction box on 1985 DT 115 and DT 140 models.

3. Disconnect the 3-pin charge coil connector and the brown wire bullet connector located inside the junction box.
4. To test the low-speed charge coil resistance, connect the tester between the brown/red and the white/black wire terminals in the 3-pin connector (**Figure 22**). Compare the low-speed charge coil resistance to the specification in **Table 6**.
5. To test the high-speed charge coil resistance, connect the tester between the red/white wire terminal in the 3-pin connector and the brown wire terminal in the single bullet connector (**Fig-

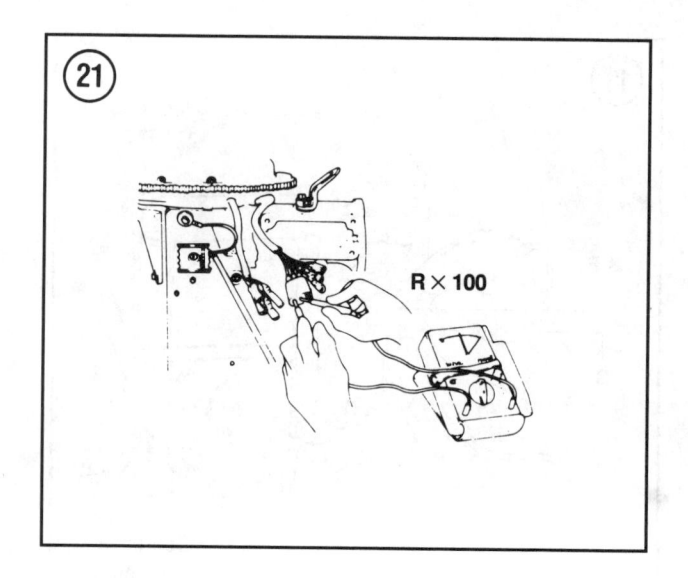

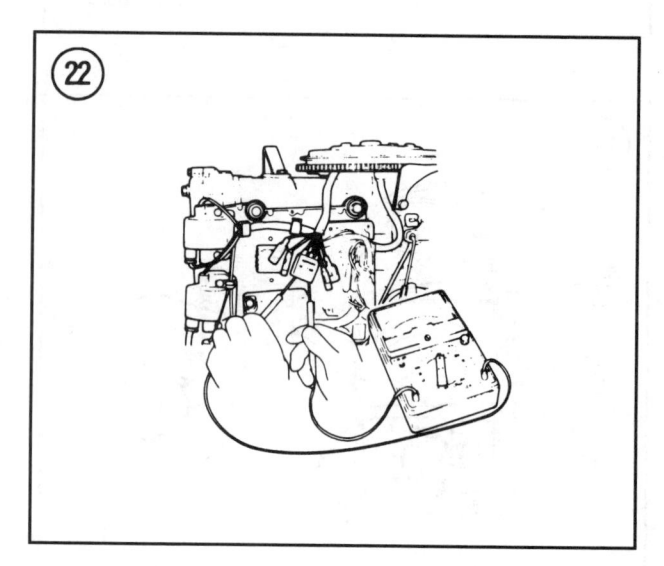

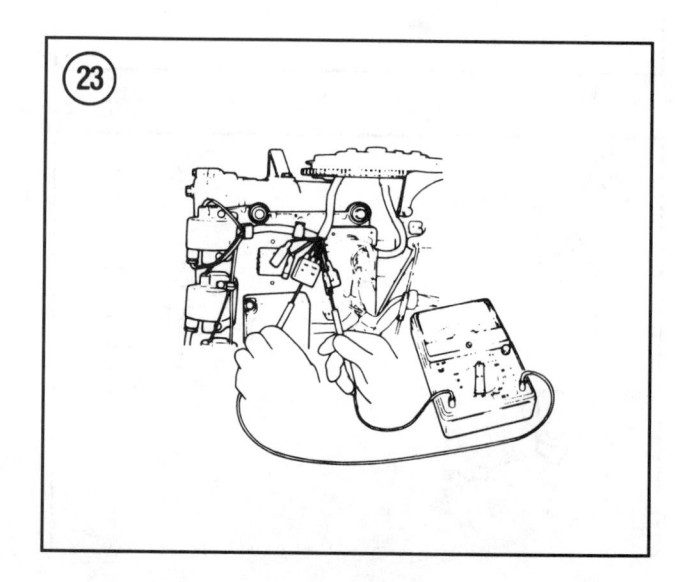

TROUBLESHOOTING

(24)

Scale: R × 10

(25)

(26)

White/Green

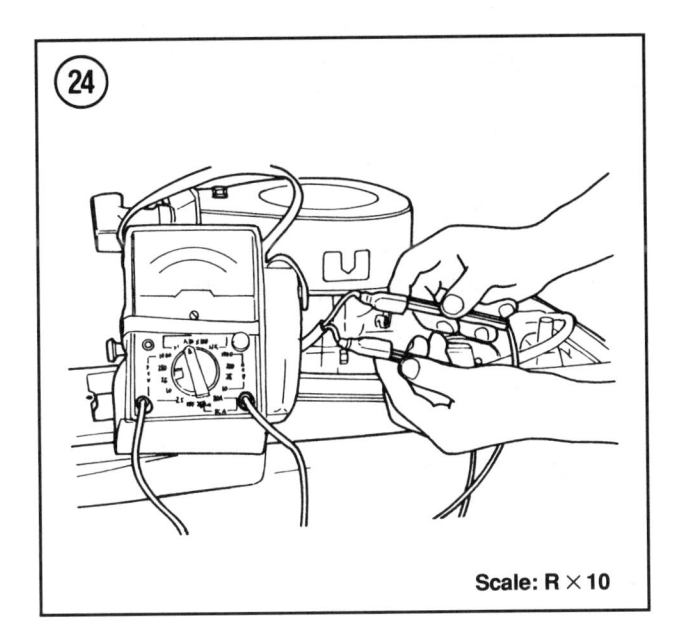

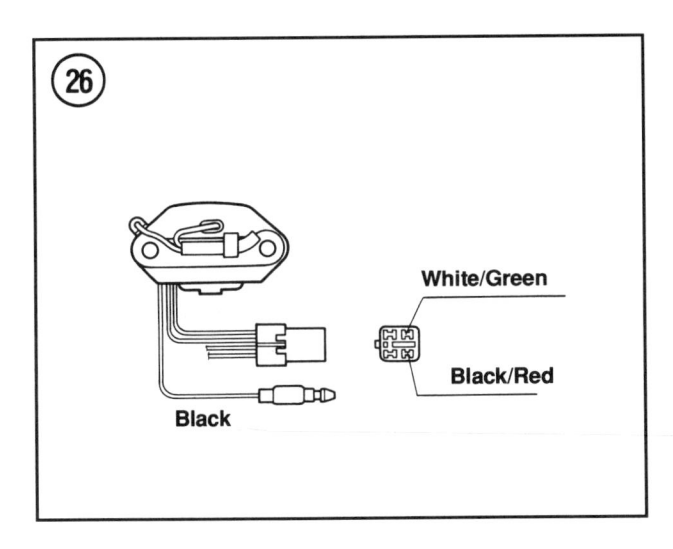

Black/Red

Black

ure 23). Compare the high-speed charge coil resistance to the specification in **Table 6**.

6. Replace the charge coil if the resistance is not as specified.

Pulser Coil Resistance Test

DT 4; 1987-on DT 6, DT 8

1. Remove the engine cover.
2. Disconnect the white/red and black wires at the connectors between the stator plate and CDI unit.
3. Connect the tester between the white/red and black wires (**Figure 24**). Compare the resistance reading to the specifications in **Table 6**. Replace the pulser coil if the resistance is not as specified. See Chapter Seven.

1985-1987 DT 6, DT 8

1. Remove the engine cover.
2. Disconnect the 3-pin connector between the stator plate and the CDI unit.
3. Connect the tester between the red/white wire terminal of the 3-pin connector and a good engine ground (**Figure 25**). Compare the resistance reading to the specifications in **Table 6**. Replace the pulser coil if the resistance is not as specified. See Chapter Seven.

1989-on DT 15

Two pulser coils are located under the flywheel 180° apart.

1. Remove the engine cover.
2. Disconnect the negative battery cable on models so equipped.
3. Disconnect the 4-pin connector and the black ground wire (**Figure 26**).
4. To test the No. 1 pulser coil, connect the tester between the black/red wire terminal in the 4-pin connector and the black ground wire.

5. To test the No. 2 pulser coil, connect the tester between the white/green wire terminal in the 4-pin connector and the black ground wire.
6. Compare the resistance reading to the specifications in **Table 6**. Replace pulser coil(s) if the resistance is not as specified. See Chapter Seven.

1988-on DT 8, DT 9.9

Two pulser coils are located 180° apart on the stator plate.
1. Remove the engine cover.
2. Disconnect the negative battery cable on models so equipped.
3. Disconnect the white/green wire (No. 1 pulser coil) or the red/black wire (No. 2 pulser coil) at the bullet connector.
4. Connect the tester between the white/green (No. 1 pulser coil) wire or the red/black (No. 2 pulser coil) wire and a good engine ground. Refer to **Figure 27**. Compare the resistance reading to the specifications in **Table 6**. Replace the pulser coil(s) if the resistance is not as specified. See Chapter Seven.

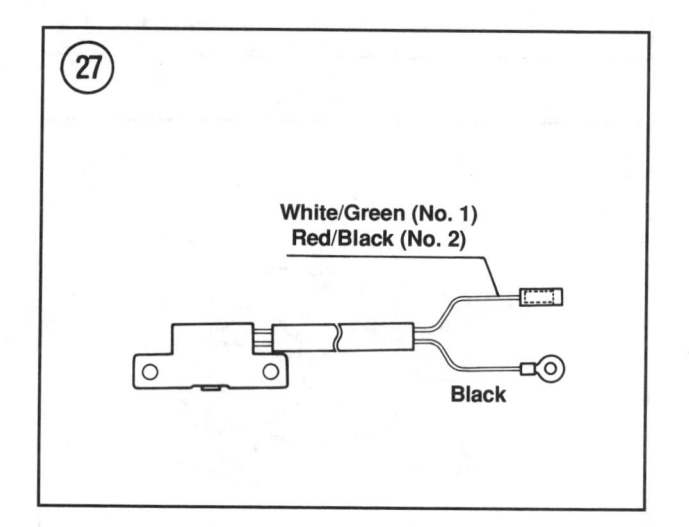

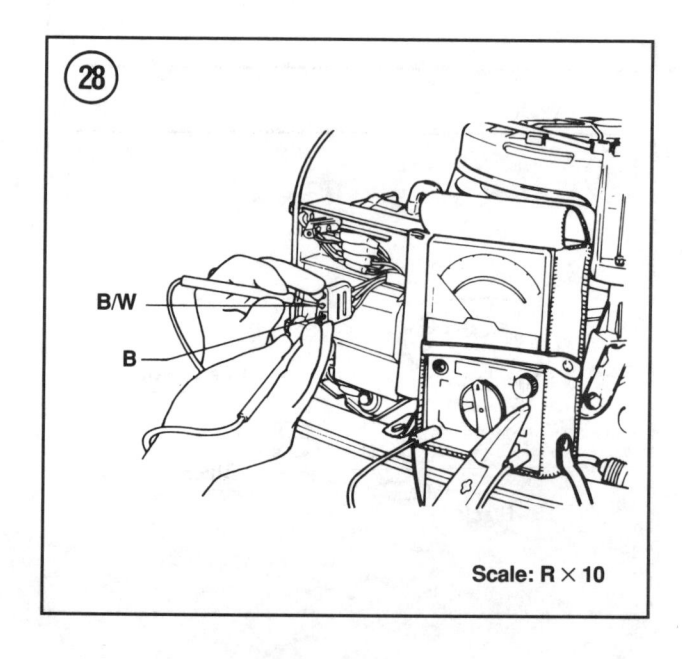

DT 20, DT 25, DT 30 (2-cylinder)

One pulser coil is located on the stator assembly.
1. Remove the engine cover.
2. Disconnect the negative battery cable on models so equipped.

> **NOTE**
> *The pulser coil wire connectors are located inside the connector junction box. See **Figure 28**.*

3. Disconnect the 3-pin connector (**Figure 28**).
4. Connect the tester between the black/white and the black wire terminal as shown in **Figure 28**. Compare the pulser coil resistance to the specification in **Table 6**. Replace the pulser coil if the resistance is not as specified. See Chapter Seven.

DT 35, DT 40

1. Remove the engine cover.
2. Disconnect the negative battery cable on models so equipped.
3. Disconnect the white/red and black pulser coil wires at the bullet connectors located inside the CDI unit mounting bracket.
4. Connect the tester between the white/red and black wires (**Figure 29**). Compare the resistance reading to the specifications in **Table 6**. Replace the pulser coil if the resistance is not as specified. See Chapter Seven.

TROUBLESHOOTING

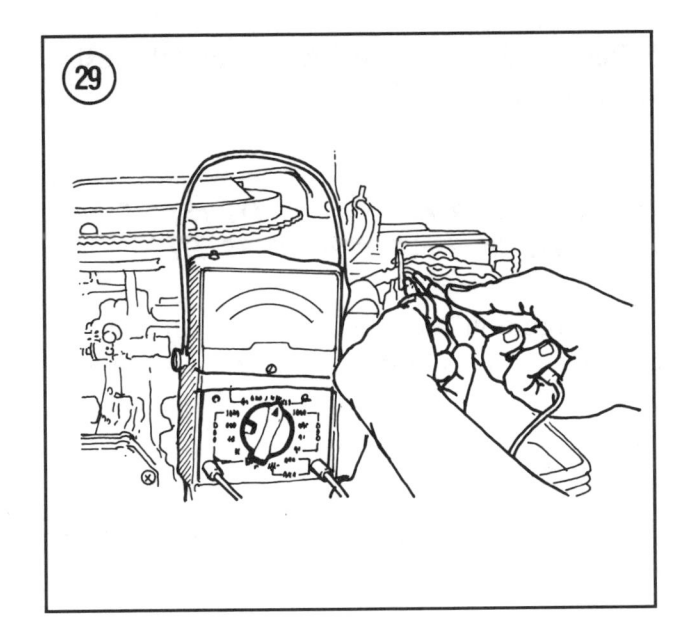

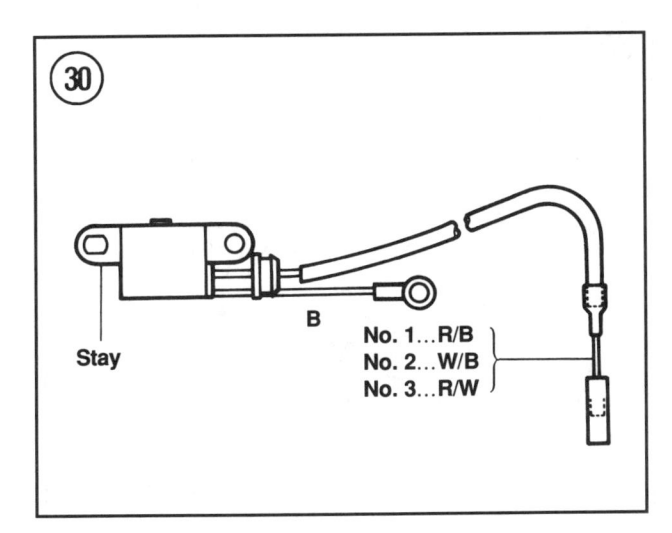

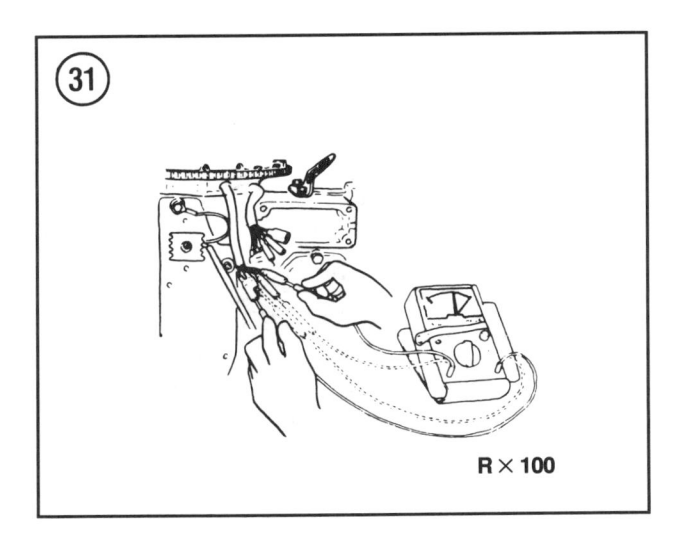

1985 DT 55, DT 65

Three pulser coils are mounted on the stator plate assembly.

1. Remove the engine cover.
2. Disconnect the negative battery cable.
3. Disconnect the pulser coils at the bullet connector.
4. Test the pulser coils as follows (**Figure 30**):
   a. No. 1—Connect the tester between the red/black wire and the black ground wire.
   b. No. 2—Connect the tester between the white/black wire and the black ground wire.
   c. No 3.—Connect the tester between the red/white wire and the black ground wire.
5. Compare the pulser coil(s) resistance to the specifications in **Table 6**.
6. Connect the tester between each pulser coil colored wire and the metal mounting stay (**Figure 30**). No continuity should be noted.
7. Replace the pulser coil(s) if not as specified. See Chapter Seven.

1985-1987 DT 75, DT 85

1. Remove the engine cover.
2. Disconnect the negative battery cable.
3. Disconnect the pulser coil wires located inside the connector junction box.
4. Connect the tester between the green and yellow/red wires, then between the pink and white/red wires (**Figure 31**). Compare the resistance reading to the specifications in **Table 6**. Replace the pulser coil(s) if the resistance is not as specified.

1985 DT 115, DT 140

Two pulser coil are mounted on the timer base assembly. One coil operates cylinders 1 and 2, and the remaining coil operates cylinders 3 and 4.

1. Remove the engine cover.
2. Disconnect the negative battery cable.

3. Disconnect the pulser coil wires at the bullet connectors located inside the connector junction box.

4. Connect the tester between the green and pink wires, then between the yellow/red and white/red wires (**Figure 31**). Compare the resistance readings to the specifications in **Table 6**. Replace the pulser coil(s) if resistance is not as specified. See Chapter Seven.

Ignition Coil Resistance Tests

DT 4

The ignition coil must be disconnected from the ignition system before testing. See Chapter Seven.

1. To check the primary winding resistance, proceed as follows:
   a. Connect the tester between the orange primary terminal and ground as shown in **Figure 32**.
   b. Compare the reading to the specification in **Table 5**.
2. To check the secondary winding resistance, connect the tester between the orange primary terminal and the spark plug terminal as shown in **Figure 33**. Compare the reading to the specification in **Table 5**.

DT 6; 1985-1987 DT 8

The ignition coil and CDI unit are a combined assembly on these models. Only ignition coil secondary windings can be tested using an ohmmeter.

1. Remove the engine cover.
2. Disconnect the spark plug wires from the spark plugs.
3. Connect the tester between the 2 spark plug terminals as shown in **Figure 34**.
4. Compare the resistance reading to the specification in **Table 5**.

5. Replace the ignition coil/CDI unit if the resistance is not as specified.

1985-1987 DT 9.9; 1985-1988 DT 15; DT 20, DT 25, DT 30 (2-cylinder); DT 35, DT 40

The ignition coils and the CDI unit are a combined unit assembly on these models. Only the ignition coil secondary windings can be tested using an ohmmeter.

1. Remove the engine cover.
2. Disconnect the spark plug leads from the spark plugs.

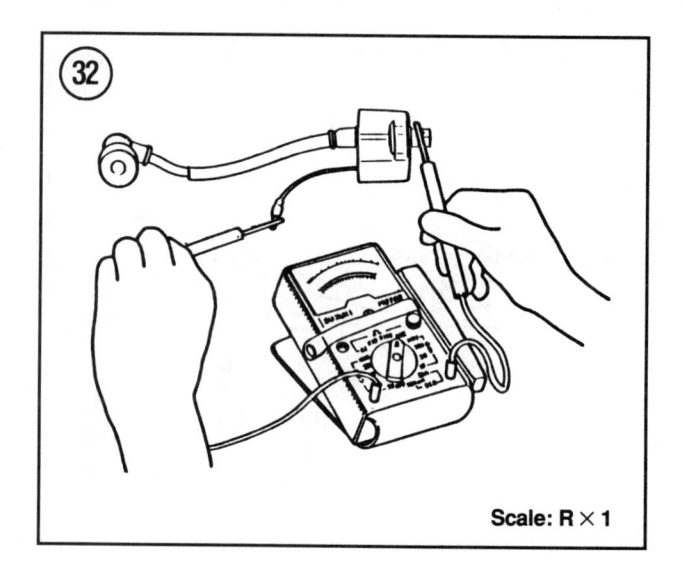

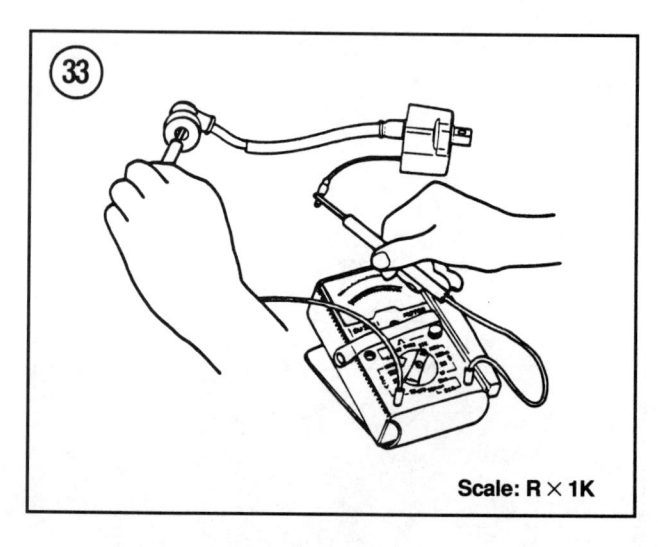

TROUBLESHOOTING

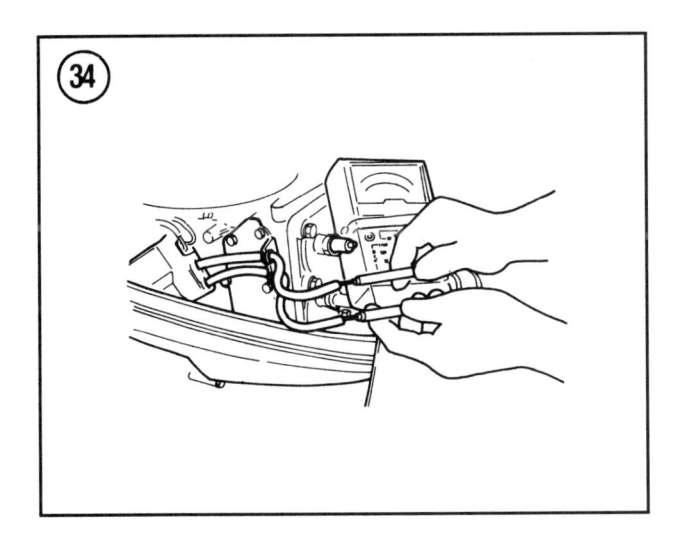

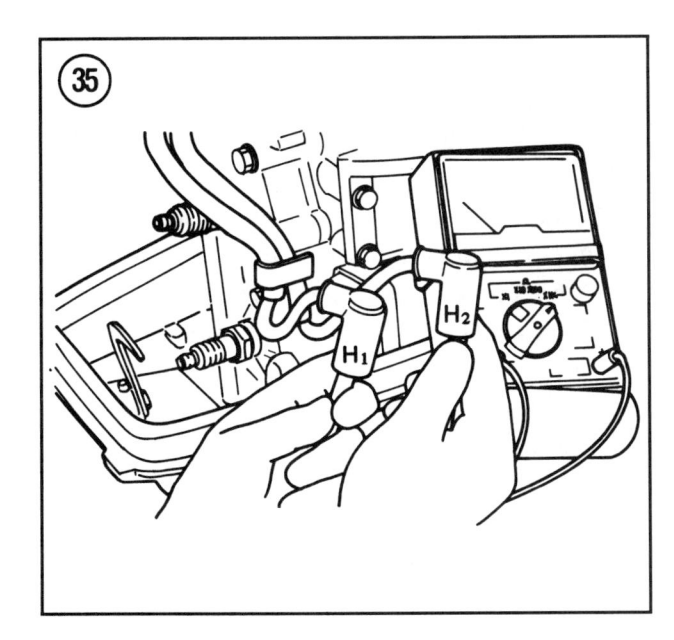

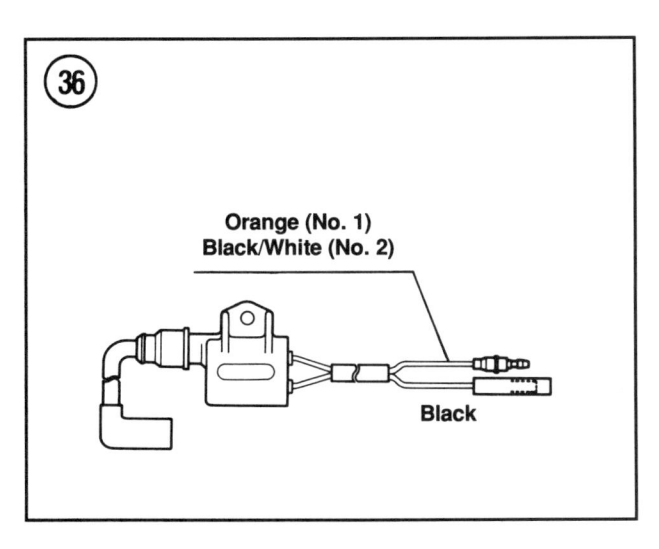

3. Connect the tester between the spark plug terminals as shown in **Figure 35**.

4. Compare the reading to the specifications in **Table 5**.

1989-on DT 15

The ignition coils must be disconnected from the ignition system before performing resistance tests. See Chapter Seven.

1. To test the primary winding resistance, connect the tester between the black ground wire and the orange primary wire on the No. 1 coil or the gray primary wire on the No. 2 coil. Compare the reading to the specification in **Table 5**.

2. To test secondary winding resistance, connect the tester between the spark plug terminal and the orange primary wire (No. 1 coil) or the gray primary wire (No. 2 coil).

3. Replace the ignition coil if primary or secondary resistance is not as specified. See Chapter Seven.

1988-on DT 8, DT 9.9

The ignition coils must be disconnected from the ignition system before testing. See Chapter Seven.

1. To test primary winding resistance, connect the tester between the orange primary wire (No. 1 coil) or the black/white primary wire (No. 2 coil) and the black ground wire. See **Figure 36**.

2. Compare the reading to the specifications in **Table 5**.

3. To test secondary winding resistance, connect the tester between the coil primary wire and the spark plug terminal.

4. Compare the reading to the specification in **Table 5**.

5. Replace the ignition coil if primary or secondary resistance is not as specified. See Chapter Seven.

1985 DT 55, DT 65

The ignition coils must be disconnected from the ignition system before testing. See Chapter Seven.

1. To test primary winding resistance, connect the tester between the black coil ground wire and the orange primary wire (No. 1 coil), blue primary wire (No. 2 coil) or the light green primary wire (No. 3 coil).
2. Compare the resistance reading to the specifications in **Table 5**.
3. To test secondary winding resistance, connect the tester between the coil primary wire and the spark plug terminal.
4. Compare the reading to the specifications in **Table 5**.
5. Replace the ignition coil(s) if primary or secondary resistance is not as specified. See Chapter Seven.

1985-1987 DT 75, DT 85; 1985 DT 115, DT 140

The ignition coils must be removed from the ignition system before testing. See Chapter Seven.

1. To test primary winding resistance, connect the tester between the primary wire and the coil as shown in **Figure 37**.
2. Compare the reading to the specifications in **Table 5**.
3. To test secondary winding resistance, connect the tester between the spark plug terminal and coil as shown in **Figure 38**.
4. Compare the reading to the specifications in **Table 5**.
5. Replace the ignition coil(s) if primary or secondary resistance is not as specified.

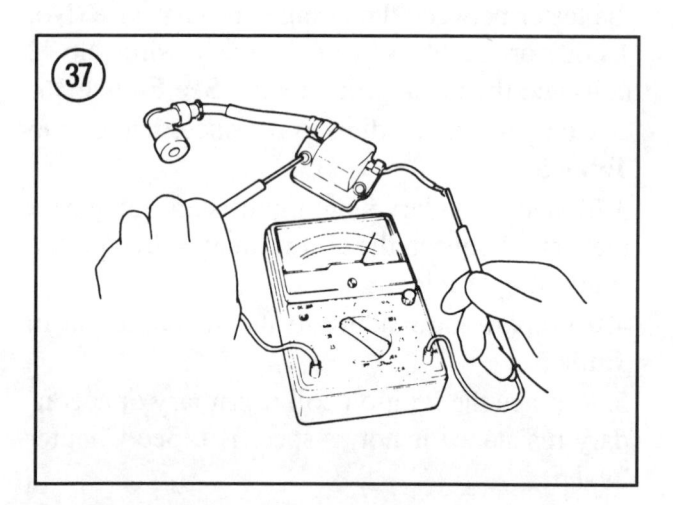

Throttle Switch Test (1985 DT 55, DT 65)

A throttle switch is used on 1985 DT 55 and DT 65 models to provide a signal to the CDI unit. This signal is used to help determine ignition timing advance. To test the switch, proceed as follows:

1. Disconnect the throttle switch wires at the bullet connectors. See **Figure 39**.
2. Connect an ohmmeter between the switch black wire (B) and the light green/red wire (Lg/R).
3. Depress the lever on the idle switch (1). No continuity should be noted.
4. Release the lever on the idle switch (1). Zero resistance should be noted.
5. Connect the ohmmeter between the brown/yellow wire (Br/Y) and the black wire (B).

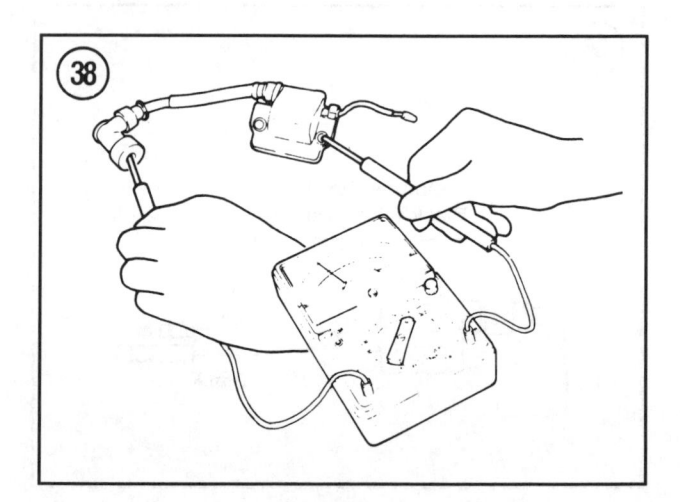

TROUBLESHOOTING

6. Depress the lever on the acceleration switch (2). No continuity should be noted.
7. Release the acceleration switch (2) lever. Zero resistance should be noted.
8. Replace the throttle switch if resistance is not as specified.

INTEGRATED CIRCUIT (IC) AND MICRO LINK IGNITION SYSTEMS

The manufacturer recommends using a Stevens Model CD-77 or a suitable peak reading voltmeter (PRV) to test the IC and Micro Link ignition systems. The Stevens Model CD-77 is available from Suzuki Motor Company (part No. 99954-53873) or from Stevens Instrument Company, P.O. Box 193, Waukegan, IL, 60079-9375.

The Model CD-77 is equipped with 2 test leads, one red and one black. The test leads have a probe type end that can be slipped under the connector sleeves or directly into an open connector. A plug-in clip connector is also provided with the tester for attaching the black test lead onto an engine ground. The tester has 3 voltage scales; 0-5 volts, 0-50 volts and 0-500 volts. The voltage range is selected by turning the knob on the right side of the meter. The selector on the left side of the meter is used to select the sensor polarity of the voltage to be measured.

During testing, if the meter needle swings hard against the right-hand side of the scale, *immediately* disconnect the test leads or discontinue the test to prevent damage to the tester. Recheck the meter switch settings or the connections.

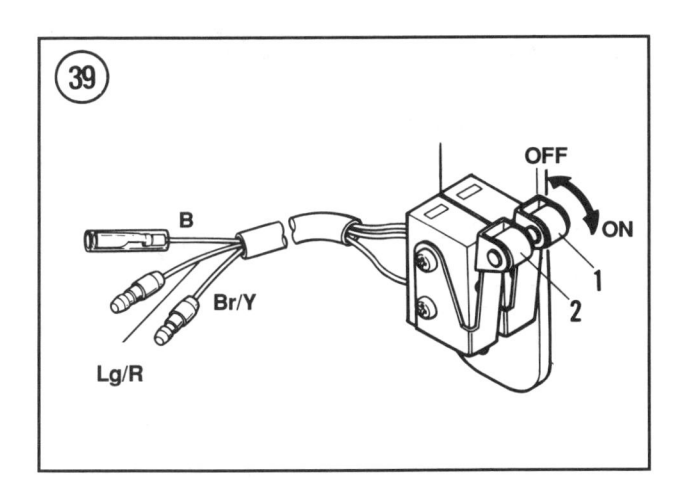

The manufacturer recommends repeating peak output tests 2-3 times to ensure consistent test results.

Slow cranking speed, caused by a weak battery, faulty starter motor or other starting system problem will result in invalid peak output readings. Make sure the battery is fully charged and the starting system is operating properly. Prior to performing peak output tests, remove all the spark plugs to ensure consistent cranking speeds.

To perform resistance tests, the manufacturer recommends using Suzuki Pocket Tester 09900-25002, or a suitable equivalent ohmmeter. An ohmmeter, although useful, is not always a good indicator of ignition system condition. This is primarily because resistance tests do not simulate actual operating conditions. For example, the power source in most ohmmeters is only 6-9 volts. A CDI charge coil, however, commonly produces 100-300 volts during normal operation. Such high voltage can cause coil insulation leakage that can not be detected with an ohmmeter.

Resistance tests should be performed with the engine cold. Resistance increases approximately 10 ohms per each degree of temperature increase. Ohmmeter tests on hot components will indicate increased resistance and may result in unnecessary parts replacement without solving the basic problem.

Peak Output Tests
1989-on DT 25; 1988-on DT 30 (3-cylinder)

Remove all the spark plugs before performing output tests. Crank the engine using the manual rewind starter during peak output tests. Pull the starter rope sharply and consistently to ensure accurate test results. Unless specified otherwise, leave all ignition components connected during voltage output tests.

CDI unit output

1. Set the meter polarity switch to POS and the voltage switch to 500.
2. Connect the black test lead to the CDI unit black ground wire or a good engine ground.
3. Connect the red test lead to the orange CDI unit output wire (Test 1, **Figure 40**).
4. Crank the engine using the rewind starter.
5. Repeat test with the red test lead connected to the blue CDI unit output wire, then the gray CDI unit output wire.
6. CDI unit output voltage should be 105 volts or more at each output wire. Replace the CDI unit if the voltage is not as specified. See Chapter Seven.

Charge coil output

Refer to Test 2, **Figure 40**.
1. Set the meter polarity switch to POS and the voltage switch to 500.
2. Connect the red test lead to the green charge coil wire and the black test lead to the black/red charge coil wire.
3. Crank the engine with the rewind starter while noting the meter.
4. Charge coil output should be 120 volts or more.
5. Replace the charge coil if the output voltage is not as specified. See Chapter Seven.

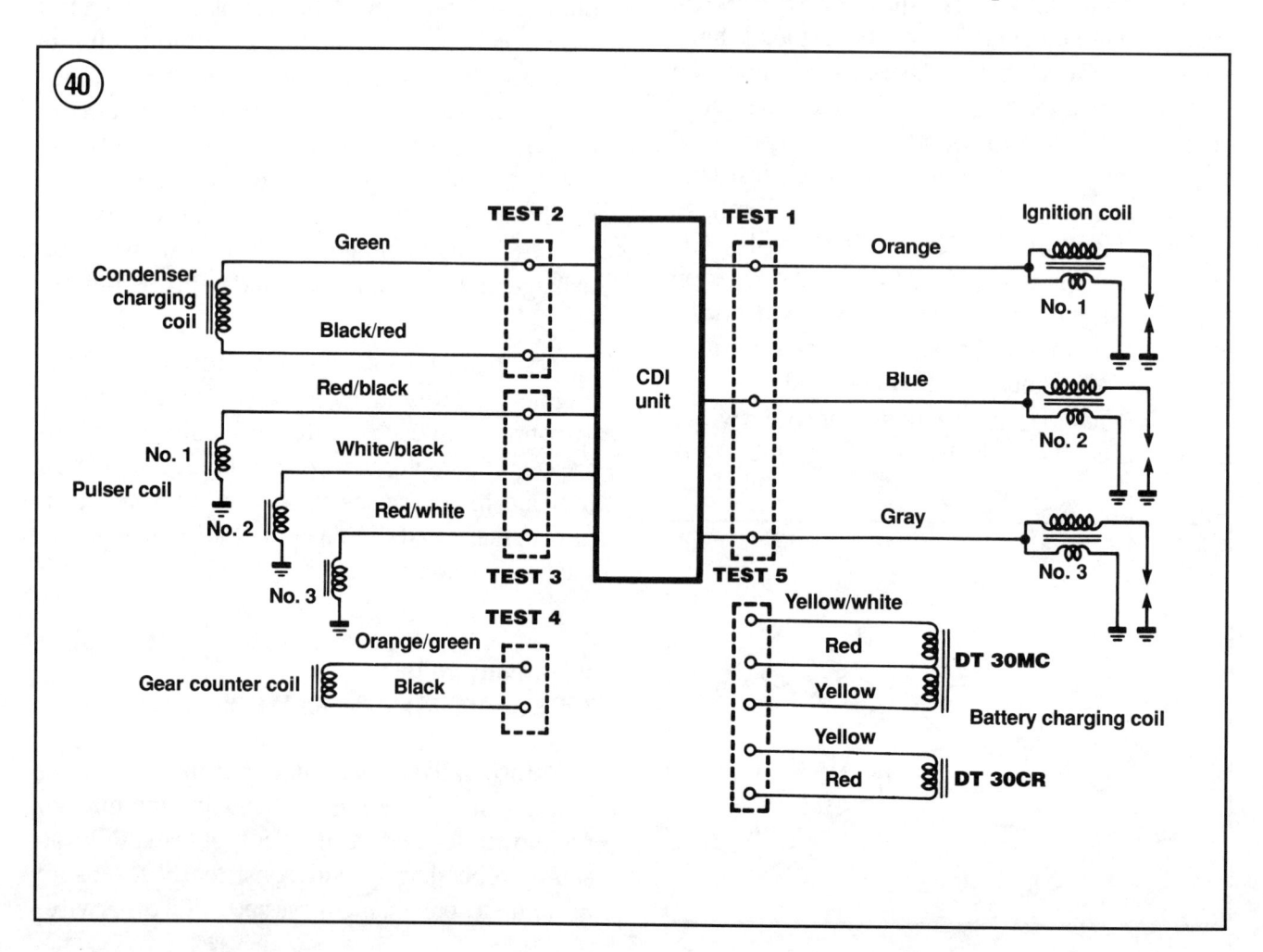

TROUBLESHOOTING

Pulser coil output

Refer to Test 3, **Figure 40**.
1. Set the meter polarity switch to SEN and the voltage switch to 50.
2. Connect the black test lead to a good engine ground and the red test lead to the No. 1 pulser coil wire (red/black).
3. Crank the engine with the rewind starter while noting the meter.
4. Repeat the test with the red test lead connected to the No. 2 pulser coil wire (white/black), then the No. 3 pulser coil wire (red/white).
5. Pulser coil output voltage should be 2.4 volts or more at each test connection. Replace pulser coil(s) if the output voltage is not as specified. See Chapter Seven.

Gear counter coil output

Refer to Test 4, **Figure 40**.
1. Set the meter polarity switch to SEN and the voltage switch to 50.
2. Connect the black test lead to black counter coil wire and the red test lead to the orange/green counter coil wire.
3. Crank the engine with the rewind starter while noting the meter.
4. Counter coil output should be 2.5 volts or more. Replace the counter coil if the voltage is not as specified. See Chapter Seven.

Peak Output Tests
1986-on DT 55, DT 65

Remove all the spark plugs before performing output tests. Crank the engine using the electric starter to take the voltage readings. Unless specified otherwise, leave all ignition components connected during voltage output testing.

> *CAUTION*
> *Do not crank the engine for more than 20 seconds when performing peak output tests. Allow the starter motor to cool for approximately 2-3 minutes between tests.*

CDI unit output

Refer to Test 1, **Figure 41**.
1. Set the meter polarity switch to POS and the voltage switch to 500.
2. Connect the black test lead to a good engine ground and the red test lead to the orange CDI unit output wire.
3. Crank the engine while noting the meter.
4. Repeat the test with the red test lead connected to the light blue CDI unit output wire, then the gray CDI output wire.
5. CDI unit output voltage at each test connection should be 120 volts or more on 1986-1988 models or 190 volts or more on 1989-on models. Replace the CDI unit if output voltage is not as specified. See Chapter Seven.

Charge coil output

Refer to Test 2, **Figure 41**.
1. Set the meter polarity switch to POS and the voltage switch to 500.
2A. 1986-1988:
   a. Connect the black test lead to a good engine ground.
   b. Connect the red test lead to the green charge coil wire.
   c. Crank the engine while noting the meter. The voltage output at the green wire should be 130 volts or more.
   d. Move the red test lead to the black/red charge coil wire.
   e. Set the meter voltage switch to 50.
   f. Crank the engine while noting the meter. Voltage output at the black/red wire should be 15 volts or more.
   g. Replace the charge coil if the voltage output is not as specified.
2B. 1989-on:

a. Set the meter polarity switch to POS and the voltage switch to 500.
b. Connect the red test lead to the green charge coil wire and the black test lead to a good engine ground.
c. Crank the engine while noting the meter.
d. Replace the charge coil if voltage output is not 190 volts or more.

Pulser coil output

Refer to Test 3, **Figure 41**.
1. Set the meter polarity switch to SEN and the voltage switch to 5.
2. Connect the black test lead to a good engine ground.
3. Connect the red test lead to the No. 1 pulser coil wire (red/black). Crank the engine while noting the meter.
4. Move the red test lead to the No. 2 pulser coil wire (white/black). Crank the engine while noting the meter.
5. Move the red test lead to the No. 3 pulser coil wire (red/white). Crank the engine while noting the meter.
6. Pulser coil output should be 1.7 volts or more on 1986-1988 models or 4 volts or more on 1989-on models. Replace any pulser coil(s) if the voltage output is not as specified. See Chapter Seven.

Counter coil output

The gear counter coil is used on 1989-on DT 55 and DT 65 models. Refer to *Gear Counter Coil Resistance Test* in this chapter to test.

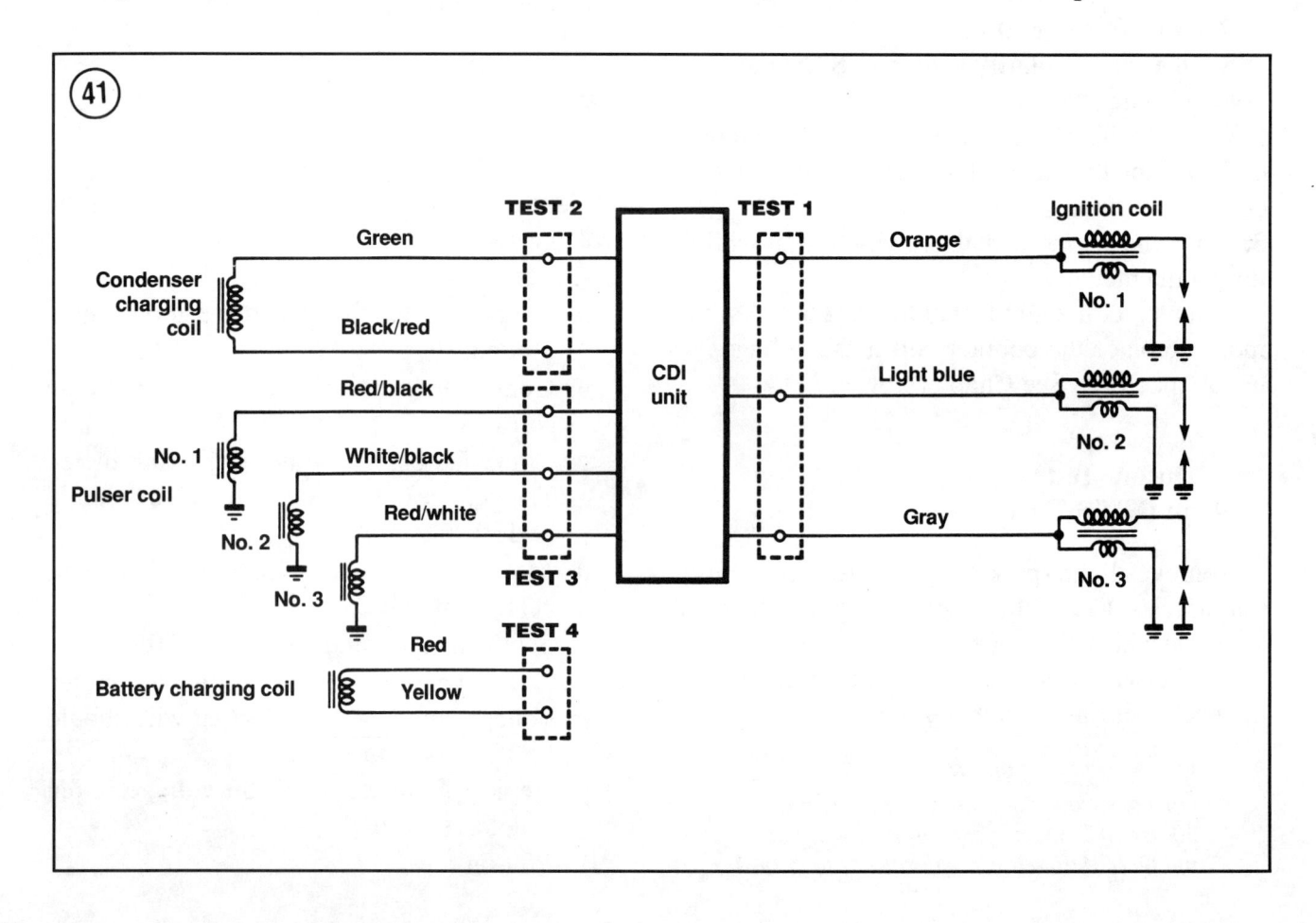

TROUBLESHOOTING

Battery charging coil output

Refer to Test 4, **Figure 41** for this procedure.

1. Set the meter polarity to POS or NEG and the voltage switch to 50.
2. Disconnect the rectifier from the battery charging coil output wires.
3. Connect the red test lead to the red charging coil wire and the black test lead to the yellow charging coil wire.
4. Crank the engine while noting the meter. Replace the battery charging coil if voltage output is not 4 volts or more on 1986-1988 models or 16 volts or more on 1989-on models.

Peak Output Tests
1988-on DT 75, DT 85

Remove all spark plugs before performing voltage output tests. Crank the engine using the electric starter to take the voltage readings. Unless specified otherwise, leave all ignition components connected during testing.

CAUTION
Do not crank the engine for more than 20 seconds when performing peak output tests. Allow the starter motor to cool for approximately 2-3 minutes between tests.

CDI unit output test

Refer to Test 1, **Figure 42** for this procedure.

1. Set the meter polarity to POS and the voltage switch to 500.
2. Connect the black test lead to a good engine ground.
3. Connect the red test lead to the orange CDI unit output wire.
4. Crank the engine while noting the meter.

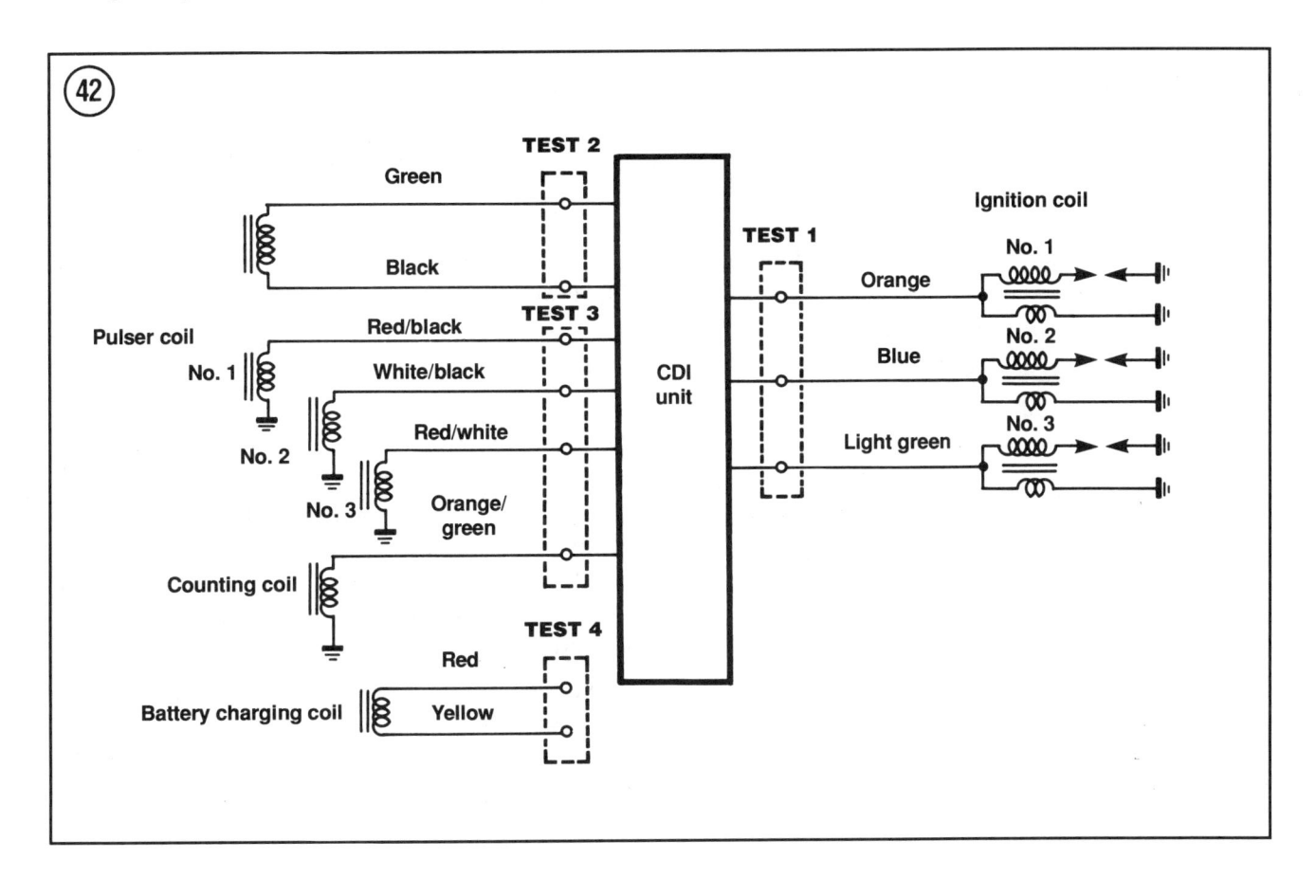

5. Move the red test lead to the blue CDI unit output wire. Repeat Step 4.
6. Move the red test lead to the light green CDI output wire. Repeat Step 4.
7. CDI unit output voltage should be 96 volts or more at each test connection.
8. Replace the CDI unit if the output voltage is not as specified. See Chapter Seven.

Charge coil output test

Refer to Test 2, **Figure 42** for this procedure.
1. Set the meter polarity switch to POS and the voltage switch to 500.
2. Connect the red test lead to the green charge coil wire and the black test lead to the black charge coil wire.
3. Crank the engine while noting the meter.
4. Replace the charge coil if the output voltage is not 107 volts or more. See Chapter Seven.

Pulser coil output

Refer to Test 3, **Figure 42** for this procedure.
1. Set the meter polarity switch to SEN and the voltage switch to 5.
2. Connect the black test lead to a good engine ground.
3. Connect the red test lead as follows:
   a. No. 1 pulser coil—red/black wire.
   b. No. 2 pulser coil—white/black wire.
   c. No. 3 pulser coil—red/white wire.
4. Crank the engine at each test connection. Pulser coil output should be 2.4 volts or more.
5. Replace the pulser coil(s) if the voltage output is not as specified. See Chapter Seven.

Counter coil output

Refer to **Figure 42** for this procedure.
1. Set the meter polarity switch to SEN and the voltage switch to 5.
2. Connect the red test lead to the orange/green counter coil wire and the black test lead to a good engine ground.
3. Crank the engine while noting the meter.
4. Replace the counter coil if the voltage output is not as specified. See Chapter Seven.

**Peak Voltage Output Tests
1986-on DT 115, DT 140**

Remove all spark plugs before performing voltage output tests. Crank the engine using the electric starter to take the voltage readings. Unless specified otherwise, leave all ignition components connected during testing.

*CAUTION
Do not crank the engine for more than 20 seconds when performing peak output tests. Allow the starter motor to cool for approximately 2-3 minutes between tests.*

CDI unit output test

Refer to Test 1, **Figure 43** for this procedure.
1. Set the meter polarity switch to POS and the voltage switch to 500.
2. Connect the black test lead to the CDI unit black ground wire or a good engine ground.
3. Connect the red test lead to the orange CDI unit output wire.
4. Crank the engine while noting the meter.
5. Move the red test lead to the blue CDI unit output wire and repeat Step 4.
6. Move the red test lead to the gray CDI unit output wire and repeat Step 4.
7. Move the red test lead to the light green CDI unit output wire and repeat Step 4.
8. CDI unit output voltage at each test connection should be 100 volts or more on 1986-1987 models, 160 volts or more on 1988 models and 144 volts or more on 1989-on models. Replace the CDI unit if output voltage is not as specified. See Chapter Seven.

TROUBLESHOOTING

Charge coil output

Refer to Test 2, **Figure 43** for this procedure.

1. Set the meter polarity switch to POS and the voltage switch to 500.
2. Disconnect the charge coil wires from the CDI unit.
3A. 1986-1988—Connect the black test lead to the black/red charge coil wire. Connect the red test lead to the No. 1 charge coil green wire.
3B. 1989-on—Connect the black test lead to the green charge coil wire and the red test lead to the charge coil brown wire.
4. Crank the engine while noting the meter.
5A. 1986-1988—Move the red test lead to the No. 2 charge coil brown wire and repeat Step 3.
6. Charge coil output should be 75 volts or more on 1986-1987 models, 200 volts or more on 1988 models and 148 volts or more on 1989-on models. Replace the charge coil(s) if output voltage is not as specified. See Chapter Seven.

Pulser coil output

Refer to Test 3, **Figure 43** for this procedure.
1. Set the meter polarity switch to SEN and the voltage switch to 50.
2. Disconnect the pulser coil(s) wires from the CDI unit.
3. Connect the black test lead to the black pulser coil ground wire.
3. Connect the red test lead to the No. 1 (red/green) pulser coil output wire.
4. Crank the engine while noting the meter.
5. Move the red test lead alternately to the following wires:
   a. No. 2 pulser coil output wire (white/black).
   b. No. 3 pulser coil output wire (red/white).
   c. No. 4 pulser coil output wire (white/green).

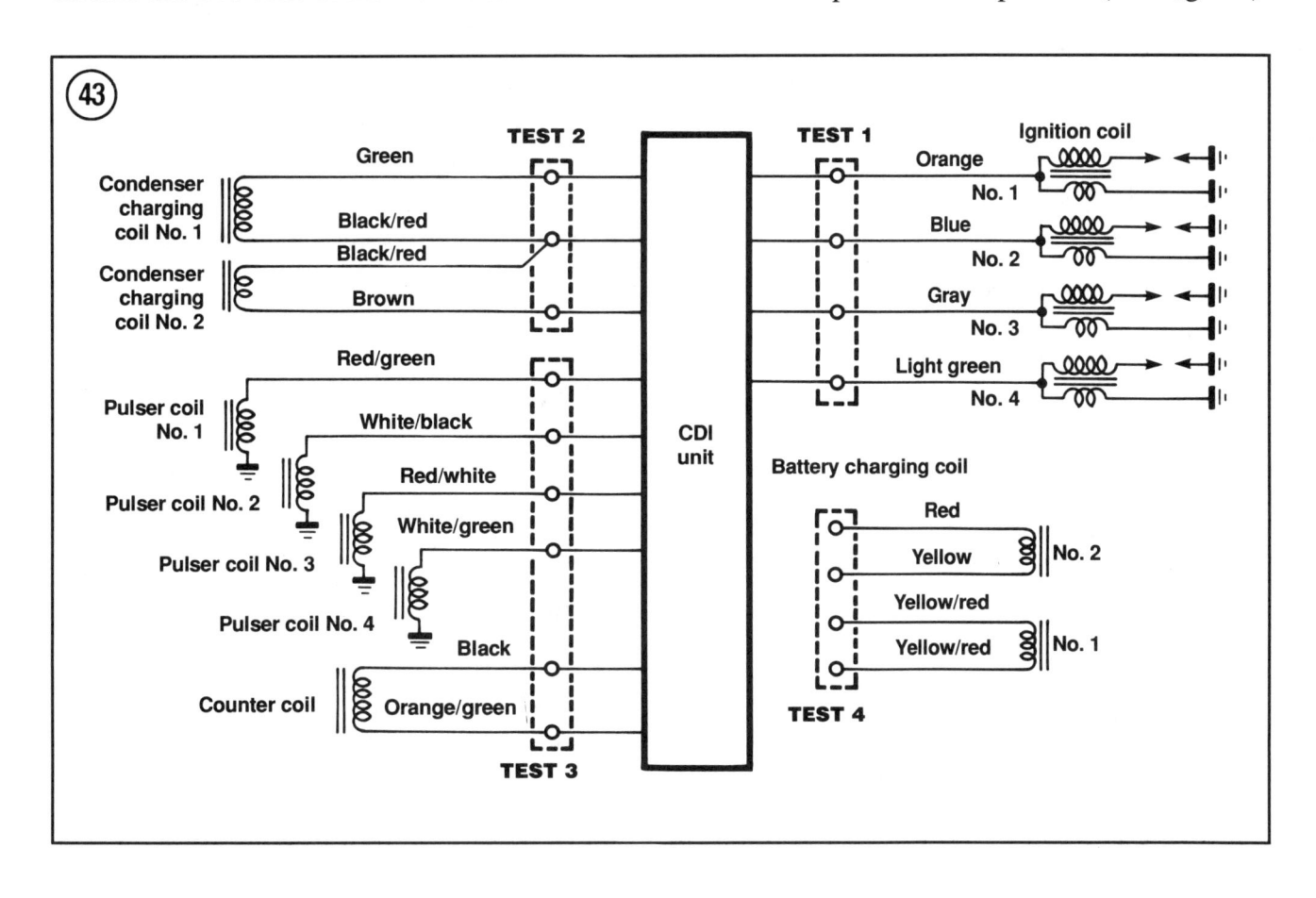

6. Repeat Step 4 at each test connection.

7. Pulser coil output should be 3.0 volts or more on 1986-1987 models, 5.5 volts or more on 1988 models and 4.4 volts or more on 1989-on models. Replace the pulser coil(s) if the output voltage is not as specified. See Chapter Seven.

Counter coil output

Refer to Test 3, **Figure 43** for this procedure.
1. Set the meter polarity switch to SEN and the voltage switch to 50.
2. Disconnect the counter coil wires from the CDI unit.
3. Connect the black test lead as follows:
   a. 1986-1988—Counter coil black wire.
   b. 1989-on—Counter coil black/green wire.
3. Connect the red test lead to the counter coil orange/green wire.
4. Crank the engine while noting the meter.
5. Counter coil output should be 3.0 volts or more on 1986-1987 models, 7.5 volts or more on 1988 models and 3.2 volts or more on 1989-on models. Replace the gear counter coil if the output voltage is not as specified. See Chapter Seven.

Battery charging coil output

Refer to Test 4, **Figure 43** for this procedure.
1. Set the meter polarity switch to POS and the voltage switch to 50.
2. Disconnect the rectifier from the battery charging coil.
3A. 1986-1988:
   a. Connect the black test lead to the yellow charging coil wire and the red test lead to the red charging coil wire.
   b. Crank the engine while noting the meter.
   c. Connect the test leads to the charging coil yellow/red leads. Crank the engine while noting the meter.
3B. 1989-on:
   a. Connect the black test lead to the charging coil yellow wire and the red test lead to the charging coil red wire.
   b. Crank the engine while noting the meter.
4. Battery charging coil output should be 3.0 volts or more on 1986-1987 models; on 1988 models, 9 volts or more on the No. 1 charging coil (yellow/red wires), 10 volts or more on the No. 2 charging coil (red and yellow wires); on 1989-on models, 4.9 volts or more. Replace the charging coil if voltage output is not as specified. See Chapter Seven.

I.C. power source coil output (1989-on models)

1. Set the meter polarity switch to POS and the voltage switch to 50.
2. Disconnect the power source wires from the CDI unit.
3. Connect the black test lead to the black power source coil wire.
4. Connect the red test lead to the power source coil black wire with the red sleeve.
5. Crank the engine while noting the meter.
6. Power source coil output should be 15.2 volts or more. Replace the coil if voltage output is not as specified.

Peak Voltage Output Tests V4 Models

Remove all spark plugs before performing voltage output tests. Crank the engine using the electric starter to take the voltage readings. Unless specified otherwise, leave all ignition components connected during testing.

CAUTION
Do not crank the engine for more than 20 seconds when performing peak output tests. Allow the starter motor to cool for approximately 2-3 minutes between tests.

TROUBLESHOOTING

CDI unit output

Refer to **Figure 44** for this procedure.

1. Set the meter polarity switch to POS and the voltage switch to 500.
2. Connect the black test lead to a good engine ground.
3. Connect the red test lead to the orange CDI unit output wire.
4. Crank the engine while noting the meter.
5. Move the red test lead to the blue CDI unit wire and repeat Step 4.
6. Move the red test lead to the gray CDI unit wire and repeat Step 4.

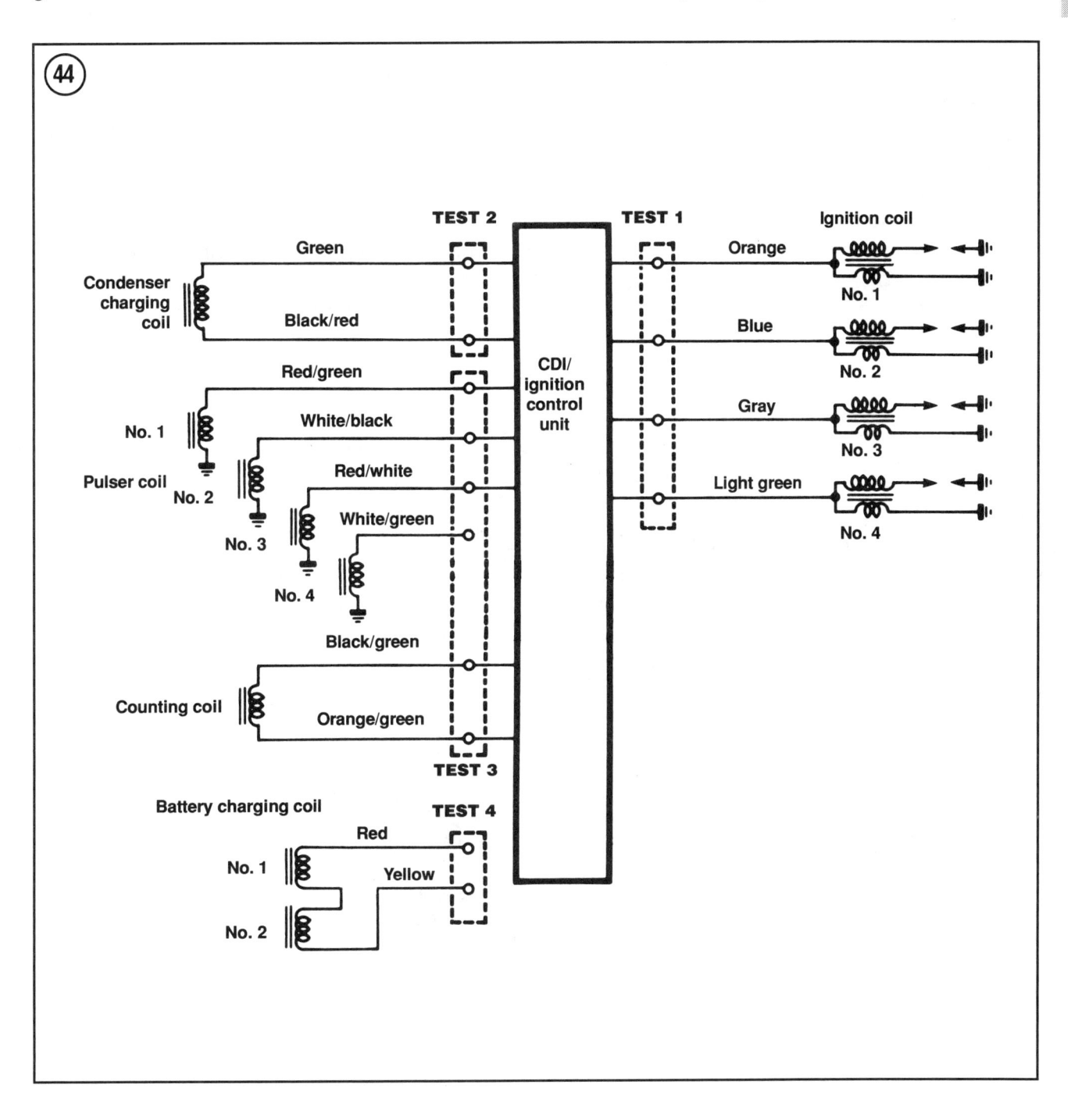

7. Move the red test lead to the light green CDI unit wire and repeat Step 4.
8. CDI unit output should be 102 volts or more at each test connection. Replace the CDI unit if output voltage is not as specified. See Chapter Seven.

Charge coil output

Refer to **Figure 44** and **Figure 45** for this procedure.
1. Set the meter polarity switch to POS and the voltage switch to 500.
2. Connect the black test lead to the green charge coil wire and the red test lead to the black/red charge coil wire.
3. Crank the engine while noting the meter.
4. Charge coil output should be 300 volts or more. Replace the charge coil if voltage output is not as specified. See Chapter Seven.

Pulser coil output

Refer to **Figure 44** and **Figure 45** for this procedure.
1. Set the meter polarity switch to SEN and the voltage switch to 50.
2. Connect the black test lead to a good engine ground.
3. Connect the red test lead to the No. 1 pulser coil output wire (red/green).
4. Crank the engine while noting the meter.

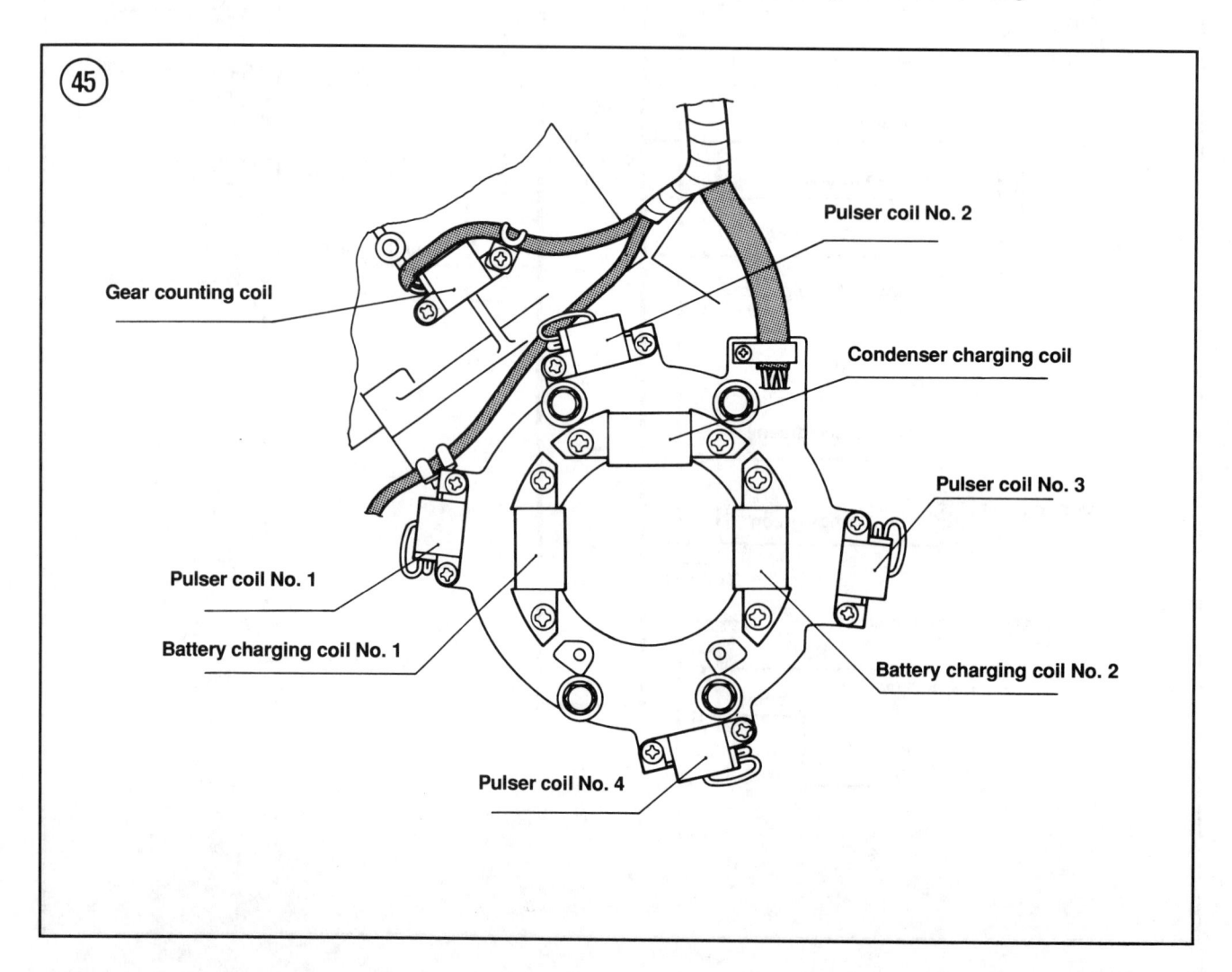

Figure 45

- Gear counting coil
- Pulser coil No. 2
- Condenser charging coil
- Pulser coil No. 3
- Pulser coil No. 1
- Battery charging coil No. 1
- Battery charging coil No. 2
- Pulser coil No. 4

TROUBLESHOOTING

5. Connect the red test lead alternately to the following wires and repeat Step 4 at each connection.

    a. No. 2 pulser coil (white/black).
    b. No. 3 pulser coil (red/white).
    c. No. 4 pulser coil (white/green).

6. Pulser coil output at each test connection should be 3.0 volts or more. Replace the pulser coil(s) if voltage output is not as specified. See Chapter Seven.

Counter coil output

Refer to **Figure 44** and **Figure 45** for this procedure.

1. Set the meter polarity switch to SEN and the voltage switch to 50.
2. Connect the black test lead to the black/green counter coil wire.
3. Connect the red test lead to the orange/green counter coil wire.
4. Crank the engine while noting the meter.
5. Counter coil output should be 4.0 volts or more. Replace the counter coil if voltage output is not as specified.

Battery charging coil output

Refer to **Figure 44** and **Figure 45** for this procedure.

1. Set the meter polarity switch to POS and the voltage switch to 50.
2. Disconnect the charge coil from the rectifier/regulator.
3. Connect the black test lead to the red charging coil wire and the red test lead to the yellow charging coil wire.
4. Crank the engine while noting the meter.
5. Battery charging coil output should be 4.8 volts or more. Replace the charging coil if voltage output is not as specified.

Peak Voltage Output Tests
V6 Models

Remove all spark plugs before performing voltage output tests. Crank the engine using the electric starter to take the voltage readings. Unless specified otherwise, leave all ignition components connected during testing.

CAUTION
Do not crank the engine for more than 20 seconds when performing peak output tests. Allow the starter motor to cool for approximately 2-3 minutes between tests.

CDI unit output test

Refer to **Figure 46** for this procedure.

1. Set the meter polarity switch to POS and the voltage switch to 500.
2. Connect the black test lead to a good engine ground.
3. Connect the red test lead to the CDI unit orange output wire.
4. Crank the engine while noting the meter.
5. Connect the red test lead alternately to the following CDI unit output wires and repeat Step 4 at each test connection.

    a. Blue wire.
    b. Gray wire.
    c. Light green wire.
    d. White/green wire.
    e. Blue/yellow wire.

6. CDI unit output should be 165 volts or more on 1986-1987 models, 220 volts or more on 1988 models, 135 volts or more on 1989-on DT 150, DT 175 and DT 200 models and 170 volts or more on 1989-on DT 150SS and DT 200V models. Replace the CDI unit if output voltage is not as specified. See Chapter Seven.

Charge coil output

Refer to **Figure 46** and **Figure 47** for this procedure.

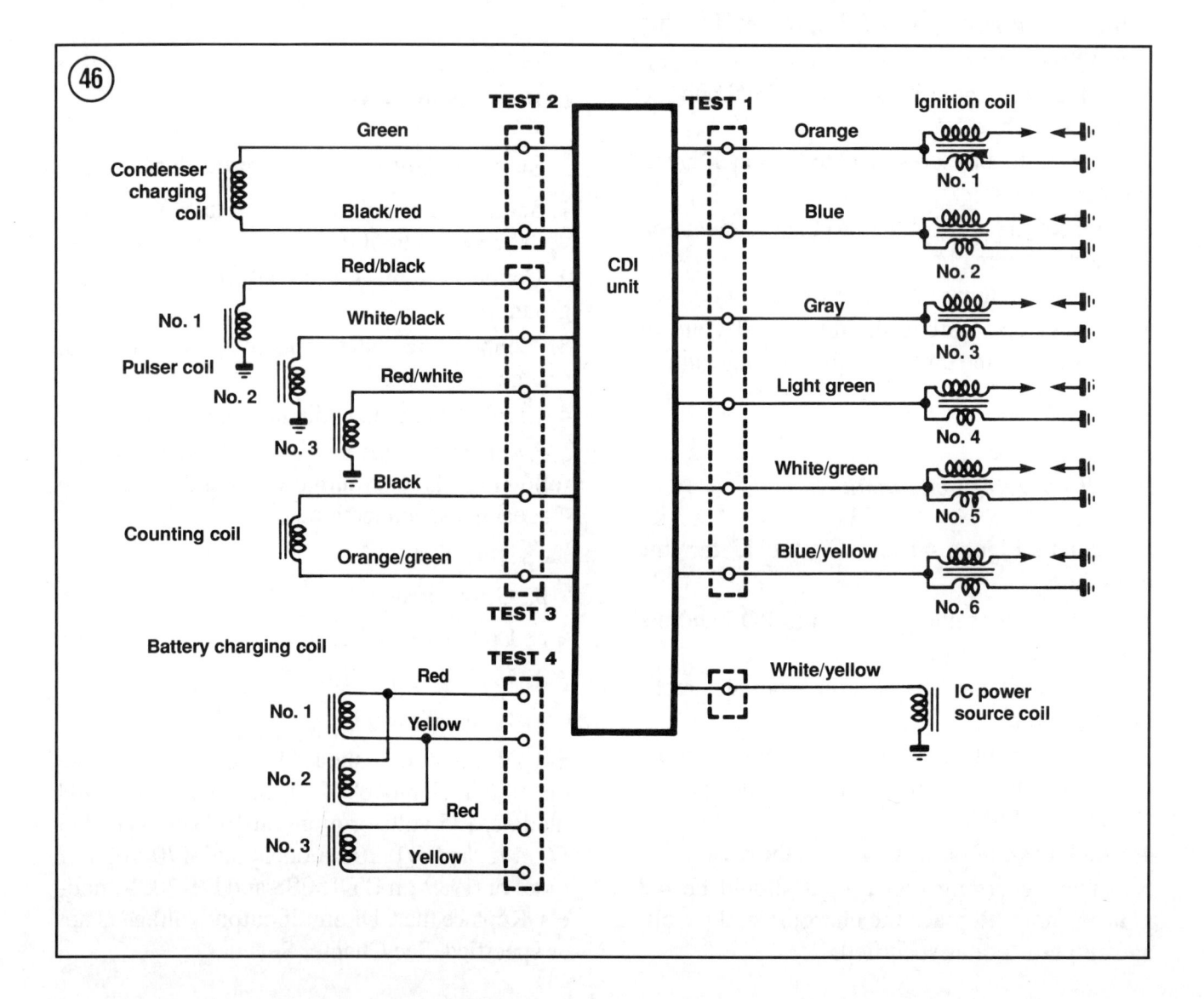

1. Set the meter polarity switch to POS and the voltage switch to 500.
2. Disconnect the charge coil green and black red wires from the CDI unit.
3. Connect the black test lead to the black/red wire on 1986-1988 models and the green wire on 1989-on models.
4. Connect the red test lead to the green wire on 1986-1988 models and the black/red wire on 1989-on models.
5. Crank the engine while noting the meter.
6. Charge coil output should be 135 volts or more on 1986-1987 models, 230 volts on 1988 models, 135 volts or more on 1989-on DT 150, DT 175 and DT 200 models and 170 volts or more on 1989-on DT 150SS and DT 200V models. Replace the charge coil if voltage output is not as specified. See Chapter Seven.

Pulser coil output

Refer to **Figure 46** and **Figure 47** for this procedure.

![Figure 46: Wiring diagram showing CDI unit connections to condenser charging coil (Green, Black/red), pulser coil No.1/No.2/No.3 (Red/black, White/black, Red/white), counting coil (Black, Orange/green), battery charging coil No.1/No.2/No.3 (Red, Yellow), ignition coils No.1-No.6 (Orange, Blue, Gray, Light green, White/green, Blue/yellow), and IC power source coil (White/yellow). Test points labeled TEST 1, TEST 2, TEST 3, TEST 4.]

TROUBLESHOOTING

1. Set the meter polarity switch to SEN and the voltage switch to 5.
2. Disconnect the pulser coil red/black, white/black and red/white wires from the CDI unit.
3. Connect the black test lead to the pulser coil black ground wire or a good engine ground.
4. Connect the red test lead to the No. 1 pulser coil wire (red/black).
5. Crank the engine while noting the meter.
6. Move the red test lead to the No. 2 pulser coil wire (red/black) and repeat Step 5.
7. Move the red test lead to the No. 3 pulser coil wire (red/white) and repeat Step 5.
8. Pulser coil output should be 2 volts or more on 1986-1987 models, 5 volts or more on 1988 models and 2.6 volts or more on 1989-on models. Replace the pulser coil(s) if voltage output is not as specified. See Chapter Seven.

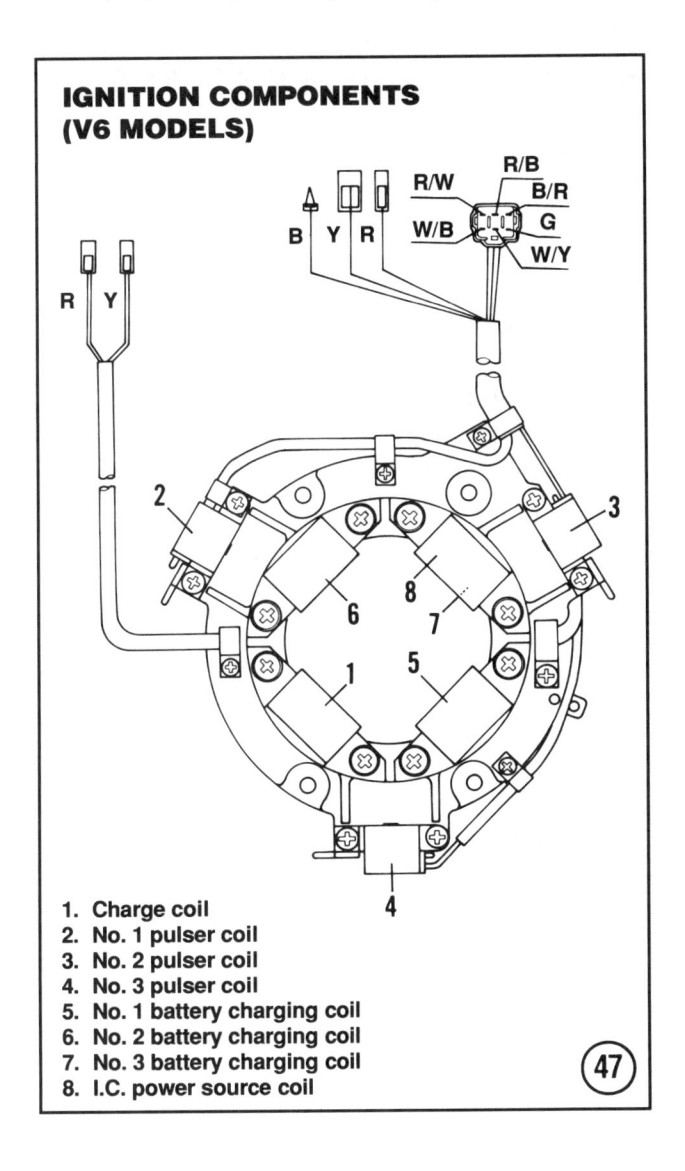

Counter coil output

Refer to **Figure 46** and **Figure 47** for this procedure.
1. Set the meter polarity switch to SEN and the voltage switch to 5.
2. Disconnect the counter coil orange/green and black (or black/green) wires from the CDI unit.
3. Connect the black test lead to the counter coil black wire on 1986-1988 models or the black/green wire on 1989-on models.
4. Connect the red test lead to the counter coil orange/green wire.
5. Crank the engine while noting the meter.
6. Counter coil output should be 0.7 volt on 1986-1987 models, 2 volts or more on 1988 models and 2.4 volts or more on 1989-on models. Replace the counter coil if voltage output is not as specified. See Chapter Seven.

I.C. power source coil output (1988-on models)

Refer to **Figure 46** and **Figure 47** for this procedure.
1. Set the meter polarity switch to POS and the voltage switch to 500 on 1986-1987 models and 50 on 1988-on models.
2. Disconnect the I.C. coil white/yellow wire from the CDI unit.
3. Connect the black test lead to a good engine ground.
4. Connect the red test lead to the I.C. coil white/yellow wire.
5. Crank the engine while noting the meter.
6. I.C. source coil output should be 60 volts or more on 1986-1987 models and 6.5 volts or more on 1988-on models.

IGNITION COMPONENTS (V6 MODELS)

1. Charge coil
2. No. 1 pulser coil
3. No. 2 pulser coil
4. No. 3 pulser coil
5. No. 1 battery charging coil
6. No. 2 battery charging coil
7. No. 3 battery charging coil
8. I.C. power source coil

Battery charging coil output

Refer to **Figure 46** and **Figure 47** for this procedure.

1. Set the meter polarity switch to POS and the voltage switch to 50.
2. Disconnect the charge coil yellow and red wires from the rectifier or rectifier/regulator.
3. Connect the black test lead to the yellow charging coil wire on 1986-1988 models or the red charging coil wire on 1989-on models.
4. Connect the red test lead to the red wire on 1986-1988 models or the yellow wire on 1989-on models.
5. Crank the engine while noting the meter.
6. Repeat Step 3, Step 4 and Step 5 on the No. 2 and No. 3 battery charging coils.
7. Battery charging coil output should be 1.7 volts or more on 1986-1987 models, 12 volts or more on the No. 1 and No. 2 coils on 1988 models, 8 volts or more on the No. 3 coil on 1988 models and 8 volts or more on 1989-on models.

Resistance Tests

Use Suzuki Pocket Tester 09900-25002 or an equivalent ohmmeter to perform resistance tests on Integrated Circuit and Micro Link ignition system components.

Charge Coil Resistance Test

1989-on DT 25; 1988-on DT 30

Refer to **Figure 48** for this procedure.

1. Remove the engine cover.
2. Disconnect the negative battery cable on models so equipped.
3. Disconnect the charge coil connector (**Figure 48**).
4. Connect the tester between the black/red and green charge coil wire terminals.
5. Compare the reading to the specifications in **Table 6**. Replace charge coil if resistance is not as specified.

1986-on DT 55, DT 65

Refer to *Charge Coil Resistance Test* under PEI ignition systems in this chapter.

1988-on DT 75, DT 85

1. Remove the engine cover.
2. Disconnect the negative battery cable.
3. Disconnect the black and green charge coil wires from the CDI unit connectors.
4. Connect the tester between the black and green wire.
5. Compare the reading to the specifications in **Table 6**. Replace the charge coil if the resistance is not as specified.

1986-on DT 115, DT 140

Refer to **Figure 49** for this procedure.

1. Remove the engine cover.
2. Disconnect the negative battery cable.
3. Disconnect the black/red and green wires (No. 1 charge coil) and the black/red and brown wires (No. 2 charge coil) at the bullet connectors.

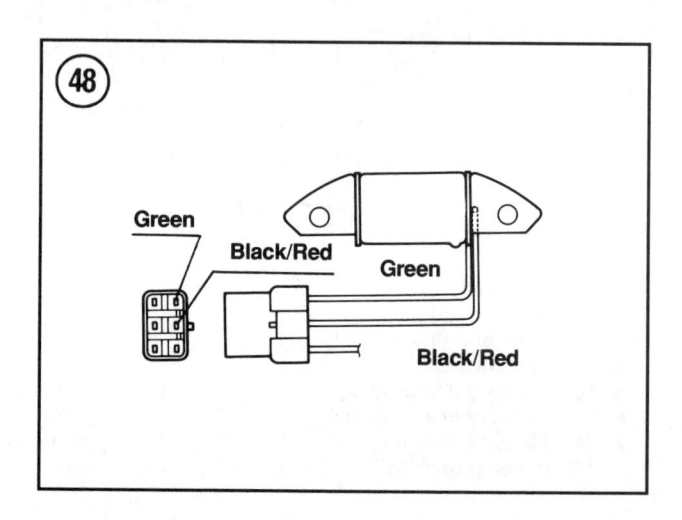

TROUBLESHOOTING

4. Connect the tester between the No. 1 charge coil wires.
5. Repeat step 4 on the No. 2 charge coil wires.
6. Compare the readings to the specifications in **Table 6**. Replace the charge coil(s) if the resistance is not as specified.

V4 models

Refer to **Figure 50** for this procedure.
1. Remove the engine cover.
2. Disconnect the negative battery cable.

3. Disconnect the 6 pin connector between the stator plate and the CDI unit.
4. Connect the tester between the charge coil black/red and green wire terminals.
5. Compare reading to the specifications in **Table 6**. Replace the charge coil if the resistance is not as specified. See Chapter Seven.

V6 models

Refer to **Figure 51** for this procedure.
1. Remove the engine cover.

49

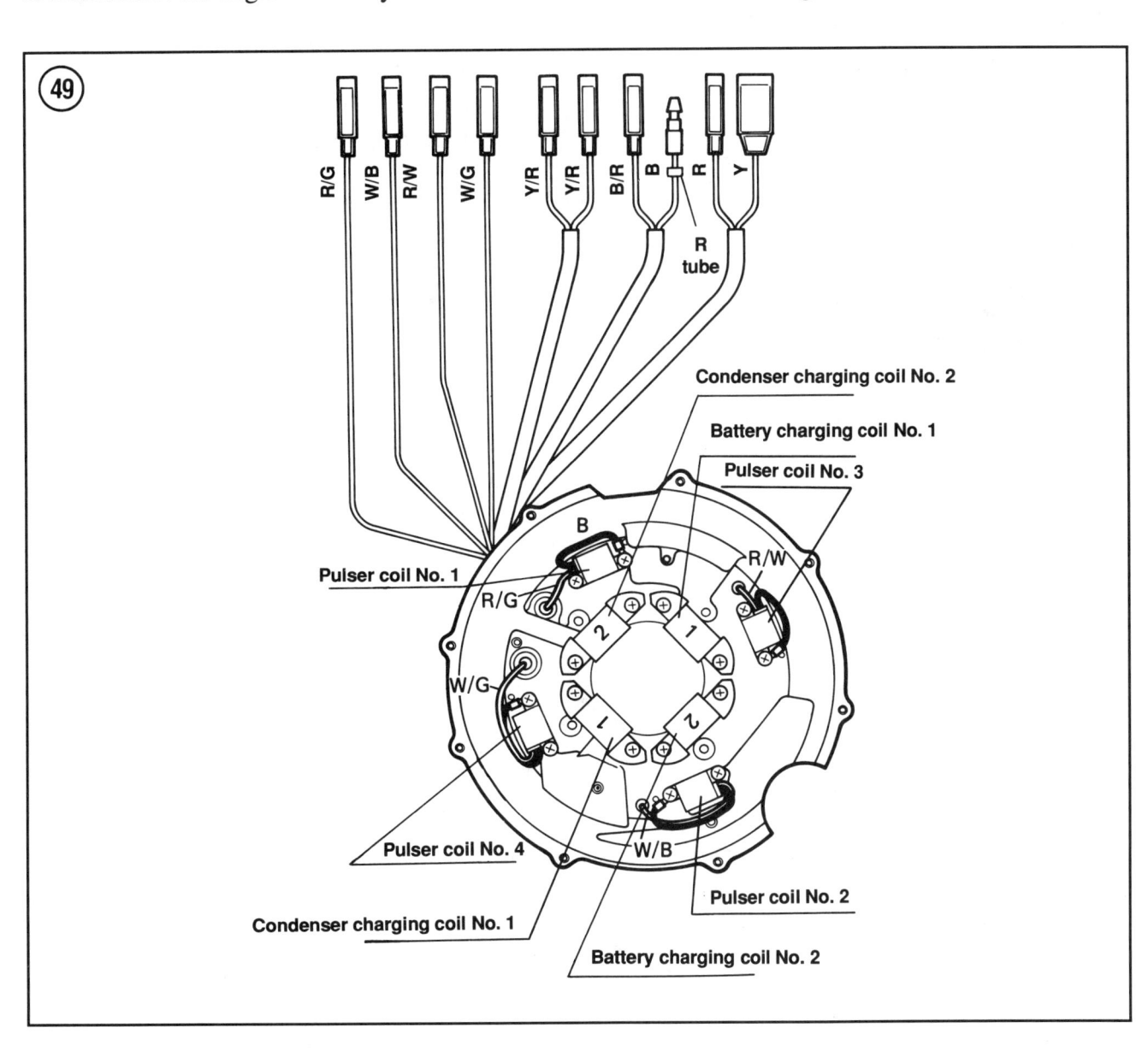

Condenser charging coil No. 2
Battery charging coil No. 1
Pulser coil No. 3
Pulser coil No. 1
Pulser coil No. 4
Condenser charging coil No. 1
Pulser coil No. 2
Battery charging coil No. 2

2. Disconnect the negative battery cable.

3. Disconnect the 6-pin connector between the stator plate and CDI unit.

4. Connect the tester between the black/red and green charge coil wires.

5. Compare the reading to the specifications in **Table 6**. Replace the charge coil if the resistance is not as specified. See Chapter Seven.

Pulser Coil Resistance Tests

1989-on DT 25; 1988-on DT 30; 1986-on DT 55, DT 65; 1988-on DT 75, DT 85

1. Remove the engine cover.
2. Disconnect the negative battery cable.
3A. DT 25, DT 30, DT 75, DT 85:

   a. Disconnect the No. 1 pulser coil red/black wire (**Figure 52**) at the bullet connector.
   b. Connect the tester between the red/black wire and a good engine ground. Note resistance reading.
   c. Repeat test at each remaining pulser coil: No. 2—white/black wire; No. 3—red/white wire.

3B. DT 55 and DT 65:

   a. Disconnect the No. 1 pulser coil red/black wire and the black ground wire (**Figure 53**).
   b. Connect the tester between the red/black wire and the black ground wire. Note the resistance reading.
   c. Repeat the test between the remaining pulser coils black ground wire and the

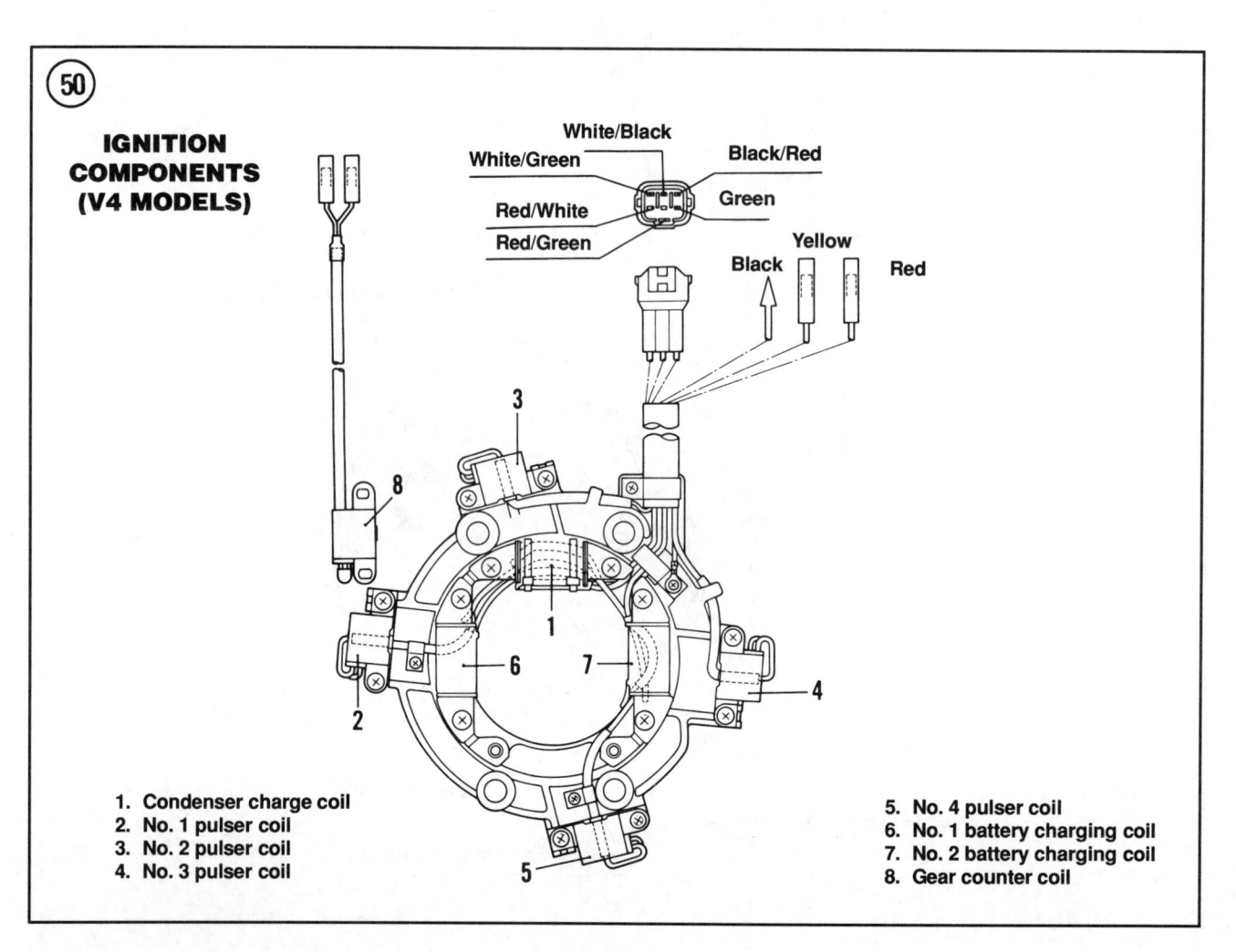

(50) IGNITION COMPONENTS (V4 MODELS)

White/Black
White/Green
Black/Red
Red/White
Green
Red/Green
Black
Yellow
Red

1. Condenser charge coil
2. No. 1 pulser coil
3. No. 2 pulser coil
4. No. 3 pulser coil
5. No. 4 pulser coil
6. No. 1 battery charging coil
7. No. 2 battery charging coil
8. Gear counter coil

TROUBLESHOOTING

white/black wire (No. 2) and red/white (No. 3).

4. Compare the readings with the specifications in **Table 6**. Replace the pulser coil(s) if the resistance is not as specified. See Chapter Seven.

5. DT 55 and DT 65: Connect the meter between each pulser coil mounting stay (**Figure 53**) and color coded wire. Replace the pulser coil(s) if any continuity is noted. See Chapter Seven.

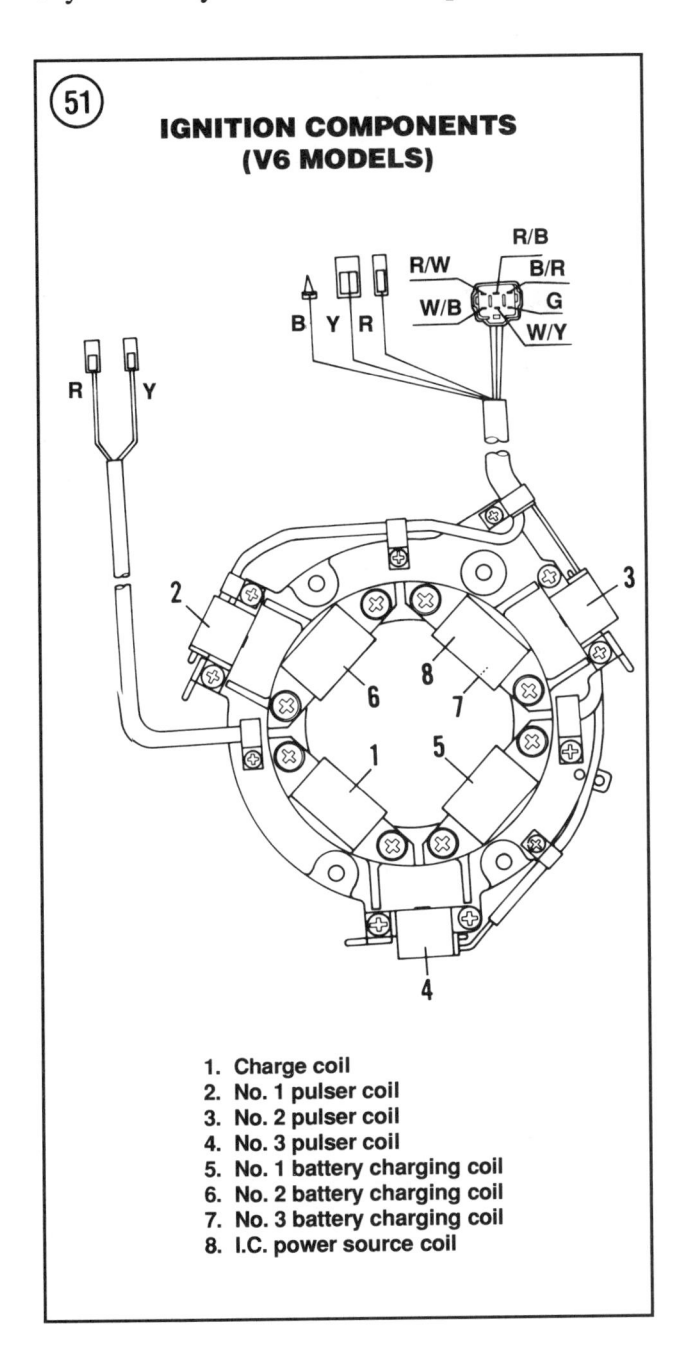

1986-on DT 115, DT 140

Refer to **Figure 54** for this procedure.

1. Remove the engine cover.
2. Disconnect the negative battery cable.
3. Disconnect the No. 1 pulser coil red/green wire.
4. Connect the tester between a good engine ground and the red/green wire. Note the resistance reading.
5. Repeat Step 3 and Step 4 on the No. 2 pulser coil white/black wire, No. 3 pulser coil red/white wire and No. 4 pulser coil white/green wire.
6. Compare readings to the specifications in **Table 6**. Replace the pulser coil(s) if the resistance is not as specified. See Chapter Seven.

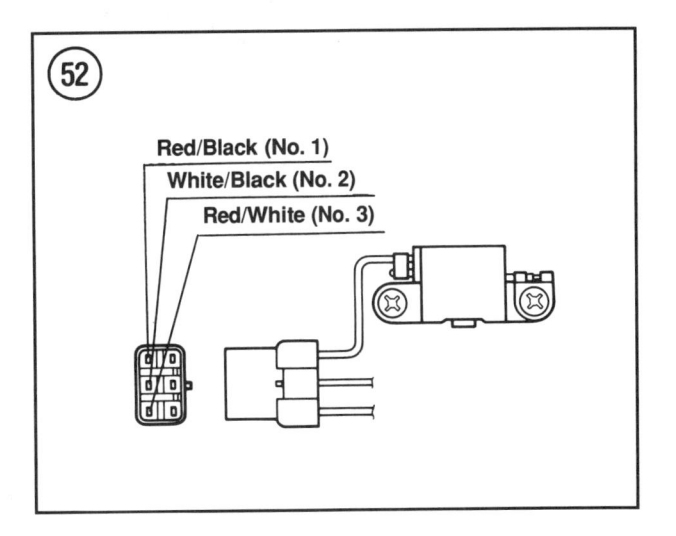

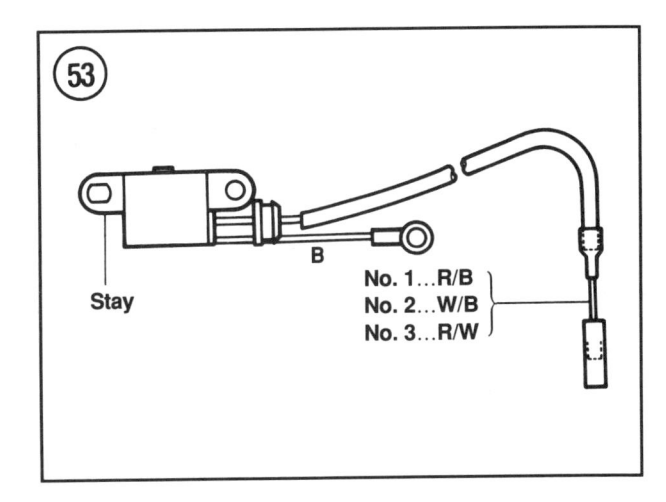

(51) IGNITION COMPONENTS (V6 MODELS)

1. Charge coil
2. No. 1 pulser coil
3. No. 2 pulser coil
4. No. 3 pulser coil
5. No. 1 battery charging coil
6. No. 2 battery charging coil
7. No. 3 battery charging coil
8. I.C. power source coil

(52)
- Red/Black (No. 1)
- White/Black (No. 2)
- Red/White (No. 3)

(53)
Stay
- No. 1...R/B
- No. 2...W/B
- No. 3...R/W

CHAPTER THREE

V4 and V6

Refer to **Figure 50** (V4) or **Figure 51** (V6) for this procedure.

1. Remove the engine cover.
2. Disconnect the negative battery cable.
3. Disconnect the connector between the stator and CDI unit (**Figure 50** or **Figure 51**).
4. V4—Connect the tester between a good engine ground and the following terminals in the connector:
   a. No. 1—red/green.
   b. No. 2—white/black.
   c. No. 3—red/white.
   d. No. 4—white/green.
5. Note the resistance reading at each connection and compare to the specifications in **Table 6**.
6. V6—Connect the tester between a good engine ground and the following terminals in the connector:
   a. No. 1—red/black.
   b. No. 2—white/black.
   c. No. 3—red/white.

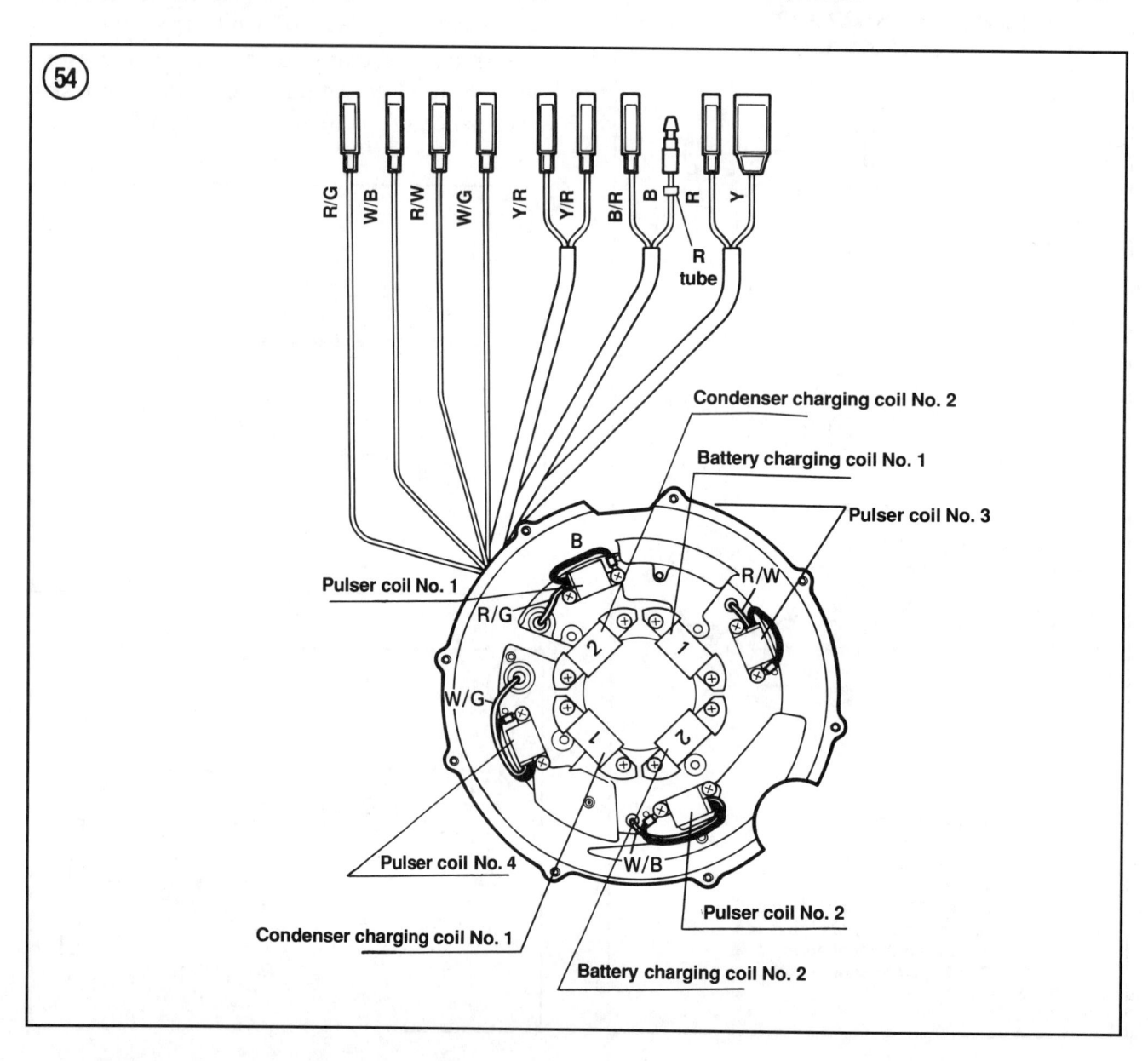

54

- Condenser charging coil No. 2
- Battery charging coil No. 1
- Pulser coil No. 3
- Pulser coil No. 1
- Pulser coil No. 4
- Condenser charging coil No. 1
- Pulser coil No. 2
- Battery charging coil No. 2

TROUBLESHOOTING

7. Note the resistance readings at each connection and compare to the specifications in **Table 6**.
8. Replace the pulser coil(s) if the resistance is not as specified. See Chapter Seven.

Ignition Coil Resistance Tests
(1989-on DT 25; 1988-on DT 30; 1986-on DT 55, DT 65; 1988-on DT 75, DT 85; 1986-on DT 115, DT 140; V4; V6)

The ignition coils must be disconnected from the ignition system before testing. See Chapter Seven.

1. To test primary winding resistance, connect the tester between both primary wire connectors as shown in **Figure 55**. Compare the reading to the specifications in **Table 5**.
2. To test secondary winding resistance, connect the tester between the spark plug terminal and the coil primary wire (not the black ground wire). Compare the reading to the specifications in **Table 5**.
3. Repeat Step 1 and Step 2 for the remaining ignition coils.
4. Replace the ignition coil(s) if resistance is not as specified.

Gear Counter Coil Resistance Test
(All Models)

The gear counter coil is mounted along the outer periphery of the flywheel on models so equipped. The counter coil provides engine speed and crankshaft position information to be used in determining the ignition timing advance.

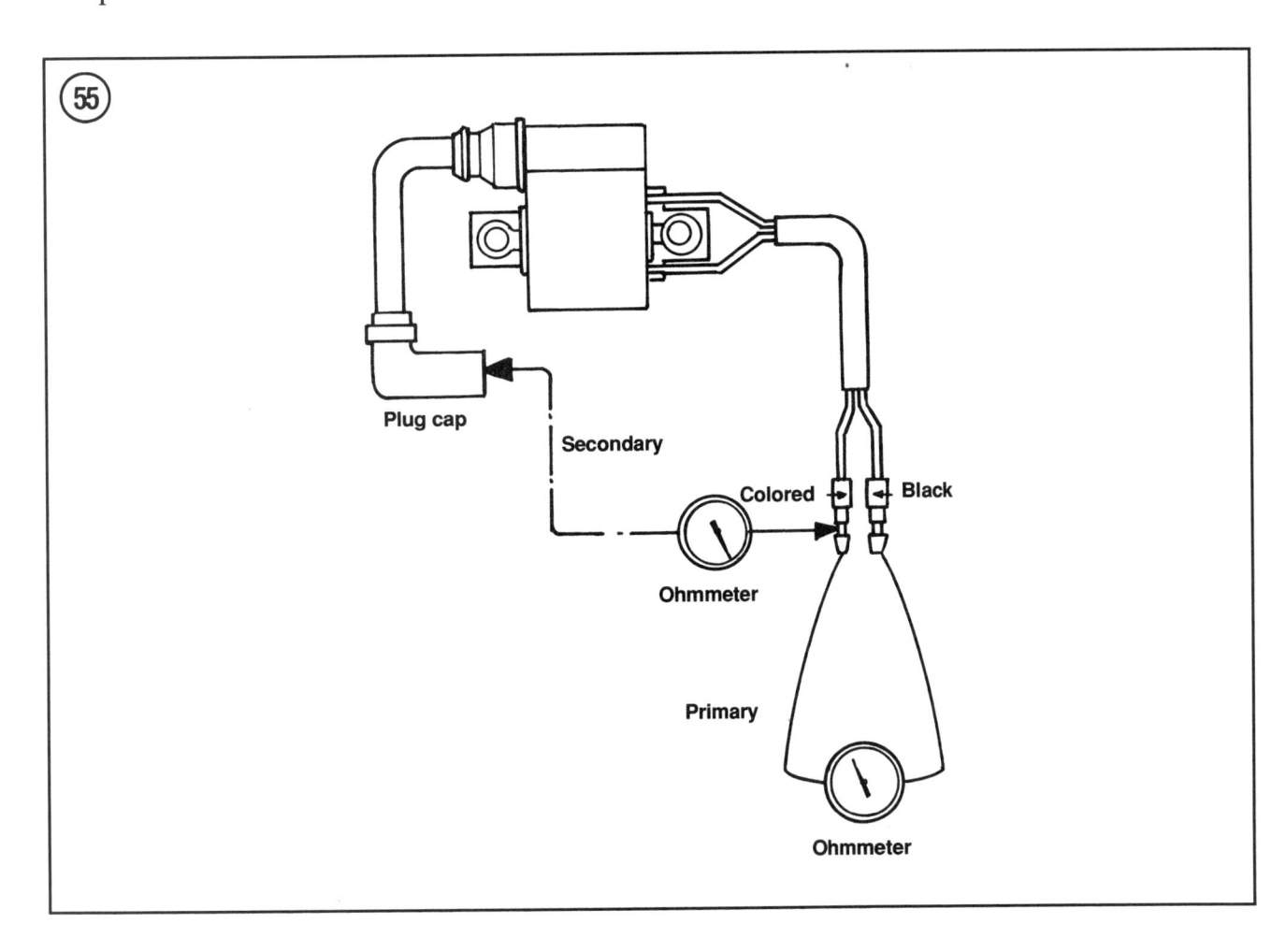

The counter coil must be disconnected from the ignition system before testing resistance.

1. Disconnect the counter coil orange/green and black or black/green wires.
2. Connect the tester between the disconnected wires.
3. Compare the reading to the specifications in **Table 7**. Replace the counter coil if the resistance is not as specified.

I.C. Power Source Coil Resistance Test (All Models)

Refer to **Figure 51** for this procedure.
1. Remove the engine cover.
2. Disconnect the negative battery cable.
3. Disconnect the I.C. source coil white/yellow wire.
4. Connect the tester between the white/yellow wire and a good engine ground. Replace the source coil if the resistance is not within 210-310 ohms on 1986-1988 models or 140-210 ohms on 1989-on models.

Idle Speed Adjustment Switch

NOTE
*A defective or maladjusted throttle valve sensor may cause the idle speed switch to be inoperative or ignition timing at idle speed to be incorrect. Prior to testing idle speed switch, make sure throttle valve sensor is operating properly and correctly adjusted. See **Throttle Valve Sensor** in this chapter.*

The engine speed should change approximately 50 rpm with each position of the idle speed switch. To test switch, connect a suitable timing light to the number 1 spark plug lead. With engine at idle speed, note ignition timing while moving the idle speed switch through each position. If the timing does not change with each switch position, the switch is defective and must be replaced.

Throttle Switch Test

A throttle switch is used on early models equipped with IC ignition to provide a signal to the CDI unit used to determine ignition timing advance. To test the switch, proceed as follows:

1. Disconnect the throttle switch wires at the bullet connectors. See **Figure 56**.
2. Connect an ohmmeter between the switch black wire (B) and the light green/red wire (Lg/R).
3. Depress the lever on the idle switch (1). No continuity should be noted.
4. Release the lever on the idle switch (1). Zero resistance should be noted.
5. Connect the ohmmeter between the brown/yellow wire (Br/Y) and the black wire.
6. Depress the lever on the acceleration switch (2). No continuity should be noted.
7. Release the acceleration switch (2) lever. Zero resistance should be noted.
8. Replace the throttle switch if resistance is not as specified.

Throttle Valve Sensor

A throttle valve sensor is used on late models equipped with IC and Micro Link ignition to provide a signal to the integrated circuit or the Micro Link computer. The signal is used to de-

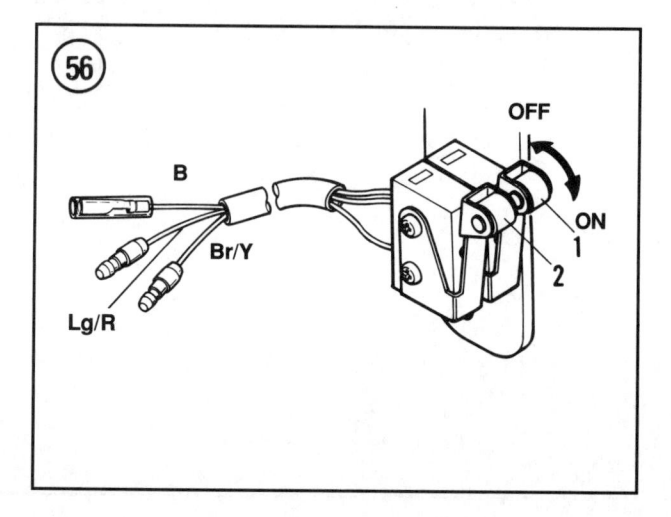

TROUBLESHOOTING

termine ignition timing advance. The throttle valve sensor is mounted on the side of the bottom carburetor on DT 25, DT 30, DT 115 and DT 140 models and the side of the center carburetor on DT 55, DT 65, DT 75 and DT 85 models. On all models, the sensor is engaged with the carburetor throttle valve shaft to directly register throttle opening.

1989-on DT 25, DT 55, DT 65, DT 115, DT 140; 1988-on DT 30, DT 75, DT 85

The following special tools are required for this procedure:

a. Small nonmagnetic screwdriver (insulated electrical type).
b. Digital voltmeter.
c. Battery (9 volt minimum).
d. Throttle valve sensor test lead (part 09930-89530) or suitable jumper leads.

1. On models prior to 1991, turn the idle speed switch to the lowest idle setting. On models after 1990, back out idle speed screw so throttle valves are completely closed.

NOTE
*Throttle valves **must** be fully closed for proper throttle sensor testing and adjustment.*

2. Disconnect the sensor 3-wire connector.
3. Connect sensor test lead (part 09930-89530) to the sensor connector.

NOTE
*If the throttle sensor test lead (part 09930-89530) is not available, connect suitable jumper leads to the sensor connector as shown in **Figure 57**.*

4. Connect test lead black/red wire to the positive battery terminal. See **Figure 57**.
5. Connect test lead black wire to the negative battery terminal. See **Figure 57**.
6. Connect voltmeter positive lead to the light green/red test lead wire and voltmeter negative lead to the negative battery terminal. See **Figure 57**.

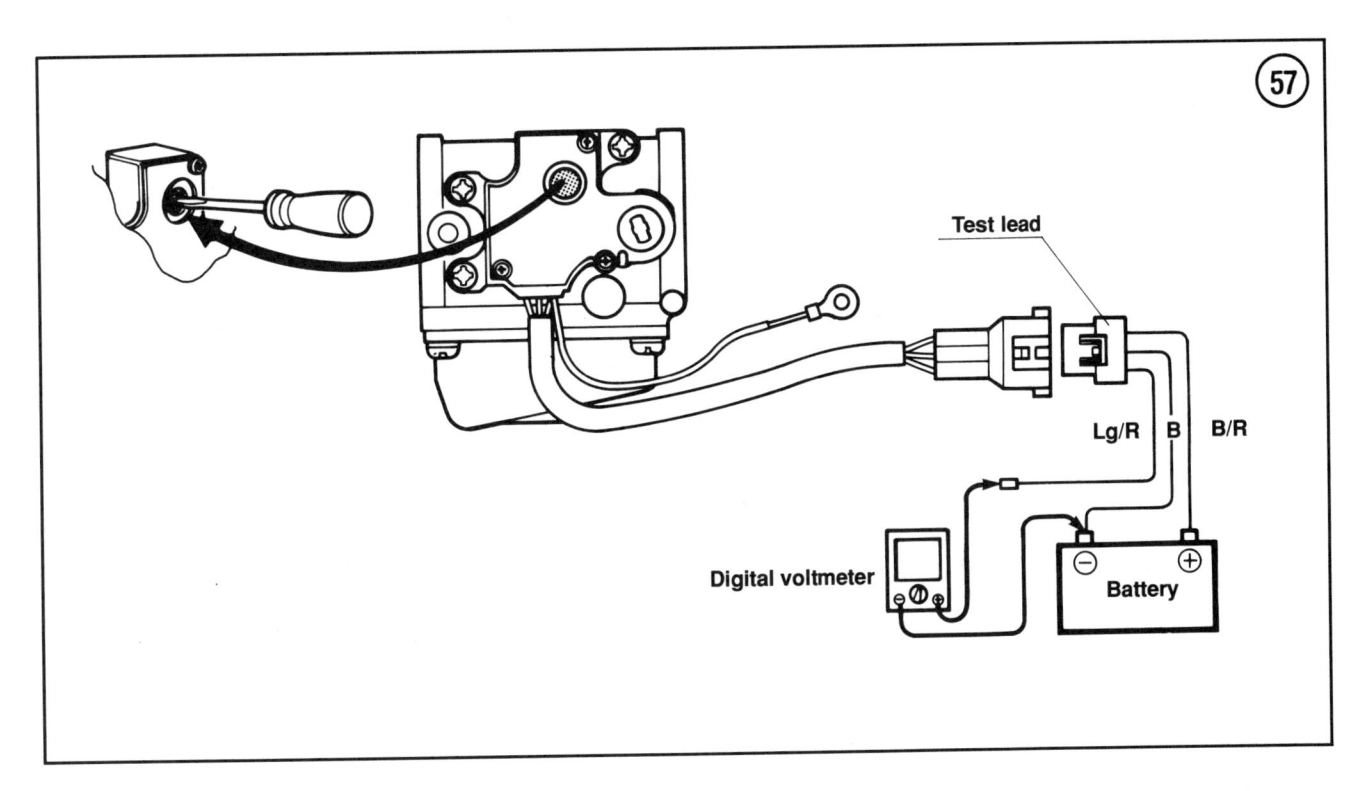

57

NOTE
Use only a nonmagnetic, insulated screwdriver to adjust sensor in Step 7, or voltage reading will not be valid.

7. Throttle sensor voltage at closed throttle should be 0.45-0.55 volt. If not, remove adjusting screw cover and turn screw (**Figure 57**) clockwise to increase the voltage or counterclockwise to decrease the voltage.

8. Advance the throttle to the wide-open position. Sensor voltage at wide-open throttle should be 2.7 volts or more. Wide-open voltage is not adjustable—if 2.7 volts or more can not be obtained, replace throttle sensor.

NOTE
The correct throttle sensor voltages are essential for proper outboard operation. If the correct closed-throttle voltage can not be obtained by turning the sensor adjusting screw, loosen the sensor mounting screws and shift position of the sensor on the carburetor. If the correct voltage still can not be obtained, replace the sensor.

9. Disconnect the test leads, voltmeter and reconnect the sensor connector to the outboard wiring harness. Reinstall sensor adjusting screw cover.

V4 and 1987-on V6

1. Remove the oil pump control rod from the oil pump lever.
2. Open and close throttle valves to check for free movement. Make sure throttle valves can be fully closed.
3. Calibrate an ohmmeter on the R × 100 scale.
4. Set the idle speed switch to the slowest position on models prior to 1991. On models after 1990, back out the idle speed screw so throttle valves are fully closed.
5. Disconnect the throttle sensor connector.
6. Connect the ohmmeter red (+) lead to the sensor light green/red wire terminal and the black (−) lead to the sensor black wire terminal.

7. The sensor resistance values at fully-closed throttle should be as follows:
 a. V4—225-275 ohms.
 b. 1987-1988 V6—200-250 ohms.
 c. 1989-on V6—225-275 ohms.
8. If the resistance is not as specified, perform the following:
 a. Make sure the throttle valves are fully closed.
 b. Loosen the sensor mounting screws and rotate the sensor as shown in **Figure 58** to obtain the specified resistance value.
 c. Retighten sensor mounting screws.
9. Disconnect the ohmmeter and reconnect the sensor to the outboard wiring harness.

DT 225 EFI

On EFI models, the throttle valve sensor is adjusted at the factory and the sensor mounting screws are secured with thread locking compound. The sensor should not require further adjustment.

FUEL SYSTEM (CARBURETTED MODELS)

NOTE
This procedure relates to all models that are equipped with carburetor(s). Fuel injected models are not covered due to

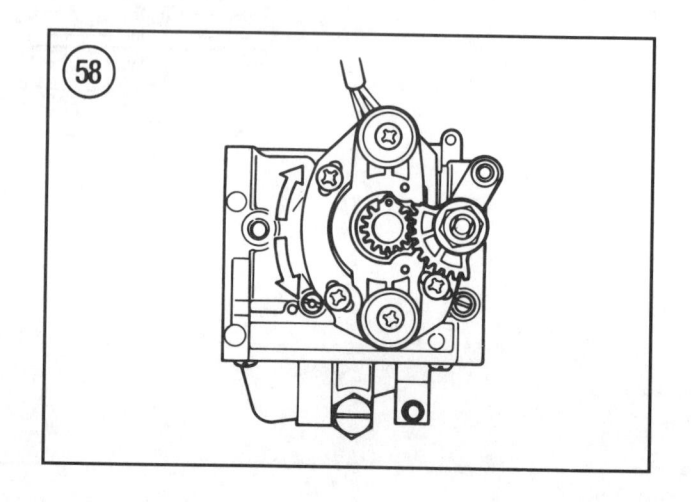

TROUBLESHOOTING

their complexity and the related electronic control systems. Refer to Chapter Six for a basic description of the EFI system.

Many outboard owners automatically assume that the carburetor(s) is at fault when the engine does not run properly. While fuel system problems are not uncommon, carburetor adjustment is seldom the answer. In many cases, adjusting the carburetor only compounds the problem by making the engine run worse.

Fuel system troubleshooting should start at the fuel tank and work all the way through the system, reserving the carburetor(s) as the final point. The majority of fuel system problems result from an empty fuel tank, sour fuel, a plugged fuel filter or a malfunctioning fuel pump. **Table 3** provides a series of symptoms and causes that can be useful in localizing fuel system problems. Chapter Six contains inspection and overhaul procedures for the fuel system components. Chapter Twelve contains inspection and service procedures for oil injection system components.

Troubleshooting

As a first step, check the fuel flow. Remove the fuel tank cap and look into the tank. If there is fuel present, disconnect and ground the spark plug lead(s) as a safety precaution. Disconnect the fuel line at the carburetor (**Figure 59**, typical) and place it in a suitable container to catch any discharged fuel. See if fuel flows freely from the line when the primer bulb is squeezed.

If there is no fuel flow from the line:

a. The fuel petcock may be shut off or blocked by rust or foreign matter.
b. The fuel line may be plugged or kinked.
c. A primer bulb check valve may be defective.
d. The fuel pump may be defective.

If a good fuel flow is present, crank the engine 10-12 revolutions to check fuel pump operation. A pump that is operating satisfactorily will deliver a good, constant flow of fuel from the line. If the amount of flow varies from pulse to pulse, the fuel pump is probably failing. Discard the spent fuel properly—do not throw it overboard into the water.

Carburetor chokes can also present problems. A choke that sticks open will cause a hard starting problem; one that sticks closed will result in a flooding condition.

During a hot engine shut-down, the fuel bowl temperature can rise above 200° F, causing the fuel inside the float bowl to boil. While outboard carburetors are vented to atmosphere to prevent this problem, there is a possibility that some fuel will percolate over the high-speed nozzle.

A leaking inlet needle and seat or a defective float will allow an excessive amount of fuel into the intake manifold. Pressure in the fuel line after the engine is shut down forces fuel past the leaking needle and seat. This raises the fuel bowl level, allowing fuel to overflow into the manifold.

Excessive fuel consumption may not necessarily mean an engine or fuel system problem. Marine growth on the boat's hull, a bent or otherwise damaged propeller or a fuel line leak can cause an increase in fuel consumption. These areas should all be checked *before* blaming the carburetor.

ENGINE TEMPERATURE AND OVERHEATING

Proper engine temperature is critical to good engine operation. An engine that runs too hot will be damaged internally. One that operates too cool will not run smoothly or efficiently resulting in poor fuel economy.

A variety of problems can cause engine overheating. Some of the most commonly encountered are a defective thermostat, a low output or defective water pump, damaged or mispositioned water passage restrictors or even engine flashing in the cylinder head casting water discharge passage that was not removed during engine manufacture.

Troubleshooting

Engine temperature can be checked with the use of Markal Thermomelt Stiks available at your local marine dealer. This heat-sensitive stick looks like a large crayon (**Figure 60**) and will melt on contact with a metal surface at the specific temperature indicated on the stick label.

Two thermomelt sticks are required to check a Suzuki outboard properly: a 125° F (52° C) stick and a 163° F (73° C) stick. The stick should not be applied to the center of the cylinder head, as this area may normally run hotter than 163° F (73° C).

The test is most efficient when carried out on a motor operating on a boat in the water. If necessary to perform the test using a test tank, run the engine at 3,000 rpm for a minimum of 5 minutes to assure that it is at operating temperature. Make sure inlet water temperature is below 80° F (26° C) and perform the test as follows.

1. Mark the cylinder water jacket with each stick. The mark will appear similar to a chalk mark. Make sure sufficient material is applied to the metal surface.
2. With the engine at operating temperature and running at idle in FORWARD gear, the 125° F stick mark should melt. If it does not melt on thermostat-equipped models, the thermostat is stuck open and the engine is running cold.
3. With the engine at operating temperature and running at full throttle in FORWARD gear, the 163° F stick mark should not melt. If it does, the power head is overheating. Look for a defective water pump or a clogged or leaking cooling system. On thermostat-equipped models, the thermostat may be stuck closed.

ENGINE

Engine problems are generally symptoms of something wrong in another system, such as ignition, fuel or starting. If properly maintained and serviced, the engine should experience no problems other than those caused by age and wear.

Overheating and Lack of Lubrication

Overheating and lack of lubrication cause the majority of engine mechanical problems. Outboard motors create a great deal of heat and are

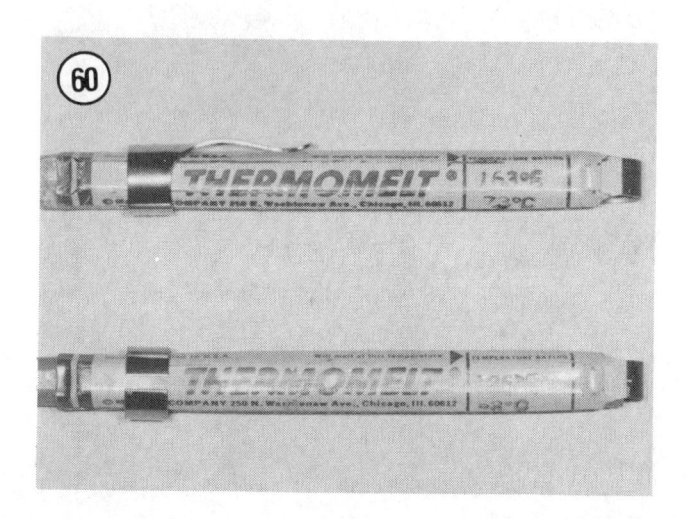

TROUBLESHOOTING

not designed to operate at a standstill for any length of time. Using a spark plug of the wrong heat range can burn a piston. Incorrect ignition timing, a defective water pump or thermostat, a propeller that is too large (over-propping) or an excessively lean fuel mixture can also cause the engine to overheat.

Preignition

Preignition is the premature burning of fuel and is caused by hot carbon spots in the combustion chamber (**Figure 61**). The fuel actually ignites before it is supposed to. Glowing deposits in the combustion chamber, inadequate cooling or overheated spark plugs can all cause preignition. This is first noticed in the form of a power loss but will eventually result in extensive damage to the internal parts of the engine because of higher combustion chamber temperatures.

Detonation

Commonly called "spark knock" or "fuel knock," detonation is the violent explosion of fuel in the combustion chamber prior to the proper time of combustion (**Figure 62**). Severe damage can result. The use of low octane gasoline is a common cause of detonation.

Even when high octane gasoline is used, detonation can still occur if the engine is improperly timed. Other causes are over-advanced ignition timing, lean fuel mixture at or near full throttle, inadequate engine cooling, cross-firing of spark plugs, spark plug(s) of the wrong heat range, excessive accumulation of deposits on piston(s) and combustion chamber(s) or the use of a propeller that is too large (over-propping).

Since outboard motors are noisy, engine knock or detonation is likely to go unnoticed by owners, especially at high engine rpm when wind noise is also present. Such inaudible detonation, as it is called, is usually the cause when engine damage occurs for no apparent reason.

Poor Idling

A poor idle can be caused by improper carburetor adjustment, incorrect timing or ignition system malfunctions. Check the gas cap vent for an obstruction.

Misfiring

Misfiring can result from a weak spark or dirty spark plug(s). Check for fuel contamination. If misfiring occurs only under heavy load, as when accelerating, it is usually caused by a defective

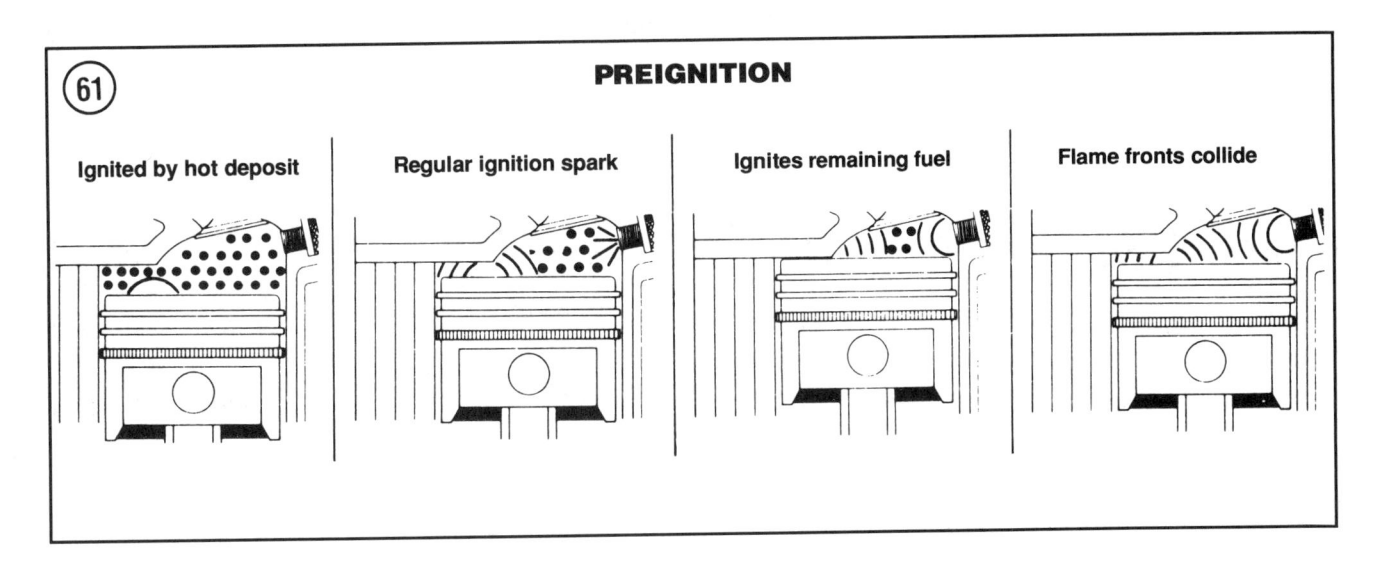

61 PREIGNITION

| Ignited by hot deposit | Regular ignition spark | Ignites remaining fuel | Flame fronts collide |

spark plug. Run the motor at night to check for spark leaks along the plug wire and under spark plug cap or use a spark leak tester.

> *WARNING*
> *Do not run engine in a dark garage to check for spark leak. There is considerable danger of carbon monoxide poisoning.*

Water Leakage in Cylinder

The fastest and easiest way to check for water leakage in a cylinder is to check the spark plug(s). Water will clean a spark plug. If one spark plug on a multi-cylinder engine is clean and the others are dirty, there is most likely a water leak in the cylinder with the clean plug.

To remove all doubt, install a dirty plug in each cylinder. Run the engine in a test tank or on the boat in water for 5-10 minutes. Shut the engine off and remove the plugs. If one plug is clean and the other dirty (or if all plugs are clean), a water leak in the cylinder(s) is the problem.

Flat Spots

If the engine seems to die momentarily when the throttle is opened and then recovers, check for a dirty main jet in the carburetor(s), water in the fuel or an excessively lean mixture.

Power Loss

Several factors can cause a lack of power and speed. Look for air leaks in the fuel line or fuel pump, a clogged fuel filter or a choke/throttle valve that does not operate properly. Check ignition timing.

A piston or cylinder that is galling, incorrect piston clearance or a worn/sticky piston ring may be responsible. Look for loose bolts, defective gaskets or leaking machined mating surfaces on the cylinder head, cylinder or crankcase. Also check the crankcase oil seal; if worn, it can allow fuel to leak between cylinders.

Piston Seizure

This is caused by one or more pistons with incorrect bore clearances, piston rings with an improper end gap, the use of an oil-fuel mixture

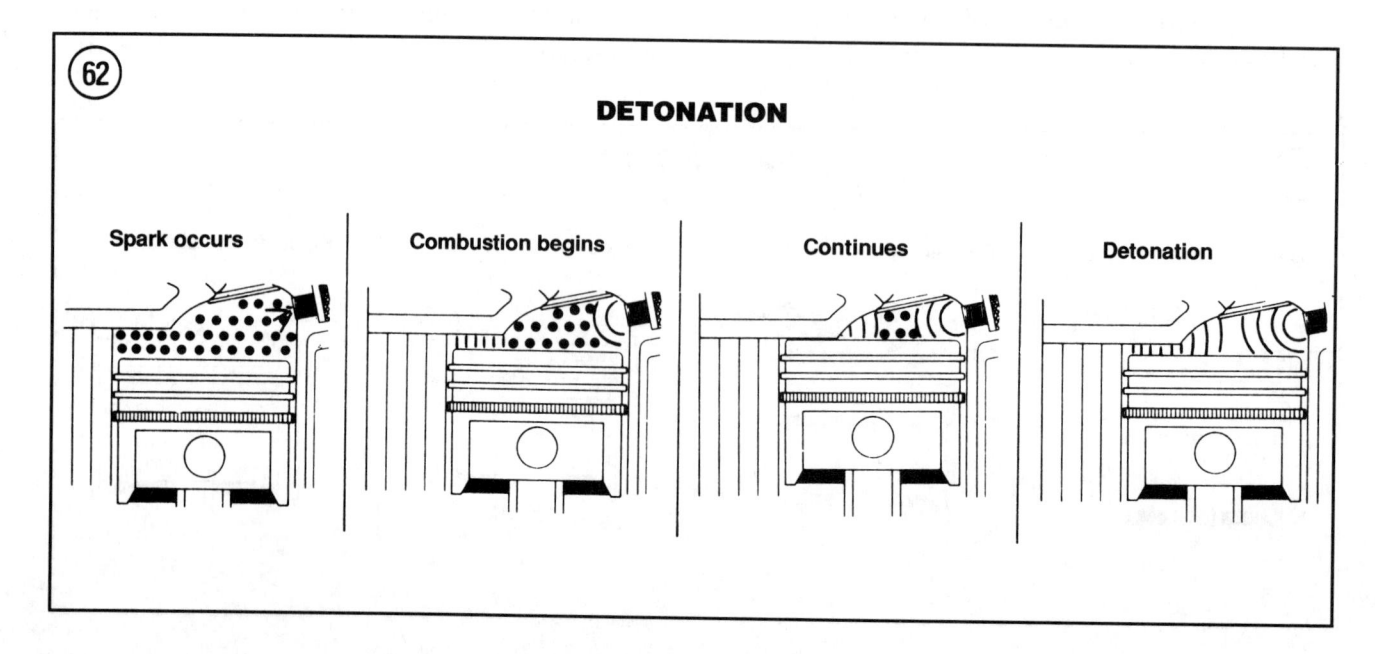

62
DETONATION

Spark occurs | Combustion begins | Continues | Detonation

TROUBLESHOOTING

containing less than 1 part oil to 50 parts of gasoline or an oil of poor quality, a spark plug of the wrong heat range or incorrect ignition timing. Overheating from any cause may result in piston seizure.

Excessive Vibration

Excessive vibration may be caused by loose motor mounts, worn bearings or a generally poor running motor.

Engine Noises

Experience is needed to diagnose accurately in this area. Noises are difficult to differentiate and even harder to describe. Deep knocking noises usually mean main bearing failure. A slapping noise generally comes from a loose piston. A light knocking noise during acceleration may be a bad connecting rod bearing. Pinging should be corrected immediately or damage to the piston will result. A compression leak at the head-to-cylinder joint will sound like a rapid on-off squeal.

Table 1 STARTER TROUBLESHOOTING

Trouble	Cause	Remedy
Pinion does not move when starter is turned on	Blown fuse	Replace fuse
	Pinion rusted to armature shaft	Remove, clean or replace as required.
	Series coil or shunt broken or shorted	Replace coil or shunt.
	Loose switch connections	Tighten connections.
	Rusted or dirty plunger	Clean plunger.
Pinion meshes with ring gear but starter does not run	Worn brushes or brush springs touching armature	Replace brushes or brush springs.
	Dirty or burned commutator	Clean or replace as required.
	Defective armature field coil	Replace armature.
	Worn or rusted armature shaft bearing	Replace bearing.
Starter motor runs at full speed before pinion meshes with ring gear	Worn pinion sleeve	Replace sleeve.
	Pinion does not stop in correct position	Replace pinion.
Pinion meshes with gear and motor starts but engine does not crank	Defective overrunning clutch	Replace overrunning clutch.
Starter motor does not stop when turned off after engine has started	Rusted or dirty plunger	Clean or replace plunger.

(continued)

Table 1 STARTER TROUBLESHOOTING (continued)

Trouble	Cause	Remedy
Starter motor has low no-speed and high-current draw	Armature may be dragging on pole shoes from bent shaft, worn bearings or loose pole shoes	Replace shaft or bearing and/or tighten pole shoes.
	Tight or dirty bearings	Loosen or clean bearings.
High current draw with no armature rotation	A direct ground switch, @ terminal or @ brushes or field connections	Replace defective parts.
	Frozen shaft bearings which prevent armature from rotating	Loosen, clean or replace bearings.
Starter motor has grounded armature or field winding	Current passes through armature first, then to ground field windings	Disconnect grounded leads, then locate any abnormal grounds in starter motor.
Starter motor fails to operate and draws no current and/or high resistance	Open circuit in fields or armature, @ connections or brushes or between brushes and commutator	Repair or adjust broken or weak brush springs, worn brushes high insulation between commutator bars or a dirty, gummy or oily commutator.
High resistance in starter motor	Low no-load speed and a low-current draw and low developed torque	Close "open" field winding on unit which has 2 or 3 circuits in starter motor (unit in which current divides as it enters, taking 2 or 3 parallel paths).
High free speed and high current draw	Shorted fields in starter motor	Install new fields and check for improved performance (fields normally have very low resistance, thus it is difficult to detect shorted fields, since difference in current draw between normal starter motor field windings would not be very great).
Excessive voltage drop	Cables too small	Install larger cables to accommodate high current draw.
High circuit resistance	Dirty connections	Clean connections.
Starter motor has grounded armature or field winding	Field and/or armature is burned or lead is thrown out of commutator due to excess leakage	Raise grounded brushes from commutator and insulate them with cardboard. Use Magneto Analyzer (part No. C-91-25213) (Selector No. 3) and test points to check between insulated terminal or starter motor and starter motor frame (remove ground connection of shunt coils on motors with this feature). If analyzer shows resistance (meter needle moves to right), there is a ground. Raise other brushes from armature and check armature and fields separately to locate ground.

(continued)

TROUBLESHOOTING

Table 1 STARTER TROUBLESHOOTING (continued)

Trouble	Cause	Remedy
Starter does not operate	Run-down battery	Check battery with hydrometer. If reading is below 1.230, recharge or replace battery.
	Poor contact @ terminals	Remove terminal clamps. Scrape terminals and clamps clean and tighten bolts securely.
	Wiring or key switch	Coat with sealer to protect against further corrosion.
Starter does not operate (continued)	Starter solenoid	Check for resistance between: (a) positive (+) terminal of battery and large input terminal of starter solenoid, (b) large wire @ top of starter motor and negative (−) terminal of battery and (c) small terminal of starter solenoid and positive battery terminal. Key switch must be in START position. Repair all defective parts. With a fully charged battery, connect a negative (−) jumper wire to upper terminal on side of starter motor and a positive (+) jumper to large lower terminal of starter motor. If motor still does not operate, remove for overhaul or replacement.
Starter turns over too slowly	Low battery or poor contact @ battery terminal	See "Starter does not operate."
	Poor contact @ starter solenoid or starter motor	Check all terminals for looseness and tighten all nuts securely.
	Starter mechanism	Disconnect positive (+) battery terminal. Rotate pinion gear in disengaged position. Pinion gear and motor should run freely by hand. If motor does not turn over easily, clean starter and replace all defective parts.
	Starter motor	See "Starter does not operate."
Starter spins freely but does not engage engine	Low battery or poor contact @ battery terminal	See "Starter does not operate."
	Poor contact @ starter solenoid or starter motor	See "Starter does not operate."
	Dirty or corroded pinion drive	Clean thoroughly and lubricate the spline underneath the pinion with Suzuki water-resistant grease (part No. 99000-25170).
Starter does not engage freely	Pinion or flywheel gear	Inspect mating gears for excessive wear. Replace all defective parts.
	Small anti-drift spring	If drive pinion interferes with flywheel gear after engine has started, inspect anti-drift spring located under pinion gear. Replace all defective parts. NOTE: If drive pinion tends to stay (continued on p. 82)
(continued)		

Table 1 STARTER TROUBLESHOOTING (continued)

Trouble	Cause	Remedy
Starter keeps on spinning after key is turned ON	Small anti-drift spring (continued)	engaged in flywheel gear when starter motor is in idle position, start motor @ 1/4 throttle to allow starter pinion gear to release flywheel ring gear instantly.
	Key not fully returned	Check that key has returned to normal ON position from START position. Replace switch if key constantly stays in START position.
	Starter solenoid	Inspect starter solenoid to see if contacts have become stuck in closed position. If starter does not stop running with small yellow lead disconnected from starter solenoid, replace starter solenoid.
	Wiring or key switch	Inspect all wires for defects. Open remote control box and inspect wiring @ switches. Repair or replace all defective parts.
Wires overheat	Battery terminals improperly connected	Check that negative marking on harness matches that of battery. If battery is connected improperly, red wire to rectifier will overheat.
	Short circuit in wiring system	Inspect all connections and wires for looseness or defects. Open remote control box and inspect wiring @ switches. Repair or replace all defective parts.
	Short circuit in choke solenoid	Check for high resistance. If blue choke wire heats rapidly when choke is used, choke solenoid may have internal short. Replace if defective.
	Short circuit in starter solenoid	If yellow starter solenoid lead overheats, there may be internal short (resistance) in starter solenoid. Replace if defective.
	Low battery voltage	Battery voltage is checked with an ampere-volt tester when battery is under a starting load. Battery must be recharged if it registers under 9.5 volts. If battery is below specified hydrometer reading of 1.230, it will not turn engine fast enough to start it.

Table 2 IGNITION TROUBLESHOOTING

Symptom	Probable cause
Engine won't start, but fuel and spark are good	Defective or dirty spark plugs. Spark plug gap set too wide. Improper spark timing. Shorted "kill" or stop button. Air leaks into fuel pump. Broken piston ring(s). Cylinder head, crankcase or cylinder sealing faulty. Worn crankcase oil seal.
(continued)	

TROUBLESHOOTING

Table 2 IGNITION TROUBLESHOOTING (continued)

Symptom	Probable cause
Engine misfires @ idle	Incorrect spark plug gap. Defective, dirty or loose spark plugs. Spark plugs of incorrect heat range. Cracked distributor cap. Leaking or broken high tension wires. Weak armature magnets. Defective coil or condenser. Defective ignition switch. Spark timing out of adjustment.
Engine misfires @ high speed	See "Engine misfires @ idle." Coil breaks down. Coil shorts through insulation. Spark plug gap too wide. Wrong type spark plugs. Too much spark advance.
Engine backfires through exhaust	Cracked spark plug insulator. Carbon path in distributor cap. Improper timing. Crossed spark plug wires.
Engine backfires through carburetor	Improper ignition timing.
Engine preignition	Spark advanced too far. Incorrect type spark plug. Burned spark plug electrodes.
Engine noises (knocking at power head)	Spark advanced too far.
Ignition coil fails	Extremely high voltage. Moisture formation. Excessive heat from engine.
Spark plugs burn and foul	Incorrect type plug. Fuel mixture too rich. Inferior grade of gasoline. Overheated engine. Excessive carbon in combustion chambers.
Ignition causing high fuel consumption	Incorrect spark timing. Leaking high tension wires. Incorrect spark plug gap. Fouled spark plugs. Incorrect spark advance. Weak ignition coil. Preignition.

Table 3 FUEL SYSTEM TROUBLESHOOTING (CARBURETTED MODELS)

Symptom	Probable cause
No fuel @ carburetor	No gas in tank.
	Air vent in gas cap not open.
	Air vent in gas cap clogged.
	Fuel tank sitting on fuel line.
	Fuel line fittings not properly connected to engine or fuel tank.
	Air leak @ fuel connection.
	Fuel pickup clogged.
	Defective fuel pump.
Flooding @ carburetor	Choke out of adjustment.
	High float level.
	Float stuck.
	Excessive fuel pump pressure.
	Float saturated beyond buoyancy.
Rough operation	Dirt or water in fuel.
	Reed valve open or broken.
	Incorrect fuel level in carburetor bowl.
	Carburetor loose @ mounting flange.
	Throttle shutter not closing completely.
	Throttle shutter valve installed incorrectly.
	Carburetor backdraft jets plugged (if so equipped).
Carburetor spit-back misfires @ high speed	Chipped or broken reed valve(s).
	Dirty carburetor.
	Lean carburetor adjustment.
	Restriction in fuel system.
	Low fuel pump pressure.
Engine backfires	Poor quality fuel.
	Air-fuel mixture too rich or too lean.
	Improperly adjusted carburetor.
Engine preignition	Excessive oil in fuel.
	Inferior grade of gasoline.
	Lean carburetor mixture.
Spark plugs burn and foul	Fuel mixture too rich.
	Inferior grade of gasoline.
High gas consumption: Flooding or leaking	Cracked carburetor casting.
	Leaks @ line connections.
	Defective carburetor bowl gasket.
	High float level.
	Plugged vent hole in cover.
	Loose needle and seat.
	Defective needle valve seat gasket.
	Worn needle valve and seat.
	Foreign matter clogging needle valve.

(continued)

TROUBLESHOOTING

Table 3 FUEL SYSTEM TROUBLESHOOTING (CARBURETTED MODELS) (continued)

Symptom	Probable cause
Flooding or leaking (continued)	Worn float pin or bracket.
	Float binding in bowl.
	High fuel pump pressure.
Overrich mixture	Choke lever stuck.
	High float level.
	High fuel pump pressure.
Abnormal speeds	Carburetor out of adjustment.
	Too much oil in fuel.

Table 4 LIGHTING/BATTERY CHARGING COIL RESISTANCE SPECIFICATIONS

Model	Ohms
DT 2	NA
DT 4	0.5-1.1
DT 6	0.37-0.45
DT 8	0.1-0.4
DT 9.9	0.1-0.4
DT 25, DT 30 (1985)	NA
DT 20, DT 25, DT 30 (1986-on)	0.2-0.6
DT 35, DT 40 (1987-on)	0.2-0.6
DT 40 (1985-1986)	NA
DT 55, DT 65	0.2-0.6
DT 75, DT 85	
1985-1986	
Red to yellow	0.54-0.66
Red to red/yellow	0.36-0.44
1987	
Red to yellow	0.4-0.8
Red to red/yellow	0.2-0.6
1988-on	0.25-0.35
DT 115, DT 140	
1985	
Red to yellow	0.4-0.8
Red to red/yellow	0.2-0.6
1986-on	0.1-0.3
V4	0.4-0.6
V6	
1986-1988	
No. 1 coil	0.05-0.2
No. 2 & No. 3 coil	0.1-0.4
1989-on	0.2-0.4

NA = Information not available from Suzuki.

Table 5 IGNITION COIL RESISTANCE SPECIFICATIONS

Model	Primary (ohms)	Secondary (ohms)
DT 2		
1985-1989	0.9-1.22	5,200-6,900
1990-on	0.8-1.2	7,000-10,000
DT 4	0.1-0.2	1,500-2,300

(continued)

Table 5 IGNITION COIL RESISTANCE SPECIFICATIONS (continued)

Model	Primary (ohms)	Secondary (ohms)
DT 6		
1985-1986	—	1,350-1,830
1987-on	—	1,272-1,908
DT 8	0.1-0.4	2,700-4,100
DT 9.9		
1985-1987	240-306	2,000-3,000
1988-on	0.1-0.4	2,700-4,100
DT 20, DT 25, DT 30		
(2-cylinder)		
1985	—	2,136-3,204
1986-on	—	2,100-3,200
DT 25, DT 30		
(3-cylinder)	0.1-0.4	1,900-2,700
DT 35	0.2-0.5	3,100-3,200
DT 40		
1985-1986	NA	2,300-3,100
1987-on	0.2-0.5	3,100-3,200
DT 55, DT 65		
1985-1988	0.5-0.8	4,700-7,000
1989-on	0.1-0.3	6,000-9,000
DT 75, DT 85		
1985-1986	0.21-0.29	2,130-2,880
1987	0.2-0.5	4,700-7,000
1988-on	0.2-0.3	1,800-2,800 +10,000*
DT 115, DT 140		
1985	0.2-0.5	4,700-7,000
1986-1988	0.4-0.7	5,000-8,000
1989-on	0.2-0.4	6,500-9,500
V4	0.15-0.25	2,600-3,800 +10,000*
V6		
1986-1987		
DT 150, DT 200	0.3-0.5	1,400-2,100
DT 150 SS	0.05-0.15	1,100-1,600
1988		
DT 150, DT 175, DT 200	0.1-0.3	2,600-3,800
DT 150 SS	0.05-0.15	1,100-1,600
1989-on		
DT 150 SS	0.05-0.15	3,300-4,900 +10,000*
DT 150, DT 175, DT 200	0.1-0.3	2,600-3,800 +10,000*
1990-on DT 225	0.3-0.15	3,300-4,900 +10,000*

NA = Information not available from Suzuki.
* = Spark plug cap resistance.

Table 6 CHARGE AND PULSER COIL RESISTANCE

Model	Charge coil resistance (ohms)	Pulser coil resistance (ohms)
DT 4	15-45	12-18
DT 6, DT 8		
1985-1986	257-314	20.5-25.1
1987	230-340	70-100
1988-on	230-340	170-250

(continued)

TROUBLESHOOTING

Table 6 CHARGE AND PULSER COIL RESISTANCE (continued)

Model	Charge coil resistance (ohms)	Pulser coil resistance (ohms)
DT 6, DT 8 (cont.)		
1988-on	230-340	170-250
DT 9.9		
1985-1987	200-310	NA
1988-on	230-340	170-250
DT 15		
1985-1988	200-310	NA
1989-on	170-250	260-380
DT 20	100-160	40-60
DT 25, DT 30 (1985)	102-154	27.9-41.9
DT 25 (1986-1988),		
DT 30 (1986-1987)	100-160	40-60
DT 25 (1989-on),		
DT 30 (1988-on)	170-250	170-250
DT 40 (1985-1986)	230-280	175-210
DT 35 (1987-1989),		
DT 40 (1987-on)	200-300	160-230
DT 55, DT 65 (1985-on)	190-270	170-250
DT 75, DT 85		
1985-1986		
Low speed	680-831	320-391
High speed	114-140	
1987		
Low speed	600-900	290-420
High speed	100-150	
1988-on	170-250	160-240
DT 115, DT 140		
1985		
Low speed	600-900	290-240
High speed	100-150	
1987-on	170-250	170-250
V4	180-270	160-230
DT 150, DT 175,		
DT 200		
1986-1988	180-260	160-240
1989-on	245-370	160-240
DT 225		
Green-Black/Red	62-92	160-240
White/Red-White/Blue	385-575	

TABLE 7 GEAR COUNTER COIL COIL RESISTANCE

Model	Counter coil resistance (ohms)
1989-on DT 25, 1988-on DT 30,	
1988 DT 75, DT 85,	
1986-on V6	160-240
1989-on DT 75, DT 85	150-260
1989-on DT 55, DT 65,	
1986-on DT 115, DT 140	170-250
V4	160-230

Chapter Four

Lubrication, Maintenance and Tune-up

The modern outboard motor delivers more power and performance than ever before, with higher compression ratios, new and improved electrical systems and other design advances. Proper lubrication, maintenance and tune-ups have thus become increasingly important as ways in which you can maintain a high level of performance, extend engine life and extract the maximum economy of operation.

You can do your own lubrication, maintenance and tune-ups if you follow the correct procedures and use common sense. The following information is based on recommendations from Suzuki that will help you keep your outboard motor operating at its peak performance level.

Table 1 provides a complete model history. **Tables 1-7** are at the end of the chapter.

LUBRICATION

Proper Fuel Selection

Two-stroke engines are lubricated by mixing oil with the fuel. The various components of the engine are lubricated as the fuel-oil mixture passes through the crankcase and cylinders. Since outboard fuel serves the dual function of producing ignition and distributing the lubrication, the use of low-octane marine white gasoline should be avoided. Such gasoline also has a tendency to cause ring sticking and port plugging.

All Suzuki outboards will use any gasoline with a minimum posted pump octane rating of 85, that works satisfactorily in an automotive engine. Unleaded gasoline is preferable to

LUBRICATION, MAINTENANCE AND TUNE-UP

leaded gasoline, as it offers longer spark plug life.

Sour Fuel

Fuel should *not* be stored for more than 60 days even under ideal conditions. Gasoline forms gum and varnish deposits as it ages. Gasoline tends to lose its potency after standing for long periods and condensation may contaminate it with water. Such fuel will cause starting and running problems. A good grade of gasoline stabilizer and conditioner additive may be used to prevent gum and varnish formation during storage or prolonged periods of non-use but it is always better to drain the tank in such cases. Always use *fresh* gasoline when mixing fuel for your outboard. This will help to keep you from getting stranded on the water with no power.

Gasohol

Some gasolines sold for marine use now contain alcohol, although this fact may not be advertised. A mixture of 10 percent ethyl alcohol and 90 percent unleaded gasoline is called gasohol. While Suzuki does *not* recommend gasohol for use in its outboards, testing to date has found that it causes no major deterioration of the fuel system or its component parts when consumed immediately after purchase.

Fuels with an alcohol content tend to slowly absorb moisture from the air. When the moisture content of the fuel reaches approximately one half of one percent, it combines with the alcohol and separates from the fuel. This separation does not normally occur when gasohol is used in an automobile, as the tank is generally emptied within a few days after filling it.

The problem does occur in marine use, however, because boats often remain idle between start-ups for days or even weeks. This length of time permits separation to take place. The alcohol-water mixture settles at the bottom of the fuel tank. Since outboard motors will not run on this mixture, it is necessary to drain the fuel tank, flush out the fuel system with clean gasoline and then remove, clean and reinstall the spark plug(s) before the engine can be started.

Continued use of fuels containing alcohol can "melt" the fuel level indicator lens in portable fuel tanks. Many late-model replacement tanks now contain an alcohol-resistant lens.

The major danger of using gasohol in an outboard motor is that a shot of the water-alcohol mixture may be picked up and sent to one of the carburetors, or fuel injectors, of a multi-cylinder engine. Since this mixture contains no oil, it will wash oil off the bore of any cylinder it enters. The other carburetor, or fuel injector, receiving good fuel-oil mixture will keep the engine running while the cylinder receiving the water-alcohol mixture can suffer internal damage.

The problem of unlabeled gasohol has become so prevalent around the United States that Miller Tools (32615 Park Lane, Garden City, MI 48135) now offers an Alcohol Detection Kit (part No. C-4846) so that owners and mechanics can determine the quality of fuel being used.

The kit cannot differentiate between types of alcohol (ethanol, methanol, etc.) nor is it considered to be absolutely accurate from a scientific standpoint, but it is accurate enough to determine whether or not there is sufficient alcohol in the fuel to cause the user to take precautions.

Recommended Fuel Mixture (Oil Injected Models)

The 1985-on DT 40, 1988-on DT 8 and DT 9.9, 1989-on DT 15, 1989-on DT 25, 1988-on DT 30, all 3- and 4-cylinder inline models and all V4 and V6 models covered in this manual are equipped with a mechanical oil injection system. A crankshaft-driven injection pump automatically injects oil into the intake manifold at a variable ratio from approximately 120:1 at idle to approximately 50:1 at full throttle. A sensor monitoring the injection systems sounds when the oil tank requires replenishment. To replenish the system, lift the lid on the engine cover. Reach inside, remove the oil tank cap and fill the tank with Suzuki CCI 50:1 Outboard Oil. See Chapter Twelve for system operation and service procedures.

CAUTION
Do not, under any circumstances, use multigrade or other high detergent automotive oils or oils containing metallic additives. Such oils are harmful to 2-stroke engines. Since they do not mix properly with gasoline, do not burn as 2-stroke oils do and leave an ash residue, their use may result in piston scoring, bearing failure or other engine damage.

NOTE
In order to comply with any applicable Suzuki warranty on 1990 and later models, you must use Suzuki Outboard motor oil or NMMA Certified TC-WII outboard motor oil. This type of motor oil is also highly recommended for use in all outboard motors covered in this book. This type of motor oil has shown no tendency to gel or cause filtration problems. The use of the normally recommended ashless oils can gel which will lead to filter blockage in the oil injection system. This blockage can cause insufficient lubrication to the cylinders and can result in engine seizure.

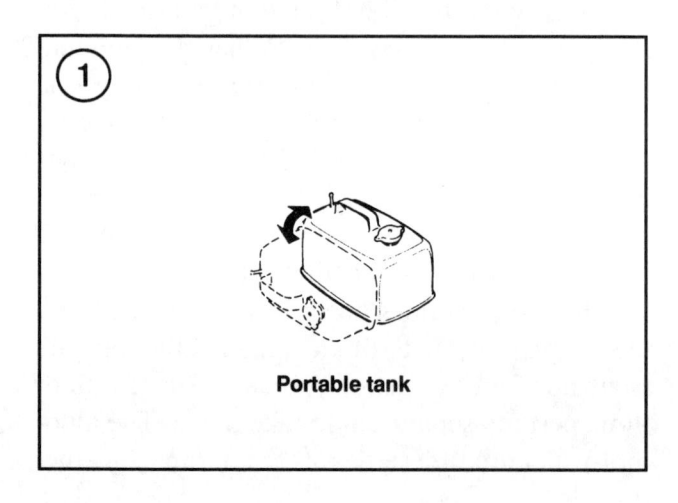

Recommended Fuel Mixture (Models Without Oil Injection)

On models that do not have oil injected engines, use the specified gasoline and mix with Suzuki CCI 50:1 Outboard Oil in the following ratios:

CAUTION
Do not, under any circumstances, use multigrade or other high detergent automotive oils or oils containing metallic additives. Such oils are harmful to 2-stroke engines. Since they do not mix properly with gasoline, do not burn as 2-stroke oils do and leave an ash residue, their use may result in piston scoring, bearing failure or other engine damage.

1. During the break-in period, thoroughly mix 26 ounces of Suzuki CCI 50:1 Outboard Oil with each 5-7/8 gallons of gasoline in your 6 gallon Suzuki fuel tank (or 17 ounces with each 3-7/8 gallons in a 4 gallon tank). This provides the recommended 30:1 mixture.

2. After engine break-in, mix 16 ounces with each 5-7/8 gallons of gasoline in your 6 gallon Suzuki fuel tank (or 10 ounces with each 3-7/8 gallons in a 4 gallon tank). This provides a 50:1 mixture.

CAUTION
*There are a number of oil products on the market which specify use at 100:1. They are **not** NMMA TC-WII approved and should **not** be used.*

NOTE
In order to comply with any applicable Suzuki warranty on 1990 and later models, you must use Suzuki Outboard motor oil or NMMA Certified TC-W II outboard motor oil. This type of motor oil is also highly recommended for use in all outboard motors covered in this book. This type of motor oil has shown

①

Portable tank

LUBRICATION, MAINTENANCE AND TUNE-UP

no tendency to gel or cause filtration problems. The use of the normally recommended ashless oils can gel which will lead to filter blockage in the oil injection system. This blockage can cause insufficient lubrication to the cylinders and can result in engine seizure.

If Suzuki CCI 50:1 Outboard Oil is not available, any high-quality 2-stroke oil intended for outboard use may be substituted provided the oil meets NMMA rating TC-WII and specifies so on the container. Follow the manufacturer's mixing instructions on the container but do *not* exceed a 50:1 ratio (30:1 during break-in). *Never* use more than one brand or type of 2-stroke oil when mixing a batch of fuel and oil. The different brands of oil may not be compatible with each other, resulting in a lack of lubrication or excessive carbon deposits.

Correct Fuel Mixing

Mix the fuel and oil outdoors or in a well-ventilated indoor location. Mix the fuel directly in the remote tank.

WARNING
*Serious fire hazards always exist around gasoline. Do **not** allow anyone to smoke where fuel is being mixed. Always have a fire extinguisher rated for gasoline fires (Class B) close by. Never use gasoline near heat, sparks or open flame.*

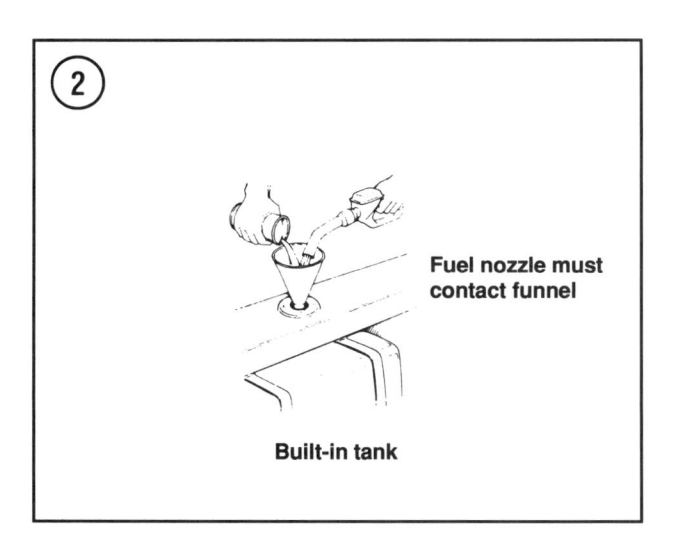

NOTE
Always mix fresh gasoline. Mix only the amount that you feel will be used for that day's outing. Gasoline loses its potency after sitting for a period of time. The oil loses some of its lubricating ability when mixed with the gasoline and not used for a period of time.

Using less than the specified amount of oil can result in insufficient lubrication and serious engine damage. Using more oil than specified causes spark plug fouling which will make engine starting difficult, if not impossible. It will also cause erratic carburetion, excessive smoking and rapid carbon accumulation on the piston crown and cylinder head which can cause preignition.

Cleanliness is of prime importance. Even a very small particle of dirt can cause carburetion problems. Always use fresh gasoline. Gum and varnish deposits tend to form in gasoline stored in a tank for any length of time. Use of sour fuel can result in carburetor problems and spark plug fouling.

Above 32° F (0° C)

Measure the required amounts of gasoline and Suzuki CCI 50:1 Outboard Oil accurately. Pour the specified amount of oil into the portable tank and add one-half of the gasoline to be mixed. Install the tank filler cap and mix the fuel by tipping the tank on its side and back to an upright position several times (**Figure 1**). Remove the tank cap and add the balance of the gasoline, then thoroughly mix again.

If a built-in tank is used, insert a large filter funnel in the tank filler neck. Slowly pour the Suzuki CCI 50:1 Outboard Oil into the funnel at the same time the tank is being filled with gasoline (**Figure 2**).

Below 32° F (0° C)

Measure the required amounts of gasoline and Suzuki CCI 50:1 Outboard Oil accurately. Pour about one gallon of gasoline in the tank and then add the required amount of oil. Install the tank filler cap and shake the tank to mix the fuel and oil thoroughly. Remove the cap and add the balance of the gasoline, then thoroughly mix again.

If a built-in tank is used, insert a large filter funnel in the tank filler neck. Mix the required amount of Suzuki CCI 50:1 Outboard Oil with one gallon of gasoline in a separate container. Slowly pour the mixture into the funnel at the same time the tank is being filled with gasoline.

Consistent Fuel Mixtures

The carburetor idle adjustment is sensitive to fuel mixture variations which result from the use of different oils and gasolines or from inaccurate measuring and mixing. This may require readjustment of the idle needle. To prevent the necessity for constant readjustment of the carburetor from one batch of fuel to the next, always be consistent. Prepare each batch of fuel exactly the same as previous ones.

Pre-mixed fuels sold at some marinas are not recommended for use in Suzuki outboards, since the quality and consistency of pre-mixed fuels can vary greatly. The possibility of engine damage resulting from use of an incorrect fuel mixture outweighs the convenience offered by pre-mixed fuel.

Gearcase Lubrication

Change the gearcase lubricant after the first 10 hours of operation and replace at 50 hour intervals or at least once per season. Use Suzuki Outboard Motor Gear Oil. If this is not available, use a high quality non-corrosive E.P. 90 outboard gear lubricant.

CAUTION
Do not use regular automotive grease in the gearcase. Its expansion and foam characteristics are not suitable for marine use.

Gearcase Lubricant Check

To assure a correct level check, the engine must be in the upright position and not run for at least 2 hours before performing this procedure. Refer to **Figure 3** for DT 2 models or **Figure 4** for all other models.

1. Remove the engine cover and disconnect the spark plug lead(s) to prevent accidental starting of the engine.
2A. On DT 2 models, locate and loosen (but do not remove) the gearcase drain plug.
2B. On all other models, locate and loosen (but do not remove) the gearcase drain plug (2, **Figure 4**).

CAUTION
If water is noted in Step 3, retighten the drain plug securely and pressure test the

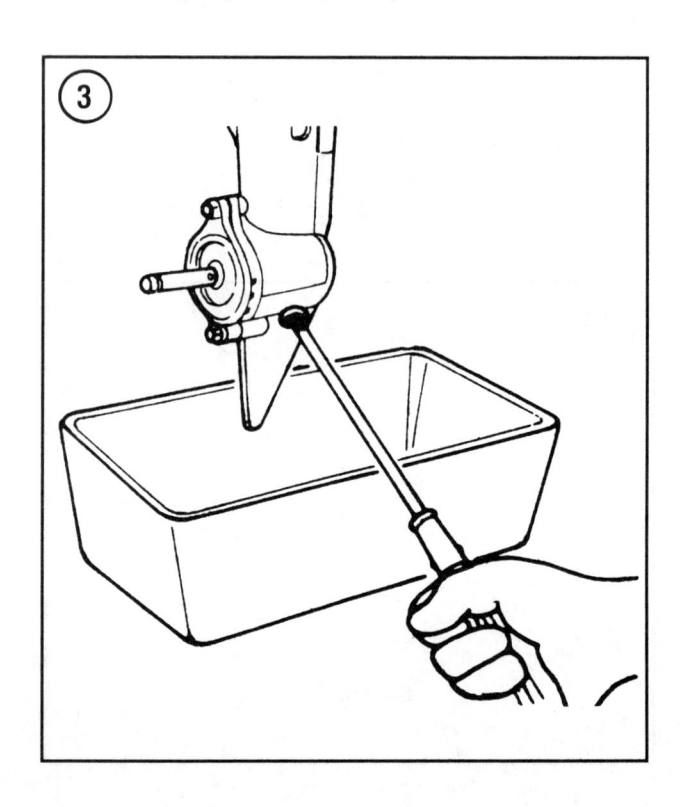

LUBRICATION, MAINTENANCE AND TUNE-UP

gearcase to determine if a seal has failed or if the water is simply condensation in the gearcase. See Chapter Nine.

3. Allow a small amount of lubricant to drain. If there is water in the gearcase, it will drain out before the lubricant. Retighten the drain plug securely.

NOTE
If lubricant level is low, proceed to Step 4. If the level is correct, reconnect the spark plug lead(s) and install the engine cover.

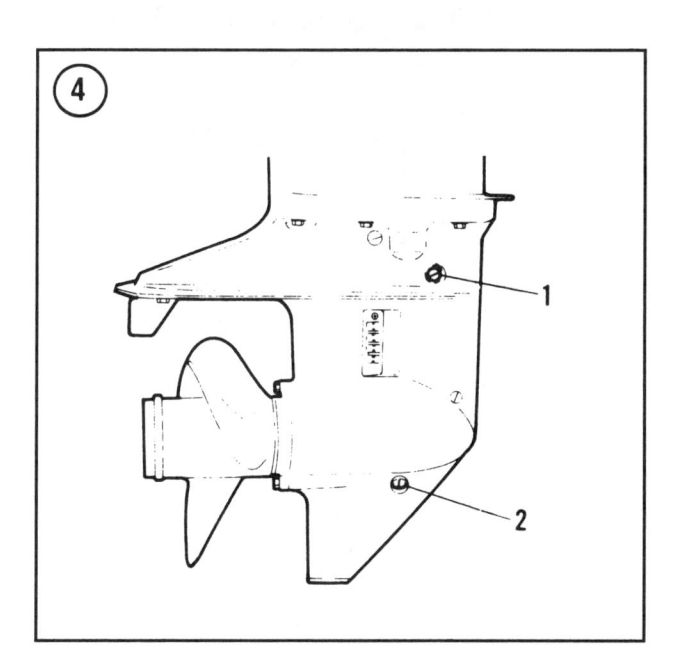

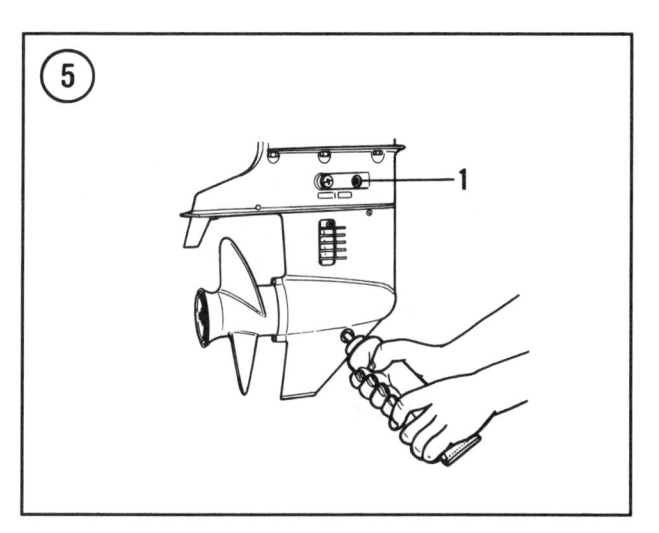

4. On all models except DT 2, remove the vent plug (1, **Figure 4**). Do not lose the accompanying washer. The lubricant should be level with the bottom of the vent plug hole.

CAUTION
Never lubricate the gearcase without first removing the vent plug, as the injected lubricant displaces air which must be allowed to escape. The gearcase cannot be completely filled otherwise.

5. On all models except DT 2, if the lubricant level is low, remove the drain plug (2, **Figure 4**).

NOTE
On DT 2 gearcases, the lubricant will flow out of the drain hole when the gearcase is full.

6A. On DT 2 models, perform the following:
 a. Inject lubricant into the fill/drain hole (**Figure 3**) until excess fluid flows out the fill/drain hole.
 b. Be sure the washers are in place under the head of the fill/drain plug, so that water will not leak past the threads into the housing. Install the fill/drain plug and tighten securely.

6B. On all other models, perform the following:
 a. Inject lubricant into the drain hole (**Figure 5**) until excess fluid flows out the vent plug hole.
 b. Be sure the washers are in place under the head of each plug. Install vent plug, then the drain plug and tighten securely.

7. Wipe any excess lubricant off the gearcase exterior.

NOTE
If the gearcase has been disassembled, recheck the lubricant level after 2-3 hours of operation. Top up if necessary.

8. On all models except DT 2, after refilling the gearcase, remove the vent plug and washer (1, **Figure 4**). Let gearcase stand upright for a minimum of 1/2 hour, then recheck the lubricant

Gearcase Lubricant Change

Refer to **Figure 3** or **Figure 4** for this procedure.

1. Remove the engine cover and disconnect the spark plug lead(s) to prevent accidental starting of the engine.
2. Place a container under the fill/drain or drain plug.
3A. On DT 2 models, locate and remove the gearcase drain plug.
3B. On all other models, locate and remove the gearcase drain plug (2, **Figure 4**). Remove the vent plug (1, **Figure 4**).

NOTE
If the lubricant is creamy in color or metallic particles are found in Step 5, remove and disassemble the gearcase to determine and correct the cause of the problem.

4. Drain all lubricant from the gearcase.
5. Wipe a small amount of lubricant on a finger and rub the finger and thumb together. Check for the presence of metallic particles in the lubricant. Note the color of the lubricant. A white or creamy color indicates water in the lubricant. Check the drain container for signs of water separation from the lubricant. If water is present, it will rise to the top of the grease.
6. Perform Steps 6-8 of *Gearcase Lubricant Check* in this chapter.

Other Lubrication Points

Refer to **Figures 6-11**, typical, and **Table 2** for other lubricant points and frequency of lubrication. Use water-resistant grease for all grease fittings and Suzuki Outboard Motor Gear Oil for topping up or changing the gearcase lubricant.

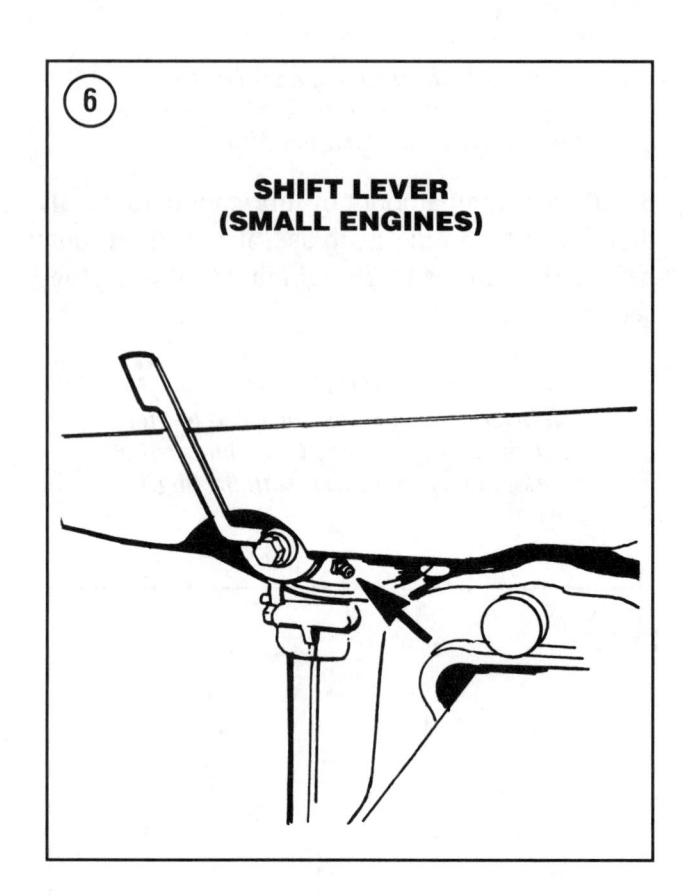

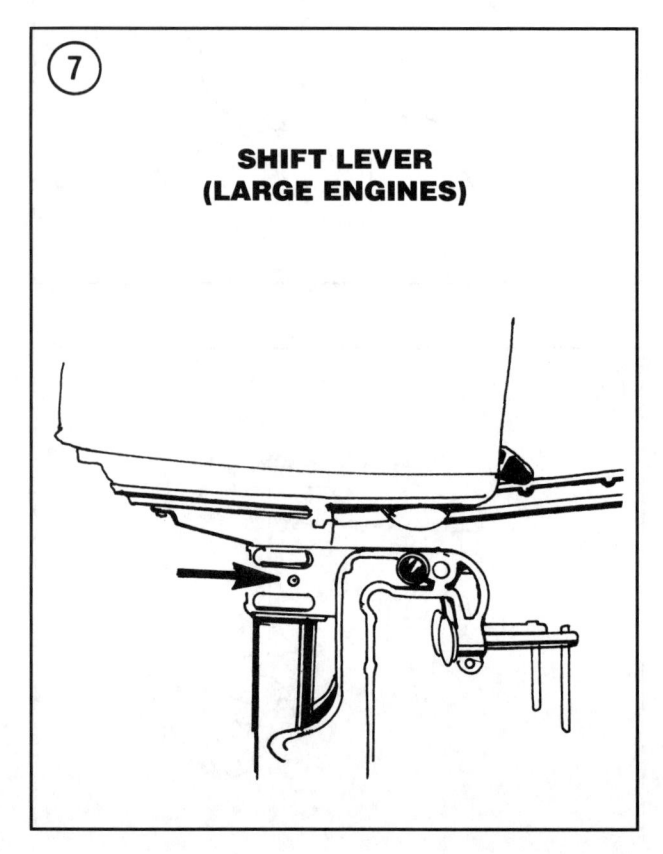

LUBRICATION, MAINTENANCE AND TUNE-UP

In addition to these lubrication points, some motors may also have grease fittings provided at critical points where bearing surfaces are not externally exposed. These fittings should be lubricated at least once each season with an automotive type grease gun and water-resistant grease.

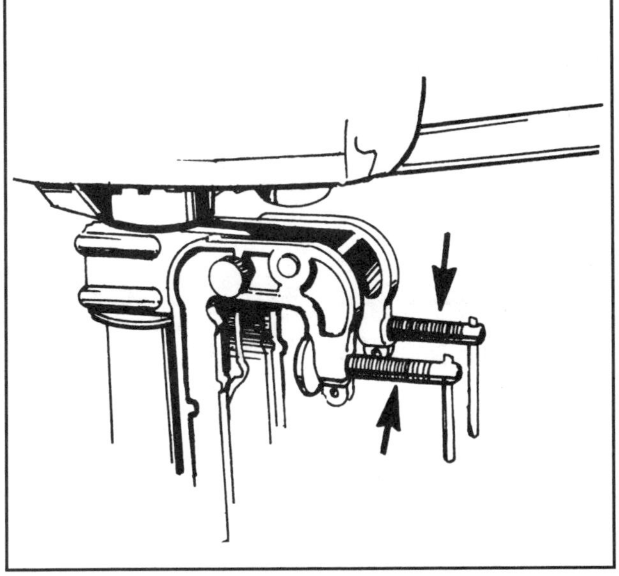

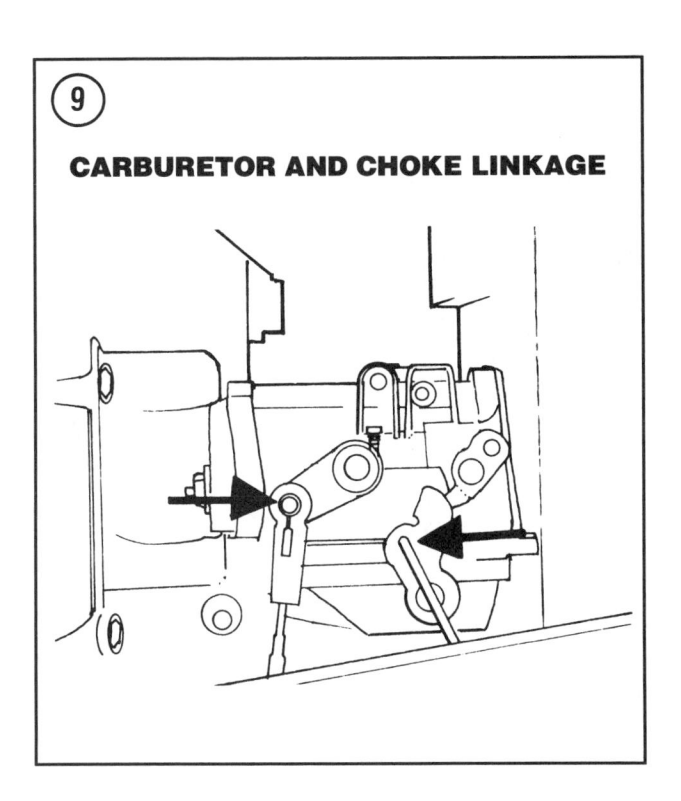

CAUTION
When lubricating the steering cable on models so equipped, make sure its core is fully retracted into the cable housing. Lubricating the cable while extended can cause a hydraulic lock to occur.

Saltwater Corrosion of Gearcase Bearing Cage or Spool

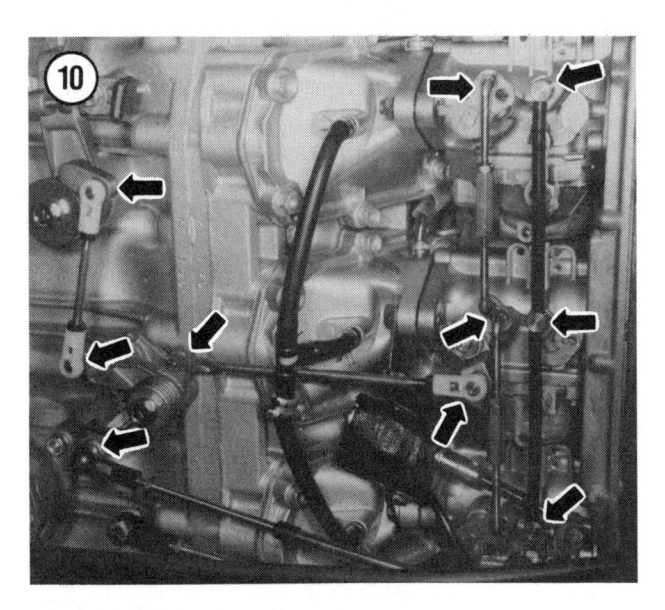

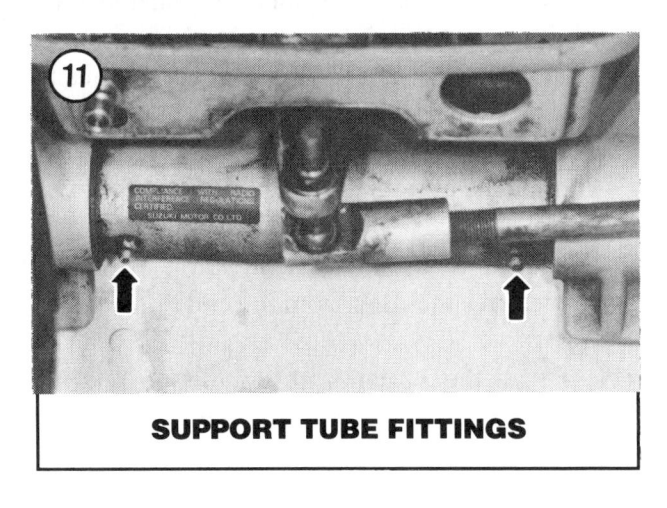

Saltwater corrosion that is allowed to build up unchecked can eventually split the gearcase and destroy the lower unit. If the motor is used in saltwater, remove the propeller shaft bearing housing (**Figure 12**, typical) at least once a year after the initial 10-hour inspection. Clean all corrosive deposits and dried-up lubricant from each end of the housing.

Install *new* O-rings on the bearing housing and wipe the outer diameter of the housing (A and B, **Figure 12**) with Suzuki water-resistant grease. Apply a thin coat of Suzuki silicone seal to the gearcase and bearing housing mating surfaces and install the housing. Lubricate the propeller shaft splines with Suzuki water-resistant grease and reinstall the propeller.

STORAGE

The major consideration in preparing an outboard motor for storage is to protect it from rust, corrosion and dirt. Suzuki recommends the following procedure.

1. Remove the engine cover.
2. Operate the motor in a test tank or attach a flush device (**Figure 13**). Start the engine and run at fast idle until warmed up.
3. Shift the engine into NEUTRAL. Disconnect the fuel line and run the engine at fast idle while pouring about 2 ounces of Suzuki CCI 50:1 Outboard Oil into the carburetor, or fuel injection air intake until the engine stalls out.
4. Remove spark plug(s) as described in this chapter. Pour about one ounce of Suzuki CCI 50:1 Outboard Oil into each spark plug hole. Slowly rotate flywheel by hand several times to distribute the oil throughout the cylinder(s). Reinstall the spark plugs.
5. Service the portable fuel tank filter as follows:
   a. Remove the bolts and washers securing the adaptor and remove the adaptor.
   b. Remove the filter from the end of the adaptor.
   c. Clean the screen with solvent to remove any particles. Refer to **Figure 14** or **Figure 15**.
   d. Reinstall filter screen to the tank adapter.
   e. Install the adaptor and the bolts and washers. Tighten the bolts securely.
6. Service the engine fuel filter as described in this chapter.
7. Drain and refill the gearcase as described in this chapter. Check the condition of vent, fill or fill/drain plug gaskets. Replace as required.
8. Refer to **Figures 6-11** and **Table 2** as appropriate and lubricate motor at all specified points. See **Table 3** for recommended types of lubricants.

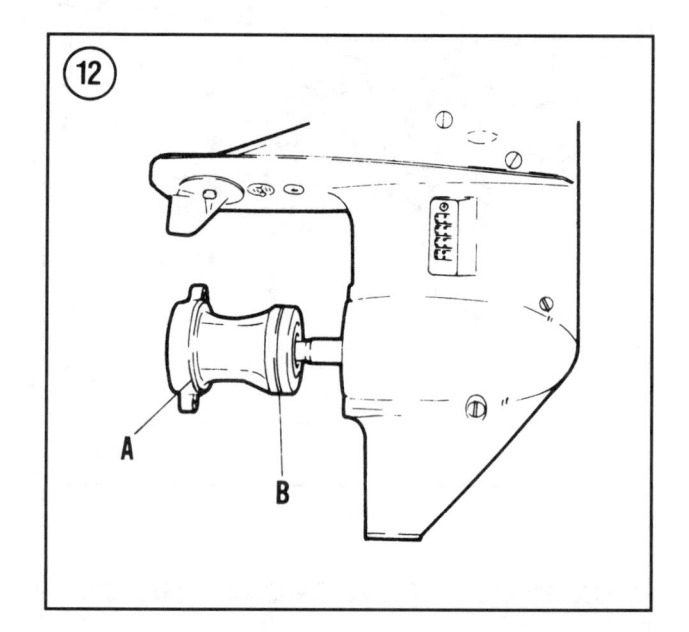

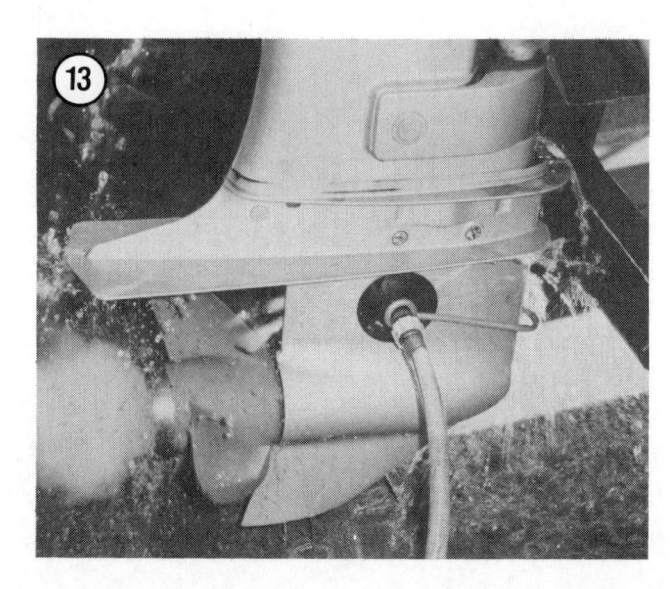

LUBRICATION, MAINTENANCE AND TUNE-UP

9. Clean the motor, including all accessible power head parts. Coat with a good marine-type wax. Install the engine cover.

10. Remove the propeller and lubricate propeller shaft splines with Suzuki water-resistant grease. Reinstall the propeller.

11. Store the motor upright in a dry and well-ventilated area.

12. Service the battery (if so equipped) as follows:

   a. Disconnect the negative battery cable, then the positive battery cable.
   b. Remove all grease, corrosion and dirt from the battery surface. 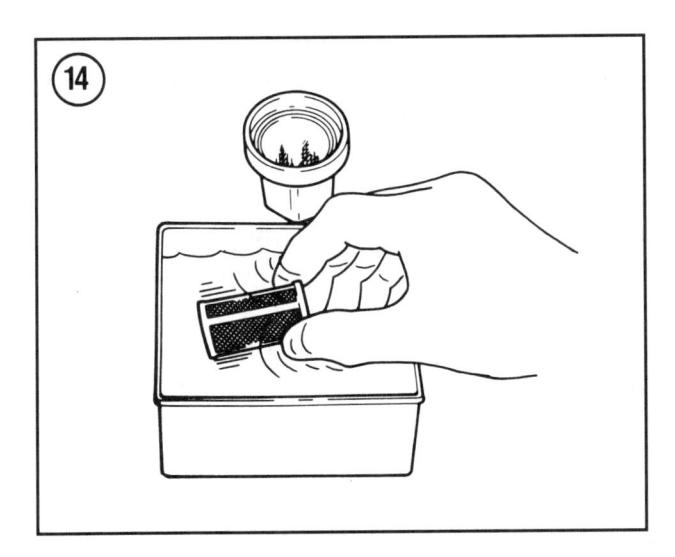 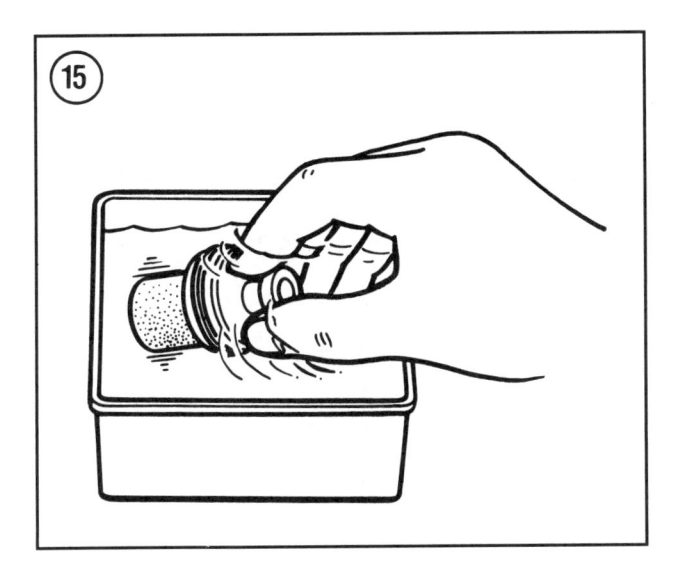
   c. Check the electrolyte level in each battery cell and top up with distilled water, if necessary. Fluid level in each cell should not be higher than 3/16 in. above the perforated baffles. Gently shake the battery to mix the water into the electrolyte.
   d. Lubricate the cable terminal bolts with grease or petroleum jelly.

   CAUTION
   A discharged battery can be damaged by freezing.

   e. With the battery in a fully charged condition (specific gravity 1.260-1.275), store in a dry place where the temperature will not drop below freezing.
   f. Recharge the battery every 45 days or whenever the specific gravity drops below 1.230. Before charging, add sufficient distilled water to cover the top of the plates, but not more than 3/16 in. above the perforated baffles. The charge rate should not exceed 6 amps. Stop charging when the specific gravity reaches 1.260 at 80° F (27° C).
   g. Before placing the battery back into service after storage, remove the excess grease from the terminals, leaving a small amount on. Make sure the battery is fully charged before returning to service.

13. Cover the motor with a tarp, blanket or heavy plastic drop cloth. Place this cover over the motor mainly as a dust cover—do not wrap it tightly, especially any plastic material, as it may trap moisture causing condensation. Leave room for air to circulate around the motor.

COMPLETE SUBMERSION

An outboard motor which has been lost overboard should be recovered as quickly as possible. If the motor was running when submerged, disassemble and clean it immediately—any delay will result in rust and corrosion of internal components once it has been removed from the water. If the motor was not running and appears to be undamaged mechanically with no abrasive dirt or silt inside, take the following emergency steps immediately.

1. Wash the outside of the motor with clean water to remove weeds, mud and other debris.

2. Remove the engine cover.

3. If recovered from saltwater, flush the motor completely with freshwater. If compressed air is available, blow off all water from exterior surfaces.

4. Remove the spark plug(s) as described in this chapter.

CAUTION
Do not force the motor if it does not turn over freely by hand in Step 5. This may be an indication of internal damage such as a bent connecting rod or broken piston.

5. Drain as much water as possible from the power head by placing the motor in a horizontal position. Manually rotate the flywheel with the spark plug hole(s) facing downward to expel the water within the cylinder(s).

6. Dry and reinstall the spark plug(s).

7. Dry all ignition components. Use compressed air if available and spray all components with an aerosol electrical contact cleaner. This will help evaporate any remaining moisture.

8. Drain the fuel lines and carburetor(s).

9. On models with an integral fuel tank, drain the tank and flush with fresh gasoline until all water has been removed.

CAUTION
If there is a possibility that sand may have entered the power head, do not try to start the motor or severe internal damage may occur.

10. Try starting the motor with a tank of fresh fuel. If the motor will start, let it run at least one hour to eliminate any water remaining inside.

CAUTION
If it is not possible to disassemble and clean the motor immediately in Step 11, resubmerge the power head in water to prevent rust and corrosion formation until it can be properly serviced.

11. If the motor will not start in Step 10, try to diagnose the cause as fuel, electrical or mechanical and correct the problem. If the engine cannot be started within 2 hours, disassemble, clean and oil all parts thoroughly as soon as possible.

ANTI-CORROSION MAINTENANCE

1. Flush the cooling system with freshwater as described in this chapter after each time the motor is used in saltwater. Completely wash exterior with freshwater.

2. Dry the exterior of the motor with compressed air to remove all moisture from all crevasses. Apply a marine primer over any paint nicks and scratches. Use only Suzuki touch-up paints available at your dealer. Do not use paints containing mercury or copper. Do *not* paint sacrificial anodes.

3. Spray the power head and all electrical connections with a good quality corrosion and rust preventative.

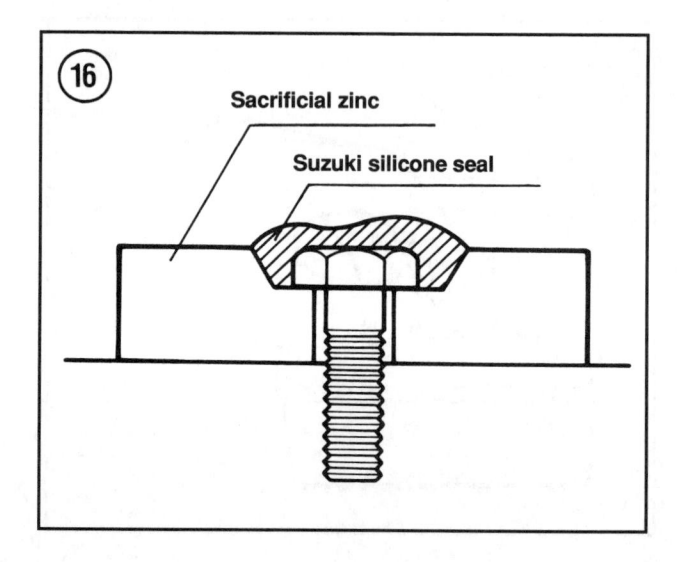

LUBRICATION, MAINTENANCE AND TUNE-UP

4. If used consistently in saltwater, reduce lubrication intervals in **Table 2** by one-half.

GALVANIC PROTECTION AND BONDING WIRES

Corrosion occurs when two or more dissimilar metals are connected to form a continuous electrical path and are submersed in a conductive solution, such as water (especially brackish, polluted or saltwater). Different metals posses different voltage potentials. When dissimilar metals are submersed in water, electrical current flows between the metals causing the metal that is the most chemically active (less noble) to erode. This process is called galvanic corrosion. See Chapter One.

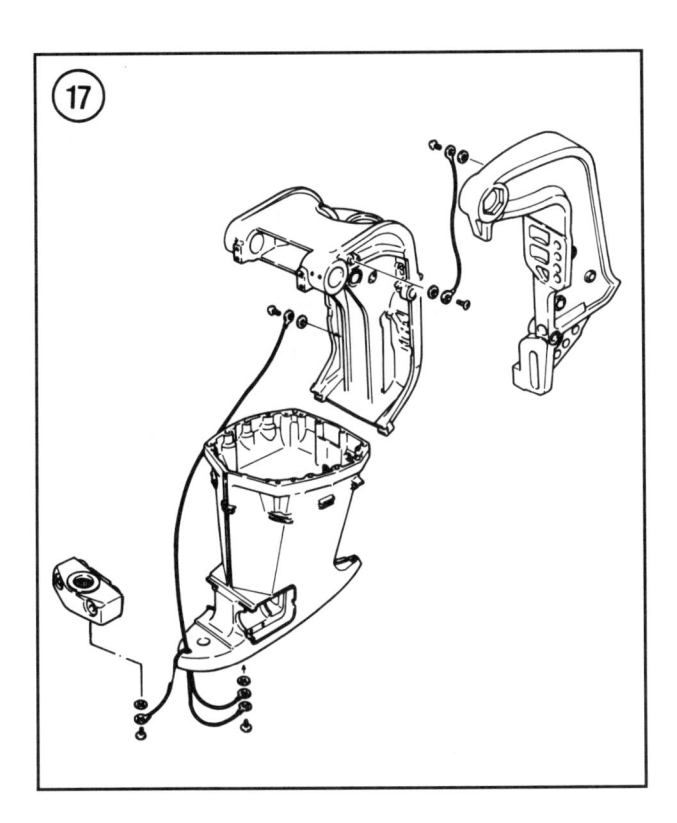

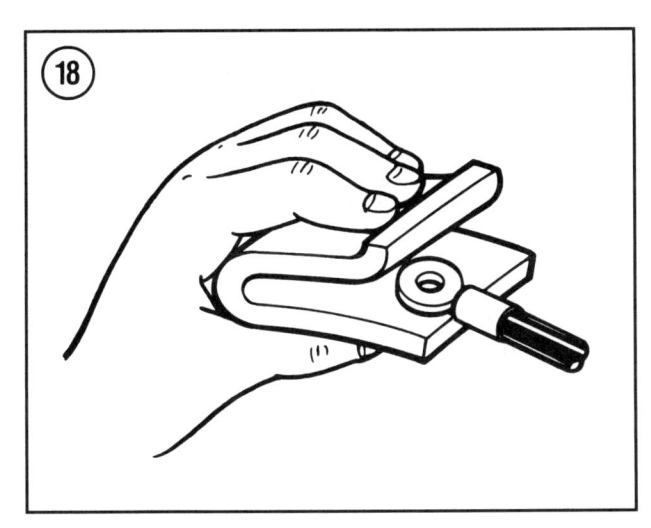

To protect the outboard motor from this galvanic action, pieces of sacrifical zinc (anodes) are installed on the outside of the motor and within some internal water passages. The zinc protects the more critical components of the outboard by sacrificing (corroding) itself.

As the zinc anodes are corroded away, they require periodic replacement. Although no specific replacement interval is provided, they should be replaced when corroded to half their original size. When replacing the zinc anode, cover the securing bolt with Suzuki Silocone Seal (part 99000-31120) (**Figure 16**). Make sure the surface to which the anode is mounted is scraped down to bare metal and the anode mounting screws are fastened tightly. The anode will not do its job unless it makes a good electrical contact.

Additional corrosion protection can be obtained by a continuity circuit. The bonding circuit assures that electrical continuity will be maintained between all components on the outboard that will be underwater. It consists of a series of ground wires that serve as bridges to connect components. Electrically speaking, the ground wires tie the individual components into one large component. **Figure 17** shows a typical layout of the bonding system.

Whenever a bonding wire is reattached to a component, clean the terminal connector with cleaning solvent and clean cloth (**Figure 18**). To prevent galvanic activity between the bonding wire and the component it is attached to, place an insulating washer under the bonding wire terminal.

ENGINE FLUSHING

Periodic engine flushing will prevent salt or silt deposits from accumulating within the engine's water passageways. This procedure should also be performed whenever an outboard motor is operated in salt water or polluted water.

Keep the engine in an upright position during and after flushing. This prevents water from passing into the power head through the drive shaft housing and exhaust ports during the flushing procedure. This also eliminates the possibility of residual water being trapped in the drive shaft housing or other passageways.

1. Attach a flushing device according to the manufacturer's instructions.
2. Connect a garden hose between a water tap and the flushing device.
3. Open the water tap partially—do not use full pressure.
4. Shift into NEUTRAL, then start engine. Keep engine at idle speed.
5. Adjust the water flow so that there is a slight loss of water around the rubber cups of the flushing device (**Figure 13**).
6. Check the engine to make sure that water is being discharged from the "tell-tale" nozzle. If not, stop the engine immediately and determine the cause of the problem.

CAUTION
Flush the engine for at least 5 minutes if used in saltwater.

7. Flush the engine until the discharged water is clear. Stop the engine.
8. Close the water tap and remove the flushing device from the gearcase.

TUNE-UP

A tune-up consists of a series of inspections, adjustments and parts replacements to compensate for normal wear and deterioration of outboard engine components. Regular tune-ups are important for optimum power, performance and economy. Suzuki recommends that its outboards be serviced every 6 months or 50 hours of operation, whichever comes first. If subjected to limited use, the engine should be tuned at least once a year.

Since proper outboard engine operation depends upon a number of interrelated system functions, a tune-up consisting of only one or two corrections will seldom give lasting results. For best results, a thorough and systematic procedure of analysis and correction is necessary.

Prior to performing a tune-up, it is a good idea to flush the engine as described in this chapter and check for satisfactory water pump operation.

NOTE
It is a good idea to start the engine after each one of the tune-up procedures is completed and make sure it runs okay. If for some reason, the procedure was not done correctly or a faulty new part(s) was installed, you can then concentrate on that specific procedure and part(s) and correct the problem. If you wait until all of the tune-up procedures are completed and then the motor runs poorly or does not start at all, then you have to narrow it down to which one of the procedures or parts is causing the problem.

The tune-up sequence recommended by Suzuki includes the following:
  a. Compression check.
  b. Spark plug service.
  c. Gearcase and water pump check.
  d. Fuel system service.
  e. Ignition system service.
  f. Battery, starter and relay/solenoid check (if so equipped).
  g. Wiring harness check.
  h. Timing, synchronization and adjustment.
  i. Performance test (on boat).

Anytime the fuel or ignition systems are adjusted or defective parts replaced, the engine timing, synchronization and adjustment *must* be checked. These procedures are described in

LUBRICATION, MAINTENANCE AND TUNE-UP

Chapter Five. Perform the timing, synchronization and adjustment procedure for your engine *before* running the performance test.

Compression Check

An accurate cylinder compression check gives a good idea of the condition of the basic working parts of the engine. It is also an important first step in any tune-up, as an engine with low or unequal compression between cylinders *cannot* be satisfactorily tuned. Any compression problem discovered during this check must be corrected before continuing with the tune-up procedure.

1. With the engine warm, disconnect the spark plug wire(s) and remove the plug(s) as described in this chapter.
2. Ground the spark plug wire(s) to the engine to disable the ignition system.
3. Connect the compression tester to the top spark plug hole according to manufacturer's instructions (**Figure 19**).
4. With the throttle set to the wide-open position, crank the engine through at least 4 compression strokes. Record the gauge reading.
5. Repeat Step 3 and Step 4 on each cylinder of multi-cylinder engines.

The actual readings are not as important as the differences in readings when interpreting the results. A variation of more than 10-15 psi between 2 cylinders indicates a problem with the lower reading cylinder, such as worn or sticking piston rings or scored pistons or cylinders. In such cases, pour a tablespoon of engine oil into the suspect cylinder and repeat Step 3 and Step 4. If the compression is raised significantly (by 10 psi in an older engine), the rings are worn and should be replaced.

Many outboard engines are plagued by hard starting and generally poor running for which there seems to be no good cause. Carburetion and ignition check out satisfactorily and a compression test may show that everything is well in the engine's upper end. With everything apparently pointing to a sound engine, owners focus on the propeller as the cause of the problem and start swapping props, often with disastrous results.

What a compression test does *not* show is lack of primary compression. In a 2-stroke engine, the crankcase must be alternately under high pressure and low pressure. After the piston closes the intake port, further downward movement of the piston causes the entrapped mixture to be pressurized so that it can rush quickly into the cylinder when the scavenging ports are opened. Upward piston movement creates a lower pressure in the crankcase, enabling fuel-air mixture to pass in from the carburetor.

When the crankshaft seals or case gaskets leak, the crankcase cannot hold pressure and proper engine operation becomes impossible. Any other source of leakage, such as defective cylinder base gaskets or a porous or cracked crankcase casting, will result in the same conditions.

If the power head shows signs of overheating (discolored or scorched paint) but the compression test turns up nothing abnormal, check the cylinder(s) visually through the transfer ports for possible scoring. A cylinder can be slightly scored and still deliver a relatively good com-

SPARK PLUG ANALYSIS
(CONVENTIONAL GAP SPARK PLUGS)

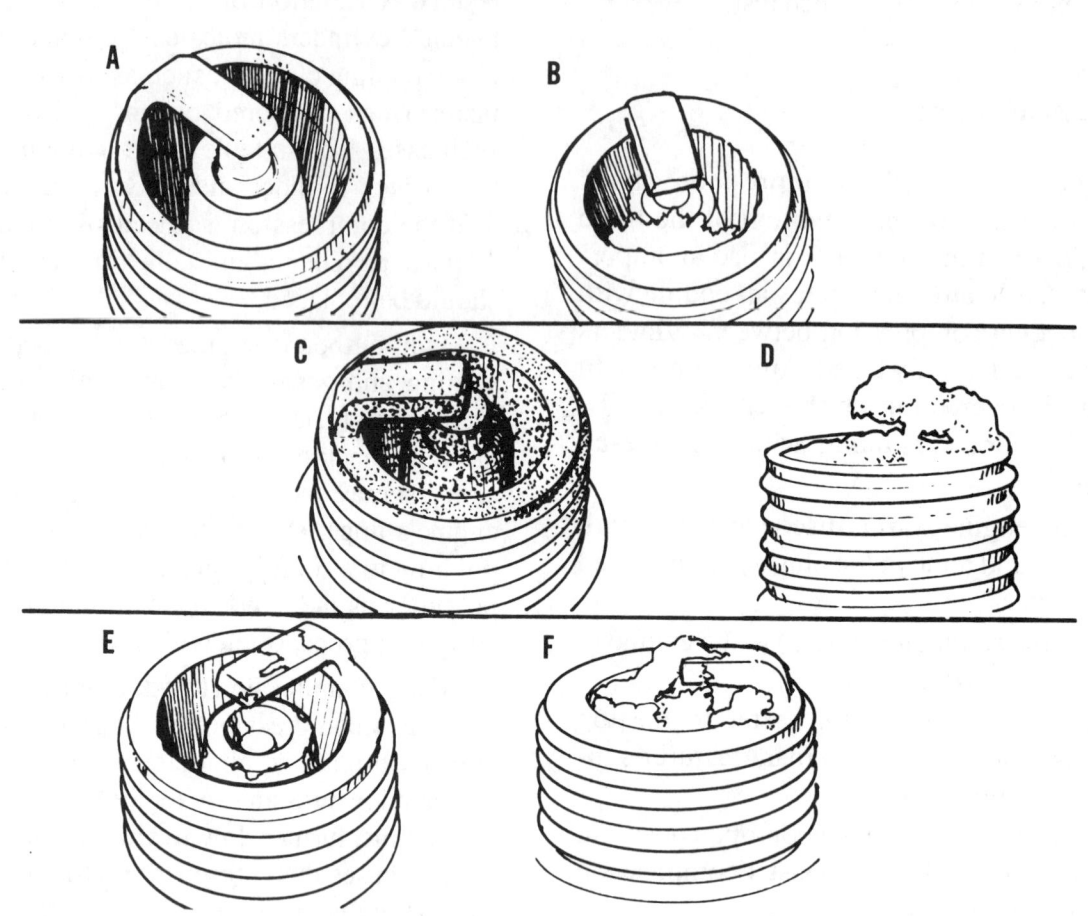

A. **NORMAL**—Light tan to gray color of insulator indicates correct heat range. Few deposits are present and the electrodes are not burned.

B. **CORE BRIDGING**—These defects are caused by excessive combustion chamber deposits striking and adhering to the firing end of the plug. In this case, they wedge or fuse between the electrode and core nose. They originate from the piston and cylinder head surfaces. Deposits are formed by one or more of the following:
 1. Excessive carbon in cylinder.
 2. Use of non-recommended oils.
 3. Immediate high-speed operation after prolonged trolling.
 4. Improper fuel-oil ratio.

C. **WET FOULING**—Damp or wet, black carbon coating over entire firing end of plug. Forms sludge in some engines. Caused by one or more of the following:
 1. Spark plug heat range too cold.
 2. Prolonged trolling.
 3. Low-speed carburetor adjustment too rich.
 4. Improper fuel-oil ratio.
 5. Induction manifold bleed-off passage obstructed.
 6. Worn or defective breaker points.

D. **GAP BRIDGING**—Similar to core bridging, except the combustion particles are wedged or fused between the electrodes. Causes are the same.

E. **OVERHEATING**—Badly worn electrodes and premature gap wear are indicative of this problem, along with a gray or white "blistered" appearance on the insulator. Caused by one or more of the following:
 1. Spark plug heat range too hot.
 2. Incorrect propeller usage, causing engine to lug.
 3. Worn or defective water pump.
 4. Restricted water intake or restriction somewhere in the cooling system.

F. **ASH DEPOSITS OR LEAD FOULING**—Ash deposits are light brown to white in color and result from use of fuel or oil additives. Lead fouling produces a yellowish brown discoloration and can be avoided by using unleaded fuels.

LUBRICATION, MAINTENANCE AND TUNE-UP

pression reading. In such a case, it is also a good idea to double-check the water pump operation as a possible cause for overheating.

Correct Spark Plug Heat Range

Suzuki outboards are equipped with NGK spark plugs selected for average use conditions. Under adverse use conditions, the recommended spark plug may foul or overheat. In such cases, check the ignition and fuel systems to make sure they are operating correctly. If no defect is found, replace the spark plug with one of a hotter or colder heat range as required.

The proper spark plug is important in obtaining maximum performance and reliability. The condition of a used spark plug can tell a trained mechanic a lot about engine condition and carburetion.

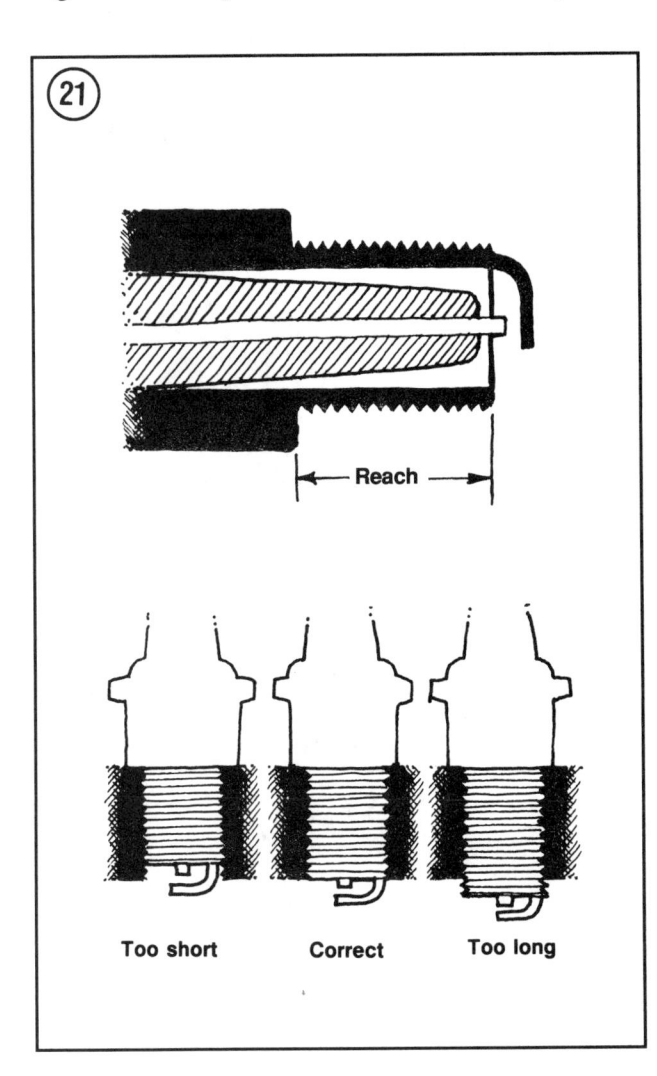

Select a plug of the heat range designed for the loads and conditions under which the motor will be run. Use of incorrect heat ranges can cause a seized piston, scored cylinder wall, or damaged piston crown.

In general, use a hot plug for low speeds and low temperatures. Use a cold plug for high speeds, high engine loads and high temperatures. The plug should operate hot enough to burn off unwanted deposits, but not so hot that they burn themselves or cause preignition. A spark plug of the correct heat range will show a light tan color on the portion of the insulator within the cylinder after the plug has been in service. See **Figure 20**.

The reach (length) of a plug is also important. A longer than normal plug could interfere with the piston, causing severe damage. Refer to **Figure 21**.

Table 4 contains the recommended spark plugs for all models covered in this book. **Table 5** provides a cross-reference for use when NGK spark plugs are not available.

Spark Plug Removal

CAUTION
Whenever a spark plug is removed, dirt around it can fall into the plug hole and into the cylinder. This can cause engine damage that is expensive to repair.

1. Blow out any foreign matter from around the spark plug(s) with compressed air. Use a compressor if you have one. If you do not, use a can of compressed inert gas, available from photo stores.
2. Disconnect the spark plug wire(s) by twisting the wire boot back and forth on the plug insulator while pulling outward. Pulling on the wire instead of the boot may cause internal damage to the wire.

NOTE
If the plug is difficult to remove, apply penetrating oil, like WD-40 or Liquid Wrench, around the base of the plug and let it soak in about 10-20 minutes.

3. Remove the plug(s) with an appropriate size spark plug socket or box-end wrench. Keep the plugs in order so you know which cylinder they came from.

4. Inspect the plug carefully. Look for a broken center porcelain, excessively eroded electrodes, and excessive carbon or oil fouling. Compare its condition with **Figure 20**. Spark plug condition indicates engine condition and can warn of developing trouble.

5. Check each plug for make and heat range. All should be of the same make and number or heat range.

6. Suzuki recommends the plugs be cleaned and reused if in good condition; however, such plugs seldom last very long. New plugs are inexpensive and far more reliable.

Spark Plug Gapping

New or cleaned plugs should be carefully gapped to ensure a reliable, consistent spark. Use a special spark plug tool with a wire gauge. See **Figure 22** for one common type.

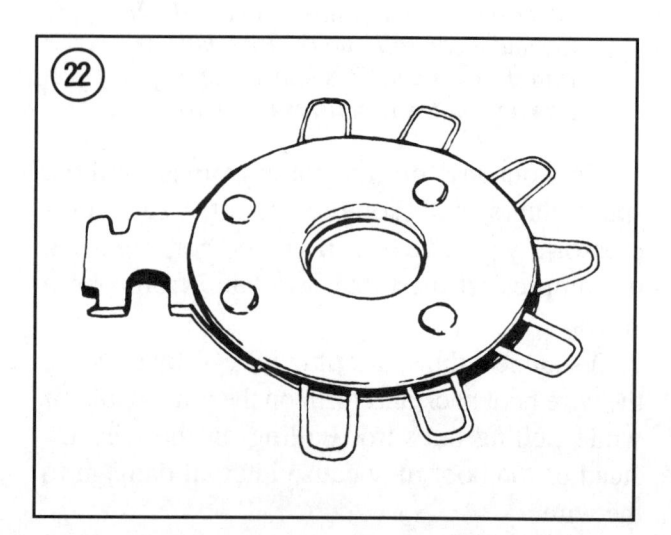

1. Remove the plugs and gaskets from the boxes. Install the gaskets.

NOTE
Some plug brands may have small end pieces that must be screwed on before the plugs can be used.

2. Insert the appropriate size wire gauge between the electrodes. See **Table 4**. If the gap is correct, there will be a slight drag as the wire is pulled through. If there is no drag or if the wire will not pull through, bend the side electrode with the gapping tool (**Figure 23**) to change the gap. Remeasure with the wire gauge.

CAUTION
Never try to close the electrode gap by tapping the spark plug on a solid surface. This can damage the plug internally. Always use the gapping and adjusting tool to open or close the gap.

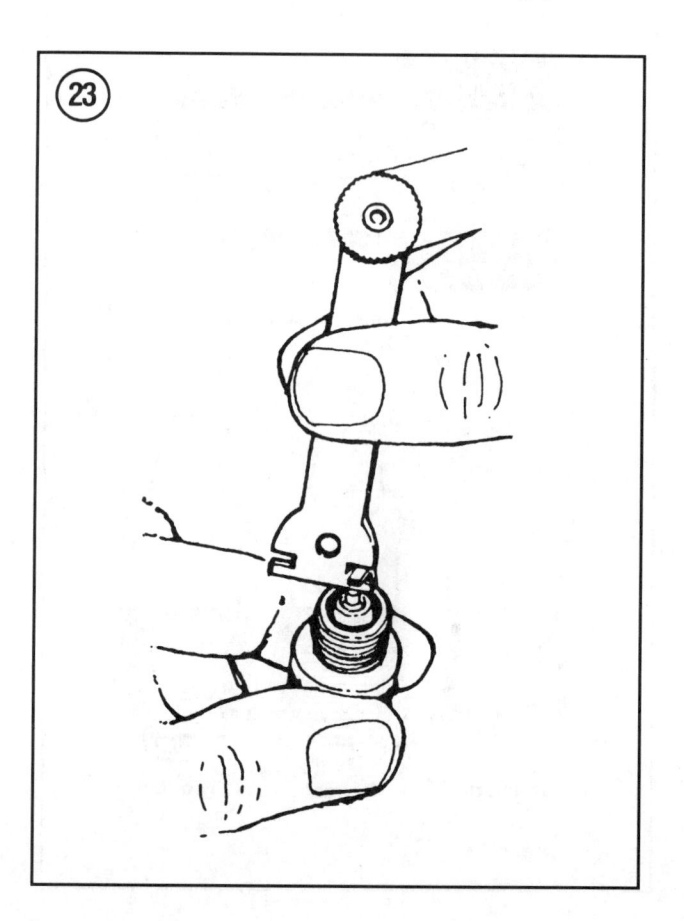

LUBRICATION, MAINTENANCE AND TUNE-UP

3. Check the spark plug hole threads and clean with an appropriate size spark plug chaser (**Figure 24**), if necessary, before installing the plug. This will remove any corrosion, carbon build-up or minor flaws in the threads. Coat the chaser threads with grease to catch chips or foreign matter. Use care to avoid cross-threading. Wipe cylinder head seats clean before installing the new plugs.

4. Apply a thin film of anti-seize compound to the spark plug threads.

Spark Plug Installation

Improper installation of spark plugs is one of the most common causes of poor spark plug performance in outboard engines. The gasket on the plug must be fully compressed against a clean plug seat in order for heat transfer to take place effectively. This requires close attention to proper tightening during installation.

1. Screw each plug in by hand until it seats. Very little effort is required. If force is necessary, the plug may be cross-threaded. Unscrew it and try again.
2. Tighten the spark plugs. If you have a torque wrench, tighten to 10-15 ft.-lb. If not, seat the plug finger-tight on the gasket, then tighten an additional 1/4 turn with a wrench.
3. Inspect each spark plug wire before reconnecting it to its cylinder. If insulation is damaged or deteriorated, install a new plug wire. Push wire boot onto plug terminal and make sure it seats fully.

Gearcase and Water Pump Check

A faulty water pump or one that performs below specifications can result in extensive engine damage. Thus, it is a good idea to replace the water pump impeller, seals and gaskets once a year or whenever the gearcase is removed for service. See Chapter Nine.

Fuel Lines

1. Visually check all fuel lines for kinks, leaks, deterioration or other damage.
2. Disconnect fuel lines and blow out with compressed air to dislodge any contamination or foreign material.
3. Reinstall the fuel lines and make sure they are on tight.

Engine Fuel Filter Service (Carburetted Models)

NOTE
On fuel injected models, the fuel filter replacement is covered in Chapter Six.

DT 2 models have an integral fuel tank and utilize gravity fuel feed instead of a fuel pump. A filter screen is attached to the fuel line fitting on the bottom of the tank. A filter screen is also installed in the carburetor inlet fuel fitting.

Other models may use one of the following:
  a. An inline filter.
  b. A canister filter.

Integral tank filter

Refer to **Figure 25** for this procedure.
1. Remove the fuel tank, if necessary, to provide working room. See Chapter Six.
2. Loosen the fuel petcock clamp. Remove the petcock from the hose.
3. Remove the filter screen from the petcock.
4. Clean the screen in gasoline. If excessively dirty or contaminated with water, discard and install a new screen.
5. Installation is the reverse of removal.

Inline filter

1. Slide each hose retaining clamp off the filter nipple with a pair of pliers and disconnect the hoses from the filter (**Figure 26**).
2. Clean the filter assembly with solvent to remove any particles. If excessively dirty or contaminated with water, discard and install a new filter.
3. Reinstall the hoses on the filter nipples. Make sure embossed arrow on filter points in the direction of fuel flow toward the carburetor (1, **Figure 27**).
4. Slide the retaining clamps on each hose over the nipple to assure a leak-free connection.
5. Check fuel filter installation for leakage by priming fuel system with fuel line primer bulb.

Canister filter

Refer to **Figure 28** for this procedure.
1. Unscrew the cup from the filter housing.
2. Remove the filter from the end of the adaptor.
3. Clean the filter screen with solvent to remove any particles (**Figure 29**). If the filter screen is severely clogged or dirty, replace it.
4. Inspect the filter screen for damage and replace if necessary.
5. Install a new O-ring seal in the filter cup.
6. Install the filter and the filter cup. Tighten the cup securely.

LUBRICATION, MAINTENANCE AND TUNE-UP

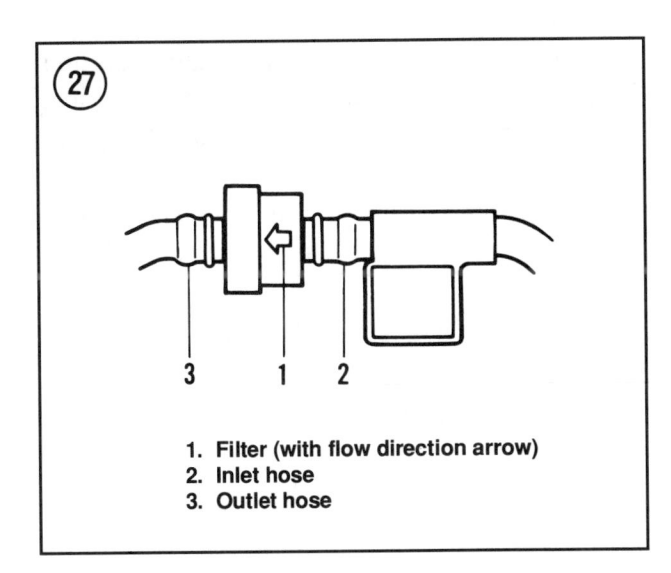

7. Check fuel filter installation for leakage by priming fuel system with fuel line primer bulb.

Fuel Pump

The fuel pump does not generally require service during a tune-up. However, if the engine has more than 100 hours on it since the fuel pump was last serviced, it is a good idea to remove and disassemble the pump, inspect each part carefully for wear or damage and reassemble it with a new diaphragm. See Chapter Six.

Fuel pump diaphragms are fragile and one that is defective often produces symptoms that are diagnosed as an ignition system problem. A common malfunction results from a tiny pinhole or crack in the diaphragm caused by an engine backfire. This defect allows gasoline to enter the crankcase and wet-foul the spark plug(s) at idle speed, causing hard starting and engine stall at low rpm. The problem disappears at higher speeds, as fuel quantity is limited. Since the plug(s) is not fouled by excess fuel at higher speeds, it fires normally.

Breaker Point Ignition System Service (1985-1989 DT 2)

The condition and gap of the breaker points will greatly affect engine operation. Burned or badly oxidized points will allow little or no current to pass. A gap that is too wide will not allow the coil to build up sufficient voltage and will result in a weak spark. An excessive point gap will allow the points to open before the primary current reaches its maximum. Point gap too narrow retards ignition timing.

While slightly pitted points can be dressed with a file, this should be done only as a temporary measure, as the points may arc after filing. Oxidized, dirty or oily points can be cleaned with alcohol but new points are inexpensive and always preferable for efficient engine operation.

27

1. Filter (with flow direction arrow)
2. Inlet hose
3. Outlet hose

The condenser absorbs the surge of high voltage from the coil and prevents current from arcing across the points when they open. Condensers can be tested as described in Chapter Three, but are also inexpensive and should be replaced as a matter of course whenever new breaker points are installed.

NOTE
Breaker points must be adjusted correctly. An error in gap of 0.0015 in. will change engine timing by as much as one degree.

The breaker point set is installed on the stator base under the flywheel. The point gap is set to 0.012-0.016 in. (0.3-0.4 mm). After establishing the breaker point gap, ignition timing should be checked. See Chapter Five.

CAUTION
*Always rotate the crankshaft in a **clockwise** direction in the following procedures. If rotated in a counterclockwise direction, the water pump impeller may be damaged.*

Breaker Point Replacement

1. Disconnect the negative battery cable, if so equipped.
2. Remove the engine cover.
3. Remove the flywheel. See Chapter Eight.
4. Remove the screw holding the breaker point set to the stator base (A, **Figure 30**).
5. Disconnect the coil and condenser leads (B, **Figure 30**) at the breaker point set. Remove the breaker point set.
6. Remove the screw holding the condenser to the stator base (C, **Figure 30**). Remove the condenser.
7. Install a new breaker point set on the stator base. Make sure the pivot point on the bottom of the point set engages the hole in the stator base. Install but do not tighten the hold-down screw.

8. Install a new condenser on the stator base and tighten attaching screw securely, then connect the coil and condenser leads to the breaker point set.
9. Squeeze the felt lubrication wick (D, **Figure 30**) to see if it is dry. If dry, lubricate with 1-2 drops of SAE 30 engine oil.
10. Adjust the breaker point gap as described in this chapter.
11. Install the flywheel. See Chapter Eight.
12. Install the engine cover.
13. Connect the negative battery cable, if so equipped.

Breaker Point Adjustment

The following procedure is used when new breaker points have been installed. Slots are provided in the flywheel rotor for checking and adjusting the point gap without flywheel rotor removal.

1. Install the flywheel nut on the crankshaft and rotate the stator base to the wide-open throttle position.
2. Place a wrench on the flywheel nut and rotate the crankshaft *clockwise* until the breaker point rubbing block rests on a high point on the cam (the points will be wide open).

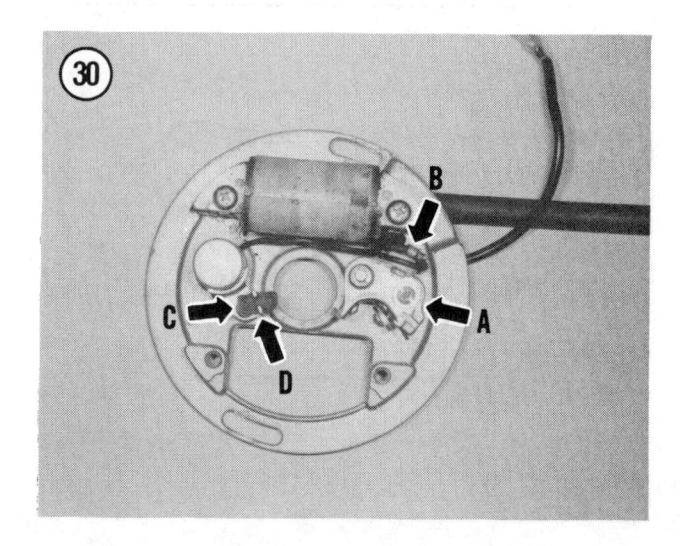

LUBRICATION, MAINTENANCE AND TUNE-UP

3. Loosen the point set hold-down screw (A, **Figure 31**). Insert a screwdriver in the adjusting notch (B, **Figure 31**) and move the point set base to obtain a gap of 0.012-0.016 in. when measured with a flat feeler gauge. The gap is correct when the feeler gauge offers a slight drag as it is slipped between the points. When the gap is correct, tighten the hold-down screw securely and recheck the point gap.

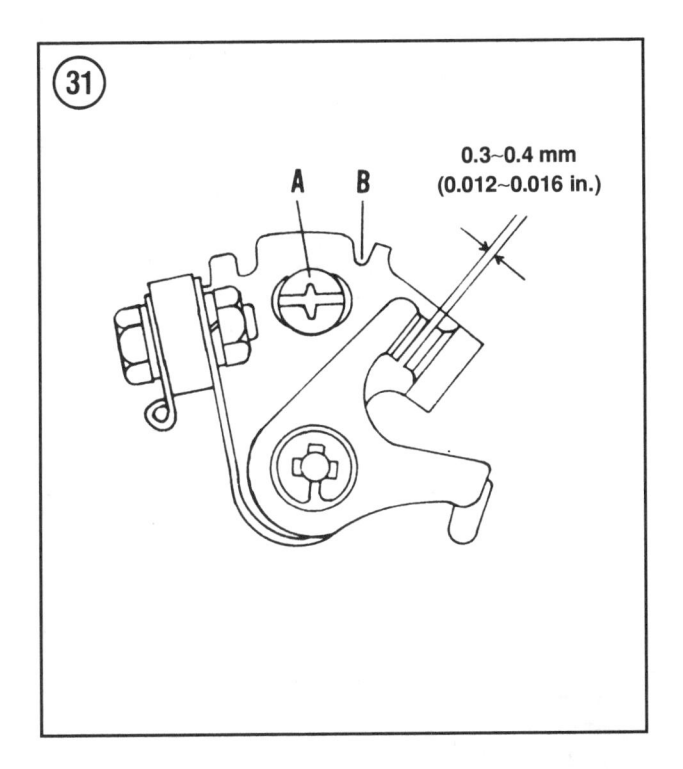

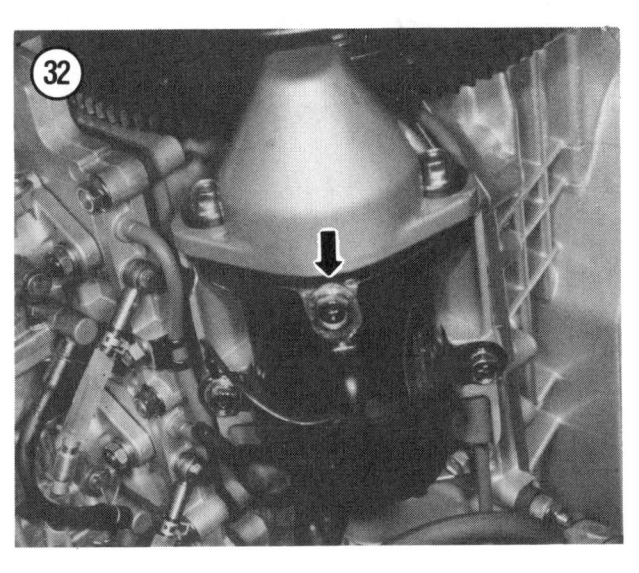

Battery and Starter Motor Check (Electric Start Models Only)

1. Check the battery's state of charge. If necessary, charge the battery prior to performing this test. See Chapter Seven.
2. Connect a voltmeter between the starter motor positive terminal (**Figure 32**) and ground.
3. Turn ignition switch to START and check voltmeter scale.
4A. If voltage exceeds 9.5 volts and the starter motor does not operate, replace the motor.
4B. If voltage is less than 9.5 volts, recheck battery and all electrical connections for corrosion and tightness. Charge the battery, if necessary, and repeat procedure.

Starter Relay Resistance Check (Electric Start Models Only)

The starter relay is attached to the starter motor housing. Refer to **Figure 33**, typical for this procedure.

Starter Relay Testing

1. Remove the starter relay as described in Chapter Seven.
2. Connect an ohmmeter between the large terminals on top of the starter relay. There should be no continuity (infinite resistance).
3. Connect the starter relay black and yellow/red (or yellow/green) electrical leads to a fully charged 12 volt battery.
4. Connect an ohmmeter between the large terminals on top of the starter relay (**Figure 34**). There should be continuity (low resistance).
5. Disconnect the electrical leads from the battery.
6. If the starter relay fails either of these tests, the relay is faulty and must be replaced.

CHAPTER FOUR

Choke Solenoid Resistance Check
(1989-on DT 25; 1988-on DT 30; 1985-on DT 55, DT 65; 1985-on DT 75, DT 85; 1985-on DT 115, DT 140 Electric Start Models Only)

NOTE
The DT 40 E is equipped with a choke solenoid but Suzuki does not provide any specifications for this model.

The choke solenoid is located on the starboard side of the engine. It may be positioned at the base of the choke linkage (**Figure 35**) or at the top of the linkage (**Figure 36**).

1. Disconnect the orange choke solenoid lead at the bullet connector.
2. Connect an ohmmeter between the solenoid terminal and a good engine ground.
3. If the reading obtained in Step 2 is not within the specifications listed in **Table 6**, replace the choke solenoid.

Wiring Harness Check

1. Check the wiring harness for signs of frayed or chafed insulation.
2. Check for loose connections between the wires and terminal ends.

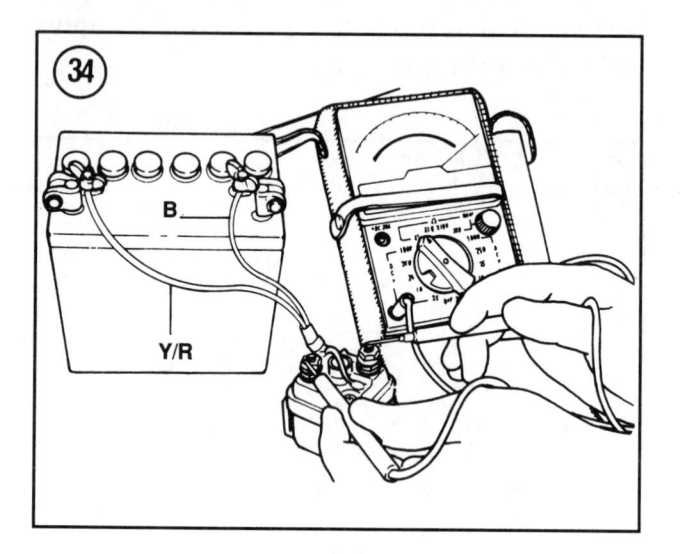

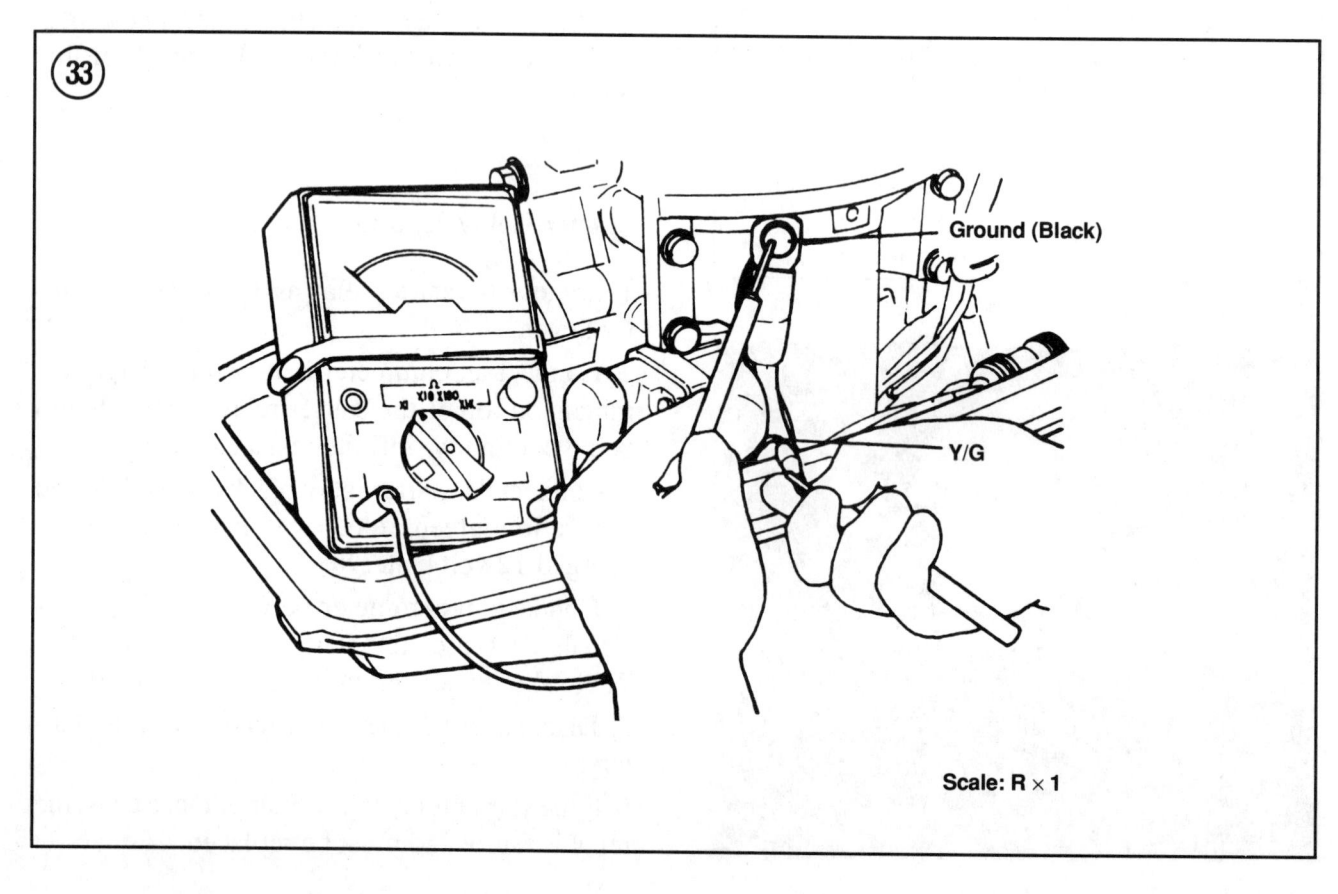

LUBRICATION, MAINTENANCE AND TUNE-UP

3. Make sure all electrical connectors are free of corrosion and are tight.

4. If the harness is suspected of contributing to electrical malfunctions, check all wires within the harness for continuity and resistance between harness connection and terminal end. Repair or replace as required.

Engine Synchronization and Adjustment

See Chapter Five.

Performance Test (On Boat)

Before performance testing the engine, make sure that the boat bottom is cleaned of all marine growth and that there is no evidence of a "hook" or "rocker" (**Figure 37**) on the bottom. Any of these conditions will reduce performance considerably.

The boat should be performance tested with an average load and with the motor tilted at an angle that will allow the boat to ride on an even keep. If equipped with an adjustable trim tab, it should be properly adjusted to allow the boat to steer in either direction with equal ease.

Check engine rpm at full throttle. If not within the maximum rpm range for the motor as specified in **Table 7**, check the propeller pitch. A high pitch propeller will reduce rpm while a lower pitch prop will increase it.

Readjust the idle mixture and speed under actual operation conditions as required to obtain the best low-speed engine performance.

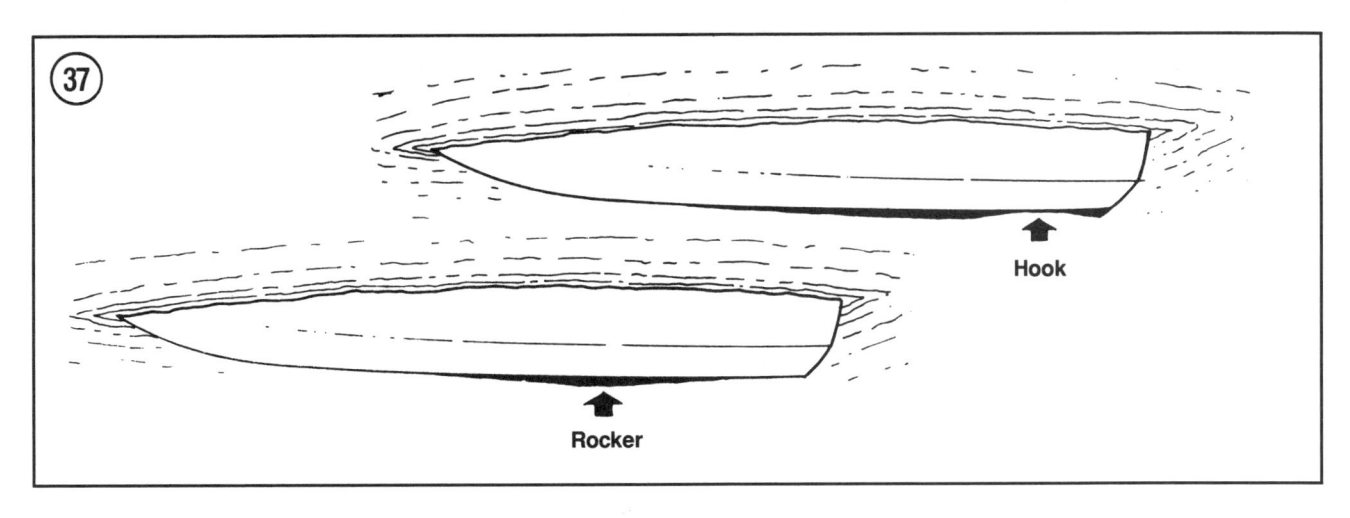

Table 1 MODEL HISTORY

	1985	1986	1987	1988	1989	1990	1991
DT 2	X	X	X	X	X	X	X
DT 4	X	X	X	X	X	X	X
DT 6	X	X	X	X	X	X	X
DT 8	X	X	X	X	X	X	X
DT 8 SAIL					X	X	X
DT 9.9	X	X	X	X	X	X	X
DT 9.9 SAIL				X	X	X	X
DT 15	X	X	X	X	X	X	X
DT 20		X	X	X			
DT 25	X	X	X	X	X	X	X
DT 30	X	X	X		X	X	X
DT 35				X	X		
DT 40	X	X	X	X	X	X	X
DT 55	X	X	X	X	X	X	X
DT 65	X	X	X	X	X	X	X
DT 75	X	X	X	X	X	X	X
DT 85	X	X	X	X	X	X	X
DT 90					X	X	X
DT 100					X	X	X
DT 100 SUPER FOUR						X	X
DT 115	X	X	X	X	X	X	X
DT 140	X	X	X	X	X	X	X
DT 150			X	X	X	X	X
DT 150 SUPER SIX		X	X	X	X	X	X
DT 175			X	X	X	X	X
DT 200		X	X	X	X	X	X
DT 200 EXANTE				X	X	X	X
DT 225						X	X

Table 2 MAINTENANCE SCHEDULE*

At first 10 hours	Change gearcase lubricant
Every 10 hours	Retighten bolts and nuts
	Check wire harness connections
	Check idle speed (DT 2-DT 15, DT 40 and 1991 DT 25-ST 200)
	Check and adjust carburetors (DT 20-DT 200)
	Check propeller for damage
	Lubricate propeller shaft splines
	Lubricate jet drive bearings
	Check fuel lines for leakage
	Check intake manifold hose for deterioration
	Lubricate steering handle
	Check neutral start interlock switch operation
	Check emergency switch operation (if so equipped)
	Check and adjust remote control linkage (if so equipped)
	Check engine key and choke operation (if so equipped)
	Check starter button and choke operation (if so equipped)

(continued)

LUBRICATION, MAINTENANCE AND TUNE-UP

Table 2 MAINTENANCE SCHEDULE* (continued)

Every 50 hours	Clean and regap spark plugs
	Decarbonize the piston(s), cylinder and cylinder head
	Change gearcase lubricant
	Check fuel strainer or filter
	Check steering handle preload
	Check starter rope condition
	Check tilt mechanism preload
	Lubricate jet drive bearings
Every 100 hours	Check and adjust ignition timing
	Check water pump impeller
Every week	Check oil injection lines (if so equipped)
Every month	Lubricate carburetor and choke linkage
	Lubricate clamp screws
Every 3 months	Lubricate swivel bracket
	Lubricate support tube
	Lubricate shift lever
Once each season	Change starter rope
	Check fuel tank condition
	Replace water pump impeller

* Not all items apply to all engines. Perform only those pertaining to your engine. Lubricate jet drive bearings after each operation.

Table 3 RECOMMENDED LUBRICANTS, SEALANTS AND ADHESIVES

Type	Part No.
Water-resistant grease	99000-25170
Outboard Motor Gear Oil	99000-22540
Suzuki CCI 50:1 Outboard Oil	99105-00153
Super Grease "A"	99000-25030
Silicone Seal	99000-31120
Bond No. 4	99000-31030
Cemedine 366E	99000-31090
Thread Lock 1342	99000-32050
Thread Lock Super 1333B	99000-32020
DEXRON automatic transmission fluid	—
Jet Grease	99954-53885

Table 4 RECOMMENDED SPARK PLUGS

	NGK part No.	Gap (in.)
DT 2		
1985-1990	BR4H	0.020-0.024
1991	BR5HS	0.024-0.028
DT 4		
1985-1989	BP6HS	0.024-0.028
1990-on	BP5HS	0.024-0.028
(continued)		

Table 4 RECOMMENDED SPARK PLUGS (continued)

	NGK part No.	Gap (in.)
DT 6		
1985-1986	BP6HS, BPR6HS	0.031-0.035
1987-on	BR6HS-10	0.035-0.039
DT 8		
1985-1987	BPR6HS	0.035-0.039
1988-on	B6HS-10	0.035-0.040
DT 9.9		
1985-1987	BR7HS-10	0.035-0.039
1988-on	B6HS-10	0.035-0.040
DT 15	BR7HS-10	0.035-0.039
DT 20, DT 25, DT 30		
(2-cylinder)	BR7HS	0.035-0.039
DT 25, DT 30 (3-cylinder)	B7HS-10	0.035-0.039
DT 35, DT 40	B8HS	0.031-0.035
DT 55, DT 65	B8HS	0.031-0.035
DT 75, DT 85	B8HS	0.031-0.035
DT 115, DT 140		
1985-1990	BR8HS	0.031-0.035
1991	BR8HCS	0.031-0.035
V4	BR8HS-10	0.035 0.039
V6		
1986		
DT 150, DT 200	BR8HS-10	0.035 0.039
DT 150 SS	B8HS	0.024 0.028
1987-1988		
DT 150, DT 175, DT 200	B8HS-10	0.035 0.039
DT 150 SS, DT 200 AT	B8HS	0.024 0.028
1989-on		
DT 150-DT 200	BR8HS-10	0.035-0.039
1990-on DT 225	BR8HS-10	0.035-0.039

Table 5 SPARK PLUG CROSS-REFERENCE CHART*

NGK	Champion	AC
B4HS, BR5HS	L81, L88A	44F, 44FF
B6HS, BR6HS	L9J, QL7J, RL7J	42F, 42FF
BP6HS, BPR6HS	RL12Y, RL87Y, L66Y	42FS, 43FS, R43FS
B7HS, BR7HS	L5, L7J	M42FF, S42FR
BR7HS-10	—	—
B8HS, BR8HS	L4J, RL4J, L78, RL78	S41FR, S40FR, M41FF
BR8HS-10	—	—

* The cross-referenced spark plugs are not exact replacements for original plug heat range and should be used only as a temporary replacement.

Table 6 CHOKE SOLENOID RESISTANCE SPECIFICATIONS

Model	Ohms
DT 25, DT 30 (3-cylinder)	2.8-4.2
DT 55, DT 65 (1988-on)	3.5-5.1

(continued)

LUBRICATION, MAINTENANCE AND TUNE-UP

Table 6 CHOKE SOLENOID RESISTANCE SPECIFICATIONS (continued)

Model	Ohms
DT 75, DT 85	
1985-1986	4.1-4.5
1987-on	3.5-5.1
DT 115, DT 140	
1985	4.1-4.5
1986-on	2.8-4.2

Model DT 40 E is equipped with a choke solenoid but Suzuki does not provide service specifications for it.

Table 7 ENGINE OPERATING RANGE

Model	Idle Speed (rpm)	Recommended Maximum (rpm)
DT 2	800-900	4200-4800
DT 4	852-900	4500-5500
DT 6	600-650	5200-5700
DT 8	600-650	5300-5700
DT 9.9	600-650	5300-5700
DT 15	600-650	5300-5700
DT 20	650-700	4800-5500
DT 25	650-700	4800-5500
DT 30	650-700	4800-5500
DT 35	650-700	4800-5500
DT 40	650-700	4800-5500
DT 55	650-700	4800-5500
DT 65	650-700	4800-5500
DT 75	600-700	4800-5500
DT 85	600-700	4800-5500
DT 90	650-700	5000-5600
DT 100	650-700	5000-5600
DT 115	600-700	4800-5500
DT 140	600-700	4800-5500
DT 150	600-700	4800-5500
DT 175	600-700	4800-5500
DT 200	600-700	5000-5600
DT 225	600-700	5000-5600

Chapter Five

Timing, Synchronization and Adjustment

Procedures for timing, synchronization and adjustment on Suzuki outboards differ according to model and type of ignition system. This chapter is divided into self-contained sections dealing with particular models/ignition systems for fast and easy reference. Each section specifies the appropriate procedure and sequence to be followed and provides the necessary tune-up data. Read the general information at the beginning of the chapter and then select the section pertaining to your specific model and year.

Ignition timing adjustment and synchronization procedures are provided for models equipped with breaker-point and Suzuki PEI ignition systems. On models equipped with Integrated Circuit (IC) and Micro Link ignition, the ignition timing advance is controlled electronically with no means for mechanical spark advance adjustment. The IC and Micro Link systems process information provided by various sensors and switches and automatically adjust timing advance according to engine speed, throttle opening and engine temperature. On these models, if a timing advance malfunction occurs, the ignition system must be tested as described in Chapter Three.

Table 1 is at the end of this chapter.

ENGINE TIMING AND SYNCHRONIZATION

Ignition timing advance and throttle opening must be synchronized to occur at the proper time to obtain optimum outboard motor performance. On 1985-on DT 6, 1985-1987 DT 8, 1987-1989 DT 35, 1985-on DT 40 and 1985 DT 115 and DT 140 models, synchronization is the process of timing the carburetor operation to the ignition timing advance. On all other models, there is no mechanical interface between these 2 systems.

Engines equipped with a breaker point ignition are static-timed by measuring the amount of piston travel relative to breaker point opening and closing.

Correct timing thus depends upon a correct breaker point gap setting. If the breaker point gap is correct, but piston travel or timing mark align-

TIMING, SYNCHRONIZATION AND ADJUSTMENT

ment is incorrect, the magneto stator position is readjusted as required.

Synchronization is automatic on models with a breaker point ignition once the point gap and piston travel or timing mark alignment are correct.

Some models listed in the following procedures, equipped with the electronic ignition, are static-timed by aligning timing marks on the throttle cam or stopper with marks on the flywheel. Initial timing and timing advance are both set in this manner, then timing is checked with a timing light.

On 1985-on DT 6, 1985-1987 DT 8, 1987-1989 DT 35, 1985-on DT 40 and 1985 DT 115 and DT 140 models, the magneto stator base is mechanically linked to the throttle. As the throttle is opened, the ignition timing is advanced. Timing advance is checked with a timing light (**Figure 1**).

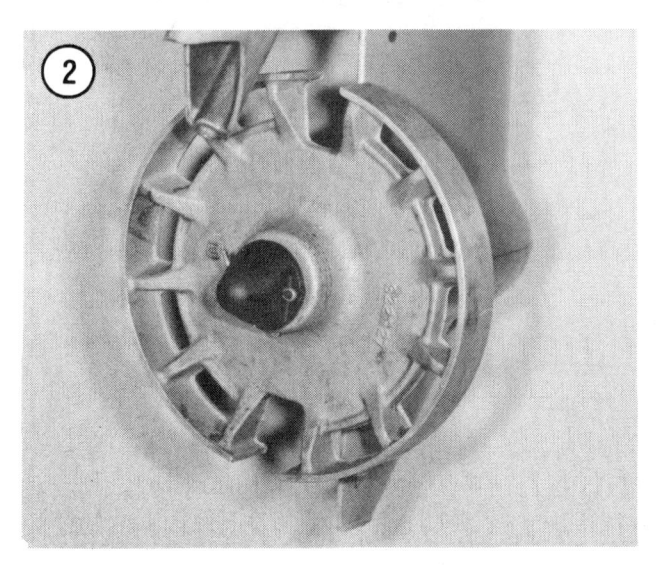

Required Equipment

Static timing of an engine with a breaker point ignition requires the use of a timing gauge (dial indicator) and timing tester (ohmmeter or self-powered test lamp) to set ignition timing properly.

CAUTION
Never operate the engine without water circulating through the gearcase to the power head. If the engine is run in a dry condition, it will damage the water pump and the gearcase and can cause severe engine damage.

Some form of water supply is required whenever the engine is operated during the procedure. The use of a test tank and test wheel is the recommended method. While the procedure may be carried out with the boat in the water, checking engine timing while speeding across open water is neither easy nor safe.

Dynamic engine timing uses a stroboscopic timing light connected to the No. 1 spark plug wire (**Figure 1**). As the engine is cranked or operated, the light flashes each time the spark plug fires. When the light is pointed at the moving flywheel, the mark on the flywheel appears to stand still. The flywheel mark should align with the stationary timing mark on the power head.

Suzuki recommends that a test wheel (**Figure 2**) be substituted for the propeller to put a load on the propeller shaft and prevent engine damage from excessive rpm. Suzuki test wheel recommendations and part numbers are given in **Table 1**.

A portable tachometer connected to the engine is used to determine engine speed during idle and high-speed adjustments.

TIMING ADJUSTMENT

Static Timing Adjustment (DT 2)

1. Remove the engine cover.
2. Remove the fuel tank. See Chapter Six.
3. Remove the rewind starter. See Chapter Ten.
4. Check and adjust breaker point gap as required. See Chapter Four.
5. Disconnect the black stator lead at the bullet connector.
6. Remove the spark plug. See Chapter Four.
7. Install timing gauge (part No. 09931-00112) or a suitable dial indicator in spark plug hole (**Figure 3**, typical).
8. Rotate flywheel *clockwise* until timing gauge shows that the piston has reached top dead center (TDC). This is the point at which the indicator needle reverses its direction of movement as the flywheel is rotated.
9. At this point, reset timing gauge indicator to zero.
10. Connect the timing tester (part No. 09900-27003) between the disconnected black stator lead and a good engine ground (**Figure 4**). Turn the tester switch to the ON position.

NOTE
If an ohmmeter or test lamp is used instead of the timing tester, the meter needle will deflect or the lamp will light in Step 11.

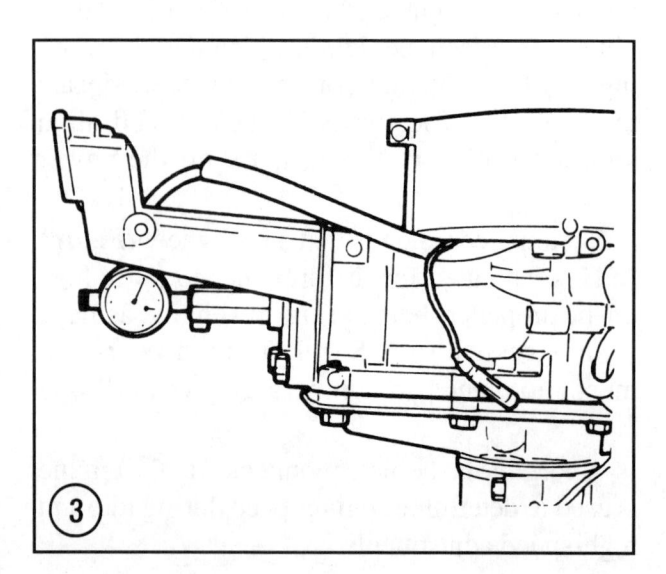

11. Slowly rotate the flywheel rotor *clockwise* until the timing tester buzzes. Stop the engine rotation as this indicates that the points have just opened.
12. Check the timing gauge to determine the amount of piston travel. If the reading is 0.032 in. (0.804 mm), the timing is properly adjusted.

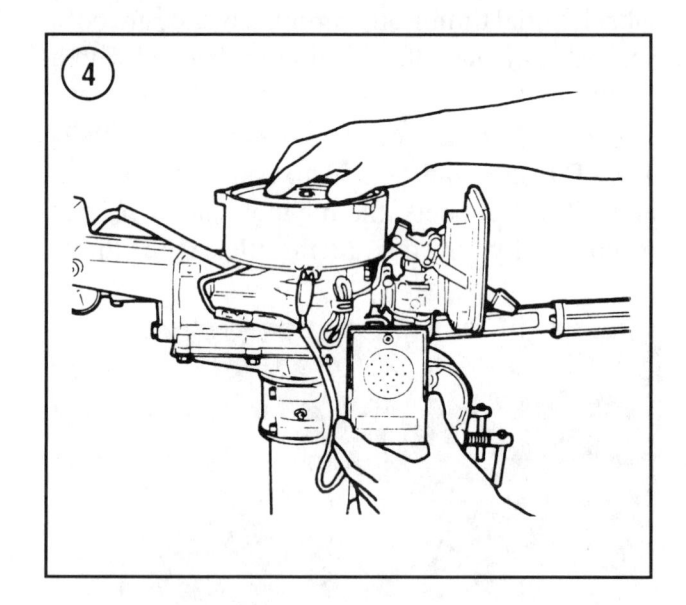

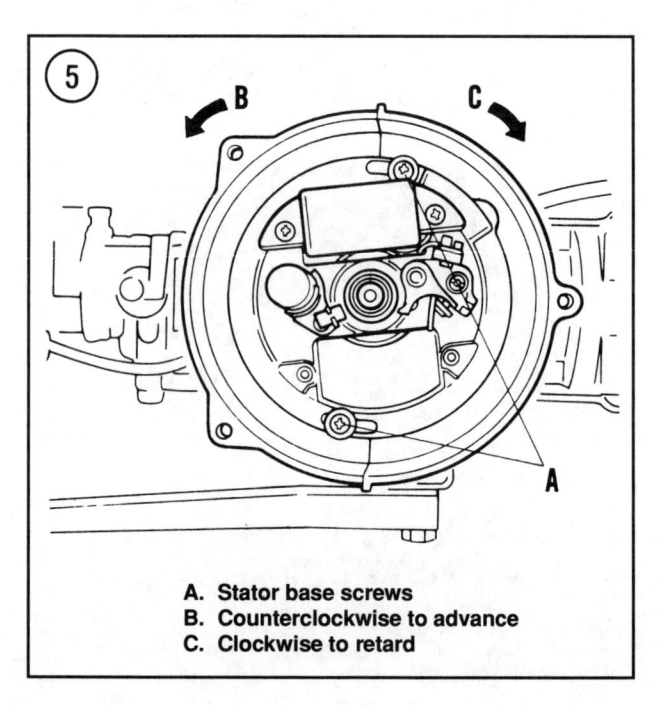

③

④

⑤

A. Stator base screws
B. Counterclockwise to advance
C. Clockwise to retard

TIMING, SYNCHRONIZATION AND ADJUSTMENT

13. If the timing gauge does not read as specified in Step 12, remove the flywheel rotor. See Chapter Eight.
14. Loosen the stator base screws (A, **Figure 5**). Rotate the stator *clockwise* to retard the timing or *counterclockwise* to advance the timing as required to bring the piston travel into specifications. Temporarily reinstall the flywheel rotor and repeat Step 11 and Step 12 to check piston travel. Repeat this step until the piston travel is correct.
15. When piston travel is correct, tighten stator base screws securely. Remove all test equipment and reinstall the flywheel rotor, rewind starter, fuel tank and engine cover.

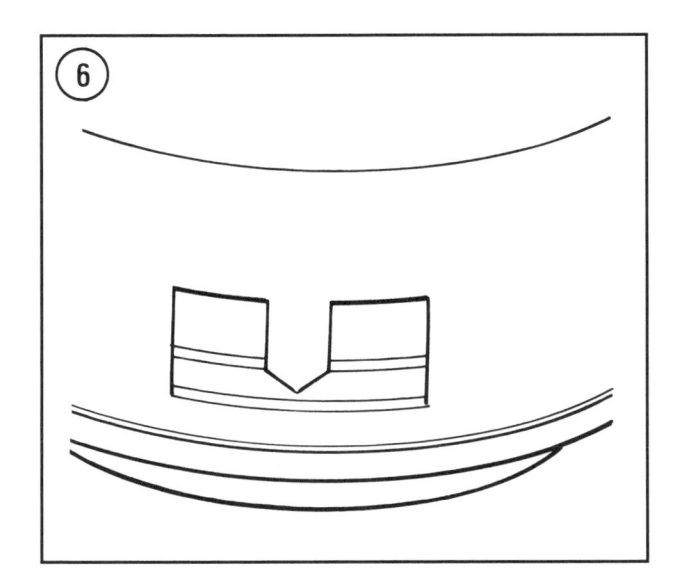

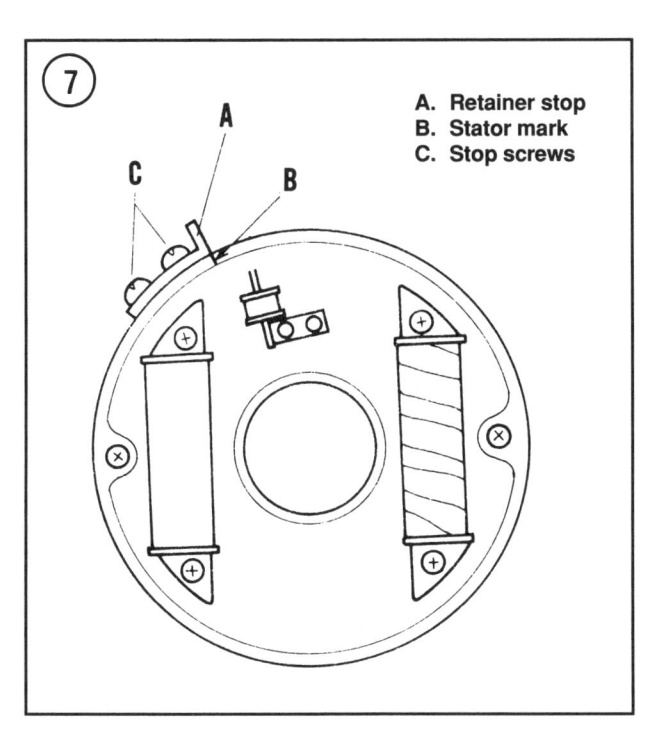

Dynamic Timing Adjustment (DT 4)

The engine should be in a test tank or on the boat in the water for this procedure.
1. Remove the engine cover.
2. Connect a timing light according to manufacturer's instructions.
3. Start the engine and allow it to warm up for approximately 5 minutes.
4. With the engine idling in NEUTRAL, point the timing light at the recoil starter timing mark area. The pointer should align with the flywheel timing mark BTDC 7° at 1,000 rpm (**Figure 6**).
5. Repeat Step 4 with the engine in NEUTRAL and increase engine speed to 5,000 rpm. The pointer should align with the flywheel timing mark BTDC 25°
6. If the timing marks do not align as specified in Step 4 or Step 5, test the CDI unit as described in Chapter Three.

Static Timing Adjustment (DT 6; 1985-1987 DT 8)

1. Remove the engine cover.
2. Rotate the twist grip to the closed throttle position.
3. Loosen the retainer stop screws. Align the end of the retainer stop with the flywheel timing mark and tighten the stop screws (**Figure 7**). This sets ignition timing to TDC ±2°.
4. Rotate the twist grip to the wide-open throttle position. The throttle arm should just contact the throttle stop.
5. If throttle arm adjustment is necessary, disconnect the stator rod from the stator base. Loosen

CHAPTER FIVE

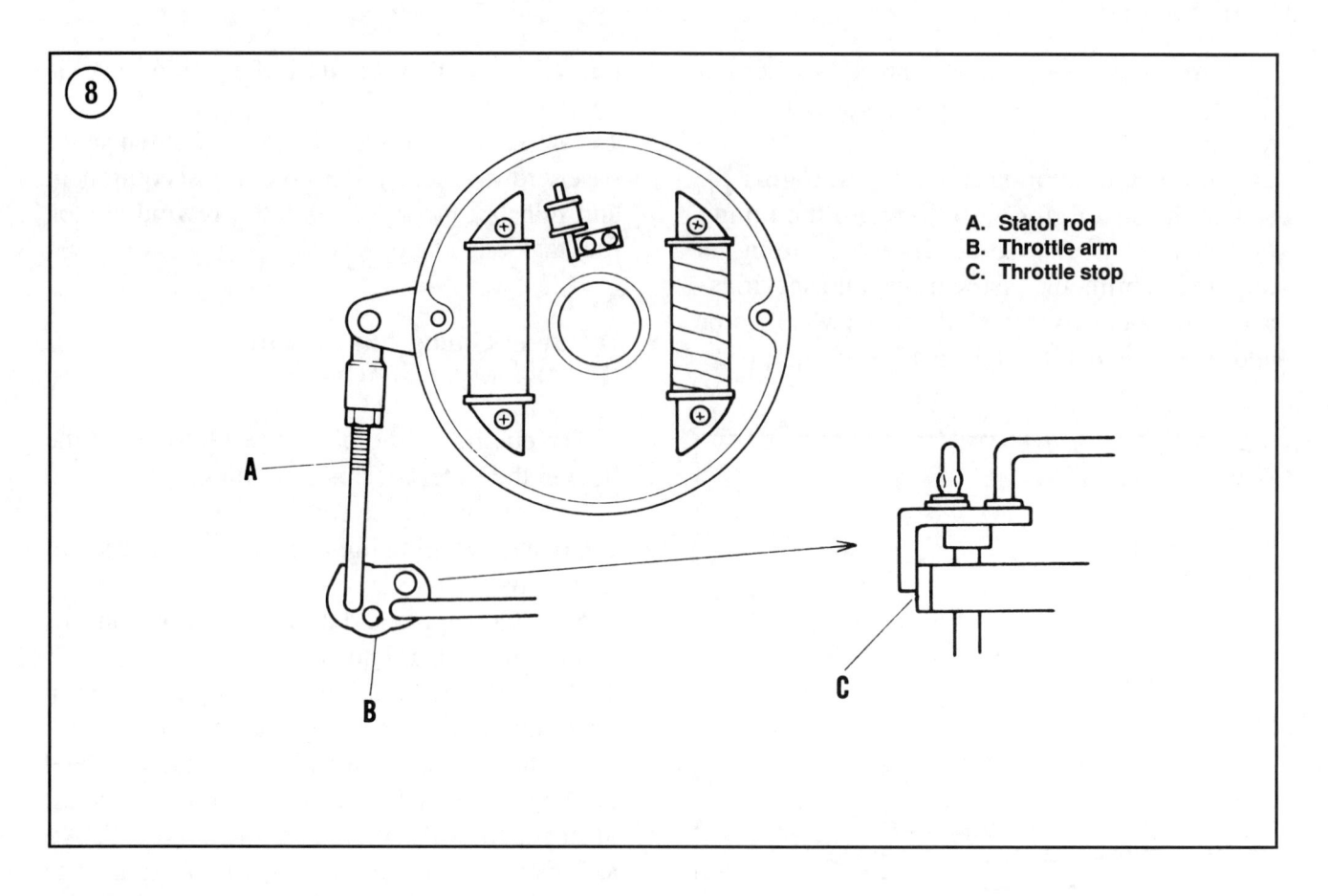

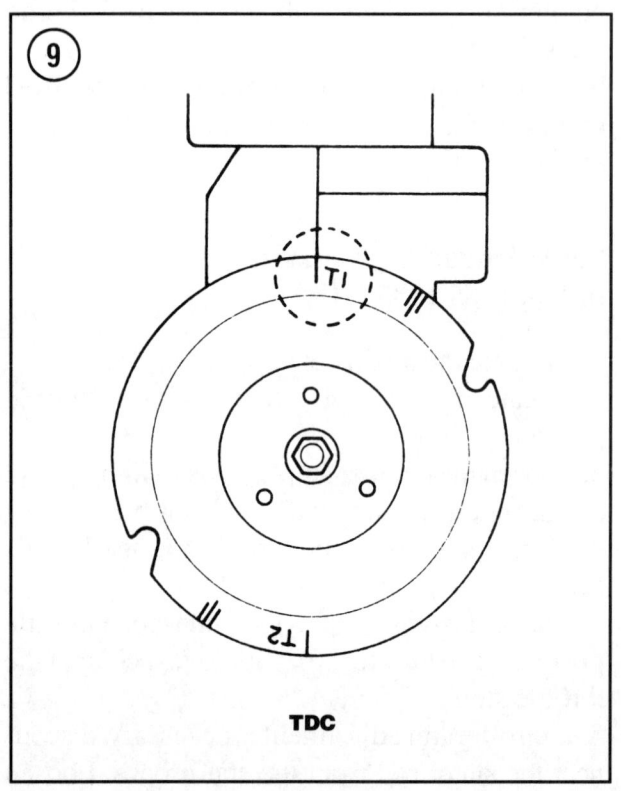

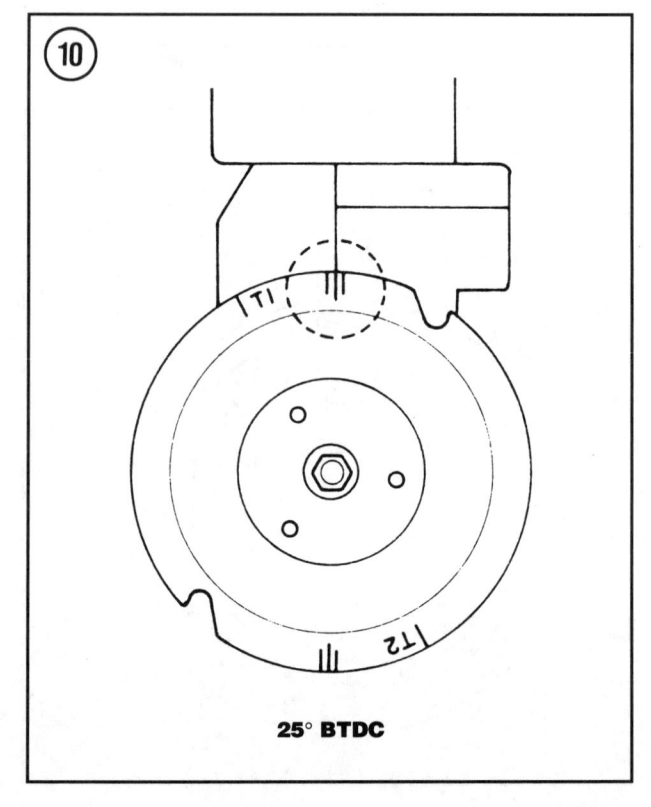

⑧
A. Stator rod
B. Throttle arm
C. Throttle stop

⑨
TDC

⑩
25° BTDC

TIMING, SYNCHRONIZATION AND ADJUSTMENT

the connector locknut and rotate the connector to adjust the stator rod to the proper length. Reconnect stator rod to stator base (**Figure 8**). This sets maximum advance to 25° BTDC ±2°.

6. Open and close the throttle with the twist grip. The flywheel mark and retainer stop should align with the throttle closed. The throttle arm should just touch its stop with the throttle open.

Dynamic Timing Check (DT 6; 1985-1987 DT 8)

The engine should be in a test tank or on the boat in the water for this procedure.

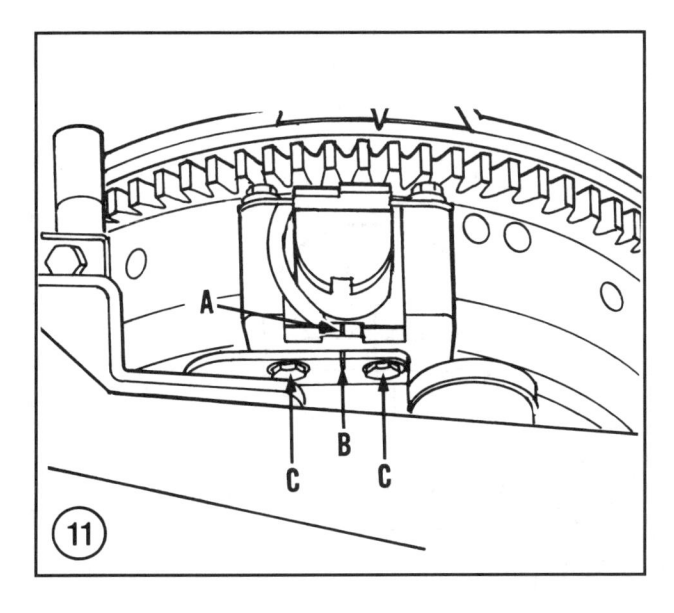

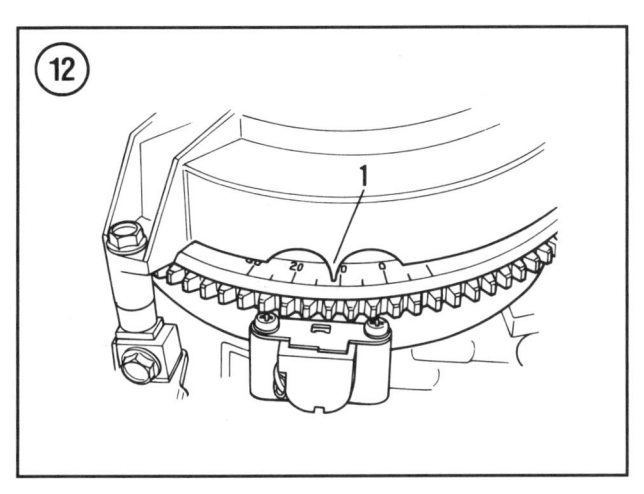

1. Remove the engine cover.
2. Connect a timing light according to manufacturer's instructions.
3. Start the engine and allow it to warm up for approximately 5 minutes.
4. With the engine idling in NEUTRAL, point the timing light at the cylinder case center line. It should align with the flywheel timing mark (**Figure 9**).
5. Repeat Step 4 with the engine in NEUTRAL and the throttle wide open. The cylinder case center line should align with one of the series of 3 flywheel marks shown in **Figure 10**.
6. If the timing marks do not align as specified in Step 4 or Step 5, repeat the *Static Timing Adjustment* in this chapter.

Dynamic Timing Check (1988-on DT 8, DT 9.9)

The engine should be in a test tank or on the boat in the water for this procedure.

NOTE
There is no static timing check for these models.

1. Remove the engine cover.
2. Make sure the stator's match mark (A, **Figure 11**) is aligned with the stopper's match mark (B, **Figure 11**).
3. If alignment is incorrect, loosen the mounting bolts (C, **Figure 11**) and move the stopper until alignment is correct. Tighten the bolts securely.
4. Connect a timing light according to manufacturer's instructions.
5. Start the engine and allow it to warm up for approximately 5 minutes.
6. With the engine idling in NEUTRAL, point the timing light at the magneto cover timing mark (1, **Figure 12**). It should align with one of the following timing marks (**Figure 12**):
   a. DT 8 C models: ATDC 11° ±2° at 600 rpm.
   b. DT 9.9 MC/CE/CN models: ATDC 12° ±2° at 600 rpm.

CHAPTER FIVE

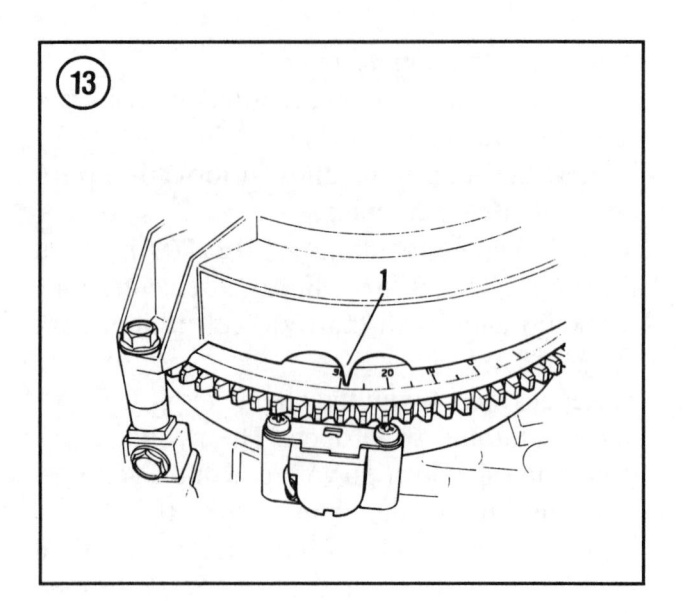

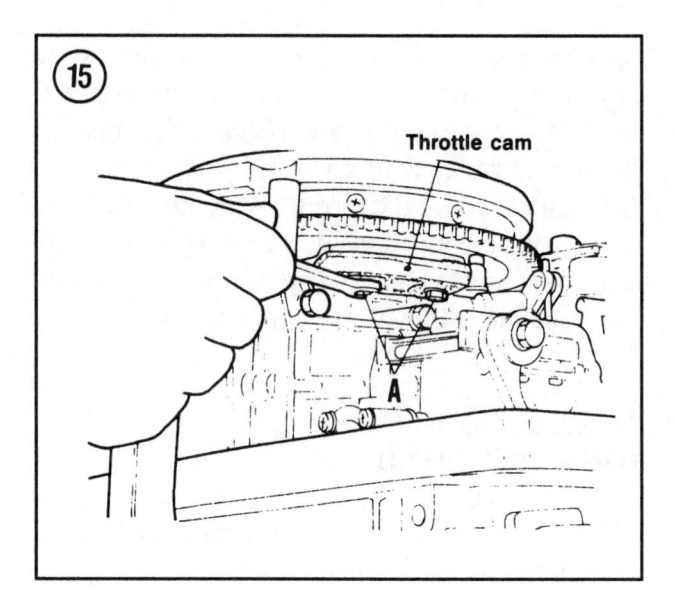

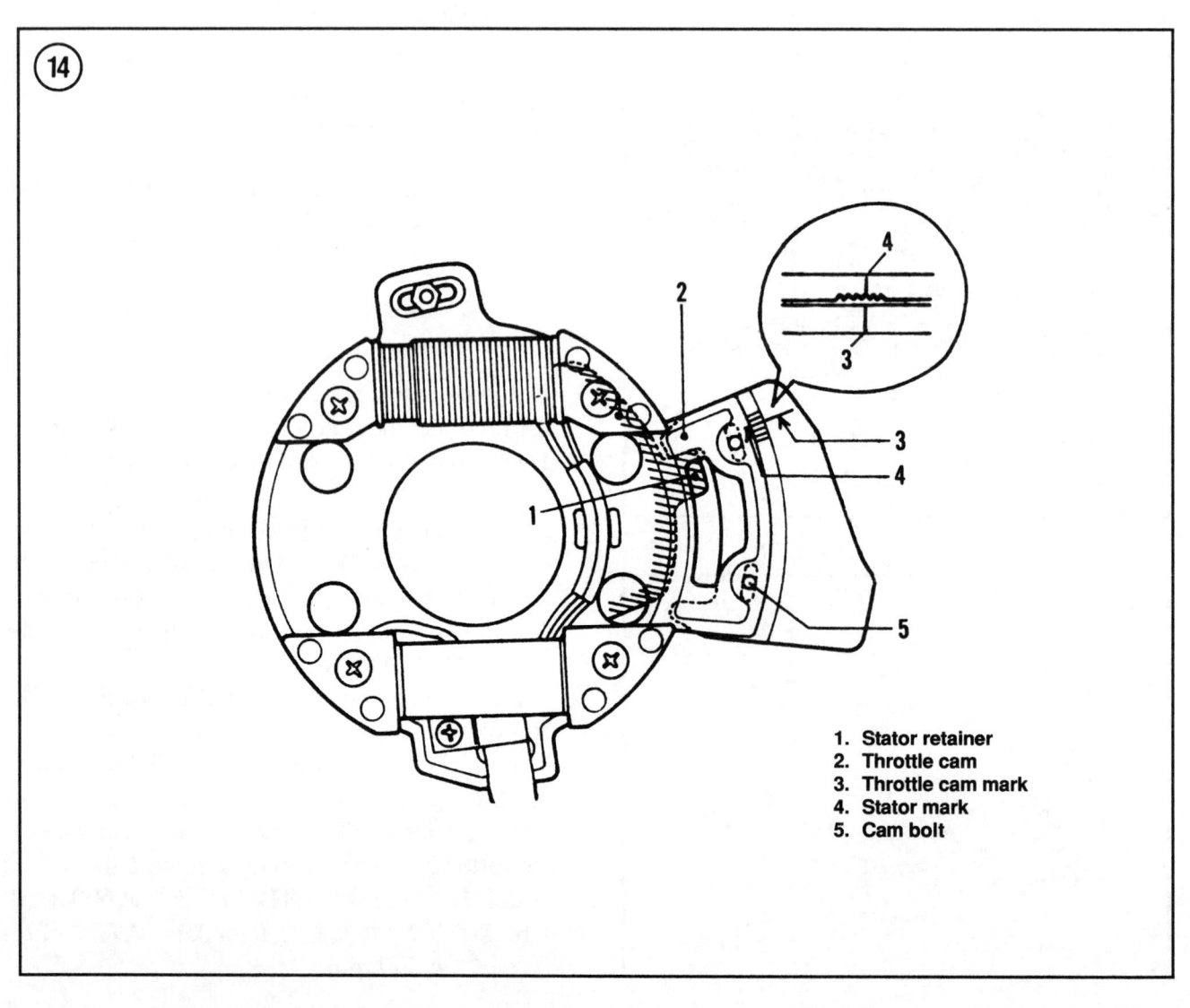

⑬

⑮
Throttle cam

⑭

1. Stator retainer
2. Throttle cam
3. Throttle cam mark
4. Stator mark
5. Cam bolt

TIMING, SYNCHRONIZATION AND ADJUSTMENT

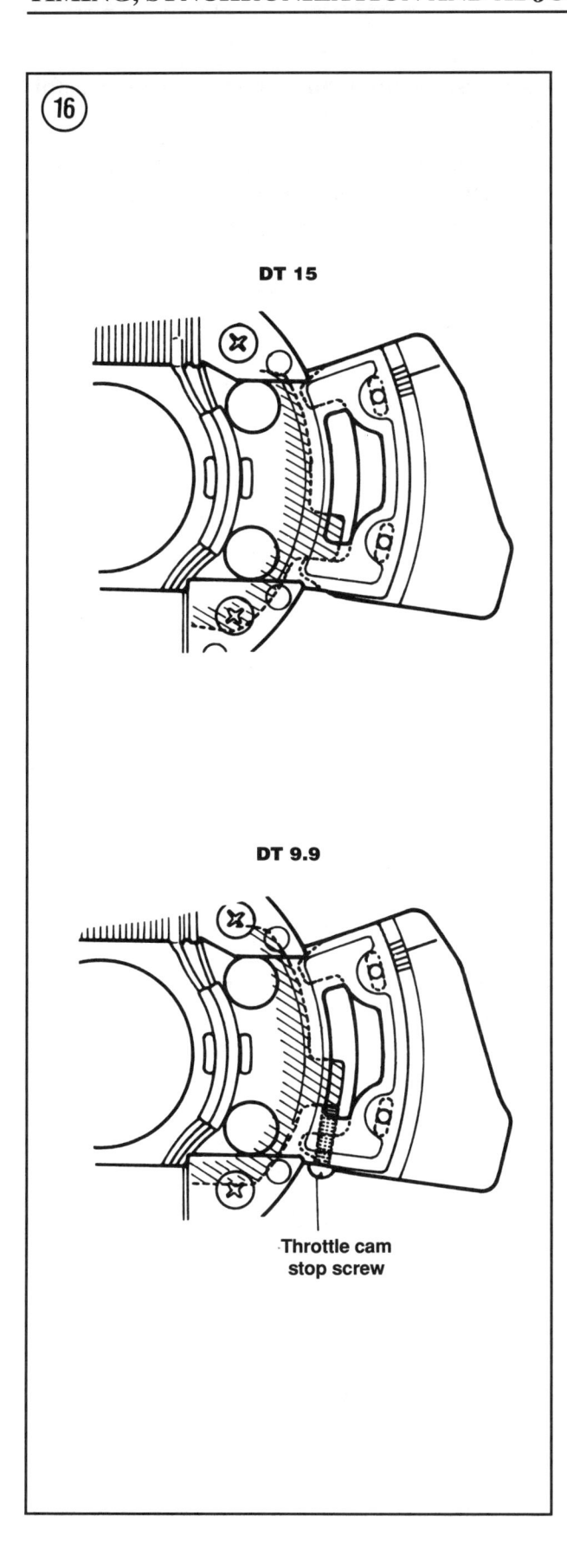

7. With the engine in NEUTRAL, increase engine speed to 2,000 rpm or more. Point the timing light at the magneto cover timing mark (1, **Figure 13**). It should align with one of the following timing marks (**Figure 13**):

   a. DT 8 C models: BTDC 18° ±2° at 2,000 rpm.
   b. DT 9.9 MC/CE/CN models: BTDC 28° ±2° at 2,000 rpm.

8. If the ignition timing is incorrect in either Step 6 or Step 7, have the CDI unit tested by a Suzuki dealer.

9. Shut the engine off and install the engine cover.

Static Timing Adjustment (1985-1987 DT 9.9; 1985-on DT 15)

1. Remove the engine cover.
2. Rotate the twist grip to the closed throttle position.
3. Rotate the stator until the retainer stop contacts the throttle cam (**Figure 14**).
4. If the cam and stator timing marks do not align, loosen the throttle cam bolts (A, **Figure 15**).
5. Reposition the throttle cam as required to align the cam and stator timing marks (**Figure 16**), then tighten the cam bolts (A, **Figure 15**). On DT 9.9 models, also tighten the throttle cam stop screw. At this point, ignition timing is 2° ATDC ±2°.
6. Rotate the twist grip to the wide-open throttle position. The magneto stop should just contact the throttle cam stop screw (DT 9.9) or throttle cam (DT 15). Maximum advance is 18.5° BTDC ±2° (DT 9.9) or 25° BTDC ±2° (DT 15) at 5,000 rpm.
7. Open and close the throttle with the twist grip. The cam and stator timing marks should align with the throttle closed. The magneto stop should just touch the throttle cam stop screw (DT 9.9) or throttle cam (DT 15) with the throttle wide open.

Dynamic Timing Check
(1985-1987 DT 9.9; 1985-on DT 15)

The engine should be in a test tank or on the boat in the water for this procedure.
1. Remove the engine cover.
2. Connect a timing light according to manufacturer's instructions.
3. Start the engine and allow it to warm up for approximately 5 minutes.
4. With the engine idling in NEUTRAL, point the timing light at the rewind starter cover timing mark. It should align with the 2° ATDC ±2° timing mark on the flywheel (A, **Figure 17**).
5. Repeat Step 4 with the engine in NEUTRAL and the throttle wide open. The rewind starter cover timing mark should align with the 18.5° BTDC ±2° mark on DT 9.9 engines or the 25° BTDC ±2° mark on DT 15 engines (B, **Figure 17**).
6. If the timing marks do not align as specified in Step 4 or Step 5, repeat the *Static Timing Adjustment* in this chapter.
7. If the ignition timing is still incorrect, have the CDI unit tested by a Suzuki dealer.
8. Shut the engine off and install the engine cover.

Static Timing Adjustment
(1985-1987 DT 20, DT 30; 1985-1988 DT 25)

1. Remove the engine cover.
2. Adjust the maximum advance position as follows:
   a. Align the 25° BTDC match mark (1, **Figure 18**) on the stator housing with the match mark on the engine (2, **Figure 18**).
   b. If alignment is incorrect, loosen the stator mounting bolts (5, **Figure 18**) and move the stator housing until alignment is correct.
   c. Contact the inner end of the stator stopper (3, **Figure 18**) with the cylinder projection (4, **Figure 18**).
   d. Tighten the bolts (5, **Figure 18**) securely.

3. Adjust the minimum advance position as follows:
   a. Align the 2° ATDC match mark (1, **Figure 19**) on the stator housing with the match mark on the engine (2, **Figure 19**).

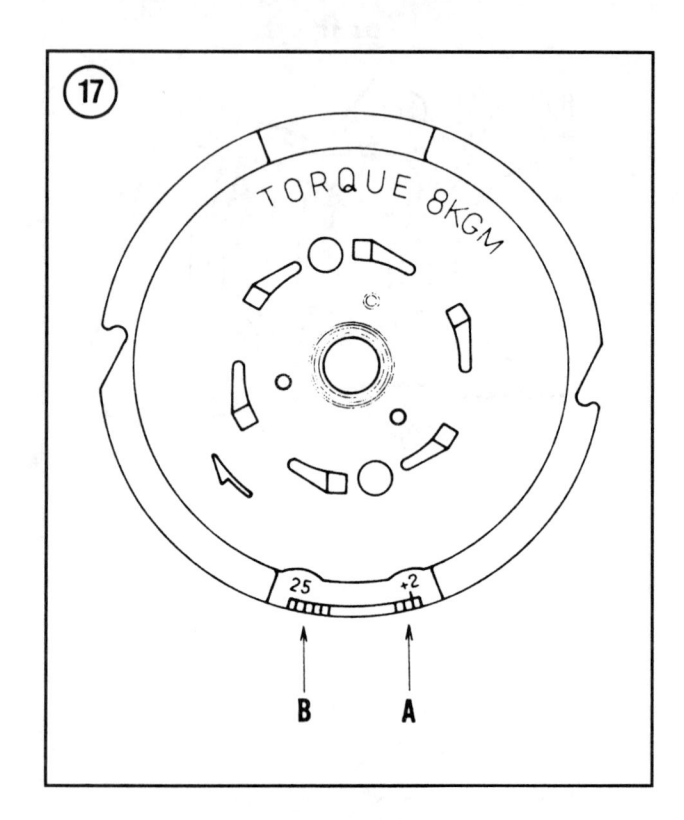

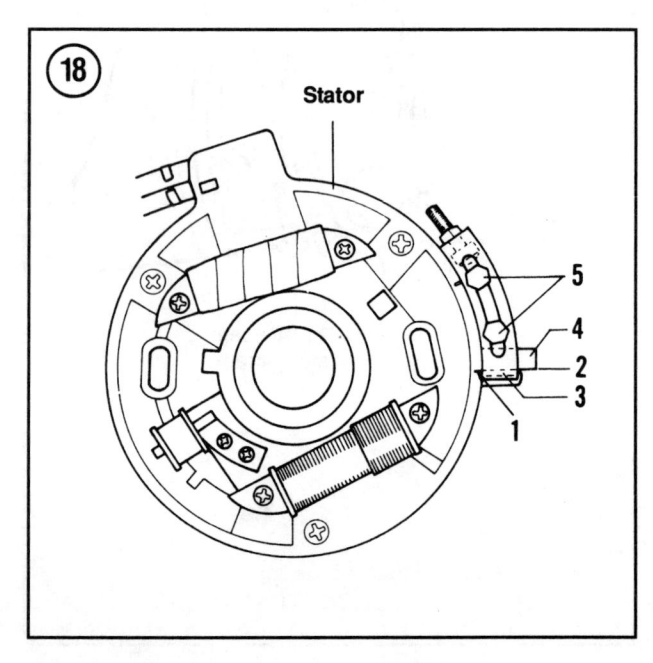

TIMING, SYNCHRONIZATION AND ADJUSTMENT

b. If alignment is incorrect, loosen the stator mounting bolts and move the stator housing until alignment is correct.

c. Loosen the locknut (5, **Figure 19**) and turn the adjustment screw until its tip comes in contact with the cylinder projection (4, **Figure 19**). Tighten the locknut on the adjustment screw.

d. Tighten the stator mounting bolts securely.

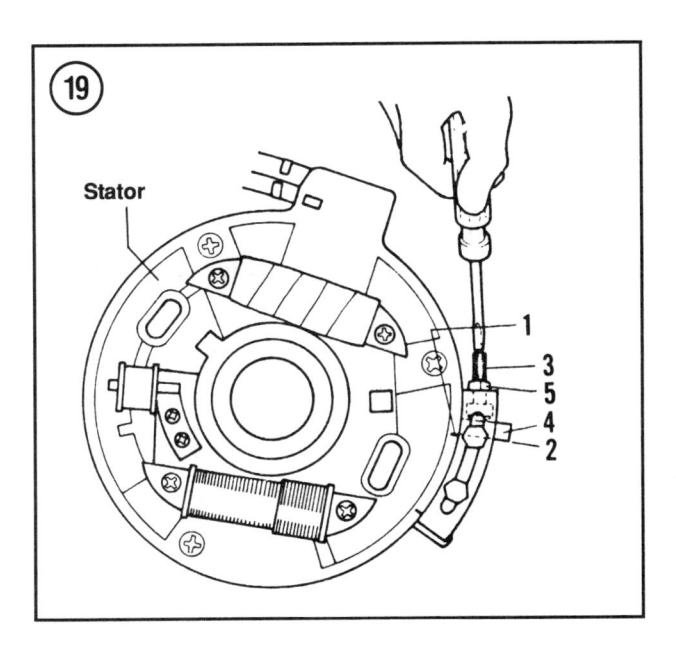

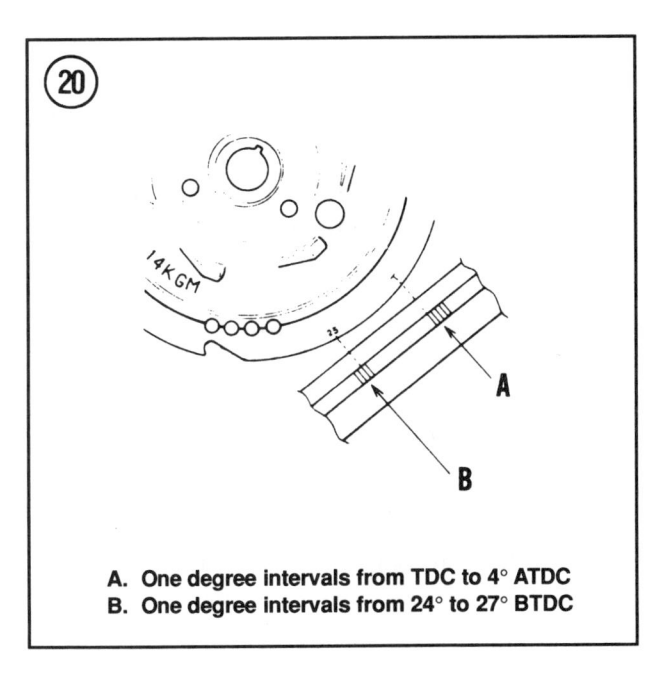

Dynamic Timing Adjustment
(1985-1987 DT 20, DT 30; 1985-1988 DT 25)

The engine should be in a test tank or on the boat in the water for this procedure.

1. Remove the engine cover.
2. Connect a timing light and tachometer according to manufacturer's instructions.
3. Start the engine and allow it to warm up for approximately 5 minutes.
4. Run the engine at 1,000 rpm in NEUTRAL and point the timing light at the rewind starter timing pointer. It should align with one of the marks (preferably the center mark) in the series of 5 flywheel timing marks (A, **Figure 20**).
5. If the timing pointer does not align as specified in Step 4, shut the engine off. Loosen the adjusting screw locknut and turn the screw *clockwise* to advance or *counterclockwise* to retard the timing (**Figure 21**).
6. Restart the engine and repeat Step 4. If the timing pointer and marks now align, shut the engine off and tighten the locknut. If they still

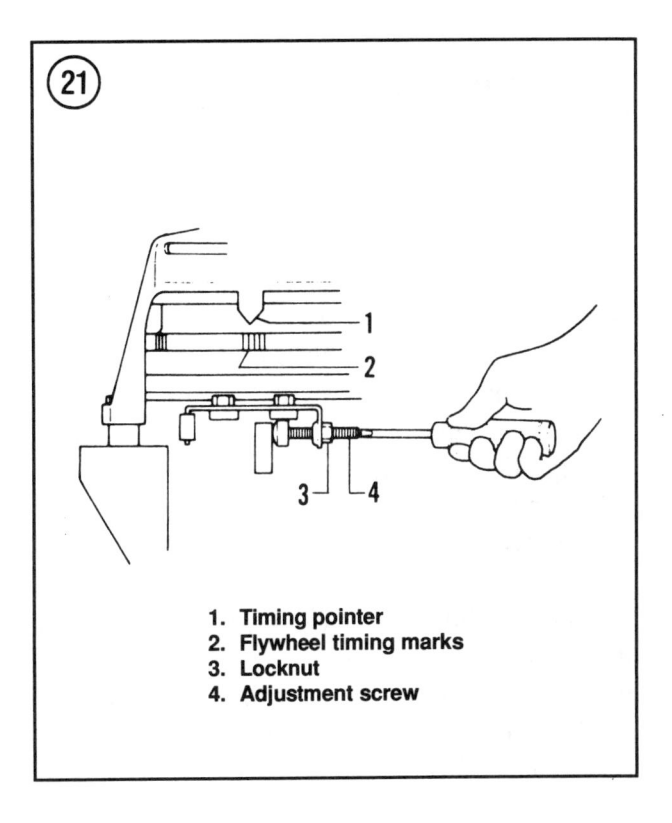

⑲
Stator

⑳
A. One degree intervals from TDC to 4° ATDC
B. One degree intervals from 24° to 27° BTDC

㉑
1. Timing pointer
2. Flywheel timing marks
3. Locknut
4. Adjustment screw

are not in alignment, repeat adjustment of the screw as required.

7. Repeat Step 4 with the engine in NEUTRAL and the throttle wide open. The timing pointer should align with the 25° BTDC mark on the flywheel (B, **Figure 20**).

8. If the timing pointer does not align as specified in Step 7, shut the engine off and proceed as follows:
   a. Disconnect the stator rod from the stator base.
   b. Loosen the connector locknut and rotate the connector to adjust the stator rod to the proper length.
   c. Reconnect the stator rod to the stator base.
   d. Start the engine and repeat Step 7.
   e. If the timing pointer and mark now align, shut the engine off and tighten the connector locknut.
   f. If the pointer and mark still do not align, repeat this step as required.

9. Shut the engine off and install the engine cover.

**Static Timing Adjustment
(1987-1989 DT 35; 1985-on DT 40)**

*NOTE
Steps 1-8 are for maximum advance setting.*

1. Remove the engine cover.
2. Shift the engine into NEUTRAL.
3. Fully open the throttle grip (or remote control lever) and secure the grip or lever in this position.
4. Align the match mark "A" (A, **Figure 22**) on the stator assembly with the cylinder projection (C, **Figure 22**). Maintain the stator in this position.
5. Loosen the screws (9, **Figure 22**) on the throttle cam (2, **Figure 22**).
6. Turn the throttle cam in the direction indicated by the arrow (S) on the cam until it contacts the crankcase stopper (E, **Figure 22**). Securely tighten the screws (9) securing the throttle cam in this position.
7. Loosen screw (10, **Figure 22**) on the carburetor, open the carburetor link in the direction indicated by the arrow (T, **Figure 22**) until it hits the stop.
8. Move the roller (U, **Figure 22**) in the direction indicated by the arrow (V, **Figure 22**) until it contacts the throttle cam (2). Tighten the screw (10) securely.

*NOTE
Steps 9-13 are for maximum retard setting.*

9. Place the shift lever (or remote control lever) in the FORWARD 1st notch.
10. Align the match mark (B, **Figure 22**) on the stator assembly with the cylinder projection (C, **Figure 22**). Maintain the stator in this position.
11. Loosen the locknut (6, **Figure 22**) and turn the adjust bolt (5, **Figure 22**) so that the adjust bolt cap (4, **Figure 22**) contacts the crankcase stopper (D, **Figure 22**). Tighten the locknut (6) securely.
12. Move the throttle limiter (1, **Figure 22**) in the direction (W, **Figure 22**) until it touches the undercover stopper (F, **Figure 22**). Check that the carburetor can be fully opened in this condition.
13. If the carburetor cannot be fully opened, loosen the locknut (8, **Figure 22**) and adjust the rod (7, **Figure 22**) to the standard length of 4.25 in. (108 mm). Tighten the nut (8).

**Dynamic Timing Adjustment
(1987-1989 DT 35; 1985-on DT 40)**

The engine should be in a test tank or on the boat in the water for this adjustment.

1. Remove the engine cover.
2. Connect a timing light and tachometer according to manufacturer's instructions.
3. Disconnect the throttle rod (11, **Figure 22**).

TIMING, SYNCHRONIZATION AND ADJUSTMENT

4. Start and run the engine until it reaches operating temperature (about 5 minutes).

NOTE
Steps 5-9 are for advanced ignition timing.

5. Shift the engine into FORWARD.

6. Increase engine speed to 1000 rpm and move the throttle limiter (1, **Figure 23**) to the full advance position (W, **Figure 23**).

7. Run the engine at 1000 rpm and point the timing light at the rewind starter timing pointer. It should align with the 18.5° BTDC mark on flywheel (**Figure 24**).

8. If the timing pointer does not align as specified in Step 7, shut the engine off. Loosen the 2 screws (1, **Figure 25**) and move the throttle cam in either direction to adjust the timing. Tighten the 2 screws and repeat Step 6 and Step 7. Readjust if necessary until advanced timing is correct.

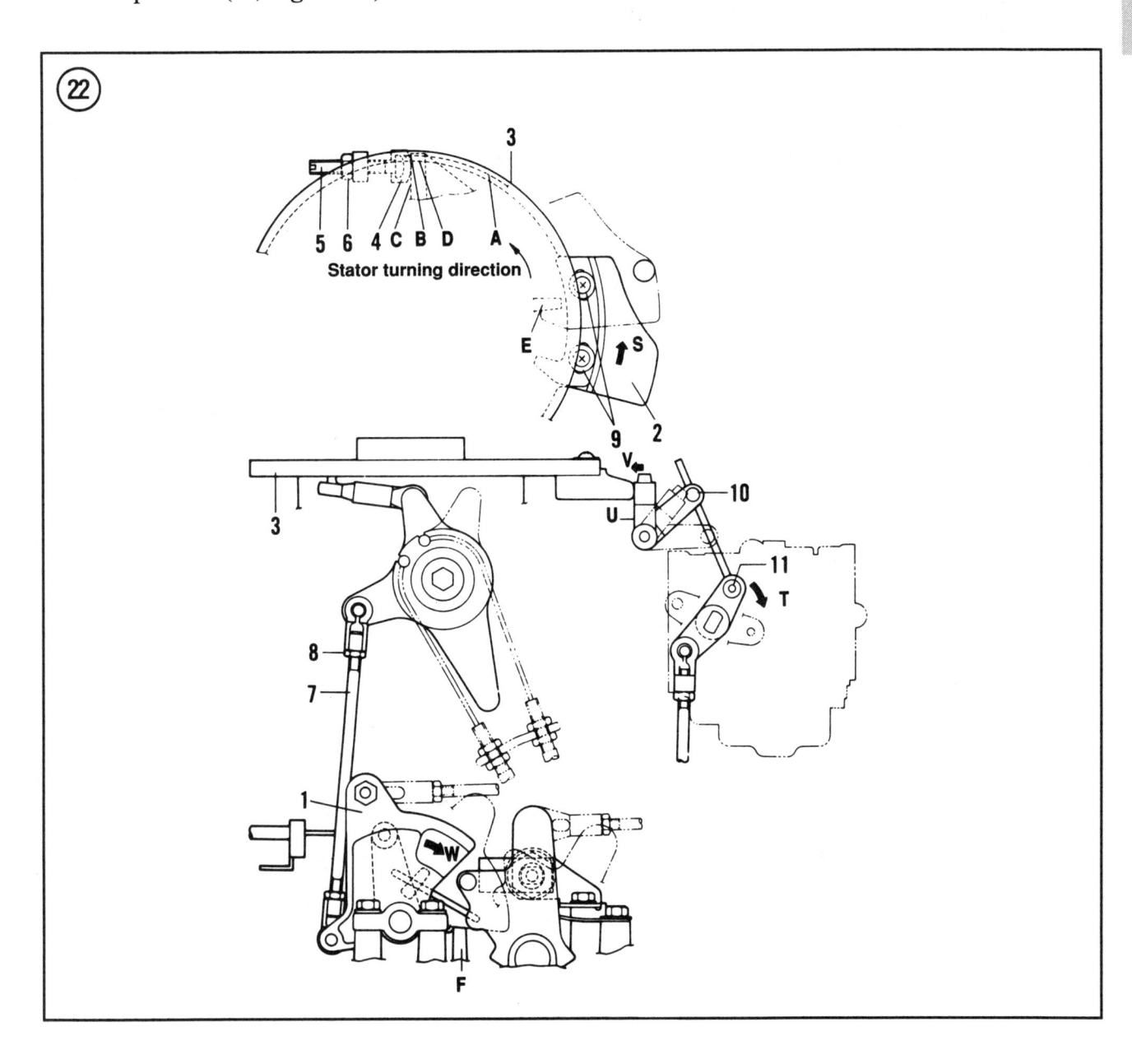

9. Connect the throttle rod (3, **Figure 26**) to its original position.

NOTE
Steps 10-12 are for fully retarded ignition timing.

10. With the engine still shifted in the FORWARD position, increase engine speed to 1,000 rpm.
11. Run the engine at 1,000 rpm and point the timing light at the rewind starter timing pointer (1, **Figure 27**). It should align with the timing marks within the range of 2° ±1° ATDC on flywheel (2, **Figure 27**).
12. If the timing pointer does not align as specified in Step 11, shut the engine off. Loosen the locknut (3, **Figure 27**) and turn the screw (4, **Figure 27**) *clockwise* to advance or *counterclockwise* to retard the timing. Tighten the locknut securely.
13. Shut the engine off and install the engine cover.

Static Timing Adjustment (1985-1987 DT 75, DT 85)

1. Place the throttle in the full advance position and align the mark (A, **Figure 28A**) on the timer with the mark (B) on the upper seal housing.
2. Loosen the 2 screws (C, **Figure 28A**) on the maximum advance stop plate (D), then position the stop plate to lightly contact the spark advance lever (E).
3. Securely tighten the 2 screws (C, **Figure 28A**).

Dynamic Timing Adjustment (1985-1987 DT 75, DT 85)

The outboard motor must be in a test tank or on the boat in the water for this adjustment.
1. Connect a timing light and tachometer according to their manufacturer's instructions.
2. Start the motor and run at fast idle until warmed to normal operating temperature.

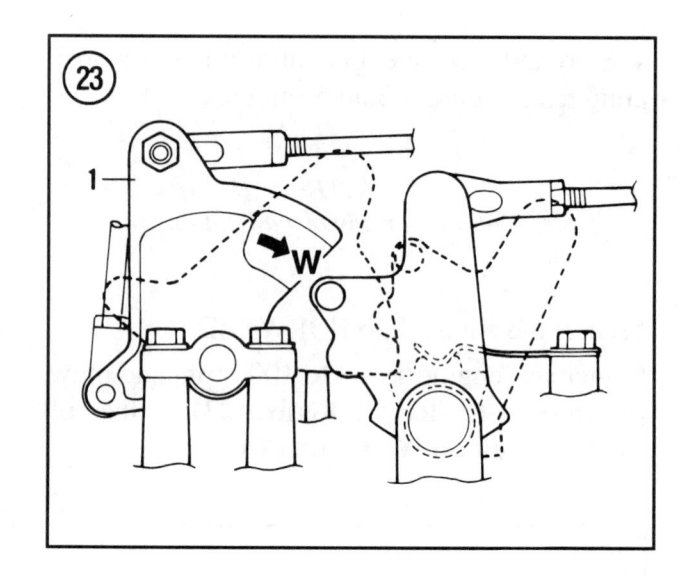

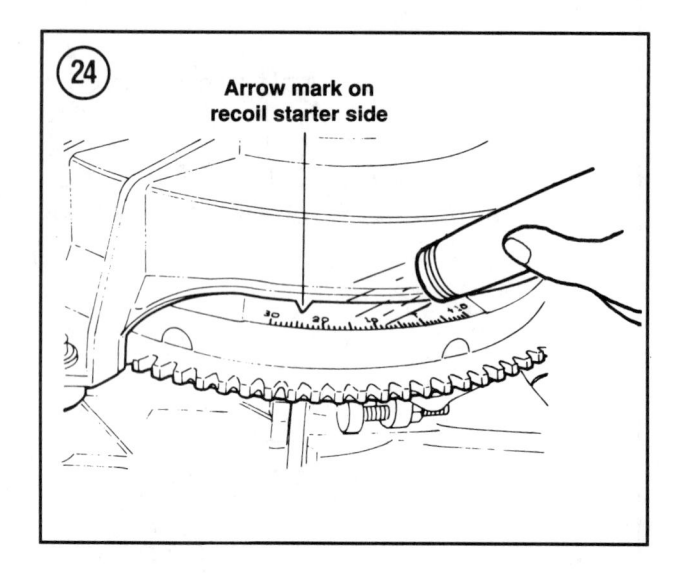

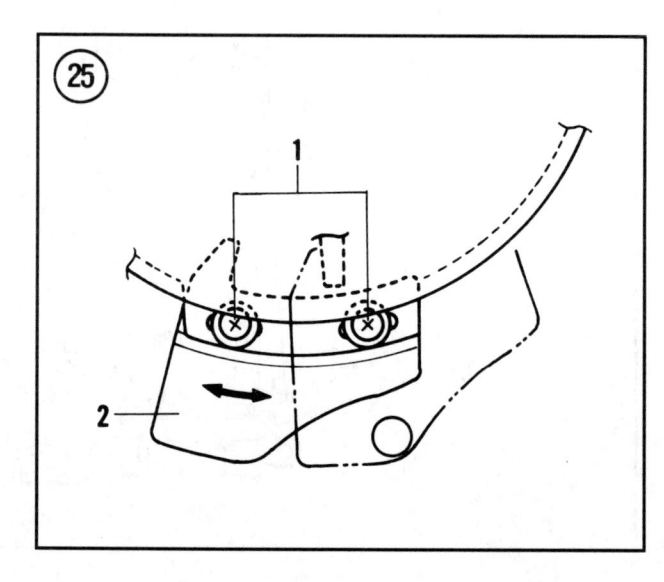

TIMING, SYNCHRONIZATION AND ADJUSTMENT

3. Adjust the idle stop screw so the motor idles at 1000 rpm in NEUTRAL.

4. Place the remote control in the fully closed-throttle position. Point the timing light at the timing pointer. Full retarded ignition timing should be:
   a. 1985-1986 models—7° ATDC at 1000 rpm.
   b. 1987 models—3° ATDC at 1000 rpm.

5. To adjust full retard timing, turn adjusting screw (F, **Figure 28A**) as necessary to obtain the specified timing.

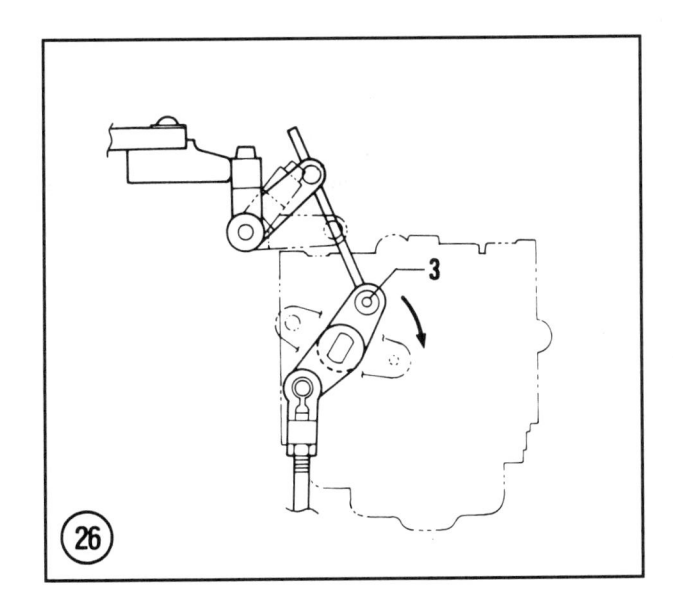

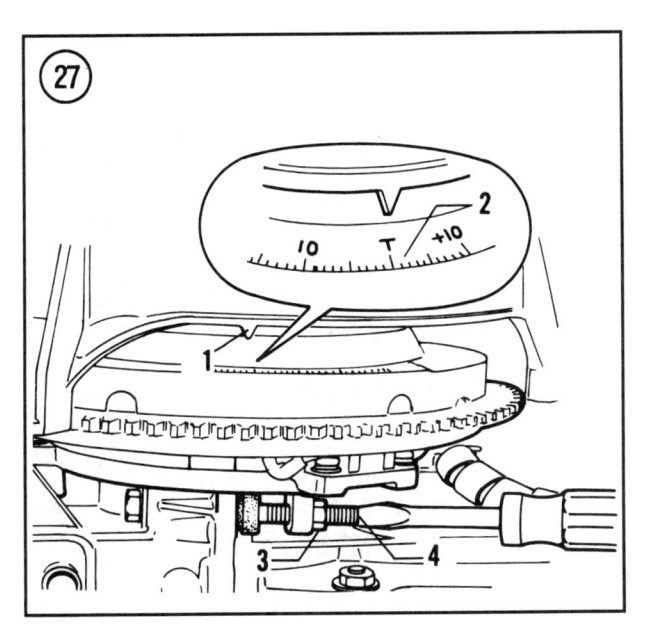

6. To check maximum advance ignition timing, run the motor at wide-open throttle while checking the timing with the timing light.

7. Maximum timing advance should be 21.5 BTDC on all models.

8. If adjustment is necessary, throttle back to idle and stop the motor. Loosen the 2 screws (C, **Figure 28A**) and reposition the maximum advance stop plate (D) as required to obtain the specified timing. Retighten the screws (C, **Figure 28A**), then recheck the timing at wide-open throttle.

9. Next, disconnect the carburetor link (G, **Figure 28A**). Then, rotate the speed control lever (H, **Figure 28A**) toward the maximum speed position until the lever (H) contacts the maximum speed stop on the crankcase.

10. Place the carburetor throttle valves in the fully-open position, then vary the length of the

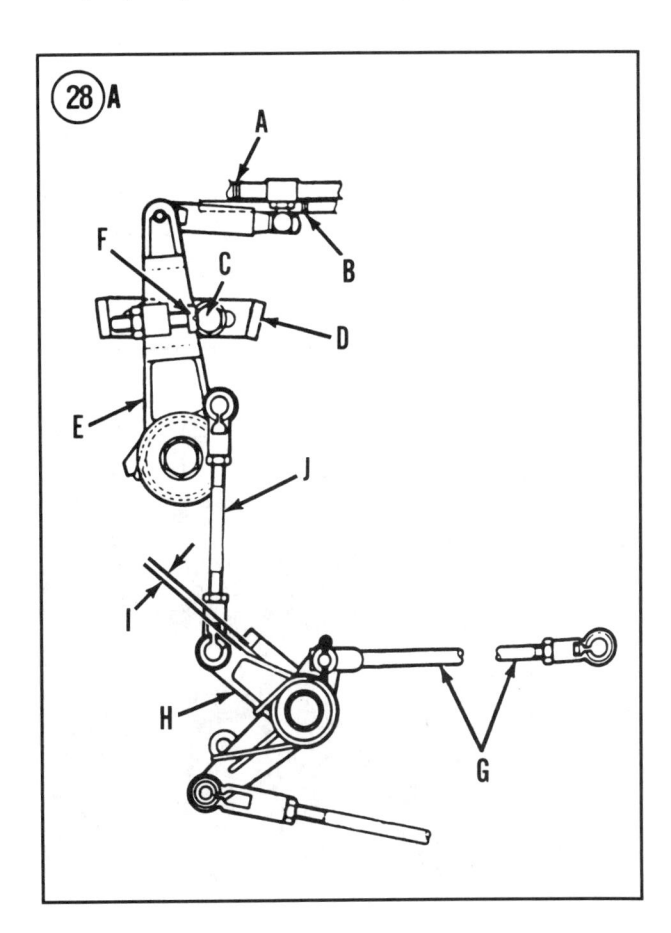

130 CHAPTER FIVE

carburetor link (G, **Figure 28A**) until the ball joint connector will just attach.

11. Place the speed control lever in the fully-retarded position. Start the motor and adjust the idle speed to 650 rpm in FORWARD gear. Then, check the clearance (I, **Figure 28A**) between the 2 levers. The clearance should be 0-1 mm (0-0.39 in.). Adjust the length of the link (J, **Figure 28A**) as necessary to adjust the clearance.

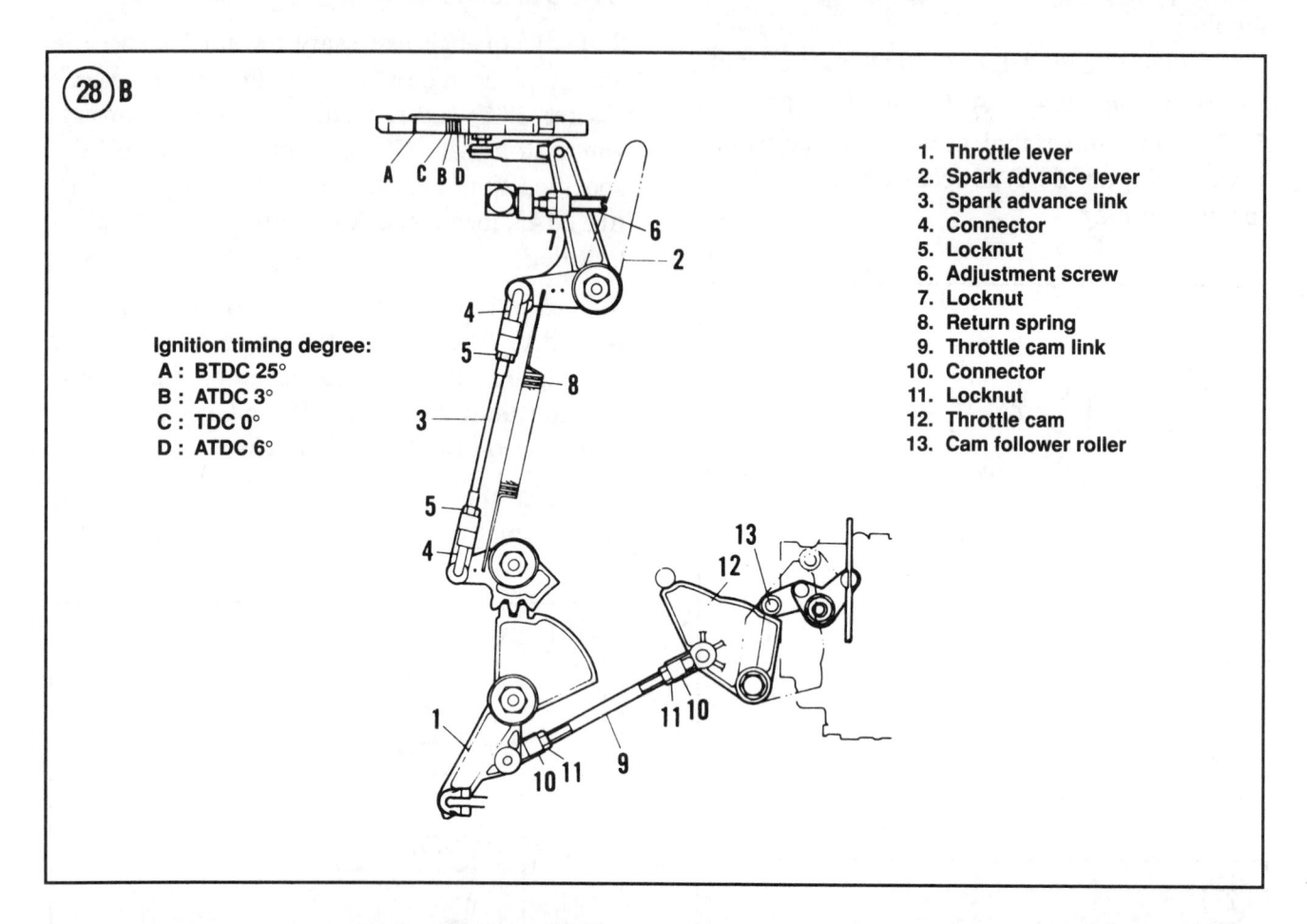

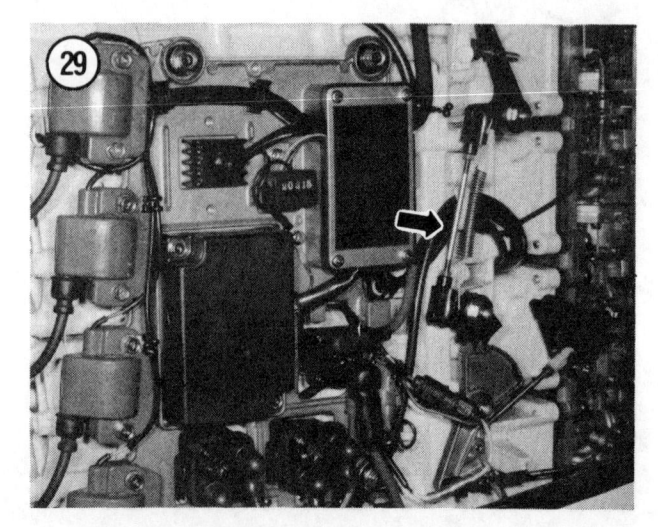

28 B

Ignition timing degree:
A : BTDC 25°
B : ATDC 3°
C : TDC 0°
D : ATDC 6°

1. Throttle lever
2. Spark advance lever
3. Spark advance link
4. Connector
5. Locknut
6. Adjustment screw
7. Locknut
8. Return spring
9. Throttle cam link
10. Connector
11. Locknut
12. Throttle cam
13. Cam follower roller

TIMING, SYNCHRONIZATION AND ADJUSTMENT

Static Timing Adjustment (1985 DT 115, DT 140)

These engines must be static-timed whenever components have been replaced. Timing should be checked during a tune-up with the *Dynamic Timing Adjustment* in this chapter. Refer to **Figure 28B** for this procedure.

1. Remove the engine cover.
2. Disconnect the throttle and shift lever cables.
3. Move the throttle lever against the wide-open throttle stop and hold in that position.
4. Check the 25° BTDC timing mark on the timer base. It should align with the timing mark stamped on the cylinder housing.
5. If the timing marks do not align as specified in Step 4, proceed as follows:
    a. Disconnect the link between the throttle and spark advance levers (**Figure 29**).
    b. Loosen the connector locknuts and rotate the connectors to lengthen or shorten the link as required.
    c. Reconnect the link and repeat Step 3 and Step 4.
    d. Repeat this step as required until the timing marks align.
    e. Tighten the connector locknuts.
6. Loosen the locknut on the throttle cam adjustment screw.
7. Rotate the throttle lever *counterclockwise* as far as possible without stretching the return spring. Hold throttle lever in this position and turn throttle cam adjustment screw as required until the 3° ATDC mark on the timer base aligns with the cylinder housing timing mark. Tighten the locknut.
8. Rotate the throttle lever until the TDC mark on the timer base aligns with the cylinder housing timing mark. Hold throttle lever in this position.
9. Loosen the throttle lever link locknuts and rotate the connectors as required until the throttle cam mark aligns with the center of the carburetor cam follower roller (**Figure 30**). Tighten the locknuts securely.
10. Rotate the throttle lever several times to make sure that it moves freely. There should be at least 0.118 in. clearance between the wire harness clamp and return spring.
11. Reconnect the shift and throttle cables.
12. Perform the *Dynamic Timing Adjustment* described in this chapter.

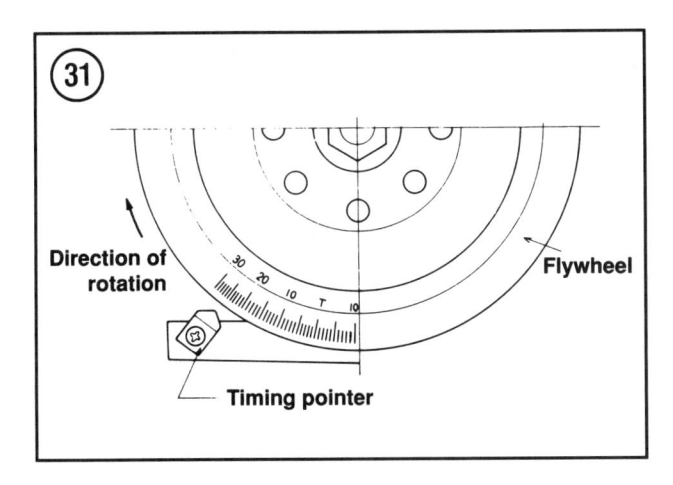

Dynamic Timing Adjustment (1985 DT 115, DT 140)

The engine should be in a test tank or on the boat in the water for this adjustment.

1. Remove the engine cover.
2. Connect a timing light and tachometer according to manufacturer's instructions.
3. Start and run the engine until it reaches operating temperature (about 5 minutes).
4. With the engine idling in NEUTRAL, move the remote control lever to the fully closed throttle position.
5. Point the timing light at the timing pointer (**Figure 31**). It should align with the 6° ATDC mark on the flywheel at 900 rpm or less.
6. If the timing marks do not align as specified in Step 5, loosen the throttle adjustment screw locknut and turn the adjustment screw as required to align the marks.

7. Disconnect the throttle cam link and move the remote control lever to the wide-open throttle position.

NOTE
Maximum advance is specified as 23° BTDC ±1° at 5,000 rpm. When checked at 2,000 rpm in Step 8, there should be a time lag of 2° BTDC, resulting in a reading of 25° BTDC.

8. Point the timing light at the timing pointer (**Figure 31**). It should align with the 25° BTDC mark on the flywheel.
9. If the timing marks do not align as specified in Step 8, loosen the throttle stop screw locknut and adjust the stop screw as required to align the marks. Tighten the locknut.
10. Once maximum advance timing is correct, connect the throttle cam link.
11. Shut the engine off and install the engine cover.

THROTTLE ADJUSTMENT

The manufacturer provides throttle adjustment information on the following models only.

1989-on DT 25, DT 30

Refer to **Figure 32** for this procedure.
1. Remove the engine cover.
2. Loosen the throttle lever adjusting screws (1) on the No. 1 and No. 3 cylinder carburetors.
3. Turn the plates (2) *counterclockwise* to make sure the throttles are completely closed.
4. Hold the plates (2) in this position and securely tighten both screws (1).
5. Loosen the locknuts on the throttle link lever rod (3).
6. Turn the throttle link lever rod (3) in either direction to achieve an *initial* length of 3.0 in. (75.5 mm) between the center of the holes in each connector.

7. At the FULL CLOSED throttle position, the throttle lever arm (4) must contact the stop boss (5) on the engine crankcase.
8. With the throttle lever arm in the position noted in Step 7, there must be 0.02-0.06 in. (0.5-1.5 mm) of clearance at point "A."
9. If the clearance is incorrect, readjust the throttle link lever rod to achieve the correct clearance.
10. Tighten the throttle link lever rod locknuts securely.
11. Check the synchronization of the carburetors as described in Chapter Six.
12. Install the engine cover.

1985-on DT 55, DT 65

Full-close adjustment—DT 55 (engine serial numbers 501001-502859) and DT 65 (engine serial numbers 501001-502959)

Refer to **Figure 33** for this procedure.
1. Remove the engine cover.

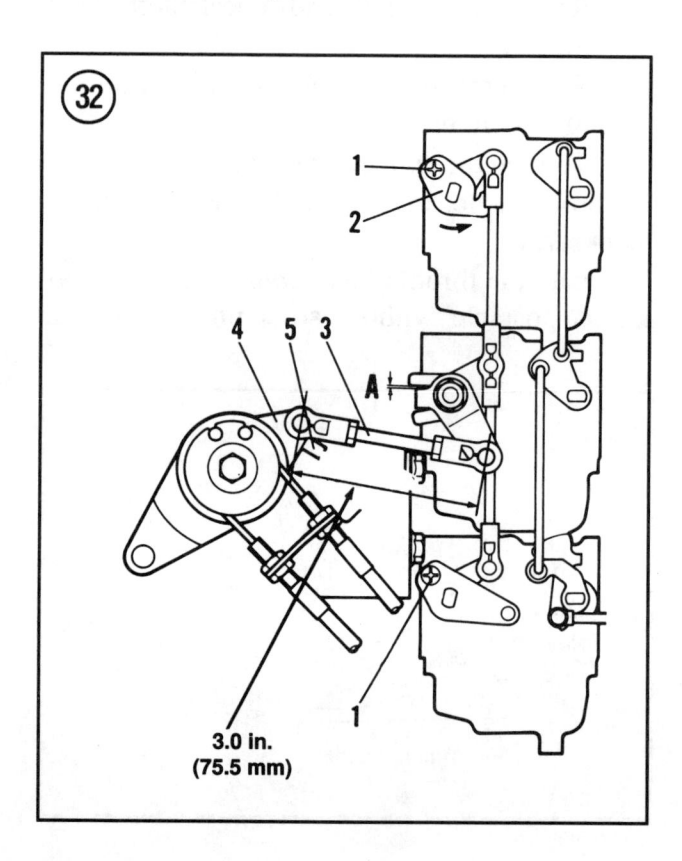

TIMING, SYNCHRONIZATION AND ADJUSTMENT

2. Disconnect the throttle rod (1) from the throttle control lever (2).

3. Loosen the set screw (4) on the throttle lever (3) on the No. 1 and No. 3 cylinder carburetors.

4. Turn the throttle valves until they are both completely closed.

5. Hold the throttle levers in this position and securely tighten both set screws (4). Do not allow the throttle lever (3) to move while tightening the set screw or the adjustment will be incorrect.

6. Check the adjustment by operating the lever (5) on the No. 2 carburetor and make sure the throttle valves on the No. 1 and No. 3 carburetors operate together.

7. If they do not work together, repeat Steps 3-5 until they work correctly.

8. Do not attach the throttle rod (1) onto the throttle control lever at this time. Adjust the throttle rod length as described in this chapter.

Full-close adjustment—DT 55 (engine serial numbers 502860-on) and DT 65 (engine serial numbers 502960-on)

Refer to **Figure 34** for this procedure.

1. Remove the engine cover.

2. Disconnect the throttle rod (1) from the throttle control lever (2).

3. Loosen the set screws (4) on the throttle lever (3) on the No. 1 and No. 3 cylinder carburetors. When the set screw is loosened, the return spring

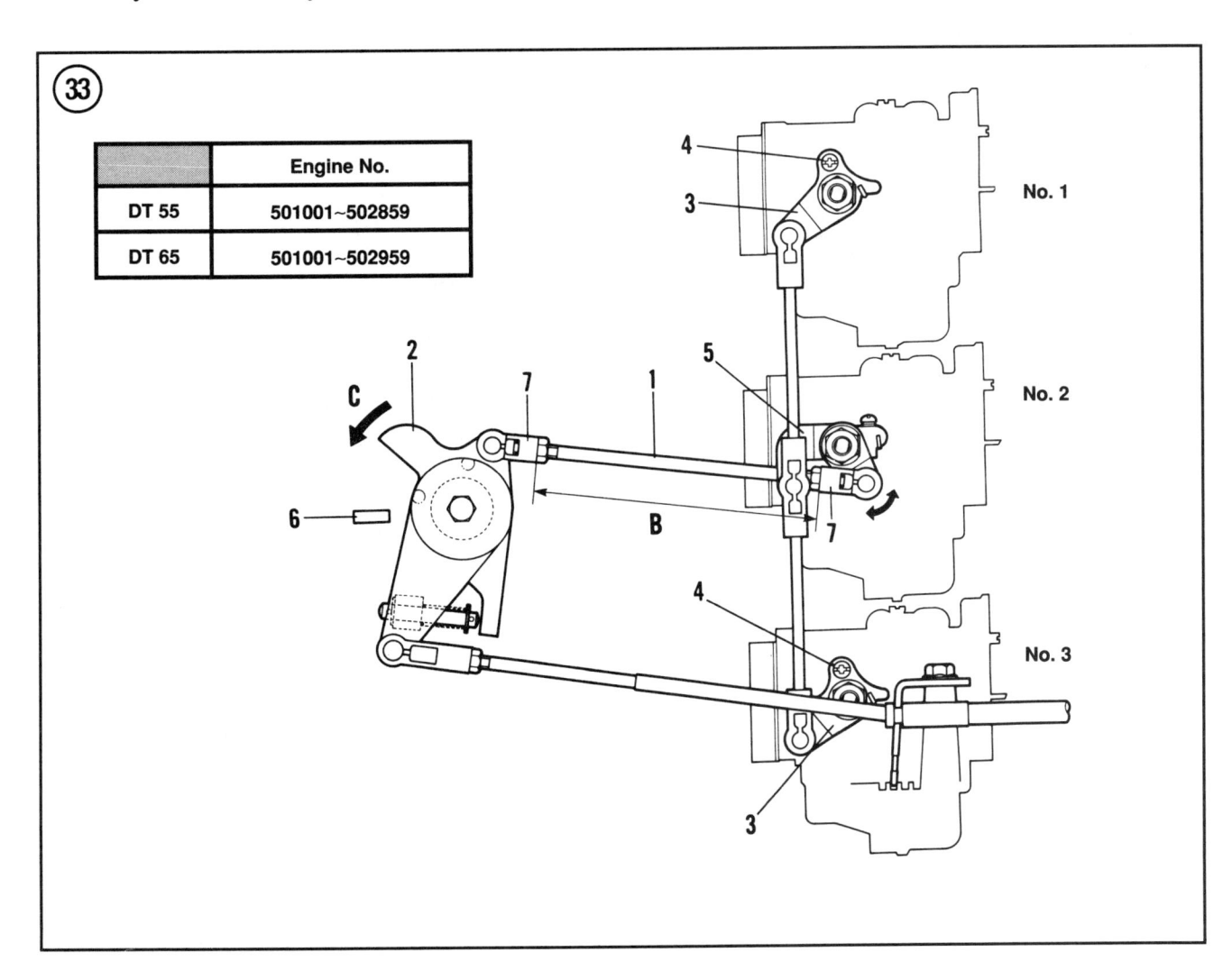

	Engine No.
DT 55	501001~502859
DT 65	501001~502959

㉝

CHAPTER FIVE

will move the throttle valve to the fully-closed position.

4. Operating the lever (5) on the No. 2 carburetor several times by more than 30° of movement as shown by arrow (A) will remove all play between all 3 carburetors.

5. Allow the No. 2 carburetor to move back to the fully-closed position. At this point all 3 carburetors should be in the full-closed position.

6. Apply Thread lock 1342 to the threads of all 4 set screws (4) and tighten securely. Do not allow the throttle lever (3) to move while tightening the set screw or the adjustment will be incorrect.

7. Check the adjustment by operating the lever (5) on the No. 2 carburetor and make sure the throttle valves on the No. 1 and No. 3 carburetors operate together.

8. If they do not work together, repeat Steps 3-6 until they work correctly.

9. Do not attach the throttle rod (1) onto the throttle control lever at this time. Adjust the throttle rod length as described in this chapter.

Throttle rod length adjustment

Refer to **Figure 33** and **Figure 34** for this procedure.

1. Loosen the locknuts on each end of the throttle rod (1).

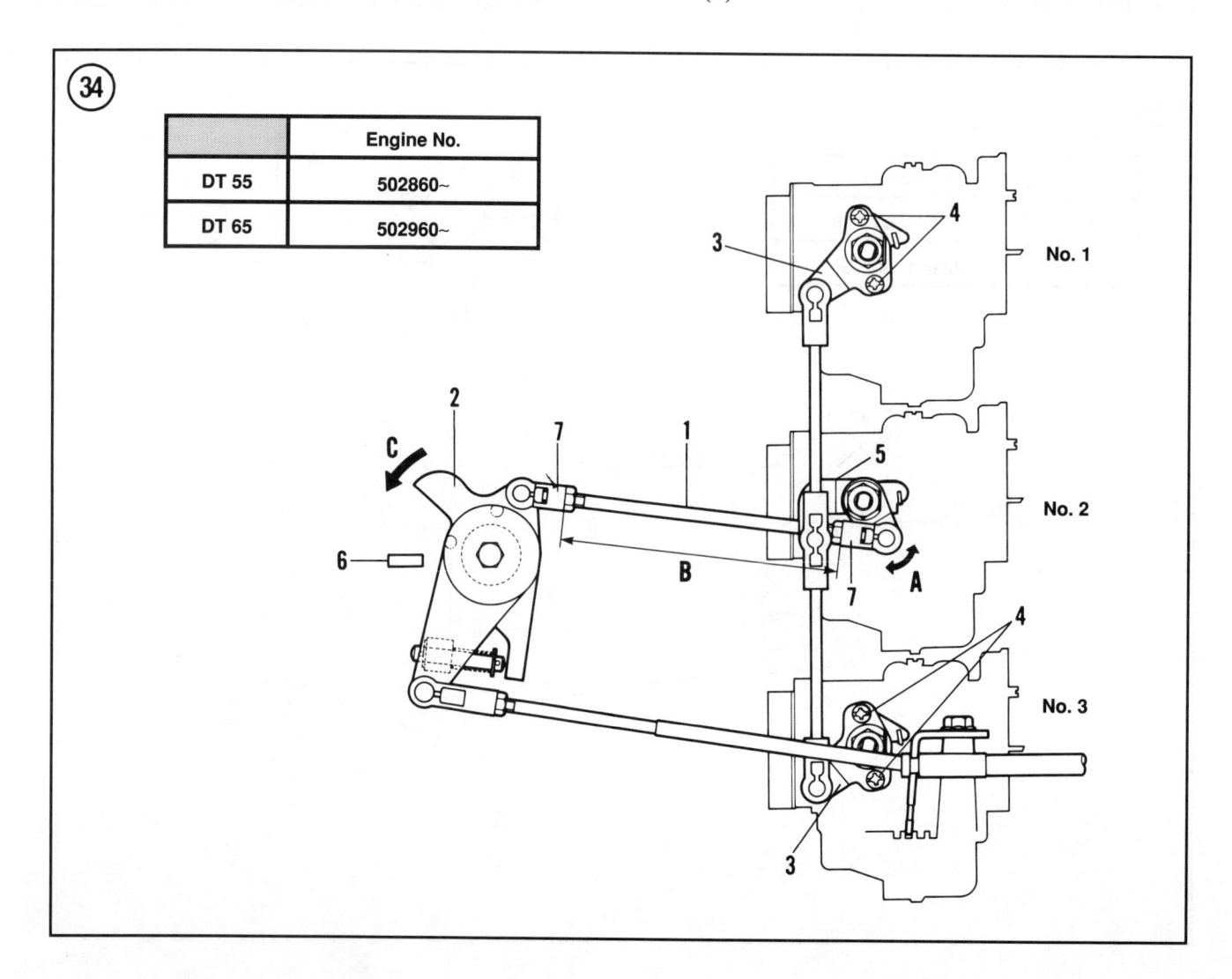

	Engine No.
DT 55	502860~
DT 65	502960~

TIMING, SYNCHRONIZATION AND ADJUSTMENT

2. Rotate the throttle rod in either direction to achieve the following dimension (B):
  a. DT 55: 4.5 in. (114 mm).
  b. DT 65: 4.25 in. (108 mm).
3. Attach the throttle rod (1) onto the throttle control lever (2). Do not tighten the locknuts at this time.
4. Move the control lever (2) in the direction of arrow (C) until the control lever comes in contact with the stopper (6).
5. At this point the throttle valves should be in the fully-open position or within 1-2° from the fully-open position. If necessary, readjust the throttle rod (1) length so the control lever (2) contacts the stopper (6) when the carburetor throttle valves are fully open.

CAUTION
Make sure a gap does not exist between the control lever (2) and the stopper (6) when the throttle valves are in the fully open position, or the throttle valve(s) or carburetor(s) may be damaged during full-throttle operation.

6. Securely tighten the throttle rod locknuts.
7. Install the engine cover.

1986-on DT 115, DT 140

Refer to **Figure 35** for this procedure.
1. Remove the engine cover.
2. Loosen the locknuts on the throttle rod (1).
3. Rotate the throttle rod in either direction to achieve a dimension of 6.3 in. (160 mm) between the center holes in the connectors (3). The connectors (3) must be at the same angle after adjusting.
4. Push the throttle control lever (5) *counterclockwise* until the throttle valves are in the completely-open position.

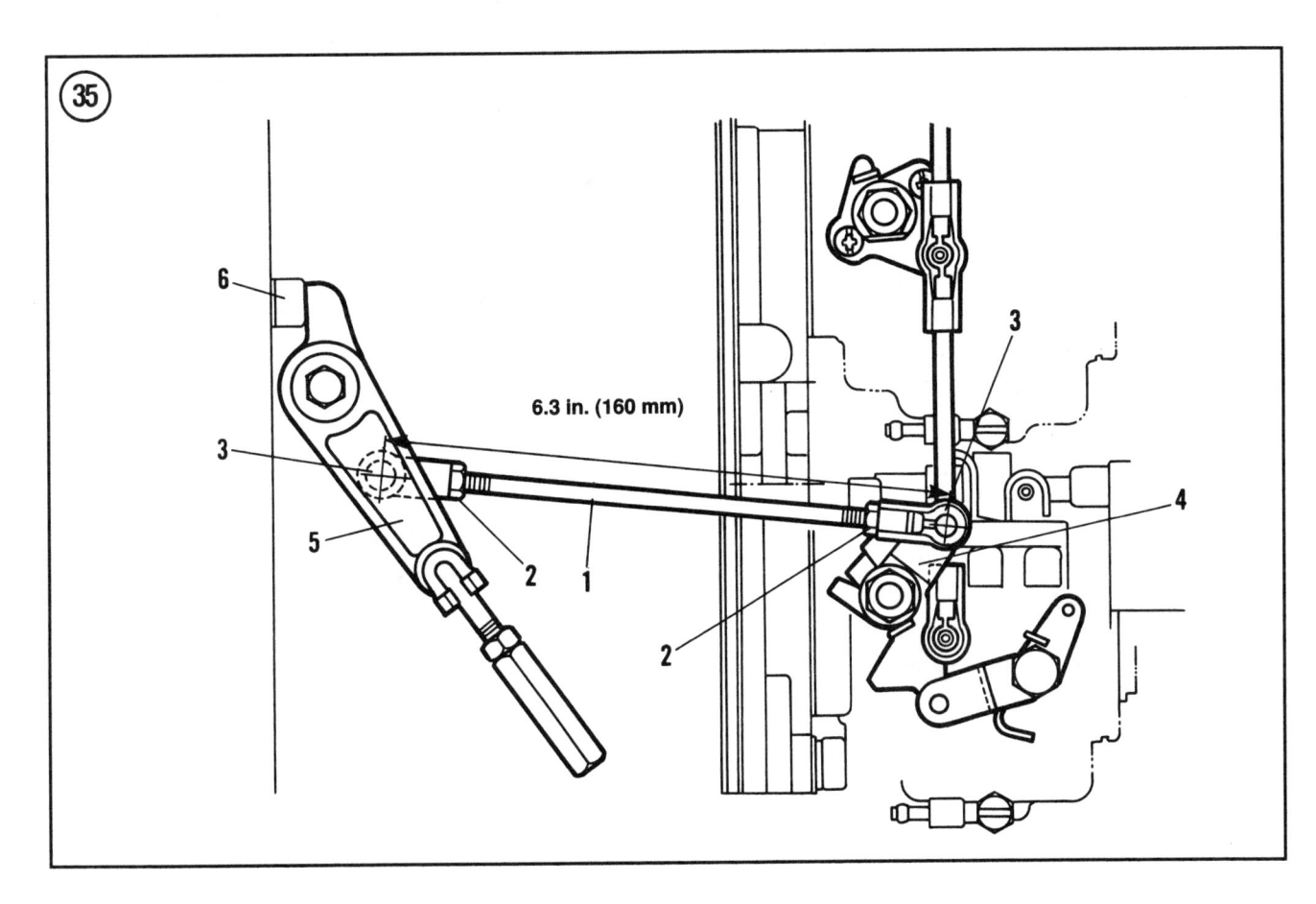

5. At this point there must be zero clearance between the control lever (5) and the stopper (6) on the crankcase.
6. If necessary, readjust the throttle rod length to accomplish this condition.

> **CAUTION**
> Make sure a gap does not exist between the control lever (5) and the stopper (6) when the throttle valves are in the fully open position, or the throttle valve(s) or carburetor(s) may be damaged during full-throttle operation.

7. Securely tighten the throttle rod (1) locknuts.
8. Install the engine cover.

V4 Models

Refer to **Figure 36** for this procedure.
1. Remove the engine cover.
2. Check the length of the throttle linkage rod (1). The correct length is 4.7 in. (120 mm).
3. If incorrect, loosen the locknuts and adjust the length of the linkage rod (1) as required. Tighten the locknuts.
4. Loosen the throttle lever adjusting screws (2) on the top carburetor.
5. Lightly push the throttle lever (3) *clockwise* until both throttle valves are completely closed. Tighten the adjusting screws securely.
6. Move the linkage several times and make sure both throttle valves close at the same time. Readjust if necessary.
7. Repeat for the other bank of carburetors.
8. Install the engine cover.

V6 Models
(Carburetted Models)

Refer to **Figure 37** for this procedure.
1. Remove the engine cover.
2. Check that the throttle linkage rod (1) moves freely with no binding.
3. Loosen the throttle lever adjusting screws (2) on the top and center carburetors.
4. Lightly push the throttle lever (3) on the center carburetor *clockwise* until both the bottom and center carburetor throttle valves are completely closed. Tighten the adjusting screws securely on the center carburetor.
5. Lightly push the throttle lever (3) on the top carburetor clockwise until all 3 carburetor throttle valves are completely closed. Tighten the adjusting screws securely on the top carburetor.
6. Move the linkage several times and make sure all throttle valves close at the same time. Readjust if necessary.
7. Repeat for the other bank of carburetors.

THROTTLE VALVE SENSOR ADJUSTMENT

On models equipped with Integrated Circuit (IC) and Micro Link ignition systems, a throttle switch (early models) or a throttle valve sensor (late models) is used to provide throttle position information to the IC or Micro Link module. The

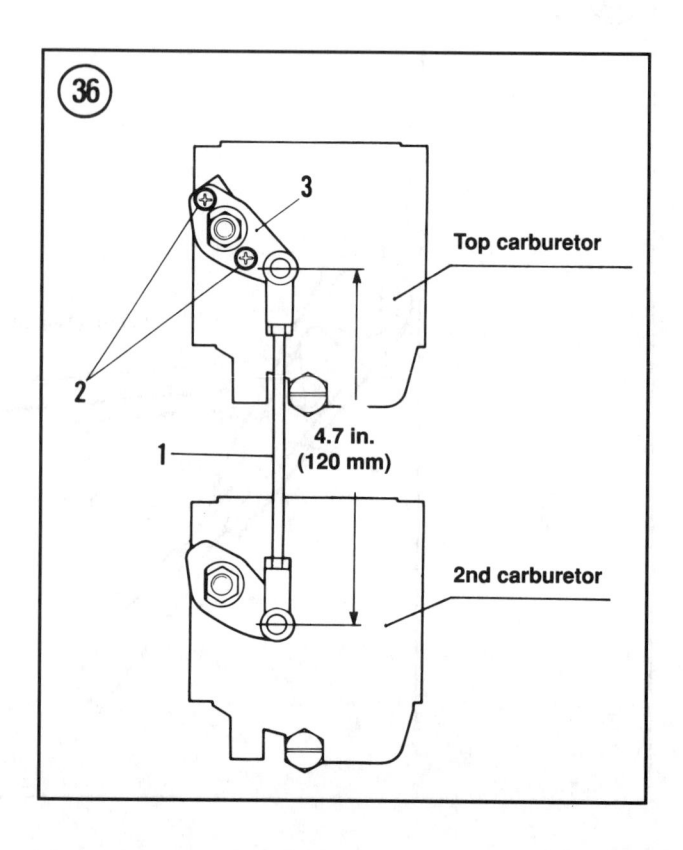

TIMING, SYNCHRONIZATION AND ADJUSTMENT

module processes throttle position, engine speed and engine temperature to determine the optimum ignition timing for all operating conditions.

The throttle valve sensor is mounted on the side of the bottom carburetor on DT 25, DT 30, DT 115 and DT 140 models and the side of the center carburetor on DT 55, DT 65, DT 75 and DT 85 models. On all models, the sensor is engaged with the carburetor throttle valve shaft to directly register throttle opening.

No adjustment is required on early models equipped with a throttle position switch. Refer to Chapter Three for throttle switch and throttle sensor troubleshooting procedures.

The following procedures apply to later models equipped with a throttle valve sensor.

1989-on DT 25, DT 55, DT 65, DT 115, DT 140; 1988-on DT 30, DT 75, DT 85

The following special tools are required for this procedure:
a. Small nonmagnetic screwdriver (insulated electrical type).
b. Digital voltmeter.
c. Battery (9 volt minimum).
d. Throttle valve sensor test lead (part 09930-89530) or suitable jumper leads.

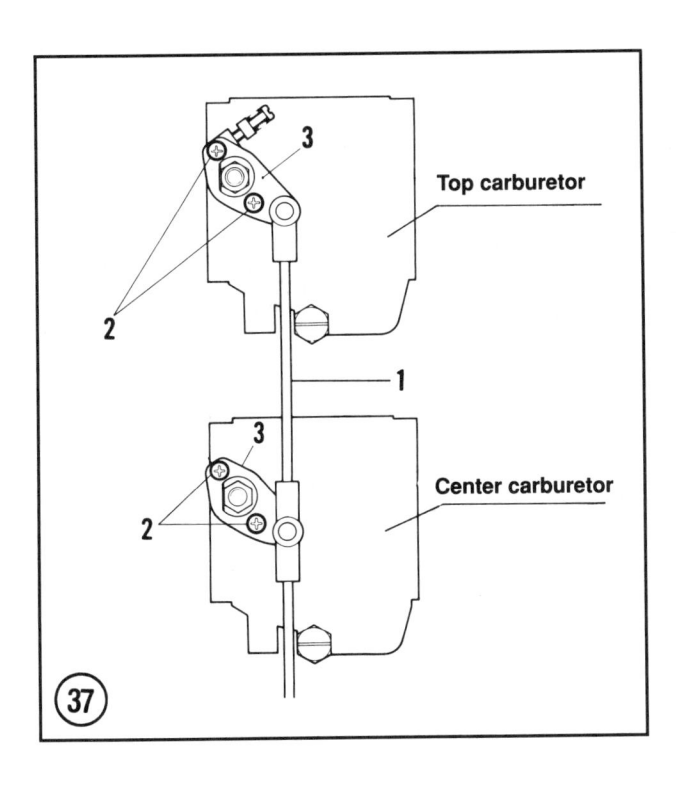

1. On models prior to 1991, turn the idle speed switch to the lowest idle setting. On models after 1990, back out idle speed screw so throttle valves are completely closed.

NOTE
*Throttle valves **must** be fully closed for proper throttle sensor adjustment.*

2. Disconnect the sensor 3-wire connector.
3. Connect sensor test lead (part 09930-89530) to the sensor connector.

NOTE
*If the throttle sensor test lead (part 09930-89530) is not available, connect suitable jumper leads to the sensor connector as shown in **Figure 38**.*

4. Connect test lead black/red wire to the positive battery terminal. See **Figure 38**.
5. Connect test lead black wire to the negative battery terminal. See **Figure 38**.
6. Connect voltmeter positive lead to the light green/red test lead wire and voltmeter negative lead to the negative battery terminal. See **Figure 38**.

NOTE
Use only a nonmagnetic, insulated screwdriver to adjust sensor in Step 7, or voltage reading will not be valid.

7. Throttle sensor voltage at closed throttle should be 0.45-0.55 volt. If not, remove adjusting screw cover and turn screw (**Figure 38**) clockwise to increase the voltage or counterclockwise to decrease the voltage.
8. Advance the throttle to the wide-open position. Sensor voltage at wide-open throttle should be 2.7 volts or more. Wide-open voltage is not adjustable—if 2.7 volts or more can not be obtained, replace throttle sensor.

NOTE
The correct throttle sensor voltages are essential for proper outboard operation. If the correct closed-throttle voltage can not be obtained by turning the sensor adjusting screw, loosen the sensor mounting screws and shift position of the sensor on the carburetor. If the correct voltage still can not be obtained, replace the sensor.

9. Disconnect the test leads, voltmeter and reconnect the sensor connector to the outboard wiring harness. Reinstall sensor adjusting screw cover.

V4 and 1987-on V6

1. Remove the oil pump control rod from the oil pump lever.

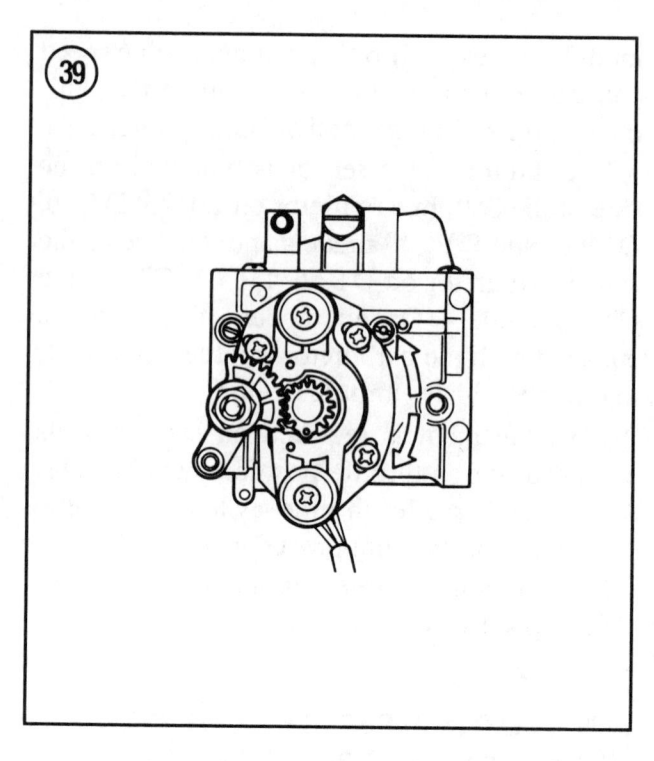

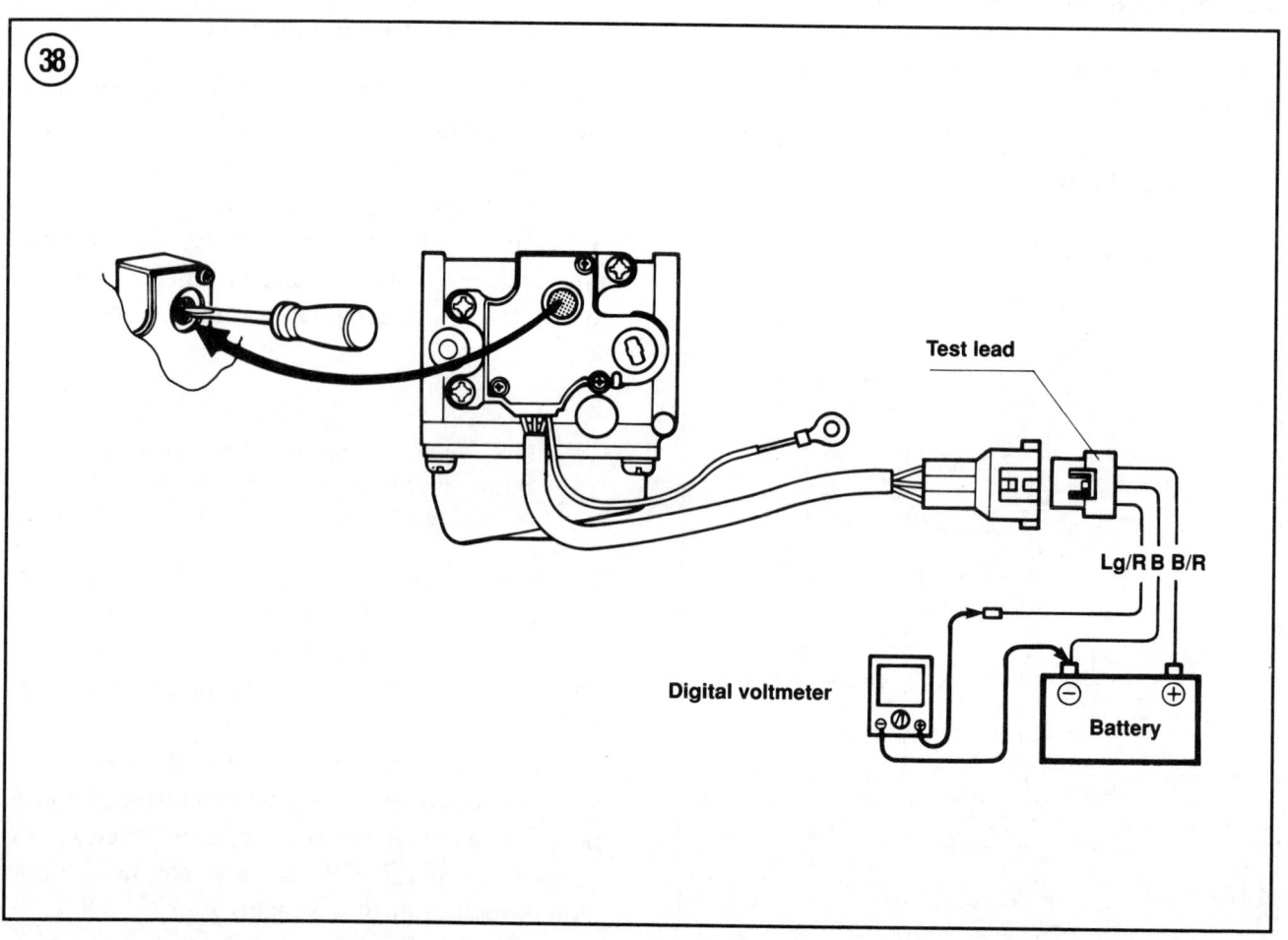

TIMING, SYNCHRONIZATION AND ADJUSTMENT

2. Open and close throttle valves to check for free movement. Make sure throttle valves can be fully closed.

3. Calibrate an ohmmeter on the R × 100 scale.

4. Set the idle speed switch to the slowest position on models prior to 1991. On models after 1990, back out the idle speed screw so throttle valves are fully closed.

5. Disconnect the throttle sensor connector.

6. Connect the ohmmeter red (+) lead to the sensor light green/red wire terminal and the black (–) lead to the sensor black wire terminal.

7. The sensor resistance values at fully-closed throttle should be as follows:
   a. V4—225-275 ohms.
   b. 1987-1988 V6—200-250 ohms.
   c. 1989-on V6—225-275 ohms.

8. If the resistance is not as specified, perform the following:
   a. Make sure the throttle valves are fully closed.
   b. Loosen the sensor mounting screws and rotate the sensor as shown in **Figure 39** to obtain the specified resistance value.
   c. Retighten sensor mounting screws.

9. Disconnect the ohmmeter and reconnect the sensor to the outboard wiring harness.

DT 225 EFI

On EFI models, the throttle valve sensor is adjusted at the factory and the sensor mounting screws are secured with thread locking compound. The sensor should not require further adjustment.

Table 1 TEST WHEEL RECOMMENDATIONS*

Model	Test Wheel Part Number
DT 2	09914-79970
DT 4	09914-79850
DT 6, DT 8	09914-79810
DT 9.9	09914-79840
DT 15	09914-79840
DT 20, DT 25, DT 30 (2-cylinder)	09914-79830
DT 25, DT 30 (3-cylinder)	09914-79831
DT 35, DT 40	09914-79980
DT 55, DT 65	09914-79980
DT 75, DT 85	
1985-1987	09914-79510
1988-on	09914-79511
DT 115, DT 140	09914-79420

* These are the only recommendations provided by Suzuki.

Chapter Six

Fuel System

The majority of the procedures in this chapter relate to the DT 2 through DT 200 models that are equipped with carburetor(s). It covers the removal, overhaul, installation and adjustment procedures for fuel pumps, carburetors, reed valves, fuel tanks and connecting lines for all carburetted models.

For the DT 225 V-6 models that are equipped with the fuel injection system, refer to the end of this chapter for a description of the various components of the fuel injection system. Service to the fuel injection system is limited for the home mechanic and these procedures are covered in that section.

Tables 1-3 are at the end of the chapter.

LOSS OF HIGH SPEED RPM (ALL DT 15)

Suzuki has determined that all DT 15 models may exhibit a high speed surge resulting from the lack of fuel. This problem was covered in Suzuki Service Bulletins No. SB89-16 dated September 1989 and No. SB90-06 dated November 1989.

There is a carburetor/fuel pump modification kit available to solve this problem. In some cases, this modification will not completely solve the problem and a modification must be made to either the reed plate assembly or to the carburetor gasket. If the outboard motor is still under any applicable warranty, return the motor to the Suzuki dealer and have the problem solved at no charge.

OFF IDLE LEAN CONDITION (1988 DT 30 AND 1989 DT 65)

Suzuki has determined that the 1988 DT 30 and 1989 DT 65 models may exhibit a poor idle and/or off idle stalling problem. These problems were covered in Suzuki Service Bulletins No. SB90-02 dated October 1989 and No. SB90-04 dated November 1989. On the DT 65 models, the

FUEL SYSTEM

engines affected are serial No. 908870 to serial No. 910130. All engines are affected on DT 30 models.

There is a carburetor modification kit available along with additional work to be performed on either the reed valve plate or to the gear counter coil, depending on model. If the outboard motor is still under any applicable warranty, return the motor to the Suzuki dealer and have the problem solved at no charge.

FUEL PUMP AND FILTER

All DT 2 outboards have an integral fuel tank that uses a gravity flow fuel system; they require no fuel pump.

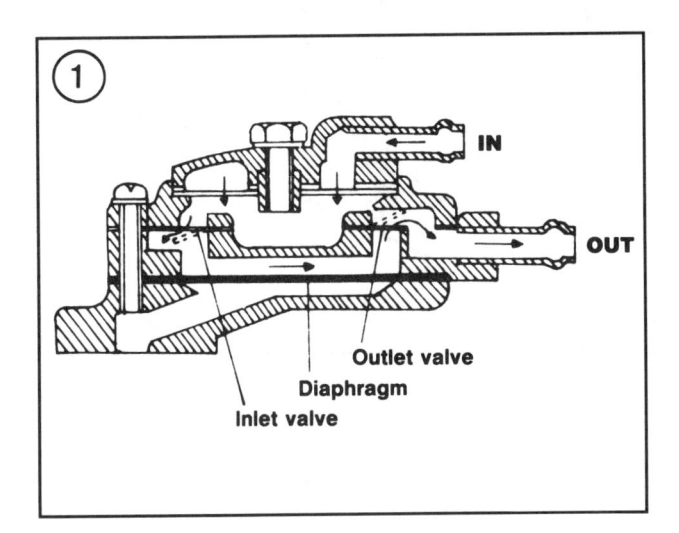

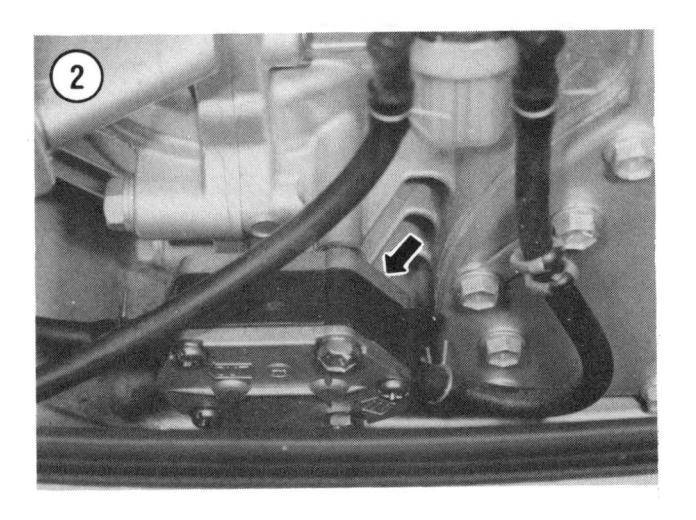

The diaphragm-type fuel pump used on all other models operates by crankcase pressure. Pressure pulsations created by movement of the pistons reach the fuel pump through a passageway between the crankcase and pump.

Upward piston motion creates a low pressure on the pump diaphragm. This low pressure opens the inlet check valve in the pump, drawing fuel from the line into the pump. At the same time, the low pressure draws the air-fuel mixture from the carburetor into the crankcase.

Downward piston motion creates a high pressure on the pump diaphragm. This pressure closes the inlet check valve and opens the outlet check valve, forcing the fuel into the carburetor and drawing the air-fuel mixture from the crankcase into the cylinder for combustion. **Figure 1** shows the operational sequence of a typical Suzuki fuel pump.

Since this type of fuel pump cannot create sufficient pressure to draw fuel from the tank during cranking, fuel is transferred from the fuel tank to the carburetor(s) for starting by operating the primer bulb installed in the fuel line.

Suzuki outboards use either a self-contained, remote fuel pump mounted on the power head or a combination fuel pump and carburetor. Pump design is extremely simple and reliable in operation. Diaphragm failures are the most common problem, although the use of dirty or improper fuel-oil mixtures can cause check valve problems.

The fuel system is also equipped with a separate fuel filter.

Remote Fuel Pump Removal/Installation

Refer to **Figure 2** or **Figure 3**, typical for this procedure.

1. Compress the fuel line fitting clamp with hose clamp pliers, slide the clamp off the fitting and disconnect the fuel line at the pump. Repeat this

step to disconnect the remaining fuel line(s) from the pump. Plug the lines to prevent leakage.

2. Remove the screws holding the fuel pump assembly to the power head. Remove the fuel pump.

3. On models so equipped, remove the fuel pump insulator (**Figure 4**).

4. Clean all gasket residue from the engine mounting pad. Work carefully to avoid gouging or damaging the mounting surface.

5. Installation is the reverse of removal, noting the following.

6. Use a new mounting gasket.

7. If so equipped, install the insulator (A, **Figure 4**) along with a new O-ring in the insulator (B, **Figure 4**).

8. Install a new O-ring in the fuel pump body (**Figure 5**).

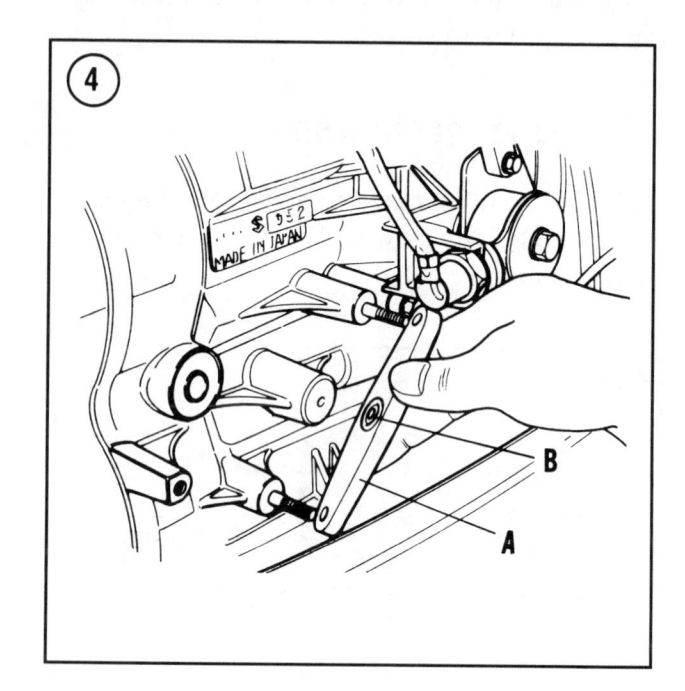

Carburetor Fuel Pump
Removal/Disassembly/Assembly/Installation

Refer to **Figure 6** for this procedure.

1. Disconnect the fuel line at the inlet cover. Plug the line to prevent leakage.

2. Remove the screws (**Figure 7**) holding the fuel pump components to the side of the carburetor.

3. Remove the pump cover, gaskets, diaphragms and pump body as an assembly.

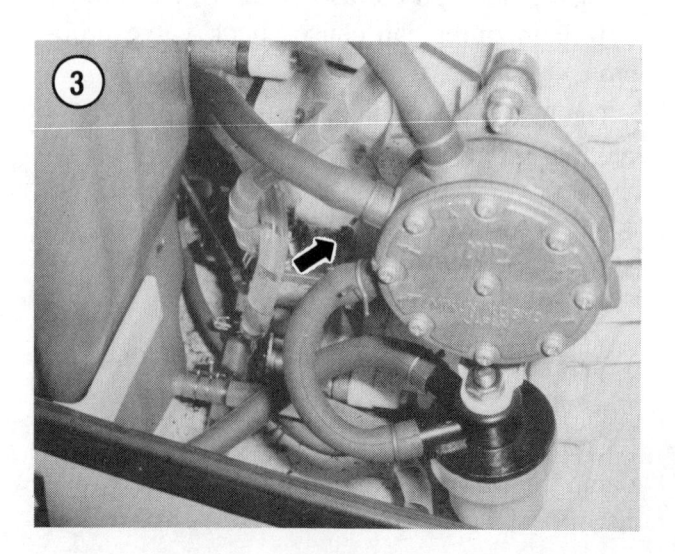

4. Separate the pump cover, diaphragm and valve assembly and fuel pump body. Peel gaskets from cover and body and discard.

5. Clean and inspect the components as described in this chapter.

6. Assemble components with new gaskets in reverse of order given in Step 4.

7. Installation is the reverse of removal.

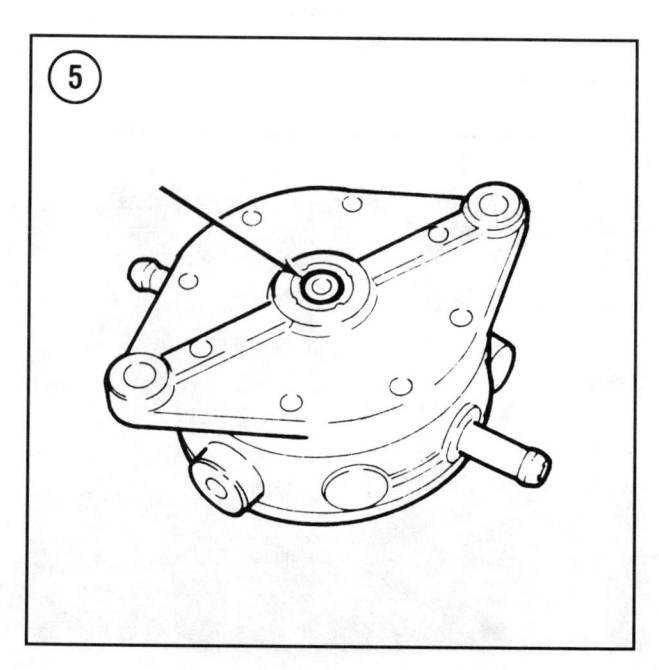

FUEL SYSTEM

Remote Fuel Pump Disassembly/Assembly

There are several variations of the fuel pump design used among the different models. All operate essentially the same, but there slight variations of parts within the fuel pump. Note the location of the these parts and any spring location during disassembly.

Refer to the following illustrations for this procedure:

a. **Figure 8**: DT 4; DT 6; 1985-1987 DT 8.
b. **Figure 9**: DT 20, DT 25, DT 30, DT 35 and DT 40.
c. **Figure 10**: 1985-on DT 55 and DT 65.
d. **Figure 11**: 1985-1989 DT 75, DT 85 and 1985 DT 115, DT 140.

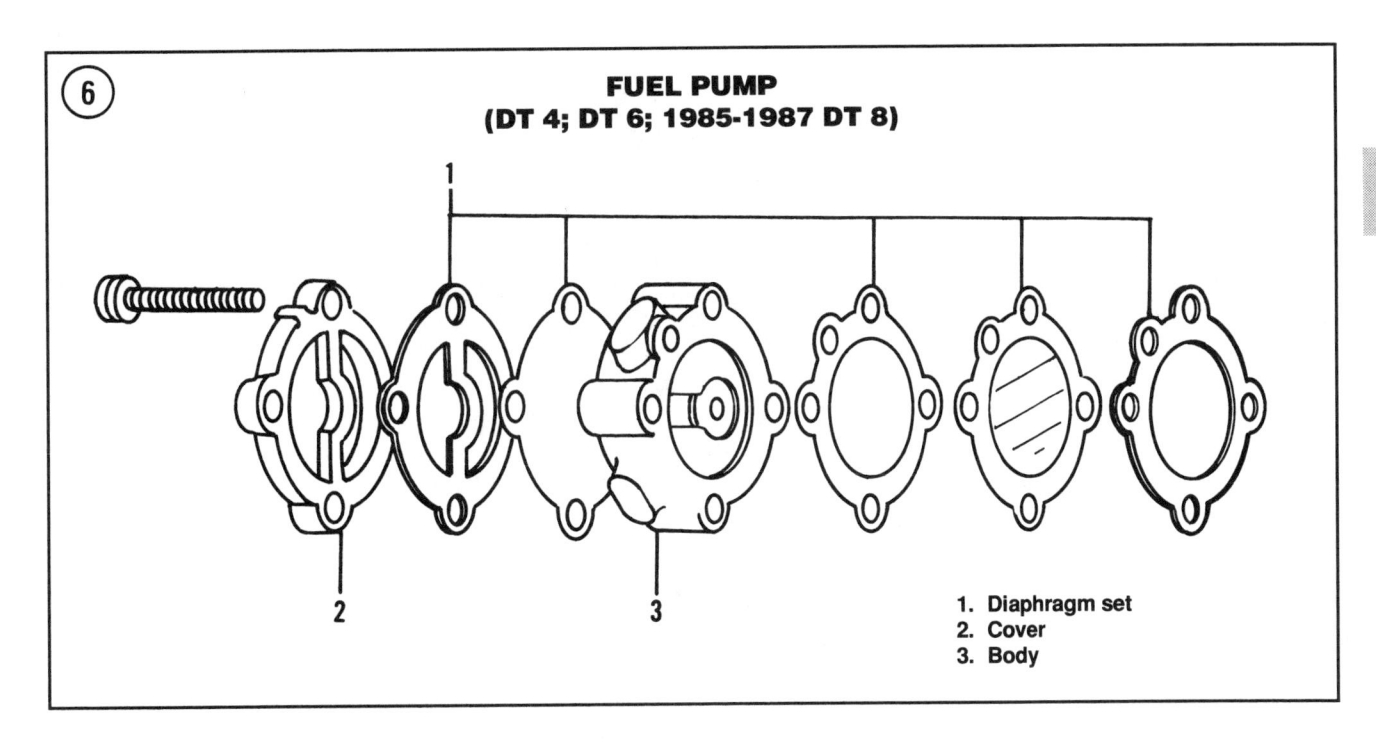

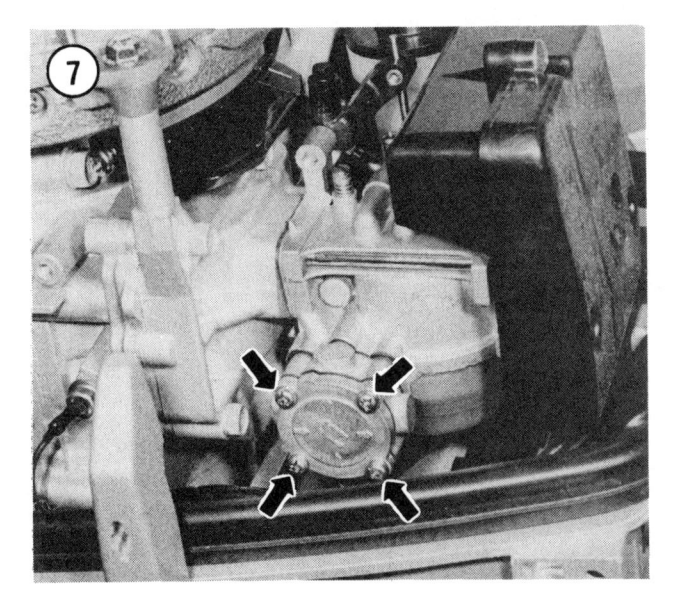

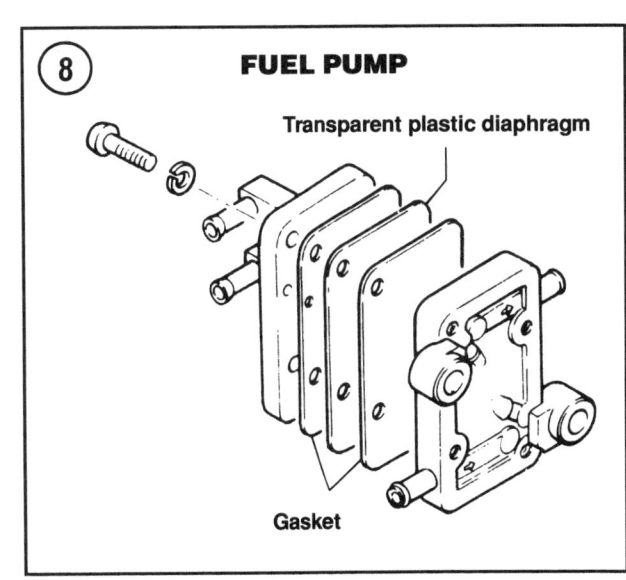

FUEL PUMP (DT 4; DT 6; 1985-1987 DT 8)

1. Diaphragm set
2. Cover
3. Body

FUEL PUMP

Transparent plastic diaphragm

Gasket

CHAPTER SIX

⑨ FUEL PUMP AND FILTER
(DT 20, DT 25, DT 30, DT 35 AND DT 40)

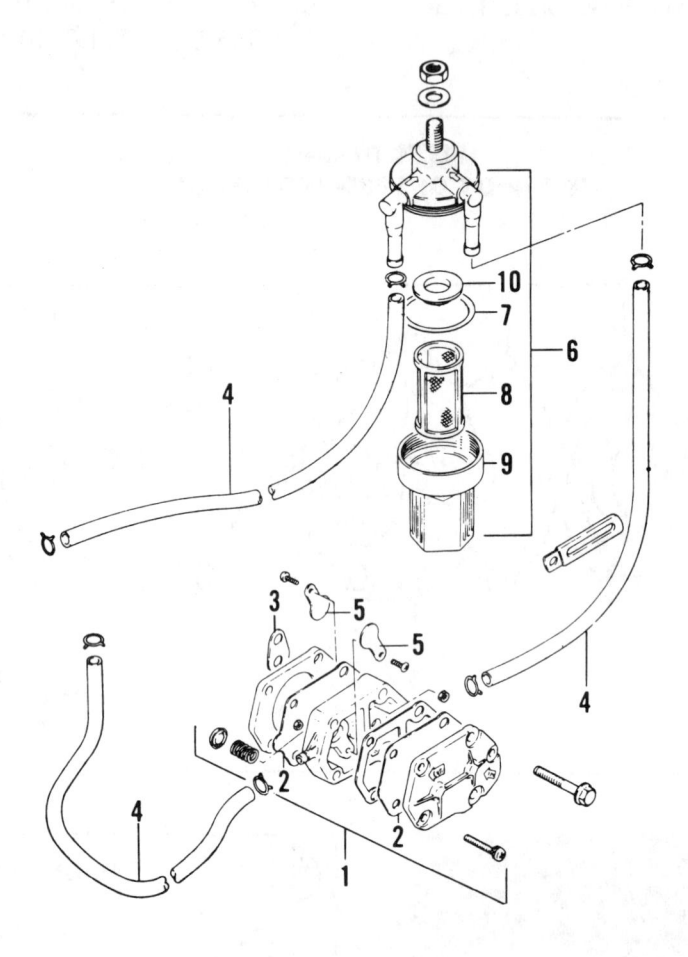

1. Fuel pump assembly
2. Diaphragm set
3. Gasket
4. Fuel hose
5. Valve
6. Filter assembly
7. O-ring
8. Strainer
9. Cup
10. Gasket

FUEL SYSTEM

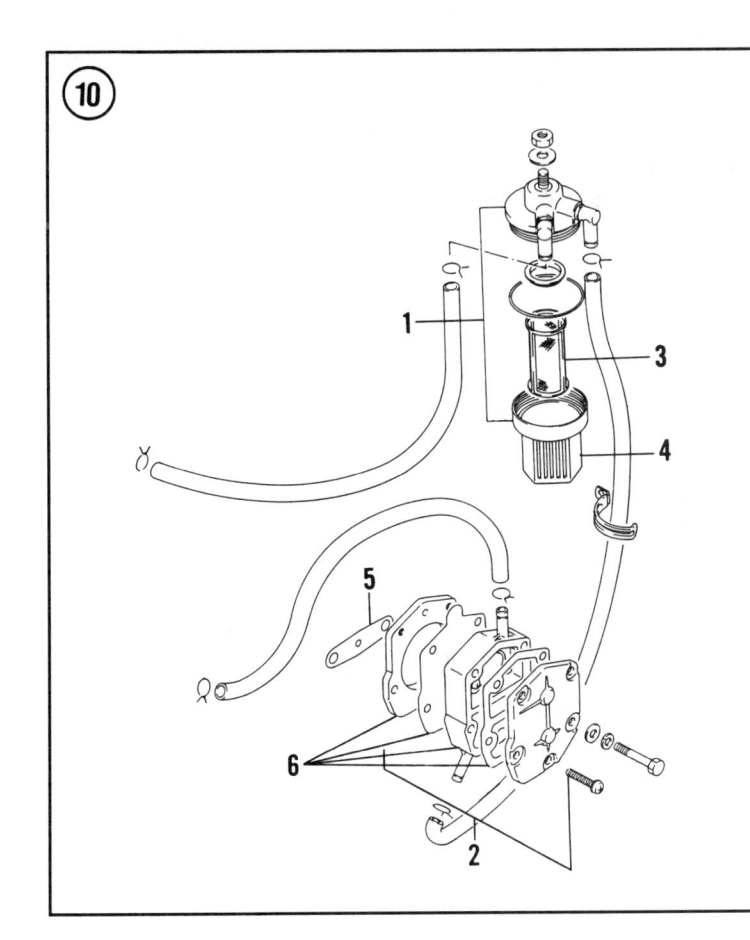

⑩ FUEL PUMP AND FILTER (1985-ON DT 55 AND DT 65)

1. Fuel filter assembly
2. Fuel pump assembly
3. Strainer
4. Cup
5. Gasket
6. Diaphragm set

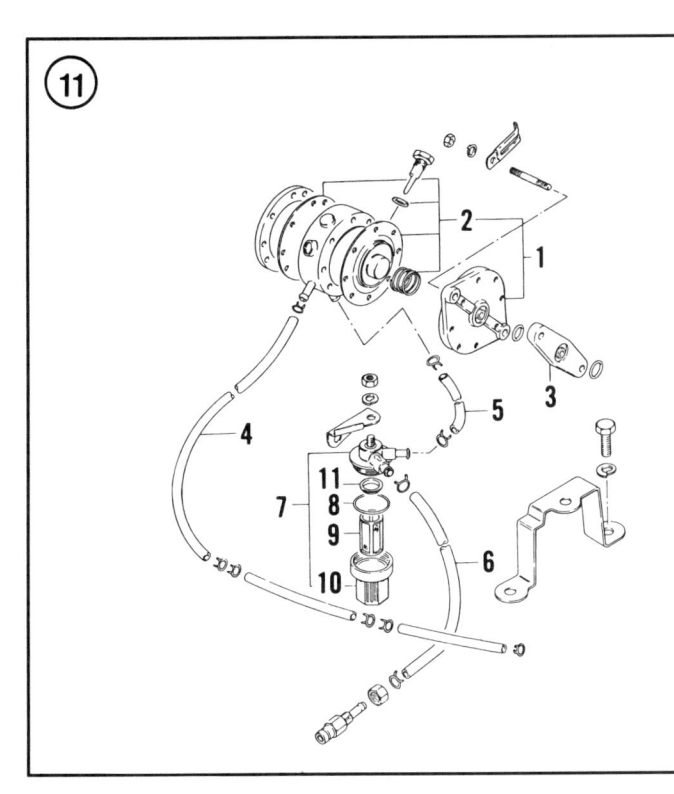

⑪ FUEL PUMP AND FILTER (1985-1989 DT 75 AND DT 85; 1985 DT 115 AND DT 140)

1. Fuel pump assembly
2. Diaphragm set
3. Insulator
4. Fuel hose
5. Fuel hose
6. Fuel hose
7. Fuel filter assembly
8. O-ring
9. Strainer
10. Cup
11. Gasket

CHAPTER SIX

⑫

**FUEL PUMP AND FILTER
(1986-ON DT 115 AND DT 140)**

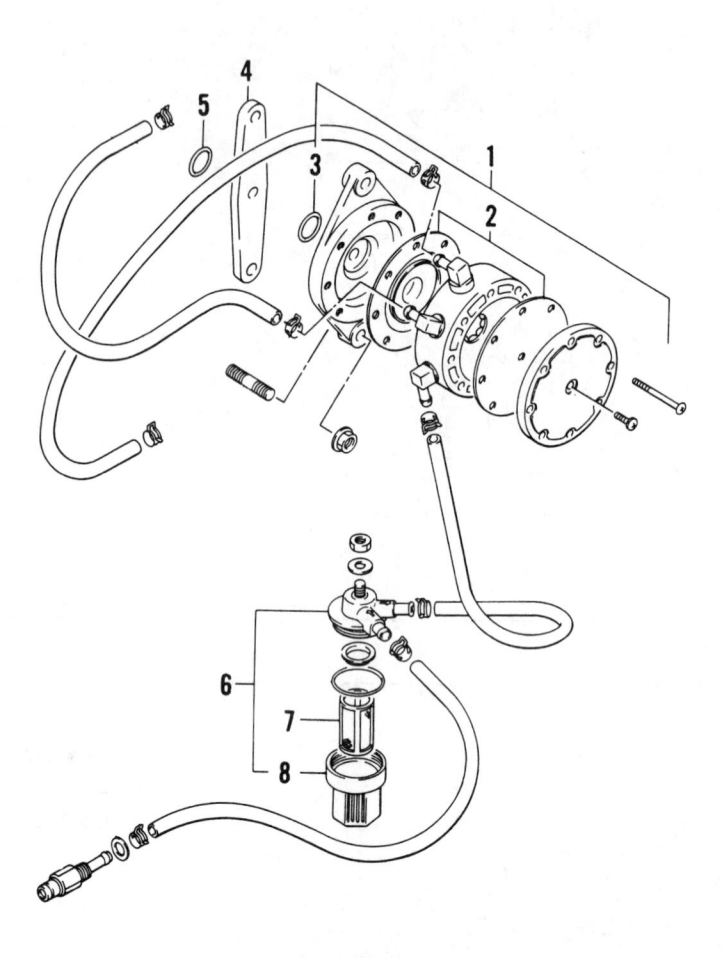

1. Fuel pump assembly
2. Diaphragm set
3. O-ring
4. Insulator
5. O-ring
6. Fuel filter assembly
7. Strainer
8. Cup

FUEL SYSTEM

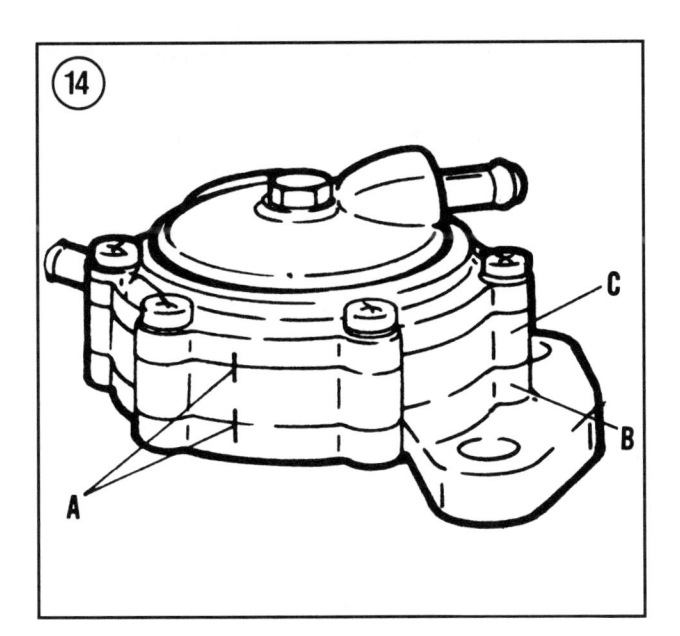

147

e. **Figure 12**: 1986-on DT 115 and DT 140.

f. **Figure 13**: V4 and V6.

1. On all models except DT 115, DT 140, V4 and V6, etch a match mark (A, **Figure 14**) across the side of the pump to provide an alignment reference for the upper and lower halves.

2. Remove the screws and washers (models so equipped) securing the fuel pump assembly together.

3. Separate the pump halves. Remove and discard all gaskets.

4. On models so equipped, remove the check valves. Some pumps use a flap-type check valve diaphragm; others use metal reed-type valves secured by a small bolt and nut.

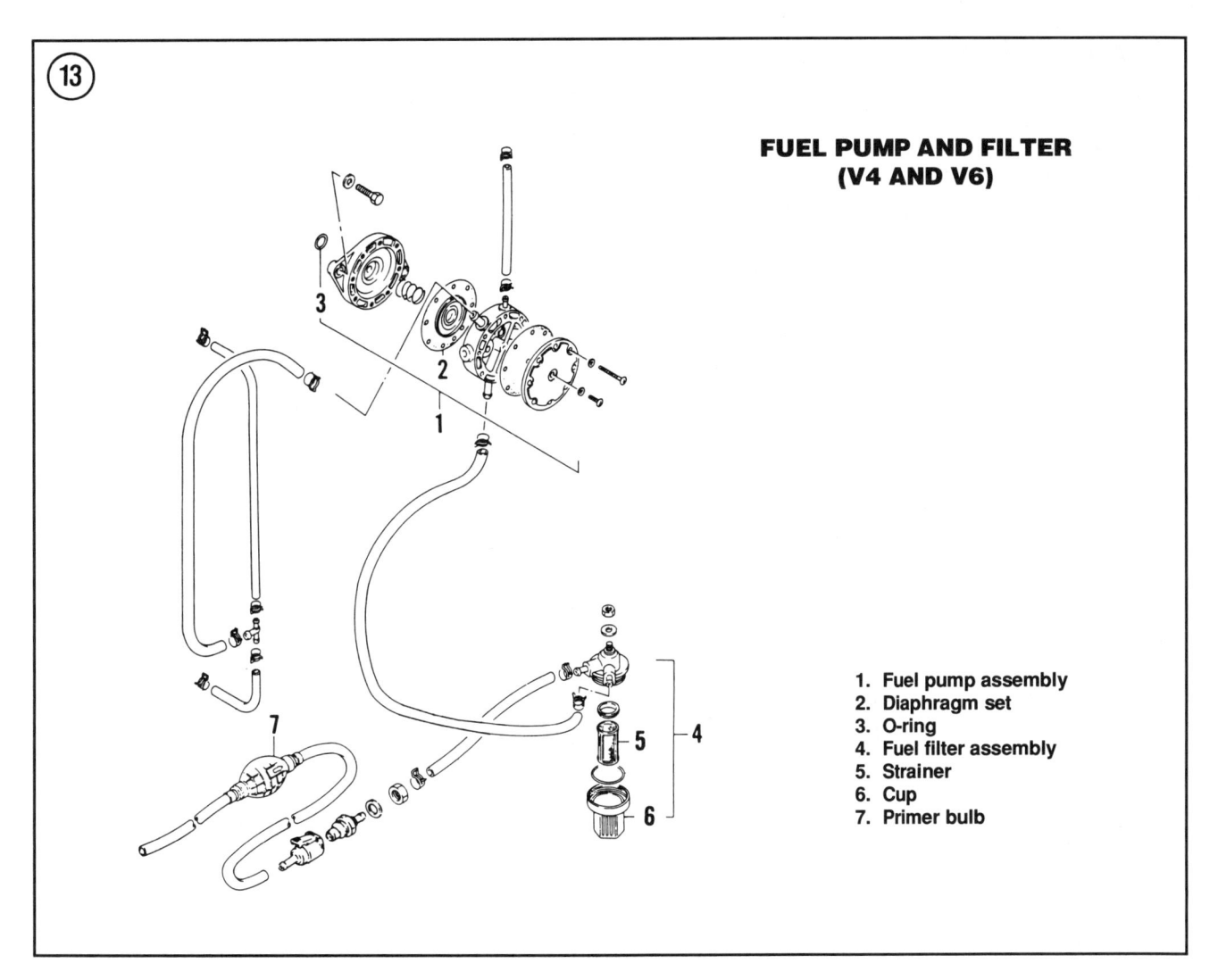

FUEL PUMP AND FILTER (V4 AND V6)

1. Fuel pump assembly
2. Diaphragm set
3. O-ring
4. Fuel filter assembly
5. Strainer
6. Cup
7. Primer bulb

5. On models so equipped, remove the spring.
6. Remove the diaphragm or diaphragm set.
7. Assembly is the reverse of disassembly, noting the following.
8. Refer to the illustration relating to your specific model for gasket and diaphragm positioning.
9A. On all models except DT 115, DT 140, V4 and V6, align the etch marks made in Step 1 and tighten pump body screws securely.
9B. On DT 115, DT 140, V4 and V6, align the match mark on the upper and lower halves and tighten pump body screws securely.
10. Inspect all fuel hoses for cracks or deterioration and replace as necessary.

Cleaning and Inspection (All Models)

1. Clean all metal parts in solvent and blow dry with compressed air.
2. Hold diaphragm up to a strong light source and check for pin holes, breaks or excessive stretching. Replace if any defects are found.
3. Check condition of fuel pump reeds (models so equipped) or check valve diaphragm. Install new reeds or diaphragm if there are signs of warpage, tension or curling of the reeds or check valve ends.
4. Check pump body and cover condition. Replace pump body or cover as required if cracks or rough gasket mating surfaces are found.

Fuel Filter

If you feel there is a restriction in the fuel flow, check the fuel filter screen for possible contaminants.

1. Unscrew the cup from the filter housing.
2. Remove the filter from the end of the adaptor.
3. Clean the filter screen with solvent to remove any particles (**Figure 15**). If the filter screen is severely clogged or dirty, replace it.
4. Inspect the filter screen for damage and replace if necessary.
5. Install a new O-ring seal in the filter cup.
6. Install the filter and the filter cup. Tighten the cup securely.

CARBURETOR REMOVAL/INSTALLATION

Suzuki outboards use a variety of Mikuni carburetors. All operate essentially the same, but housing shape and design varies slightly according to engine size.

All carburetors use a fixed main jet and require no high-speed adjustment. Poor fuel quality or

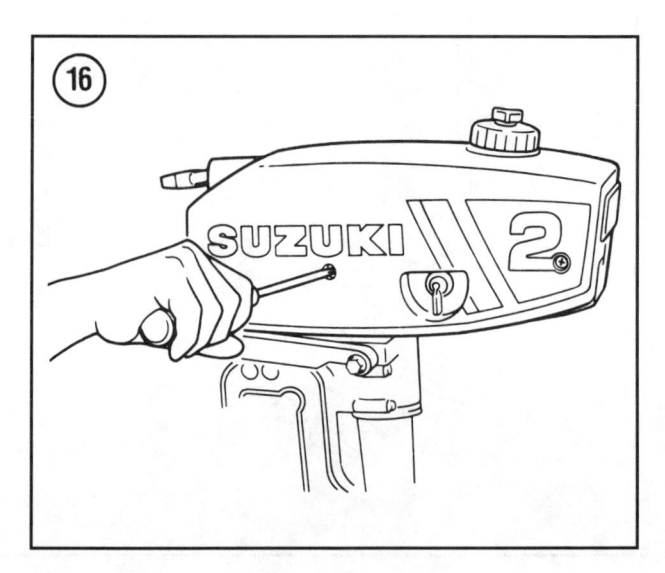

FUEL SYSTEM

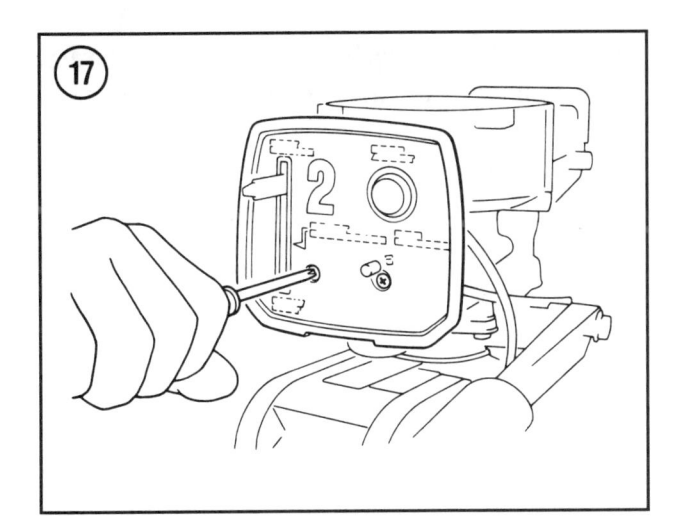

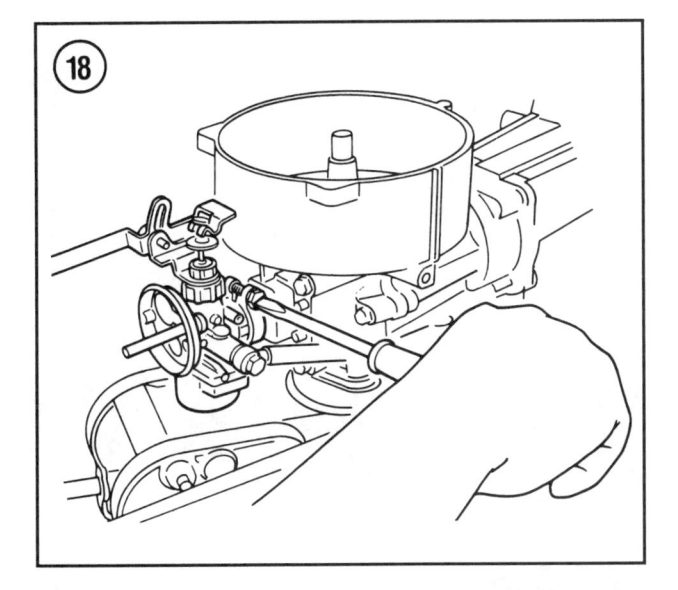

use in high-altitude areas may make rejetting necessary. See your Suzuki dealer for jet recommendations.

When installing a carburetor, make sure that the mounting nuts are securely tightened. A carburetor that is loose will cause a lean-running condition.

Before removing and disassembling any carburetor, be sure you have the correct overhaul kit, the proper tools and a sufficient quantity of fresh cleaning solvent.

DT 2 Models

1. Remove the right and left engine covers (**Figure 16**).
2. Loosen the choke knob setscrew. Remove the knob from the control panel.
3. Remove the throttle lever knob.
4. Remove the 2 screws holding the control panel to the carburetor (**Figure 17**).
5. Slide the fuel line clamp from the carburetor fuel fitting with hose clamp pliers and disconnect the fuel line at the carburetor. Plug the line to prevent leakage.
6. Loosen the clamp screw holding the carburetor to the crankcase (**Figure 18**). Remove the carburetor.
7. Remove and discard the O-ring inside the carburetor throat.
8. Installation is the reverse of removal, noting the following.
9. Install a new O-ring in the carburetor throat.
10. Tighten the clamp screw securely to prevent an air leak.

All Other Models

1. Remove the engine cover.
2. Remove the air silencer and gasket, if so equipped (**Figure 19**, typical).
3. Slide the fuel line clamp from the carburetor fuel fitting with hose clamp pliers and disconnect the fuel line at each carburetor to be removed

(**Figure 20**, typical). Plug the line(s) to prevent leakage.

4. Disconnect the throttle and choke linkage at each carburetor to be removed (**Figure 21**, typical).

5. Remove the choke rod knob, if so equipped.

6A. Clamp screw attachment—Loosen the clamp screw holding the carburetor to the intake manifold (**Figure 18**). Remove the carburetor. Remove and discard the O-ring inside the carburetor throat.

6B. Hex nut attachment—Remove the 2 hex nuts holding the carburetor to the intake manifold (**Figure 22**). Remove the carburetor from the manifold, pulling the choke rod (if so equipped) through the lower support grommet. Remove and discard the carburetor gasket.

7. Installation is the reverse of removal, noting the following.

8. Use a new mounting gasket or install a new O-ring in the carburetor throat as required.

9. Tighten the hex nuts or clamp screw securely to prevent an air leak. Adjust the carburetor as described in this chapter.

CARBURETOR OVERHAUL/ADJUSTMENT

Work slowly and carefully, follow the disassembly procedures and refer to the exploded drawing of your carburetor when necessary. When referring to the exploded drawing, note that not all carburetors will use all of the components shown.

Do not apply excessive force at any time.

It is not necessary to disassemble the carburetor linkage or remove the throttle cam or other external components. Remove the throttle or choke plate only if it is damaged or binds. Carburetor specifications are provided in **Table 1**.

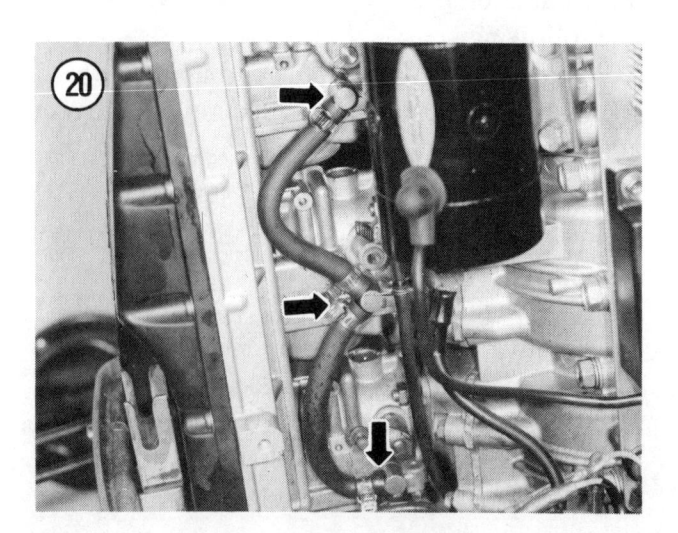

FUEL SYSTEM

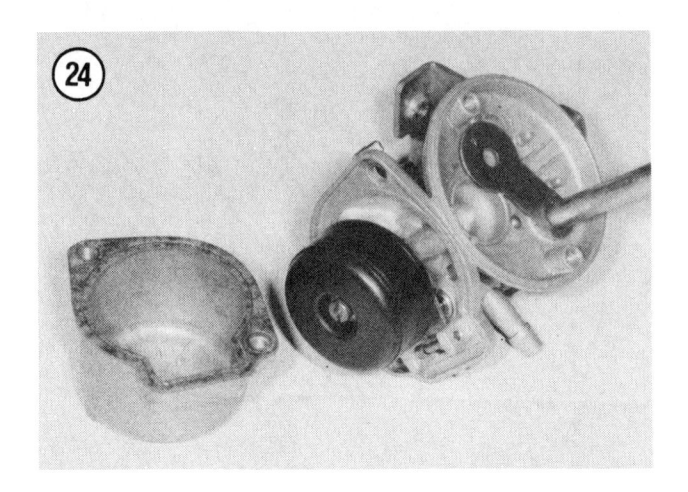

THROTTLE PLUNGER CARBURETOR (DT 2)

Overhaul

Refer to **Figure 23** for this procedure.

1. Remove the fuel bowl attaching screws. Separate the fuel bowl from carburetor body (**Figure 24**). Remove and discard the fuel bowl gasket.
2. Remove the float hinge pin from float arm. Remove the float arm and needle valve from fuel bowl (**Figure 25**).

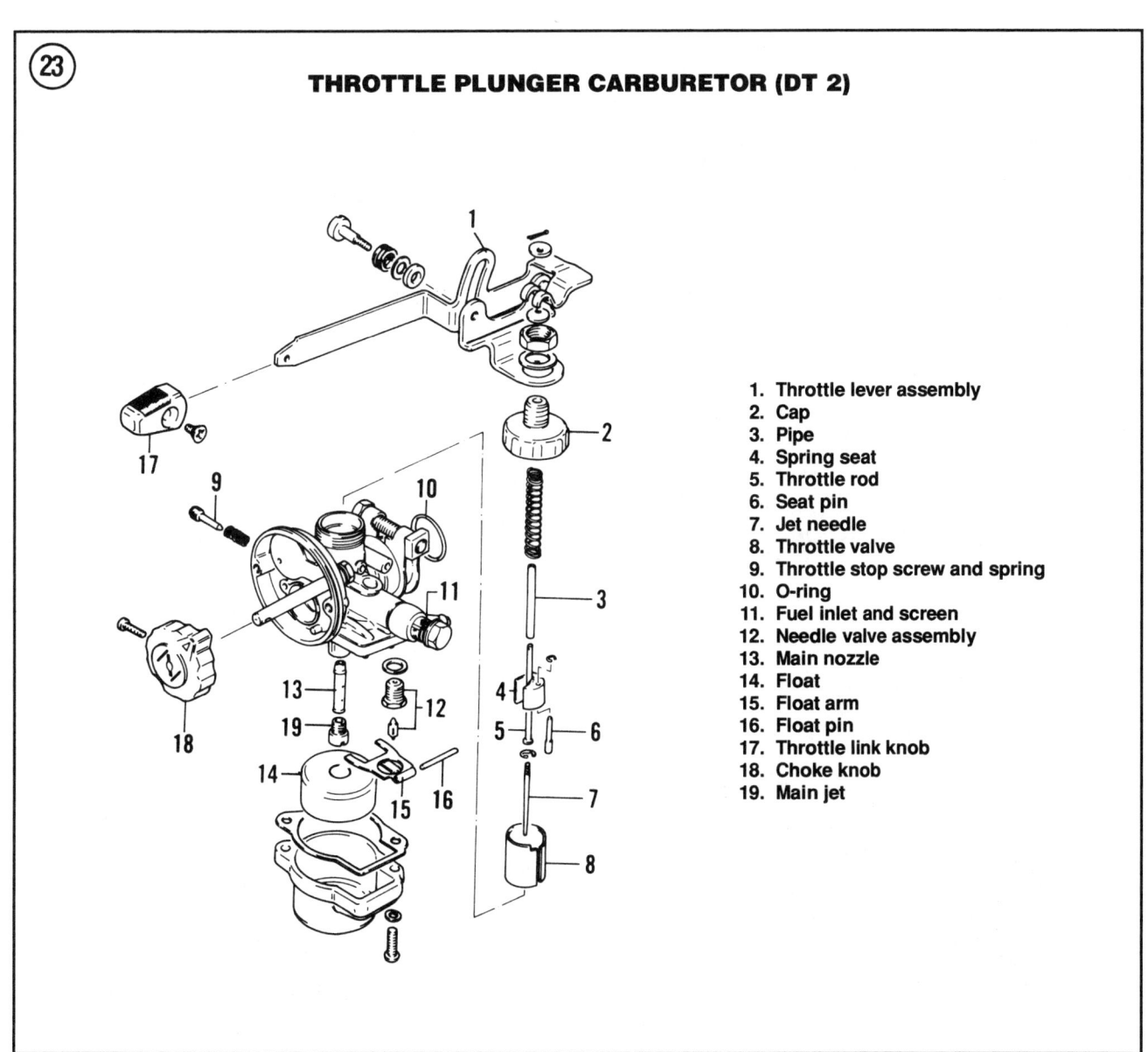

THROTTLE PLUNGER CARBURETOR (DT 2)

1. Throttle lever assembly
2. Cap
3. Pipe
4. Spring seat
5. Throttle rod
6. Seat pin
7. Jet needle
8. Throttle valve
9. Throttle stop screw and spring
10. O-ring
11. Fuel inlet and screen
12. Needle valve assembly
13. Main nozzle
14. Float
15. Float arm
16. Float pin
17. Throttle link knob
18. Choke knob
19. Main jet

3. Remove the main nozzle and the main jet assembly (A, **Figure 26**).

4. Remove the needle seat and gasket (B, **Figure 26**) with a wide-blade screwdriver. Discard gasket.

5. Loosen the throttle stop screw and unscrew the retainer cap. Remove the cap with bracket and throttle valve assembly (**Figure 27**).

6. Disconnect the throttle plunger from needle valve. Remove the valve, retainer and spring. Make sure the tiny E-clip is intact on the jet needle valve (**Figure 28**).

7. Lightly seat the throttle stop screw, counting the number of turns required for reinstallation reference. Back the screw out and remove from the carburetor with spring (**Figure 29**).

8. Remove the fuel inlet fitting and filter screen.

9. Clean and inspect all parts as described in this chapter.

10. Assembly is the reverse of disassembly, noting the following.

11. Install the throttle stop screw and spring. Lightly seat screw, then back out the number of turns noted during removal.

12. Reassemble the throttle valve assembly. The jet needle retainer should be positioned with slot aligned as shown in **Figure 30**.

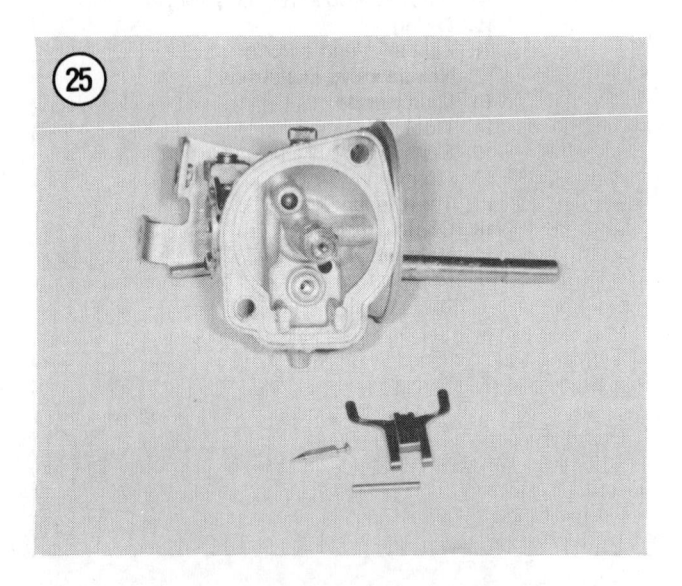

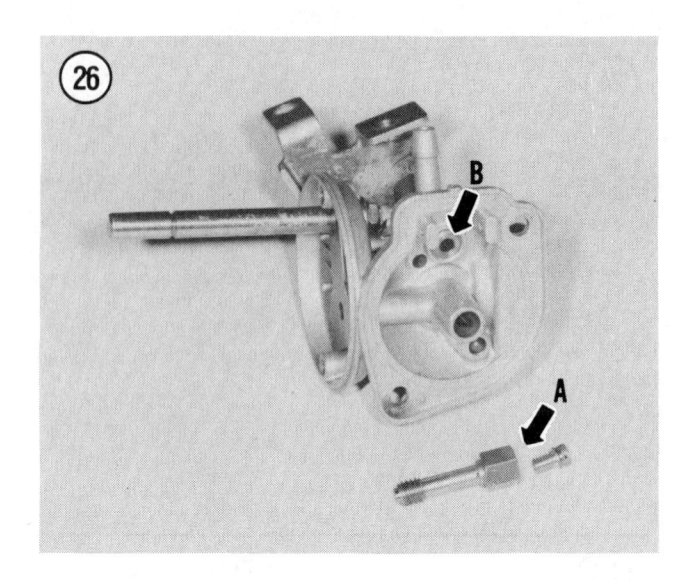

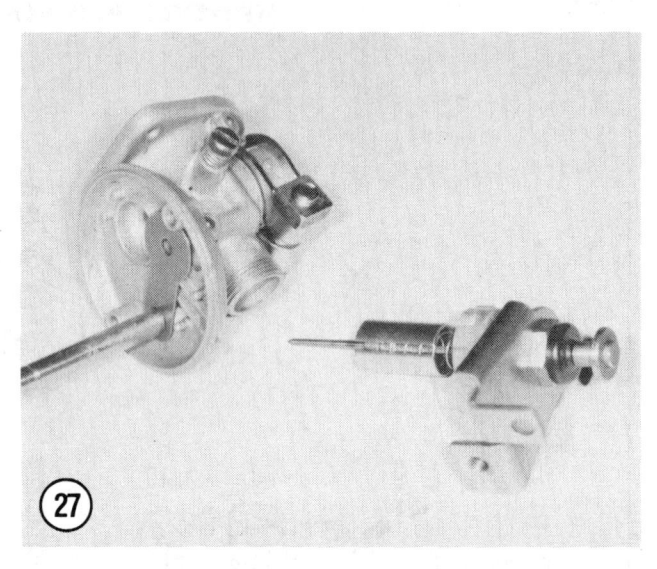

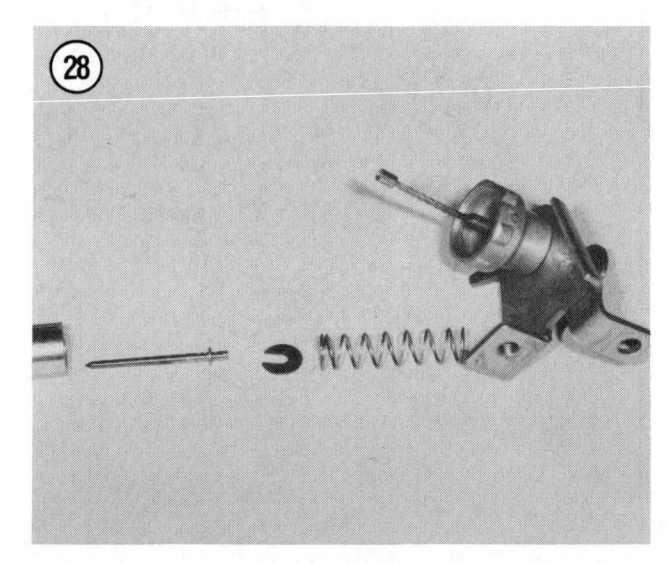

FUEL SYSTEM

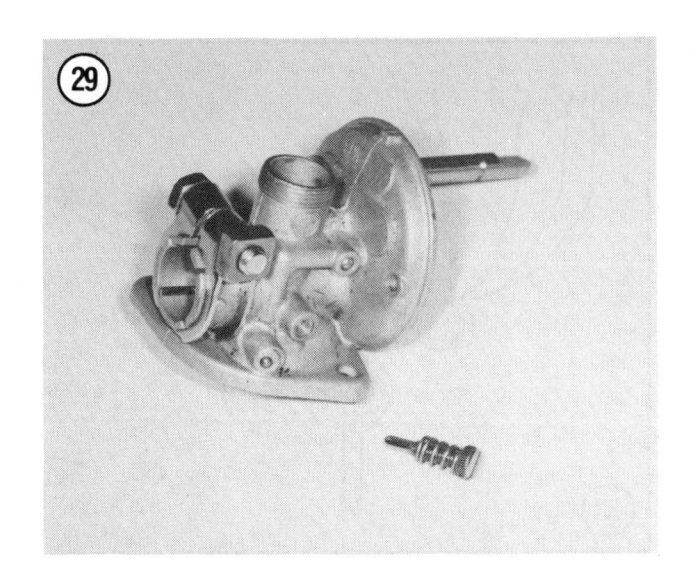

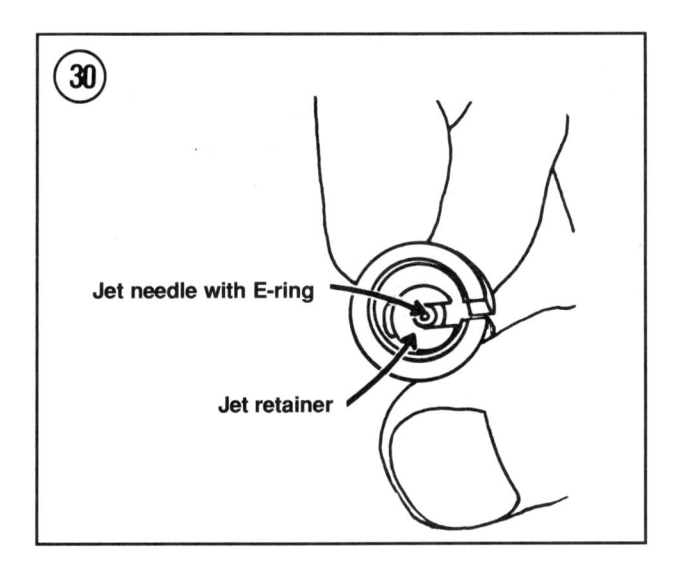

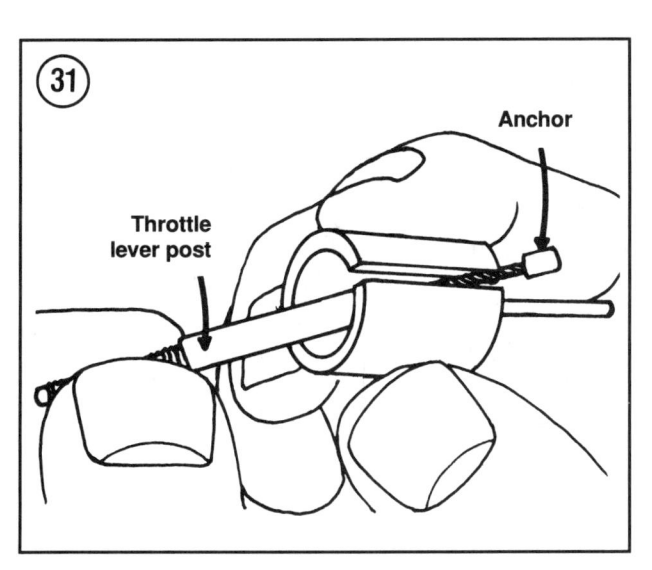

13. Install the throttle lever post in the throttle valve with anchor in pocket at bottom of valve body (**Figure 31**).

14. Check the float adjustment as described in this chapter.

Float Adjustment

1. Invert the carburetor body and lower float until adjusting tab just touches inlet needle. Hold the float in this position and measure the distance between the carburetor body (without gasket) and the bottom of the float (**Figure 32**).

CAUTION
*Bend the float adjustment arm or tang carefully when adjustment is required—do **not** press down on the float. Downward pressure on the float will press the inlet needle tip into its seat and can damage the tip surface.*

2. If the float level is not within specifications (**Table 1**), adjust the float by bending the adjusting tab (**Figure 33**) as required.

Mid-range Jet Needle Adjustment

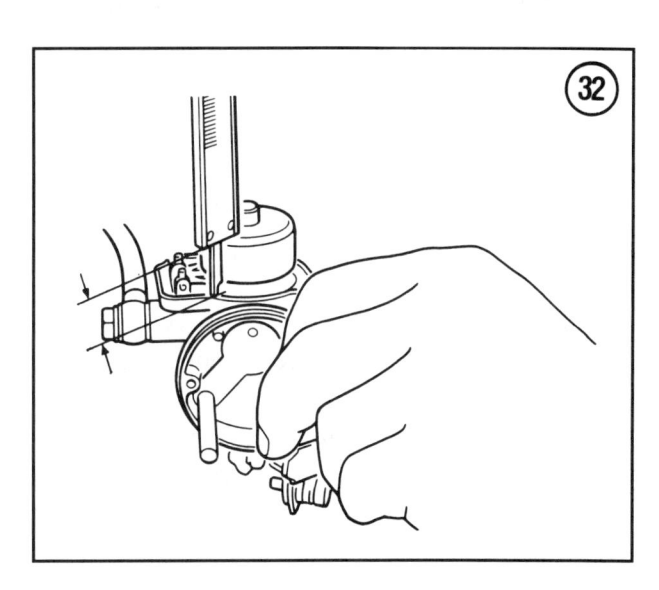

The position of the E-ring on the jet needle determines the proper low- and medium-speed range mixture (**Figure 34**). The standard setting (the 3rd groove from the top) should be suitable for most operating conditions.

If a too-rich or too-lean condition results from extreme changes in elevation, temperature or humidity, the E-ring position can be changed. The closer the E-ring is installed to the top of the needle, the leaner the mixture. Remember that it is always preferable to run a slightly richer mixture than one that is too lean.

Idle Speed Adjustment

The engine should be installed in a test tank or on the boat in the water for this procedure.
1. Remove the engine cover.
2. Check and adjust the ignition timing. See Chapter Five.
3. Connect a tachometer according to manufacturer's instructions.
4. Start the engine and run for 5-10 minutes to bring it to normal operating temperature.
5. Shift the engine into FORWARD.
6. If idle speed is not 800-900 rpm, adjust the throttle stop screw (A, **Figure 35**) as required to bring idle speed within specifications.

CENTERBOWL CARBURETOR (DT 4, DT 6; 1985-1987 DT 8)

Overhaul

Several variations of this carburetor design have been used on Suzuki DT 4, DT 6, 1985-1987 DT 8 outboards. All operate essentially the same, varying primarily in calibration and linkage design.

The needle valve seat is not serviceable. If worn or damaged, the entire carburetor must be replaced.

Refer to **Figure 36**, typical, for this procedure.
1. Remove the hex head bolt and washer from the float bowl. Remove the float bowl and O-ring from the fuel bowl. Discard the O-ring.

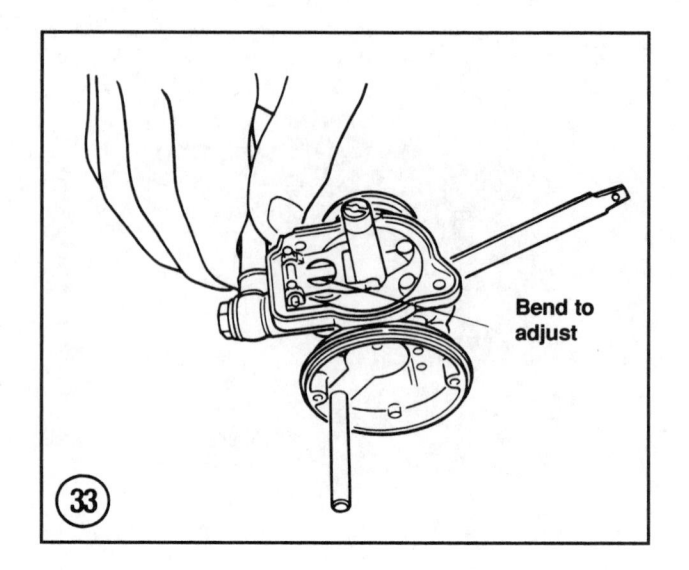

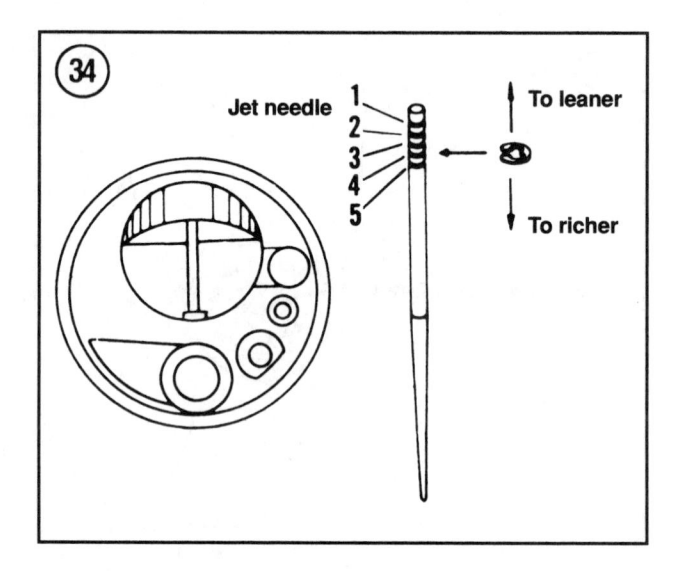

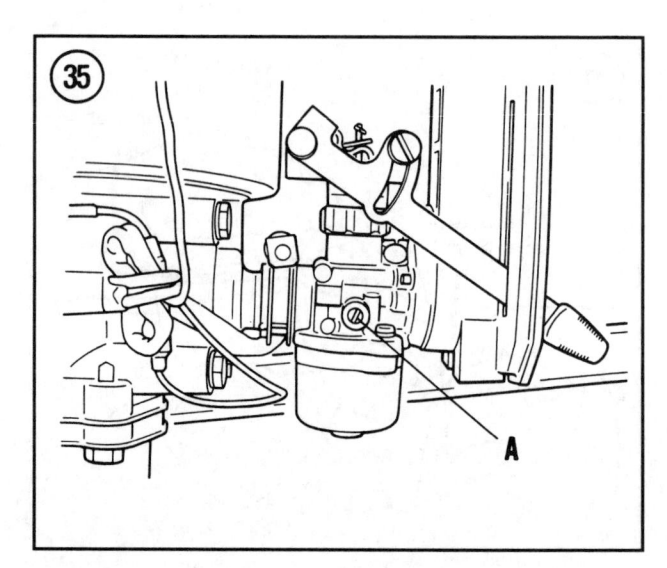

FUEL SYSTEM

36

**CENTERBOWL CARBURETOR
(DT 4, DT 6, 1985-1987 DT 8)**

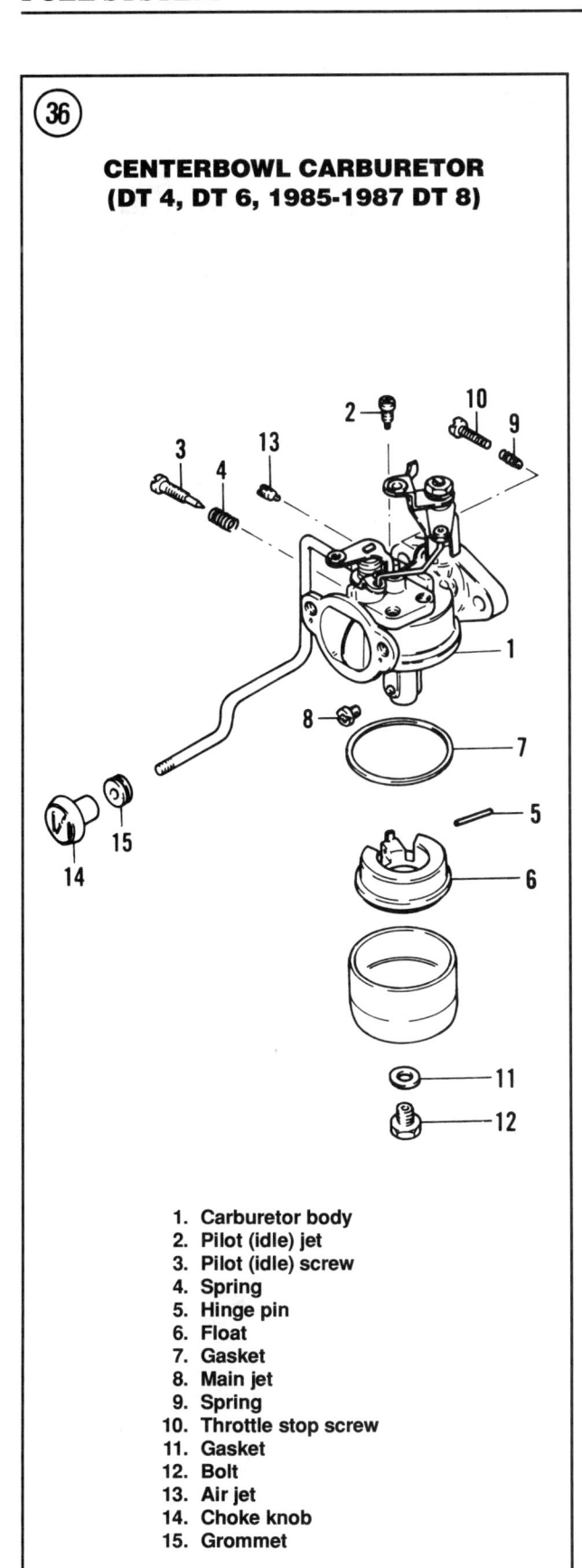

1. Carburetor body
2. Pilot (idle) jet
3. Pilot (idle) screw
4. Spring
5. Hinge pin
6. Float
7. Gasket
8. Main jet
9. Spring
10. Throttle stop screw
11. Gasket
12. Bolt
13. Air jet
14. Choke knob
15. Grommet

2. Slide the float hinge pin to one side and remove the float and pin assembly from the carburetor.

CAUTION
The inlet needle is permanently installed in the valve seat on some models. Do not try to remove needle individually in Step 3 unless it comes out easily.

3. Remove the inlet needle.
4. Remove the main jet with a jet remover or wide-blade screwdriver.
5. Remove the pilot (idle) jet with a suitable removal tool.
6. Remove the air jet with a suitable removal tool.
7. Lightly seat the pilot (idle) screw, counting the number of turns required for reinstallation reference. Back screw out and remove from carburetor with spring.
8. Clean and inspect all parts as described in this chapter.
9. Assembly is the reverse of disassembly, noting the following.
10. Install the pilot (idle) screw and spring. Lightly seat screw, then back out the number of turns noted during removal.
11. Check the float adjustment as described in this chapter.

Float Adjustment

1. Invert the carburetor body and lower float until the adjusting tab just touches inlet needle. Measure the distance between the carburetor body (without gasket) and the bottom of the float, as shown in A, **Figure 37**.

CAUTION
*Bend the float adjustment arm or tang carefully when adjustment is required—do **not** press down on the float. Downward pressure on the float will press the inlet needle tip into its seat and can damage the tip surface.*

2. If the float level is not within specifications (**Table 1**), adjust the float by bending the adjusting tab (B, **Figure 37**) as required.

Idle Speed Adjustment

The engine should be installed in a test tank or on the boat in the water for this procedure.

1. Remove the engine cover.
2. Check and adjust the ignition timing. See Chapter Five.
3. Connect a tachometer according to manufacturer's instructions.
4. Start the engine and run for 5-10 minutes to bring it to normal operating temperature.
5. Lightly seat the pilot (idle) air screw, then back it out gradually until the engine speed stops increasing. Refer to **Figure 38** for typical variations in air screw location.
6. Shift the engine into FORWARD.
7. If the idle speed is not 850-900 rpm (DT 4) or 600-650 rpm (all others), adjust the throttle stop screw as required to bring idle speed within specifications. See **Figure 39** for typical variations in throttle stop screw location.

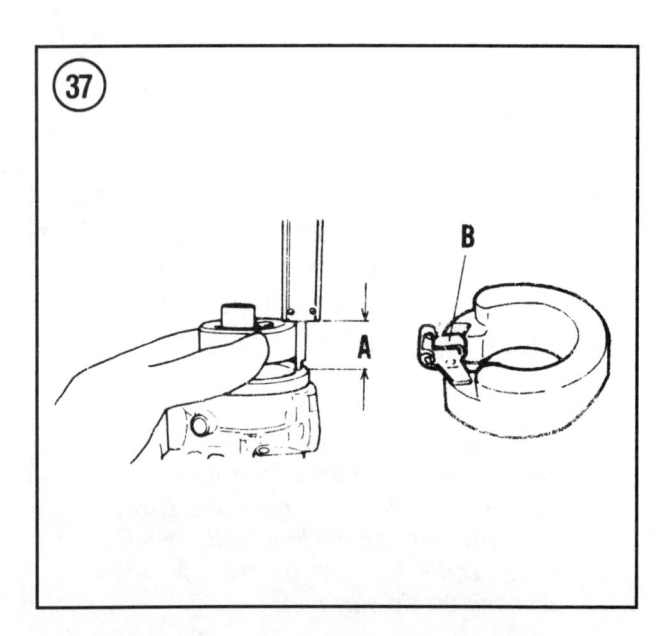

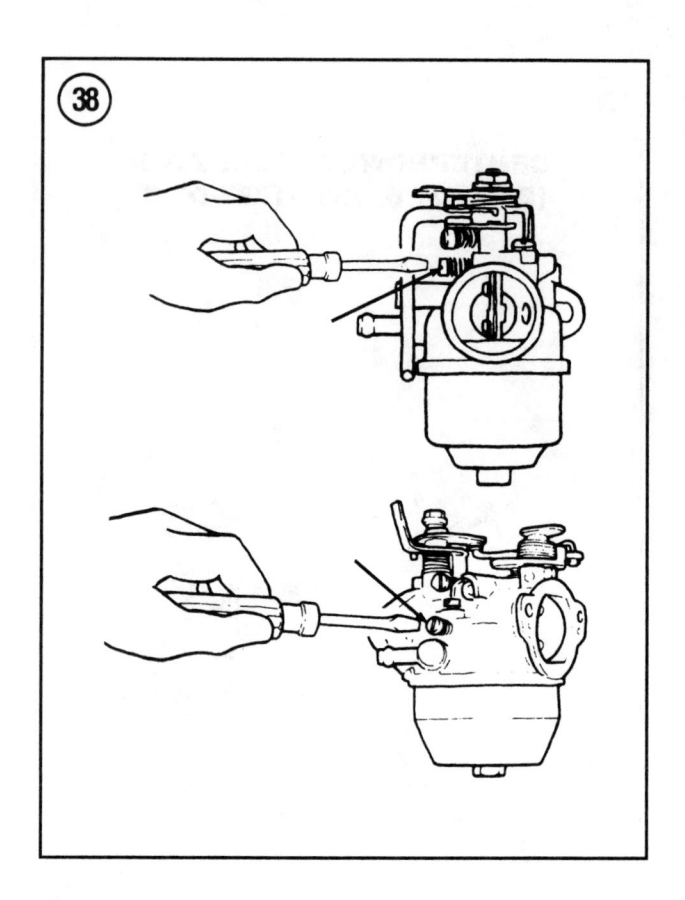

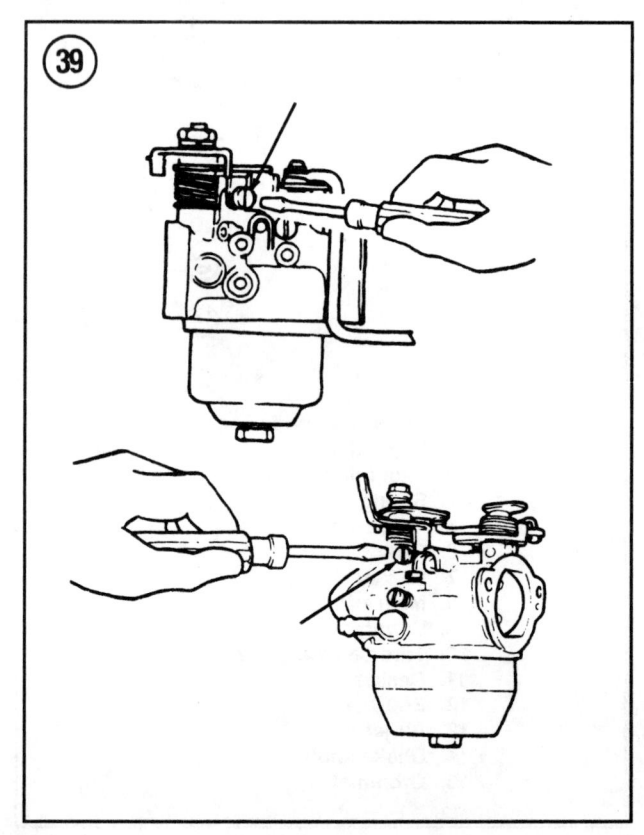

FUEL SYSTEM

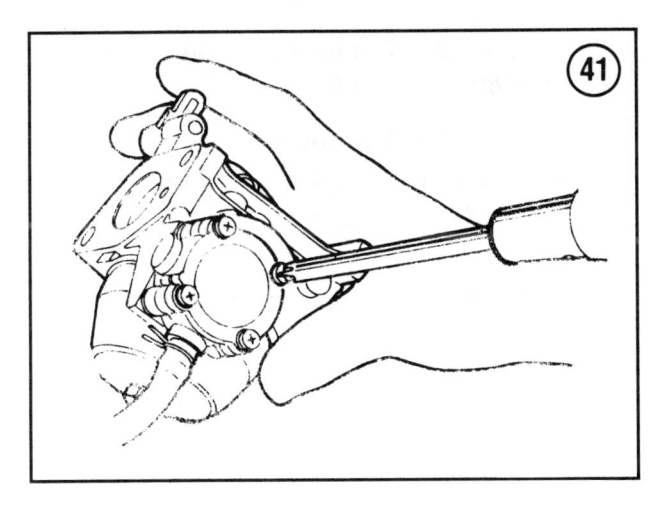

INTEGRAL FUEL PUMP CARBURETOR (1988-ON DT 8; 1985-ON DT 9.9, DT 15)

Refer to **Figure 40** for this procedure.

1. Remove the fuel pump components as described in this chapter (**Figure 41**).

2. Remove the hex head bolt and washer from the float bowl. Remove the float bowl and O-ring from the fuel bowl. Discard the O-ring.

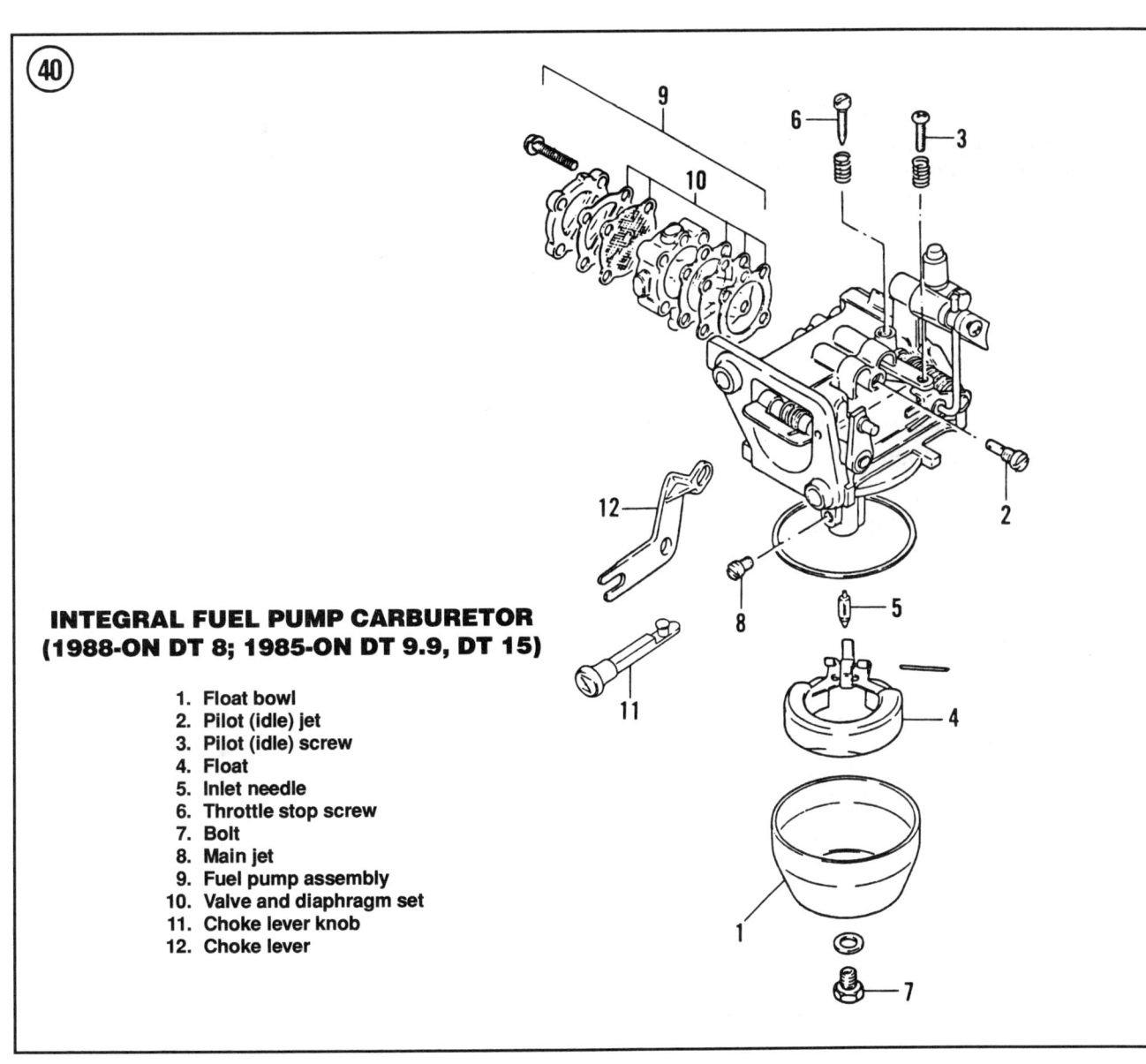

INTEGRAL FUEL PUMP CARBURETOR (1988-ON DT 8; 1985-ON DT 9.9, DT 15)

1. Float bowl
2. Pilot (idle) jet
3. Pilot (idle) screw
4. Float
5. Inlet needle
6. Throttle stop screw
7. Bolt
8. Main jet
9. Fuel pump assembly
10. Valve and diaphragm set
11. Choke lever knob
12. Choke lever

3. Slide the float hinge pin to one side and remove the float and pin assembly from the carburetor.

NOTE
The needle valve seat is permanently installed. If needle valve seat is damaged, replace the carburetor body.

4. Remove the inlet needle.
5. Remove the main jet with a jet remover or wide-blade screwdriver.
6. Remove the pilot (idle) jet with a suitable removal tool.

NOTE
Before removing the pilot screw, record the number of turns necessary until the screw lightly seats. Record the number of turns as the screw must be reinstalled to the exact same setting.

7. Unscrew the pilot (idle) screw and spring.
8. Clean and inspect all parts as described in this chapter.
9. Assembly is the reverse of disassembly, noting the following.
10. Install the pilot (idle) screw and spring. Lightly seat screw, then back out the number of turns recorded during removal.
11. Check the float adjustment as described in this chapter.

Float Adjustment

1. Invert the carburetor body and lower the float until adjusting tab just touches inlet needle. Hold the float in this position and measure the distance between the carburetor body (without gasket) and the bottom of the float, as shown in **Figure 42**.

CAUTION
*Bend the float adjustment arm or tang carefully when adjustment is required—do **not** press down on the float. Downward pressure on the float will press the inlet needle tip into its seat and can damage the tip surface.*

2. If the float level is not within specifications (**Table 1**), adjust float by bending the adjusting tab (**Figure 43**) as required.

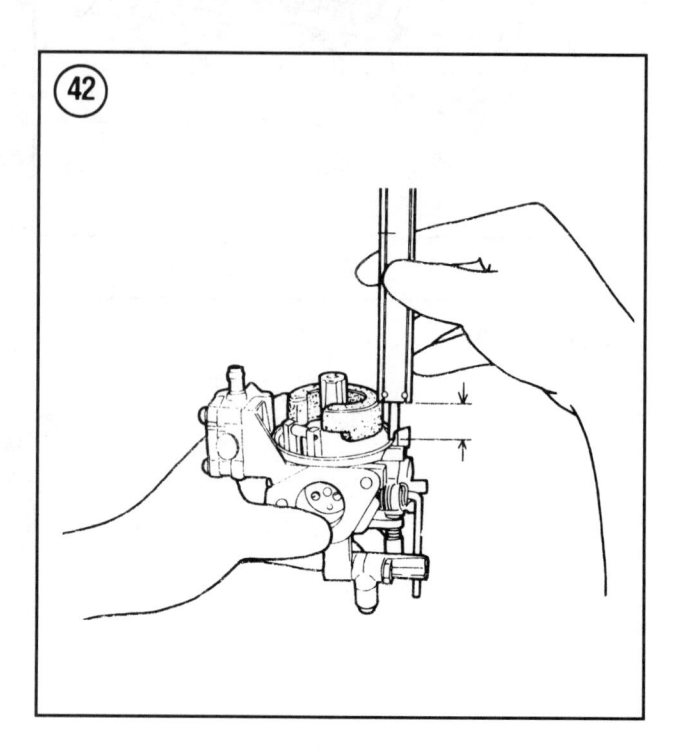

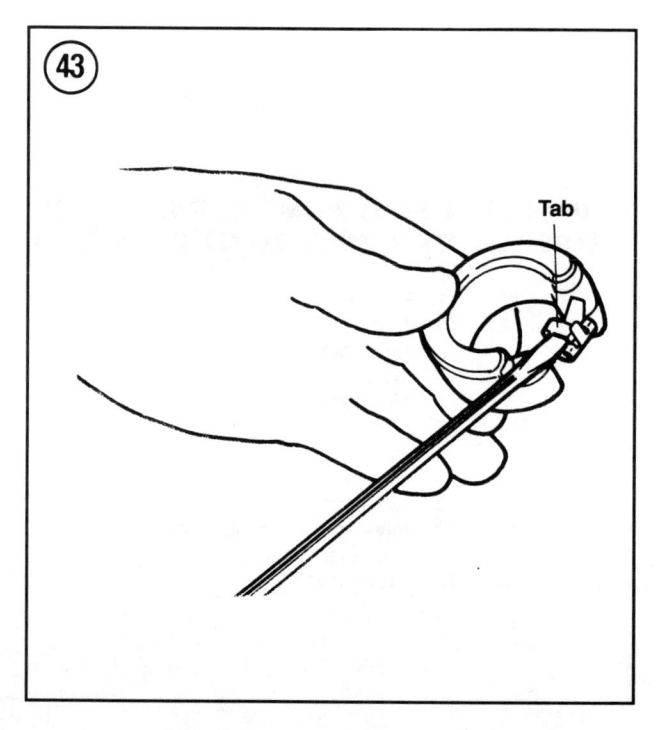

FUEL SYSTEM

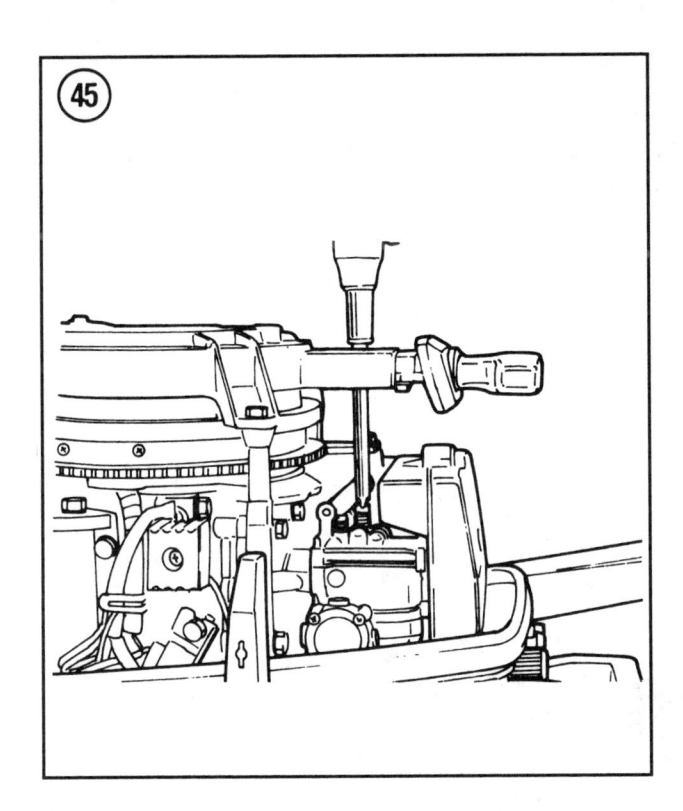

Idle Speed Adjustment

The engine should be installed in a test tank or on the boat in the water for this procedure.

1. Remove the engine cover.
2. Check and adjust the ignition timing. See Chapter Five.
3. Connect a tachometer according to manufacturer's instructions.
4. Start the engine and run for 5-10 minutes to bring it to normal operating temperature.
5. Lightly seat the pilot (idle) air screw, then back it out 1 1/4-1 3/4 turns (**Figure 44**).
6. Shift the engine into FORWARD.
7. Rotate the throttle lever to the closed position (full retard).
8. If idle speed is not 600-650 rpm, adjust the throttle stop screw (**Figure 45**) as required to bring idle speed within specifications.

CHAPTER SIX

**CENTERBOWL SINGLE THROAT CARBURETOR
(DT 20 THROUGH DT 85; DT 115, DT 140)**

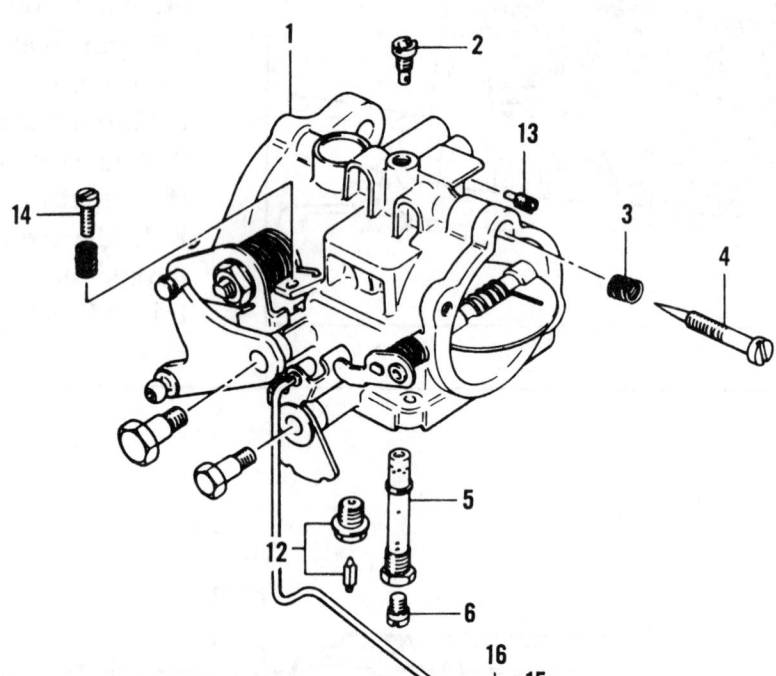

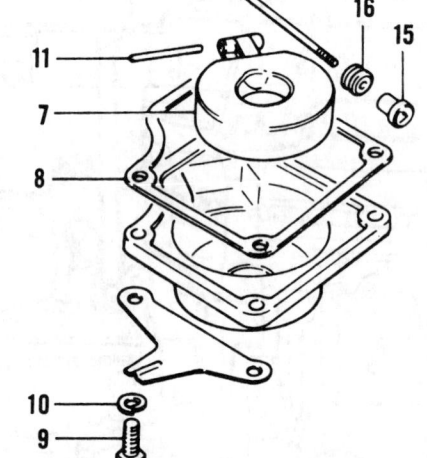

1. Carburetor body
2. Pilot (idle) jet
3. Air screw spring
4. Idle air screw
5. Main nozzle
6. Main jet
7. Float
8. Gasket
9. Screw
10. Washer
11. Hinge pin
12. Needle valve and seat
13. Air jet
14. Throttle stop screw and spring
15. Choke knob
16. Grommet

FUEL SYSTEM

CENTERBOWL SINGLE-THROAT CARBURETOR (DT 20 THROUGH DT 85; DT 115, DT 140)

Overhaul

Several variations of this carburetor design have been used among these models. All operate essentially the same, varying primarily in calibration and linkage connections. Refer to **Figure 46**, typical for this procedure.

1. Remove the float bowl attaching screws. Remove the float bowl and gasket from the fuel bowl. Discard the gasket.
2. Slide the float hinge pin to one side and remove the float and pin assembly from the carburetor.
3. Remove the inlet needle. Remove the needle valve seat with an appropriate size socket wrench (on models with removable seat).
4. Remove the main jet and nozzle with a jet remover or wide-blade screwdriver. Remove and discard the nozzle O-ring.
5. Remove the pilot (idle) jet with a suitable removal tool.
6. Remove the air jet (if so equipped) with a suitable removal tool.

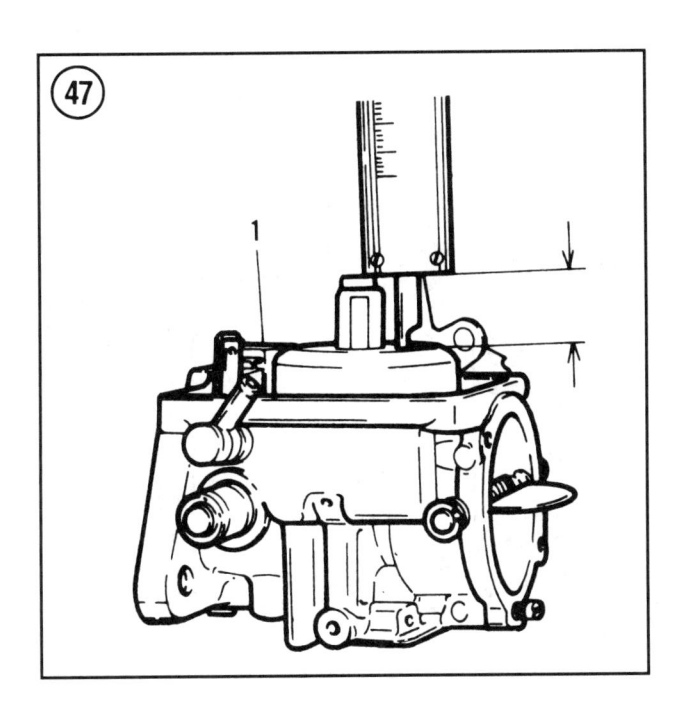

NOTE
Before removing the pilot screw, record the number of turns necessary until the screw lightly seats. Record the number of turns as the screw must be reinstalled to the exact same setting.

7. Unscrew the pilot (idle) screw and spring.
8. Clean and inspect all parts as described in this chapter.
9. Assembly is the reverse of disassembly, noting the following.
10. Install the pilot (idle) screw and spring. Lightly seat screw, then back out the number of turns recorded during removal.
11. Check the float adjustment as described in this chapter.

Float Adjustment

1. Invert the carburetor body and lower the float until adjusting tab just touches inlet needle. Hold the float in this position and measure the distance between the carburetor body and the bottom of the float, as shown in **Figure 47**, on all models except 1989-on DT 25 and 1988-on DT 30. On 1989-on DT 25 and 1988-on DT 30, measure from float bowl mating surface (gasket removed) to bottom of float.

CAUTION
*Bend the float adjustment arm or tang carefully when adjustment is required—do **not** press down on the float. Downward pressure on the float will press the inlet needle tip into its seat and can damage the tip surface.*

2. If the float level is not within specifications (**Table 1**), adjust float by bending the adjusting tab (1, **Figure 47**) as required.

NOTE
After adjusting the float level, the float should be approximately level with the carburetor body-to-float bowl mating surface.

Idle Speed Adjustment

The engine should be installed in a test tank or on the boat in the water for this procedure.

1985-1987 DT 20, DT 30; 1985-1988 DT 25; 1991 DT 25, DT 30

1. Remove the engine cover.
2. Check and adjust the ignition timing. See Chapter Five.
3. Connect a tachometer according to manufacturer's instructions.
4. Start the engine and run for 5-10 minutes to bring it to normal operating temperature.
5. Lightly seat the pilot (idle) air screw (**Figure 48**). Refer to **Table 2** and back it out the number of turns specified.
6. Shift the engine into FORWARD.

NOTE
Make sure the choke valve is in the full-open position.

7. Rotate the throttle lever to the closed position (full retard).
8. If idle speed is not 600-650 rpm, adjust the throttle stop screw (A, **Figure 49**) as required to bring the idle speed within specifications.
9. Shut the engine off.
10. Disconnect the tachometer and install the engine cover.

1985-on DT 35, DT 40

1. Remove the engine cover.
2. Check and adjust ignition timing. See Chapter Five.
3. Connect a tachometer according to manufacturer's instructions.
4. Start the engine and run for 5-10 minutes to bring it to normal operating temperature.
5. Lightly seat the pilot (idle) air screw (**Figure 50**). Refer to **Table 2** and back it out the number of turns specified.

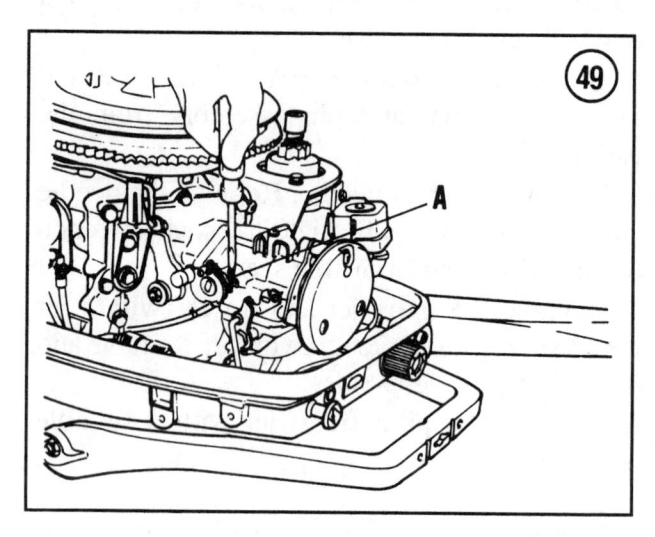

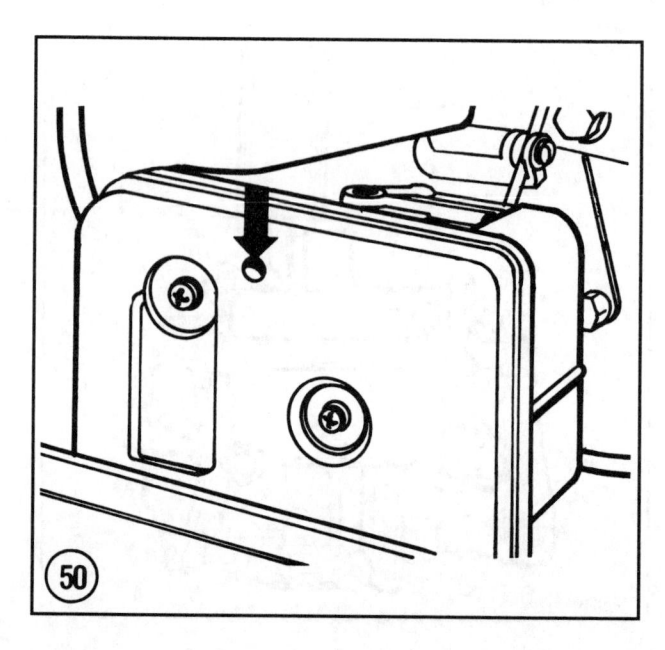

FUEL SYSTEM

6. Shift the engine into the FORWARD 1st notch.

NOTE
Make sure the choke valve is in the full-open position.

7. If idle speed is not 650-700 rpm, adjust the throttle stop screw (**Figure 51**) as required to bring the idle speed within specifications.

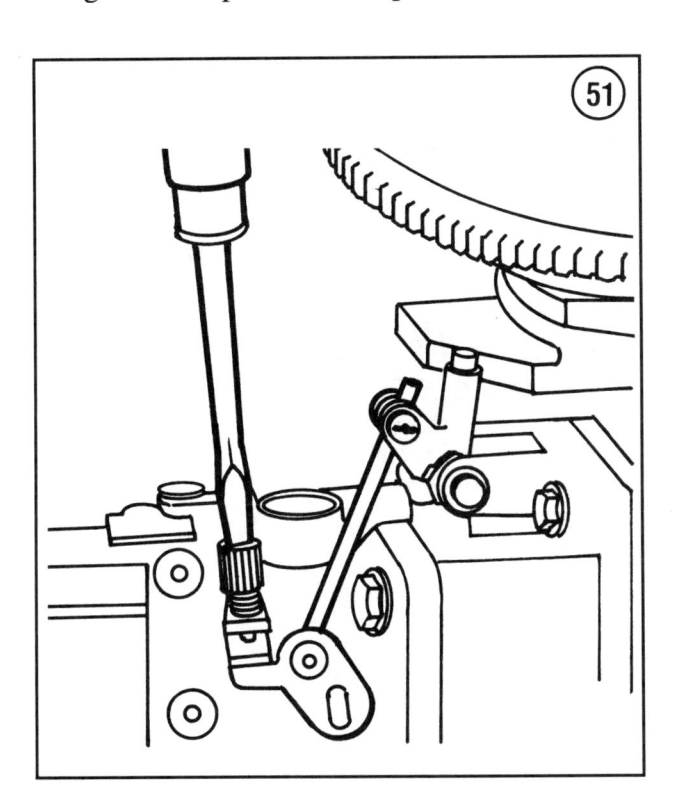

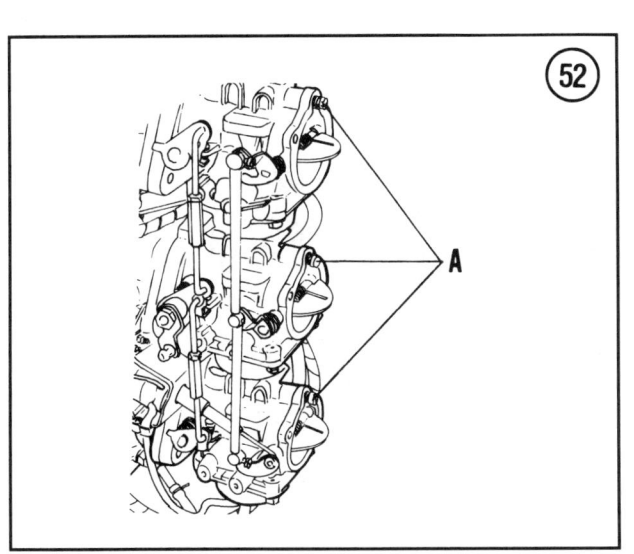

Figure 51

Figure 52

8. Shut the engine off.
9. Disconnect the tachometer and install the engine cover.

1985-1987 DT 75, DT 85; 1991 DT 75, DT 85

1. Remove the engine cover.
2. Check and adjust ignition timing. See Chapter Five.
3. Connect a tachometer according to manufacturer's instructions.
4. Start the engine and run for 5-10 minutes to bring it to normal operating temperature.
5. Lightly seat each pilot (idle) air screw (A, **Figure 52**). Refer to **Table 2** and back it out the number of turns specified.
6. Shift the engine into FORWARD.

NOTE
Make sure the choke valve is in the full-open position.

NOTE
Do not adjust the throttle stop screws to make the adjustment in Step 7.

7. If idle speed is not 600-700 rpm, loosen the timing adjust bolt locknut and adjust the bolt (**Figure 53**) as required to bring idle speed within

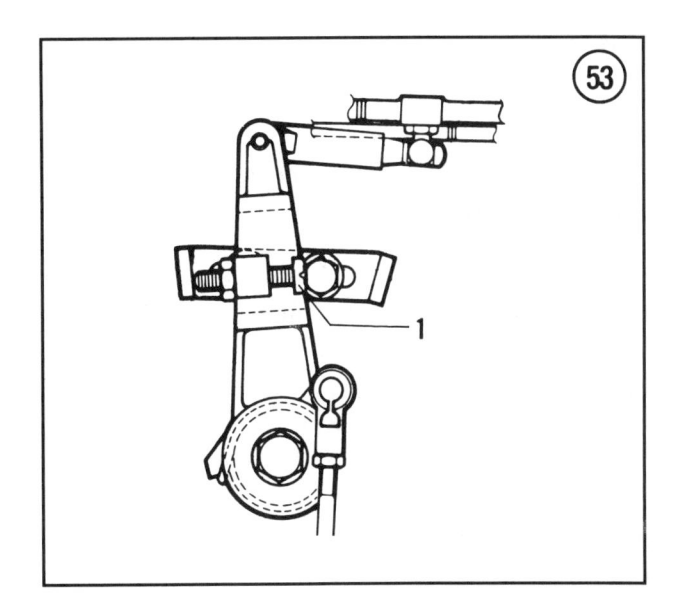

Figure 53

specifications. When idle speed is correct, tighten locknut.
8. Shut the engine off.
9. Disconnect the tachometer and install the engine cover.

1985 DT 115, DT 140; 1991 DT 115, DT 140

1. Remove the engine cover.
2. Check and adjust ignition timing. See Chapter Five.
3. Connect a tachometer according to manufacturer's instructions.
4. Start the engine and run for 5-10 minutes to bring it to normal operating temperature.
5. Lightly seat each pilot (idle) air screw (A, **Figure 54**). Refer to **Table 2** and back it out the number of turns specified.
6. Shift the engine into the FORWARD 1st notch (approx 34°).

NOTE
Make sure the choke valve is in the full-open position.

7. If idle speed is not 600-700 rpm, adjust the throttle stop screw as required to bring the idle speed within specifications.
8. Shut the engine off.
9. Disconnect the tachometer and install the engine cover.

1989-1990 DT 25; 1988-1990 DT 30; 1985-on DT 55, DT 65; 1988-1990 DT 75, DT 85; 1986-on DT 115, DT 140

On these models, initial idle speed has been factory-adjusted at the optimum value. The idle speed can be adjusted during operation by turning the idle speed switch.
1. Remove the engine cover.
2. Check and adjust the ignition timing. See Chapter Five.
3. Connect a tachometer according to manufacturer's instructions.

4. Start the engine and run for 5-10 minutes to bring it to normal operating temperature.
5A. On 1989-1990 DT 25, 1988-1990 DT 30 models, lightly seat the pilot (idle) air screw (**Figure 55**). Refer to **Table 2** and back it out the number of turns specified.
5B. On 1985-1990 DT 55, DT 65 and 1986-1990 DT 115, DT 140 models, lightly seat the pilot (idle) air screw (**Figure 56**). Refer to **Table 2** and back it out the number of turns specified.
6A. On 1989-1990 DT 25, 1988-1990 DT 30 models, place the clutch lever or remote control lever just into FORWARD position.

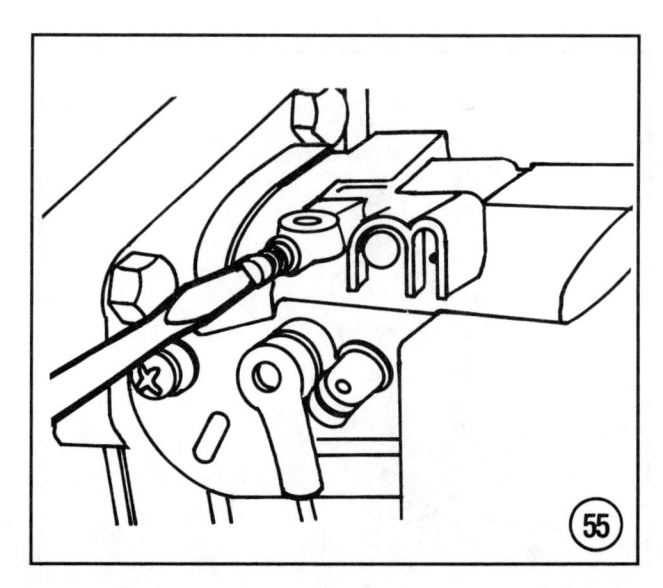

FUEL SYSTEM

6B. On 1985-1990 DT 55, DT 65 and 1986-1990 DT 115, DT 140 models, place the remote control lever just into FORWARD position.

NOTE
Make sure the choke valve is in the full-open position.

7A. On 1989-1990 DT 25, 1988-1990 DT 30 models, if idle speed is not 650-700 rpm, turn the idle speed adjusting knob (**Figure 57**) to bring the idle speed within specifications.

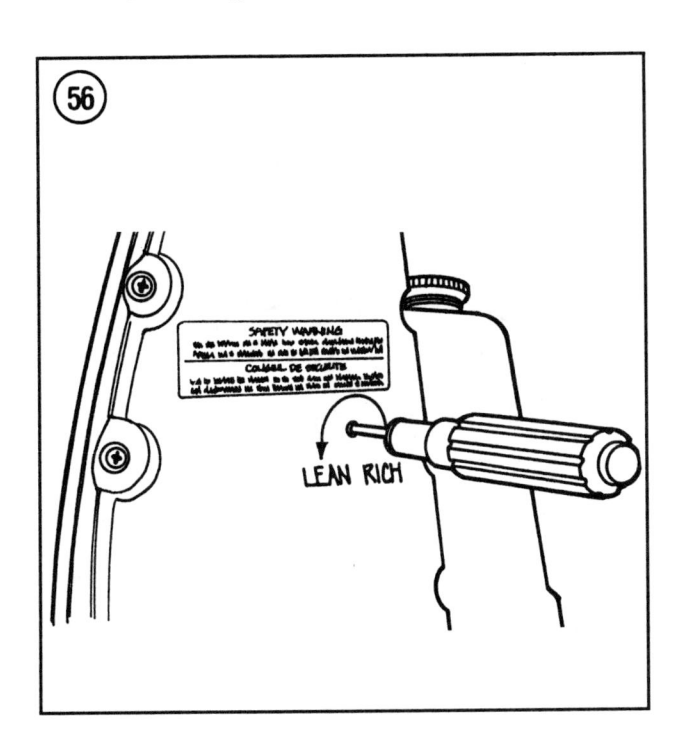

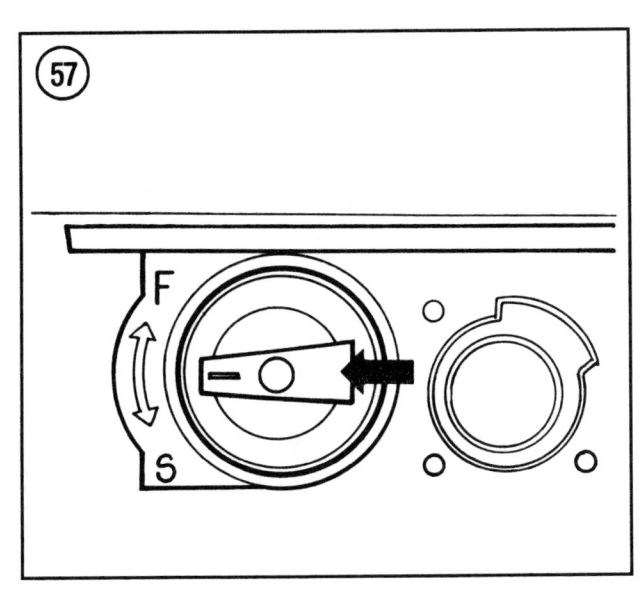

7B. On 1985-1990 DT 55, DT 65 models, if idle speed is not 650-700 rpm, turn the idle speed adjusting knob to bring the idle speed within specifications.

7C. On 1986-1990 DT 115, DT 140 models, if idle speed is not 600-700 rpm, turn the idle speed adjusting knob (**Figure 58**) to bring the idle speed within specifications.

8. Shut the engine off.

9. Disconnect the tachometer and install the engine cover.

Carburetor Balancing (DT 75 and DT 85)

Carburetors on DT 75 and DT 85 engines should be balanced whenever they are installed after servicing. This procedure requires the use of a carburetor balancer (part No. 09913-13121) and tachometer.

1. Connect a tachometer according to manufacturer's instructions.
2. Disconnect the intake manifold line at the top carburetor. Plug the line to prevent leakage.
3. Connect the right balance line to the top carburetor fitting (**Figure 59**).

4. Start the engine and run it at a steady 1,000 rpm.

5. Turn the air screw at the bottom of the balancer line which is connected to the engine (**Figure 60**) until the steel ball inside the tube rests on the center line (**Figure 61**).

6. Disconnect the balancer line at the top carburetor intake manifold fitting. Remove the intake manifold line at the center carburetor. Connect the second balancer line in its place. Connect the third balancer line to the bottom carburetor intake manifold fitting. Repeat Step 4 until the steel balls in the second and third balancer tubes rest on the center line.

7. Once a balancer tube has been adjusted for each carburetor, shut the engine off and connect the balancer lines to the carburetors as follows:

   a. Connect the first balancer tube to the top carburetor intake manifold fitting.

   b. Connect the second tube to the center carburetor fitting.

   c. Connect the third tube to the bottom carburetor fitting.

8. Restart the engine and run at 1,000 rpm. Make sure the steel ball in the top balancer tube rests on the center line (the others will not).

9. Loosen the throttle rod locknuts, rotate the throttle rods until the steel balls in the second and third balancer tubes rest on the center line (**Figure 62**) then tighten the locknuts (**Figure 63**).

10. Shut the engine off. Disconnect the balancer from the intake manifold and reconnect the intake hose(s).

11. Remove the tachometer.

CENTERBOWL DUAL-THROAT CARBURETOR (V4 AND V6)

Overhaul

Refer to **Figure 64**, typical for this procedure.

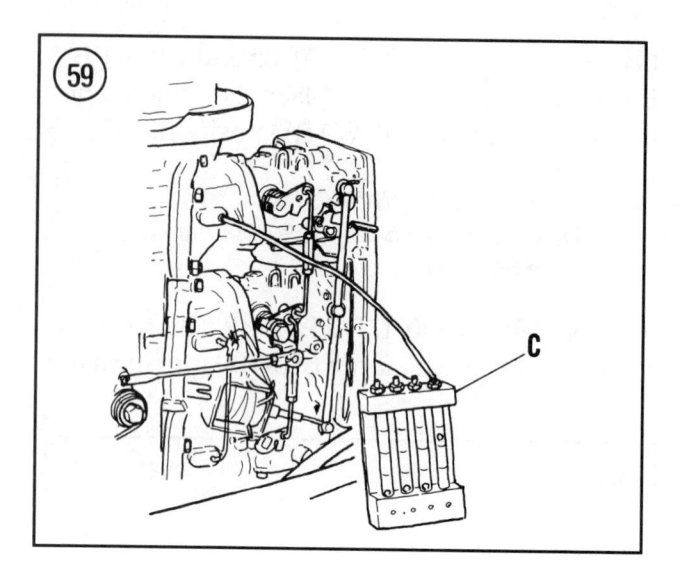

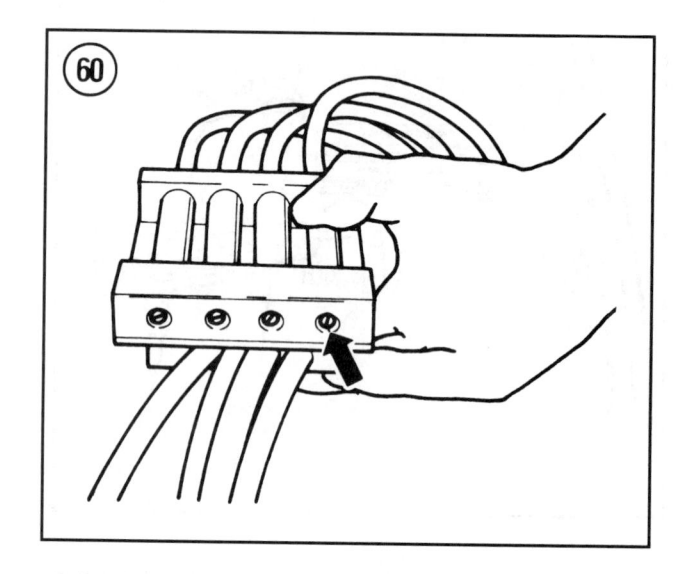

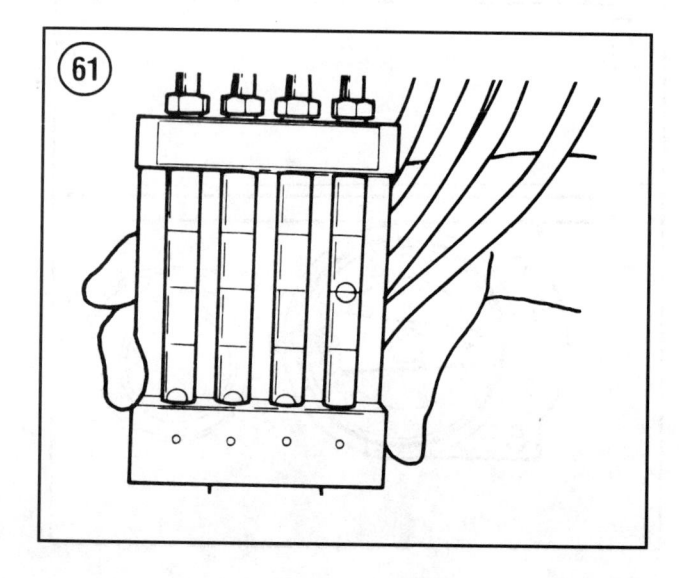

FUEL SYSTEM

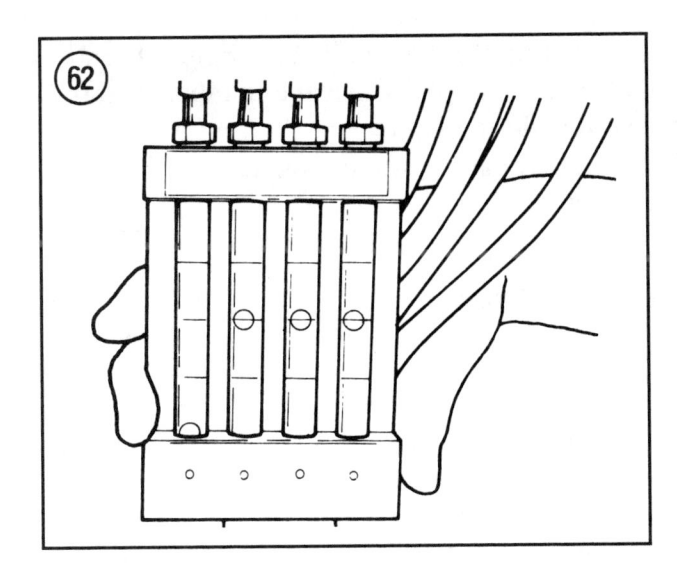

1. Remove the float bowl attaching screws. Remove the float bowl and gasket from the fuel bowl. Discard the gasket.

2. Slide the float hinge pin to one side and remove the float and pin assembly from the carburetor.

3. Remove the inlet needle. Remove the needle valve seat with an appropriate size socket wrench.

4. From the carburetor body, remove the following:

   a. Remove the pilot (idle) jet with a suitable removal tool.

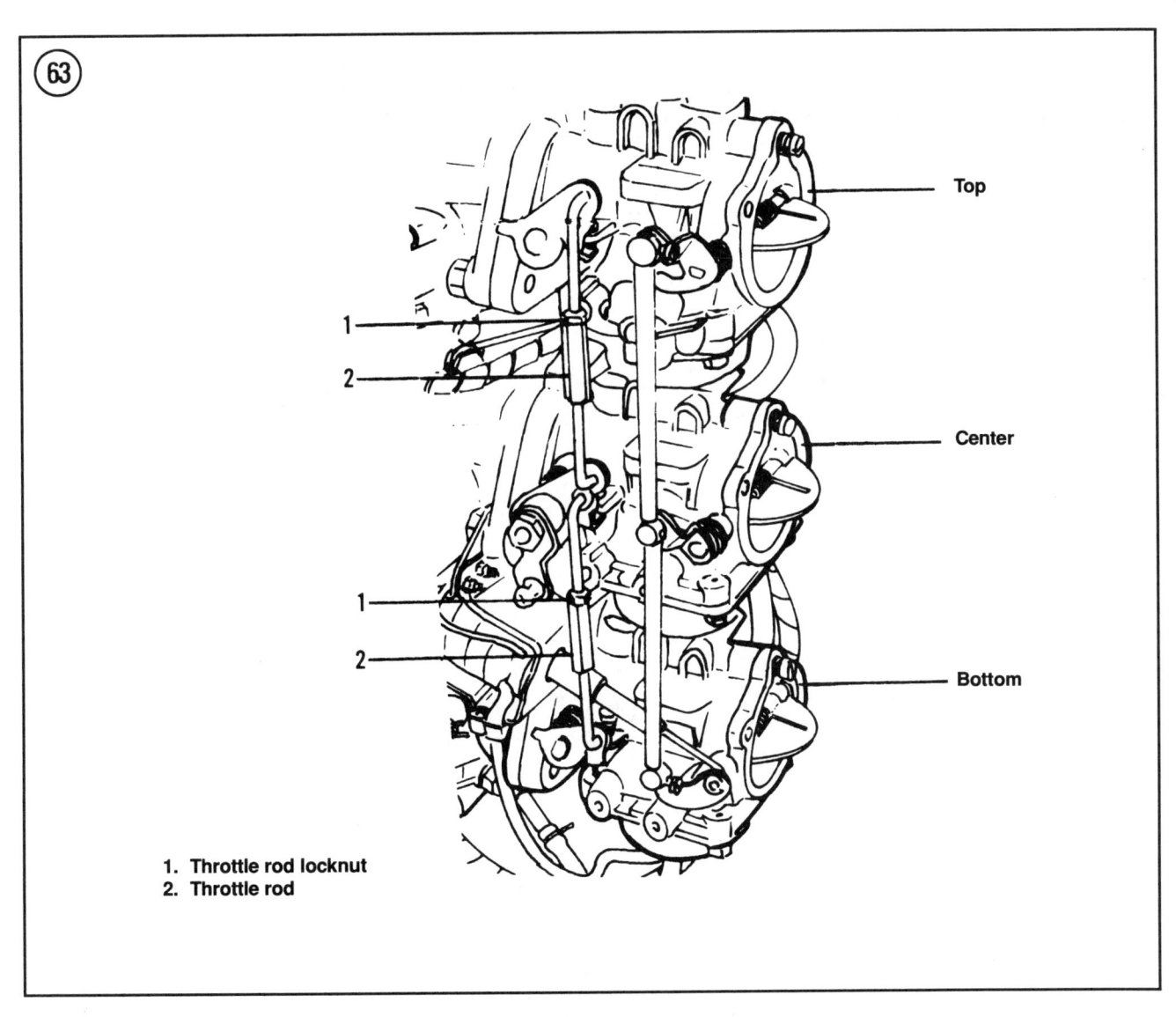

1. Throttle rod locknut
2. Throttle rod

NOTE
Before removing the pilot screw, record the number of turns necessary until the screw lightly seats. Record the number of turns as the screw must be reinstalled to the exact same setting.

b. Unscrew the pilot (idle) screw and spring.
c. Remove both high speed nozzles.

5. From the float bowl, remove the following:
   a. Remove the main jet holder from each side of the float bowl.
   b. Remove the main jet from each main jet holder with a jet remover or wide-blade screwdriver. Remove and discard O-ring on each holder.

6. Clean and inspect all parts as described in this chapter.
7. Assembly is the reverse of disassembly, noting the following.
8. Install pilot (idle) screw and spring. Lightly seat screw, then back out the number of turns recorded during removal.
9. Check float adjustment as described in this chapter.

Fuel Level Test

This test requires the use of 2 Suzuki special tools. They are the Fuel Level Gauge Adaptor (part No. 09913-18711) and Fuel Level Gauge (part No. 09932-28211).

1. Remove the engine cover.
2. Connect a tachometer according to manufacturer's instructions.
3. Remove the left-hand main jet holder from the carburetor float bowl.
4. Remove the main jet and O-ring seal from the main jet holder.

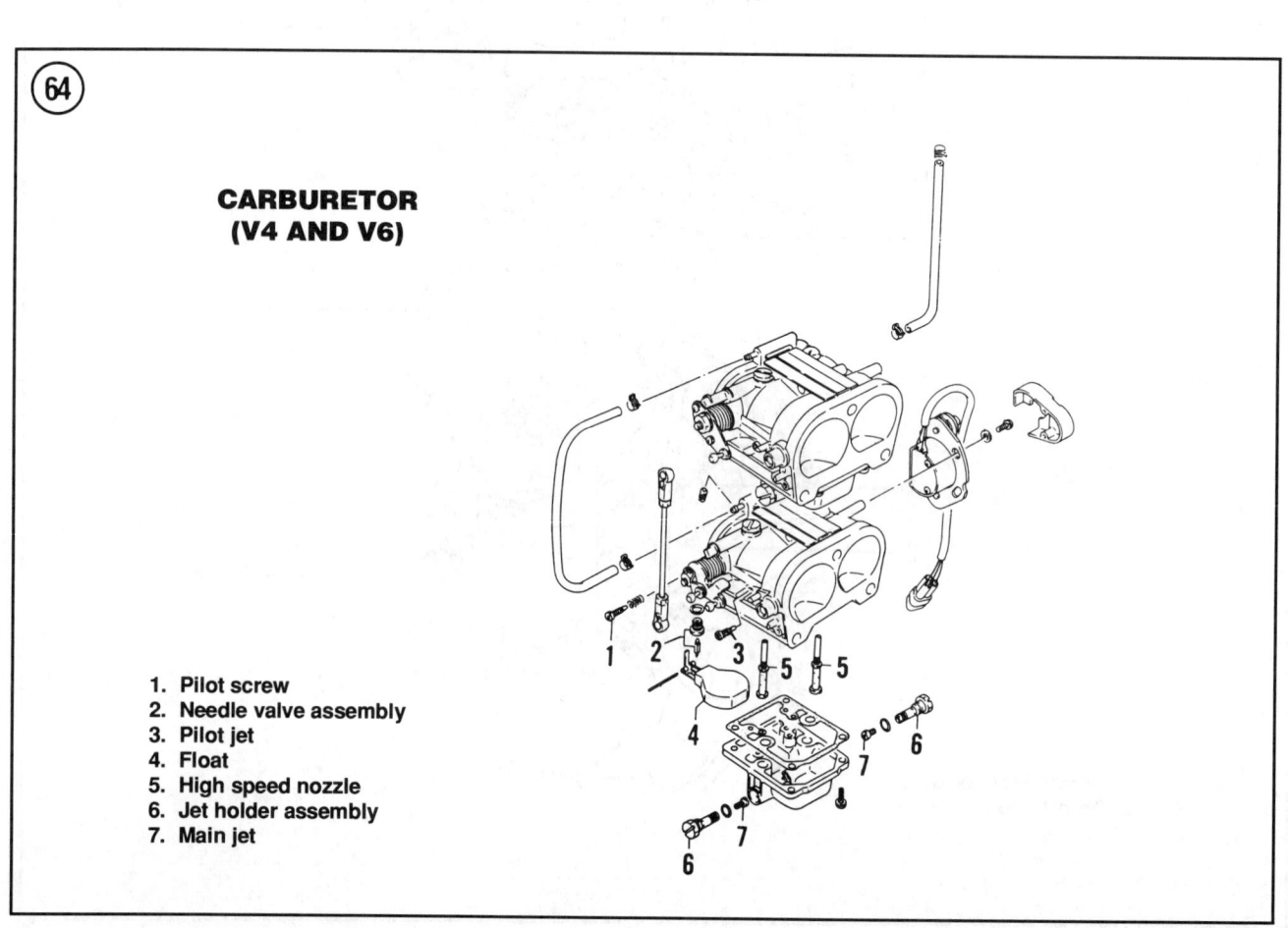

CARBURETOR (V4 AND V6)

1. Pilot screw
2. Needle valve assembly
3. Pilot jet
4. Float
5. High speed nozzle
6. Jet holder assembly
7. Main jet

FUEL SYSTEM

5. Install the main jet (A, **Figure 65**) and O-ring seal (B, **Figure 65**) onto the adapter.

6. Install the fuel level gauge to the adaptor and install this assembly into the carburetor float bowl.

7. Start the engine and let it idle at 600-700 rpm. Readjust idle speed if necessary as described in this chapter.

8. The fuel in the fuel level gauge should be 21.0-23.0 mm (0.83-0.91 in.). See Dimension A, **Figure 66**.

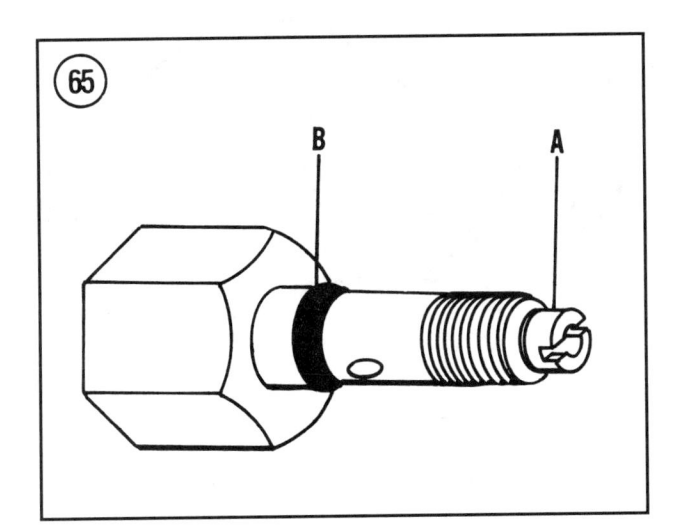

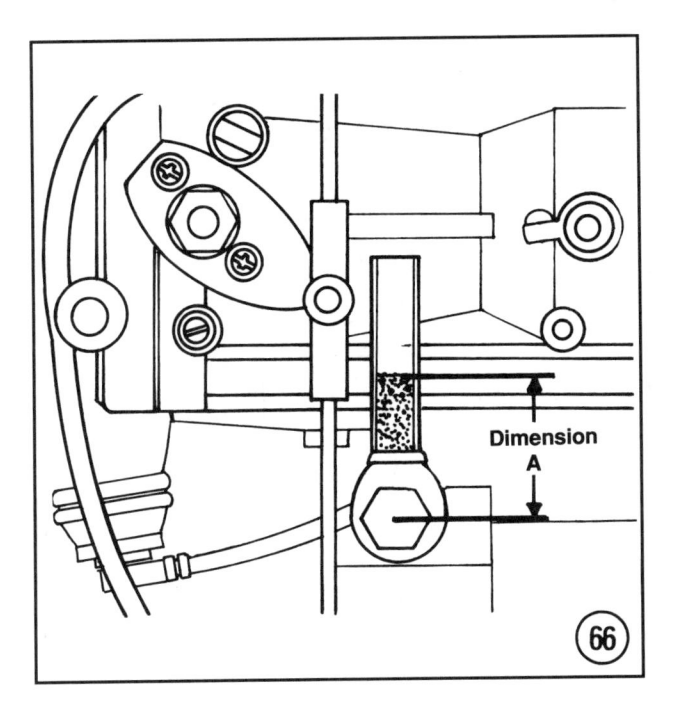

9. If the fuel level is incorrect, adjust the fuel level by adjusting the float level tab as described in this chapter.

10. Remove the main jet and O-ring seal from the adapter.

11. Reinstall the main jet and O-ring seal onto the main jet holder and install into the carburetor float bowl.

12. Repeat this procedure for the other carburetor(s).

13. Disconnect the tachometer and install the engine cover.

Float Adjustment

1. Invert the carburetor body and lower the float until adjusting tab just touches inlet needle. Hold float in this position and measure the distance between the carburetor body (without gasket) and the bottom of the float, as shown in **Figure 67**.

CAUTION
*Bend the float adjustment arm or tang carefully when adjustment is required—do **not** press down on the float. Downward pressure on the float will press the inlet needle tip into its seat and can damage the tip surface.*

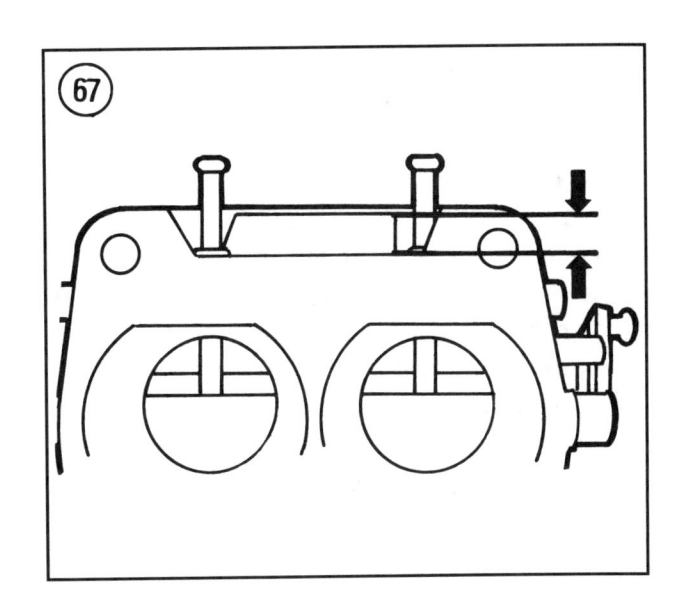

2. If the float level is not within specifications (**Table 1**), adjust float by bending the adjusting tab (**Figure 68**) as required.

Idle Speed Adjustment

The engine should be installed in a test tank or on the boat in the water for this procedure.

On models prior to 1991, minor idle speed adjustments can be made by turning the idle speed switch. On 1991 models, the idle speed switch is eliminated and the idle speed is adjusted by the idle speed screw.

1. Remove the engine cover.
2. Check and adjust the ignition timing. See Chapter Five.
3. Connect a tachometer according to manufacturer's instructions.
4. Start the engine and run for 5-10 minutes to bring it to normal operating temperature.
5. Lightly seat the pilot (idle) air screw. Refer to **Table 2** and back it out the number of turns specified.
6. Shift the remote control lever into FORWARD gear 1st notch.

NOTE
Make sure the choke valve is in the full-open position.

7A. On V4 models, turn the idle speed adjusting switch to position No. 5 (slow). This should result in an idle speed of 600-650 rpm. The engine must maintain this speed for 3 minutes.
7B. On V6 models, turn the idle speed adjusting switch to achieve a speed of 600-700 rpm.
8. On V4 models, if idle speed is less than 600 rpm, or the engine will not maintain this speed for 3 minutes, perform the following:
   a. Adjust the top carburetor pilot screw to maintain the 600-650 rpm.
   b. Recheck the throttle valve sensor resistance and reset if necessary. Refer to Chapter Six.
9. Shut the engine off.
10. Disconnect the tachometer and install the engine cover.

Throttle Linkage Synchronizing

The throttle linkage must be synchronized to ensure that all throttle valves completely close at the same time.

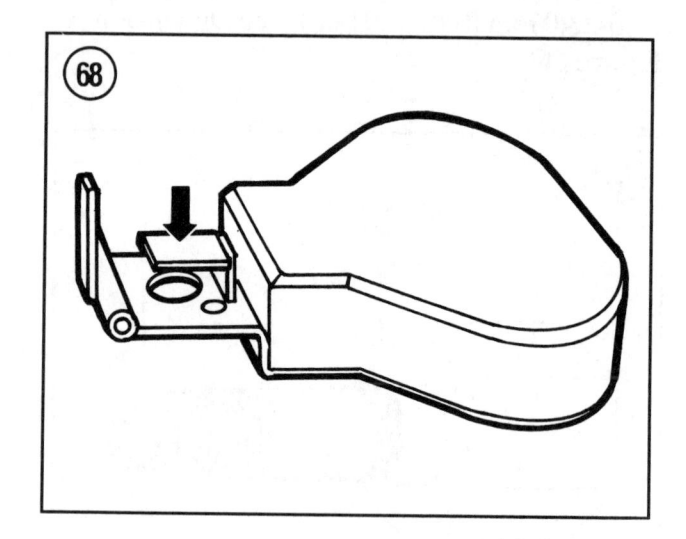

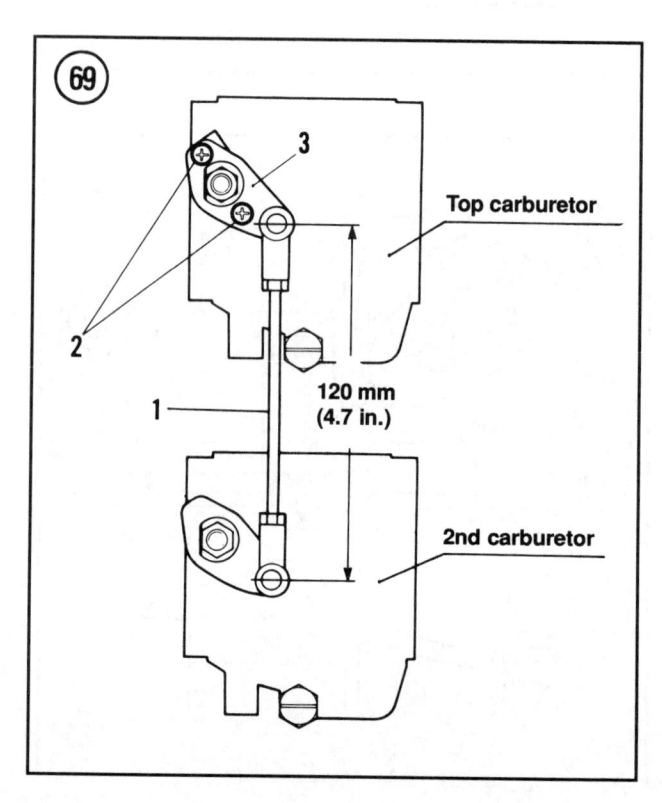

FUEL SYSTEM

V4 models

1. Check the length of the throttle linkage rod (1, **Figure 69**). The correct length is 120 mm (4.7 in.). If incorrect, loosen the locknuts and turn the linkage rod in either direction until the dimension is correct. Tighten the locknuts.
2. Loosen the throttle lever adjusting screws (2, **Figure 69**) on the top carburetor.
3. Lightly push the throttle lever (3, **Figure 69**) clockwise until both throttle valves are completely closed. Tighten the adjusting screws securely.
4. Move the linkage several times and make sure both throttle valves are synchronized in the completely closed position. Readjust if necessary.
5. Repeat for the other bank of carburetors.

V6 models

1. Check that the throttle linkage rod (1, **Figure 70**) moves freely with no binding.
2. Loosen the throttle lever adjusting screws (2, **Figure 70**) on the top and center carburetors.
3. Lightly push the throttle lever (3, **Figure 70**) on the center carburetor clockwise until both the bottom and center carburetor throttle valves are completely closed. Tighten the adjusting screws securely on the center carburetor.
4. Lightly push the throttle lever (3, **Figure 70**) on the top carburetor clockwise until all 3 carburetor throttle valves are completely closed. Tighten the adjusting screws securely on the top carburetor.
5. Move the linkage several times and make sure all 3 throttle valves are synchronized in the completely closed position. Readjust if necessary.
6. Repeat for the other bank of carburetors.

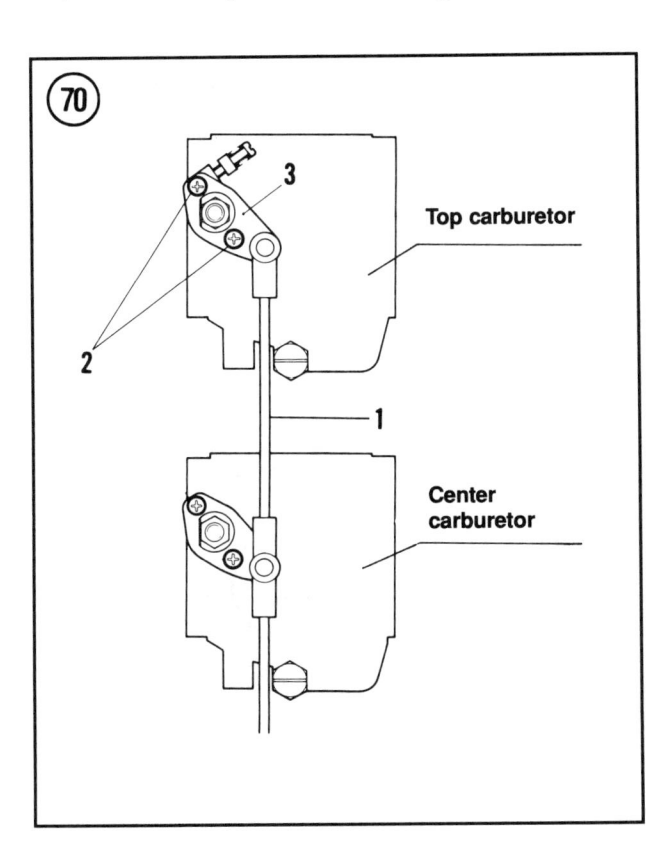

CARBURETOR CLEANING AND INSPECTION

WARNING
Carburetor cleaner is extremely caustic and can cause permanent eye damage. Always wear eye protection when using any type of carburetor cleaner.

Carburetors are best cleaned by completely disassembling them and cleaning the fuel and air orifices with an aerosol carburetor cleaner. Never use a wire to clean out jets or orifices; such a process could enlarge the passage which would adversely affect the air-to-fuel ratio.

Outboard carburetors have much smaller air and fuel passages than automotive carburetors. For this reason, soaking the carburetor parts in an automotive type carburetor cleaner is not recommended. The exterior of nearly all outboard carburetors is usually coated with a corrosion-protective clear coating. These caustic liquid cleaners will remove the protective coatings from the outside of the carburetor body. The dissolved coating could plug one or more of the air or fuel passages within the carburetor. Also, if the cleaner was used previously, there will be sediment held in suspension within the solution. The sediment could also plug a passage.

Clean the carburetor parts in a good grade of fresh solvent and thoroughly dry with compressed air. Many good aerosol carburetor cleaners (i.e. Zep Choke and Carburetor Cleaner) can help remove any residue not removed with the solvent. Thoroughly rinse off all parts with clean water and dry with compressed air. If you do not have access to compressed air, place the cleaned parts on a piece of newspaper and allow to dry.

1. Clean all parts, except rubber or plastic parts, with a good grade of aerosol carburetor cleaner or cleaning solvent.

NOTE
*A special carburetor cleaner is **not** usually necessary to clean a carburetor unless it is very dirty or corroded. A good grade of parts cleaning solvent will usually clean most carburetors sufficiently.*

CAUTION
Do not put non-metallic parts such as floats, gaskets and O-rings in special carburetor cleaner as these components will be damaged. Clean these components in common solvent or kerosene.

2. Remove all parts from the cleaner and wash thoroughly in soap and water. Rinse with clean water and dry thoroughly.

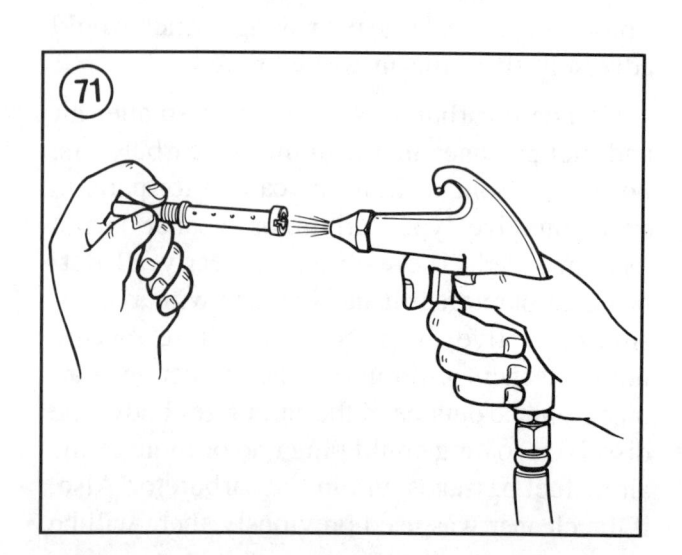

CAUTION
If compressed air is not available, allow the parts to air dry or use a clean lint-free cloth. Do not use paper towels to dry carburetor parts, as small paper particles may plug openings in the carburetor body or jets.

3. Blow out the jets with compressed air. *Do not use a piece of wire to clean them as minor gouges in the jet can alter flow rate and upset the fuel/air mixture.* If compressed air is not available, use a piece of straw from a broom to clean the jets.

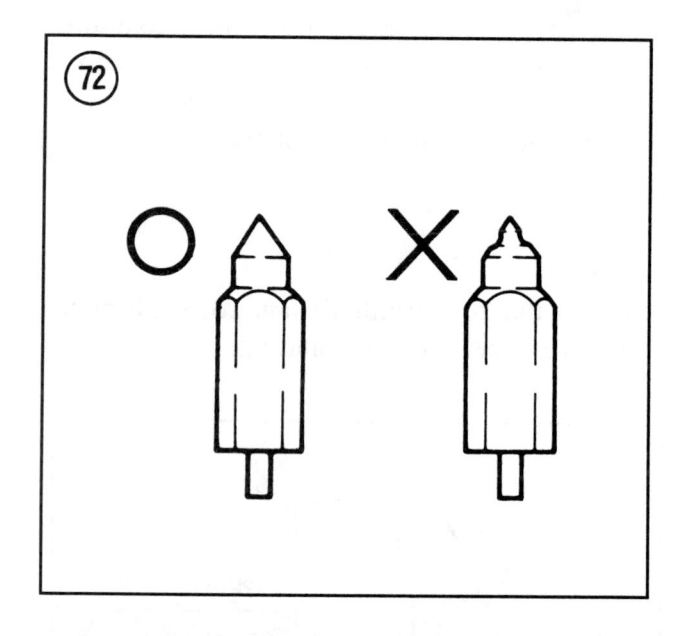

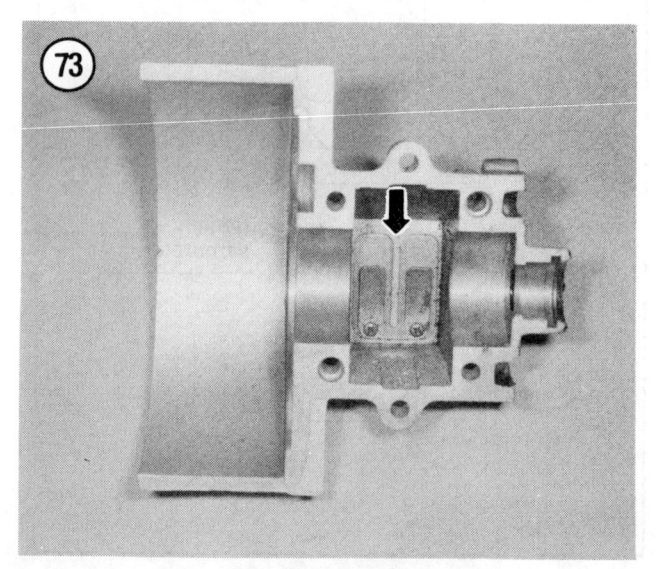

FUEL SYSTEM

4. On V4 and V6 models, make sure the small openings in the high speed nozzle (**Figure 71**) are clean and open.

5. Inspect the tip of the inlet needle tip (**Figure 72**) for wear or damage. Replace the valve and seat as a set.

6. O-ring seals tend to become hardened after prolonged use and therefore lose their ability to seal properly. Inspect all O-rings and replace if necessary.

7. Remove the O-ring gasket from the float bowl and install a new gasket.

8. Check the float for leaks. Fill the float bowl with water and push the float down. There should be no sign of bubbles. Replace the float if it leaks.

9. Check the throttle valve for scratches or other damage that would allow it to stick open during engine operation.

10. Check the throttle and choke shafts for excessive wear or play. The throttle and choke valves must move freely without binding. Replace carburetor if any of these defects are noted. Make sure the screws securing the throttle and choke valves are tight. Tighten securely if necessary.

11. Blow out all jets and passages in the carburetor bodies with compressed air. Clean out if they are plugged in any way.

12. Inspect the float pivot pin posts for cracks or damage. If any damage is noted, replace the carburetor assembly.

REED VALVE ASSEMBLY

On DT 2 models, the reed valve assembly is an integral part of the front crankcase half (**Figure 73**). On all other models, the reed valve assembly is mounted between the intake manifold and the crankcase (**Figure 74**).

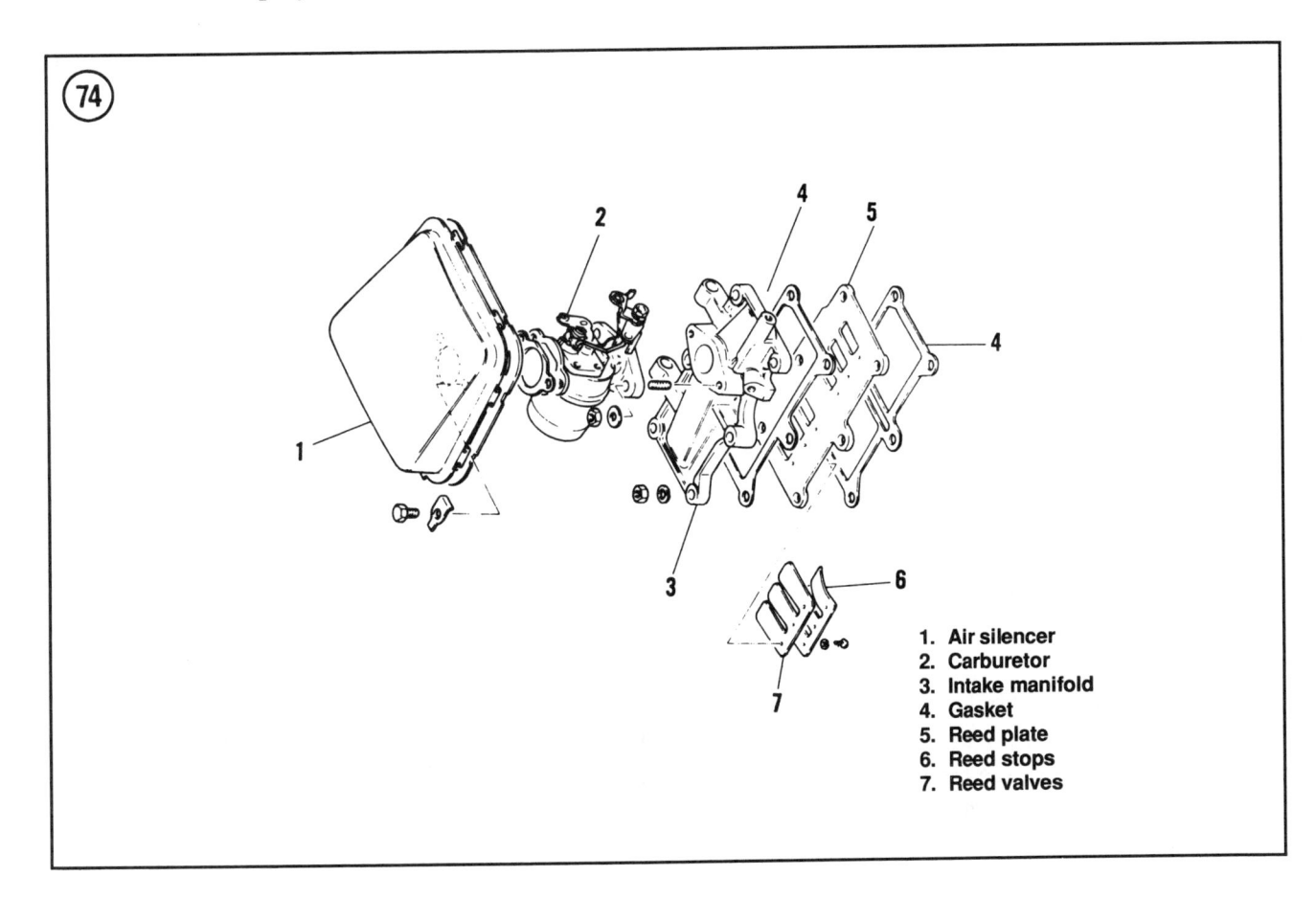

74

1. Air silencer
2. Carburetor
3. Intake manifold
4. Gasket
5. Reed plate
6. Reed stops
7. Reed valves

Reed valves control the passage of air-fuel mixture into the crankcase by opening and closing as crankcase pressure changes. When crankcase pressure is high, the reeds maintain contact with the reed plate to which they are attached. As crankcase pressure drops on the compression stroke, the reeds move away from the plate and allow air-fuel mixture to pass. Reed travel is limited by the reed stop. As crankcase pressure increases, the reeds return to the reed plate.

Figure 75 and **Figure 76** show two typical reed valve arrangements.

Removal/Installation (DT 2)

The reed valves are attached to the inside of the front crankcase half on this model. Access to the reed valves for inspection and service requires that the engine be removed and the crankcase separated. See Chapter Eight.

Removal/Installation (All Other Models)

1. Remove the carburetor as described in this chapter.
2. Disconnect any hoses connected to the intake manifold.

NOTE
On some models, the reed valve assembly is secured to the crankcase by one or more separate fasteners. With this design, the intake manifold must be removed first, then the reed valve assembly fastener(s) (Figure 77) removed before the assembly and gasket can be removed.

3. Remove the fasteners holding the intake manifold(s) to the crankcase cover. Remove the intake manifold(s), gasket(s), reed valve assembly(ies) and gasket(s) from the crankcase. Discard the gaskets.

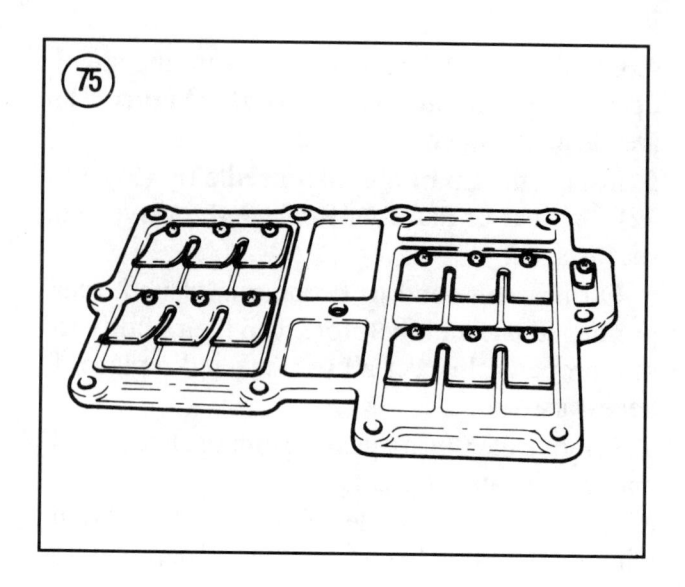

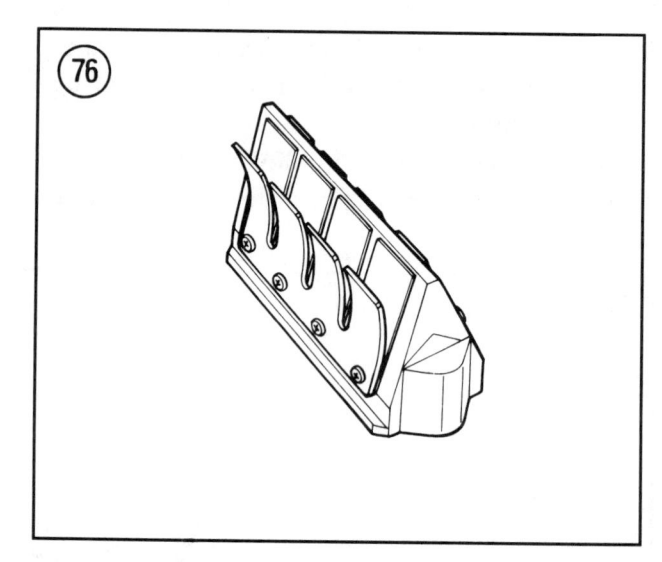

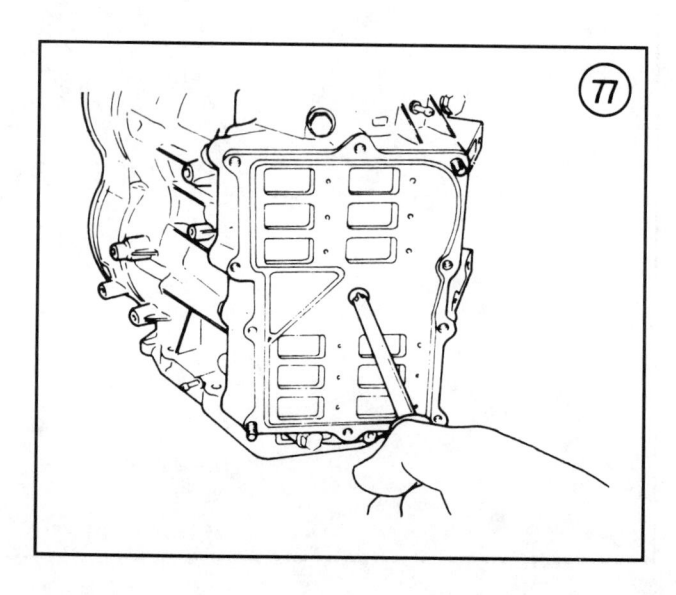

FUEL SYSTEM

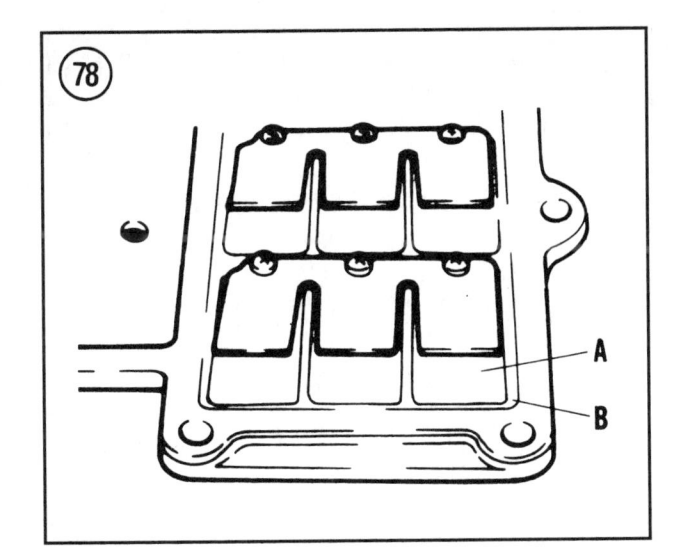

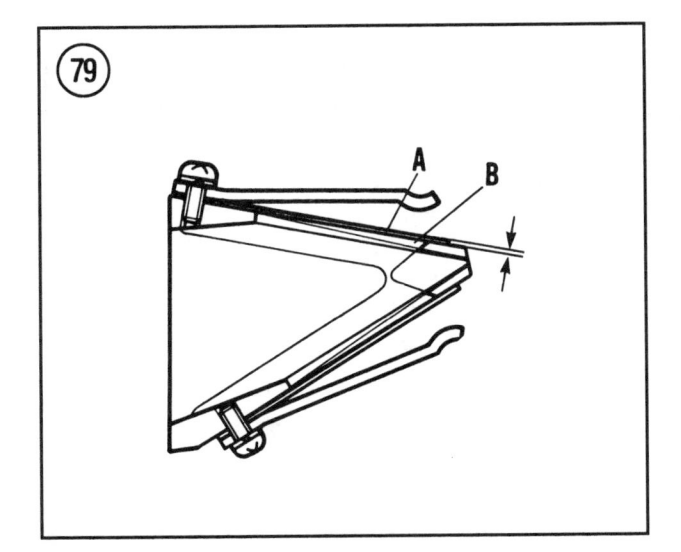

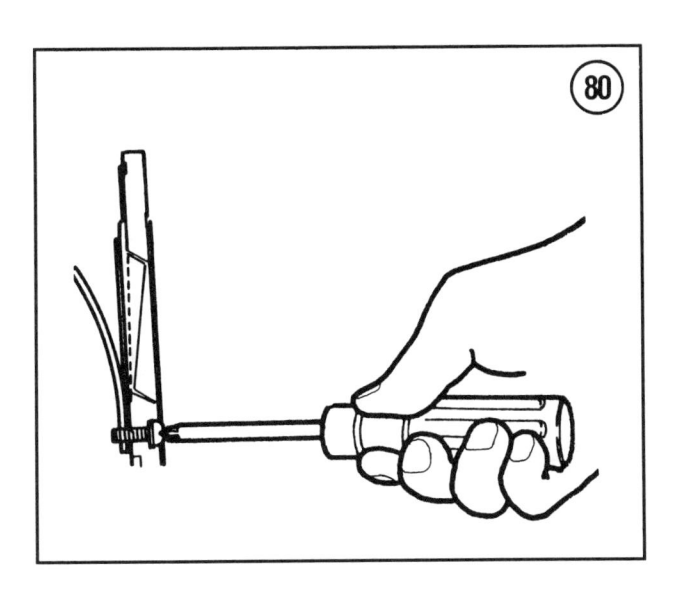

4. Clean all mating surfaces of gasket or sealant residue.

5. Installation is the reverse of removal. Use new gaskets and make sure that reed assembly(ies) faces the crankcase.

Inspection

1. Check reeds for cracking or other damage. Replace if any defects are noted.

2. Refer to **Figure 78** or **Figure 79** for typical reed valve layouts. Reeds (A) should lie flat on the valve seat (B) with no preload.

3. To check flatness, gently push each reed petal out. Constant resistance should be felt with no audible noise.

4. Check the clearance between the reed and the valve seat with a flat feeler gauge. If greater than 0.008 in. (0.20 mm), replace the reed set as described in this chapter.

5. Measure the distance between the reed stop and valve seat and compare to **Table 3**. If not within specifications, check the valve seat for warpage and replace as required. If valve seat is not warped, replace the reed stop assembly as described in this chapter.

Reed and Reed Stop Replacement

1. Remove the screws holding the reed stop and reeds to the valve seat (**Figure 80**).

2. Remove the reed stop and reeds.

3. Place a new reed on the valve seat and check for flatness.

4. Center the reed over the valve seat openings.

5. Wipe reed stop screw threads with Thread Lock No. 1342. Install reed stop and tighten screws securely.

6. Check reed tension and opening. See *Inspection* in this chapter.

FUEL TANK

The DT 2 and DT 4 models have an integral fuel tank. The DT 2 transports fuel from the tank to the carburetor by gravity feed.

Since the DT 4 can also be operated from a remote fuel tank, it has a fuel pump installed between the integral tank and carburetor (**Figure 81**). When running the engine on its integral tank, the fuel petcock should be in a horizontal position. To run the engine on a remote tank, connect the remote tank line to the fuel line socket and turn the integral tank petcock to a vertical position.

Removal/Installation (DT 2)

1. Remove the right and left engine covers.
2. Make sure the fuel petcock lever is in the OFF position that is marked with an "S."
3. Disconnect the fuel line at the petcock (**Figure 82**). Plug the end of the line to prevent leakage.
4. Remove the front and rear screws holding the fuel tank to the support plate (**Figure 83**).
5. Remove the tank and petcock assembly from the engine.
6. If petcock removal is required, loosen the hose clamp and pull the petcock and hose from the fuel tank (**Figure 84**).
7. Installation is the reverse of removal.

Removal/Installation (DT 4)

Refer to **Figure 85** for this procedure.
1. Remove the engine cover.
2. Make sure the fuel petcock lever is in the OFF position marked "B."
3. Disconnect the fuel line at the petcock. Plug the end of the line to prevent leakage.
4. Remove the bolts and retaining brackets securing the fuel tank to the engine. Remove the fuel tank.

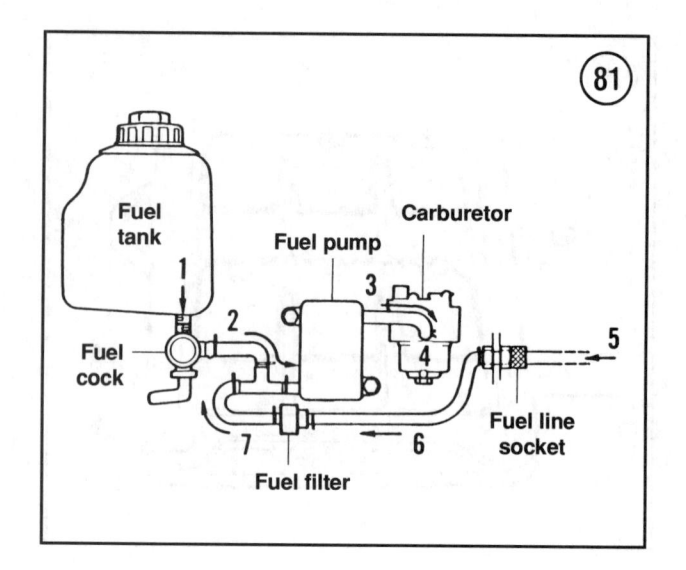

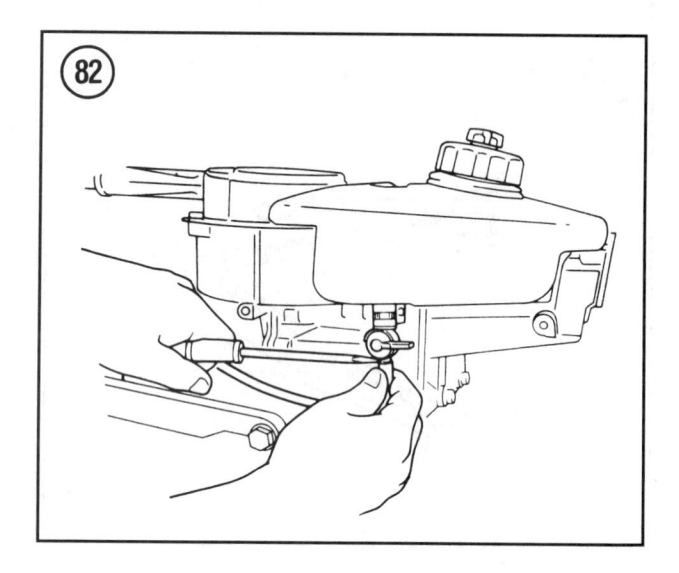

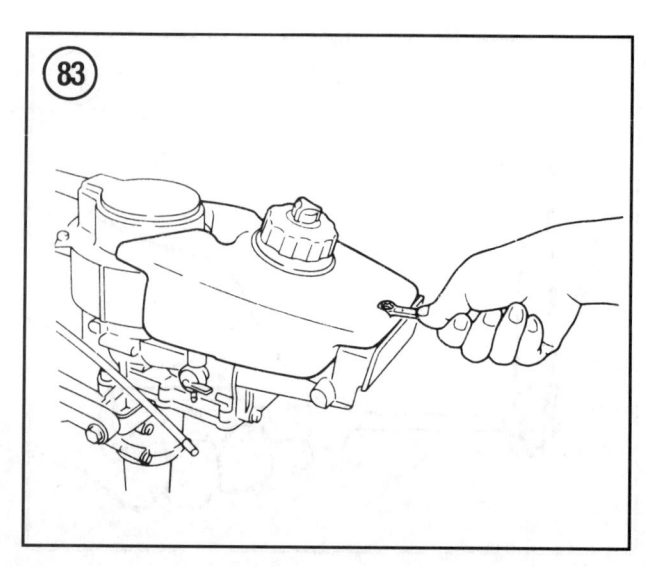

FUEL SYSTEM

5. Installation is the reverse of removal.

Portable Fuel Tank

Suzuki offers 2 types of portable or remote fuel tanks. **Figure 86** (recreational) and **Figure 87** (commercial) show the components of the fuel tanks, including the primer bulb assembly.

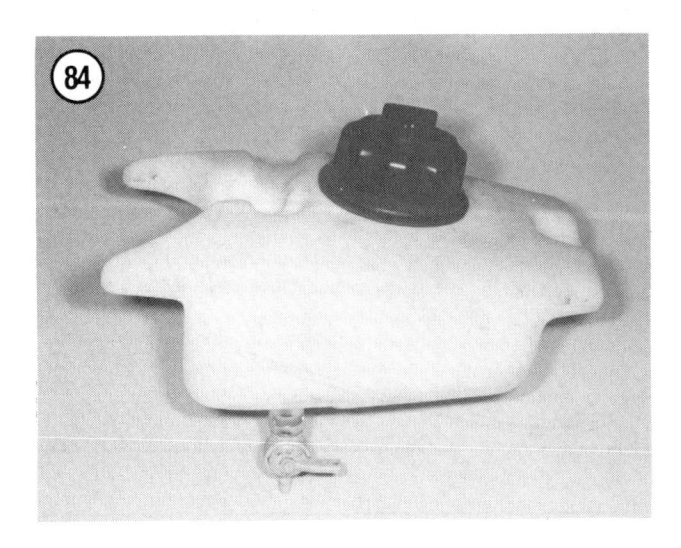

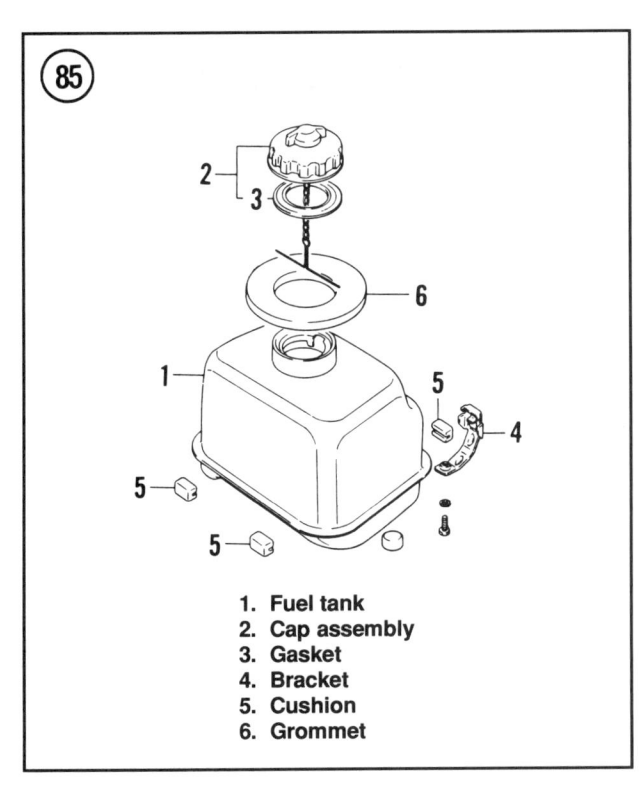

85

1. Fuel tank
2. Cap assembly
3. Gasket
4. Bracket
5. Cushion
6. Grommet

When some oils are mixed with gasoline and stored in a warm place, a bacterial substance will form. This colorless substance covers the fuel pickup, restricting flow through the fuel system. Bacterial formation can be prevented by using a good quality fuel conditioner on a regular basis. If present, bacteria can be removed with a good marine engine cleaner.

To remove any dirt or water that may have entered the tank during refilling and to prevent the build-up of gum and varnish, clean the inside of the tank once each season by flushing with clean lead-free gasoline or kerosene.

Check the inside and outside of the tank for signs of rust, leakage or corrosion. Replace as required. Do not attempt to patch the tank with automotive fuel tank repair materials. Portable marine fuel tanks are subject to much greater pressure and vacuum conditions than automotive fuel tanks.

Portable Fuel Tank Filter Cleaning

If you feel there is a restriction in the fuel flow, check the fuel tank adapter filter screen for possible contaminants.

1. Remove the bolts and washers securing the adaptor and remove the adaptor.
2. Remove the filter from the end of the adaptor.
3. Inspect the hose and screen for damage, replace if necessary.
4. Clean the screen with solvent to remove any particles. Refer to **Figure 88** or **Figure 89**.
5. Install all parts removed.

Portable Fuel Tank Fuel Gauge Inspection

If the fuel gauge indicator does not indicate the correct amount of fuel within the fuel tank, carefully bend or straighten the arm (**Figure 90**) until the indication is correct.

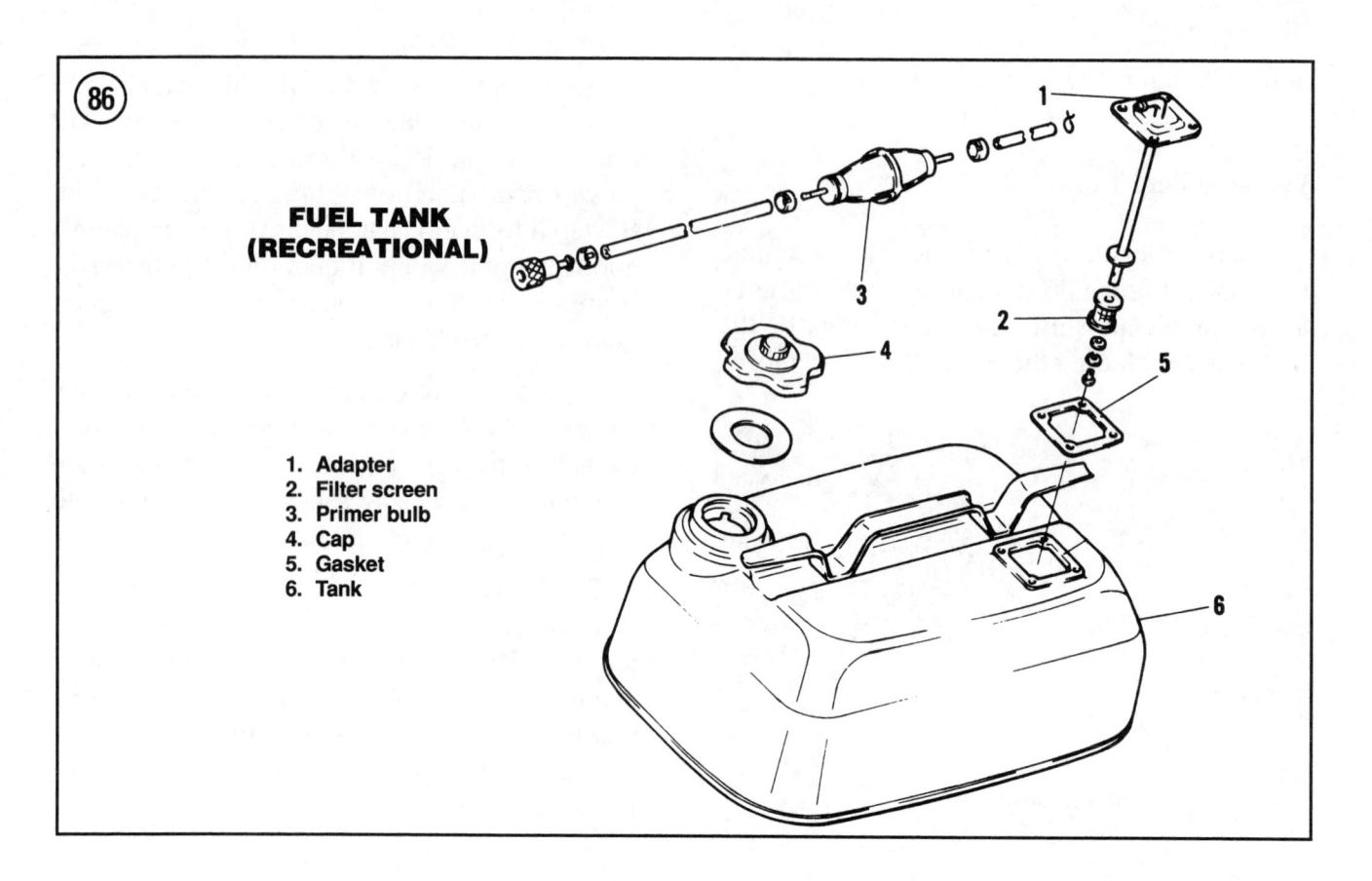

178   CHAPTER SIX

86 FUEL TANK (RECREATIONAL)

1. Adapter
2. Filter screen
3. Primer bulb
4. Cap
5. Gasket
6. Tank

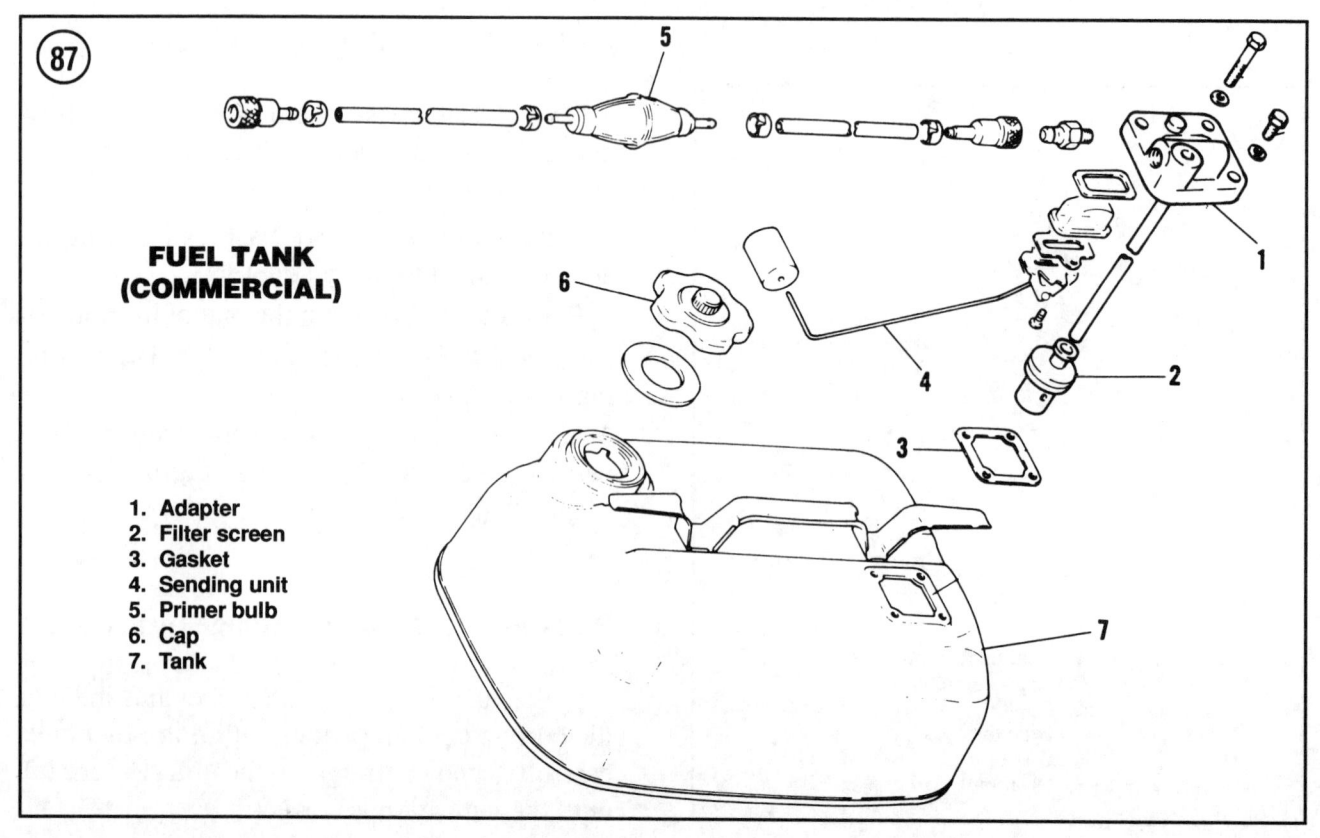

87 FUEL TANK (COMMERCIAL)

1. Adapter
2. Filter screen
3. Gasket
4. Sending unit
5. Primer bulb
6. Cap
7. Tank

FUEL SYSTEM

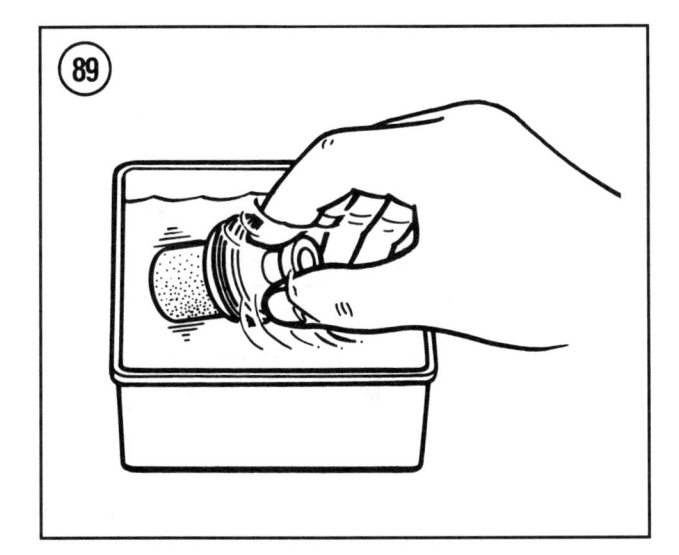

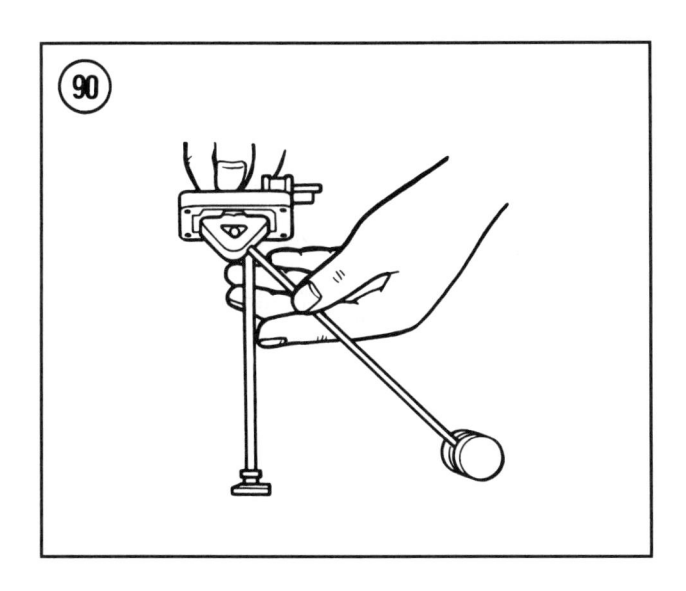

FUEL LINE, CONNECTOR AND PRIMER BULB

When priming the engine, the primer bulb should gradually become firm. If it does not become firm or if it stays firm even when disconnected, the check valve inside the primer bulb is malfunctioning.

WARNING
Leaking gasoline presents a real fire danger that may lead to loss of life and damage or total destruction of the boat. Always replace any fuel lines or connectors immediately after any fuel leakage is observed.

The fuel line and primer bulb should be checked periodically for cracks, breaks, restrictions and chafing. The bulb should be checked periodically for proper operation. Make sure it is installed correctly within the fuel line with the fuel flow arrow pointing in the correct direction (**Figure 91**). Make sure all fuel line connections are tight and securely clamped.

The fuel line connectors have internal O-ring seals (4, **Figure 92**). If these seals become hard or start to deteriorate, the connectors must be replaced as the O-rings cannot be replaced.

FUEL INJECTION SYSTEM (DT 225 V6)

This section includes service procedures for parts of the fuel injection system that should be attempted by the home mechanic. Also, typical of complex electronic systems, most parts cannot be repaired, only replaced. This section describes how the fuel injection system works, how to maintain it and how to replace some of the parts. The majority of the components must be serviced by a Suzuki dealer, either due to any applicable warranty or because they require expensive, complicated electronic troubleshooting equipment and a thorough knowledge of the fuel injection system. Some of these components are

180   CHAPTER SIX

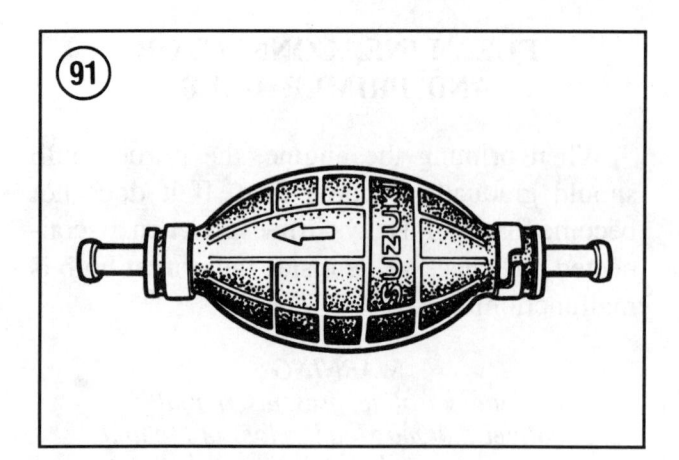

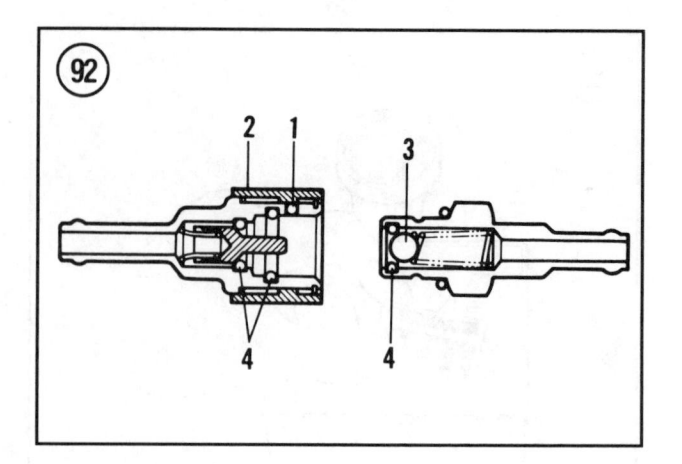

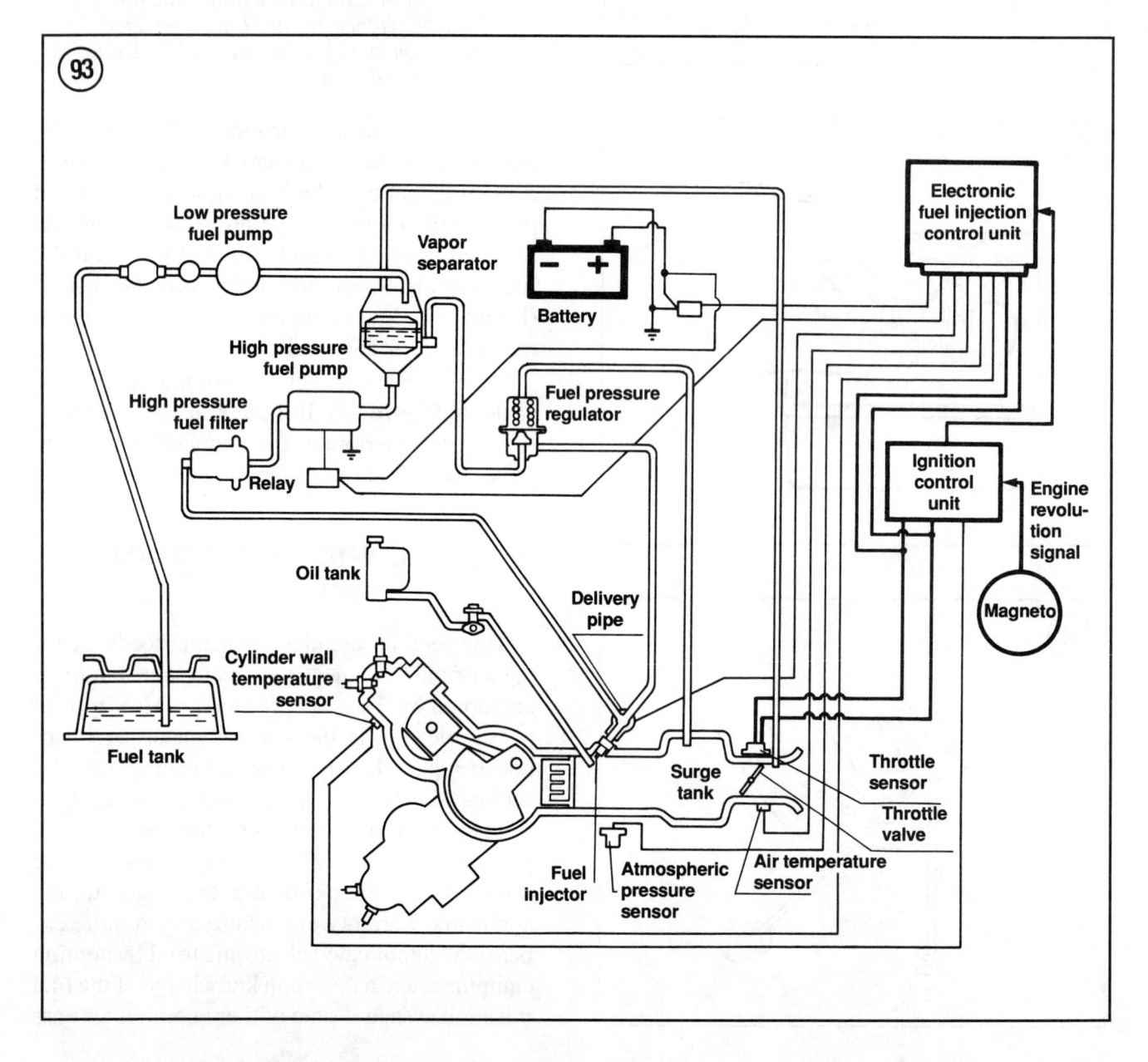

FUEL SYSTEM

very expensive and could be damaged by someone unfamiliar with the test equipment or the components.

> *CAUTION*
> *Servicing the electronic fuel injection requires special precautions to prevent damage to the expensive electronic control units. Common electrical system service procedures acceptable on other fuel systems may cause damage to several parts of the fuel injection system. Be sure to read **Fuel Injection Precautions** in this chapter.*

Fuel Injection System Description

The fuel injection system monitors the engine's current conditions by various sensors within the system. The system receives its primary information from the throttle sensor and the revolution sensor. Secondary information consists of atmospheric pressure, cylinder wall temperature and intake air temperature. These sensors send information on the engine conditions to the micro-computer via electronic signals. With these electronic inputs, the computer determines the optimum amount of fuel needed for the engine's current condition and calculates corresponding injection time duration. The fuel injector receives this duration signal and injects fuel into the intake manifold at the precise amount for maximum power and efficiency.

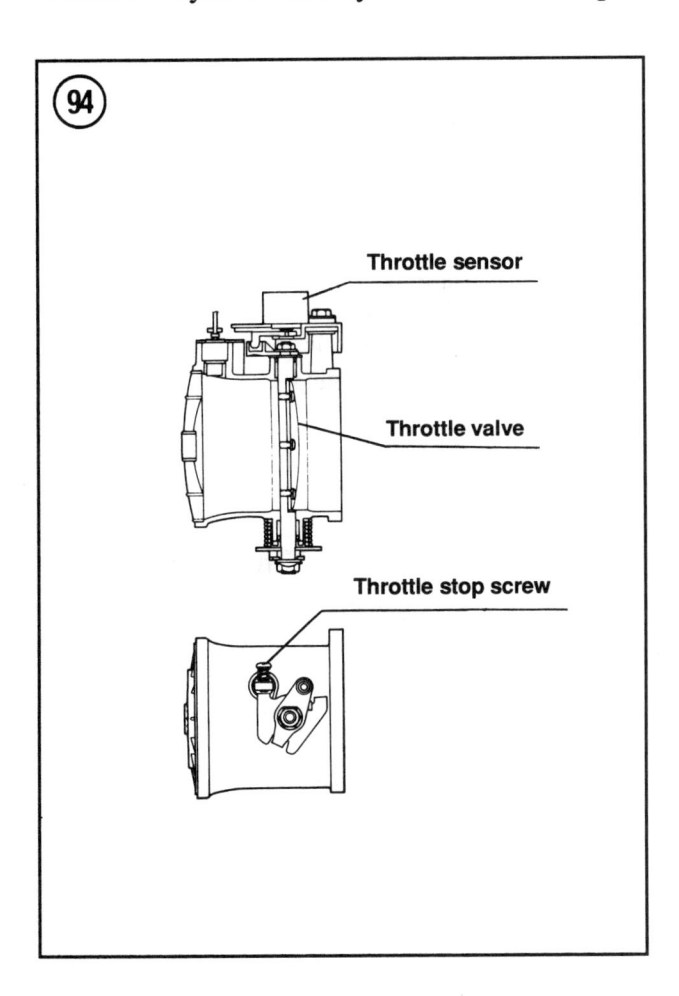

The fuel injection system is split into three different basic systems. They are the Intake System, the Fuel System and the Control System. **Figure 93** is a basic layout of the fuel injection system.

Fuel Injection System Components

This brief description of the fuel injection system will help familiarize you with the system and describe the function of each component. An understanding of the function of the fuel injection system components and their relation to one another is a valuable aid for pinpointing a basic source of fuel injection problems. **Figure 93** is a basic layout of the fuel injection system.

Throttle body

The throttle body (**Figure 94**) consists of the throttle valve, throttle sensor (senses the throttle valve angle), throttle stop screw and other miscellaneous small components.

Vapor separator

Within the vapor separator (**Figure 95**) is a float that maintains a constant fuel level. When the fuel level drops, the fuel sent from the low pressure fuel pump flows into the vapor separa-

182 CHAPTER SIX

tor. The fuel vapor contained in the fuel is separated from the liquid fuel and bled to the throttle body through the pipe provided.

High pressure fuel pump

The fuel pump is an in-line type (connected within the fuel line) and is operated electrically. The pump consists of the pump spacer, rotor and rollers (**Figure 96**). As the rotor rotates, centrifugal force pushes the rollers out and the rollers revolve along with the pump spacer's wall. During rotation, the space normally formed by these parts alternates from larger to smaller (**Figure 97**). This change in shape variation creates fuel suction and creates pressure feed.

There is a built-in pressure check valve to keep residual pressure in the fuel line after the engine is shut off. This maintains fuel pressure for the next start up.

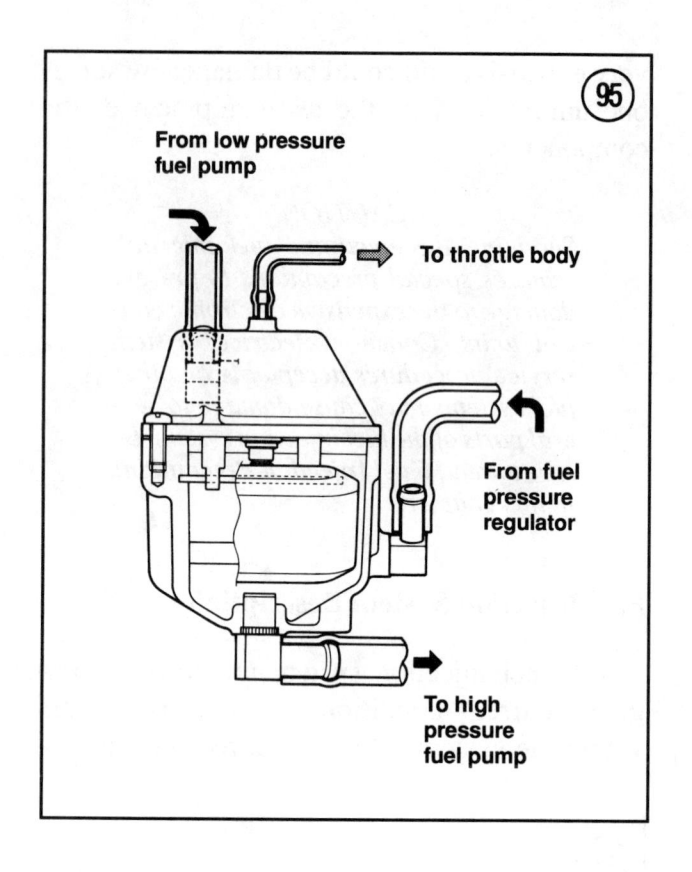

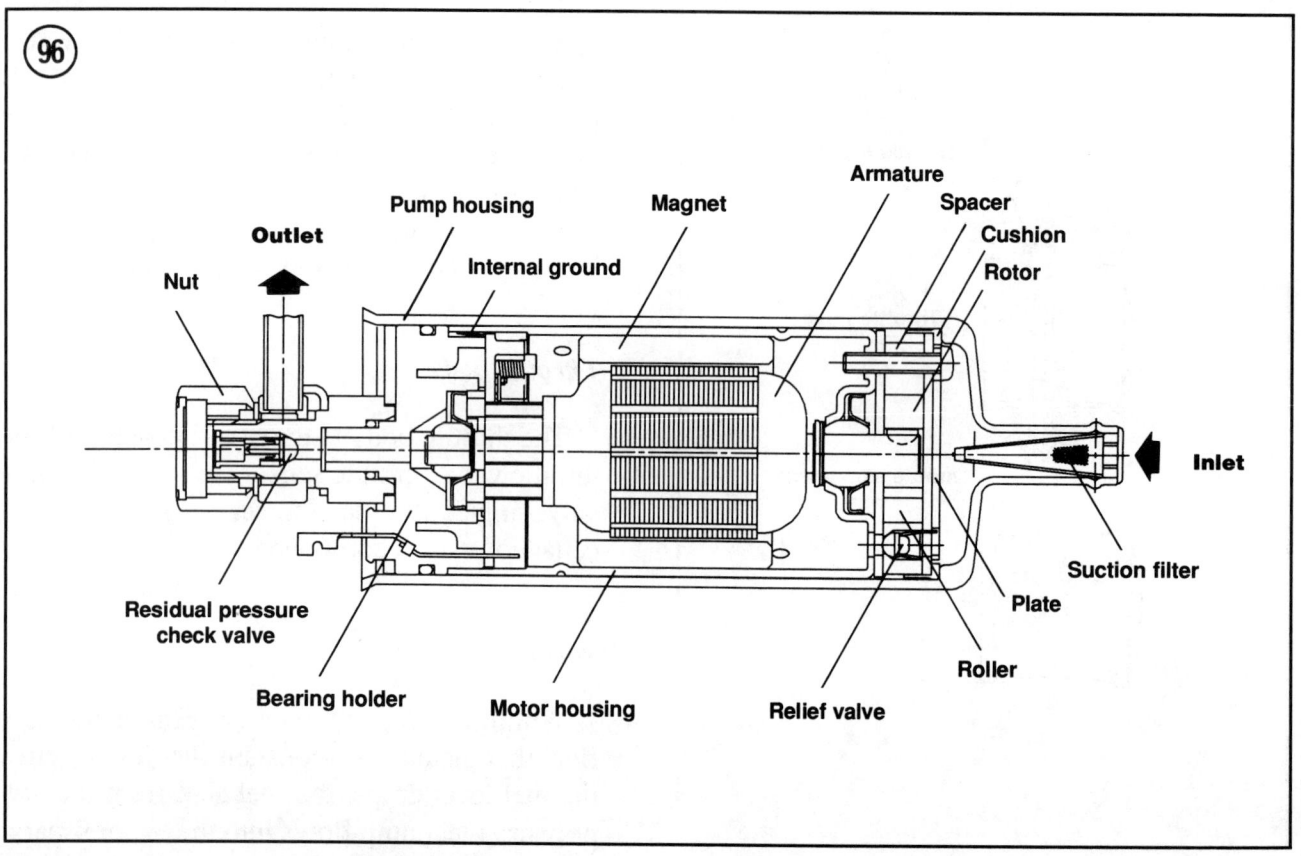

FUEL SYSTEM

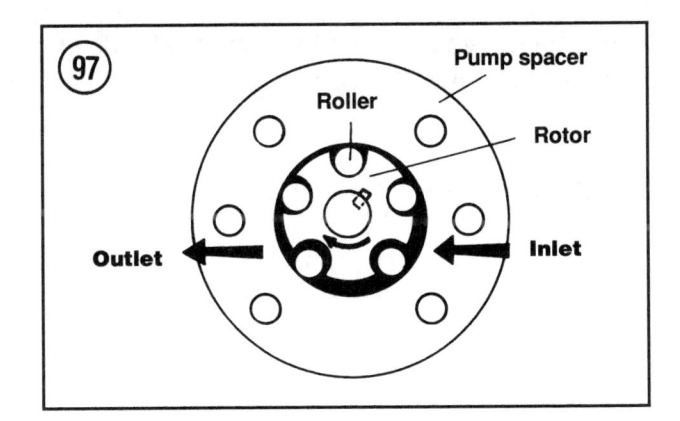

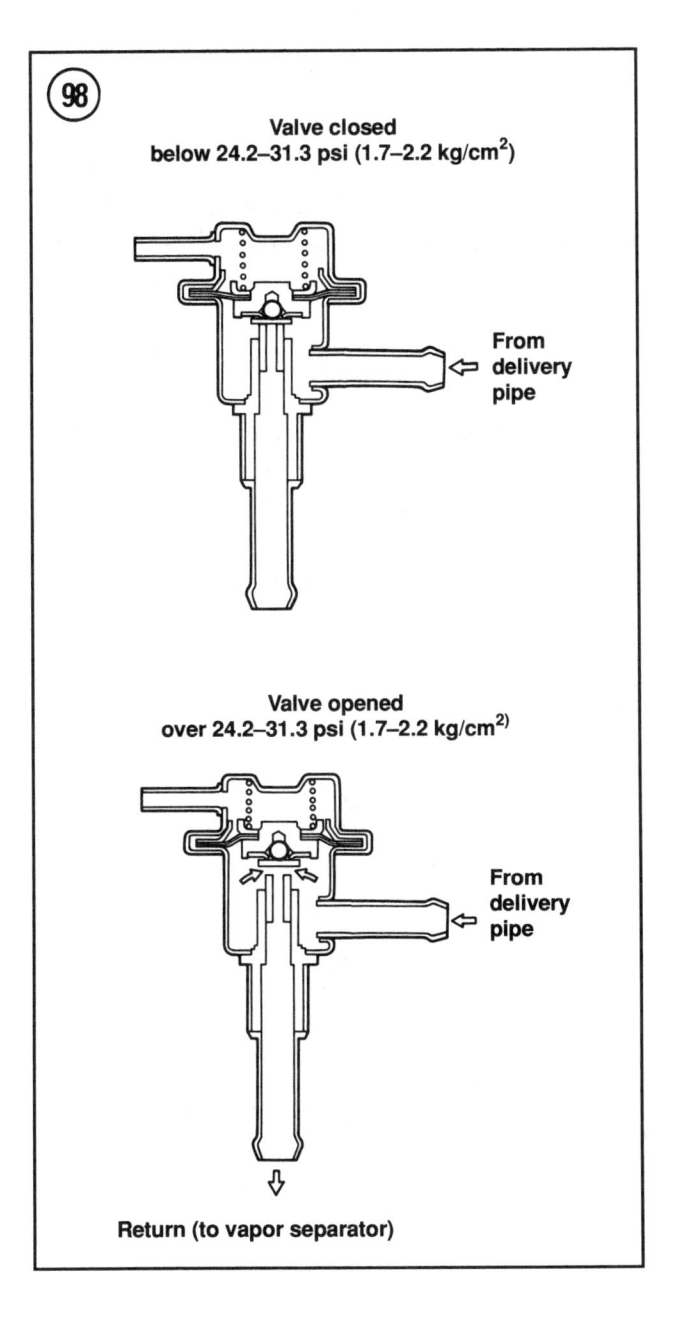

A relief valve is provided to prevent overpressure in the high pressure side of the fuel line.

High pressure fuel filter

The filter is high pressure resistant and filters very fine particles from the fuel using a special element with an average filtering mesh of 25 microns.

Fuel pressure regulator

The pressure regulator (**Figure 98**) maintains a constant fuel pressure relative to manifold pressure. The regulator diaphragm chamber is connected with the surge tank to keep the pressure balanced. The fuel pressure is adjusted by the regulator to be constantly higher than surge tank pressure by 24.2-31.3 psi (1.7-2.2 kg/cm^2).

When the fuel pressure differential between the regulator and the surge tank begins to exceed 24.2-31.3 psi (1.7-2.2 kg/cm^2), the diaphragm is pushed up and allows the fuel to flow through the return pipe to the vapor separator. Upon returning to the vapor separator, the excess fuel pressure is relieved, thus maintaining a constant fuel pressure within the system.

Fuel injectors

The fuel injectors are solenoid-actuated constant-stroke pintle type consisting of a solenoid, plunger, needle valve and the housing (**Figure 99**). When electrical current is applied to the solenoid coil the valve is lifted and the fuel is injected into the intake manifold.

The fuel injector's opening is fixed and fuel pressure is constant at all times. The amount of fuel injected is controlled by the length of time the injector is open which is controlled by the fuel injection control unit.

Air temperature sensor

The air temperature sensor is mounted inside the throttle body where it measures the air temperature flowing through the body opening. The sensor provides the temperature information as a resistance signal—as the temperature lowers, the thermistor's resistance rises and vise-versa. With the different air temperature the intake air mass also changes thus the amount of fuel that is to be injected must also vary. This information is constantly being monitored by the fuel injection control unit which in turn compensates the fuel injection command signal to the fuel injectors.

Cylinder wall temperature sensor

The cylinder wall temperature sensor is mounted on top of the cylinder where it measures the temperature of the cylinder wall. The sensor provides the temperature information as a resistance signal—as the temperature lowers, the thermistor's resistance rises and vise-versa. This information is constantly being monitored by the fuel injection control unit which in turn compensates the fuel injection command signal to the fuel injectors.

Throttle valve sensor

The throttle valve sensor is mounted on the upper portion of the throttle body (**Figure 94**). The sensor provides throttle valve opening angle information for both the ignition control unit and the electronic fuel injection unit. The ignition control and fuel injection units use the throttle position information to help determine optimum timing advance and fuel injector opening duration.

Atmospheric pressure sensor

The atmospheric pressure sensor is mounted on the upper portion of the crankcase. The sensor detects atmospheric pressure and sends that signal to the electronic fuel injection unit in the form of a voltage signal. This information helps determine the fuel injectors opening duration.

Fuel Injection Control Unit Precautions

CAUTION
Servicing the electronic fuel injection system requires special precautions to prevent damage to the expensive fuel injection control unit. Common electrical system service procedures acceptable on other fuel systems may cause damage to several parts of the fuel injection system.

1. Unless otherwise specified in a procedure, do not start the motor while any electrical connectors are disconnected. Do not disconnect the battery cables or any electrical connector while the ignition switch is ON. The fuel injection control unit will be damaged and will require replacement.
2. Before disconnecting any electrical connectors, turn the ignition switch OFF.
3. When repairs are completed, do not try to start the engine without double checking to make sure all fuel injection electrical connectors are con-

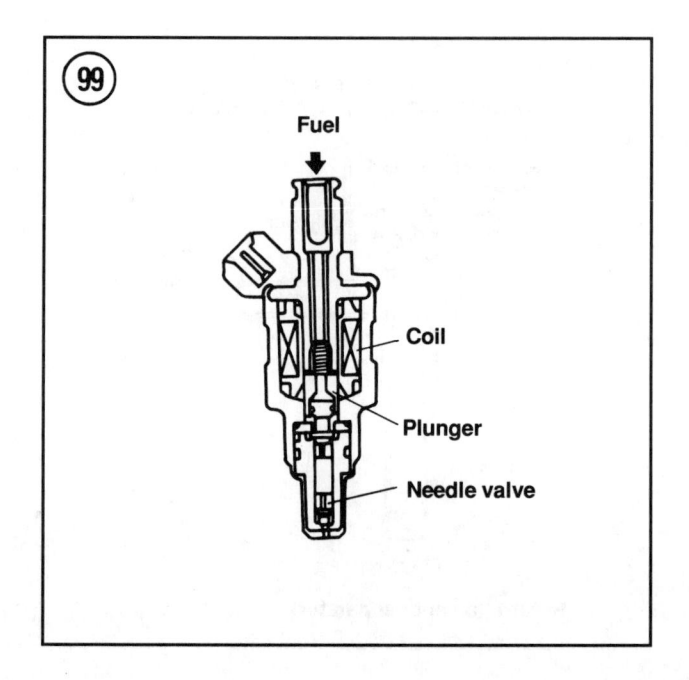

FUEL SYSTEM

nected; faulty connectors may cause damage to the control unit and its related components.

4. Do not disconnect the battery while the engine is running.

5. Do not apply anything other than a 12 volt battery to the electrical system. The battery must be removed before attaching a battery charger.

6. Use only the specified Suzuki test equipment suggested in the test procedures to prevent damaging the sensitive circuits in the EFI system.

Fuel System Precautions

1. The fuel system is pressurized so wear eye protection whenever working on the fuel system, especially when disconnecting fuel lines.

2. Do not add any lubricants, preservatives or additives to the gasoline as fuel system corrosion or clogging may result.

Depressurizing The Fuel System

The fuel system is pressurized to 24.2-31.3 psi (1.7-2.2 kg/cm^2) while the engine is running. This pressure is maintained for some time after the engine is shut off and *must be depressurized* prior to working on, or disconnecting, any component of the fuel system.

1. Disconnect the fuel supply hose from the fuel tank.

2. Start the engine and allow it to idle until it stops running. When the engine stops running the fuel is exhausted and the pressure is relieved.

3. After working on the system, reconnect the fuel supply hose to the fuel tank.

Vapor Separator Removal/Installation

1. Depressurize the fuel system as described in this chapter.

2. Remove the engine cover.

3. Remove the bolts securing the oil tank. Disconnect the oil level sensor leads and move the oil tank to one side. It is not necessary to remove the oil tank completely, just move it out of the way.

4. Release the hose clamp and disconnect the fuel hose that runs from the vapor separator to the fuel pressure regulator.

5. Release the hose clamp and disconnect the fuel hose that runs from the vapor separator to the low pressure fuel pump.

6. Remove the bolts securing the vapor separator to the engine and move it away from the engine.

7. Release the hose clamp and disconnect the fuel hose that runs from the vapor separator to the throttle body.

8. Release the hose clamp and disconnect the fuel hose that runs from the vapor separator to the high pressure fuel pump.

9. Remove the vapor separator.

10. Installation is the reverse of removal, noting the following.

11. Tighten the bolts securing the vapor separator to 3-5 ft.-lb. (4-7 N•m).

12. Make sure all fuel hoses and hose clamps are in good condition, replace if necessary.

13. Push all fuel hoses onto their respective fitting until they stop and install the hose clamp securely.

14. Make sure there are no kinks or sharp bends in any of the fuel hoses.

15. Start the engine and check for fuel leaks. Correct any problem immediately.

High Pressure Fuel Filter Removal/Installation

1. Depressurize the fuel system as described in this chapter.

2. Remove the engine cover.

3. Release the hose clamp and disconnect the fuel hose that runs from the high pressure fuel filter to the high pressure fuel pump.

4. Release the hose clamp and disconnect the fuel hose that runs from the high pressure fuel filter to the delivery pipe.

5. Remove the bolts securing the vapor separator to the engine and remove it.

6. Installation is the reverse of removal, noting the following.

7. Tighten the bolts securely.

8. Make sure all fuel hoses and hose clamps are in good condition, replace if necessary.

9. Push all fuel hoses onto their respective fitting until they stop and install the hose clamp securely.

10. Make sure there are no kinks or sharp bends in any of the fuel hoses.

11. Start the engine and check for fuel leaks. Correct any problem immediately.

Table 1 CARBURETOR SPECIFICATIONS

DT 2	
Type	
1985-1989	VM-11-10
1990-on	VM-11-18
Main jet	
1985-1989	No. 95
1990-on	No. 90
Main air jet	
1985-1989	—
1990-on	2.1 mm
Jet needle	3E6-3
Needle jet	
1985-1989	2.0 mm
1990-on	C-0
Float level	0.75-0.83 in.
DT 4	
Type	BV18-15
Main jet	No. 97.5
Pilot idle jet	No. 45
Pilot air jet	No. 1.5
Main air jet	1.6 mm
Float level	0.47-0.55 in.
DT 6	
Type	BV18-14
Main jet (1985-1986)	
Short shaft	No. 115
Long shaft	No. 120
Main jet (1987-on)	
S type	No. 100
L and UL type	No. 115
Main air jet	
1985-1986	1.8 mm
1987-on	
S type	1.3 mm
L and UL type	1.6 mm
(continued)	

Table 1 CARBURETOR SPECIFICATIONS (continued)

DT 6 (continued)	
Pilot (idle) jet	
1985-1987	No. 20
1989-on	No. 40
Pilot (idle) outlet	2.0 mm
Float level	0.51-0.59 in.
DT 8	
Type	
1985-1987	BV24-18
1988-on	B24-18
Main jet	
1985-1986	
Short shaft	No. 95
Long shaft	No. 100
1987	
S type	102.5 mm
L and UL type	105 mm
1988-on	137.5 mm
Pilot (idle) jet	
1985-1987	No. 50
1988-on	No. 65
Pilot (idle) outlet	
1985-1987	0.7 mm
1988-on	1.2 mm
Main air jet	
1985-1987	1.2 mm
1988-on	1.8 mm
Float level	
1985-1987	0.87-0.94 in.
1988-on	0.9-1.0 in.
DT 9.9	
Type	
1985-1987	BV24-15
1988-on	B24-18
Main jet	
1985-1987	No. 110
1988-on	No. 137.5
Pilot (idle) jet	
1985-1987	No. 52.5
1988-on	No. 65
Air jet	
1985-1987	1.4 mm
1988-on	1.2 mm
Float level	0.91-0.98 in.
DT 15	
Type	BV24-18
Main jet	No. 122.5
Main air jet	1.2 mm
Pilot (idle) jet	
1985-1988	No. 57.5
1989-on	No. 65
Pilot (idle) air jet	1.2 mm
Float level	0.91-0.98 in.

(continued)

Table 1 CARBURETOR SPECIFICATIONS (continued)

DT 20	
Type	BV28-22
Main jet	No. 20
Pilot (idle) jet	No. 67.5
Air jet	1.5, 2.0 mm
Float level	0.45-0.53 in.
DT 25 (2-cylinder)	
Type	BV32-24
Main jet	No. 142.5
Pilot (idle) jet	No. 90
Pilot (idle) air jet	1.5 mm
Main air jet	1.4 mm
Float level	
1985-1986	0.39-0.47 in.
1987-1988	0.43-0.47 in.
DT 25 (3-cylinder)	
Type	B26-20
Main jet	
1989	No. 122.5
1990	No. 125
1991	No. 120
Pilot (idle) jet	
1989	No. 67.5
1990	No. 72.5
1991	No. 70.5
Pilot (idle) air jet	1.2 mm
Main air jet	1.6 mm
Float level	0.47-0.55 in.
DT 30	
Type	B26-20
Main jet	
1988-1990	No. 125
1991	No. 120
Pilot (idle) jet	
1988	No. 70
1989	No. 67.5
1990	No. 70
1991	No. 62.5
Pilot (idle) air jet	1.2 mm
Main air jet	1.6 mm
Float level	0.47-0.55 in.
DT 35	
Type	B40-32
Main jet	No. 190
Pilot (idle) jet	No. 77.5
Pilot (idle) air jet	2.0 mm
Main air jet	1.3 mm
Float level	0.69-0.75 in.
DT 40	
Type	B40-32
Main jet	No. 180
Pilot (idle) jet	No. 70
Pilot (idle) air jet	2.0 mm
Main air jet	1.3 mm

(continued)

FUEL SYSTEM

Table 1 CARBURETOR SPECIFICATIONS (continued)

DT 40 (continued)	
Float level	
1985-1986	0.65-0.73 in.
1987-on	0.69-0.75 in.
DT 55	
Type	B32-24
Main jet	
1985-1988	No. 117.5
1989-on	No. 120
Pilot (idle) jet	
1985	No. 70
1986-1987	No. 75
1988-on	No. 77.5
Pilot (idle) air jet	
1985	2.0 mm
1986-1987	1.2 mm
1988-on	2.0 mm
Main air jet	1.4 mm
Float level	
1985	0.43-0.73 in.
1986	0.67-0.75 in.
1987-on	0.39-0.47 in.
DT 65	
Type	B40-32
Main jet	
1985-1987	No. 147.5
1988-1989	No. 155
1990-on	No. 160*
Pilot (idle) jet	
1985	No. 75
1986-1988	No. 77.5
1989-on	No. 80
Pilot (idle) air jet	2.0 mm
Main air jet	
1986-1987	1.4 mm
1989-on	1.5 mm
Float level	
1985-1988	0.39-0.47 in.
1989-on	0.67-0.75 in.
DT 75	
Type	B32-28
Main jet	
1985-1986	No. 140
1987	No. 135
1988-on	No. 140
Pilot (idle) jet	
1985-1986	No. 80
1987-on	No. 87.5
Pilot (idle) air jet	2.0 mm
Main air jet	
1985-1986	1.2 mm
1987-on	1.3 mm
Float level	
1985-1986	0.49-0.57 in.
1987-on	0.39-0.47 in.

(continued)

Table 1 CARBURETOR SPECIFICATIONS (continued)

DT 85	
Type	B40-32
Main jet	
1985-1986	No. 162.5
1987-on	No. 157.5
Pilot (idle) jet	
1985-1986	No. 75
1987-on	No. 80
Pilot (idle) air jet	1.5 mm
Main air jet	1.2 mm
1985-1986	1.2 mm
1987-on	1.3 mm
Float level	
1985-1986	0.69-0.77 in.
1987-on	0.67-0.75 in.
DT 115	
Type	B32-28
Main jet	No. 135
Pilot (idle) jet	
1985	No. 75
1986-1988	No. 82.5
1989	No. 65
1990-on	No. 70
Pilot (idle) air jet	
1985	1.4 mm
1986-on	2.0 mm
Main air jet	
1985-1986	1.0 mm
1983-on	1.2 mm
Float level	
1985	0.39-0.42 in.
1986-on	0.39-0.47 in.
DT 140	
Type	B40-32
Main jet	
1985	No. 160
1986-1988	
Top carburetor	No. 157.5
3rd carburetor	No. 162.5
2nd and 4th carburetor	No. 155
1989-on	No. 150
Pilot (idle) jet	
1985-1988	No. 80
1989-on	No. 85
Pilot (idle) air jet	
1985	1.3 mm
1986-on	2.0 mm
Main air jet	1.2 mm
Float level	
1985	0.67-0.75 in.
1986-on	0.39-0.47 in.
DT 90	
Type	BW36-24 × 2
Main jet	No. 132.5
Pilot (idle) jet	No. 90
Pilot (idle) air jet	1.1 mm

(continued)

FUEL SYSTEM

Table 1 CARBURETOR SPECIFICATIONS (continued)

DT 90 (continued)	
Main air jet	1.5 mm
Float level	0.37-0.45 in.
Fuel level	0.83-0.91 in.
DT 100, DT 100 S	
Type	BW40-32 × 2
Main jet	No. 165
Pilot (idle) jet	No. 77.5
Pilot (idle) air jet	1.0 mm
Main air jet	2.0 mm
Float level	0.37-0.45 in.
Fuel level	0.83-0.91 in.
DT 150, DT 150 SS (1986)	
Type	BW40-28
Main jet	
DT 150	No. 157.5
DT 150 SS	No. 147.5
Pilot (idle) jet	No. 77.5
Pilot (idle) air jet	1.2 mm
Main air jet	2.0 mm
Float level	0.37-0.45 in.
Fuel level	0.83-0.91 in.
DT 150, DT 150 SS (1987-on)	
Type	BW36-24
Main jet	No. 132.5
Pilot (idle) jet	No. 77.5
Pilot (idle) air jet	0.8 mm
Main air jet	1.5 mm
Float level	0.37-0.45 in.
Fuel level	0.83-0.91 in.
DT 175 (1987-on)	
Type	BW40-28
Main jet	No. 142.5
Pilot (idle) jet	No. 90
Pilot (idle) air jet	1.0 mm
Main air jet	1.2 mm
Float level	0.37-0.45 in.
Fuel level	0.83-0.91 in.
DT 200 (1986)	
Type	BW40-32
Main jet	No. 167.5
Pilot (idle) jet	No. 82.5
Pilot (idle) air jet	1.1 mm
Main air jet	2.0 mm
Float level	0.37-0.45 in.
Fuel level	0.83-0.91 in.
DT 200, DT 200 AE (1987-on)	
Type	BW40-32
Main jet	No. 162.5
Pilot (idle) jet	No. 92.5
Pilot (idle) air jet	1.0 mm
Main air jet	1.3 mm
Float level	0.37-0.45 in.
Fuel level	0.83-0.91 in.

* Main jet changed from No. 155 to No. 160 from frame No. 909777-on.

Table 2 IDLE AIR SCREW ADJUSTMENT

Model	Turns out from lightly seated position
DT 2	
1985-1989	NA
1990	1 1/4-1 3/4
DT 4	1-1 1/2
DT 6	
1985-1986	1 3/4-2 1/4
1987-on	
S-type	7/8-1 3/8
L and UL type	1-1 1/2
DT 8	
1985-1986	3/4- 1 1/4
1987	1/2-1
1988-on	1 3/4-2 1/4
DT 9.9	
1985-1987	1 1/4-1 3/4
1988-on	1 3/4-2 1/4
DT 15	
1985-1987	1 1/4-1 3/4
1988-on	1 1/2-2
DT 20	1 3/4-2 1/4
DT 25 (2-cylinder)	
1985	3/4-1 1/4
1986-1988	1 1/4-1 3/4
DT 30 (2-cylinder)	
1985	1 1/4-1 3/4
1986-1987	1 1/2-2
DT 25, DT 30 (3-cylinder)	
1988 (DT 30)	1 1/2-2
1989 (DT 25 and DT 30)	
MC	1 1/4-1 3/4
LE	1 1/2-2
1990	1 1/2-2
1991	1-1 1/2
DT 35	1 1/2-2
DT 40	
1985-1986	1 1/8-1 5/8
1987-on	1 1/2-2
DT 55	
1985	1 1/8-1 5/8
1986-1989	1 1/4-1 3/4
1990-on	1-1 1/2
DT 65	
1985	1 1/2-2
1986-1989	1 1/4-1 3/4
1990-on	1-1 1/2
DT 75	
1985-1986	1-1 1/2
1987	1 3/4-2 1/4
1988-on	1 3/4-2 1/4
DT 75, DT 85	
1985-1986	1-1 1/2
1987	1 3/4-2 1/4
1988-on	1 5/8-2 1/8

(continued)

Table 2 IDLE AIR SCREW ADJUSTMENT (continued)

Model	Turns out from lightly seated position
DT 90	1 1/8-1 5/8
DT 100	1 1/8-1 5/8
DT 115	
1985	1 1/2
1986-1988	1 1/4-1 3/4
1989	7/8
1990-on	1-1 1/2
DT 140	
1985	1 3/8
1986-1988	1-1 1/2
1989-on	1 1/8-1 5/8
DT 150	
1986	1 1/4-1 3/4
1987-1988	1-1 1/2
1989-on	1 1/2-2
DT 150 SS	1 1/4-1 3/4
DT 175	
1987-1988	1 1/2-2
1989-on	1 1/4-1 3/4
DT 200	
1986	1 1/2-2
1987-on	1 1/4-1 3/4
DT 200 AE	1 1/4-1 3/4

Table 3 REED STOP OPENING

Model	Opening (in.)
DT 2	0.16
DT 4	0.19-0.20
DT 6	0.16-0.18
DT 8	
1985-1987	0.16-0.18
1988-on	0.31
DT 9.9	
1985-1987	0.09-0.10
1988-on	0.31
DT 15	0.22-0.23
DT 20	0.24-0.25
DT 25	
1985-1988	0.24-0.25
1989-on	0.15
DT 30	
1985-1987	0.24-0.25
1988-on	0.15
DT 35	0.24-0.25
DT 40	
1985-1986	0.30-0.31
1988-on	0.24-0.25
DT 55, DT 65	
1985	0.30-0.31
1986-on	0.37-0.38
DT 75, DT 85	0.30-0.31

(continued)

Table 3 REED STOP OPENING (continued)

Model	Opening (in.)
DT 90, DT 100	0.413
DT 115, DT 140	0.30-0.31
DT 150	
1986	0.29
1987-1989	0.24
1990-on	0.29
DT 150SS	
1986	0.24
1987-1989	0.29
1990-on	0.24
DT 175, DT 200	0.29
DT 225 EFI	0.26

Chapter Seven

Electrical Systems

This chapter provides service procedures for the battery, starter motor (if so equipped) and each ignition system used on Suzuki outboard motors. Wiring diagrams are included at the end of the book.

Table 1 and **Table 2** are at the end of the chapter.

ELECTRICAL CONNECTORS

Some models are equipped with special electrical connectors that require care when disconnecting them to avoid damage to the connector and to the electrical wires.

On 2 or 3 pin electrical connectors, *push* the tab and then pull the connector apart as shown in **Figure 1**.

On 4 or 6 pin electrical connectors, *lift* the tab and then pull the connector apart as shown in **Figure 2**.

When connecting, correctly align the connectors, then push them together until you hear a "click." After the click, make sure the connector is fully engaged and is locked together.

BATTERY

Since batteries used in marine applications endure far more rigorous treatment than those used in an automotive electrical system, they are constructed differently. Marine batteries have a thicker exterior case to cushion the plates inside during tight turns and rough water. Thicker plates are also used, with each one individually fastened within the case to prevent premature failure. Spill-proof caps on the battery cells prevent electrolyte from spilling into the bilges.

Automotive batteries are not designed to be run down and recharged repeatedly. For this reason, they should *only* be used in an emergency

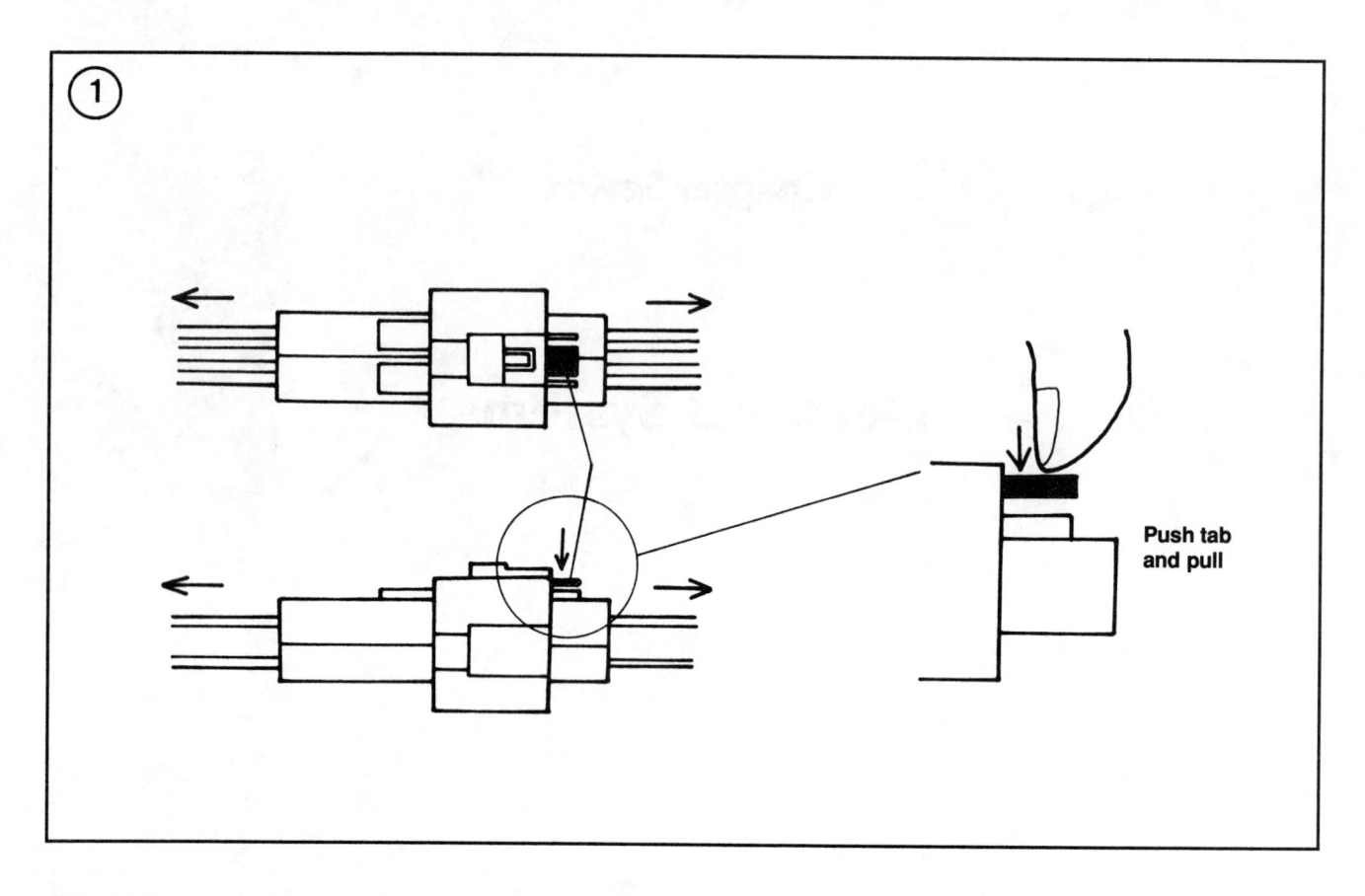

CHAPTER SEVEN

①

Push tab
and pull

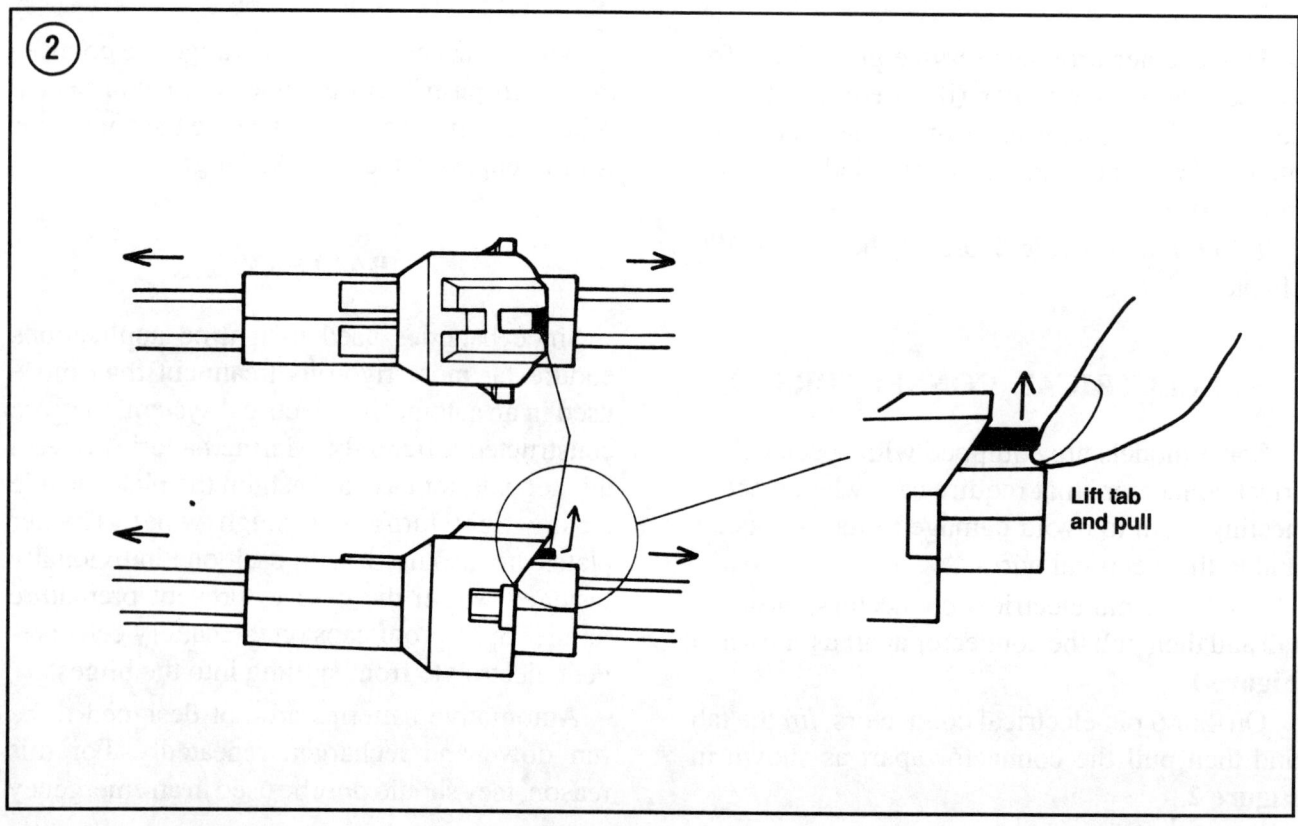

②

Lift tab
and pull

ELECTRICAL SYSTEMS

situation when a suitable marine battery is not available.

Suzuki recommends that any battery used to crank DT 9.9 through DT 40 motors have a *minimum* rating of 35 amp hours. DT 55-DT 225 motors require a battery with a *minimum* rating of 70 amp hours.

CAUTION
*Sealed or maintenance-free batteries are **not** recommended for use with the unregulated charging systems used on some Suzuki outboards. Excessive charging during continued high-speed operation will cause the electrolyte to boil, resulting in its loss. Since water cannot be added to such batteries, such overcharging will ruin the battery.*

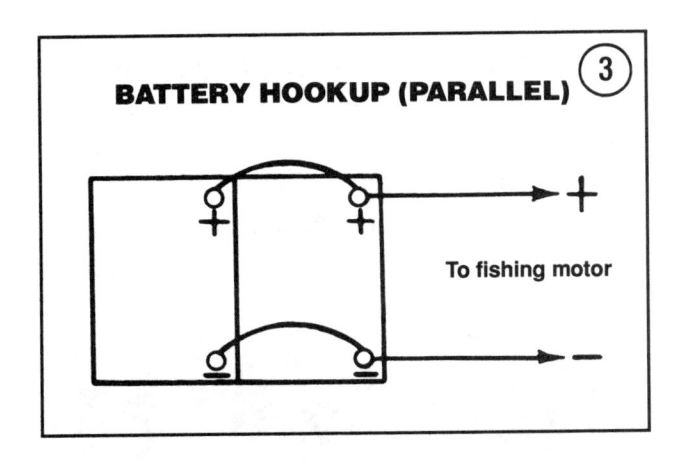

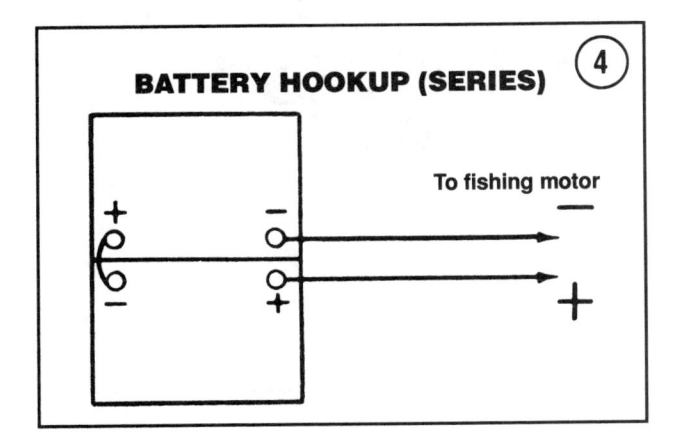

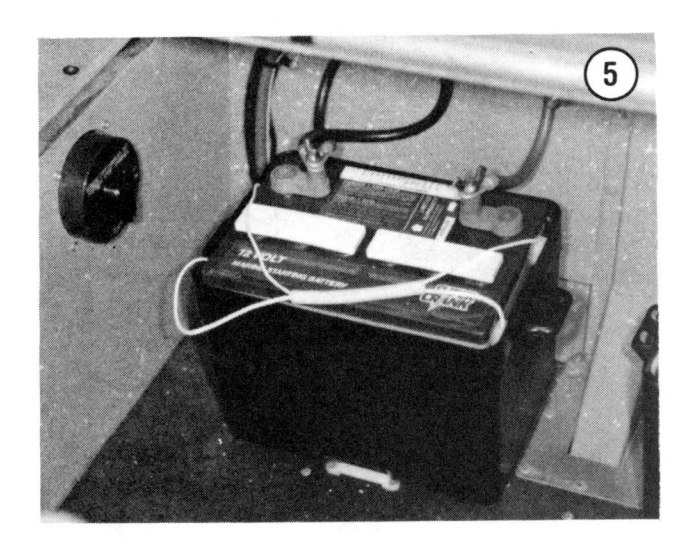

Separate batteries may be used to provide power for any accessories such as lighting, fish finders, depth finder, etc. To determine the required capacity of such batteries, calculate the average current draw rate of the accessories and refer to **Table 1**. Batteries may be wired in parallel to double the ampere hour capacity while maintaining a 12-volt system (**Figure 3**). For accessories which require 24 volts, batteries may be wired in series (**Figure 4**) but only accessories specifically requiring 24 volts should be connected to the system. Whether wired in parallel or in series, disconnect and charge the batteries individually.

Battery Installation in Aluminum Boats

If a battery is not properly secured and grounded when installed in an aluminum boat, it may contact the hull and short to ground. This will burn out remote control cables, tiller handle cables or wiring harnesses.

The following preventive steps should be taken when installing a battery in a metal boat.

1. Choose a location as far as practical from the fuel tank while providing access for maintenance.

2. Install the battery in a plastic battery box with cover and tie-down strap (**Figure 5**).

3. If a covered container is not used, cover the positive battery terminal with a non-conductive shield or boot (**Figure 6**).

4. Make sure the battery is secured inside the battery box and that the box is fastened in position with the tie-down strap.

Care and Inspection

1. Remove the battery container cover (**Figure 5**) or hold-down (**Figure 6**).
2. Disconnect the negative (–) battery cable. Disconnect the positive (+) battery cable.

NOTE
*Some batteries have a built-in carry strap (**Figure 7**) for use in Step 3.*

3. Attach a battery carry strap to the terminal posts. Remove the battery from the battery tray or container.
4. Check the entire battery case for cracks or electrolyte leakage.
5. Inspect the battery tray or container for corrosion and clean if necessary with a solution of baking soda and water.

NOTE
Keep cleaning solution out of the battery cells in Step 6 or the electrolyte will be seriously weakened.

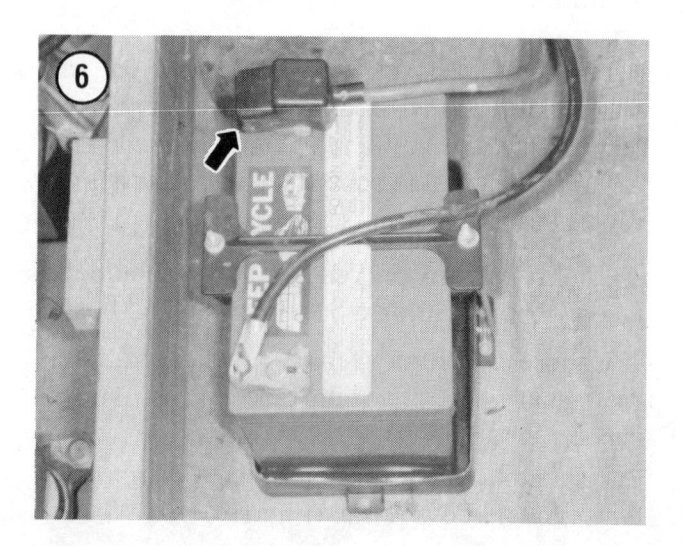

6. Clean the top of the battery with a stiff bristle brush using the baking soda and water solution (**Figure 8**). Rinse the battery case with clear water and wipe dry with a clean cloth or paper towel.
7. Position the battery in the battery tray or container.
8. Clean the battery cable clamps with a stiff wire brush or one of the many tools made for this purpose (**Figure 9**). The same tool is used for cleaning the battery posts (**Figure 10**).
9. Reconnect the positive battery cable, then the negative cable.

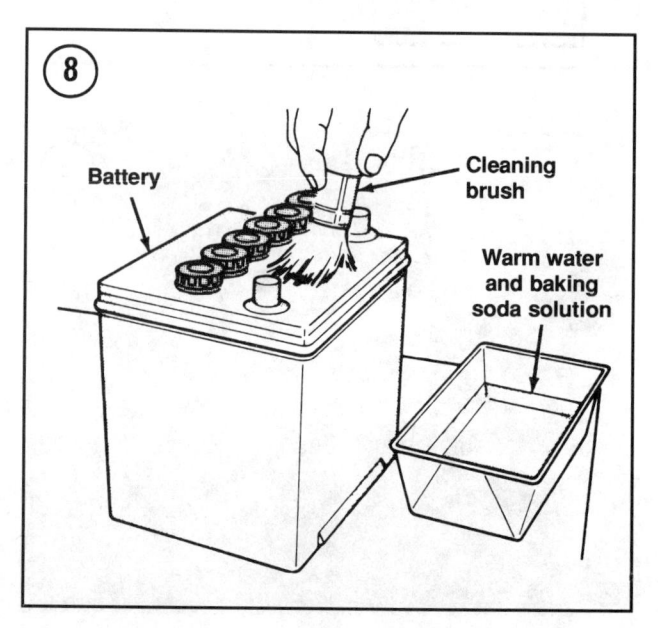

ELECTRICAL SYSTEMS

CAUTION
Be sure the battery cables are connected to their proper terminals. Connecting the battery backwards will reverse the polarity and damage the rectifier and some components of the electronic ignition system.

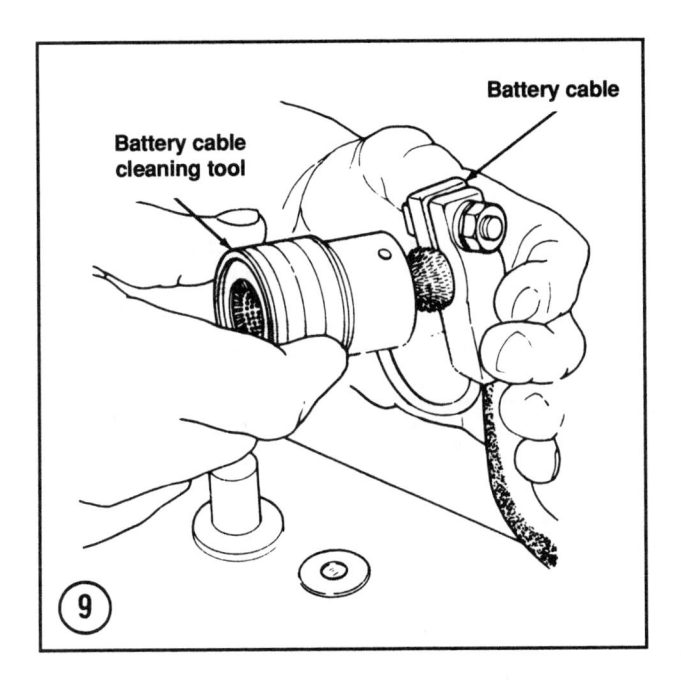

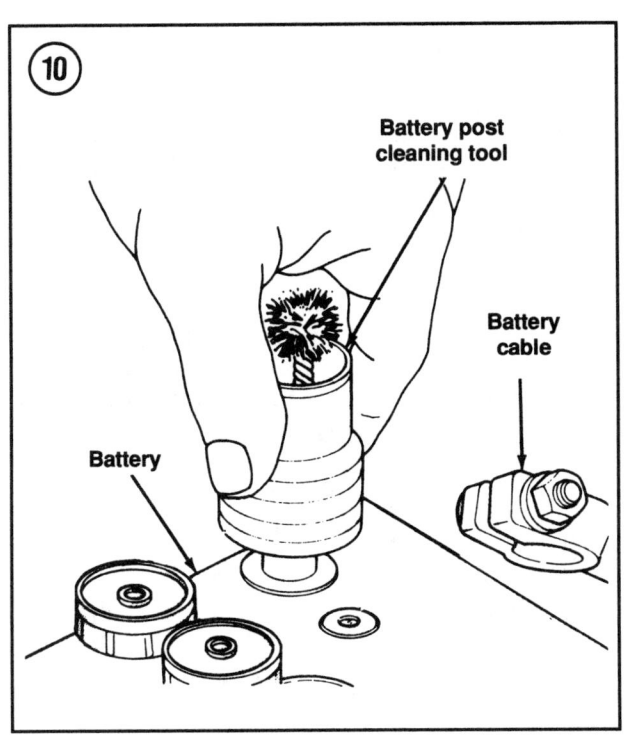

10. Tighten the battery connections and coat with a petroleum jelly such as Vaseline or a light mineral grease.

NOTE
Do not overfill the battery cells in Step 11. The electrolyte expands due to heat from charging and will overflow if the level is more than 3/16 in. above the battery plates.

11. Remove the filler caps and check the electrolyte level. Add distilled water, if necessary, to bring the level up to 3/16 in. above the plates in the battery case (**Figure 11**).

NOTE
If distilled water has been added, reinstall the filler caps, unsecure the battery and gently shake the battery for several minutes to mix the existing electrolyte with the new water. Properly secure the battery.

Testing

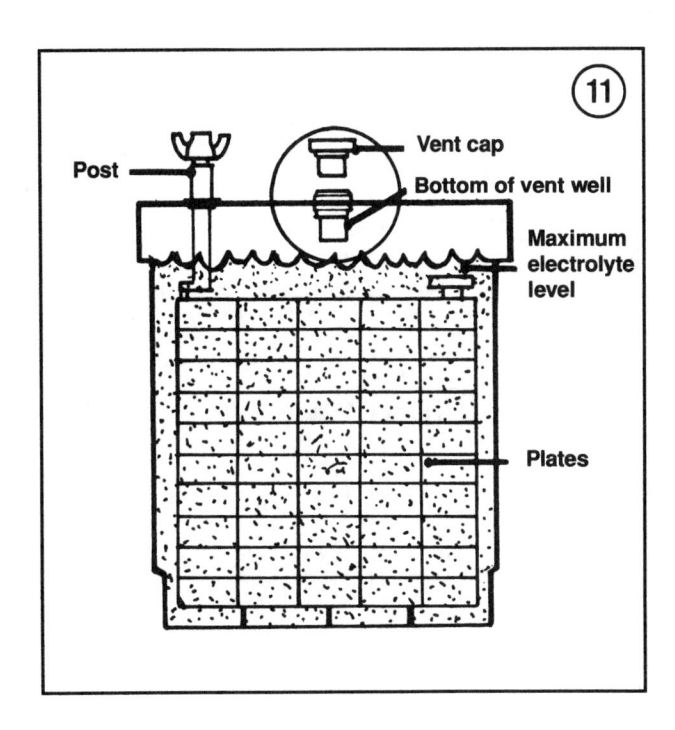

Hydrometer testing is the best way to check battery state of charge. Use a hydrometer with numbered graduations from 1.100-1.300 rather than one with just color-coded bands. To use the hydrometer, squeeze the rubber ball, insert the tip in a cell and release the ball (**Figure 12**).

NOTE
Do not attempt to test a battery with a hydrometer immediately after adding water to the cells. Charge the battery for 15-20 minutes at a rate high enough to cause vigorous gassing and allow the water and electrolyte to mix thoroughly.

Draw enough electrolyte to suspend the weighted float inside the hydrometer. When using a temperature-compensated hydrometer, release the electrolyte and repeat this process several times to make sure the thermometer has adjusted to the electrolyte temperature before taking the reading.

Hold the hydrometer vertically and note the number in line with the surface of the electrolyte (**Figure 13**). This is the specific gravity for that cell. Return the electrolyte to the cell from which it came.

The specific gravity of the electrolyte in each battery cell is an excellent indicator of that cell's condition. A fully charged cell will read 1.260 or more at 68° F (20° C). A cell that is 75 percent charged will read from 1.220-1.230 while one with a 50 percent charge reads from 1.170-1.180.

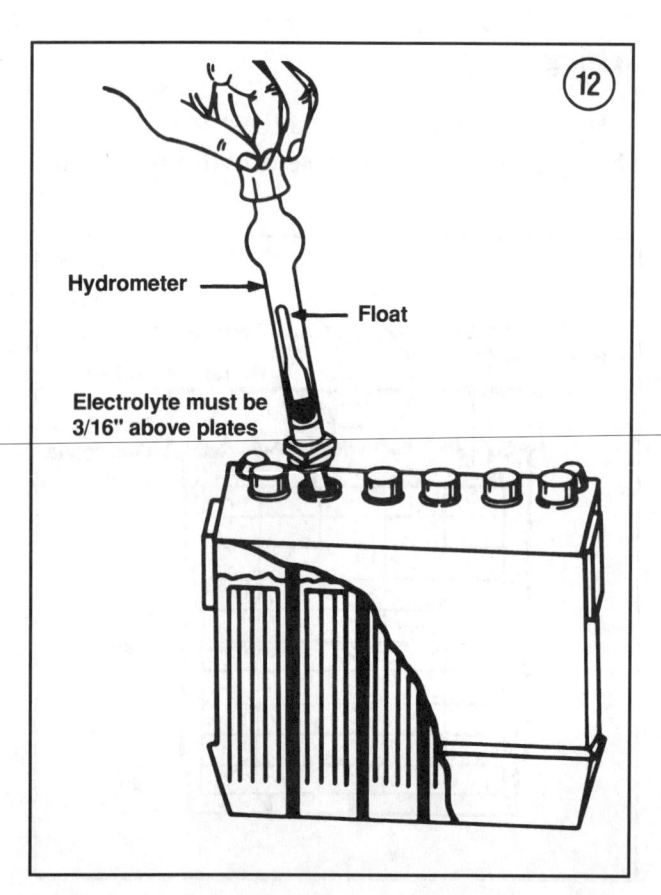

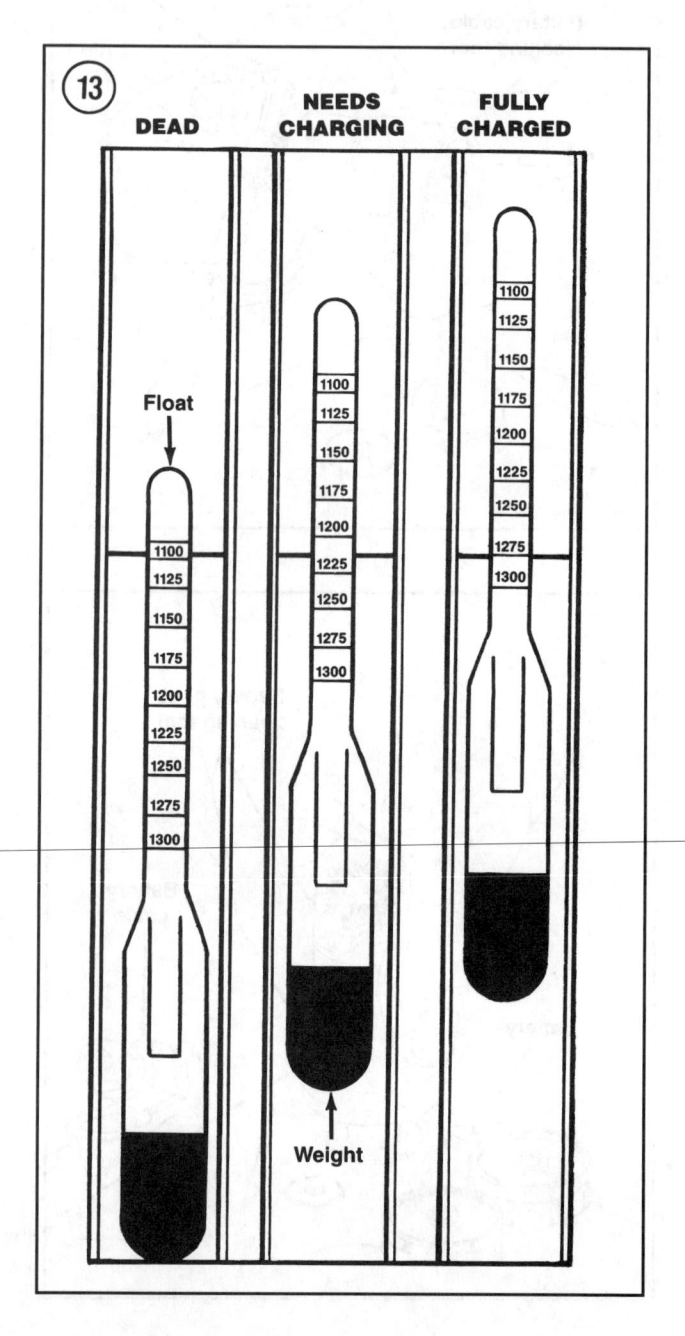

ELECTRICAL SYSTEMS

if the cell tests below 1.120, the battery must be recharged and one that reads 1.140 or below is dead. Charging is also necessary if the specific gravity varies more than 0.050 from cell to cell.

NOTE
If a temperature-compensated hydrometer is not used, add 0.004 to the specific gravity reading for every 10° above 80° F (27° C). For every 10° below 80° F (27° C), subtract 0.004.

Storage

Wet cell batteries slowly discharge when stored. They discharge faster when warm than when cold. See **Table 2**. Before storing a battery for the season, perform the following.

1. Clean the case with a solution of baking soda and water. Rinse with clear water and wipe dry.
2. Charge the battery to a fully charged condition (no change in specific gravity when 3 readings are taken 1 hour apart).
3. Install the filler caps and make sure they are on tight.
4. Coat the battery posts with a petroleum jelly such as Vaseline or a light mineral grease.
5. Store the battery in as cool and dry a place as possible.

Charging

A good state of charge should be maintained in batteries used for starting. Check the battery with a voltmeter as shown in **Figure 14**. Any battery that cannot deliver at least 9.6 volts under a starting load should be recharged. If recharging does not bring it up to strength or if it does not hold the charge, replace the battery.

WARNING
During the charging process, highly explosive hydrogen gas is released from the battery. The battery should be charged only in a well-ventilated area away from any open flames (including pilot lights on home gas appliances). Do not allow any smoking in the area. Never check the charge by arcing (connecting pliers or other metal objects) across the terminals; the resulting spark can ignite the hydrogen gas.

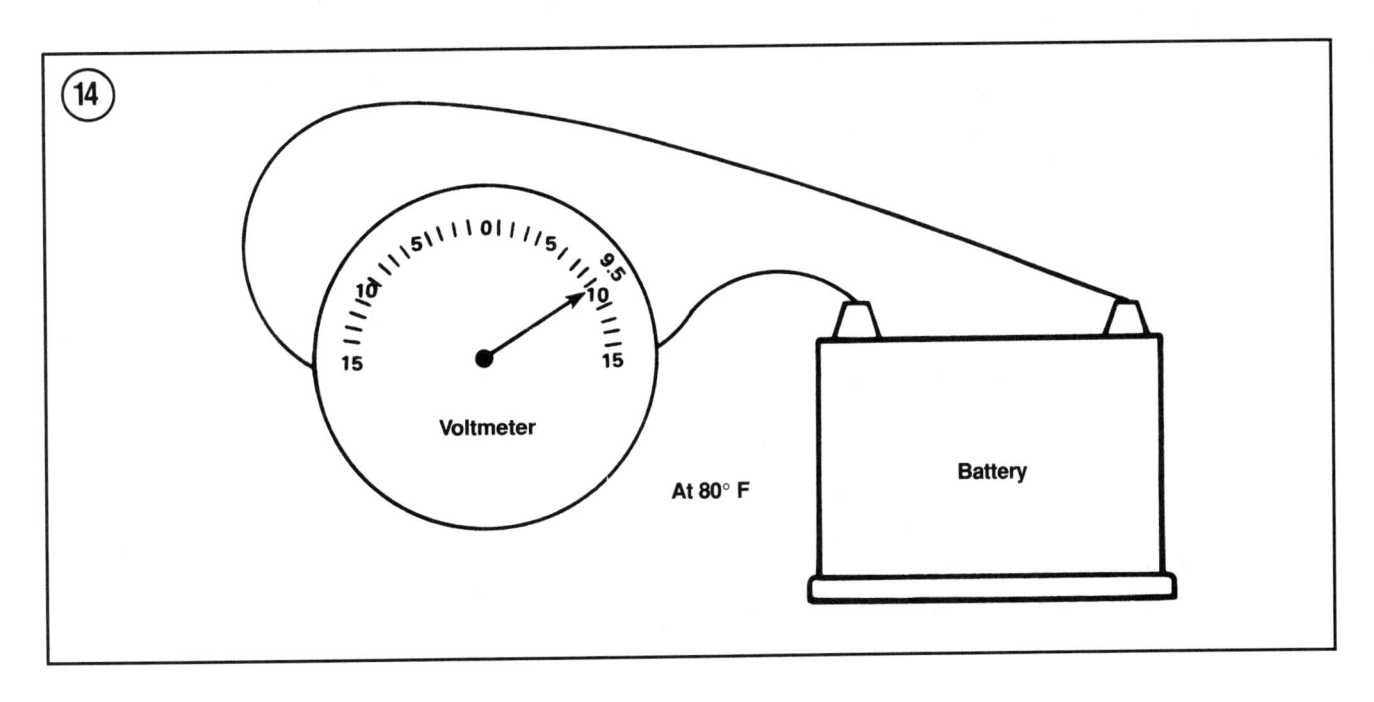

The battery does not have to removed from the boat for charging; but it is a recommended safety procedure since a charging battery gives off highly explosive hydrogen gas. In many boats, the area around the battery is not well ventilated and the gas may remain in the area for hours after the charging process has been completed. Sparks or flames occurring near the battery can cause it to explode, spraying battery acid over a wide area. Also, the corrosive mist that is emitted during the charging process will corrode all surrounding surfaces.

For this reason, it is important that you observe the following precautions:

a. Do not allow anyone to smoke around batteries that are charging or have been recently charged.
b. Do not break a live circuit at the battery terminals and cause an electrical arc that can ignite the hydrogen gas.

1. Disconnect the negative (–) battery cable first, then the positive (+) battery cable.
2. Make sure the electrolyte is fully topped up.
3. Connect the charger to the battery—negative to negative, positive to positive.
4. If the charger output is variable, select a 4 amp setting. Set the voltage regulator to 12 volts and plug the charger in. If the battery is severely discharged, allow it to charge for at least 8 hours. Batteries that are not as badly discharged require less charging time. **Table 2** gives approximate charge rates for batteries used primarily for cranking. Check the charging progress with the hydrometer.

Jump Starting

If the battery becomes severely discharged, it is possible to start and run an engine by jump starting it from another battery. If the proper procedure is not followed however, jump starting can be dangerous. Check the electrolyte level before jump starting any battery. If it is not visible or if it appears to be frozen, *do not* attempt to jump start the battery.

WARNING
Use extreme caution when connecting a booster battery to the battery that is discharged to avoid personal injury or damage to the system.

1. Connect the jumper cables in the order and sequence shown in **Figure 15**.

WARNING
An electrical arc may occur when the final connection is made. This could cause an explosion if it occurs near either battery. For this reason, the final connection should be made to a good ground away from the battery and not to the battery itself.

2. Check that all jumper cables are out of the way of moving engine parts.

CAUTION
Running the engine at wide-open throttle may cause damage to the electrical system.

3. Start the engine. Once it starts, run it at a moderate speed.
4. Remove the jumper cables in the exact reverse order shown in **Figure 15**. Remove the cables from point 4, then 3, 2 and 1.

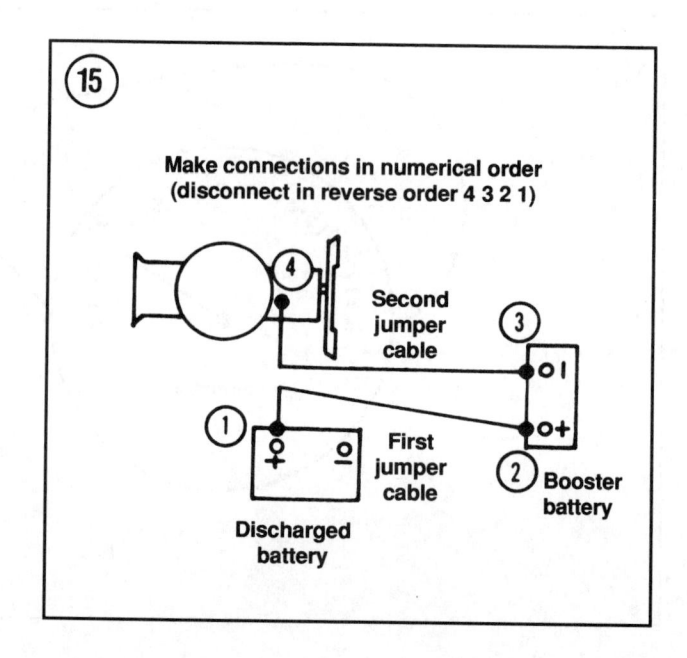

ELECTRICAL SYSTEMS

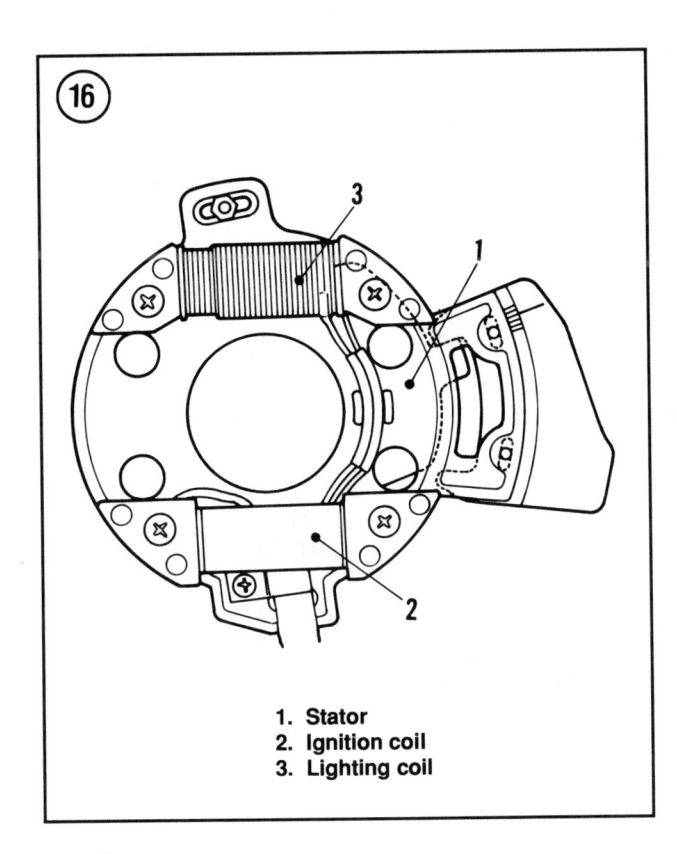

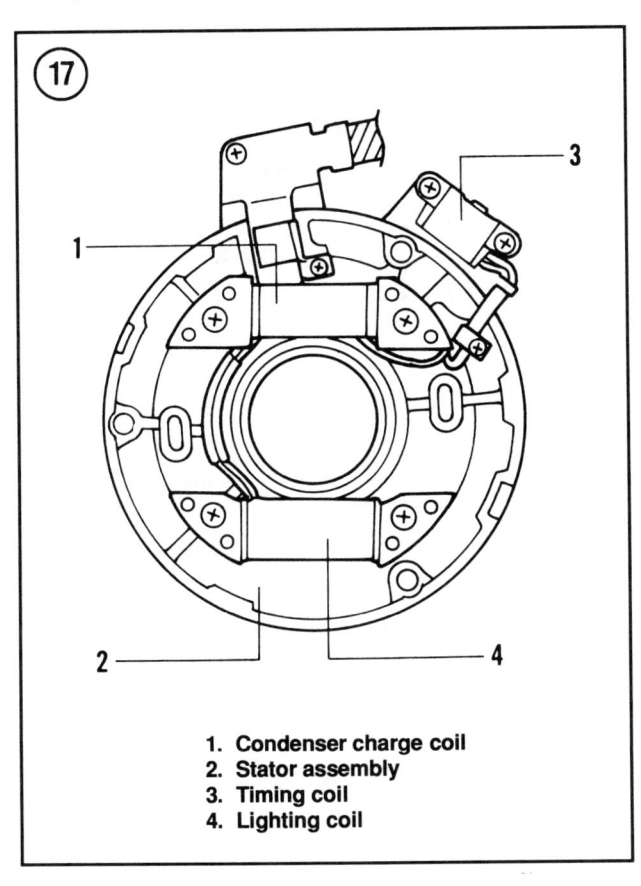

16

1. Stator
2. Ignition coil
3. Lighting coil

17

1. Condenser charge coil
2. Stator assembly
3. Timing coil
4. Lighting coil

New Battery Installation

When replacing the old battery with a new one, be sure to charge it completely (specific gravity, 1.260-1.280) before installing it in the boat. Failure to do so, or using the battery with a low electrolyte level will permanently damage the battery. When purchasing a new battery, be sure to purchase the correct battery capacity for your specific engine and boat.

NOTE
*Recycle your old battery. When you replace the old battery, be sure to turn in the old battery at that time. The lead plates and the plastic case can be recycled. Most outboard dealers will accept your old battery in trade when you purchase a new one, but if they will not, many automotive supply stores certainly will. **Never** place an old battery in your household trash since it is illegal, in most states, to place any acid or lead (heavy metal) contents in landfills. There is also the danger of the battery being crushed in the trash truck and spraying acid on the truck operator.*

LIGHTING SYSTEM
(1985-1986 DT 9.9, DT 15, DT 40)

An AC lighting system is optional on 1985-1986 DT 9.9, DT 15 and DT 40 models with manual start. This system is used to power lights used for boating or fishing at night. When the engine is running at approximately 4,500-5,000 rpm, the system will deliver 80 watts of 12-volt AC current.

The AC lighting system consists of an AC lighting coil on the magneto stator base, permanent magnets located in the flywheel rim and connecting wiring between the lighting coil and a plug-in socket. Refer to **Figure 16** for DT 9.9 and DT 15 models or **Figure 17** for DT 40 models.

Rotation of the flywheel magnets past the lighting coil creates alternating current. This cur-

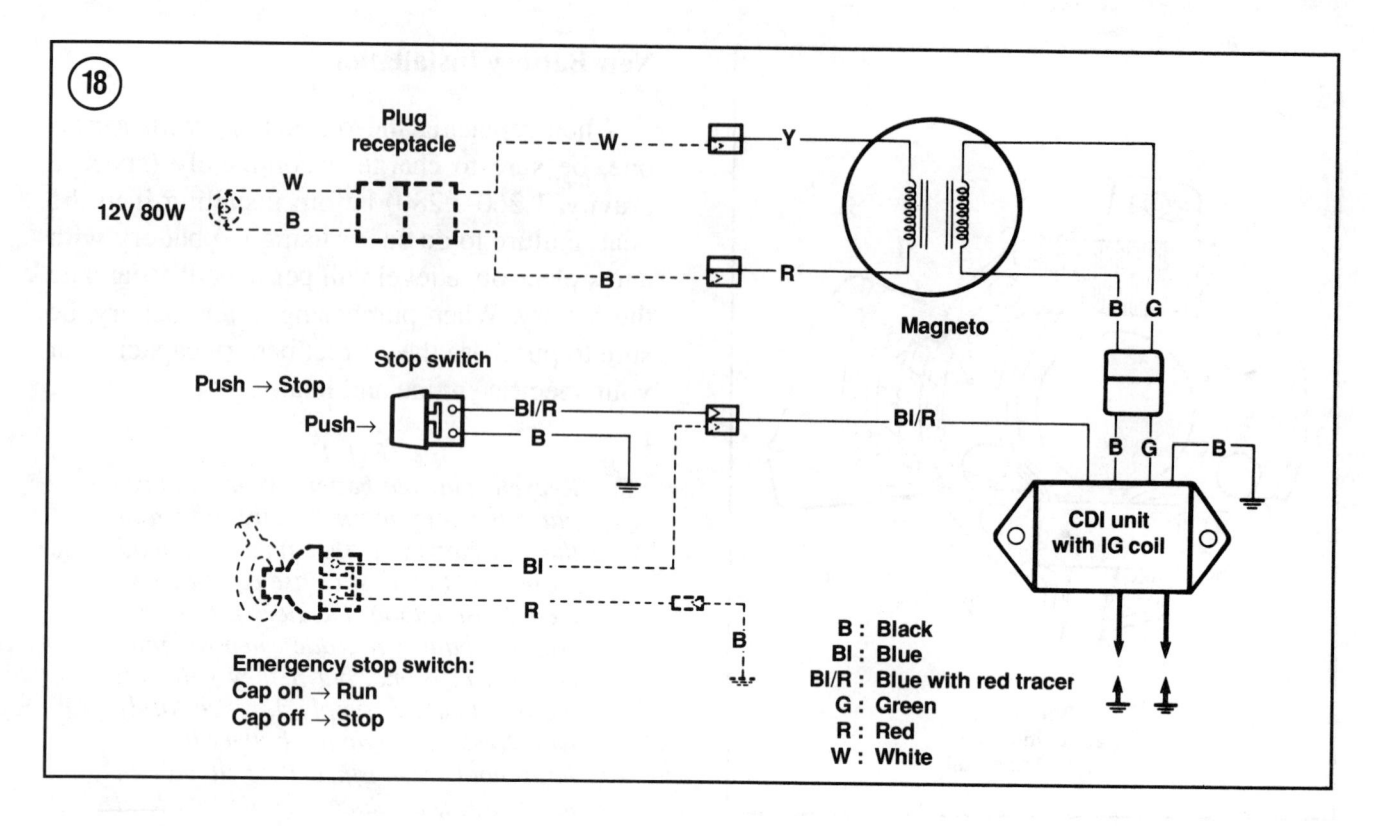

CHAPTER SEVEN

⑱

Plug receptacle

12V 80W

Stop switch
Push → Stop
Push →

Emergency stop switch:
Cap on → Run
Cap off → Stop

Magneto

CDI unit with IG coil

B : Black
Bl : Blue
Bl/R : Blue with red tracer
G : Green
R : Red
W : White

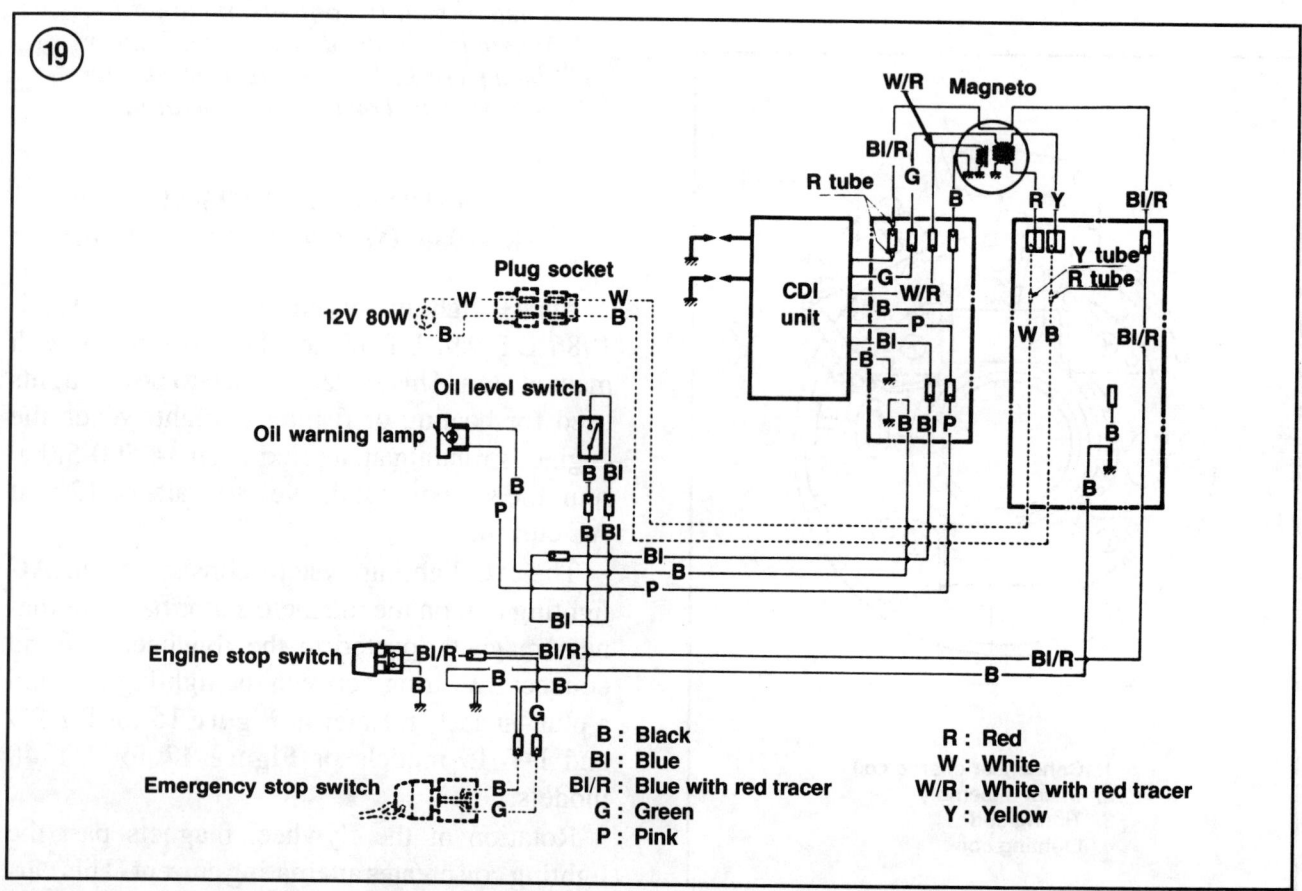

⑲

W/R Magneto
Bl/R
R tube
Y tube
R tube

Plug socket
12V 80W

CDI unit

Oil level switch
Oil warning lamp

Engine stop switch

Emergency stop switch

B : Black
Bl : Blue
Bl/R : Blue with red tracer
G : Green
P : Pink

R : Red
W : White
W/R : White with red tracer
Y : Yellow

ELECTRICAL SYSTEMS

rent is sent to the plug-in socket installed on the engine cover to power accessories. Schematics of a typical AC lighting circuit and DC battery charging circuit are shown in **Figure 18** for DT 9.9 and DT 15 models or **Figure 19** for DT 40 models.

Lighting Coil Replacement

Refer to **Figure 20** for DT 9.9 and DT 15 models or **Figure 21** for DT 40 models for this procedure.

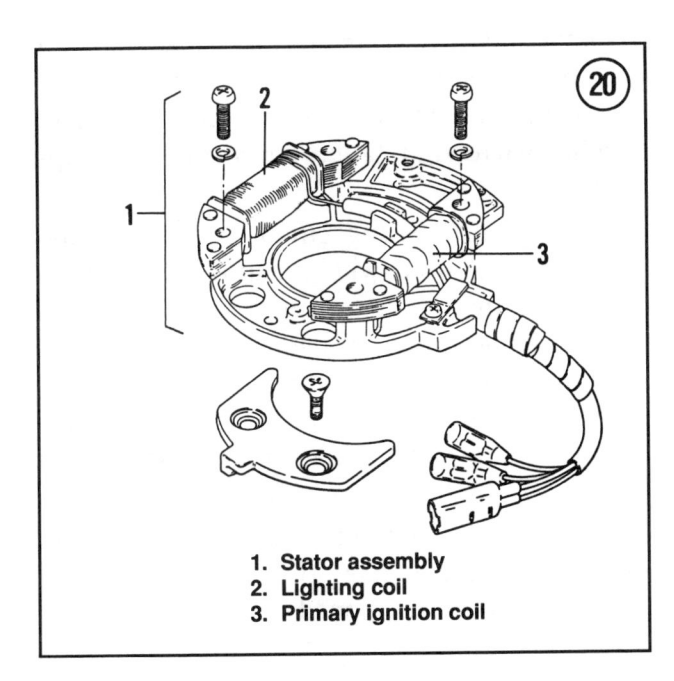

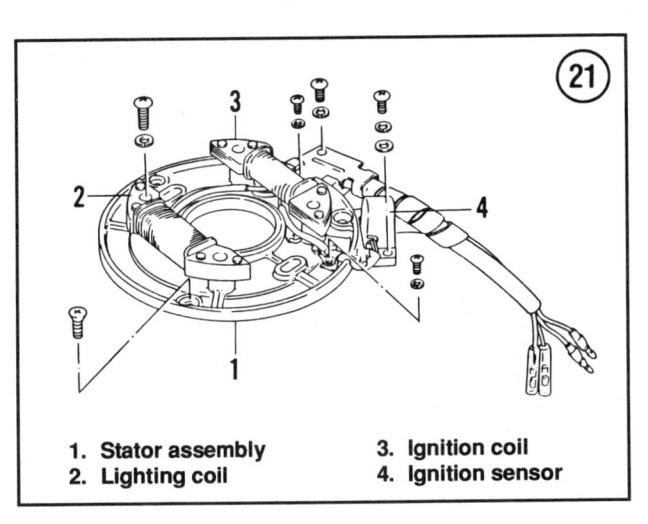

1. Stator assembly
2. Lighting coil
3. Primary ignition coil

1. Stator assembly
2. Lighting coil
3. Ignition coil
4. Ignition sensor

1. Remove the engine cover.
2. Remove the flywheel. See Chapter Eight.
3. Disconnect the one white and one black lighting coil leads at the quick-disconnect terminals.
4. Remove the screw and clamp securing the electrical harness to the stator base.
5. Remove the lighting coil mounting screws and lockwashers. Remove the coil from the stator base.
6. Installation is the reverse of removal, noting the following.
7. Make sure both electrical connectors are free of corrosion and tight.
8. Be sure to route electrical wires so they do not contact or interfere with any moving components.

BATTERY CHARGING SYSTEM

A battery charging system is standard on all electric start models.

The battery charging system consists of charge coil(s) (or an AC lighting coil on some models), permanent magnets located in the flywheel rim, a rectifier (voltage regulator/rectifier on 1986-on DT 115, DT 140 and all V4 and V6 models), a battery and connecting wiring with a 20 or 25 amp fuse.

On all DT 75, DT 85 models and on 1985 DT 115 and DT 140 models, the optional 15 amp system is similar to all other systems except a rectifier/regulator unit is used in place of the standard rectifier.

Rotation of the flywheel magnets past the charging coils, or AC lighting coil(s), creates alternating current. This current is sent to the rectifier, or voltage regulator/rectifier, where it is converted into direct current and then supplied to the battery or electrical accessories through a fuse.

A malfunction in the battery charging system will result in an undercharged battery. Perform the following visual inspection to determine the cause of the problem. If the visual inspection proves satisfactory, test the lighting coil and rectifier, or voltage regulator/rectifier. See Chapter Three.

1. Check the fuse in the wire between the rectifier, or voltage regulator/rectifier and the battery.
2. Make sure that the battery cables are connected properly. The red cable must be connected to the positive (+) battery terminal. If polarity is reversed, check for a damaged rectifier or voltage regulator/rectifier.
3. Inspect the battery terminals for loose or corroded connections. Tighten or clean as described in this chapter.
4. Inspect the physical condition of the battery. Look for bulges or cracks in the case, leaking electrolyte or corrosion build-up.
5. Carefully check the wiring between the charging coil(s) or the lighting coil and the battery for chafing, deterioration or other damage.
6. Check the circuit wiring for corroded, loose or disconnected electrical connections. Remove any corrosion, clean, tighten or connect as required.
7. Determine if the accessory load on the battery is greater than the battery capacity.

Battery Charging Coil Replacement

See *Lighting Coil Replacement* in this chapter for all models except 1985-1987 DT 75, DT 85 and 1985 DT 115, DT 140. On 1985-1987 DT 75, DT 85 and 1985 DT 115, DT 140 models with an alternator stator, the entire stator must be replaced as described in this chapter if one or more of the charge or lighting coils are defective.

Rectifier, or Voltage Regulator/Rectifier Replacement

1. Disconnect the negative battery cable.
2. Disconnect the electrical connectors from the rectifier, or voltage regulator/rectifier unit.
3. Remove the fastener holding the unit to the power head or electrical component mounting bracket. Remove the unit.
4. On models so equipped, remove the black ground lead screw. Disconnect the ground lead from the power head.
5. Installation is the reverse of removal, noting the following.
6. Make sure all electrical connectors are free of corrosion and are tight.

ELECTRIC STARTING SYSTEMS

Outboards covered in this manual either use a rope-operated mechanical (rewind) starting system or an electric (starter motor) starting system. Mechanical rewind starters are covered in Chapter Ten.

The electric starting circuit consists of the battery, an ignition switch, neutral start switch, the starter motor, starter relay and connecting wiring. While control of the neutral start switch is a function of the ignition system on most engines, some engines may also have a mechanical starter interlock device to prevent starting in gear.

An optional switch connected into the ignition system shuts the engine off in case of an emergency. Outboards equipped with a remote con-

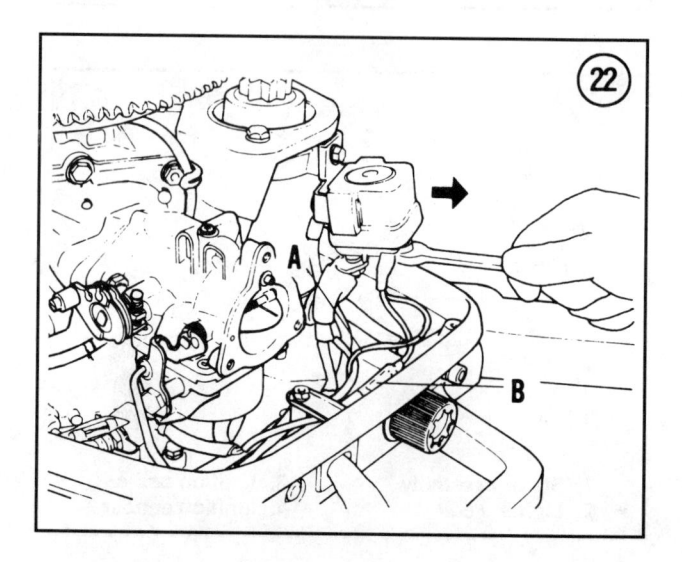

ELECTRICAL SYSTEMS

trol have the emergency switch located on the remote box.

Starting system operation and troubleshooting procedures are described in Chapter Three.

STARTER MOTOR

Marine starter motors are very similar in design and operation to those found on automotive engines. They use an inertia-type drive in which external spiral splines on the armature shaft mate with internal splines on the drive assembly.

The starter motor produces very high torque but only for a brief period of time, due to heat buildup. Never operate the starter motor continuously for more than 10 seconds. Let the motor cool for at least 2 minutes before operating it again.

If the starter motor does not turn over, check the battery and all connecting wiring for loose or corroded connections. If this does not solve the problem, refer to Chapter Three. Except for brush replacement, service to the starter motor is limited to replacement with a new or rebuilt unit.

Suzuki outboards use a variety of starter motors, manufactured primarily by Hitachi. The starter motors are equipped with either 2, 3 or 4 brushes depending on model.

Starter Motor Removal/Installation (Attached Starter Relay)

1. Disconnect the negative battery cable.
2. Remove the engine cover.
3. Disconnect the starter relay electrical lead at its bullet connector. Remove the screw holding the black relay ground lead. See A and B, **Figure 22**.
4. Disconnect the battery and starter cables from the relay terminals.
5. Remove the relay (**Figure 22**).
6. On models so equipped, remove the clamp holding the starter motor to the base of the mounting bracket.
7. Remove the mounting bolts holding the starter motor to its mounting bracket (**Figure 23**). Remove the starter motor.

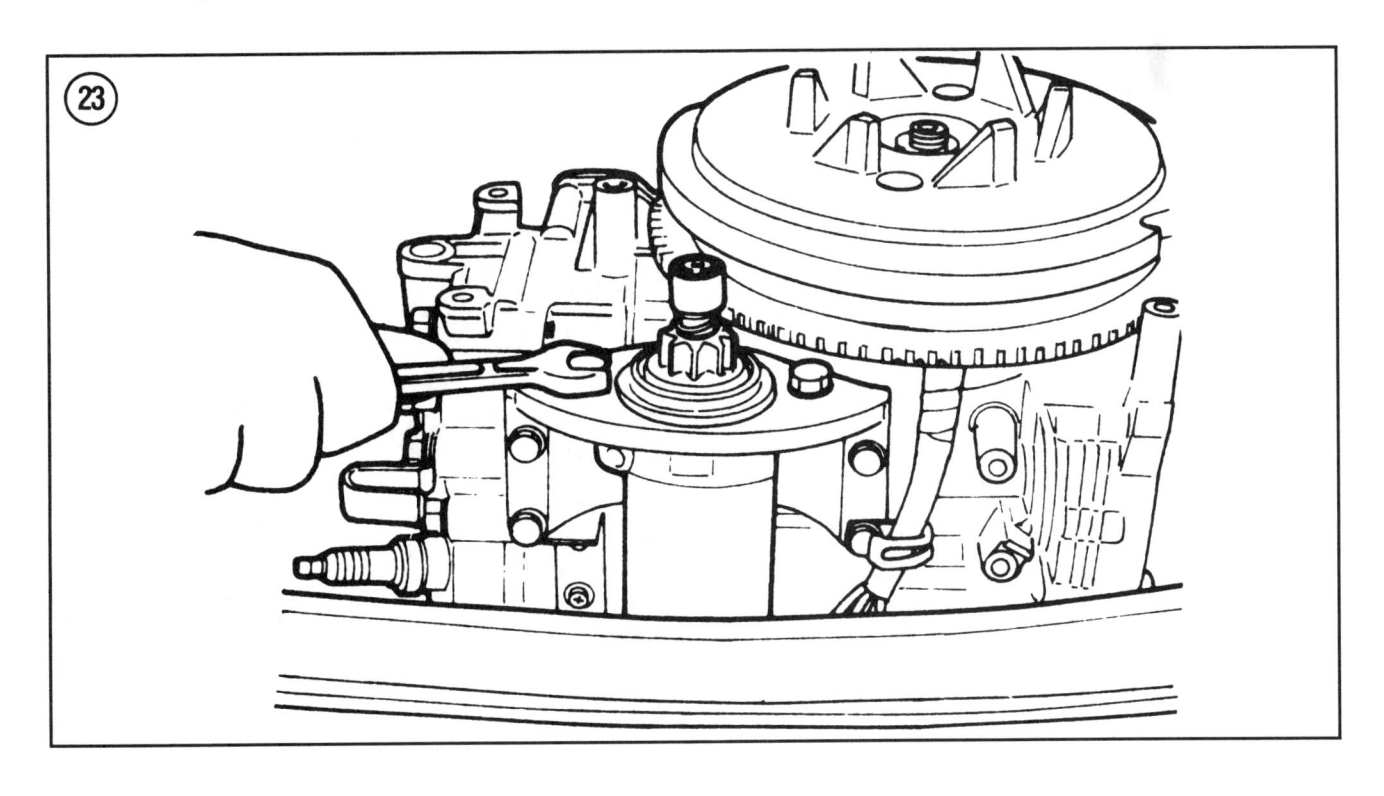

8. Installation is the reverse of removal, noting the following.
9. Make sure all electrical connectors are free of corrosion and are tight.

Starter Motor Removal/Installation (Remote Starter Relay)

1. Disconnect the negative battery cable.
2. Remove the engine cover.
3. Disconnect the starter cables from their terminals (**Figure 24**, typical).
4. On models so equipped, remove the clamp holding the starter motor to the base of the mounting bracket (**Figure 25**, typical).
5. Remove the 2 vertical mounting bolts holding the starter motor to its mounting bracket (**Figure 26**).
6. Lower the starter motor from the mounting bracket and remove from the engine.
7. Installation is the reverse of removal, noting the following.
8. Make sure all electrical connectors are free of corrosion and are tight.

Brush Replacement (DT 8 through DT 40 With 2-Brush Starter)

Refer to **Figure 27** for a typical 2-brush design for this procedure. Always replace all brushes in complete sets.

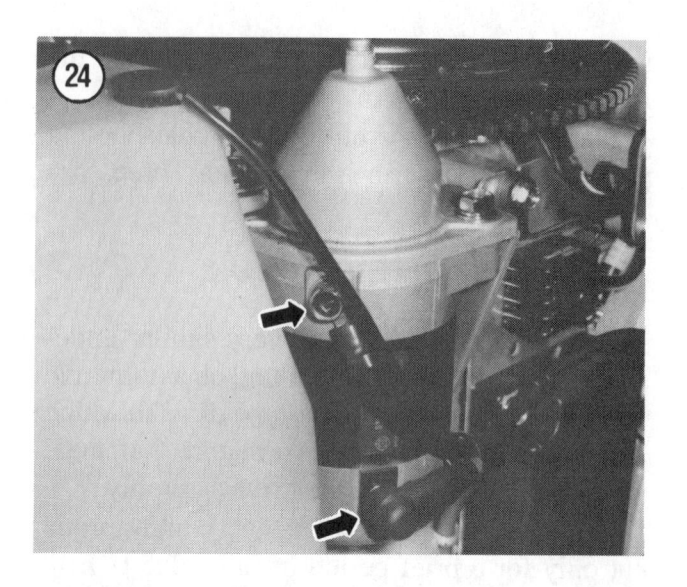

1. Remove the starter as described in this chapter.
2A. On 1988-on DT 8, 1985-on DT 9.9 and 1985-on DT 15 models, perform the following:
   a. Remove the terminal nut (**Figure 28**) and remove the starter cable.
   b. Remove the 2 through-bolts from the starter (**Figure 29**).
   c. Lightly tap on brush holder cover with a rubber mallet until it breaks free of starter housing. Remove brush holder cover, taking care not to lose the brush springs (**Figure 30**).

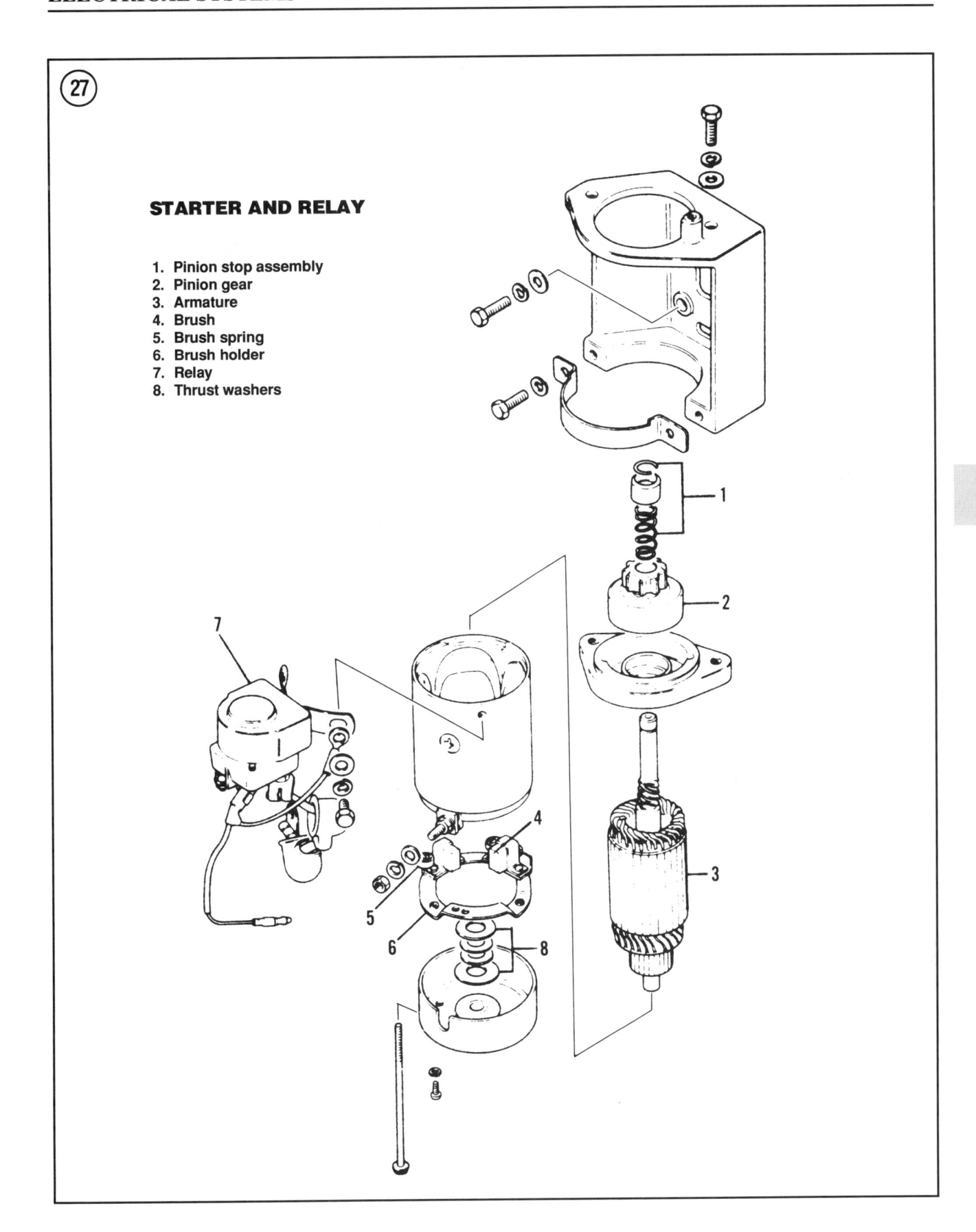

ELECTRICAL SYSTEMS

STARTER AND RELAY

1. Pinion stop assembly
2. Pinion gear
3. Armature
4. Brush
5. Brush spring
6. Brush holder
7. Relay
8. Thrust washers

CHAPTER SEVEN

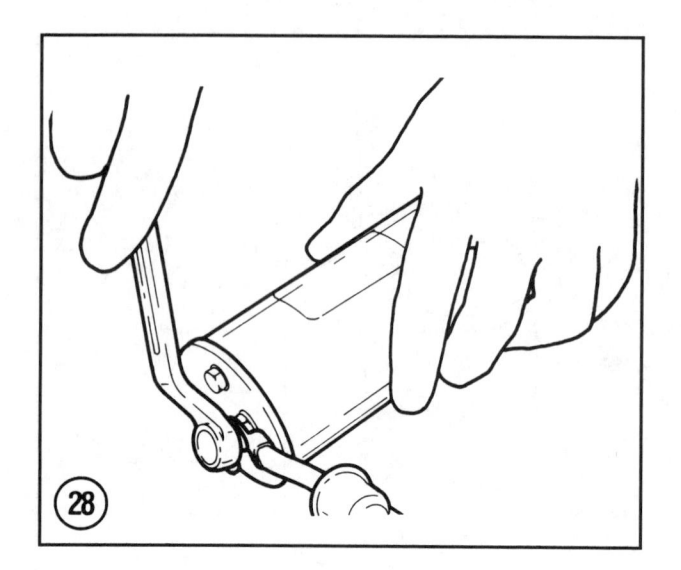

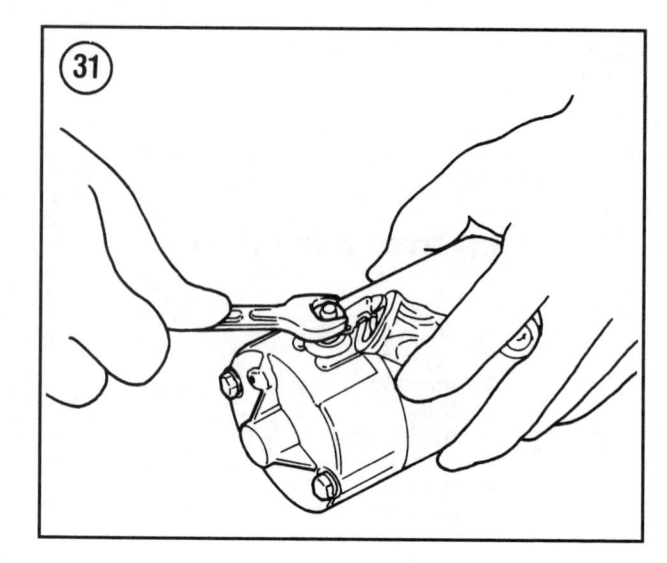

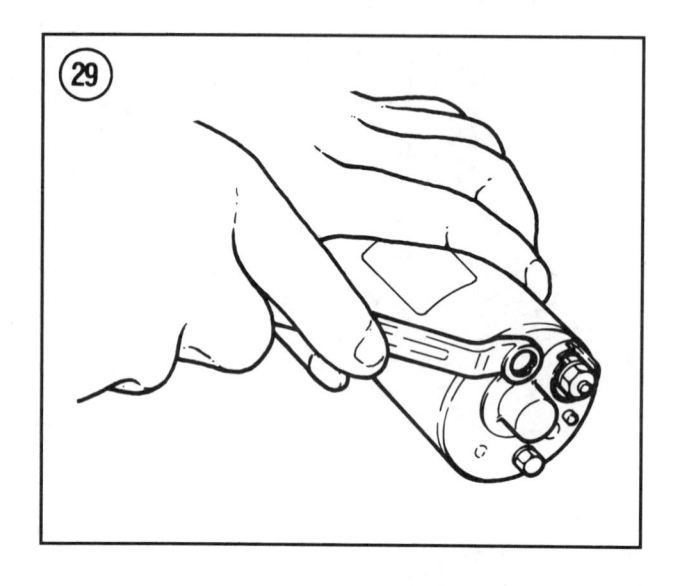

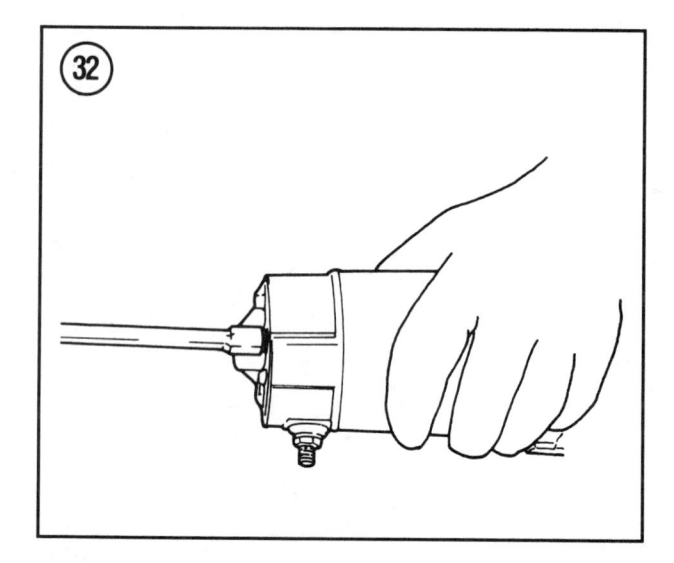

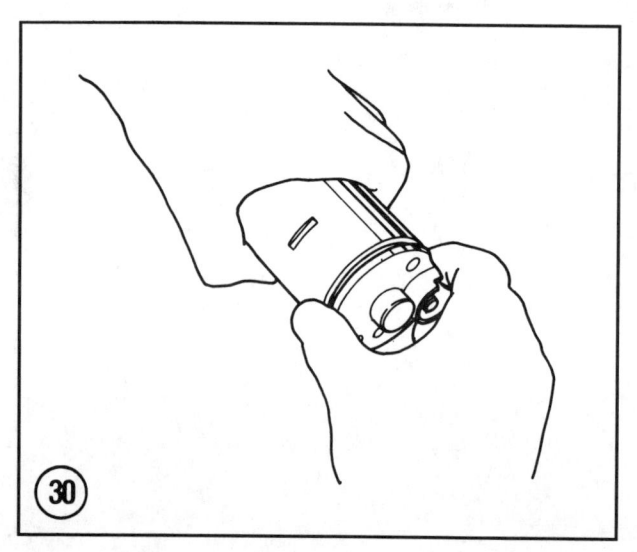

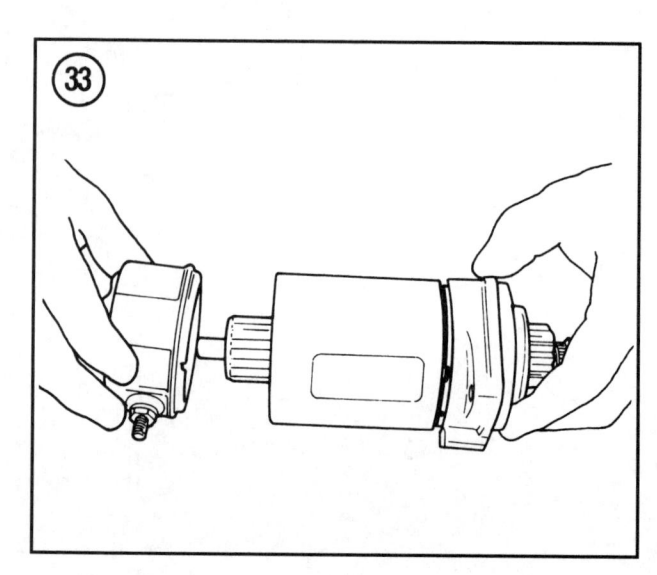

ELECTRICAL SYSTEMS

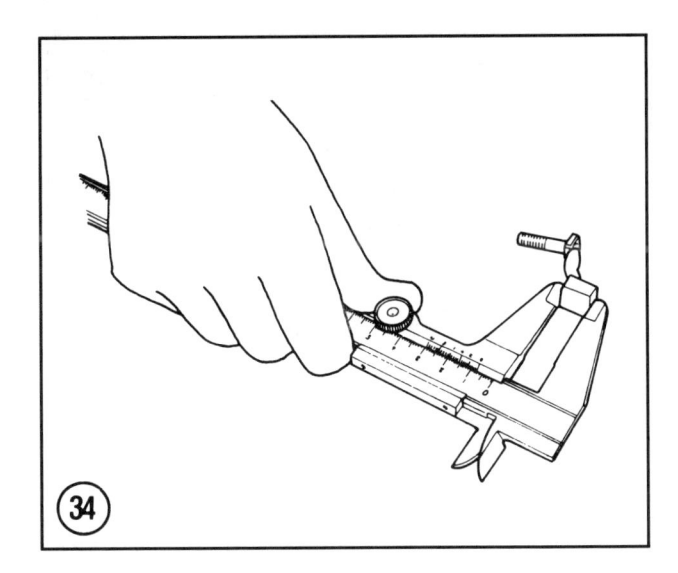

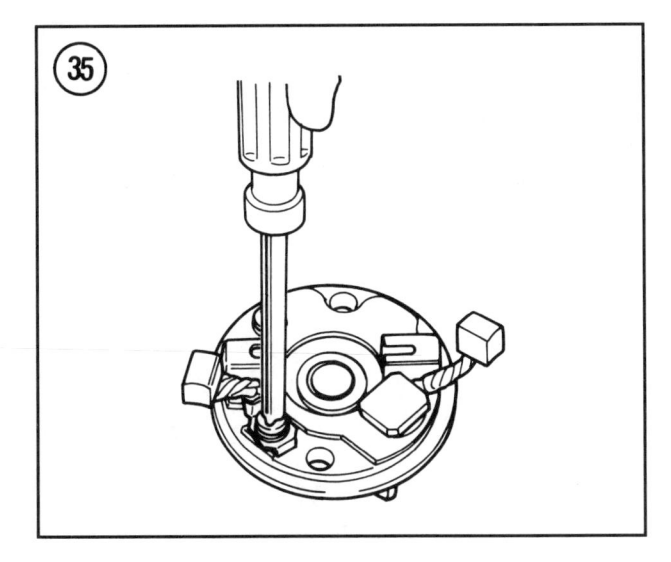

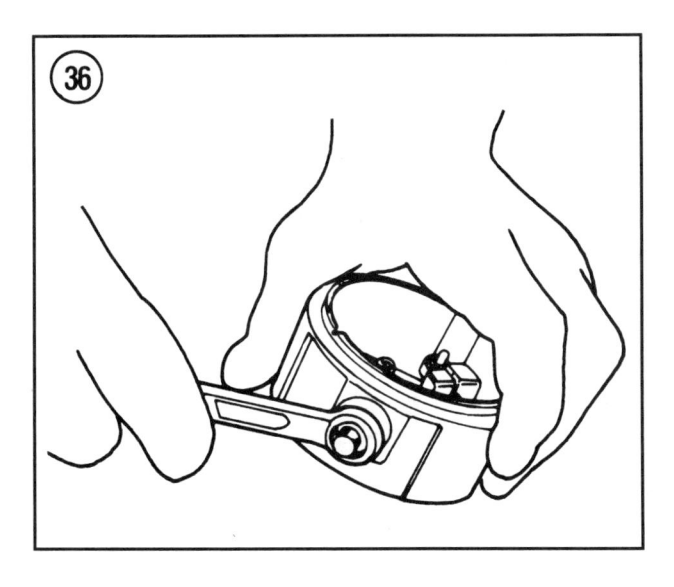

d. Remove the brush holder cover thrust washer from the armature.

2B. On all other models, perform the following:
   a. Remove the terminal nut (**Figure 31**) and remove the starter cable.
   b. Remove the 2 through-bolts from the starter (**Figure 32**).
   c. Lightly tap on brush holder cover with a rubber mallet until it breaks free of starter housing. Remove brush holder cover, taking care not to lose the brush springs (**Figure 33**).
   d. Remove the brush holder cover thrust washer from the armature.

NOTE
If corrosion causes the brushes to stick during Step 3, replace the brush holder plate.

3. Check brush spring tension by pulling spring back and releasing it. Replace the spring if it does not snap the brush firmly into position.
4. Remove the brushes and springs from the brush holder plate.
5. Inspect the brushes and replace all brushes if any are pitted or oil-soaked.
6. Measure each brush with a vernier caliper (**Figure 34**). Replace brushes if worn to the following dimension or less:
   a. 1988-on DT 8, 1985-on DT 9.9 and 1985-on DT 15: 0.18 in. (4.5 mm).
   b. All other models: 0.35 in. (9 mm).

7A. On 1988-on DT 8, 1985-on DT 9.9 and 1985-on DT 15 models, remove the screw or nut securing the brushes (**Figure 35**).
7B. On all other models, perform the following:
   a. Remove the positive terminal nut, insulators and O-ring (**Figure 36**).
   b. Remove the screws holding the brush holder to the end cap (**Figure 37**).
   c. Remove the brush holder plate from the end cap.
8. Connect an ohmmeter between the negative and positive brush holders on the brush holder

plate (**Figure 38**). Replace the brush holder plate if the meter shows continuity.

9. Install a new positive terminal and brush assembly to the brush holder.

10. Install a new ground brush to the brush holder plate.

11. Fit the springs and brushes into their respective brush holders.

12. Coat brush holder cover bore with water-resistant grease.

13. Press the brushes into the holders and use a narrow strip of flexible metal or plastic as shown in **Figure 39** to keep them in place.

14. Install the brush holder cover thrust washer(s) onto the armature.

15. Partially fit brush holder cover in place, removing the temporary brush retainer as the brushes slip over the commutator.

16. Align the notch in the brush holder cover (B) with the alignment mark (A) on the field coil case. Refer to **Figure 40** for 1988-on DT 8, 1985-on DT 9.9 and 1985-on DT 15 models or **Figure 41** for all other models.

17. Install through-bolts and tighten securely.

18. Install the starter cable and the terminal nut.

19. Install the starter as described in this chapter.

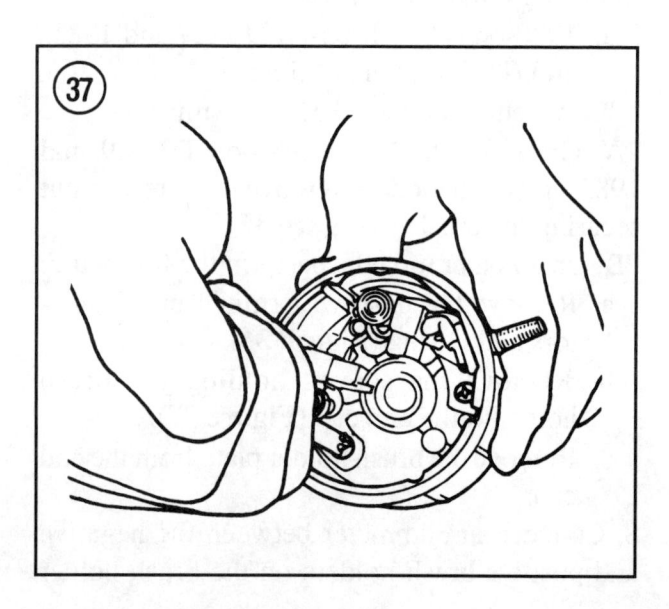

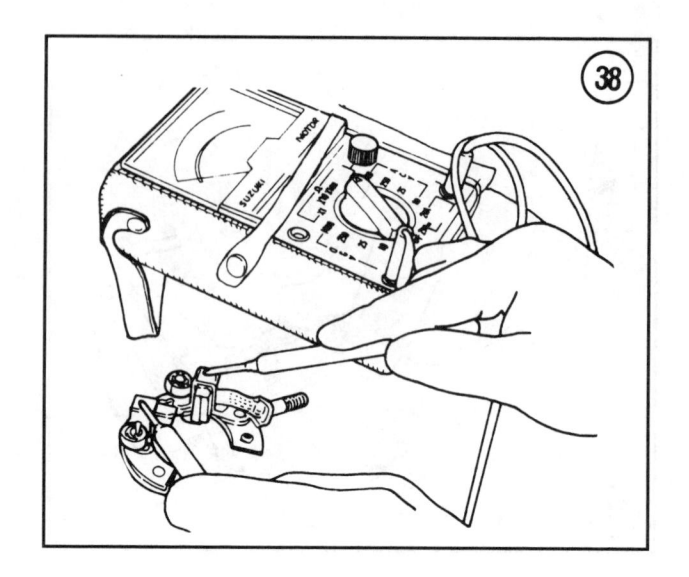

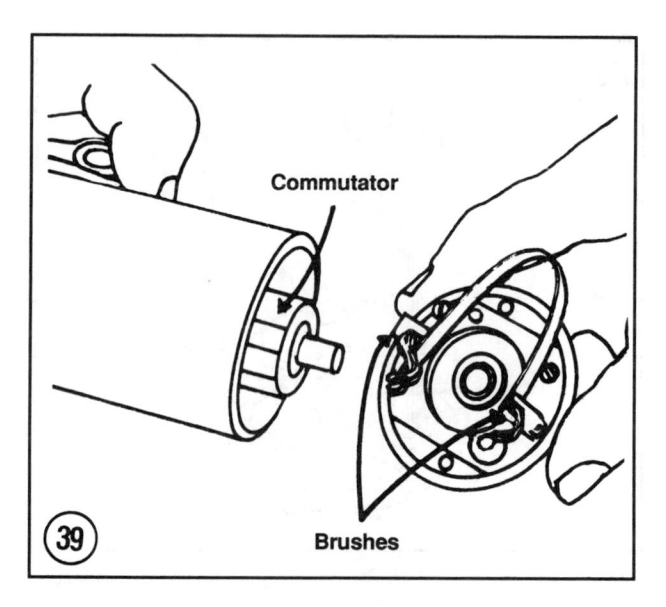

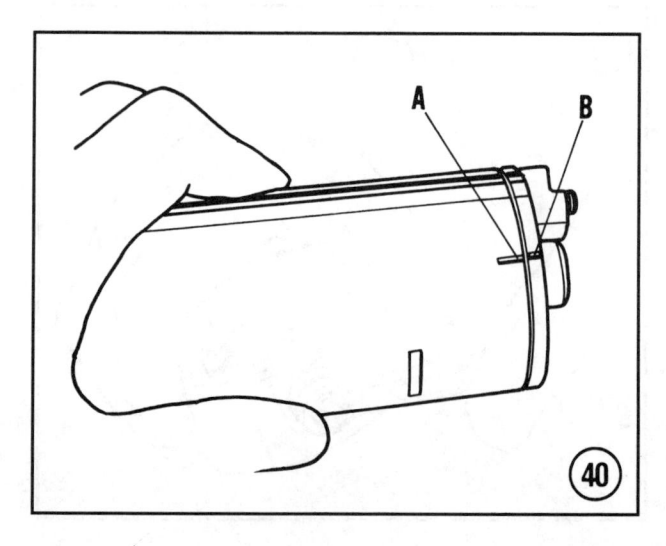

ELECTRICAL SYSTEMS

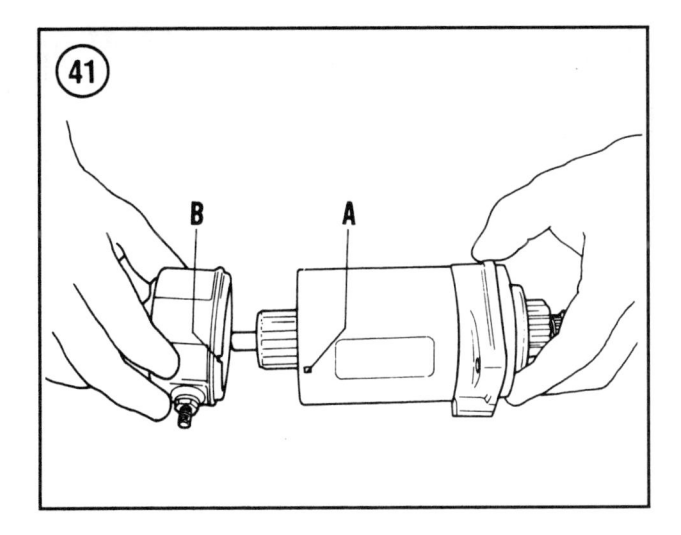

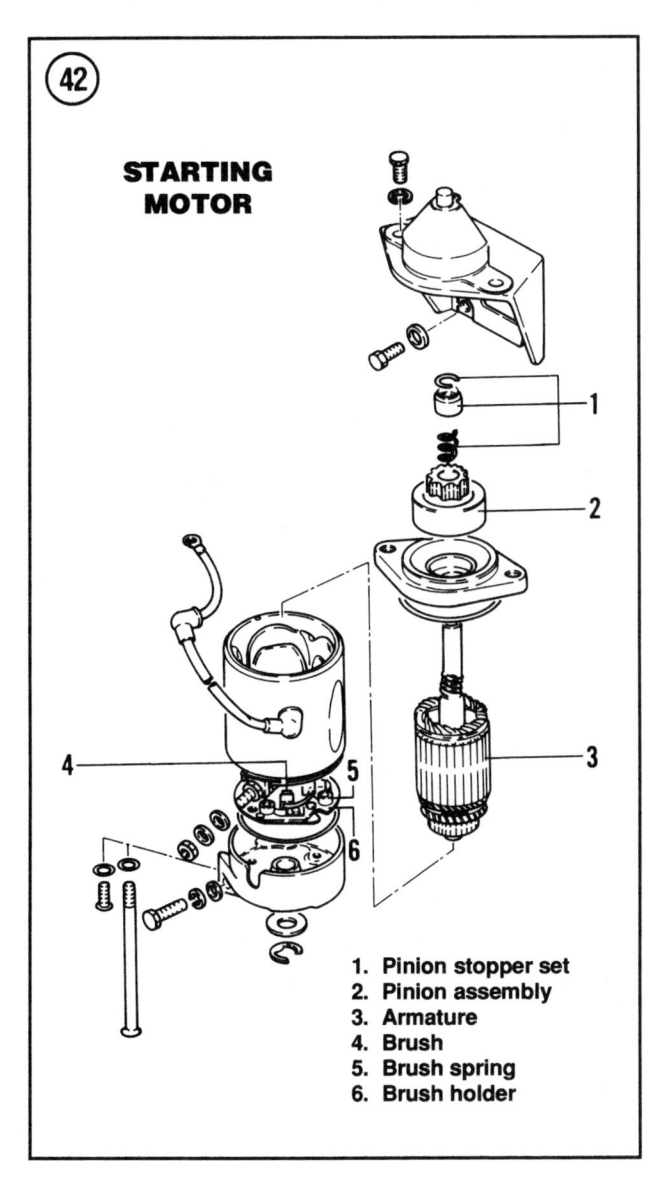

Brush Replacement
(1985-on DT 75, DT 85 and
1985 DT 115 With 3- and 4-Brush Starter)

Refer to **Figure 42** for a typical 3-brush design for this procedure. Always replace all brushes in complete sets.

1. Remove the starter as described in this chapter.
2. Remove the terminal nut (**Figure 43**) and remove the starter cable.
3. Scribe alignment marks between the brush holder cover and the field coil case and between the pinion bearing carrier and the field coil case. These marks will be used during assembly.
4. Remove the 2 through-bolts and washers from the starter (**Figure 44**).

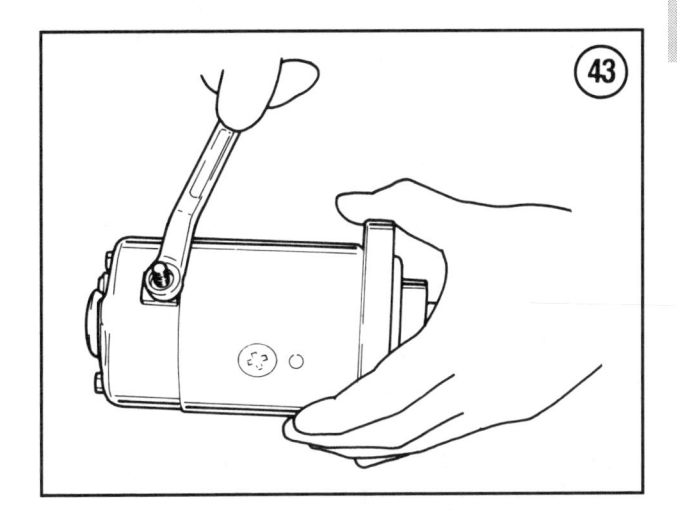

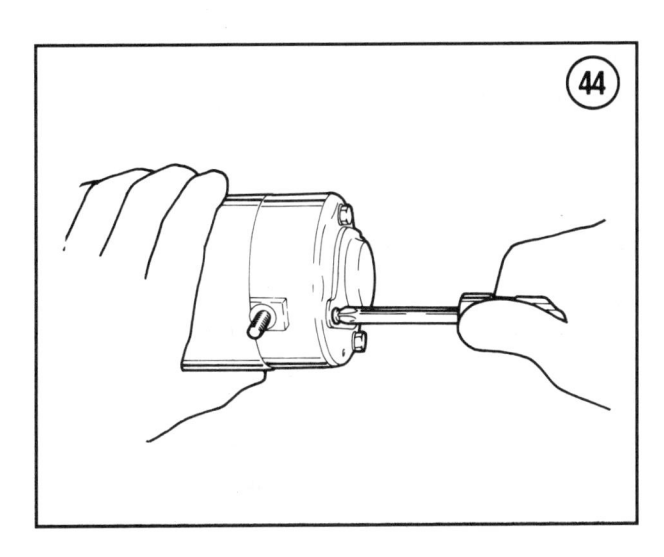

STARTING MOTOR

1. Pinion stopper set
2. Pinion assembly
3. Armature
4. Brush
5. Brush spring
6. Brush holder

5. Remove the E-clip and thrust washer (**Figure 45**) from the brush holder cover.
6. Remove the 2 screws securing the brush holder cover (**Figure 46**).
7. Lightly tap on brush holder cover with a rubber mallet until it breaks free of starter housing. Remove brush holder cover.
8. Withdraw the armature assembly from the field coil case (**Figure 47**).

NOTE
If corrosion causes the brushes to stick during Step 9, replace the brush holder plate.

9. Check brush spring tension by pulling spring back and releasing it. Replace the spring if it does not snap the brush firmly into position.
10. Unhook the springs and remove the brushes from the brush holder plate (**Figure 48**).
11. Remove the brush holder plate.
12. Inspect the brushes as follows:
   a. Replace all brushes if any are pitted or oil-soaked.
   b. Measure each brush with a vernier caliper (**Figure 49**). Replace brushes if worn to 0.45 in. (11.5 mm) or less.
13. To remove the brushes, unsolder them from the electrical lead either on the brush holder or the field coil assembly. Resolder new brushes onto the electrical leads.
14. Connect an ohmmeter between the negative and positive brush holders on the brush holder plate (**Figure 50**). Replace the brush holder plate if the meter shows continuity.
15. Fit the springs and brushes into their respective brush holders.
16. Install the armature assembly into the field coil case (**Figure 47**). Refer to alignment marks made in Step 3 to align the pinion bearing carrier with the field coil case.
17. Coat brush holder cover bore with water-resistant grease and install the brush holder cover. Refer to the alignment marks made in Step 3 to align the brush holder cover to the field coil case.

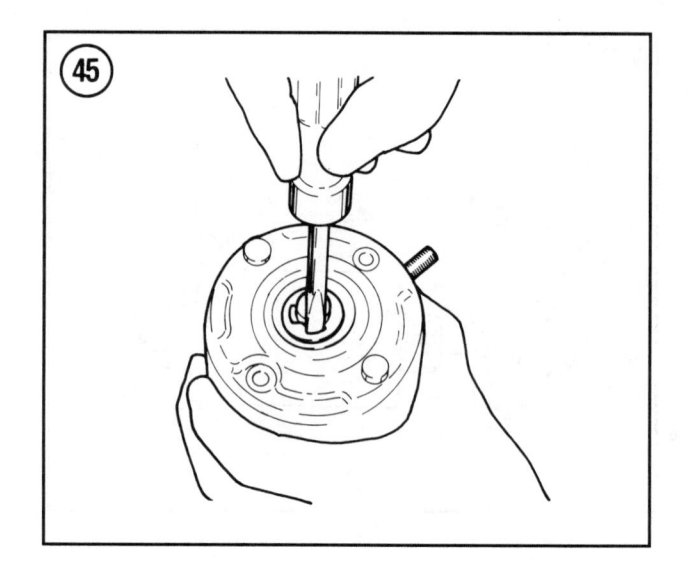

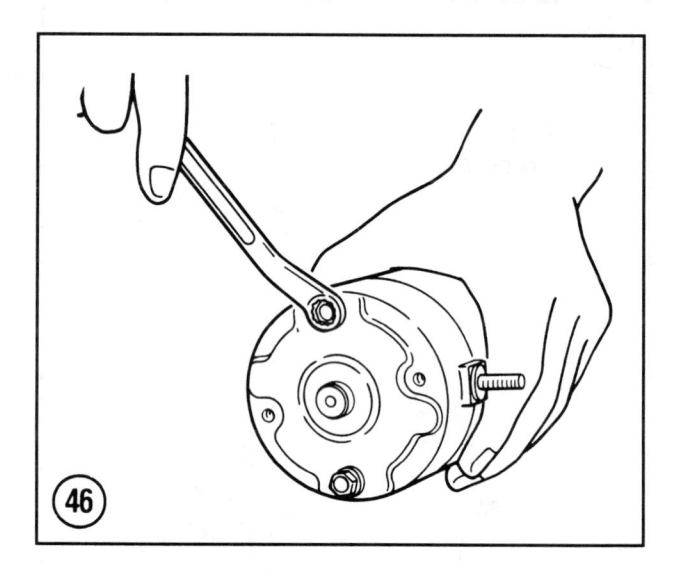

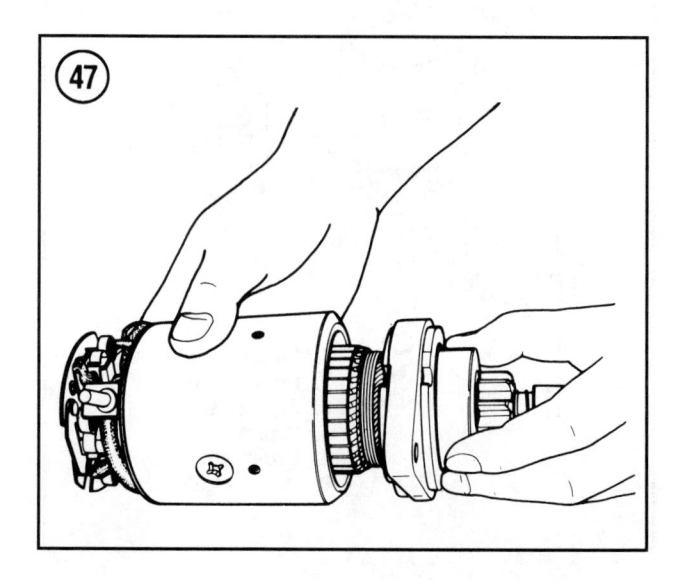

ELECTRICAL SYSTEMS

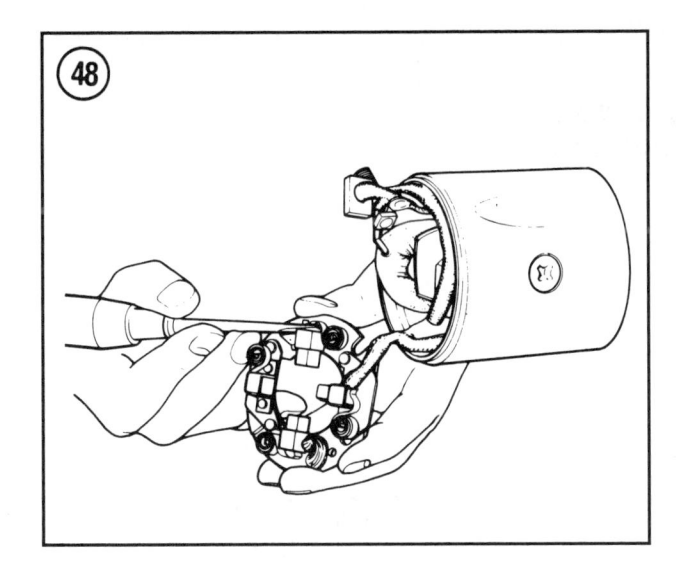

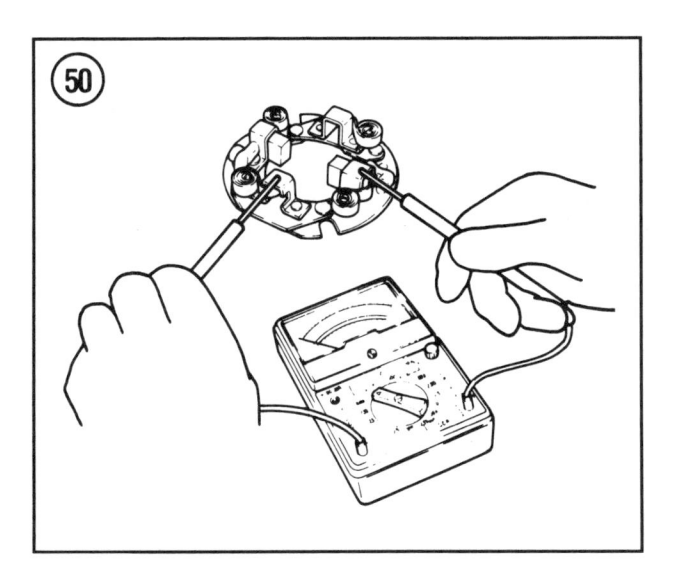

18. Install the 2 screws securing the brush holder cover.

19. Install thrust washer and E-clip onto the brush holder cover.

20. Install the 2 through-bolts and washers into the starter and tighten securely.

21. Install the starter cable and the terminal nut.

22. Install the starter as described in this chapter.

Brush Replacement
(1989-on DT 25; 1988-on DT 30; 1985-on DT 35, DT 40; 1985-on DT 55, DT 65; 1986-on DT 115, DT 140; V4 and V6 With 2- and 3-Brush Starter)

Refer to **Figure 51** for the 2-brush design or **Figure 52** for the typical 3-brush design for this procedure. Always replace all brushes in complete sets.

1. Remove the starter as described in this chapter.
2. If still attached, remove the terminal nut (**Figure 53**) and remove the starter cable.
3. Remove the 2 through-bolts from the starter (**Figure 54**).
4. Lightly tap on brush holder cover with a rubber mallet until it breaks free of starter housing. Remove brush holder cover, taking care not to lose the brush springs (**Figure 55**).
5. On models so equipped, remove the brush holder cover thrust washer from the armature.

NOTE
If corrosion causes the brushes to stick during Step 6, replace the brush holder plate.

6. Check brush spring tension by pulling spring back and releasing it. Replace the spring if it does not snap the brush firmly into position.
7. Remove the nut from the brush holder cover and remove the insulators and O-ring seal (**Figure 56**).
8A. On 1989-on DT 25, 1988-on DT 30, 1985-on DT 35, DT 40 models, remove the screws

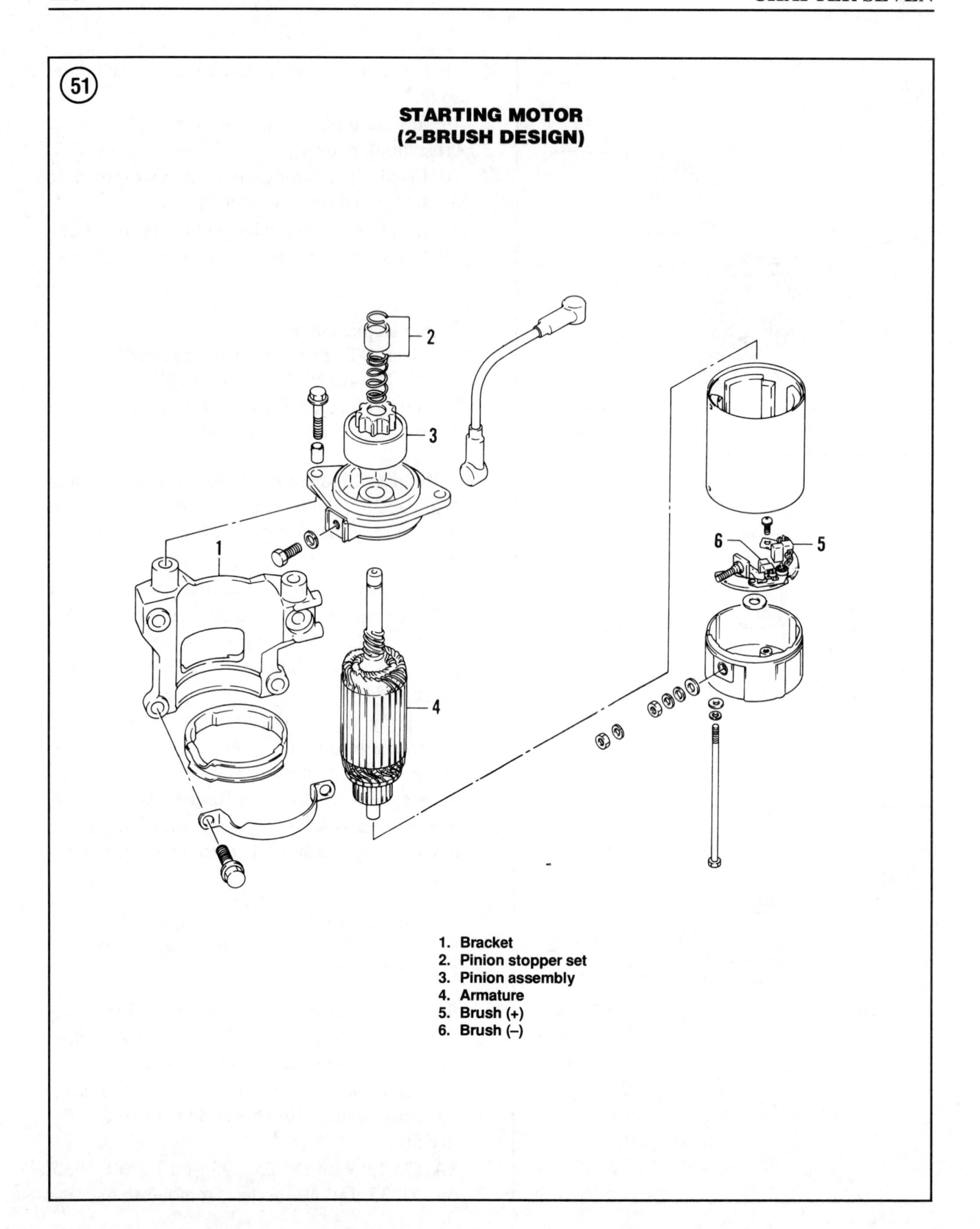

(51)

STARTING MOTOR (2-BRUSH DESIGN)

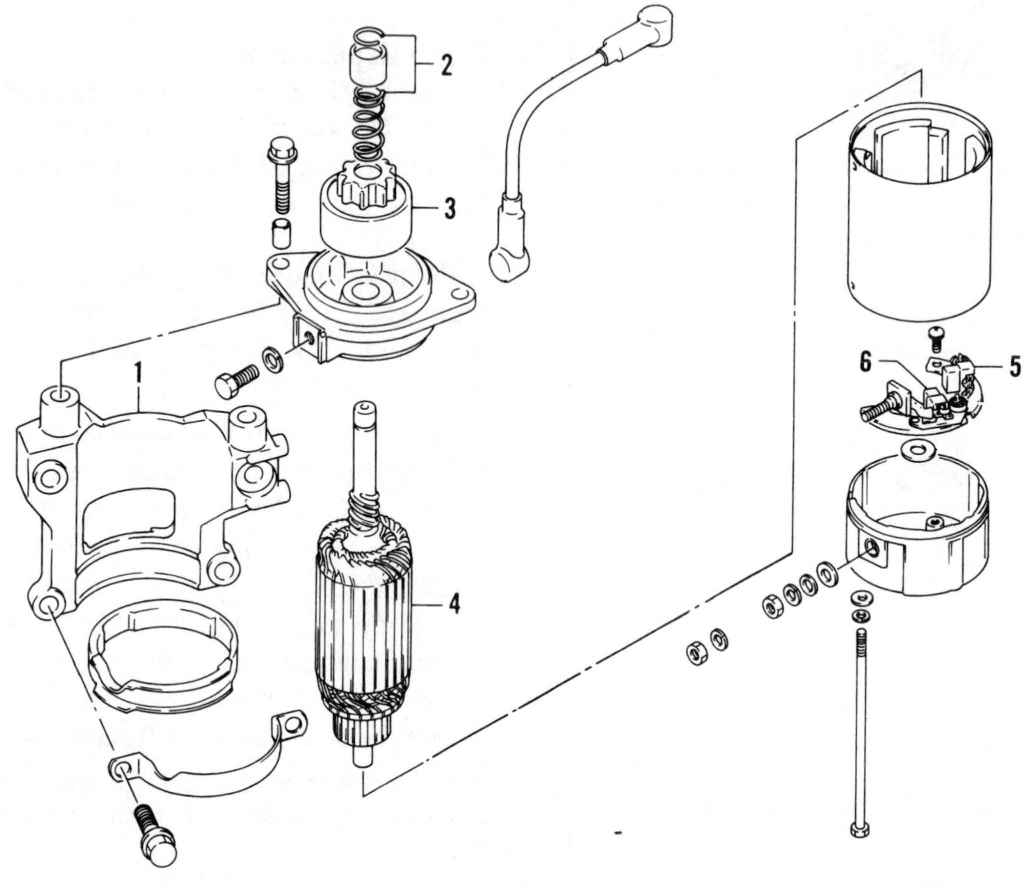

1. Bracket
2. Pinion stopper set
3. Pinion assembly
4. Armature
5. Brush (+)
6. Brush (–)

Electrical Systems

217

⑤²

**STARTING MOTOR
(3-BRUSH DESIGN)**

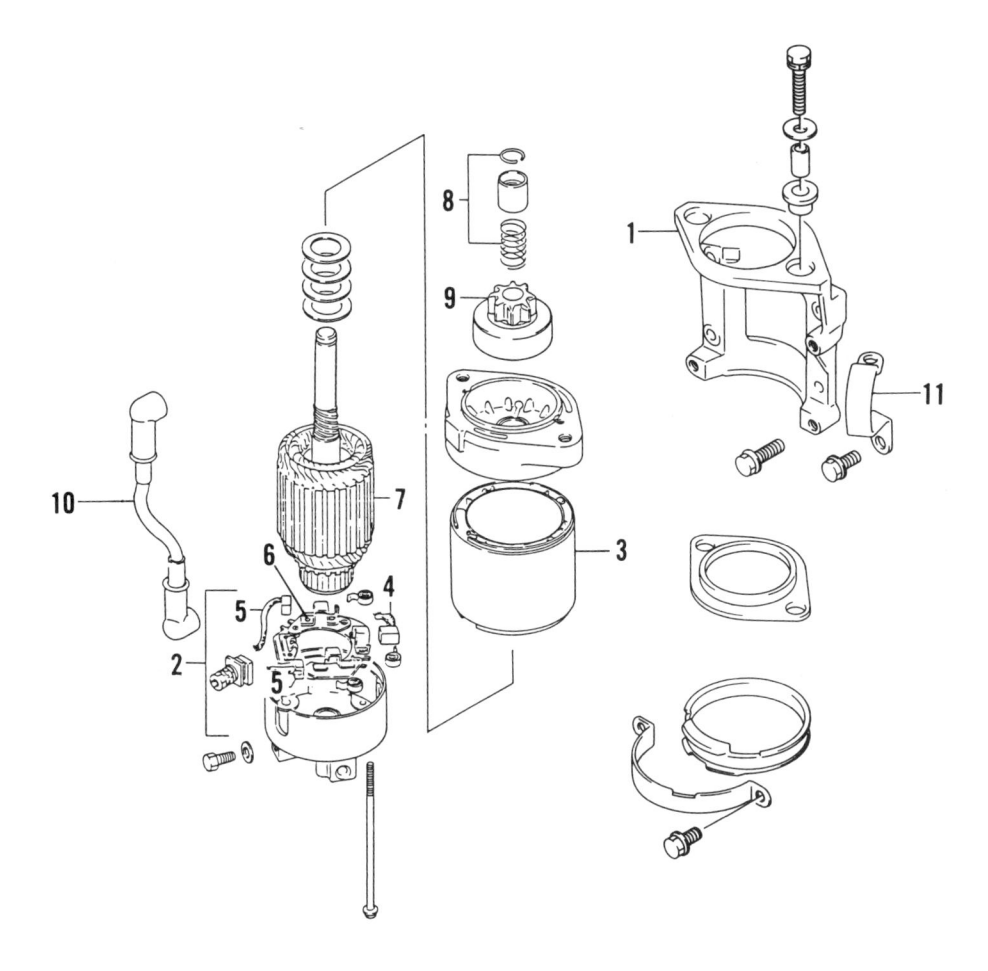

1. Bracket
2. Brush holder cover assembly
3. Field coil case
4. Brush (−)
5. Brush (+)
6. Brush holder assembly
7. Armature
8. Pinion stopper set
9. Pinion assembly
10. Lead wire (+)
11. Engine hook

7

(**Figure 57**) securing the brush holder plate and remove the plate.

8B. On all other models, remove the screws (**Figure 58**) securing the brush holder plate and remove the plate.

9. Remove the brushes and springs from the brush holder plate.

10. Inspect the brushes and replace all brushes if any are pitted or oil-soaked.

11. Measure each brush with a vernier caliper (**Figure 59**). Replace brushes if worn to the following dimension or less:
   a. 1989-on DT 25, 1988-on DT 30, 1985-on DT 35, DT 40: 0.35 in. (9.0 mm).
   b. All other models: 0.45 in. (11.5 mm).

12. Remove the brushes from the brush holder.

13. Connect an ohmmeter between the negative and positive brush holders on the brush holder plate (**Figure 60**). Replace the brush holder plate if the meter shows continuity.

14. Install the brushes into the brush holder.

15. Fit the springs and brushes into their respective brush holders.

16. Coat brush holder cover bore with water-resistant grease

17A. On 1989-on DT 25, 1988-on DT 30, 1985-on DT 35, DT 40 models, install the brush holder plate and install the screws (**Figure 57**) securing the plate. Tighten the screws securely.

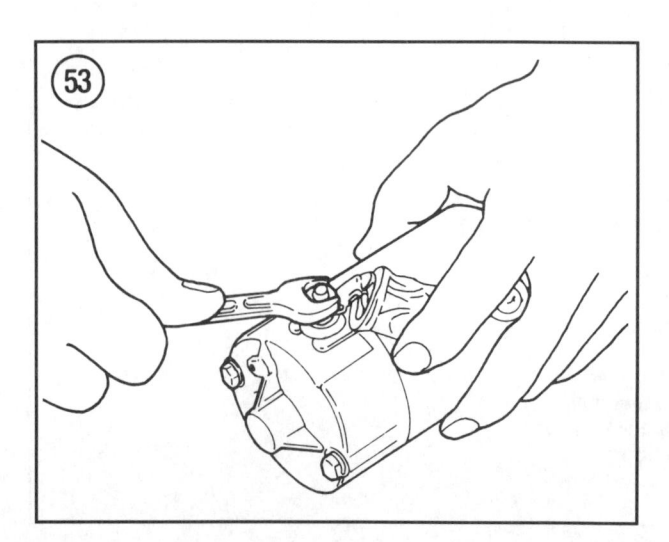

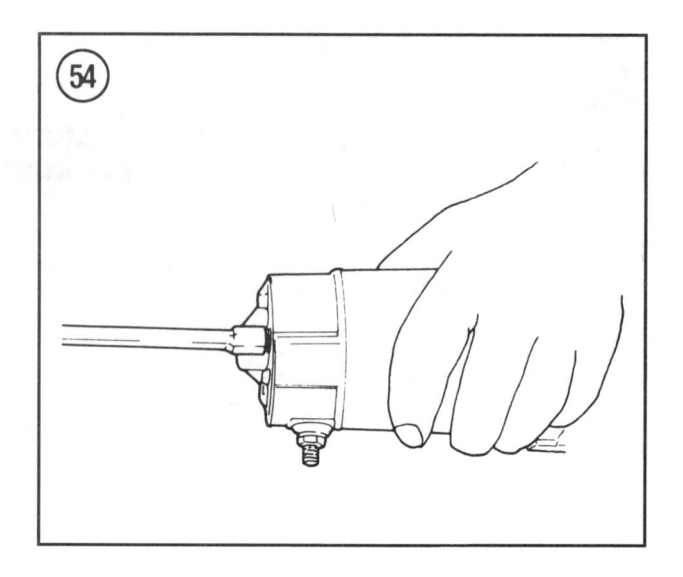

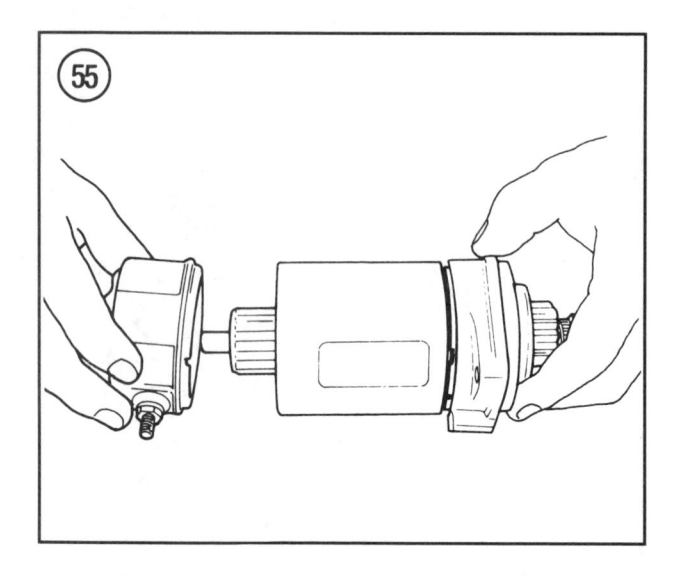

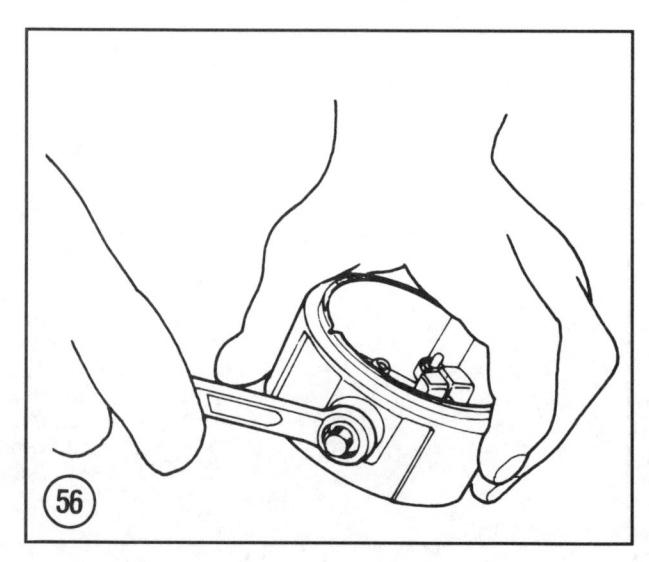

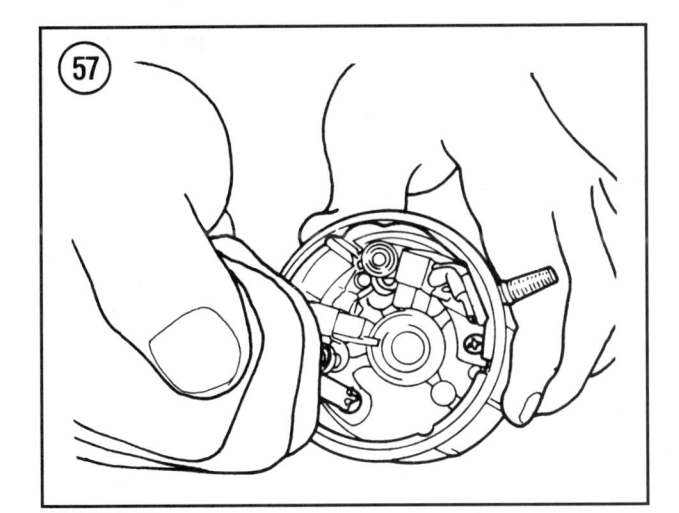

ELECTRICAL SYSTEMS

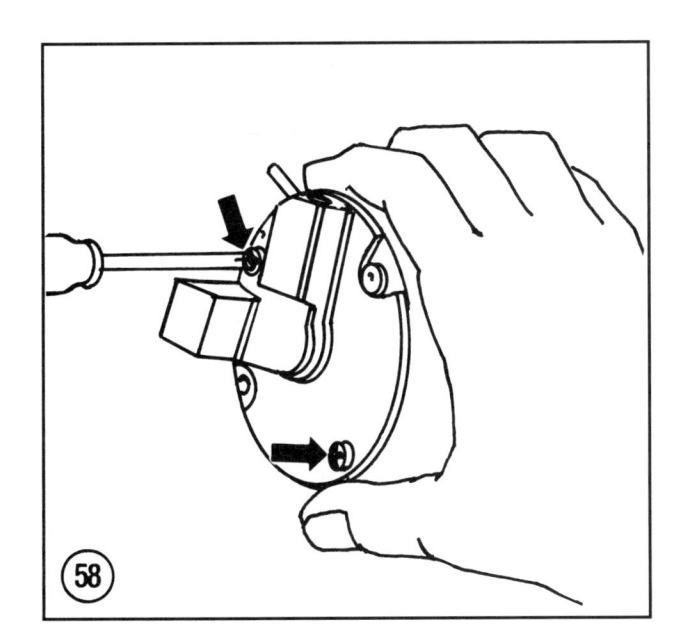

17B. On all other models, install the brush holder plate and install the screws (**Figure 58**) securing the plate. Tighten the screws securely.

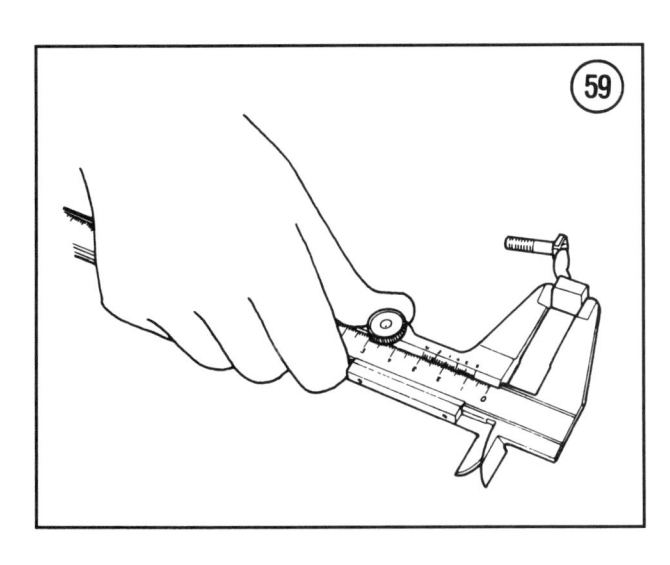

18. Install the O-ring seal and insulators onto the threaded stud (**Figure 56**).

19. On models so equipped, install the brush holder cover thrust washer onto the armature.

20. Install the brush holder cover.

21. Align the notch in the brush holder cover (B) with the alignment mark (A) on the field coil case (**Figure 61**).

22. Install through-bolts and tighten securely.

23. If removed from the starter, install the starter cable and the terminal nut.

24. Install the starter as described in this chapter.

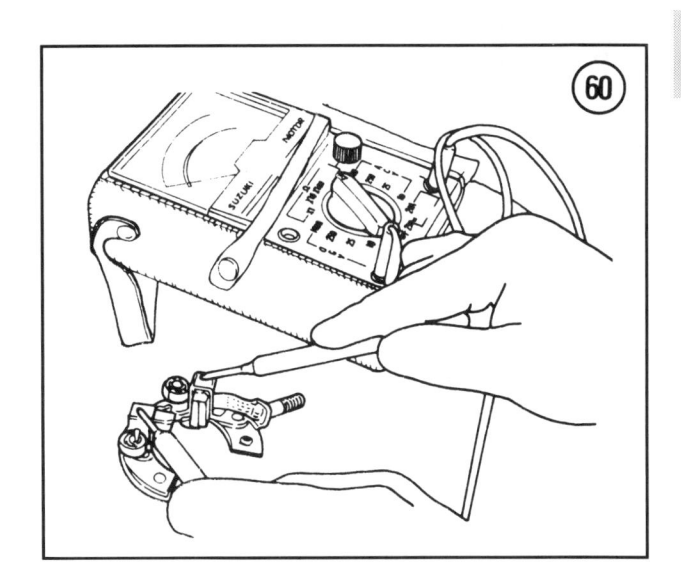

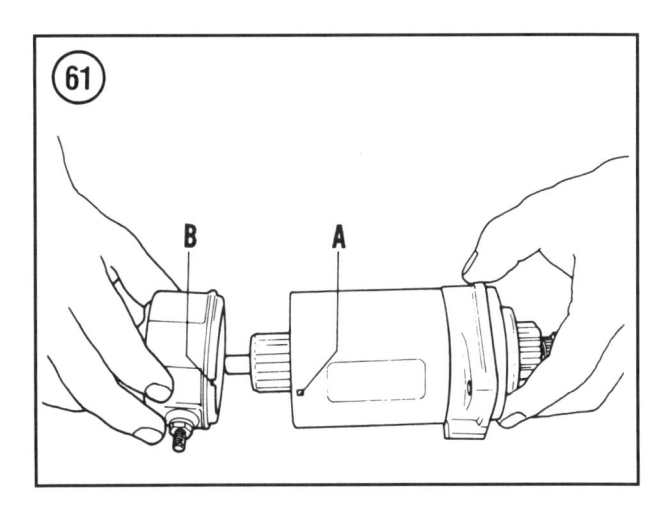

CHAPTER SEVEN

Starter Relay Replacement

1. Disconnect the negative battery cable.
2. Remove the engine cover.
3. Disconnect the electrical leads from the starter relay lead at its bullet connector. Remove the screw holding the black relay ground lead.
4. Disconnect the battery and starter cables from the relay terminals.
5. Remove the relay.
6. Installation is the reverse of removal, noting the following.
7. Make sure all electrical connectors are free of corrosion and are tight.

Starter Relay Testing

1. Remove the starter relay as described in this chapter.
2. Connect an ohmmeter's test leads to the large terminals on top of the starter relay. There should be no continuity (infinite resistance).
3. Connect the starter relay black and yellow/red (or yellow/green) electrical leads to a fully charged 12 volt battery (**Figure 62**).
4. Connect an ohmmeter's test leads to the large terminals on top of the starter relay. There should be continuity (low resistance).
5. Disconnect the electrical leads from the battery.
6. If the starter relay fails either of these tests, the relay is faulty and must be replaced.

MAGNETO BREAKER POINT IGNITION (1985-1989 DT 2)

The DT 2 uses a magneto ignition with a combined primary/secondary ignition coil, a condenser and one set of breaker points.

The ignition coil, condenser and breaker point set are mounted on the stator base under the flywheel. **Figure 63** shows the stator base with primary/secondary ignition coil (A), condenser (B) and breaker point set (C).

Troubleshooting and test procedures are given in Chapter Three.

Operation

As the flywheel rotates, magnets around its outer diameter create a current that flows through the closed breaker points into the ignition coil primary windings. This flow of current through the coil primary winding builds a strong magnetic field. When the cam opens the point set, the magnetic field collapses, inducing a high voltage (approximately 18,000 volts) in the coil secondary winding; this voltage is sent to the spark plug. The condenser absorbs any residual

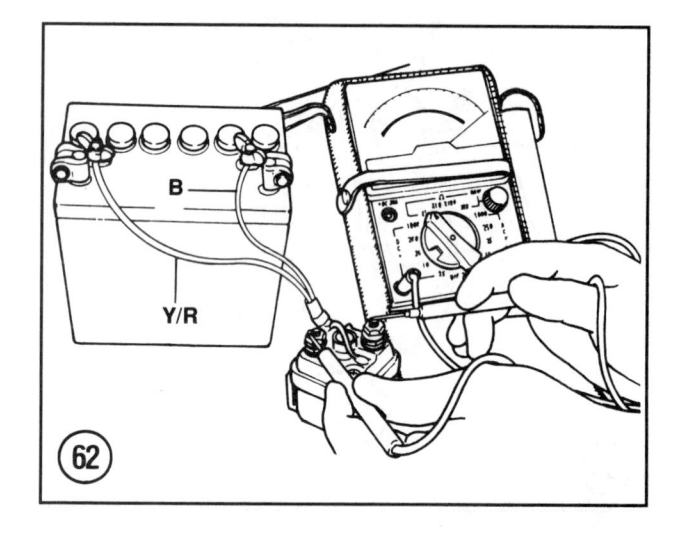

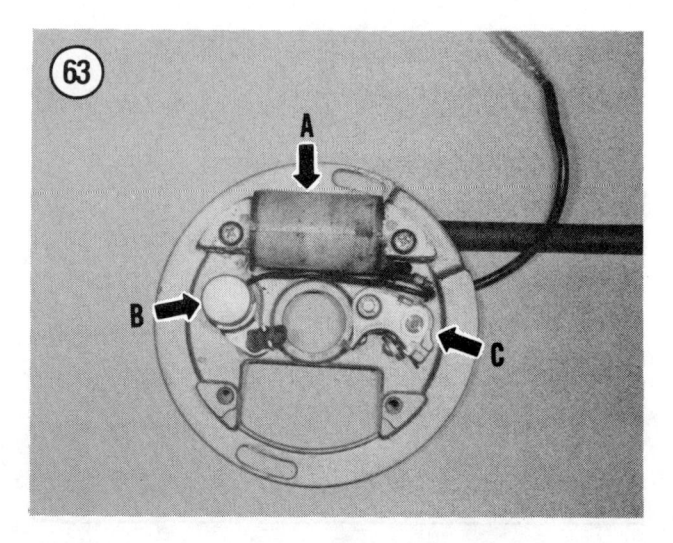

ELECTRICAL SYSTEMS

current remaining in the primary windings while the breaker points are open. This eliminates arcing at the points and produces a stronger spark at the plug. The breaker points close and the flywheel continues to rotate, duplicating the sequence for the next firing cycle.

Stator Base
Removal/Installation

Refer to **Figure 64** for this procedure.

1. Remove the engine cover.
2. Remove the fuel tank. See Chapter Six.
3. Remove the rewind starter assembly. See Chapter Ten.
4. Remove the flywheel (2). See Chapter Eight.
5. Disconnect the spark plug lead (3) from the spark plug.
6. Remove the spark plug cap from the spark plug lead (**Figure 65**).
7. Disconnect the stator lead wires at their bullet connectors.
8. Remove the screws holding the stator to the magneto housing (**Figure 66A**). Remove the stator base.
9. Installation is the reverse of removal, noting the following.

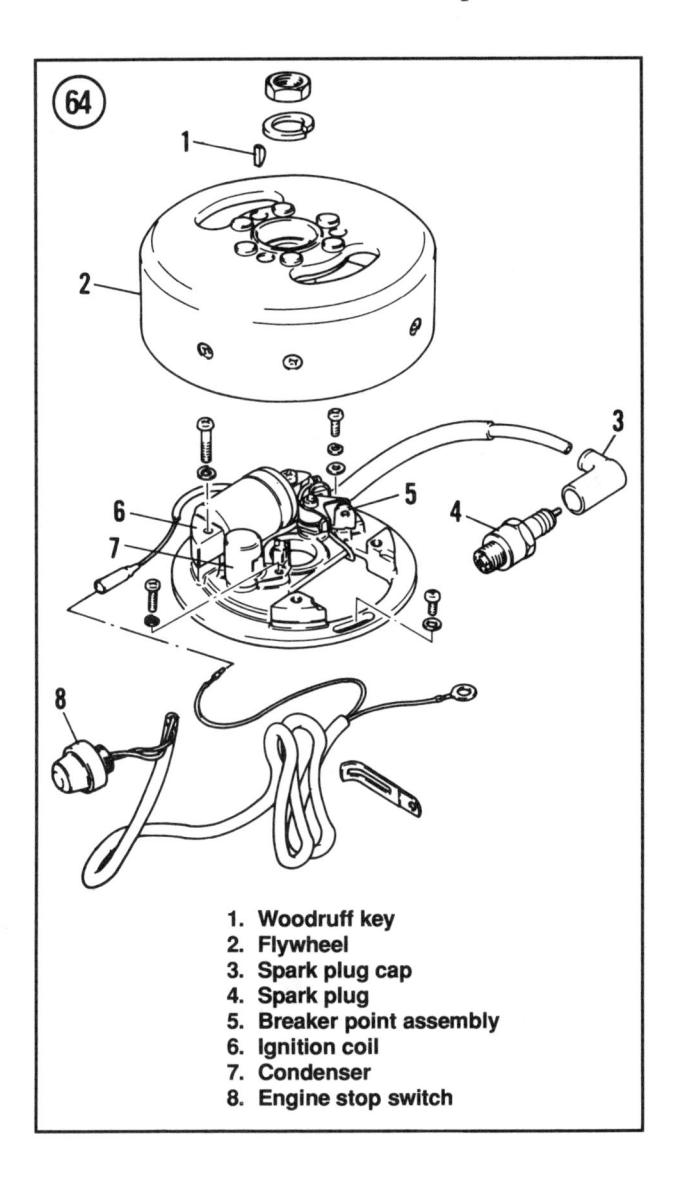

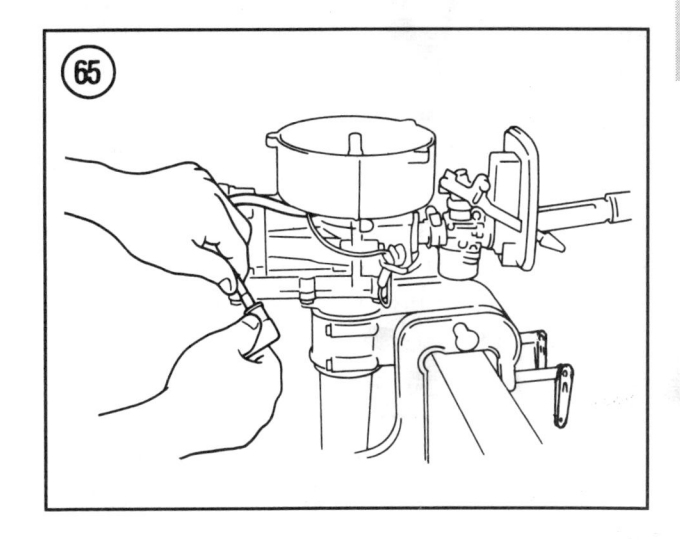

1. Woodruff key
2. Flywheel
3. Spark plug cap
4. Spark plug
5. Breaker point assembly
6. Ignition coil
7. Condenser
8. Engine stop switch

10. Make sure the stator mounting screws are tightened securely.
11. Make sure all electrical connectors are free of corrosion and are tight.

Breaker Point and Condenser Replacement

See *Tune-up*, Chapter Four.

Primary/Secondary Ignition Coil Removal/Installation

Refer to **Figure 64** for this procedure.
1. Remove the stator base as described in this chapter.
2. Disconnect the coil lead wires at the breaker point set and ground.
3. Remove the spark plug cap (3) from the plug lead.
4. Remove the 2 screws and washers holding the coil (6) to the stator base.
5. Pull the spark plug lead through the stator base grommet and remove the coil.
6. Installation is the reverse of removal, noting the following.
7. Insert the spark plug lead through the stator base cutout before installing coil mounting screws.
8. Make sure all electrical connectors are free of corrosion and are tight.

BREAKERLESS TRANSISTORIZED IGNITION (1990-ON DT 2)

A breakerless, transistorized ignition system is used on 1990-on DT 2 models. The ignition coil, transistors and electronic circuitry are contained in a one-piece igniter unit mounted outside the flywheel. See **Figure 66B**. Refer to Chapter Three for troubleshooting procedures.

Igniter Unit Removal/Installation

Refer to **Figure 66B** for this procedure.
1. Remove the engine cover.
2. Disconnect the negative battery cable.
3. Remove the fuel tank. See Chapter Six.
4. Remove the rewind starter. See Chapter Ten.
5. Disconnect the spark plug lead from the spark plug.
6. Disconnect the igniter unit primary lead at the bullet connector.
7. Remove the 2 screws securing the igniter unit to the power head. Lift off the igniter unit.
8. Installation is the reverse of removal. Note the following.
9. Using a feeler gage, set the air gap (A, **Figure 66B**) between the legs of the igniter unit and flywheel magnet to 0.016 in., then securely tighten igniter unit mounting fasteners.

Stator Base Removal/Installation

1. Remove the engine cover.
2. Remove the fuel tank. See Chapter Six.
3. Remove the rewind starter. See Chapter Ten.
4. Remove the igniter unit. See *Igniter Unit Removal/Installation* in this chapter.

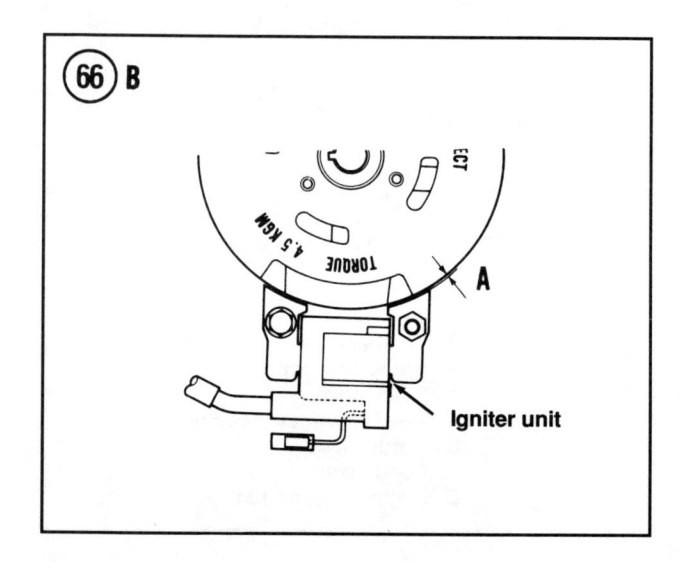

ELECTRICAL SYSTEMS

5. Remove the flywheel. See Chapter Eight.

6. Remove the screws holding the stator to the magneto housing (**Figure 66A**). Lift the stator base off the power head.

7. Installation is the reverse of removal. Make sure stator base mounting screws are securely tightened. Make sure electrical connections are clean and tight.

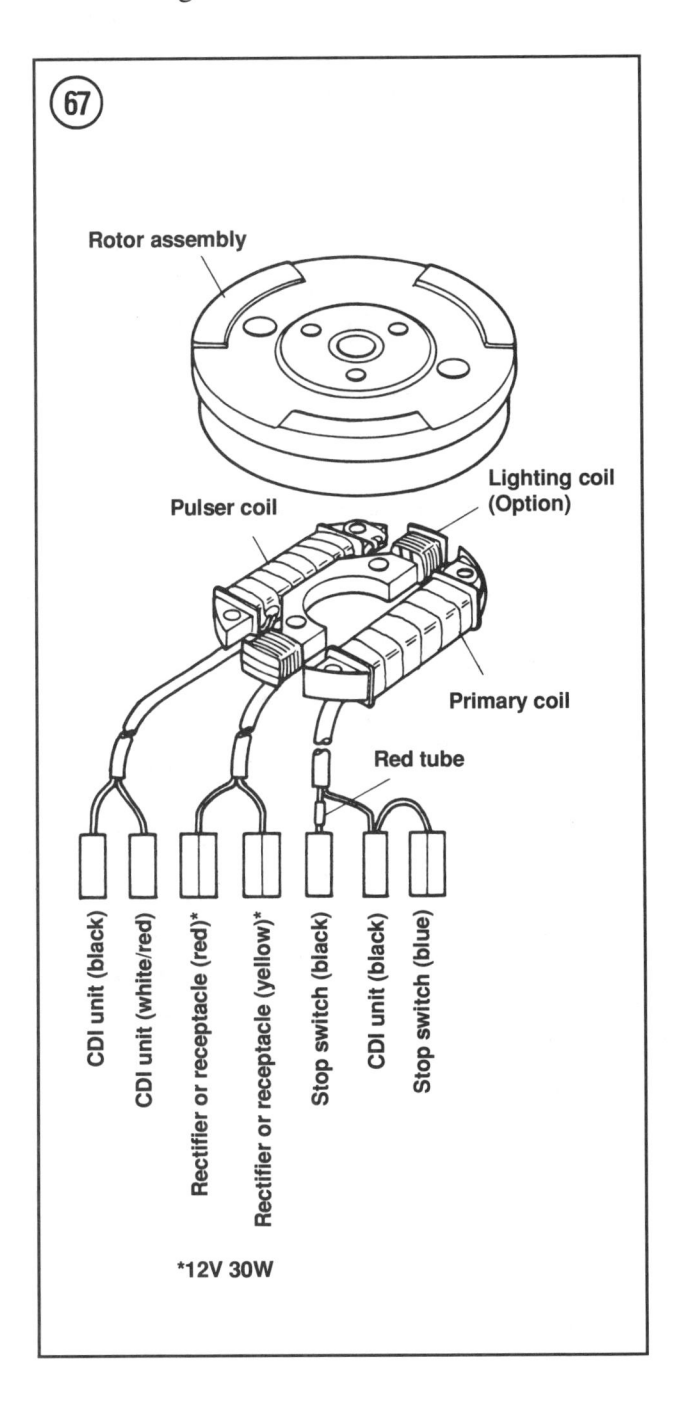

PEI IGNITION
(DT 4; DT 6; DT 8, DT 9.9; DT 15; 1985-1987 DT 20, DT 30; 1985-1988 DT 25; 1987-1989 DT 35; 1985-ON DT 40; 1985-1987 DT 75, DT 85; 1985 DT 115, DT 140)

The Suzuki PEI (Pointless Electronic Ignition) system is a magneto CDI (capacitor discharge ignition) system. Several variations have been used on the models covered in this manual:

1. The DT 4 models have an ignition which uses a single primary coil, 1 pulser coil, an optional lighting coil and 1 ignition coil. Refer to **Figure 67**. The CDI unit electronically advances ignition timing when firing the spark plugs. The CDI unit and secondary ignition coils are separate components.

2. The DT 6 and 1985-1987 DT 8 models use a single condenser charge coil, a single pulser coil and a lighting coil (**Figure 68**) which supplies current to a combined CDI unit and single secondary coil. Ignition advance is mechanical, with the stator base and throttle interlocked.

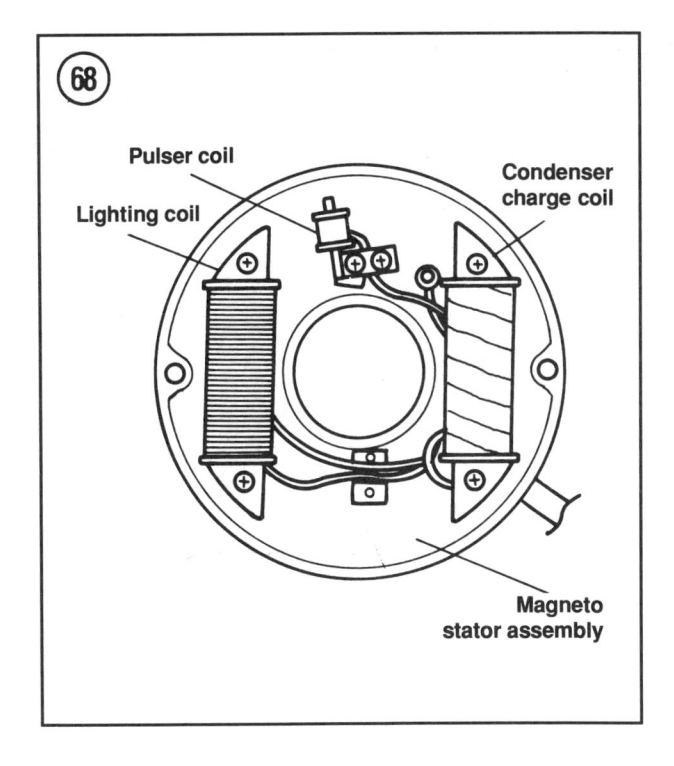

3. The 1988-on DT 8 and DT 9.9 and 1989-on DT 15 models have an ignition which uses a single condenser charging coil, 2 pulser coils and 2 ignition coils, one for each cylinder. Refer to **Figure 69** for 1988-on DT 8, DT 9.9 models or **Figure 70** for 1985-on DT 9.9, DT 15. The CDI unit electronically advances ignition timing when firing the spark plugs. The CDI unit and secondary ignition coils are separate components.

4. The 1985-1987 DT 9.9 and 1985-1988 DT 15 models use a single charge coil which supplies current to a combined CDI unit and single secondary coil (**Figure 71**). Ignition advance is mechanical, with the stator base and throttle interlocked. The CDI unit and secondary ignition coil are separate components.

5. The 1985-1987 DT 20, DT 30 and 1985-1988 DT 25 models use a single charge coil and a single trigger coil (**Figure 72**) which supplies current to a combined CDI unit and single secondary coil. Ignition advance is mechanical, with the stator base and throttle interlocked.

6. The 1987-1989 DT 35 and 1985-on DT 40 models use a single charge coil and a single trigger coil (**Figure 73**) which supplies current to a combined CDI unit and single secondary coil. Ignition advance is mechanical, with the stator base and throttle interlocked.

7. The 1985-1987 DT 75, DT 85 and 1985 DT 115, DT 140 models use an alternator stator with 2 charge coils and 8 (3-cylinder) or 10 (4-cylinder) battery charging coils. **Figure 74** shows the 3-cylinder stator assembly. An aluminum timer

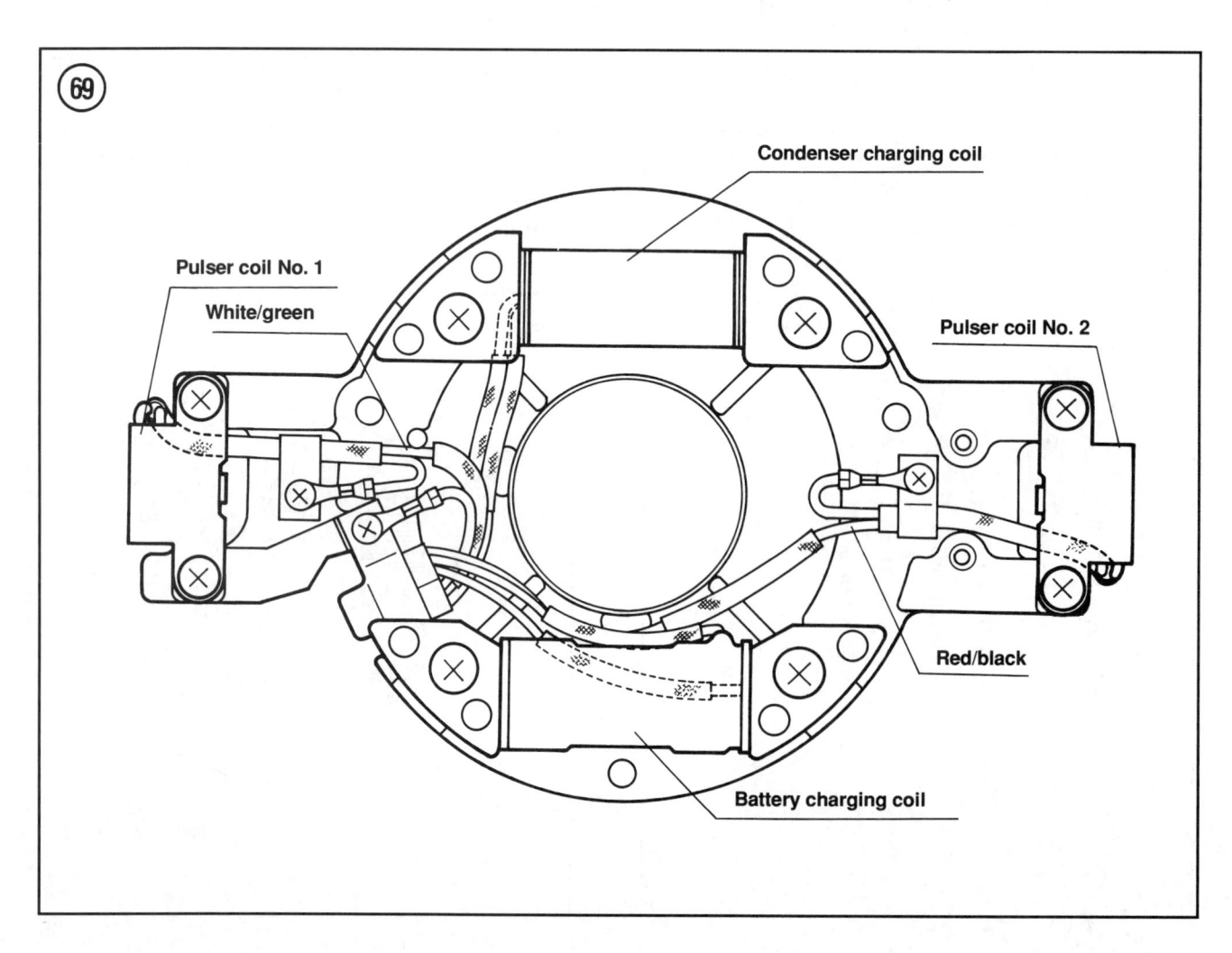

ELECTRICAL SYSTEMS

225

⑦

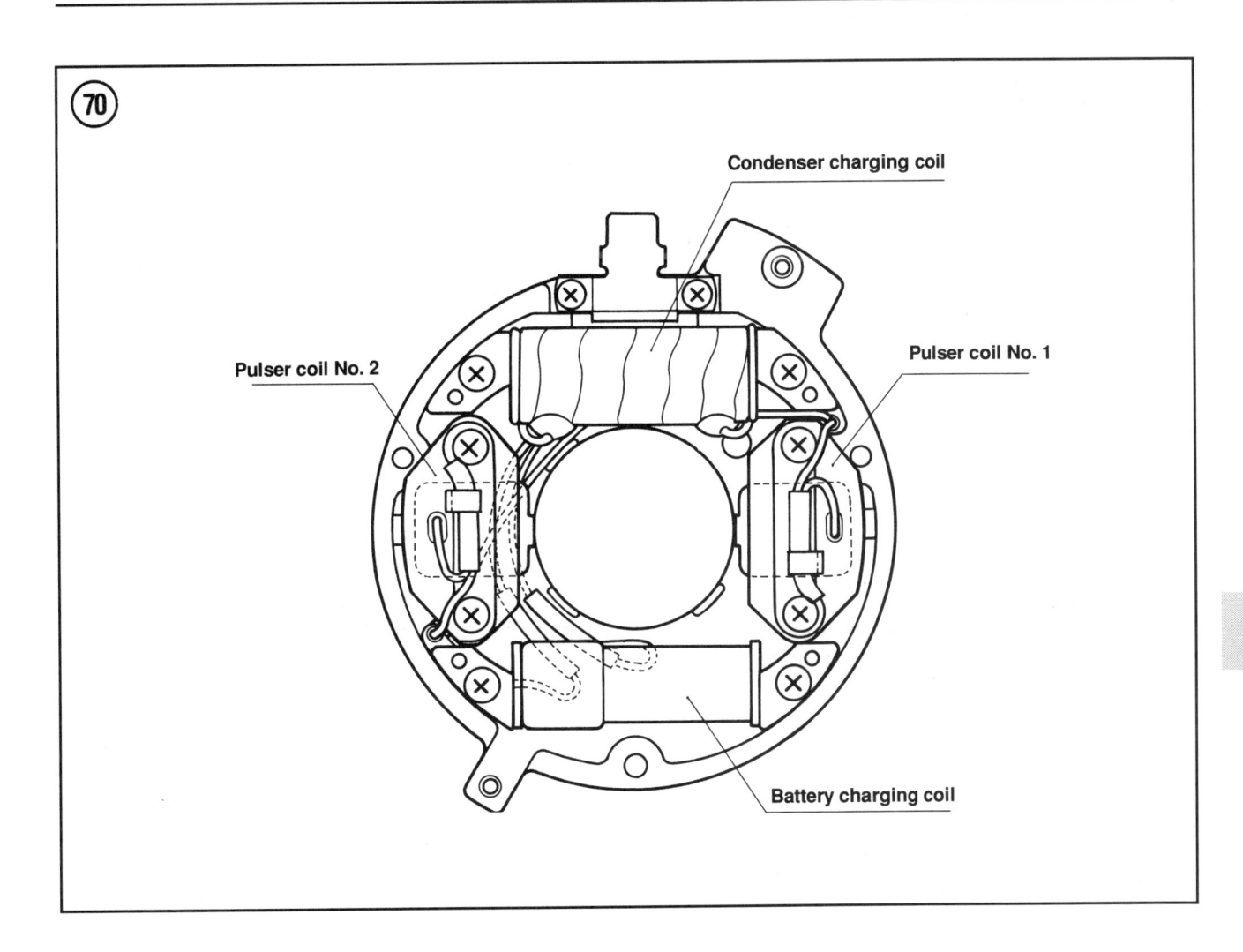

- Condenser charging coil
- Pulser coil No. 2
- Pulser coil No. 1
- Battery charging coil

⑦¹

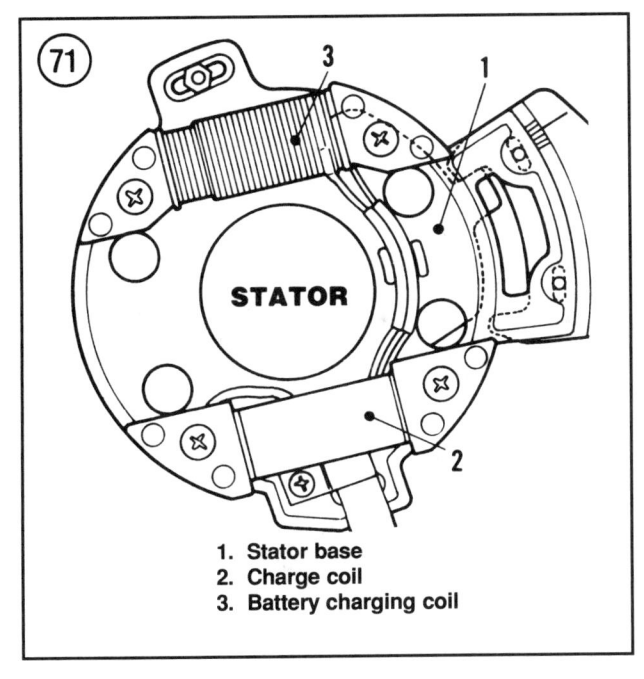

STATOR

1. Stator base
2. Charge coil
3. Battery charging coil

⑦²

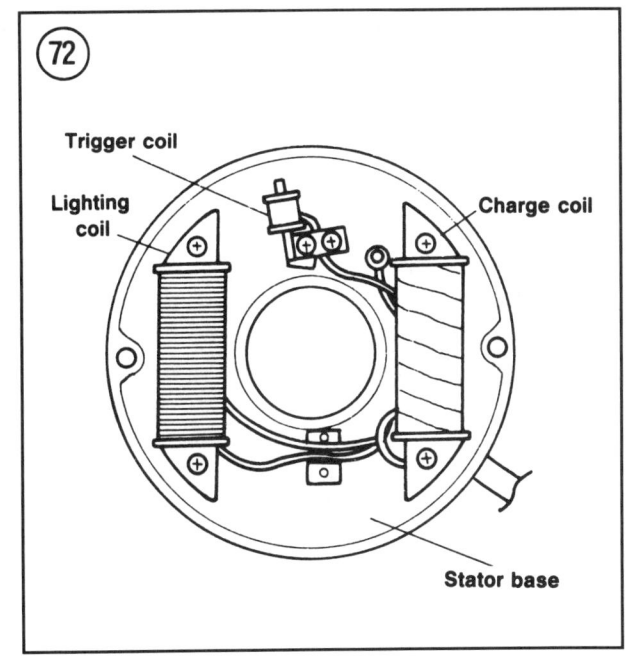

- Trigger coil
- Lighting coil
- Charge coil
- Stator base

7

base located underneath the stator contains 2 trigger coils enclosed in iron to help build up the magnetic field and prevent interference from external sources. With 3-cylinder engines, one trigger coil controls the No. 1 and No. 3 cylinders (**Figure 75**). The other trigger coil controls the No. 2 cylinder. On 4-cylinder engines, one trigger coil controls the No. 1 and No. 2 cylinders; the second trigger coil controls the No. 3 and No. 4 cylinders (**Figure 76**). Ignition advance is mechanical.

Ignition timing should be adjusted (Chapter Five) whenever a component is replaced.

Operation

The outer rim of the flywheel contains a series of magnets which create a magnetic field during rotation. This magnetic field cuts through the charge coil windings and produces an alternating current of positive and negative wave forms. This current is sent to the CDI unit where it is changed into direct current by an internal rectifier and stored in a capacitor.

On models without a trigger (pulser) coil, the CDI capacitor is charged by positive wave forms. Negative wave forms cause an electronic switch in the CDI unit to close, discharging the stored voltage into the ignition coil where it is stepped up to a higher voltage and sent to the spark plugs.

On models with a trigger (pulser) coil, the rotation of a timing magnet in the flywheel hub

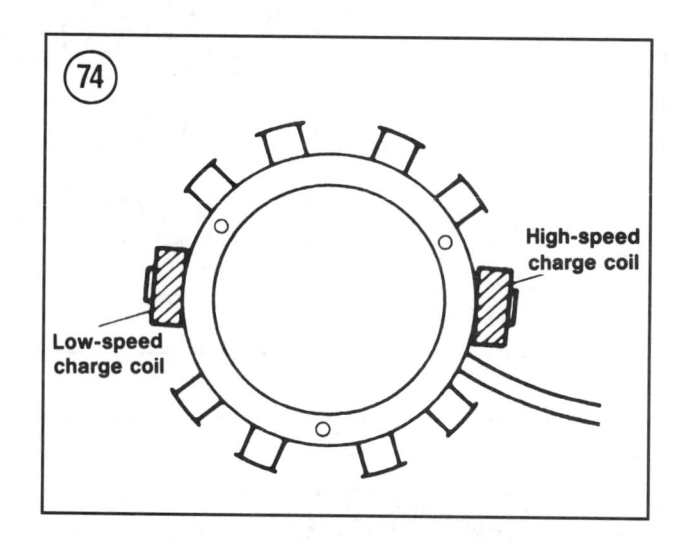

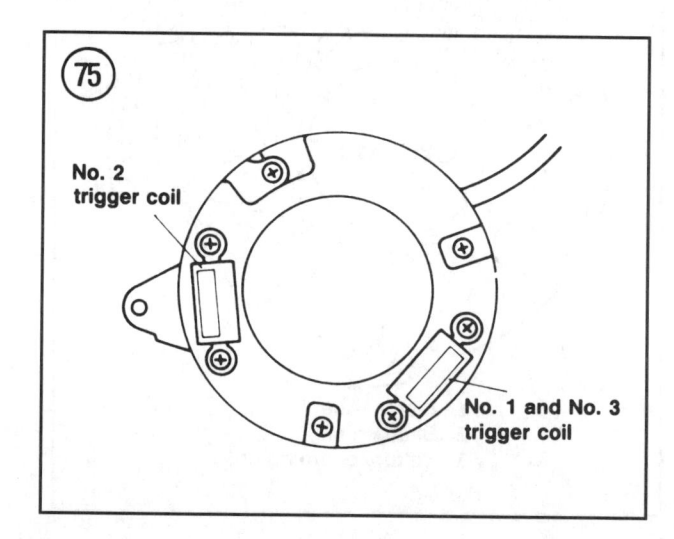

ELECTRICAL SYSTEMS

76

No. 3 and No. 4 trigger coil

No. 1 and No. 2 trigger coil

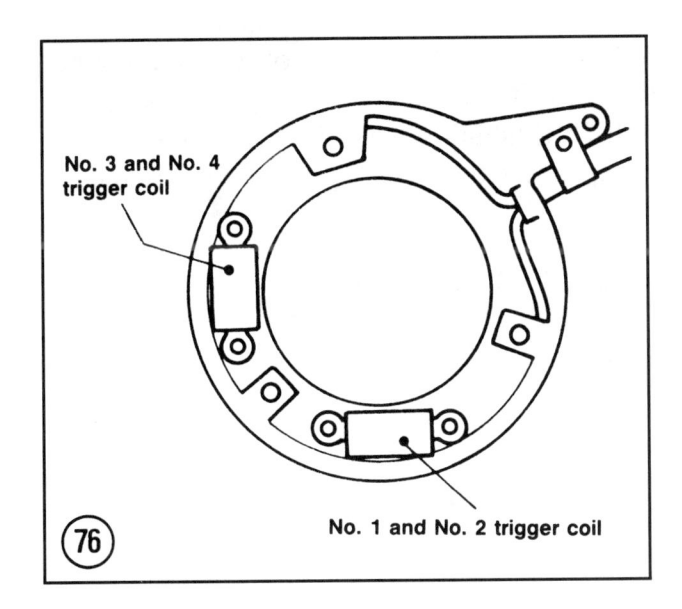

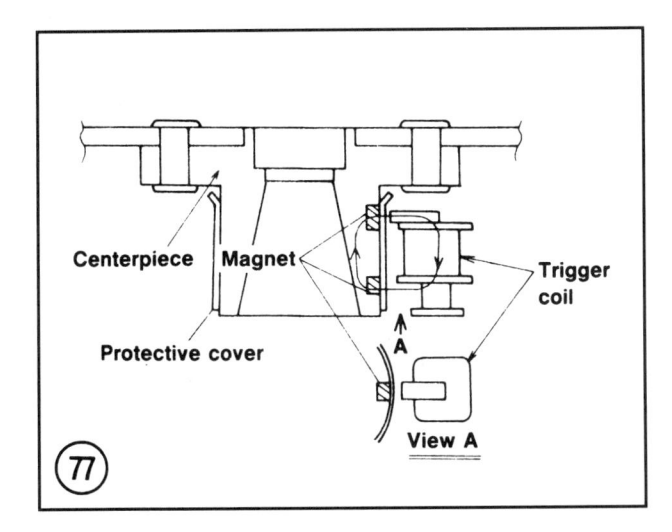

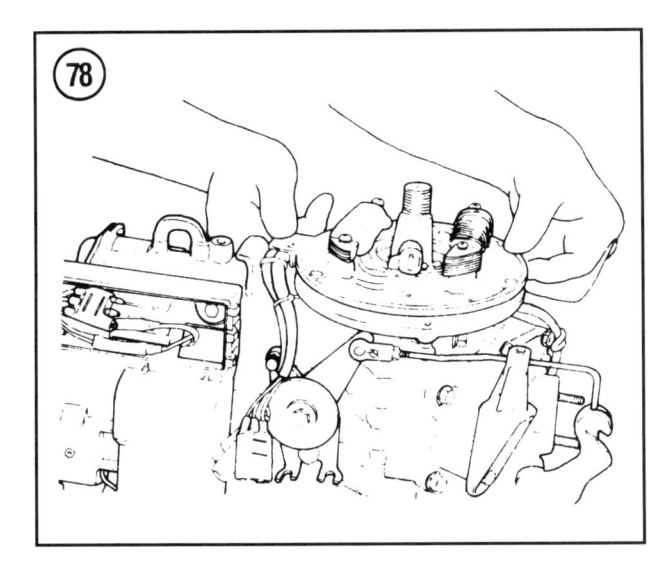

77

Centerpiece | Magnet | Trigger coil
Protective cover
View A

78

past the trigger coil on the stator base creates a magnetic field (**Figure 77**). As the flywheel continues to rotate, this magnetic field collapses, inducing a small voltage pulse in the trigger coil. This pulse causes an electronic switch in the CDI module to close, discharging the stored voltage into the ignition coil where it is stepped up to a higher voltage and sent to the spark plugs.

Independent ignitions use a separate ignition coil to fire each cylinder; simultaneous ignitions use a single coil to fire both cylinders. The simultaneous ignition system is also called a "waste spark" system. When the piston in one cylinder is at TDC, the other is at BDTC. If both spark plugs fire at the same time, the spark in the cylinder with the piston at TDC is "used" while the spark in the other cylinder is "wasted."

Depressing the stop switch shorts the charge coil to ground and shuts the engine off. On some models, the charge coil is shorted to ground through a low voltage engine stop circuit to prevent a possible voltage leak.

Stator Base/Alternator Stator Removal/Installation

1. Disconnect the negative battery cable, if so equipped.
2. Remove the engine cover.
3. Remove the flywheel. See Chapter Eight.
4. Disconnect the stator leads at their connectors. On some models, the stator lead connections are housed in the electrical junction box.
5. Disconnect the stator leads at their connectors.
6A. Stator base—Remove the screws holding the stator base to the upper oil seal housing. Remove the stator base (**Figure 78**, typical).
6B. Alternator stator—Remove the screws holding the alternator stator to the timer base. Remove the alternator stator (**Figure 79**, typical).
7. Installation is the reverse of removal, noting the following.
8. Make sure all electrical connectors are free of corrosion and are tight.

Timer Base Removal/Installation
(1988-on DT 8, DT 9.9; 1985-1987 DT 75, DT 85)

1. Remove the alternator stator as described in this chapter.
2. Disconnect the timer base link from the spark advance lever.
3. Remove the junction box cover and disconnect the trigger coil leads.
4. Remove the timer base retaining screws (**Figure 80**). Remove the timer base.
5. Installation is the reverse of removal, noting the following.
6. Make sure all electrical connectors are free of corrosion and are tight.

Charge Coil or Lighting/ Charging Coil Replacement
(1985-1987 DT 75, DT 85 and 1985 DT 115, DT 140)

The alternator stator is replaced as an assembly if any coil is defective. The individual coils cannot be replaced separately.

Charge Coil, Lighting Coil, Battery Charging Coil or Trigger Coil Replacement
(Except 1985-1987 DT 75, DT 85; 1985 DT 115, DT 140)

1. Remove the stator base as described in this chapter.
2. Remove the 2 screws holding the defective coil to the stator base.
3. Unwrap the tie straps or wiring harness cover to separate the stator leads. On some models, PVC tubing may be used to form a wiring harness cover. If so, carefully slit the tubing to remove the coil lead(s).
4. Remove the defective coil from the stator base, pulling its lead(s) through the base cutout.
5. Installation is the reverse of removal, noting the following.
6. If PVC tubing was used for a wiring harness cover, secure the leads together with tie straps.

Trigger Coil Replacement
(1985-1987 DT 75, DT 85 and 1985 DT 115, DT 140)

See *Charge Coil, Lighting Coil, Battery Charging Coil or Trigger Coil Replacement (Except 1985-1987 DT 75, DT 85 and 1985 DT 115, DT 140)* in this chapter.

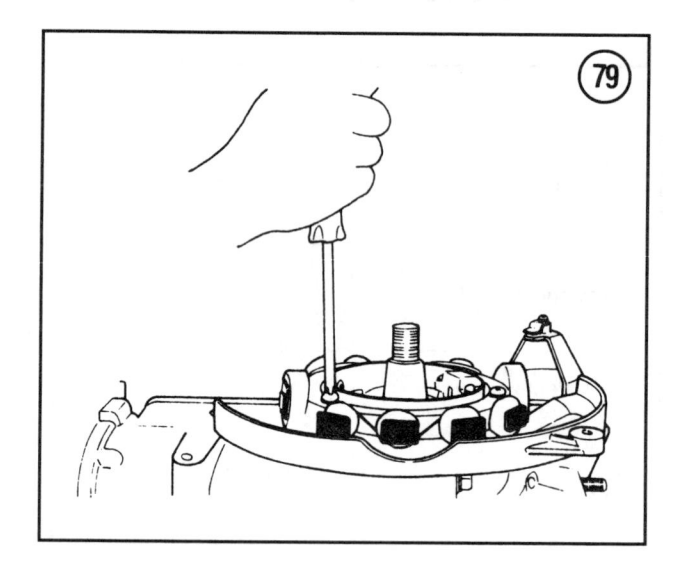

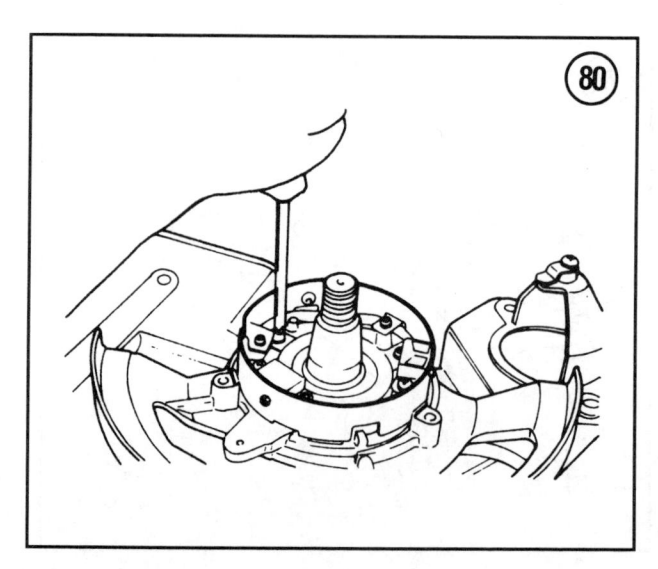

ELECTRICAL SYSTEMS

INTEGRATED CIRCUIT (I.C.) IGNITION SYSTEM
(1989-ON DT 25; 1988-ON DT 30; 1986-ON DT 55, DT 65; 1988-ON DT 75, DT 85; 1986-ON DT 115, DT 140; 1985-1988 V6)

The Suzuki I.C. electronic ignition system is a magneto CDI (capacitor discharge ignition) system. Because of the integrated circuit feature, it has the ability to monitor the degree of throttle opening as well as engine rpm to determine the ideal spark timing for best possible performance (**Figure 81**, typical). This creates improved acceleration by maintaining optimum carburetor and ignition synchronization which enables the engine to run smoother. This system also creates easier start-ups, crisper acceleration and improved top-end performance.

There are several variations to the system used on these models but all operate in the same manner. There is a single charge coil and a pulser coil for each cylinder which supply current to the CDI unit (**Figure 82**, typical). The CDI unit in turn triggers each ignition coil to fire the spark plugs as required. Ignition advance is electronic.

Ignition timing adjustment is not necessary on these models with the exception of adjusting the throttle valve sensor that is covered in Chapter Five.

Operation

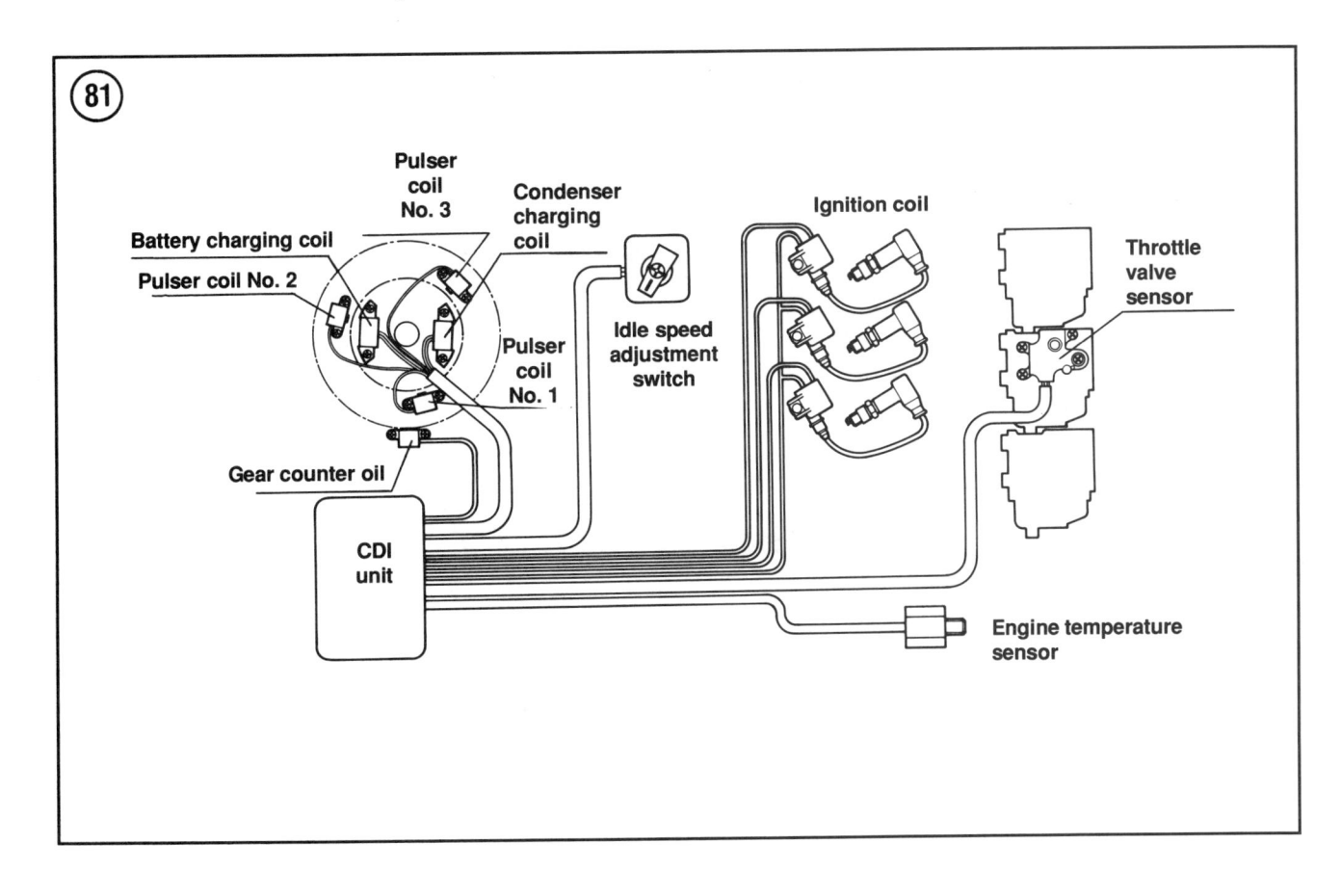

The IC ignition system operates the same as a conventional capacitor discharge ignition system. The magnets in the outer rim of the flywheel induce an alternating current (AC) into the condenser charge coils (located under the flywheel) as the flywheel rotates. This current is sent to the CDI unit where it is rectified (converted to direct current) and stored as DC voltage in an internal capacitor. As the flywheel continues to rotate, magnets on the flywheel induce current into the pulser coils. This current is routed to the CDI unit where it causes an electronic switch inside the CDI unit to close, allowing the voltage stored in the capacitor to flow into the ignition coils. The ignition coils greatly amplify this voltage to create the spark at the spark plug gaps. Depressing the stop switch or turning off the ignition switch shorts the charge coil output to ground, shutting off the engine.

IC ignition is different from conventional CDI in that it monitors information input from various sensors and switches to electronically calculate the optimum ignition timing for all operating conditions. Throttle opening position data is provided by the throttle position switch (early models) or the throttle position sensor (late models). Engine speed and crankshaft position information is provided by the gear counter coil mounted outside the flywheel.

The IC system also monitors the oil level, cooling system and overspeed caution systems. If one or more caution systems indicate a malfunction, the IC system activates the appropriate warning buzzer and lamp, and automatically reduces the engine speed to a predetermined level.

Stator Base/Alternator Stator Removal/Installation

1. Disconnect the negative battery cable.
2. Remove the engine cover.

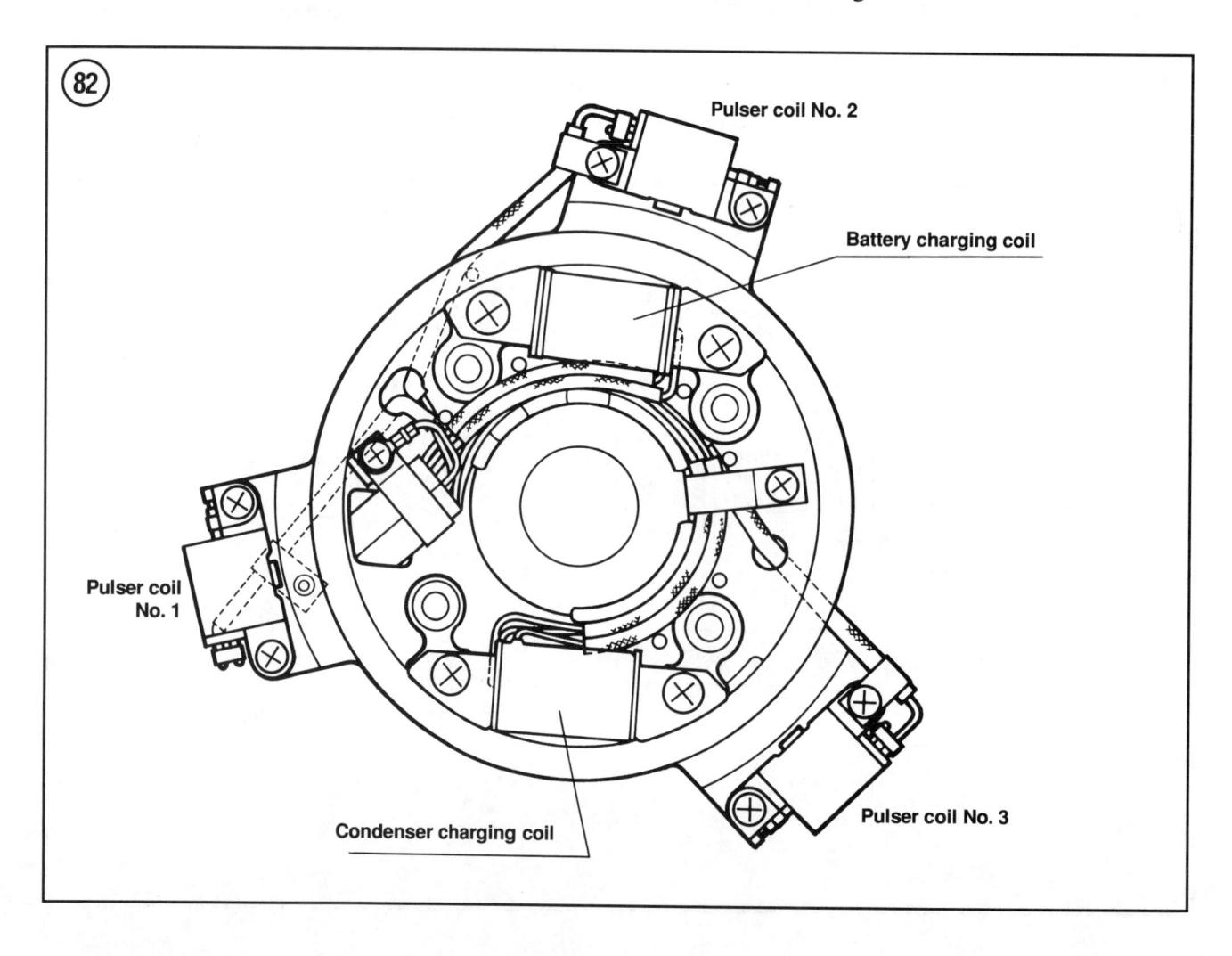

82

Pulser coil No. 2

Battery charging coil

Pulser coil No. 1

Condenser charging coil

Pulser coil No. 3

ELECTRICAL SYSTEMS

3. Remove the flywheel. See Chapter Eight.

4. Disconnect the stator leads at their connectors.

5. Remove the screws or bolts (**Figure 83**) holding the stator base to the upper oil seal housing. Remove the stator base and plate.

6. Installation is the reverse of removal, noting the following.

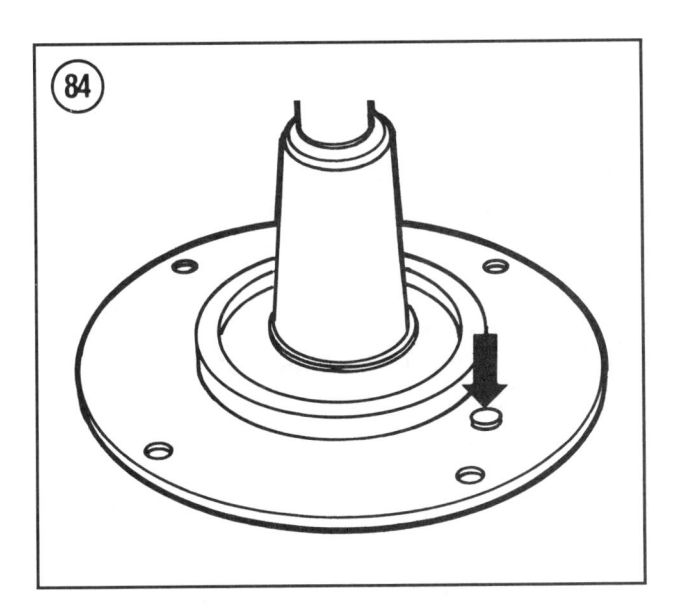

7. On models so equipped, be sure to install the locating dowel (**Figure 84**) into the plate.

8. Apply Thread Lock 1342 (part No. 99000-32050) to the screws or bolts securing the stator base prior to installation, then tighten securely.

9. Make sure all electrical connectors are free of corrosion and are tight.

Condenser Charging Coil, Gear Counter Coil, Battery Charging Coil or Pulser Coil Replacement

NOTE
A special Suzuki tool is required to install and locate the coils and to adjust the air gap clearance on 1986-on DT 55, DT 65 and 1986-on DT 115, DT 140 models.

1. Disconnect the negative battery cable, if so equipped.

2. Remove the engine cover.

3. Remove the flywheel/rotor. See Chapter Eight.

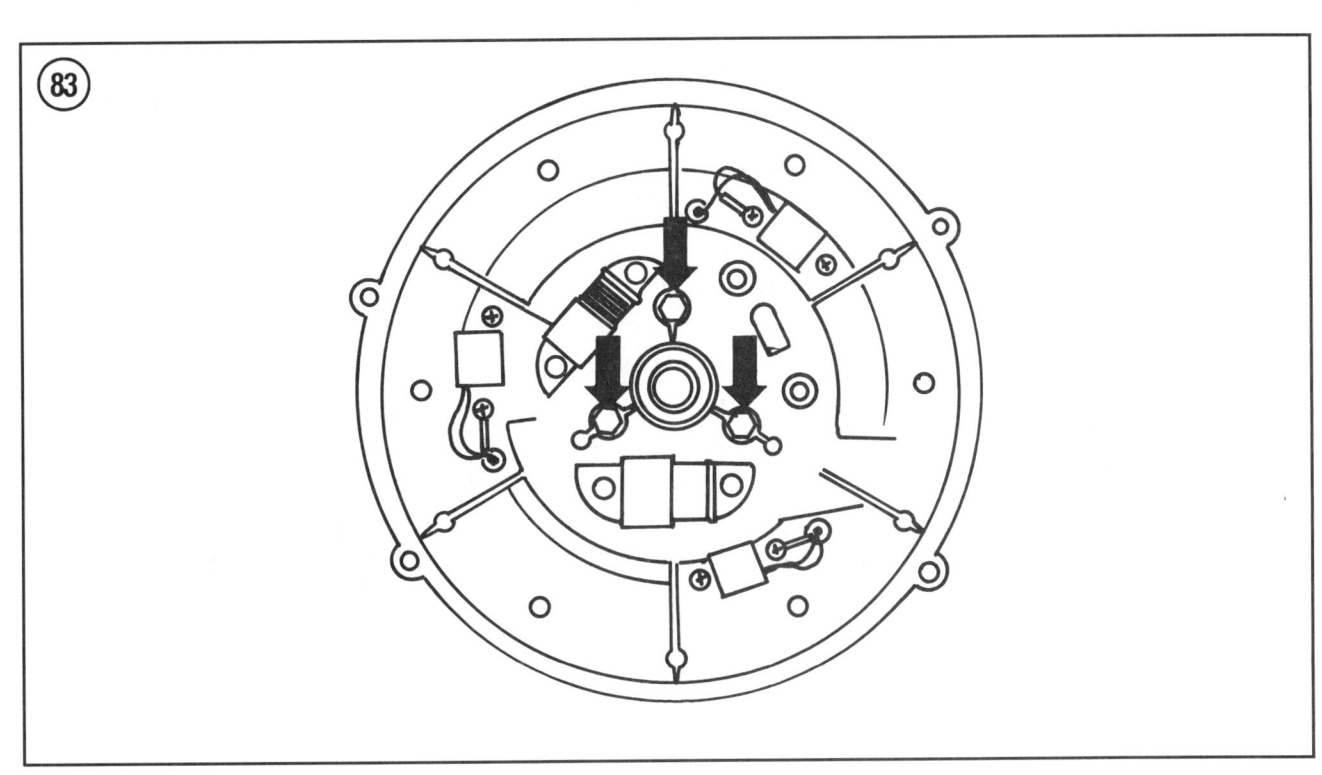

4. Remove the 2 screws and washers holding the defective coil to the stator base.

5. Unwrap the tie straps or wiring harness cover to separate the stator leads. On some models, PVC tubing may have been used to form a wiring harness cover. If so, carefully slit the tubing to remove the coil lead(s).

6. Remove the defective coil from the stator base, pulling its lead(s) through the base cutout.

NOTE
Suzuki does not provide information for the pulser coil air gap for the 1988-on DT 75, DT 85 models.

7A. 1989-on DT 25 and 1988-on DT 30 models— Refer to **Figure 82** and correctly install and locate the pulser coil(s) clearance as follows:

  a. Install the pulser coil(s) in their correct location(s).

  b. Apply Thread Lock 1342 (part No. 99000-32050) to the screws securing the pulser coil(s) prior to installation, then install the screws and washers finger-tight at this time.

  c. Install the flywheel/rotor onto the crankshaft. Install the center mounting bolt just sufficiently to hold the flywheel/rotor in place.

  d. Place a 0.75 mm (0.03 in.) flat feeler gauge between the pulser and the flywheel/rotor surface (**Figure 85**).

  e. Push the pulser coil so that the pulser bar is up against the feeler gauge and tighten both screws securely with an angled screwdriver. Tighten the screws sufficiently to hold the coil in place. This will achieve an air gap of 0.75 mm (0.03 in.). Remove the feeler gauge.

  f. Remove the flywheel/rotor straight up and off of the crankshaft. Do not move the pulser coil(s) during flywheel/rotor removal.

  g. Tighten the pulser coil(s) screws securely, then reinstall the flywheel/rotor and recheck the clearance, readjust if necessary.

7B. DT 55, DT 65 and 1986-on DT 115, DT 140 models— Refer to **Figure 86** for DT 55, DT 65 models or **Figure 87** for 1986-on DT 115, DT 140 models and correctly locate and install the pulser coil(s) clearance as follows:

  a. Install the special tool, Pulser Coil Locating Jig, onto the end of the crankshaft and secure it with the bolt. For DT 55, DT 65 models, use locating tool (part No. 09931-89410) or for 1986-on DT 115, DT 140 models, use locating tool (part No. 09931-89420).

  b. Install the pulser coil(s) in their correct location(s). Place the coil so that the pulser bar is up against the locating jig surface. This will provide a final clearance between the pulser bar and the flywheel/rotor of 0.03 in. (0.75 mm) when the flywheel/rotor is installed.

  c. Apply Thread Lock 1342 (part No. 99000-32050) to the screws securing the pulser coil(s) prior to installation, then install and tighten securely.

  d. Remove the special tool from the crankshaft.

8. Adjust the gear counter coil air gap as follows:

  a. Install the gear counter coil in its correct location.

  b. Apply Thread Lock 1342 (part No. 99000-32050) to the threads of the coil mounting screws. Install the screws finger tight.

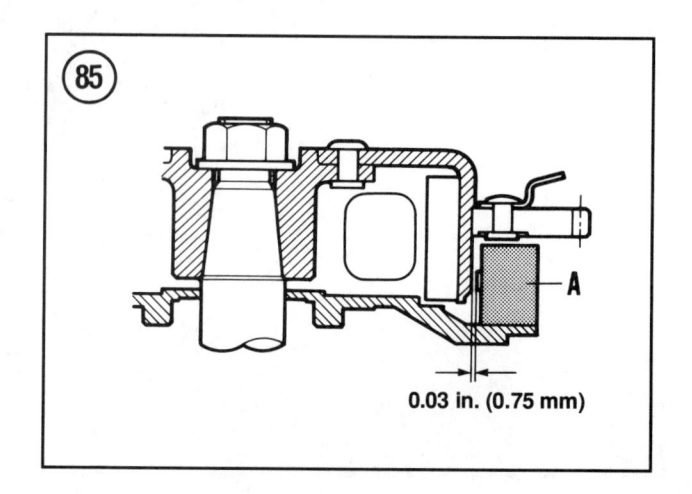

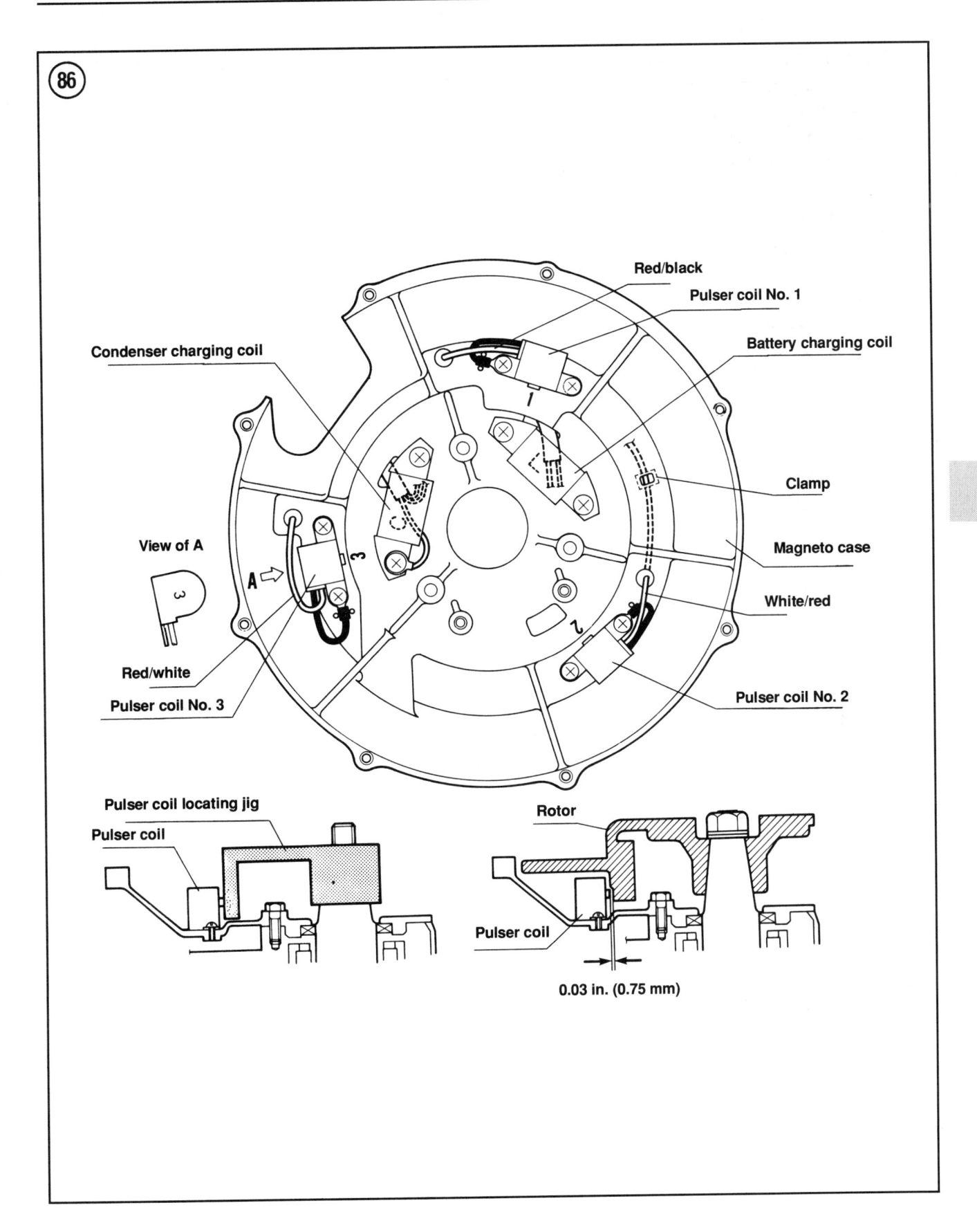

ELECTRICAL SYSTEMS

86

- Condenser charging coil
- View of A
- Red/white
- Pulser coil No. 3
- Red/black
- Pulser coil No. 1
- Battery charging coil
- Clamp
- Magneto case
- White/red
- Pulser coil No. 2
- Pulser coil locating jig
- Pulser coil
- Rotor
- Pulser coil
- 0.03 in. (0.75 mm)

234

CHAPTER SEVEN

c. Place a 0.02 in. (0.5 mm) flat feeler gage between the gear counting coil and the flywheel gear teeth (**Figure 88**).
d. Push the counter coil against the feeler gage and securely tighten the screws.

9. The remainder of reassembly is the reverse of removal, noting the following.
10. If PVC tubing was used for a wiring harness cover, secure the leads together with tie straps.
11. Apply Thread Lock 1342 (part No. 99000-32050) to the screws or bolts securing the stator base prior to installation, then tighten securely.
12. Make sure all electrical connectors are free of corrosion and are tight.

MICRO LINK IGNITION SYSTEM (V4 AND V6)

The Suzuki Micro Link electronic ignition system is a magneto CDI (capacitor discharge ignition) system and a microcomputer. Because of the microcomputer's integrated circuit feature, it has the ability to determine the crankshaft

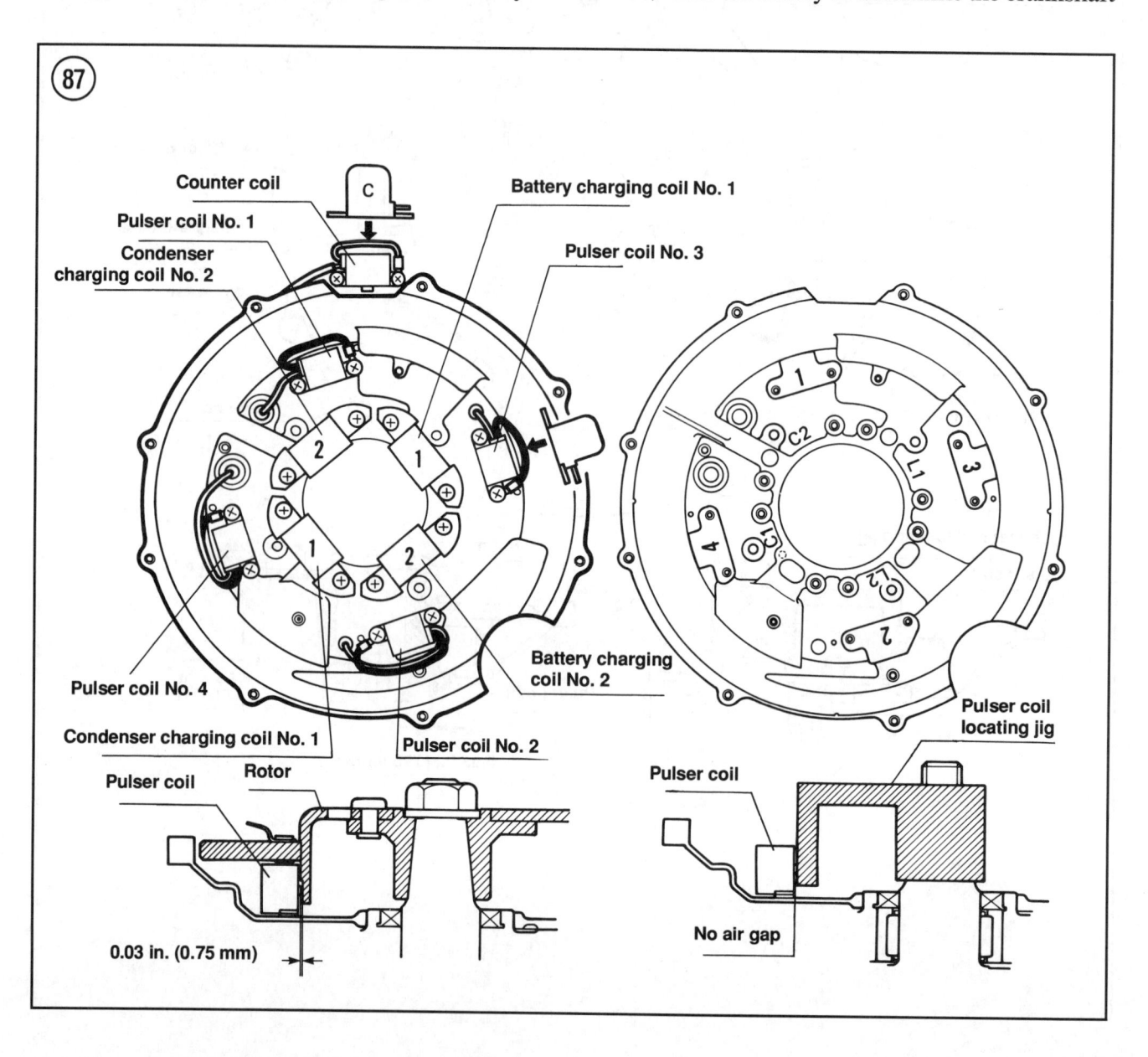

87

- Counter coil
- Pulser coil No. 1
- Condenser charging coil No. 2
- Battery charging coil No. 1
- Pulser coil No. 3
- Pulser coil No. 4
- Condenser charging coil No. 1
- Pulser coil No. 2
- Battery charging coil No. 2
- Pulser coil
- Rotor
- 0.03 in. (0.75 mm)
- Pulser coil
- Pulser coil locating jig
- No air gap

ELECTRICAL SYSTEMS

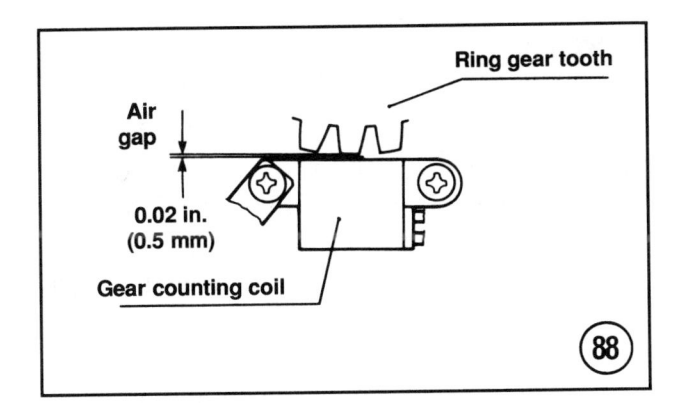

Air gap 0.02 in. (0.5 mm)
Ring gear tooth
Gear counting coil

Figure 88

angle accurately, ignition timing control for engine starting, trolling timing control, neutral revolution control, ignition advance map control for normal operating range, engine rpm rev limiter control, oil level warning reset function and tachometer pulse signal conversion. The basic system layout is shown in **Figure 89**, typical and the microcomputer is powered by the battery and not the magneto. This system creates improved acceleration by maintaining optimum carburetor

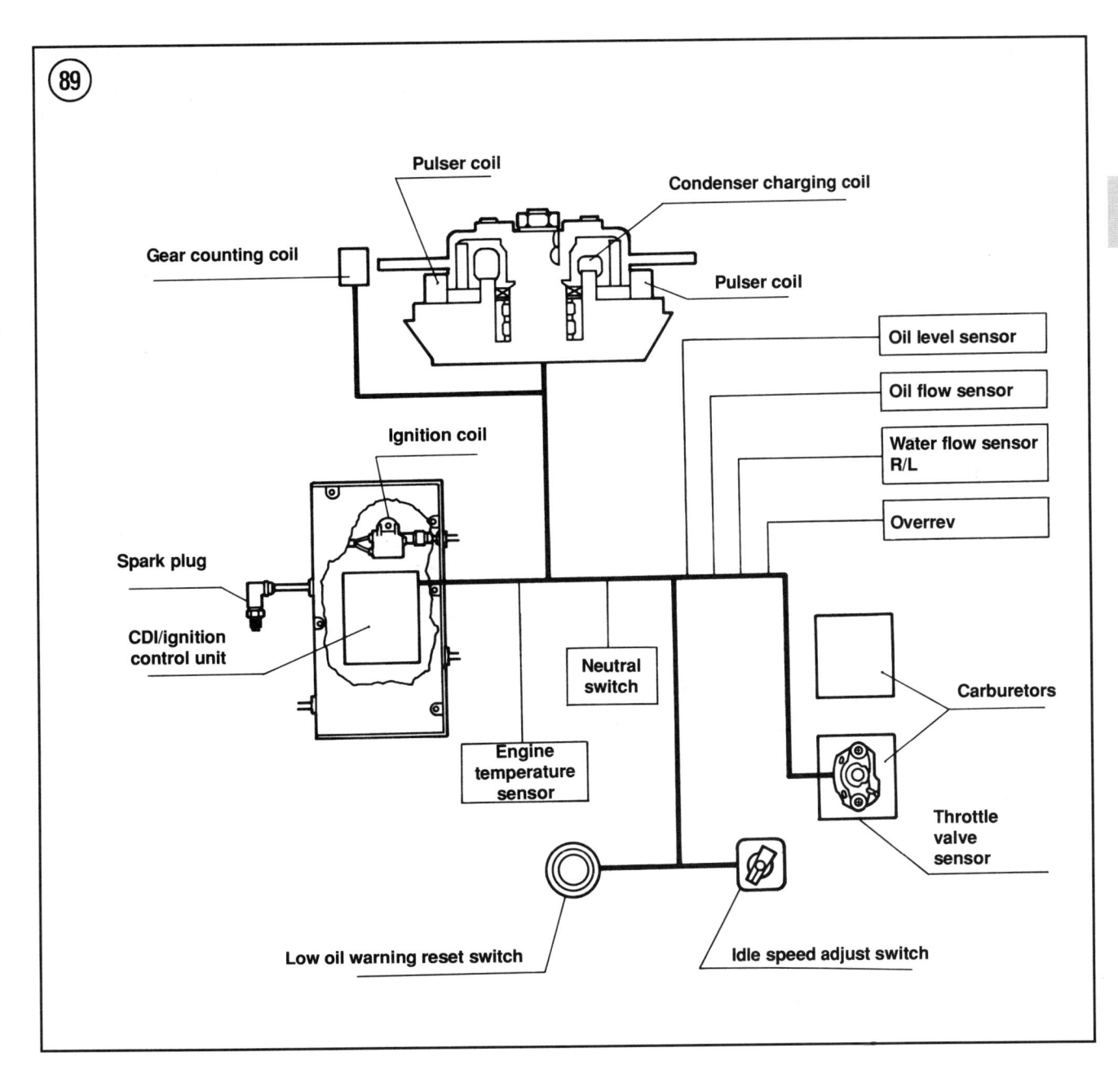

Figure 89 — System layout showing: Pulser coil, Condenser charging coil, Gear counting coil, Pulser coil, Oil level sensor, Oil flow sensor, Water flow sensor R/L, Overrev, Ignition coil, Spark plug, CDI/ignition control unit, Neutral switch, Carburetors, Engine temperature sensor, Throttle valve sensor, Low oil warning reset switch, Idle speed adjust switch

and ignition synchronization which enables the engine to run smoother. This system also creates easy start-ups, crisp acceleration and improved top-end performance.

There are several variations to the system used on these models but all operate in the same manner. There is a single charge coil and pulse coil for each cylinder which supply current to the CDI unit (**Figure 90**, typical). The CDI unit in turn triggers each ignition coil to fire the spark plugs as required. Ignition advance is electronic.

Ignition timing adjustment is not necessary on these models with the exception of adjusting the throttle valve sensor that is covered in Chapter Five.

Operation

Magnets in the outer rim of the flywheel induce an alternating current (AC) into the condenser charge coil (located under the flywheel) as the flywheel rotates. This current is sent to the CDI unit where it is rectified (converted to direct current) and stored as DC voltage in an internal capacitor. As the flywheel continues to rotate, magnets on the flywheel induce current into the pulser coils. This current is routed to the CDI unit where it causes an electronic switch inside the CDI unit to close, allowing the voltage stored in the capacitor to flow into the ignition coils. The ignition coils greatly amplify this voltage to

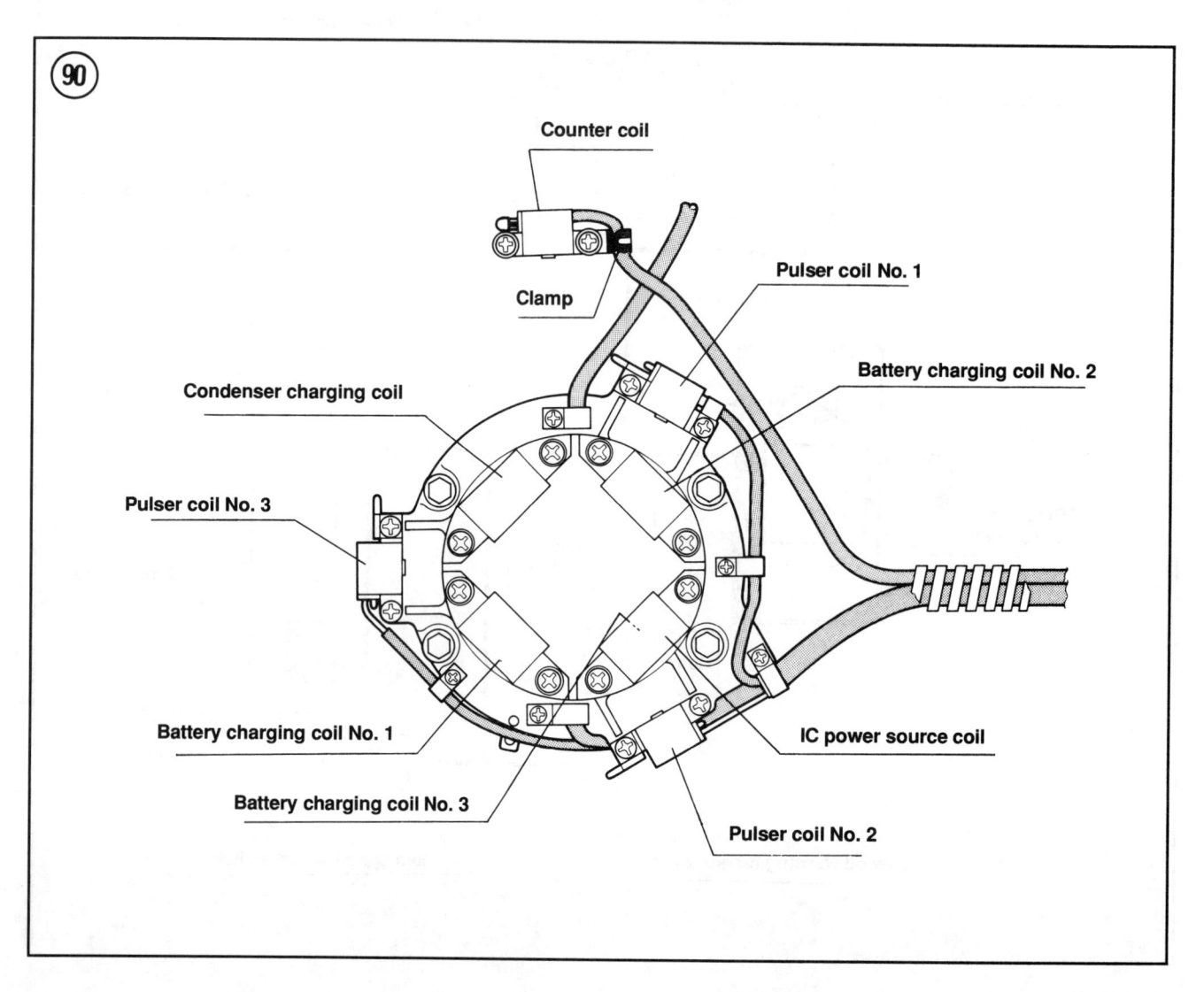

ELECTRICAL SYSTEMS

create the spark at the spark plug gaps. Depressing the stop switch or turning off the ignition switch shorts the charge coil output to ground, shutting off the engine.

The Micro Link microcomputer processes information from various sensors and switches including throttle valve opening, engine speed, engine temperature and shift lever position, then calculates the optimum ignition timing for all operating conditions.

The Micro Link system also monitors oil level, oil flow, water temperature, water flow and overspeed caution systems. If one or more cautions systems detect a malfunction, the microcomputer activates the appropriate warning buzzer and lamp, and automatically reduces engine speed to a predetermined level.

Stator Base/Alternator Stator Removal/Installation

1. Disconnect the negative battery cable.
2. Remove the engine cover.
3. Remove the flywheel. See Chapter Eight.
4. Disconnect the stator leads at their connectors.
5. Remove the bolts and washers holding the stator base to the upper oil seal housing.
6. Remove the bolt and washers securing the gear counter coil.
7. Remove the stator base and gear counter assembly.
8. Installation is the reverse of removal, noting the following.
9. Apply Thread Lock 1342 (part No. 99000-32050) to the screws or bolts securing the stator base prior to installation, then tighten securely.
10. Make sure all electrical connectors are free of corrosion and are tight.

Condenser Charging Coil, Gear Counter Coil, Battery Charging Coil or Pulser Coil Replacement

NOTE
A special Suzuki tool is required to install and locate the coils and to adjust the air gap clearance.

1. Disconnect the negative battery cable.
2. Remove the engine cover.
3. Remove the flywheel/rotor. See Chapter Eight.
4. Remove the 2 screws and washers holding the defective coil(s) to the stator base.
5. Unwrap the tie straps or wiring harness cover to separate the stator leads. On some models, PVC tubing may have been used to form a wiring harness cover. If so, carefully slit the tubing to remove the coil lead(s).
6. Remove the defective coil(s) from the stator base, pulling its lead(s) through the base cutout.
7. Refer to **Figure 91** for V4 models or **Figure 92** for V6 models and correctly locate and install the pulser coil(s) clearance as follows:
   a. Install the special tool, Pulse Coil Locating Jig (part No. 09931-88710) onto the end of the crankshaft and secure it with the bolt.
   b. Install the pulser coil(s) in their correct location(s). Place the coil so that the pulser bar is up against the locating jig surface. This will provide a final clearance between the pulser bar and the flywheel/rotor of 0.03 in. (0.75 mm) when the flywheel/rotor is installed.
   c. Apply Thread Lock 1342 (part No. 99000-32050) to the screws securing the pulser coil(s) prior to installation, then install and tighten securely.
   d. Remove the special tool from the crankshaft.
8. Adjust the gear counter coil air gap as follows:
   a. Install the gear counter coil in its correct location.

238   CHAPTER SEVEN

b. Apply Thread Lock 1342 (part No. 99000-32050) to the screws securing the gear counter coil prior to installation, and install the screws finger-tight at this time.

c. Place a 0.02 in. (0.5 mm) flat feeler gauge between the gear counting coil and the flywheel/rotor gear teeth (**Figure 88**).

d. Push the gear counter coil up against the feeler gauge and tighten both screws securely. This will achieve an air gap of 0.02 in. (0.5 mm).

9. Installation is the reverse of removal, noting the following.

10. If PVC tubing was used for a wiring harness cover, secure the leads together with tie straps.

11. Apply Thread Lock 1342 (part No. 99000-32050) to the screws or bolts securing the stator base prior to installation, then tighten securely.

12. Make sure all electrical connectors are free of corrosion and are tight.

IGNITION COIL (ALL MODELS)

Individual Ignition Coil Removal/Installation

1. Disconnect the negative battery cable.
2. Remove the engine cover.

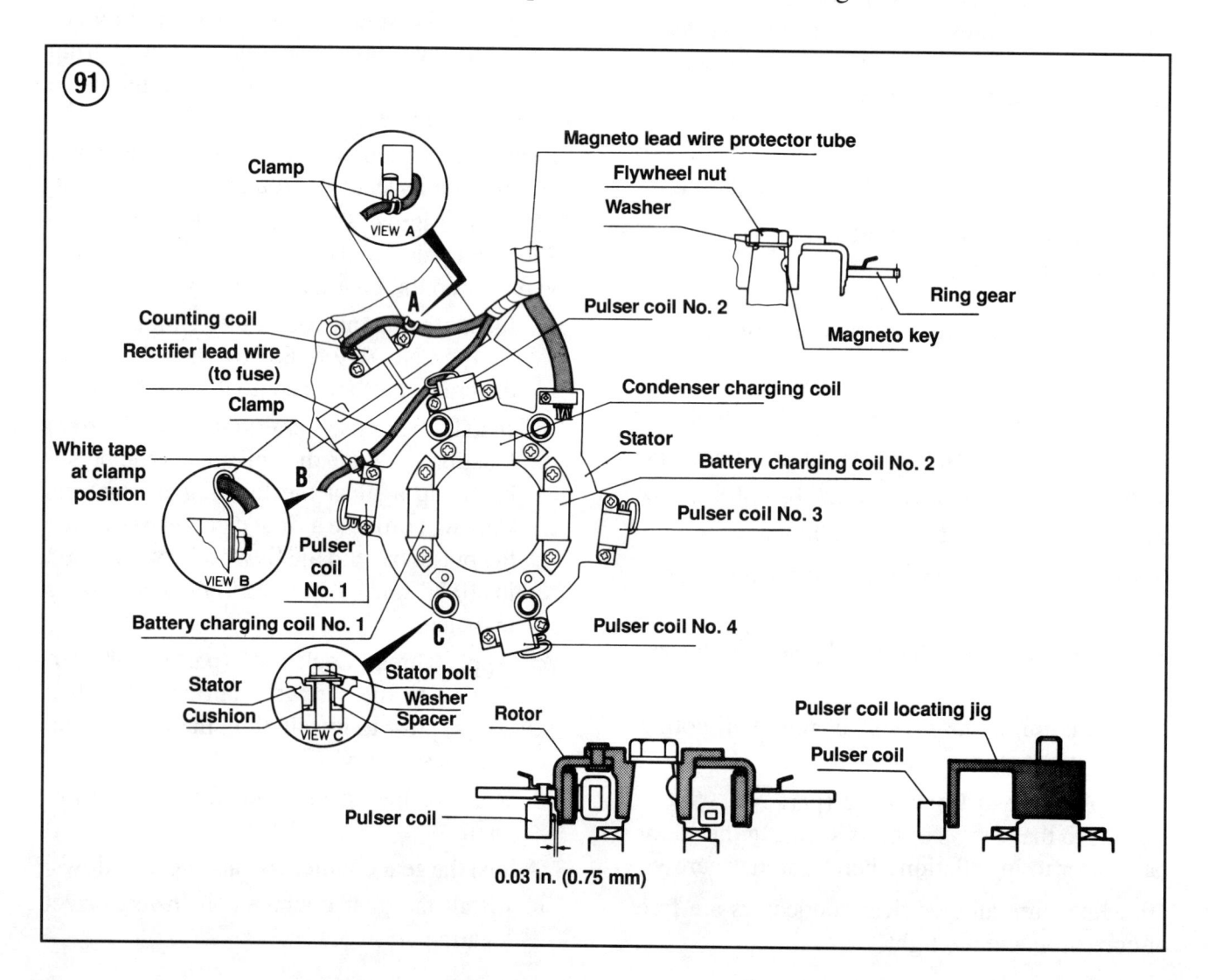

ELECTRICAL SYSTEMS

3. On DT 55, DT 65, 1986-on DT 115, DT 140, V4 and V6 models, remove the cover from the electric junction box.

4. Disconnect the coil primary leads at their bullet connectors.

5. Disconnect the coil secondary lead from the spark plug.

6. Remove the coil mounting screws and lockwashers. Remove the coil(s).

7. On DT 55, DT 65, 1986-on DT 115, DT 140, V4 and V6 models, carefully remove the rubber grommet that surrounds the secondary lead where it exits the electric junction box.

8. Installation is the reverse of removal, noting the following.

9. On DT 55, DT 65, 1986-on DT 115, DT 140, V4 and V6 models, be sure to install the rubber grommet into the receptacle in the electric junc-

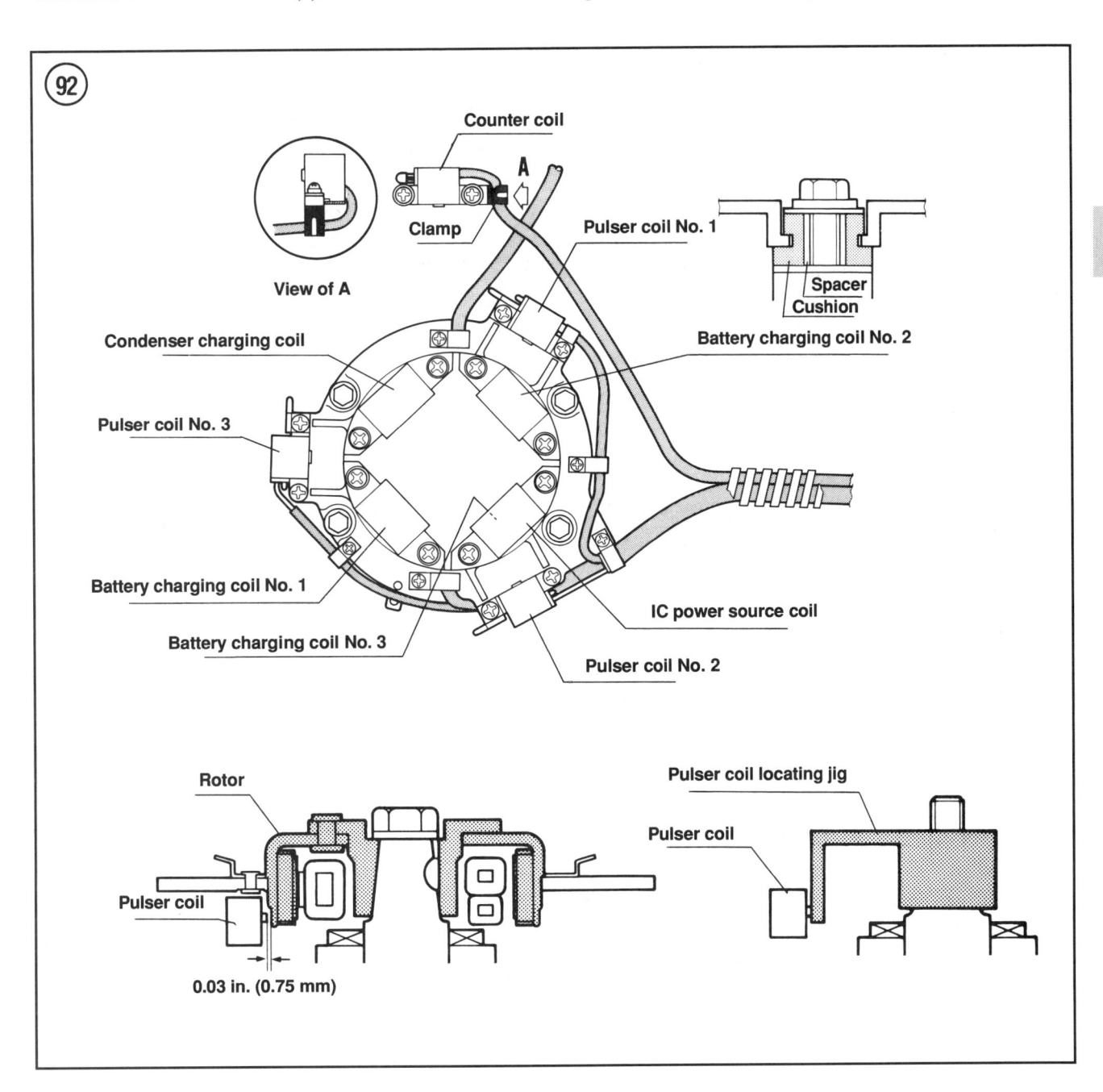

tion box. This is necessary to keep moisture from entering the electric junction box.

10. Be sure to reinstall the black ground lead under one of the coil mounting screws.

11. Make sure all electrical connectors are free of corrosion and are tight.

Combined Ignition Coil/CDI Unit Removal/Installation

The following models are equipped with a combined ignition coil/CDI unit:
  a. 1985-on DT 6 and DT 40.
  b. 1985-1987 DT 8, DT 9.9, DT 20 and DT 30.
  c. 1985-1988 DT 15 and DT 25.
  d. 1987-1989 DT 35.

1. Disconnect the negative battery cable, if so equipped.
2. Remove the engine cover.
3. Disconnect the secondary wires at the spark plugs.
4A. On 1985-1987 DT 20, DT 30 and 1985-1988 DT 25, perform the following:
  a. Remove the junction box cover and disconnect the 3-wire yellow connector and the 1-wire red lead wire connectors (**Figure 93**).
  b. Remove the 3 bolts holding the electric component assembly to the power head (**Figure 94**).
4B. On all others models, perform the following:
  a. Disconnect the CDI/coil leads at their bullet connectors.
  b. Remove the screws holding the CDI/coil unit to the power head.
5A. On 1985-on DT 20, DT 25, DT 30, perform the following:
  a. Remove the electric component assembly.
  b. From the rear of the electric component assembly, remove the band holding the CDI/coil unit in place. Remove the CDI/coil unit from the component assembly.

5B. On all other models, remove the CDI/coil unit.

6. Installation is the reverse of removal, note the following.
7. Be sure to reinstall the black ground lead under one of the mounting screws.
8. Make sure all electrical connectors are free of corrosion and are tight.

CDI UNIT (ALL MODELS)

Combined Ignition Coil/CDI Unit Removal/Installation

See preceding procedure under *Ignition Coil*.

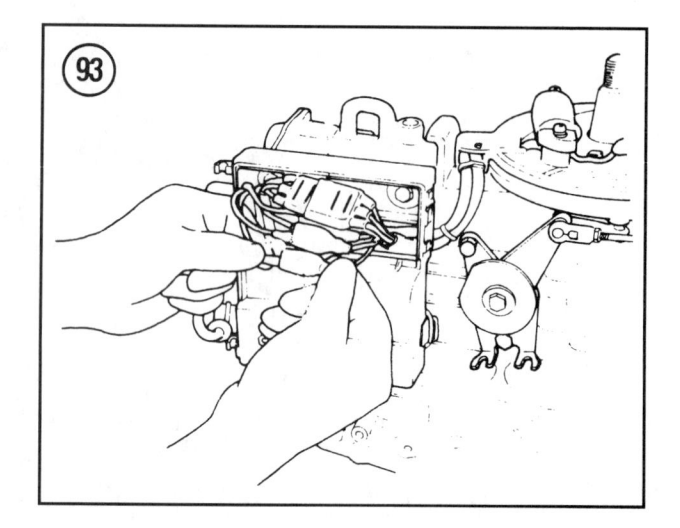

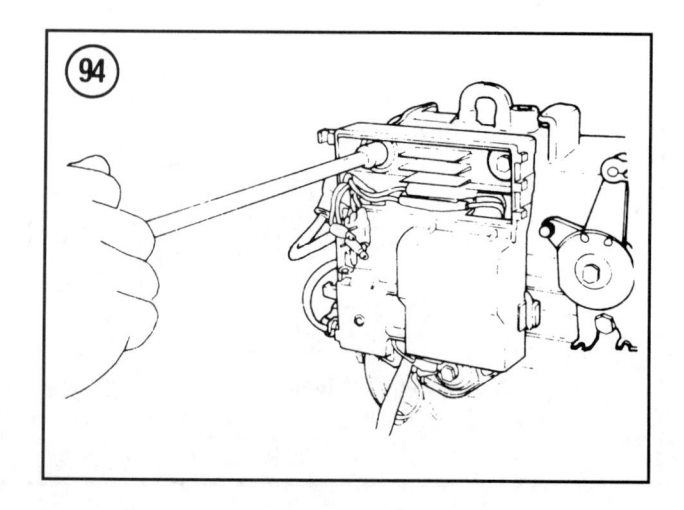

ELECTRICAL SYSTEMS

Individual CDI Unit Removal/Installation

NOTE
Note the location and routing of all electrical wires prior to disconnecting any of the CDI units. Make a drawing or take a Polaroid picture of the assembly prior to disconnecting any of the connectors.

1. Disconnect the negative battery cable, if so equipped.
2. Remove the engine cover.
3A. On 1988-on DT 8 and DT 9.9 models, perform the following:
   a. Disconnect the CDI unit leads at their bullet connectors.
   b. Remove the CDI unit ground lead screw (A, **Figure 95**).
   c. Remove the CDI unit mounting screw (B, **Figure 95**).
   d. Remove the CDI unit (C, **Figure 95**).
3B. On 1989-on DT 15 models, perform the following:
   a. Disconnect the CDI unit leads at their bullet connectors.
   b. Remove the CDI unit ground lead screw.
   c. Remove the CDI unit mounting screws (**Figure 96**) and cable clamp. Note the stop on the CDI unit in relation to the clamp. This clamp is used as a stopper to prevent the CDI unit from moving in one direction.
   d. Remove the CDI unit.
3C. On 1988-on DT 25 and DT 30 models, perform the following:
   a. Remove the cover from the CDI enclosure.
   b. Disconnect the CDI unit leads at their bullet connectors.
   c. Remove the CDI unit ground lead screw (**Figure 97**).
   d. Remove the bands securing the CDI unit and remove the CDI unit.
3D. On 1985-on DT 55, DT 65 and 1986-on DT 115, DT 140 models, perform the following:
   a. Remove the cover from the electric parts holder.
   b. Disconnect the CDI unit leads at their bullet connectors.
   c. Remove the CDI unit ground lead bolt and washer. Refer to **Figure 98** for 1985-on DT 55 and DT 65 or **Figure 99** for 1986-on DT 115 and DT 140 models.
   d. On 1985-on DT 55 and DT 65, unhook the CDI unit mounting bands and remove the CDI unit.
   e. On 1986-on DT 115, DT 140 models, remove the CDI unit mounting screws and remove the CDI unit.

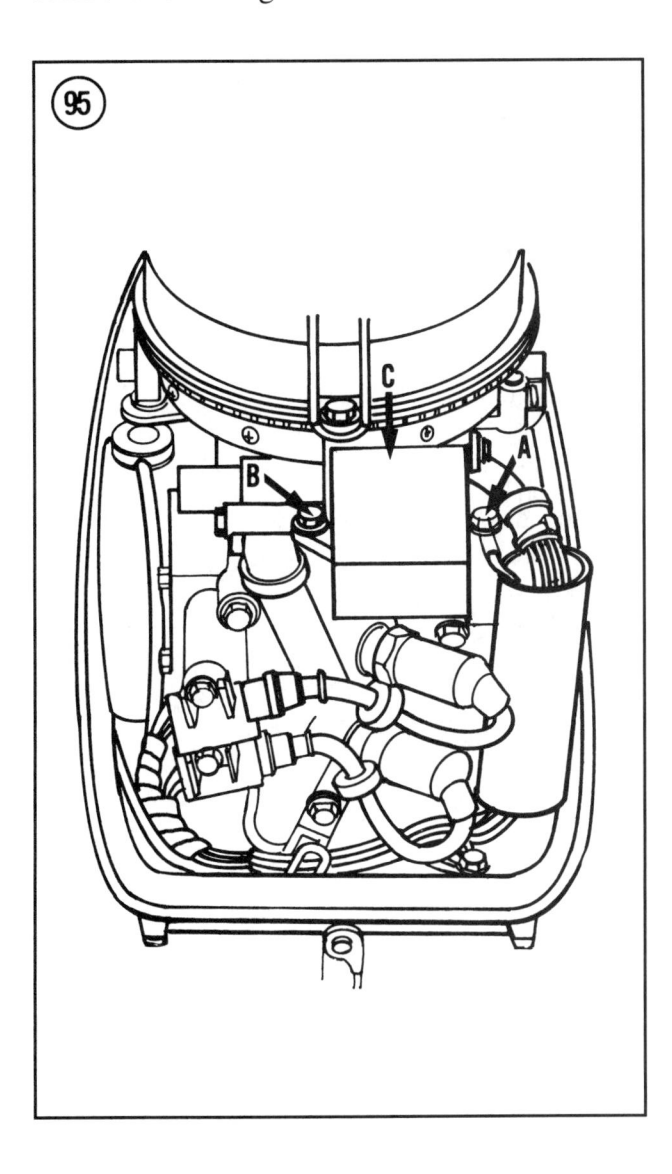

3E. On 1985-1986 DT 75 and DT 85 models, perform the following:
   a. Remove the cover from the junction box.
   b. Disconnect the CDI unit leads at their bullet connectors.
   c. Remove the CDI unit ground lead screw.
   d. Remove the CDI unit mounting screws (**Figure 100**) and remove the CDI unit.

3F. On 1987-on DT 75 and DT 85 models, perform the following:
   a. Remove the cover from the electric parts holder.
   b. Disconnect the CDI unit leads at their bullet connectors.
   c. Remove the CDI unit ground lead bolt and washer (**Figure 101**).
   d. Unhook the CDI unit mounting bands and remove the CDI unit.

3G. On 1985 DT 115 and DT 140 models, perform the following:
   a. Remove the cover from the junction box (A, **Figure 102**).
   b. Disconnect the CDI unit leads at their bullet connectors.
   c. Remove the CDI unit ground lead screw then remove the CDI unit mounting screw. Remove the CDI unit (B, **Figure 102**).

NOTE
On V4 and V6 models, the CDI unit is mounted on the backside of the electric parts holder.

3H. On V4 and V6 models, perform the following:
   a. Remove the cover from the electric parts holder.
   b. Remove the bolts securing the electric parts holder and partially pull the electric parts holder out to gain access to the backside.
   c. Disconnect the CDI unit leads at their bullet connectors.
   d. Remove the CDI unit ground lead bolt and washer (**Figure 103**).

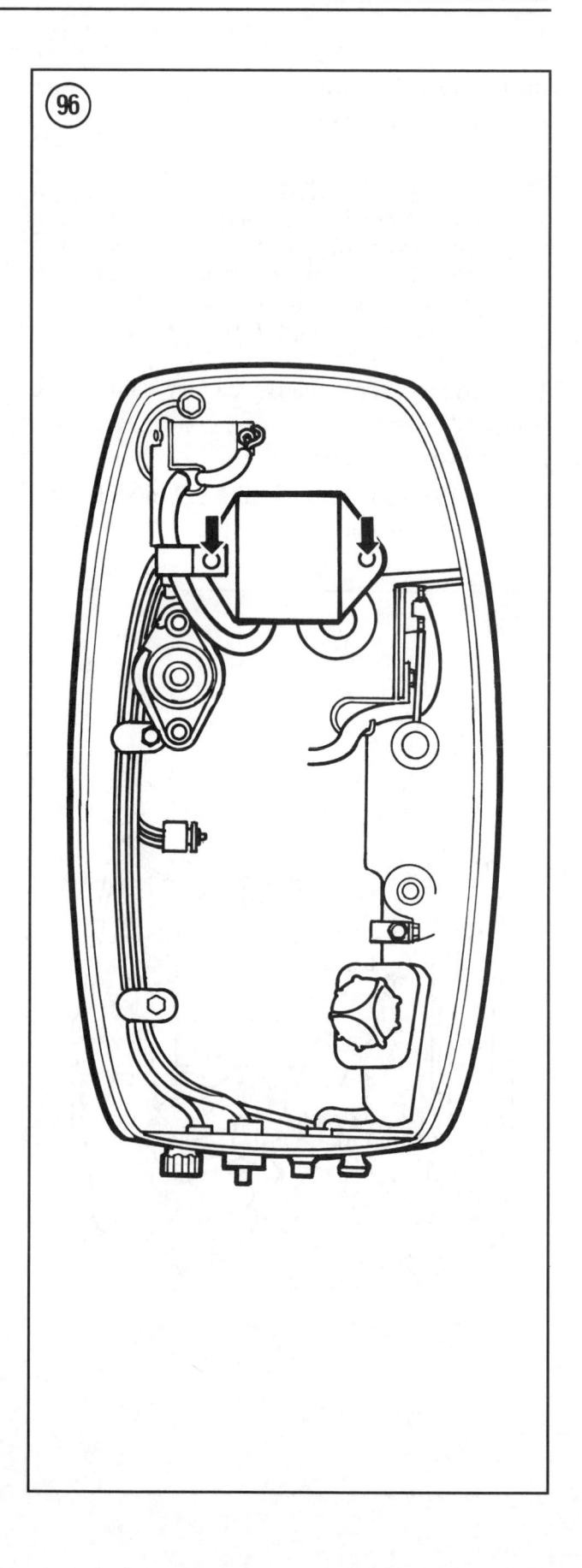

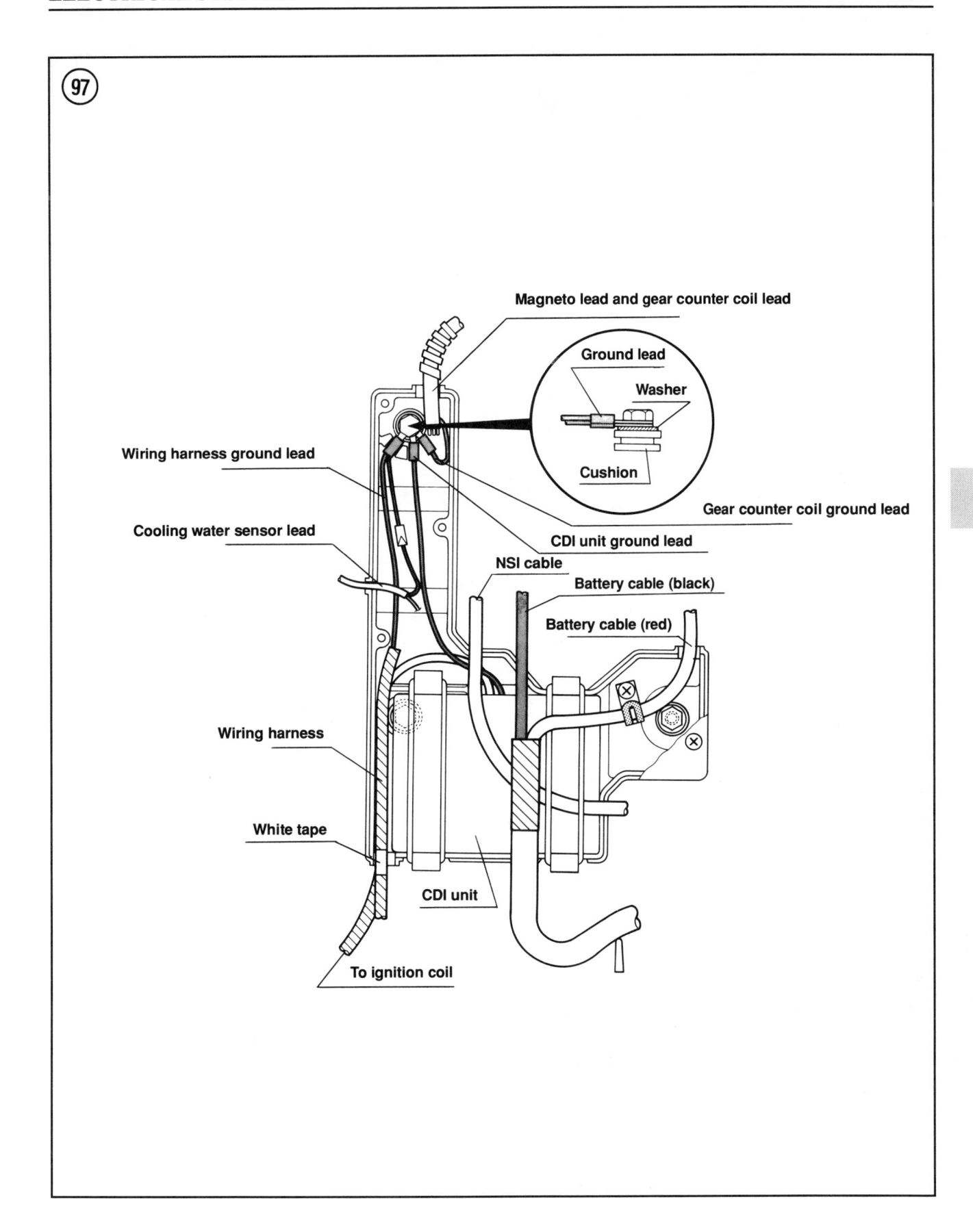

ELECTRICAL SYSTEMS

97

- Magneto lead and gear counter coil lead
- Ground lead
- Washer
- Cushion
- Wiring harness ground lead
- Cooling water sensor lead
- Gear counter coil ground lead
- CDI unit ground lead
- NSI cable
- Battery cable (black)
- Battery cable (red)
- Wiring harness
- White tape
- CDI unit
- To ignition coil

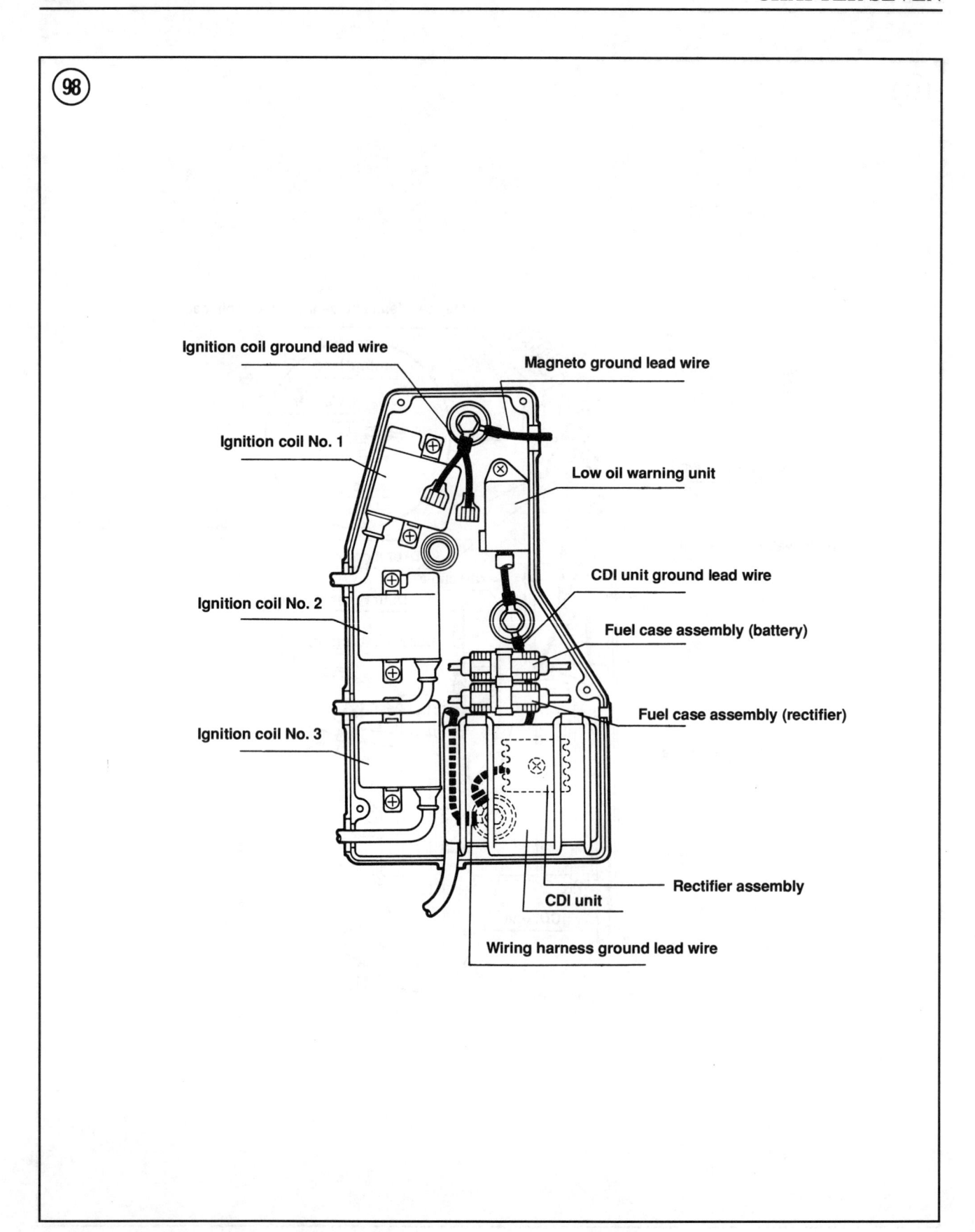

CHAPTER SEVEN

98

- Ignition coil ground lead wire
- Magneto ground lead wire
- Ignition coil No. 1
- Low oil warning unit
- Ignition coil No. 2
- CDI unit ground lead wire
- Fuel case assembly (battery)
- Fuel case assembly (rectifier)
- Ignition coil No. 3
- Rectifier assembly
- CDI unit
- Wiring harness ground lead wire

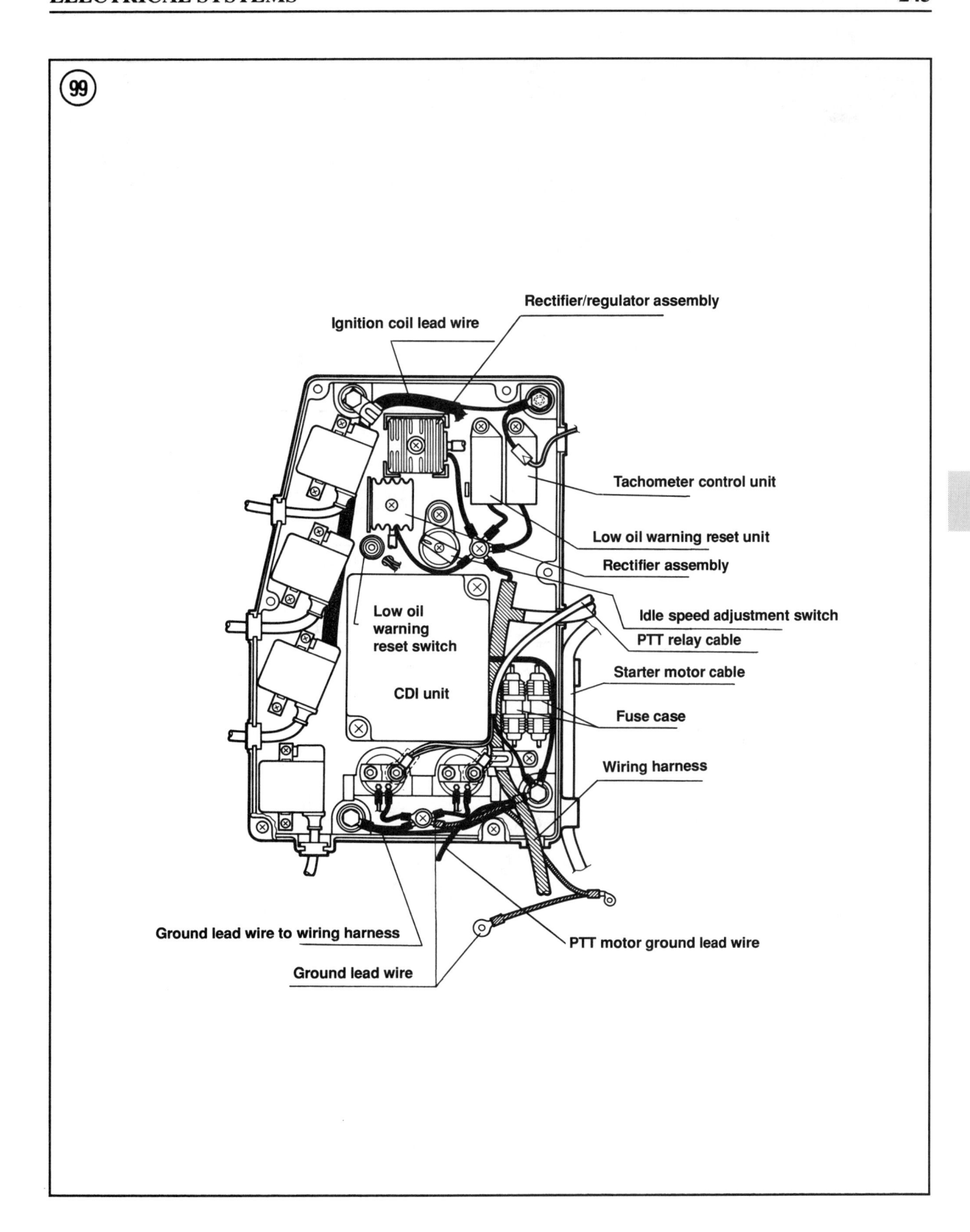

ELECTRICAL SYSTEMS

⑨⑨

Ignition coil lead wire
Rectifier/regulator assembly
Tachometer control unit
Low oil warning reset unit
Rectifier assembly
Idle speed adjustment switch
PTT relay cable
Starter motor cable
Fuse case
Wiring harness
Low oil warning reset switch
CDI unit
Ground lead wire to wiring harness
PTT motor ground lead wire
Ground lead wire

246

CHAPTER SEVEN

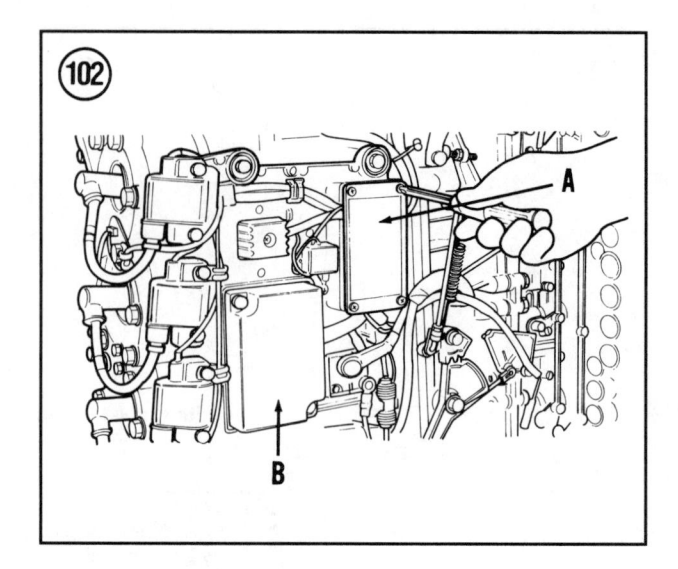

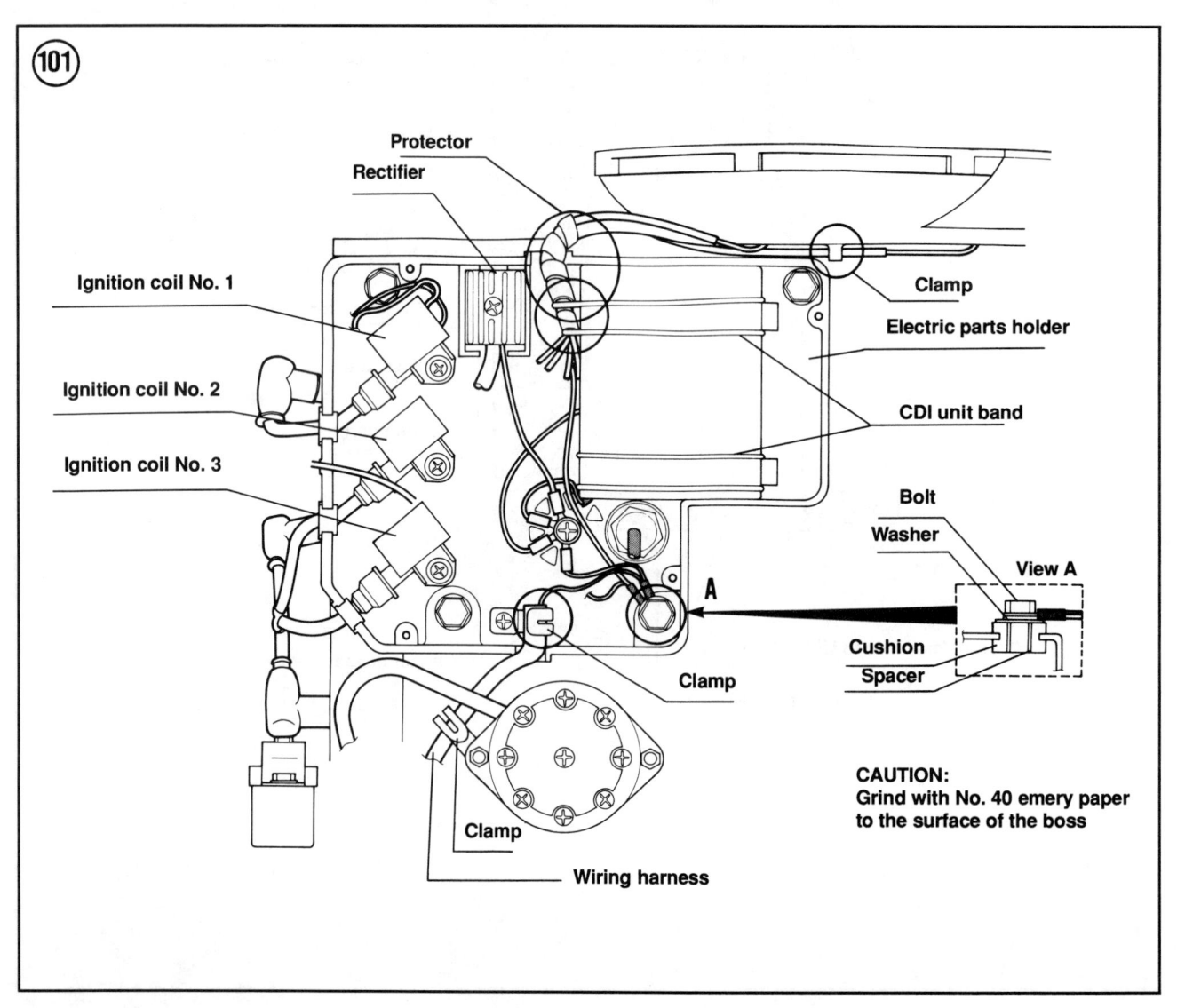

100

102

101

- Ignition coil No. 1
- Ignition coil No. 2
- Ignition coil No. 3
- Protector
- Rectifier
- Clamp
- Electric parts holder
- CDI unit band
- Bolt
- Washer
- View A
- Cushion
- Spacer
- A
- Clamp
- Clamp
- Wiring harness

CAUTION:
Grind with No. 40 emery paper to the surface of the boss

ELECTRICAL SYSTEMS

e. Remove the CDI unit mounting bolts (**Figure 103**) and remove the CDI unit.

4. Installation is the reverse of removal, noting the following.

5. Make sure all electrical connectors are free of corrosion and are tight.

6. On models so equipped, be sure to install the rubber grommet into the receptacle in the electric parts holder. This is necessary to keep moisture from entering the electric parts holder.

7. Be sure to reinstall the ground lead under the bolt or screw.

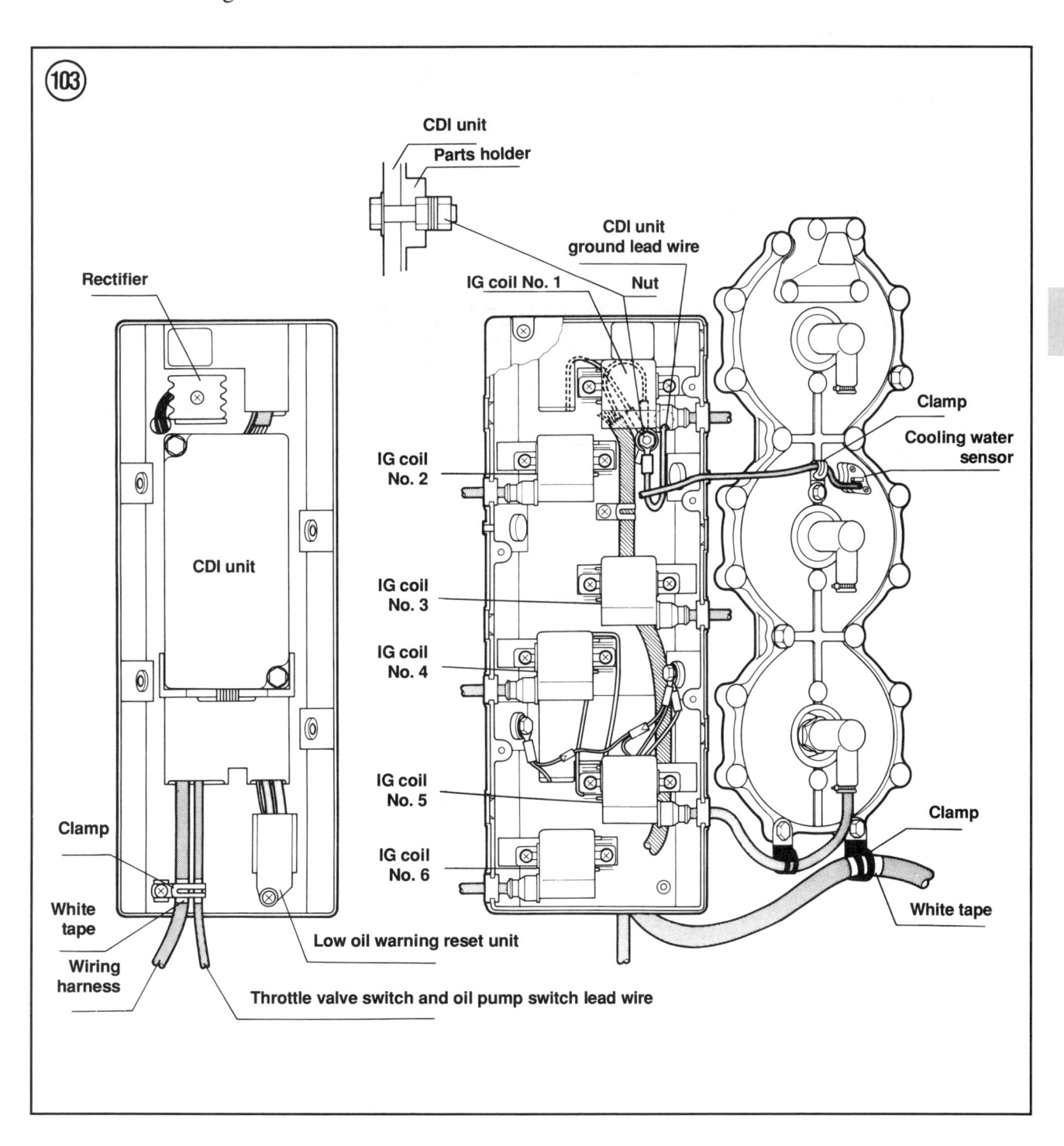

Table 1 BATTERY CAPACITY (HOURS)

Accessory draw	80 amp-hour battery provides continuous power for:	Approximate recharge time
5 amps	13.5 hours	16 hours
15 amps	3.5 hours	13 hours
25 amps	1.8 hours	12 hours

Accessory draw	105 amp-hour battery provides continuous power for:	Approximate recharge time
5 amps	15.8 hours	16 hours
15 amps	4.2 hours	13 hours
25 amps	2.4 hours	12 hours

Table 2 SELF-DISCHARGE RATE

Temperature	Approximate allowable self-discharge per day for first 10 days (specific gravity)
100° F (37.8° C)	0.0025 points
80° F (26.7° C)	0.0010 points
50° F (10.0° C)	0.0003 points

Chapter Eight

Power Head

This chapter covers the basic repair of Suzuki outboard power heads. The procedures involved are similar from model to model, with minor differences. Some procedures require the use of special tools, which can be purchased from a dealer. Certain tools may also be fabricated by a machinist, often at substantial savings. Power head stands are available from specialty shops such as Bob Kerr's Marine Tool Co. (P.O. Box 1135, Winter Garden, FL 32787).

Work on the power head requires considerable mechanical ability. You should carefully consider your own capabilities before attempting any operation involving major disassembly of the engine.

Much of the labor charge for dealer repairs involves the removal and disassembly of other parts to reach the defective component. Even if you decide not to tackle the entire power head overhaul after studying the text and illustrations in this chapter, it can be cheaper to perform the preliminary operations yourself and then take the power head to your dealer. Since many marine dealers have lengthy waiting lists for service (especially during the spring and summer season), this practice can reduce the time your unit is in the shop. If you have done much of the preliminary work, your repairs can be scheduled and performed much quicker.

Repairs go much faster and easier if your motor is clean before you begin work. There are special cleaners for washing the motor and related parts. Just spray or brush on the cleaning solution following the manufacturer's instructions, let it stand the specified length of time, then rinse it away with a garden hose. If compressed air is available, apply low air pressure and blow away water residue from the crevices of the engine. Clean all oily or greasy parts with fresh solvent as you remove them.

WARNING
Never use gasoline as a cleaning agent. It presents an extreme fire hazard. Be sure to work in a well-ventilated area when using cleaning solvents. Keep a fire extinguisher rated for gasoline and oil fires nearby in case of emergency.

Once you have decided to do the job yourself, read this chapter thoroughly until you have a good idea of what is involved in completing the overhaul satisfactorily. Make arrangements to buy or rent any special tools necessary and obtain replacement parts before you start. It is frustrating and time-consuming to start an overhaul and then be unable to complete it because the necessary tools or parts are not at hand.

Before beginning the job, re-read Chapter Two of this manual. You will do a better job with this information fresh in your mind.

Remember that new engine break-in procedures should be followed after an engine has been overhauled. Refer to your owner's manual for specific instructions.

Since this chapter covers a large range of models over a lengthy time period, the procedures are somewhat generalized to accommodate all models. Where individual differences occur, they are specifically pointed out. The power heads shown in the accompanying pictures are current designs. While it is possible that the components shown in the pictures may not be identical to those being serviced, the step-by-step procedures may be used with all models covered in this manual.

Tables 1-3 are at the end of the chapter.

ENGINE SERIAL NUMBER

Suzuki outboards are identified by engine serial number and model number. These numbers are stamped on a plate riveted to the port side stern bracket or to the starboard side of the support plate.

This information identifies the outboard and indicates if there are unique parts or if internal changes have been made during the model run. The serial and model numbers should be used when ordering any replacement parts for your outboard.

FASTENERS AND TORQUE

Always replace a worn or damaged fastener with one of the same size, type and torque requirement.

Power head tightening torques are listed in **Table 1**. Where a specification is not provided for a given bolt or nut, use the standard bolt and nut torque according to fastener size as listed at the end of **Table 1**.

Retighten the cylinder head bolts after the engine has been run for 15 minutes and allowed to cool. It is a good idea to retorque them again after 10 hours of operation. To retighten the

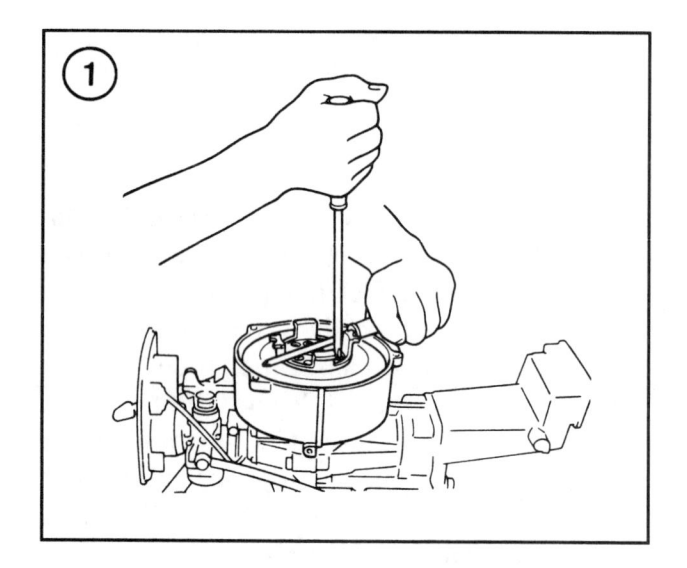

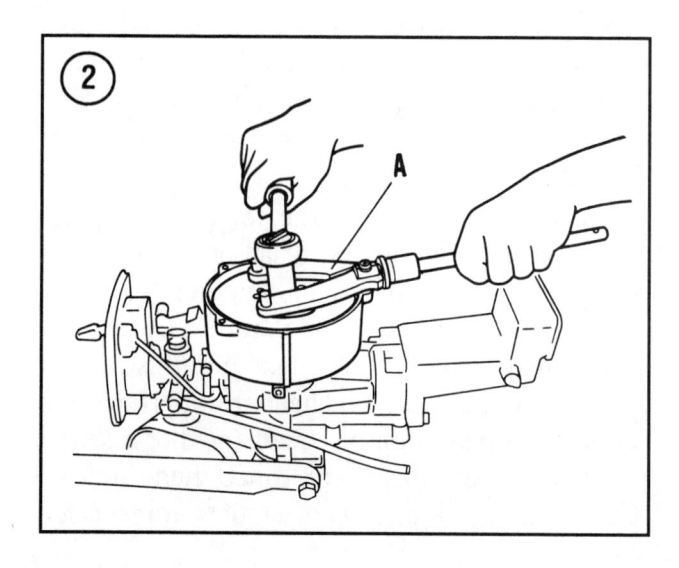

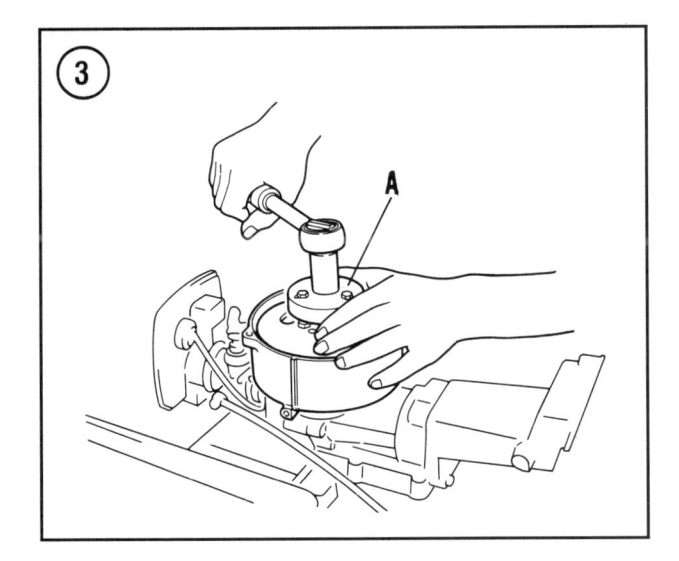

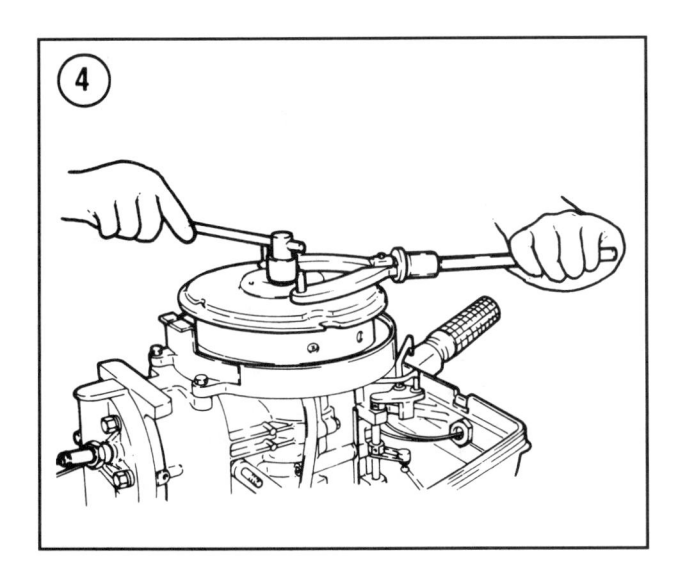

POWER HEAD

power head mounting fasteners properly, back them out one turn and then retighten to the proper torque specifications.

When spark plugs are reinstalled after an overhaul, tighten to the specified torque. Warm the engine to normal operating temperature, let it cool down and retorque the plugs.

FLYWHEEL

Removal/Installation (DT 2 and DT 4)

The following Suzuki special tools are required for this procedure:

a. Flywheel holder, part No. 09930-40113 on DT 4 and 1985-1989 DT 2 or part No. 09930-48720 on 1990-on DT 2.

b. Flywheel rotor remover, part No. 09930-30713.

1. Remove the engine cover.
2. On DT 2 models, remove the fuel tank. See Chapter Six.
3. Disconnect the spark plug lead to prevent accidental starting of the engine.
4. Remove the rewind starter. See Chapter Ten.
5. Remove the bolt holding the starter rewind cup (**Figure 1**). Remove the cup.
6A. On DT 2 models, perform the following:
   a. Install flywheel holder to hold the flywheel while loosening the flywheel nut (**Figure 2**).
   b. Remove the flywheel nut and install flywheel puller (A, **Figure 3**) to the flywheel with the puller bolts.
6B. On DT 4 models, perform the following:
   a. Install flywheel holder to hold the flywheel while loosening the flywheel nut (**Figure 4**).
   b. Remove the flywheel nut and install flywheel puller to the flywheel with the puller bolts (**Figure 5**).

CAUTION
Do not strike puller screw with excessive force in Step 7 or crankshaft and/or bearing damage may result.

7. Hold puller body with puller handle and tighten center screw. If flywheel does not pop from the crankshaft taper, lightly tap the puller center screw with a brass hammer (**Figure 6**).
8. Remove puller from flywheel. Remove flywheel from crankshaft. Remove flywheel key from crankshaft if it does not come off with the flywheel.
9. Inspect flywheel carefully as described in this chapter.
10. Install flywheel key, flywheel, starter cup and bolt. Tighten flywheel nut to specifications (**Table 1**).
11. Install the rewind starter. See Chapter Ten.
12. Reconnect the spark plug lead.
13. On DT 2 models, install the fuel tank. See Chapter Six.
14. Install the engine cover.

Removal/Installation (DT 6; 1985-1987 DT 8)

The following Suzuki special tools are required for this procedure:
   a. Flywheel holder, part No. 09930-40113.
   b. Flywheel rotor remover, part No. 09930-30713.

1. Remove the engine cover.
2. Disconnect the negative battery cable and the spark plug leads to prevent accidental starting of the engine.
3. Remove the overhead starter. See Chapter Ten.
4. Install the flywheel holder (**Figure 7**) onto the flywheel.
5. Hold the flywheel stationary and, using a suitable size socket, remove the flywheel nut.
6. Install the flywheel rotor remover (**Figure 8**) onto the flywheel and secure it with the remover bolts.

CAUTION
Do not strike remover plate center bolt with excessive force in Step 7 or crankshaft and/or bearing damage may result.

7. Hold onto the flywheel holder and tighten the center bolt. If the flywheel does not pop from the crankshaft taper, lightly tap the puller center bolt with a brass hammer.
8. Remove the flywheel and special tool from the crankshaft. Remove flywheel key from the crankshaft if it does not come off with the flywheel.

POWER HEAD

9. Remove the puller bolts and remove the flywheel remover from the flywheel.
10. Thoroughly inspect flywheel as described in this chapter.
11. Inspect the crankshaft and flywheel tapers. They must be perfectly clean and free of oil. Clean the tapered surfaces with a lint-free cloth and solvent, then blow dry with compressed air.
12. Installation is the reverse of removal, noting the following.

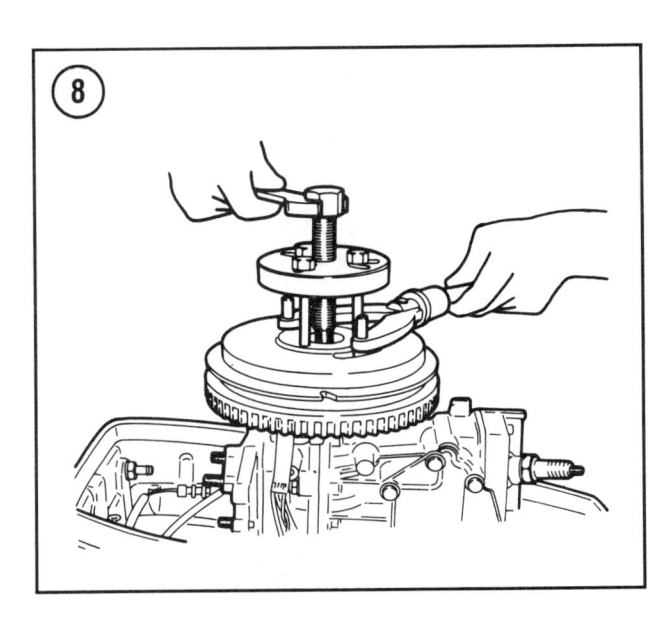

13. Install flywheel key with outer edge of key parallel to the crankshaft centerline.
14. Install the flywheel on the crankshaft.
15. Use the same tool set-up used for removal to keep the flywheel from turning while tightening the flywheel nut. Install flywheel nut and tighten to specifications (**Table 1**).

Removal/Installation (1988-on DT 8, DT 9.9)

The following Suzuki special tools are required for this procedure:
  a. Flywheel holder, part No. 09930-48720.
  b. Flywheel holder attachment bolts, part No. 09930-49210.
  c. Flywheel rotor remover, part No. 09930-39411.
  d. Flywheel puller bolts, part No. 09930-39210.

1. Remove the engine cover.
2. Disconnect the negative battery cable and the spark plug leads to prevent accidental starting of the engine.
3. Remove the rewind starter. See Chapter Ten.
4. On electric start models, remove the starter motor and bracket. See Chapter Seven.
5. Install the attachment bolts, washer and nuts (B, **Figure 9**) to the flywheel holder.
6. Install the flywheel holder (A, **Figure 9**) onto the flywheel.
7. Hold the flywheel stationary and using a suitable size socket, remove the flywheel nut.
8. Remove the flywheel holder (A, **Figure 9**) assembly from the flywheel.
9. Install the flywheel rotor remover (C, **Figure 10**) onto the flywheel and secure it with the special remover bolts (D, **Figure 10**).

CAUTION
Do not strike remover plate center bolt with excessive force in Step 10 or crankshaft and/or bearing damage may result.

10. Hold puller body with puller handle and tighten the center bolt. If the flywheel does not pop from the crankshaft taper, lightly tap the puller center bolt with a brass hammer.

11. Remove the flywheel and special tool from the crankshaft. Remove flywheel key from the crankshaft if it does not come off with the flywheel.

12. Remove the special puller bolts and remove the flywheel remover plate from the flywheel.

13. Thoroughly inspect flywheel as described in this chapter.

14. Inspect the crankshaft and flywheel tapers. They must be perfectly clean and free of oil. Clean the tapered surfaces with a lint-free cloth and solvent, then blow dry with compressed air.

15. Installation is the reverse of removal, noting the following.

16. Install flywheel key with outer edge of key parallel to the crankshaft centerline.

17. Install the flywheel on the crankshaft.

18. Use the same tool set-up used for removal to keep the flywheel from turning while tightening the flywheel nut. Install flywheel nut and tighten to specifications (**Table 1**).

Removal/Installation
(1985-1987 DT 9.9; 1985-on DT 15)

The following Suzuki special tools are required for this procedure:

  a. Flywheel holder, part No. 09930-49310.
  b. Flywheel remover plate, part No. 09930-30713.

1. Remove the engine cover.

2. Disconnect the negative battery cable and the spark plug leads to prevent accidental starting of the engine.

3. On models so equipped, remove the rewind starter (**Figure 11**). See Chapter Ten.

4. On electric start models, remove the starter motor (**Figure 12**). See Chapter Seven.

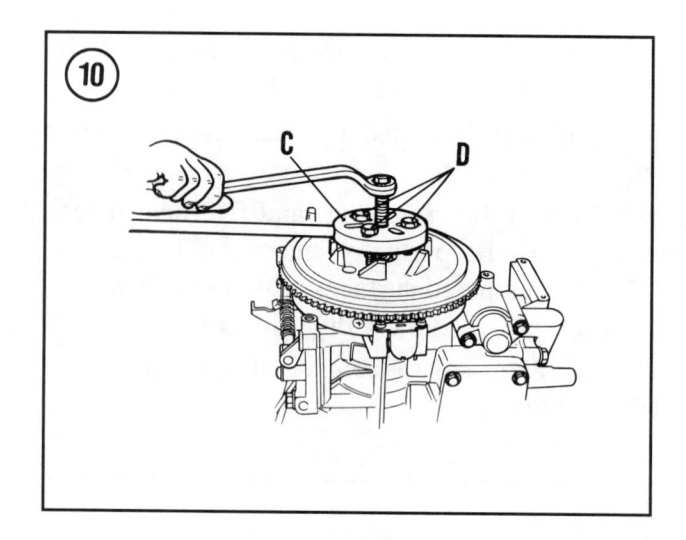

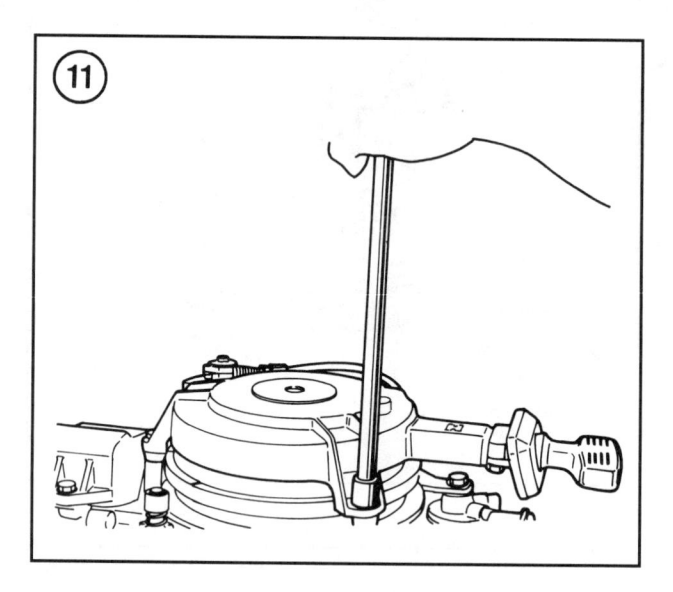

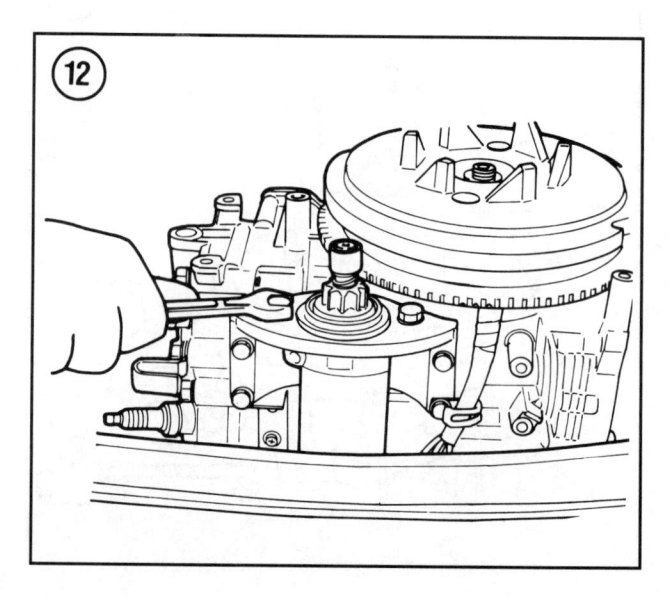

POWER HEAD

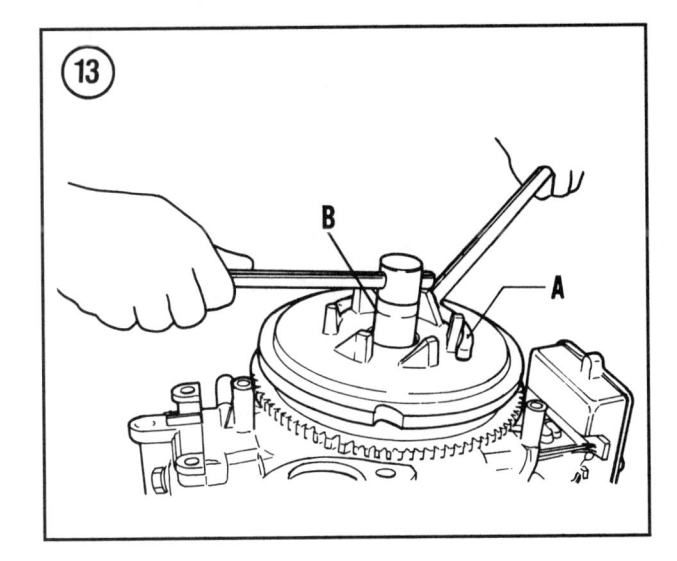

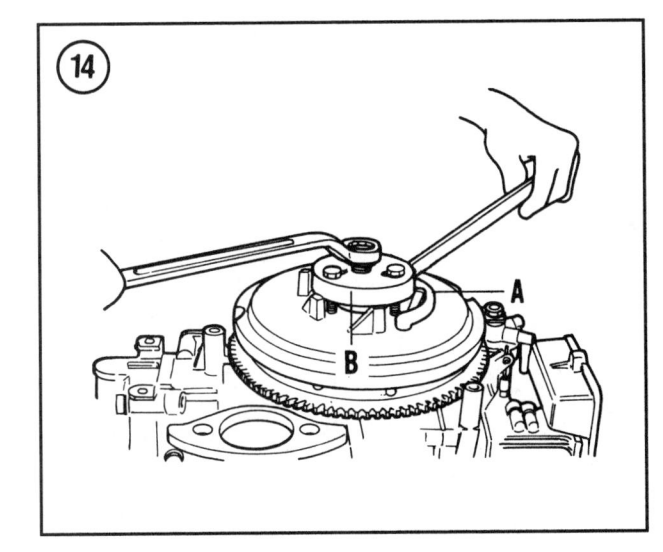

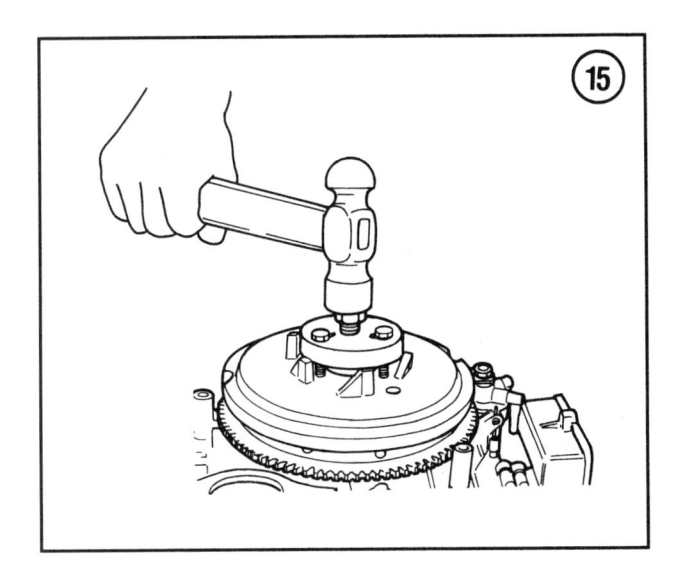

5. Install the flywheel holder (A, **Figure 13**) onto the flywheel and lock it in place on the raised bosses.

6 Hold the flywheel stationary and using a suitable size socket (B, **Figure 13**) remove the flywheel nut.

7. Leave the flywheel holder (A, **Figure 14**) in place on the flywheel.

8. Install the flywheel remover plate (B, **Figure 14**) onto the flywheel and secure it with the special puller bolts

CAUTION
Do not strike remover plate center bolt with excessive force in Step 9 or crankshaft and/or bearing damage may result.

9. Hold puller body with puller handle and tighten the center bolt. If the flywheel does not pop from the crankshaft taper, lightly tap the puller center bolt with a brass hammer (**Figure 15**).

10. Remove the flywheel and special tool from the crankshaft. Remove flywheel key from the crankshaft if it does not come off with the flywheel.

11. Remove the special puller bolts and remove the flywheel remover from the flywheel.

12. Thoroughly inspect flywheel as described in this chapter.

13. Inspect the crankshaft and flywheel tapers. They must be perfectly dry and free of oil. Clean the tapered surfaces with a lint-free cloth and solvent, then blow dry with compressed air.

14. Installation is the reverse of removal, noting the following.

15. Install flywheel key with outer edge of key parallel to the crankshaft centerline.

16. Install the flywheel on the crankshaft.

17. Use the same tool set-up used for removal to keep the flywheel from turning while tightening the flywheel nut. Install flywheel nut and tighten to specifications (**Table 1**).

**Removal/Installation
(1986-1988 DT 20; 1985-1988 DT 25;
1985-1987 DT 30; 1987-1989 DT 35;
1985-on DT 40, DT 55, DT 65)**

The following Suzuki special tools are required for this procedure:
   a. Flywheel holder, part No. 09930-39520.
   b. Flywheel remover plate, part No. 09930-49410.
   c. Flywheel puller bolts, part No. 09930-39420.

1. Remove the engine cover.
2. Disconnect the negative battery cable and the spark plug leads to prevent accidental starting of the engine.
3. On models so equipped, remove the rewind starter (**Figure 16**). See Chapter Ten.
4. On electric start models, remove the starter motor and bracket (**Figure 17**). See Chapter Seven.
5. Install the flywheel holder (A, **Figure 18**) onto the flywheel and secure with bolts.
6. Insert the handle (B, **Figure 18**) from the flywheel holder plate set of tools into the flywheel holder.
7. Hold the flywheel stationary and using a suitable size socket, remove the flywheel nut.
8. Remove the flywheel holder (A, **Figure 18**) from the flywheel.
9. Install the flywheel remover plate (A, **Figure 19**) onto the flywheel and secure it with the special puller bolts (B, **Figure 19**).
10. Insert the handle into the flywheel remover plate.

*CAUTION
Do not strike remover plate center bolt with excessive force in Step 11 or crankshaft and/or bearing damage may result.*

11. Hold puller body with puller handle and tighten the center bolt. If the flywheel does not pop from the crankshaft taper, lightly tap the

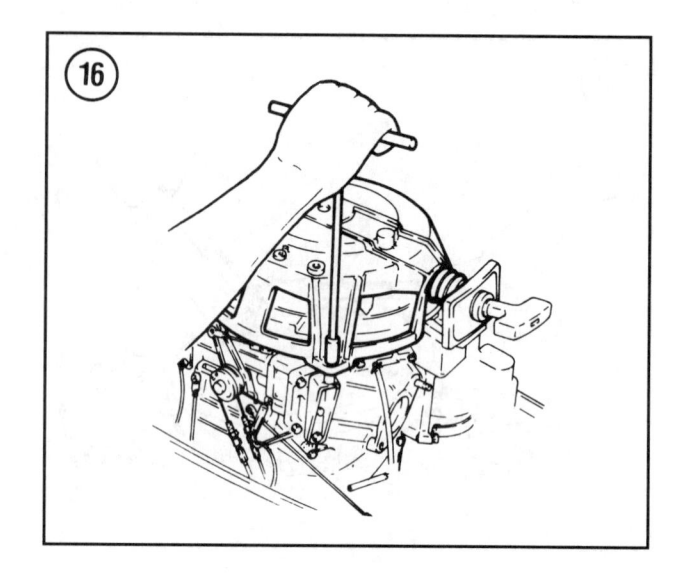

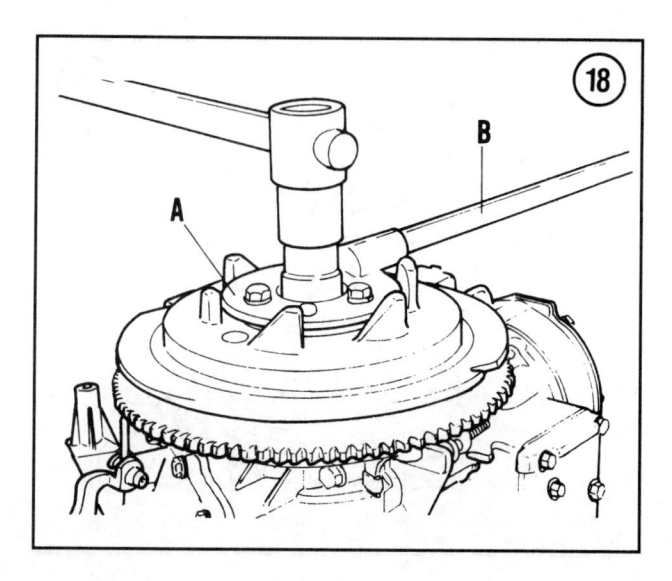

POWER HEAD

puller center bolt with a brass hammer (**Figure 20**).

12. Remove the flywheel and special tool from the crankshaft. Remove flywheel key from the crankshaft if it does not come off with the flywheel.

13. Remove the special puller bolts and remove the flywheel remover plate from the flywheel.

14. Thoroughly inspect flywheel as described in this chapter.

15. Inspect the crankshaft and flywheel tapers. They must be perfectly dry and free of oil. Clean the tapered surfaces with a lint-free cloth and solvent, then blow dry with compressed air.

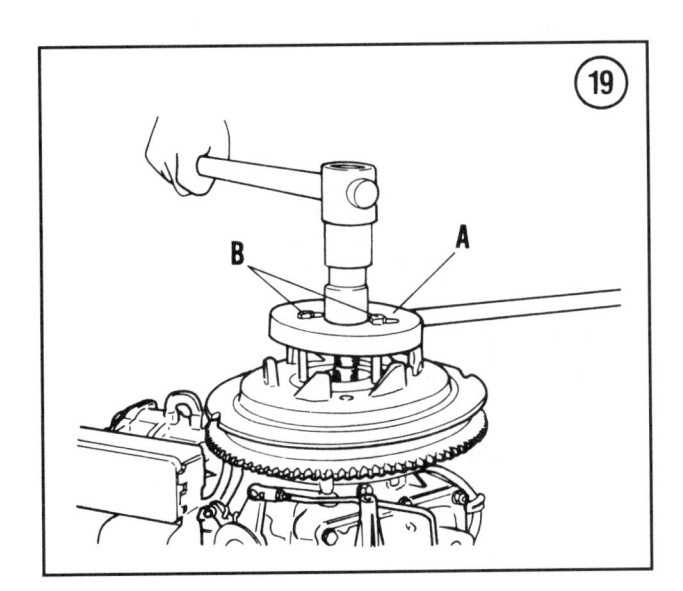

16. Installation is the reverse of removal, noting the following.

17. Install flywheel key with outer edge of key parallel to the crankshaft centerline.

18. Install the flywheel on the crankshaft.

19. Use the same tool set-up used for removal to keep the flywheel from turning while tightening the flywheel nut. Install flywheel nut and tighten to specifications (**Table 1**).

Removal/Installation
(1989-on DT 25; 1988-on DT 30)

The following Suzuki special tools are required for this procedure:

   a. Flywheel holder, part No. 09930-48720.
   b. Flywheel rotor remover, part No. 09930-39411.
   c. Flywheel remover bolts, part No. 09930-39420.

1. Remove the engine cover.
2. Disconnect the negative battery cable and the spark plug leads to prevent accidental starting of the engine.
3. On models so equipped, remove the rewind starter. See Chapter Ten. Remove the starter pulley from the flywheel.
4. Remove the gear counter coil from its mounting bracket.
5. On electric start models, remove the starter motor. See Chapter Seven.
6. Install the flywheel holder (A, **Figure 21**) onto the flywheel and secure with bolts.
7. Hold the flywheel stationary and using a suitable size socket (**Figure 21**), remove the flywheel nut.
8. Remove the flywheel holder (A, **Figure 21**) from the flywheel.
9. Install the flywheel rotor remover (A, **Figure 22**) onto the flywheel and secure it with the special puller bolts (B, **Figure 22**).

CAUTION
Do not strike remover plate center bolt with excessive force in Step 10 or crankshaft and/or bearing damage may result.

10. Hold flywheel rotor remover handle and tighten the center bolt. If the flywheel does not pop from the crankshaft taper, lightly tap the puller center bolt with a brass hammer.

11. Remove the flywheel and special tool from the crankshaft. Remove flywheel key from the crankshaft if it does not come off with the flywheel.

12. Remove the special puller bolts and remove the flywheel rotor remover from the flywheel.

13. Thoroughly inspect flywheel as described in this chapter.

14. Inspect the crankshaft and flywheel tapers. They must be perfectly dry and free of oil. Clean the tapered surfaces with a lint-free cloth and solvent, then blow dry with compressed air.

15. Installation is the reverse of removal, noting the following.

16. Install flywheel key with outer edge of key parallel to the crankshaft centerline.

17. Install the flywheel on the crankshaft.

18. Use the same tool set-up used for removal to keep the flywheel from turning while tightening the flywheel nut. Install flywheel nut and tighten to specifications (**Table 1**).

Removal/Installation (DT 75, DT 85; 1985 DT 115)

The following Suzuki special tools are required for this procedure:

a. Flywheel holder, part No. 09930-49410.

b. Flywheel remover plate, part No. 09930-39410.

c. Flywheel holder plate bolts, part No. 09930-39420.

1. Remove the engine cover.

2. Disconnect the negative battery cable and the spark plug leads to prevent accidental starting of the engine.

3. Remove the bolts securing the engine hook (**Figure 23**) and remove the hook.

4. Using the engine hook bolts, secure the flywheel holder (A, **Figure 24**) to the block. Make sure the flywheel holder is meshed properly with the teeth on the flywheel.

5. With the flywheel held stationary, use a suitable size socket and remove the flywheel nut and washer.

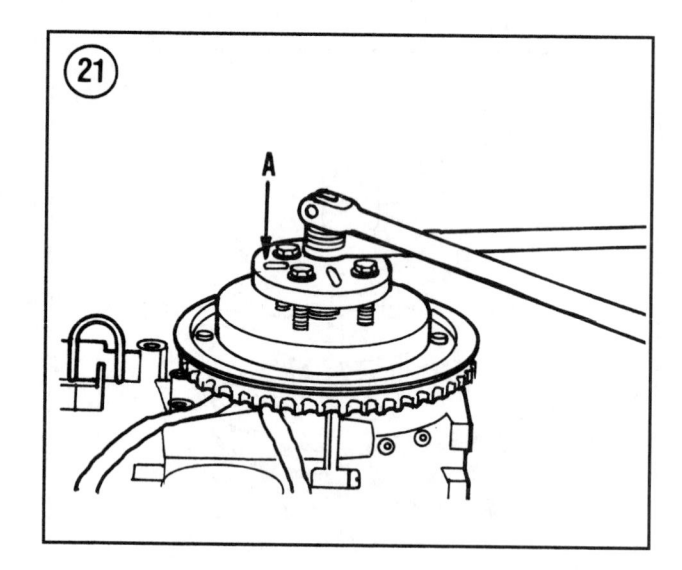

POWER HEAD

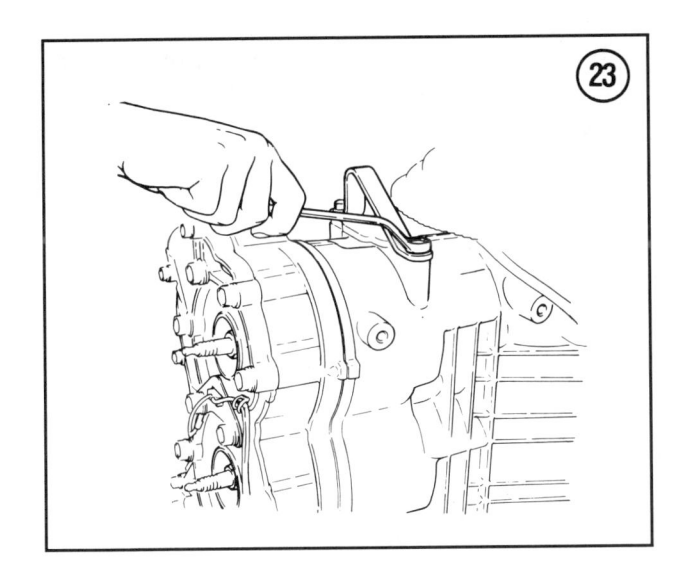

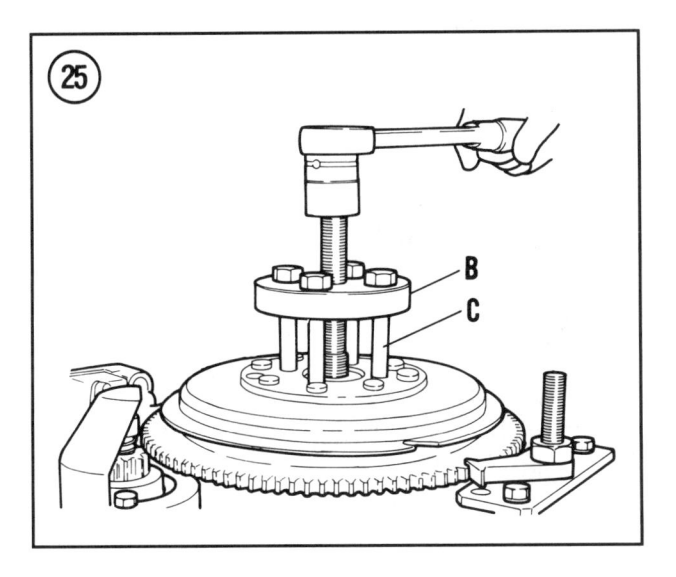

6. Install the flywheel remover plate (B, **Figure 25**) onto the flywheel and secure it with the special bolts (C, **Figure 25**).

CAUTION
Do not strike remover plate center bolt with excessive force in Step 7 or crankshaft and/or bearing damage may result.

7. Tighten the center bolt. If the flywheel does not pop from the crankshaft taper, lightly tap the puller center bolt with a brass hammer.
8. Remove the flywheel and special tool from the crankshaft. Remove flywheel key from the crankshaft if it does not come off with the flywheel.
9. Remove the special remover bolts and remove the flywheel remover plate from the flywheel.
10. Thoroughly inspect flywheel as described in this chapter.
11. Inspect the crankshaft and flywheel tapers. They must be perfectly dry and free of oil. Clean the tapered surfaces with a lint-free cloth and solvent, then blow dry with compressed air.
12. Installation is the reverse of removal, noting the following.
13. Install flywheel key with outer edge of key parallel to the crankshaft centerline.
14. Install the flywheel on the crankshaft.
15. Use the same tool set-up used for removal to keep the flywheel from turning while tightening the flywheel nut. Install flywheel nut and tighten to specifications (**Table 1**).

Removal/Installation (1986-on DT 115, DT 140; V4; V6)

The following Suzuki special tools are required for this procedure:
  a. Flywheel holder, part No. 09930-48720.
  b. Flywheel rotor remover, part No. 09930-39411.
  c. Flywheel remover bolts, part No. 09930-39420.

1. Remove the engine cover.
2. Disconnect the negative battery cable and spark plug leads to prevent accidental starting of the engine.
3. Remove the flywheel cover.
4. Install the flywheel holder (A, **Figure 26**) onto the flywheel and secure with bolts.
5. Hold the flywheel stationary and using a suitable size socket (B, **Figure 26**), remove the flywheel nut and washer.
6. Remove the flywheel holder (A, **Figure 26**) from the flywheel.
7. Install the flywheel rotor remover (A, **Figure 27**) onto the flywheel and secure it with the special remover bolts (B, **Figure 27**).

CAUTION
Do not strike remover plate center bolt with excessive force in Step 8 or crankshaft and/or bearing damage may result.

8. Hold flywheel rotor remover handle and tighten the center bolt (C, **Figure 27**). If the flywheel does not pop from the crankshaft taper, lightly tap the puller center bolt with a brass hammer.
9. Remove the flywheel and special tool from the crankshaft. Remove flywheel key from the crankshaft if it does not come off with the flywheel.
10. Remove the special remover bolts and remove the flywheel rotor remover from the flywheel.
11. Thoroughly inspect flywheel as described in this chapter.
12. Inspect the crankshaft and flywheel tapers. They must be perfectly dry and free of oil. Clean the tapered surfaces with a lint-free cloth and solvent, then blow dry with compressed air.
13. Installation is the reverse of removal, noting the following.
14. Install flywheel key with outer edge of key parallel to the crankshaft centerline.
15. Install the flywheel on the crankshaft.

16. Use the same tool set-up used for removal to keep the flywheel from turning while tightening the flywheel nut. Install flywheel nut and tighten to specifications (**Table 1**).

POWER HEAD

Inspection

1. Check the flywheel carefully for cracks or fractures.

WARNING
A cracked or chipped flywheel must be replaced. A damaged flywheel may fly apart at high rpm, throwing metal fragments over a large area. Do not attempt to repair a damaged flywheel.

2. Check tapered bore of flywheel and crankshaft taper for signs of fretting or working.

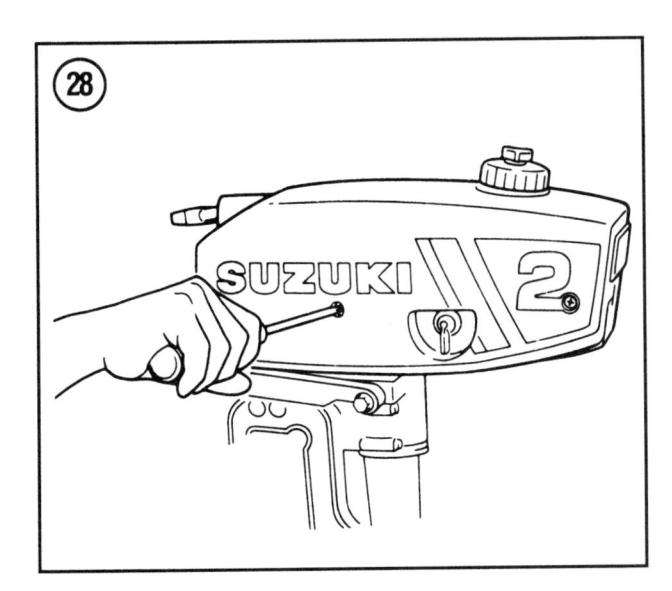

3. On electric start models, check the flywheel teeth for excessive wear or damage.
4. Check crankshaft and flywheel nut threads for wear or damage.
5. Replace flywheel, crankshaft and/or flywheel nut as required.

POWER HEAD

When removing any power head, it is a good idea to make a sketch or take an instant picture of the location, routing and positioning of electrical wiring, brackets and J-clamps for reassembly reference. Take notes as you remove wires, washers and engine grounds so they may be reinstalled in their correct position. Unless specified otherwise, install lockwashers on the engine side of the electrical lead to assure a good ground.

CAUTION
After overhauling an oil-injected engine, the first 30 gallons (5 full tanks) of fuel used should be a 50:1 fuel-oil mixture (see Chapter Four) in addition to the lubricant supplied by the injection pump. Mark the power head oil tank level and make sure the injection system works properly (oil level diminishes) before switching over to plain gasoline at the end of the 30 gallon break-in period.

Removal/Installation (DT 2)

1. Remove the starboard and port engine covers (**Figure 28**).
2. Remove the fuel tank. See Chapter Six.
3. Remove the flywheel as described in this chapter.
4. Disconnect the spark plug lead.
5. Disconnect the 2 stator lead wires.
6. Remove the screws holding the stator (**Figure 29**). Remove the stator assembly.
7. Remove the choke knob (**Figure 30**).

8. Remove the throttle link knob and the control panel (**Figure 31**).

9. Remove the carburetor and fuel shut-off valve. See Chapter Six.

10. Remove the 6 bolts holding the power head to the drive shaft housing (**Figure 32**). Remove the power head.

11. Remove and discard the power head mounting gasket.

12. Clean the power head mounting and drive shaft housing gasket surfaces of all gasket residue.

13. Installation is the reverse of removal, noting the following.

14. Use a new power head gasket.

15. Lightly coat drive shaft splines with water-resistant grease (part No. 99000-25610) or equivalent.

16. Rotate propeller as required to align drive shaft and crankshaft splines.

17. Coat power head attaching screw threads with Thread Lock 1342 (part No. 99000-32050) and tighten to specifications (**Table 1**).

18. Tighten all fasteners to specifications (**Table 1**).

19. Perform engine synchronization and linkage adjustments. See Chapter Five.

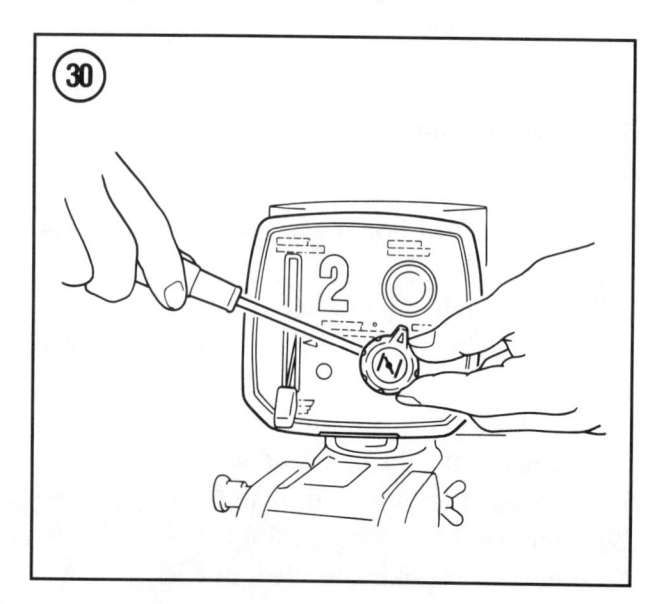

Removal/Installation (DT 4)

1. Remove the engine cover.
2. Disconnect the spark plug lead to prevent accidental starting of the engine.
3. Remove the fuel tank. See Chapter Six.
4. Remove the overhead starter. See Chapter Ten.
5. Remove the carburetor and fuel pump. See Chapter Six.

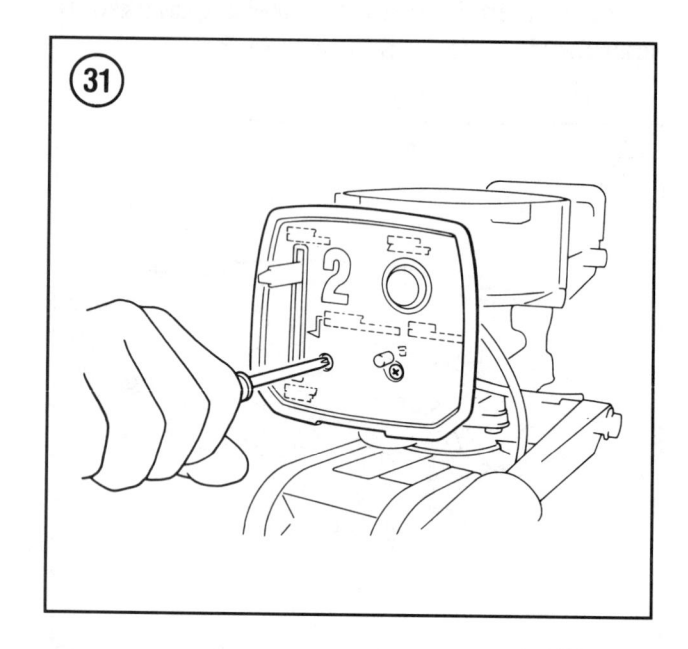

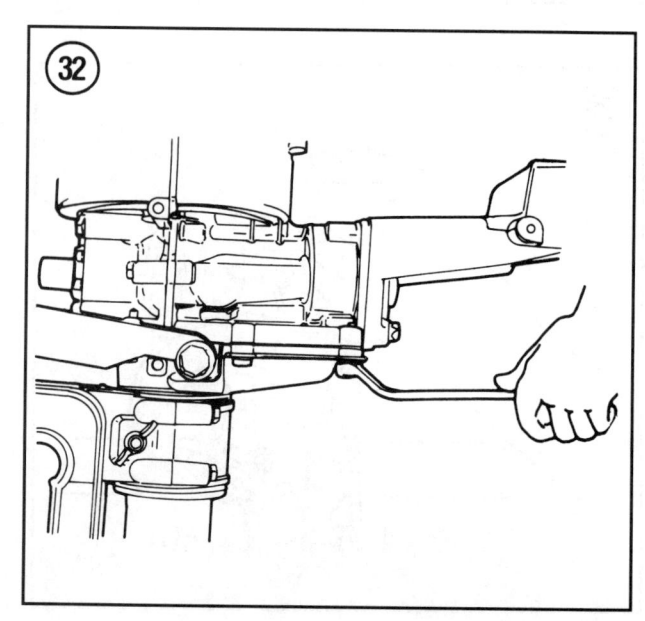

POWER HEAD

6. Disconnect the stator and CDI unit electrical leads. Remove the ignition coil and the CDI unit. See Chapter Seven.
7. Remove the flywheel as described in this chapter.
8. Remove the screws holding the stator to the power head. Remove the stator.

NOTE
At this point, there should be no linkage, ground leads or other electrical wiring connecting the power head to the support plate. Recheck to make sure that nothing will hamper power head removal.

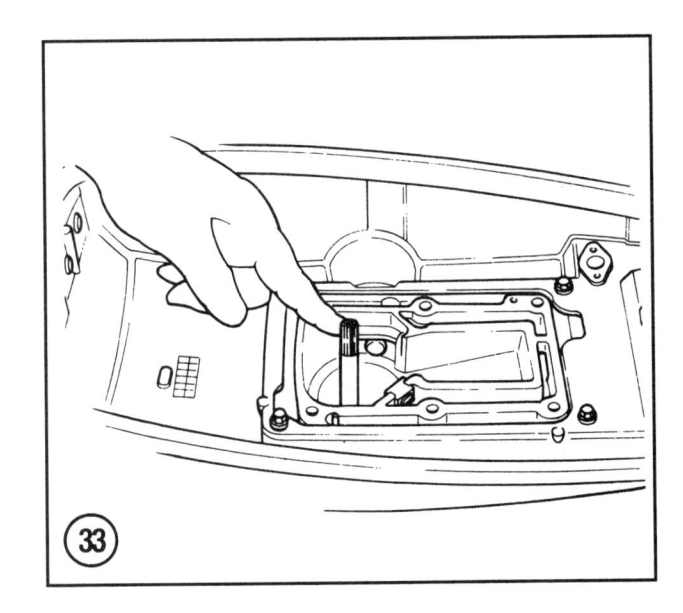

9. Remove the 6 bolts holding the power head to the drive shaft housing. Remove the power head and place on a clean workbench.
10. Remove and discard the power head mounting gasket.
11. Clean the power head mounting and drive shaft housing gasket surfaces of all gasket residue.
12. Installation is the reverse of removal, noting the following.
13. Lightly coat the drive shaft splines with water-resistant grease (part No. 99000-25160) or equivalent (**Figure 33**).
14. Coat power head attaching screw threads with Silicone Seal (part No. 99000-31120) or equivalent.
15. Tighten all fasteners to specifications (**Table 1**).
16. Perform engine synchronization and linkage adjustments. See Chapter Five.

Removal/Installation (DT 6; 1985-1987 DT 8)

1. Remove the engine cover.
2. On models so equipped, disconnect the battery negative lead.
3. Disconnect the spark plug leads to prevent accidental starting of the engine.
4. Remove the carburetor silencer cover (**Figure 34**).
5. Disconnect the fuel line at the carburetor and remove the choke knob.
6. Remove the carburetor, fuel pump and fuel filter. See Chapter Six.
7. Remove the CDI unit and ignition coils. See Chapter Seven.
8. Remove the rewind starter. See Chapter Ten.
9. Loosen the throttle cable locknuts (**Figure 35**) and remove the cable.

10. Remove the flywheel as described in this chapter.
11. Disconnect the stator leads. Remove the stator screws (**Figure 36**) and remove the stator.
12. Remove the 4 screws holding the upper oil seal housing to the power head (**Figure 37**).

NOTE
At this point, there should be no linkage, ground leads or other electrical wiring connecting the power head to the support plate. Recheck to make sure that nothing will hamper power head removal.

13. Remove the 6 bolts holding the power head to the drive shaft housing (**Figure 38**).
14. Remove the power head from the drive shaft housing and remove the support plate. Place the power head on a clean workbench.
15. Remove and discard the power head mounting gasket.
16. Clean the power head mounting and drive shaft housing gasket surfaces of all gasket residue.
17. Installation is the reverse of removal, noting the following.
18. Install a new power head mounting gasket.
19. Lightly coat the drive shaft splines with water-resistant grease (part No. 99000-25160 or equivalent).
20. Rotate the propeller as required to align the crankshaft and drive shaft splines.

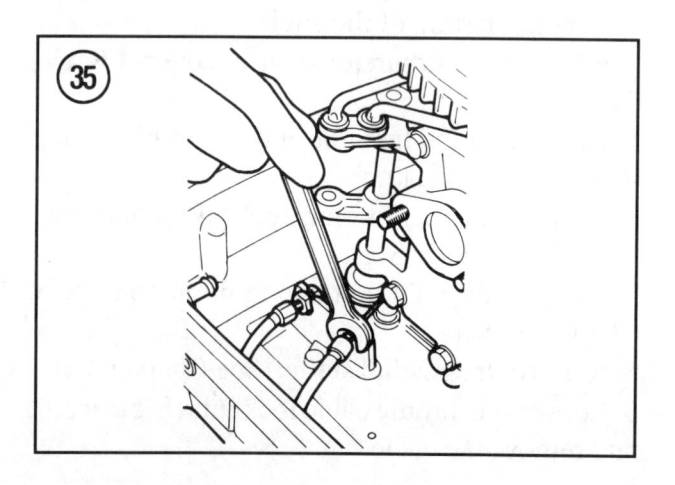

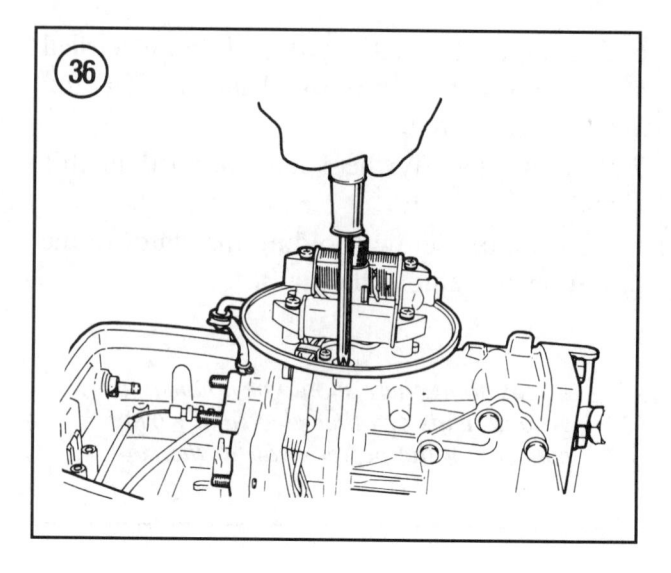

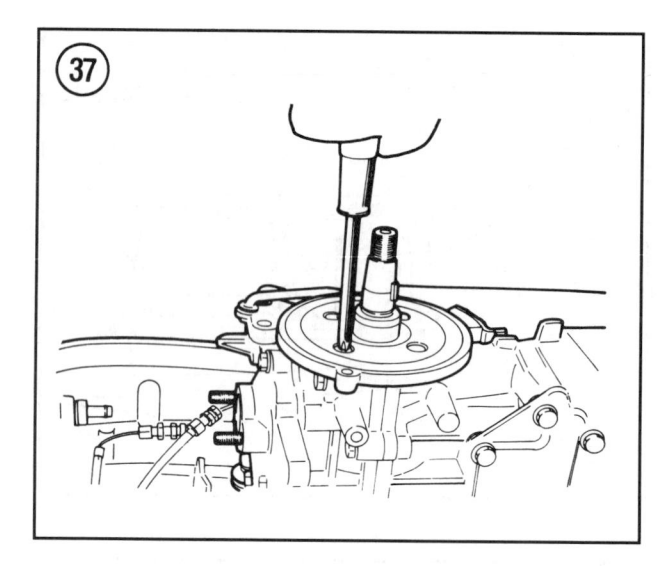

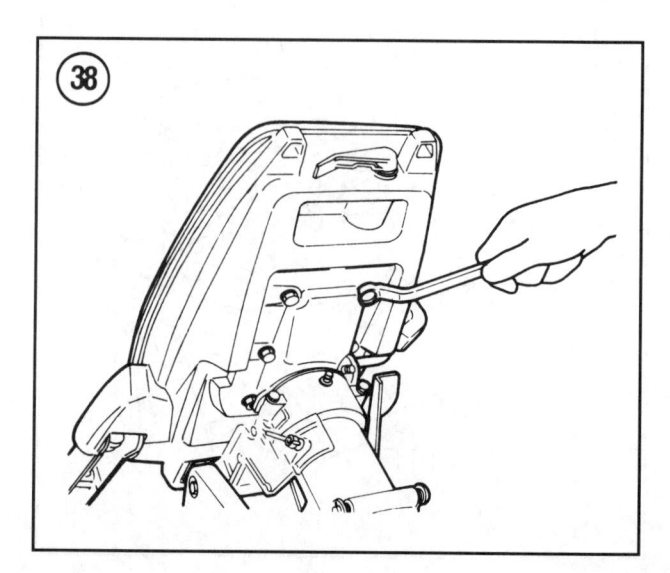

POWER HEAD

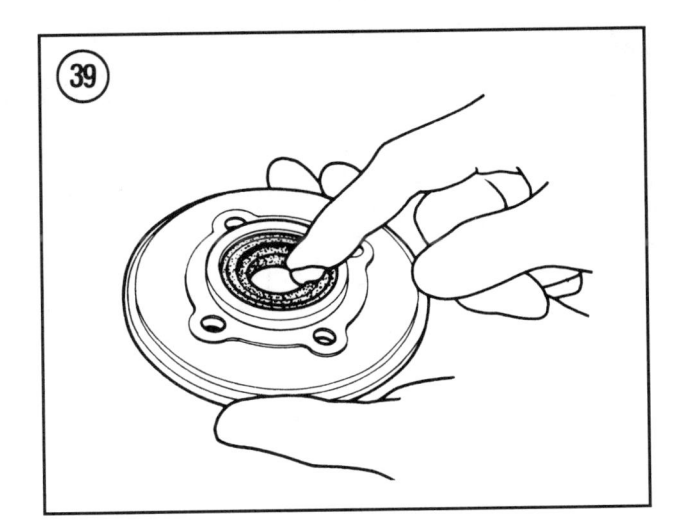

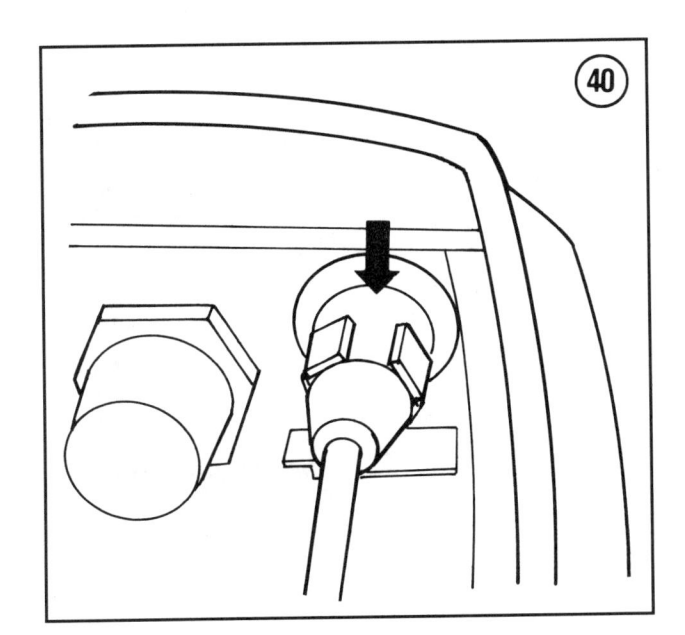

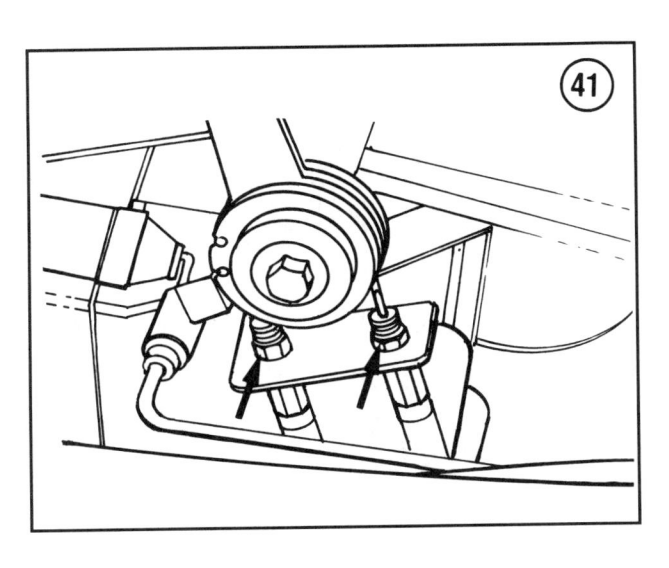

21. Install a new upper oil seal housing seal with its lip facing down toward the power head. Lubricate the seal lips with Super Grease A (part No. 990000-25010) (**Figure 39**).
22. Coat power head attaching screw threads with water resistant grease or equivalent.
23. Tighten all fasteners to specifications (**Table 1**).
24. Perform engine synchronization and linkage adjustments. See Chapter Five.

Removal/Installation (1988-on DT 8, DT 9.9)

1. Remove the engine cover.
2. On models equipped, disconnect the negative battery cable.
3. Remove the rewind starter. See Chapter Ten.
4. Disconnect the spark plug wires and wiring harness connector from the power head.
5. Remove the E-ring and take off the idle speed control knob (**Figure 40**).
6. Loosen the throttle cable locknuts (**Figure 41**) and remove the cables.
7. Remove the oil tank (A, **Figure 42**). See Chapter Twelve.
8. Remove the carburetor silencer cover (B, **Figure 42**).
9. Remove the carburetor and choke knob (C, **Figure 42**). See Chapter Six.

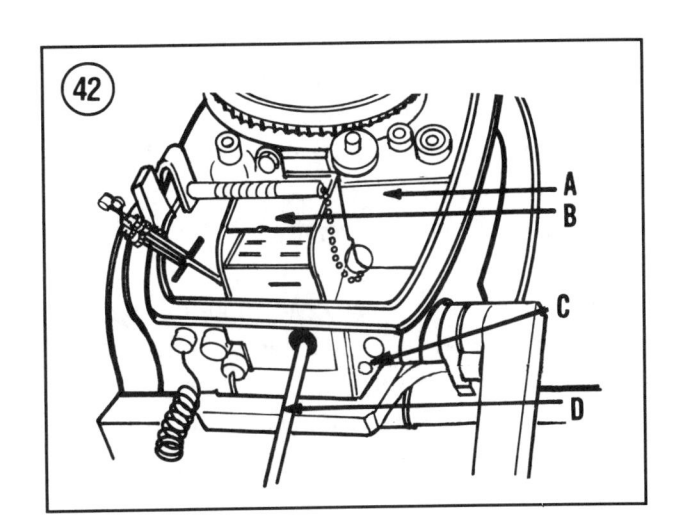

CHAPTER EIGHT

NOTE
The lower carburetor bolt can be accessed by removing cap in the lower cover and inserting wrench through hole (D, Figure 42).

10. Disconnect the oil hose from the power head.

NOTE
At this point, there should be no linkage, ground leads or other electrical wiring connecting the power head to the support plate. Recheck to make sure that nothing will hamper power head removal.

11. Remove the 4 bolts and 2 nuts securing the power head (**Figure 43**) to the drive shaft housing.
12. Lift the power head from the drive shaft housing and remove the support plate. Place the power head on a clean workbench.
13. Clean all gasket residue from the power head mounting and drive shaft housing support plate surfaces.
14. Remove the bolts securing the exhaust tube to the power head and remove the exhaust tube.
15. Remove the CDI unit, rectifier, ignition coils, starter relay, starter motor, neutral switch and spark plugs. See Chapter Seven.
16. Remove the throttle cable bracket.
17. Remove the flywheel as described in this chapter.
18. Disconnect the stator leads. Remove the 3 screws securing the stator and remove the stator.
19. Remove the 3 screws holding the upper retainer and spacer to the power head. Remove the upper retainer, spacer and gaskets.
20. Installation is the reverse of removal, noting the following.
21. Install a new power head mounting gasket (A, **Figure 44**). Make sure the locating dowels are in place (B, **Figure 44**).
22. Lightly coat the drive shaft splines with water-resistant grease (part No. 99000-25160 or equivalent).

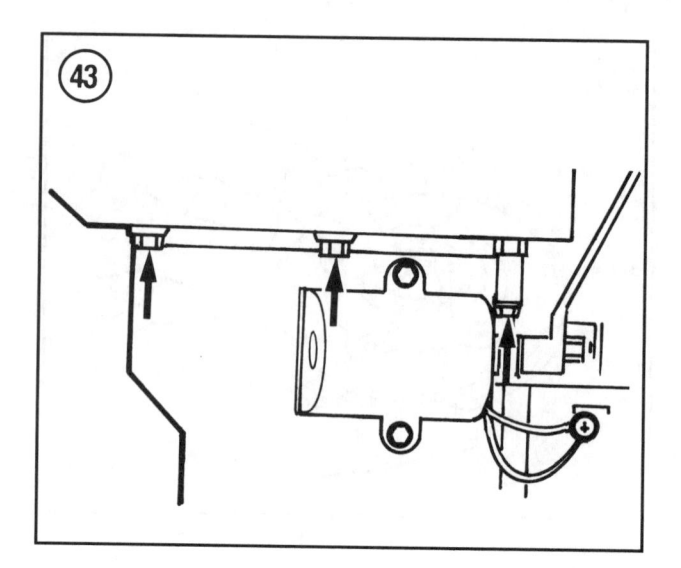

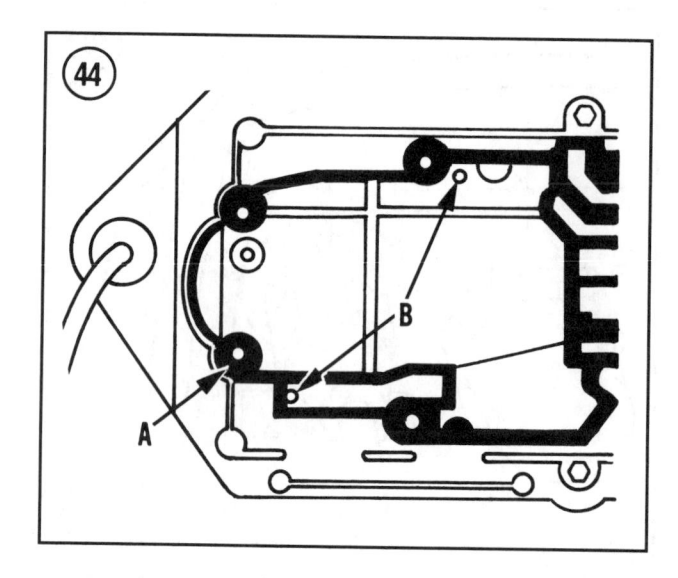

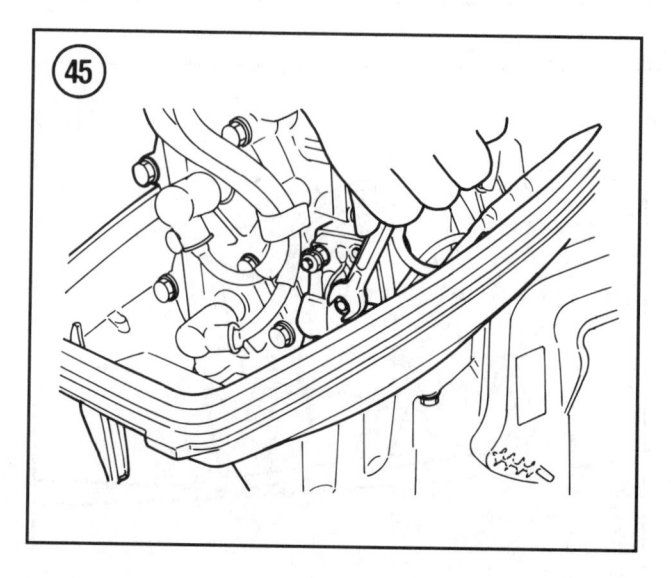

POWER HEAD

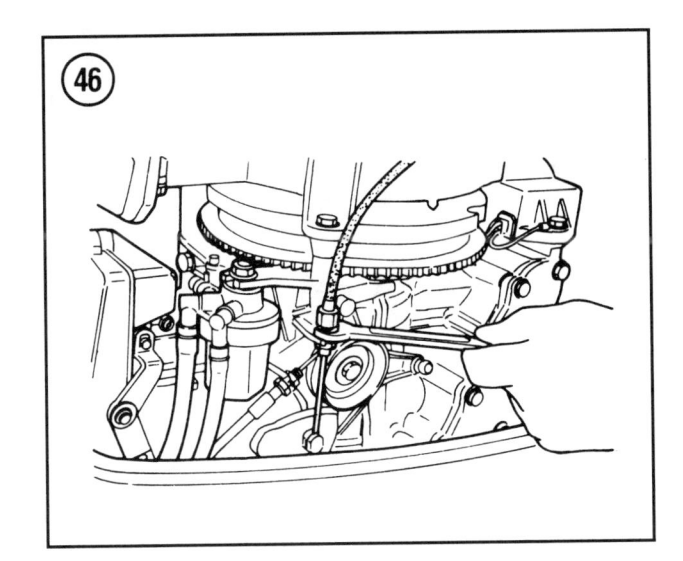

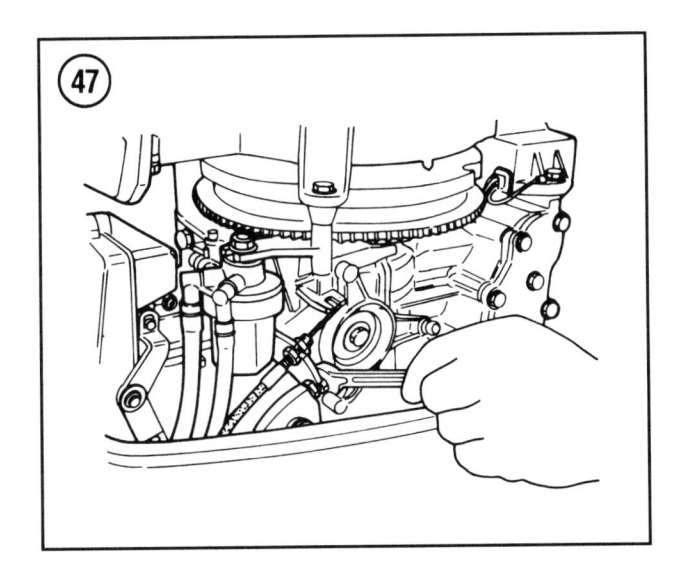

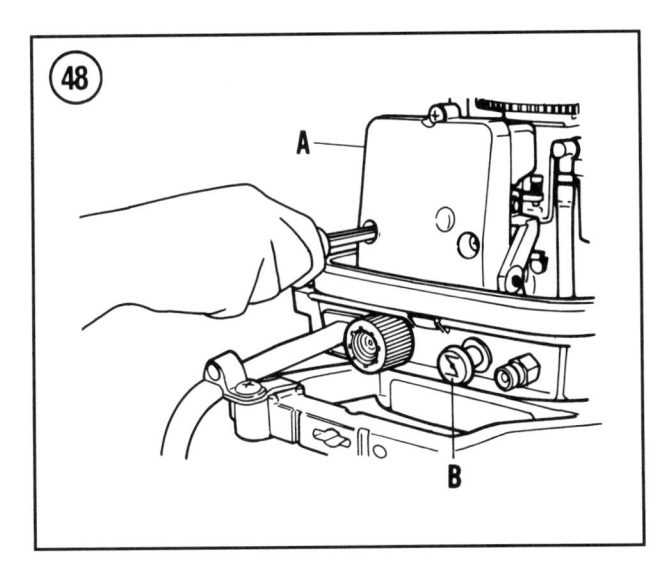

23. Rotate propeller as required to align crankshaft and drive shaft splines.
24. Coat power head attaching screw threads with water-resistant grease or equivalent.
25. Tighten all fasteners to specifications (**Table 1**).
26. Make sure all electrical connections are free of corrosion and are tight.
27. Perform engine synchronization and linkage adjustments. See Chapter Five.

Removal/Installation
(1985-1987 DT 9.9; DT 15)

1. Remove the engine cover.
2. Disconnect the spark plug leads. Remove the spark plugs.
3. On electric start models, perform the following:
   a. Remove the nuts and disconnect the electrical cables from the battery and starter (**Figure 45**).
   b. Disconnect the yellow/green wire from the neutral switch.
4. Loosen the neutral starter interlock locknut and disconnect the interlock cable from the throttle limiter (**Figure 46**).
5. Loosen the throttle cable locknuts and disconnect the cable from the control lever (**Figure 47**).
6. Remove the screws securing the silencer cover (A, **Figure 48**) and remove the cover.
7. Pull off the choke lever knob (B, **Figure 48**).
8. Remove the rewind starter assembly. See Chapter Ten.
9. Remove the CDI unit. See Chapter Seven.
10. Disconnect the brown electrical wire and remove the neutral start switch (**Figure 49**).
11. Remove the red and yellow electrical wires from the stator and remove the rectifier (**Figure 50**).
12. Remove the electric starter. See Chapter Seven.
13. Remove the flywheel as described in this chapter.

14. Disconnect the stator connector and remove the throttle control lever (**Figure 51**).

15. Disconnect the hoses from the fuel filter and remove the fuel filter. Plug the ends of the fuel hoses to prevent the loss of fuel.

16. Remove the bolts and remove the silencer case and carburetor (**Figure 52**).

17. Remove the fuel pump. See Chapter Six.

18. Remove the nut on the backside of the starter switch assembly (**Figure 53**).

19. Unclamp the starter cable clamp and remove the grommet from the cable (**Figure 54**). Pull the starter cable out.

20. Remove the screws holding the stator to the power head (**Figure 55**). Remove the stator assembly.

21. Remove the 6 bolts holding the power head to the drive shaft housing (**Figure 56**).

NOTE
At this point, there should be no hoses, wires or linkage connecting the power head to the drive shaft housing. Recheck this to make sure nothing will hamper power head removal.

22. Remove the power head from the drive shaft housing and place it on a clean workbench.

23. Remove the 4 bolts holding the starter motor bracket to the power head. Remove the starter motor bracket.

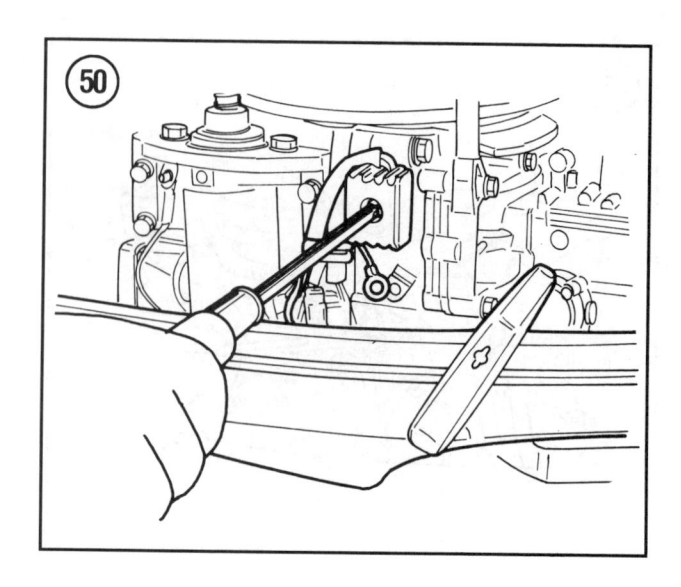

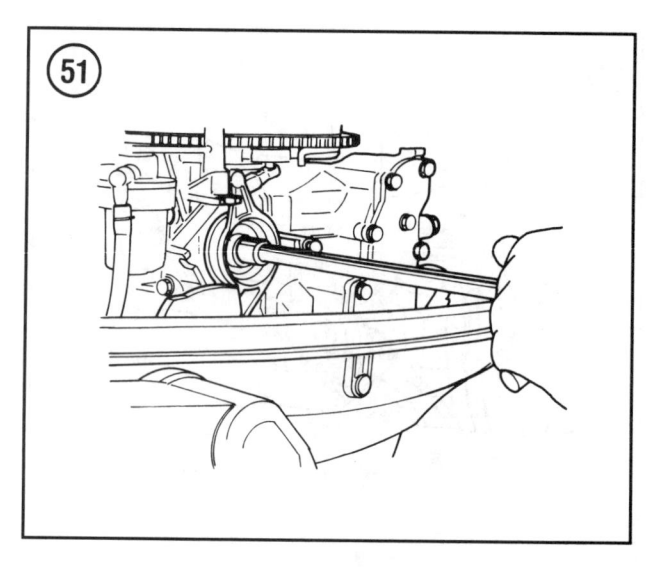

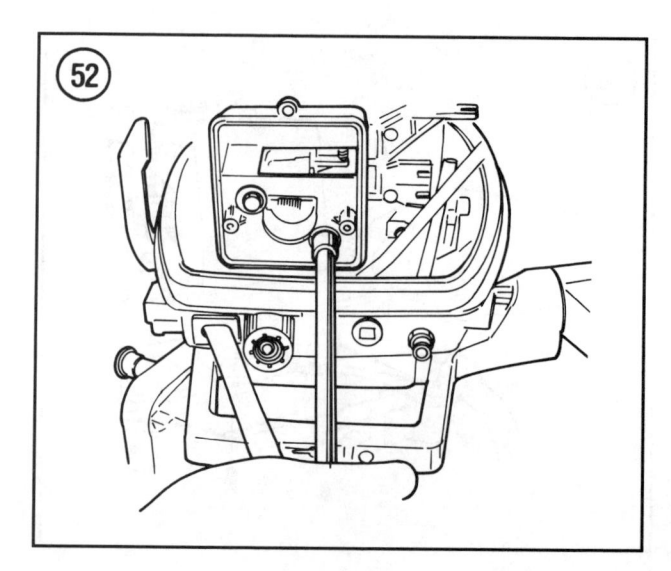

POWER HEAD

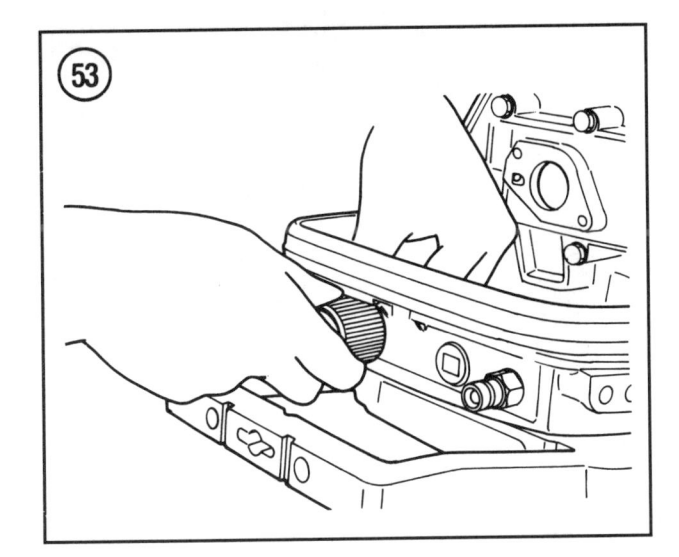

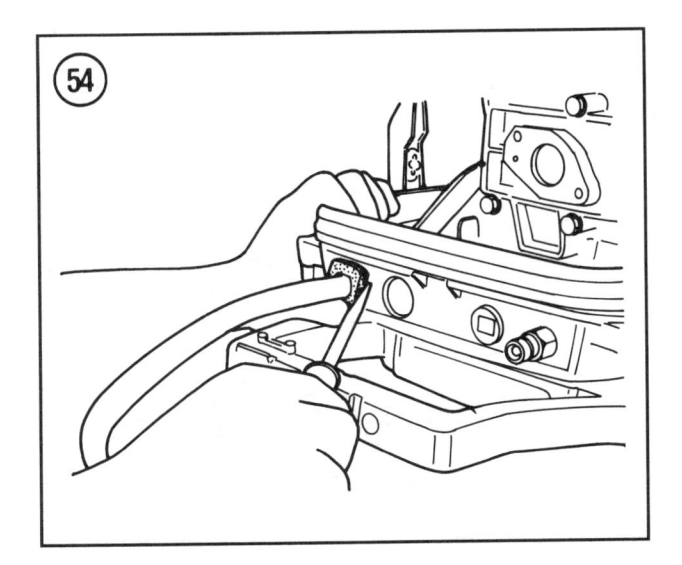

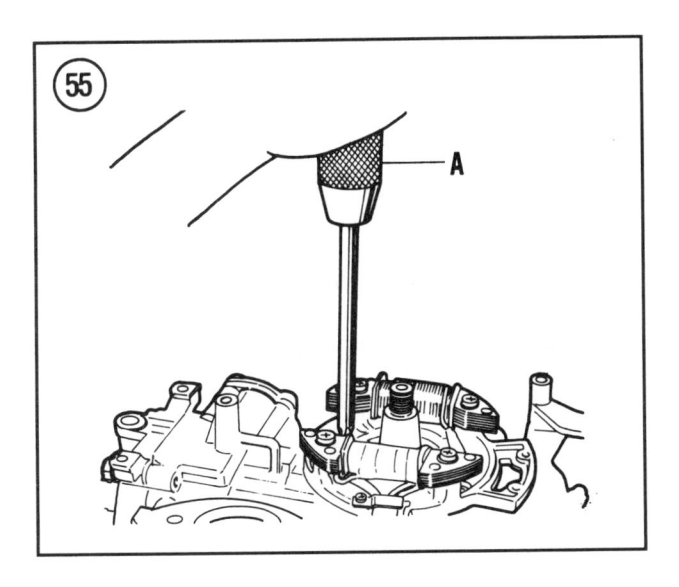

24. Remove the bolts and lockwashers securing the exhaust tube and remove it from the base of the power head.

25. Clean all gasket residue from the power head and drive shaft engine holder surfaces.

26. Installation is the reverse of removal, noting the following.

27. Use a new power head gasket.

28. Lightly coat drive shaft splines with water-resistant grease (part No. 99000-25610) or equivalent.

29. Make sure the locating dowels in the drive shaft support plate are in position.

30. Rotate the propeller as required to align crankshaft and drive shaft splines.

31. Coat the power head attaching bolt threads with Silicone Seal (part No. 99000-31120) or equivalent.

32. Tighten all fasteners to specifications (**Table 1**).

33. Make sure all electrical connections are free of corrosion and are tight.

34. Perform engine synchronization and linkage adjustments. See Chapter Five.

Removal/Installation
(DT 20, DT 25, DT 30—2-cylinder)

1. Remove the engine cover.
2. Disconnect the negative battery cable.

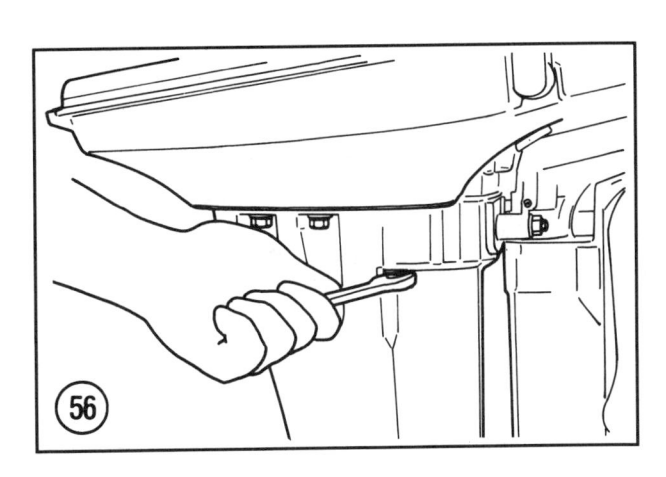

3. Remove the rewind starter assembly. See Chapter Ten.

4. Remove the 3 screws and remove the carburetor silencer.

5. Remove the junction box cover. Disconnect all electrical leads (**Figure 57**). Remove the starter relay, starter motor and motor bracket. See Chapter Seven.

6. Remove the clamp holding the battery cable. Remove the battery cable grommet and pull the cable from the lower support (**Figure 58**).

7. Remove the choke knob.

8. Disconnect the fuel line connections and remove the carburetor. See Chapter Six.

9. Remove the clip and slide the fuel hose from the under cover (**Figure 59**).

10. Remove the 2 bolts and disconnect the electrical wire veil (**Figure 60**).

11. Disconnect the blue/red wire connector (**Figure 61**).

12. Disconnect the throttle rod from the control lever (**Figure 62**).

13. Loosen the throttle cable locknuts and pull the cable ends (A, **Figure 63**) from the control lever.

NOTE
At this point, there should be no hoses, wires or linkage connecting the power head to the drive shaft housing. Recheck

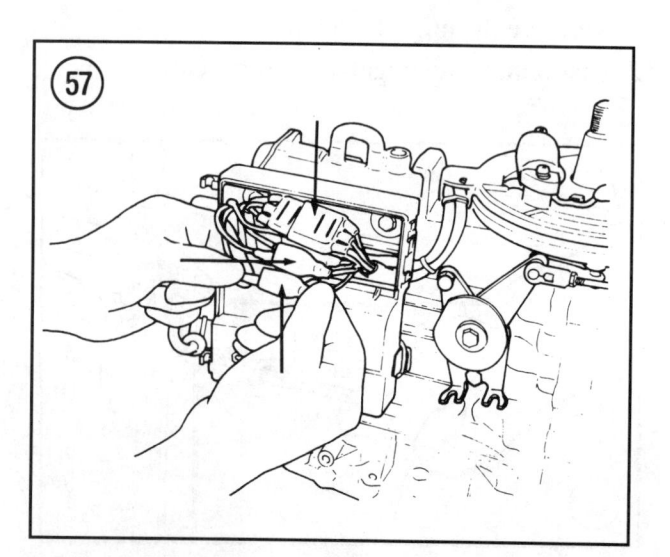

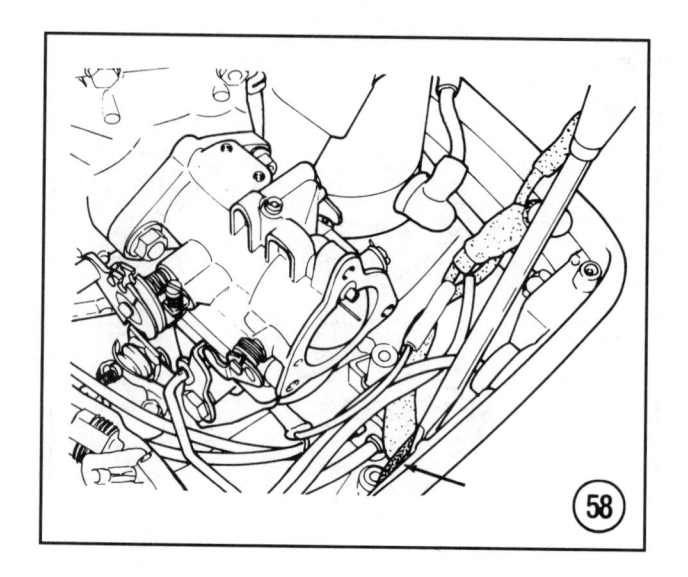

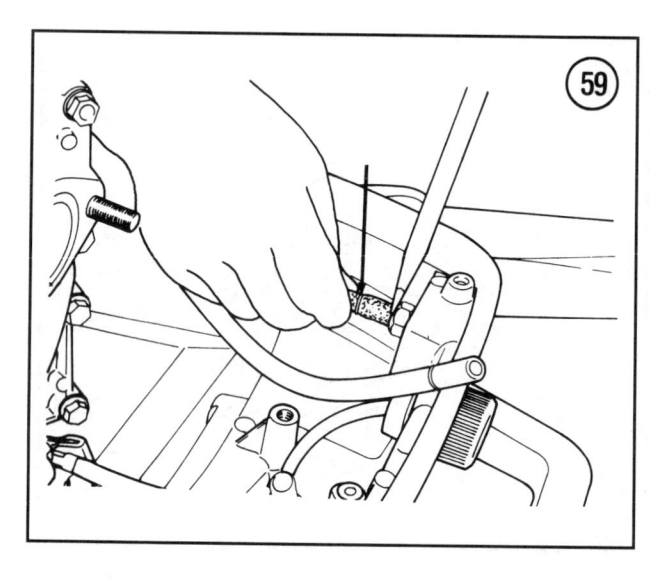

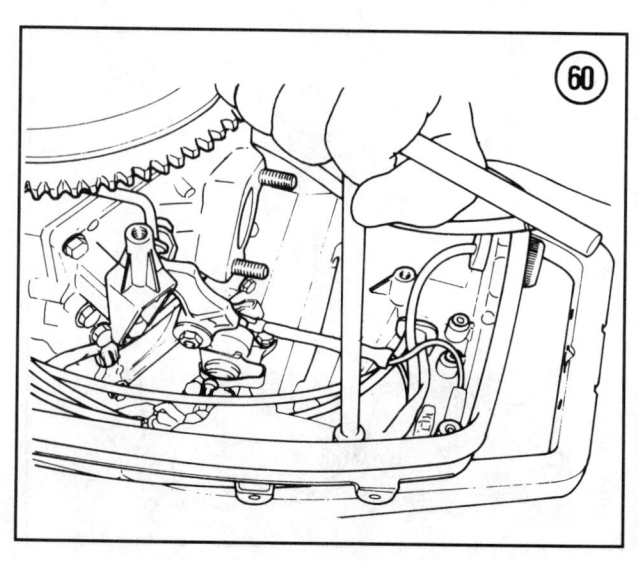

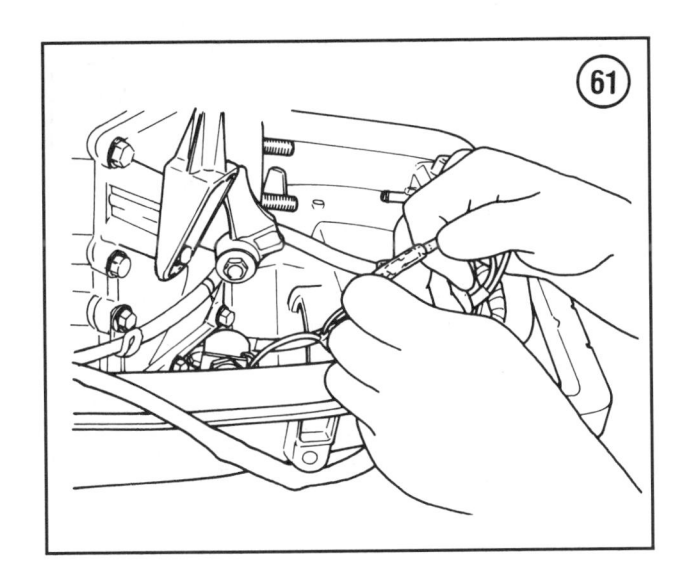

POWER HEAD

this to make sure nothing will hamper power head removal.

14. Remove the 8 bolts holding the power head to the drive shaft housing (**Figure 64**).

15. Remove the power head from the drive shaft housing and place it on a clean workbench.

16. Remove the 5 bolts and 2 dowel pins (A, **Figure 65**) holding the exhaust tube and lower oil seal housing to the power head and remove the assembly.

17. Clean all gasket residue from the power head and drive shaft housing surfaces.

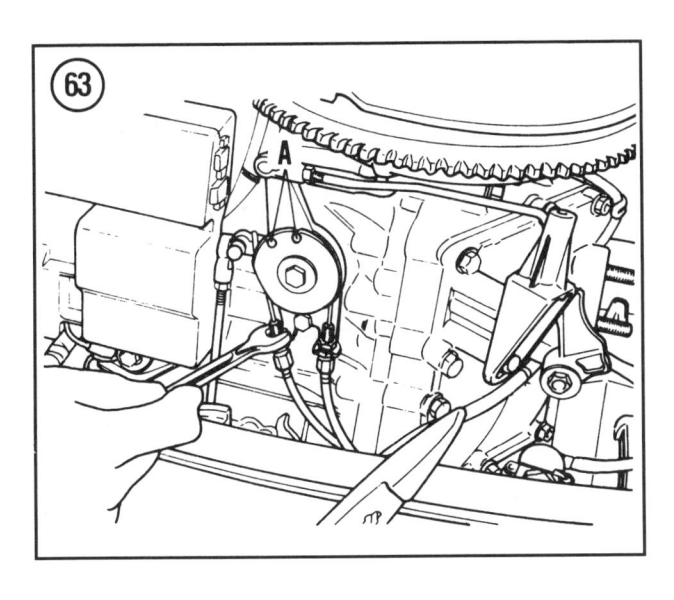

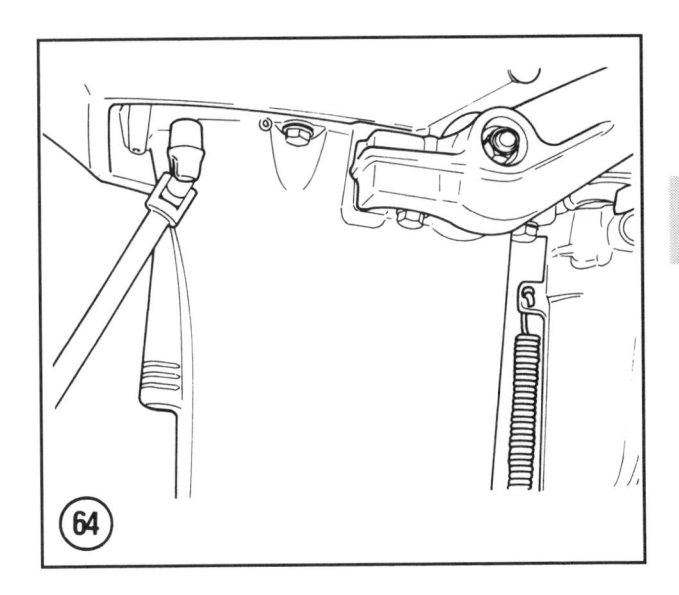

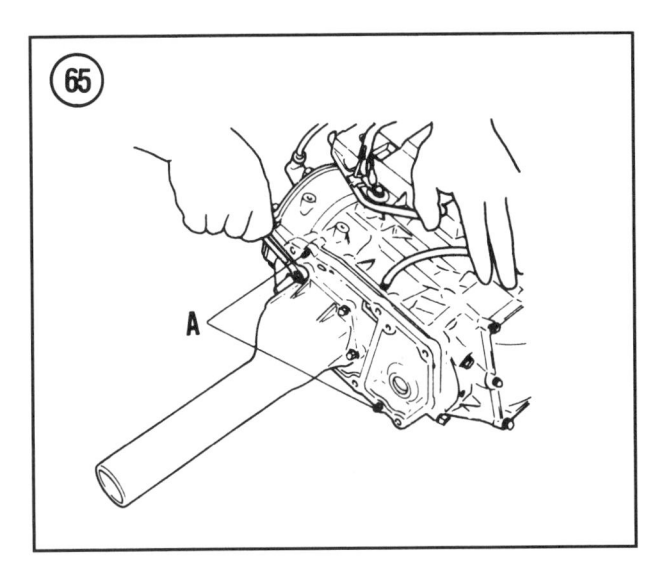

18. Remove the fuel filter and fuel pump. See Chapter Six.
19. Remove the flywheel as described in this chapter.
20. Remove the screws holding the stator assembly to the power head. Remove the stator assembly (**Figure 66**).
21. Remove the 4 screws holding the stator retainer and upper oil seal housing.
22. Disconnect the stator retainer-to-throttle control lever connector (**Figure 67**). Remove the assembly and discard the gasket.
23. Remove the electrical parts holder (**Figure 68**), the throttle control lever (**Figure 69**), throttle cable stay (**Figure 70**) and the rewind starter brace (**Figure 71**).
24. Disconnect the lubrication hose (**Figure 72**).
25. Installation is the reverse of removal, noting the following.
26. Install the lubrication hose with the check valve arrow pointing toward the top of the power head (**Figure 73**).
27. Install a new power head gasket.
28. Install a new upper oil seal housing seal with its lip facing the power head. Lubricate seal lips with Super Grease A (part No. 99000-25030) or equivalent.
29. Install a new lower oil seal housing seal with its lip facing away from the power head. Lubri-

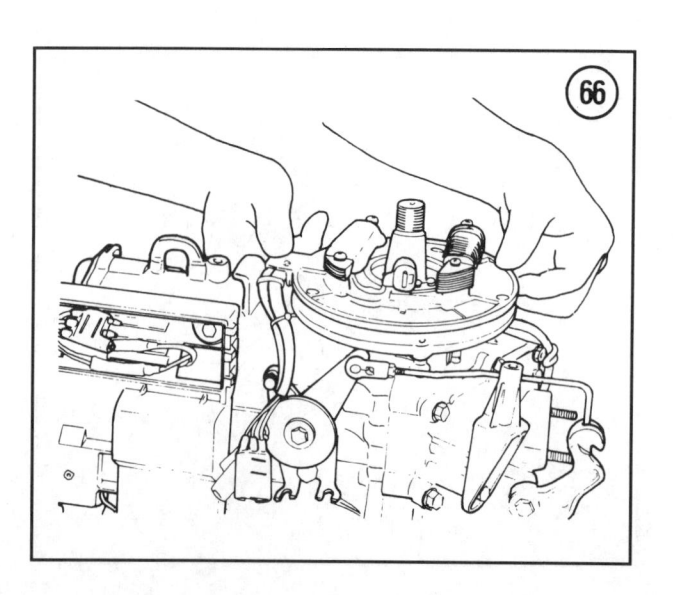

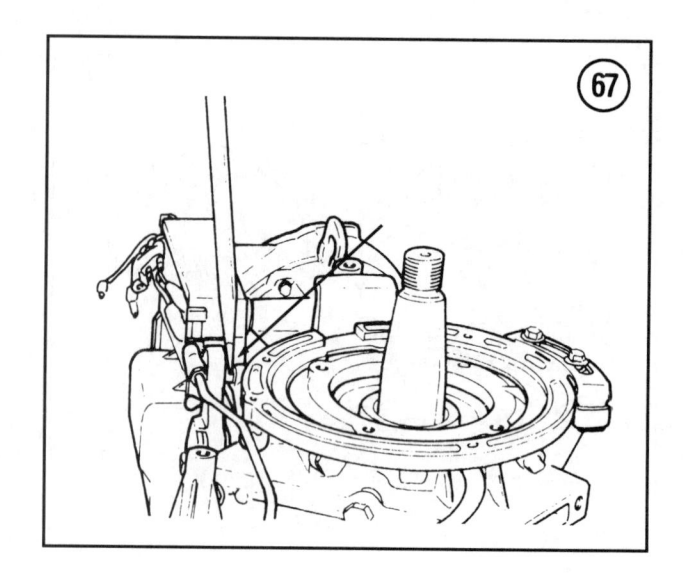

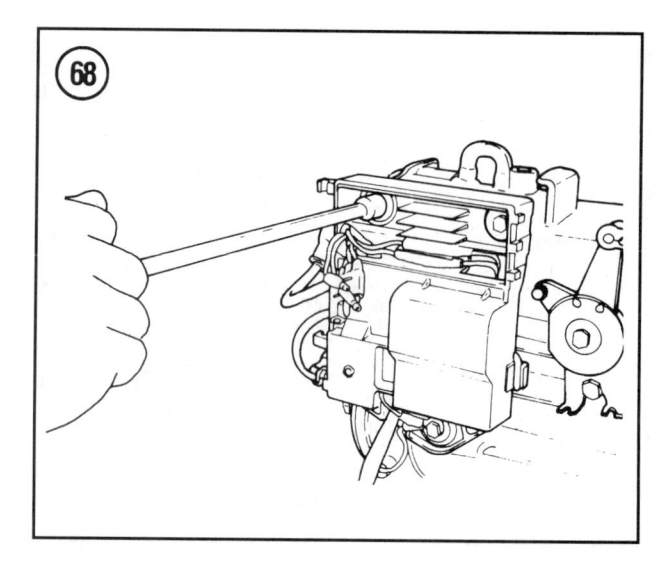

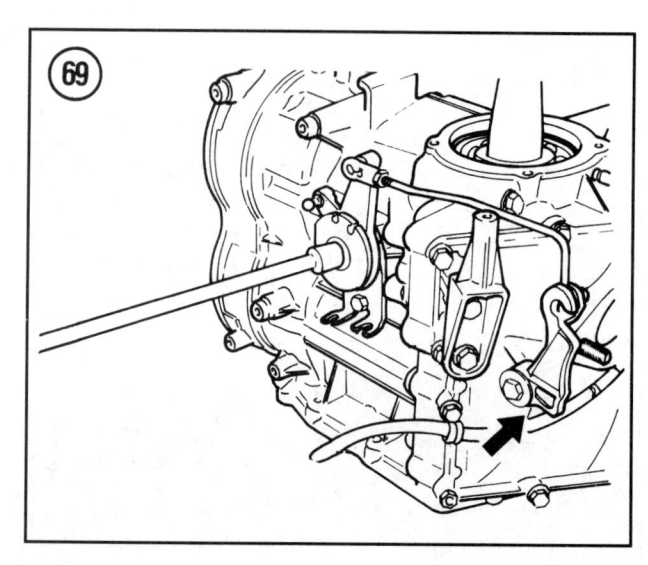

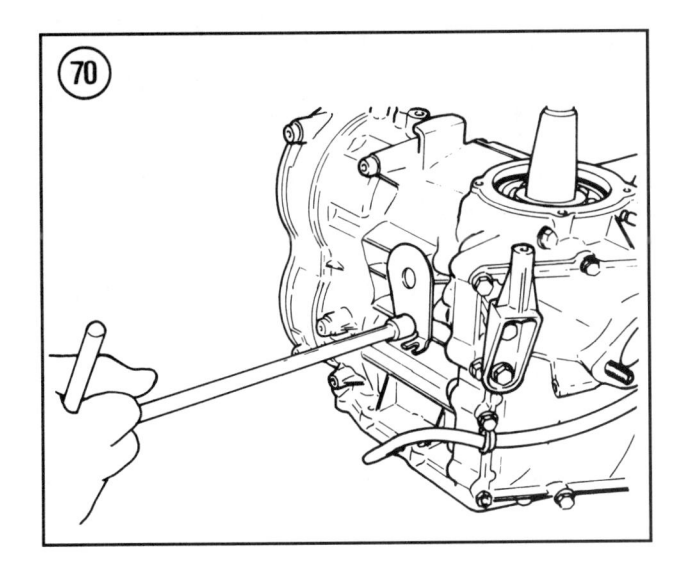

Power Head

cate seal lips with Silicone Seal (part No. 99000-31120) or equivalent (**Figure 74**).

30. Lightly coat the drive shaft splines with water-resistant grease (part No. 99000-25160) or equivalent.

31. Make sure the locating dowels in the drive shaft support plate are in position.

32. Rotate the propeller as required to align crankshaft and drive shaft splines.

33. Coat the power head attaching bolt threads with Silicone Seal (part No. 99000-31120) or equivalent.

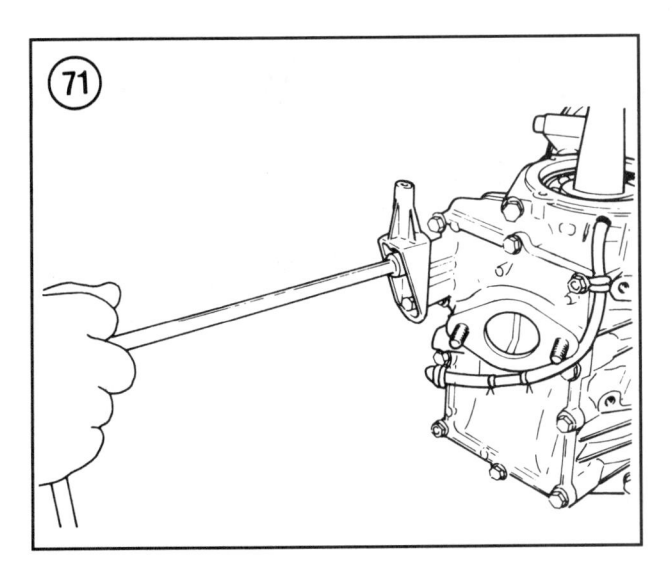

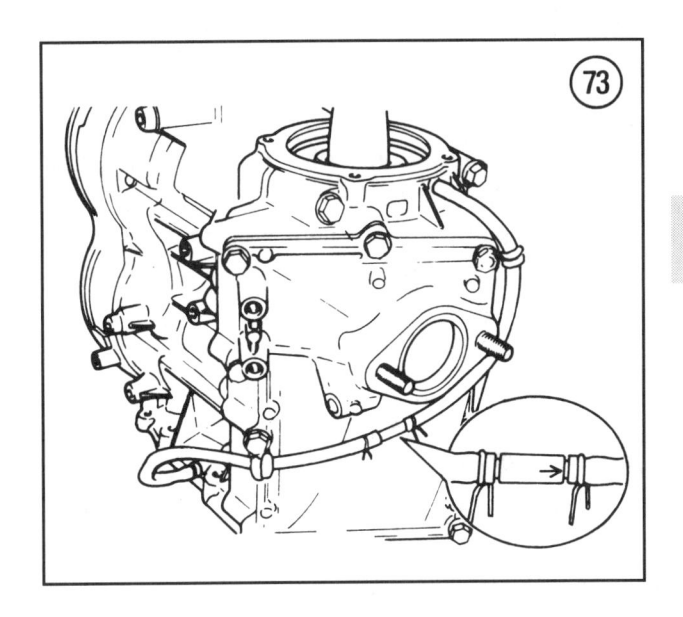

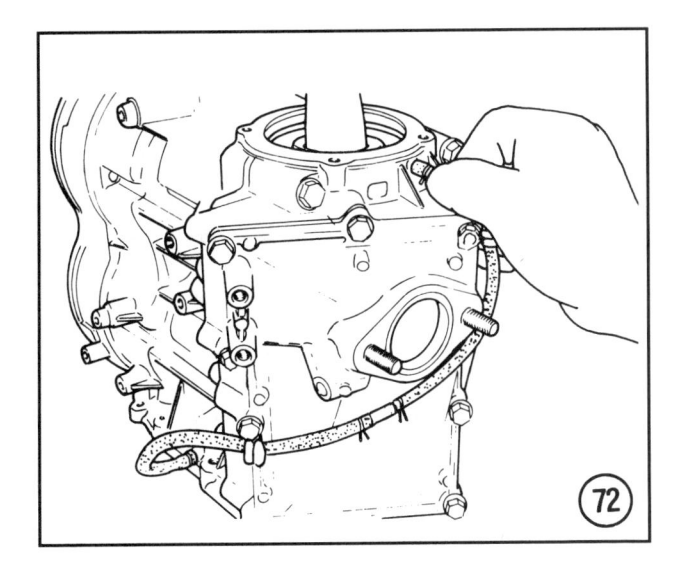

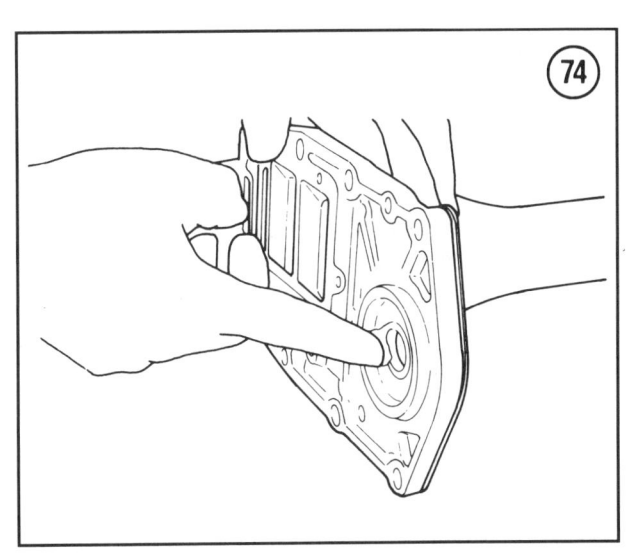

34. Tighten all fasteners to specifications (**Table 1**).

35. Make sure all electrical connections are free of corrosion and are tight.

36. Perform engine synchronization and linkage adjustments. See Chapter Five.

**Removal/Installation
(DT 25, DT 30—3-cylinder)**

1. Remove the engine cover.
2. Disconnect the negative battery cable.
3. Disconnect the spark plug leads. Remove the spark plugs.
4. Remove the rewind starter assembly. See Chapter Ten.
5. Remove the cover from the electrical parts cover and the rectifier.
6. Disconnect the rewind starter cable from the neutral start interlock lever.
7. Disconnect the battery cables from the starter motor and relay.
8. Remove the CDI unit. See Chapter Seven.
9. Disconnect all electrical leads and remove the electrical parts holder.
10. Remove the oil tank. See Chapter Twelve.
11. Refer to **Figure 75** and disconnect the following electrical leads:
    a. Throttle valve sensor (A).
    b. Choke solenoid (B).
    c. Warning lamp (C).
    d. Idle speed adjustment switch on models prior to 1991(D).
12. Disconnect the fuel hose (E, **Figure 75**) from the engine under cover.
13. Remove the wiring harness stopper (A, **Figure 76**) and disconnect the ignition coil lead wires.
14. Remove the CDI unit (B, **Figure 76**) and the wiring harness assembly (C, **Figure 76**).
15. Remove the silencer cover.
16. Disconnect the choke lever (**Figure 77**).

*NOTE
At this point, there should be no hoses, wires or linkage connecting the power head to the drive shaft housing. Recheck this to make sure nothing will hamper power head removal.*

17. Remove the 8 bolts (**Figure 78**) securing the power head to the drive shaft housing.

18. Remove the power head from the drive shaft housing and place it on a clean workbench.

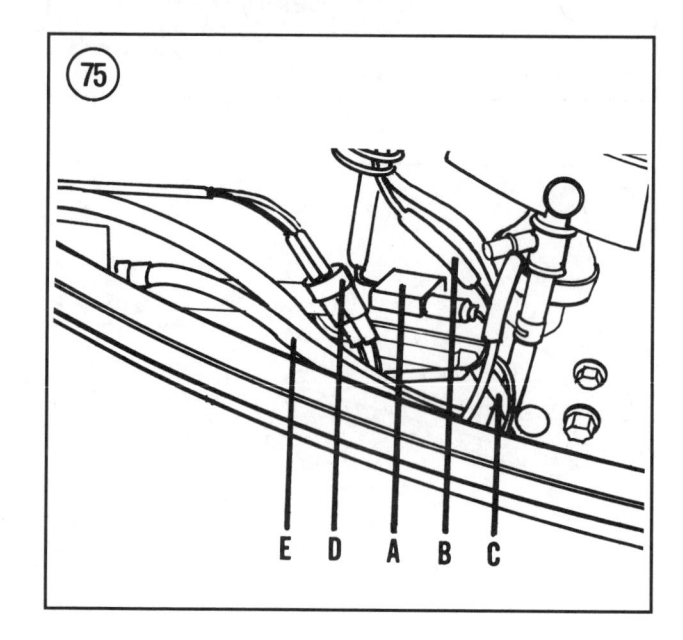

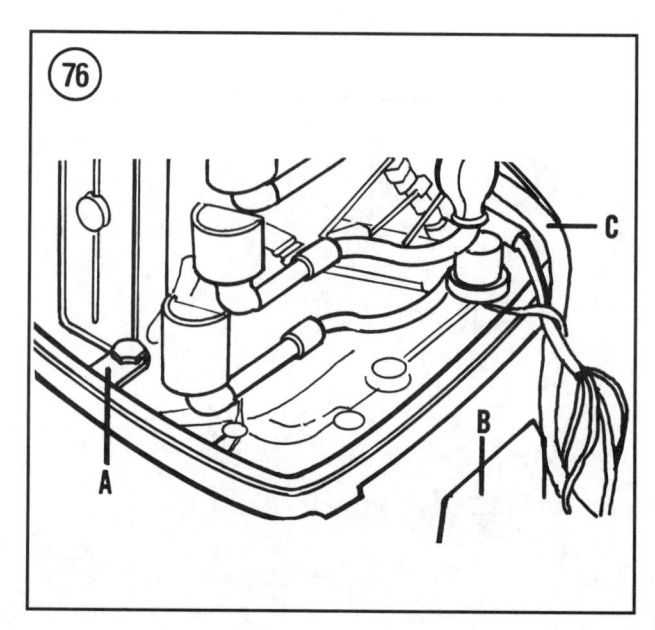

POWER HEAD

19. Remove the bolts and nuts holding the exhaust tube and lower oil seal housing to the power head and remove the assembly.
20. Disconnect the fuel line connections and remove the carburetors. See Chapter Six.
21. Remove the throttle control lever.
22. Disconnect the starter motor and starter relay electrical leads. Remove the starter motor assembly. See Chapter Seven.
23. Remove the fuel filter and fuel pump. See Chapter Six.
24. Remove the oil pump and oil hoses. See Chapter Twelve.
25. Remove the choke solenoid.

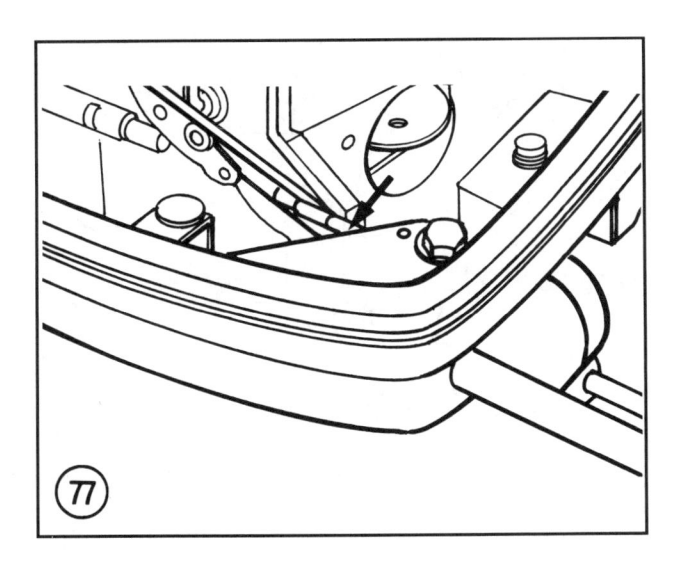

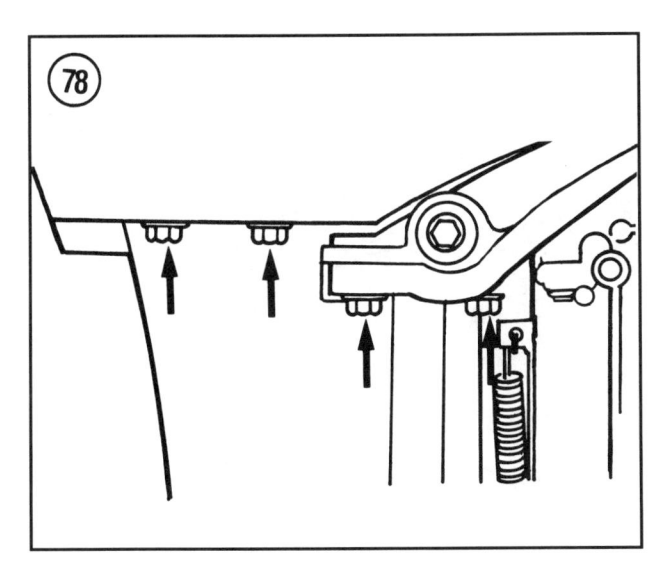

CAUTION
In Step 26, do not pry too hard as the gasket sealing surfaces may be damaged.

26. Remove the bolts securing the inlet case. Insert a flat-bladed screwdriver into the pry point slots indicated by an "O" and gently pry the inlet case away from the crankcase. Remove the case and gasket.
27. Remove the ignition coils. See Chapter Seven.
28. Remove the engine temperature sensor.
29. Remove the flywheel as described in this chapter.
30. Remove the screws holding the stator assembly to the power head. Remove the stator assembly.
31. Installation is the reverse of removal, noting the following.
32. Use a new power head gasket.
33. Install a new upper oil seal housing seal with its lip facing the power head. Lubricate seal lips with Super Grease A (part No. 99000-25030) or equivalent.
34. Fill the cavity between the lower oil seal and backup oil seal in the lower oil seal housing with water-resistant grease (part No. 99000-25160) or equivalent.
35. Lightly coat the drive shaft splines with water-resistant grease.
36. Make sure the locating dowels in the drive shaft support plate are in position, if so equipped.
37. Rotate the propeller as required to align crankshaft and drive shaft splines.
38. Coat the power head attaching bolt threads with Silicone Seal (part No. 99000-31120) or equivalent.
39. Tighten all fasteners to specifications (**Table 1**).
40. Make sure all electrical connections are free of corrosion and are tight.
41. Perform engine synchronization and linkage adjustments. See Chapter Five.

Removal/Installation (DT 35, DT 40)

1. Remove the engine cover.
2. Disconnect the negative battery cable.
3. Disconnect the spark plug leads. Remove the spark plugs.
4. Disconnect the neutral start interlock cable from the throttle limiter (**Figure 79**).
5. Remove the oil tank. See Chapter Twelve.
6. Remove the rewind starter assembly. See Chapter Ten.
7. Remove the 2 screws and remove the silencer cover and case.
8. Disconnect the oil pump control rod from the carburetor.
9. Remove the choke knob.
10. Disconnect the fuel line connections and remove the carburetor (**Figure 80**). See Chapter Six.
11. Disconnect the throttle control link rods and remove the throttle lever (**Figure 81**).
12. Disconnect the starter motor, starter relay and neutral switch electrical leads. Remove the starter motor assembly. See Chapter Seven.
13. Loosen the bolt in the electrical parts holder and disconnect the black ground wire. Disconnect all electrical leads.
14. Disconnect the hoses from the fuel filter and remove the fuel filter. Plug the ends of the fuel hoses to prevent the loss of fuel.
15. Remove the flywheel as described in this chapter.

NOTE
At this point, there should be no hoses, wires or linkage connecting the power head to the drive shaft housing. Recheck this to make sure nothing will hamper power head removal.

16. Remove the bolts securing the power head to the drive shaft housing.
17. Remove the power head from the drive shaft housing and place it on a clean workbench.
18. Remove the 6 bolts holding the exhaust tube and lower oil seal housing to the power head and remove the assembly.
19. Remove the screws holding the stator assembly to the power head. Remove the stator assembly (**Figure 82**).
20. Remove the upper oil seal housing (**Figure 83**).
21. Make sure all electrical wires are disconnected within the electrical part holder. Remove the bolts and remove the holder.
22. Remove the ignition coil and CDI unit.

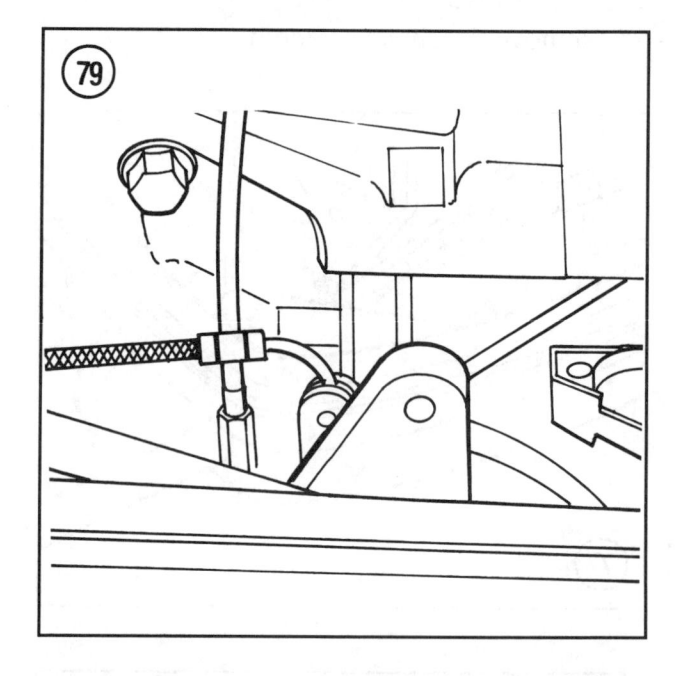

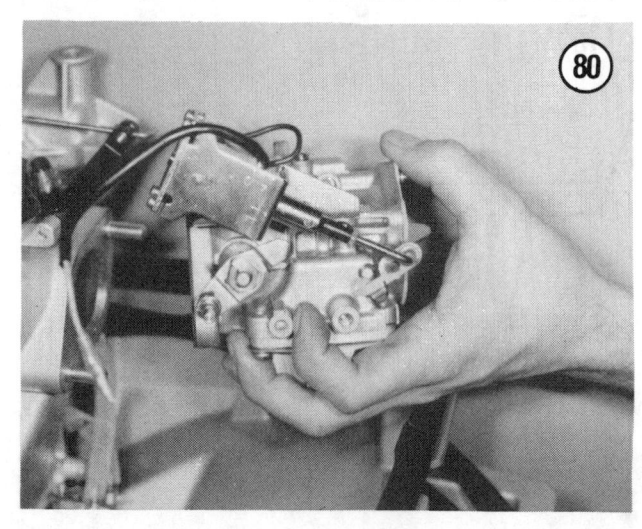

POWER HEAD

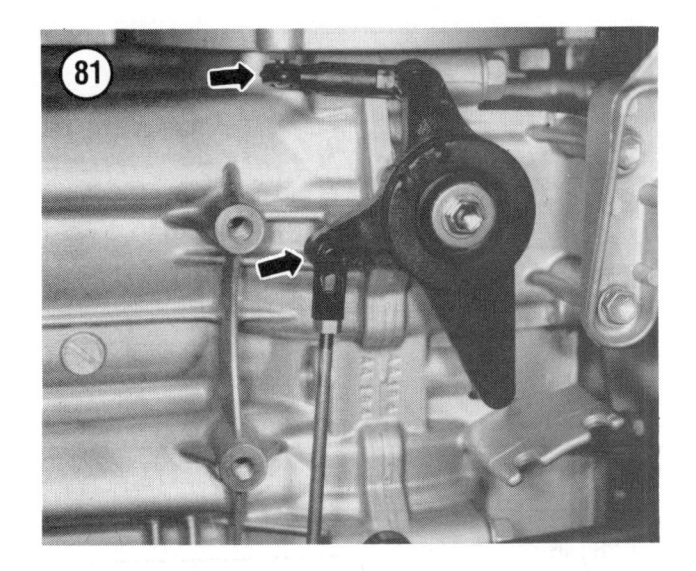

23. Remove the fuel pump and fuel filter.
24. Remove the oil pump and driven gear assembly. See Chapter Twelve.
25. Loosen the clip and remove the oil hose.
26. Remove the throttle control lever and the throttle cam follower lever.
27. Remove the bolts securing the rewind starter base on each side.
28. Remove the engine hook.
29. Installation is the reverse of removal, noting the following.
30. Use a new power head gasket.
31. Install a new upper oil seal housing seal with its lip facing the power head. Lubricate seal lips with Super Grease A (part No. 99000-25030) or equivalent.
32. Fill the cavity between the lower oil seal and backup oil seal in the lower oil seal housing with water-resistant grease (part No. 99000-25160) or equivalent.
33. Lightly coat the drive shaft splines with water-resistant grease.
34. Make sure the locating dowels in the drive shaft support plate are in position, if so equipped.
35. Rotate the propeller as required to align crankshaft and drive shaft splines.
36. Coat the power head attaching bolt threads with Silicone Seal (part No. 99000-31120) or equivalent.
37. Tighten all fasteners to specifications (**Table 1**).
38. Make sure all electrical connections are free of corrosion and are tight.
39. Perform engine synchronization and linkage adjustments. See Chapter Five.

Removal/Installation (DT 55, DT 65)

1. Remove the engine cover.
2. Disconnect the negative battery cable.
3. Disconnect the spark plug leads. Remove the spark plugs.
4. Remove the oil tank. See Chapter Twelve.

5. Remove the power trim and tilt motor relay cover, then remove the relays from the cylinder (**Figure 84**).
6. Remove the electrical parts holder cover and disconnect all electrical connectors within it.
7. Remove the CDI unit. See Chapter Seven.
8. Disconnect the electrical leads from the starter motor and relay.
9. Remove the flywheel as described in this chapter.
10. Disconnect the fuel hose from the connector (A, **Figure 85**). Plug the end of the fuel hose to prevent the loss of fuel.
11. Remove the clutch link (B, **Figure 85**) from the throttle control arm and the clutch shaft side arm.
12. Disconnect the water outlet hose (C, **Figure 85**) from the engine lower side cover.
13. Remove the 8 bolts (A, **Figure 86**) securing the lower under cover (B, **Figure 86**) on each side and remove both under covers (B).

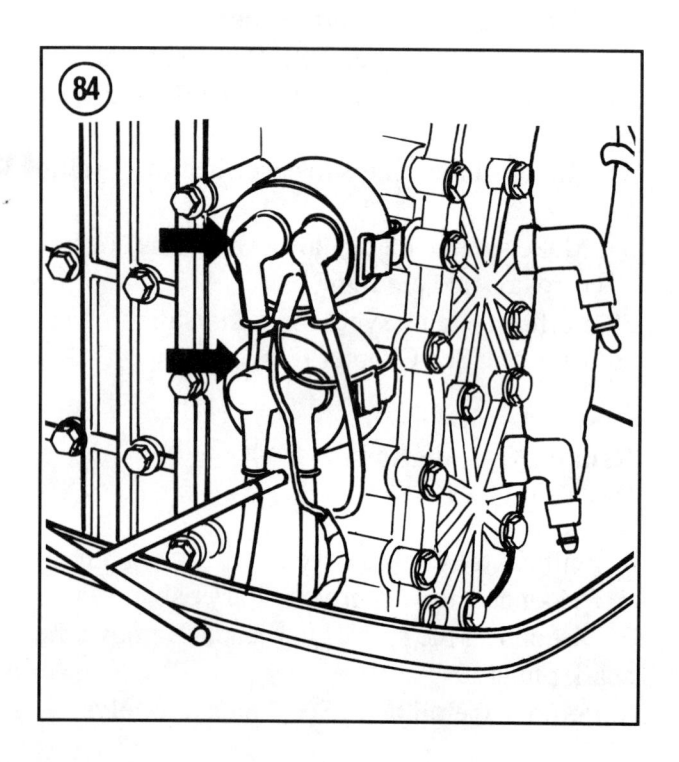

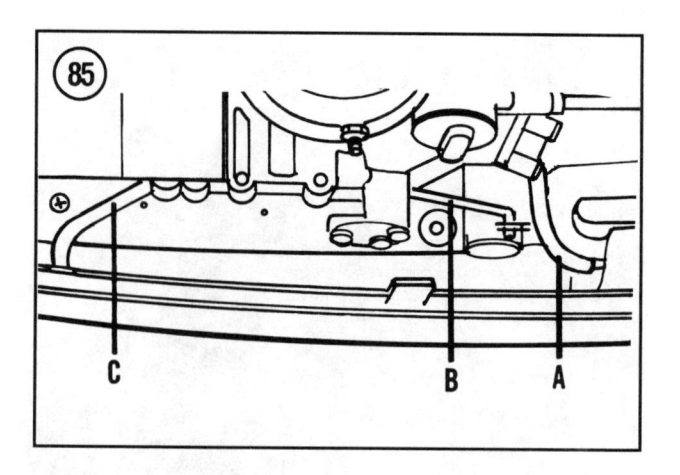

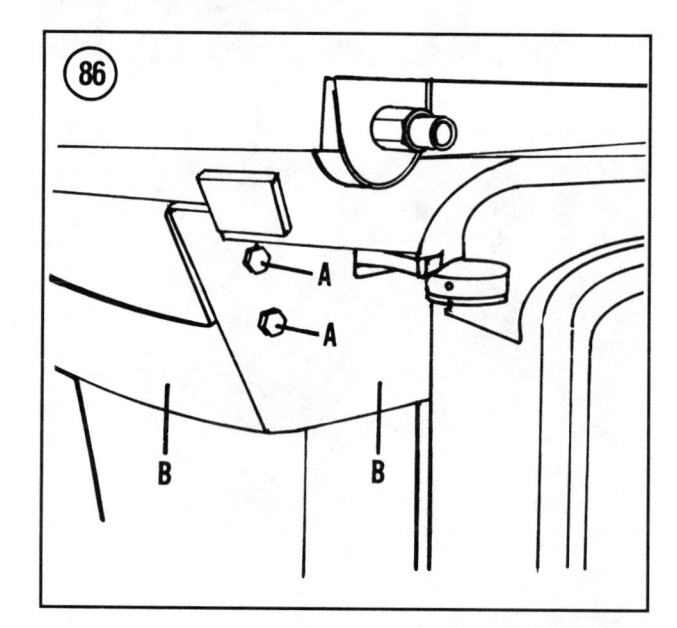

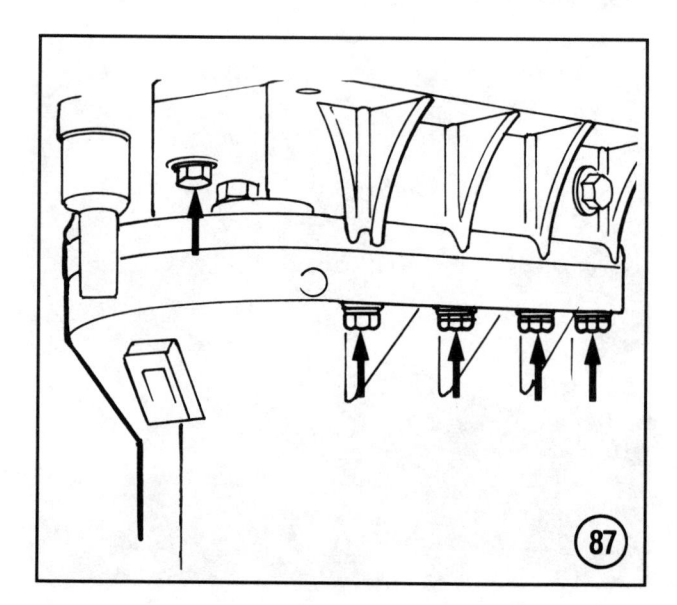

NOTE
At this point, there should be no hoses, wires or linkage connecting the power

POWER HEAD

head to the drive shaft housing. Recheck this to make sure nothing will hamper power head removal.

14. Remove the 9 bolts securing the power head to the drive shaft housing (**Figure 87**).

15. Remove the power head from the drive shaft housing and place it on a clean workbench.

16. Remove the bolts holding the exhaust tube and lower oil seal housing to the power head and remove the assembly.

17. Remove the ignition coils (A, **Figure 88**). See Chapter Seven.

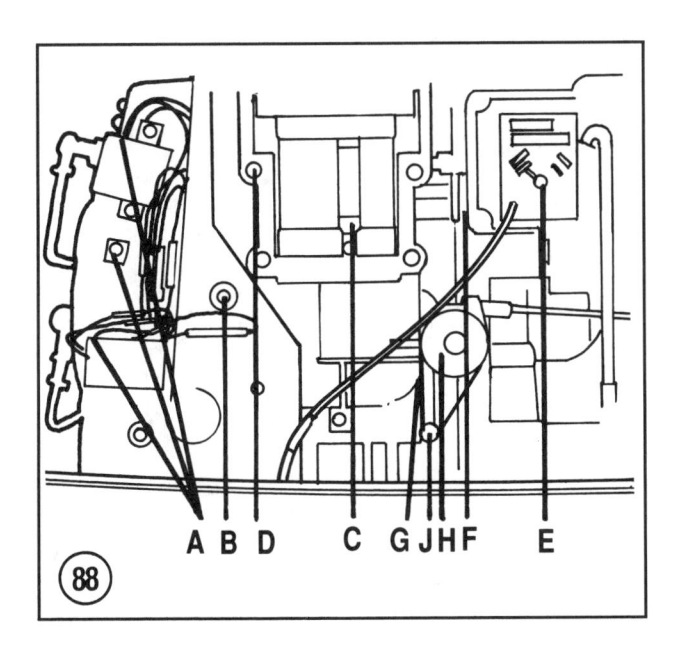

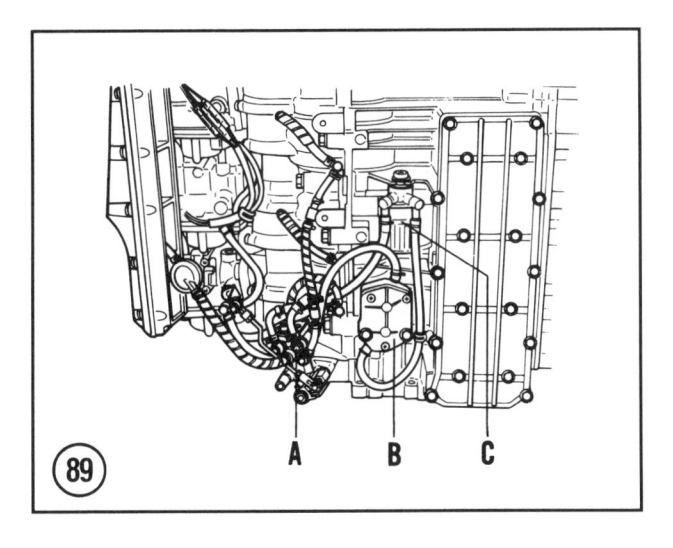

18. Remove the electrical parts holder (B, **Figure 88**).

19. Remove the starter (C, **Figure 88**). See Chapter Seven. Remove the starter bracket (D, **Figure 88**).

20. Remove the idle speed adjusting switch (E, **Figure 88**) on models prior to 1991, and bracket (F, **Figure 88**).

21. Remove the throttle control arm (G, **Figure 88**) and lever (H, **Figure 88**).

22. Remove the clutch control arm (J, **Figure 88**).

23. Remove the screws holding the stator assembly to the power head. Remove the stator assembly.

24. Remove the oil pump (A, **Figure 89**), oil pump retainer and the driven gear. See Chapter Twelve.

25. Remove the fuel pump (B, **Figure 89**) and fuel filter (C, **Figure 89**).

26. Remove the carburetor assemblies and choke solenoid (**Figure 90**).

27. Remove the engine hook.

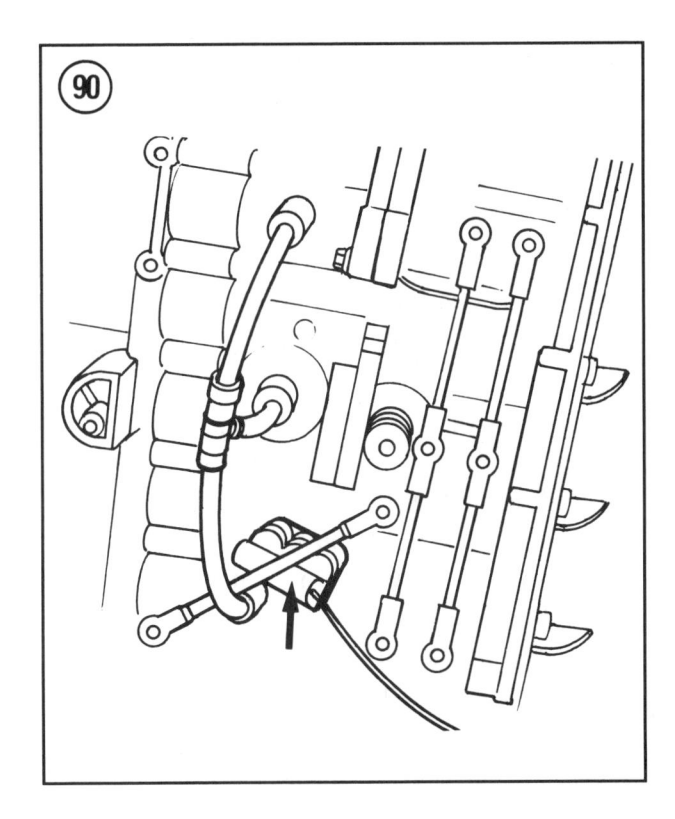

28. Installation is the reverse of removal, noting the following.

29. Use a new power head gasket.

30. Install a new upper oil seal housing seal with its lip facing the power head. Lubricate seal lips with Super Grease A (part No. 99000-25030) or equivalent.

31. Fill the cavity between the lower oil seal and backup oil seal in the lower oil seal housing with water-resistant grease (part No. 99000-25160) or equivalent.

32. Lightly coat the drive shaft splines with water-resistant grease.

33. Make sure the locating dowels in the drive shaft support plate are in position, if so equipped.

34. Rotate the propeller as required to align crankshaft and drive shaft splines.

35. Coat the power head attaching bolt threads with Silicone Seal (part No. 99000-31120) or equivalent.

36. Tighten all fasteners to specifications (**Table 1**).

37. Make sure all electrical connections are free of corrosion and are tight.

38. Perform engine synchronization and linkage adjustments. See Chapter Five.

Removal/Installation (DT 75, DT 85)

1. Remove the engine top cover.

2. Remove the lower front and rear engine covers.

3. Disconnect the negative battery cable.

4. Remove the battery cable clamp, disconnect the cable grommet (**Figure 91**) from the lower support housing and remove the cables.

5. Disconnect the spark plug leads. Remove the spark plugs.

6. Disconnect the fuel hose from the connector (**Figure 92**). Plug the end of the fuel hose to prevent the loss of fuel.

7. Remove the fuel filter. See Chapter Six.

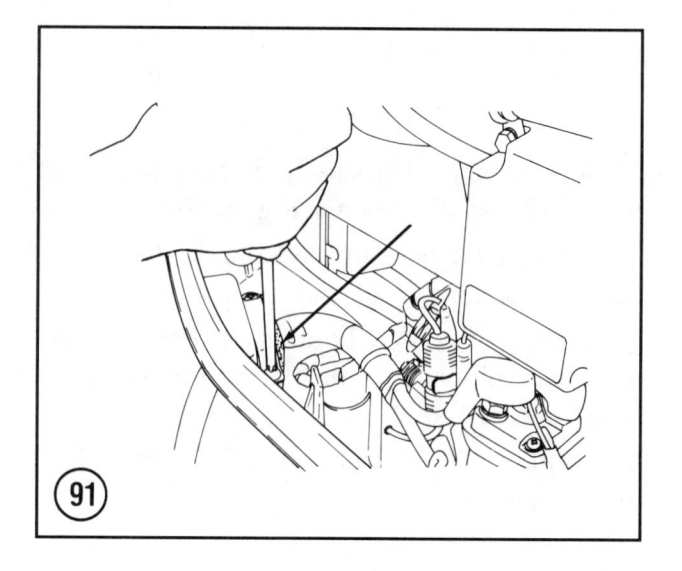

POWER HEAD

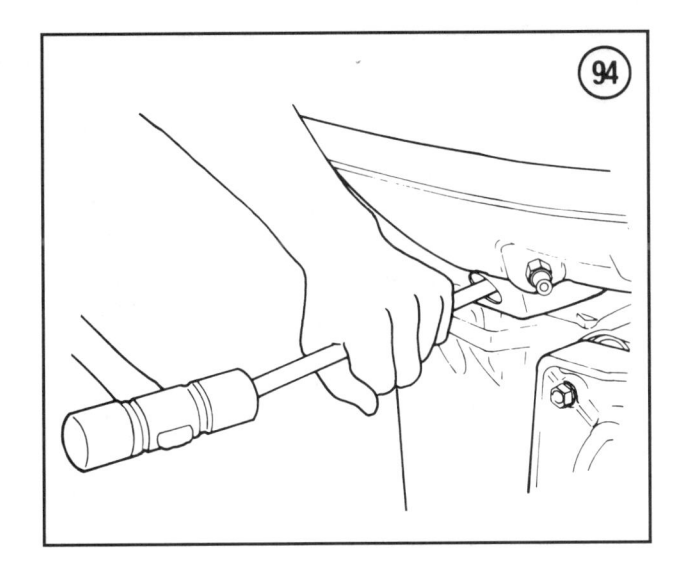

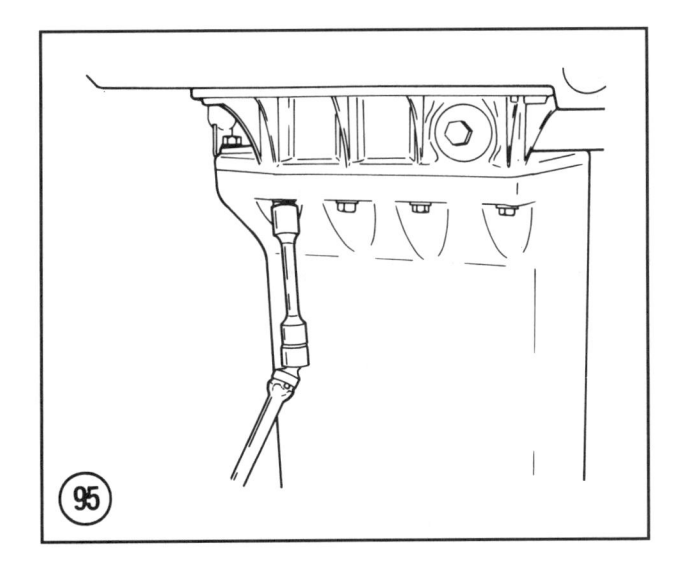

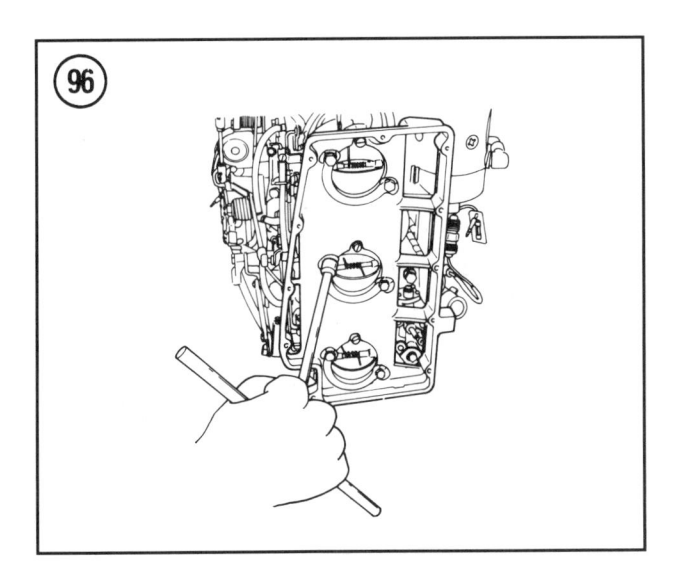

8. Remove the grommet on the starboard side of the drive shaft housing.

9. Insert an appropriate size socket wrench into the opening and remove the clutch shaft nuts (**Figure 93**).

10. Disconnect the clutch rod and shaft with a punch and hammer (**Figure 94**).

NOTE
At this point, there should be no hoses, wires or linkage connecting the power head to the drive shaft housing. Recheck this to make sure nothing will hamper power head removal.

11. Remove the 8 bolts and 1 nut (**Figure 95**) holding the power head to the drive shaft housing.

WARNING
If a hoist is not available for use in Step 12, have an assistant help with power head removal to avoid possible serious personal injury due to the weight of the power head.

12. Attach a hoist to the engine hooks and remove the power head from the drive shaft housing.

13. Check the power head base for alignment dowel pins. If any came off with the power head, remove and reinstall in the drive shaft housing mounting flange.

14. Place the power head on a clean workbench or suitable power head stand.

15. Remove the oil tank. See Chapter Twelve.

16. Remove the carburetor silencer cover and straighten the case lockwashers. Remove the bolts (**Figure 96**) and remove the case.

17. Disconnect the throttle control rods (**Figure 97**).

18. Disconnect the orange wire and disconnect the oil pump control rods (**Figure 98**).

19. Remove the carburetors and choke solenoid. See Chapter Six.

20. Remove the fuel pump (**Figure 99**) and insulator assembly. See Chapter Six.

21. Remove the power trim solenoids. See Chapter Eleven.

22. Remove the oil injection check valves, pump (**Figure 100**) and driven gear retainer. See Chapter Twelve.

23. Remove the starter motor and relay. See Chapter Seven.

24. Remove the junction box cover and disconnect all electrical leads.

25. Remove the starter motor relay (**Figure 101**).

26. Remove the cover from the junction box.

27. Disconnect the yellow, red and white electrical leads and remove the rectifier.

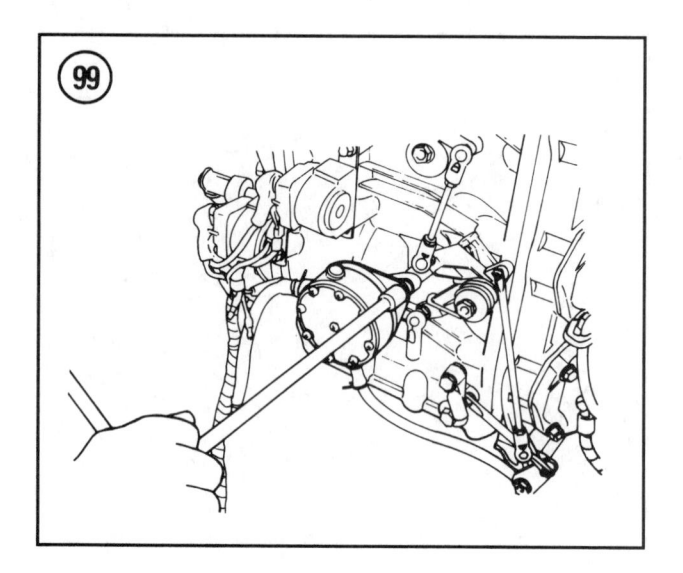

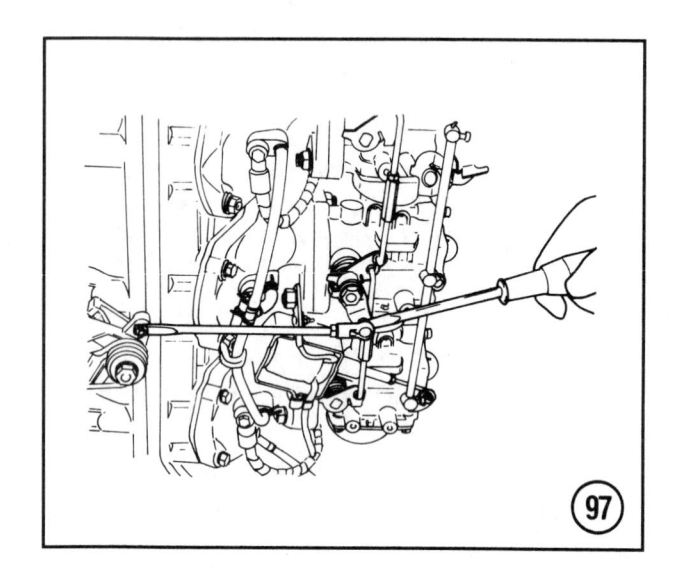

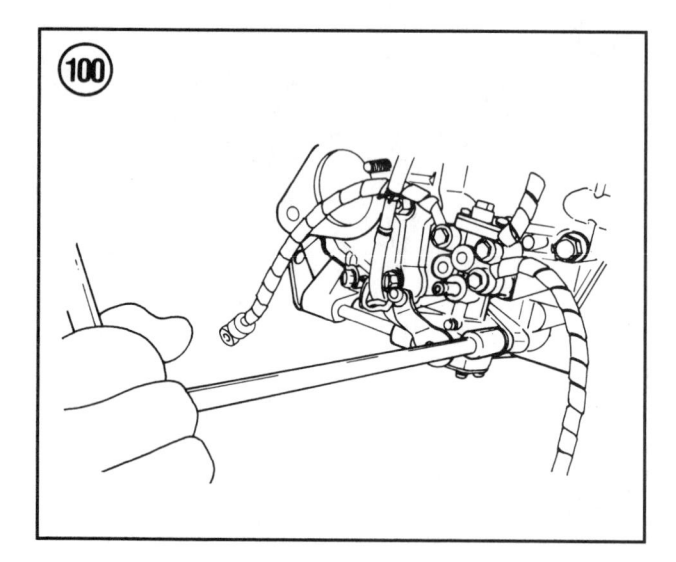

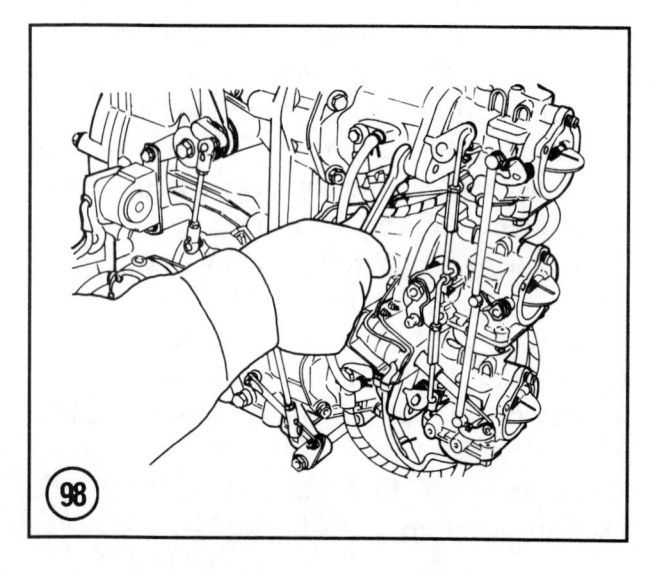

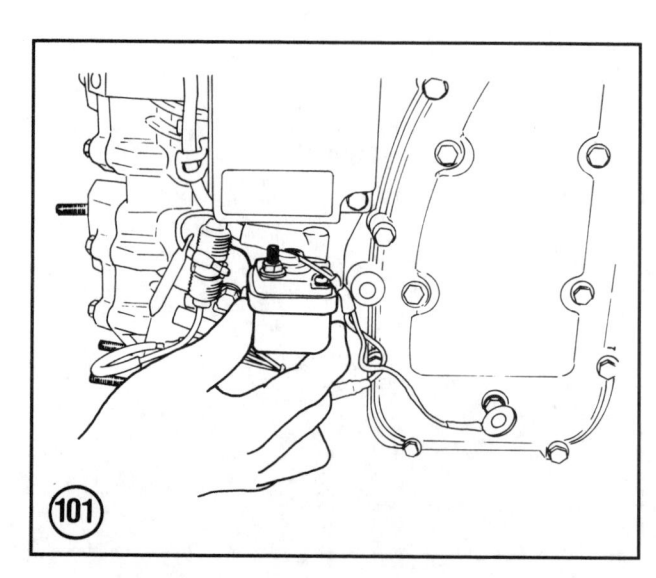

Power Head

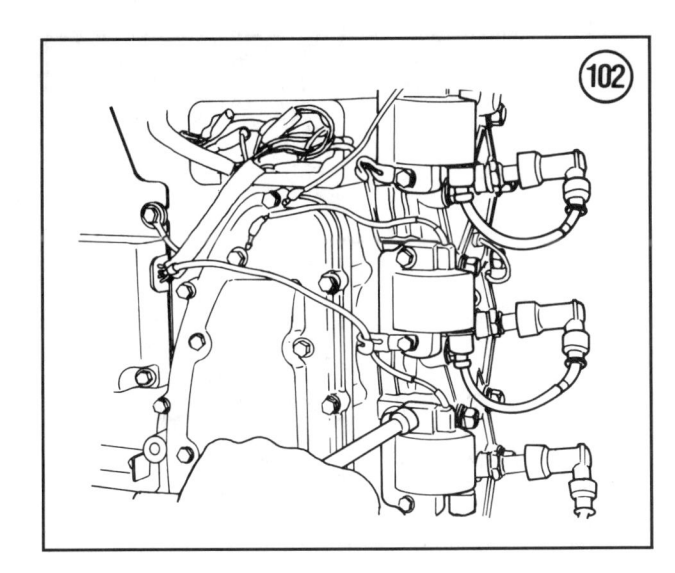

28. Remove the ignition coils (**Figure 102**).

29. Disconnect the 2-pin and 3-pin electrical connectors, then disconnect the pink and green electrical leads. Remove the CDI unit (**Figure 103**).

30. Remove the electric parts holder, then remove the wiring harness (**Figure 104**).

31. Disconnect the spark advance lever connector and remove the lever (**Figure 105**).

32. Remove the throttle control lever and link (**Figure 106**).

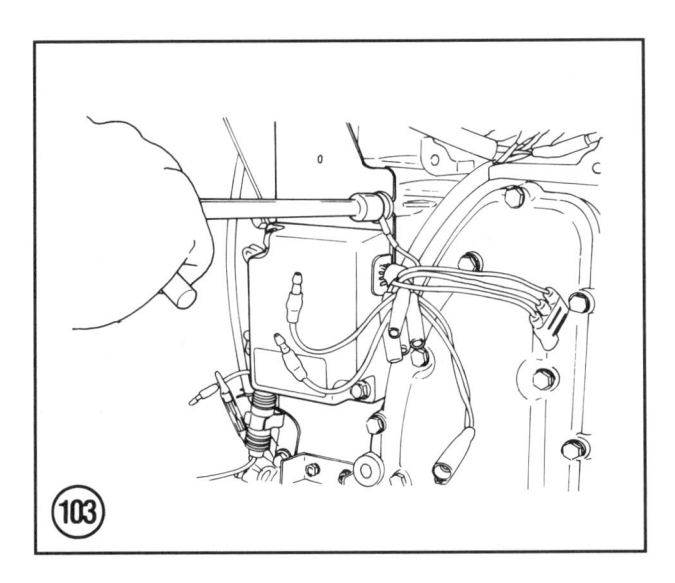

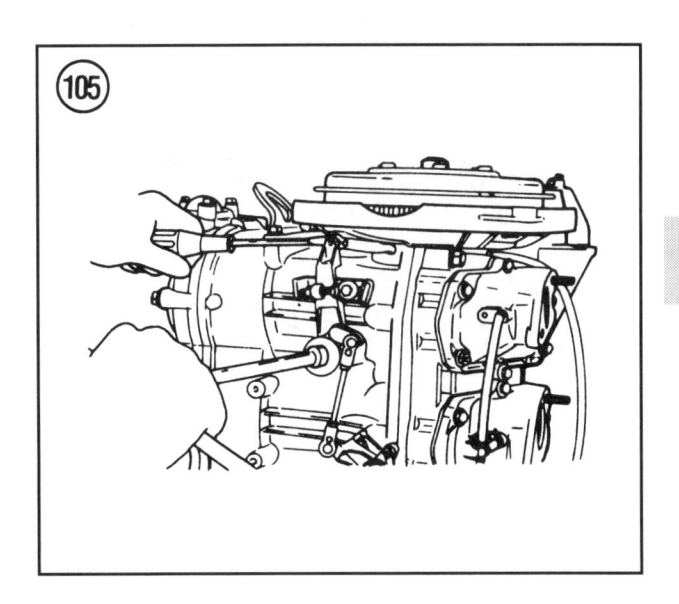

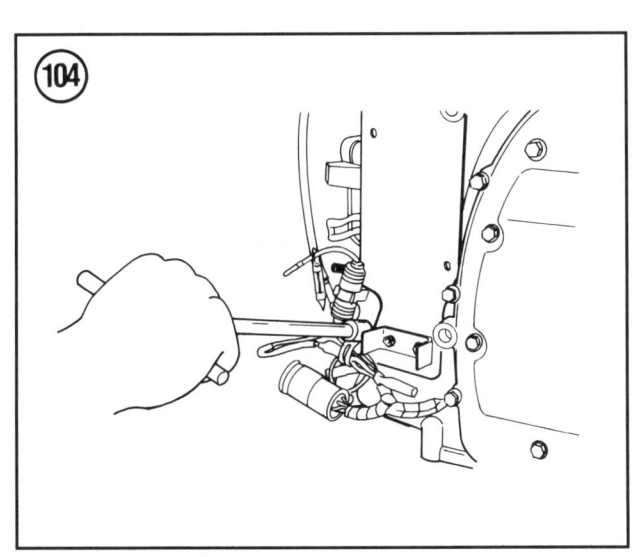

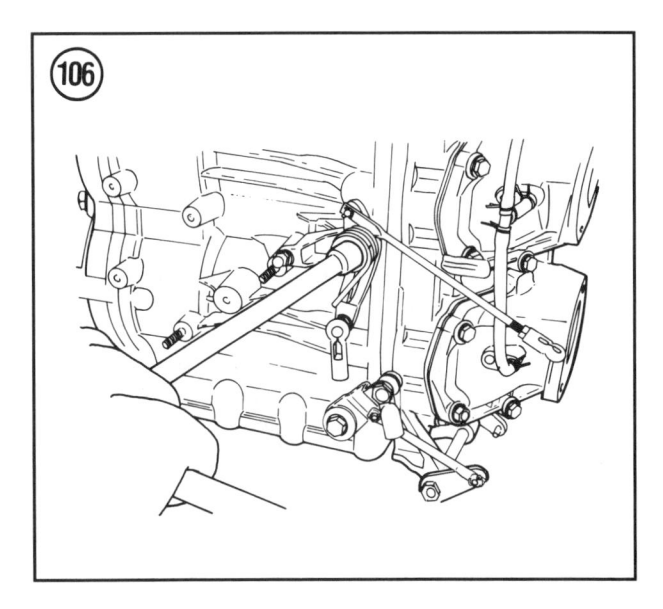

33. Remove the spark advance adjustment plate (**Figure 107**).
34. Remove the clutch control arm (**Figure 108**).
35. Remove the flywheel as described in this chapter.
36. Remove the screws holding the stator assembly to the power head (**Figure 109**). Remove the stator assembly.
37. Remove the screws holding the timer base and sensor assembly (**Figure 110**). Remove the assembly.
38. Remove the upper oil seal housing (**Figure 111**).
39. Remove the starter mounting bracket.

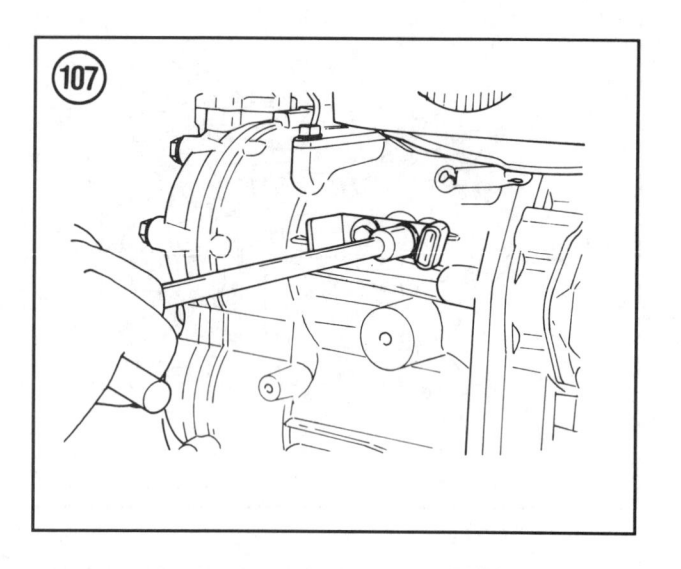

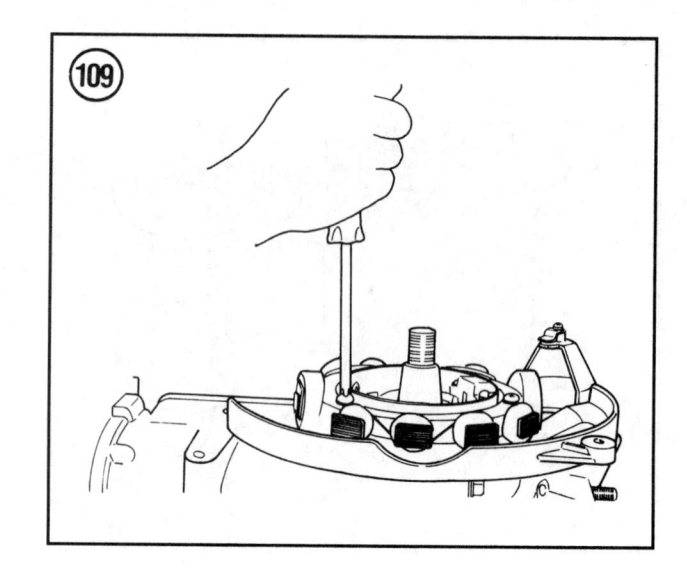

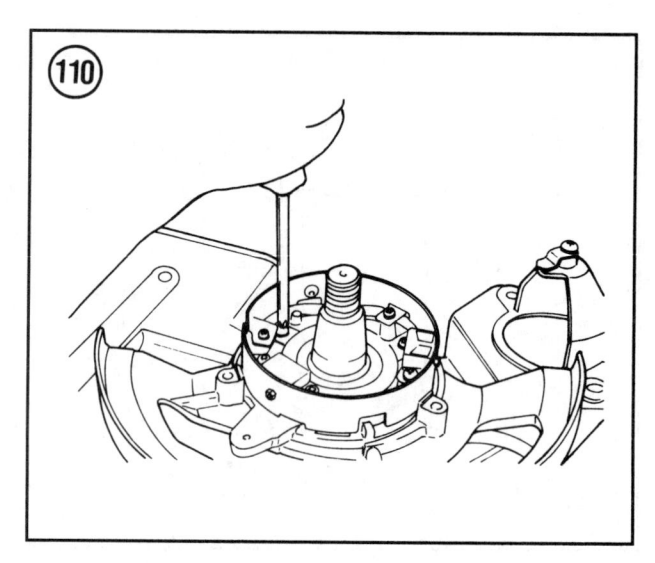

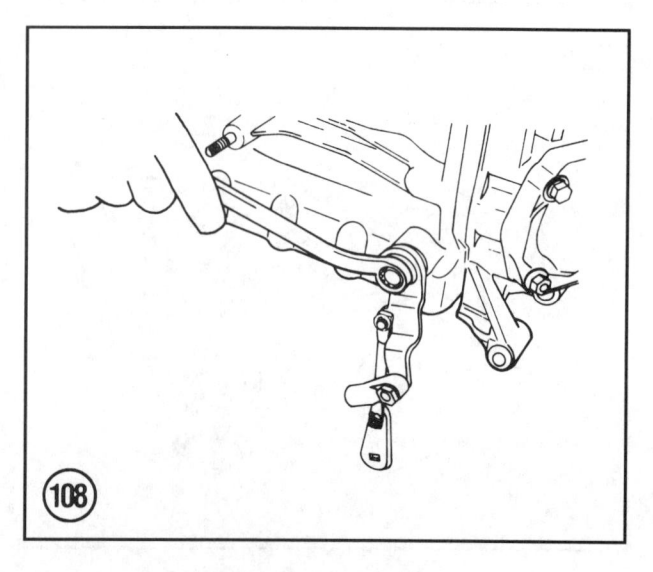

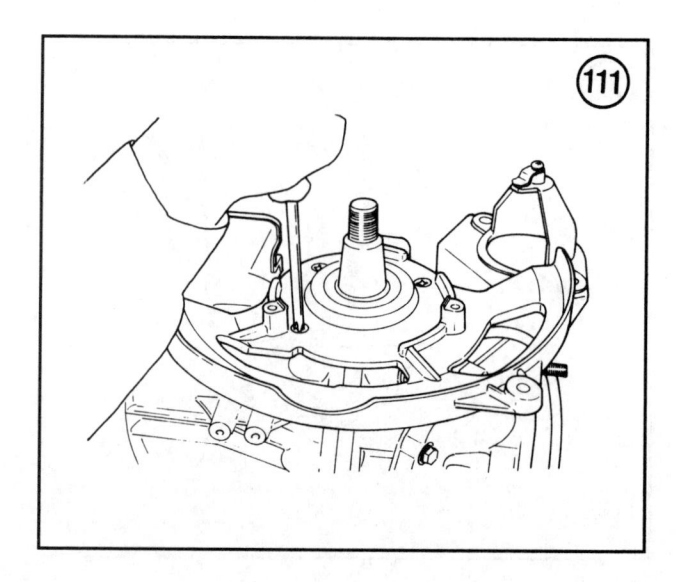

POWER HEAD

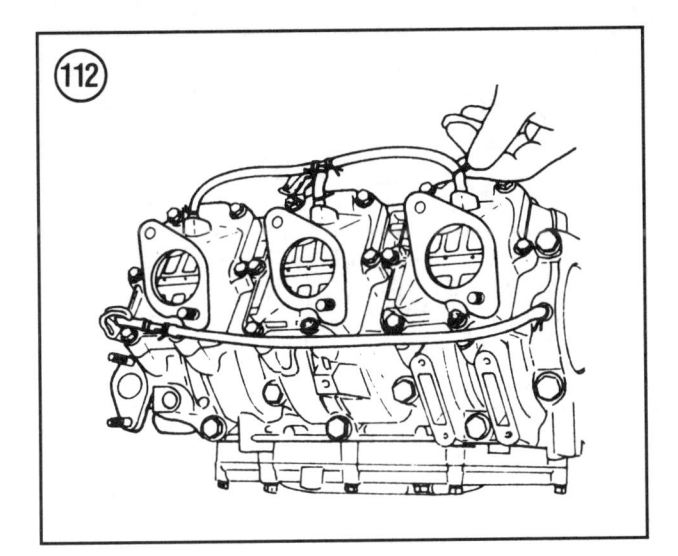

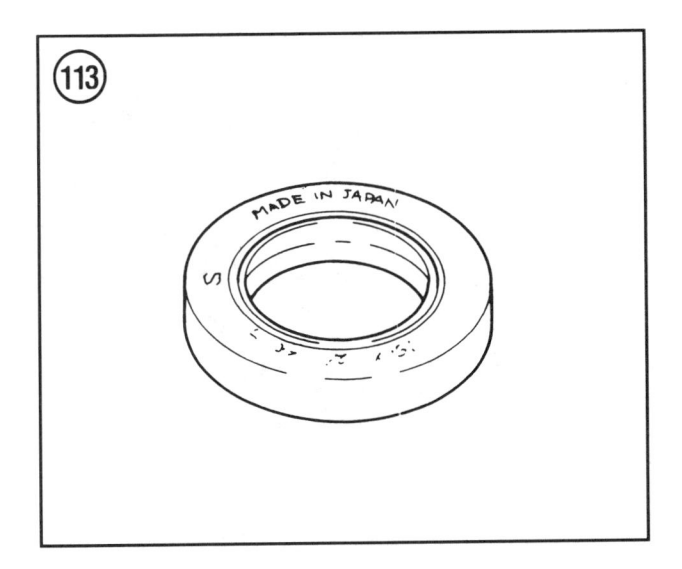

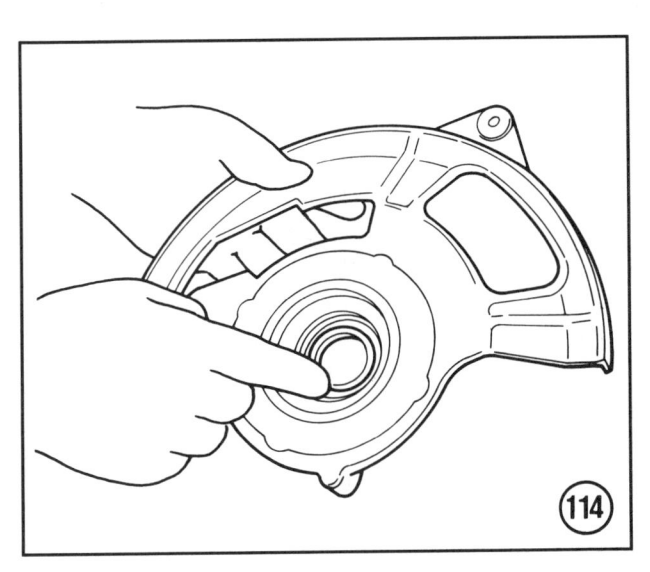

40. Note how the lubrication hoses are connected (**Figure 112**), then disconnect and remove them from the power head brackets.

41. Installation is the reverse of removal, noting the following.

42. Use a new power head gasket.

43. Install a new upper oil seal housing seal with its lip facing the power head (**Figure 113**). Lubricate seal lips with Super Grease A (part No. 99000-25030) or equivalent.

44. Fill the cavity between the lower oil seal and backup oil seal in the lower oil seal housing with water-resistant grease (part No. 99000-25160) or equivalent (**Figure 114**).

45. Lightly coat the drive shaft splines with water-resistant grease.

46. Make sure the locating dowels in the drive shaft support plate are in position, if so equipped.

47. Rotate the propeller as required to align crankshaft and drive shaft splines.

48. Coat the power head attaching bolt threads with Silicone Seal (part No. 99000-31120) or equivalent.

49. Tighten all fasteners to specifications (**Table 1**).

50. Install the coils in sequence according to the number on the secondary lead. The coil with the lead marked "1" should be installed to the No. 1 cylinder, etc. (**Figure 115**).

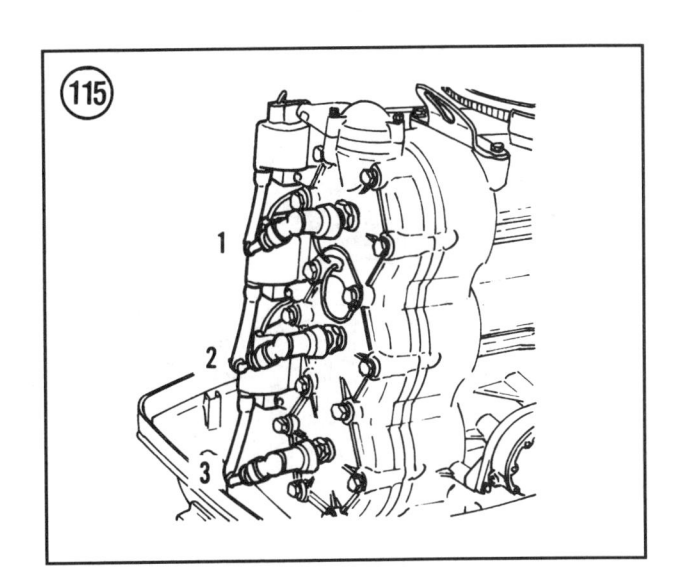

51. Make sure all electrical connections are free of corrosion and are tight.

52. Perform engine synchronization and linkage adjustments. See Chapter Five.

Removal/Installation (1985 DT 115, DT 140)

1. Remove the engine top cover.

2. Remove the 4 bolts holding the lower engine holder covers. Lower the covers.

3. Disconnect the negative battery cable.

4. Disconnect the clutch and throttle cables (**Figure 116**).

5. Disconnect the electrical coupler between the remote control cable and power trim unit (**Figure 117**).

6. Disconnect the spark plug leads. Remove the spark plugs.

7. Remove the battery cables (**Figure 118**). Disconnect the electrical connector (A, **Figure 118**).

8. Remove the oil tank. See Chapter Twelve.

9. Disconnect the fuel hose from the connector (**Figure 119**). Plug the end of the fuel hose to prevent the loss of fuel.

10. Disconnect the water inspection hose at the lower support housing (**Figure 120**).

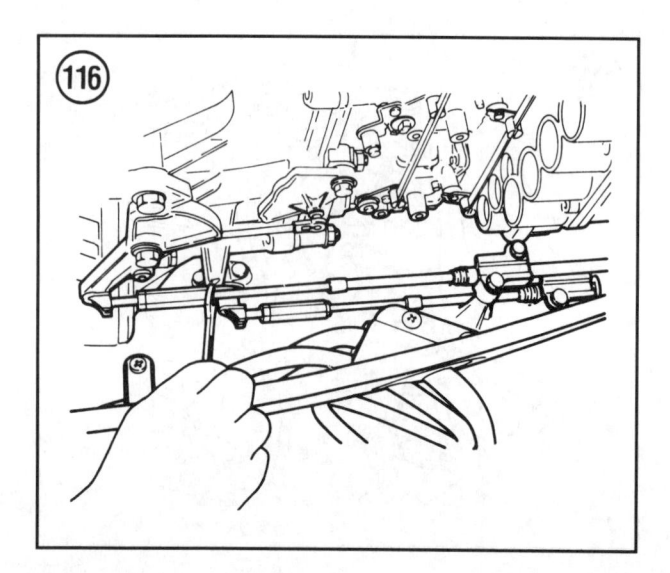

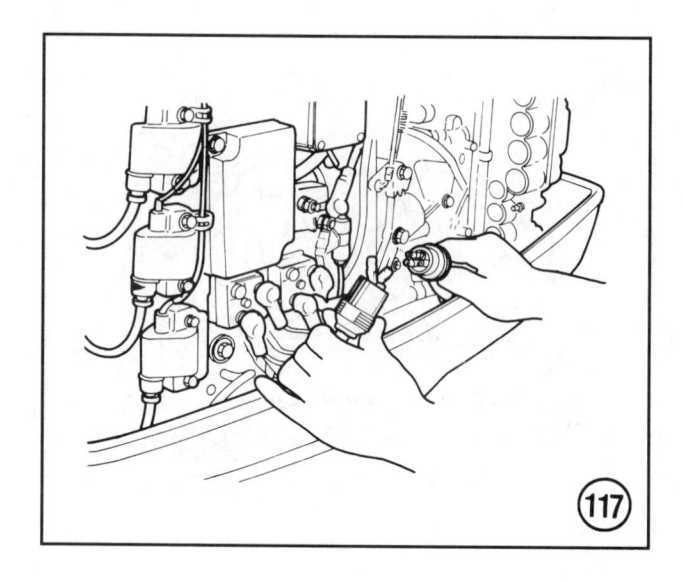

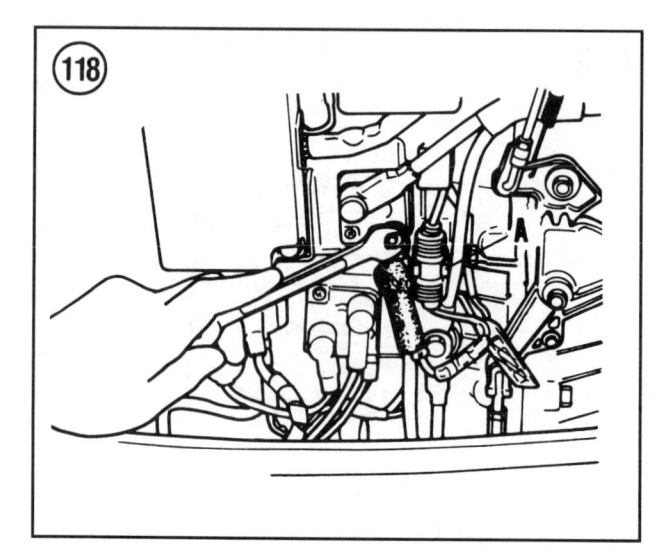

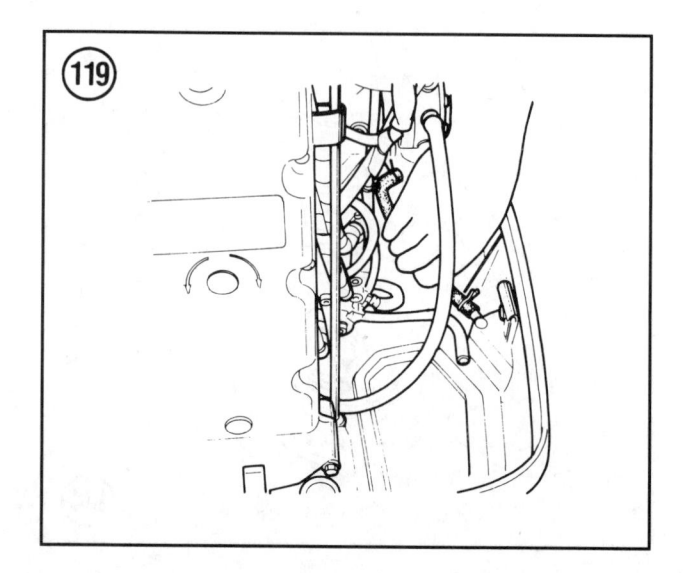

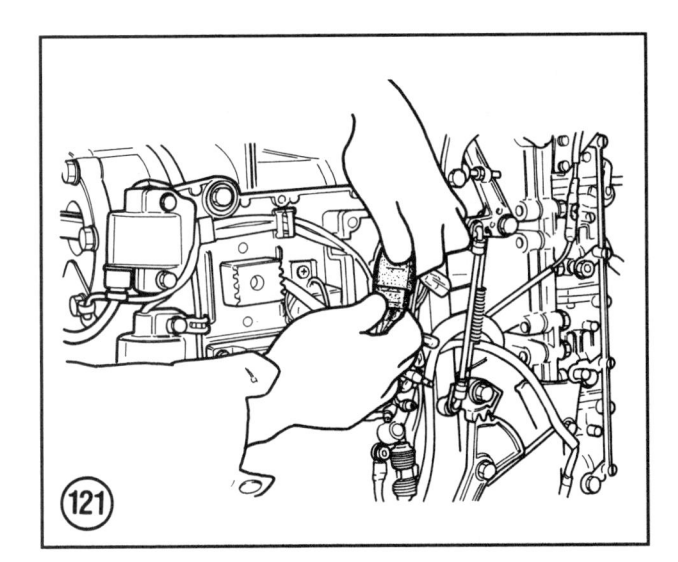

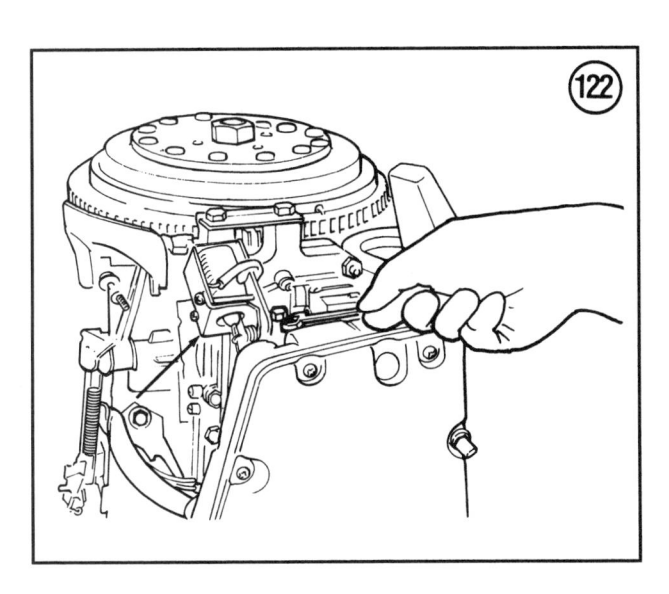

POWER HEAD

NOTE
At this point, there should be no hoses, wires or linkage connecting the power head to the drive shaft housing. Recheck this to make sure nothing will hamper power head removal.

11. Remove the 8 bolts and 2 nuts holding the power head to the drive shaft housing.

WARNING
If a hoist is not available for use in Step 12, have an assistant help with power head removal to avoid possible serious personal injury.

12. Attach a hoist to the engine hooks and remove the power head from the drive shaft housing.
13. Check the power head base for alignment dowel pins. If any came off with the power head, remove and reinstall in the drive shaft housing mounting flange.
14. Place the power head on a clean workbench or suitable power head stand.
15. Remove the starter motor. See Chapter Eight.
16. Remove the junction box cover and disconnect all electrical leads (**Figure 121**).
17. Remove the choke solenoid (**Figure 122**).
18. Remove the bolts securing the electrical parts holder (**Figure 123**) and remove the holder.

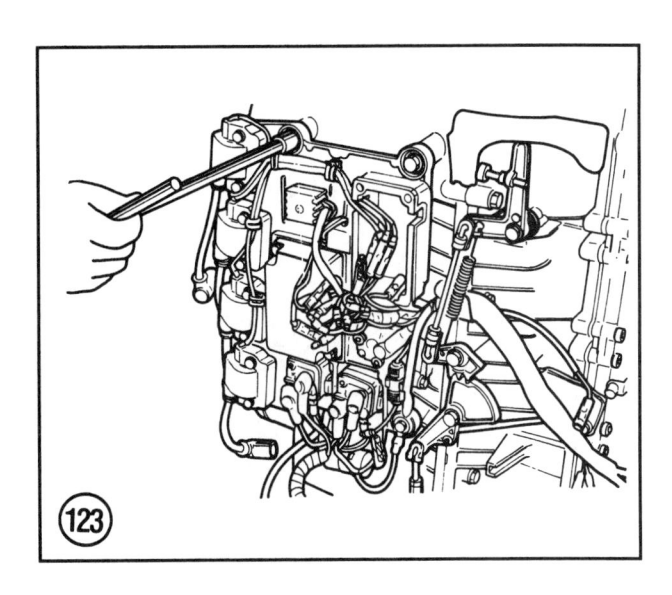

19. Remove the flywheel as described in this chapter.

20. Remove the 3 screws holding the stator and timer base assembly (**Figure 124**). Remove the assembly.

21. Remove the 4 nuts holding the link rods and throttle cam assembly (**Figure 125**). Remove the assembly.

22. Remove the carburetor silencer cover and gasket (**Figure 126**). Remove the silencer case and gasket (**Figure 127**).

23. Remove the starter bracket.

24. Remove the fuel filter and the fuel pump/insulator assembly (**Figure 128**). See Chapter Six.

25. Disconnect the oil injection pump control rod (**Figure 129**).

26. Remove the oil injection pump and driven gear retainer (**Figure 130**). See Chapter Twelve.

27. Remove the upper oil seal housing.

28. Installation is the reverse of removal, noting the following.

29. Use a new power head gasket.

30. Install a new upper oil seal housing seal with its lip facing the power head. Lubricate seal lips with Super Grease A (part No. 99000-25030) or equivalent.

31. Fill the cavity between the lower oil seal and backup oil seal in the lower oil seal housing with

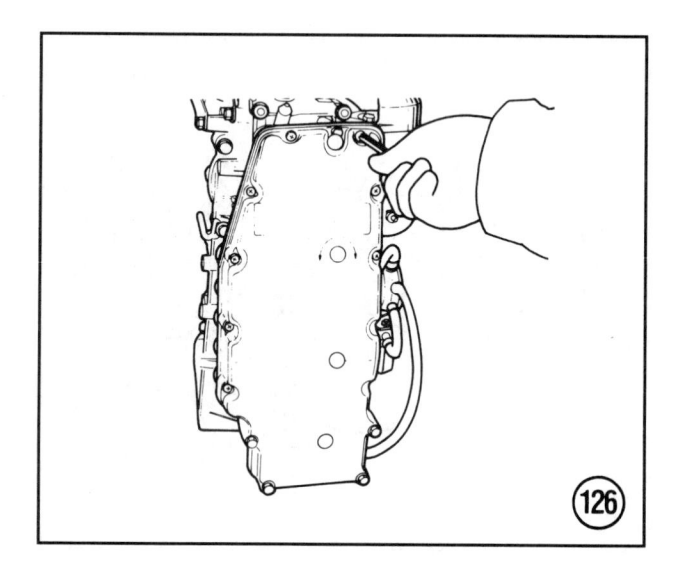

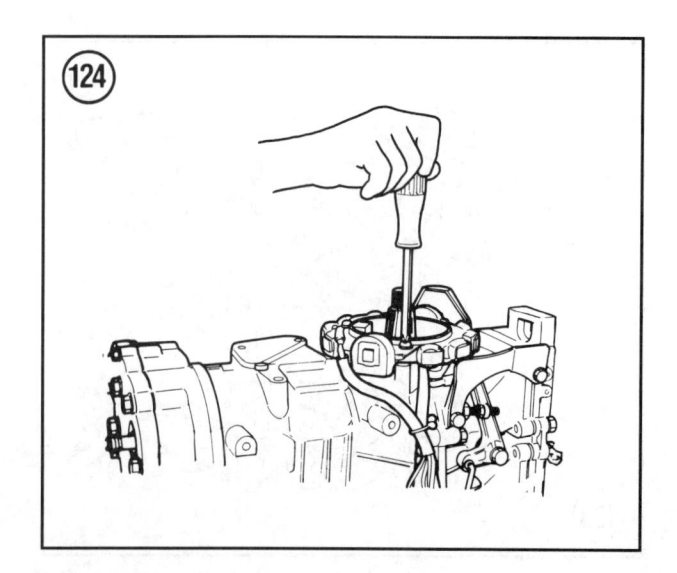

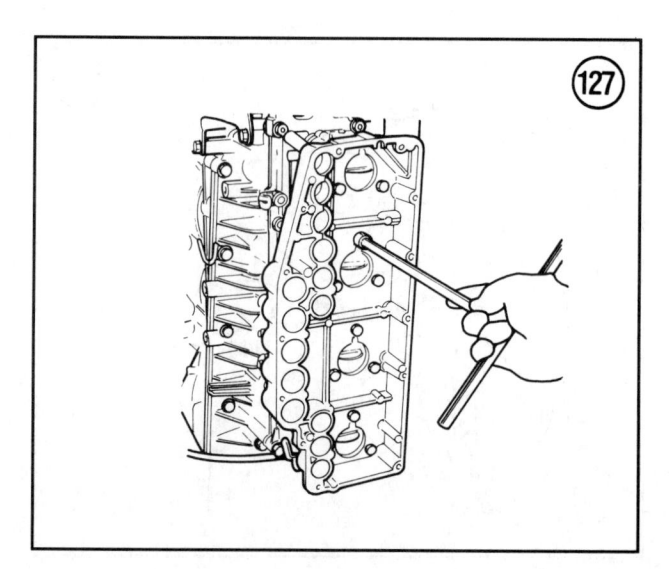

POWER HEAD

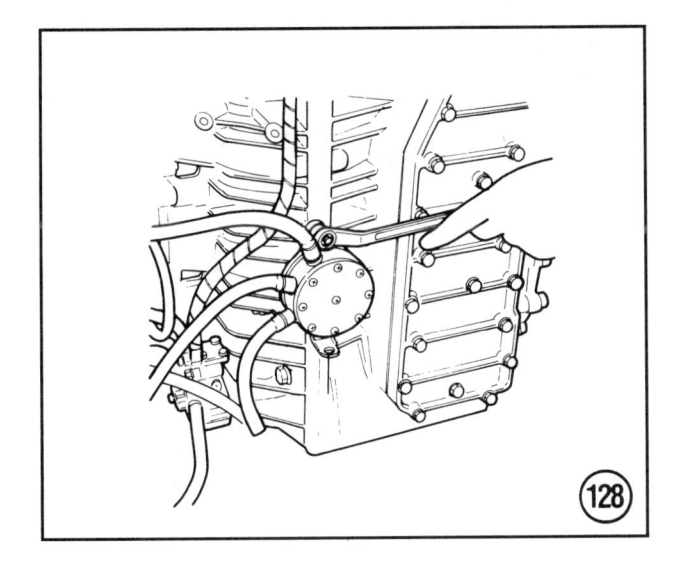

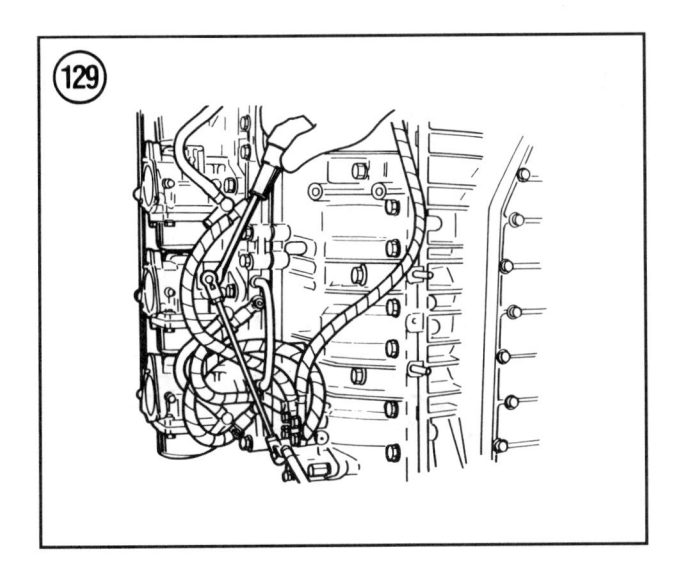

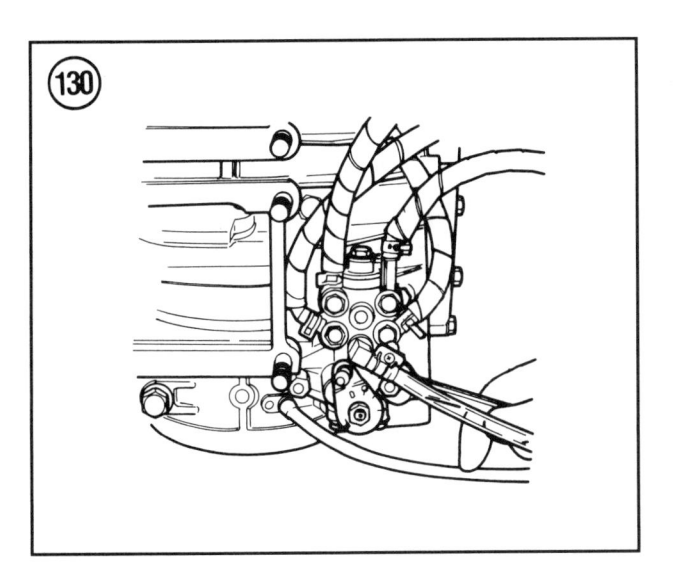

(128)

(129)

(130)

water-resistant grease (part No. 99000-25160) or equivalent.

32. Lightly coat the drive shaft splines with water-resistant grease.

33. Make sure the locating dowels in the drive shaft support plate are in position, if so equipped.

34. Rotate the propeller as required to align crankshaft and drive shaft splines.

35. Coat the power head attaching bolt and nut threads with Thread Lock 1341 (part No. 99000-32050) or equivalent.

36. Tighten all fasteners to specifications (**Table 1**).

37. Install the coils in sequence according to the number on the secondary lead. The coil with the lead marked "1" should be installed to the No. 1 cylinder, etc. (**Figure 131**).

38. Perform engine synchronization and linkage adjustments. See Chapter Five.

Removal/Installation (1986-on DT 115, DT 140)

1. Remove the engine top cover.
2. Disconnect the negative battery cable.
3. Disconnect the oil level switch lead wires and the oil hose, then remove the oil injection tank. See Chapter Twelve.

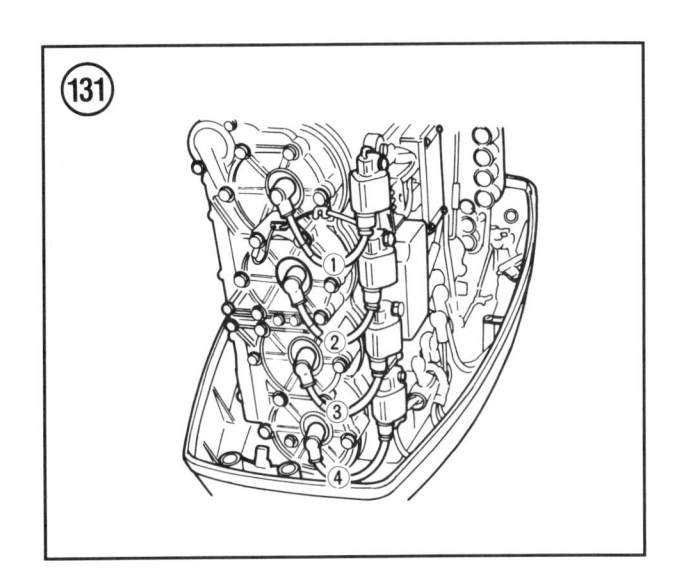

(131)

4. Remove the idle speed adjustment switch (prior to 1991) and electrical parts holder cover.
5. Disconnect the battery leads from the starter and starter switch.
6. Disconnect the power trim and tilt electrical leads.
7. Remove the bolts securing the electrical parts holder and remove the assembly.
8. Disconnect the fuel hose from the fuel filter (**Figure 132**). Plug the end of the fuel hose to prevent the loss of fuel.
9. Remove the 4 bolts holding the lower engine covers. Lower the covers.
10. Disconnect the water outlet hose at the lower support housing (**Figure 133**).

NOTE
At this point, there should be no hoses, wires or linkage connecting the power head to the drive shaft housing. Recheck this to make sure nothing will hamper power head removal.

11. Remove the 8 bolts and 2 nuts holding the power head to the drive shaft housing (**Figure 134**).

WARNING
If a hoist is not available for use in Step 12, have an assistant help with power head removal to avoid possible serious personal injury.

12. Attach a hoist to the engine hooks and remove the power head from the drive shaft housing.
13. Check the power head base for alignment dowel pins. If any came off with the power head, remove and reinstall in the drive shaft housing mounting flange.
14. Place the power head on a clean workbench or suitable power head stand.
15. Remove the flywheel as described in this chapter.
16. Remove the 3 screws holding the stator and counter coil assembly (**Figure 135**). Remove the assembly.

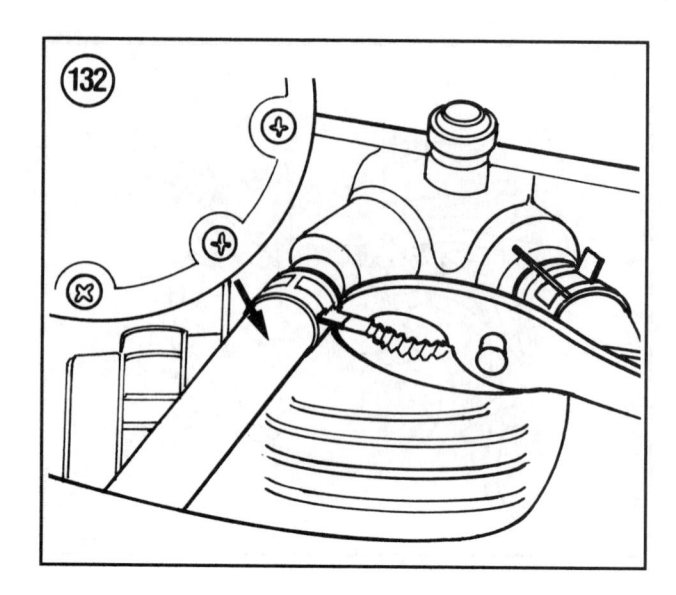

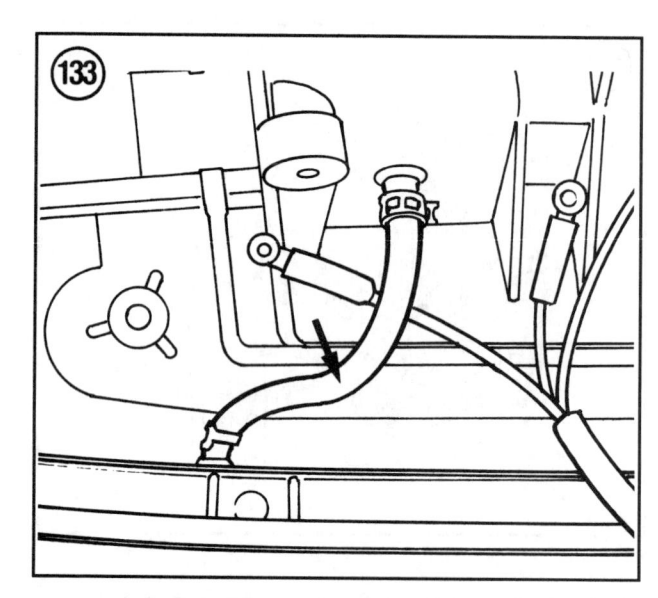

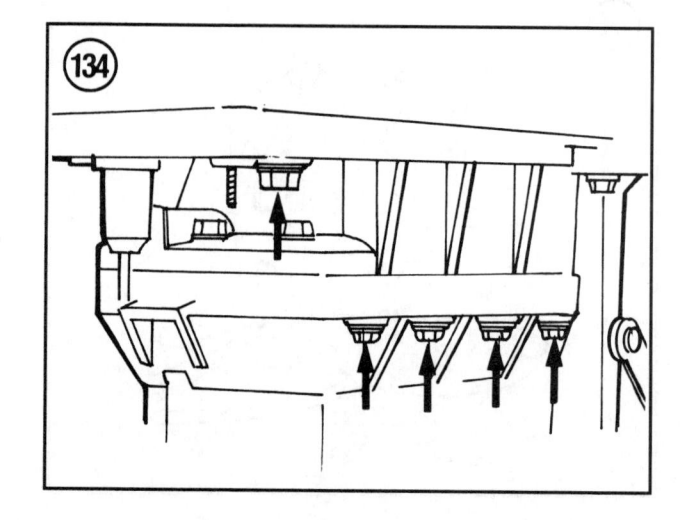

POWER HEAD

17. Remove the starter motor. See Chapter Seven.

18. Remove the starter bracket.

19. Remove the fuel filter and the fuel pump/insulator assembly. See Chapter Six.

20. Disconnect the oil injection pump control rod. Remove the oil injection pump and driven gear retainer. See Chapter Twelve.

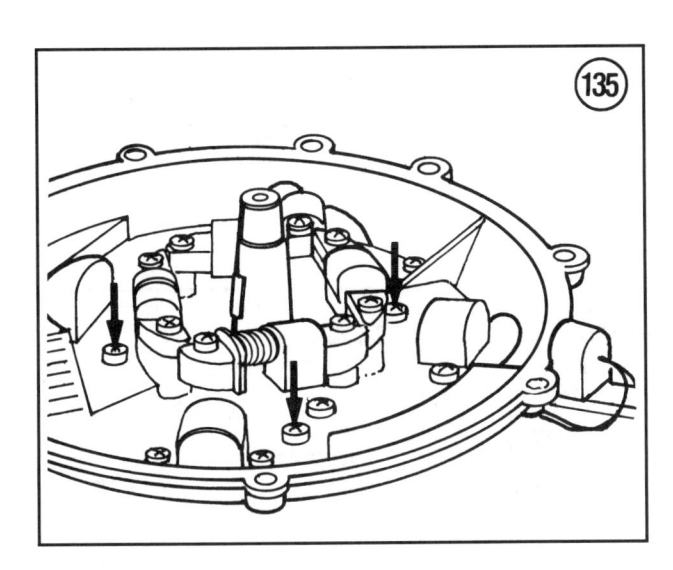

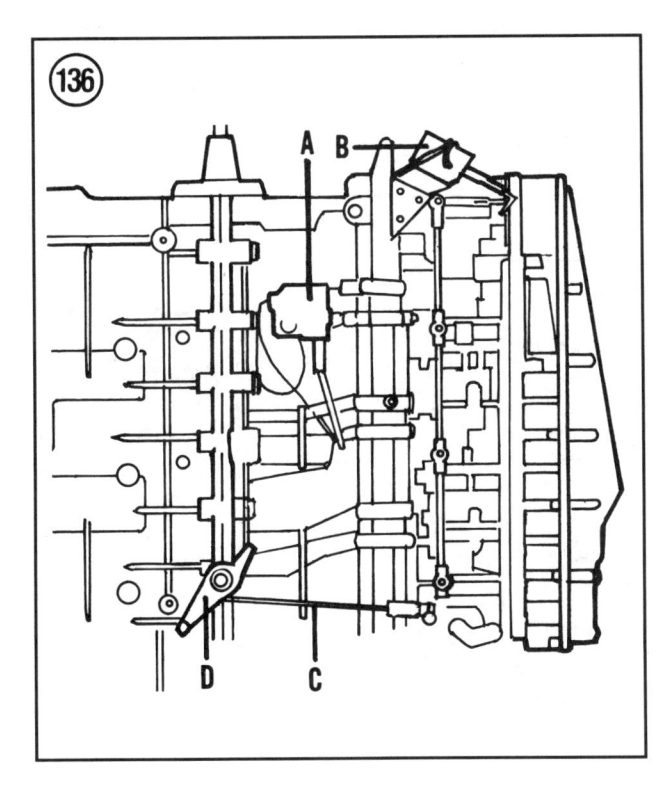

21. Remove the starter motor relay (A, **Figure 136**).

22. Remove the screws and bolts and remove the choke solenoid (B, **Figure 136**).

23. Remove the throttle lever rod (C, **Figure 136**) from the carburetor.

24. Remove the throttle control lever (D, **Figure 136**).

25. Remove the carburetor silencer cover and gasket. Remove the silencer case and gasket.

26. Remove the carburetors. See Chapter Six.

27. Remove the inlet case and reed valve assembly.

28. Installation is the reverse of removal, noting the following.

29. Use a new power head gasket.

30. Install a new upper oil seal housing seal with its lip facing the power head. Lubricate seal lips with Super Grease A (part No. 99000-25030) or equivalent.

31. Fill the cavity between the lower oil seal and backup oil seal in the lower oil seal housing with water-resistant grease (part No. 99000-25160) or equivalent.

32. Lightly coat the drive shaft splines with water-resistant grease.

33. Make sure the locating dowels in the drive shaft support plate are in position.

34. Rotate the propeller as required to align crankshaft and drive shaft splines.

35. Coat the power head attaching bolt and nut threads with Thread Lock 1341 (part No. 99000-32050) or equivalent.

36. Tighten all fasteners to specifications (**Table 1**).

37. Install the coils in sequence according to the number on the secondary lead. The coil with the lead marked "1" should be installed to the No. 1 cylinder, etc.

38. Perform engine synchronization and linkage adjustments. See Chapter Five.

Removal/Installation (V4)

1. Remove the engine top cover.
2. Remove the relay cover under the starter.
3. Remove the sub-cable from the starter (A, **Figure 137**).
4. Disconnect the starter relay from the relay holder (B, **Figure 137**).
5. Remove the relay holder and bolt (C, **Figure 137**).
6. Disconnect the left-hand bolt on the starter motor securing band and remove the ground wire (D, **Figure 137**).
7. Remove the flywheel cover.
8. Remove the bolts (**Figure 138**) securing the power trim and tilt relays and remove the relays and their holder.
9. Remove the cover from the electrical parts holder.
10. Disconnect all electrical wire leads within the parts holder.
11. Remove the bolts (**Figure 139**) securing the electrical parts holder and remove it.
12. Disconnect the ground wire from each cylinder head.
13. Remove the oil tank. See Chapter Twelve.
14. Disconnect the fuel hose from the connector on the fuel filter (**Figure 140**). Plug the end of the fuel hose to prevent the loss of fuel.
15. Refer to **Figure 141** and disconnect the throttle valve sensor lead wire, the oil sensor lead wire and the starter valve lead wire.
16. Remove the 2 bolts (**Figure 142**) holding the lower front under cover. Remove the cover.
17. Remove the 4 bolts (**Figure 143**) holding the lower rear under cover. Remove the cover.
18. Disconnect the water outlet hose.
19. Remove the 5 bolts (**Figure 144**) holding the power head to the drive shaft housing.
20. Remove the 10 additional bolts (**Figure 145**) holding the power head to the drive shaft housing.

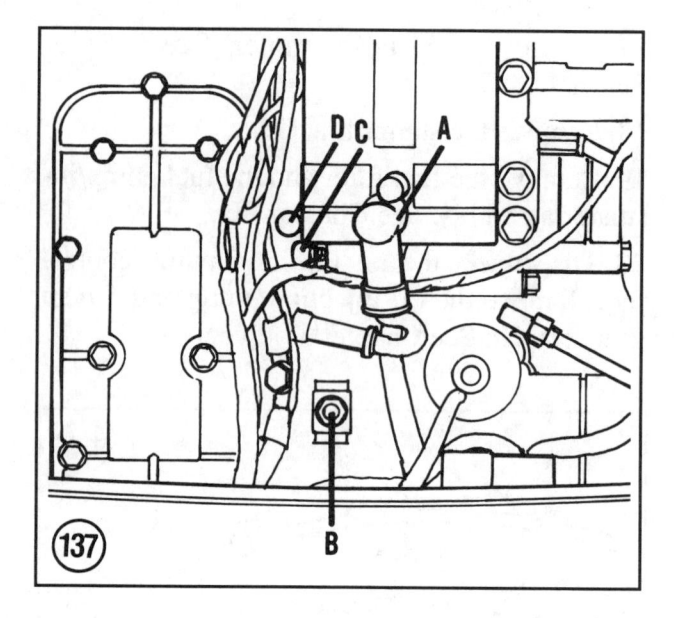

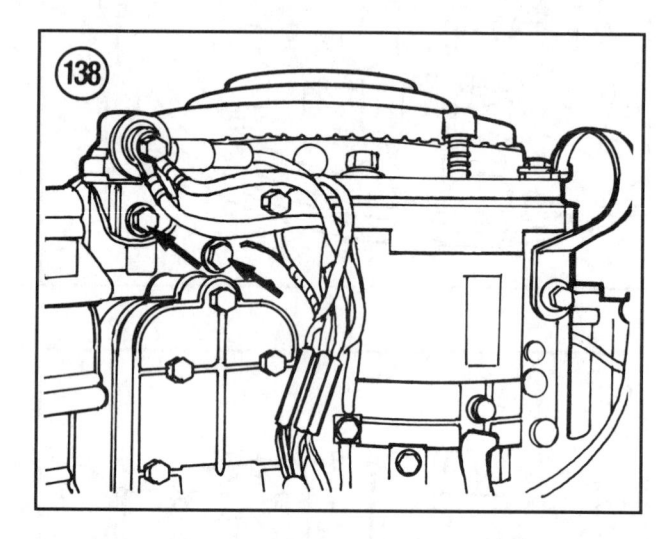

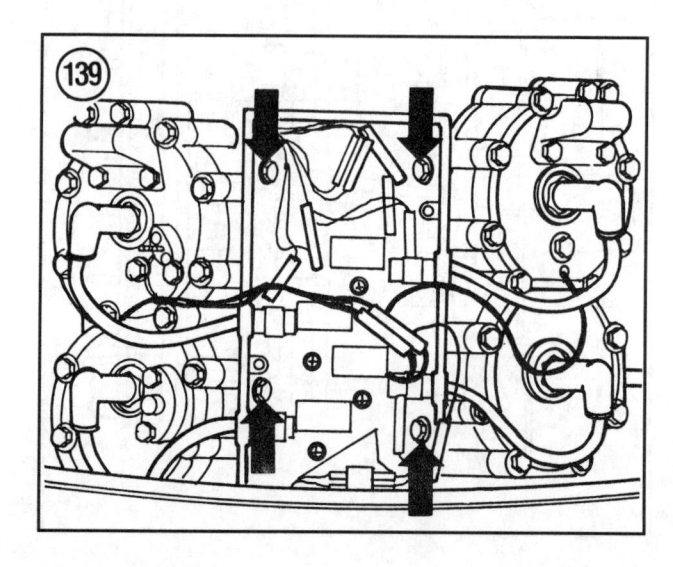

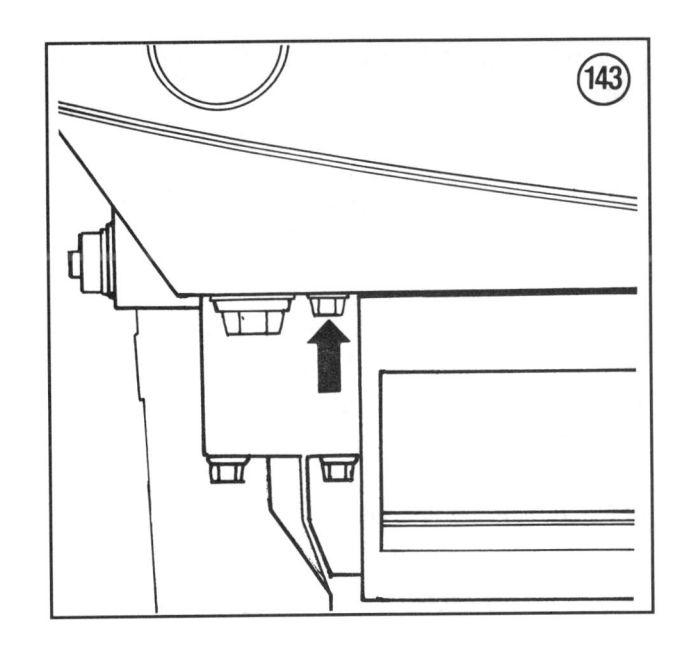

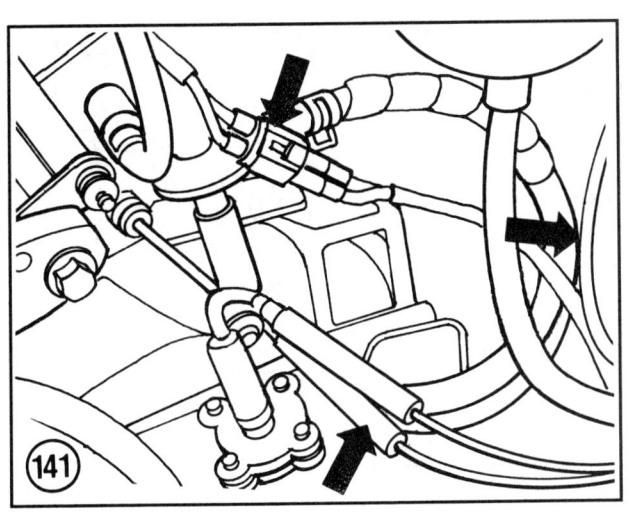

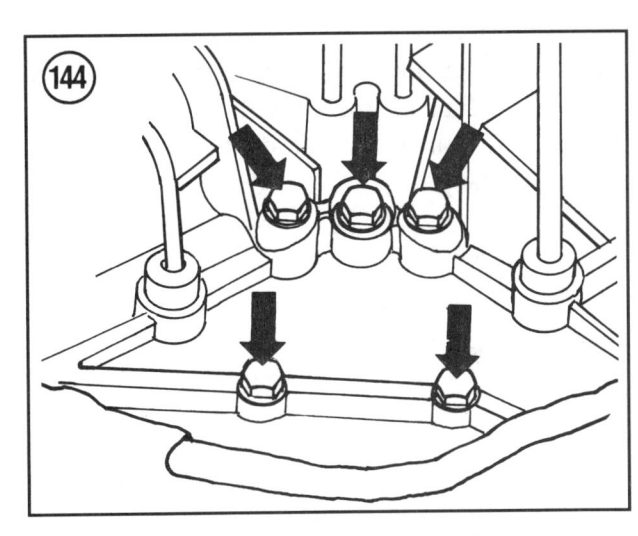

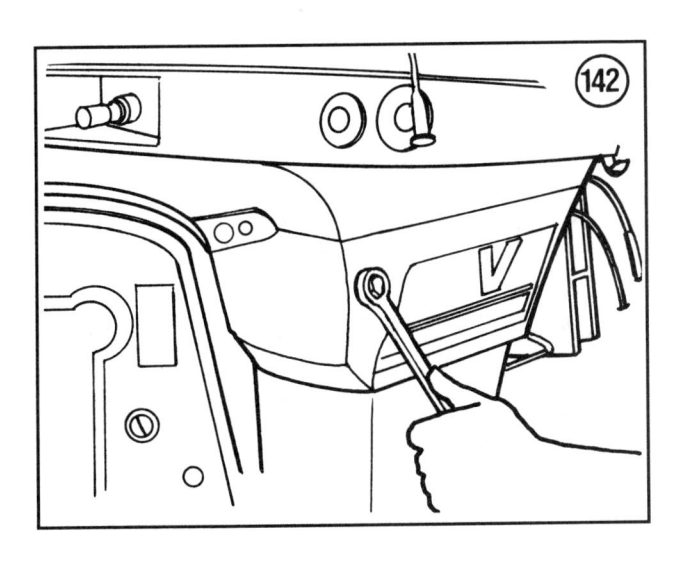

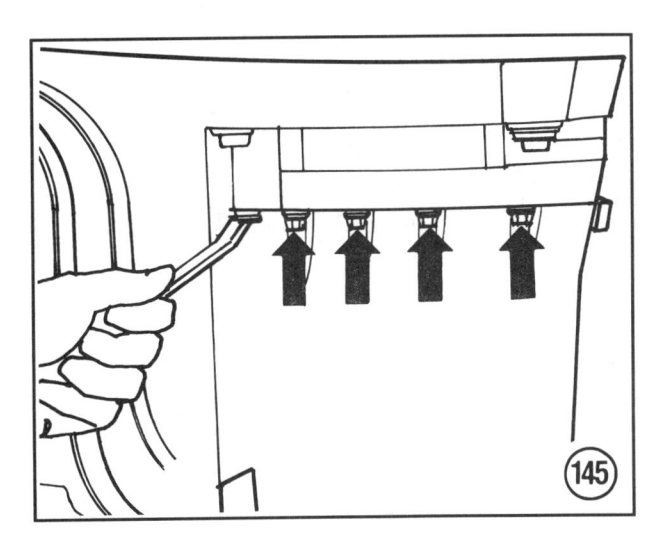

POWER HEAD

293

21. Remove the clutch connector rod pin (**Figure 146**).

WARNING
If a hoist is not available for use in Step 22, have at least 2 assistants help with power head removal to avoid possible serious personal injury.

22. Attach a hoist to the engine hooks and *partially* remove the power head from the drive shaft housing.

23. Disconnect the clutch lever rod (A, **Figure 147**) from the clutch shaft (B, **Figure 147**).

24. Disconnect the upper clutch rod (C, **Figure 147**) from the clutch shaft (B, **Figure 147**).

25. Completely remove the power head from the drive shaft housing.

26. Check the power head base for alignment dowel pins. If any came off with the power head, remove and reinstall in the drive shaft housing mounting flange.

27. Place the power head on a clean workbench or suitable power head stand.

28. Remove the flywheel as described in this chapter.

29. Remove the screws holding the stator (A, **Figure 148**) and gear counter assembly (B, **Figure 148**). Remove the assembly.

30. Remove the fuel filter and the fuel pump/insulator assembly. See Chapter Six.

31. Remove the starter valve.

32. Remove the bolts securing the engine hook and remove the hook.

33. Remove the starter motor. See Chapter Seven.

34. Remove the starter bracket.

35. Disconnect the throttle control rod (A, **Figure 149**) from the bottom carburetor.

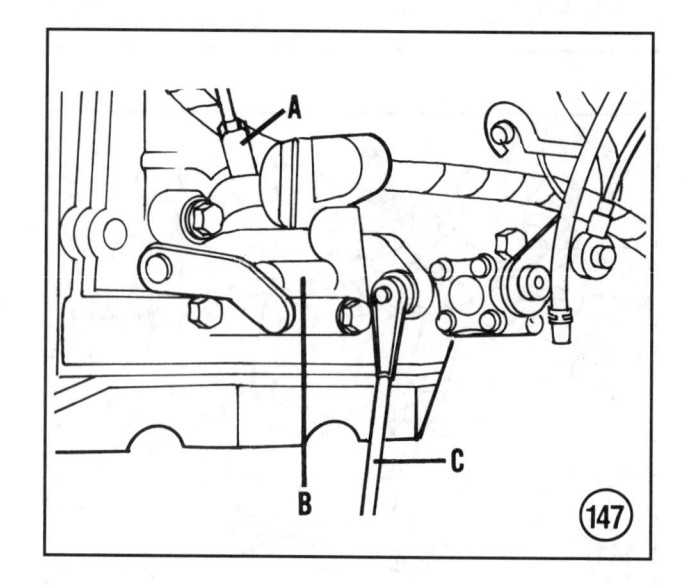

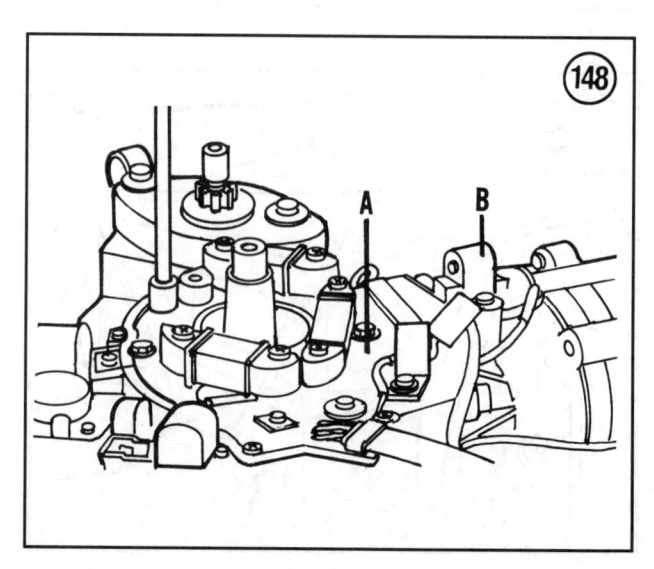

POWER HEAD

36. Remove the throttle control lever (B, **Figure 149**) and the clutch lever (C, **Figure 149**).

37. Remove the oil pump control rod (A, **Figure 150**).

38. Disconnect the oil pump hoses from the pre-atomization valves (B, **Figure 150**).

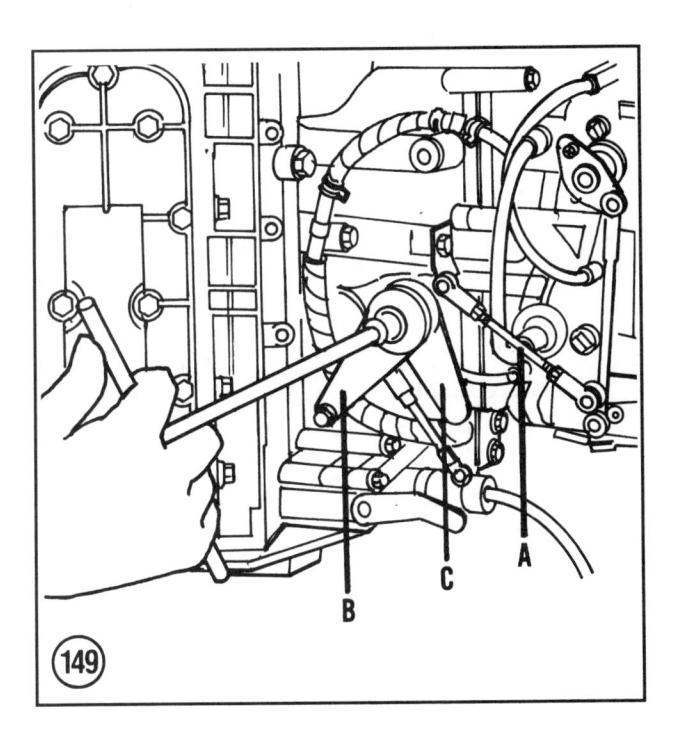

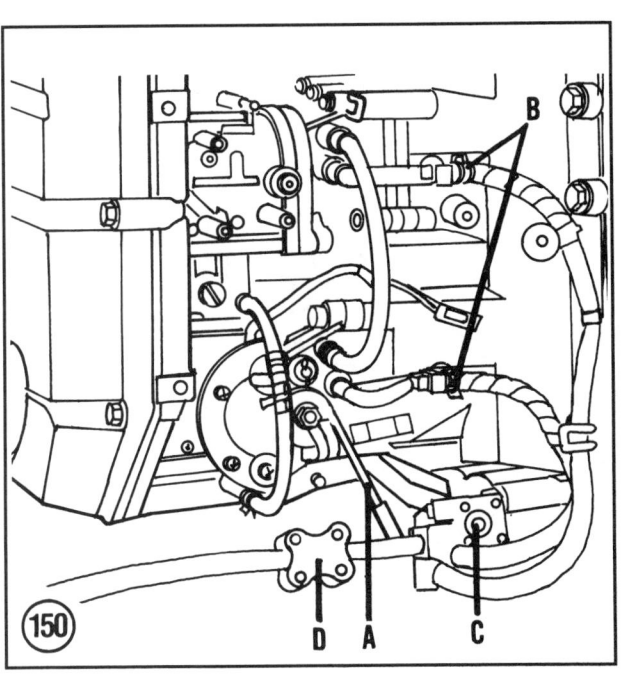

39. Remove the oil pump (C, **Figure 150**) and oil flow sensor (D, **Figure 150**). See Chapter Twelve.

40. Remove the bolts securing the exhaust cover and remove the cover (**Figure 151**).

CAUTION
In Step 41, do not pry too hard as the gasket sealing surfaces may be damaged.

41. Insert a flat-bladed screwdriver into the pry point slots indicated by an "O" and gently pry the exhaust case away from the crankcase. Remove the case, cover and gasket.

42. Installation is the reverse of removal, noting the following.

43. Use a new power head gasket.

44. Install a new upper oil seal housing seal with its lip facing the power head. Lubricate seal lips with Super Grease A (part No. 99000-25030) or equivalent.

45. Fill the cavity between the lower oil seal and backup oil seal in the lower oil seal housing with water-resistant grease (part No. 99000-25160) or equivalent.

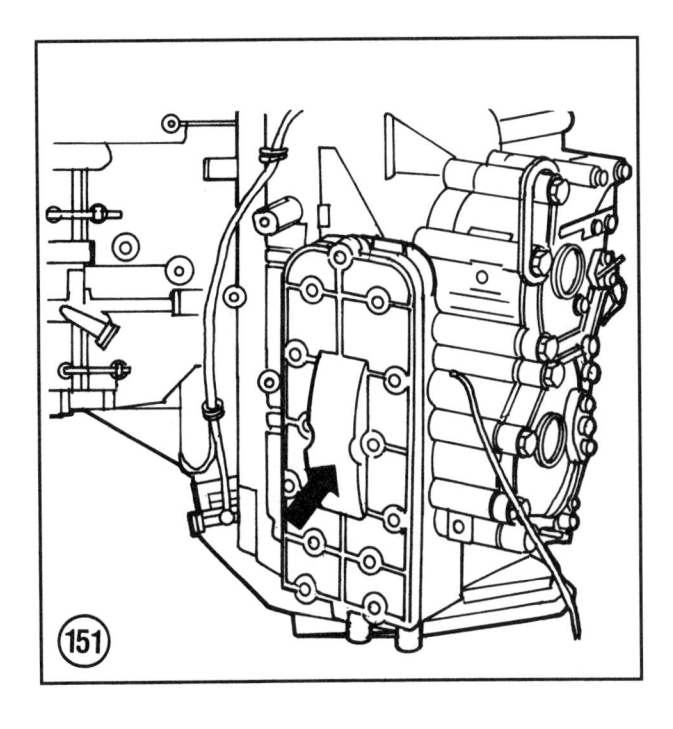

46. Lightly coat the drive shaft splines with water-resistant grease.

47. Make sure the locating dowels in the drive shaft support plate are in position, if so equipped.

48. *Partially* lower the power head into position and perform the following:
   a. Place the shift arm in the reverse position.
   b. Raise the upper clutch rod (C, **Figure 147**) and engage the pin on the clutch shaft (B, **Figure 147**) into the hole on the clutch rod.
   c. Install the clutch rod connector (A, **Figure 147**).

49. Continue to lower the power head into position and rotate the propeller as required to align crankshaft and drive shaft splines.

50. Install the clevis and a new cotter pin (**Figure 152**) between the upper and lower shift rods. Bend the ends over completely.

51. Coat the power head attaching bolt and nut threads with Suzuki Silicone Seal (part No. 99000-31120) or equivalent.

52. Tighten all fasteners to specifications (**Table 1**).

53. Install the coils in sequence according to the number on the secondary lead. The coil with the lead marked "1" should be installed to the No. 1 cylinder, etc.

54. Make sure all electrical connections are free of corrosion and are tight.

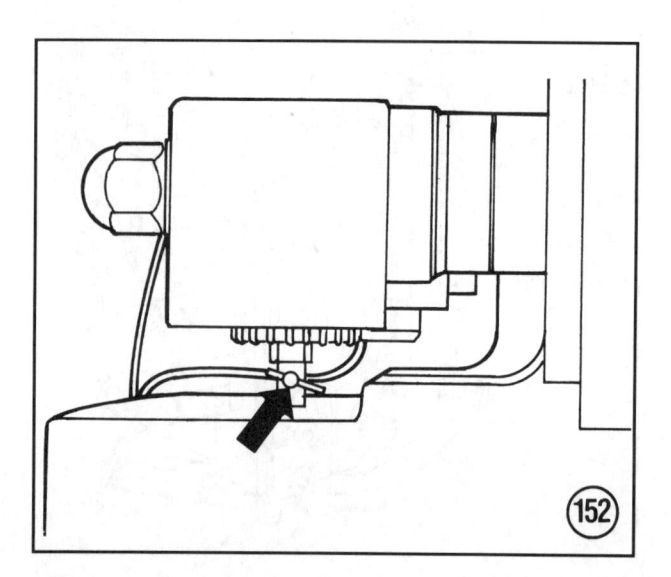

55. Perform engine synchronization and linkage adjustments. See Chapter Five.

Removal/Installation (V6)

1. Remove the engine top cover.

2. Remove the cover from the electrical parts holder on the starboard side of the engine.

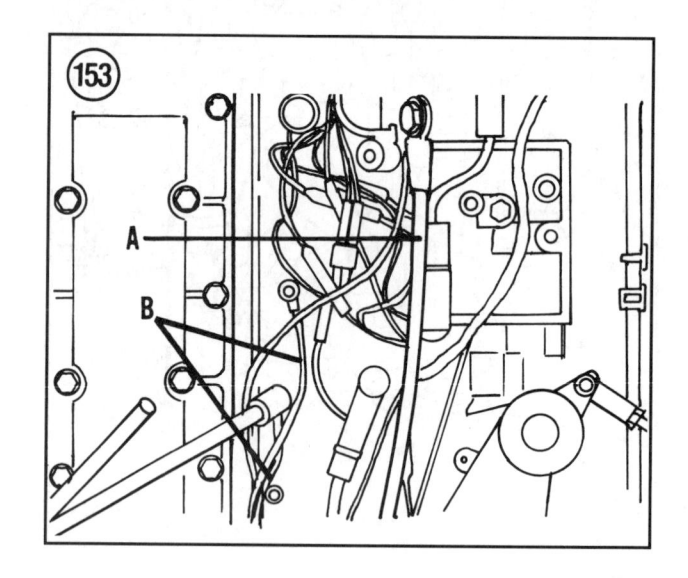

POWER HEAD

3. Disconnect the battery cable (A, **Figure 153**) and the power trim and tilt motor electrical leads (B) from the engine.

4. Disconnect the fuel hose from the connector on the fuel filter (**Figure 154**). Plug the end of the fuel hose to prevent the loss of fuel.

5. Remove the cover from the electrical parts holder located between the cylinder heads (ignition coils location). Disconnect all electrical wire leads within the parts holder, then remove the electrical parts holder.

6. Remove the 5 bolts (**Figure 155**) holding the power head to the drive shaft housing.

7. Remove the 2 bolts (**Figure 156**) holding the lower front under cover. Remove the cover.

8. Remove the 4 bolts (A, **Figure 157**) holding the lower rear under cover. Remove the cover.

9. Disconnect the water outlet hose (B, **Figure 157**).

10. Remove the 10 additional bolts (**Figure 158**) holding the power head to the drive shaft housing.

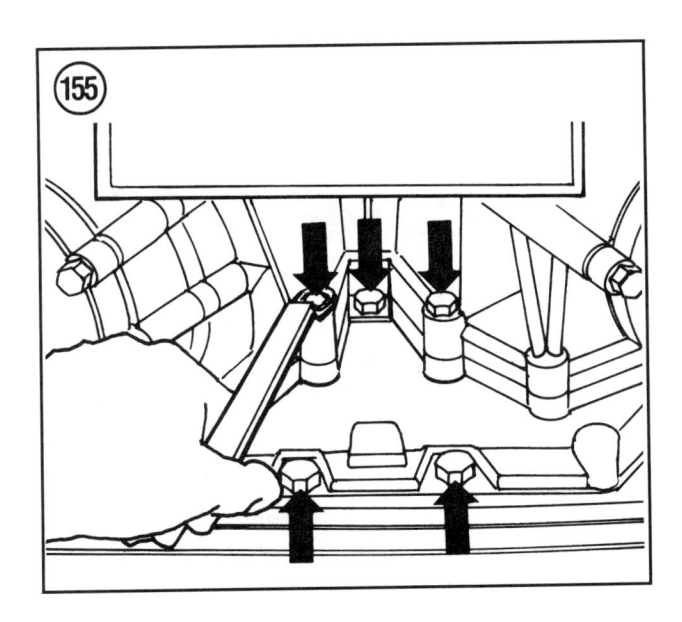

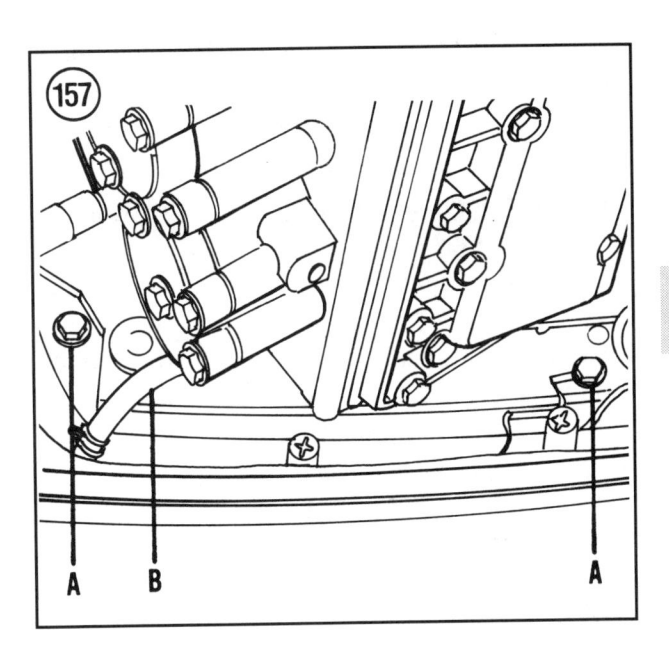

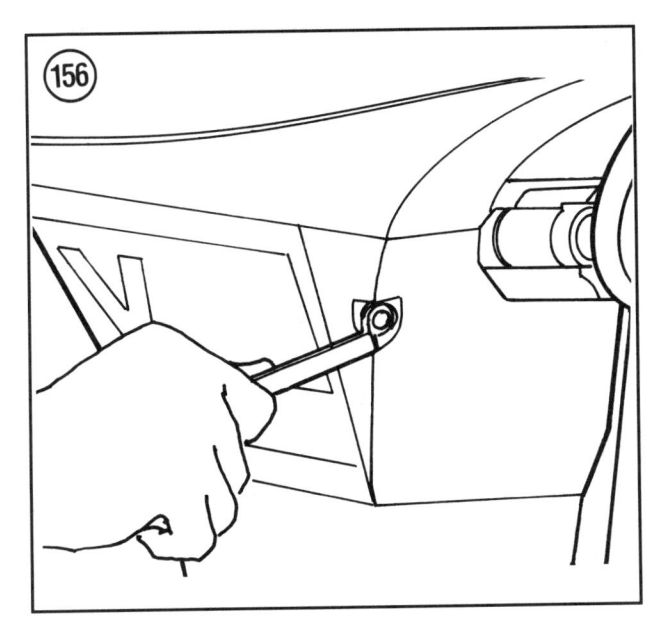

11. Remove the clutch connector rod pin (**Figure 159**).

WARNING
If a hoist is not available for use in Step 12, have at least 2 assistants help with power head removal to avoid possible serious personal injury.

12. Attach a hoist to the engine hooks and *partially* remove the power head from the drive shaft housing.

13. Disconnect the clutch lever rod connector (A, **Figure 160**) from the clutch shaft (B, **Figure 160**).

14. Disconnect the upper clutch rod (C, **Figure 160**) from the clutch shaft (B, **Figure 160**).

15. Completely remove the power head from the drive shaft housing.

16. Check the power head base for alignment dowel pins. If any came off with the power head, remove and reinstall in the drive shaft housing mounting flange.

17. Place the power head on a clean workbench or suitable power head stand.

18. Remove the oil tank. See Chapter Twelve.

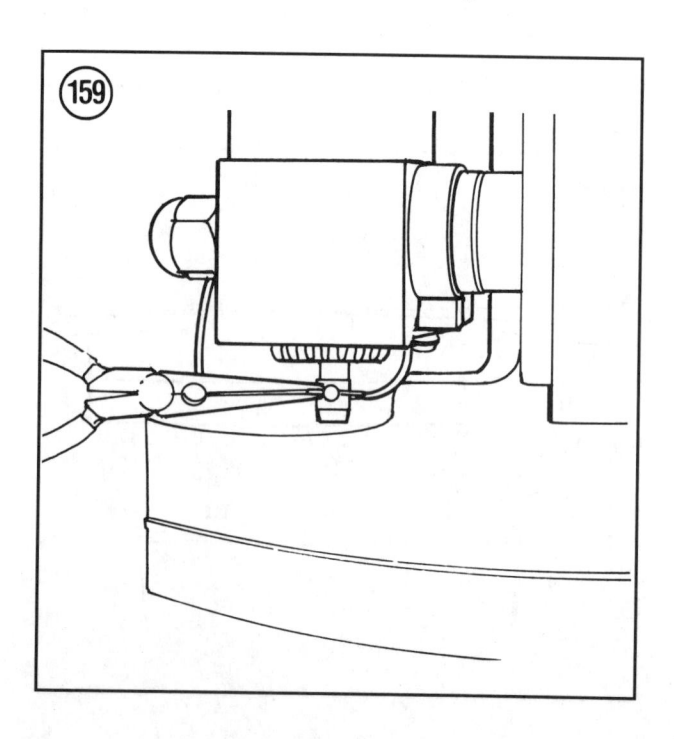

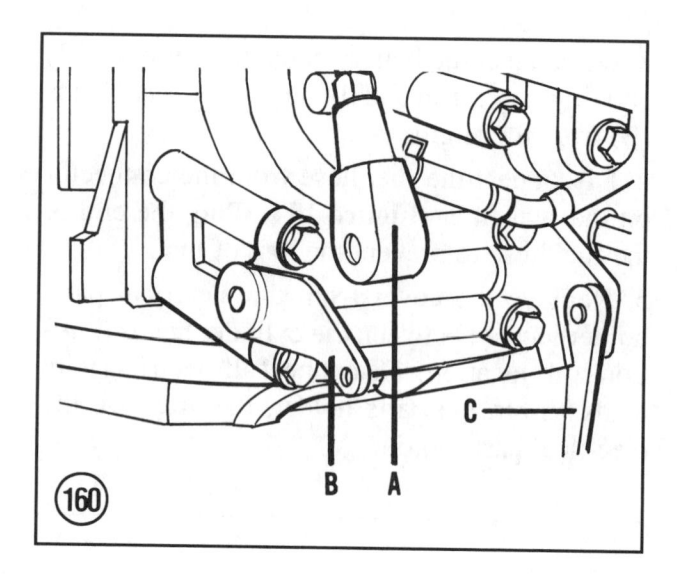

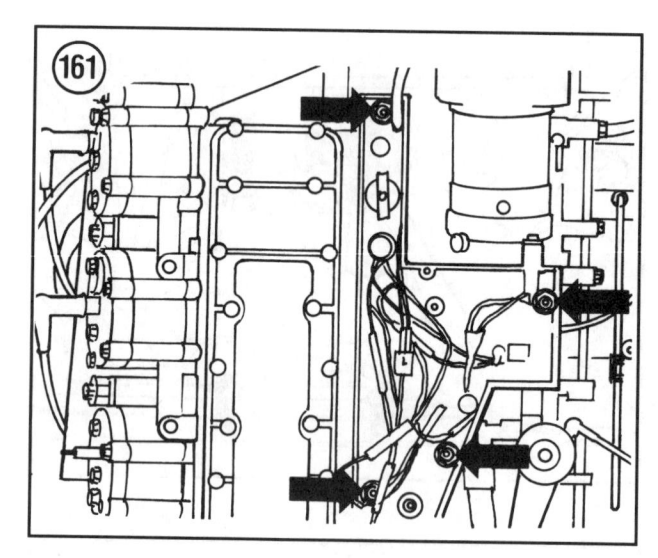

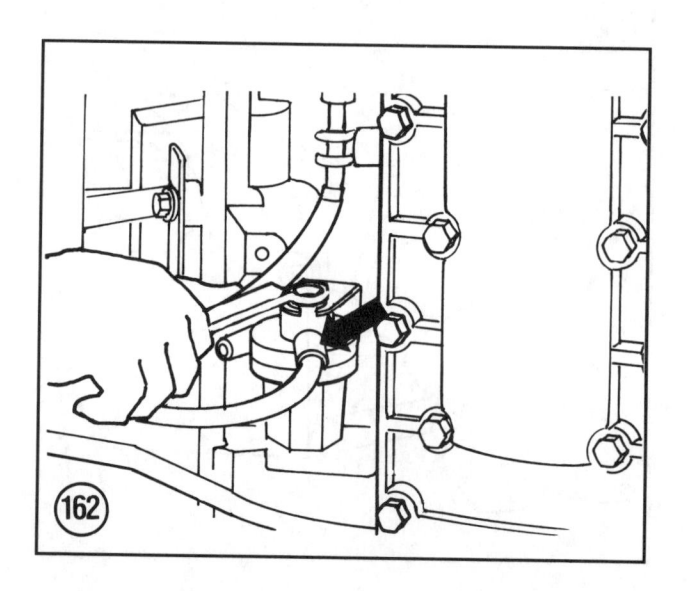

POWER HEAD

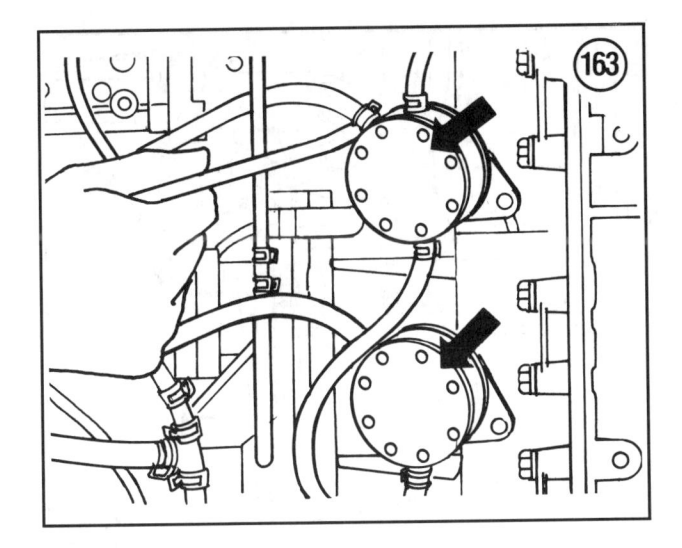

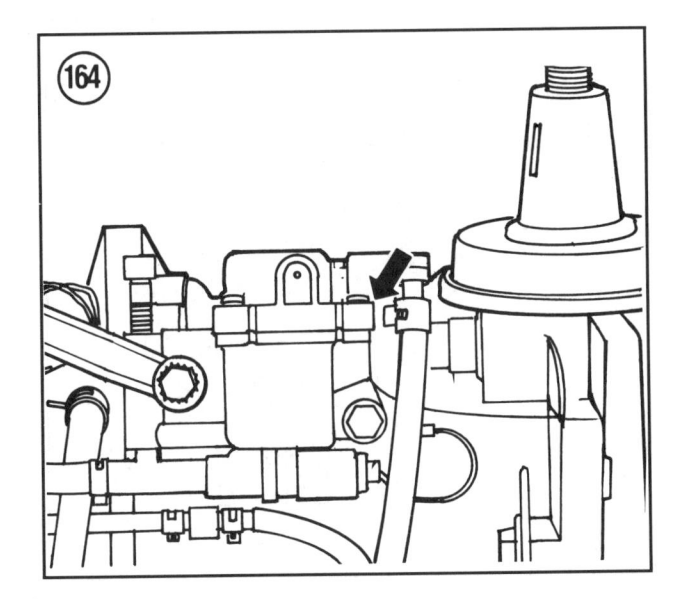

19. Remove the flywheel as described in this chapter.
20. Remove the electrical parts holder (**Figure 161**) from the starboard side of the engine.
21. Remove the screws holding the stator and gear counter assembly. Remove the assembly.
22. Remove the starter motor. See Chapter Seven.
23. Remove the starter bracket.
24. Remove the fuel filter (**Figure 162**) and the fuel pumps (**Figure 163**). See Chapter Six.
25. Remove the starter valve (**Figure 164**).
26. Disconnect the throttle control rod (**Figure 165**) from the bottom carburetor.
27. Remove the bolts securing the silencer and carburetors and remove the assembly.
28. Disconnect the oil hoses and remove the oil pump and oil flow sensor. See Chapter Twelve.
29. Remove the throttle control lever (A, **Figure 166**) and arm (B, **Figure 166**).
30. Remove the bolts securing the exhaust cover.

CAUTION
In Step 31, do not pry too hard as the gasket sealing surfaces may be damaged.

31. Insert a flat-bladed screwdriver into the pry point slots indicated by an "O" and gently pry the exhaust cover away from the crankcase. Remove the cover and plate.

32. Installation is the reverse of removal, noting the following.

33. Use a new power head gasket.

34. Lightly coat the drive shaft splines with water-resistant grease.

35. Make sure the locating dowels in the drive shaft support plate are in position, if so equipped.

36. *Partially* lower the power head into position and perform the following:

   a. Place the shift arm in the reverse position.
   b. Raise the upper clutch rod (A, **Figure 167**) and engage the pin on the clutch shaft (B, **Figure 167**) into the hole on the clutch rod.
   c. Install the clutch rod connector.

37. Continue to lower the power head into position and rotate the propeller as required to align crankshaft and drive shaft splines.

38. Install the clevis and a new cotter pin between the upper and lower shift rods. Bend the cotter pin ends over completely.

39. Coat the power head attaching bolt and nut threads with Suzuki Silicone Seal (part No. 99000-31120) or equivalent.

40. Tighten all fasteners to specifications (**Table 1**).

41. Install the coils in sequence according to the number on the secondary lead. The coil with the lead marked "1" should be installed to the No. 1 cylinder, etc.

42. Make sure all electrical connections are free of corrosion and are tight.

43. Perform engine synchronization and linkage adjustments. See Chapter Five.

Disassembly (DT 2)

1. Remove the carburetor and intake manifold assembly.

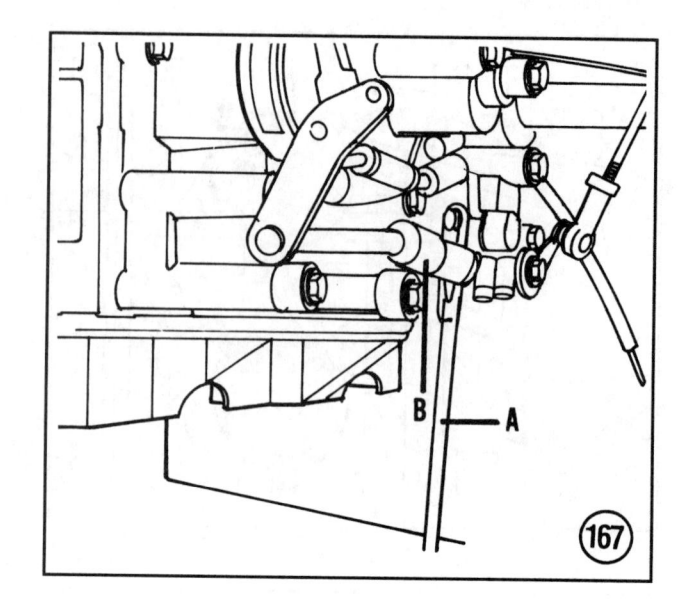

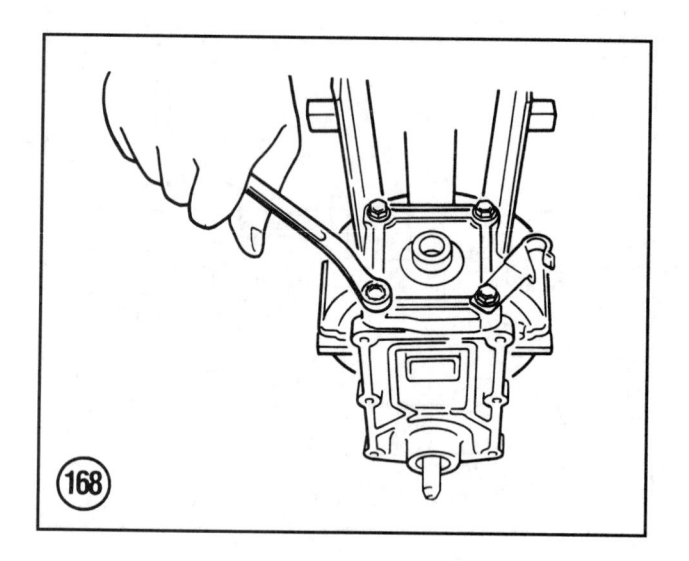

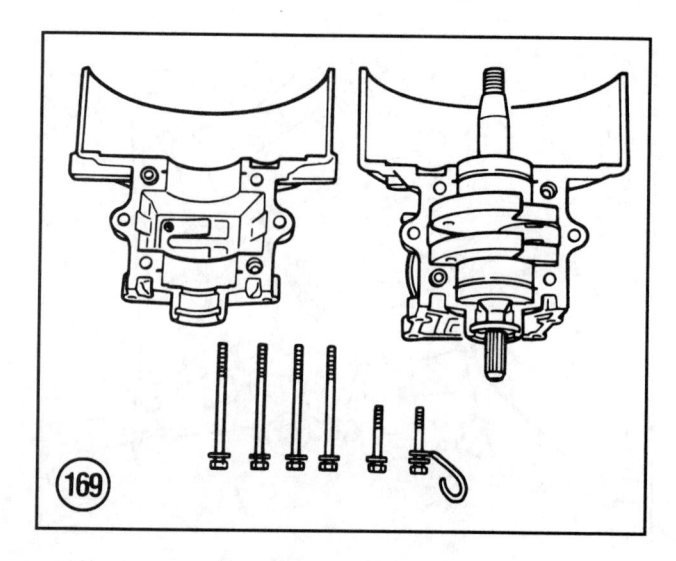

POWER HEAD

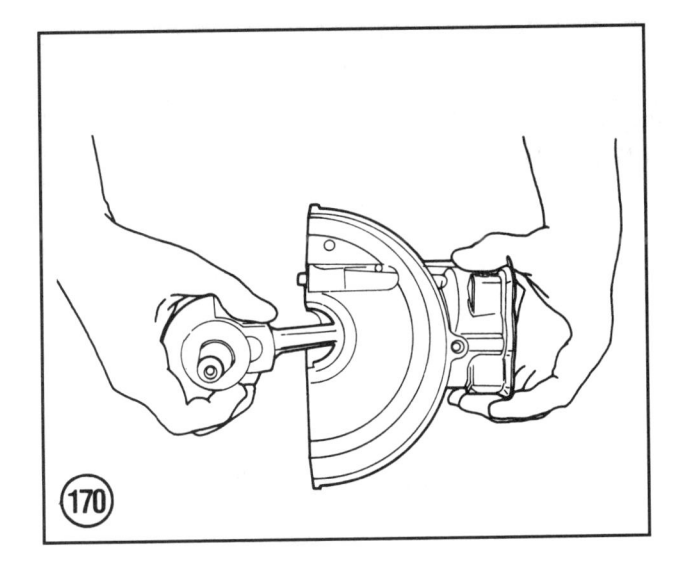

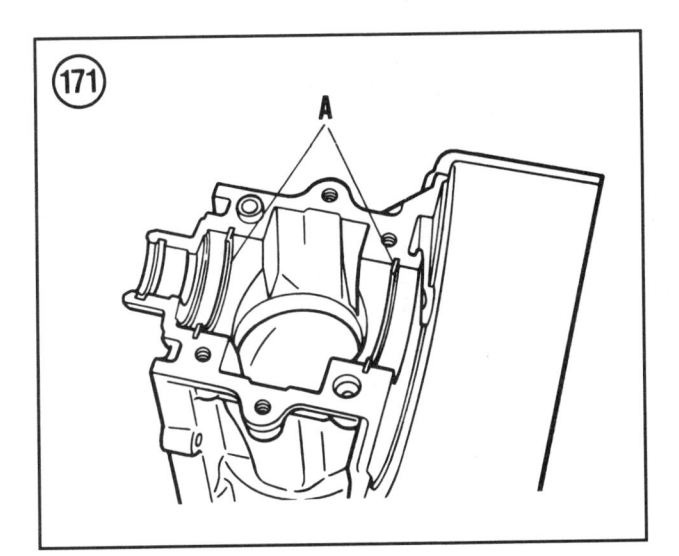

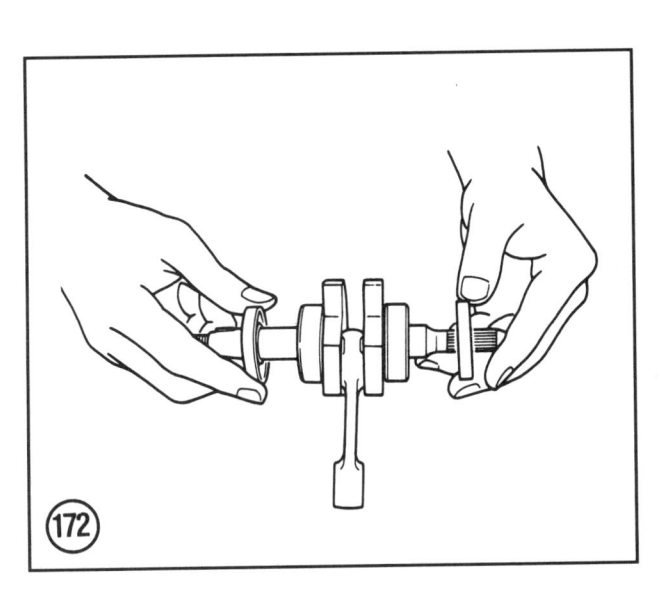

2. Loosen the 4 cylinder head nuts (**Figure 168**) in several stages to prevent head warpage. Remove the head and gasket. Discard the gasket.

3. Loosen the 6 crankcase bolts in several stages. Carefully pry apart and separate the crankcase halves (**Figure 169**).

4. Remove the O-ring from the crankshaft lower end. Hold the crankshaft in one hand and carefully pull the cylinder block off the piston (**Figure 170**).

5. Remove the crankshaft thrust rings (A, **Figure 171**).

6. If the piston is to be removed from the connecting rod:
  a. Cup one hand around the piston and carefully pry out the piston pin circlip, catching it as it comes free. Repeat this step to remove the opposite piston pin circlip. Discard the circlips as they should not be used again.
  b. Push the piston pin from the piston/connecting rod with a suitable size drift.
  c. Remove the piston and caged needle bearing from the connecting rod.

7. Remove the upper and lower crankshaft oil seals (**Figure 172**).

8. Carefully spread and remove the piston rings as shown in **Figure 173**.

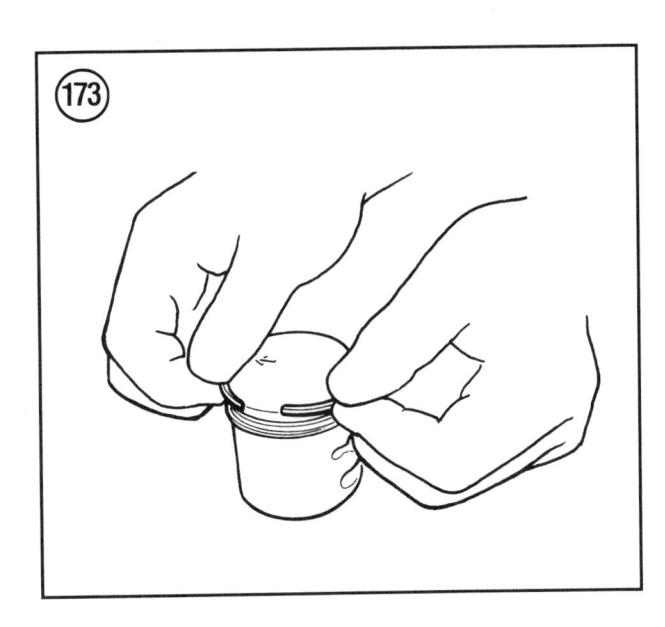

CHAPTER EIGHT

174

**POWER HEAD
(DT 6, 1985-1987 DT 8)**

1. Intake manifold
2. Gasket
3. Reed plate
4. Gasket
5. Crankcase cover
6. Cylinder block
7. Gasket
8. Cylinder head
9. Gasket
10. Reed valves
11. Gasket
12. Exhaust cover

POWER HEAD

Disassembly
(DT 4)

1. Loosen the clips and disconnect the lubrication hose from each fitting.
2. Remove the 2 mounting bolts and remove the rewind starter mounting base.
3. Remove the screws and remove the water jacket cover and gasket.
4. Loosen the 5 cylinder head bolts in several stages to prevent head warpage. Remove the head and gasket. Discard the gasket.
5. Loosen the 6 crankcase bolts in several stages. Carefully pry apart and separate the crankcase halves.
6. Carefully lift the crankshaft from the crankcase half.
7. If the piston is to be removed from the connecting rod:
   a. Cup one hand around the piston and carefully pry out the piston pin circlip, catching it as it comes free. Repeat this step to remove the opposite piston pin circlip. Discard the circlips as they should not be used again
   b. Push the piston pin from the piston/connecting rod with a suitable size drift.
   c. Remove the piston and caged needle bearing from the connecting rod.

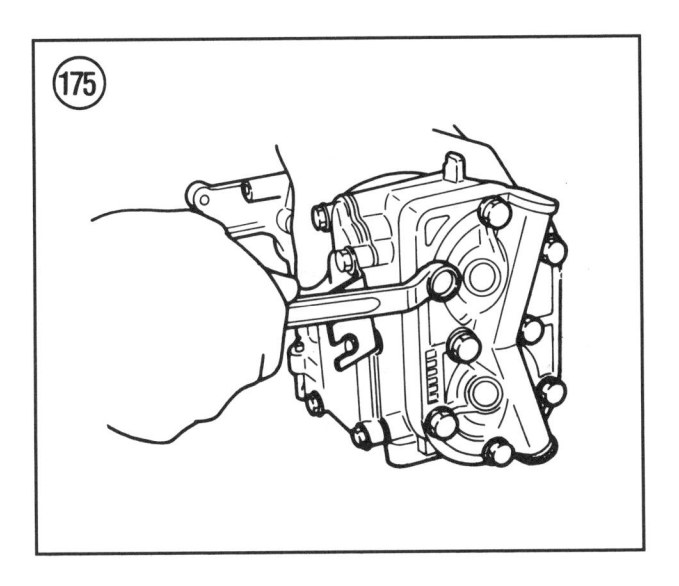

WARNING
The edges of all piston rings are very sharp. Be careful when handling them to avoid cutting fingers.

8. Carefully spread and remove the piston rings as shown in **Figure 173**.

Disassembly
(2-cylinder Engines)

A large number of bolts and screws of different lengths are used to secure the various covers and components. It is a good idea to use a cupcake tin or similar compartmented container to hold the various fasteners removed from each cover or component. This will make reassembly easier and faster.

Some power heads have pry points in the casting and are sometimes marked with a raised "O" mark adjacent to the pry point. This makes for easier removal of each component. If no pry points are provided, break the gasket seal with a wide-blade putty knife and mallet, then carefully separate the components.

DT 6, 1985-1987 DT 8

Refer to **Figure 174** for this procedure.
1. To prevent warpage of the cylinder head, loosen the cylinder head bolts 1/2 turn at a time in a crisscross pattern (**Figure 175**). Remove all bolts and lockwashers.

NOTE
*If the cylinder head is difficult to remove, use a plastic hammer or soft faced mallet and **carefully** tap around the perimeter of the cylinder head to help break it loose.*

2. Remove the cylinder head and gasket.
3. Remove the intake manifold bolts (**Figure 176**) and remove the manifold and gasket. Discard the gasket.

4. Remove the screw (**Figure 177**) holding the reed valve assembly and remove the assembly and gasket. Discard the gasket.

5. Remove the exhaust cover bolts (**Figure 178**). Remove the exhaust cover and gasket. Discard the gasket.

6. Remove the bolts and lockwashers (**Figure 179**) holding the lower oil seal housing and remove the lower oil seal housing assembly.

7. Loosen the crankcase bolts 1/2 turn at a time in a crisscross pattern (**Figure 180**) then remove all bolts and lockwashers.

NOTE
*If the crankcase is difficult to remove, use a plastic hammer or soft faced mallet and **carefully** tap around the perimeter of the crankcase to help break it loose.*

8. With the crankcase and cylinder block on a solid surface, carefully lift the crankcase off of the block.

9. Carefully lift the crankshaft assembly up and out of the cylinder block (**Figure 181**). If necessary, lightly tap on underside of crankshaft taper with a rubber mallet to break the seal.

10. Install the crankshaft assembly on a suitable power head stand, if available.

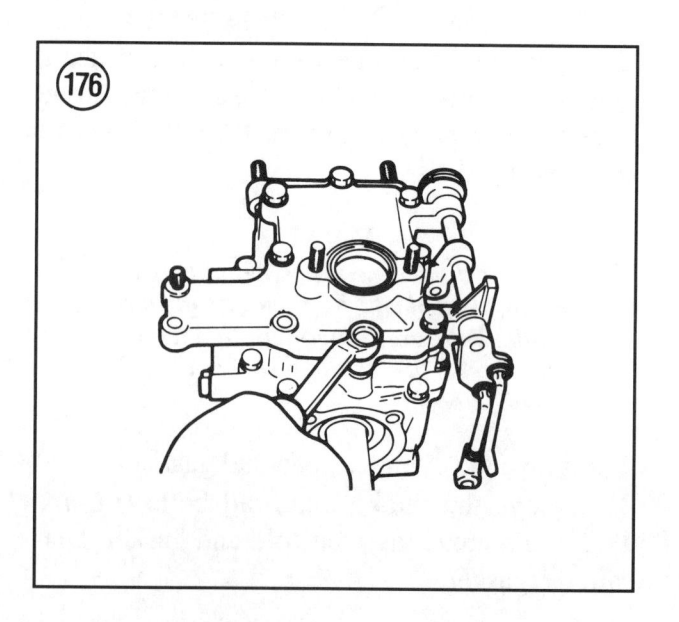

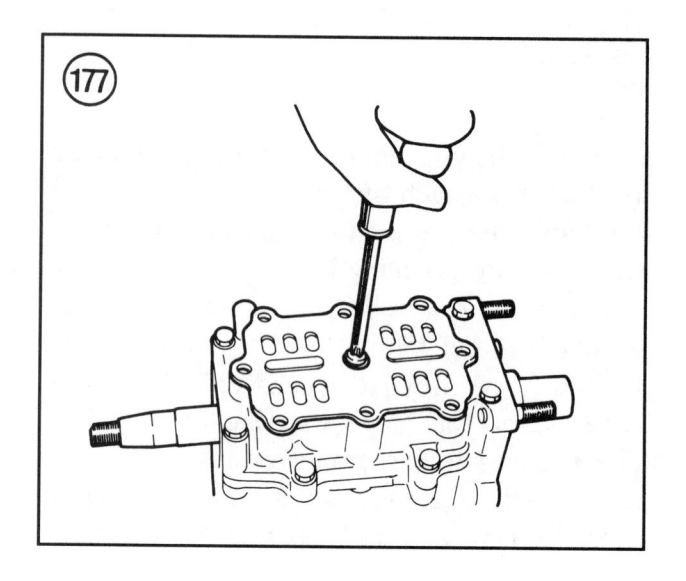

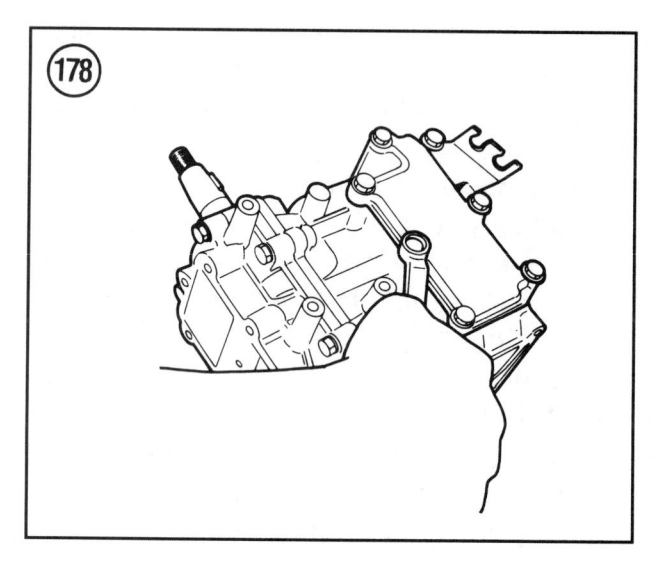

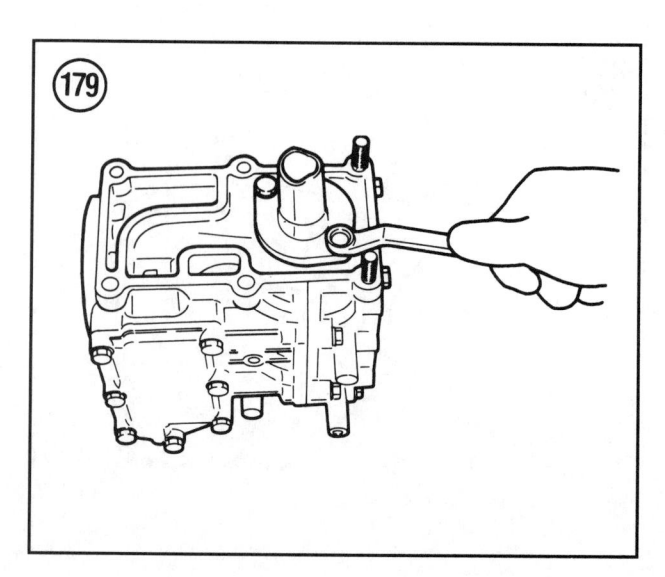

POWER HEAD

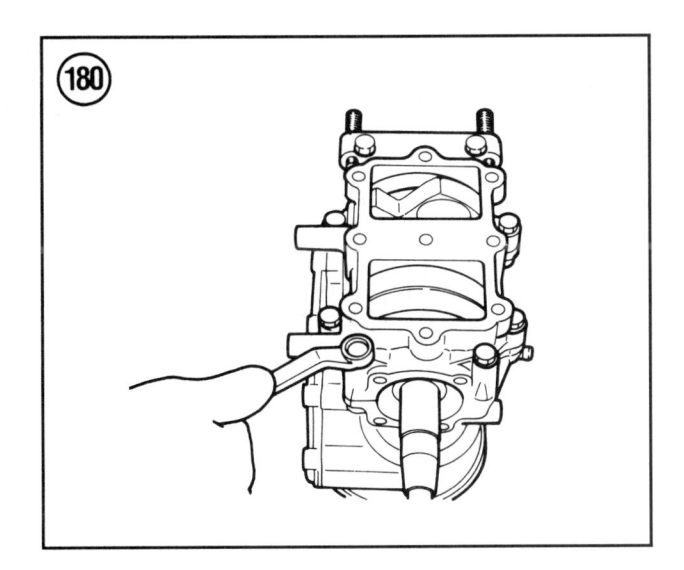

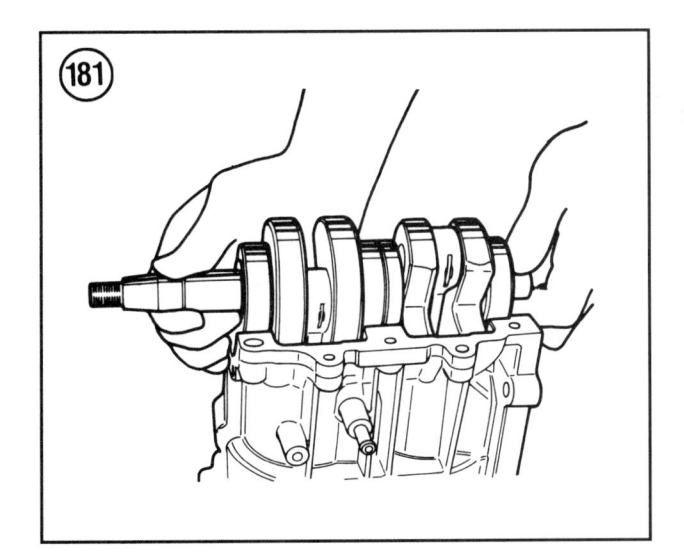

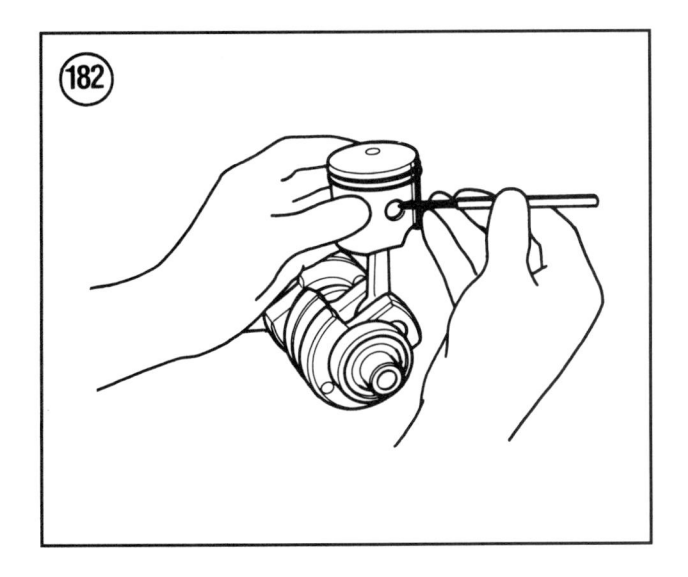

11. If the pistons are to be removed from the connecting rod:
    a. Mark the cylinder number on the top of the piston with a permanent type felt-tipped pen.
    b. Remove the piston pin circlip with a hooked tool (**Figure 182**). Repeat this step to remove the opposite piston pin circlip. Discard the circlips as they should not be used again.
    c. Push the piston pin from the piston/connecting rod with a suitable size drift (**Figure 183**).
    d. Remove the piston and caged needle bearing from the connecting rod. Keep the piston and bearing together as a set for reinstallation.
    e. Repeat for the other piston assembly.

WARNING
The edges of all piston rings are very sharp. Be careful when handling them to avoid cutting fingers.

12. Carefully spread the top ring ends with your thumbs and lift the ring up and over the piston as shown in **Figure 184**. Repeat for the lower ring and note their location in the ring grooves in the piston. The rings must be reinstalled in the same location if reused. Repeat for the other piston assembly.

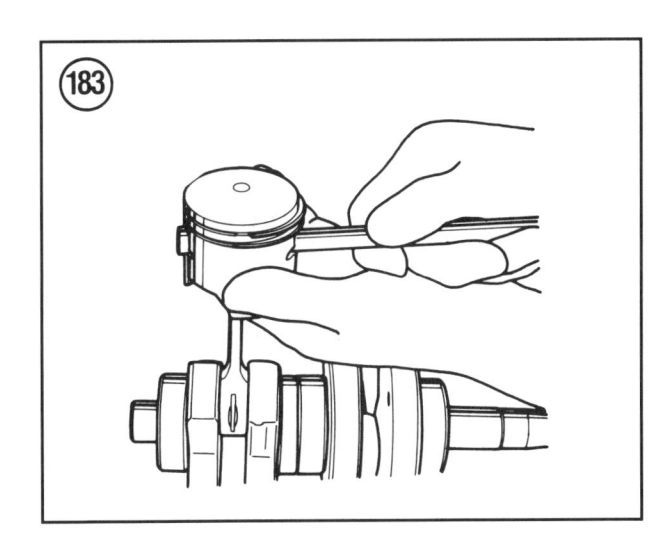

1988-on DT 8; 1985-on DT 9.9 and DT 15

Refer to **Figure 185** for 1988-on DT 8, DT 9.9 models or **Figure 186** for 1985-1987 DT 9.9 and 1985-on DT 15 models for this procedure.

1. Remove the intake manifold bolts (**Figure 187**). Remove the manifold and gasket(s). Discard the gasket(s).
2. Remove the reed valve assembly (**Figure 188**, typical). On 1985-1987 DT 9.9 and 1985-on DT 15 models, don't lose the loose piece between the reed valve assembly and intake manifold.
3. Remove the exhaust cover bolts (**Figure 189**).

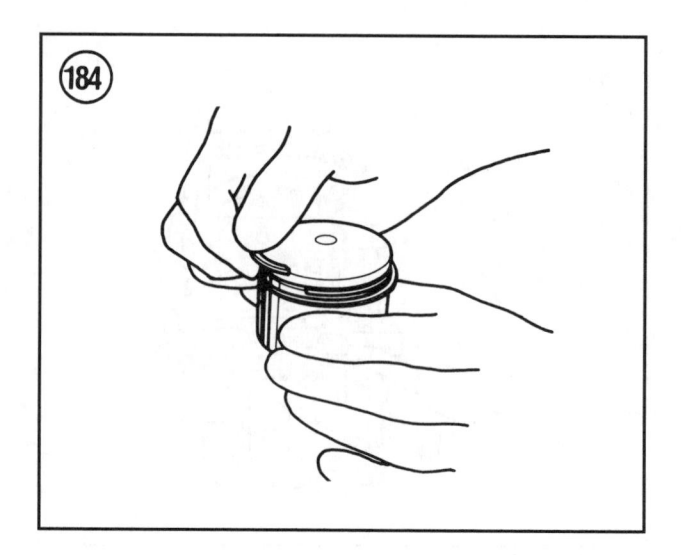

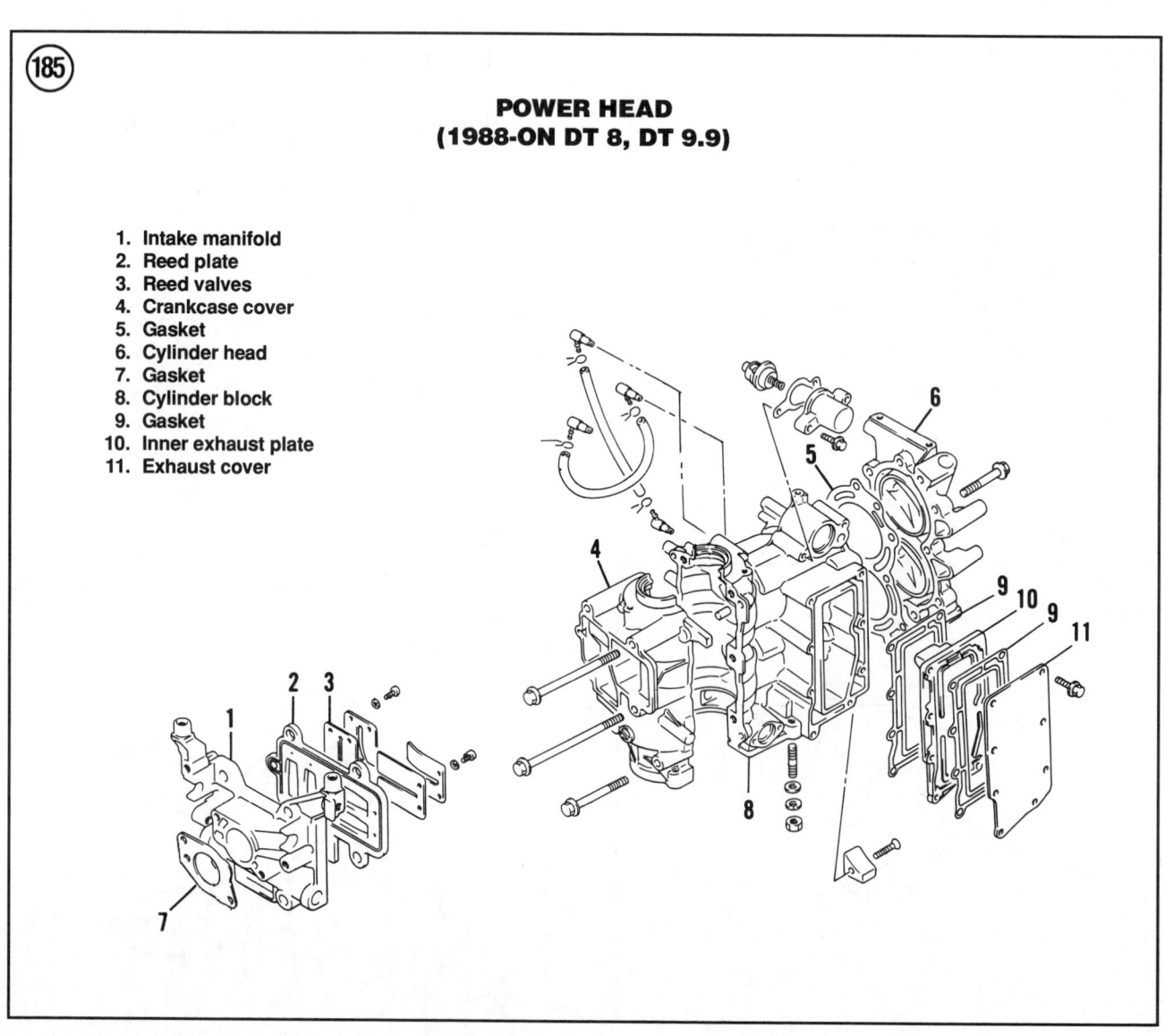

**POWER HEAD
(1988-ON DT 8, DT 9.9)**

1. Intake manifold
2. Reed plate
3. Reed valves
4. Crankcase cover
5. Gasket
6. Cylinder head
7. Gasket
8. Cylinder block
9. Gasket
10. Inner exhaust plate
11. Exhaust cover

POWER HEAD

186

**POWER HEAD
(1985-1987 DT 9.9, 1985-ON DT 15)**

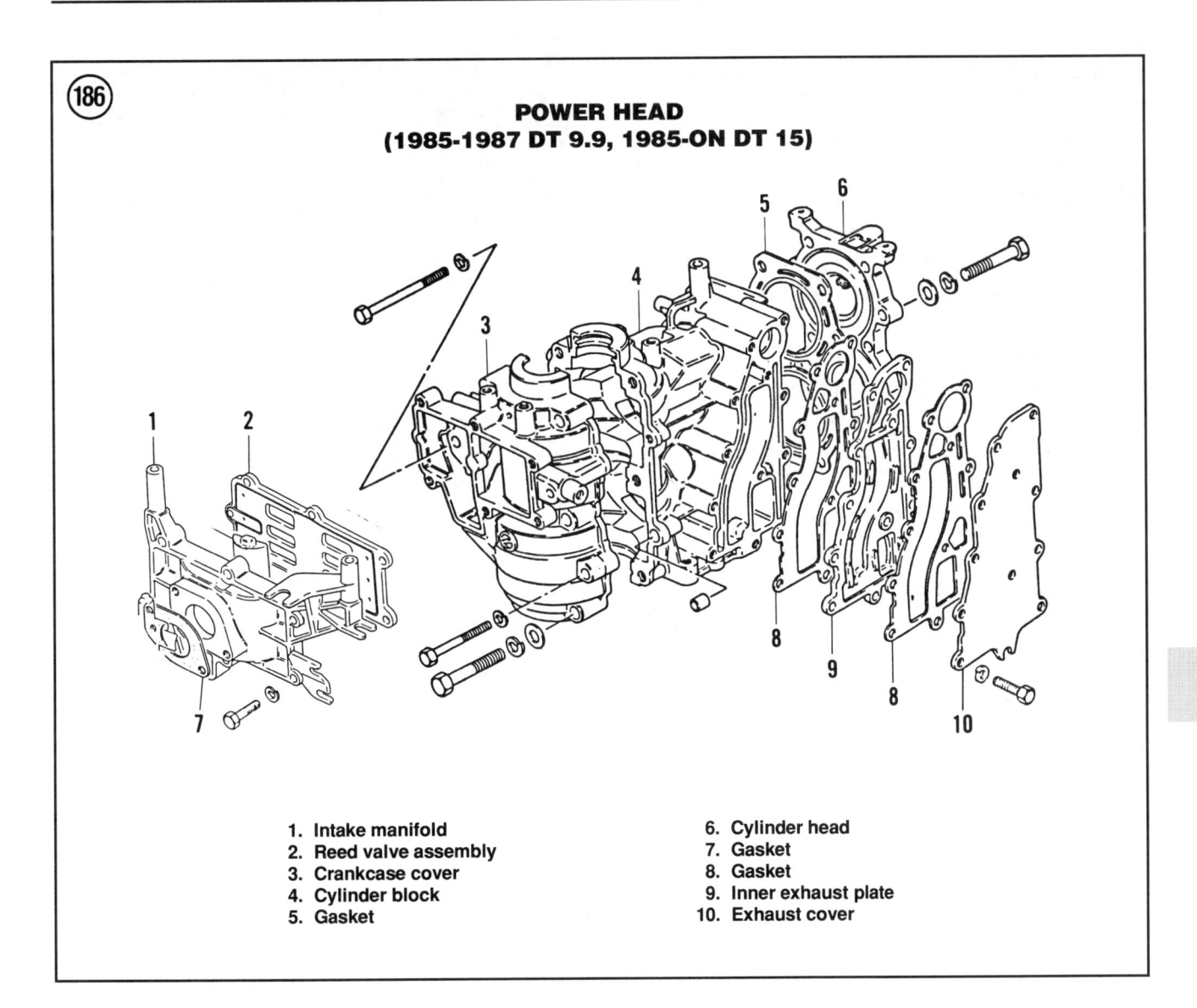

1. Intake manifold
2. Reed valve assembly
3. Crankcase cover
4. Cylinder block
5. Gasket
6. Cylinder head
7. Gasket
8. Gasket
9. Inner exhaust plate
10. Exhaust cover

187

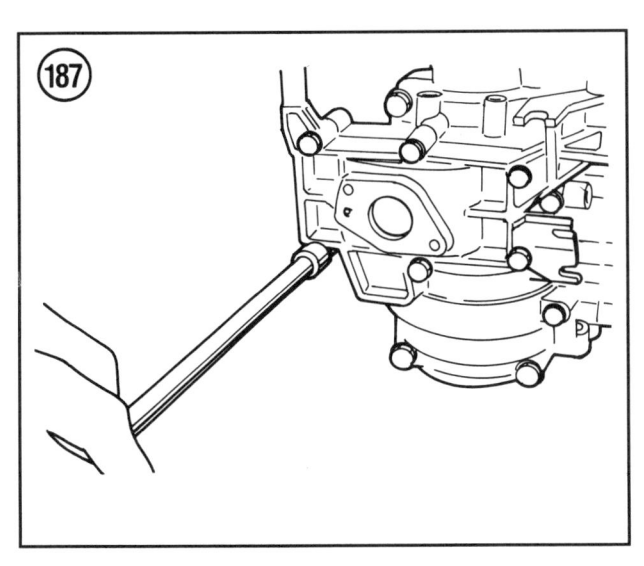

188

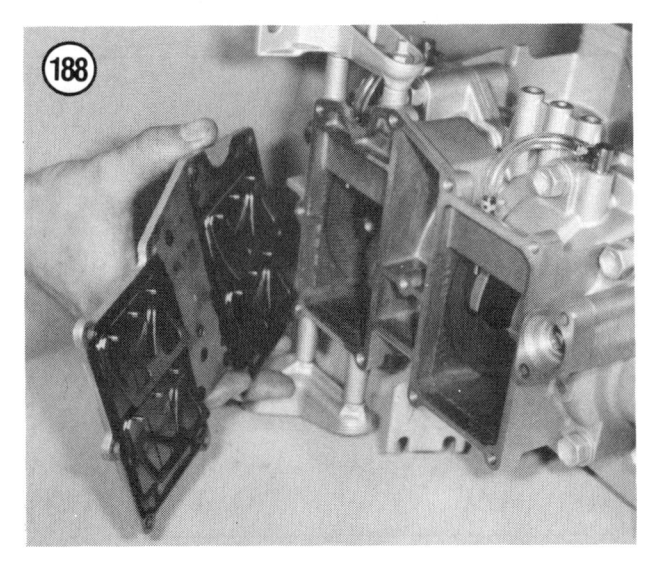

307

8

4. Insert a flat-bladed screwdriver into the pry point slots indicated by an "O" and gently pry the exhaust cover away from the crankcase (**Figure 190**). Remove the exhaust cover and gasket (**Figure 191**, typical). Discard the gasket.

5. Remove the inner exhaust plate and gasket (**Figure 192**). Discard the gasket.

6. To prevent warpage of the cylinder head, loosen the cylinder head bolts 1/2 turn at a time in a crisscross pattern (**Figure 193**, typical). Remove all bolts, noting the location of any clamps beneath the bolt heads.

NOTE
If the cylinder head is difficult to remove, use a plastic hammer or soft faced

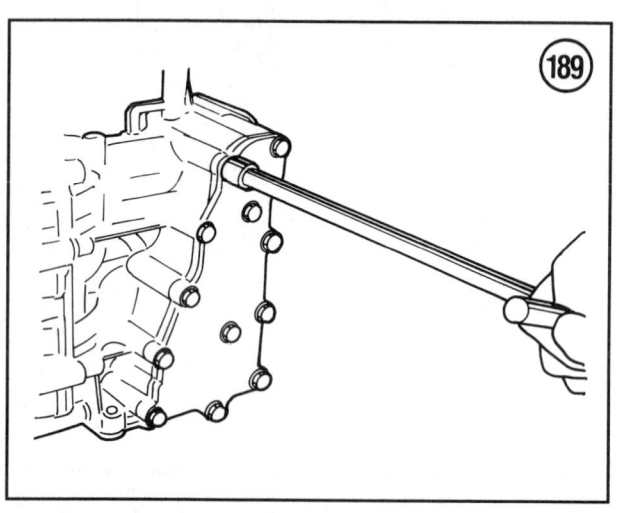

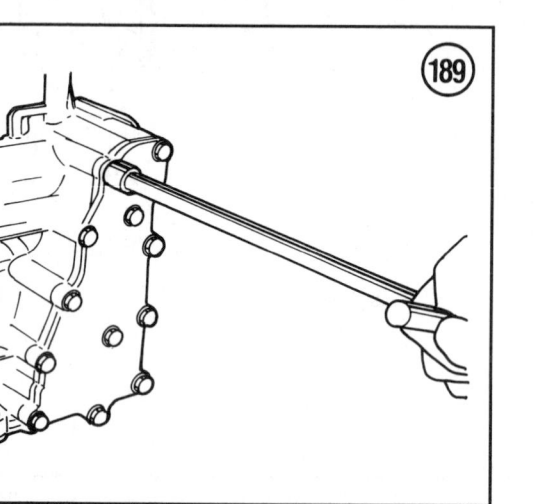

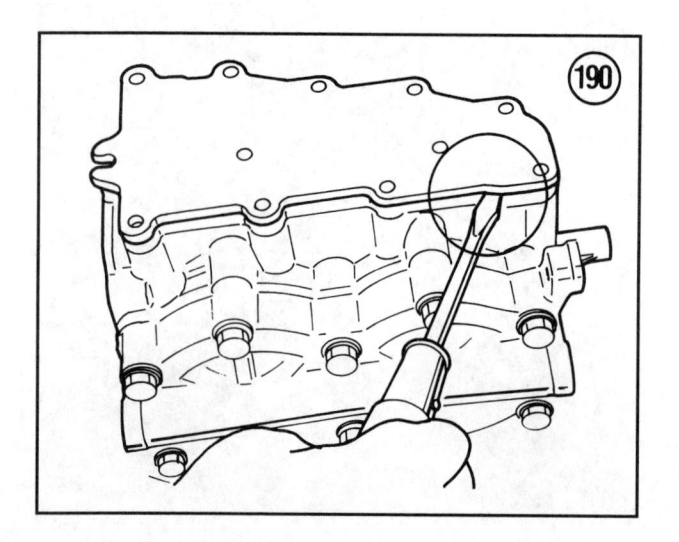

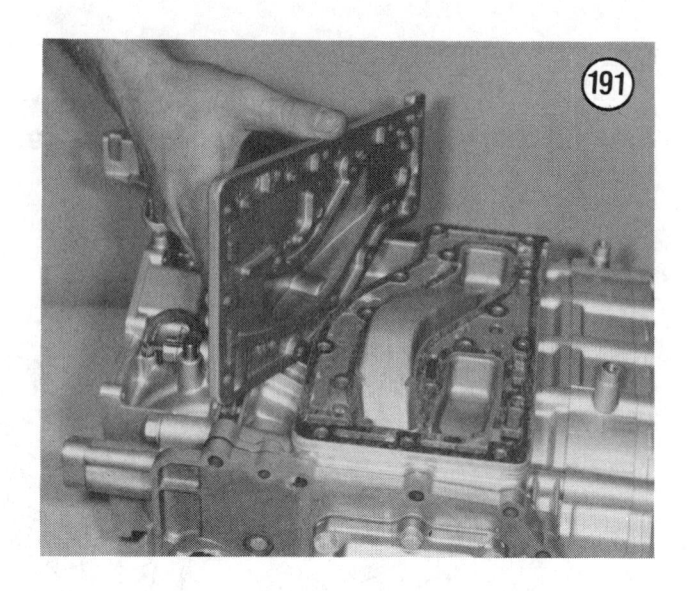

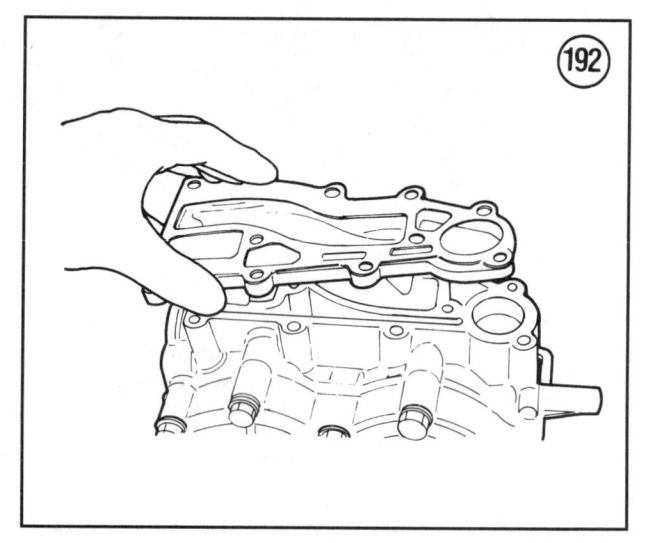

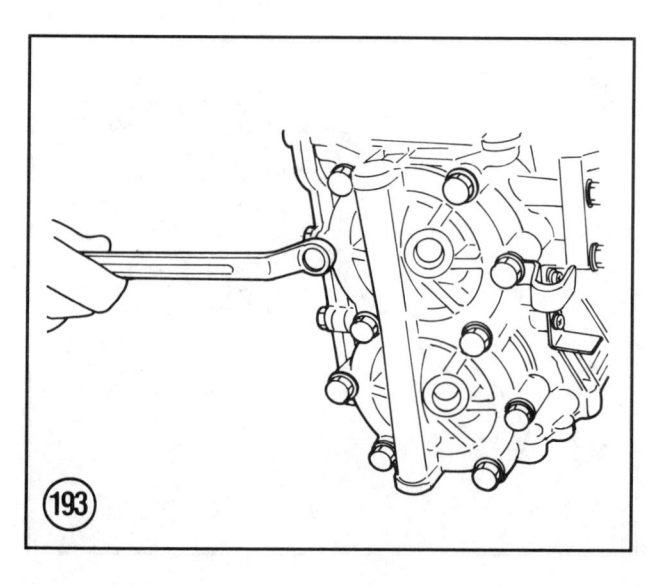

POWER HEAD

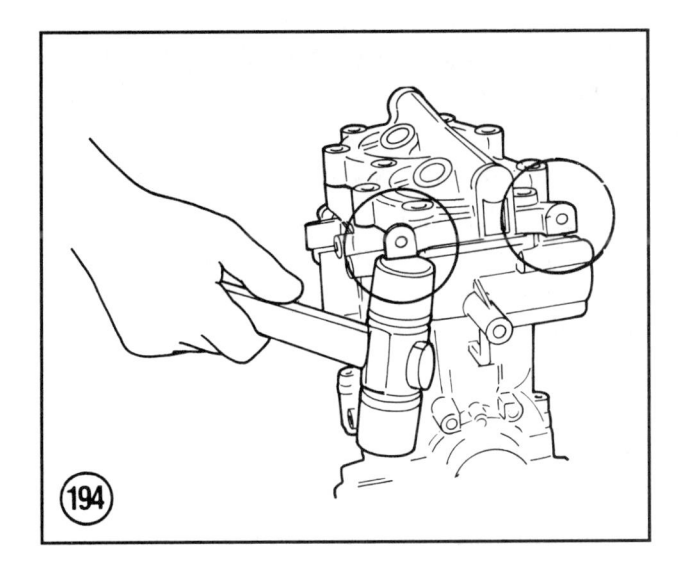

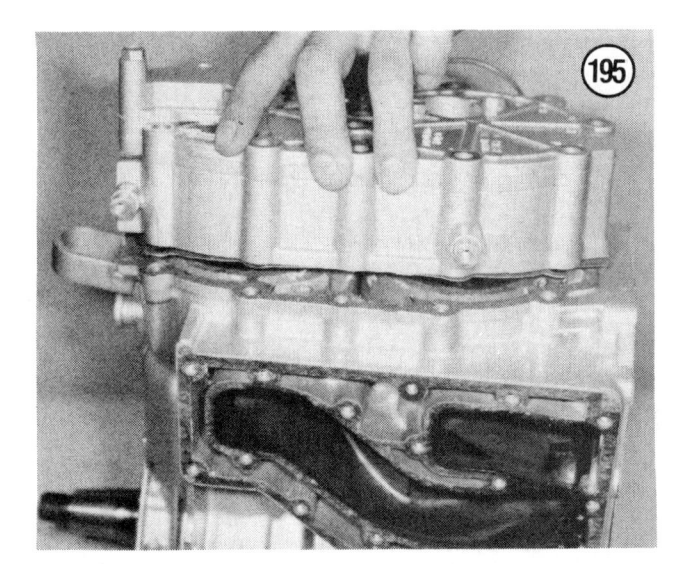

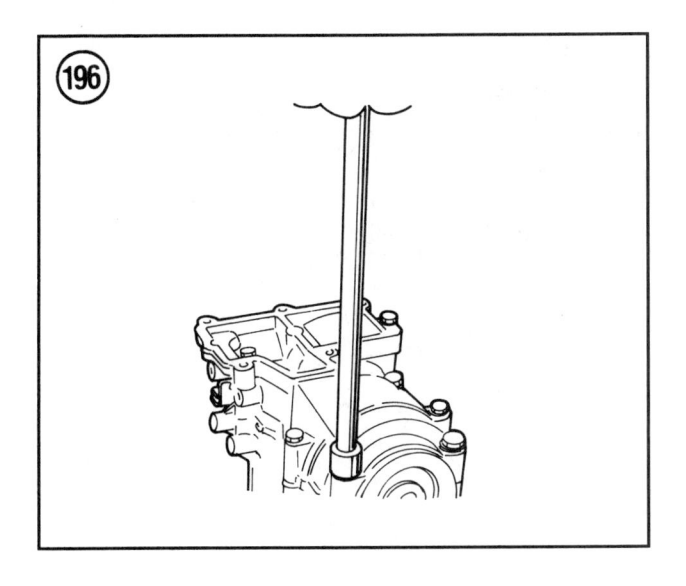

*mallet and **carefully** tap on the underside of the projections on the cylinder head as shown in **Figure 194**.*

7. Remove the cylinder head and gasket (**Figure 195**, typical). Discard the gasket.

8. Loosen the crankcase bolts 1/2 turn at a time in a crisscross pattern (**Figure 196**, typical), then remove all bolts.

9. With the crankcase and cylinder block on a solid surface, carefully lift the crankcase cover off of the block. If necessary, pry the parts apart using a suitable pry tool at the pry points provided (**Figure 197**).

10. Remove the crankcase cover from the block (**Figure 198**).

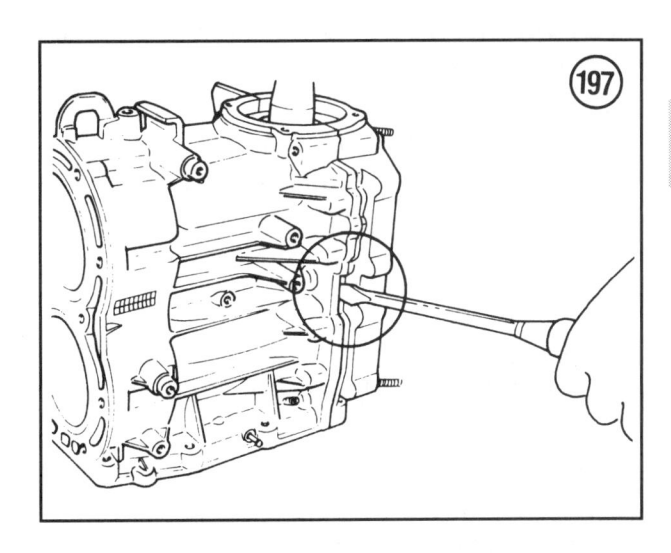

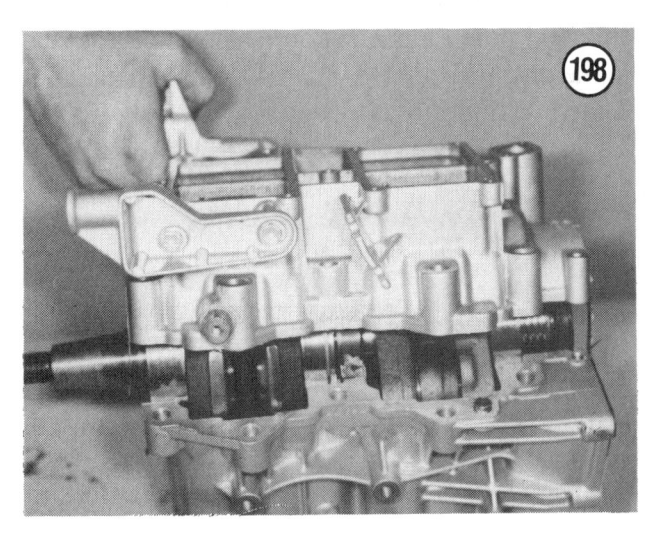

11. Carefully lift the crankshaft assembly up and out of the cylinder block (**Figure 199**). If necessary, lightly tap on underside of crankshaft taper with a rubber mallet to break the seal.

12. Install the crankshaft assembly on a suitable power head stand, if available.

13. If the pistons are to be removed from the connecting rod:
   a. Mark the cylinder number on the top of the piston with a permanent type felt-tipped pen.
   b. Remove the piston pin circlip with a suitable tool (**Figure 200**). Repeat this step to remove the opposite piston pin circlip. Discard the circlips as they should not be used again
   c. Push the piston pin from the piston/connecting rod with a suitable size drift.
   d. Remove the piston and caged needle bearing (**Figure 201**) from the connecting rod. Keep the piston and bearing together as a set for reinstallation.
   e. Repeat for the other piston assembly.

WARNING
The edges of all piston rings are very sharp. Be careful when handling them to avoid cutting fingers.

14. Carefully spread the top ring ends with your thumbs and lift the ring up and over the piston as shown in **Figure 202**. Repeat for the lower ring and note their location in the ring grooves in the piston. The rings must be reinstalled in the same location if reused.

DT 20, DT 25, DT 30 (2-cylinder); 1987-on DT 35; DT 40

Refer to **Figure 203** for DT 20, DT 25 and DT 30 models or **Figure 204** for 1987-on DT 35 and 1985-on DT 40 models for this procedure.

1. Remove the intake manifold bolts and nuts (**Figure 205**, typical). Remove the manifold and gasket(s). Discard the gasket(s).

2. Remove the screw (**Figure 206**, typical) securing the reed valve assembly and remove the assembly.

3. Remove the exhaust cover bolts (**Figure 207**).

4. Insert a flat-bladed screwdriver into the pry point slots indicated by an "O" and gently pry the exhaust cover away from the crankcase (**Figure 208**). Remove the exhaust cover and gasket. Discard the gasket.

5. Remove the inner exhaust plate and gasket (**Figure 209**). Discard the gasket.

6. On 1987-on DT 35 and 1985-on DT 40 models, perform the following:

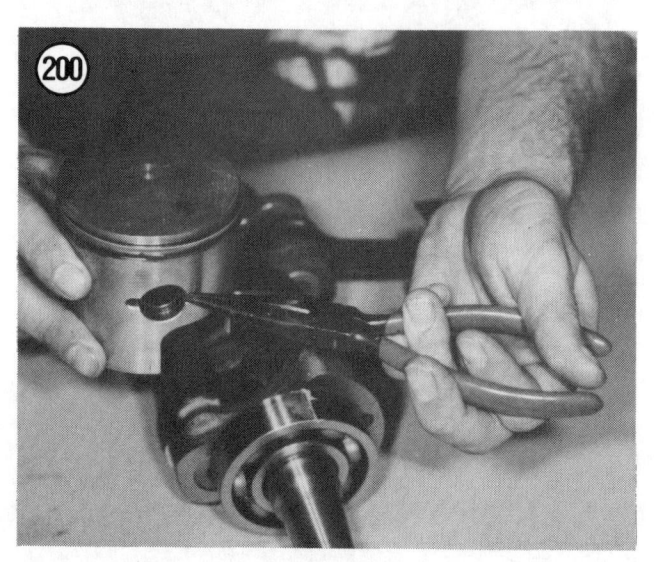

POWER HEAD

a. Remove the bolts (**Figure 210**) securing the lower oil seal housing.

b. Insert a flat-bladed screwdriver between the oil seal housing and the crankcase and gently pry the oil seal housing away from the crankcase (**Figure 211**).

c. Remove the oil seal housing (**Figure 212**).

d. Remove the bolts securing the thermostat cover (**Figure 213**) to the cylinder head and remove the cover.

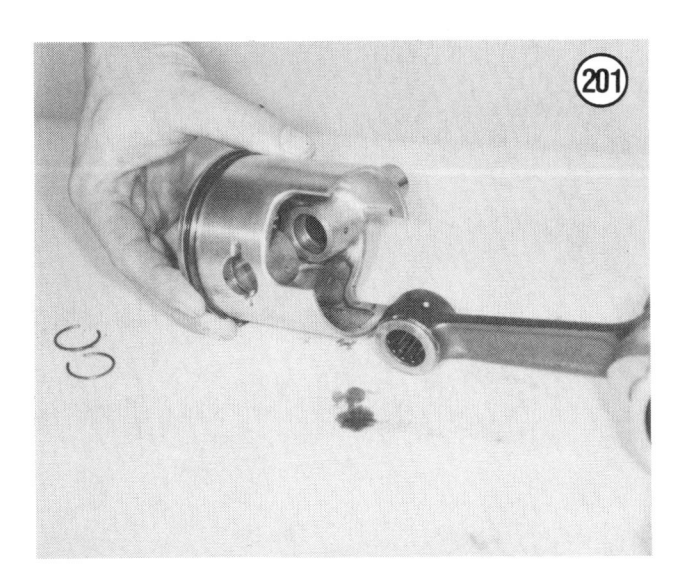

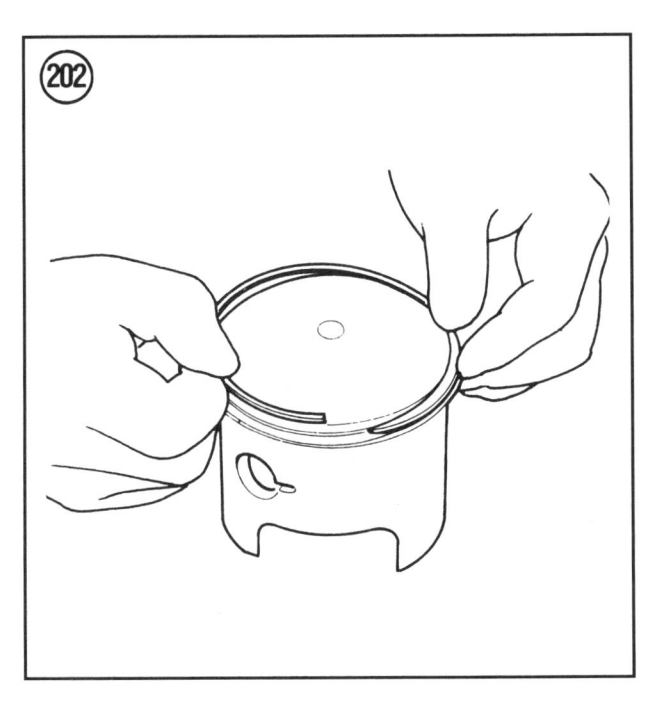

e. Remove the thermostat from the receptacle in the cylinder head (**Figure 214**).

7. To prevent warpage of the cylinder head, loosen the cylinder head bolts 1/2 turn at a time in a crisscross pattern (**Figure 215**, typical). Remove the bolts.

8. Remove the cylinder head and gasket. Discard the gasket.

NOTE
The cylinder head is a 2-piece assembly that must be disassembled.

9. Disassemble the cylinder head as follows:

a. Remove the bolts from the cylinder head (**Figure 216**, typical).

b. Insert a flat-bladed screwdriver into the pry point slots indicated by an "O" and gently pry the cylinder head cover away from the cylinder head (**Figure 217**).

10. On 1987-on DT 35 and 1985-on DT 40 models, remove the heat sensor from the cylinder head (**Figure 218**).

11. Loosen the crankcase bolts 1/2 turn at a time in a crisscross pattern (**Figure 219**), then remove all bolts.

12. With the crankcase and cylinder block on a solid surface, carefully lift the crankcase cover off of the block. If necessary, pry the cover from the block using a suitable pry tool at the pry points provided (**Figure 220**).

13. Remove the crankcase cover from the block (**Figure 198**).

14. Carefully lift the crankshaft assembly up and out of the cylinder block (**Figure 199**). If necessary, lightly tap on underside of crankshaft taper with a rubber mallet to break the seal.

15. Install the crankshaft assembly on a suitable power head stand, if available.

16. If the pistons are to be removed from the connecting rod:

a. Mark the cylinder number on the top of the piston with a permanent type felt-tipped pen.

CHAPTER EIGHT

203

**POWER HEAD
(1985-ON DT 20, 1985-1988 DT 25 AND 1985-1987 DT 30)**

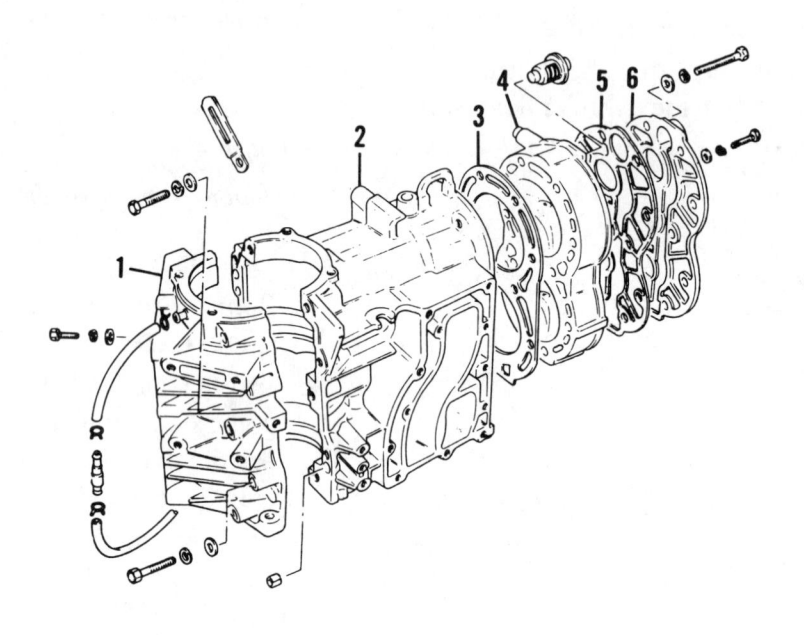

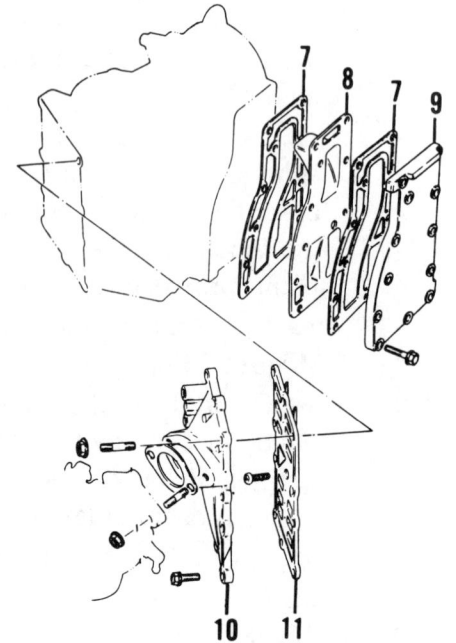

1. Crankcase cover
2. Cylinder block
3. Gasket
4. Cylinder head
5. Gasket
6. Cylinder head cover
7. Gasket
8. Inner exhaust plate
9. Exhaust cover
10. Intake manifold
11. Reed valve assembly

POWER HEAD

POWER HEAD
(1987-ON DT 35 AND 1985-ON DT 40)

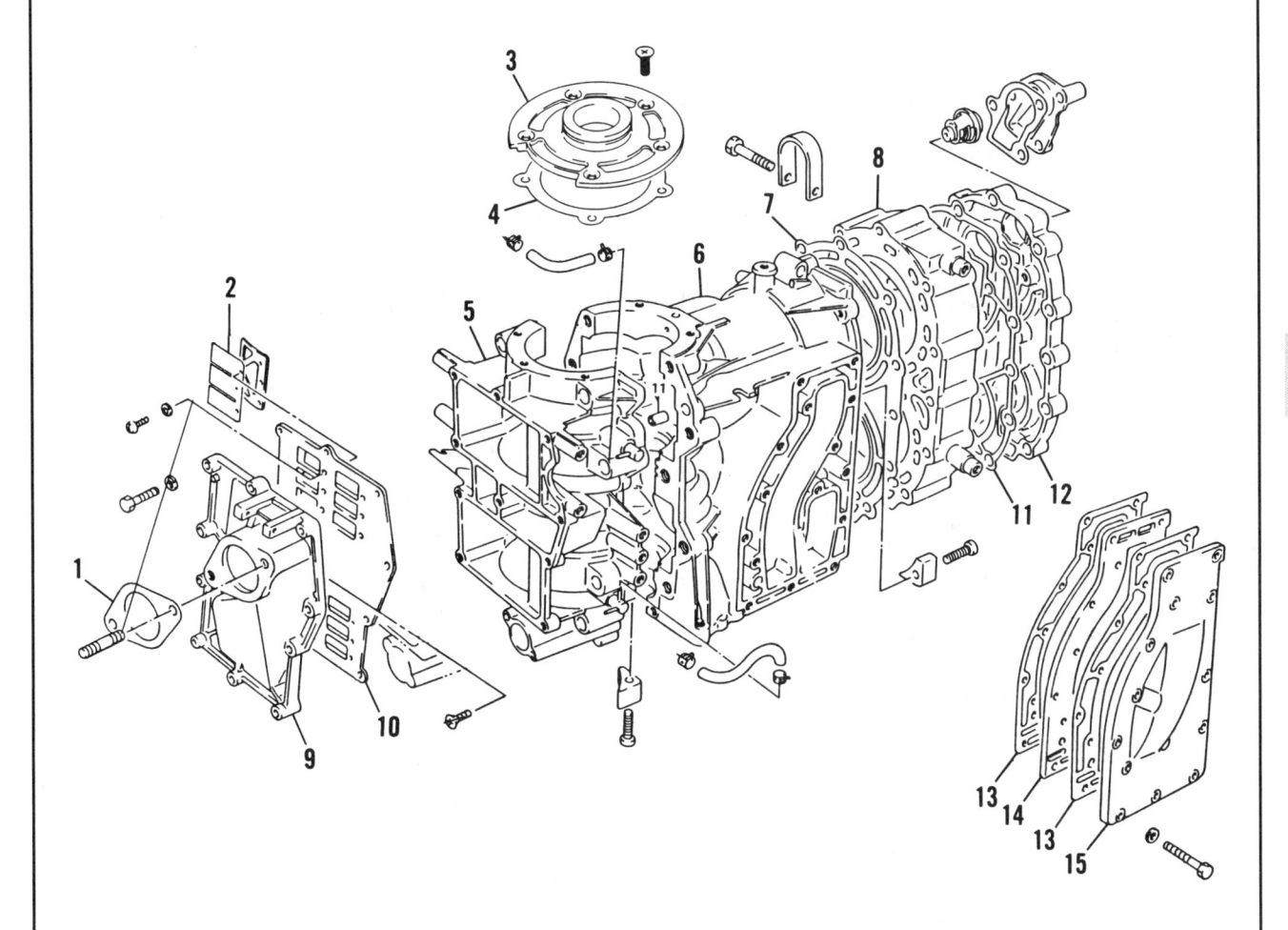

1. Gasket
2. Reed valves
3. Upper oil seal housing
4. Gasket
5. Crankcase cover
6. Cylinder block
7. Gasket
8. Cylinder head
9. Intake manifold
10. Reed plate
11. Gasket
12. Cylinder head cover
13. Gasket
14. Inner exhaust plate
15. Exhaust cover

314   **CHAPTER EIGHT**

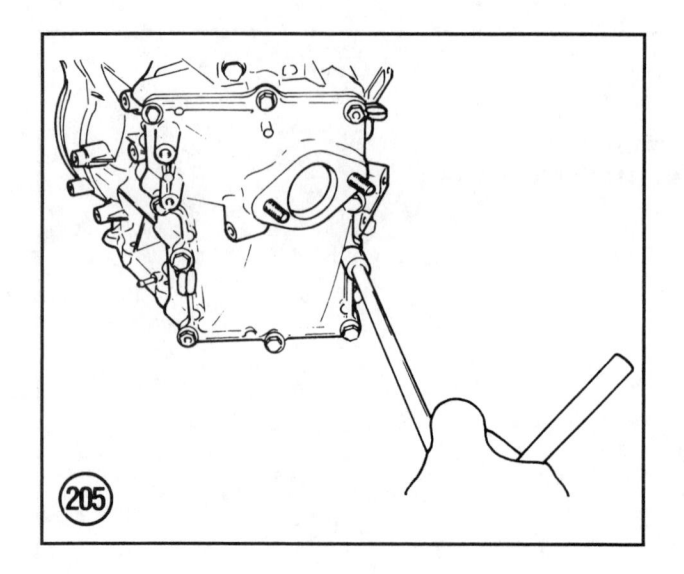

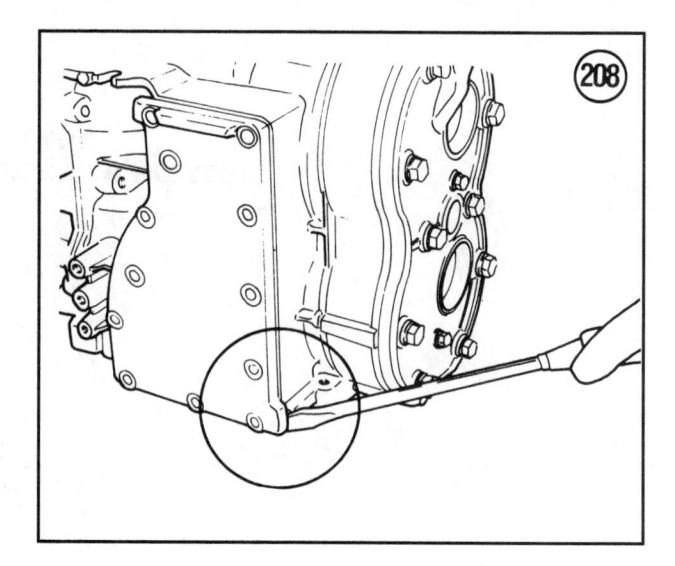

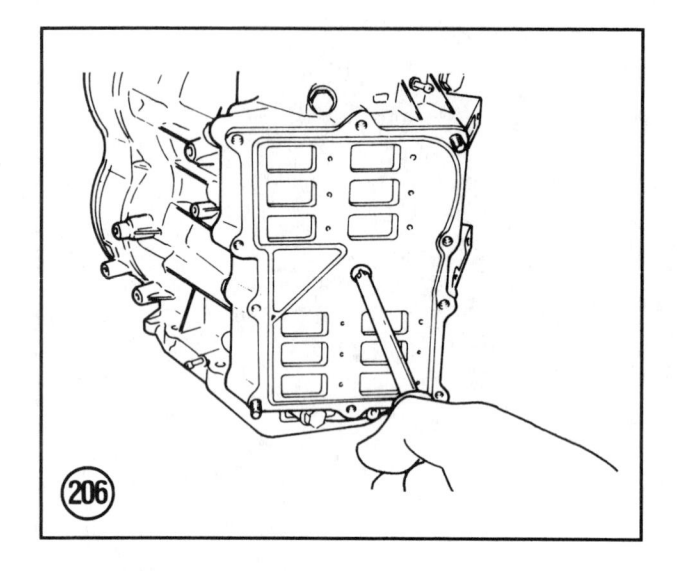

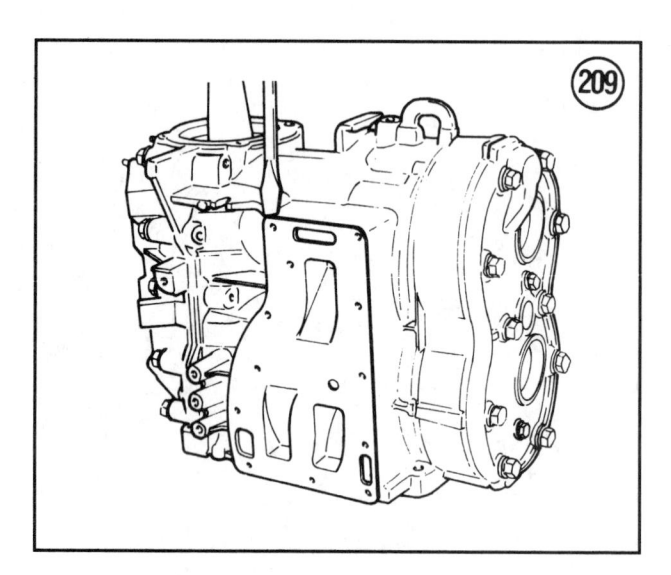

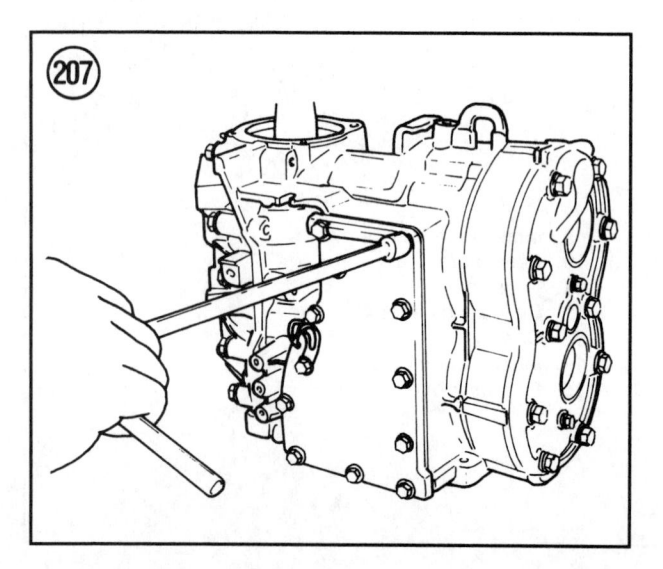

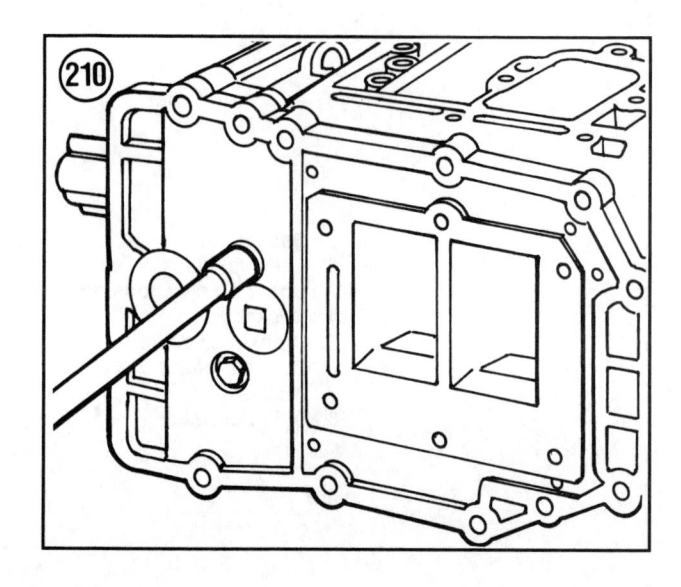

205

206

207

208

209

210

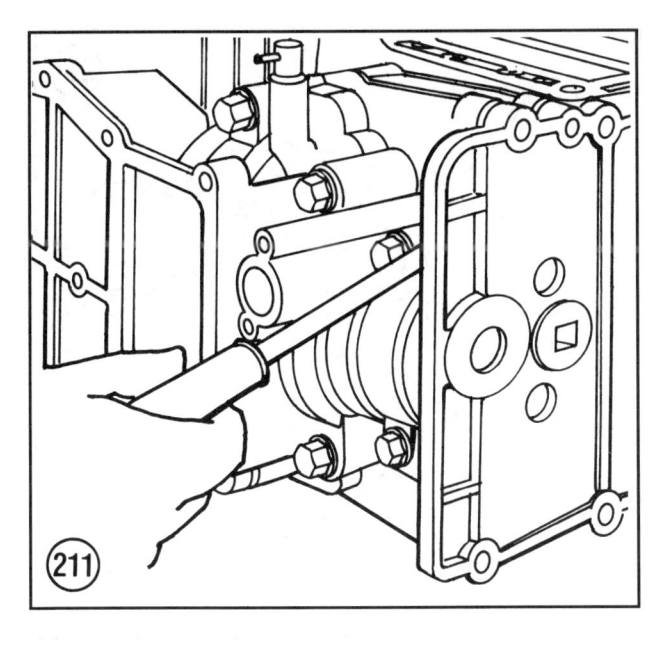

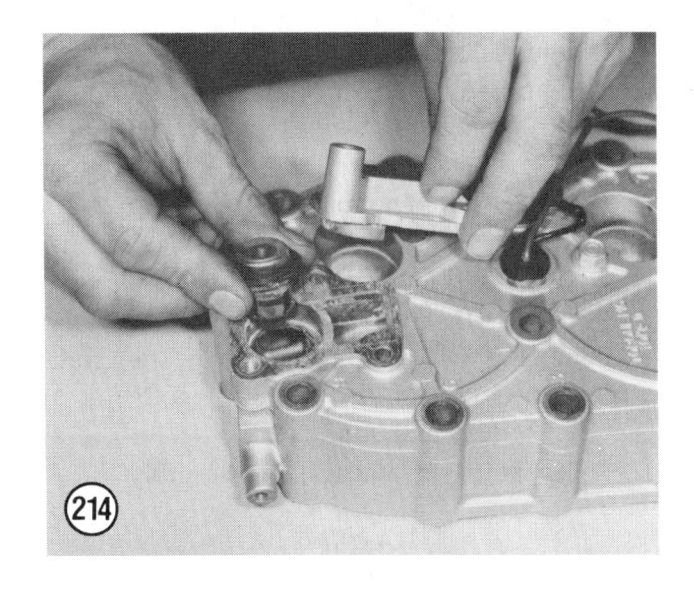

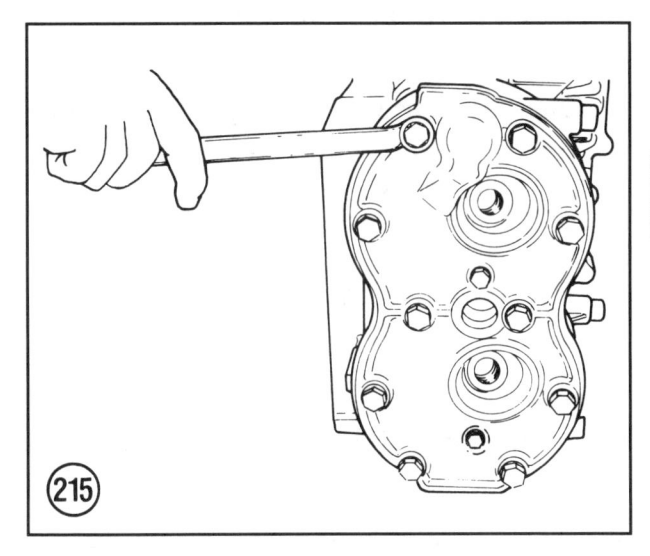

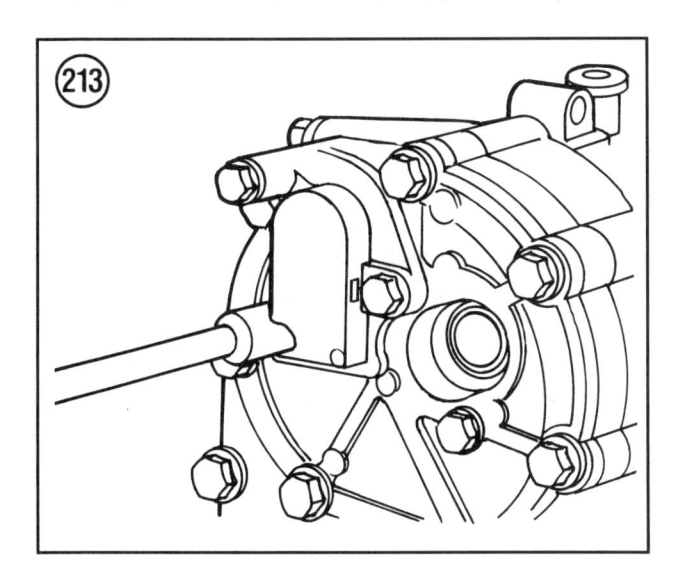

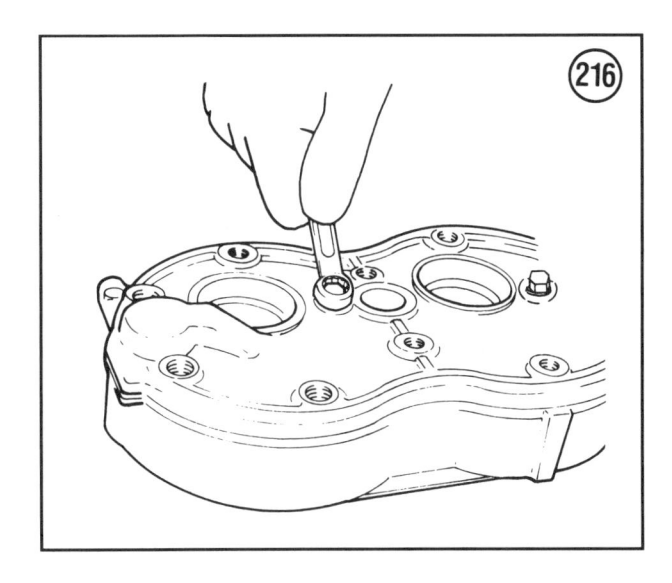

POWER HEAD

315

8

b. Remove the piston pin circlip with needlenose pliers or a hooked tool (**Figure 200**). Repeat this step to remove the opposite piston pin circlip. Discard the circlips as they should not be used again.
c. Push the piston pin from the piston/connecting rod with a suitable size drift.
d. Remove the piston and caged needle bearing (**Figure 201**) from the connecting rod. Keep the piston and bearing together as a set for reinstallation.
e. Repeat for the other piston assembly.

WARNING
The edges of all piston rings are very sharp. Be careful when handling them to avoid cutting fingers.

17. Carefully spread the top ring ends with your thumbs and lift the ring up and over the piston as shown in **Figure 202**. Repeat for the lower ring and note their location in the ring grooves in the piston. The rings must be reinstalled in the same location if reused.

Disassembly
(All 3-cylinder Engines)

A large number of bolts and screws of different lengths are used to secure the various covers

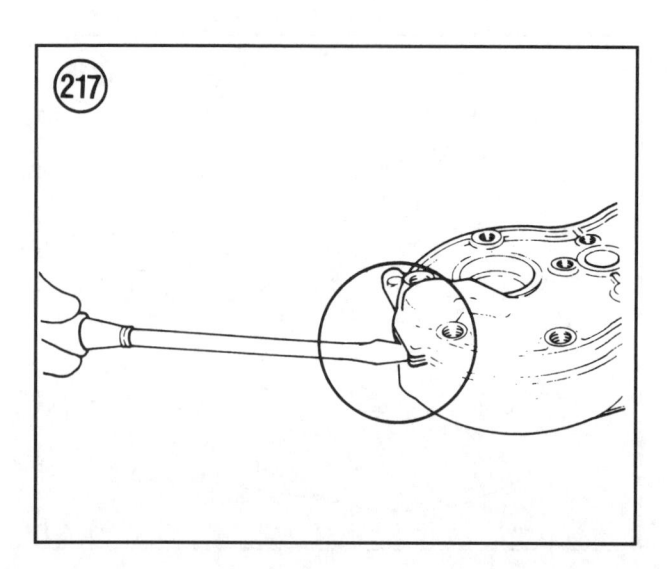

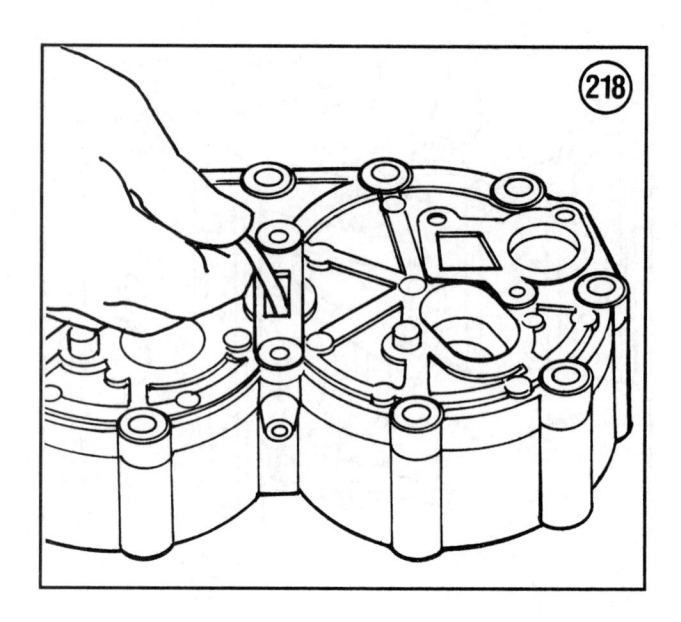

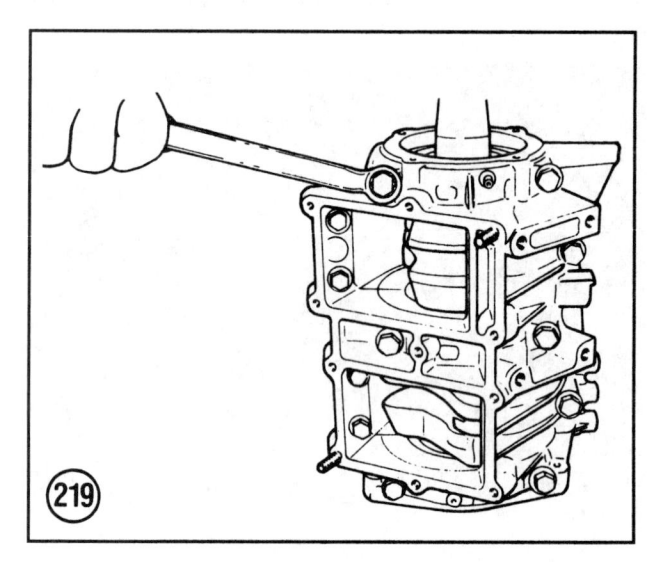

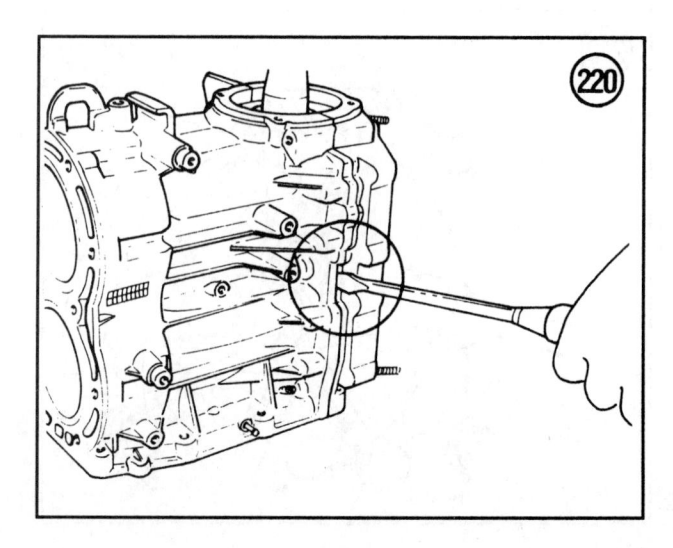

POWER HEAD

and components. It is a good idea to use a cupcake tin or similar compartmented container to hold the various fasteners removed from each cover or component. This will make reassembly easier and faster.

Some power heads have pry points in the casting for easier removal of each component. If no pry points are provided, break the gasket seal with a wide-blade putty knife and mallet, then carefully separate the components.

Refer to the following illustrations for this procedure:

a. **Figure 221**: 1989-on DT 25 and 1988-on DT 30.
b. **Figure 222**: 1985-on DT 55 and DT 65.
c. **Figure 223**: 1985-on DT 75 and DT 85.

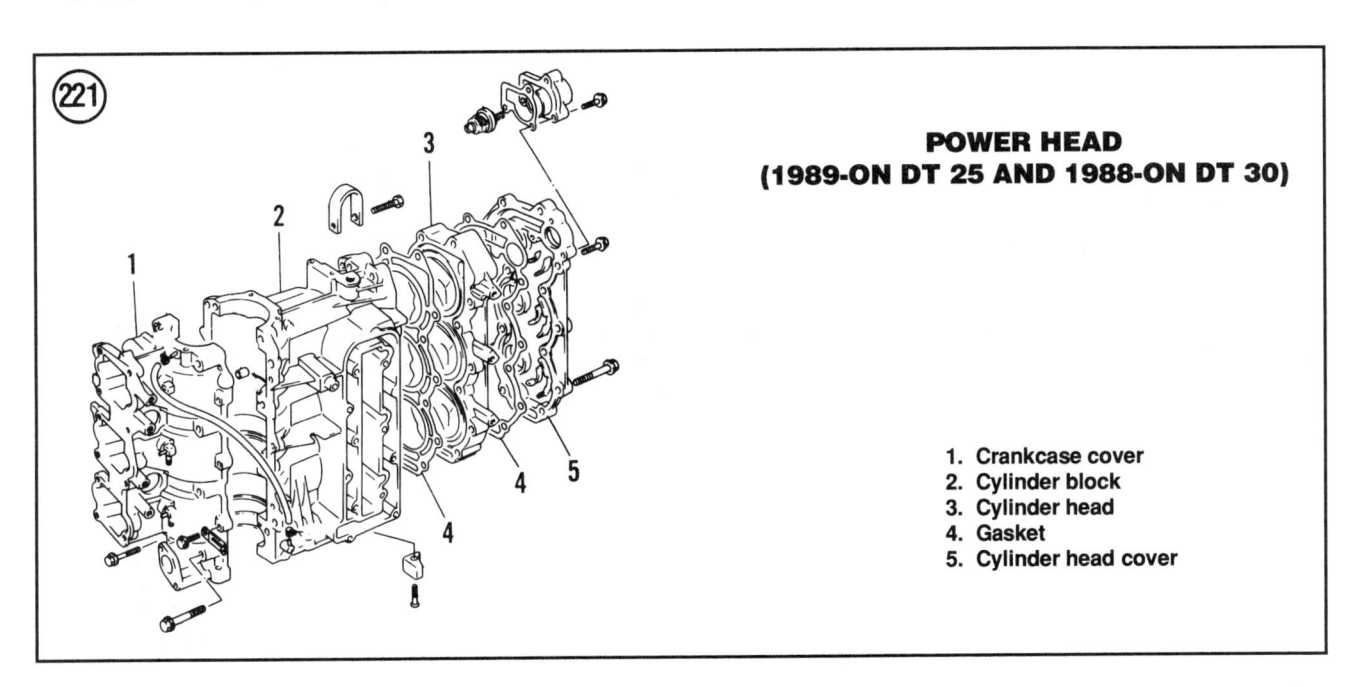

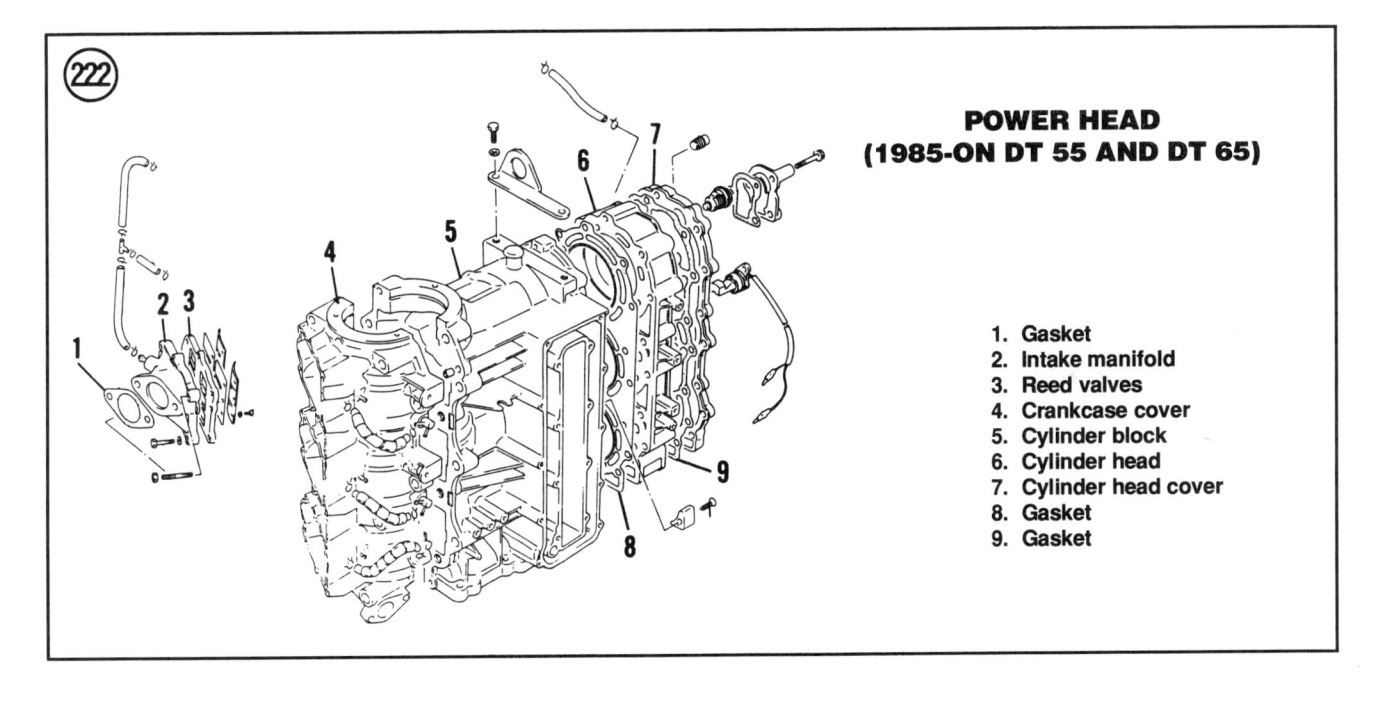

POWER HEAD (1989-ON DT 25 AND 1988-ON DT 30)

1. Crankcase cover
2. Cylinder block
3. Cylinder head
4. Gasket
5. Cylinder head cover

POWER HEAD (1985-ON DT 55 AND DT 65)

1. Gasket
2. Intake manifold
3. Reed valves
4. Crankcase cover
5. Cylinder block
6. Cylinder head
7. Cylinder head cover
8. Gasket
9. Gasket

CHAPTER EIGHT

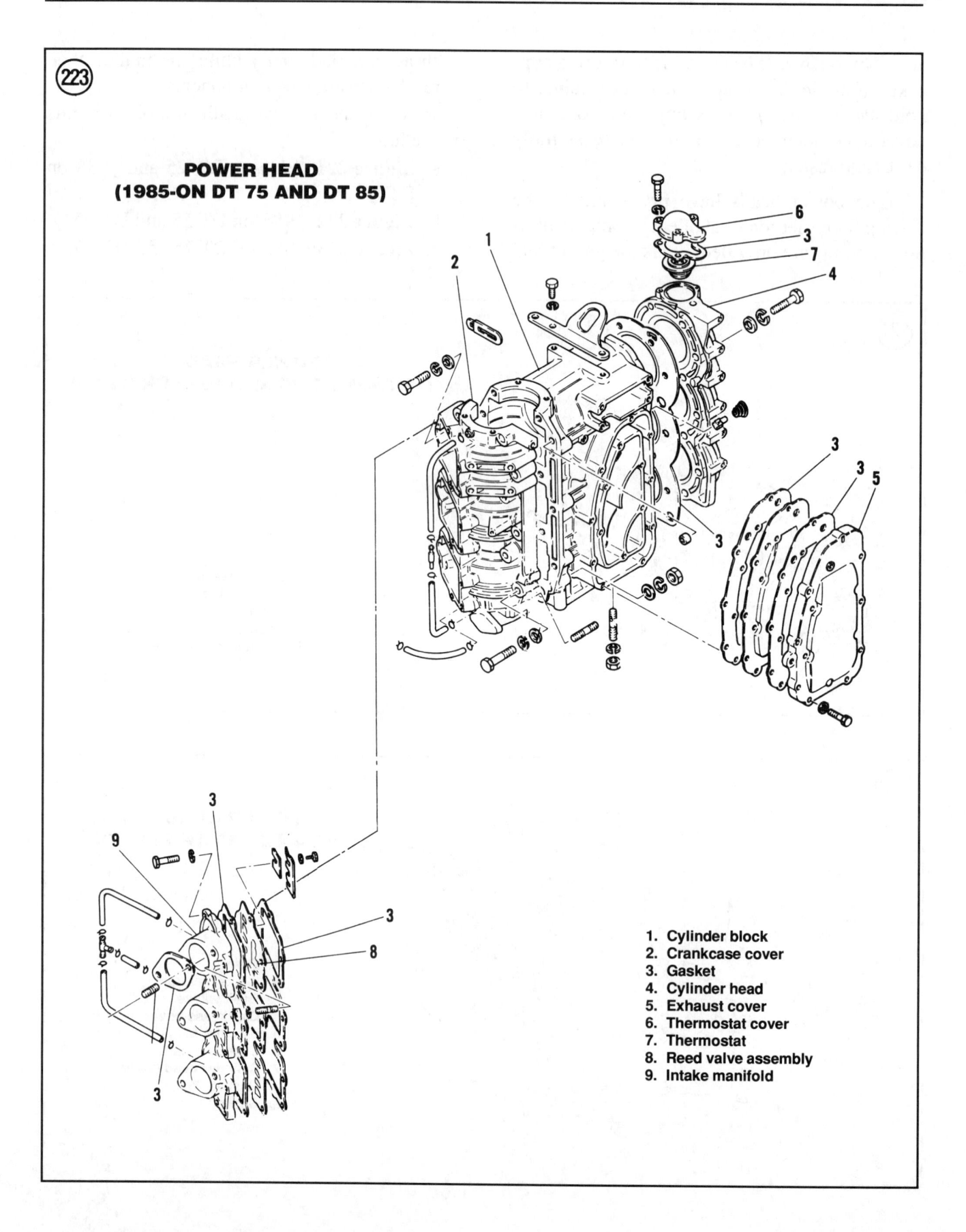

223

**POWER HEAD
(1985-ON DT 75 AND DT 85)**

1. Cylinder block
2. Crankcase cover
3. Gasket
4. Cylinder head
5. Exhaust cover
6. Thermostat cover
7. Thermostat
8. Reed valve assembly
9. Intake manifold

POWER HEAD

1. If not already removed, use pliers and slide each lubrication tube hose clamp away from the fitting enough to disconnect the tube. Pull the tube off the intake manifold (**Figure 224**).

2. Remove the intake manifold bolts (**Figure 225**, typical). Remove the manifold and the reed valve assembly.

3. Remove the screws holding the thermostat cover to the cylinder head (**Figure 226**, typical). Remove the cover, gasket and thermostat. Discard the gasket.

4. On models so equipped, remove the bolts securing the cooling water flow sensor to the cylinder head and remove the sensor assembly.

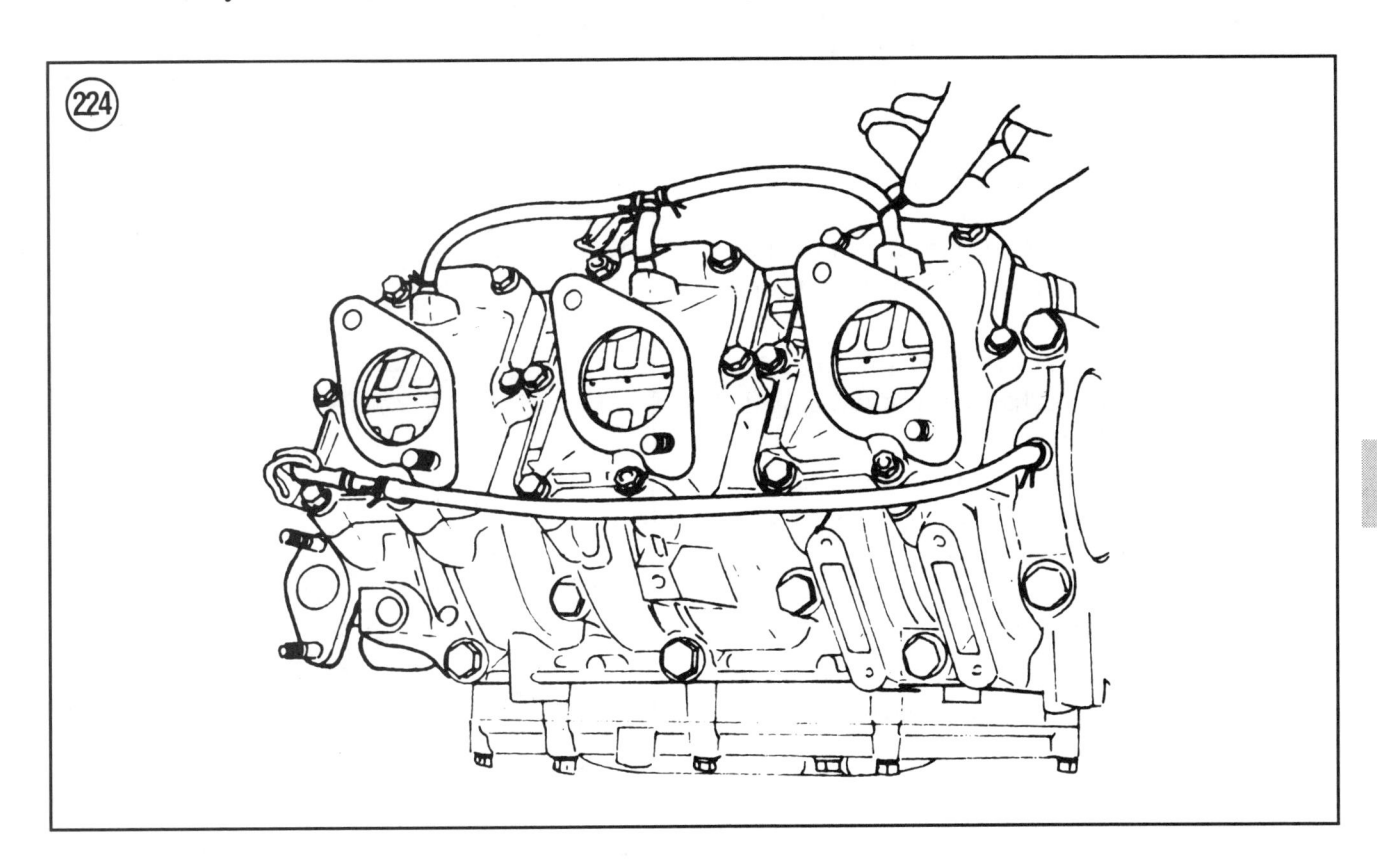

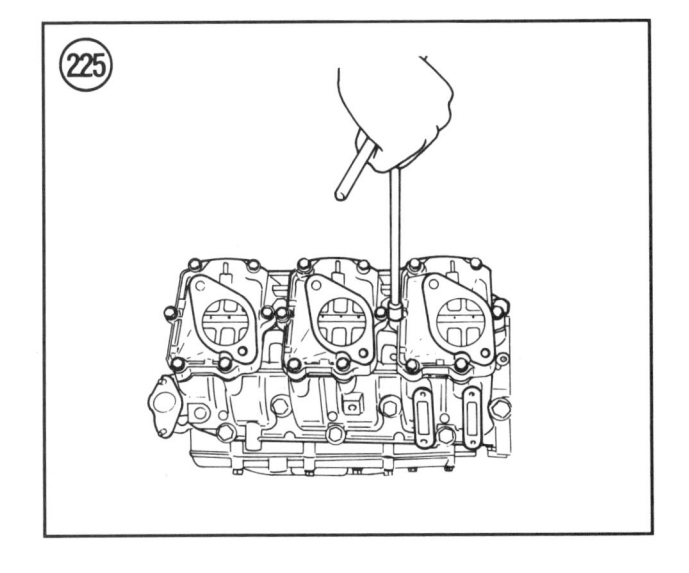

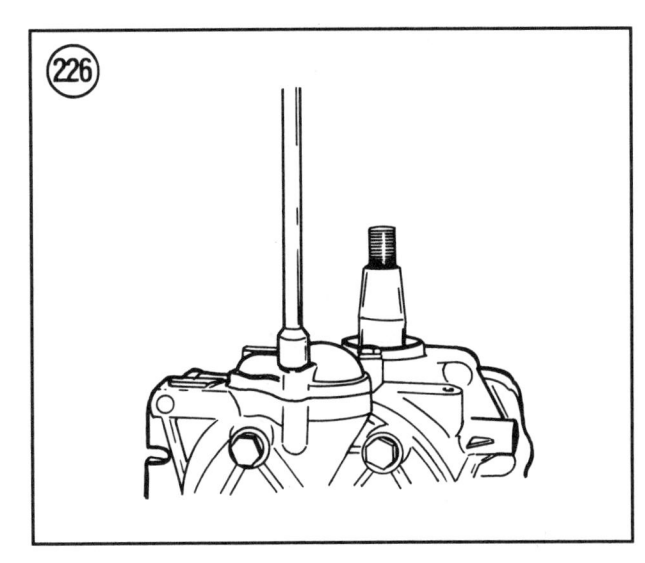

NOTE
On some models there is a number cast into the cylinder head adjacent to each cylinder head bolt hole. If your model has these numbers, loosen the bolts in reverse order of the numbers (e.g. 18, 17, 16, etc. all the way down to 1).

5. To prevent warpage of the cylinder head, loosen the cylinder head bolts 1/2 turn at a time in a crisscross pattern (**Figure 227**, typical). Remove the bolts.

6. Remove the cylinder head and gasket. Discard the gasket.

7. On models so equipped, remove the water valve (A, **Figure 228**) and the spring (B, **Figure 228**) from the cylinder block.

8. Remove the exhaust cover bolts (**Figure 229**, typical). Remove the exhaust cover and gasket. Discard the gasket.

9. Insert a flat-bladed screwdriver into the pry point slots indicated by an "O" and gently pry the inner exhaust cover plate away from the crankcase. Remove the exhaust cover and gasket. Discard the gasket.

10A. On DT 75 and DT 85 models, remove the crankcase cover bolts and nut (**Figure 230**, typical).

10B. On all other models, remove the crankcase cover bolts. Note that the bolts vary in thread diameter. Be sure to install the correct size bolt in the correct hole during assembly.

11. With the crankcase cover and cylinder block on a solid surface, carefully pry the crankcase cover and block apart using a suitable pry tool at the pry points provided.

12. Remove the crankcase cover from the cylinder block.

13. Carefully lift the crankshaft assembly up and out of the cylinder block (**Figure 231**). If neces-

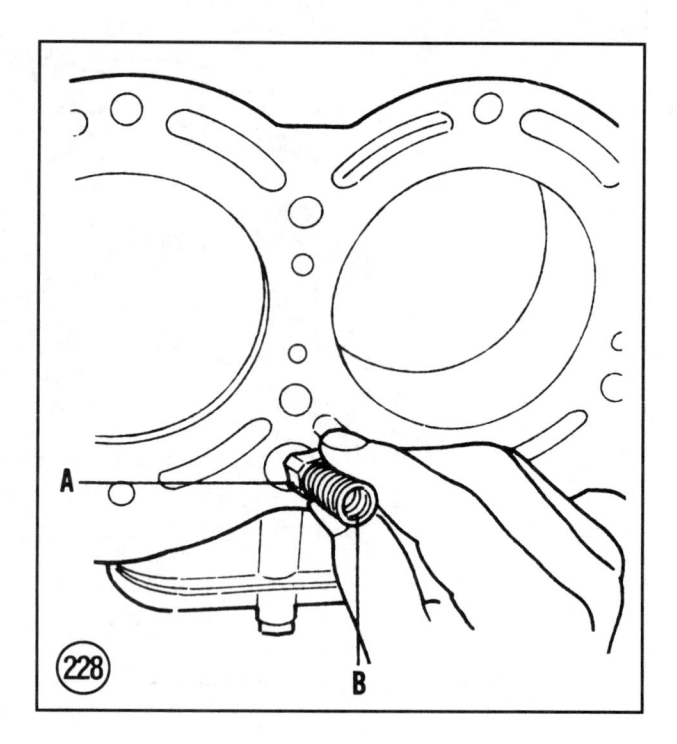

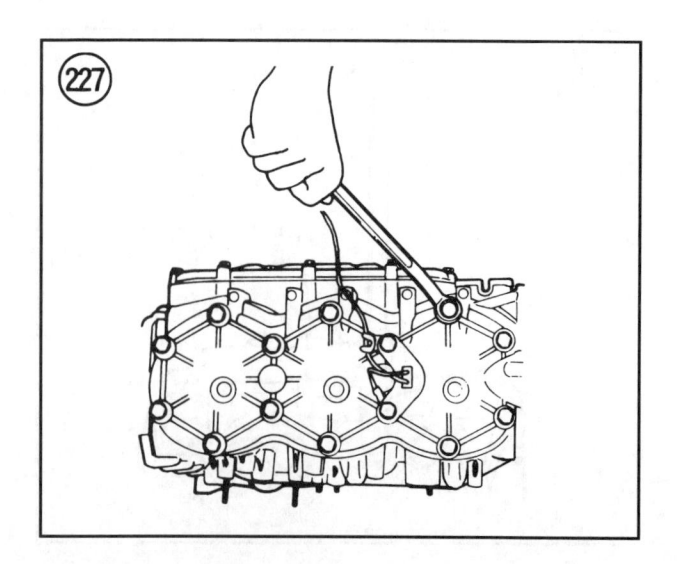

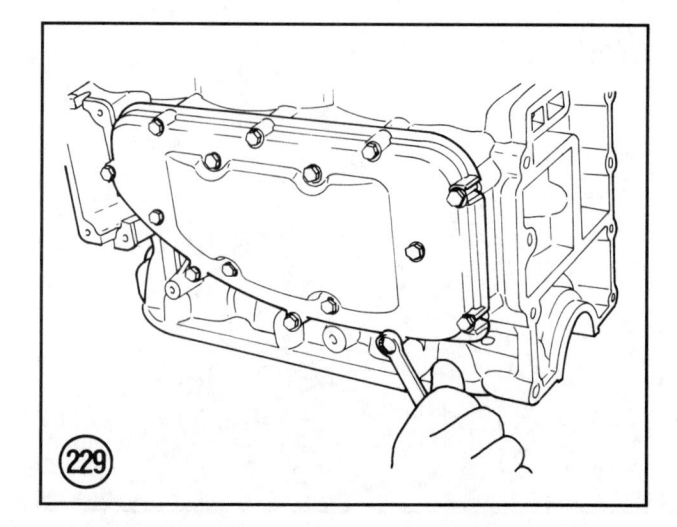

POWER HEAD

sary, lightly tap on underside of crankshaft taper with a rubber mallet to break the seal.

14. Install the crankshaft assembly on a suitable power head stand, if available.

15. Remove the upper and lower crankshaft oil seals (**Figure 232**).

16. If the pistons are to be removed from the connecting rod:

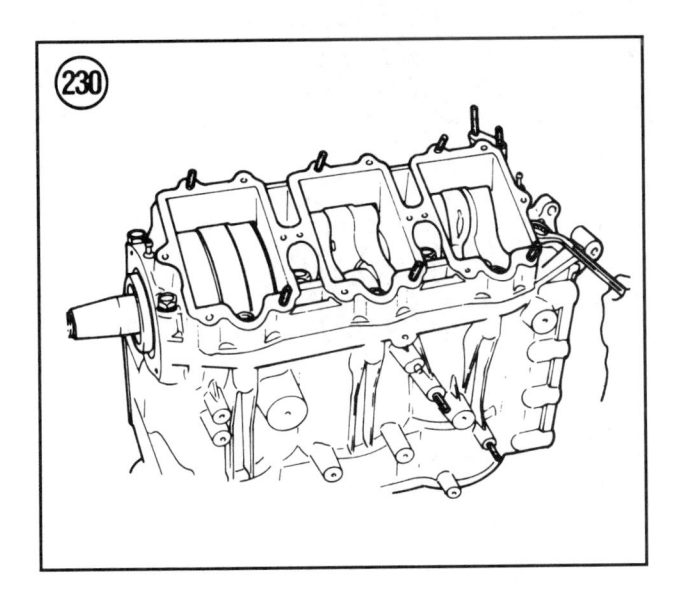

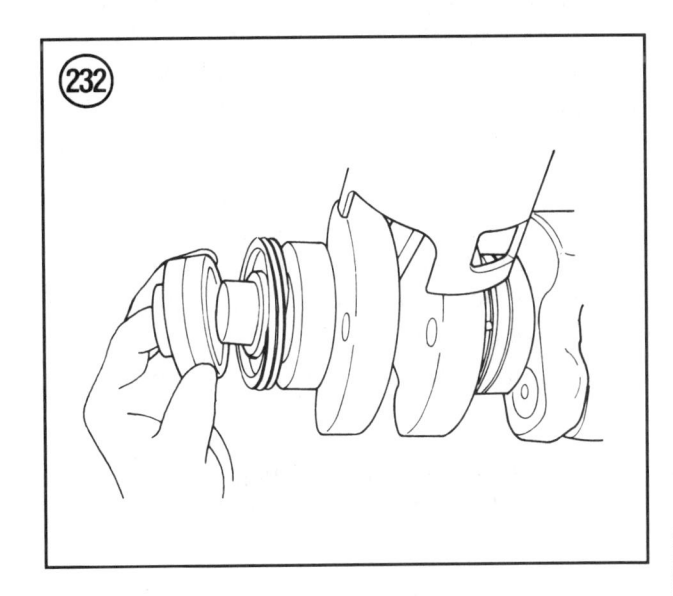

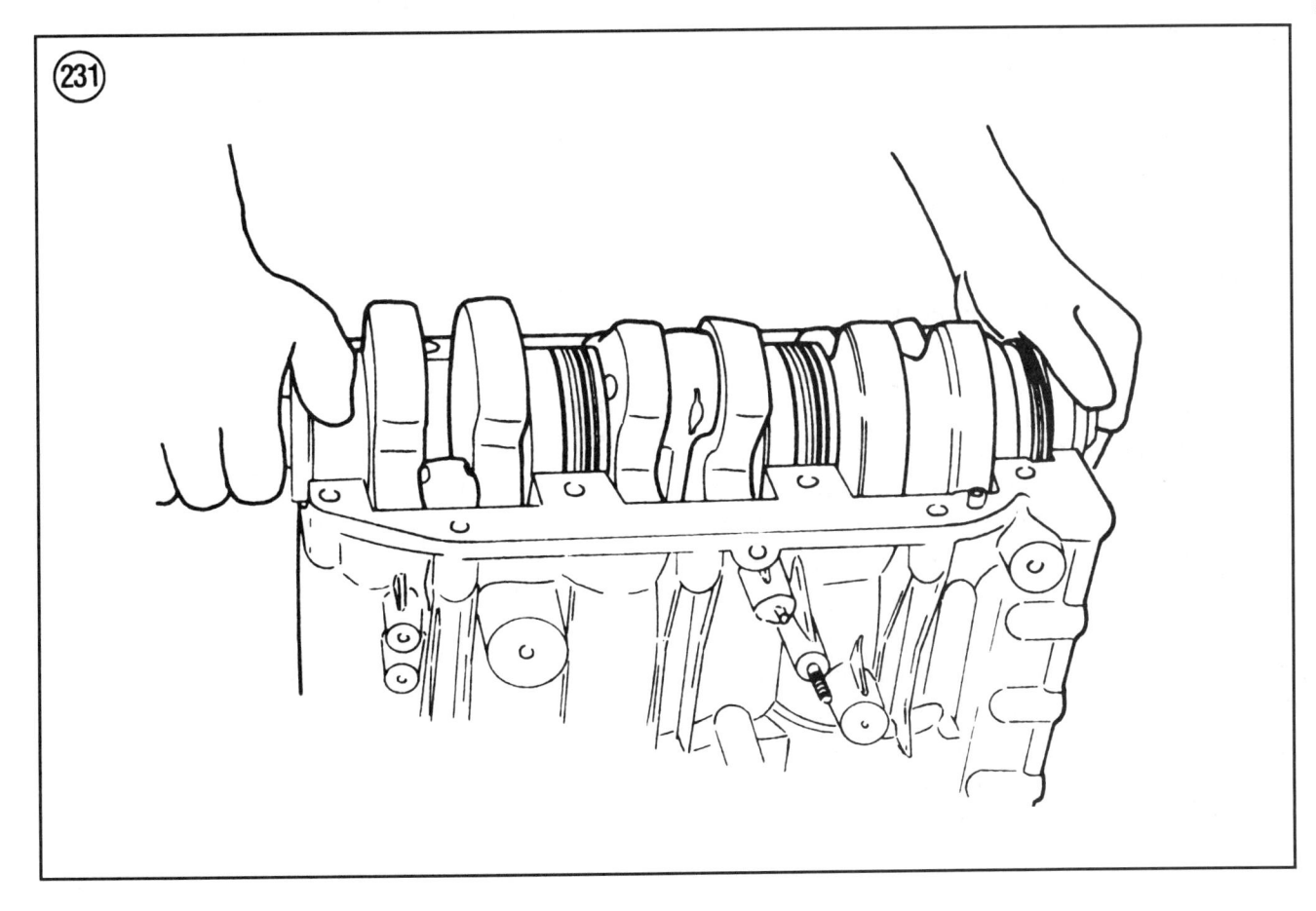

a. Mark the cylinder number on the top of the piston with a permanent type felt-tipped pen.
b. Remove the piston pin circlip with needlenose pliers or a hooked tool (**Figure 233**). Repeat this step to remove the opposite piston pin circlip. Discard the circlips as they should not be used again.
c. Push the piston pin from the piston/connecting rod with a suitable size drift.
d. Remove the piston and caged needle bearing (**Figure 234**) from the connecting rod. Keep the piston and bearing together as a set for reinstallation.
e. Repeat for the other piston assemblies.

WARNING
The edges of all piston rings are very sharp. Be careful when handling them to avoid cutting fingers.

17. Carefully spread the top ring ends with your thumbs and lift the ring up and over the piston as shown in **Figure 235**. Repeat for the lower ring and note their location in the ring grooves in the piston. The rings must be reinstalled in the same location if reused.

Disassembly
(All Inline 4-Cylinder Engines)

A large number of bolts and screws of different lengths are used to secure the various covers and components. It is a good idea to use a cupcake tin or similar compartmented container to hold the various fasteners removed from each cover or component. This will make reassembly easier and faster.

Some power heads have pry points in the casting for easier removal of each component. If no pry points are provided, break the gasket seal with a wide-blade putty knife and mallet, then carefully separate the components.

Refer to the following illustrations for this procedure:

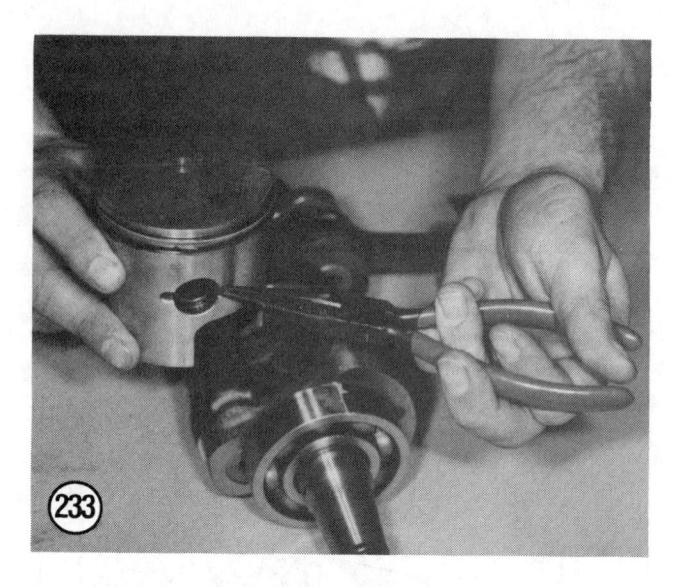

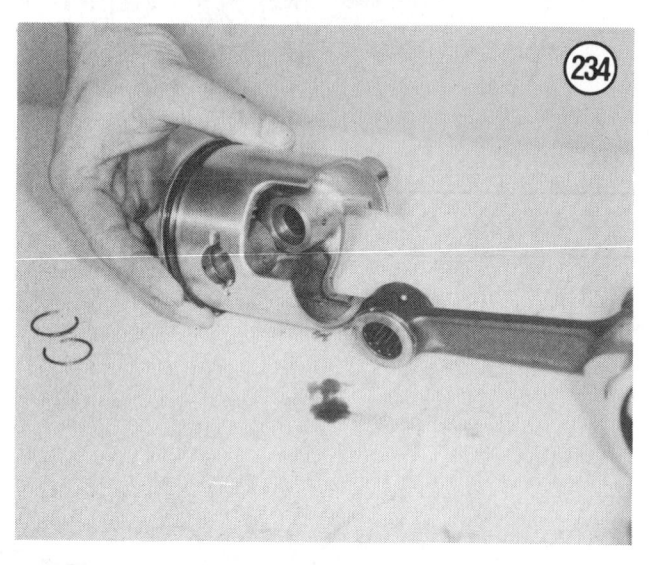

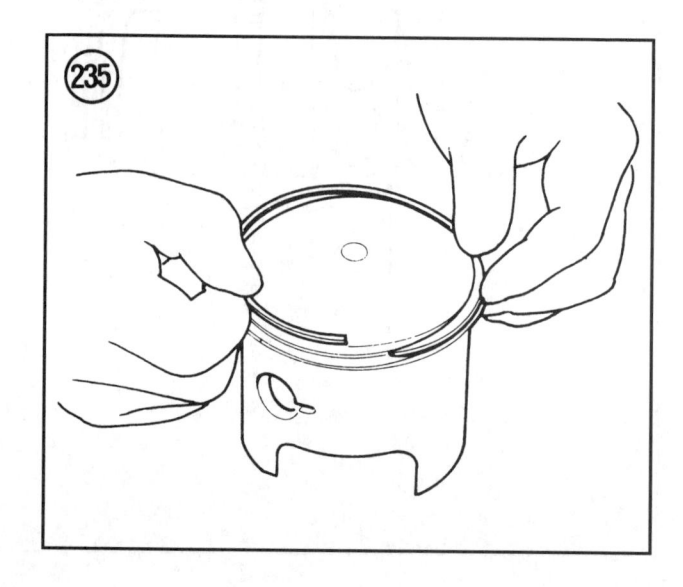

POWER HEAD

a. **Figure 236**: 1985 DT 115 and DT 140.

b. **Figure 237**: 1986-on DT 115 and DT 140.

1. Remove the intake manifold bolts (**Figure 238**, typical).

2. Insert a flat-bladed screwdriver into the pry point slots indicated by an "O" and gently pry the intake manifold away from the crankcase. Remove the intake manifold and the gasket. Discard the gasket.

3. Remove the reed valve assembly and gasket (**Figure 239**). Discard the gasket.

4. To prevent warpage of the cylinder head, loosen the cylinder head bolts 1/2 turn at a time in a crisscross pattern (**Figure 240**, typical). Remove the bolts.

236

**POWER HEAD
(1985 DT 115 AND DT 140)**

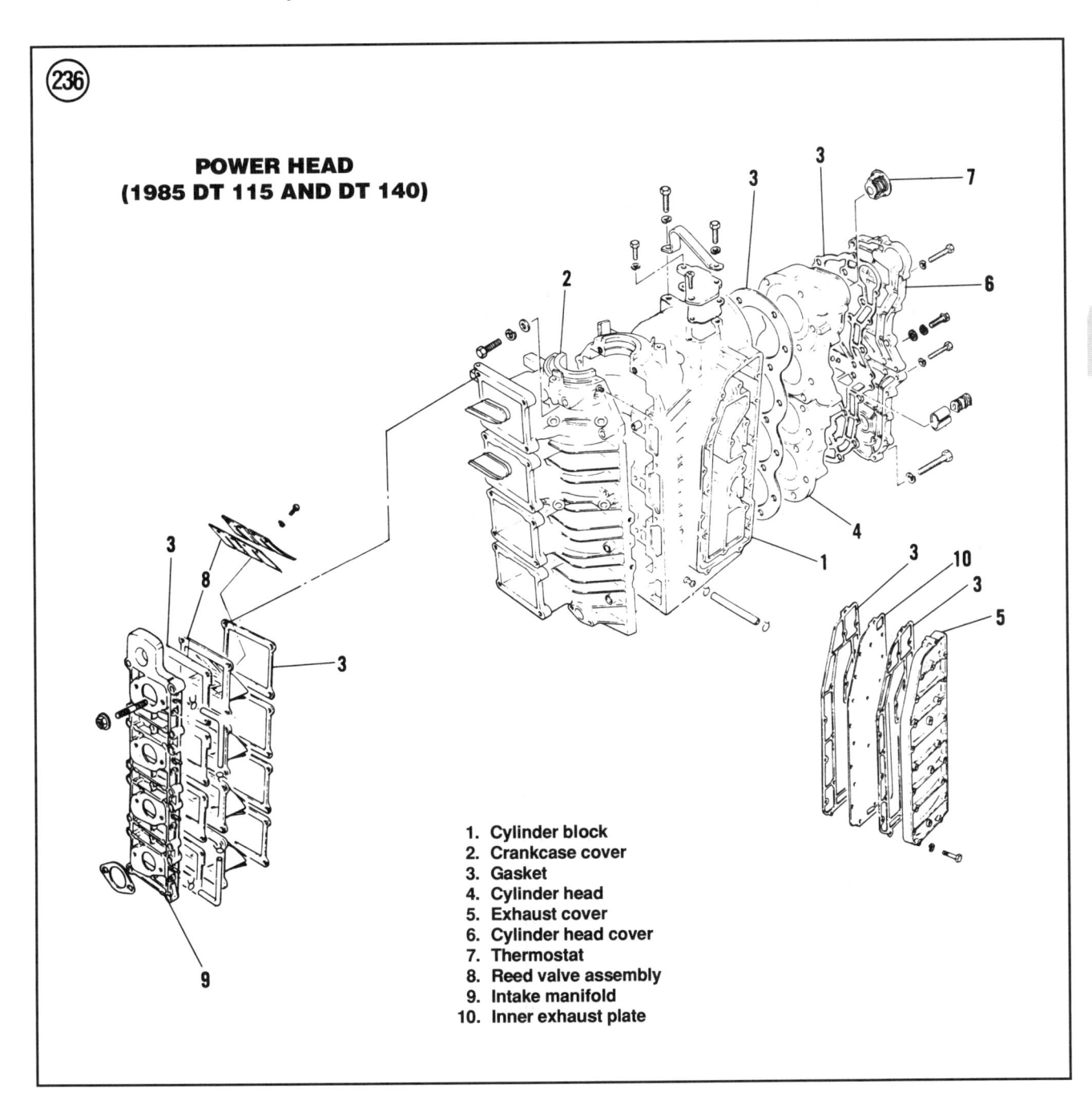

1. Cylinder block
2. Crankcase cover
3. Gasket
4. Cylinder head
5. Exhaust cover
6. Cylinder head cover
7. Thermostat
8. Reed valve assembly
9. Intake manifold
10. Inner exhaust plate

CHAPTER EIGHT

237 POWER HEAD (1986-ON DT 115 AND DT 140)

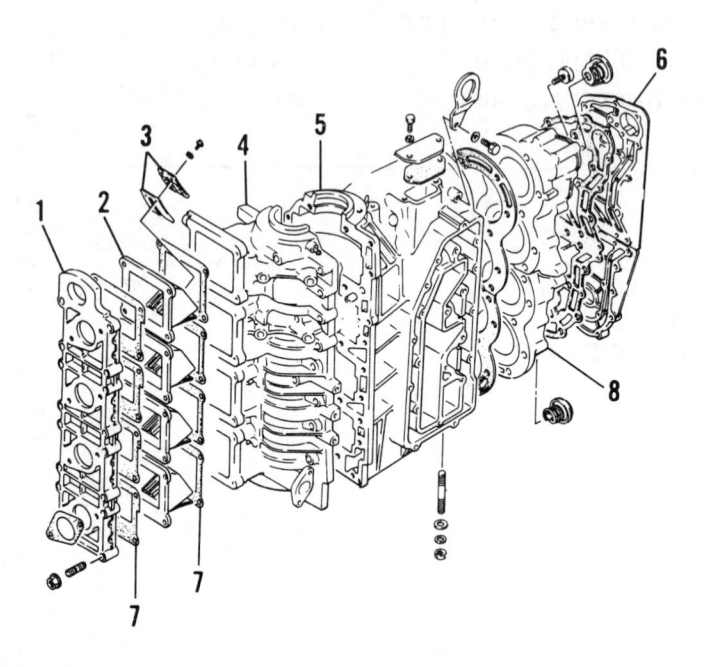

1. Intake manifold
2. Reed plate
3. Reed valve and stop
4. Crankcase cover
5. Cylinder block
6. Cylinder head cover
7. Gasket
8. Cylinder head

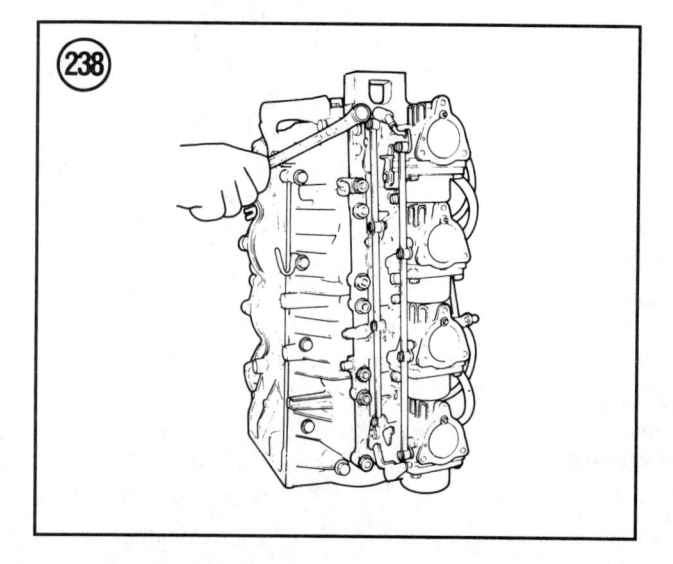

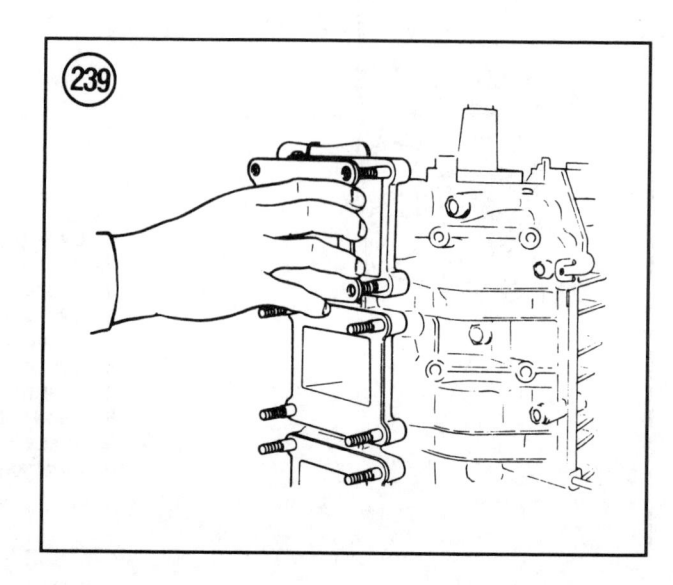

238

239

POWER HEAD

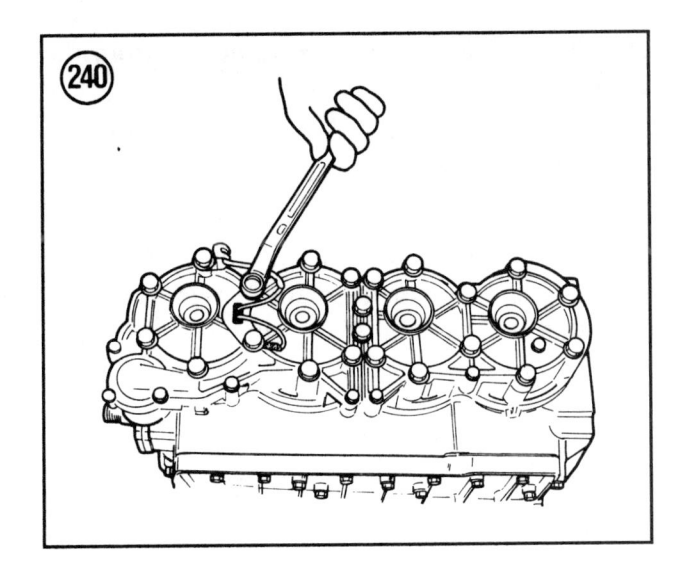

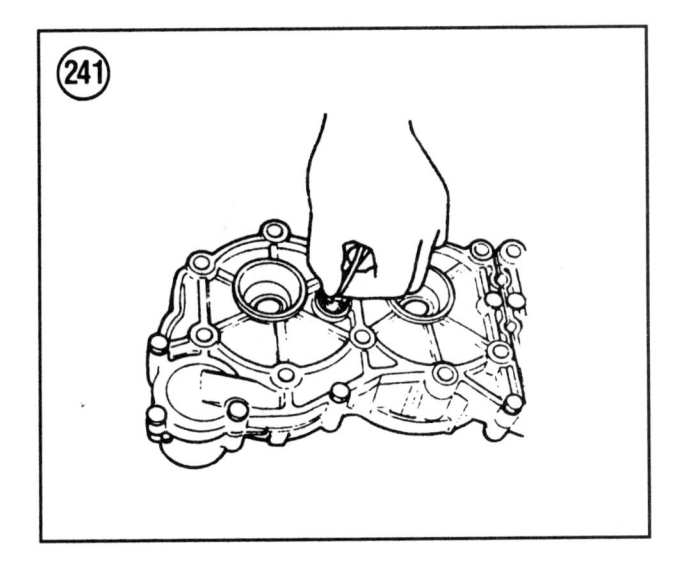

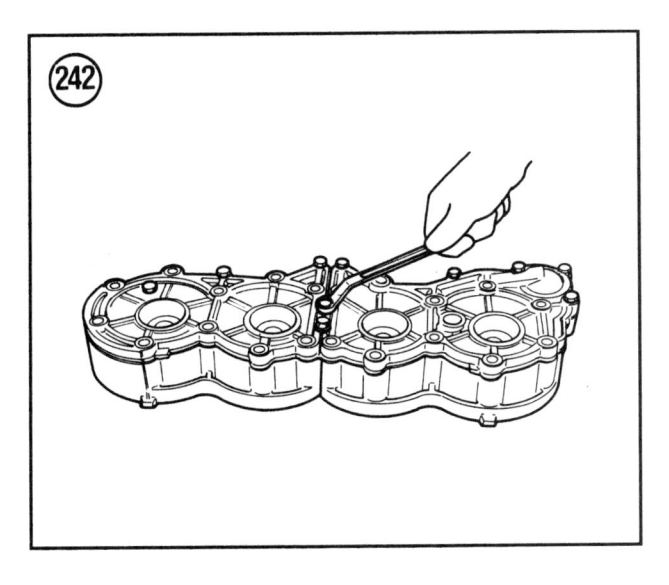

240

241

242

5. Remove the cylinder head assembly and gasket. Discard the gasket.
6. Remove the temperature switch (**Figure 241**) from the cylinder head cover.

NOTE
The cylinder head cover is a 2-piece assembly on 4-cylinder power heads.

7. Remove the bolts from each cylinder head cover (**Figure 242**, typical). Remove the cover and gasket. Discard the gasket.
8. Remove the 3 cylinder head connectors with a screwdriver (**Figure 243**).
9. Remove the thermostat from the cylinder head (A, **Figure 243**).
10. Remove the exhaust cover bolts (**Figure 244**, typical).

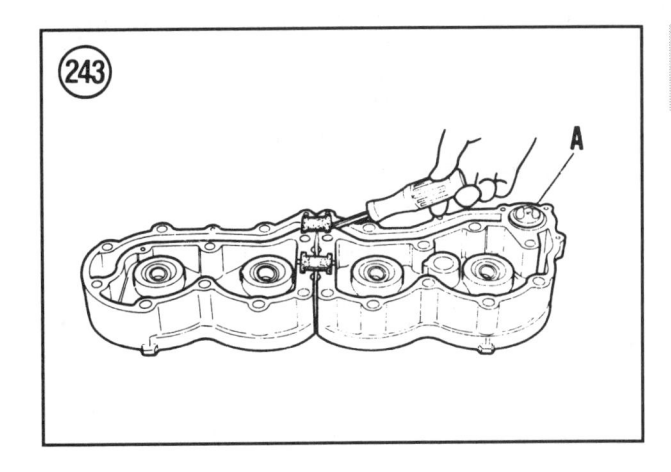

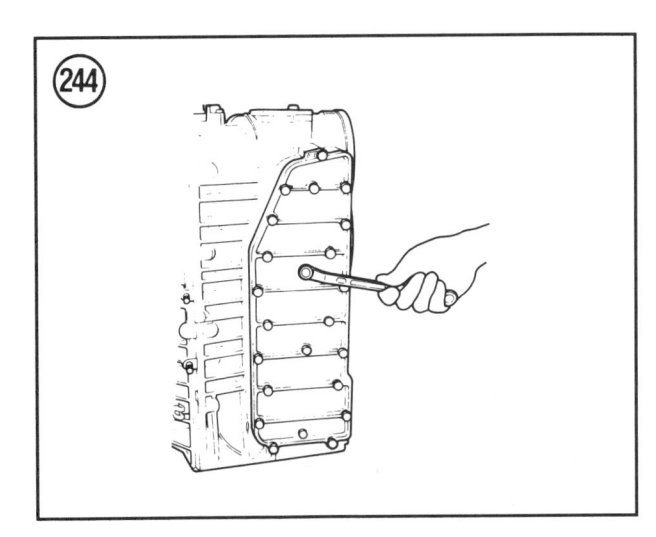

243

244

11. Insert a flat-bladed screwdriver into the pry point slots indicated by an "O" and gently pry the exhaust cover and plate away from the crankcase. Remove the exhaust cover, plate and gaskets. Discard the gaskets.

12. Remove the crankcase cover bolts (**Figure 245**, typical).

13. With the crankcase cover and cylinder block on a solid surface, carefully pry the crankcase cover and cylinder block apart using a suitable pry tool at the pry points provided.

14. Remove the crankcase cover from the cylinder block.

NOTE
*On 1985 models, the piston and connecting rod assemblies must be removed through the **top** of the cylinder block.*

15A. On 1985 models, perform the following:
  a. Mark the cylinder number on connecting rods and caps. Remove each connecting rod cap, roller bearings and bearing cage. Place bearings and cage from each rod in separate containers.
  b. Remove the crankshaft from the cylinder block. If necessary, lightly tap on crankshaft taper with a rubber mallet to break the seal.
  c. Remove the remaining connecting rod roller bearings and cages. Place in their respective container with their mating roller bearings already removed.
  d. Reinstall each rod cap to its respective connecting rod.
  e. Mark the cylinder number on the top of the piston with a permanent type felt-tipped pen.
  f. Carefully push each piston and connecting rod assembly up and out of the top of the cylinder block (**Figure 246**).

15B. On 1986-on models, carefully remove the crankshaft assembly out of the cylinder block. If necessary, lightly tap on crankshaft taper with a rubber mallet to break the seal.

16. If the pistons are to be removed from the connecting rod:
  a. On 1986-on models, mark the cylinder number on the top of the piston with a felt-tipped pen.
  b. Remove the piston pin circlip with a hooked tool or needlenose pliers. Repeat this step to remove the opposite piston pin circlip. Discard the circlips as they should not be used again

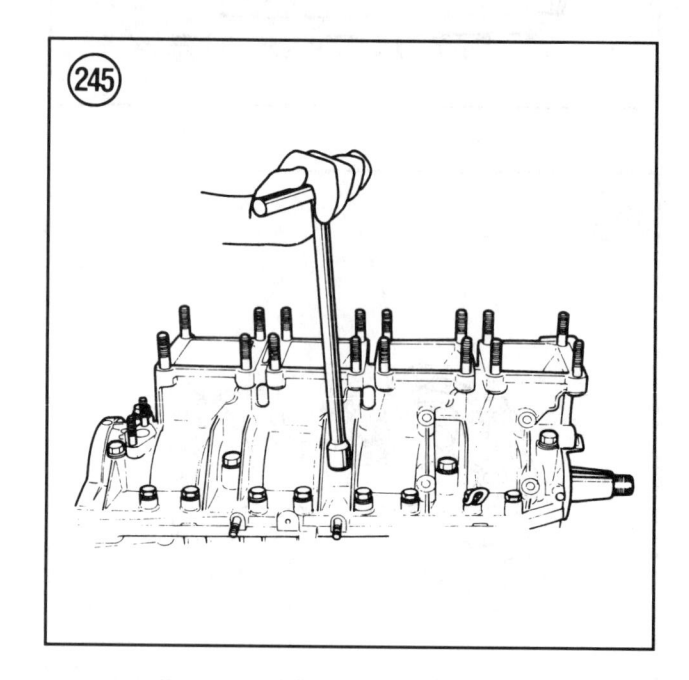

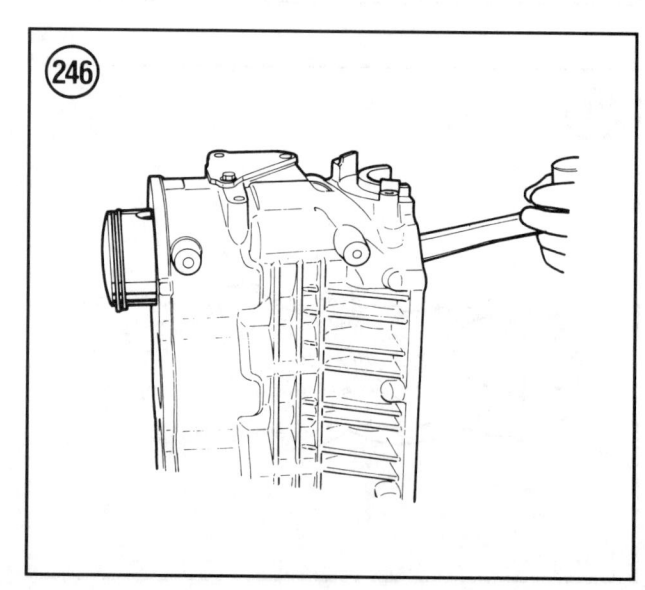

POWER HEAD

c. Push the piston pin from the piston/connecting rod with a suitable size drift.

d. Remove the piston, caged needle bearing and washers. Keep the piston and bearing together for reinstallation.

WARNING
The edges of all piston rings are very sharp. Be careful when handling them to avoid cutting fingers.

17. Carefully spread the top ring ends with your thumbs and lift the ring up and over the piston as shown in **Figure 235**. Repeat for the lower ring and note their location in the ring grooves in the piston. The rings must be reinstalled in the same location if reused.

Disassembly
(All V4 Models)

A large number of bolts and screws of different lengths are used to secure the various covers and components. It is a good idea to use a cupcake tin or similar compartmented container to hold the various fasteners removed from each cover or component. This will make reassembly easier and faster.

Some power heads have pry points in the casting for easier removal of each component. If no pry points are provided, break the gasket seal with a wide-blade putty knife and mallet, then carefully separate the components.

1. Cover the workbench surface with a flat sheet of heavy rubber or piece of clean masonite to protect the sealing surface of the silencer case. The work bench must also be level so that the power head will be stable during the disassembly procedure.

CAUTION
The complete power head assembly is heavy and also very tall and top heavy during the initial part of the disassembly procedure. When breaking bolts loose or removing a stubborn component, hold onto the power head so that it will not fall over. If necessary, have an assistant hold onto the power head during the removal of a difficult component.

2. Position the power head on the workbench with the cylinder heads facing up for easy access.

3. To prevent warpage of the cylinder head, loosen the cylinder head bolts 1/2 turn at a time in a crisscross pattern (**Figure 247**). Remove the bolts.

NOTE
*If the cylinder head is difficult to remove, use a plastic hammer or soft faced mallet and **carefully** tap on the underside of the projections on the cylinder head.*

4. Remove the cylinder head assembly and gasket from the cylinder. Discard the gasket.

5. Repeat Step 3 and Step 4 for the other cylinder head.

6. Remove the 11 bolts securing the under oil seal housing (**Figure 248**) and remove the housing.

328 CHAPTER EIGHT

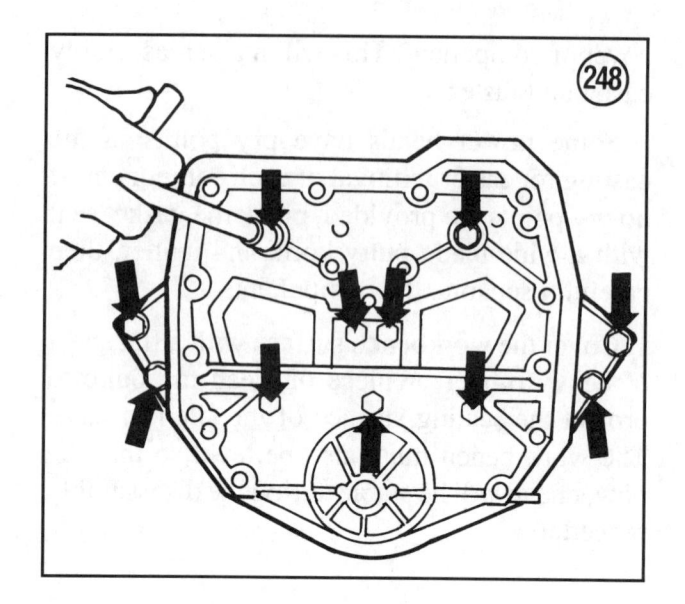

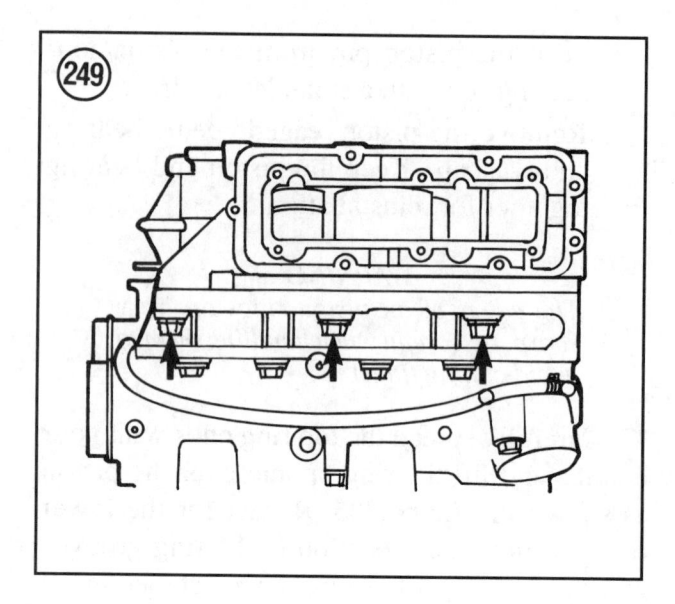

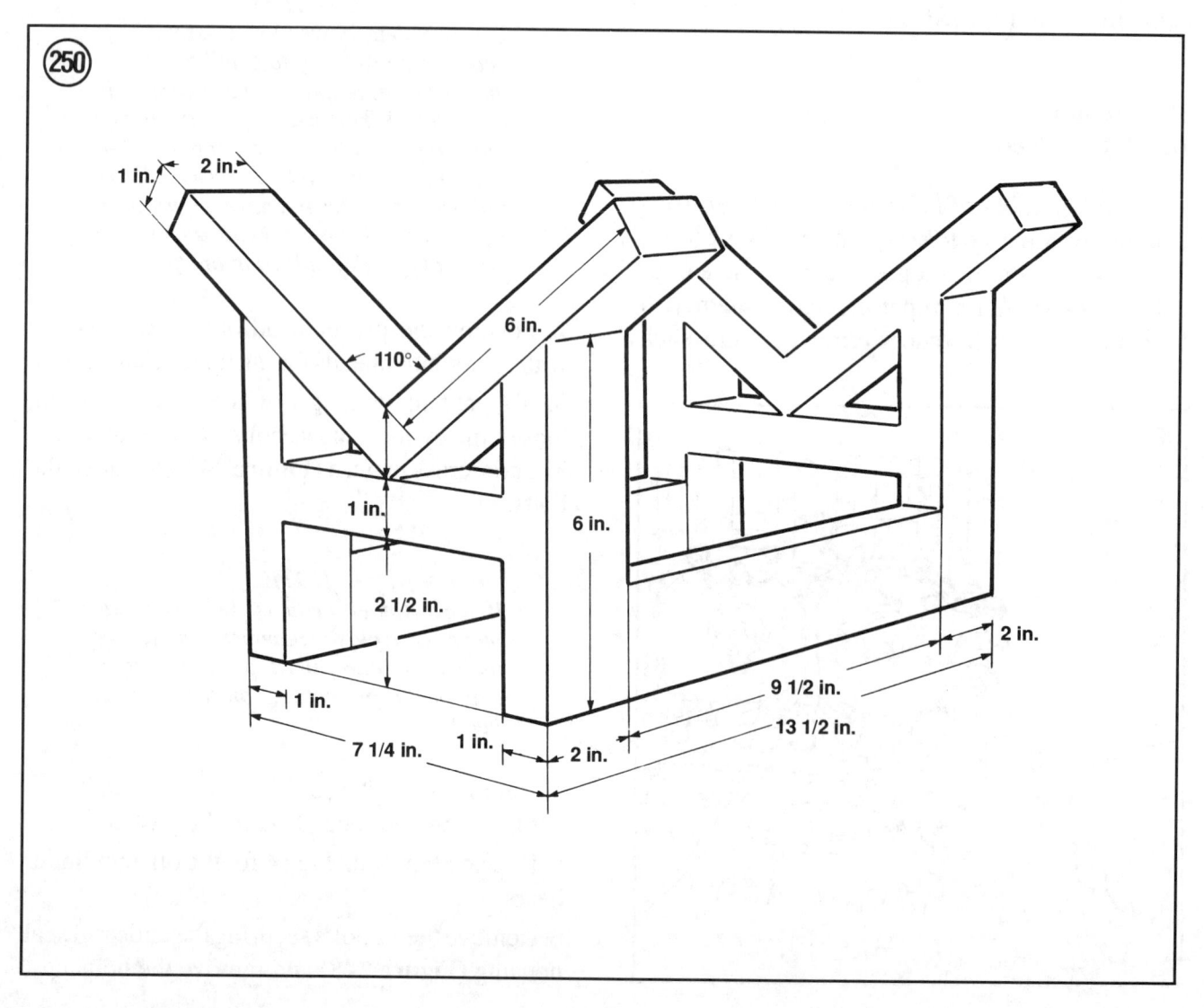

POWER HEAD

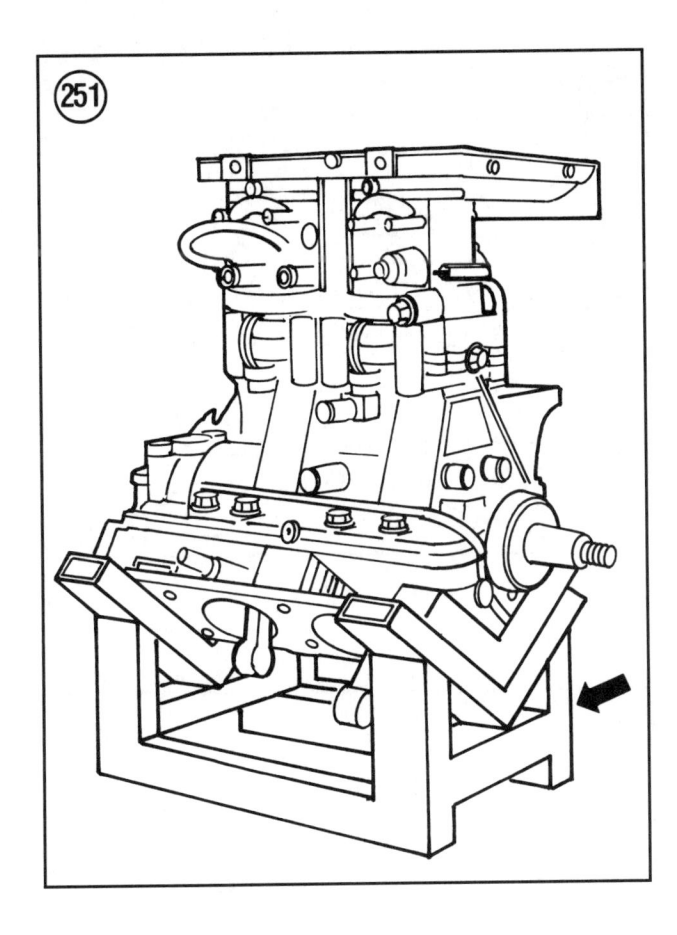

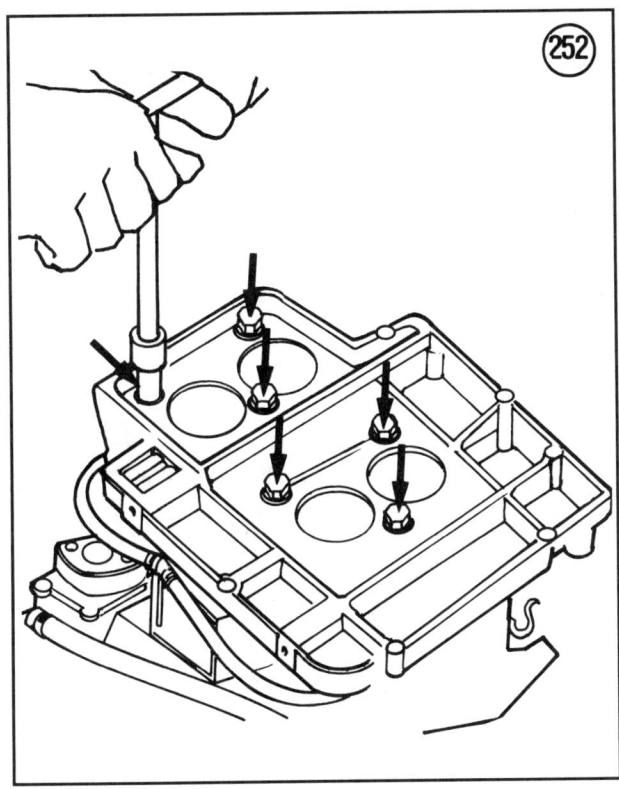

NOTE
A special Suzuki tool (part No. 09911-78730) is required to remove the cylinder bolts.

7. In the valley between the cylinders, remove the 6 bolts securing the cylinders to the crankcase.
8. Remove the nuts (**Figure 249**) securing the cylinders to the crankcase.
9. Carefully pull the cylinder straight up and off of the pistons. Repeat for the other cylinder.
10. Remove all 4 pistons from their connecting rods as follows:
   a. Mark the cylinder number on the top of each piston with a felt-tipped pen.
   b. Remove the piston pin circlip with a hooked tool or needlenose pliers. Repeat this step to remove the opposite piston pin circlip. Discard the circlips as they should not be used again.
   c. Push the piston pin from the piston/connecting rod with a suitable size drift.
   d. Remove the piston, caged needle bearing and washers. Keep the piston and bearing together for reinstallation.

NOTE
*The engine must now be turned upside down with the crankcase cover facing up. The ends of the connecting rods are now protruding through the openings in the crankcase and must be protected. Suzuki does not have a special tool for this purpose but does provide dimensions so that a stand can be fabricated from wooden 1 × 2's. Use the dimensions and angles shown in **Figure 250** and fabricate the stand. Make sure all joints are strong since it has to hold the heavy power head while removing components in the following steps.*

11. Turn the power head upside down and set it in the stand (**Figure 251**).
12. Remove the silencer case bolts (**Figure 252**) and remove the case.

13. Remove the intake manifold bolts and nuts (**Figure 253**) and remove the intake manifold and reed valve assembly. Remove the gasket and discard it.

14. Remove the clutch shaft bolts (**Figure 254**) and remove the clutch shaft assembly.

15. Remove the crankcase cover bolts (**Figure 255**). Carefully pry the crankcase cover and crankcase apart using a suitable pry tool at the pry points provided.

16. Remove the crankcase cover from the crankcase.

17. Carefully remove the crankshaft assembly from the crankcase (**Figure 256**). If necessary, lightly tap on crankshaft taper with a rubber mallet to break the seal.

Disassembly
(All V6 Models)

A large number of bolts and screws of different lengths are used to secure the various covers and components. It is a good idea to use a cupcake tin or similar compartmented container to hold the various fasteners removed from each cover or component. This will make reassembly easier and faster.

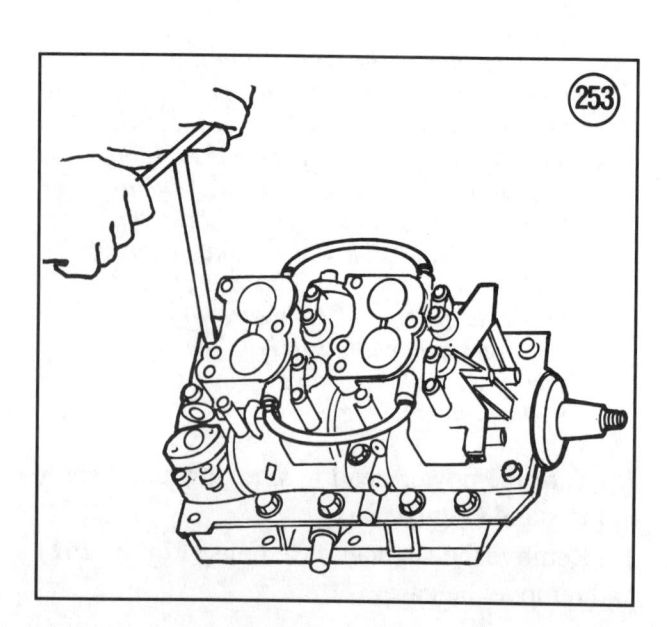

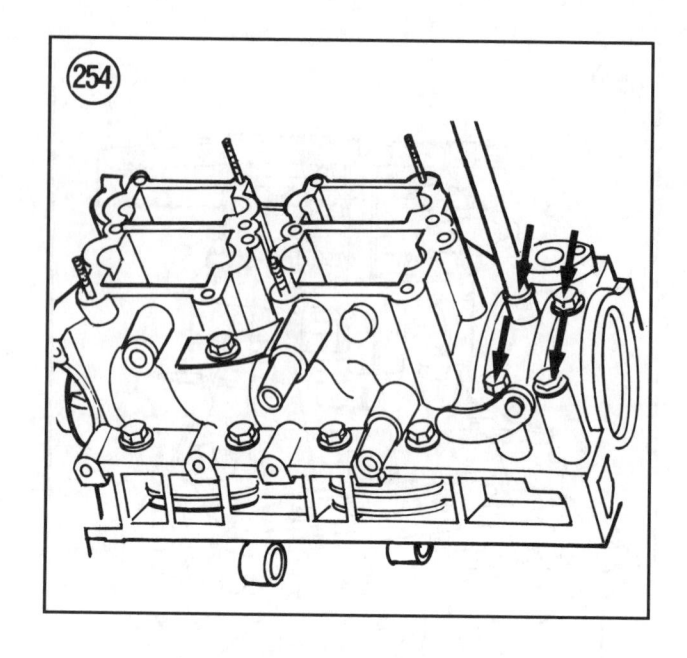

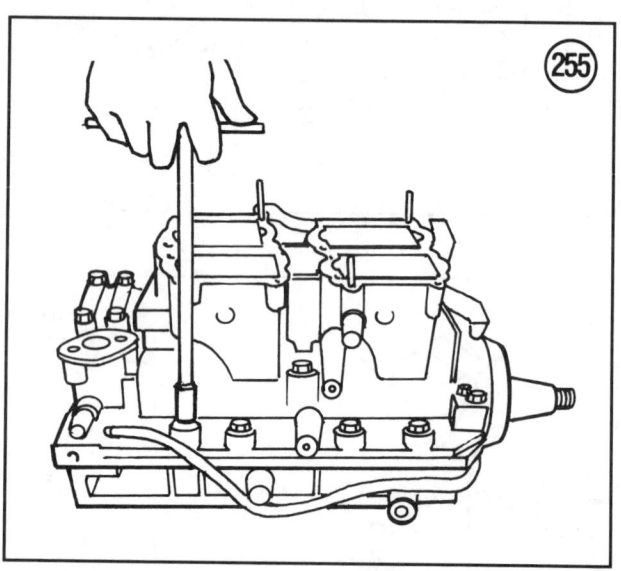

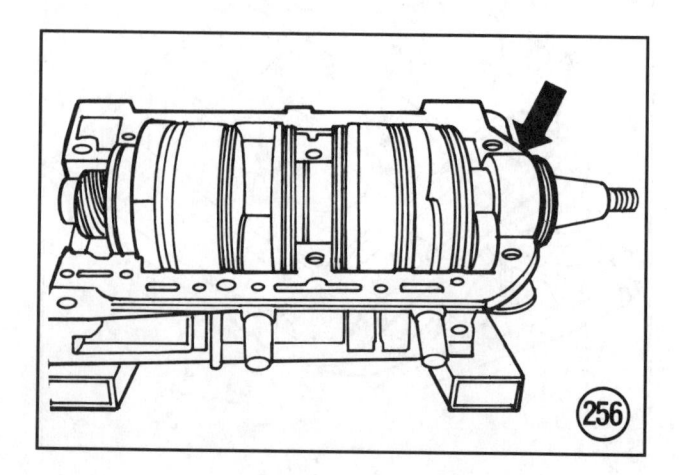

POWER HEAD

Some power heads have pry points in the casting for easier removal of each component. If no pry points are provided, break the gasket seal with a wide-blade putty knife and mallet, then carefully separate the components.

CAUTION
The complete power head assembly is heavy and also very tall and top heavy during the initial part of the disassembly procedure. When breaking bolts loose or removing a stubborn component, hold onto the power head so that it will not fall over. If necessary, have an assistant hold onto the power head during the removal of a difficult component.

1. Construct a wooden stand from 3/4 in. plywood and 2 × 4's as shown in **Figure 257**. Use 8 mm bolts and attach it to the inlet case securely.

2. Position the power unit on the workbench on the wooden stand with the cylinder heads facing up for easy access.

3. To prevent warpage of the cylinder head, loosen the cylinder head bolts 1/2 turn at a time in a crisscross pattern (**Figure 258**). Remove the bolts.

4. Insert a flat-bladed screwdriver into the pry point slots indicated by an "O" and gently pry the cylinder head away from the cylinder (**Figure 259**).

5. Remove the cylinder head assembly and gasket. Discard the gasket.

6. Repeat Steps 3-5 for the other cylinder head.

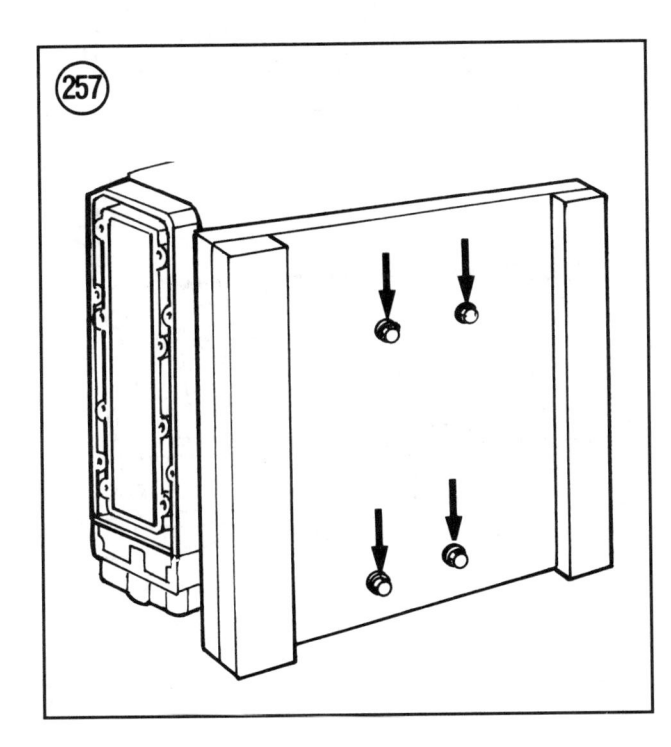

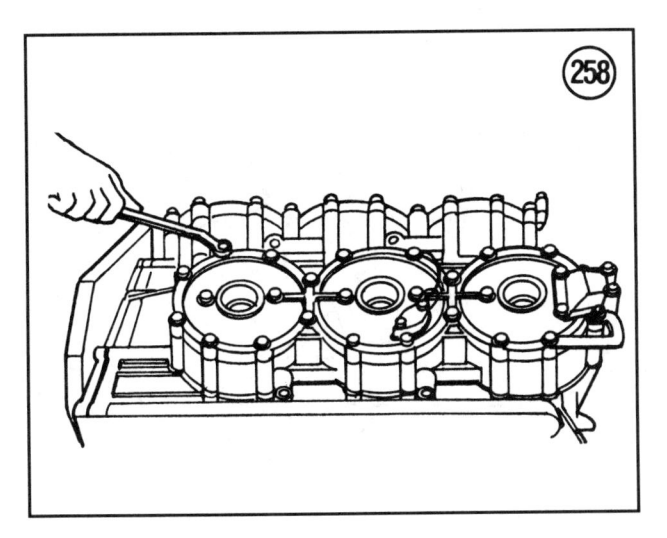

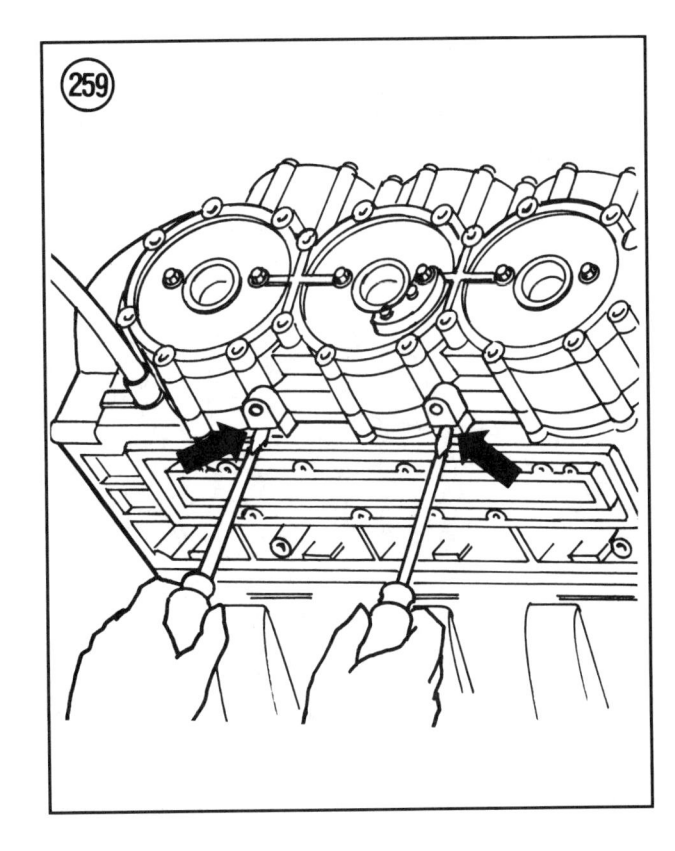

7. Remove the 11 bolts securing the under oil seal housing (**Figure 260**) and remove the housing.

NOTE
A special 8 mm hex wrench (Suzuki tool part No. 09911-78710) is required to remove the cylinder bolts.

8. In the valley between the cylinders, remove the 8 bolts securing the cylinders to the crankcase (**Figure 261**).

9. Remove the additional bolts (**Figure 262**) securing the cylinders to the side of the crankcase.

10. Insert a flat-bladed screwdriver into the pry point slots and gently pry the cylinder away from the crankcase.

11. Carefully pull the cylinder straight up and off of the pistons. Repeat for the other cylinder.

12. Remove all 4 pistons from their connecting rods as follows:

  a. Mark the cylinder number on the top of each piston with a felt-tipped pen.
  b. Remove the piston pin circlip with a hooked tool or needlenose pliers (**Figure 263**). Repeat this step to remove the opposite piston pin circlip. Discard the circlips as they should not be used again.

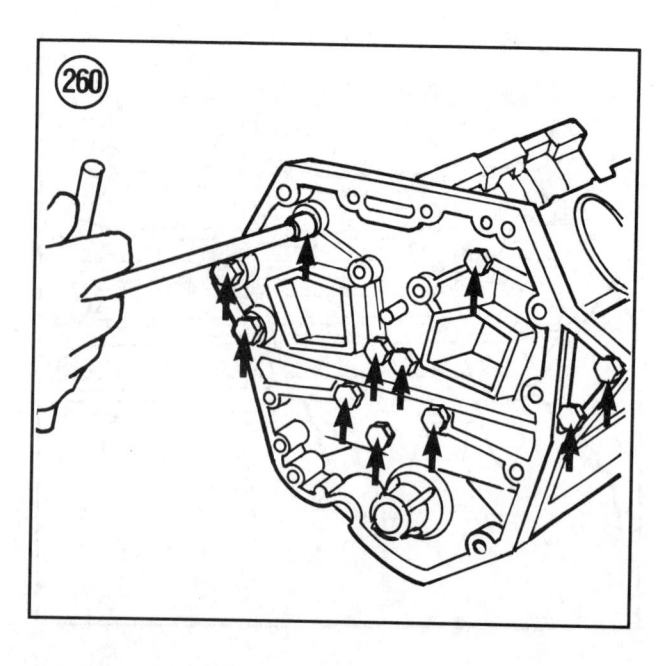

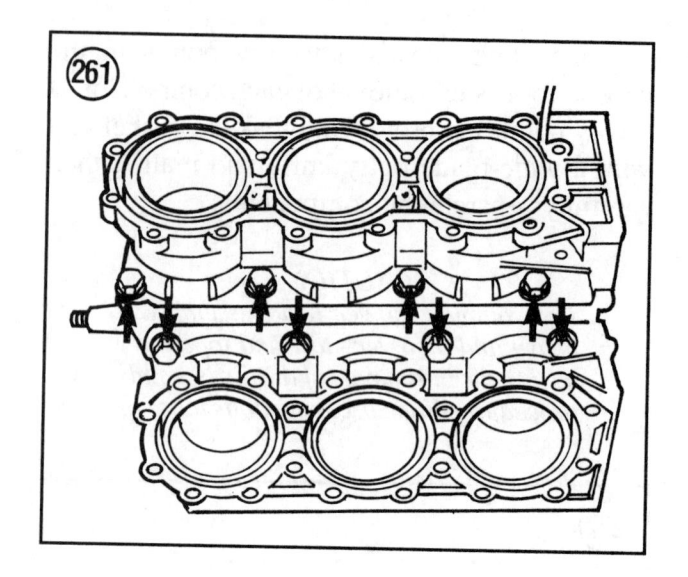

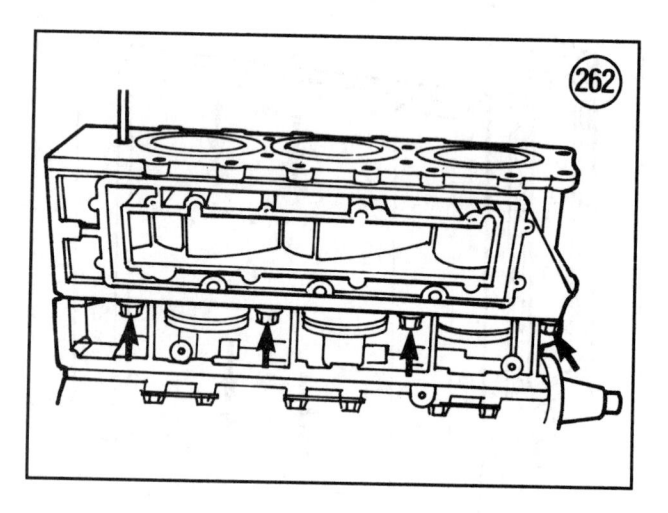

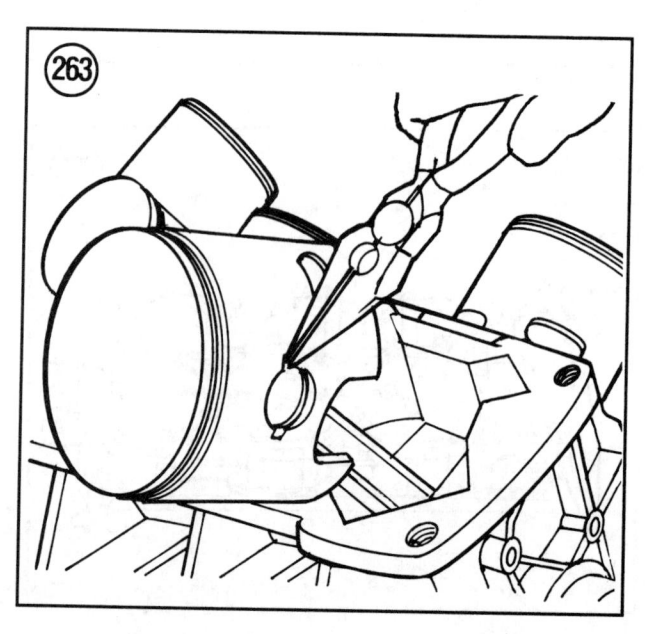

POWER HEAD

c. Push the piston pin from the piston/connecting rod with a suitable size drift.

d. Remove the piston, caged needle bearing and washers. Keep the piston and bearing together for reinstallation.

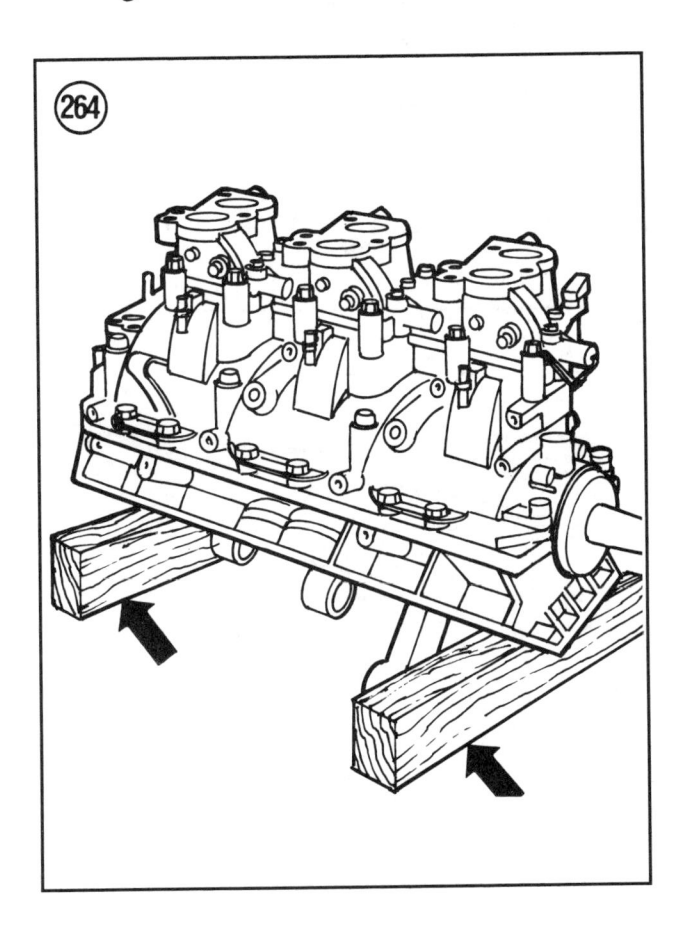

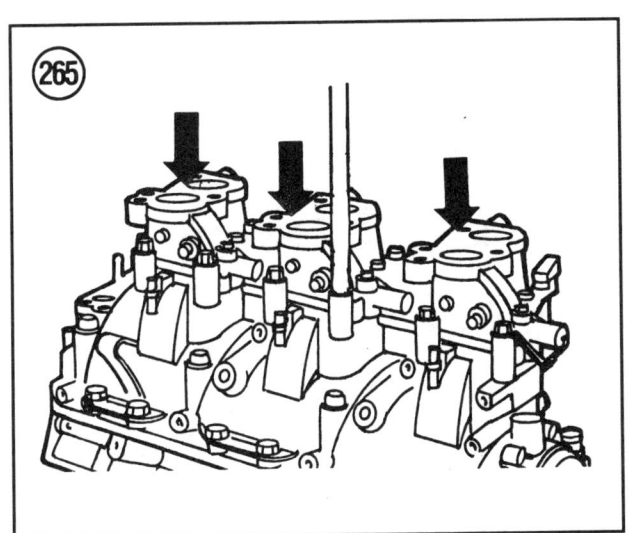

13. Remove the wood base installed in Step 1.

NOTE
*The engine must now be turned upside down with the crankcase cover facing up. The ends of the connecting rods are now protruding through the openings in the crankcase and must be protected. Suzuki does not have a special tool for this purpose. Several wooden 2 × 4's placed at the correct location will hold the power head in position as shown in **Figure 264**. Make sure the wood is properly located since it has to hold the heavy power head while removing components in the following steps.*

14. Turn the power head upside down and set it on the wooden 2 × 4's (**Figure 264**).

15. Remove the intake manifold bolts (**Figure 265**) and remove all 3 intake manifolds and reed valve assemblies. Remove the gaskets and discard them.

16. Remove the crankcase cover bolts (**Figure 266**).

17. Insert a flat-bladed screwdriver into the pry point slots indicated by an "O" and gently pry the crankcase cover away from the crankcase (**Figure 267**).

18. Remove the crankcase cover from the crankcase.

19. Carefully lift the crankshaft assembly out of the crankcase. If necessary, lightly tap on crankshaft taper with a rubber mallet to break the seal.

Cylinder Head, Cylinder Block and Crankcase Cleaning and Inspection (All Engines)

This procedure represents all single cylinder, inline and V-type power heads. The procedures apply to all types unless otherwise specified.

Suzuki outboard cylinder blocks and crankcase covers are matched and align-bored assemblies. For this reason, you should not attempt to assemble an engine with parts salvaged from other blocks. If inspection indicates that either the block or cover requires replacement, replace both as an assembly.

1. Carefully remove all gasket and sealant residue from the cylinder block and crankcase cover mating surfaces with lacquer thinner.

2. Clean the aluminum surfaces carefully to avoid nicking them. A dull putty knife can be used, but a piece of Lucite with one edge ground to a 45° angle is more efficient and will also

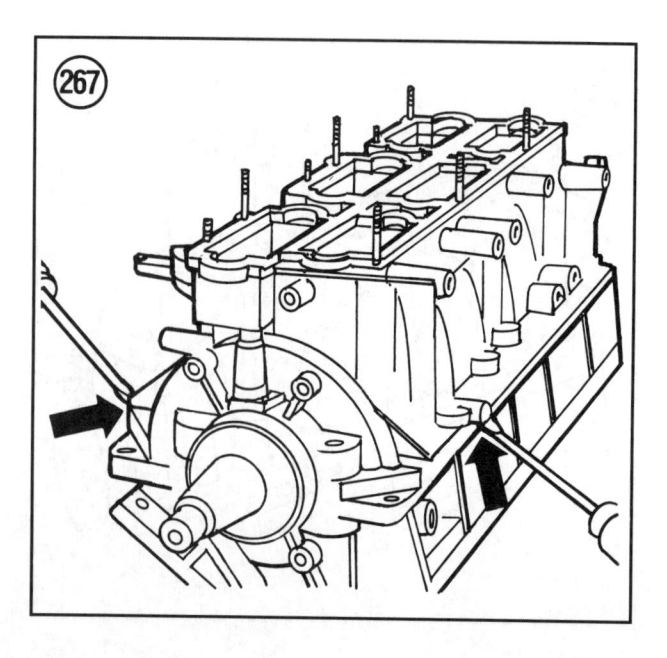

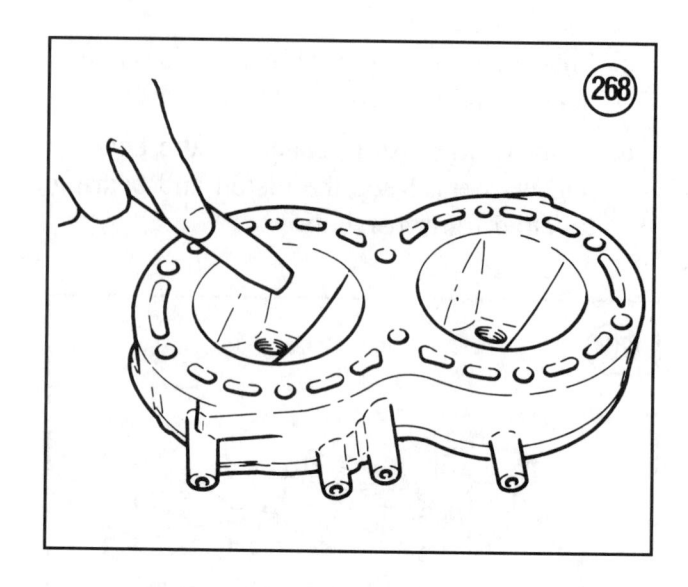

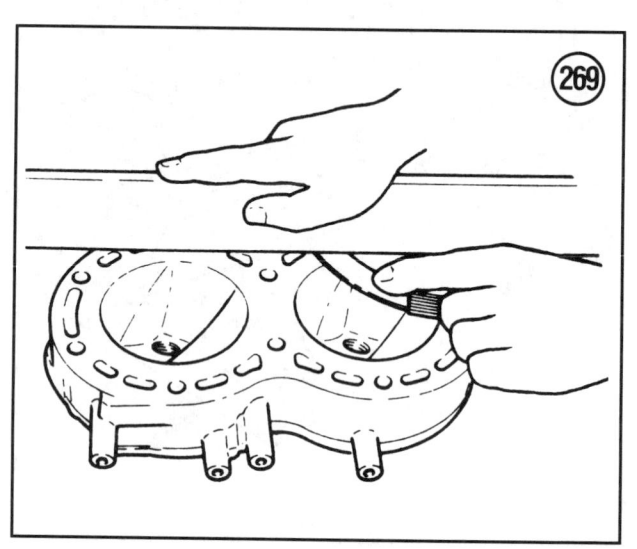

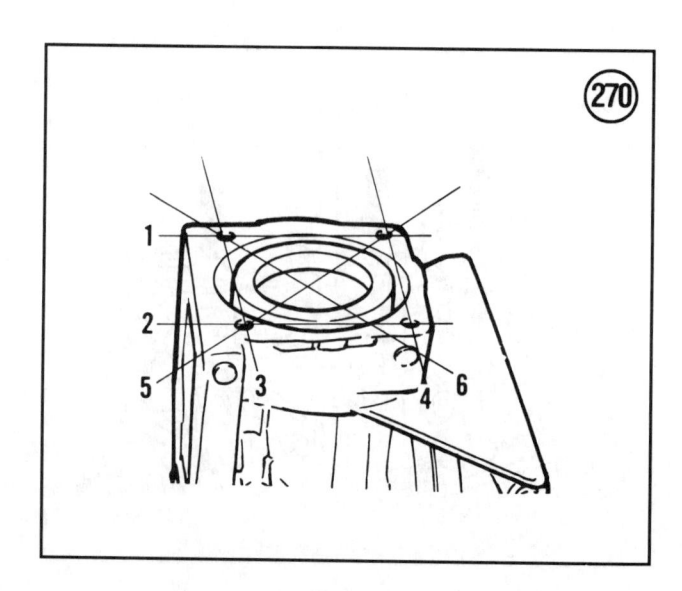

POWER HEAD

⑤

⑥

⑦

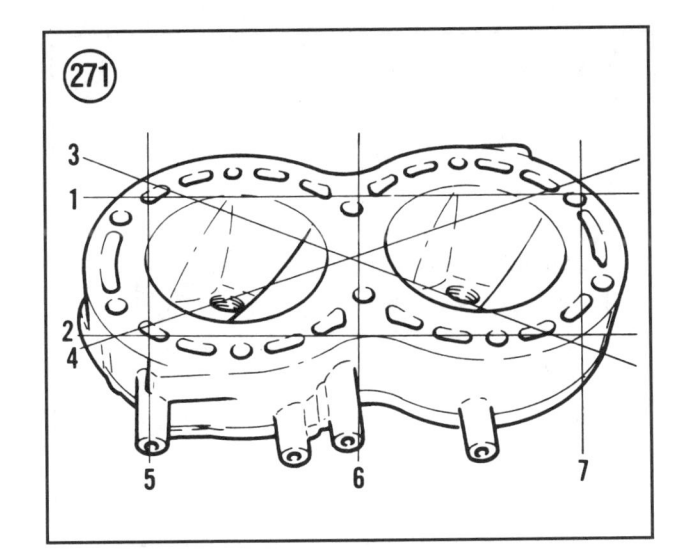

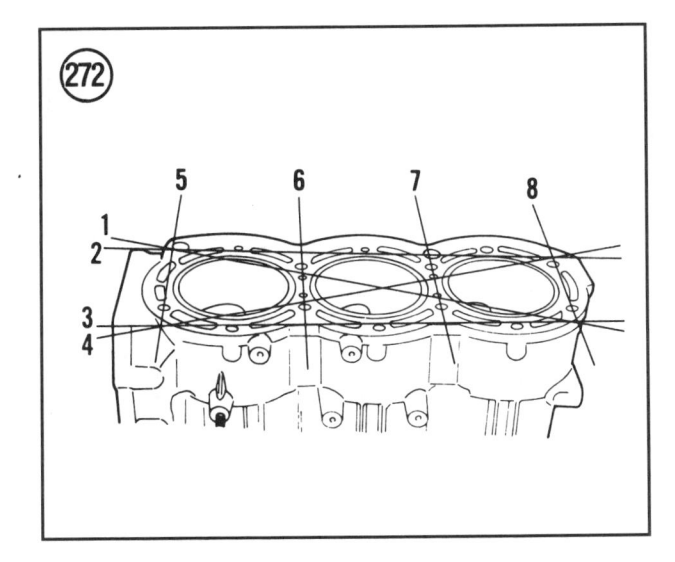

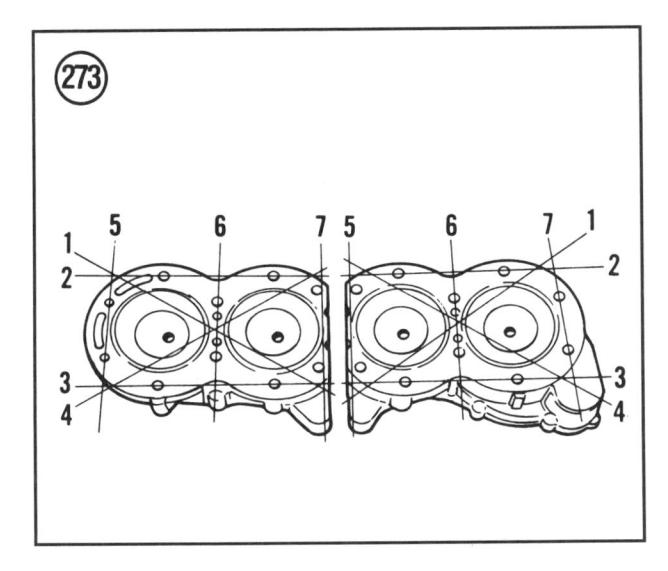

reduce the possibility of damage to the surfaces. When reassembling the crankcase cover and cylinder block, both mating surfaces must be free of all sealant residue, dirt and oil or leaks will develop.

3. Remove all carbon deposits from the combustion chambers, exhaust ports and cylinder head(s) (**Figure 268**). Use a hardwood or Lucite scraper and solvent. Be careful not to scratch or gouge the areas while cleaning.

4. Once all carbon is removed, clean the cylinder block and cylinder head thoroughly with solvent and a brush, then dry with compressed air.

5. On V-type engines, clean the under oil seal housing in solvent and dry with compressed air.

6. Carefully remove all gasket and sealant residue from all mating surfaces.

7. Check the cylinder head gasket mating surface with a straightedge and flat feeler gauge (**Figure 269**). Check for warpage in the directions shown in **Figures 270-273**.

8. If warpage exceeds 0.0012 in. (0.03 mm) in any direction, place a large sheet of No. 400 (or finer) sandpaper on a pane of plate glass or surface plate. Apply a slight amount of pressure and move the component in a figure 8 pattern. Remove the component and repeat Step 7 to recheck surface flatness.

9. Check the cylinder block-to-head surface for warpage as described in Step 7. Follow the directions shown in **Figures 270-273**. If warpage exceeds 0.0012 in. (0.03 mm) in any direction, correct as described in Step 8.

10. Check the exhaust port surfaces for warpage as described in Step 7. Follow the directions shown in **Figure 274** (2-cylinder), **Figure 275** (3-cylinder) or **Figure 276** (inline 4-cylinder, V4 and V6). If warpage exceeds 0.0012 in. (0.03 mm) in any direction, correct as described in Step 8.

11. Check the block, cylinder head(s) and cover for cracks, fractures, stripped bolt or spark plug threads or other defects.

12. Check the gasket mating surfaces for nicks, grooves, cracks or excessive distortion. Any of these defects will cause compression leakage. Replace as required.

13. Check all oil and water passages in the cylinder block and cover for obstructions. Make sure any plugs installed are properly tightened.

NOTE
With older engines, it is a good idea to have the cylinder walls lightly honed with a medium stone even if they are in good condition. This will break up any glaze that might reduce compression.

14. Check each cylinder bore for signs of aluminum transfer from the pistons to the cylinder walls. If scoring is present, but not excessive, have the cylinders honed by a dealer or qualified machine shop.

15. Measure each cylinder bore with an inside micrometer or bore gauge (**Figure 277**) at points A, B and C in **Figure 278** (1-cylinder), **Figure 279** (2-cylinder), **Figure 280** (3-cylinder), **Figure 281** (4-cylinder) or **Figure 282** (V4 and V6). Record the readings.

16. Turn the cylinder 90° and repeat the measurements. Subtract the smallest from the largest reading. If the difference between the 2 measurements exceeds 0.004 in. (0.1 mm), have all cyl-

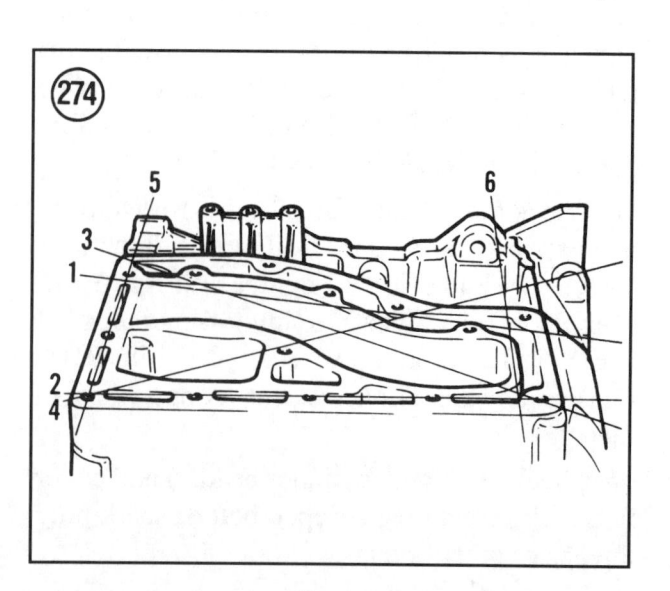

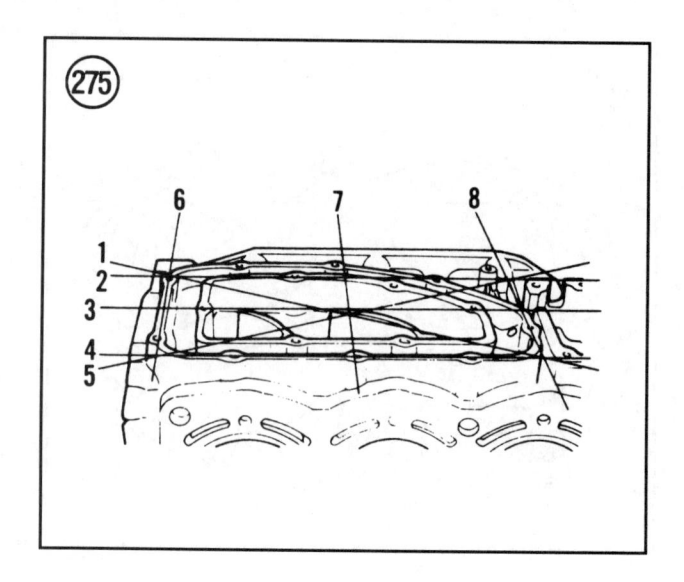

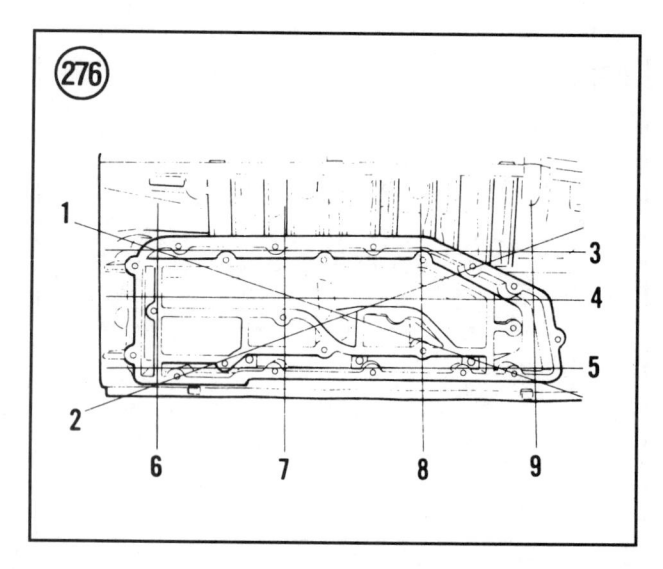

POWER HEAD

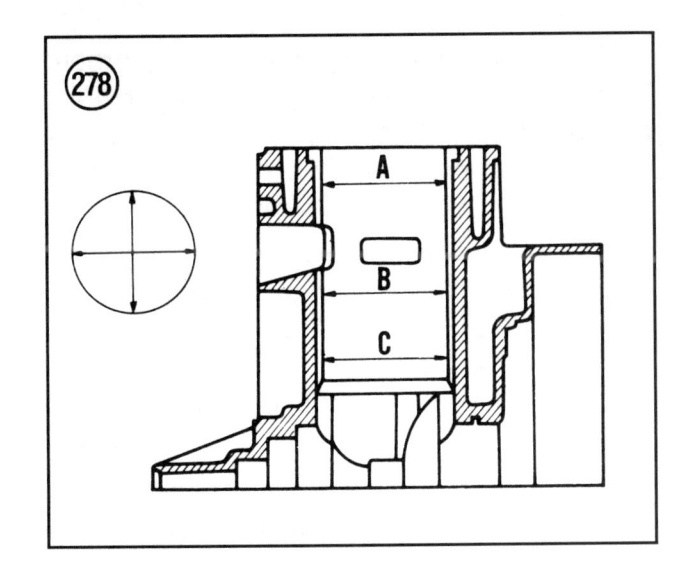

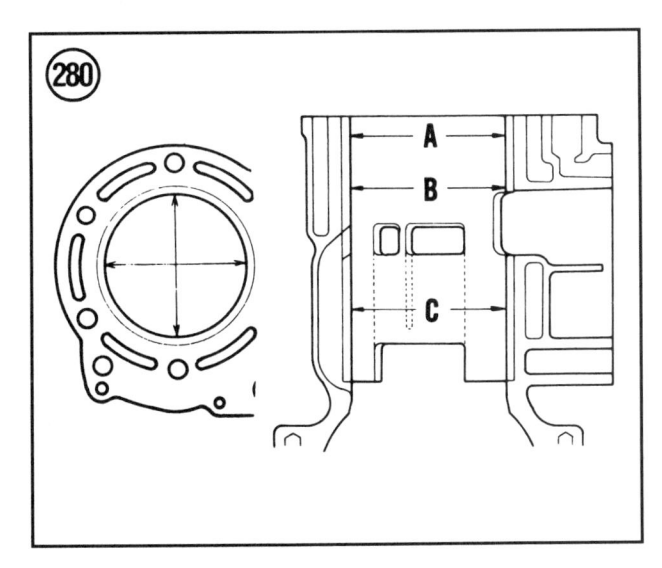

inders rebored by a dealer or qualified machine shop and install oversize pistons.

NOTE
Obtain the new pistons and measure them before having the cylinders bored. This allows for variations in piston size due to manufacturing tolerances.

Crankshaft and Connecting Rod Bearings Cleaning and Inspection (All Engines)

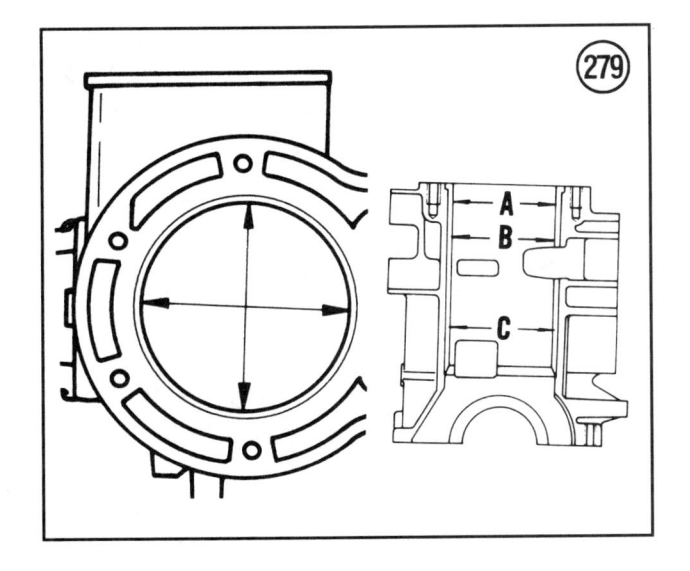

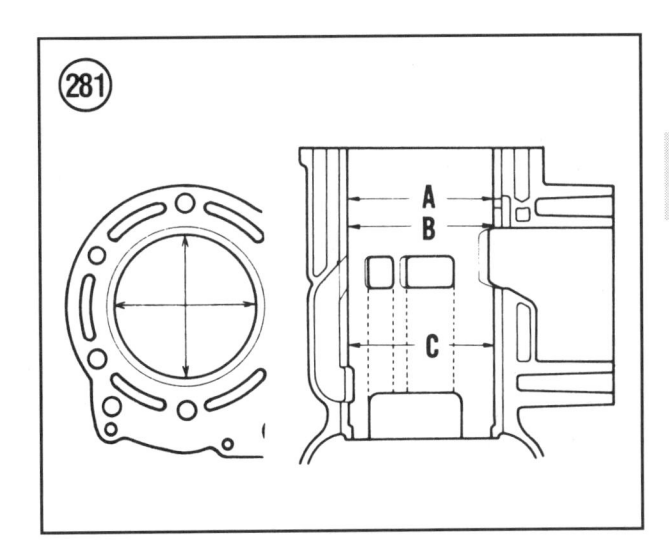

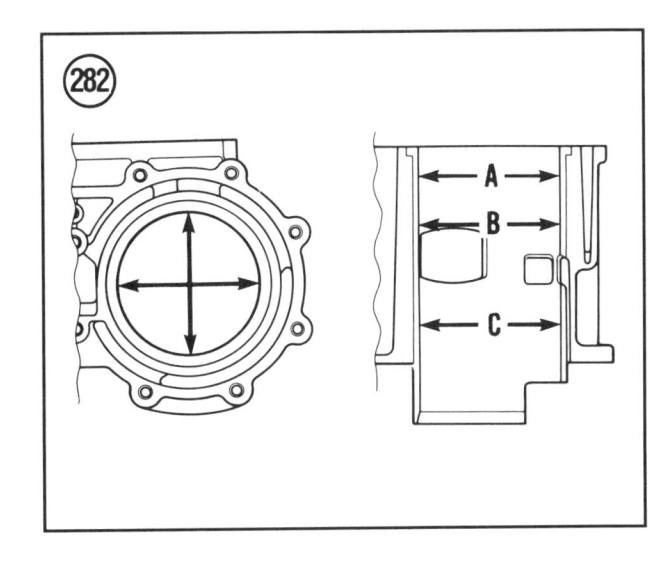

Bearings can be reused if they are in good condition. To be on the safe side however, it is a good idea to discard all removable bearings and install new ones whenever the engine is disassembled. New bearings are inexpensive compared to the cost of another overhaul caused by the use of marginal bearings.

1. Remove any sealer from outer edge of ball bearings with a scraper, then clean bearing surface with kerosene.
2. Place ball bearings in a wire basket and submerge in a suitable container of fresh kerosene or solvent. The bottom of the basket should not touch the bottom of the container.
3. Agitate basket containing bearings to loosen all grease, sludge and other contamination within the bearings.
4. Dry ball bearings with dry filtered compressed air. Be careful not to spin the bearings as they will be damaged.
5. Lubricate the dry bearings with a light coat of Suzuki Outboard Oil and inspect for rust, wear, scuffed surfaces, heat discoloration or other defects. Replace as required.
6. If caged roller bearings are to be reused, repeat Steps 2-5, cleaning one set at a time to prevent any possible mixup. Check bearings for flat spots. If one roller is defective, replace the entire bearing.

Piston Cleaning and Inspection (All Engines)

1. Check the piston(s) for scoring, cracking, cracked or worn piston pin bosses or metal damage. Replace the piston and pin as an assembly if any of these defects are noted.
2. Remove any carbon deposits on the piston crown with a hardwood or plastic scraper.

NOTE
To remove stubborn carbon deposits in Step 3, carefully scrape the ring groove with the recessed end of a broken ring. Do not use an automotive ring groove cleaning tool as it can damage the piston groove locating pin.

3. Check each piston ring groove for distortion, loose ring locating pins or excessive wear. If the flexing action of the rings has not kept the lower surface of the ring grooves free of carbon, clean with a bristle brush and solvent.
4. Immerse the pistons in a carbon removal solution to remove any carbon deposits not removed in Step 3. If the solution does not remove all of the carbon, carefully use a fine wire brush and avoid burring or rounding of the machined edges. Clean the piston skirt with crocus cloth.
5. Measure the piston diameter at right angles to the piston pin with a micrometer (**Figure 283**).

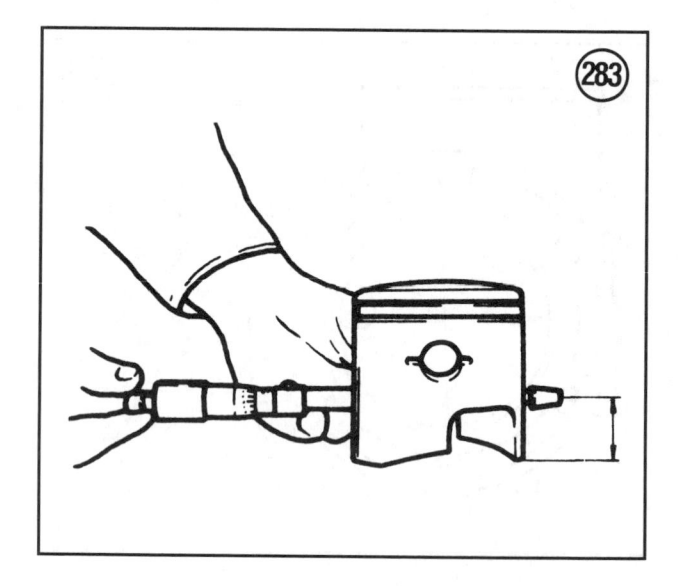

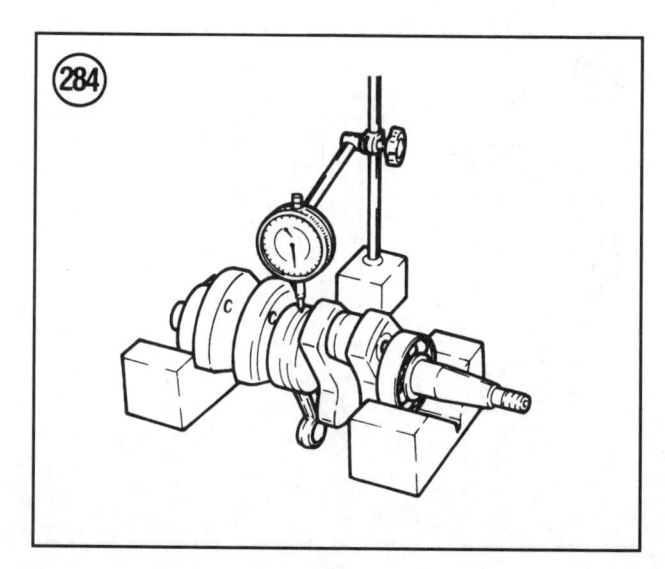

POWER HEAD

Refer to **Table 2** for exact measurement point on piston skirt according to engine.

6. Measure the piston bore diameter in the cylinder block with a bore gauge at the point indicated in **Table 2**.

7. Subtract the measurement obtained in Step 5 from the Step 6 measurement to determine the cylinder-to-piston clearance. Compare clearance to specifications (**Table 2**). If clearance exceeds specifications, have the cylinder block rebored by a dealer or machine shop and install oversize pistons.

NOTE
Obtain the new pistons and measure them before having the cylinders bored. This allows for variations in piston size to manufacturing tolerances.

Crankshaft Cleaning and Inspection (All Engines)

Only on 1985 DT 115 and DT 140 models, can the connecting rods be disassembled from the crankshaft. On all other models, if either the rod(s) or the crankshaft is defective, replace the entire assembly.

1. Clean the crankshaft thoroughly with kerosene and a brush. Blow dry with dry filtered compressed air and lubricate with a light coat of Suzuki Outboard Motor oil.

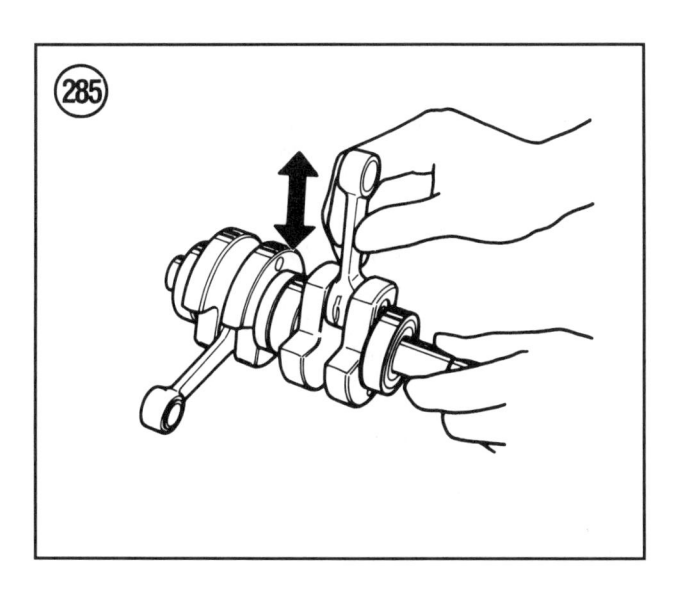

2. Check the crankshaft journals and crankpins for scratches, heat discoloration or other defects.

3. Check drive shaft splines, flywheel taper threads, keyway and oil injection pump drive gear (if so equipped) for wear or damage. Replace crankshaft as required.

4. If lower crankshaft ball bearing has not been removed, grasp inner race and try to work it back and forth. Replace bearing if excessive axial play is noted.

5. Lubricate all ball bearings with Suzuki Outboard Motor oil and rotate outer race. Replace the bearing if it sounds or feels rough or if it does not rotate smoothly.

6. Support the crankshaft on a pair of V-blocks and check runout with a dial indicator (**Figure 284**). Divide runout reading in half to determine crankshaft deflection. If deflection exceeds 0.0012 in. (0.03 mm) for DT 2 models or 0.002 in. (0.05 mm) for all others, replace the crankshaft assembly.

7. With all pistons removed and the crankshaft on V-blocks, check wear and condition of the big end of the connecting rod. Hold crankshaft steady and move connecting rod up and down with your hand (**Figure 285**). Make sure there is no rattle in the big end.

Piston and Connecting Rod Assembly (All Engines)

If the pistons were removed from the connecting rods, they must be correctly oriented when reassembling. The arrow on the piston crown (**Figure 286**, typical) must face as follows:
  a. DT 2: arrow pointing down toward the splined end of the crankshaft.
  b. DT 4: arrow up.
  c. All other models: arrow toward the exhaust port side.

1. Coat piston pin and piston pin/connecting rod small end bores with Suzuki Outboard Motor oil.

2. Check the piston dome number made during disassembly and match piston with its correct connecting rod.

3. Insert caged bearing in connecting rod small end bore. If washers are used, position on each side of bore.

4A. On V6 models, perform the following:
   a. With the piston properly oriented, fit the piston over the connecting rod small end, align the piston and small end bores.
   b. Position the piston pin with the closed off end facing up, open end down (**Figure 287**).
   c. Work the pin through the piston and connecting rod small end bores.

4B. On all other models, with the piston properly oriented, fit the piston over the connecting rod small end, align the piston and small end bores, then work the pin through the piston and connecting rod small end bores.

5. Install *new* piston pin circlips with snap ring pliers. Sharp edge of circlip should face outward and circlip opening should face directly up or down. Make sure circlips seat in piston grooves.

6. Repeat procedure for each remaining piston.

Piston Ring Installation (All Engines)

Some pistons use a keystone design top ring and a flat design bottom ring. Check new rings carefully before installation and install them in their proper grooves.

1. Check end gap of new rings before installing on piston. Place ring in cylinder bore just above the intake and exhaust ports, then square it up by inserting the bottom of an old piston. Measure the gap with a feeler gauge (**Figure 288**) and compare to specifications (**Table 3**).

2. If ring gap is excessive in Step 1, repeat the step with the ring in another cylinder. If gap is also excessive in that cylinder, discard and replace with another new ring.

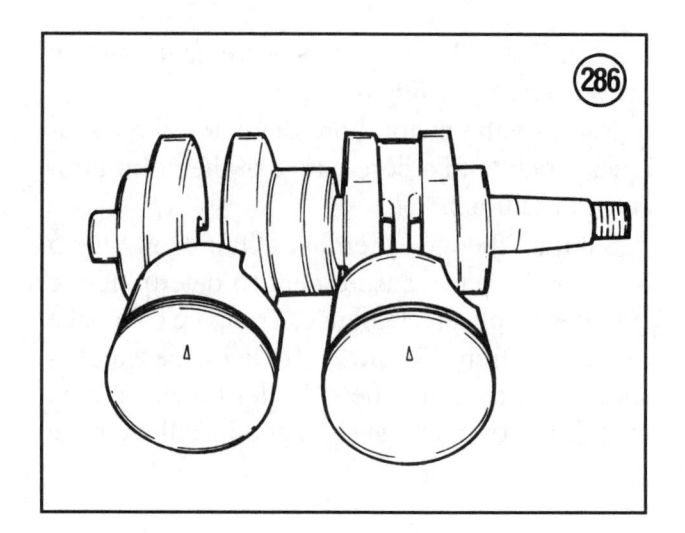

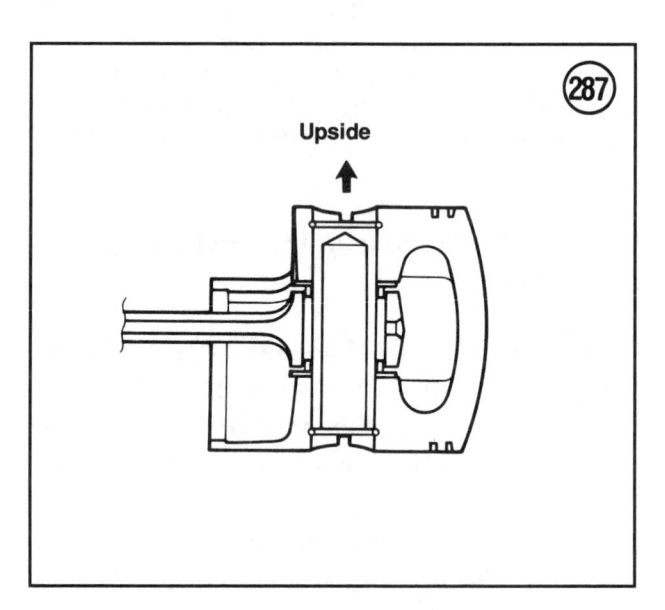

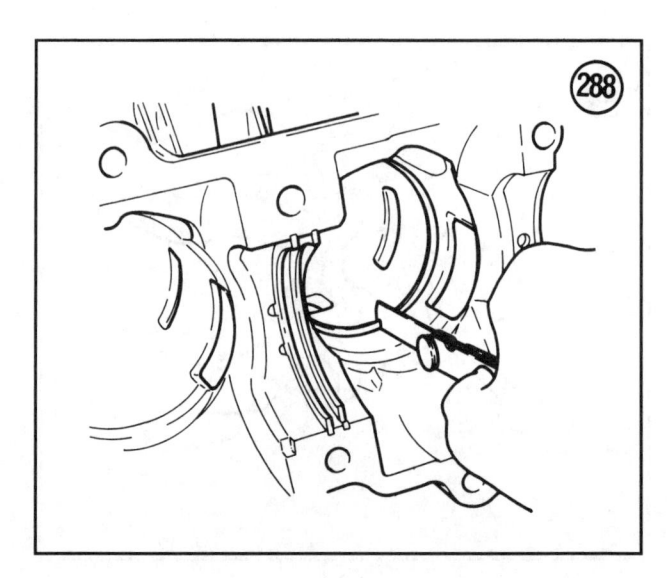

POWER HEAD

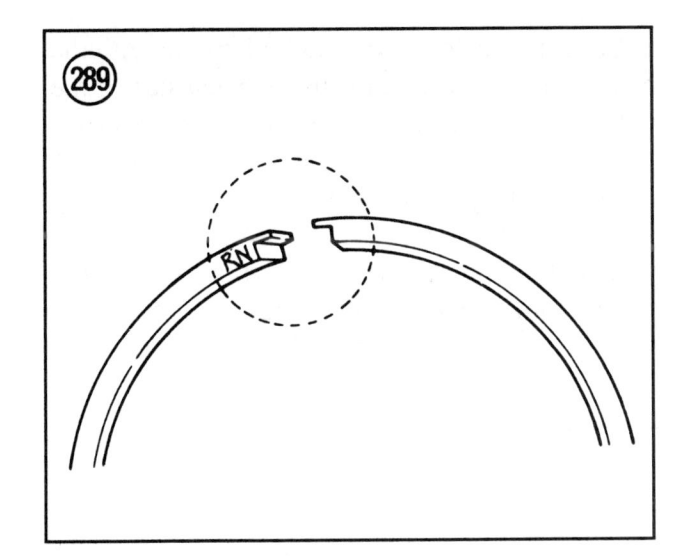

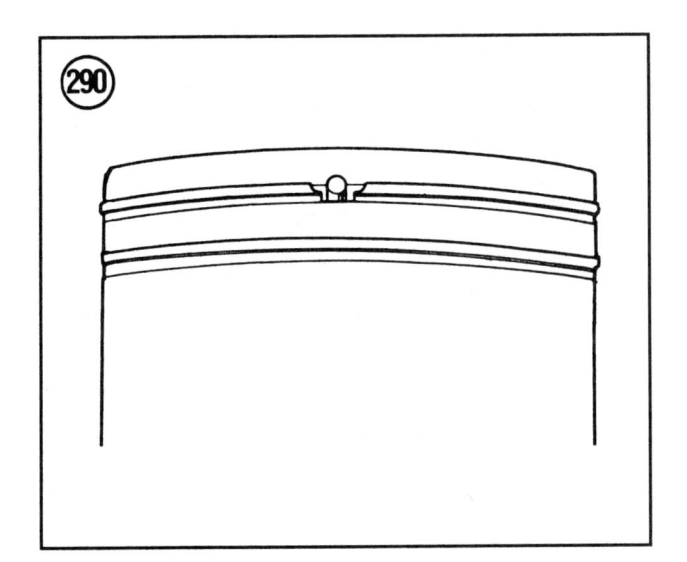

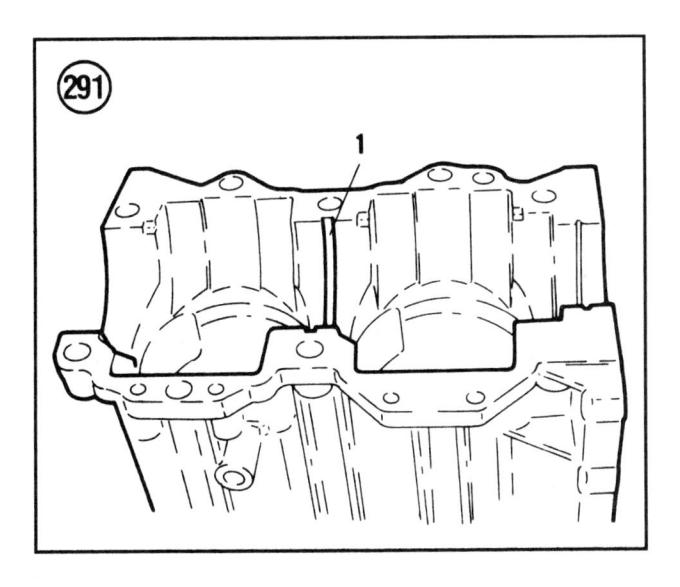

3. If ring gap is insufficient in Step 1, the ends of the ring can be filed slightly. Clean ring thoroughly and recheck gap as in Step 1.

NOTE
*Piston rings must be installed in Step 4 and Step 5 with the mark on the end of the ring facing the piston crown (**Figure 289**). On some late models, the piston rings are not marked and may be installed either way.*

4. Once the ring gaps are correctly established, spread the ring just enough to fit it over the piston head and into position. Repeat this step to install the upper ring.

5. Position each ring so that the piston groove locating pin fits in the ring gap (**Figure 290**). Proper ring positioning is necessary to minimize compression loss and prevent the ring ends from catching on the cylinder ports.

Piston, Connecting Rod and Crankshaft Installation (All Inline Models Except 1985 DT 115 and DT 140)

1. Coat the piston(s), rings and cylinder bore(s) with Suzuki Outboard Motor oil.
2. Install any labyrinth seals (**Figure 291**) or thrust rings (**Figure 292**) in the crankcase.

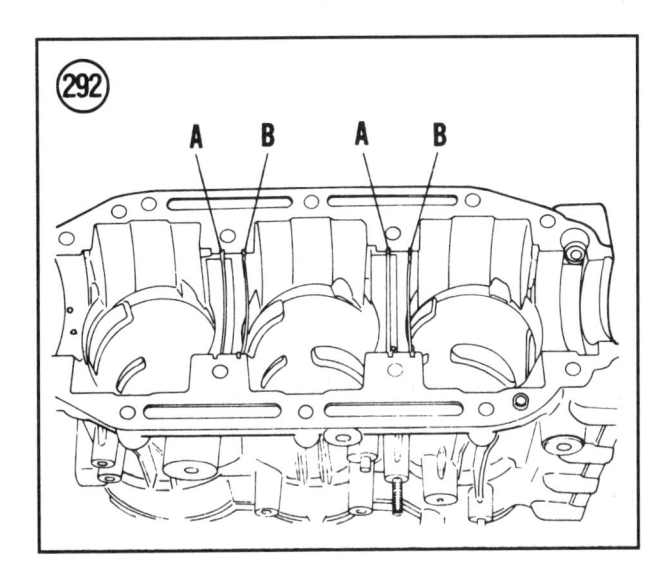

3. On 4-cylinder models, make sure the bearing locating pins are tightly installed in the crankcase.

4. With the crankcase on a clean flat surface, carefully lower the crankshaft assembly until the pistons start to enter the cylinder bore(s). Compress the rings on one piston with your fingers, making sure that the end gaps fit over the locating pins, and slowly work the piston/ring assembly into the cylinder. Repeat this step to fit each remaining piston into its bore.

5. When all piston/ring assemblies have entered the cylinder bores, apply sufficient downward pressure to seat the pistons in their bores and the crankshaft within the crankcase.

6. On 4-cylinder models, index the hole in each bearing with the locating pin in the crankcase. This alignment is necessary otherwise the crankshaft will not completely seat in the crankcase.

7. Reach through the exhaust ports and lightly depress each ring with a small screwdriver blade. The ring should snap back when pressure is released. If it does not, the ring was broken during piston installation and will have to be replaced.

8. On DT 2 models, wipe a new crankshaft O-ring with water-resistant grease (part No. 99000-25160) or equivalent and install on lower end of crankshaft. Coat the splined end of the crankshaft with water-resistant grease.

9. Make sure oil seal flanges fit into crankcase cutouts.

10. On models so equipped, rotate each bearing to position its locating pin in the crankcase cutout (**Figure 293**).

Piston and Connecting Rod Installation (1985 DT 115 and DT 140)

1. Coat the pistons, rings and cylinder bores with Suzuki Outboard Motor oil.

2. Check the piston dome number made during disassembly and match piston with its correct cylinder bore.

3. Insert the piston into its cylinder bore with the arrow on its crown facing the exhaust side of the cylinder block. Make sure the rings are properly positioned in their grooves and that the ring gaps are correctly positioned in relation to the ring groove locating pins.

4. Install a suitable ring compressor over the piston dome and rings. With the ring compressor resting on cylinder head, tighten it until the rings

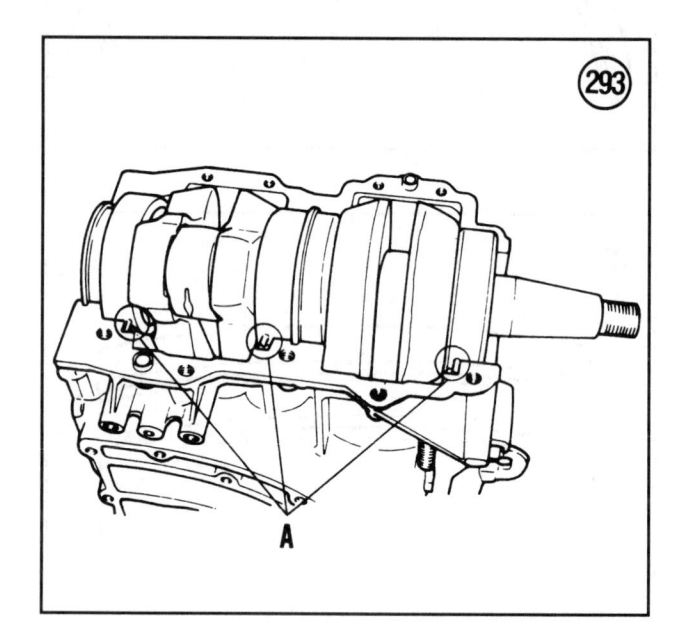

POWER HEAD

are compressed sufficiently to enter the bore (**Figure 294**).

5. Hold the connecting rod end with one hand to prevent it from scraping or scratching the cylinder bore and slowly push the piston into the cylinder.

6. Remove the ring compressor tool and repeat the procedure to install remaining pistons.

7. Reach through the exhaust ports and lightly depress each ring with a small screwdriver blade. The ring should snap back when pressure is released. If it does not, the ring was broken during piston installation and will have to be replaced.

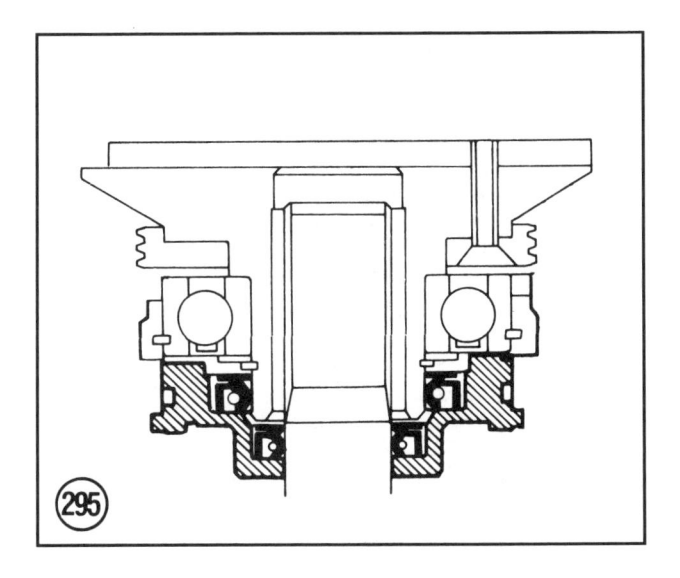

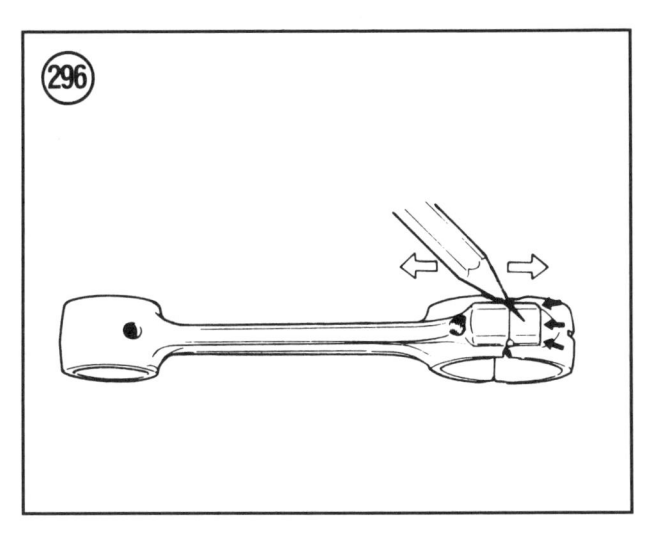

Crankshaft Installation
(1985 DT 115 and DT 140)

1. Install new seals into the lower oil seal housing as shown in **Figure 295**. Fill the cavity between the seals with water-resistant grease (part No. 99000-25160) or equivalent.

2. Install a new O-ring on the lower oil seal housing. Wipe O-ring with water-resistant grease.

3. Install lower oil seal housing on the crankshaft.

4. Remove the connecting rod caps. Coat connecting rod bearing surfaces and bearing cages with Suzuki Outboard Motor oil. Install a bearing cage in the rod.

5. Lubricate the crankshaft with Suzuki Outboard Motor oil and install into the cylinder block.

6. Position the connecting rod under the crankshaft and pull it up to the crankpin.

7. Lightly coat the exposed part of the crankpin with Suzuki Outboard Motor oil. Install the other bearing cage.

8. Install new connecting rod cap screws and tighten finger-tight.

CAUTION
The procedure detailed in Step 9 is very important to proper engine operation as it affects bearing action. If not done properly, major engine damage can result. It can also be a time-consuming and frustrating process. Work slowly and with patience. If alignment cannot be achieved, replace the connecting rod.

9. Run a dental pick or pencil point along the cap match marks to check cap offset (**Figure 296**). Rod and cap must be aligned so that the dental pick or pencil point will pass smoothly across the fracture line at each of the 3 faces indicated by the arrows in **Figure 296**. If alignment is not correct, gently tap cap with a plastic hammer. When rod and cap are properly aligned, tighten cap bolts to specifications (**Table 1**).

344 CHAPTER EIGHT

10. Rotate the crankshaft to check for binding. If the connecting rod does not float freely over the full length of the crankpin, loosen the rod cap and repeat Step 9.

11. Repeat Steps 6-10 for remaining cylinders.

Crankshaft and Connecting Rod Installation (V4 and V6)

NOTE
Be sure to locate the crankshaft with the tapered end facing up into the crankcase.

1. Coat the cylinder bores and bearing bores in the crankcase with Suzuki Outboard Motor oil.

2. If removed, install the center bearing(s) onto the crankshaft. Install the retaining clip onto the bearing so that the clip ends cover each split face of the outer race (**Figure 297**).

3. If removed, install the new oil seals as follows:
   a. Apply Suzuki Super Grease "A" (part No. 99000-25030) to the seal lip. If the existing oil seals are being reused, coat their seal with the same grease.
   b. Install the oil seals with the lip facing downward as shown in **Figure 298**.

4. Place the crankcase upside down in the wooden stand used during disassembly.

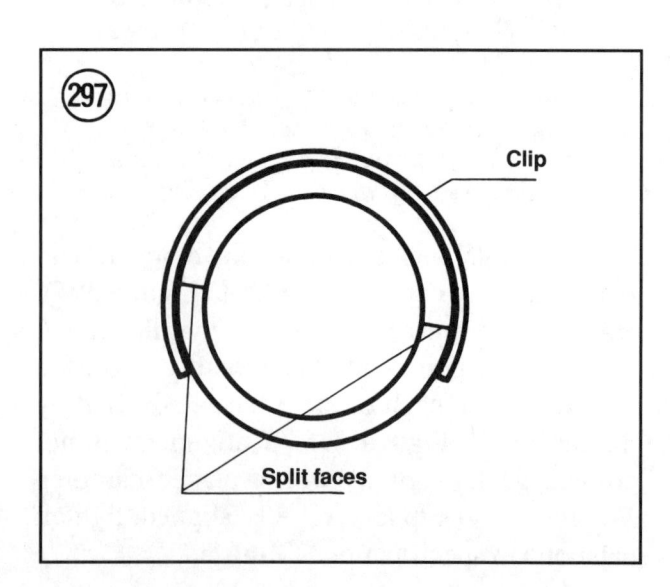

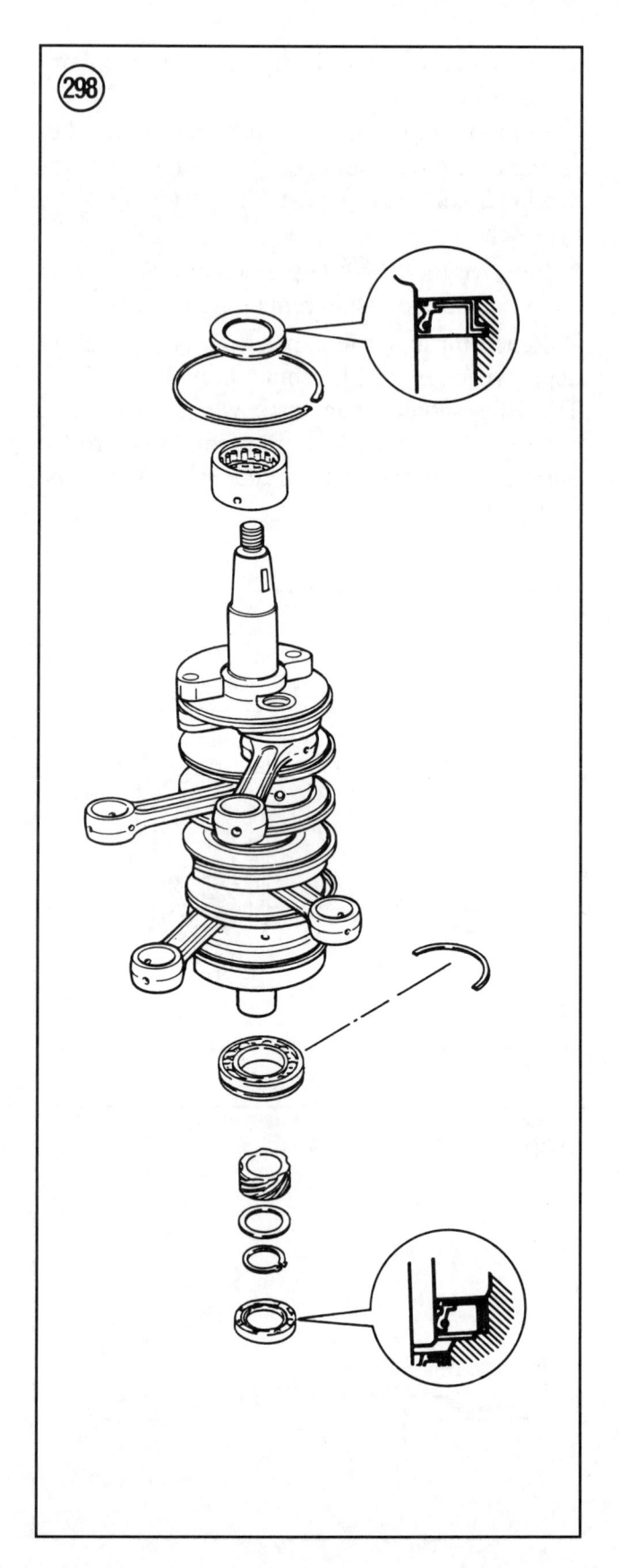

POWER HEAD

5. Make sure the bearing locating pins are tightly installed in the crankcase.

CAUTION
Wrap the ends of the connecting rods with several layers of duct tape. This will keep the connecting rod ends from scratching the cylinder walls during crankshaft installation.

6. Carefully lower the crankshaft and connecting rod assembly into the crankcase while guiding the connecting rods into their correct cylinder bores.

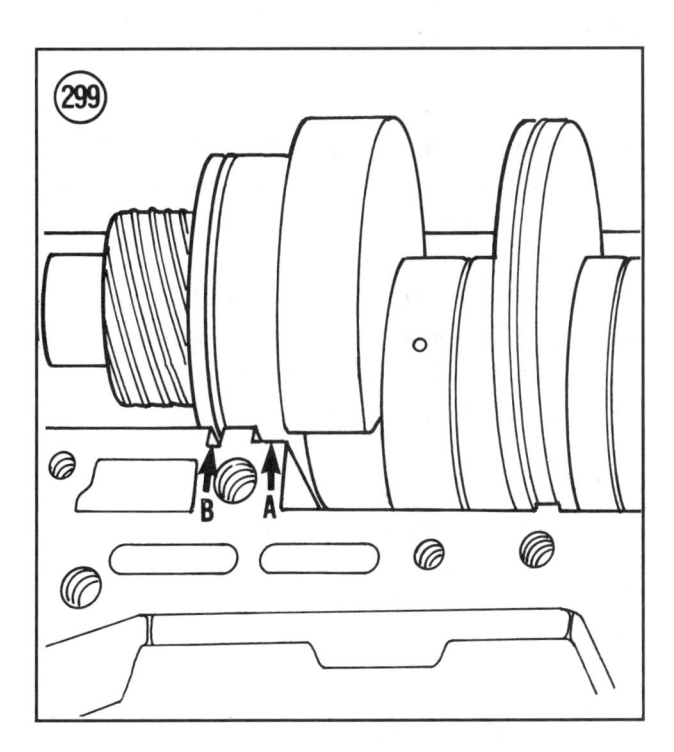

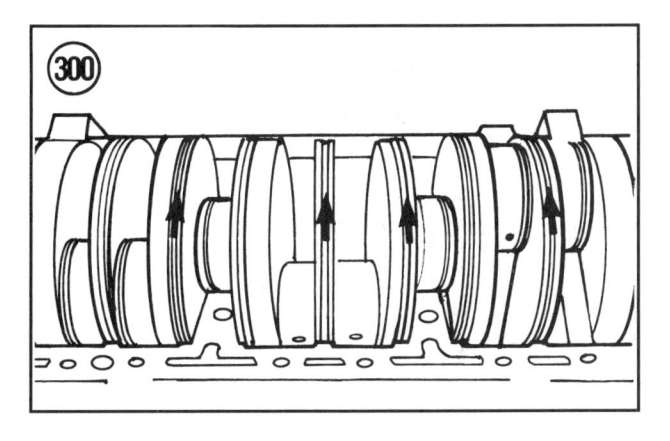

7. Index the hole in each bearing with the locating pin in the crankcase. This alignment is necessary otherwise the crankshaft will not completely seat in the crankcase. Push the crankshaft all the way down into position in the crankcase. Remove the duct tape from the ends of the connecting rods.

8. Make sure upper oil seal flanges fit into crankcase cutout.

9. Rotate the lower bearing to position its locating pin in the crankcase cutout (A, **Figure 299**).

10. Make sure the "C" ring is correctly located in the crankcase groove (B, **Figure 299**).

11. At this point the end gaps of the seal rings (**Figure 300**) must all face UP. Reposition if necessary.

12. Coat the crankshaft crankpin bearings and seal rings with Suzuki Outboard Motor oil.

Assembly (DT 2)

1. Install the piston onto the connecting rod as described in this chapter.

2. Install the crankshaft as described in this chapter.

3. Apply a liberal quantity of Suzuki Outboard Motor Oil to the connecting rod big end.

4. Apply Suzuki Bond No. 4 (part No. 99000-31030) or equivalent to the mating surfaces on both halves of the crankcase (**Figure 301**).

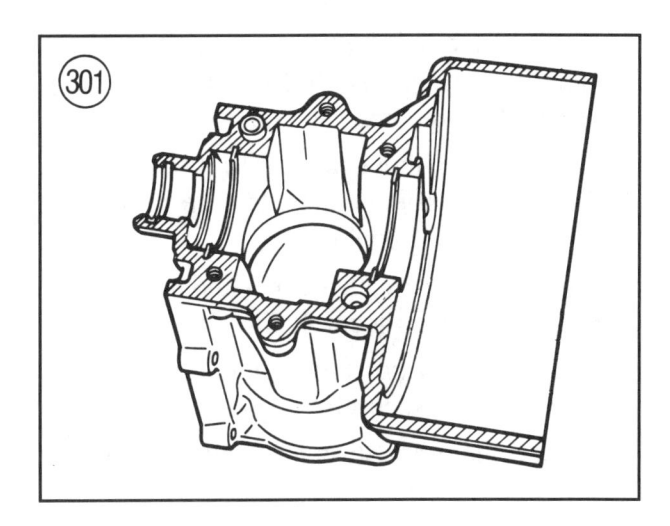

5. Install the crankcase and tighten the bolts to specifications (**Table 1**) in a diagonal pattern.

6. Rotate the crankshaft several turns to check for binding. If crankshaft does not turn easily, disassemble and correct the interference.

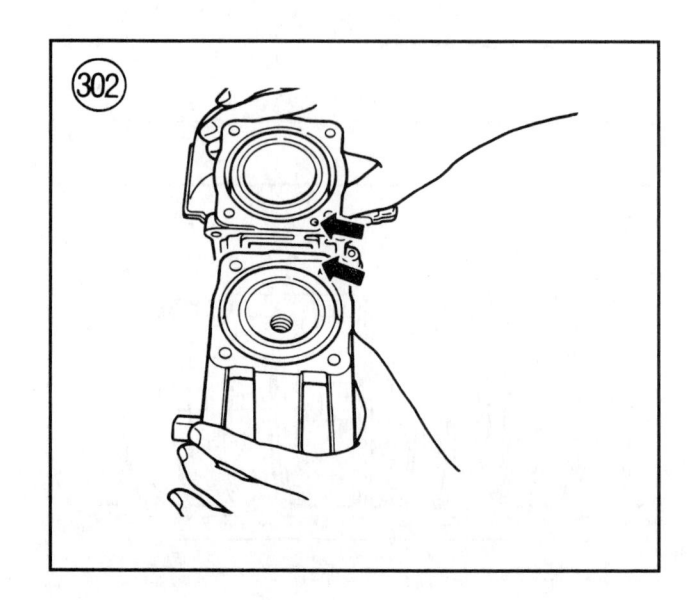

7. Install the cylinder head with a new gasket. Make sure that the water passages in the cylinder head, gasket and crankcase align (**Figure 302**). Tighten the cylinder head nuts to specifications (**Table 1**) in a diagonal pattern.

8. Install the intake manifold and carburetor assembly.

9. Install the power head as described in this chapter.

Assembly (DT 4)

1. Install the pistons onto the crankshaft as described in this chapter.

2. Install the crankshaft as described in this chapter.

3. Apply a liberal quantity of Suzuki Outboard Motor Oil to the connecting rod big end.

4. Apply Suzuki Bond No. 4 (part No. 99000-31030) or equivalent to the mating surfaces on both halves of the crankcase (**Figure 301**).

5. Install the crankcase and tighten the bolts to specifications (**Table 1**) in a diagonal pattern.

6. Rotate the crankshaft several turns to check for binding. If crankshaft does not turn easily, disassemble and correct the interference.

7. Install the cylinder head with a new gasket. Tighten the cylinder head bolts to specifications (**Table 1**) in a diagonal pattern.

8. Install the water jacket cover and new gasket. Tighten the screws securely.

9. Install the rewind starter mounting base and tighten the bolts securely.

10. Install the lubrication hose onto each fitting and tighten the clips.

Assembly

DT 6; 1985-1987 DT 8

Refer to **Figure 303** for this procedure.

1. Install the pistons onto the connecting rod as described in this chapter.

2. Install the crankshaft as described in this chapter.

3. Apply a liberal quantity of Suzuki Outboard Motor Oil to the connecting rod big end.

4. Apply Suzuki Bond No. 4 (part No. 99000-31030) or equivalent to the mating surfaces on both halves of the crankcase (**Figure 301**).

5. Install the crankcase but do *not* tighten the bolts at this time.

6. Position the reed valve assembly and new gasket so the reed valves face in toward the crankcase as shown in A, **Figure 304**. Install the screw and tighten securely.

7. Install the crankcase bolts and lockwashers. Be sure to install the clamp (A, **Figure 305**) in the correct location. Tighten the bolts to specifications (**Table 1**) in the sequence shown in **Figure 305**.

8. Rotate the crankshaft several turns to check for binding. If crankshaft does not turn easily, disassemble and correct the interference.

POWER HEAD

347

303

**POWER HEAD
(DT 6 AND 1985-1987 DT 8)**

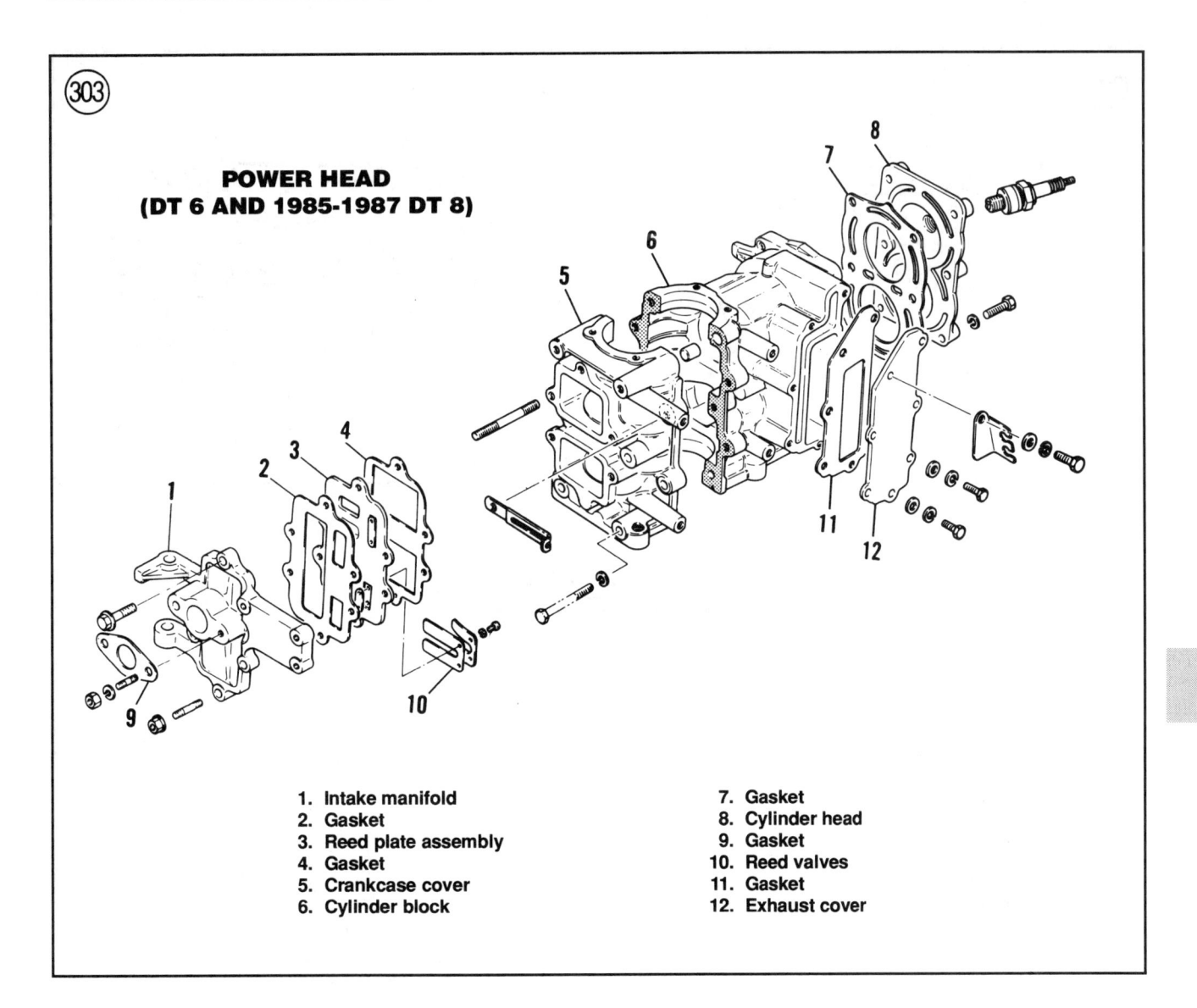

1. Intake manifold
2. Gasket
3. Reed plate assembly
4. Gasket
5. Crankcase cover
6. Cylinder block
7. Gasket
8. Cylinder head
9. Gasket
10. Reed valves
11. Gasket
12. Exhaust cover

304

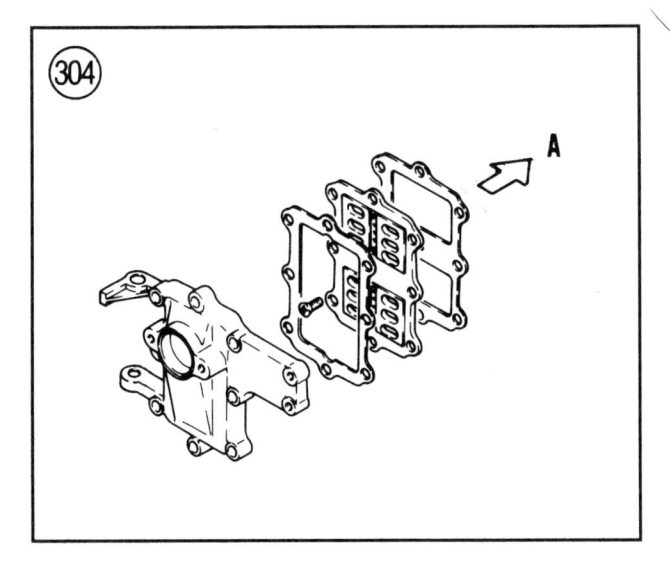

305

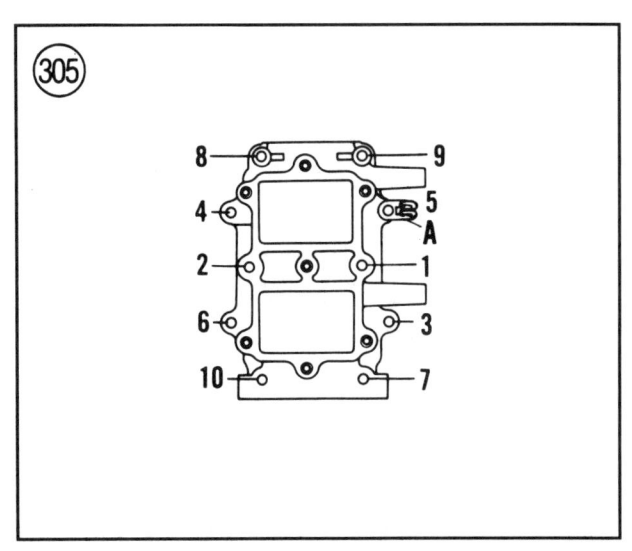

8

348 CHAPTER EIGHT

306

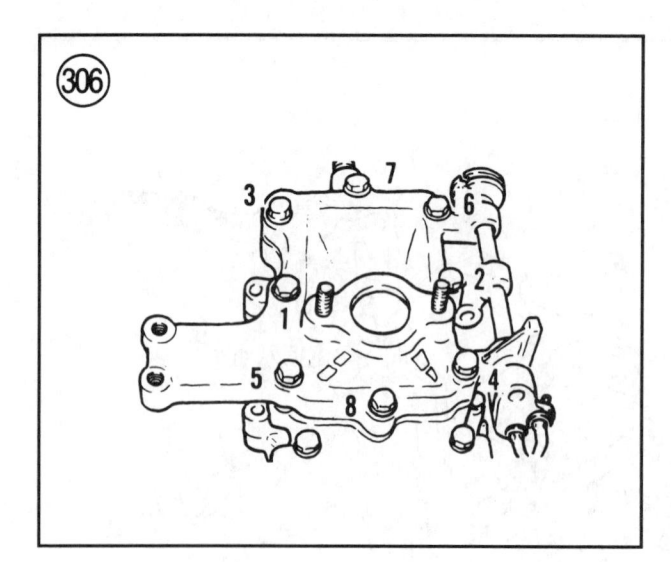

307

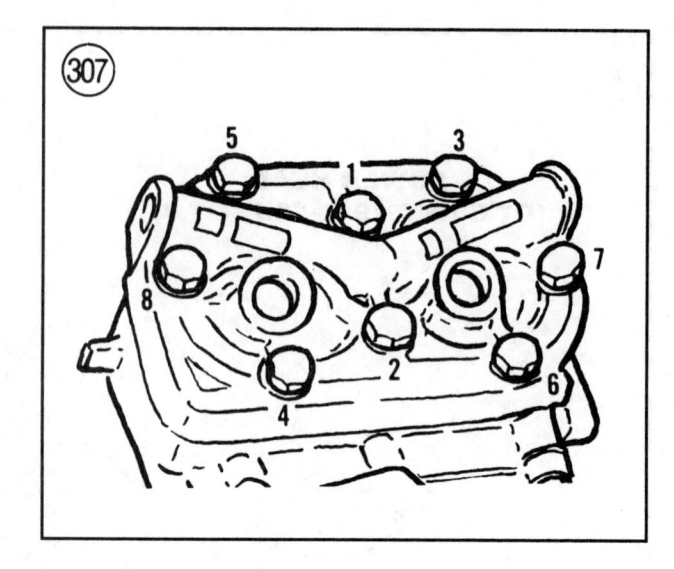

308

POWER HEAD
(1988-ON DT 8, DT 9.9)

1. Intake manifold
2. Reed plate assembly
3. Reed valves
4. Crankcase cover
5. Gasket
6. Cylinder head
7. Gasket
8. Cylinder block
9. Gasket
10. Inner exhaust plate
11. Exhaust cover

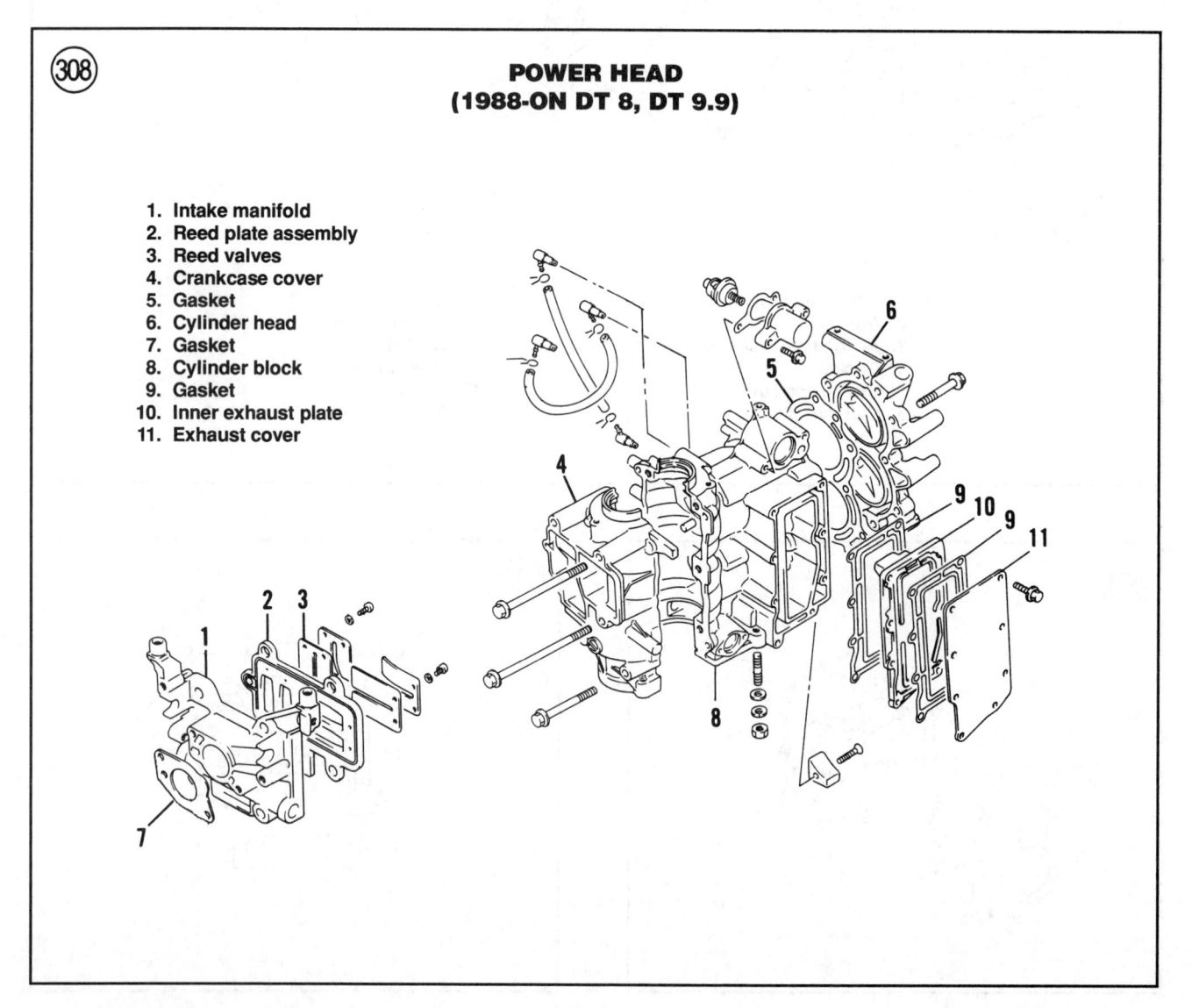

POWER HEAD

9. Install the exhaust cover and new gasket. Install the bolts and tighten securely.
10. Install the intake manifold and new gasket. Tighten the bolts in the sequence shown in (**Figure 306**).
11. Apply Suzuki Super Grease "A" (part No. 990000-25010) to the lower seal lips and install the lower seal assembly. Tighten the bolts and lockwashers securely.
12. Install the cylinder head and new gasket. Install the bolts and tighten to specifications (**Table 1**) in the sequence shown in **Figure 307**.
13. Install the power head as described in this chapter.

1988-on DT 8; 1985-on DT 9.9, DT 15

Refer to **Figure 308** for 1988-on DT 8 and DT 9.9 models or **Figure 309** for 1985-1987 DT 9.9 and 1985-on DT 15 models for this procedure.

1. Install the crankshaft assembly as described in this chapter.
2. Apply Suzuki Super "A" grease (part No. 99000-25030) or equivalent to the crankshaft lower seal lip.
3. Install the new lower oil seal onto the crankshaft so the lip faces away from the crankshaft or downward.

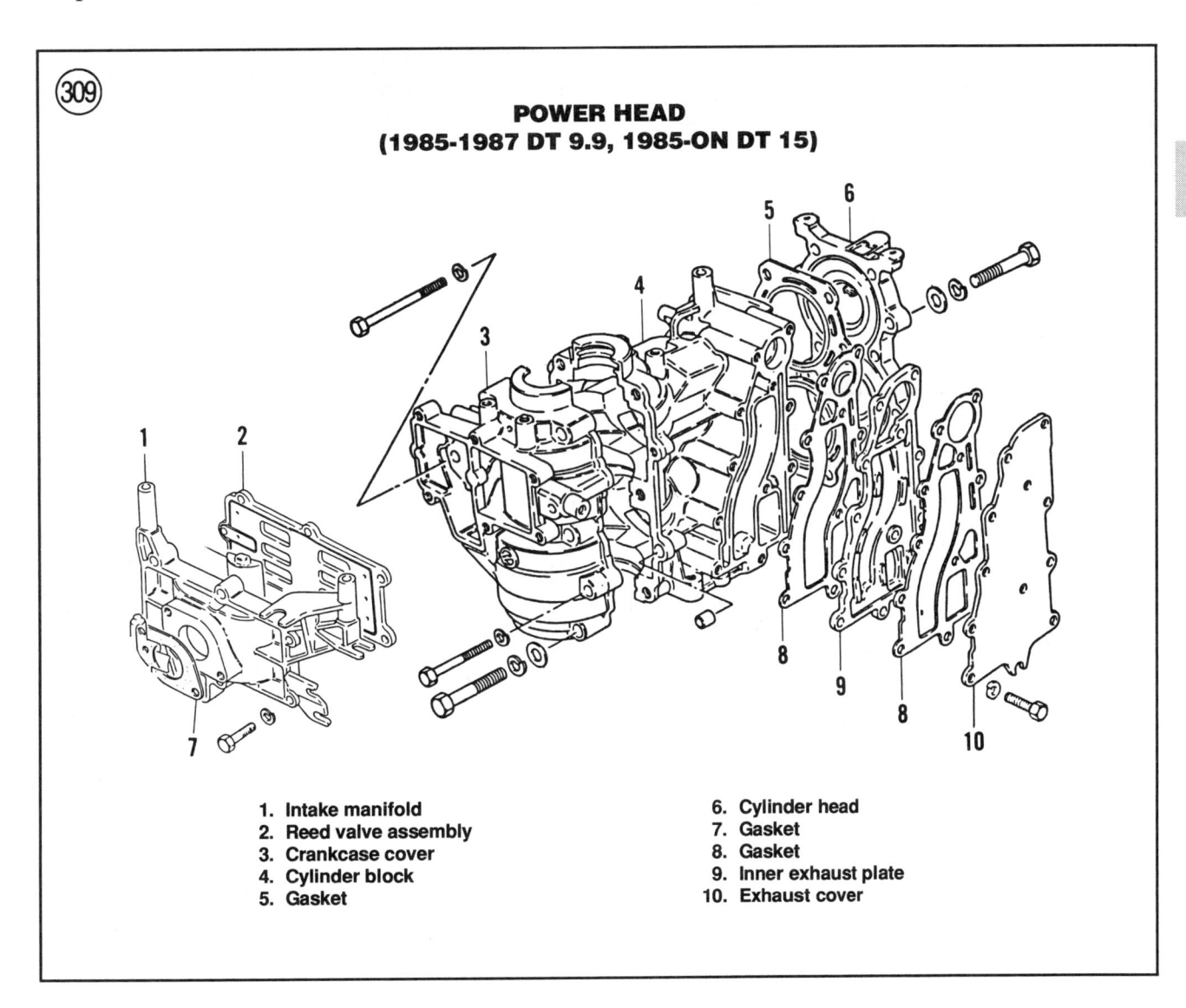

309 **POWER HEAD (1985-1987 DT 9.9, 1985-ON DT 15)**

1. Intake manifold
2. Reed valve assembly
3. Crankcase cover
4. Cylinder block
5. Gasket
6. Cylinder head
7. Gasket
8. Gasket
9. Inner exhaust plate
10. Exhaust cover

4. Apply Suzuki Bond No. 4 (part No. 99000-31030), or Suzuki Bond No. 1207B (part No. 99104-31140) or equivalent to the mating surfaces of both the crankcase cover and cylinder block (**Figure 310**).

5. Install the crankcase cover onto the cylinder block and install the bolts. Tighten the bolts to specifications (**Table 1**) following the appropriate sequence as shown in **Figure 311** for 1985-1987 DT 9.9 and 1985-on DT 15 or **Figure 312** for 1988-on DT 8 and DT 9.9 models.

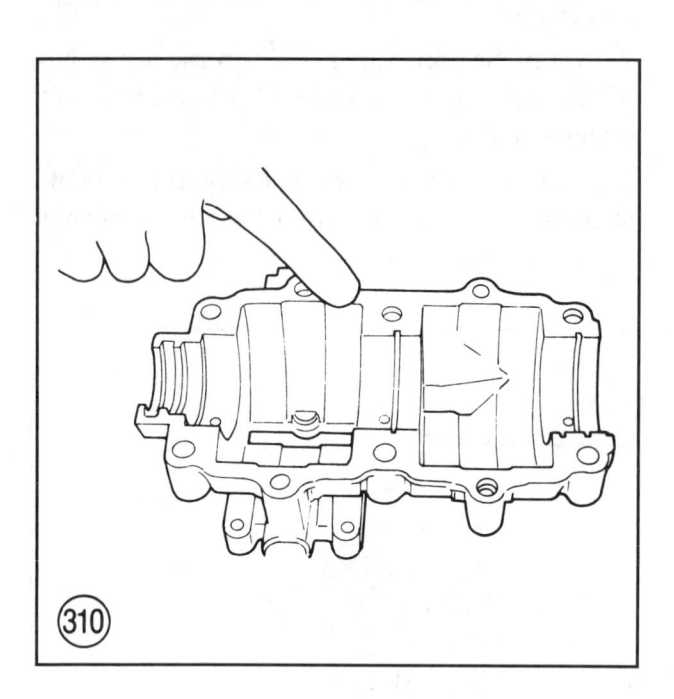

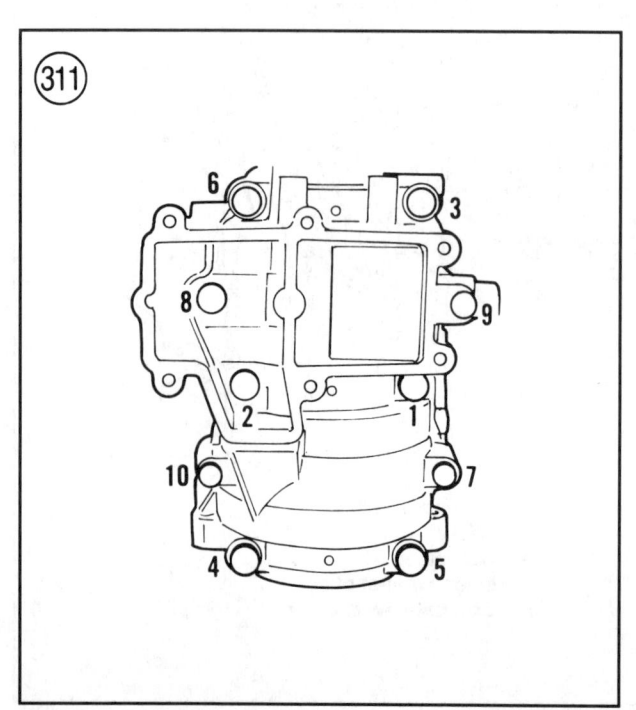

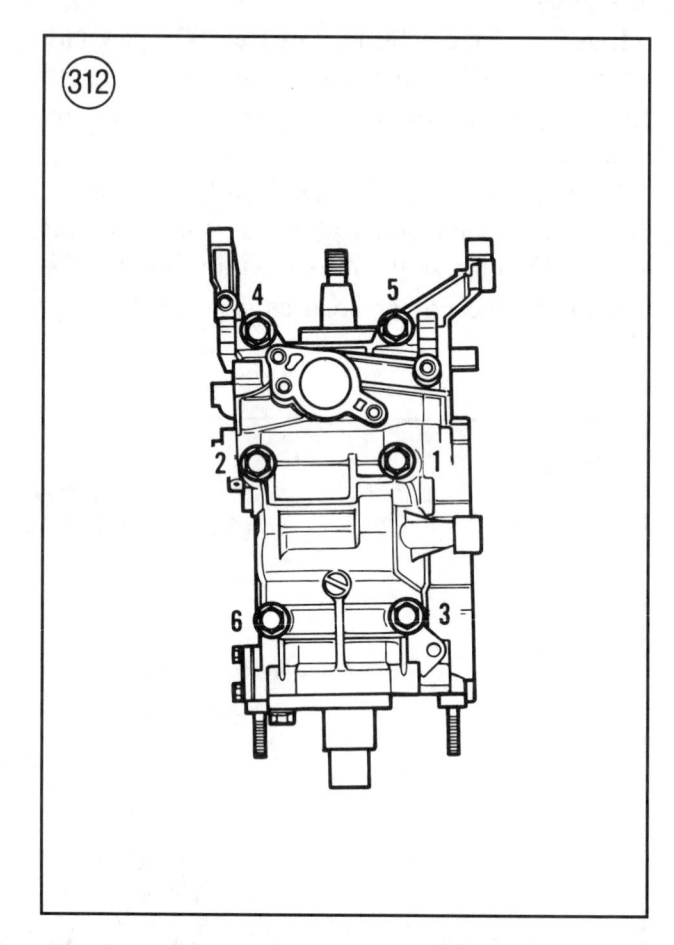

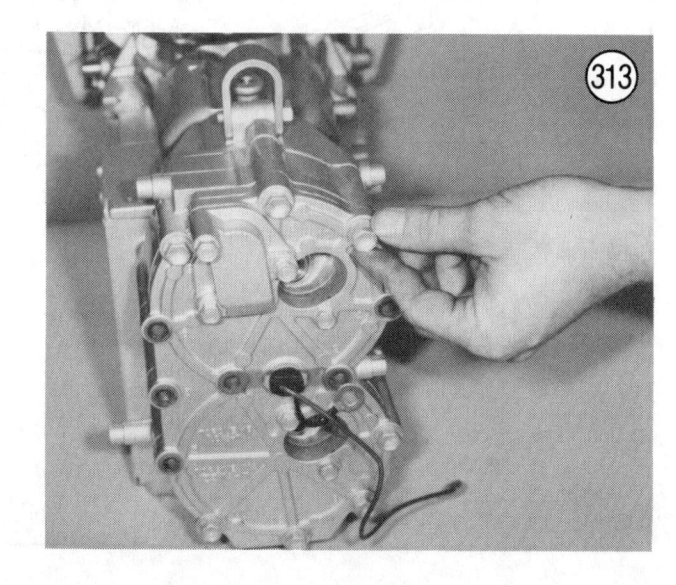

POWER HEAD

6. Rotate the crankshaft several turns to check for binding. If the crankshaft does not turn easily, disassemble and correct the interference.

7. Apply a liberal quantity of Suzuki Outboard Motor Oil to each connecting rod big end through the reed valve openings.

8. Install the cylinder head with a new gasket.

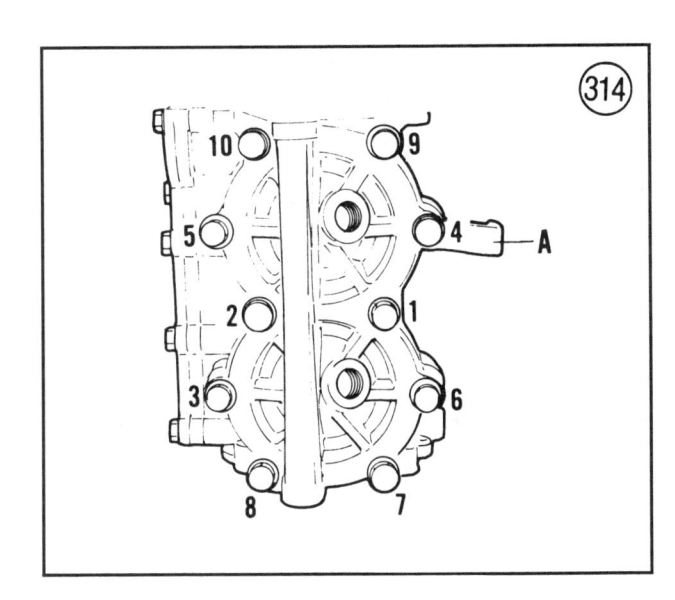

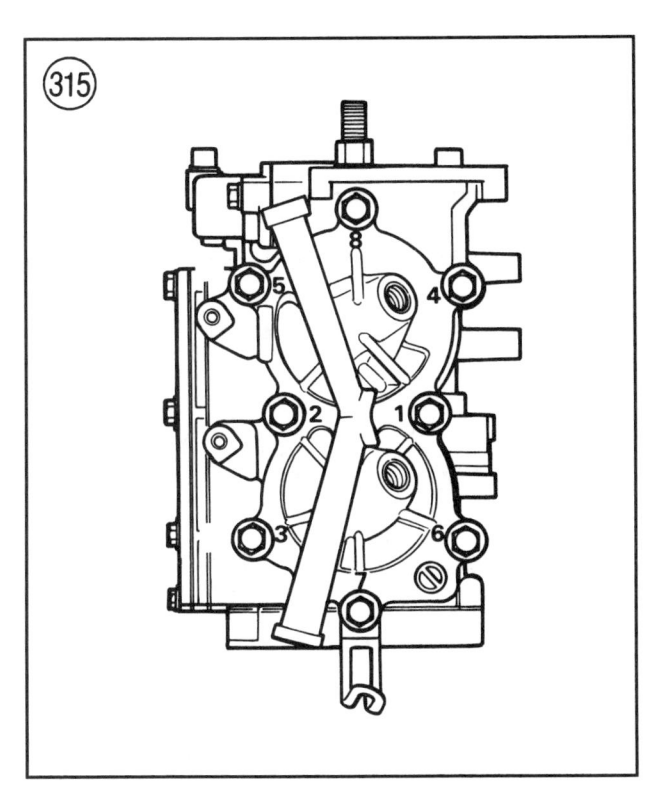

9. Install the cylinder head bolts (**Figure 313**) and any clamps to the locations noted during removal.

10. Tighten the bolts to specifications (**Table 1**) following the appropriate sequence shown in **Figure 314** for 1985-1987 DT 9.9 and 1985-on DT 15 or **Figure 315** for 1988-on DT 8 and DT 9.9 models.

11. Install the inner exhaust plate with a new gasket (**Figure 316**).

12. Install the exhaust cover with a new gasket (**Figure 317**). Install the bolts and tighten securely in a crisscross pattern.

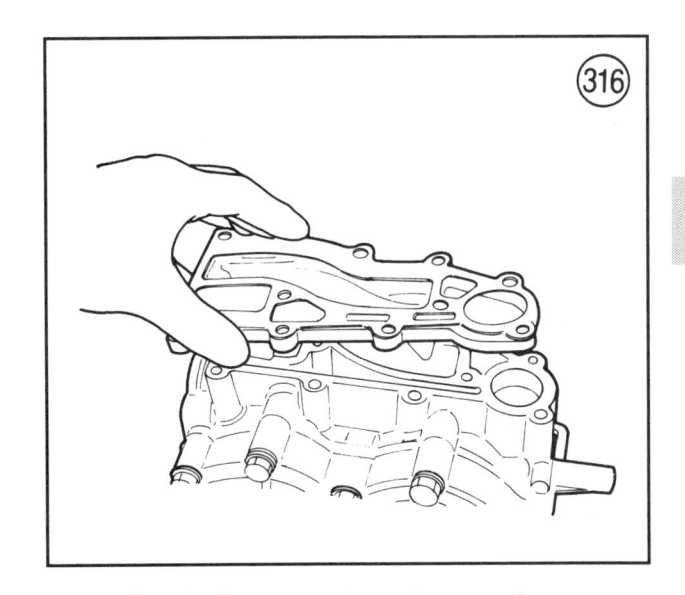

13. Install the reed valve assembly with the reed side facing the crankshaft (**Figure 318**, typical).
14. On 1985-1987 DT 9.9 and 1985-on DT 15 models, install the loose piece into the notches in the intake manifold (**Figure 319**). Hold the loose piece in place during manifold installation.
15. Install the intake manifold and new gasket. Tighten the bolts securely in a crisscross pattern.
16. Install the exhaust cover and new gasket. Tighten the bolts securely in a crisscross pattern.
17. On 1988-on DT 8 and DT 9.9 models, if removed, install the oil hose check valves, nozzles and hoses as follows:
  a. Each oil nozzle (1, **Figure 320**) has a groove and the check valves (2, **Figure 320**) *do not* have a groove.
  b. Install the oil nozzles (1) and check valves (2) into the cylinder in the locations shown in (**Figure 321**). There is a raised bar on the cylinder indicating the direction of the outlet of each part (A, **Figure 321**).
  c. Connect the oil hoses to the oil nozzles and check valves as shown in **Figure 322**.
18. Install the power head as described in this chapter.

DT 20, DT 25, DT 30 (2-cylinder); 1987-1989 DT 35; 1985-on DT 40

1. Install the crankshaft assembly as described in this chapter.

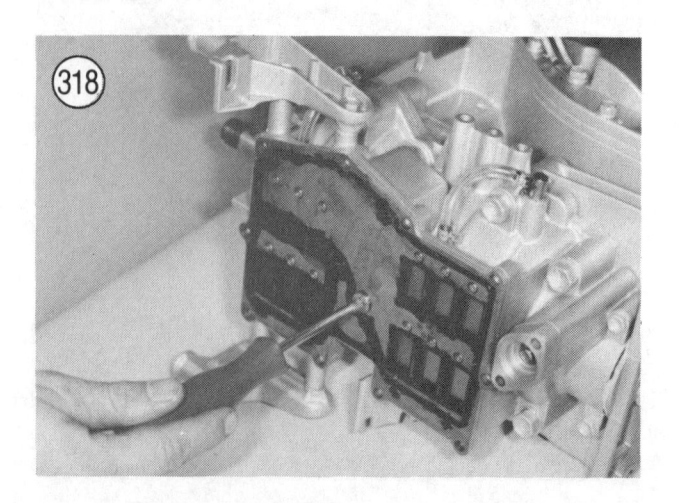

2. Apply Suzuki Super "A" grease (part No. 99000-25030) or equivalent to the crankshaft lower seal lip.

3. Install the new lower oil seal onto the crankshaft so the lip faces away from the crankshaft (down).

4. Apply Suzuki Bond No. 4 (part No. 99000-31030), or Suzuki Bond No. 1215 (part No. 99104-31110) or equivalent to the mating surfaces of both the crankcase cover and cylinder block (**Figure 310**).

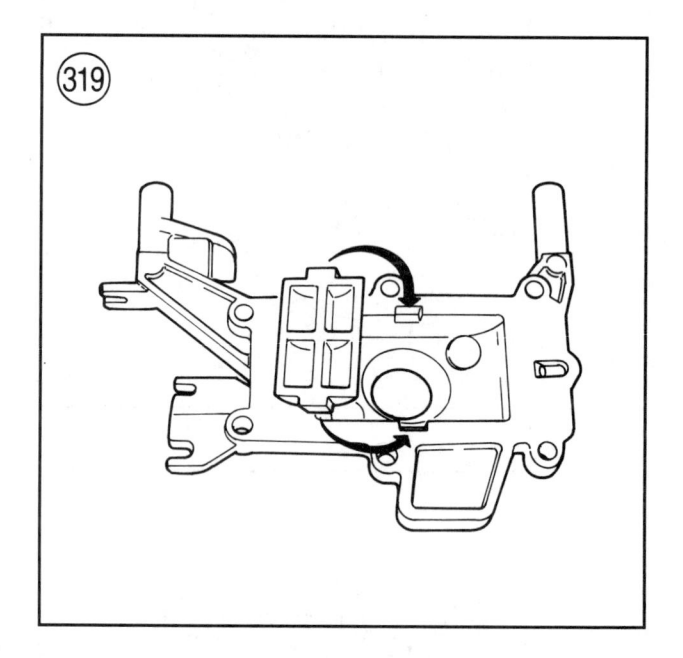

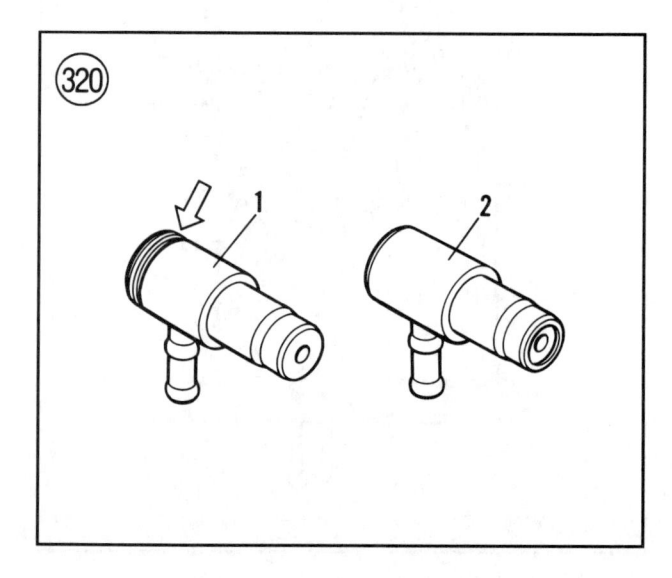

POWER HEAD

5. Install the crankcase cover to the cylinder block and tighten the bolts to specifications (**Table 1**) following the appropriate sequence as shown in **Figure 323A** for DT 20, DT 25 and DT 30 or **Figure 323B** for DT 35 and DT 40 models.

6. Rotate the crankshaft several turns to check for binding. If crankshaft does not turn easily, disassemble and correct the interference.

7. Apply a liberal quantity of Suzuki Outboard Motor Oil to each connecting rod big end through the reed valve openings.

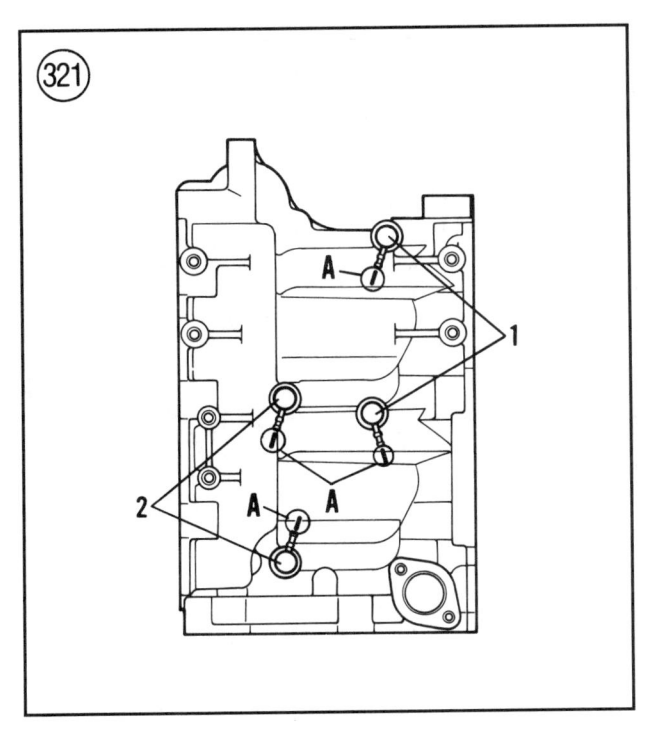

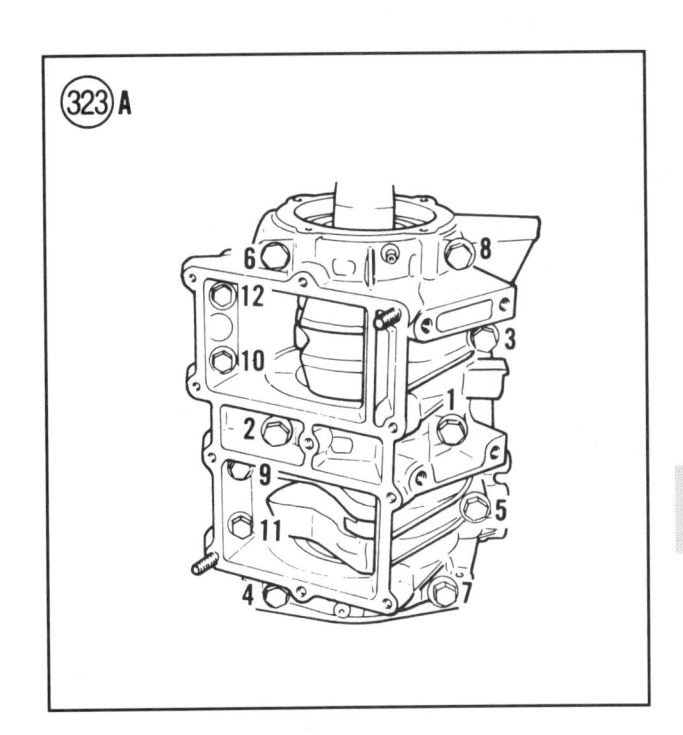

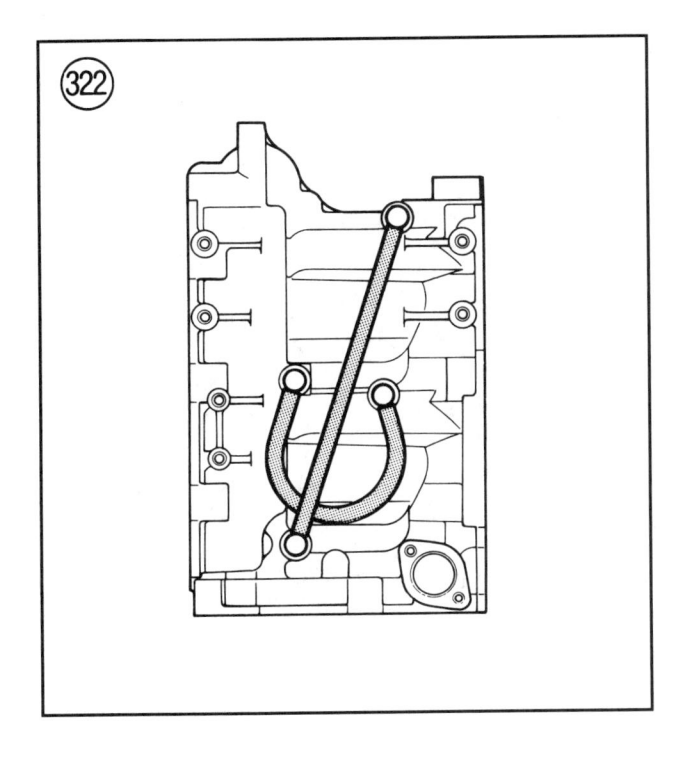

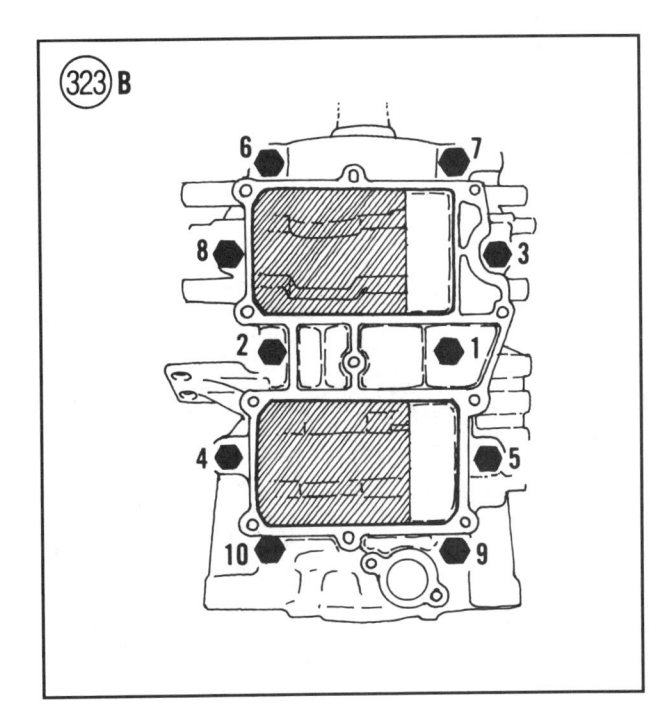

8. On 1987-on DT 35 and 1985-on DT 40 models, install the heat sensor into the cylinder head (**Figure 324**).

NOTE
On 1986-1987 DT 20, 1985-1987 DT 25 and DT 30 models, position the cylinder head gasket so the projected portion (A, Figure 325) is on the left-hand side.

9. Install the cylinder head and new gasket.
10. Install the cylinder head cover and new gasket onto the cylinder head.
11. Install the head bolts and tighten to specifications (**Table 1**) following the appropriate sequence shown in **Figure 325** for DT 20, DT 25 and DT 30 models or **Figure 326** for DT 35 and DT 40 models.
12. On DT 35 and DT 40 models, perform the following:
    a. Install the thermostat in the cylinder head cover with its air breather hole facing upward.
    b. Install the cylinder head cover on the cylinder head with a new gasket.
13. On DT 35 and DT 40 models, perform the following:
    a. Install the lower oil seal housing (**Figure 327**).
    b. Install the bolts (**Figure 328**) securing the lower oil seal housing. Tighten the bolts securely.
14. Install the inner exhaust plate and new gasket.

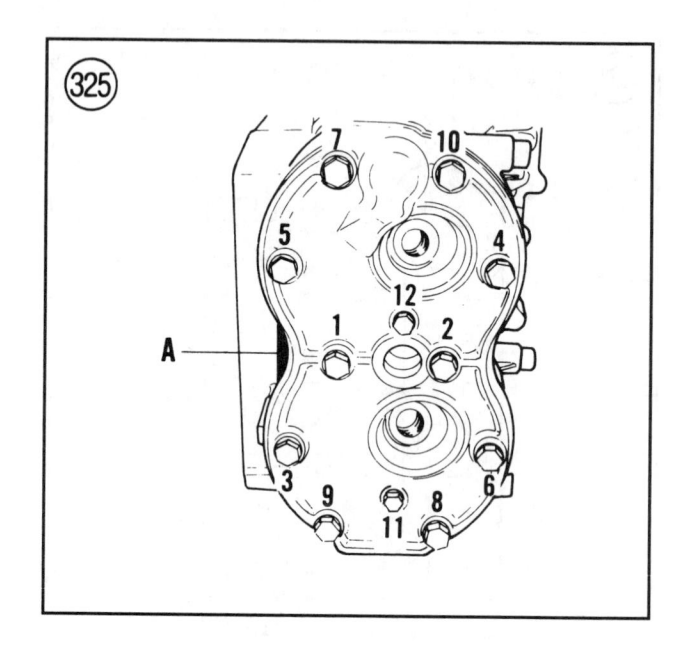

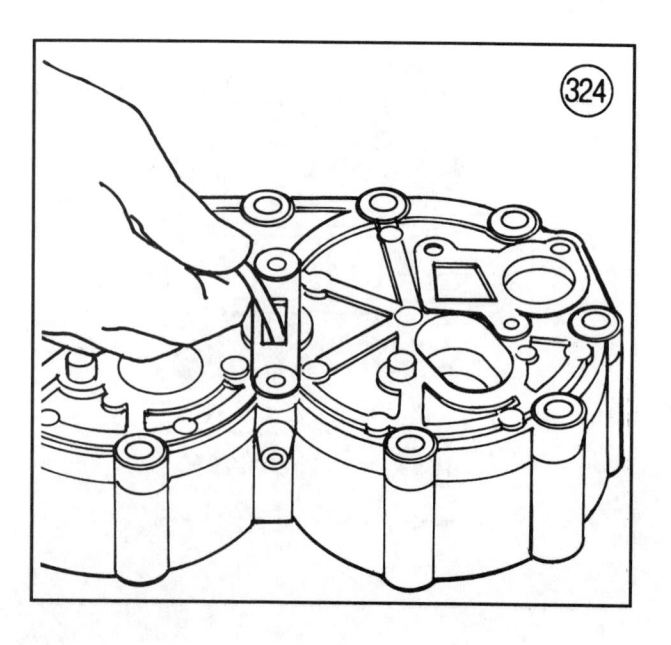

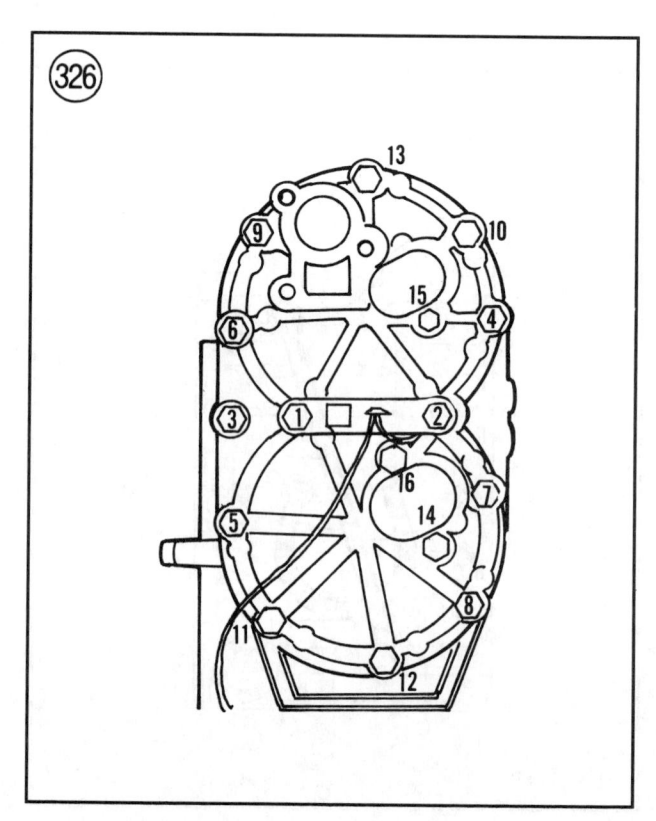

POWER HEAD

15. Install the exhaust cover and new gasket.

16. Install the exhaust cover bolts and tighten securely in a crisscross pattern.

17. Install the reed valve assembly with the reed side facing toward the crankshaft (**Figure 318**, typical).

18. Install the intake manifold with a new gasket (**Figure 329**, typical).

19. Install the intake manifold bolts and tighten in a crisscross pattern and tighten securely.

20. Install any mounting brackets removed during disassembly.

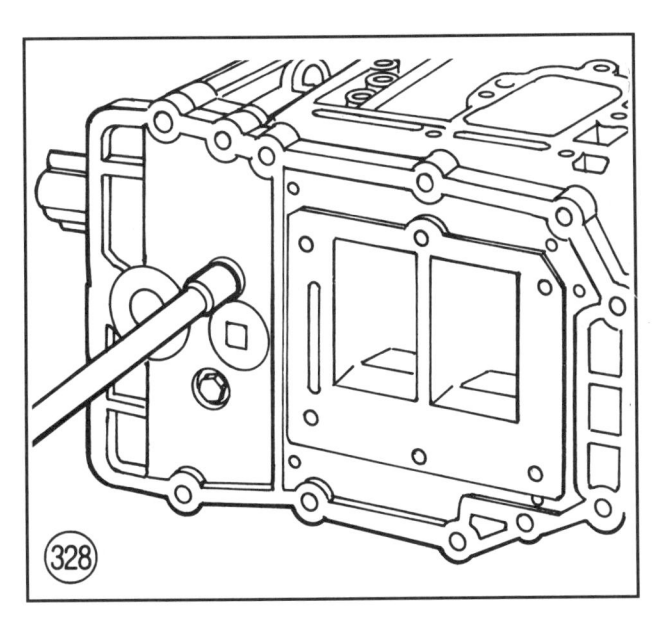

21. Install the power head as described in this chapter.

Assembly
(All 3-cylinder Engines)

Refer to the following illustrations for this procedure:
   a. **Figure 330**: 1989-on DT 25 and 1988-on DT 30.
   b. **Figure 331**: DT 55 and DT 65.
   c. **Figure 332**: DT 75 and DT 85.

1. Install the crankshaft assembly as described in this chapter.

2. Install the upper and lower crankshaft oil seals (**Figure 333**).

3. Apply Suzuki Bond No. 4 (part No. 99000-31030) or equivalent to the mating surfaces of the crankcase cover and the cylinder block (**Figure 334**).

4. Install the crankcase cover to the cylinder block.

NOTE
Note that the bolts vary in thread diameter. Be sure to install the correct size bolt in the correct hole.

5. Install the bolts and 1 nut (on DT 75 and DT 85 models) and tighten to specifications (**Table**

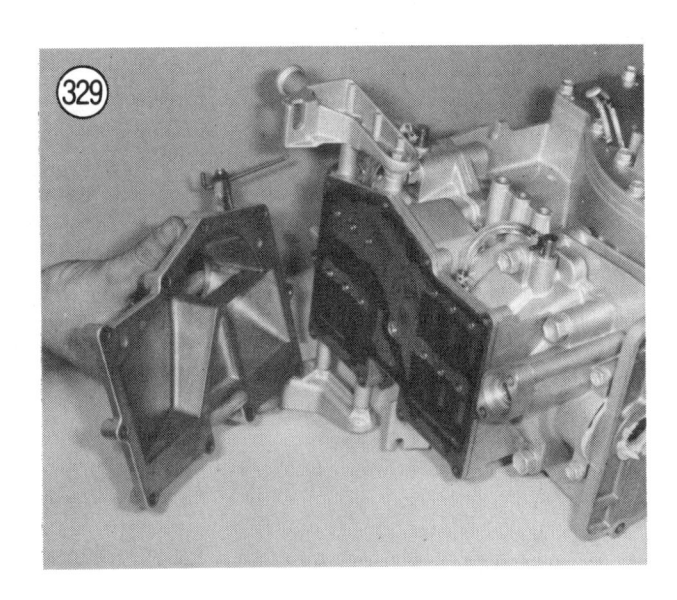

356

CHAPTER EIGHT

③③⓪

**POWER HEAD
(1989-ON DT 25 AND 1988-ON DT 30)**

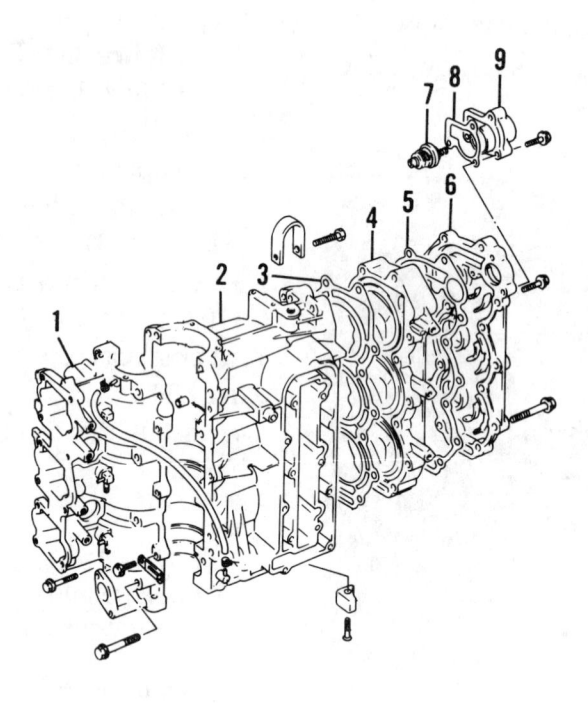

1. Crankcase cover
2. Cylinder block
3. Gasket
4. Cylinder head
5. Gasket
6. Cylinder head cover
7. Gasket
8. Thermostat
9. Thermostat cover

POWER HEAD

POWER HEAD
(1985-ON DT 55 AND DT 65)

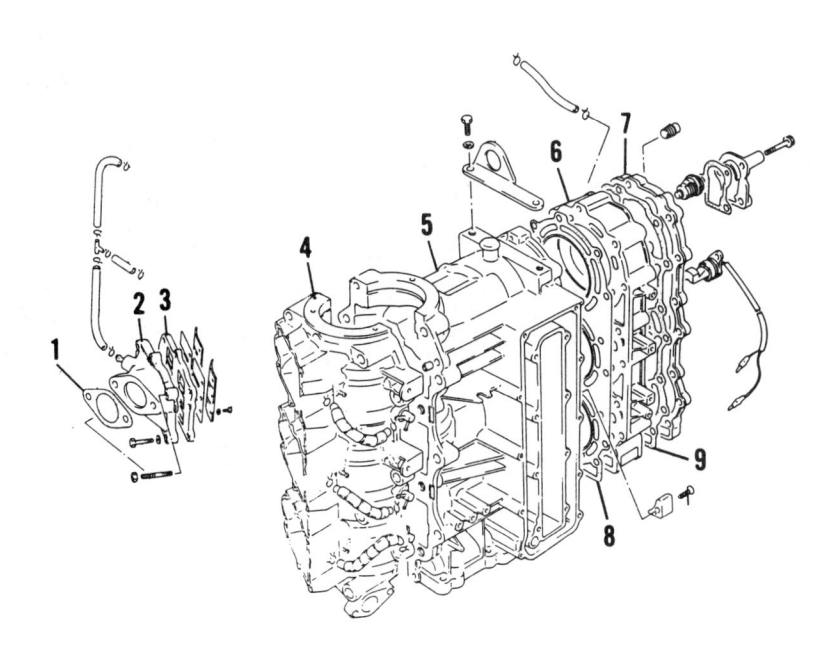

1. Gasket
2. Intake manifold
3. Reed valves
4. Crancase cover
5. Cylinder block
6. Cylinder head
7. Cylinder head cover
8. Gasket
9. Gasket

CHAPTER EIGHT

POWER HEAD (1985 DT 75 AND DT 85)

1. Cylinder block
2. Crancase cover
3. Gasket
4. Cylinder head
5. Exhaust cover
6. Thermostat cover
7. Thermostat
8. Reed valve assembly
9. Intake manifold

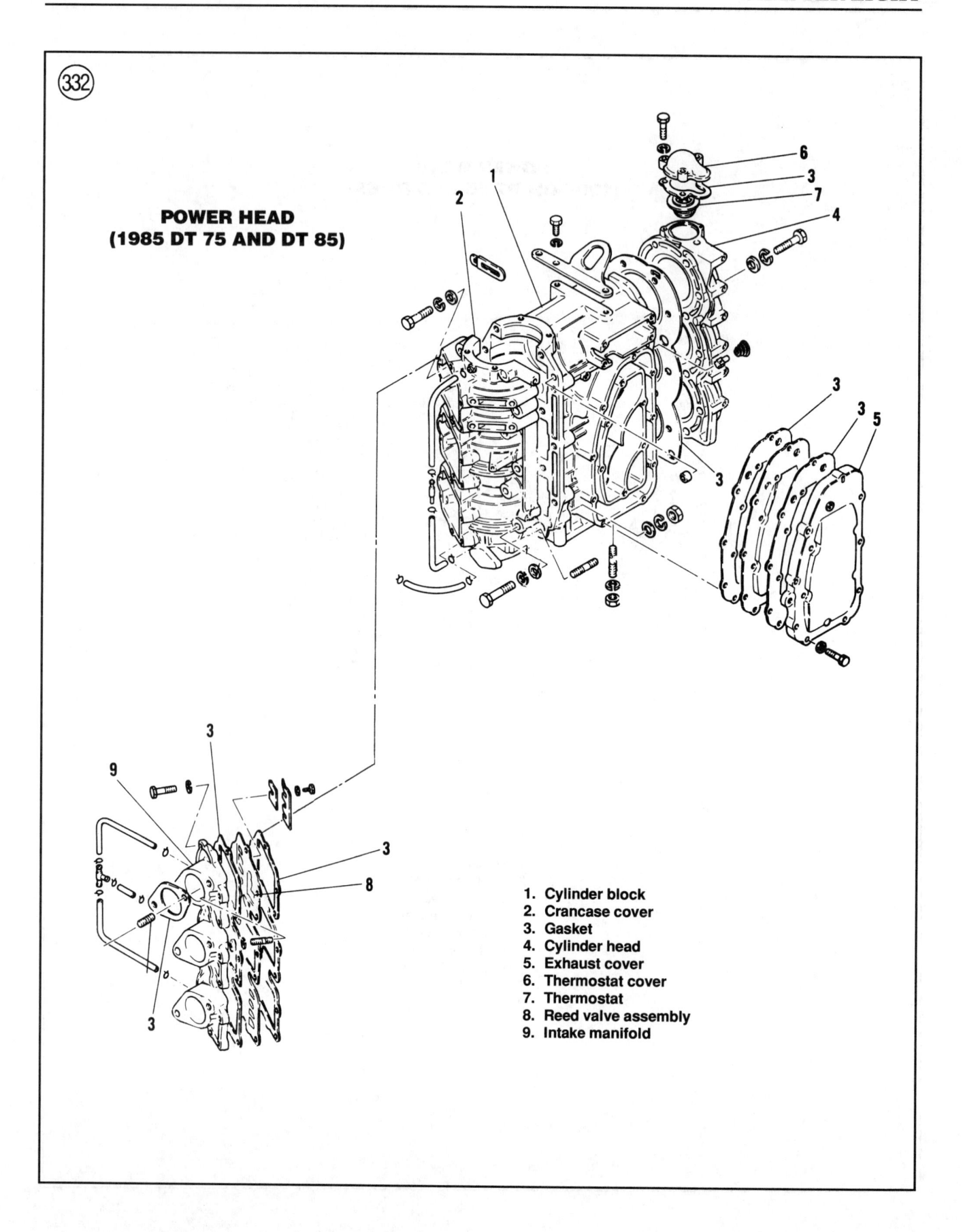

POWER HEAD

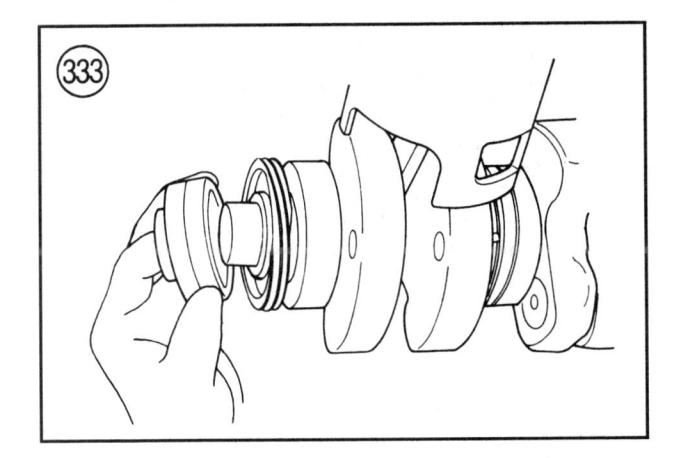

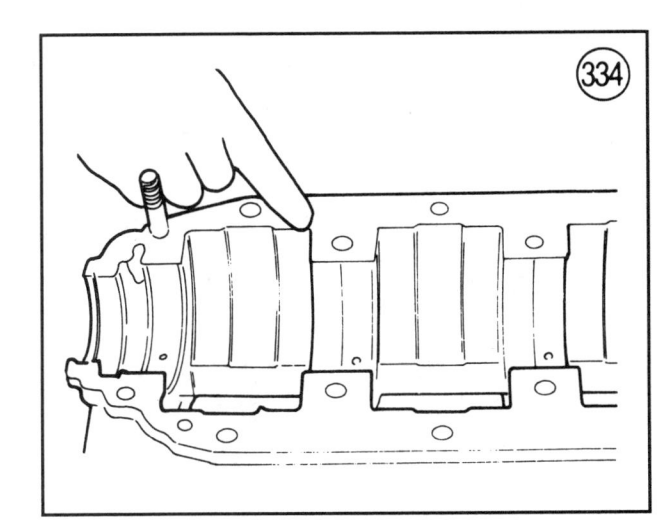

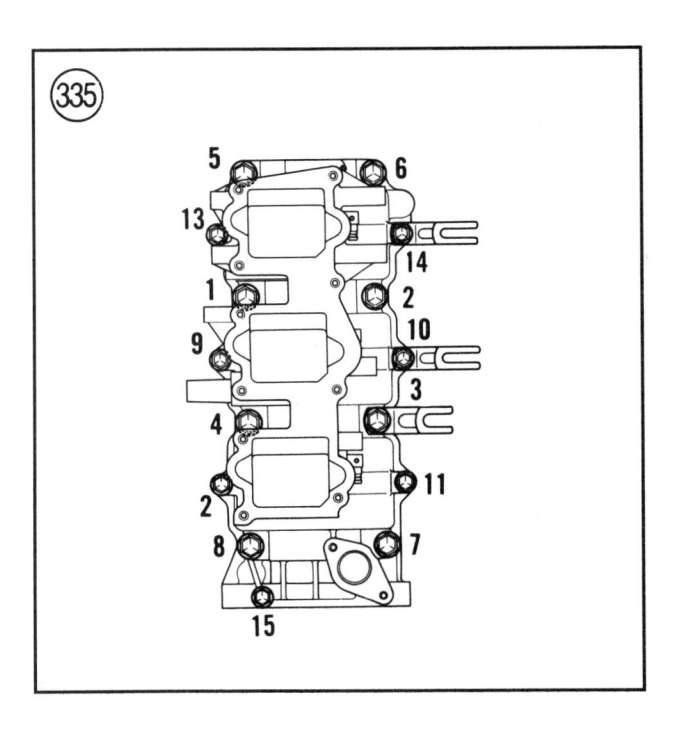

1) following the appropriate sequence as shown in the following:

   a. **Figure 335**: 1989-on DT 25 and 1988-on DT 30.

   b. **Figure 336**: DT 55 and DT 65.

   c. **Figure 337**: DT 75 and DT 85.

On models so equipped, be sure to install the clamp(s) under the correct bolt head as indicated in the illustrations.

6. Rotate the crankshaft several turns to check for binding. If crankshaft does not turn easily, disassemble and correct the interference.

7. Apply a liberal quantity of Suzuki Outboard Motor Oil to each connecting rod big end through the reed valve openings.

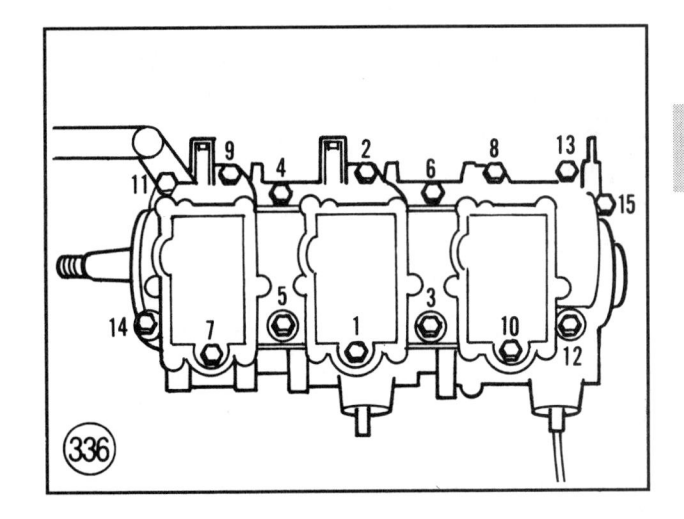

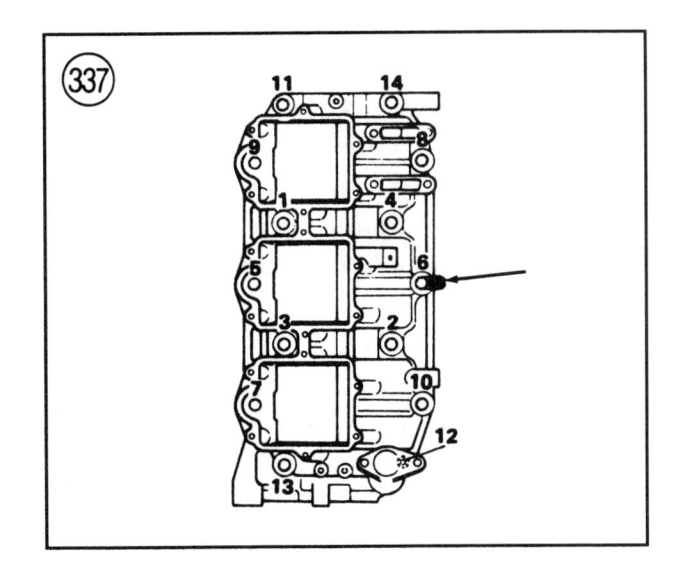

8. On DT 75 and DT 85 models, perform the following:

a. Install a new upper drive shaft seal (A, **Figure 338**) into the seal housing (B, **Figure 338**). The open side of the seal (A) should face toward the crankshaft (up).

b. Pack the open side of the upper drive shaft seal and housing with water-resistant grease (part No. 99000-25160) or equivalent.

c. Make sure the flange of the lower crankshaft seal (D, **Figure 339**) is properly positioned in the cylinder block groove (C, **Figure 339**), then install the upper drive shaft seal and housing assembly (B, **Figure 339**) into the cylinder block groove.

9. Install the exhaust cover and new gasket.

10. Install the exhaust cover bolts and tighten in a crisscross pattern to specifications (**Table 1**).

11. On models so equipped, install the water valve (A, **Figure 228**) and the spring (B, **Figure 228**) into the cylinder block.

12. Install the cylinder head and new gasket.

13A. On DT 55 and DT 65 models, refer to the number cast into the cylinder head adjacent to each cylinder head bolt hole. Tighten the bolts 1/2 turn at a time in the order of the cast numbers. Tighten the bolts to specifications (**Table 1**).

13B. On all other models, to prevent warpage of the cylinder head, tighten the cylinder head bolts 1/2 turn at a time to specifications (**Table 1**) following the sequence shown in **Figure 340** for 1989-on DT 25 and 1988-on DT 30 models or **Figure 341** for DT 75 and DT 85 models.

14. On models so equipped, install the cooling water flow sensor into the cylinder head. Install the bolts and tighten securely.

15. On DT 75 and DT 85 models, perform the following if removed:

a. Coat the thermo unit with Thermo Unit Grease (part No. 34855-95550) or equivalent and install into the cylinder head (**Figure 342**).

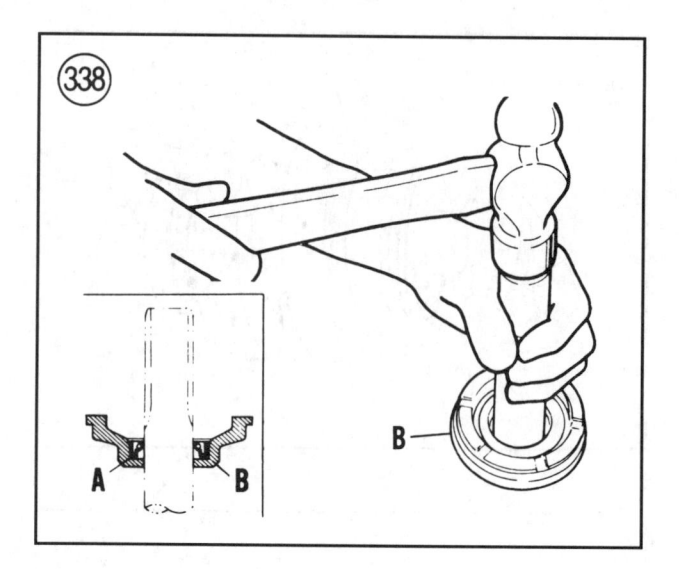

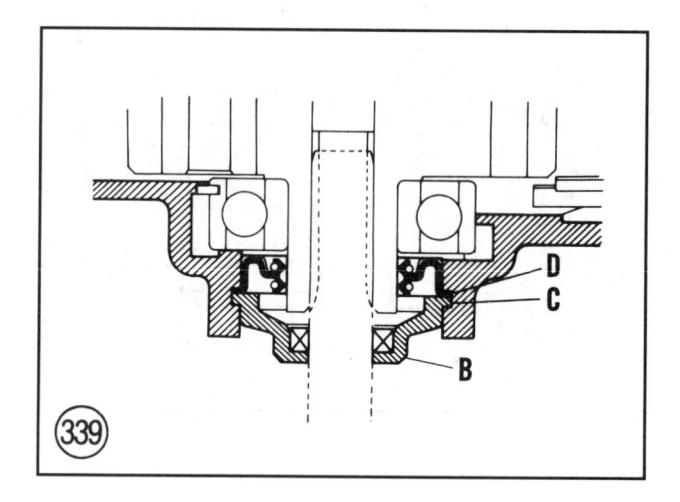

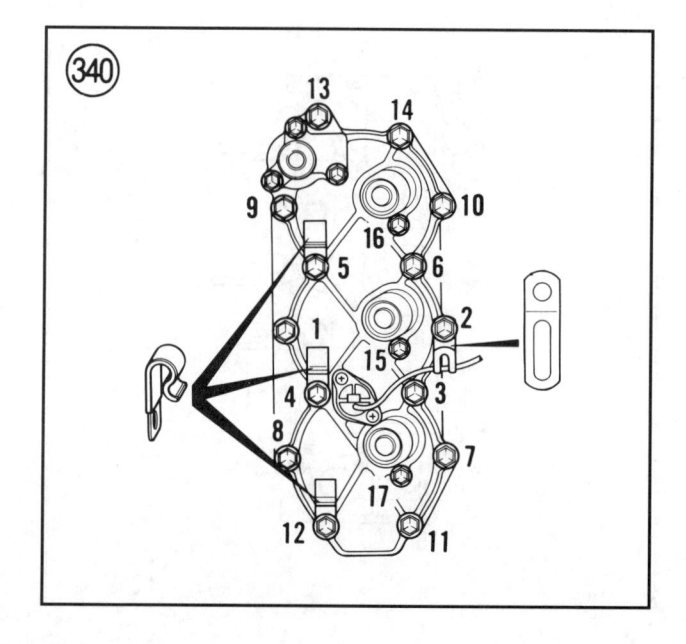

POWER HEAD

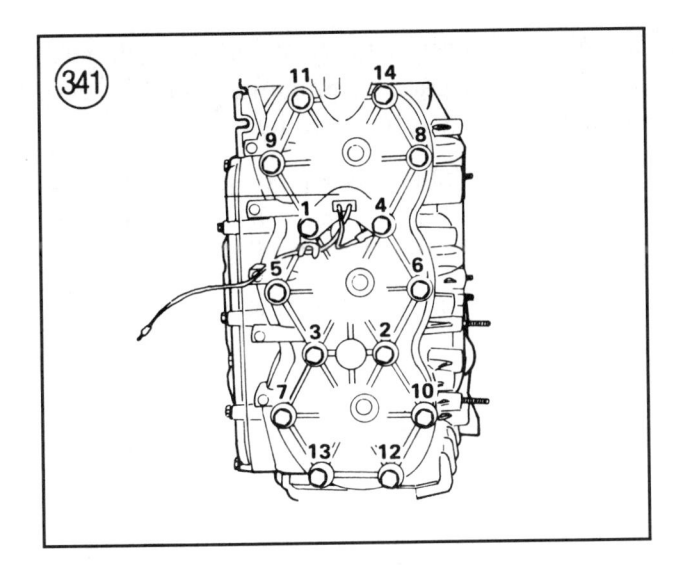

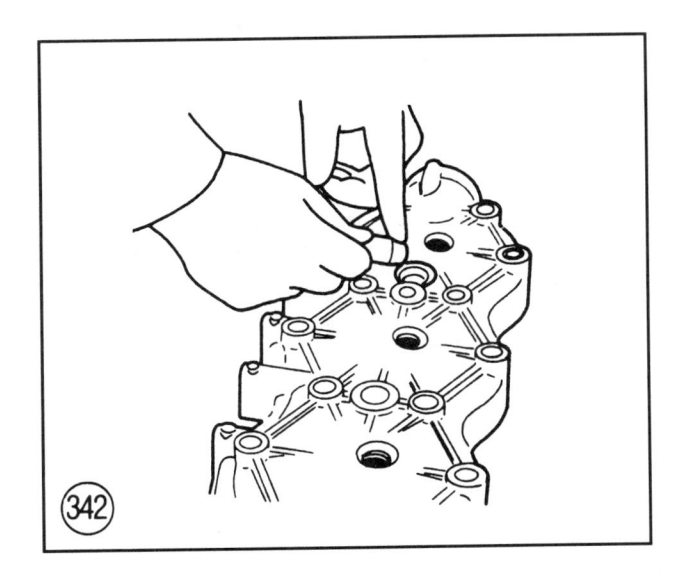

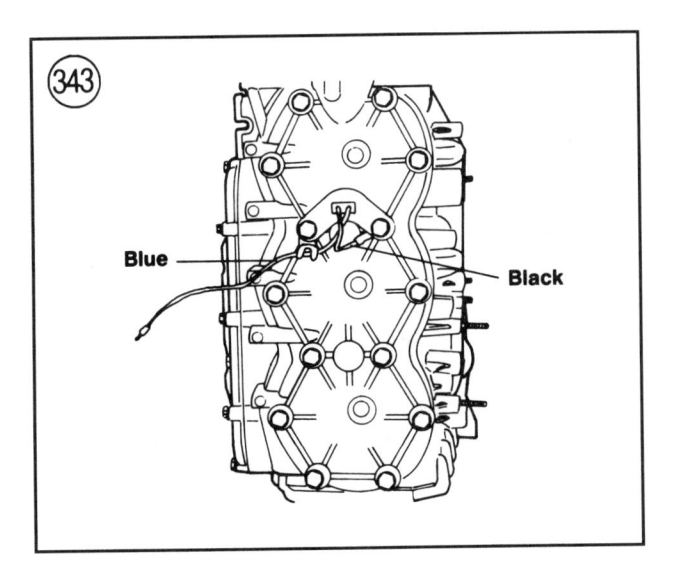

(341)

b. Install the heat switch over the thermo unit in the cylinder head. Attach the black lead under the head bolt to provide ground (**Figure 343**).

16. Install the thermostat, new gasket, cover and bolts. Tighten the bolts securely.

17. Install the reed valve assembly and intake manifold. Tighten the bolts securely in a criss-cross pattern.

18. Install the lubrication tube hoses and slide the hose clamps onto the fittings (**Figure 344**).

19. Install any mounting brackets removed during disassembly.

20. Install the power head as described in this chapter.

Assembly
(All Inline 4-Cylinder Engines)

Refer to the following illustrations for this procedure:

   a. **Figure 332**: 1985 DT 115 and DT 140.

   b. **Figure 345**: 1986-on DT 115 and DT 140.

1. Install the crankshaft assembly as described in this chapter.

2. Apply Suzuki Bond No. 4 (part No. 99000-31030), or Suzuki Bond No. 1207B (part No. 99104-31140), or equivalent to the mating surfaces of the crankcase cover and the cylinder block (**Figure 346**).

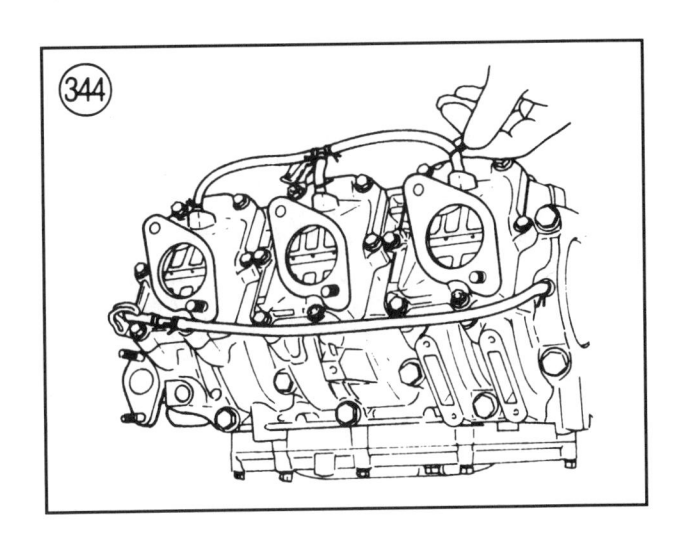

362　　CHAPTER EIGHT

3. Install the crankcase cover to the cylinder block and tighten the bolts to specifications (**Table 1**) following the appropriate sequence as shown in **Figure 347** for 1985 DT 115 and DT 140 models or **Figure 348** for 1986-on DT 115 and DT 140 models.

4. Rotate the crankshaft several turns to check for binding. If crankshaft does not turn easily, disassemble and correct the interference.

5. Apply a liberal quantity of Suzuki Outboard Motor Oil to each connecting rod big end through the reed valve openings.

6. Install the exhaust cover plate, gaskets and the exhaust cover. Install the bolts and tighten to specifications (**Table 1**) in a crisscross pattern.

7. Install the cylinder head and new gasket.

8. Install the cylinder head cover and new gasket.

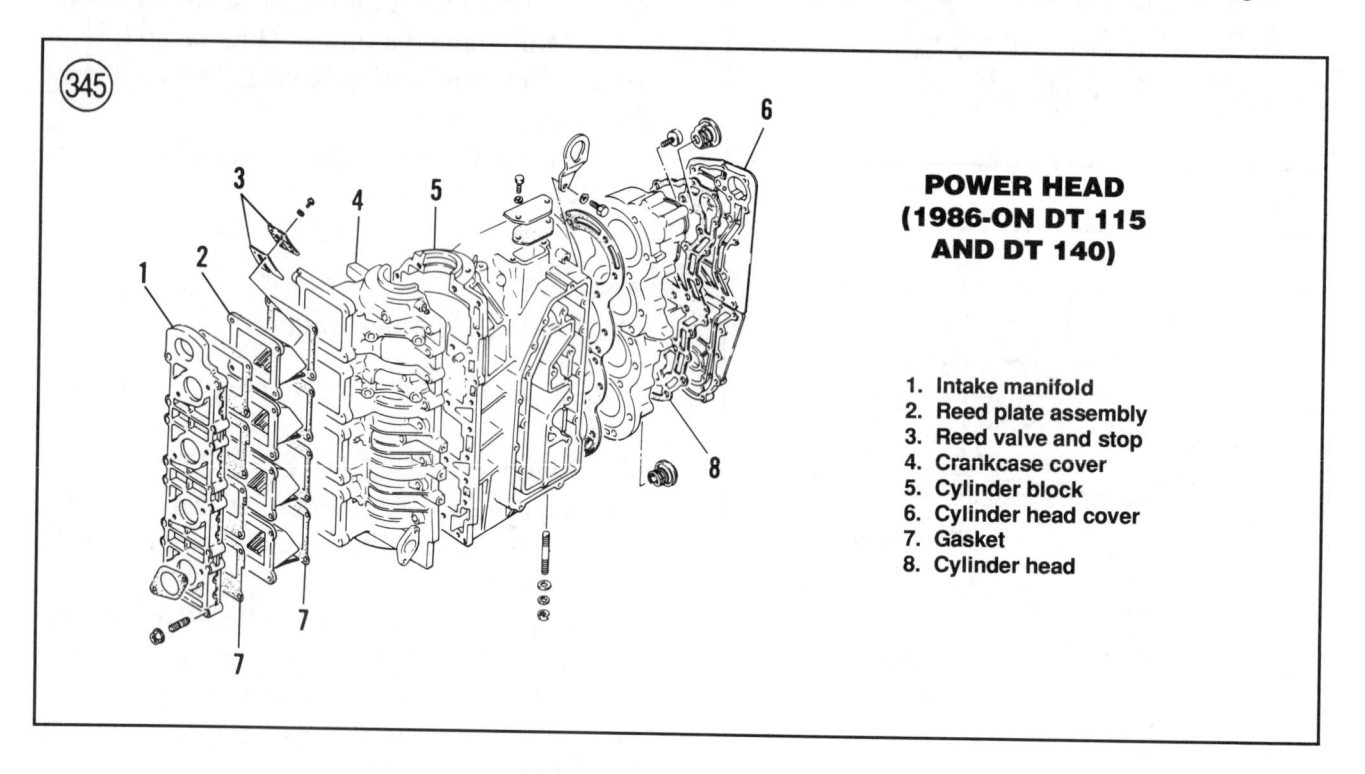

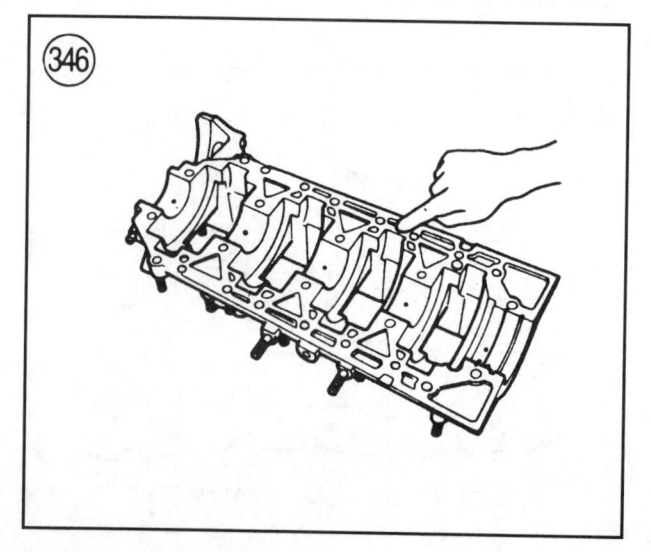

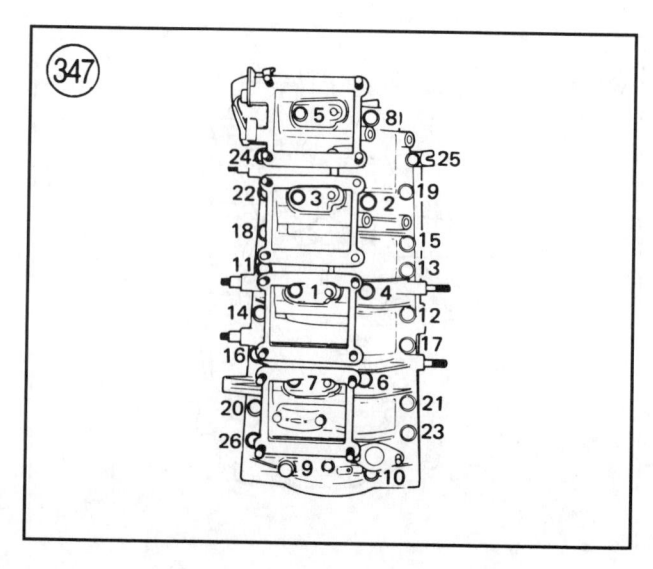

POWER HEAD (1986-ON DT 115 AND DT 140)

1. Intake manifold
2. Reed plate assembly
3. Reed valve and stop
4. Crankcase cover
5. Cylinder block
6. Cylinder head cover
7. Gasket
8. Cylinder head

POWER HEAD

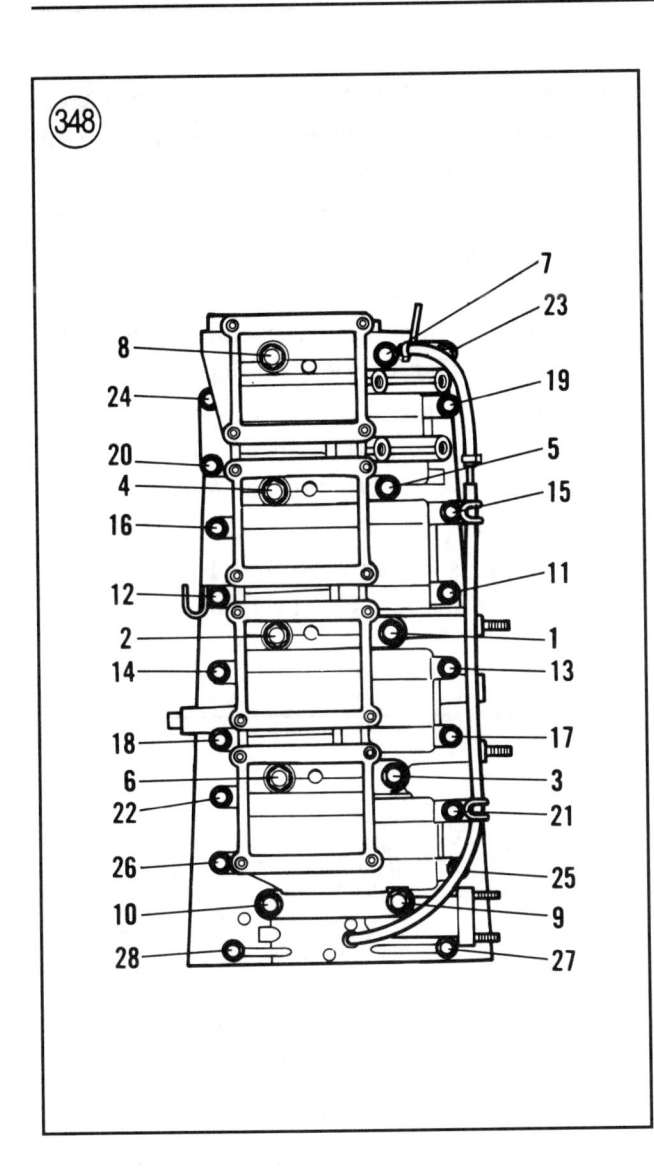

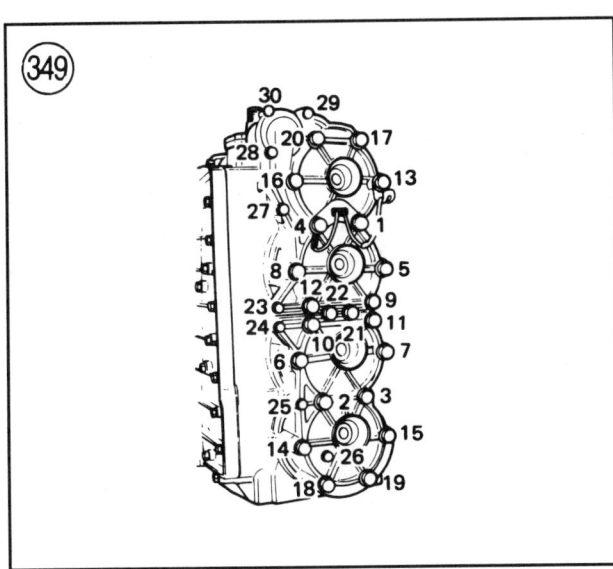

9. Install and tighten the head bolts to specifications (**Table 1**) following the sequence shown in **Figure 349** for 1985 DT 115 and DT 140 models or **Figure 350** for 1986-on DT 115 and DT 140 models.

10. Install the temperature switch in the cylinder head cover. Attach the black lead under one of the cover bolts to provide ground.

11. Install the 3 cylinder head connectors.

12. Install the thermostat in the cylinder head with the mark on its flange facing *upward*.

13A. On 1985 models, install the plugs into the crankcase. Install the small plug in the top cylinder (**Figure 351**).

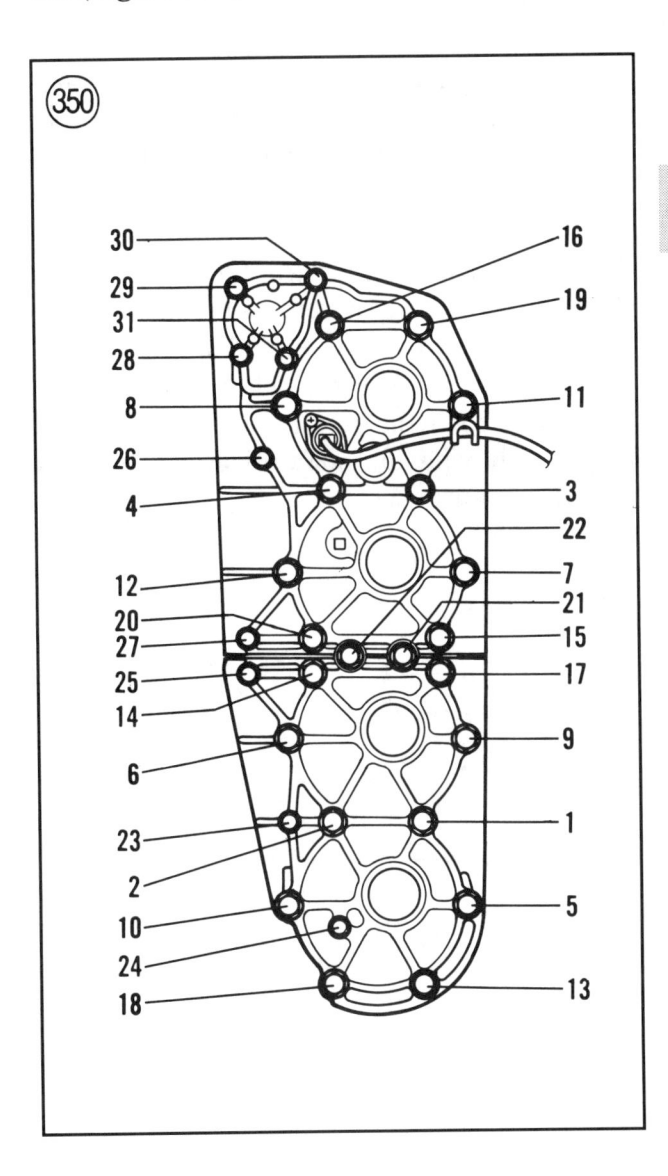

13B. On 1986-on models, install the plugs into the crankcase with the tab facing out. Install the short plug in the top cylinder.

14. Install the reed valve assemblies with new gaskets.

15. Install the intake manifold with new gaskets. Tighten the bolts securely in a crisscross pattern.

16. Install any mounting brackets removed during disassembly.

17. Install the power head as described in this chapter.

Assembly
(All V4 Models)

1. Place the crankcase in the same wooden stand used during disassembly and install the crankshaft assembly as described in this chapter.

2. Make sure the locating pins are installed in the crankcase.

3. Apply Suzuki Bond No. 1207B (part No. 99104-31140) or equivalent to the mating surfaces of the crankcase cover and the crankcase.

4. Install the crankcase cover to the cylinder block and tighten the bolts to specifications (**Table 1**) following the appropriate sequence as shown in **Figure 352**.

5. Rotate the crankshaft several turns to check for binding. If crankshaft does not turn easily, disassemble and correct the interference.

6. Apply a liberal quantity of Suzuki Outboard Motor Oil to each connecting rod big end through the reed valve openings.

7. Install the clutch shaft assembly and bolts (**Figure 353**). Tighten the bolts securely.

8. Install the intake manifold and reed valve assembly and new gasket. Install the bolts and nuts (**Figure 354**) and tighten securely in a crisscross pattern.

9. Install the silencer case and bolts (**Figure 355**). Tighten the bolts securely in a crisscross pattern.

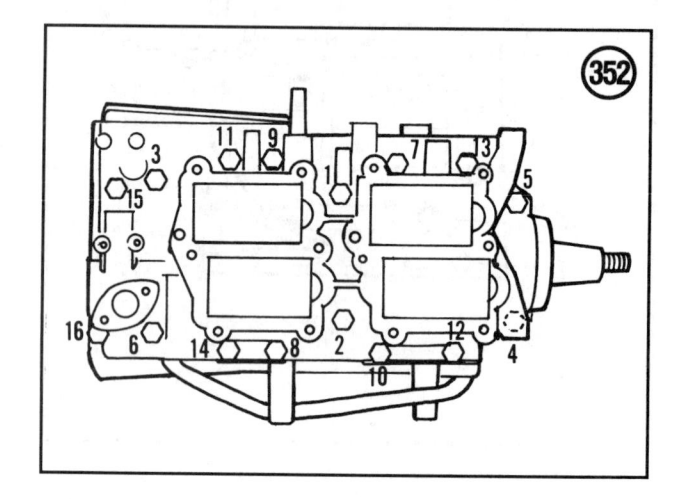

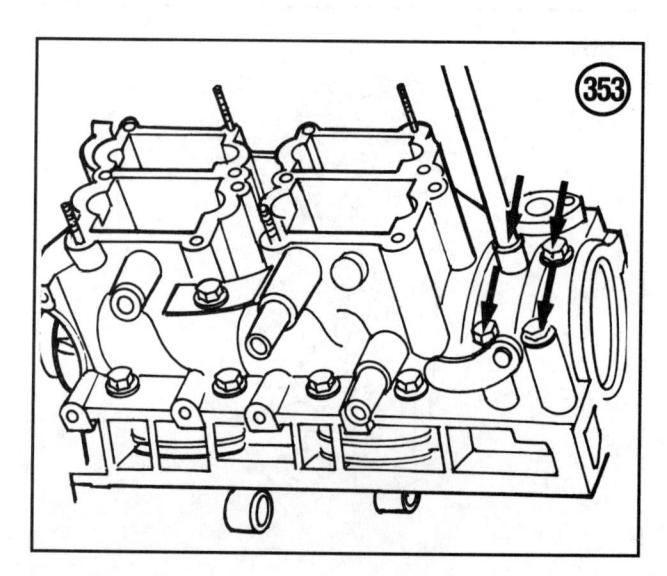

POWER HEAD

10. Turn the power head right-side up and set it on the silencer case.

CAUTION
The power head assembly is very tall and will become more top heavy as additional components are added. When tightening bolts or nuts, hold the power head so that it will not fall over. If necessary, have an assistant hold the power head during the installation of a component.

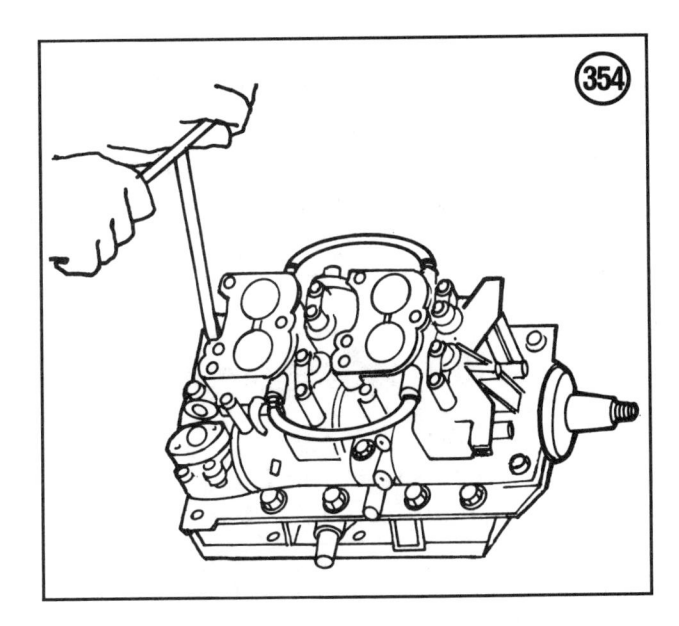

11. Install all 4 pistons onto their connecting rods as described in this chapter.

12. Coat the cylinder bores and pistons with Suzuki Outboard Motor oil.

13. Carefully install the cylinder straight down onto the pistons. Compress each piston ring as it enters the cylinder. Push the cylinder down until it is completely seated.

14. In the valley between the cylinders, install the 3 bolts securing the cylinder to the crankcase. Install the nuts securing the cylinder to the crankcase. Tighten the bolts and nuts only finger-tight at this time. The cylinder must be able to move slightly in Step 16.

15. Repeat Step 13 and Step 14 for the other cylinder.

CAUTION
*A special Suzuki tool, Cylinder/Crankcase Assembling Plate (part No. 09912-68720) is required to align the cylinder heads to the crankcase properly. Do **not** attempt to install the cylinders without this special tool as alignment will be incorrect resulting in engine damage.*

16. Install the special Suzuki tool, Cylinder/Crankcase Assembling Plate, to both cylinders and the crankcase. Install the bolts and tighten to specifications (**Table 1**) following the appropriate sequence as shown in **Figure 356**.

17. Use the same special tool used during disassembly for the bolts. Tighten the 6 bolts and 6 nuts securing the cylinders to the crankcase and tighten to the preliminary setting torque specification (**Table 1**).

18. Remove the bolts securing the special tool and remove the special tool.

19. Tighten the 6 bolts and 6 nuts securing the cylinders to the crankcase and tighten to the final setting torque specification (**Table 1**).

20. Apply water-resistant grease to the oil seal (A, **Figure 357**) and O-ring (B, **Figure 357**) in the under oil seal housing.

21. Install the under oil seal housing and bolts. Tighten the bolts in a crisscross pattern to the specifications (**Table 1**).

22. Install the cylinder head and new gasket. Install the cylinder head bolts and tighten to specification (**Table 1**) following the sequence as shown in **Figure 358**.

23. Repeat Step 22 for the other cylinder head.

Assembly (All V6 Models)

1. Correctly place the crankcase in the same wooden 2 × 4's used during disassembly and install the crankshaft assembly as described in this chapter.

2. Make sure the locating pins are installed in the crankcase.

3. Apply Suzuki Bond No. 1207B (part No. 99104-31140), or equivalent to the mating surfaces of the crankcase cover and the crankcase.

4. Install the crankcase cover to the cylinder block and tighten the bolts to specification (**Table 1**) following the appropriate sequence as shown in **Figure 359**.

5. Rotate the crankshaft several turns to check for binding. If crankshaft does not turn easily, disassemble and correct the interference.

6. Apply a liberal quantity of Suzuki Outboard Motor Oil to each connecting rod big end through the reed valve openings.

7. Install the intake manifolds and reed valve assemblies and new gaskets. Install the bolts and nuts and tighten securely in a crisscross pattern.

8. Attach the wooden stand of 3/4 in. plywood and 2 × 4's used during disassembly to the inlet case securely. Tighten the bolts securely.

9. Turn the power head right-side up and set it on the wooden stand.

CAUTION
The power head assembly is very tall and will become more top heavy as additional components are added. When tightening bolts or nuts, hold the power head so that it will not fall over. If necessary, have an assistant hold the power head during the installation of a component.

10. Install all 6 pistons onto their connecting rods as described in this chapter.

11. Coat all cylinder bores and pistons with Suzuki Outboard Motor oil.

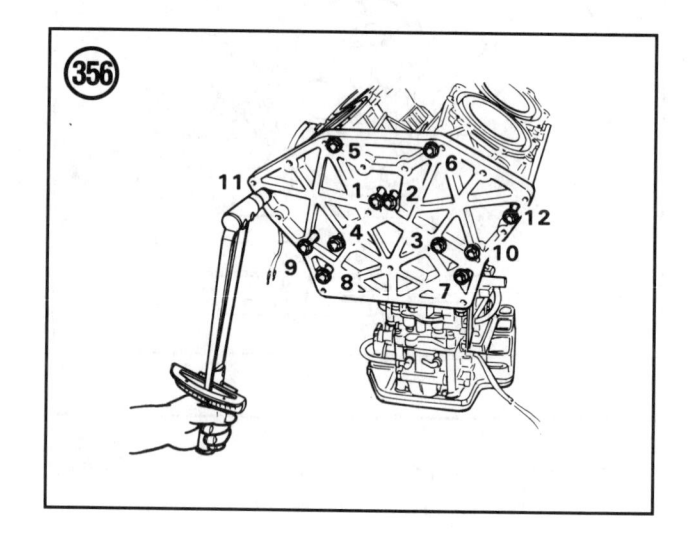

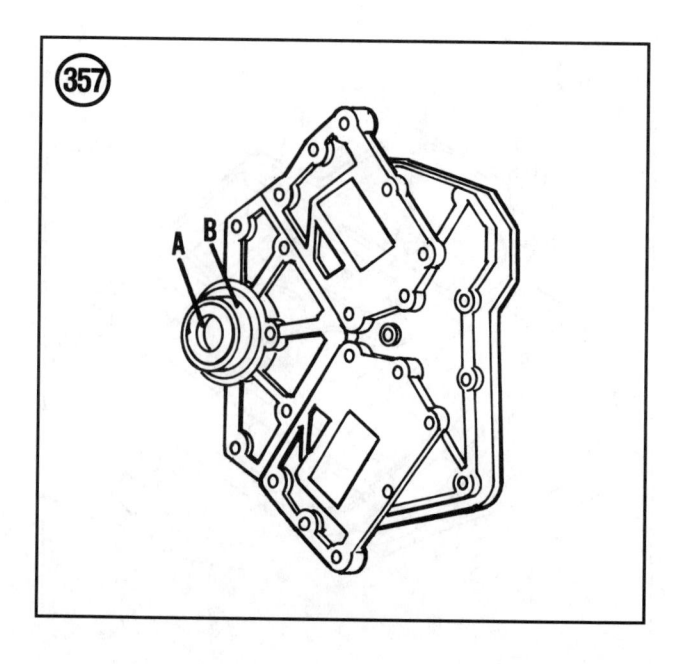

POWER HEAD

12. Carefully install the cylinder straight down onto the pistons. Compress each piston ring as it enters the cylinder. Push the cylinder down until it is completely seated.

13. In the valley between the cylinders, install the 4 bolts securing the cylinder to the crankcase. Install the 4 nuts securing the cylinder to the crankcase. Tighten the bolts and nuts only finger-tight at this time. The cylinders must be able to move slightly in Step 15.

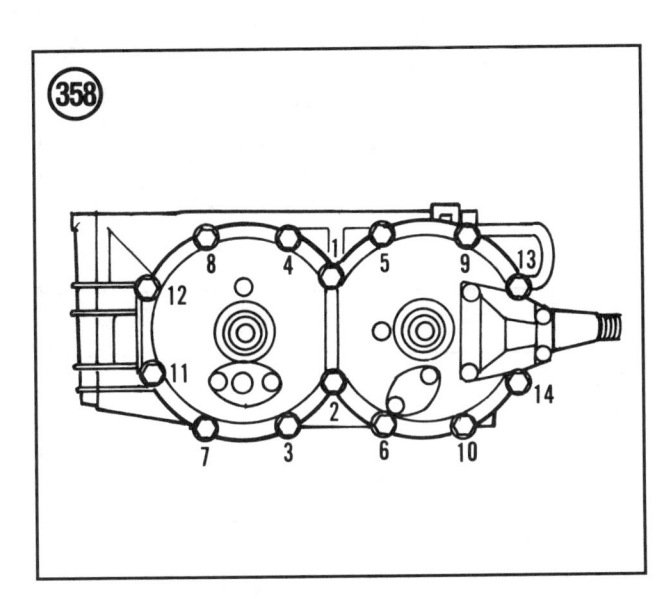

14. Repeat Step 12 and Step 13 for the other cylinder.

CAUTION
*A special Suzuki tool, Cylinder/Crankcase Assembling Plate (part No. 09912-68710) is required to align the cylinder heads to the crankcase properly. Do **not** attempt to install the cylinders without this special tool as alignment will be incorrect resulting in engine damage.*

15. Install the special Suzuki tool, Cylinder/Crankcase Assembling Plate, to both cylinders and the crankcase. Install the bolts and tighten to specifications (**Table 1**) following the appropriate sequence as shown in **Figure 360**.

16. Use the same special tool used during disassembly for the bolts. Tighten the 8 bolts and 8 nuts securing the cylinders to the crankcase and tighten to the preliminary setting torque specification (**Table 1**).

17. Remove the bolts securing the special tool and remove the special tool.

18. Tighten the 8 bolts and 8 nuts securing the cylinders to the crankcase and tighten to the final setting torque specification (**Table 1**).

19. Apply water-resistant grease to the oil seal (A, **Figure 361**) and O-ring (B, **Figure 361**) in the under oil seal housing.

20. Make sure the locating dowel (C, **Figure 361**) is in place in the under oil seal housing.

21. Install the under oil seal housing and bolts (**Figure 362**). Tighten the bolts in a crisscross pattern to specification (**Table 1**).

22. Install the cylinder head and new gasket. Install the cylinder head bolts and tighten to specification (**Table 1**) following the sequence as shown in **Figure 363**.

23. Repeat Step 22 for the other cylinder head.

24. Remove the wooden base installed in Step 8.

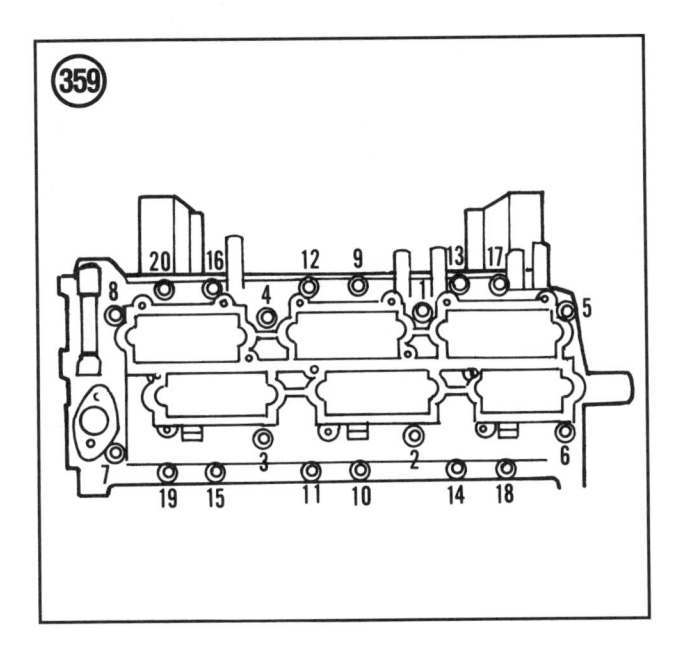

Figures 360-363 and Tables 1-3 are on the following pages.

368 CHAPTER EIGHT

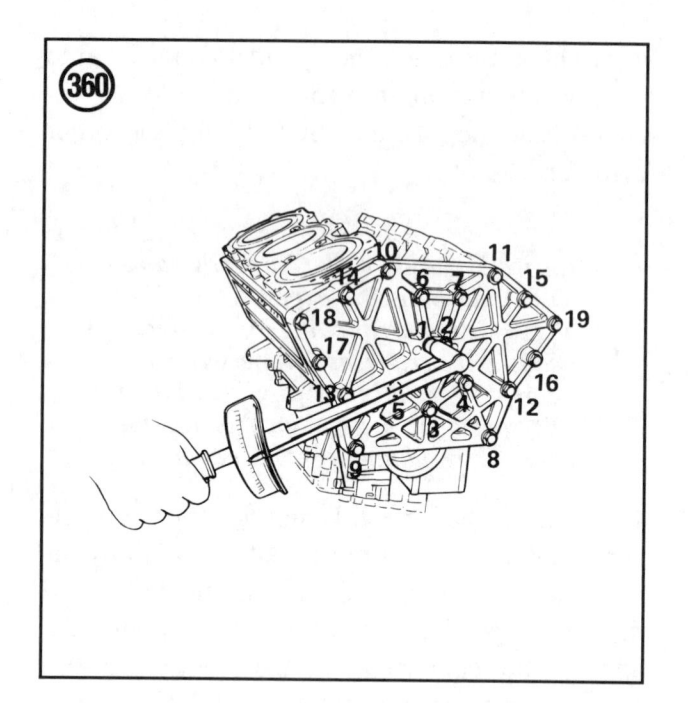

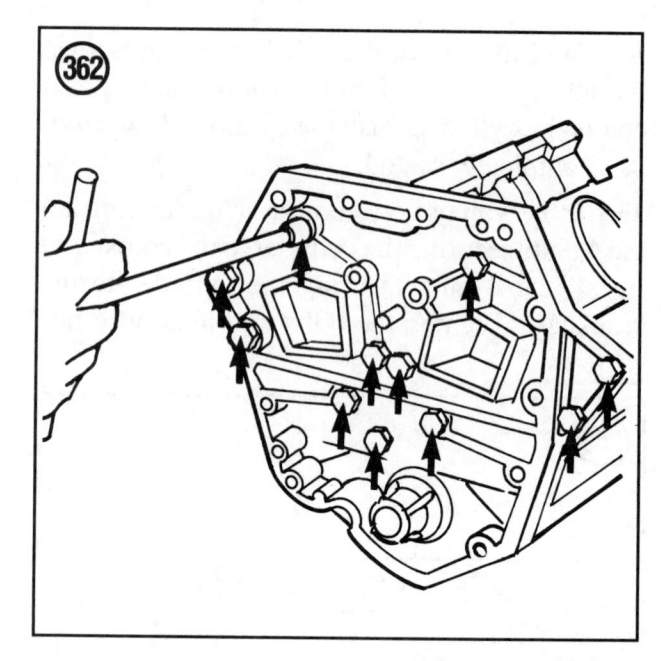

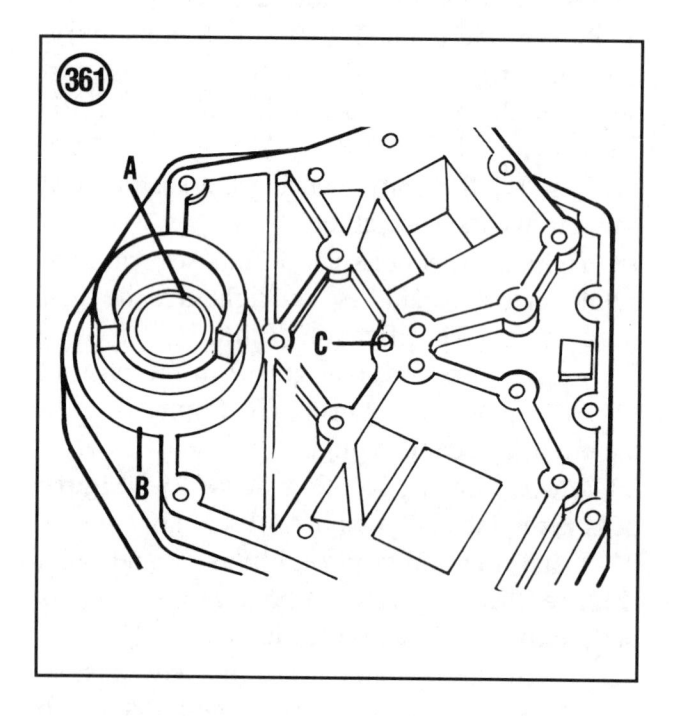

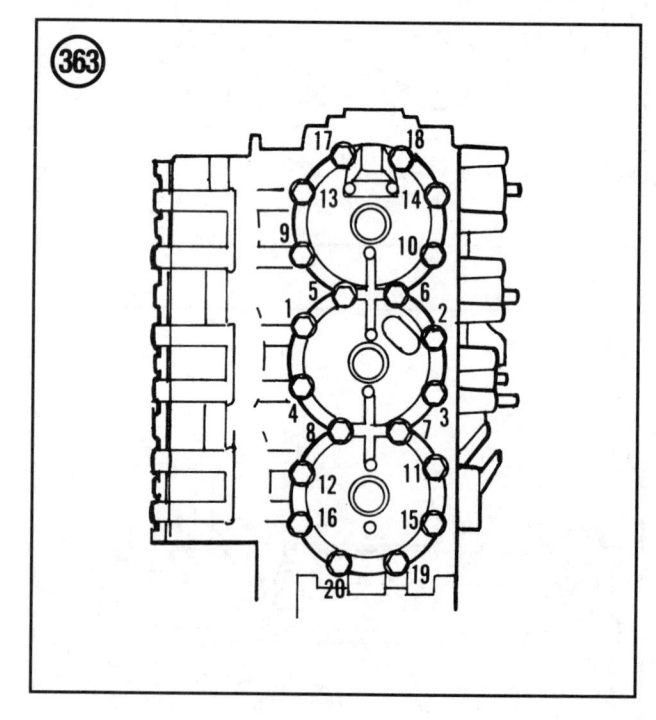

POWER HEAD

Table 1 TIGHTENING TORQUES*

Fastener	ft.-lb.	N·m
Connecting rod bolts		
DT 115, DT 140 (1985)	22-25	30-35
Crankcase bolt		
DT 2, DT 4, DT 6	6-9	8-12
DT 8 (1985-1987), DT 9.9 (1985-1987)	6-9	8-12
DT 8 (1988-1991), DT 9.9 (1988-1991), DT 15	15-19	20-26
DT 20, DT 25, DT 30, DT 35, DT 40		
6 mm	6-9	8-12
8 mm	15-19	20-26
DT 55, DT 65		
6 mm	6-9	8-12
10 mm	34-39	46-54
DT 75, DT 85	30-44	40-60
DT 115, DT 140		
8 mm	15-18	21-25
10 mm	34-39	46-54
V4		
6 mm	6-9	8-12
8 mm	15-19	20-26
10 mm	34-39	46-54
V6		
8 mm	15-19	20-26
10 mm	34-39	46-54
Cylinder bolts and nuts		
V4, V6		
Preliminary	15-19	20-26
Final	34-39	46-54
Cylinder head bolt		
DT 2	6-9	8-12
DT 4	NA	NA
DT 6, DT 8 (1985-1987)	11-14	15-20
DT 9.9 (1985-1987), DT 15	15-19	20-26
DT 8 (1988-1991), DT 9.9 (1988-1991)	15-18	21-25
DT 20, DT 25 (1985-1988), DT 30 (1985-1987)	15-19	20-26
DT 25 (1989-1991), DT 30 (1988-1991)		
6 mm	6-9	8-12
8 mm	15-18	21-25
DT 35, DT 40		
6 mm	6-9	8-12
8 mm	15-19	20-26
DT 55, DT 65		
6 mm	6-9	8-12
8 mm	15-19	20-26
DT 75, DT 85		
6 mm	6-9	8-12
10 mm	30-44	40-60
V4, V6	21-23	28-32

(continued)

Table 1 TIGHTENING TORQUES* (continued)

Fastener	ft.-lb.	N·m
Cylinder-to-crankcase assembly plate bolts		
V4	15-19	20-26
Exhaust cover screws		
DT 55-DT 225	6-9	8-12
Flywheel nut		
DT 2, DT 4	30-36	40-50
DT 6, DT 8 (1985-1987)	44-51	60-70
DT 8 (1988-1991), DT 9.9, DT 15	59-66	80-90
DT 20, DT 25, DT 30	96-110	130-150
DT 55, DT 65, DT 75, DT 85	148-155	200-210
V4, V6	184-192	250-260
Power head-to-drive shaft housing mounting bolts and nuts		
DT 2	4.5-7	6-10
DT 4, DT 6	11-14	15-20
DT 8 (1985-1987), DT 9.9 (1985-1987), DT 15	11-14	15-20
DT 8 (1988-1991), DT 9.9 (1988-1991)	13-20	18-28
DT 20, DT 25, DT 30	11-14	15-20
DT 35, DT 40, DT 55, DT 65	13-20	18-28
DT 75, DT 85	11-14	15-20
DT 115, DT 140, V4, V6		
8 mm	11-14	15-20
10 mm	25-30	34-41
Under oil seal housing bolts		
V4	21-23	28-32
V6		
8 mm	15-19	20-26
10 mm	34-39	46-54
Standard torque values		
Bolt head marked "4"		
5 mm	2-3	2-4
6 mm	3-5	4-7
8 mm	7-11	10-16
10 mm	16-25	22-35
Bolt head marked "7"		
5 mm	2-4	3-6
6 mm	6-9	8-12
8 mm	13-20	18-28
10 mm	30-44	40-60
Stainless steel bolt		
5 mm	2-3	2-4
6 mm	4-7	6-10
8 mm	11-14	15-20
10 mm	25-30	34-41

* Use standard torque values for any fasteners not specifically listed.

POWER HEAD

Table 2 CYLINDER AND PISTON SPECIFICATIONS

Engine	Piston measurement point*	Cylinder measurement point - Above exhaust port	Cylinder measurement point - Below cylinder top	Piston-to-cylinder clearance
DT 2	0.59 in. (15 mm)	0.20 in. (5 mm)	—	0.0021-0.0031 in. (0.053-0.080 mm)
DT 4	0.75 in. (19 mm)	—	0.98 in. (25 mm)	0.0020-0.0026 in. (0.052-0.067 mm)
DT 6	0.59 in. (15 mm)	0.20 in. (5 mm)	0.91 in. (23 mm)	0.0020-0.0026 in. (0.052-0.067 mm)
DT 8				
1985-1987	0.59 in. (15 mm)	0.20 in. (5 mm)	0.91 in. (23 mm)	0.0020-0.0026 in. (0.052-0.067 mm)
1988-on	0.75 in. (19 mm)	—	0.98 in. (25 mm)	0.0020-0.0026 in. (0.052-0.067 mm)
DT 9.9				
1985-1987	0.83 in. (21 mm)	—	1.10 in. (28 mm)	0.0020-0.0026 in. (0.052-0.067 mm)
1988-on	0.75 in. (19 mm)	—	0.98 in. (25 mm)	0.0020-0.0026 in. (0.052-0.067 mm)
DT 15	0.83 in. (21 mm)	—	1.10 in. (28 mm)	0.0020-0.0026 in. (0.052-0.067 mm)
DT 20, DT 25 (1985-1988), DT 30 (1985-1987)	0.94 in. (24 mm)	—	1.57 in. (40 mm)	0.0026-0.0032 in. (0.067-0.082 mm)
DT 25 (1989-on), DT 30 (1988-on)	0.90 in. (23 mm)	—	1.1 in. (28 mm)	0.0024-0.0035 in. (0.060-0.090 mm)
DT 35, DT 40	1.10 in. (28 mm)	—	1.38 in. (35 mm)	0.0039-0.0045 in. (0.099-0.114 mm)
DT 50, DT 65	1.38 in. (35 mm)	—	1.57 in. (40 mm)	0.0044-0.0050 in. (0.112-0.127 mm)
DT 75, DT 85	1.38 in. (35 mm)	—	1.57 in. (40 mm)	0.0044-0.0050 in. (0.112-0.127 mm)
DT 115, DT 140	1.0 in. (25 mm)	—	1.2 in. (30 mm)	0.0045-0.0049 in. (0.115-0.125 mm)
V4	0.20 in. (5 mm)	—	1.0 in. (25 mm)	0.0047-0.0059 in. (0.120-0.150 mm)
V6	1.2 in. (30 mm)	—	1.6 in. (40 mm)	0.0051-0.0055 in. (0.130-0.140 mm)

* Above the base of the piston skirt.

Table 3 PISTON RING END GAP

	Standard	Limit
DT 2		
1985-1989	0.0039-0.0098 in. (0.10-0.25 mm)	0.024 in. (0.6 mm)
1990-on	0.0039-0.012 in. (0.10-0.30 mm)	0.024 in. (0.6 mm)
DT 4	0.0059-0.0138 in. (0.15-0.35 mm)	0.028 in. (0.7 mm)

(continued)

Table 3 PISTON RING END GAP (continued)

	Standard	Limit
DT 6, DT 8 (1985-1987)	0.0039-0.0118 in. (0.10-0.30 mm)	0.024 in. (0.6 mm)
DT 8 (1988-on)	0.0059-0.0118 in. (0.15-0.30 mm)	0.031 in. (0.8 mm)
DT 9.9 (1985-1987), DT 15	0.0059-0.0138 in. (0.15-0.35 mm)	0.028 in. (0.7 mm)
DT 25 (1989-on), DT 30 (1988-on)	0.0059-0.0138 in. (0.15-0.35 mm)	0.031 in. (0.8 mm)
DT 150SS (1989-1990)	0.016-0.024 in. (0.4-0.6 mm)	0.039 in. (1.0 mm)
All other models	0.008-0.016 in. (0.2-0.4 mm)	0.031 in. (0.8 mm)

Chapter Nine

Gearcase

Torque is transferred from the power head's crankshaft to the gearcase by a drive shaft. A pinion gear on the drive shaft meshes with a drive gear in the gearcase to change the vertical power flow into a horizontal flow through the propeller shaft. The power head drive shaft rotates clockwise continuously when the engine is running, but propeller rotation is controlled by the gear train shifting mechanism.

On Suzuki outboards with a reverse gear, a sliding clutch engages the appropriate gear in the gearcase. This creates a direct coupling that transfers the power flow from the pinion to the propeller shaft. **Figure 1**, typical, shows the operation of the gear train.

Smaller outboards with a neutral but no reverse gear utilize a spring-loaded clutch to shift between neutral and forward gear. Gear train operation is shown disengaged in **Figure 2** or engaged **Figure 3**.

The gearcase can be removed without removing the entire outboard from the boat. This chapter contains removal, overhaul and installation procedures for the propeller, gearcase and water pump. **Table 1** is at the end of the chapter.

The gearcases covered in this chapter differ somewhat in design and construction over the years covered and thus require slightly different service procedures. This chapter is arranged in a normal disassembly/assembly sequence. When only a partial repair is required, follow the procedure(s) for your gearcase to the point where the faulty parts can be replaced, then reassemble the unit.

Since this chapter covers a wide range of models from 1985-on, the gearcases shown in the accompanying illustrations are the most common ones. While it is possible that the components shown in the pictures may not be identical to those being serviced, the step-by-step procedures may be used with all models covered in this manual.

PROPELLER

The outboards covered in this manual use variations of 2 propeller attachment designs.

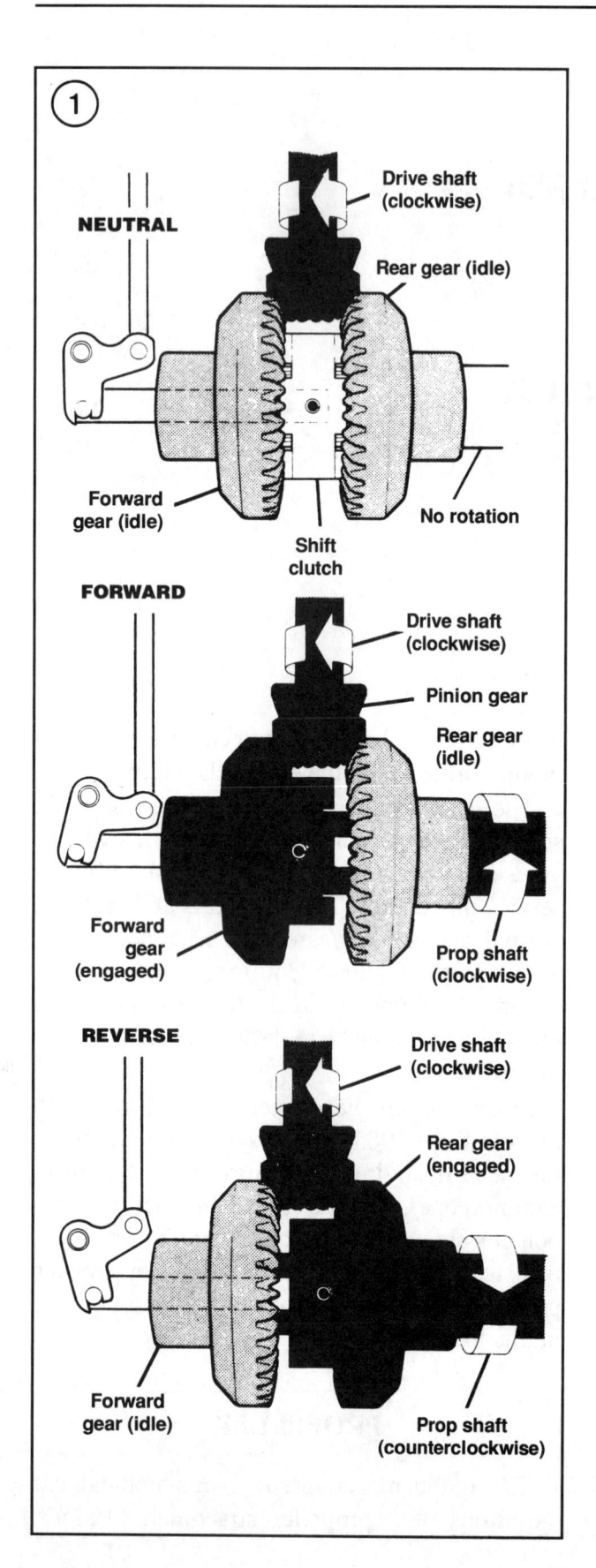

CHAPTER NINE

①

NEUTRAL

Drive shaft (clockwise)
Rear gear (idle)
Forward gear (idle)
Shift clutch
No rotation

FORWARD

Drive shaft (clockwise)
Pinion gear
Rear gear (idle)
Forward gear (engaged)
Prop shaft (clockwise)

REVERSE

Drive shaft (clockwise)
Rear gear (engaged)
Forward gear (idle)
Prop shaft (counterclockwise)

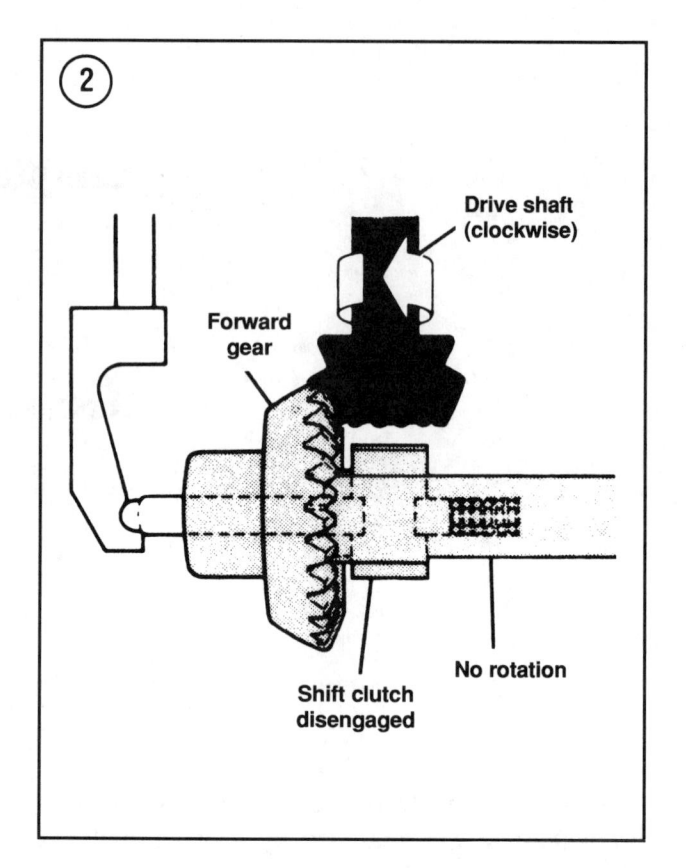

②

Drive shaft (clockwise)
Forward gear
Shift clutch disengaged
No rotation

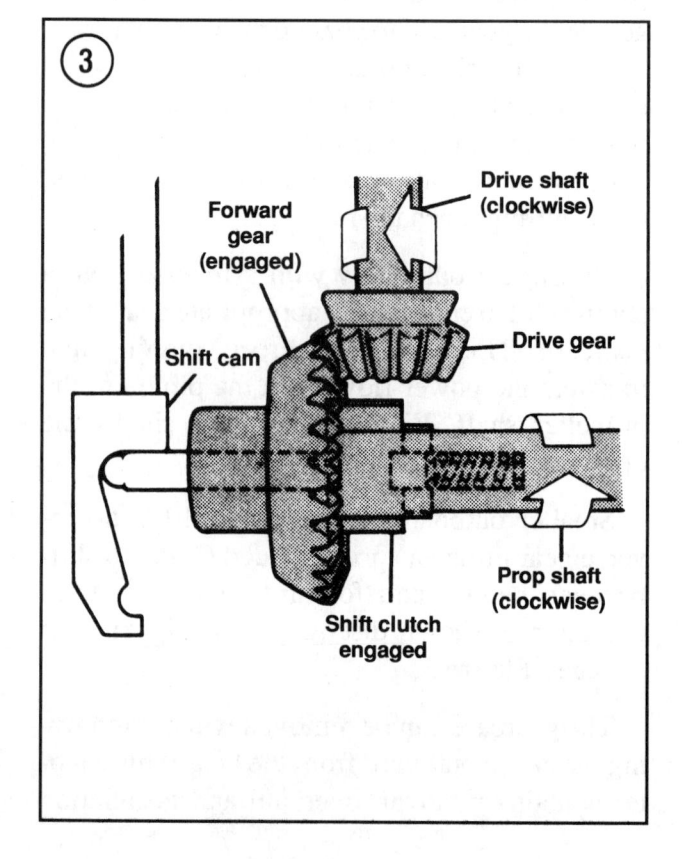

③

Forward gear (engaged)
Shift cam
Drive shaft (clockwise)
Drive gear
Shift clutch engaged
Prop shaft (clockwise)

GEARCASE

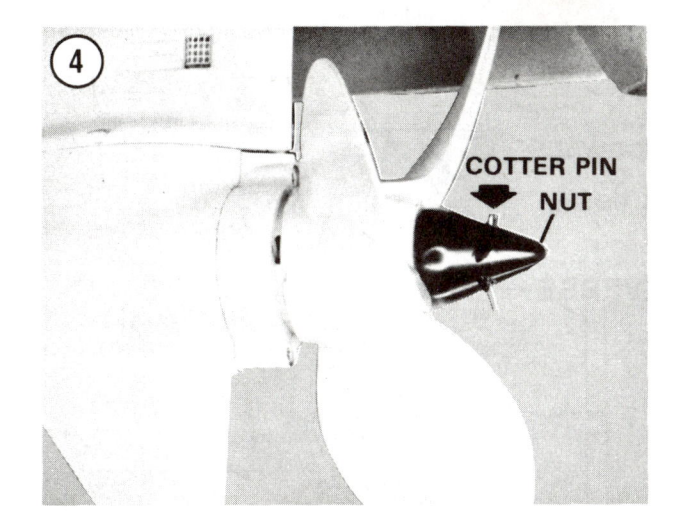

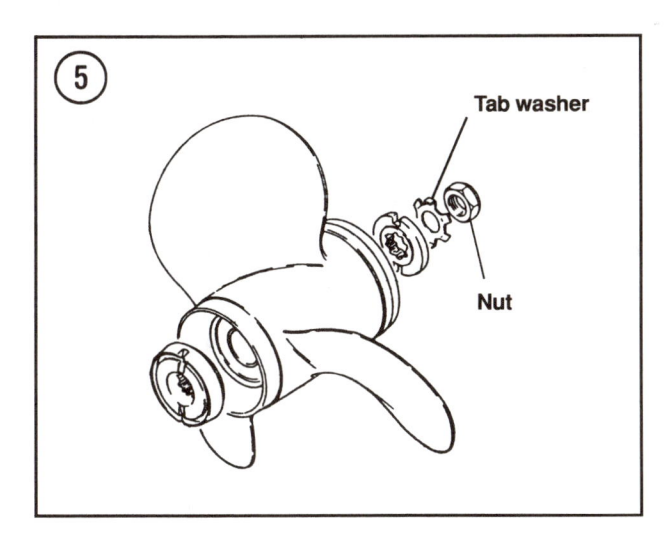

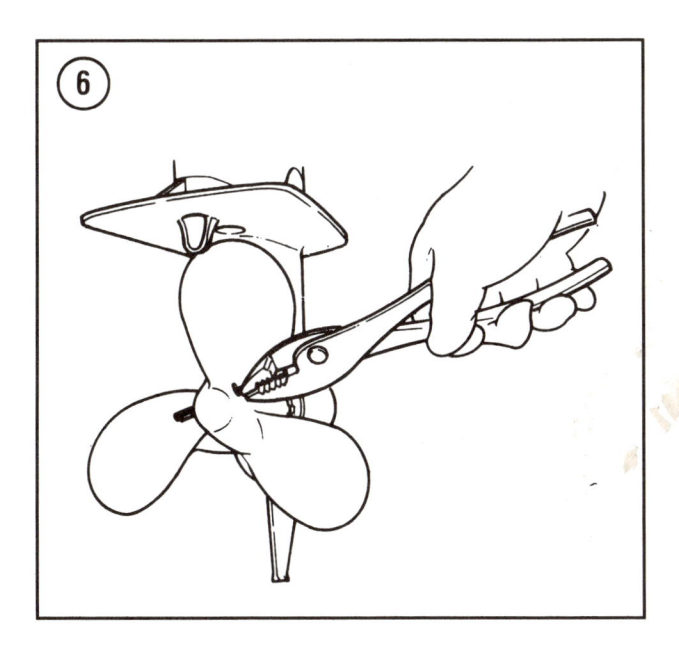

375

Smaller gearcases use a drive pin that engages a slot in the propeller hub, which is retained by a cotter pin.

The propeller on DT 2 models is retained by a cotter pin only with no retaining nut. The DT 4, DT 6 and 1985-1987 DT 8 models, use a separate hub nut retained by a cotter pin (**Figure 4**). In this design, a metal pin installed in the propeller shaft engages a recessed slot in the propeller hub. As the shaft rotates, the pin rotates the propeller. The drive pin is designed to break if the propeller hits an obstruction in the water. This design has 2 advantages. The pin absorbs the impact to prevent possible propeller damage. It also alerts the user to the fact that something is wrong, since the engine speed will increase immediately if the pin breaks.

On 1988-on DT 8 and all other larger displacement models, the propeller rides on a ratchet-type rubber bushing or a spline drive rubber hub and is retained by a castellated nut and cotter pin or a nut and tab-type lockwasher arrangement (**Figure 5**). Any underwater impact is absorbed by the propeller bushing/hub.

Removal/Installation

DT 2, DT 4, DT 6; 1985-1987 DT 8

1. Disconnect the spark plug lead(s) to prevent accidental starting of the engine.

2A. On DT 2 models, remove and discard the cotter pin (**Figure 6**). Remove the propeller from the shaft.

2B. On DT 4 and DT 6 models, perform the following:

    a. Remove and discard the cotter pin.

    b. Unscrew the propeller nut (**Figure 7**).

    c. Remove the propeller from the shaft (**Figure 8**).

3. Remove drive pin from propeller with an appropriate punch (**Figure 9**).

4. Clean propeller shaft thoroughly.

5. Inspect the pin engagement slot in the propeller hub and shaft for wear or damage. Replace if necessary.
6. Installation is the reverse of removal, noting the following.
7. Lubricate the propeller shaft with water-resistant grease (part No. 99000-25160) or equivalent.
8. Install new drive and cotter pins. Bend the ends of the cotter pin over completely.

1988-on DT 8; DT 9.9 through DT 140

1. Disconnect the spark plug leads to prevent accidental starting of the engine.
2A. On models equipped with a cotter pin, remove it. Discard the cotter pin as it must not be reused.
2B. On models equipped with a lockwasher, straighten the tab lockwasher.
3. Remove the propeller nut from the shaft (**Figure 10**).
4. Remove the tab lockwasher (if used) and propeller nut spacer from the shaft.
5. Remove the propeller (**Figure 11**) and bushing stopper (**Figure 12**) from the shaft.
6. Clean the propeller shaft splines thoroughly.
7. Installation is the reverse of removal, noting the following.
8. Lubricate propeller shaft with water-resistant grease.
9. If a tab lockwasher is used, check washer tab condition and replace as required.
10. If a cotter pin is used, install a new one and bend the ends over completely.

WATER PUMP

The water pump is mounted on top of the gearcase housing on all outboards covered in this manual except the DT 2. The DT 2 water pump is mounted in a pump case installed on the propeller shaft between the gearcase housing and propeller.

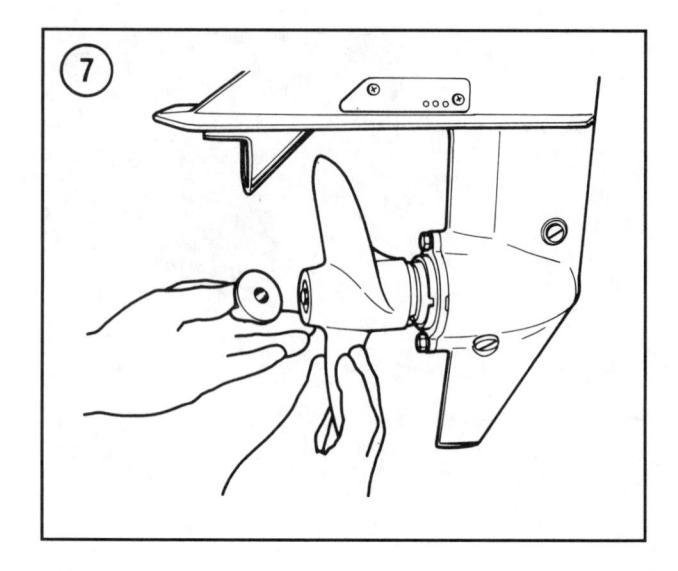

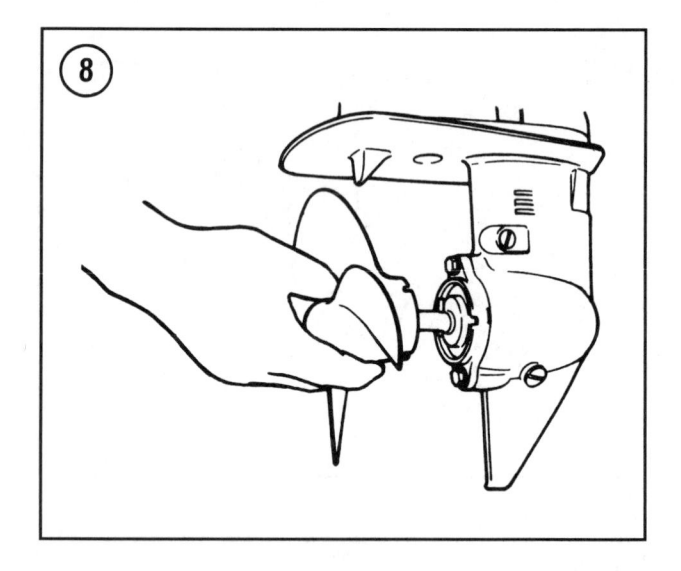

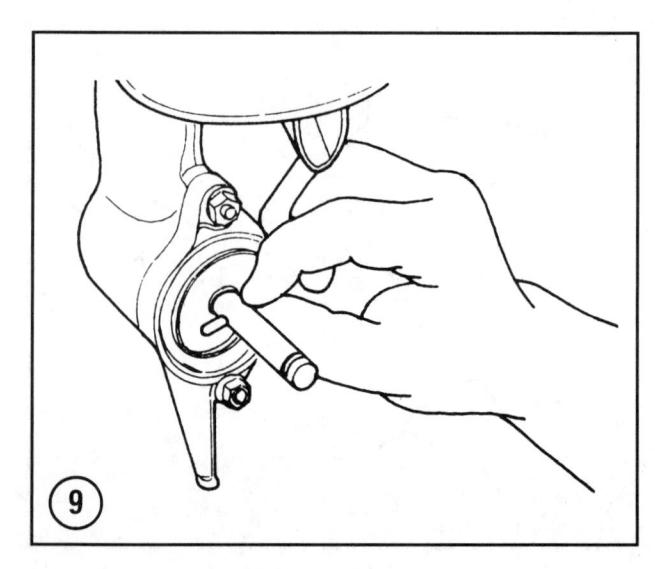

GEARCASE

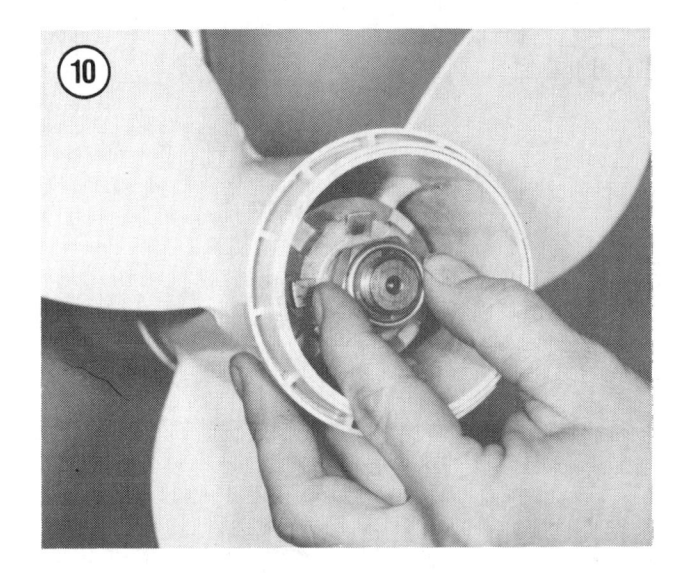

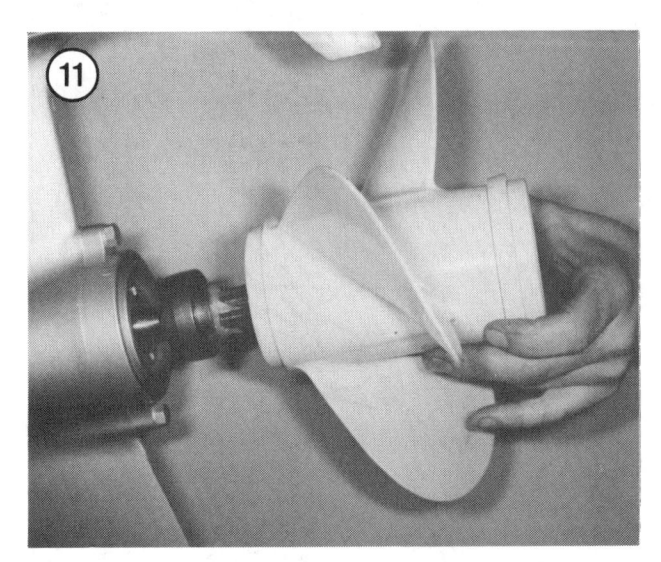

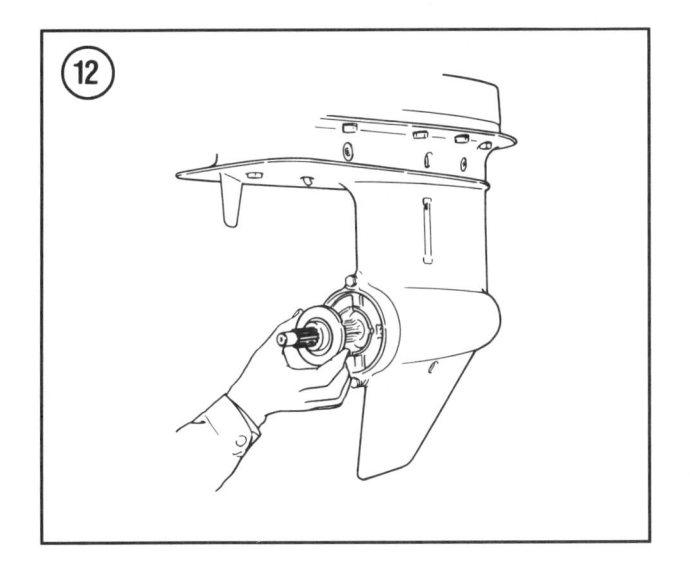

The DT 2 pump impeller is secured to the propeller shaft by a pin that fits into the propeller shaft and a similar cutout in the impeller hub. The impeller on other small displacement models is secured to the drive shaft in the same manner. On larger displacement models, a drive shaft key engages a flat on the drive shaft and a cutout in the impeller hub. As the drive shaft rotates, the impeller rotates with it. Water between the impeller blades and pump housing is pumped up to the power head through the water tube.

The offset center of the pump housing causes the impeller vanes to flex during rotation. At low speeds, the pump acts as a displacement type; at high speeds, water resistance forces the vanes to flex inward and the pump becomes a centrifugal type (**Figure 13**).

All seals and gaskets should be replaced whenever the water pump is removed. Since proper water pump operation is critical to outboard operation, it is also a good idea to install a new impeller at the same time.

Do not turn a used impeller over and reuse it. The impeller rotates in a clockwise direction with the drive shaft and the vanes gradually take a set in one direction. Turning the impeller over will cause the vanes to move in a direction opposite to that which caused the set. This will result in premature impeller failure and can cause extensive power head damage.

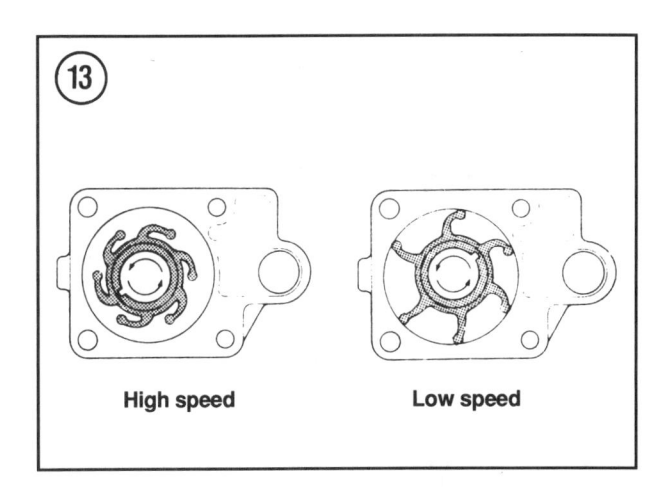

Removal and Disassembly (DT 2)

The water pump can be serviced on this model without removing the gearcase from the drive shaft housing.

1. Remove the propeller as described in this chapter.
2. Place a suitable container under the gearcase. Remove the drain screw and drain the lubricant from the unit.
3. Remove the 2 bolts holding the water pump case cover to the gearcase housing (**Figure 14**) and remove the cover.
4. Carefully pry the impeller from the water pump case, then remove the impeller drive pin from the propeller shaft (**Figure 15**).

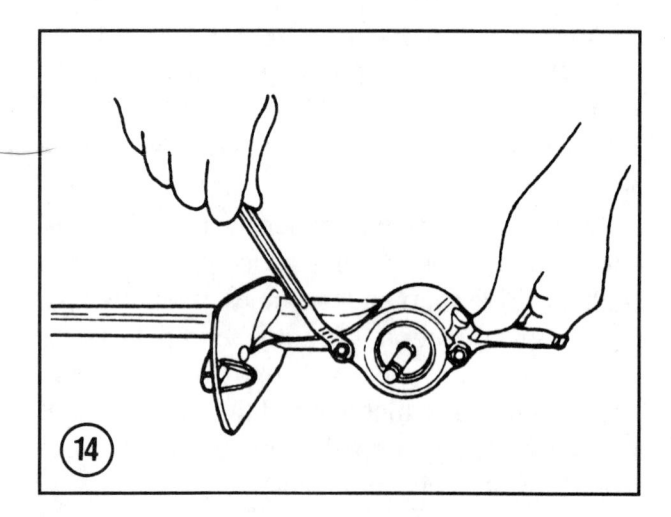

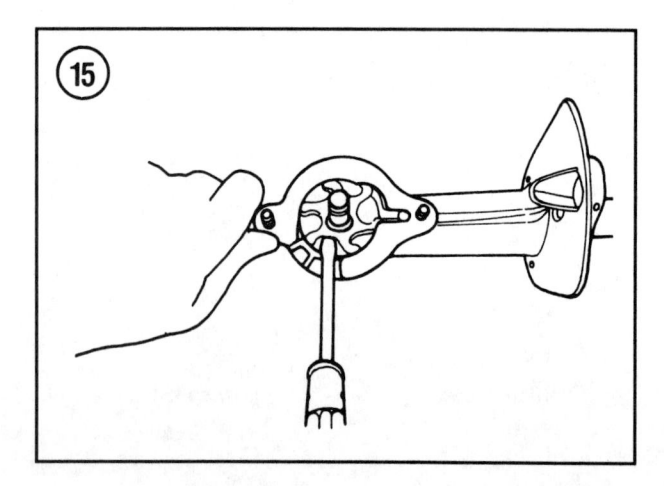

WATER PUMP (DT 4)

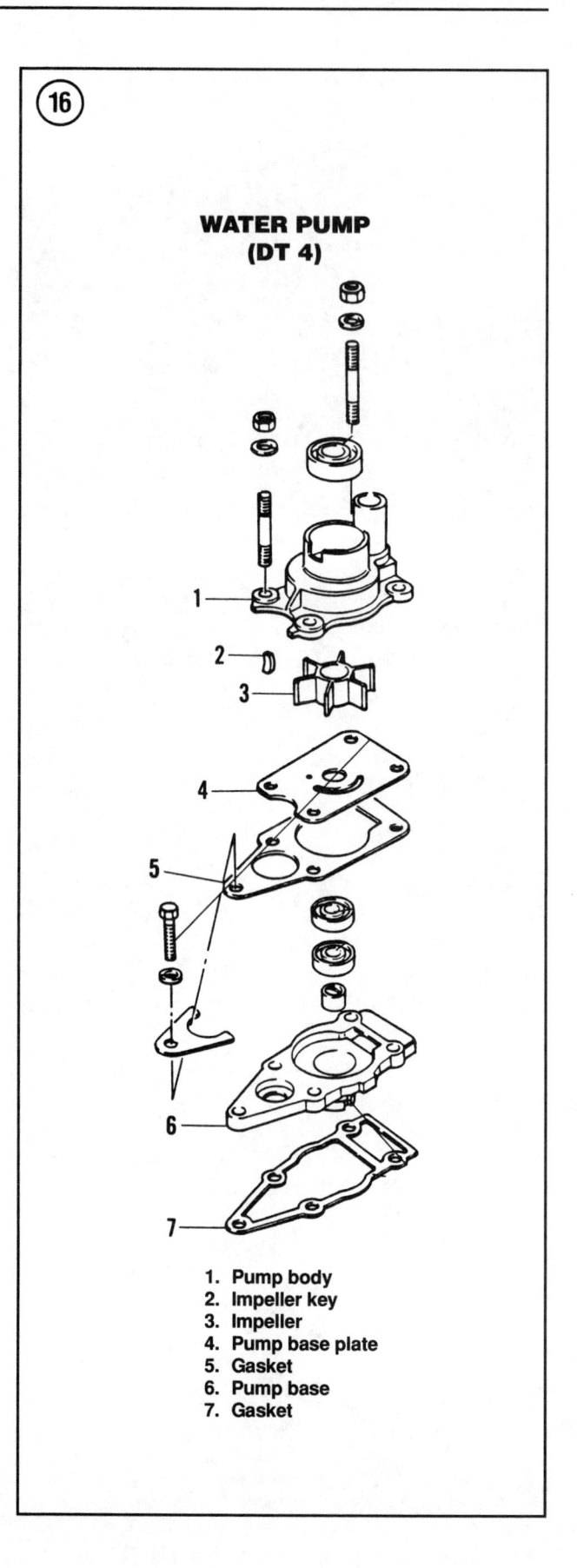

1. Pump body
2. Impeller key
3. Impeller
4. Pump base plate
5. Gasket
6. Pump base
7. Gasket

GEARCASE

Removal and Disassembly (DT 4, DT 6, 1985-1987 DT 8)

The water pump used on these models varies primarily in shape and size. All use the same basic components and are serviced in essentially the same way. It is not necessary to remove the drive shaft bearing housing to service the water pump.

Refer to the following illustrations for this procedure.

   a. **Figure 16**: DT 4.
   b. **Figure 17**: DT 6, 1985-1987 DT 8.

1. Remove the gearcase as described in this chapter.
2. Secure the gearcase in a suitable holding fixture or a vise with protective jaws. If protective jaws are not available, position the gearcase upright in the vise with the skeg between wooden blocks.
3. Remove the 4 bolts and lockwashers holding the water pump cover in place (**Figure 18**). Remove the pump cover.
4. If the impeller does not come off with the cover, slide it off the drive shaft (**Figure 19**).
5. Remove the impeller drive pin or key (A, **Figure 19**) from the drive shaft.
6. Carefully pry the pump base plate and gasket assembly free of the gearcase housing. Discard the gasket.

Removal and Disassembly (1988-on DT 8; DT 9.9 and DT 15)

The water pump used on these models varies primarily in shape and size. All use the same

---

17 WATER PUMP (DT 6, 1985-1987 DT 8)

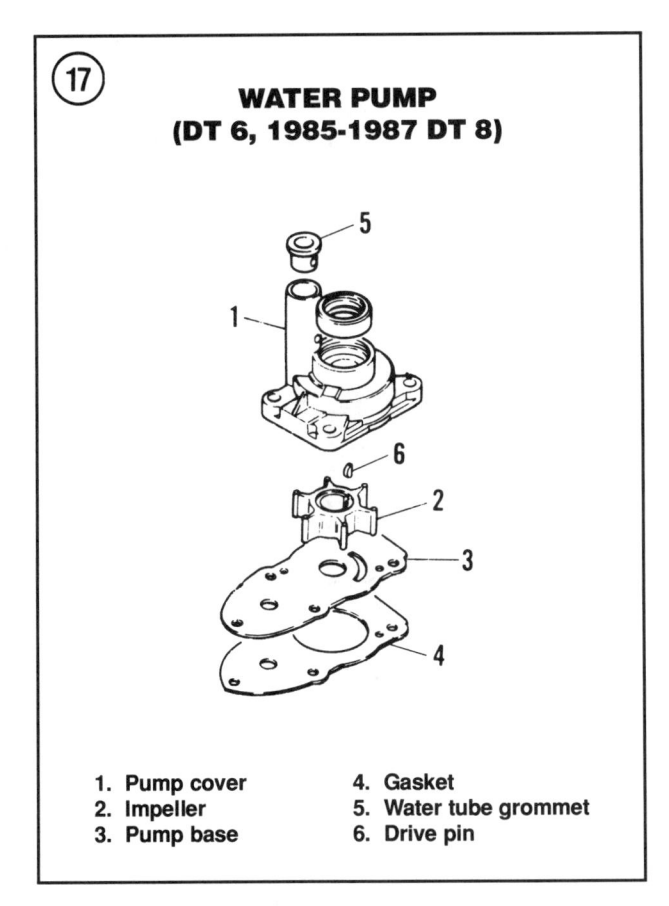

1. Pump cover
2. Impeller
3. Pump base
4. Gasket
5. Water tube grommet
6. Drive pin

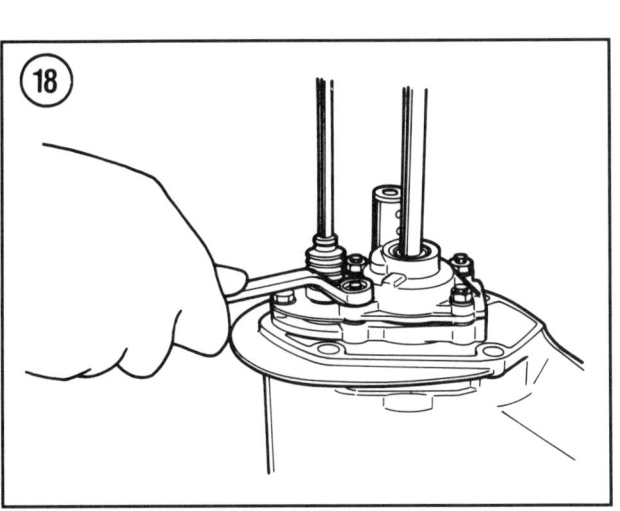

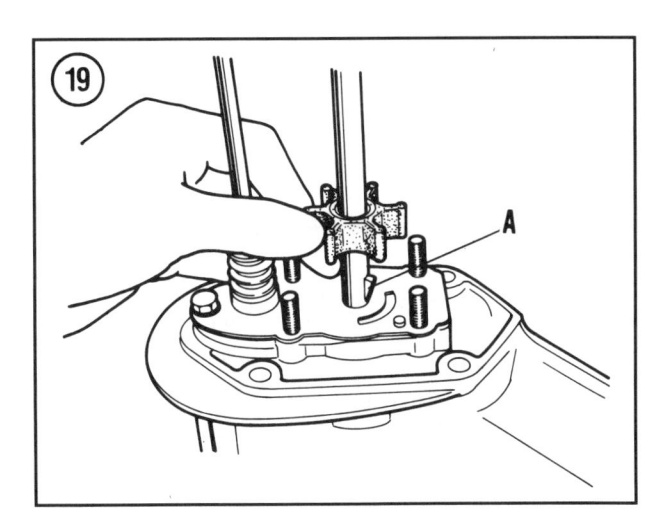

basic components and are serviced in essentially the same way. It is not necessary to remove the drive shaft bearing housing to service the water pump.

Refer to the following illustrations for this procedure:

a. **Figure 20**: 1985-1987 DT 9.9 and DT 15.
b. **Figure 21**: 1988-on DT 8 and DT 9.9.

1. Remove the gearcase as described in this chapter.
2. Secure the gearcase in a suitable holding fixture or a vise with protective jaws. If protective

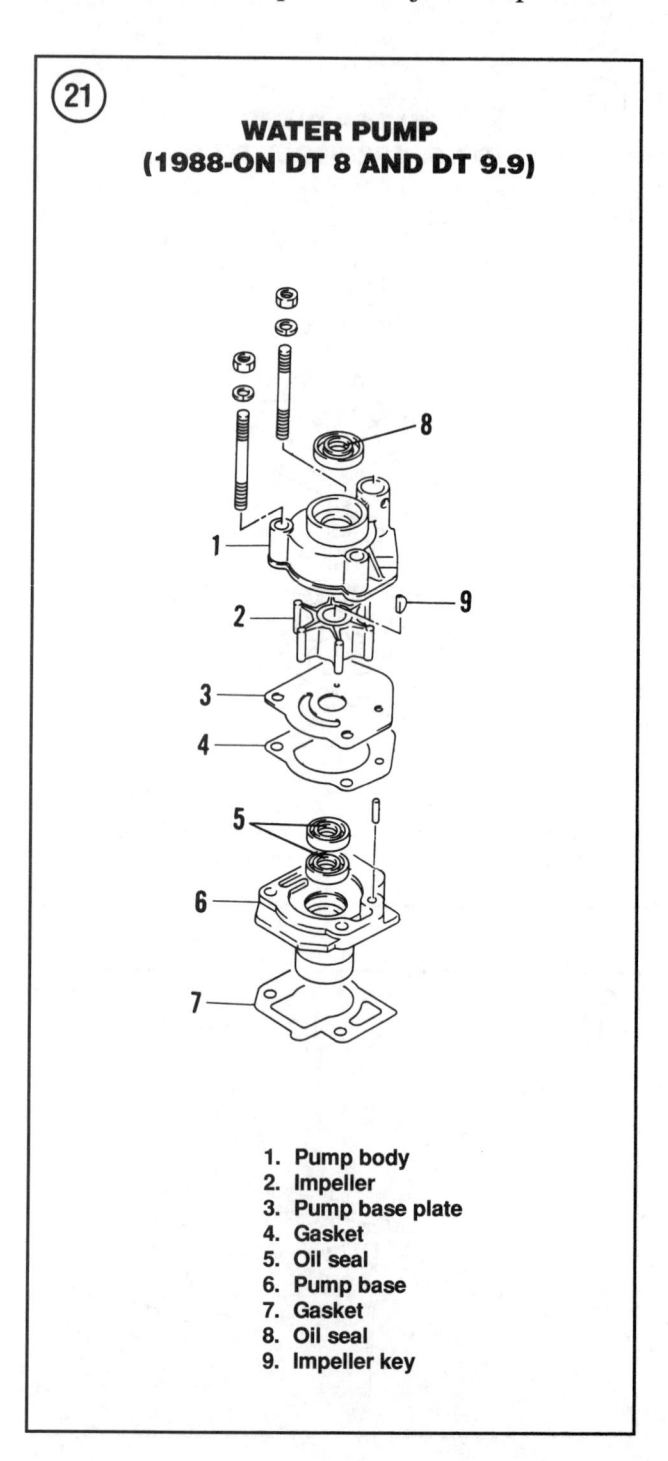

WATER PUMP (1985-1987 DT 9.9 AND DT 15)

1. Pump body
2. Water tube seal
3. Impeller
4. Pump base plate
5. Gasket
6. Water tube bushing
7. Pump body grommet
8. Impeller key
9. Pump base
10. Base gasket

WATER PUMP (1988-ON DT 8 AND DT 9.9)

1. Pump body
2. Impeller
3. Pump base plate
4. Gasket
5. Oil seal
6. Pump base
7. Gasket
8. Oil seal
9. Impeller key

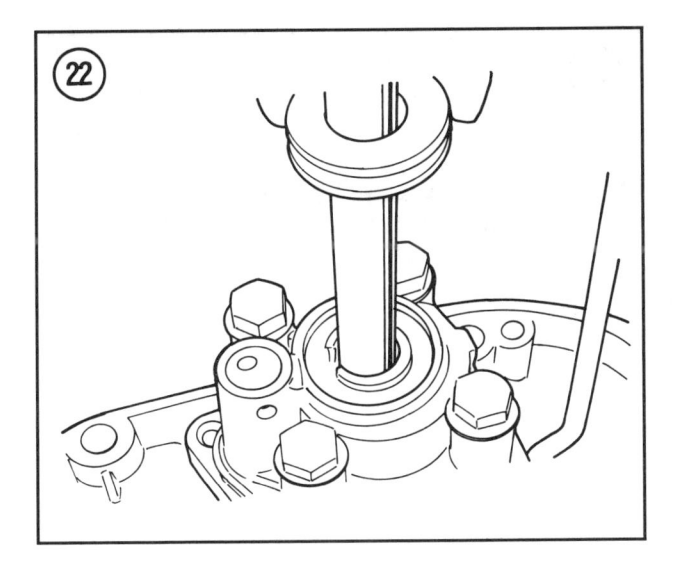

GEARCASE

381

jaws are not available, position the gearcase upright in the vise with the skeg between wooden blocks.

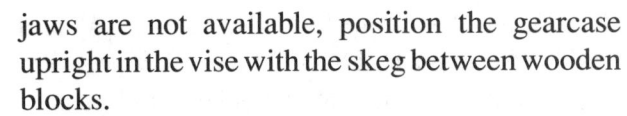

3. Remove the water pump grommet (**Figure 22**) from the pump cover.

4. Remove the fasteners holding the water pump cover (**Figure 23**). Remove the pump cover.

5. If the impeller does not come off with the cover, slide it off the drive shaft (**Figure 24**).

6. Remove the impeller drive key from the drive shaft (**Figure 25**).

7. Carefully pry the pump base plate and gasket (A, **Figure 26**) assembly free of the gearcase housing.

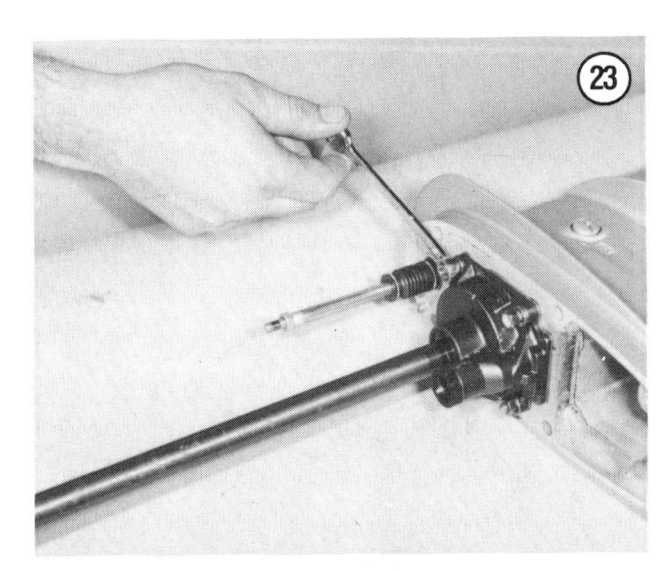

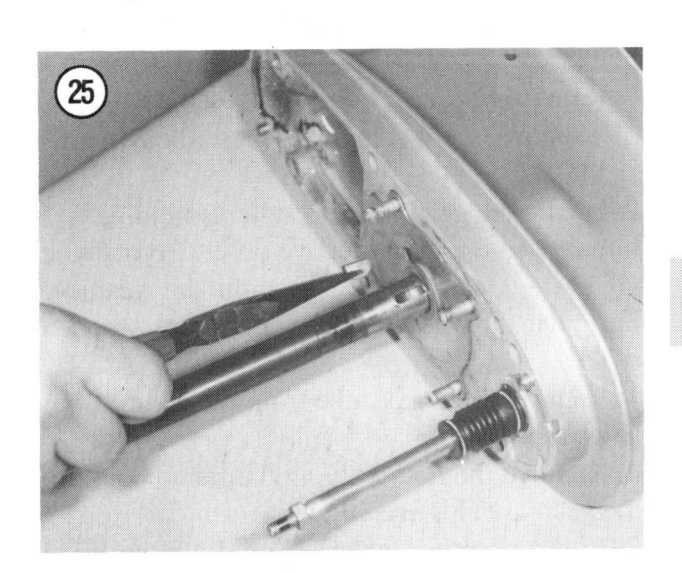

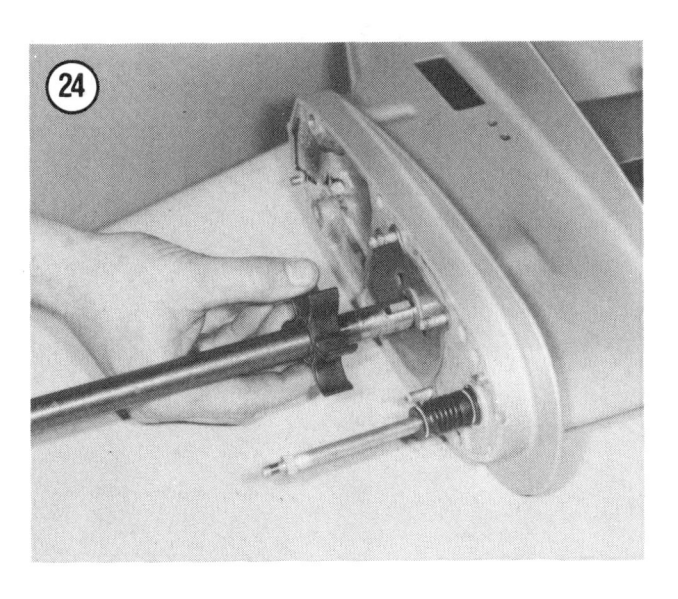

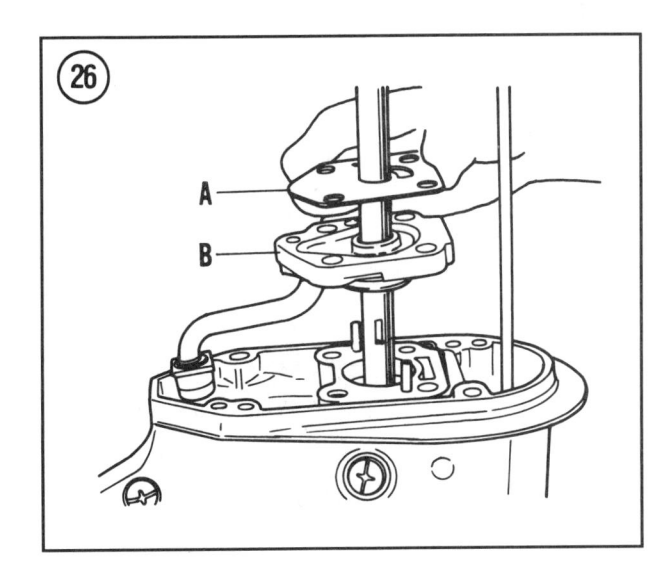

8. Remove the pump base (B, **Figure 26**) and gasket (**Figure 27**). Discard the gasket.

9. Remove the water inlet housing and gasket. Discard the gasket.

**Removal and Disassembly
(DT 20, DT 25, DT 30 [2-cylinder];
DT 35, DT 40; DT 25, DT 30 [3-cylinder];
DT 55, DT 65)**

The water pump used on these models varies primarily in shape and size. All use the same basic components and are serviced in essentially the same way. It is not necessary to remove the drive shaft bearing housing to service the water pump.

Refer to **Figure 28** for this procedure.

1. Remove the gearcase as described in this chapter.
2. Secure the gearcase in a suitable holding fixture or a vise with protective jaws. If protective jaws are not available, position the gearcase upright in the vise with the skeg between wooden blocks.
3. Remove the water tube from the pump cover (**Figure 29**).
4. Remove the nuts holding the pump cover to the gearcase (**Figure 30**).
5. Slide the pump cover up and off the drive shaft.

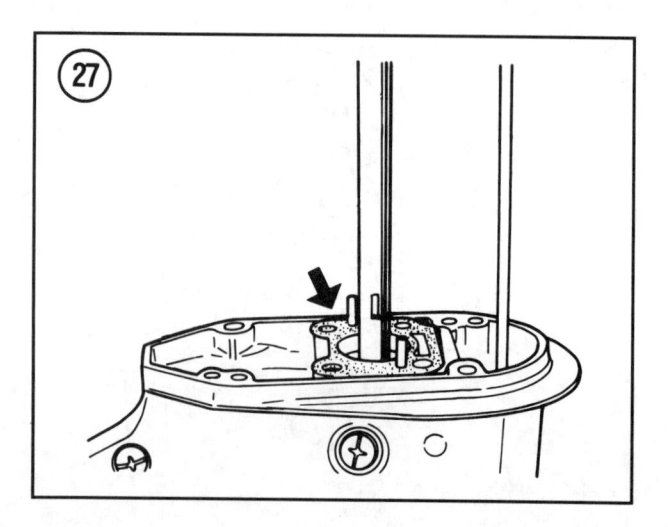

**WATER PUMP
(1985-1987 DT 20, DT 25, DT 30)**

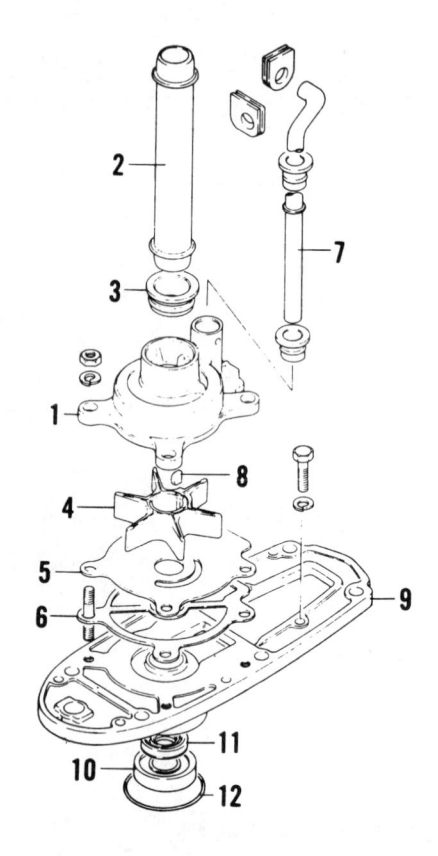

1. Pump cover
2. Tube
3. Grommet
4. Impeller
5. Pump base plate
6. Gasket
7. Tube
8. Impeller key
9. Drive shaft housing bearing
10. Bearing
11. Oil seal
12. O-ring

GEARCASE

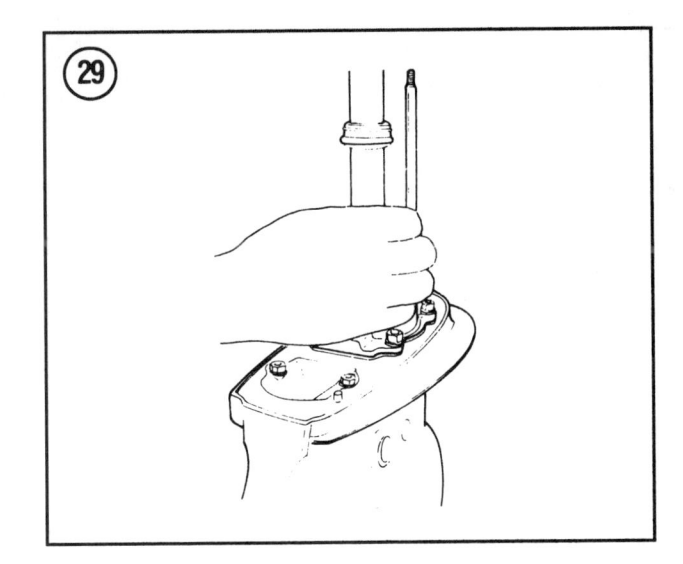

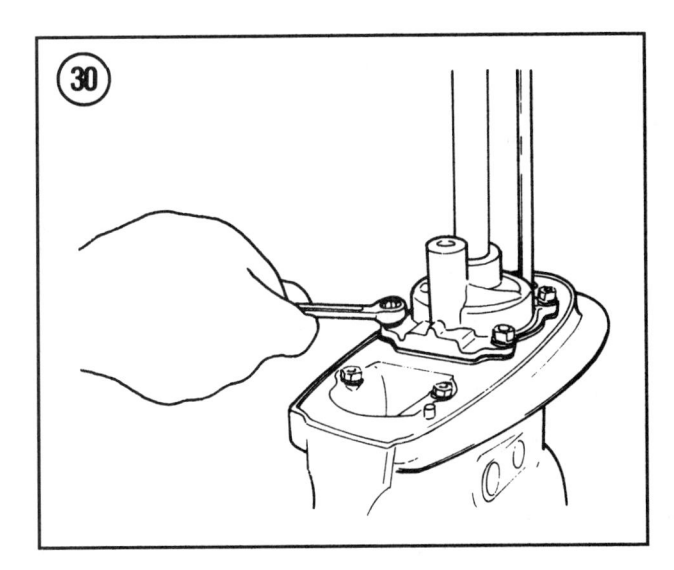

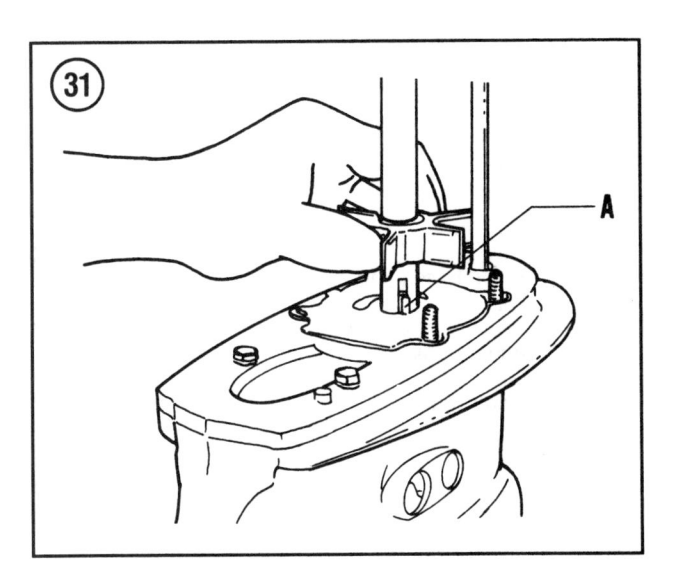

6. Remove the impeller from the pump cover. If it did not come off with the cover, slide the impeller up and off the drive shaft (**Figure 31**).

7. Remove the impeller drive key from the drive shaft.

8. Carefully pry the pump base and gasket from the gearcase housing (**Figure 32**). Remove the base and discard the gasket.

Removal and Disassembly (DT 75, DT 85)

It is not necessary to remove the drive shaft bearing housing to service the water pump.

Refer to **Figure 33** for this procedure.

1. Remove the gearcase as described in this chapter.

2. Secure the gearcase in a suitable holding fixture or a vise with protective jaws. If protective jaws are not available, position the gearcase upright in the vise with the skeg between wooden blocks.

3. Remove the water tube from the pump cover.

4. Remove the bolts (**Figure 34**) holding the pump cover to the gearcase.

5. Slide the pump cover up and off the drive shaft.

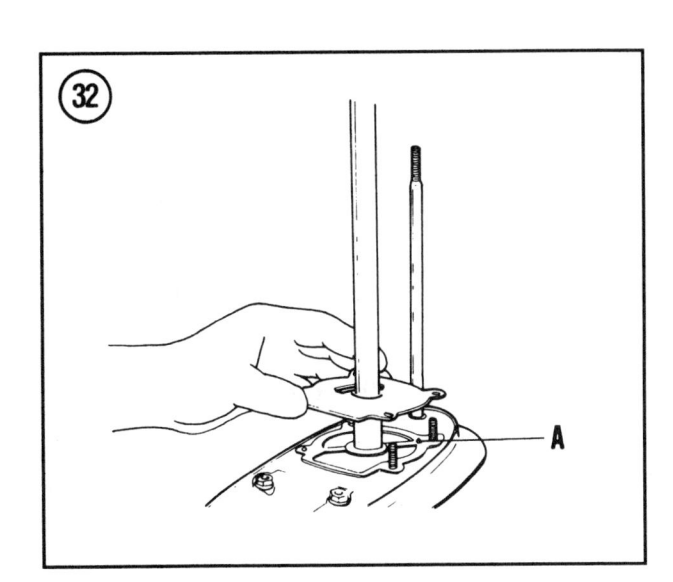

CHAPTER NINE

㉝

**WATER PUMP
(DT 75, DT 85)**

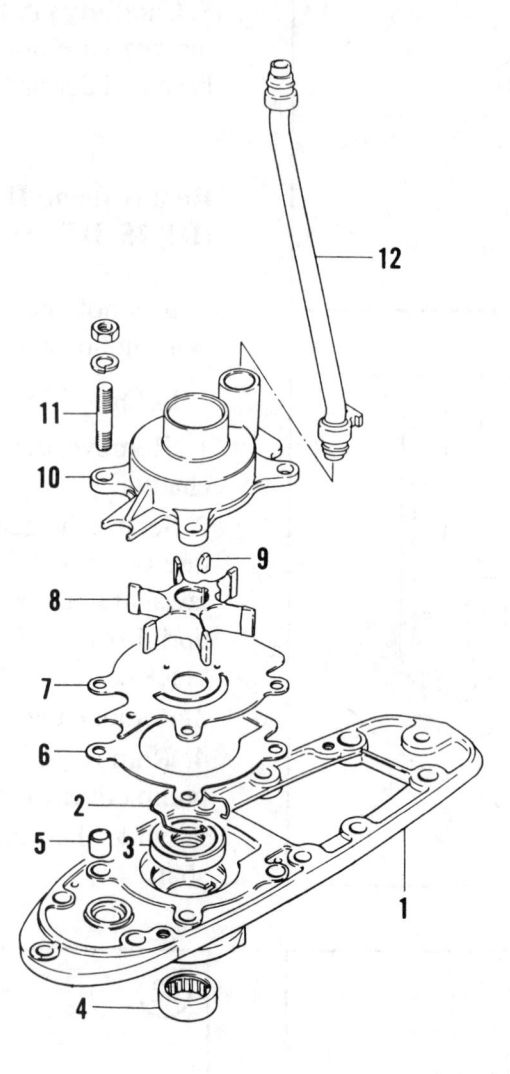

1. Drive shaft bearing housing
2. Oil seal retaining ring
3. Oil seal
4. Bearing
5. Dowel pin
6. Gasket
7. Pump base plate
8. Impeller
9. Impeller drive key
10. Pump body
11. Stud
12. Water tube

GEARCASE

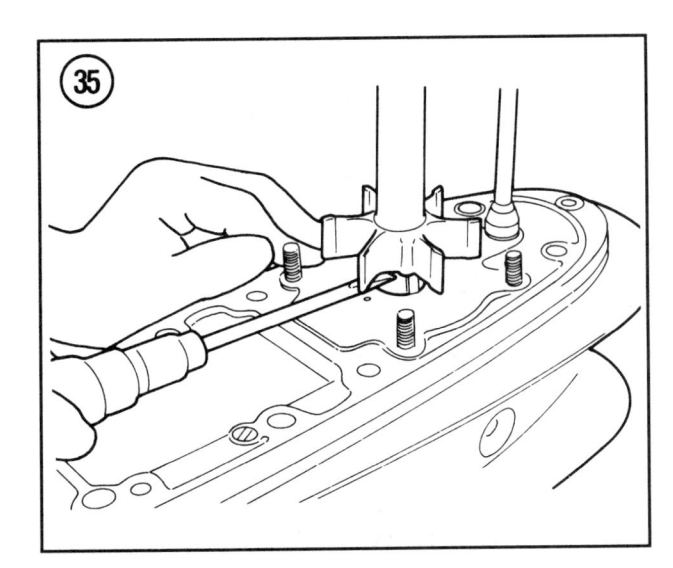

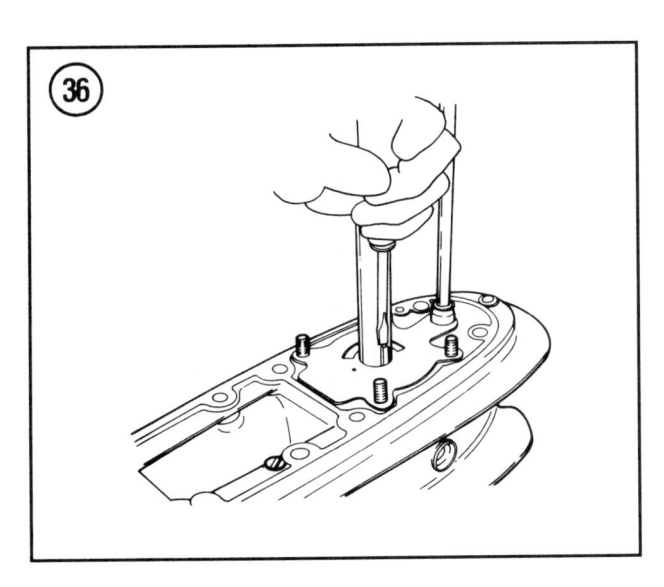

6. Remove the impeller from the pump cover. If it did not come off with the cover, slide the impeller up and off the drive shaft (**Figure 35**).

7. Remove the impeller drive key (**Figure 36**) from the drive shaft.

8. Carefully pry the pump base and gasket from the gearcase housing (**Figure 37**). Remove the base and discard the gasket.

Removal and Disassembly (DT 115, DT 140)

The water pump used on these models varies primarily in shape and size. All use the same basic components and are serviced in essentially the same way. It is not necessary to remove the drive shaft bearing housing to service the water pump.

Refer to the following illustrations for this procedure:

a. **Figure 38**: 1985 DT 115, DT 140.
b. **Figure 39**: 1986-on DT 115, DT 140.

1. Remove the gearcase as described in this chapter.

2. Secure the gearcase in a suitable holding fixture or a vise with protective jaws. If protective jaws are not available, position the gearcase

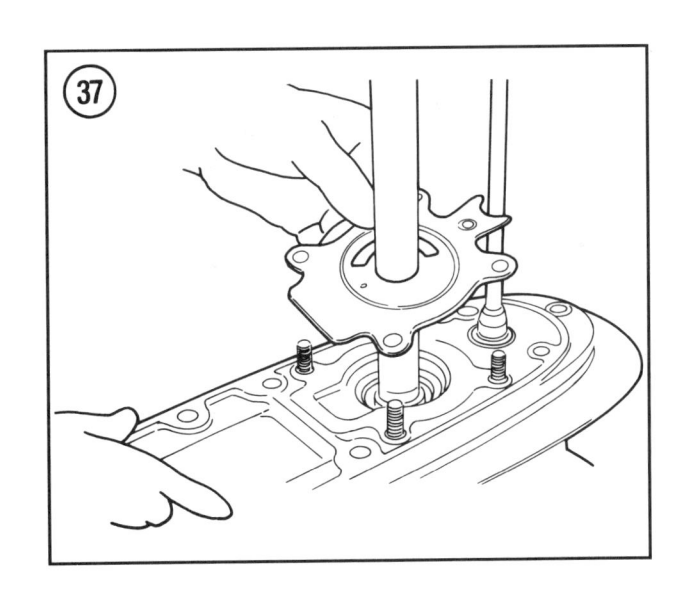

⑲ WATER PUMP (1985 DT 115, DT 140)

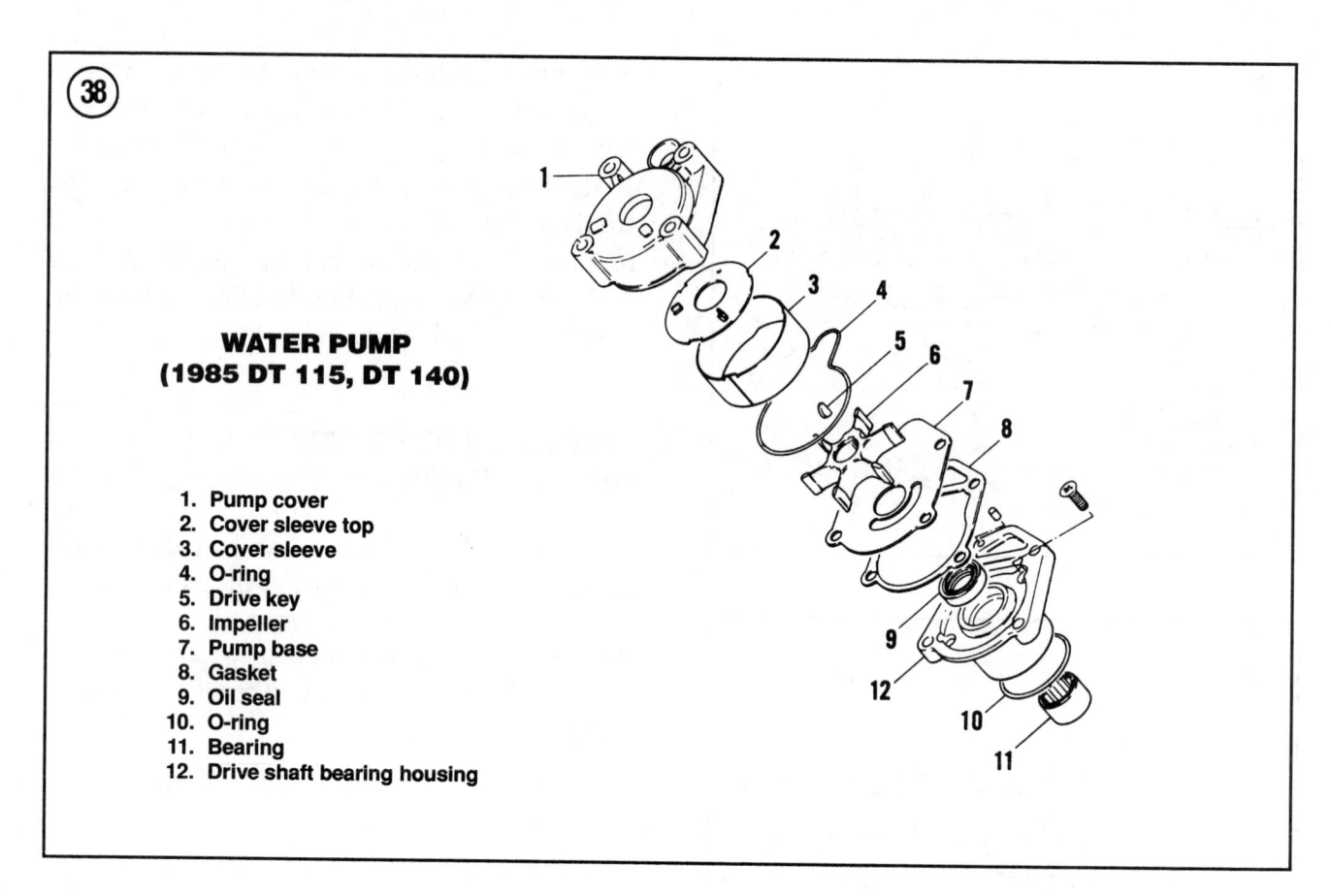

1. Pump cover
2. Cover sleeve top
3. Cover sleeve
4. O-ring
5. Drive key
6. Impeller
7. Pump base
8. Gasket
9. Oil seal
10. O-ring
11. Bearing
12. Drive shaft bearing housing

㊴ WATER PUMP (1986-ON DT 115, DT 140)

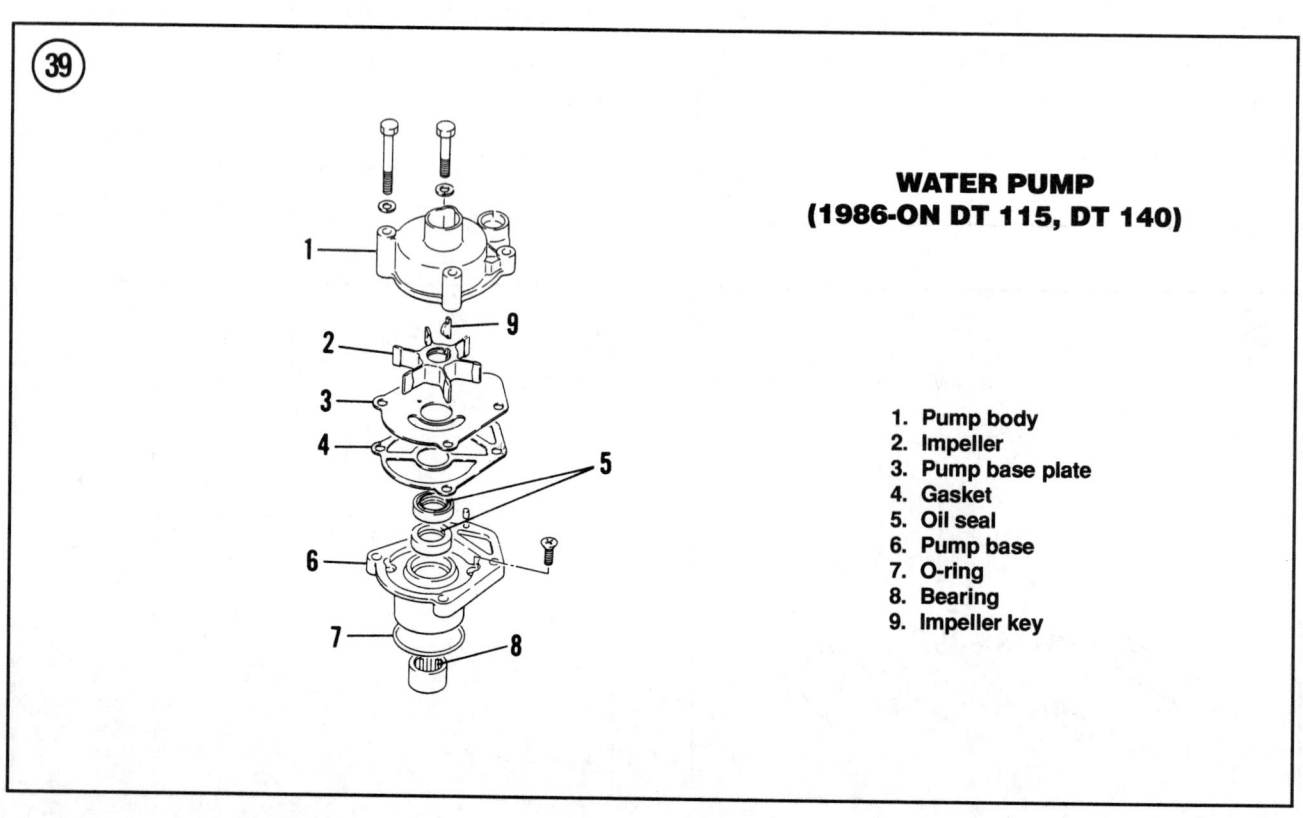

1. Pump body
2. Impeller
3. Pump base plate
4. Gasket
5. Oil seal
6. Pump base
7. O-ring
8. Bearing
9. Impeller key

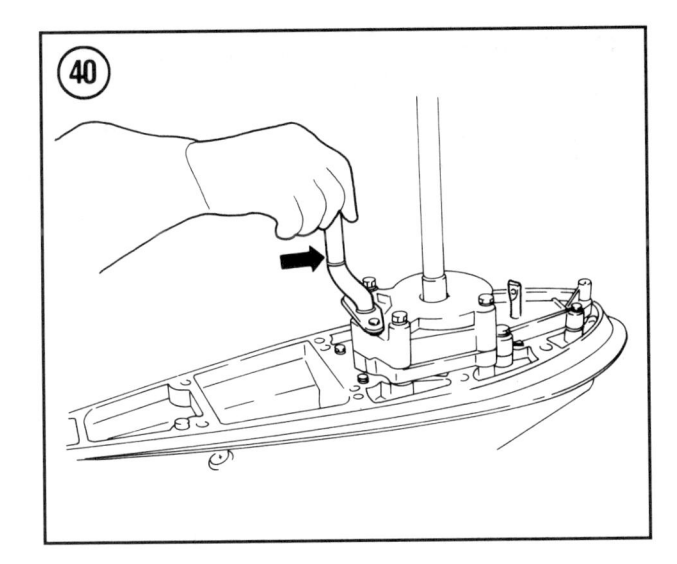

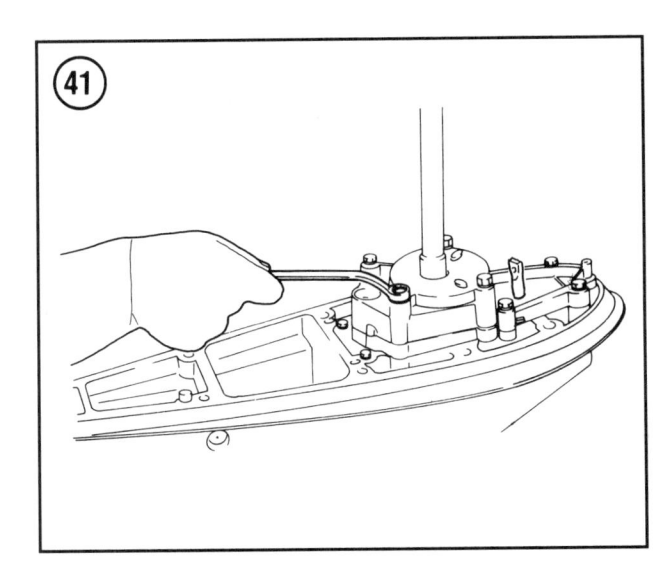

GEARCASE

upright in the vise with the skeg between wooden blocks.

3. Remove the water tube from the pump cover (**Figure 40**).

4. Remove the bolts (**Figure 41**) holding the pump cover to the gearcase.

5A. On 1985 models, slide the pump cover, cover sleeve top and cover sleeve up and off the drive shaft. Remove the O-ring.

5B. On 1986-on models, slide the pump cover up and off the drive shaft.

6. Remove the impeller from the pump cover. If it did not come off with the cover, slide the impeller up and off the drive shaft (**Figure 42**).

7. Remove the impeller drive key (**Figure 43**) from the drive shaft.

8. Carefully pry the pump base and gasket from the gearcase housing (**Figure 44**). Remove the base and discard the gasket.

Removal and Disassembly (V4, V6)

It is not necessary to remove the drive shaft bearing housing to service the water pump.

1. Remove the gearcase as described in this chapter.

2. Secure the gearcase in a suitable holding fixture or a vise with protective jaws. If protective

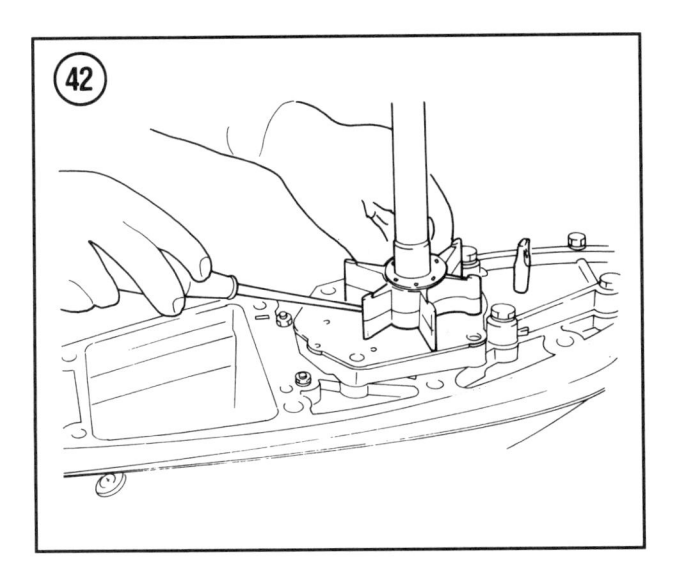

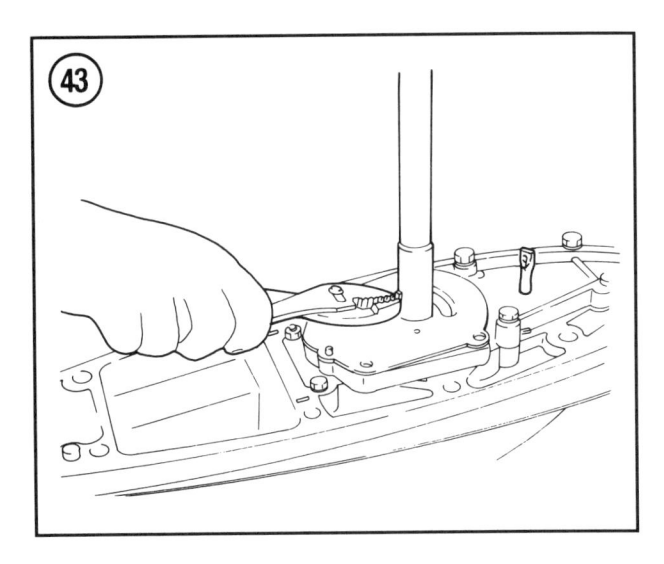

jaws are not available, position the gearcase upright in the vise with the skeg between wooden blocks.

3. Remove the water tube from the pump cover.
4. Remove the bolts (**Figure 45**) holding the pump cover to the gearcase. Slide the pump cover up and off the drive shaft.
5. Remove the impeller from the pump cover. If it did not come off with the cover, slide the impeller (A, **Figure 46**) up and off the drive shaft (B, **Figure 46**).
6. Remove the impeller drive key (A, **Figure 47**) from the drive shaft.
7. Carefully pry the pump base and gasket (B, **Figure 47**) from the gearcase housing. Remove the base and discard the gasket.

Cleaning and Inspection

When removing seals from water pump cover, note and record the direction in which each seal lip faces for proper reinstallation.

1. Remove the drive shaft rubber grommet (if so equipped) and water tube seal from the pump cover. Inspect these rubber parts for wear, hardness or deterioration and replace if necessary.
2. On models so equipped, remove and discard the water pump cover O-ring (**Figure 48**). 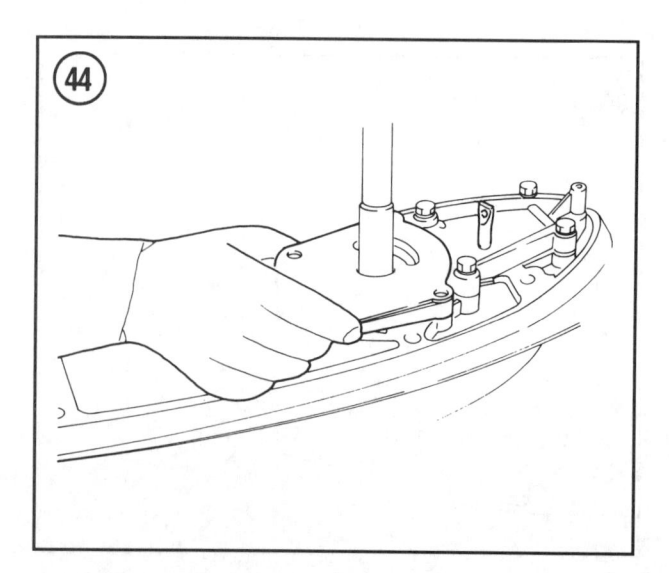
3. Check the pump cover for cracks, distortion or melting. Replace as required.
4. Clean the pump cover and base plate in solvent and blow dry with compressed air.
5. Carefully remove all gasket residue from the mating surfaces.

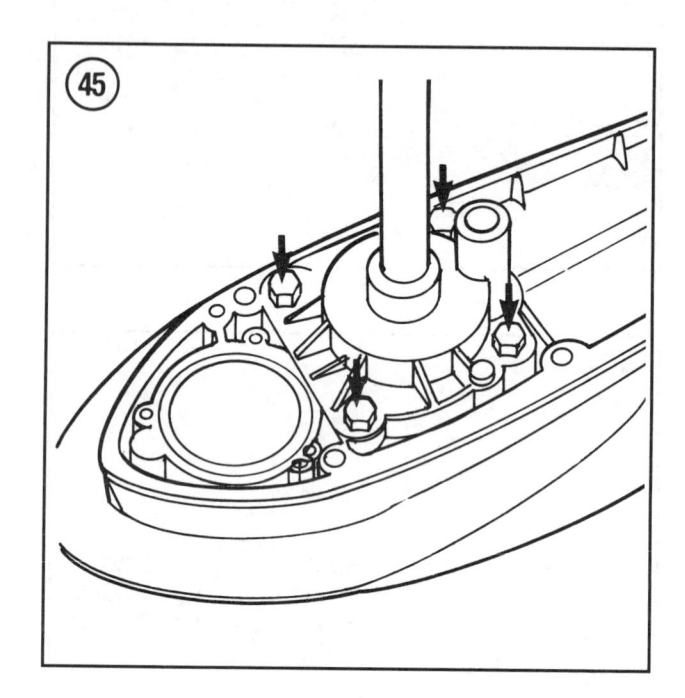

GEARCASE

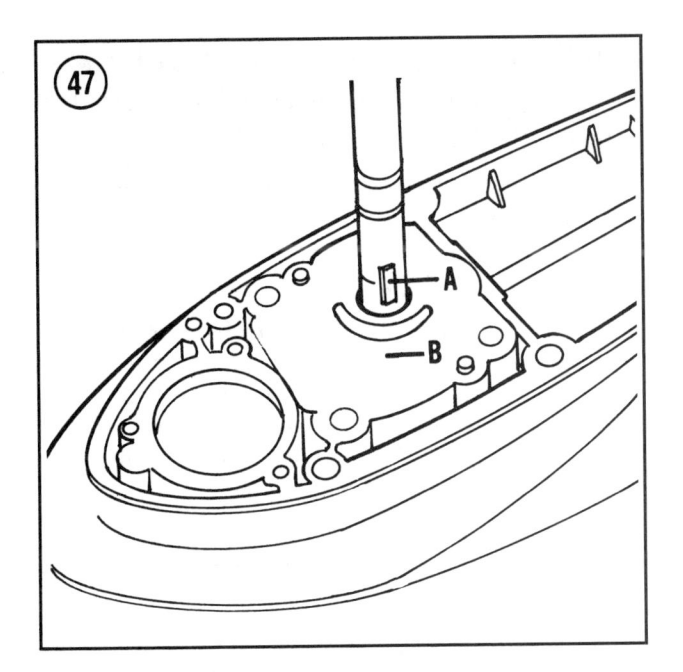

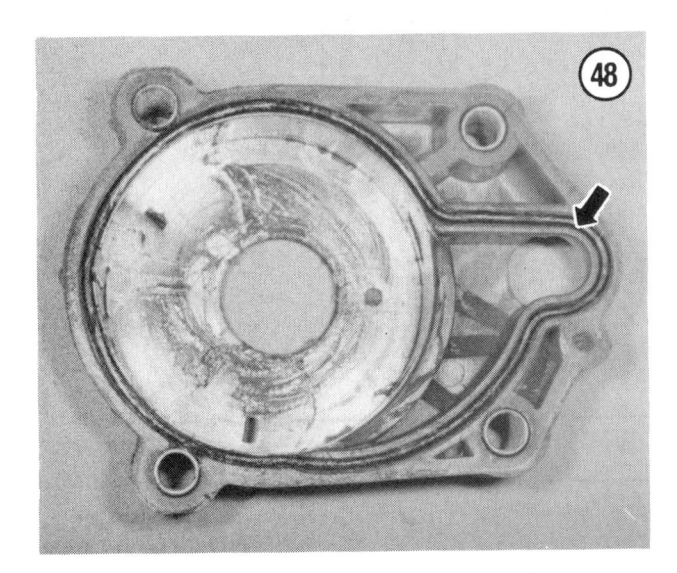

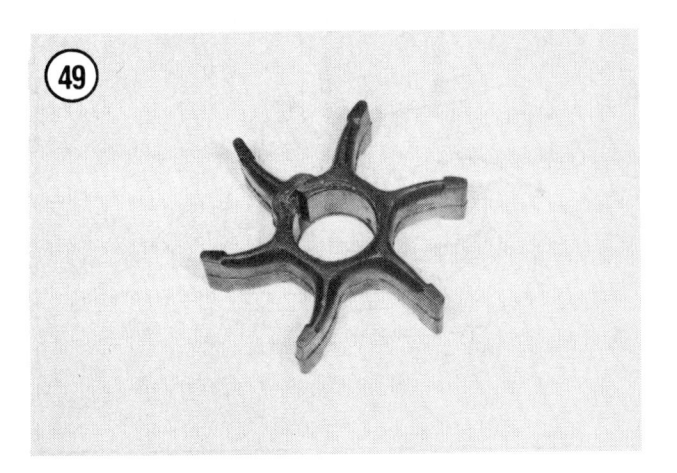

6. If the existing impeller (**Figure 49**) is to be reused, check bonding to hub. Check side seal surfaces and vane ends for cracks, tears, wear or a glazed or melted appearance. If any of these defects are noted, replace the impeller, do *not* reuse it.

Assembly and Installation (DT 2)

1. Insert impeller drive pin in propeller shaft slot.
2. Install the impeller in the pump body with a counterclockwise rotating motion. When properly installed, the impeller vanes should be positioned as shown in **Figure 50**.
3. Install the pump case cover and tighten the fasteners securely.

Assembly and Installation (DT 4, DT 6; 1985-1987 DT 8)

1. Install the pump base plate and new gasket.

CAUTION
If the original impeller is to be reused, install it in the same rotational direction as removed to avoid premature failure. The drive pin groove should be visible

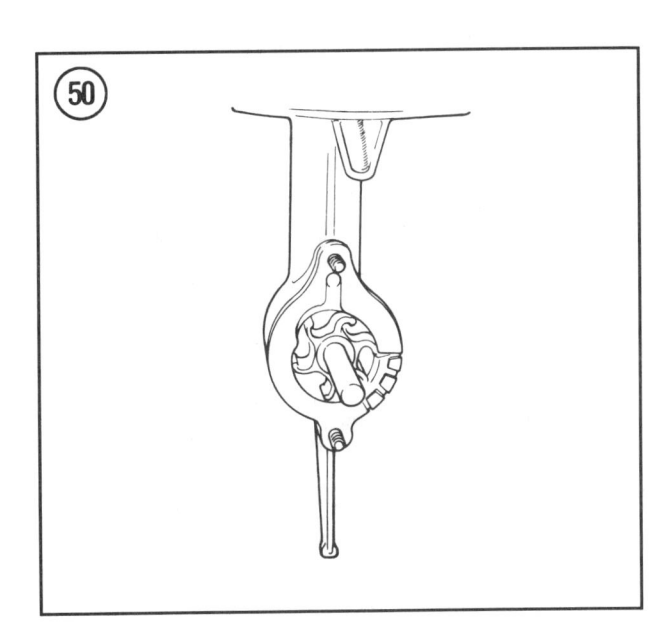

and the curl of the blades positioned as shown in **Figure 51**.

2. Install the impeller drive key into the drive shaft (A, **Figure 52**).

3. Install the impeller onto the drive shaft (**Figure 53**). Align the impeller slot with the drive shaft key.

4. Slide the water pump cover down the drive shaft (**Figure 54**).

5. Slowly rotate the drive shaft *clockwise* by hand to flex the impeller into the housing in the right direction.

CAUTION
Correct housing fastener torque is important in Step 6. Excessive torque can cause the pump to crack during operation; insufficient torque may result in leakage and exhaust induction which will cause overheating.

6. Coat the water pump cover bolt threads with Thread Lock 1342 (part No. 99000-32050). Don't forget to install the lockwashers under the bolts. Install and tighten securely.

Assembly/Installation (1988-on DT 8; DT 9.9 and DT 15)

1. Install the water inlet housing and new gasket.

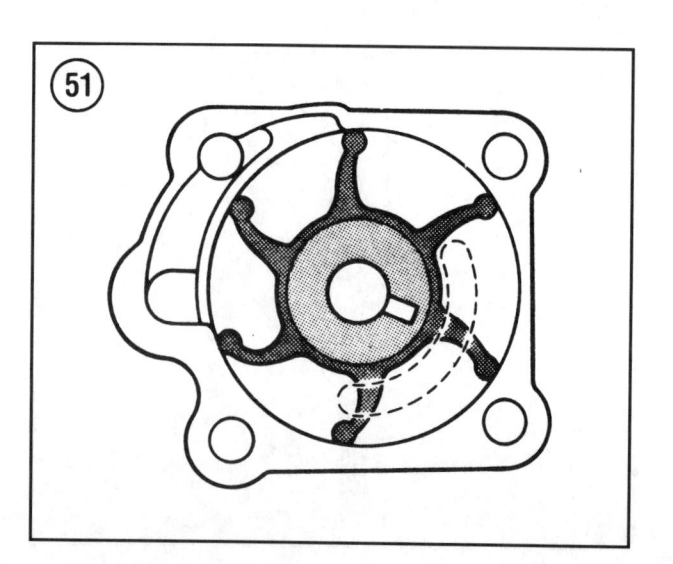

2. Install the pump base with a new gasket.

3. Install the pump base plate and new gasket.

4. Install the impeller drive key into the receptacle in the drive shaft. Make sure it seats correctly and squarely.

CAUTION
*If the original impeller is to be reused, install it in the same rotational direction as removed to avoid premature failure. The drive key groove should be visible and the curl of the blades positioned as shown in **Figure 51**.*

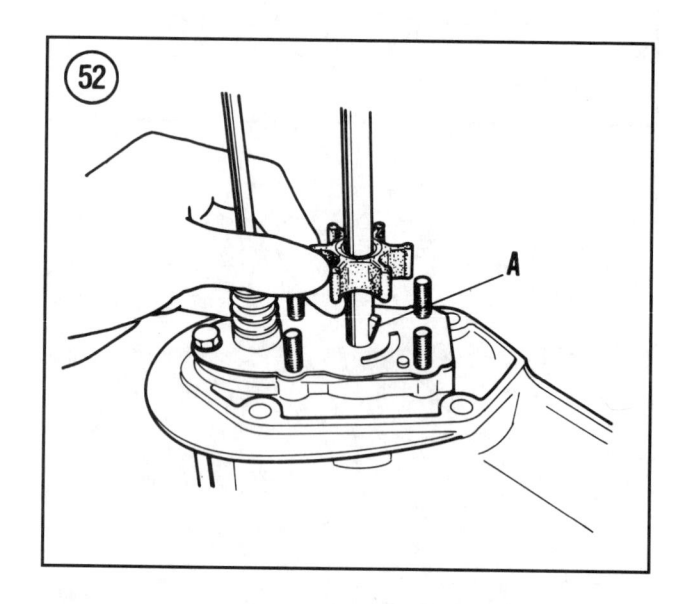

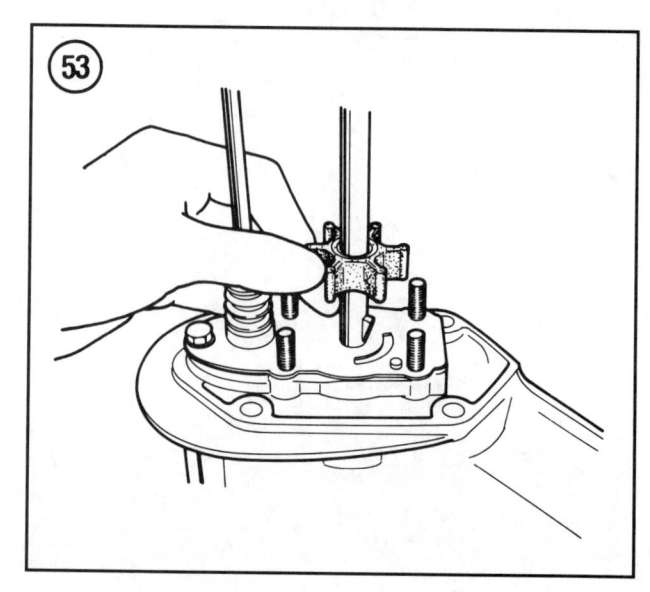

GEARCASE

5. Install the impeller in the water pump cover (**Figure 55**).

6. Slide the water pump cover down the drive shaft. Align the impeller slot with the key in the drive shaft.

7. Apply Thread Lock 1342 (part No. 99104-32050) to the water pump cover bolts prior to installation.

8. Install the water pump cover bolts and tighten to specification (**Table 1**).

9. Install the water pump grommet into the pump cover.

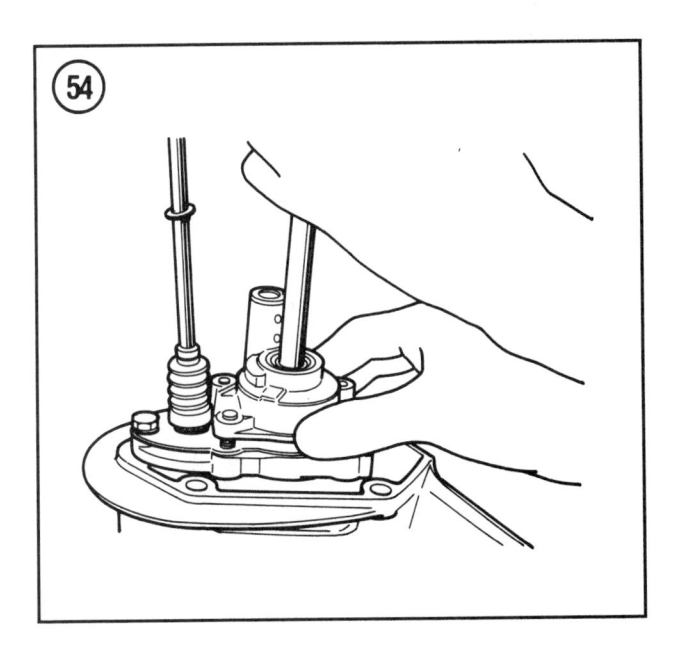

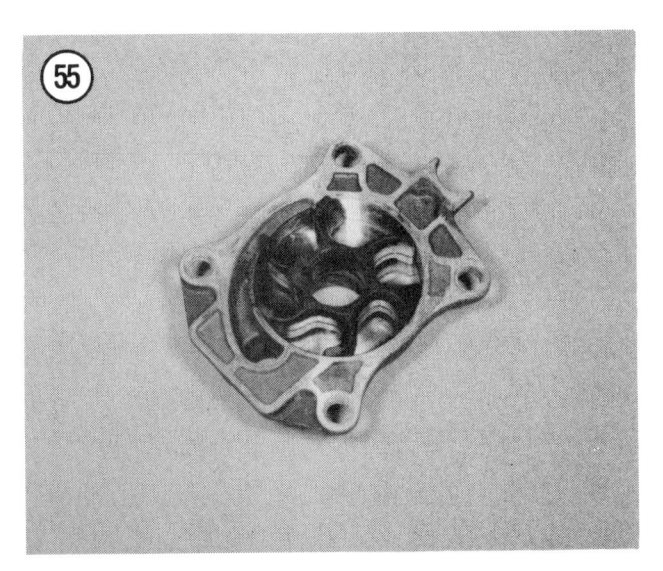

Assembly/Installation
(DT 20, DT 25, DT 30 [2-cylinder]; DT 35, DT 40; DT 25, DT 30 [3-cylinder]; DT 55, DT 65)

1. Install the pump base with a new gasket into the gearcase.

2. Install the impeller drive key into the receptacle in the drive shaft. Make sure it seats correctly and squarely.

CAUTION
If the original impeller is to be reused, install it in the same rotational direction as removed to avoid premature failure. The drive key groove should be visible and the curl of the blades positioned as shown in Figure 51.

3. Install the impeller into the water pump cover (**Figure 55**).

4. Slide the water pump cover down the drive shaft. Align the impeller slot with the key in the drive shaft.

CAUTION
Correct housing fastener torque is important in Step 5. Excessive torque can cause the pump to crack during operation; insufficient torque may result in leakage and exhaust induction which will cause overheating.

5. Apply Thread Lock 1342 (part No. 99104-32050) to the water pump cover nuts prior to installation.

6. Install the water pump cover nuts and tighten to specifications (**Table 1**).

7. Install the water tube into the pump cover.

Assembly/Installation
(DT 75, DT 85)

1. Install the pump base with a new gasket into the gearcase.

2. Install the impeller drive key into the receptacle in the drive shaft. Make sure it seats correctly and squarely.

CAUTION
If the original impeller is to be reused, install it in the same rotational direction as removed to avoid premature failure. The drive key groove should be visible and the curl of the blades positioned as shown in Figure 51.

3. Install the impeller into the water pump cover (**Figure 55**).
4. Slide the water pump cover down the drive shaft. Align the impeller slot with the key in the drive shaft.

CAUTION
Correct housing fastener torque is important in Step 5. Excessive torque can cause the pump to crack during operation; insufficient torque may result in leakage and exhaust induction which will cause overheating.

5. Apply Thread Lock 1342 (part No. 99104-32050) to the water pump cover bolts prior to installation.
6. Install the water pump cover bolts and tighten to specifications (**Table 1**).
7. Install the water tube into the pump cover.

Assembly/Installation (DT 115, DT 140)

1. Install the pump base with a new gasket into the gearcase.
2. Install the impeller drive key into the receptacle in the drive shaft. Make sure it seats correctly and squarely.

CAUTION
If the original impeller is to be reused, install it in the same rotational direction as removed to avoid premature failure. The drive key groove should be visible and the curl of the blades positioned as shown in Figure 51.

3. Install the impeller into the water pump cover (**Figure 55**).
4A. On 1985 models, perform the following:
 a. Install a new O-ring seal in the pump cover.
 b. Slide cover sleeve, cover sleeve top and the pump cover down the drive shaft.
 c. Align the impeller slot with the key in the drive shaft.
 d. Apply a water-resistant grease to the O-ring seal in the pump cover.
4B. On 1986-on models, slide the pump cover down the drive shaft. Align the impeller slot with the key in the drive shaft.
5. Apply Thread Lock 1342 (part No. 99104-32050) to the water pump cover bolts prior to installation.
6A. On 1985 models, install the water pump cover bolts and tighten to specifications (**Table 1**).
6B. On 1986-on models, install the water pump cover bolts and tighten securely.
7. Install the water tube into the pump cover.

Assembly/Installation (V4, V6)

1. Install the pump base with a new gasket into the gearcase.
2. Install the impeller drive key into the receptacle in the drive shaft. Make sure it seats correctly and squarely.

CAUTION
If the original impeller is to be reused, install it in the same rotational direction as removed to avoid premature failure. The drive key groove should be visible and the curl of the blades positioned as shown in Figure 51.

3. Install the impeller into the water pump cover (**Figure 55**).

GEARCASE

4. Make sure the locating pins (C, **Figure 46**) are in place prior to installing the pump cover.

5. Slide the pump cover down the drive shaft. Align the impeller slot with the key in the drive shaft.

6. Apply Thread Lock 1342 (part No. 99104-32050) to the water pump cover bolts prior to installation.

7. Install the water pump cover bolts and tighten securely.

8. Install the water tube into the pump cover.

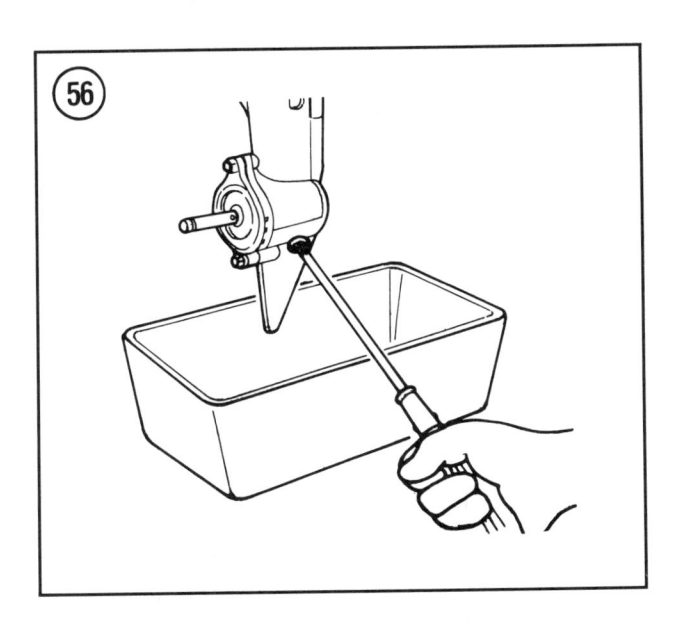

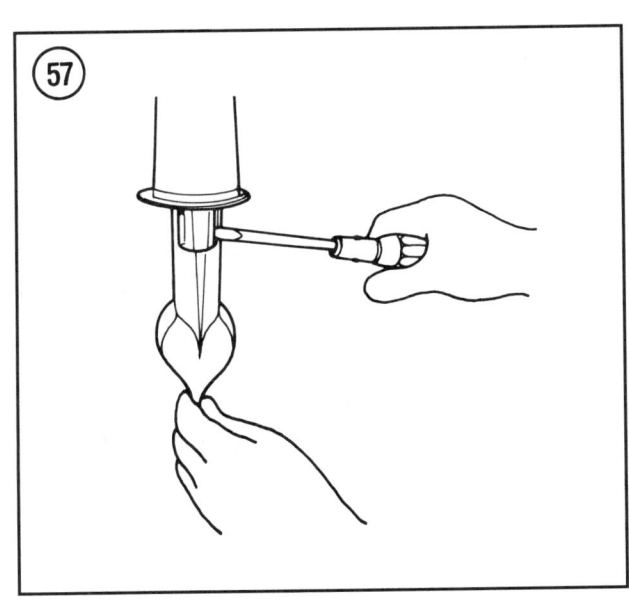

GEARCASE

Removal/Installation
(DT 2)

1. Remove the engine cover and disconnect the spark plug lead as a safety precaution to prevent any accidental starting of the engine during lower unit removal.

2. Place the shift lever in FORWARD.

3. Remove the propeller as described in this chapter.

4. Place a suitable container under the gearcase. Remove the drain screw (**Figure 56**). Drain the lubricant from the unit.

NOTE
If the lubricant is creamy in color or metallic particles are found in Step 5, the gearcase must be completely disassembled to determine and correct the cause of the problem.

5. Wipe a small amount of lubricant on a finger and rub the finger and thumb together. Check for the presence of metallic particles in the lubricant. Note the color of the lubricant. A white or creamy color indicates water in the lubricant. Check the drain container for signs of water separation from the lubricant.

6. Remove the bolt cover (**Figure 57**).

7. Remove the 2 nuts holding the gearcase to the drive shaft housing (**Figure 58**)

8. Carefully separate then remove the gearcase from the drive shaft housing.

9. Install the gearcase in an appropriate holding fixture.

CAUTION
Do not grease the top of the drive shaft in Step 10. This may excessively preload the drive shaft and crankshaft when the mounting fasteners are tightened and cause a premature failure of the power head or gearcase.

10. Lightly lubricate the drive shaft splines with water-resistant grease (part No. 99000-25160) or equivalent.

11. Apply a thin but uniform coat of Silicone Seal (part No. 99000-31120) or equivalent to the gearcase and drive shaft housing mating surfaces.

12. Wipe the drive shaft housing bolt threads with Thread Lock 1342 (part No. 99000-32050) or equivalent.

13. Position the gearcase under the drive shaft housing. Align the drive shaft splines with the crankshaft.

CAUTION
*Do **not** rotate the flywheel counterclockwise in Step 14. This can damage the water pump impeller.*

14. Seat the gearcase against the drive shaft housing, rotating the flywheel *clockwise* as required until the drive shaft and crankshaft engage.

15. Install the gearcase nuts and lockwashers and tighten securely.

16. Install the mounting nut cover.

17. Install the propeller as described in this chapter.

18. Reconnect the spark plug lead and refill the gearcase with the proper type and quantity of lubricant. See Chapter Four.

Removal/Installation (DT 4, DT 6; 1985-1987 DT 8)

1. Remove the engine cover and disconnect the spark plug lead(s) as a safety precaution to prevent any accidental starting of the engine during lower unit removal.

2. Place the shift lever in FORWARD.

3. Remove the propeller as described in this chapter.

4. On DT 6, DT 8 models, perform the following:

   a. Remove the bolts holding the shift rod adjusting cover and remove the cover (**Figure 59**).

   b. Loosen the shift rod connector bolt (**Figure 60**).

5A. On DT 4 models, remove the 2 bolts holding the gearcase to the drive shaft housing.

5B. On DT 6 and DT 8 models, remove the 4 bolts holding the gearcase to the drive shaft housing (**Figure 61**).

6. Tilt the drive shaft housing up and carefully separate it from the gearcase. Remove the gearcase from the drive shaft housing.

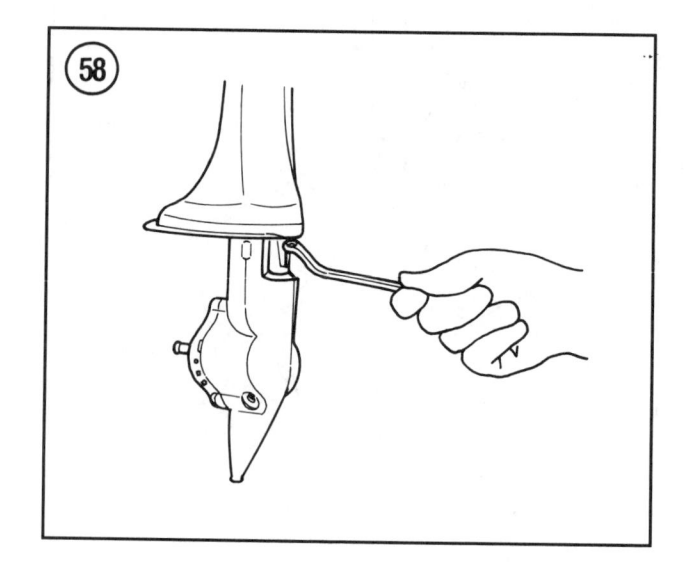

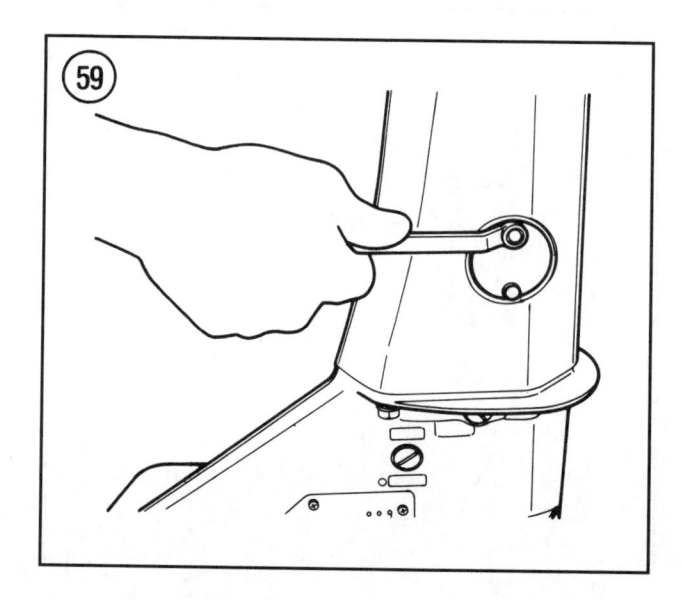

GEARCASE

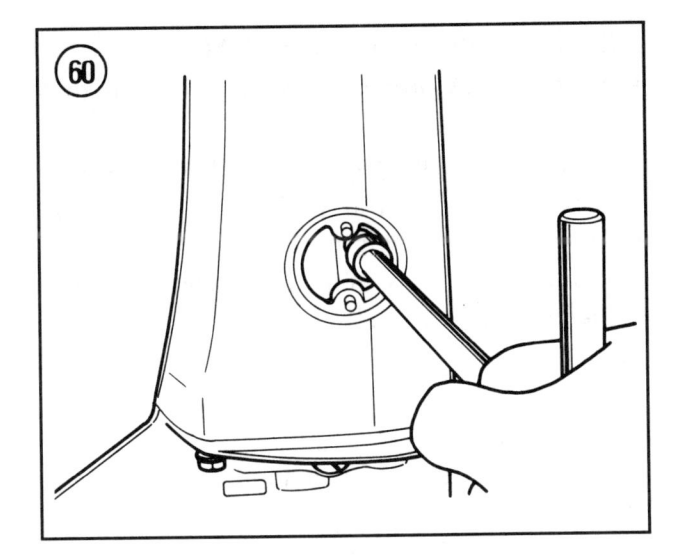

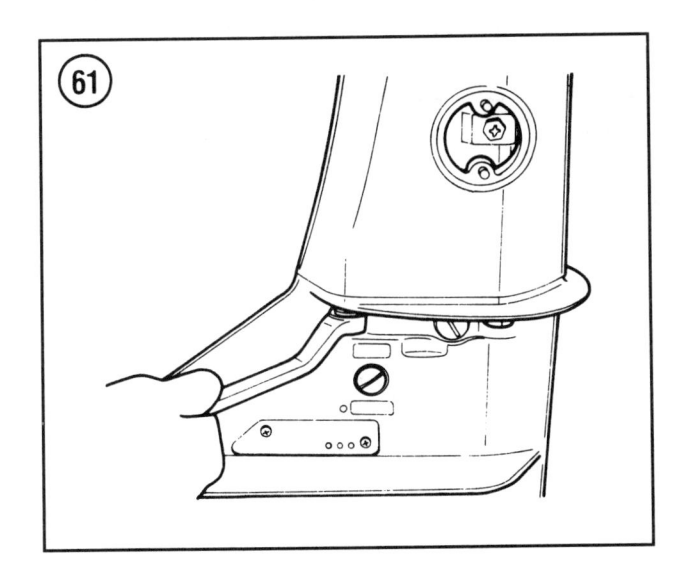

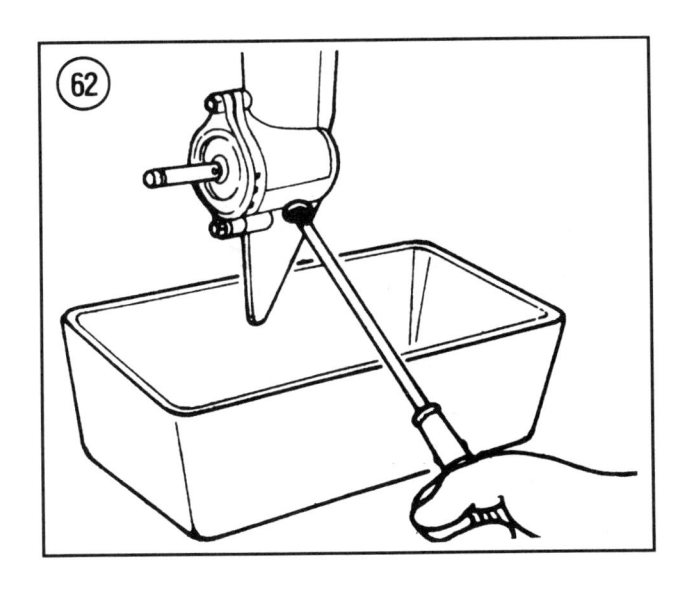

395

7. Install the gearcase in an appropriate holding fixture.

8. Place a suitable container under the gearcase. Remove the drain screw (**Figure 62**). Drain the lubricant from the unit.

NOTE
If the lubricant is creamy in color or metallic particles are found in Step 9, the gearcase must be completely disassembled to determine and correct the cause of the problem.

9. Wipe a small amount of lubricant on a finger and rub the finger and thumb together. Check for the presence of metallic particles in the lubricant. Note the color of the lubricant. A white or creamy color indicates water in the lubricant. Check the drain container for signs of water separation from the lubricant.

CAUTION
Do not grease the top of the drive shaft in Step 10. This may excessively preload the drive shaft and crankshaft when the mounting fasteners are tightened and cause a premature failure of the power head or gearcase.

10. To reinstall the gearcase, make sure the shift rod is in the NEUTRAL position and lightly lubricate the drive shaft splines with water-resistant grease (part No. 99000-25160) or equivalent.

11. Apply a thin but uniform coat of Silicone Seal (part No. 99000-31120) or equivalent to the gearcase and drive shaft housing mating surfaces.

12. Wipe the drive shaft housing bolt threads with Thread Lock 1342 (part No. 99000-32050) or equivalent.

13. Position the gearcase under the drive shaft housing. Align the drive shaft splines with the crankshaft, insert the water tube into the water pump case and fit the upper shift rod into the shift rod connector.

CAUTION
*Do **not** rotate the flywheel counterclockwise in Step 14. This can damage the water pump impeller.*

14. Seat the gearcase against the drive shaft housing, rotating the flywheel *clockwise* as required until the drive shaft and crankshaft engage.
15. Install the gearcase fasteners and tighten to specifications (**Table 1**).
16. On DT 6 and DT 8 models, perform the following:
    a. Make sure the engine shift lever is in FORWARD and install the shift rod connector bolt.
    b. Place the shift lever in NEUTRAL and make sure the propeller rotates freely, then shift back into FORWARD and make sure the propeller will only rotate clockwise. If the propeller does not rotate as indicated, loosen the shift rod connector bolt and readjust the position of the connector or rods as required.
    c. Install the shift rod adjusting cover, gasket and bolts. Tighten the bolts securely.
17. Install the propeller as described in this chapter.
18. Reconnect the spark plug lead(s) and refill the gearcase with proper type and quantity of lubricant. See Chapter Four.

Removal/Installation
(1988-on DT 8; DT 9.9 through DT 40)

The gearcase used on these models varies primarily in shape, size and component location. This is a basic procedure that covers all models listed. Your model may vary slightly from the one shown, but has the same basic components and all models listed are serviced in essentially the same way.

1. Remove the engine cover and disconnect the spark plug leads to prevent accidental starting of the engine during lower unit removal.
2. Shift the outboard into FORWARD.
3. Remove the propeller as described in this chapter.
4. Loosen the clutch rod locknut (**Figure 63**).
5. Loosen the clutch rod turnbuckle (**Figure 64**) and separate the clutch rod from the shift rod.
6. Remove the bolts and lockwashers (**Figure 65**) securing the gearcase to the drive shaft housing.
7. Carefully separate the gearcase from the drive shaft housing.
8. Install the gearcase in an appropriate holding fixture.
9. Place a suitable container under the gearcase. Remove the vent (A, **Figure 66**) and drain

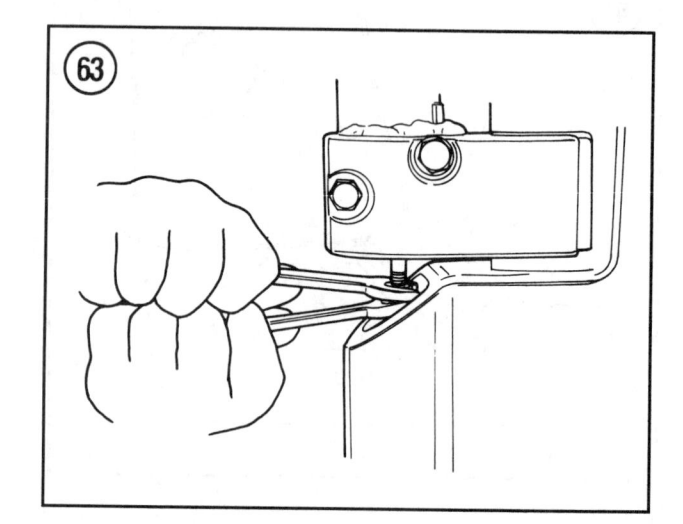

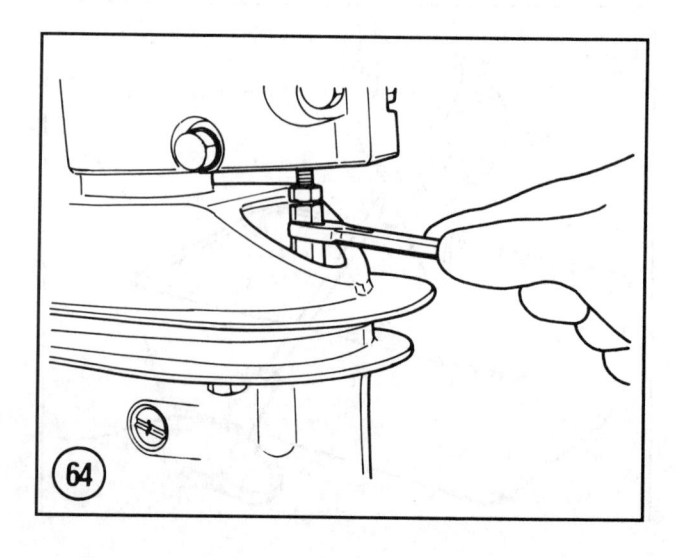

GEARCASE

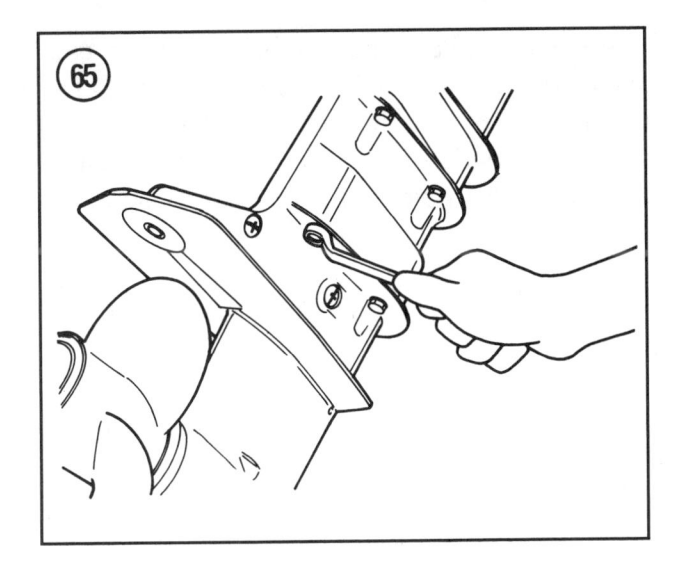

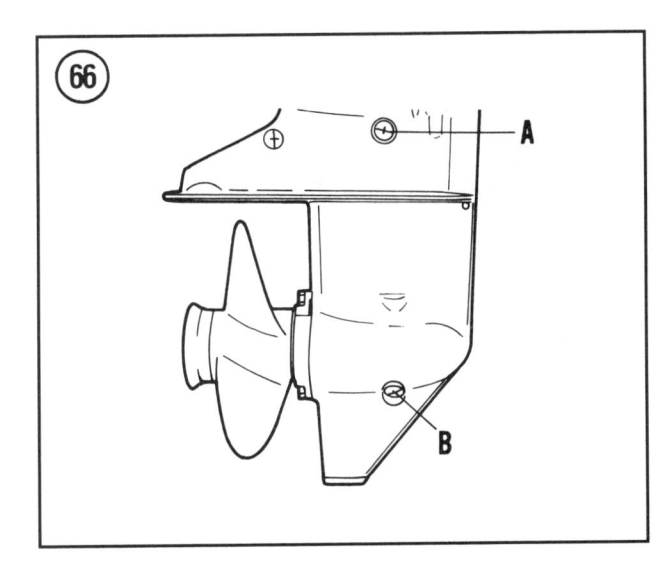

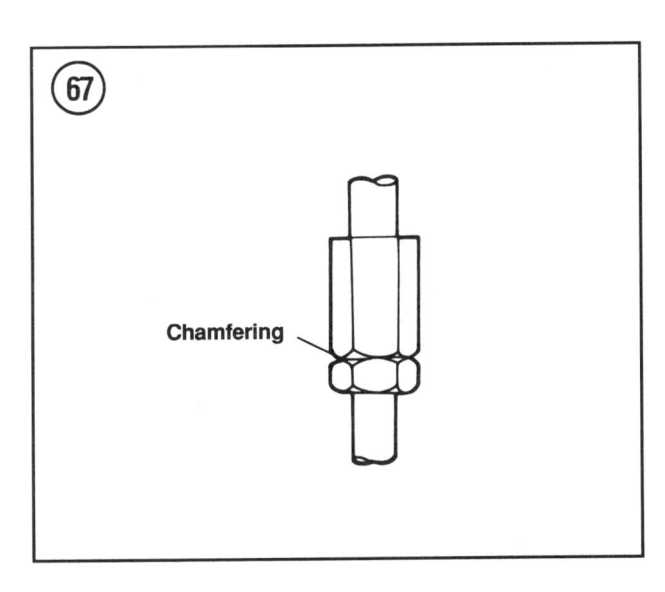

screws (B, **Figure 66**). Drain the lubricant from the unit.

NOTE
If the lubricant is creamy in color or metallic particles are found in Step 10, the gearcase must be completely disassembled to determine and correct the cause of the problem.

10. Wipe a small amount of the drained lubricant on a finger and rub the finger and thumb together. Check for the presence of metallic particles in the lubricant. Note the color of the lubricant. A white or creamy color indicates water in the lubricant. Check the drain container for signs of water separation from the lubricant.

CAUTION
Do not grease the top of the drive shaft in Step 11. This may excessively preload the drive shaft and crankshaft when the mounting fasteners are tightened and cause a premature failure of the power head or gearcase.

11. To reinstall the gearcase, pull the lower shift rod upward as far as possible to make sure it is in the FORWARD position and lightly lubricate the drive shaft splines with water-resistant grease (part No. 99000-25160) or equivalent.

12. Apply a thin but uniform coat of Silicone Seal (part No. 99000-31120) or equivalent to the gearcase and drive shaft housing mating surfaces.

13. Make sure the locating dowels are in place.

14. Position the gearcase under the drive shaft housing.

15. Align the drive shaft splines with the crankshaft, insert the water tube into the water pump case and the water pump seal tube.

NOTE
*On 1988-on models, make sure to connect the clutch rod to the shift shaft with chamfered side of the lower nut facing upward (**Figure 67**).*

16. Attach the shift rod into the clutch rod.

CAUTION
*Do not rotate the flywheel **counterclockwise** in Step 17. This can damage the water pump impeller.*

17. Seat the gearcase in place, rotating the flywheel *clockwise* as required until the drive shaft and crankshaft engage.

18. Install the gearcase bolts and lockwashers and tighten to specification (**Table 1**).

19. Make sure the engine shift lever is in FORWARD and tighten the shift rod connector or upper shift rod turnbuckle. Place the shift lever in NEUTRAL and make sure the propeller rotates freely, then shift back into FORWARD and make sure the propeller will only rotate clockwise. If the propeller does not rotate as indicated, loosen the shift rod connector or upper shift rod turnbuckle and readjust the position of the connector/turnbuckle or rods as required. When shift pattern is correct, tighten the shift rod connector locknut securely.

20. Install the propeller as described in this chapter.

21. Reconnect the spark plug leads and refill the gearcase with proper type and quantity of lubricant. See Chapter Four.

22. Install the engine cover.

Removal/Installation (DT 55, DT 65)

1. Remove the engine cover and disconnect the spark plug leads to prevent accidental starting of the engine during lower unit removal.

2. Place a container under the gearcase. Remove the vent (A, **Figure 68**) and drain plugs (B, **Figure 68**). Drain the lubricant from the unit.

NOTE
If the lubricant is creamy in color or metallic particles are found in Step 3, the gearcase must be completely disassembled to determine and correct the cause of the problem.

3. Wipe a small amount of lubricant on a finger and rub the finger and thumb together. Check for the presence of metallic particles in the lubricant. Note the color of the lubricant. A white or creamy color indicates water in the lubricant. Check the drain container for signs of water separation from the lubricant.

4. Remove the propeller as described in this chapter.

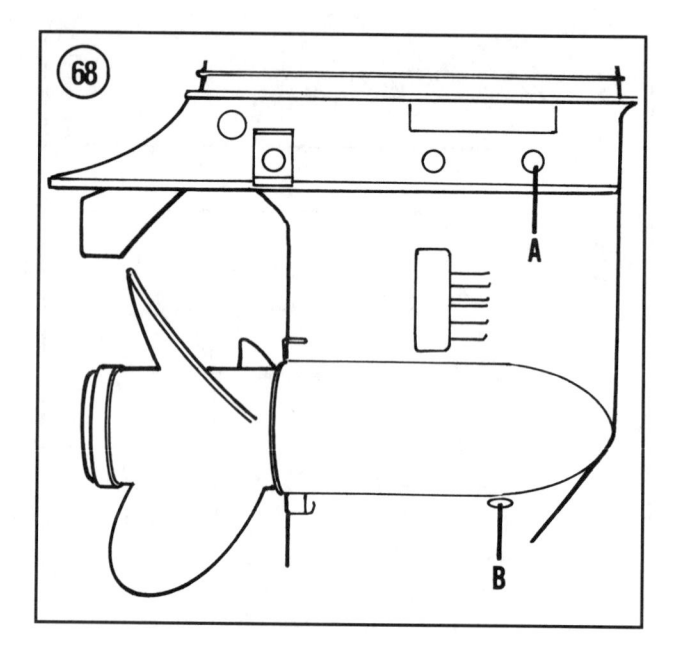

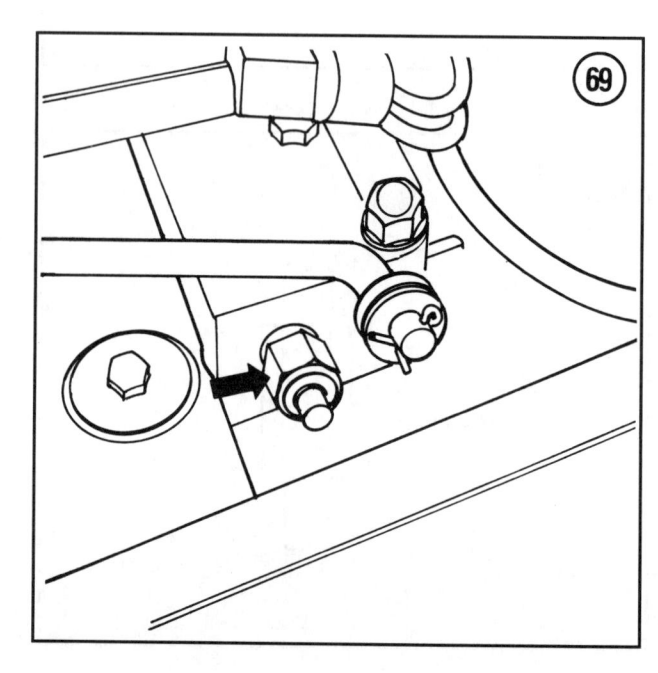

GEARCASE

5. Place the outboard in NEUTRAL.
6. Remove the nut (**Figure 69**) from the clutch shaft.
7. Disconnect the clutch shaft from the clutch rod.
8. Remove the bolts and lockwashers (**Figure 70**) securing the gearcase to the drive shaft housing.
9. Remove the gearcase from the drive shaft housing and mount it in a suitable holding fixture.

CAUTION
Do not grease the top of the drive shaft in Step 10. This may excessively preload the drive shaft and crankshaft when the mounting bolts are tightened and cause a premature failure of the power head or gearcase.

10. To reinstall the gearcase, lightly lubricate the drive shaft splines and the O-ring seal around the drive shaft bearing with water-resistant grease (part No. 99000-32160) or equivalent.
11. Apply a thin but uniform coat of Silicone Seal (part No. 99000-31120) or equivalent to the gearcase and drive shaft housing mating surfaces.
12. Make sure the locating dowels are in place.
13. Position the gearcase under the drive shaft housing.
14. Align the drive shaft splines with the crankshaft, insert the water tube into the water pump case and the water pump seal tube.
15. Guide the clutch rod into the clutch shaft.

CAUTION
*Do not rotate the flywheel **counterclockwise** in Step 16. This can damage the water pump impeller.*

16. Seat the gearcase in place, rotating the flywheel clockwise as required until the drive shaft and crankshaft engage.
17. Apply Silicone Seal (part No. 99000-31120) or equivalent to the mounting bolt threads and tighten to specifications (**Table 1**).
18. Connect the clutch shaft onto the clutch rod and install the nut. Tighten the nut securely.
19. Install the propeller as described in this chapter.
20. Reconnect the spark plug leads and refill the gearcase with proper type and quantity of lubricant. See Chapter Four.
21. Install the engine cover.

Removal/Installation (DT 75, DT 85, DT 115, DT 140, V4, V6)

1. Remove the engine cover.
2. Disconnect the spark plug leads to prevent accidental starting of the engine during lower unit removal.
3. Place a container under the gearcase. Remove the vent (A, **Figure 71**) and drain plugs (B, **Figure 71**). Drain the lubricant from the unit.

NOTE
If the lubricant is creamy in color or metallic particles are found in Step 4, the gearcase must be completely disassembled to determine and correct the cause of the problem.

4. Wipe a small amount of lubricant on a finger and rub the finger and thumb together. Check for the presence of metallic particles in the lubricant. Note the color of the lubricant. A white or creamy color indicates water in the lubricant. Check the drain container for signs of water separation from the lubricant.

5. Remove the propeller as described in this chapter.

6. Place the outboard in NEUTRAL and remove the cover at the front of the clutch rod.

7. Use needlenose pliers and remove the clutch rod connector cotter pin (**Figure 72**). Discard the cotter pin.

8. Remove the connector pin (**Figure 73**).

9A. On V6 models, remove the trim tab bolt access cap on the top surface of the drive shaft housing. Remove the bolt and lockwasher securing the trim tab and remove the trim tab.

9B. On all other models, remove the bolt and lockwasher securing the trim tab (**Figure 74**) and remove the trim tab.

10. Remove the bolts and lockwashers (**Figure 75**) securing the gearcase to the drive shaft housing.

11. Remove the gearcase from the drive shaft housing and mount it in a suitable holding fixture.

CAUTION
Do not grease the top of the drive shaft in Step 12. This may excessively preload the drive shaft and crankshaft when the mounting bolts are tightened and cause a premature failure of the power head or gearcase.

12. To reinstall the gearcase, lightly lubricate the drive shaft splines with water-resistant grease (part No. 99000-32160) or equivalent.

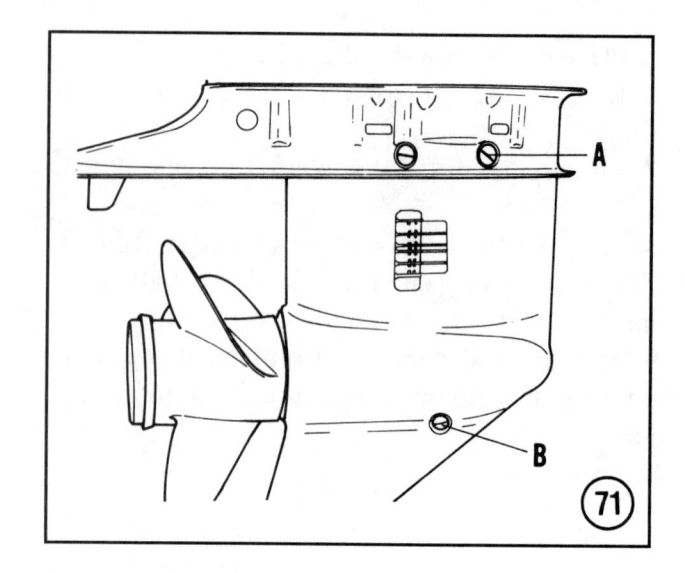

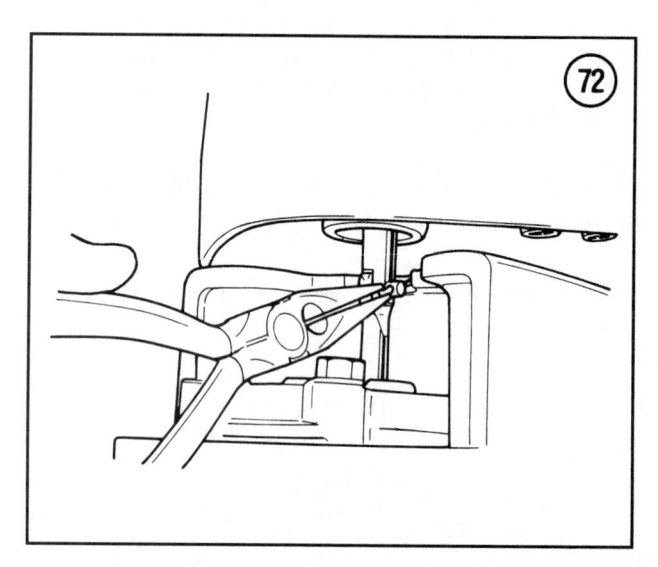

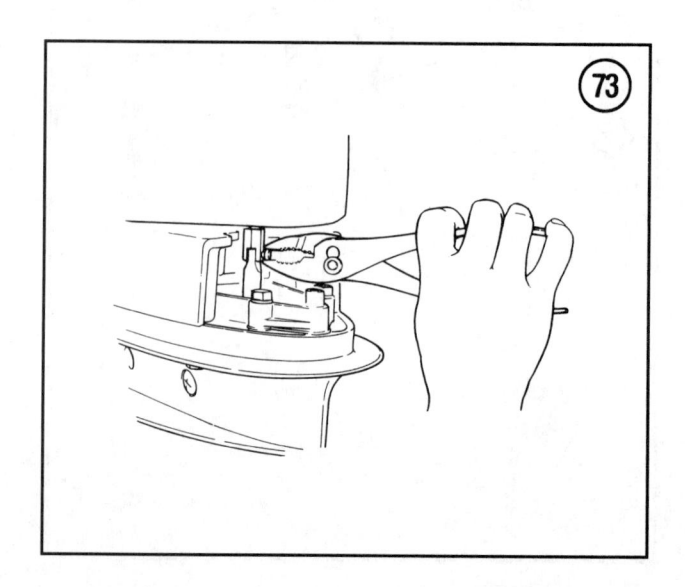

GEARCASE

13. Apply a thin but uniform coat of Silicone Seal (part No. 99000-31120) or equivalent to the gearcase and drive shaft housing mating surfaces.

14. Make sure the locating dowels are in place.

15. Position the gearcase under the drive shaft housing.

16. Align the drive shaft splines with the crankshaft, insert the water tube into the water pump case and the water pump seal tube.

17. Guide the clutch rod into the clutch rod connector.

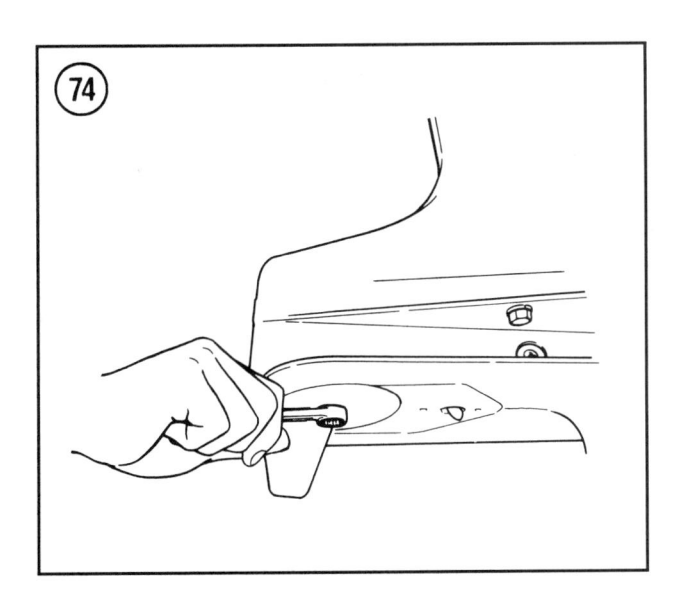

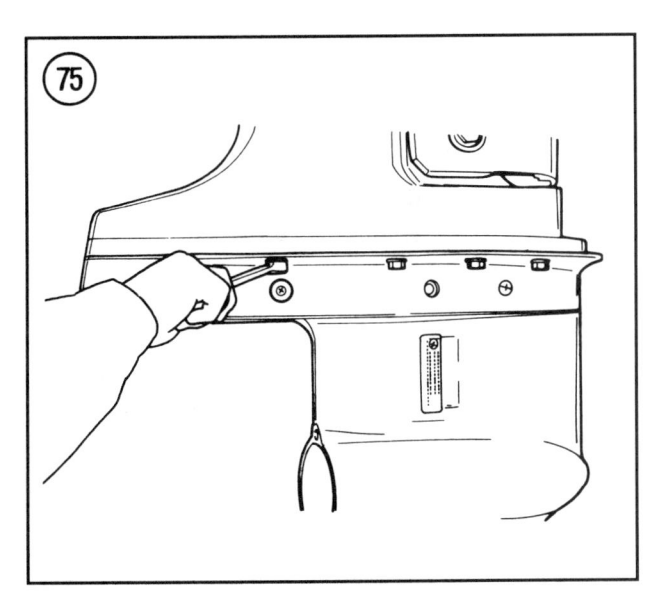

CAUTION
*Do not rotate the flywheel **counterclockwise** in Step 18. This can damage the water pump impeller.*

18. Seat the gearcase in place, rotating the flywheel clockwise as required until the drive shaft and crankshaft engage.

19A. On 1985 DT 115, DT 140 models, apply Thread Lock 1342 (part No. 99000-32050) to the mounting bolt threads and tighten securely.

19B. On all other models, apply Silicone Seal (part No. 99000-31120) or equivalent to the mounting bolt threads and tighten to specifications (**Table 1**).

20. Install the trim tab and the bolt and lockwasher. Tighten the bolt securely. On V6 models, install the trim cap over the bolt hole in the drive shaft housing.

21. Connect the clutch rod into the clutch rod connector and insert the connector pin through both parts.

22. Install a *new* cotter pin. Bend the ends over completely.

23. Install the clutch rod cover.

24. Install the propeller as described in this chapter.

25. Reconnect the spark plug leads and refill the gearcase with proper type and quantity of lubricant. See Chapter Four.

26. Install the engine cover.

Gearcase Disassembly/Assembly (DT 2)

Refer to **Figure 76** for this procedure.

1. Remove the gearcase as described in this chapter.

2. Secure the gearcase in a suitable holding fixture or a vise with protective jaws. If protective jaws are not available, position the gearcase upright with the skeg between wooden blocks in the vise.

3. Remove the water pump as described in this chapter.

CHAPTER NINE

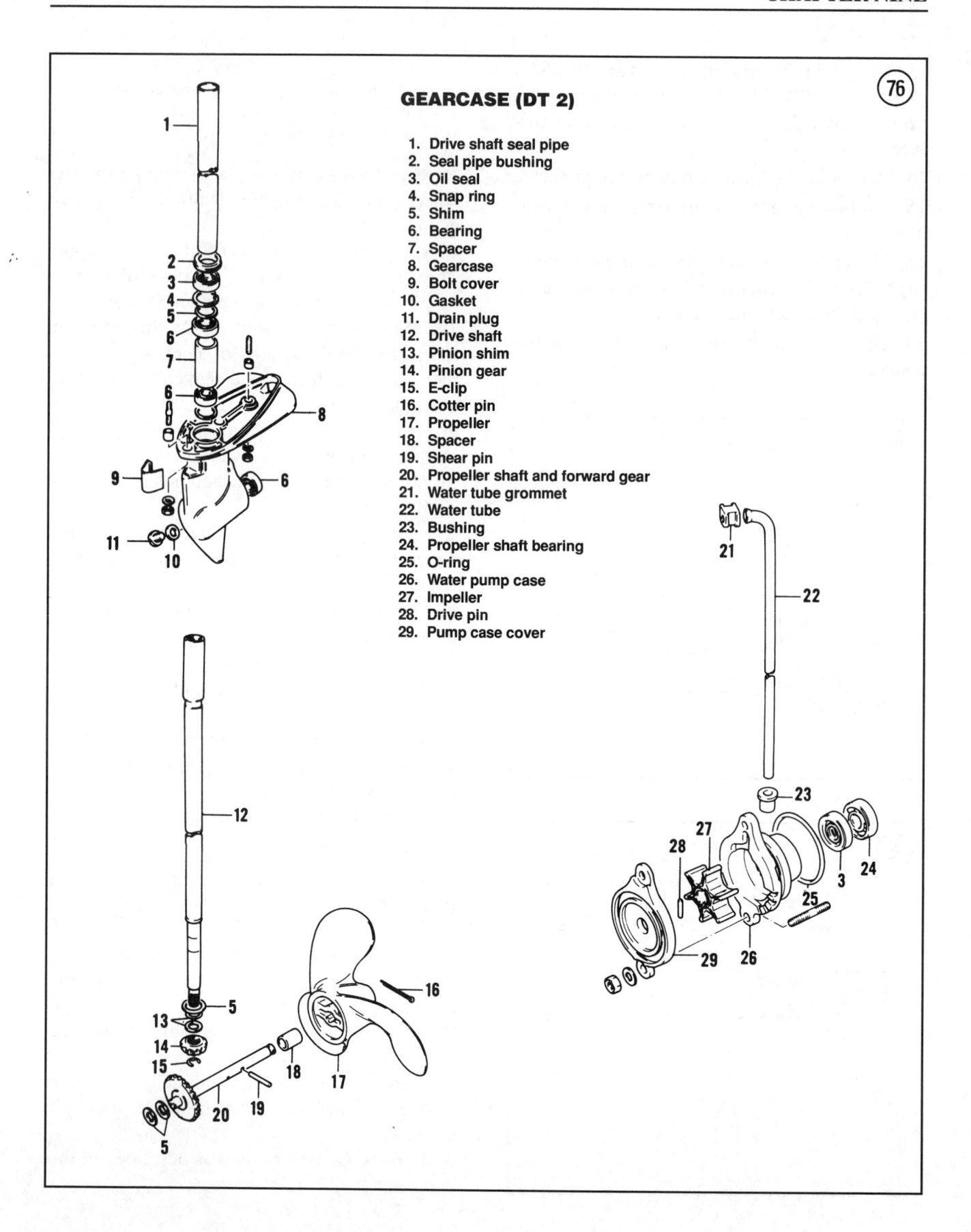

GEARCASE (DT 2)

1. Drive shaft seal pipe
2. Seal pipe bushing
3. Oil seal
4. Snap ring
5. Shim
6. Bearing
7. Spacer
8. Gearcase
9. Bolt cover
10. Gasket
11. Drain plug
12. Drive shaft
13. Pinion shim
14. Pinion gear
15. E-clip
16. Cotter pin
17. Propeller
18. Spacer
19. Shear pin
20. Propeller shaft and forward gear
21. Water tube grommet
22. Water tube
23. Bushing
24. Propeller shaft bearing
25. O-ring
26. Water pump case
27. Impeller
28. Drive pin
29. Pump case cover

GEARCASE

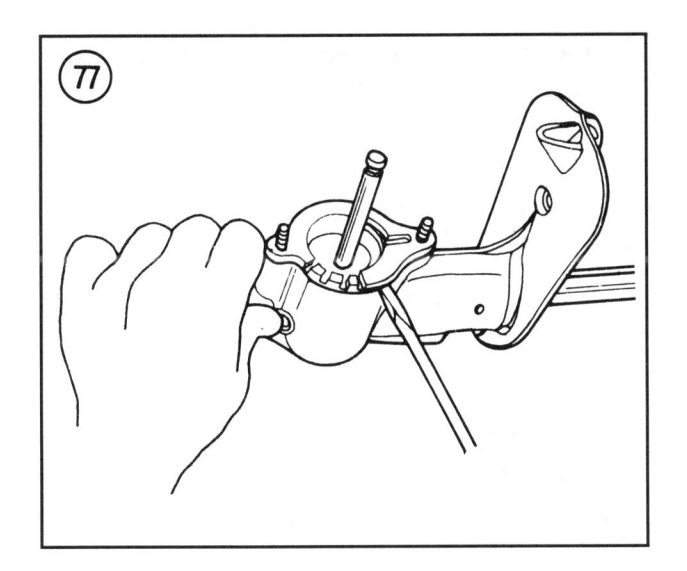

403

4. Carefully pry the water pump case free from the gearcase housing (**Figure 77**).

5. Remove the water tube and drive shaft seal pipe (**Figure 78**).

6. Use a pair of screwdrivers as shown in **Figure 79** to pry the E-clip from the pinion gear.

7. Remove the drive shaft, then reach into the gearcase housing and remove the pinion gear and shim(s) (**Figure 80**).

8. Remove the propeller shaft and gear assembly (**Figure 81**). Reach into the gearcase housing and remove the shim(s).

9. Remove the drive shaft seal pipe bushing and oil seal.

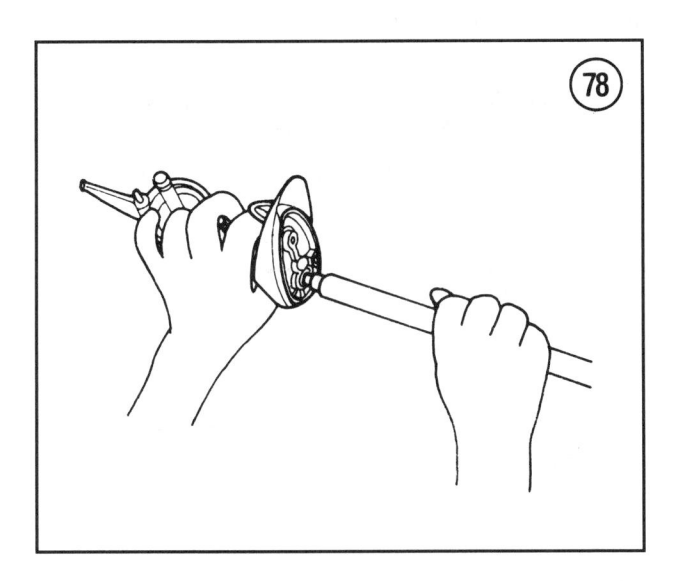

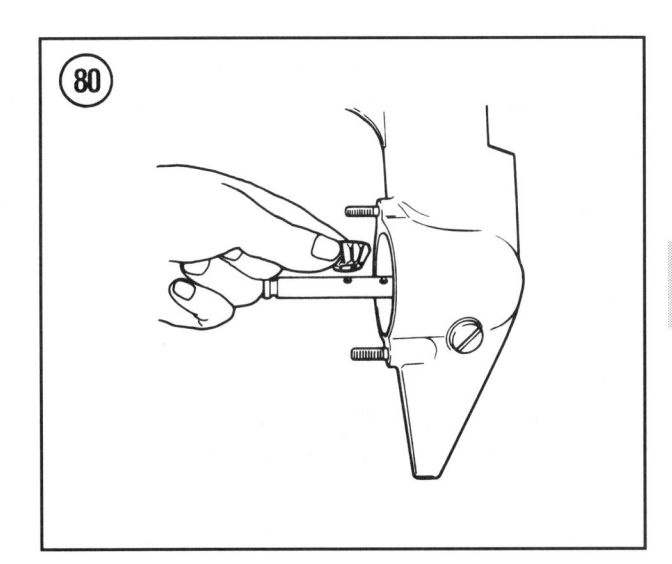

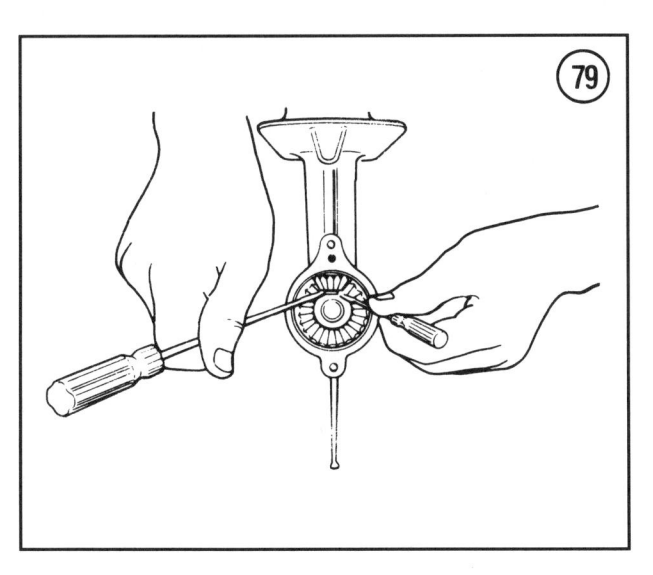

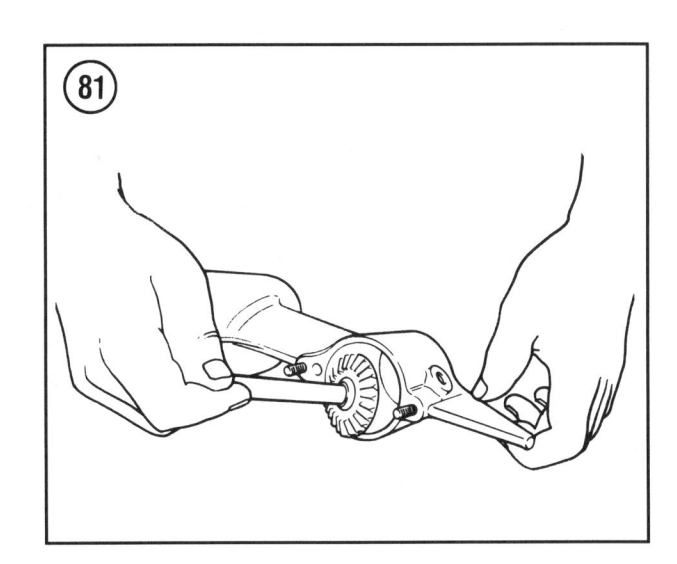

10. If inspection of the drive shaft bearings indicates replacement is required, proceed as follows:
   a. Remove the drive shaft snap ring with pliers (part No. 09900-06108) or equivalent.
   b. Remove the shim(s) from the top drive shaft bearing.
   c. Insert a bearing remover handle into the seal pipe bore.
   d. Insert the bearing remover into the gearcase bore and attach to the remover handle.
   e. Pull the 2 drive shaft bearings and spacer from the gearcase.
11. Clean and inspect all parts as described in this chapter.
12. If the drive shaft bearings are replaced:
   a. Install new bearings with the spacer between them using an appropriate bearing installer.
   b. Install the shim(s) on the top drive shaft bearing.
   c. Install the drive shaft snap ring with pliers (part No. 09900-06108) or equivalent.
   d. Temporarily install the drive shaft with the pinion gear and shims. Pull upward on drive shaft. It should not move more than 0.004 in. (0.1 mm). If movement exceeds this specification, remove the snap ring and exchange the shim(s) as required to bring the clearance within specifications (**Figure 82**).
13. Once drive shaft bearing shimming is correct, install a new oil seal and drive shaft bushing with their lips positioned as shown in A and B, **Figure 83**. Coat lips with water-resistant grease (part No. 99000-25160) or equivalent.
14. Install the propeller shaft shims and propeller shaft in the gearcase housing.
15. Position the pinion gear and shim(s) in the housing under the drive shaft bore, insert the drive shaft and engage the pinion gear. Install the E-clip on the end of the drive shaft to retain the pinion gear.
16. Install the drive shaft seal pipe and water tube.
17. Remove the water pump case bearing and oil seal.
18. Install a new bearing with an installer (part No. 09914-79510).
19. Install a new oil seal and coat its lips with water-resistant grease.
20. Remove and discard the water pump case O-ring. Lubricate a new O-ring with water-resistant grease and install on case.
21. Install the water pump as described in this chapter.
22. Install the gearcase as described in this chapter. Fill with recommended type and quantity of lubricant. See Chapter Four.
23. Check gearcase lubricant level after engine has been run. Change the lubricant after 10 hours of operation (break-in period). See Chapter Four.

Gearcase Disassembly/Assembly (DT 4)

Refer to **Figure 84** for this procedure.
1. Remove the gearcase as described in this chapter.

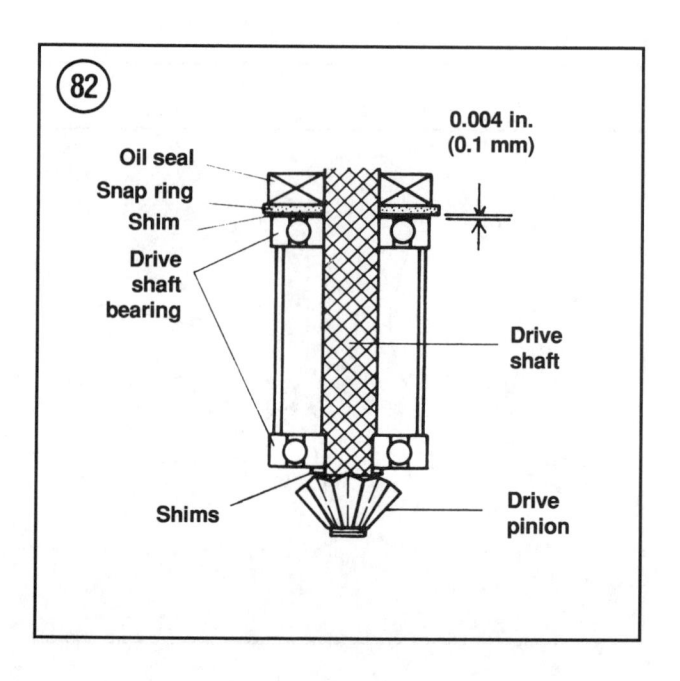

GEARCASE

2. Secure the gearcase in a suitable holding fixture or a vise with protective jaws. If protective jaws are not available, position the gearcase upright with the skeg between wooden blocks in a vise.

3. Remove the water pump as described in this chapter.

NOTE
*In Step 4 if the housing is difficult to remove, carefully pry the housing free with a screwdriver at the "O" marked points on the housing (**Figure 85**).*

4. Remove the 2 bearing housing bolts. Remove the bearing housing and propeller shaft assembly (**Figure 86**).

5. Remove the bolt securing the shift rod guide stopper (**Figure 87**) and remove the stopper.

6. Withdraw the shift rod assembly (**Figure 88**).

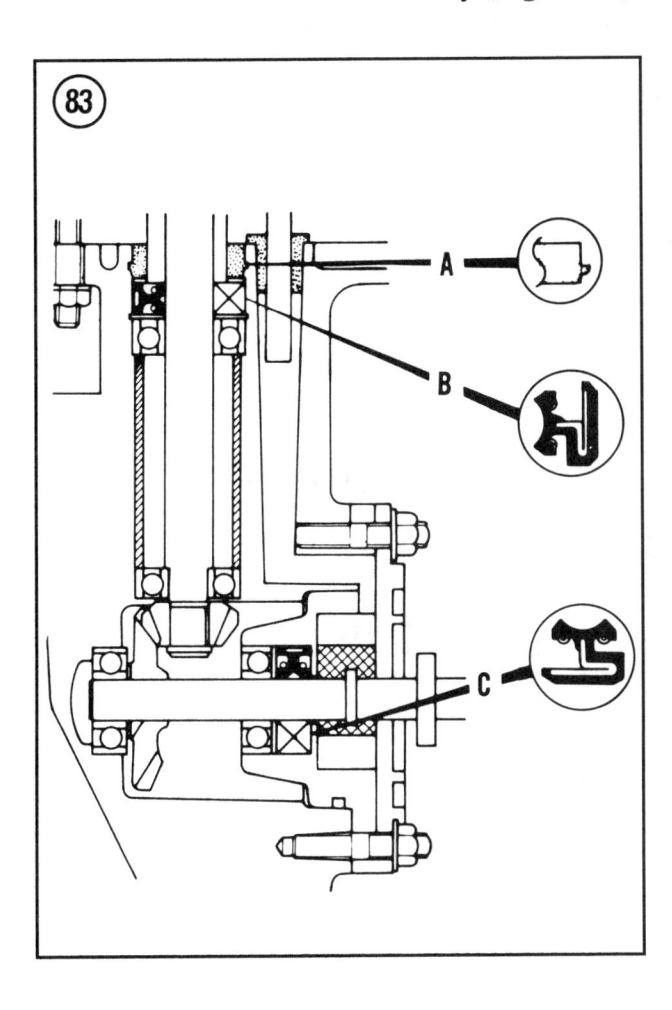

7. Carefully pry the drive shaft bearing housing free (**Figure 89**) with a screwdriver and remove the bearing housing and drive shaft.

8. Reach into the propeller shaft bore and remove the pinion gear (**Figure 90**) and the thrust washer (**Figure 91**).

9. Remove the forward gear, shims and thrust washer (**Figure 92**).

10. Remove the bolt (**Figure 93**) then remove the water filter.

11. Clean and inspect all parts as described in this chapter.

12. Install the water filter and bolt. Tighten the bolt securely.

13. Fit the forward gear shims over the shaft at the rear of the gear and install into the prop shaft bore.

14. Insert the pinion gear with shims into the prop shaft bore. Fit the gear into the drive shaft bore and mesh it with the forward gear.

15. Apply silicone seal (part No. 099000-31120) to the mating surfaces of the drive shaft bearing housing and gearcase housing mating surfaces.

16. Insert the drive shaft into the gearcase housing with a rotating motion and engage the pinion gear.

17. Lubricate the drive shaft bearing housing oil seal lip with water-resistant grease. Install the housing on the drive shaft and seat into the gearcase housing.

18. Install the shift rod assembly.

19. Install the shift rod guide stopper and bolt. Tighten the bolt securely.

20. Install the water pump as described in this chapter.

21. Check pinion gear depth and forward gear backlash as described in this chapter.

22. Insert propeller shaft into housing bore and engage the forward gear.

23. Coat bearing housing outer edges (front and rear) with water-resistant grease.

24. Carefully install bearing housing on propeller shaft.

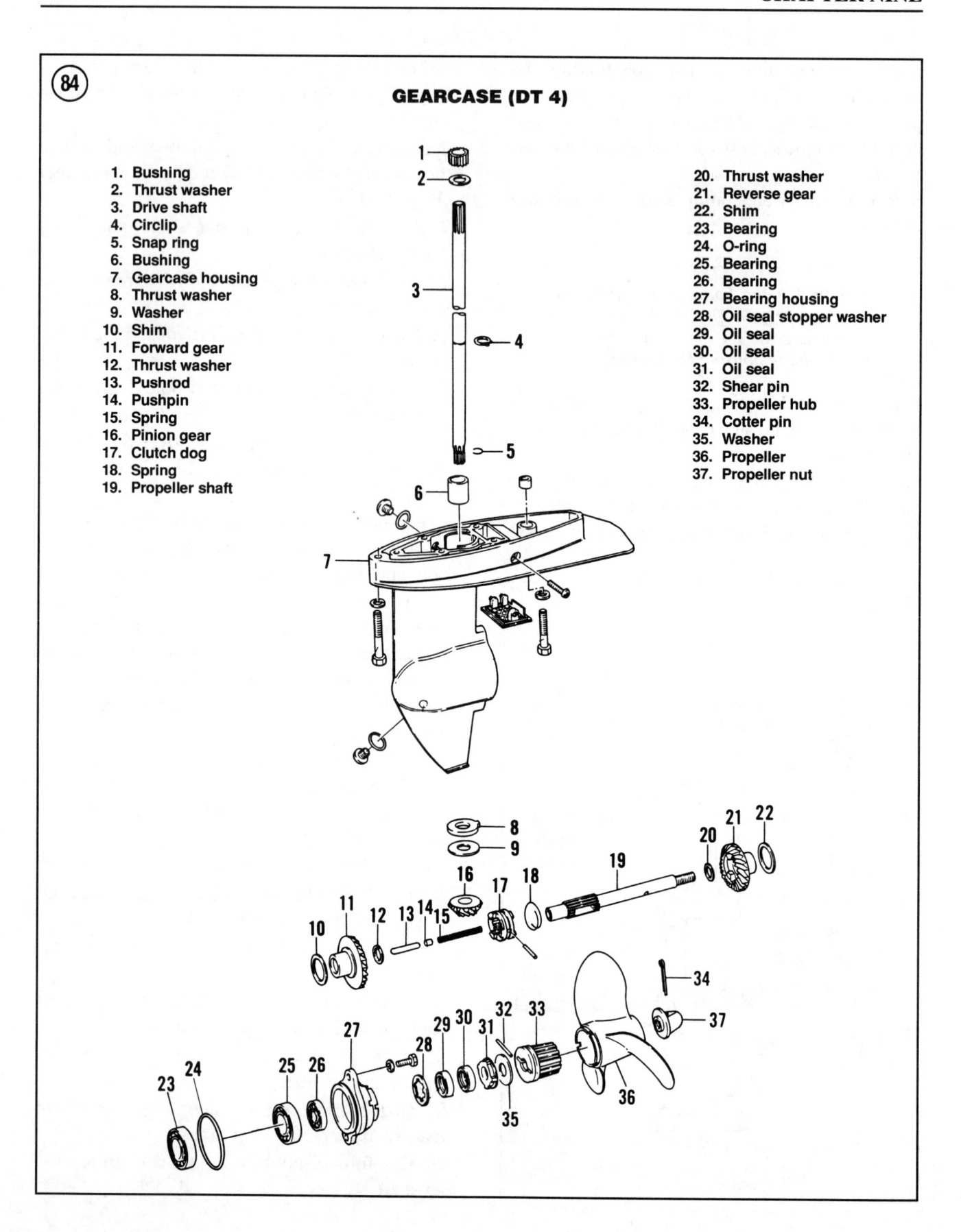

CHAPTER NINE

GEARCASE (DT 4)

1. Bushing
2. Thrust washer
3. Drive shaft
4. Circlip
5. Snap ring
6. Bushing
7. Gearcase housing
8. Thrust washer
9. Washer
10. Shim
11. Forward gear
12. Thrust washer
13. Pushrod
14. Pushpin
15. Spring
16. Pinion gear
17. Clutch dog
18. Spring
19. Propeller shaft
20. Thrust washer
21. Reverse gear
22. Shim
23. Bearing
24. O-ring
25. Bearing
26. Bearing
27. Bearing housing
28. Oil seal stopper washer
29. Oil seal
30. Oil seal
31. Oil seal
32. Shear pin
33. Propeller hub
34. Cotter pin
35. Washer
36. Propeller
37. Propeller nut

GEARCASE

25. Check propeller shaft thrust clearance as described in this chapter.

26. When proper propeller shaft thrust clearance has been established, coat the bearing housing bolt threads with Silicone Seal (part No. 99000-31120) or equivalent. Install housing (use installer part No. 09914-79610) and tighten bolts to specifications (**Table 1**).

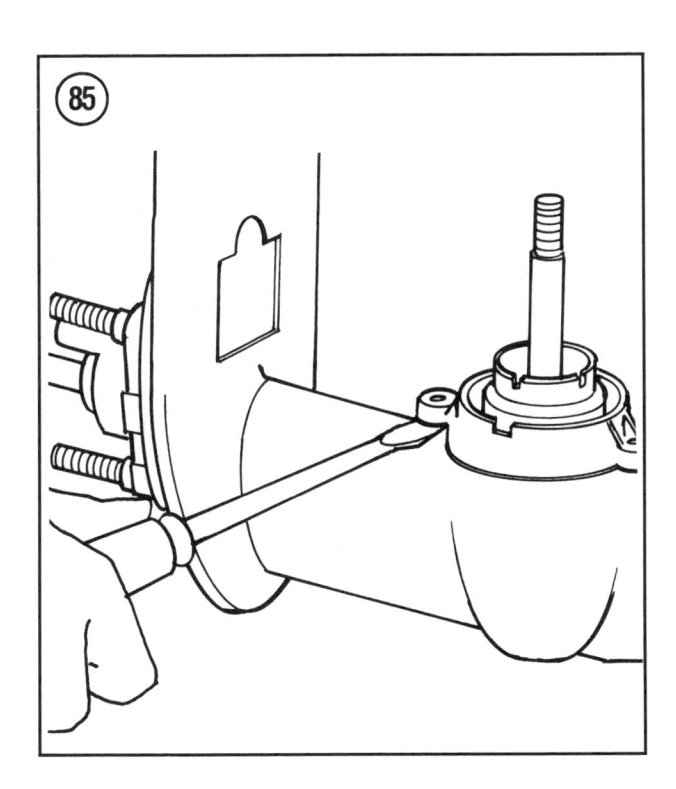

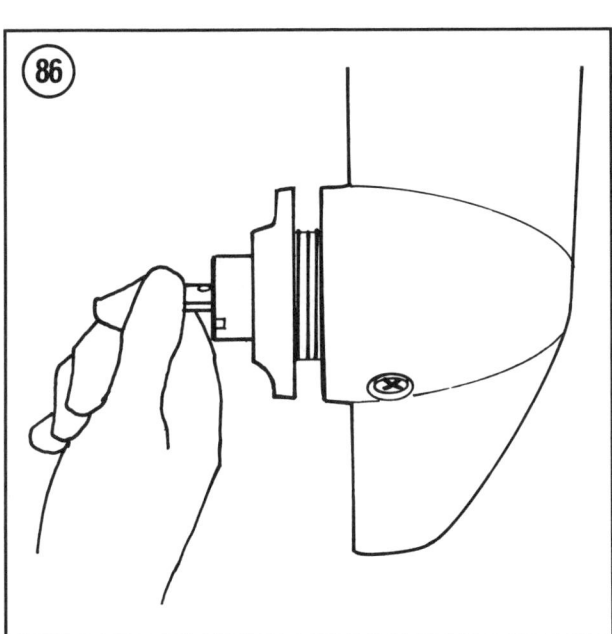

27. Install the gearcase as described in this chapter. Fill with recommended type and quantity of lubricant. See Chapter Four.

28. Check gearcase lubricant level after engine has been run. Change the lubricant after 10 hours of operation (break-in period). See Chapter Four.

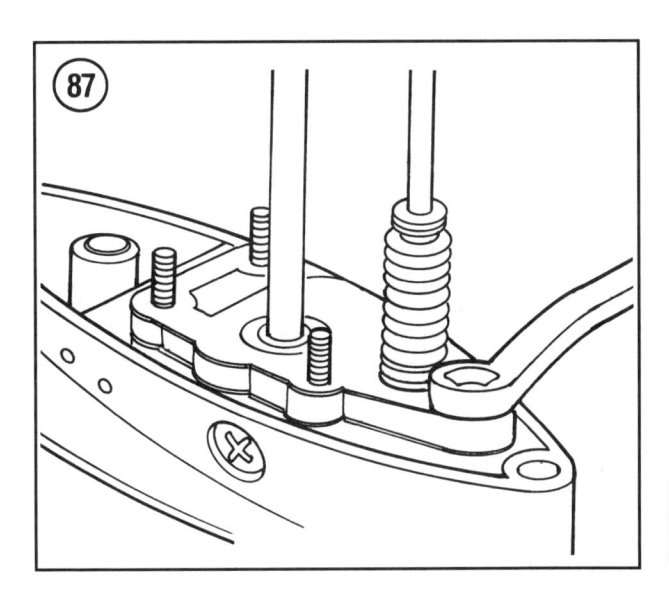

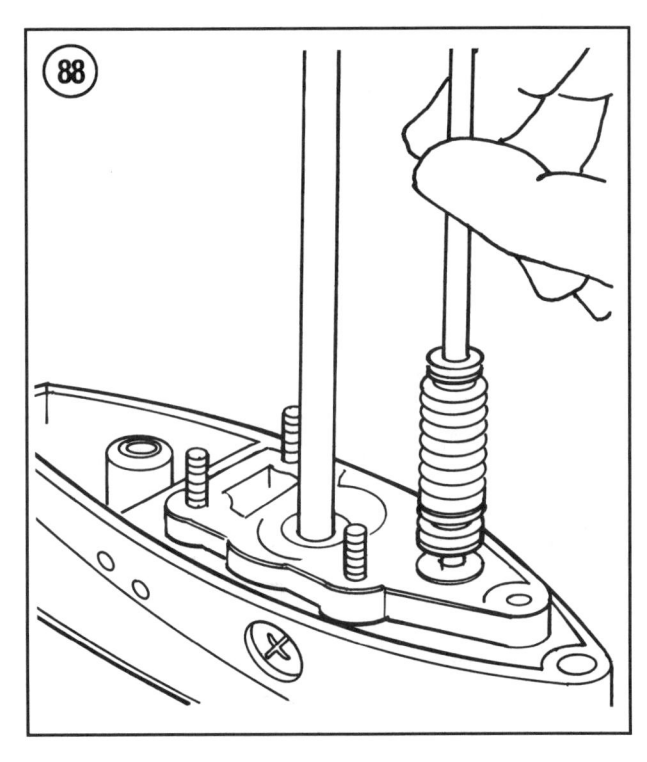

408

CHAPTER NINE

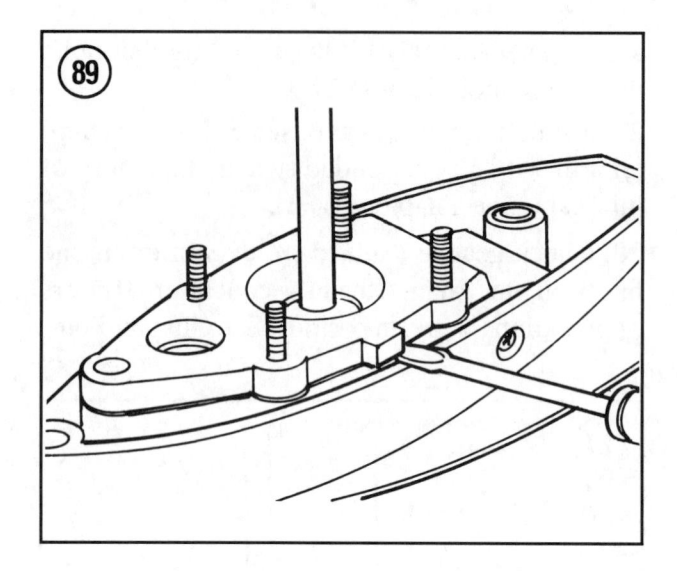

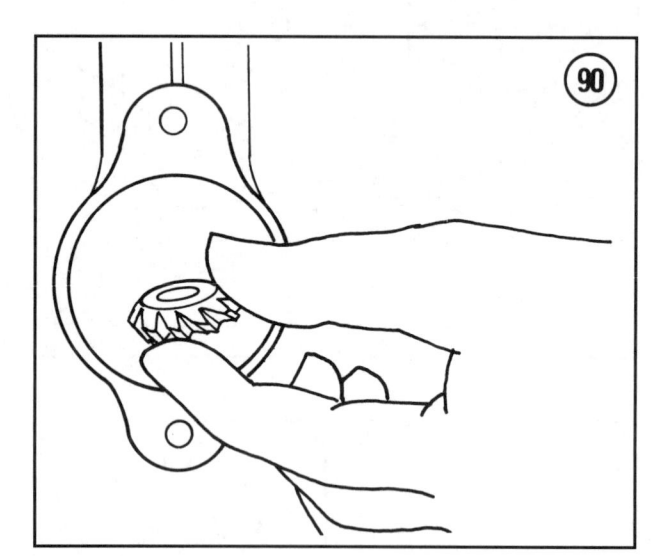

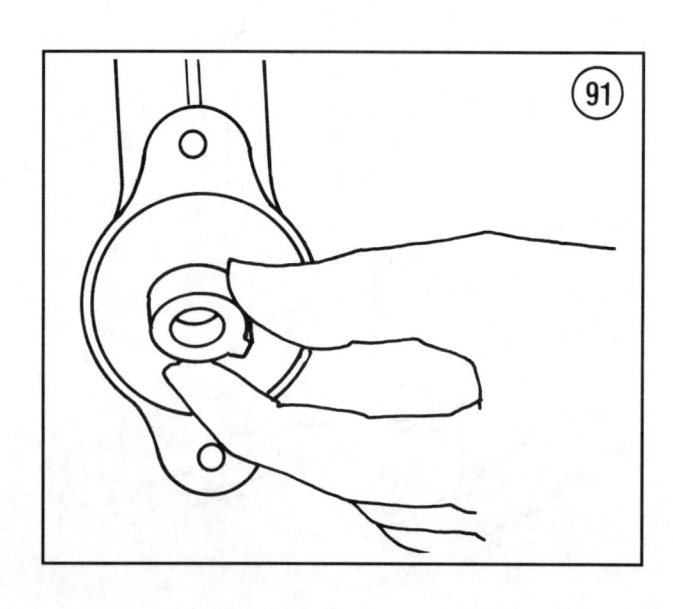

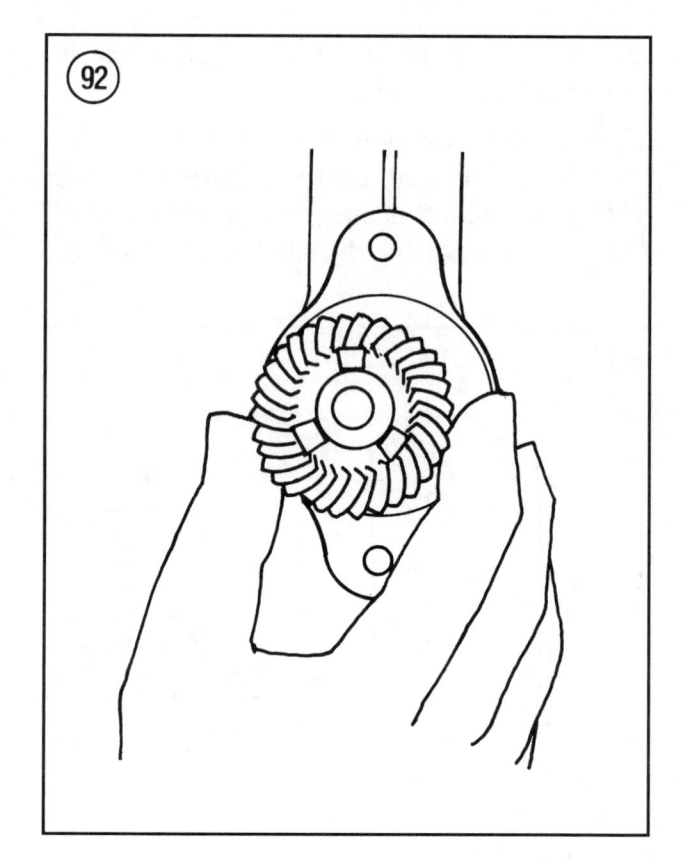

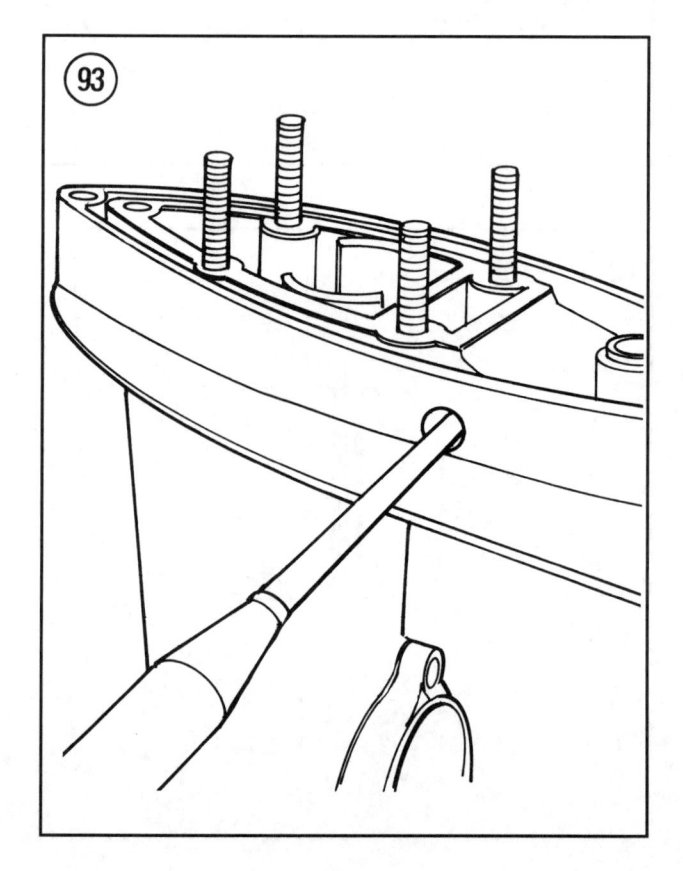

GEARCASE

Gearcase Disassembly/Assembly (DT 6; 1985-1987 DT 8)

Refer to **Figure 94** for this procedure.

1. Remove the gearcase as described in this chapter.
2. Secure the gearcase in a suitable holding fixture or a vise with protective jaws. If protective jaws are not available, position the gearcase upright with the skeg between wooden blocks in a vise.
3. Remove the water pump as described in this chapter.

NOTE
If bearing housing is corroded in prop shaft bore and cannot be removed easily

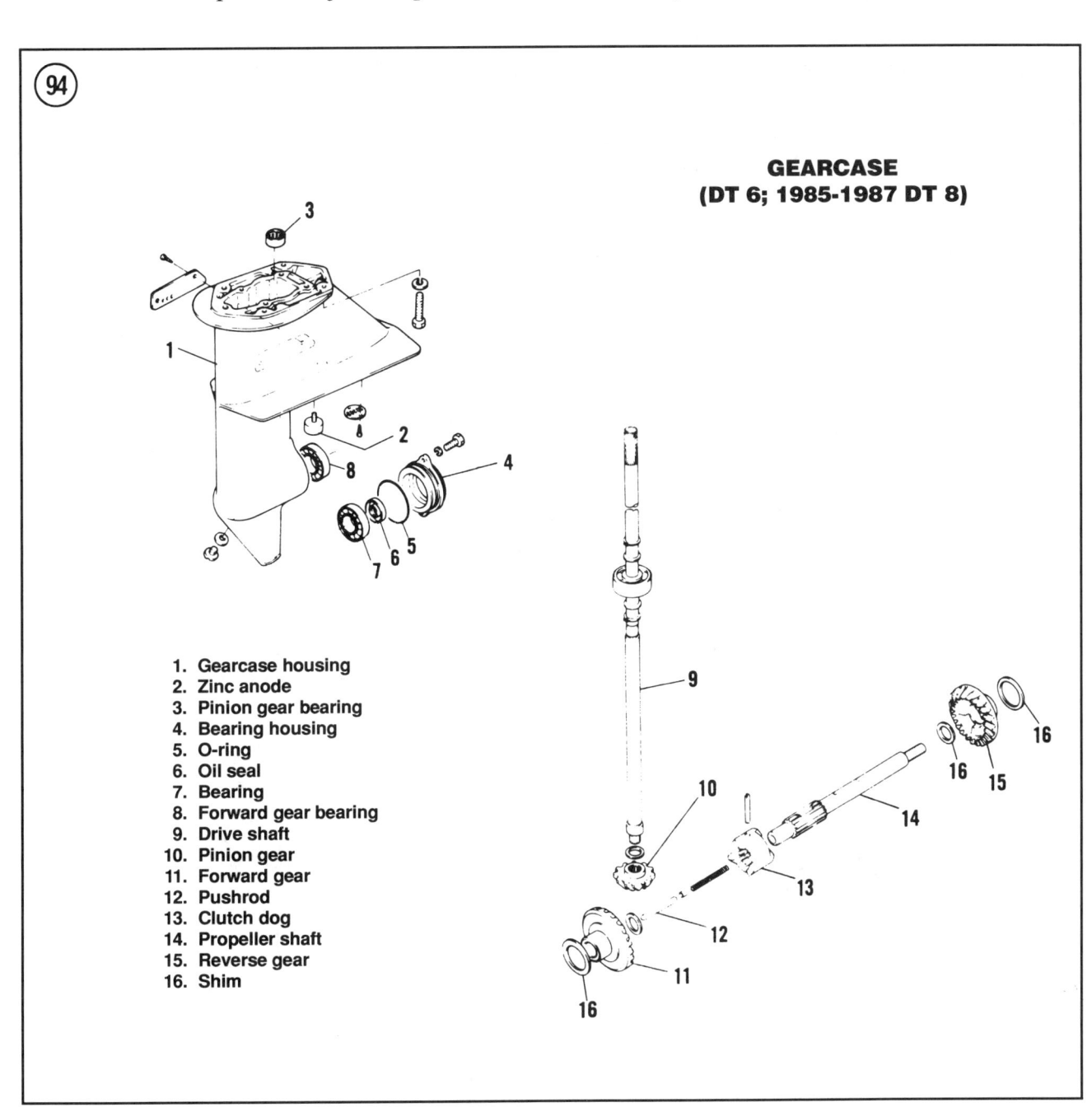

GEARCASE (DT 6; 1985-1987 DT 8)

1. Gearcase housing
2. Zinc anode
3. Pinion gear bearing
4. Bearing housing
5. O-ring
6. Oil seal
7. Bearing
8. Forward gear bearing
9. Drive shaft
10. Pinion gear
11. Forward gear
12. Pushrod
13. Clutch dog
14. Propeller shaft
15. Reverse gear
16. Shim

by hand in Step 4, tap side of housing cap with a rubber mallet to rotate cap ears, then pry bearing housing off.

4. Remove the 2 bearing housing bolts. Remove the bearing housing and propeller shaft assembly (**Figure 95**).

5. Unbolt the drive shaft bearing housing (**Figure 96**). Remove the bearing housing, the drive shaft and the shift rod assembly.

6. Remove the drive shaft bearing housing snap ring and separate the housing from the drive shaft (**Figure 97**).

7. Reach into the propeller shaft bore and remove the pinion gear (**Figure 98**), then the forward gear and shim(s) (**Figure 99**).

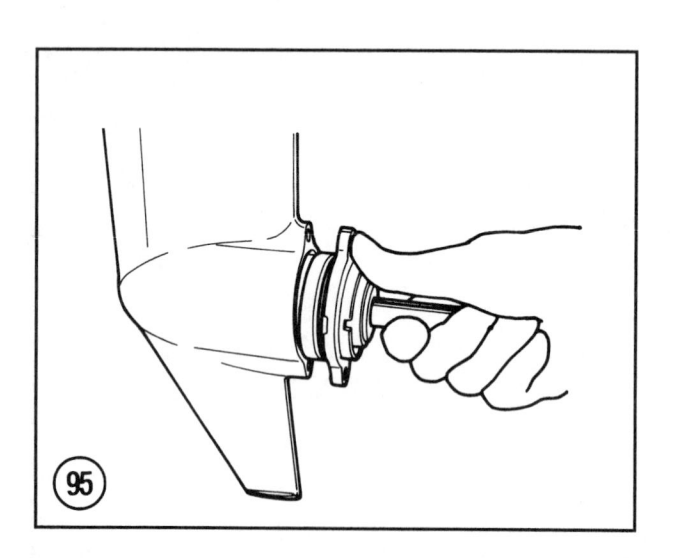

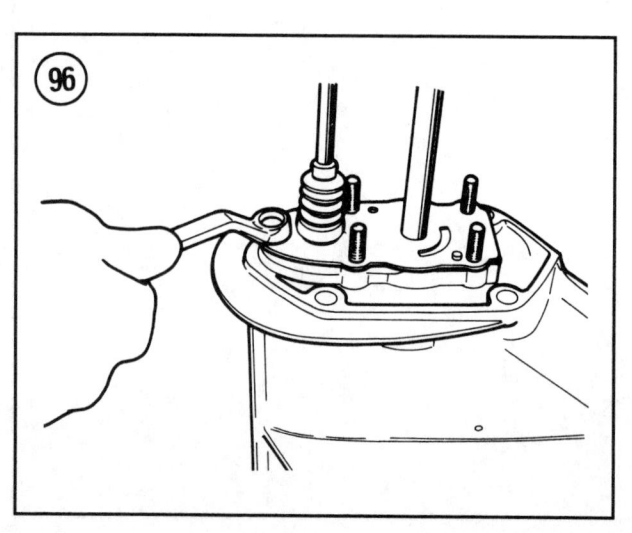

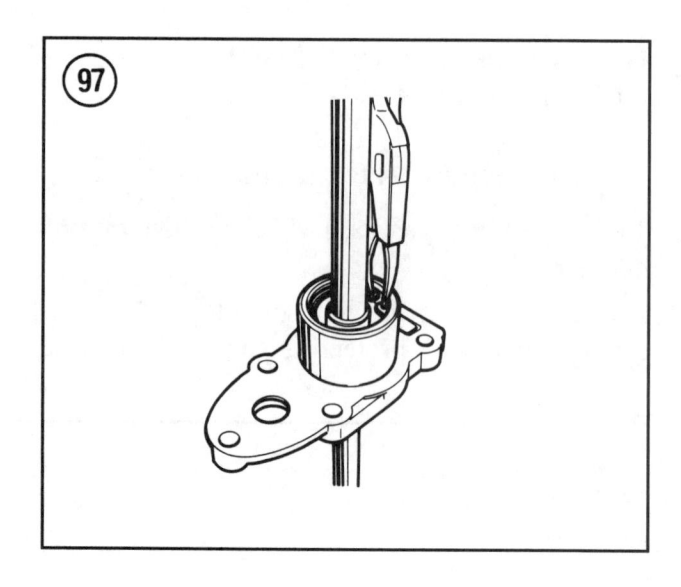

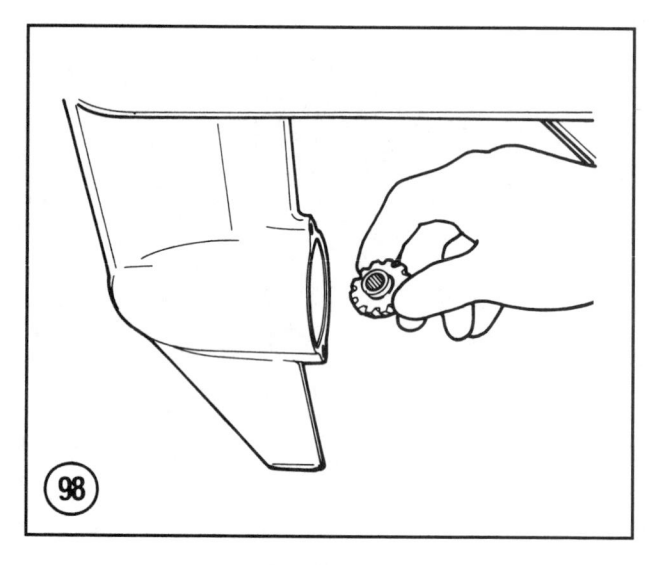

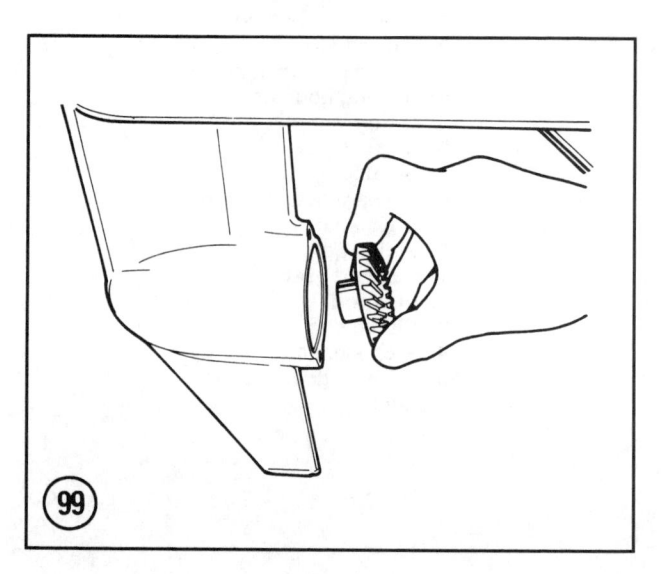

GEARCASE

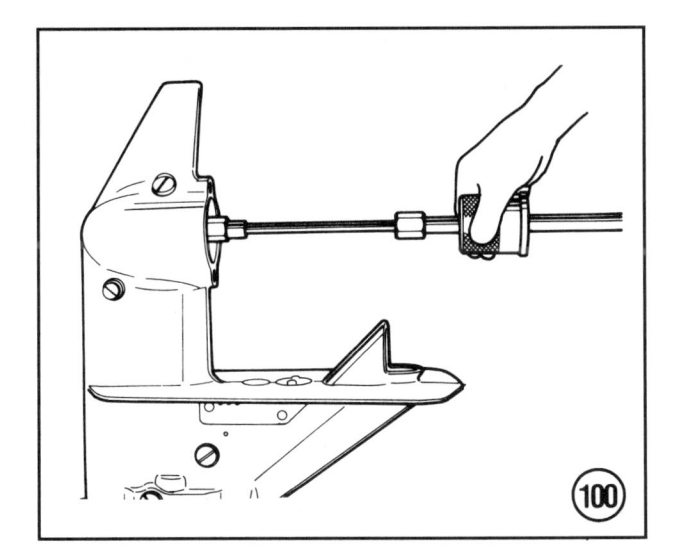

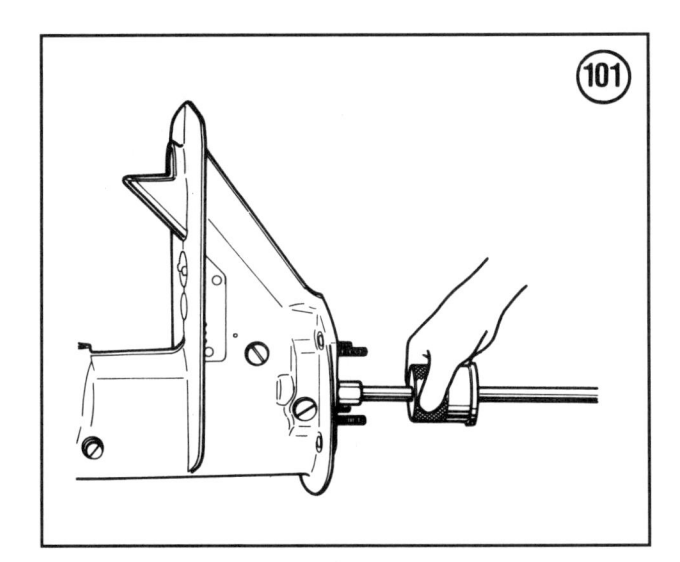

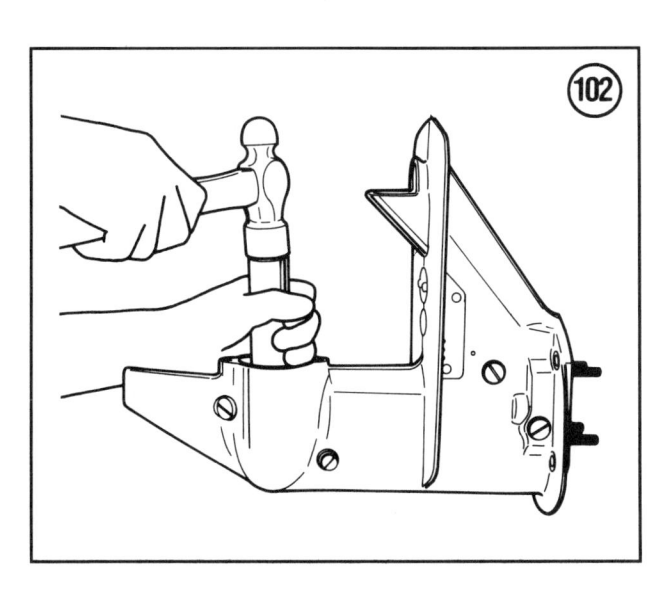

8. If necessary, attach bearing remover (part No. 09913-69911) to a slide hammer and remove the forward gear bearing (**Figure 100**).

9. Use the same remover/slide hammer combination to remove the gearcase drive shaft bearing (**Figure 101**).

10. Clean and inspect all parts as described in this chapter.

11. Install new forward gear and drive shaft bearings with installer (part No. 09914-79610) or equivalent (**Figure 102**).

12. Fit the forward gear shim(s) over the shaft at the rear of the gear and install into the prop shaft bore.

13. Insert the pinion gear with shim(s) into the prop shaft bore. Fit the gear into the drive shaft bore and mesh it with the forward gear.

14. Lubricate the drive shaft bearing housing oil seal lip with water-resistant grease. Install the housing on the drive shaft and seat into the gearcase housing. Install housing snap ring (if used) and make sure it fits properly into its groove.

15. Insert the drive shaft into the gearcase housing with a rotating motion and engage the pinion gear.

16. Lubricate the shift rod O-ring with water-resistant grease (part No. 99000-25160) or equivalent and insert shift rod assembly into gearcase (**Figure 103**).

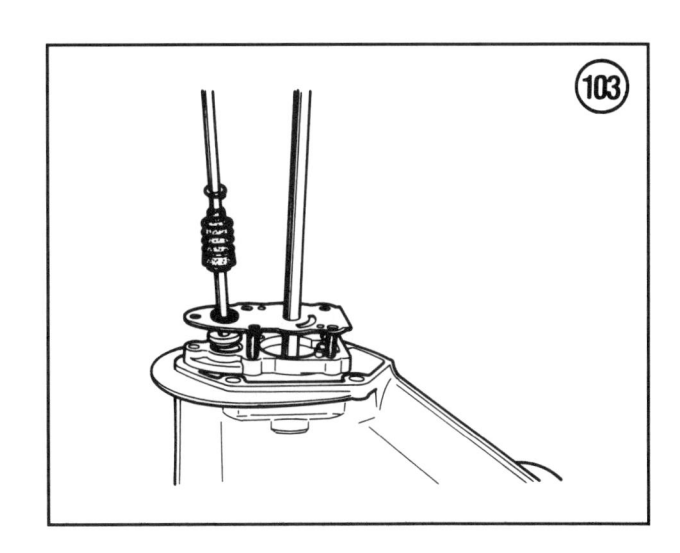

17. Install shift cam stop screw.
18. Install the water pump as described in this chapter.
19. Check pinion gear depth and forward gear backlash as described in this chapter.
20. Remove and discard the bearing housing bearing and oil seal.
21. Install a new housing bearing with installer (part No. 09914-79610) or equivalent (**Figure 104**). Coat seal lips with water-resistant grease.
22. Install a new bearing housing O-ring. Lubricate housing O-ring and end of propeller shaft pushrod with water-resistant grease.
23. Insert propeller shaft into housing bore and engage the forward gear.
24. Coat bearing housing outer edges (front and rear) with water-resistant grease.
25. Carefully install bearing housing on propeller shaft.
26. Check propeller shaft thrust clearance as described in this chapter.
27. When proper propeller shaft thrust clearance has been established, coat the bearing housing bolt threads with Silicone Seal (part No. 99000-31120) or equivalent. Install housing and tighten bolts securely.
28. Install the gearcase as described in this chapter. Fill with recommended type and quantity of lubricant. See Chapter Four.
29. Check gearcase lubricant level after engine has been run. Change the lubricant after 10 hours of operation (break-in period). See Chapter Four.

Gearcase Disassembly/Assembly (1988-on DT 8; DT 9.9, DT 15)

Refer to the following illustrations for this procedure:
   a. **Figure 105**: 1988-on DT 8, DT 9.9.
   b. **Figure 106**: 1985-1987 DT 9.9; DT 15.
1. Remove the gearcase as described in this chapter.
2. Secure the gearcase in a suitable holding fixture or a vise with protective jaws. If protective jaws are not available, position the gearcase upright with the skeg between wooden blocks in a vise.
3. Remove the 2 bearing housing bolts (**Figure 107**).

NOTE
If the bearing housing is corroded in the prop shaft bore and cannot be removed easily by hand in Step 4A, carefully tap the side of the housing cap with a rubber mallet to rotate the cap ears, then pry housing off.

4A. On 1985-1987 DT 9.9 and all DT 15 models, attach remover tool (part No. 09950-59320) or equivalent to the propeller shaft splined end and connect a slide hammer (**Figure 108**) to it.
4B. On 1988-on DT 8 and DT 9.9 models, attach remover tool (part No. 09950-59330) or equivalent to the propeller shaft splined end and connect a slide hammer (**Figure 108**) to it.
5. Using the special tool described in Step 4A or Step 4B, remove the bearing housing and propeller shaft assembly.
6. Remove the water pump as described in this chapter.
7A. On 1985-1987 DT 9.9 and all DT 15 models, perform the following:
   a. Remove the rubber exhaust seal (**Figure 109**).

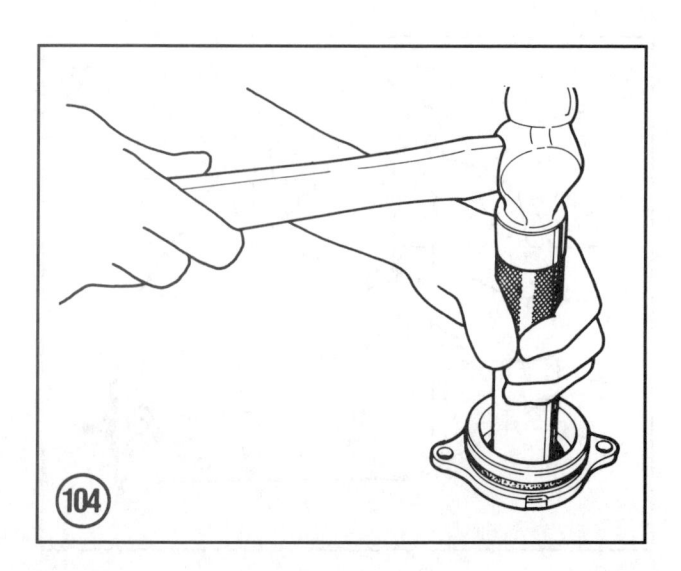

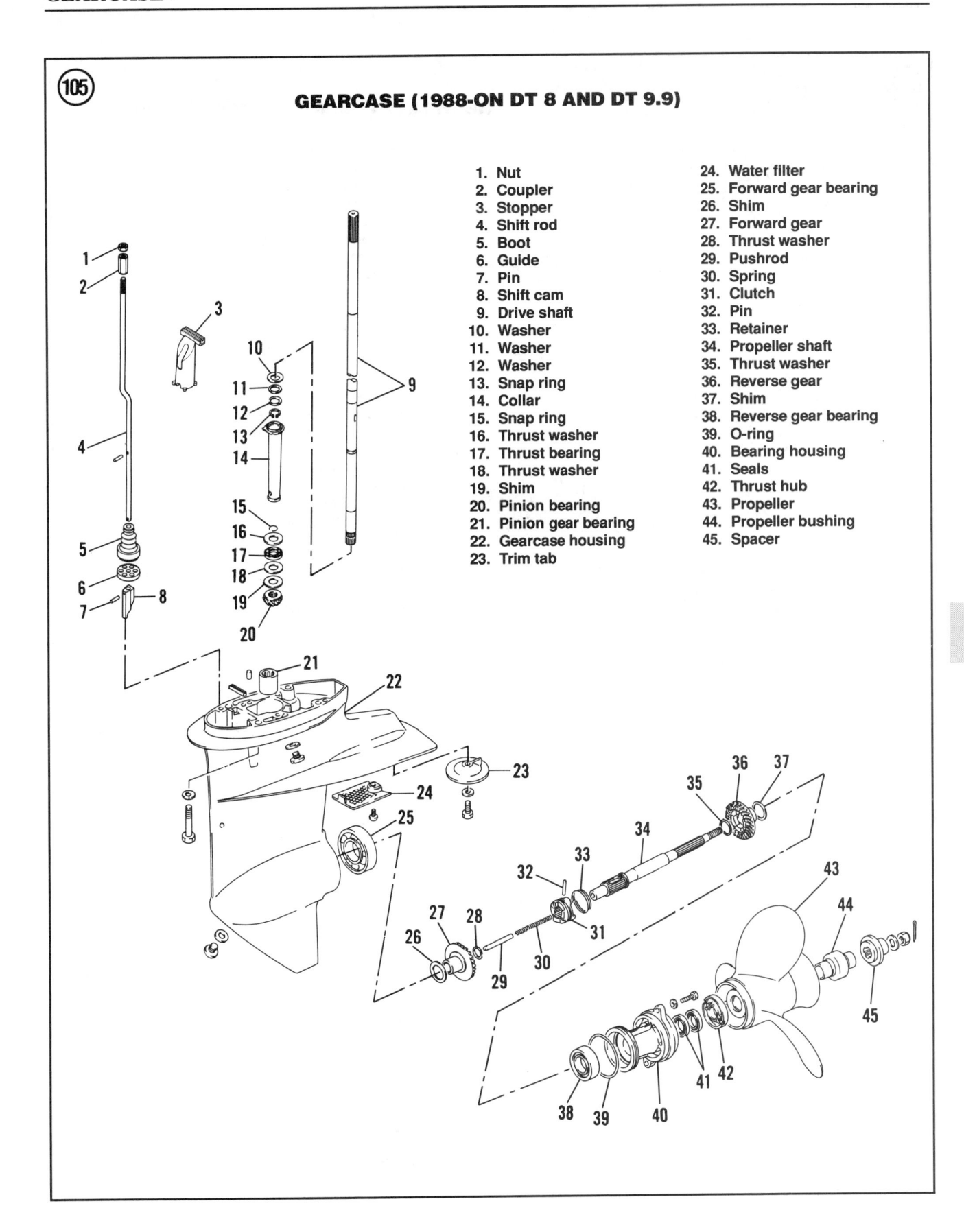

GEARCASE

413

GEARCASE (1988-ON DT 8 AND DT 9.9)

1. Nut
2. Coupler
3. Stopper
4. Shift rod
5. Boot
6. Guide
7. Pin
8. Shift cam
9. Drive shaft
10. Washer
11. Washer
12. Washer
13. Snap ring
14. Collar
15. Snap ring
16. Thrust washer
17. Thrust bearing
18. Thrust washer
19. Shim
20. Pinion bearing
21. Pinion gear bearing
22. Gearcase housing
23. Trim tab
24. Water filter
25. Forward gear bearing
26. Shim
27. Forward gear
28. Thrust washer
29. Pushrod
30. Spring
31. Clutch
32. Pin
33. Retainer
34. Propeller shaft
35. Thrust washer
36. Reverse gear
37. Shim
38. Reverse gear bearing
39. O-ring
40. Bearing housing
41. Seals
42. Thrust hub
43. Propeller
44. Propeller bushing
45. Spacer

9

CHAPTER NINE

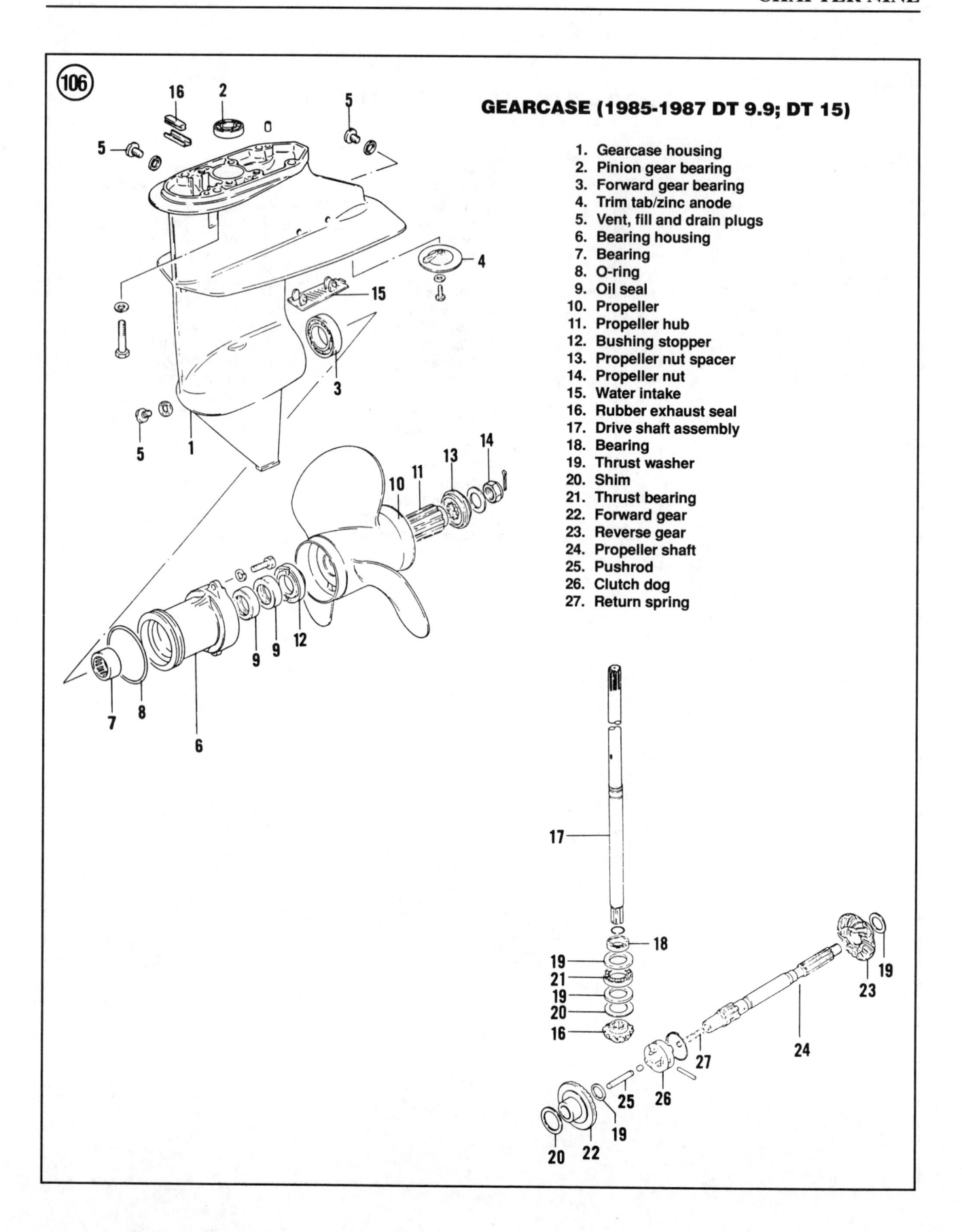

106 GEARCASE (1985-1987 DT 9.9; DT 15)

1. Gearcase housing
2. Pinion gear bearing
3. Forward gear bearing
4. Trim tab/zinc anode
5. Vent, fill and drain plugs
6. Bearing housing
7. Bearing
8. O-ring
9. Oil seal
10. Propeller
11. Propeller hub
12. Bushing stopper
13. Propeller nut spacer
14. Propeller nut
15. Water intake
16. Rubber exhaust seal
17. Drive shaft assembly
18. Bearing
19. Thrust washer
20. Shim
21. Thrust bearing
22. Forward gear
23. Reverse gear
24. Propeller shaft
25. Pushrod
26. Clutch dog
27. Return spring

GEARCASE

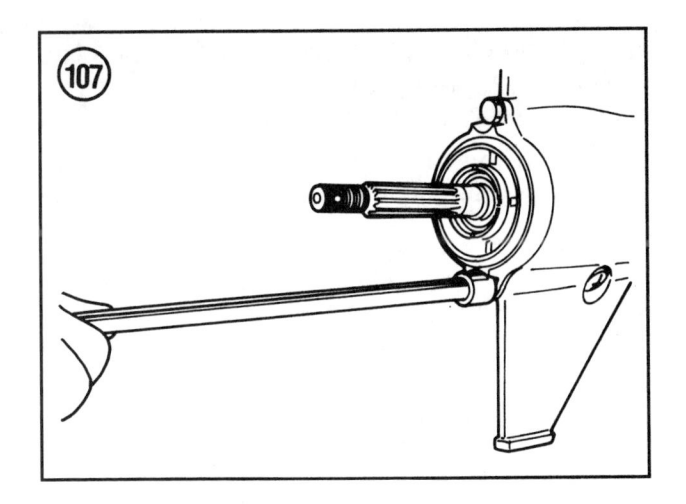

415

b. Remove the drive shaft bearing snap ring with snap ring pliers (**Figure 110**). Discard the snap ring.

c. Remove the drive shaft from the gearcase housing (**Figure 111**). The shim pack, thrust bearing and thrust washer(s) should come with it.

7B. On 1988-on DT 8 and DT 9.9 models, pull the drive shaft (A, **Figure 112**) and drive shaft housing (B, **Figure 112**) from the gearcase. Remove the thrust collar from the drive shaft.

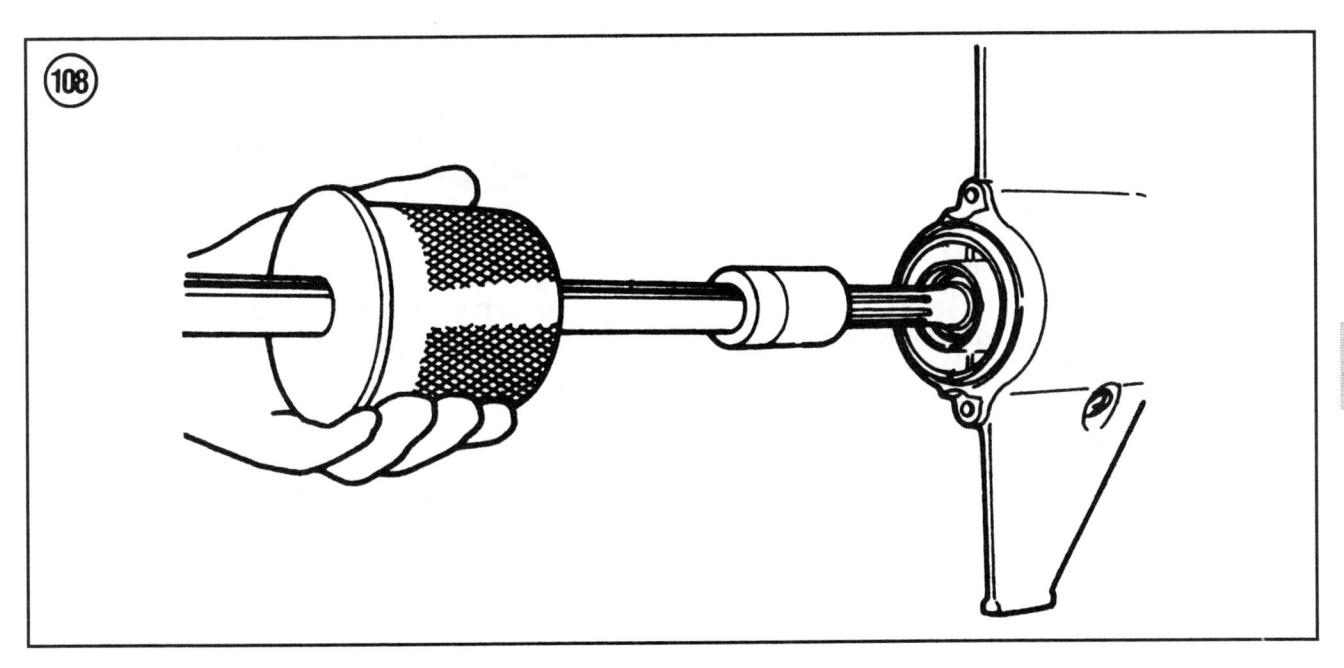

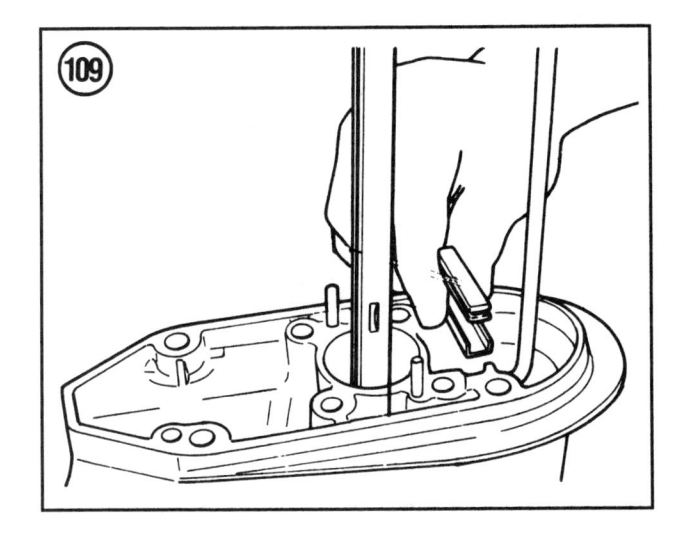

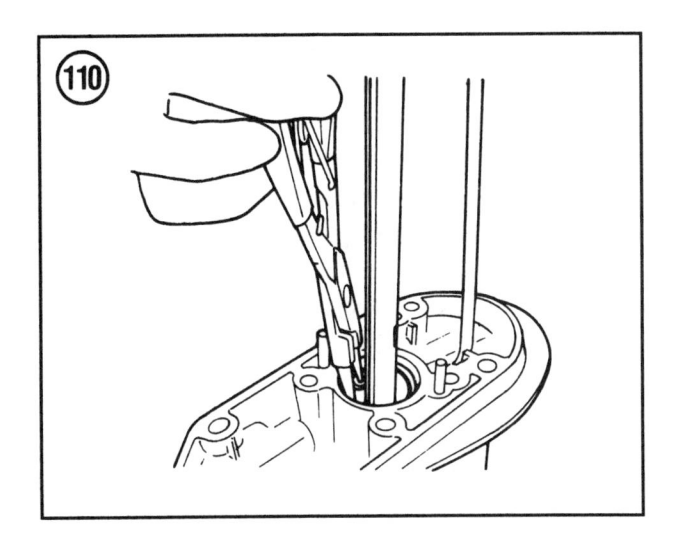

8. Reach into the propeller shaft bore and remove the pinion gear, then the forward gear and shim(s). Be sure to remove all shims and reinstall the same amount of shims.
9. Remove the shift rod collar nut then remove the shift rod assembly.
10. Disassemble the propeller shaft and clutch assembly as described in this chapter.
11. Clean and inspect all parts as described in this chapter.
12. Fit the forward gear shim(s) over the shaft at the rear of the gear and install gear/shim assembly into the prop shaft bore.
13. Insert the shift rod assembly into the gearcase bore and install the collar nut.
14. Insert the pinion gear into the prop shaft bore. Fit the gear into the drive shaft bore and mesh it with the forward gear.
15. Reassemble the bearing, shim pack and thrust washer to the drive shaft in the same order as noted during removal.
16. Insert the drive shaft into the gearcase housing with a rotating motion and engage the pinion gear.
17. On 1985-1987 DT 9.9 and all DT 15 models, perform the following:
 a. Install a new drive shaft bearing snap ring. Make sure the snap ring fits properly into its groove.
 b. Install rubber exhaust seal (**Figure 109**).
18. Check pinion gear depth and forward gear backlash as described in this chapter.
19. Install a new bearing housing O-ring. Lubricate housing O-ring with water-resistant grease.
20. Insert propeller shaft into housing bore and engage the forward gear.
21. Coat bearing housing outer edges (front and rear) with water-resistant grease.
22. Carefully install bearing housing on propeller shaft.
23. Check propeller shaft thrust clearance as described in this chapter.
24. When proper propeller shaft end play/thrust clearance has been established, coat the bearing housing bolt threads with Silicone Seal (part No. 99000-31120) or equivalent.
25A. On 1985-1987 DT 9.9 and all DT 15 models, install housing (use installer part No. 09922-59510) and tighten bolts to specification (**Table 1**).
25B. On 1988-on DT 8 and DT 9.9 models, install housing (use installer part No. 09922-59410) and tighten bolts to specification (**Table 1**).
26. Install the water pump as described in this chapter.
27. Install the gearcase as described in this chapter. Fill with recommended type and quantity of lubricant. See Chapter Four.
28. Check gearcase lubricant level after engine has been run. Change the lubricant after 10 hours of operation (break-in period). See Chapter Four.

Gearcase Disassembly/Assembly (DT 20, DT 25, DT 30 [2-cylinder]; DT 25, DT 30 [3-cylinder]; DT 35, DT 40, DT 55, DT 65)

Refer to the following illustrations for this procedure:
 a. **Figure 113**: DT 20, DT 25, DT 30 (2-cylinder); DT 25, DT 30 (3-cylinder).

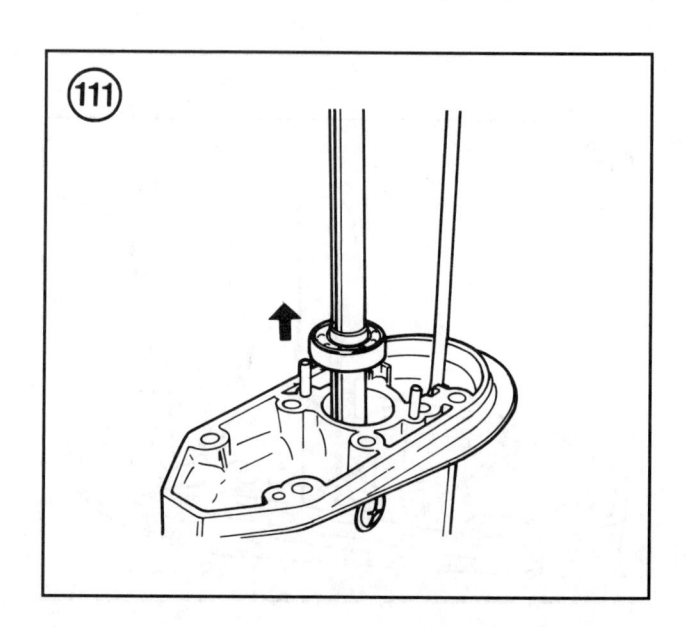

GEARCASE

b. **Figure 114**: DT 35, DT 40.

c. **Figure 115**: DT 55, DT 65.

1. Remove the gearcase as described in this chapter.

2. Secure the gearcase in a suitable holding fixture or a vise with protective jaws. If protective jaws are not available, position the gearcase upright with the skeg between wooden blocks in a vise.

3. Remove the water pump as described in this chapter.

4. Remove the 2 bolts holding the bearing housing (**Figure 116**).

5A. On DT 55 and DT 65 models, attach propeller shaft remover tool (part No. 09930-30161) (A, **Figure 117**) to the propeller shaft splined end and connect a slide hammer (B, **Figure 117**) to it.

5B. On all other models, attach propeller shaft remover tool (part No. 09950-59310) (A, **Figure 117**) to the propeller shaft splined end and connect a slide hammer (B, **Figure 117**) to it.

6. Using the special tool described in Step 5A or Step 5B, remove the bearing housing and propeller shaft assembly.

7. Fit an appropriate size box-end wrench over the pinion nut and pad the sides of the prop shaft bore to prevent distortion or damage from contact with the wrench.

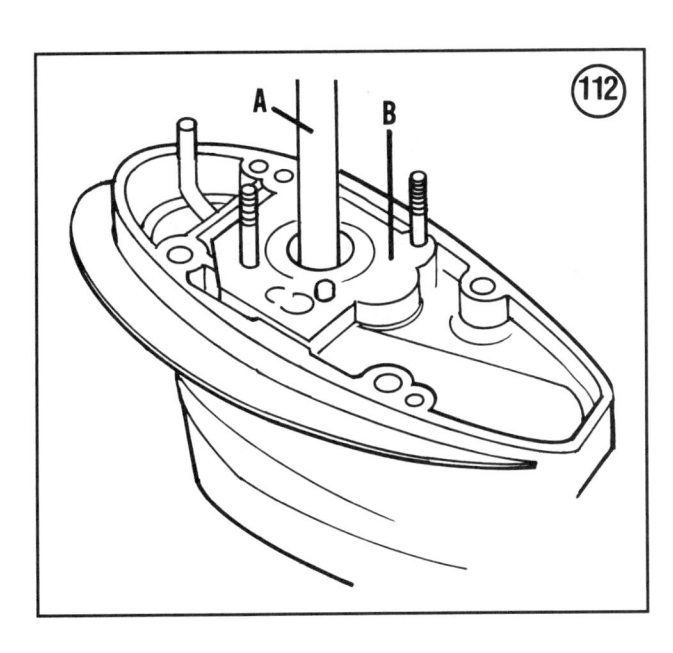

8. Holding the pinion nut with the wrench installed in Step 7, turn the drive shaft counterclockwise to loosen the pinion nut (**Figure 118**).

9. Remove the tools. Remove the pinion nut and pinion gear (**Figure 119**).

10. Remove the forward gear and shim(s) (**Figure 120**).

11. Unbolt the drive shaft bearing housing (**Figure 121**).

CAUTION
In Step 12, the bolts must be tightened evenly otherwise the bearing housing will be distorted and damaged during removal.

12. To separate the bearing housing, install two 6 mm bolts as shown in 1, **Figure 122**. Tighten the bolts evenly and alternately and force the bearing housing free.

13A. On DT 35 and DT 40 models, perform the following:

   a. Remove the bearing housing and the shift rod assembly from the gearcase as an assembly.

   b. Remove the drive shaft from the gearcase.

13B. On all other models, remove the drive shaft and bearing housing from the gearcase as an assembly.

14. On DT 20, DT 25, DT 30 (2-cylinder) and DT 25 and DT 30 (3-cylinder) models, remove the pinion gear shim(s) from the drive shaft bore.

15. On DT 20, DT 25, DT 30 (2-cylinder) and DT 25 and DT 30 (3-cylinder) models, perform the following:

   a. Remove the screw securing the shift rod guide (**Figure 123**).

   b. Withdraw the shift rod assembly (**Figure 124**).

16A. On DT 35 and DT 40 models, remove the retaining pin, then remove the preload spring collar and spring from the drive shaft.

16B. On all other models, perform the following:

CHAPTER NINE

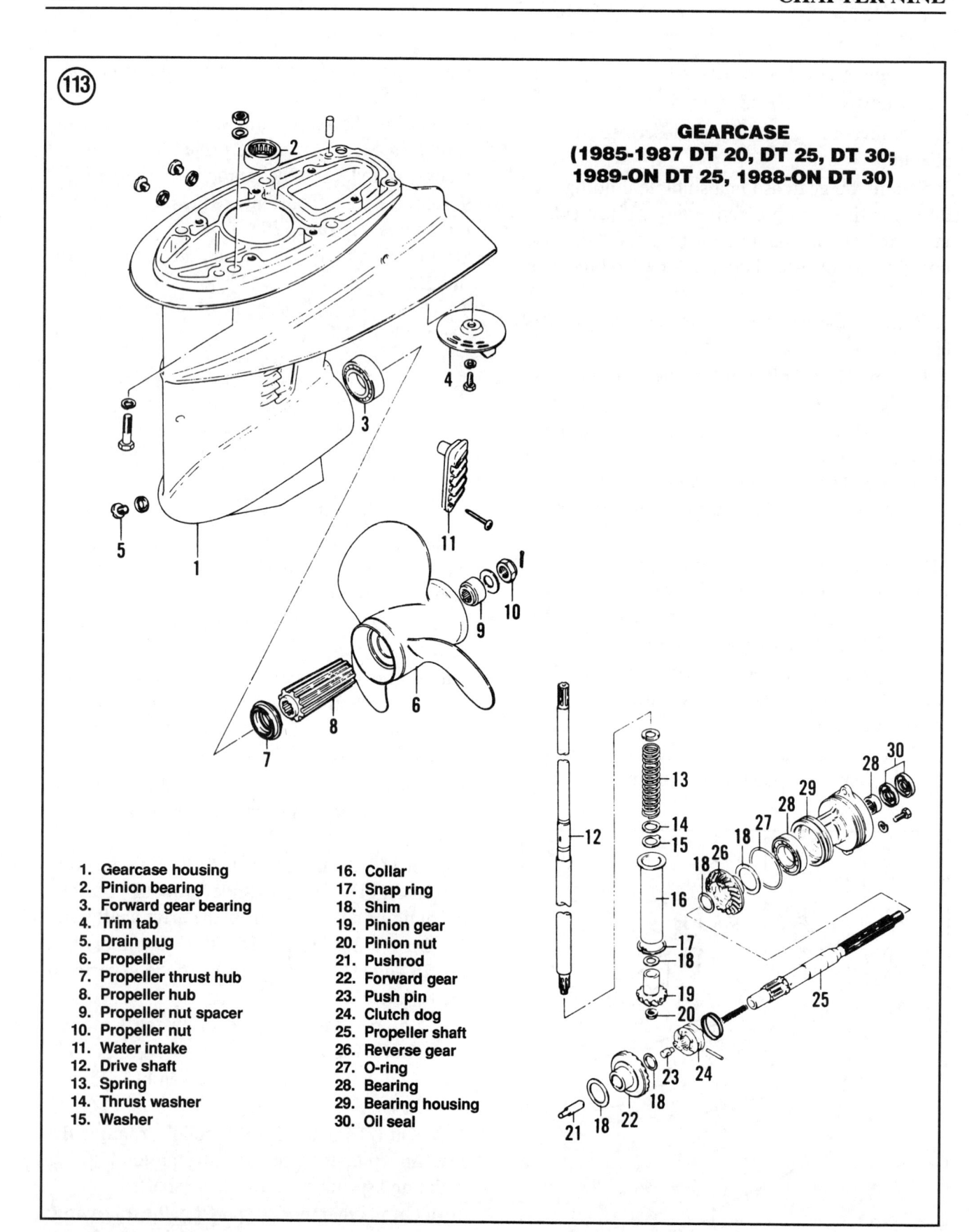

GEARCASE
(1985-1987 DT 20, DT 25, DT 30; 1989-ON DT 25, 1988-ON DT 30)

1. Gearcase housing
2. Pinion bearing
3. Forward gear bearing
4. Trim tab
5. Drain plug
6. Propeller
7. Propeller thrust hub
8. Propeller hub
9. Propeller nut spacer
10. Propeller nut
11. Water intake
12. Drive shaft
13. Spring
14. Thrust washer
15. Washer
16. Collar
17. Snap ring
18. Shim
19. Pinion gear
20. Pinion nut
21. Pushrod
22. Forward gear
23. Push pin
24. Clutch dog
25. Propeller shaft
26. Reverse gear
27. O-ring
28. Bearing
29. Bearing housing
30. Oil seal

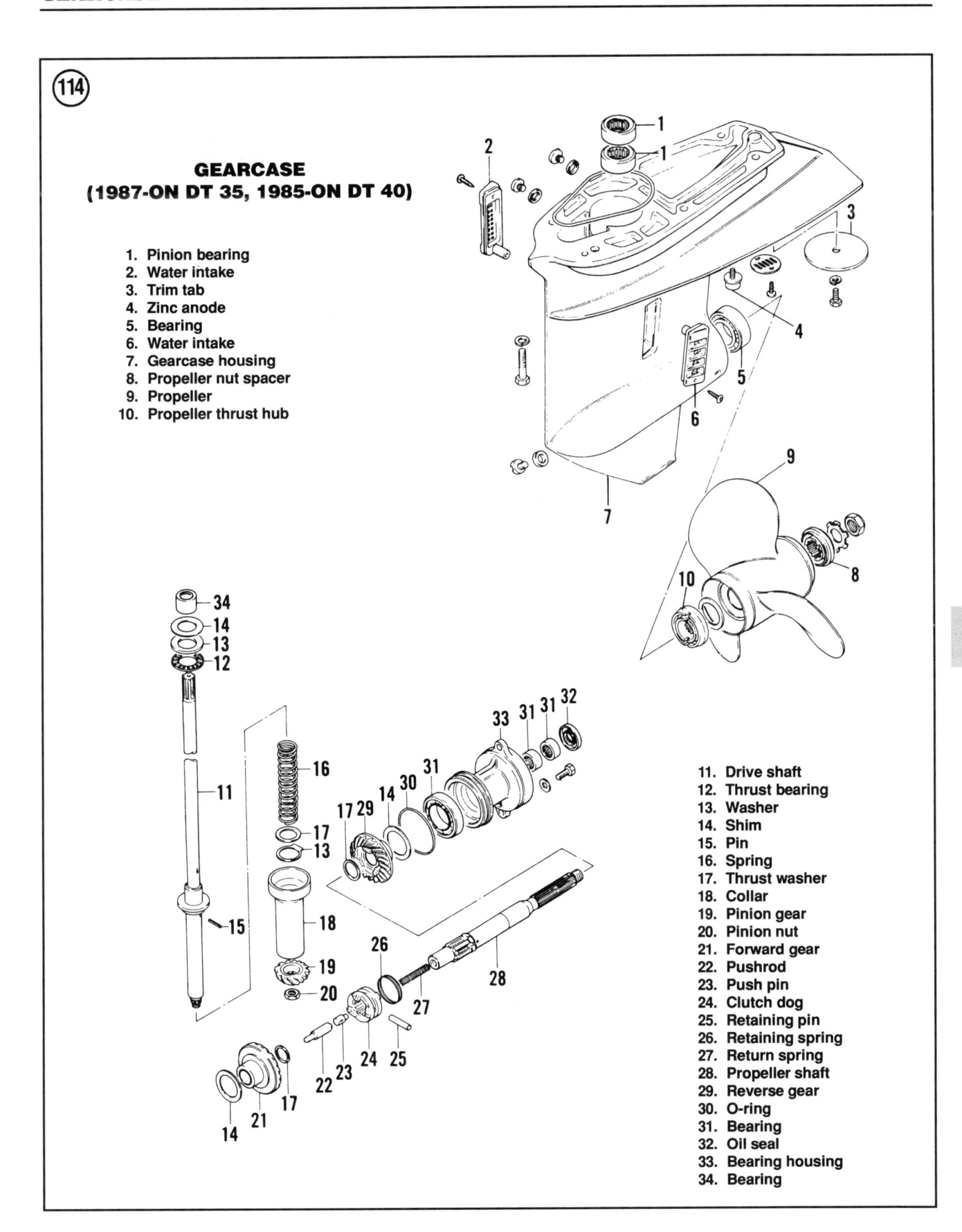

GEARCASE

419

GEARCASE
(1987-ON DT 35, 1985-ON DT 40)

1. Pinion bearing
2. Water intake
3. Trim tab
4. Zinc anode
5. Bearing
6. Water intake
7. Gearcase housing
8. Propeller nut spacer
9. Propeller
10. Propeller thrust hub
11. Drive shaft
12. Thrust bearing
13. Washer
14. Shim
15. Pin
16. Spring
17. Thrust washer
18. Collar
19. Pinion gear
20. Pinion nut
21. Forward gear
22. Pushrod
23. Push pin
24. Clutch dog
25. Retaining pin
26. Retaining spring
27. Return spring
28. Propeller shaft
29. Reverse gear
30. O-ring
31. Bearing
32. Oil seal
33. Bearing housing
34. Bearing

CHAPTER NINE

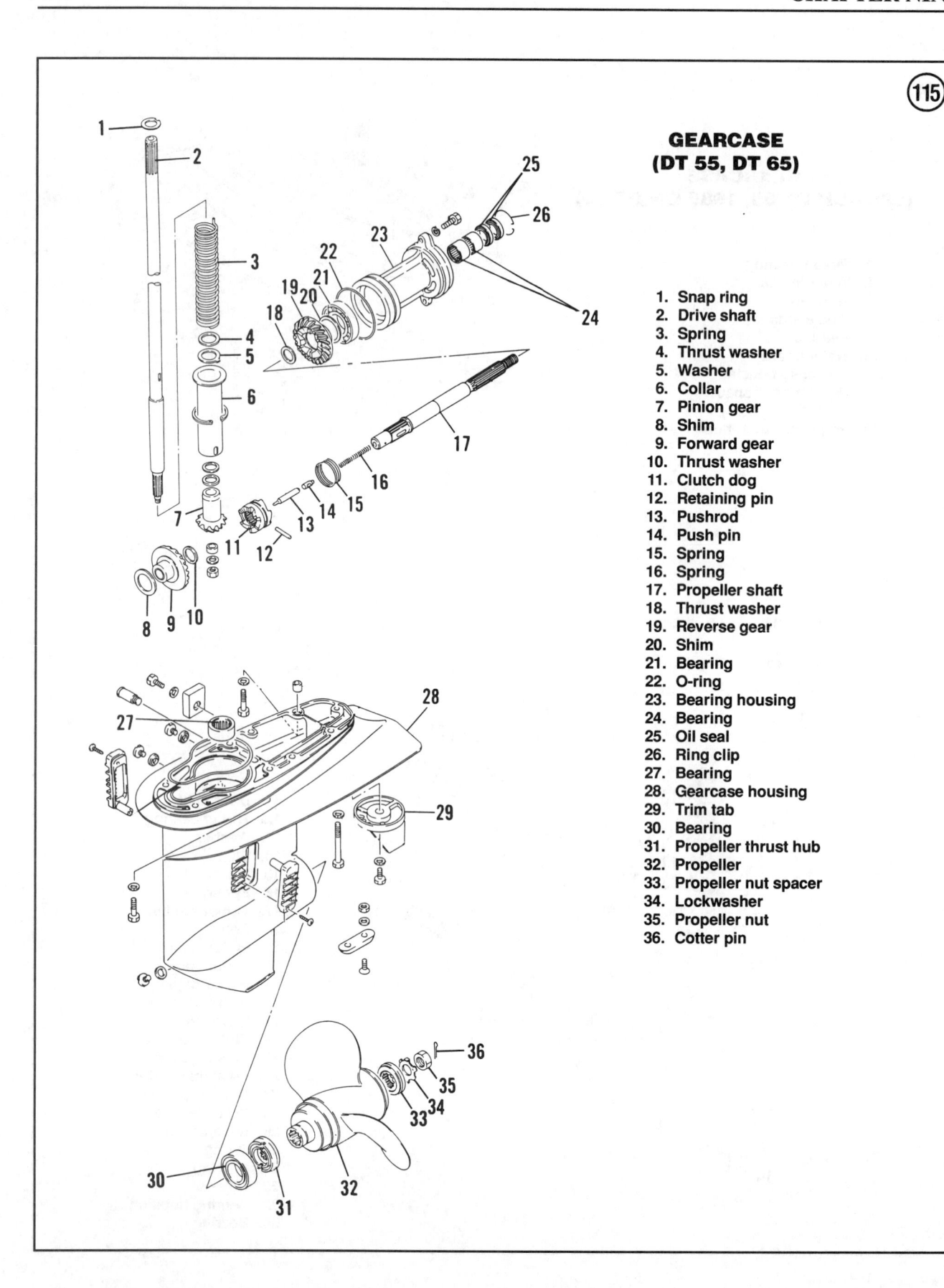

GEARCASE (DT 55, DT 65)

1. Snap ring
2. Drive shaft
3. Spring
4. Thrust washer
5. Washer
6. Collar
7. Pinion gear
8. Shim
9. Forward gear
10. Thrust washer
11. Clutch dog
12. Retaining pin
13. Pushrod
14. Push pin
15. Spring
16. Spring
17. Propeller shaft
18. Thrust washer
19. Reverse gear
20. Shim
21. Bearing
22. O-ring
23. Bearing housing
24. Bearing
25. Oil seal
26. Ring clip
27. Bearing
28. Gearcase housing
29. Trim tab
30. Bearing
31. Propeller thrust hub
32. Propeller
33. Propeller nut spacer
34. Lockwasher
35. Propeller nut
36. Cotter pin

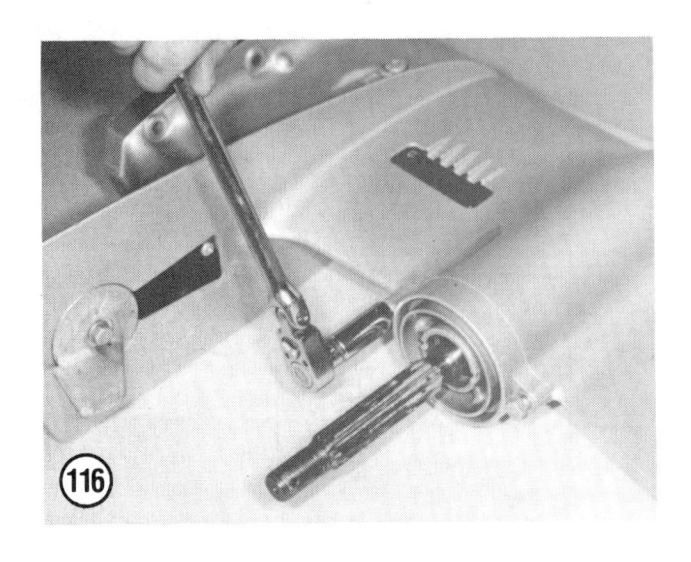

GEARCASE

421

(116)

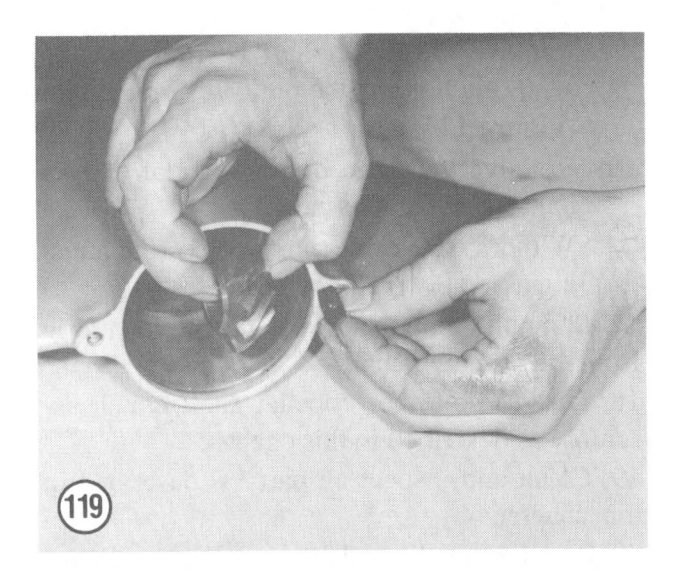

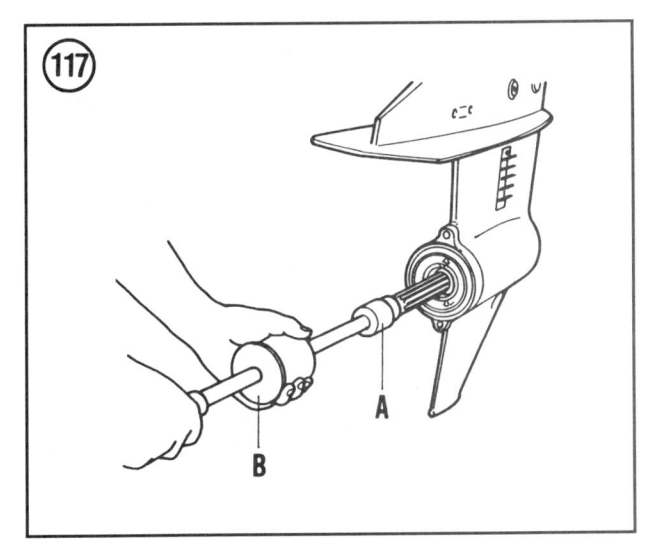

(117)

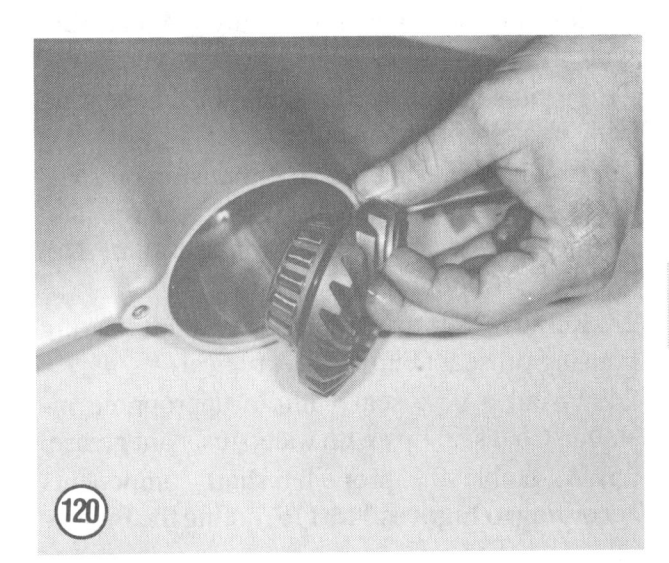

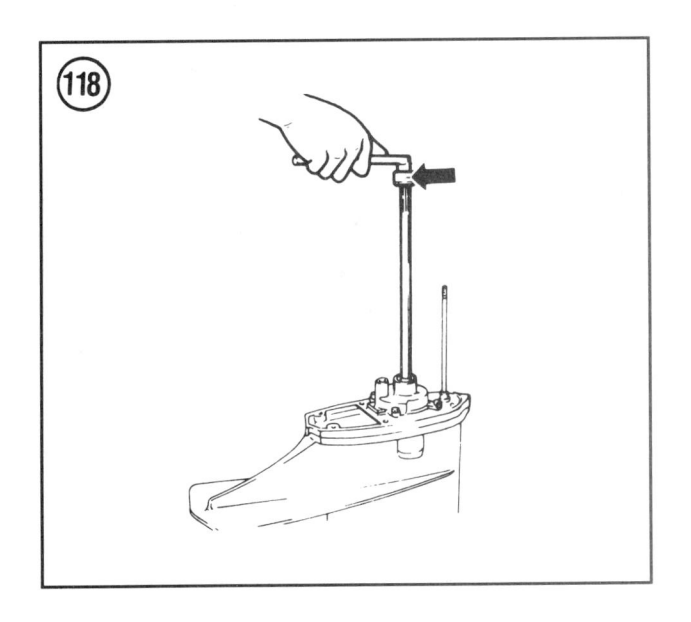

(118)

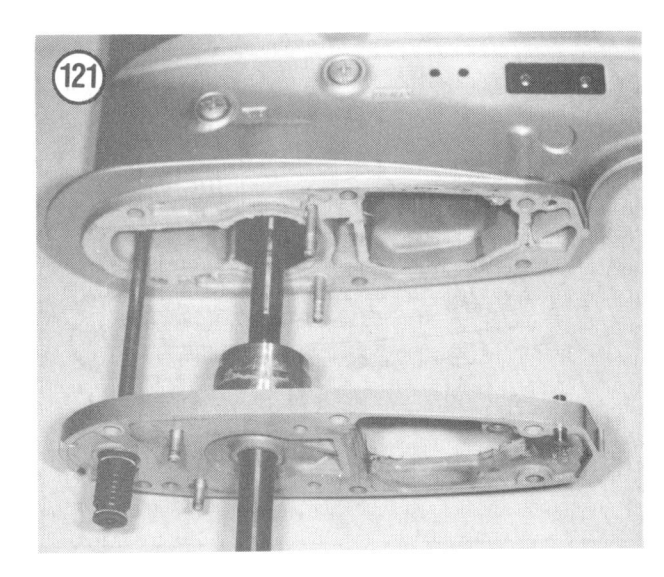

(119)

(120)

(121)

a. Remove the drive shaft bearing housing snap ring holding the preload spring collar (**Figure 125**).
b. Remove the preload spring collar or preload spring and washers.
c. Withdraw the drive shaft and bearing housing assembly from the gearcase.

17. Slide all components except the clutch dog from the propeller shaft.
18. Disassemble the propeller shaft clutch assembly as described in this chapter.
19. Clean and inspect all parts as described in this chapter.
20. Remove and discard the propeller shaft bearing housing oil seals and bearing (**Figure 126**, typical).
21. Install a new propeller shaft housing bearing with an appropriate installer.
22. Install new propeller shaft housing oil seals and a new housing O-ring. Coat seal lips and O-ring with water-resistant grease (part No. 99000-25160) or equivalent.
23. Remove and discard the drive shaft bearing housing oil seal (**Figure 127**, typical).
24. Install a new seal with an appropriate installer. Coat seal lips with water-resistant grease.
25. Assemble the propeller shaft components according to **Figures 113-115**, noting the following.

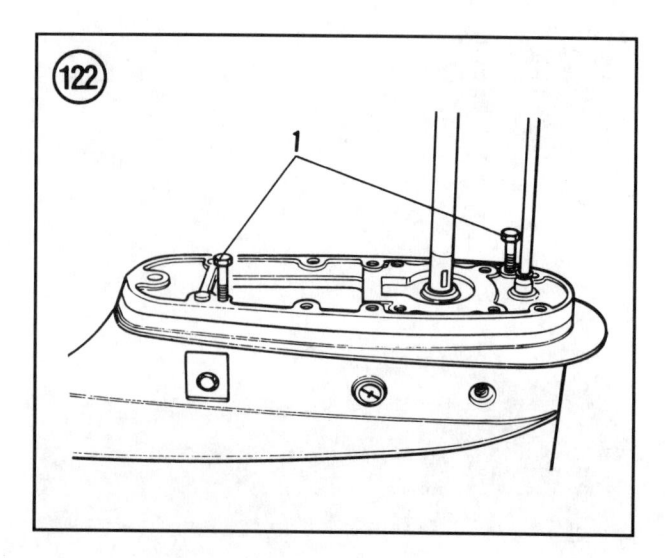

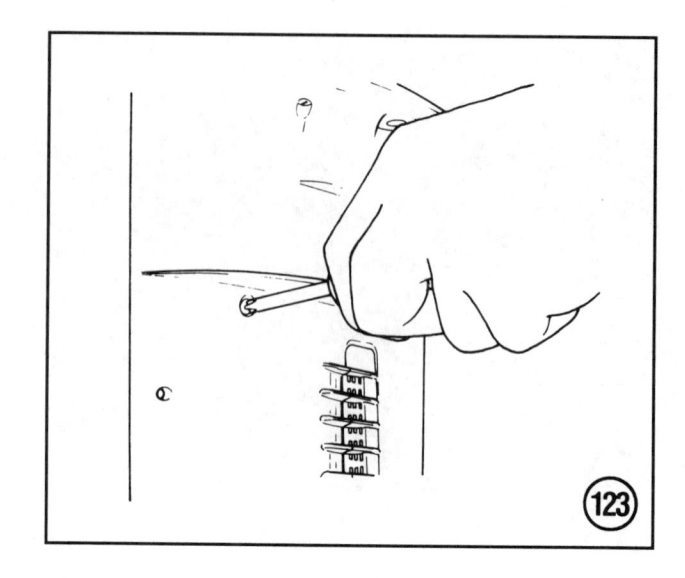

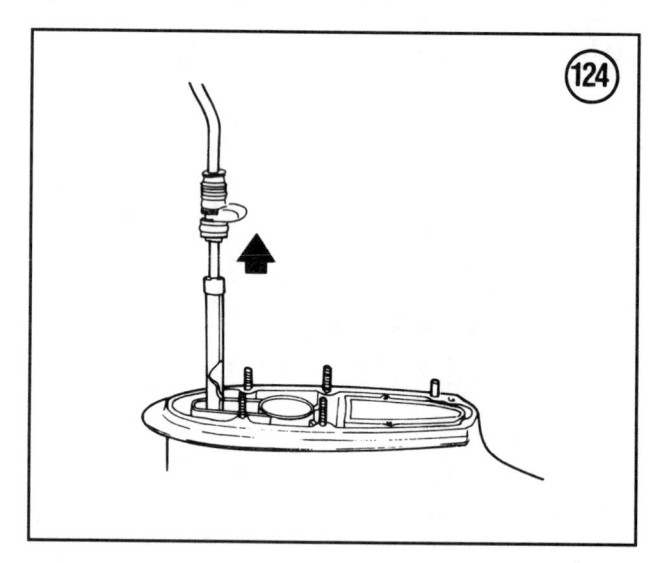

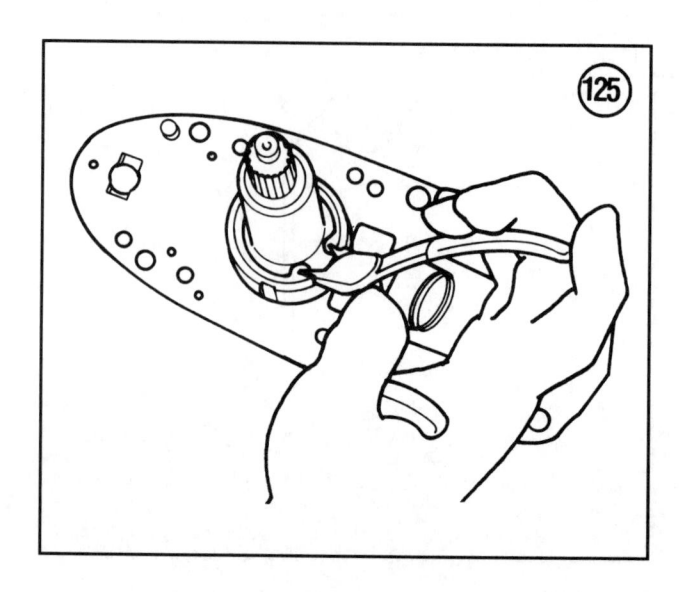

GEARCASE

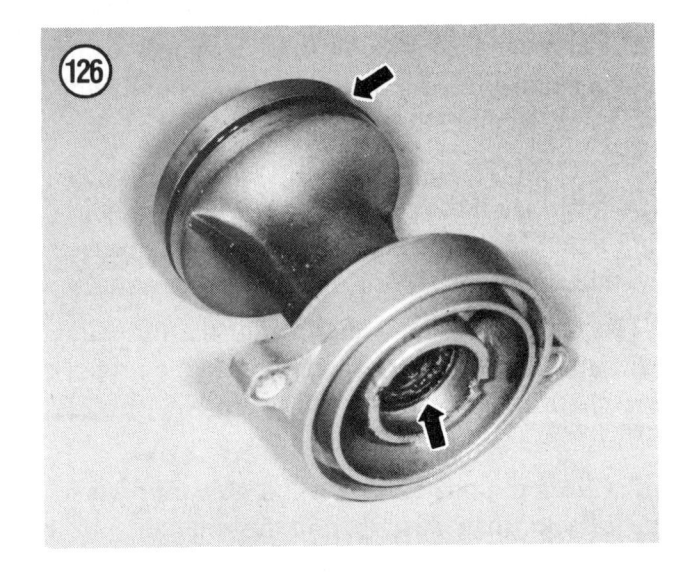

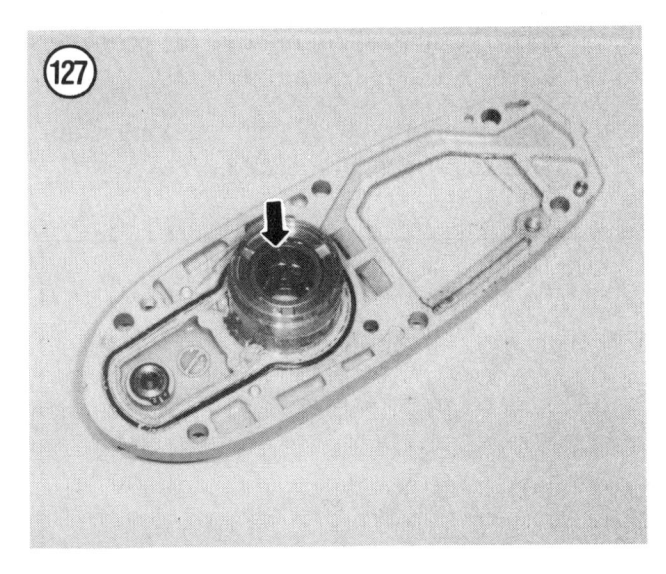

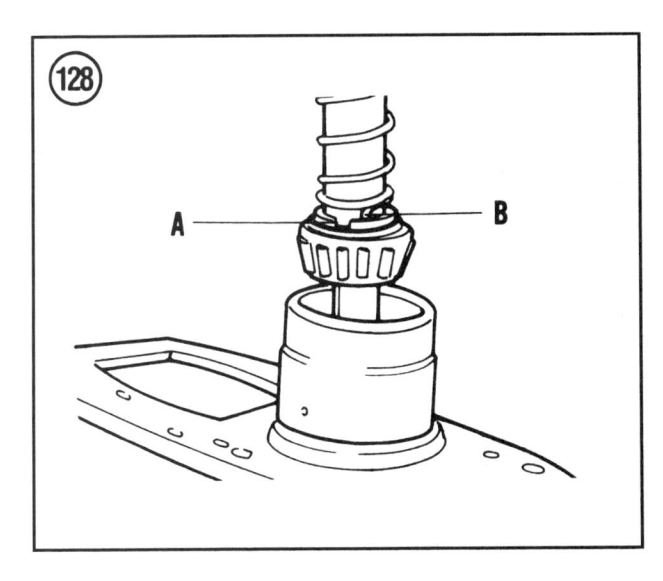

26. Lubricate all parts with Suzuki Outboard Motor Gear Oil.

27. Lightly lubricate the shift rod guide and boot with water-resistant grease. Install the shift rod assembly into the gearcase. Install the drive shaft housing screw which retains the shift rod guide.

28A. On DT 35 and DT 40 models, perform the following:

   a. Install the spring, spring collar and preload spring onto the drive shaft.
   b. Install the retaining pin and push it in all the way.
   c. Install the drive shaft into the gearcase.

28B. On all other models, perform the following:

   a. Make sure the end of the preload spring (B, **Figure 128**) fits into the washer notch provided (A, **Figure 128**).
   b. Install the drive shaft bearing housing snap ring to the preload collar.
   c. Lightly lubricate the bearing housing O-ring with water-resistant grease.
   d. Install the drive shaft and bearing housing assembly into the gearcase.

29. On DT 20, DT 25, DT 30 (2-cylinder) and DT 25 and DT 30 (3-cylinder) models, perform the following:

   a. Install the shift rod assembly.
   b. Install the screw securing the shift rod guide and tighten securely.

30. On DT 35 and DT 40 models, if removed, install the preload spring collar into the gearcase drive shaft bore.

31. Install the pinion gear shim pack into the gearcase drive shaft bore.

32. Install the forward gear and shim pack into the propeller shaft bore.

33. Install the pinion gear into the propeller shaft bore and mesh with the forward gear.

34. Lightly coat mating surfaces of drive shaft bearing housing and gearcase with Silicone Seal (part No. 99000-31120) or equivalent.

35. Holding pinion gear in place, install the drive shaft assembly into the gearcase. Rotate shaft to

align its splines with those of the pinion gear and seat gear on shaft.

36. Install and tighten the drive shaft bearing housing fasteners.

37. Coat the pinion nut threads with Thread Lock 1342. Position nut over drive shaft threads in prop shaft bore. Start nut by hand.

38. Hold pinion gear nut with an appropriate size box-end wrench and pad side of prop shaft bore to prevent damage from contact with the wrench.

39A. On DT 35, DT 40, DT 55 and DT 65 models, attach the drive shaft holder (part No. 09921-29510) on crankshaft end of drive shaft. Rotate drive shaft and tighten the pinion nut to specification (**Table 1**).

39B. On all other models, attach the drive shaft holder (part No. 09921-29610) on crankshaft end of drive shaft. Rotate drive shaft and tighten the pinion nut to specification (**Table 1**).

40. Lubricate the propeller shaft bearing housing O-ring and the shift mechanism pushrod with water-resistant grease.

41A. On DT 20, DT 25, DT 30 (2-cylinder) models, install the propeller shaft/bearing housing assembly into the propeller shaft bore. Use installer part No. 09922-59510.

41B. On all other models, install the propeller shaft/bearing housing assembly into the propeller shaft bore. Use installer part No. 09922-59610.

42. Check pinion gear depth and forward/reverse gear backlash as described in this chapter.

43. When forward and reverse gear backlash has been properly established, remove the propeller shaft/bearing housing assembly:

  a. Apply a light coat of gear marking compound to 5-6 teeth on each gear.

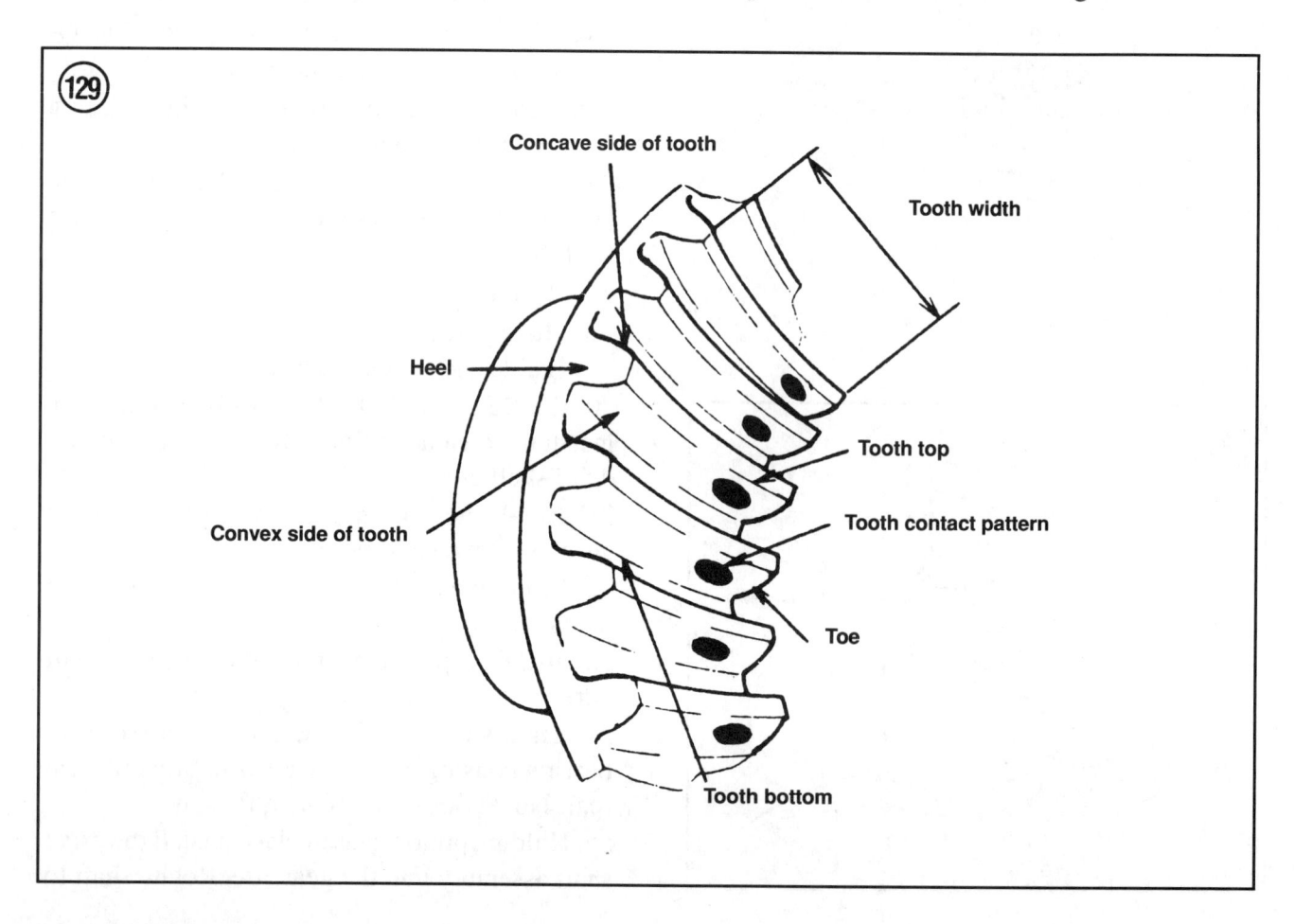

GEARCASE

b. Temporarily reinstall the propeller shaft/bearing housing assembly. Tighten the bolts snugly.
c. Prevent the propeller shaft from moving and rotate the drive shaft about 10 turns.
d. Remove the bearing housing and propeller shaft assembly. Check the tooth contact pattern.
e. If contact pattern does not look like that shown in **Figure 129**, repeat Step 42 as required.

NOTE
*If the end play is out of tolerance, maintain the front thrust washer at the standard thickness. If necessary, add a shim(s) to the **rear** thrust washer only.*

f. Temporarily reinstall the propeller shaft/bearing housing assembly. Install a dial indicator and check the propeller shaft end play by pushing/pulling the shaft. If it is not within 0.008-0.016 in. (0.2-0.4 mm), replace the *rear* thrust washer with a shim(s) as shown in **Figure 130** to bring end play within specifications.

44. When gear tooth contact pattern and propeller shaft end play are correct, coat the gearcase and bearing housing mating surfaces with Silicone Seal and reinstall assembly. Install and tighten the bearing housing bolts to specifications (**Table 1**).
45. Install the water pump as described in this chapter.
46. Install the gearcase as described in this chapter. Fill with recommended type and quantity of lubricant. See Chapter Four.
47. Check gearcase lubricant level after engine has been run. Change the lubricant after 10 hours of operation (break-in period). See Chapter Four.

Gearcase Disassembly/Assembly (DT 75, DT 85, DT 115, DT 140)

Refer to the following illustrations for this procedure:
  a. **Figure 131**: DT 75, DT 85.
  b. **Figure 132**: DT 115, DT 140.
1. Remove the gearcase as described in this chapter.

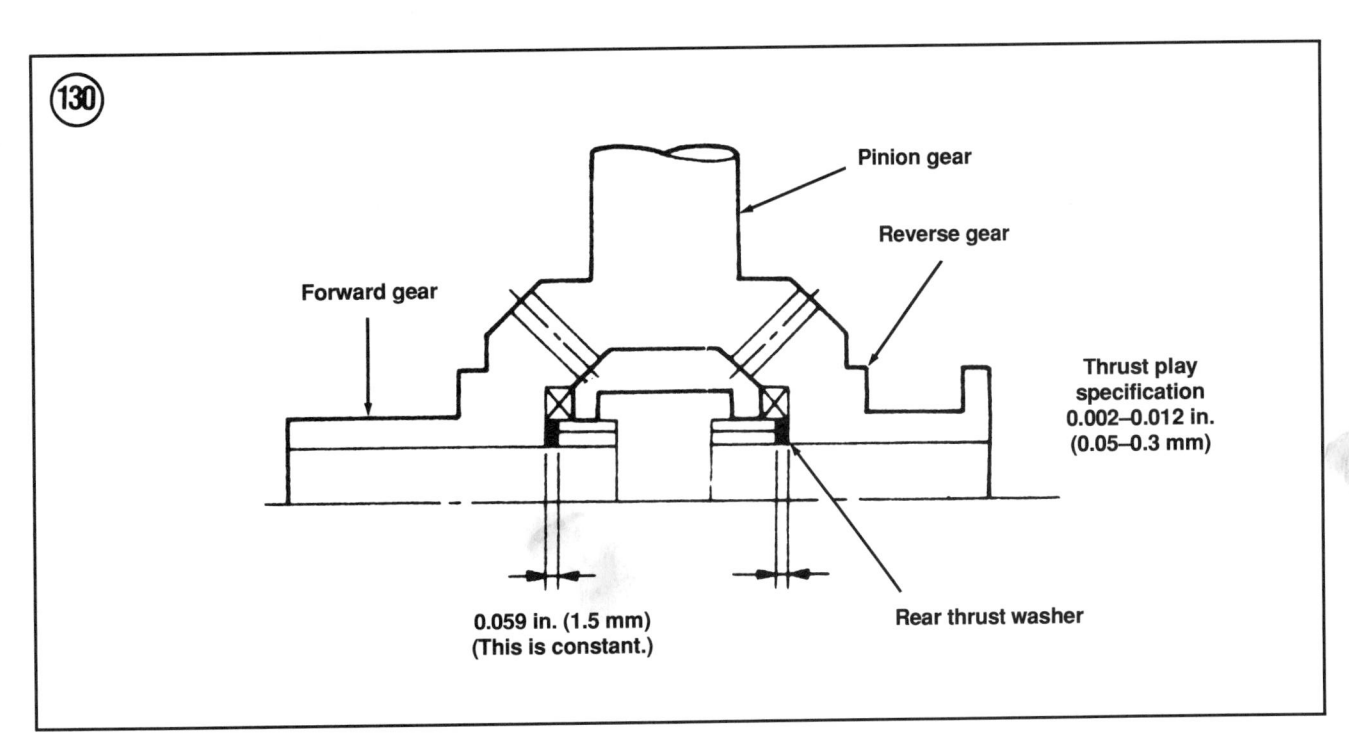

Figure 130

Forward gear — Pinion gear — Reverse gear

Thrust play specification 0.002–0.012 in. (0.05–0.3 mm)

0.059 in. (1.5 mm) (This is constant.) — Rear thrust washer

426

CHAPTER NINE

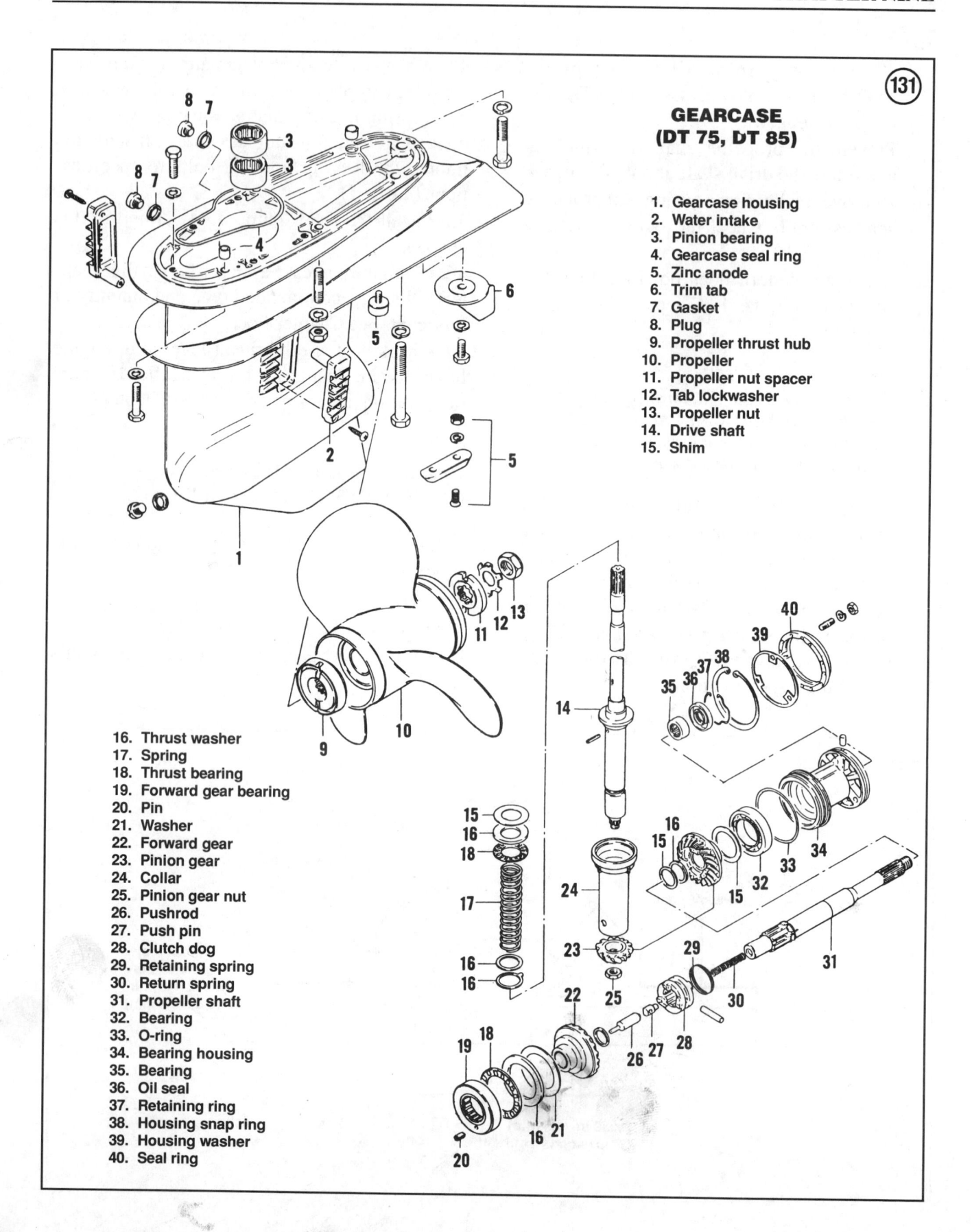

GEARCASE (DT 75, DT 85)

1. Gearcase housing
2. Water intake
3. Pinion bearing
4. Gearcase seal ring
5. Zinc anode
6. Trim tab
7. Gasket
8. Plug
9. Propeller thrust hub
10. Propeller
11. Propeller nut spacer
12. Tab lockwasher
13. Propeller nut
14. Drive shaft
15. Shim
16. Thrust washer
17. Spring
18. Thrust bearing
19. Forward gear bearing
20. Pin
21. Washer
22. Forward gear
23. Pinion gear
24. Collar
25. Pinion gear nut
26. Pushrod
27. Push pin
28. Clutch dog
29. Retaining spring
30. Return spring
31. Propeller shaft
32. Bearing
33. O-ring
34. Bearing housing
35. Bearing
36. Oil seal
37. Retaining ring
38. Housing snap ring
39. Housing washer
40. Seal ring

GEARCASE

GEARCASE (DT 115, DT 140)

1. Gearcase housing
2. Pinion bearing
3. Water intake
4. Trim tab
5. Zinc anode
6. Exhaust seal plate
7. Gasket
8. Plug
9. Propeller thrust hub
10. Propeller nut spacer
11. Tab lockwasher
12. Propeller nut
13. Propeller
14. Drive shaft
15. Drive shaft spring collar 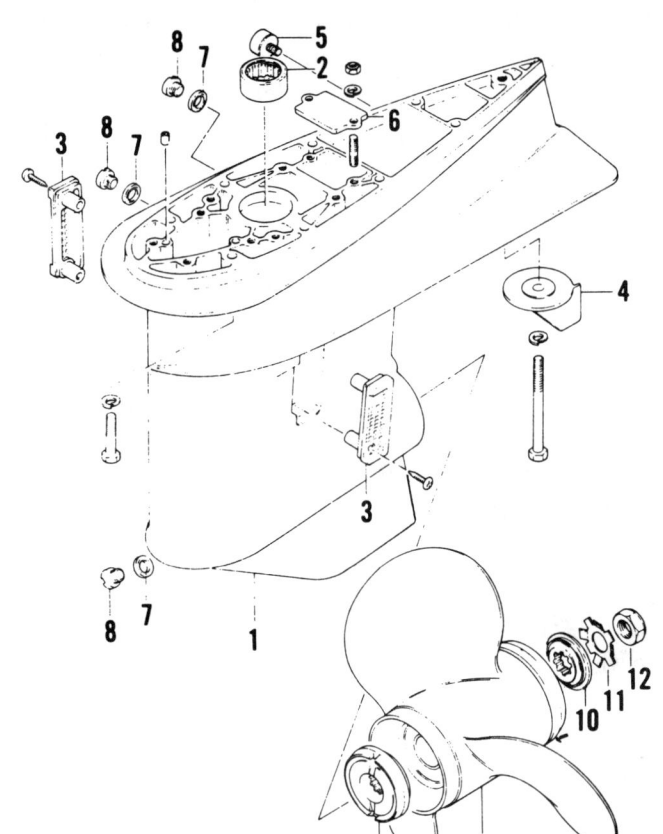 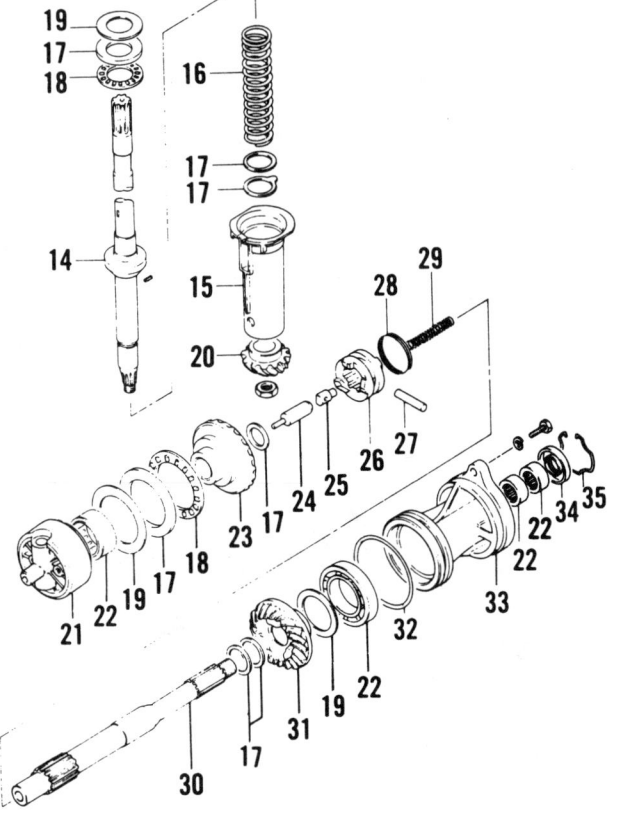
16. Spring
17. Thrust washer
18. Thrust bearing
19. Shim
20. Pinion gear
21. Forward gear bearing housing
22. Bearing
23. Forward gear
24. Pushrod
25. Push pin
26. Clutch dog
27. Retaining pin
28. Retaining spring
29. Return spring
30. Propeller shaft
31. Reverse gear
32. O-ring
33. Bearing housing
34. Oil seal
35. Retaining ring

2. Secure the gearcase in a suitable holding fixture or a vise with protective jaws. If protective jaws are not available, position the gearcase upright with the skeg between wooden blocks in a vise.

3. Remove the water pump as described in this chapter.

4A. On DT 75 and DT 85 models, remove the seal ring and housing washer (**Figure 133**).

4B. On DT 115 and DT 140 models, remove the 2 bolts holding the bearing housing (**Figure 134**).

5. On DT 75 and DT 85 models, remove the snap ring with pliers (**Figure 135**).

6A. On DT 75 and DT 85 models, perform the following:
   a. Install the bearing housing remover (part No. 09930-39410) and special bolts (part No. 09950-59520) as shown in **Figure 136**.
   b. Remove the propeller shaft/bearing housing assembly.

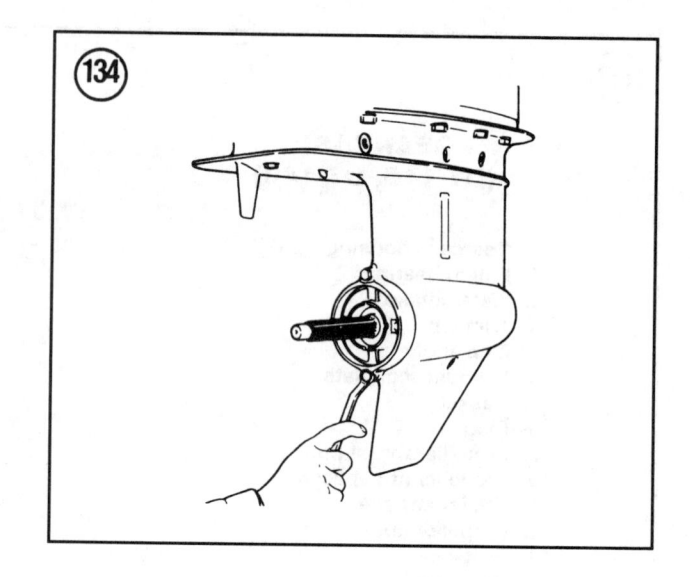

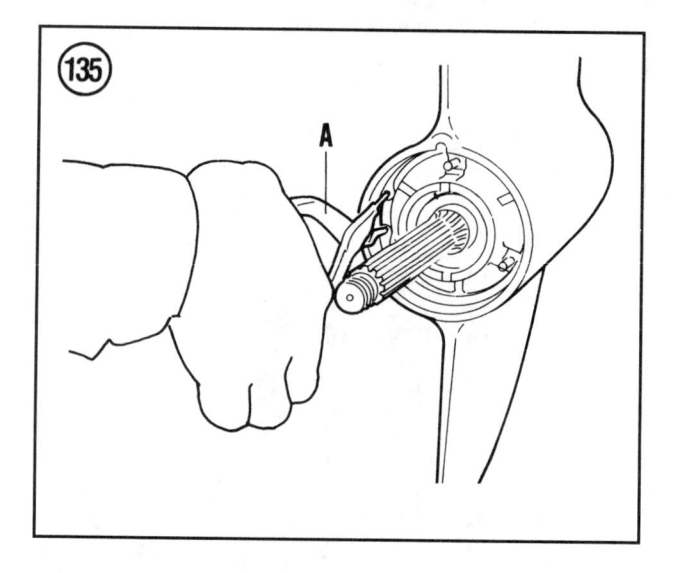

6B. On 1985 DT 115 and DT 140 models, perform the following:
   a. Install the bearing housing remover (part No. 09930-39410) and special bolts (part No. 09930-39430) as shown in **Figure 136**.
   b. Remove the propeller shaft/bearing housing assembly.

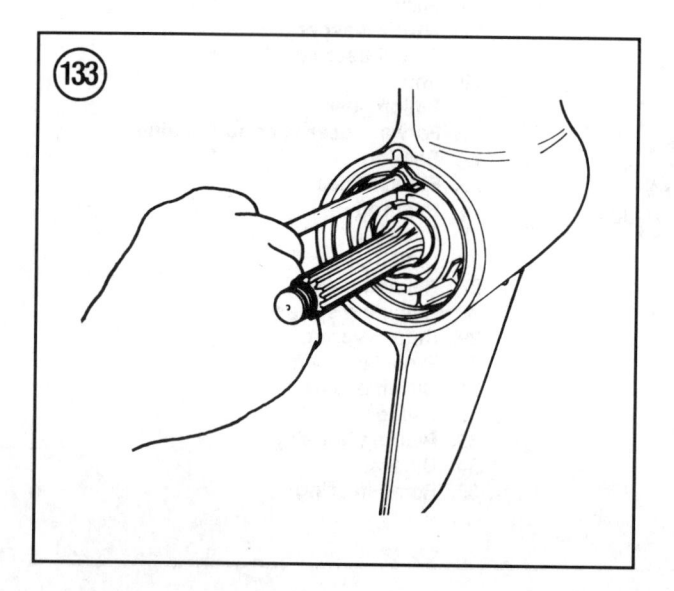

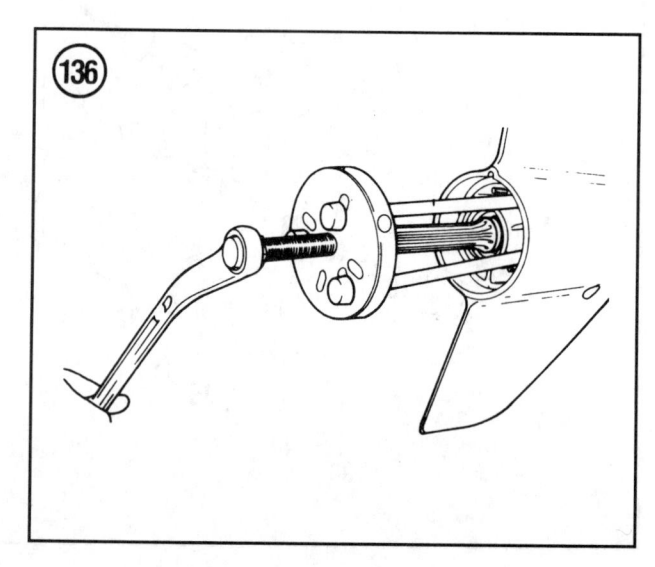

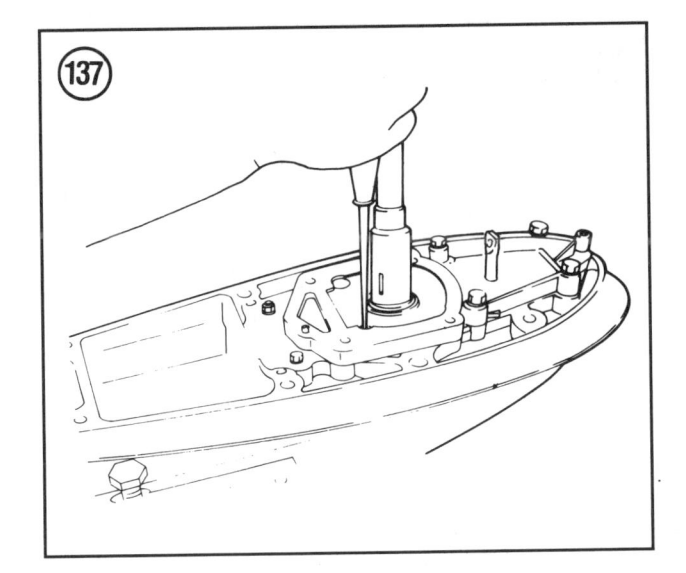

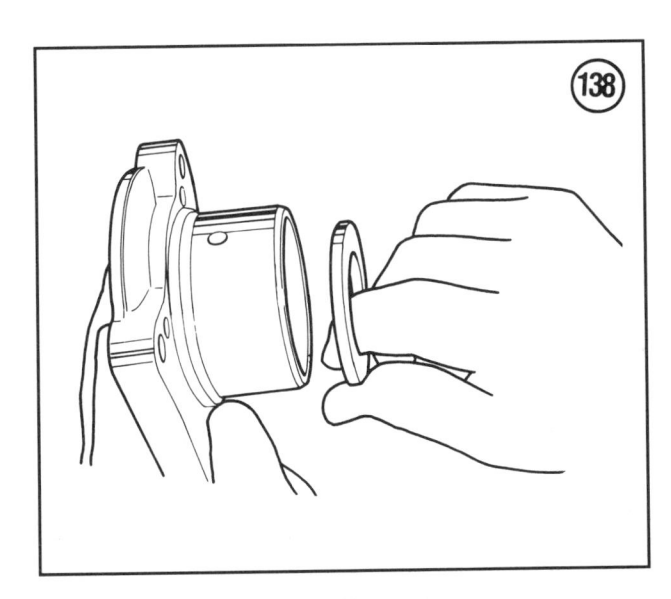

GEARCASE

6C. On 1986-on DT 115, DT 140 models, perform the following:
   a. Install the bearing housing remover (part No. 09930-39411) and special bolts (part No. 09930-39430) as shown in **Figure 136**.
   b. Remove the propeller shaft/bearing housing assembly.

7. On DT 115 and DT 140 models, perform the following:
   a. Remove the drive shaft bearing housing fasteners (**Figure 137**).
   b. Pry the drive shaft bearing housing free and slide it up and off of the drive shaft.
   c. Carefully remove the drive shaft thrust bearing and shim from the bearing housing (**Figure 138**).

8A. On DT 75 and DT 85 models, attach drive shaft holder tool (part No. 09950-79510) to the top of the drive shaft.

8B. On DT 115 and DT 140 models, attach drive shaft holder tool (part No. 09921-29410) to the top of the drive shaft.

9. Fit an appropriate size box-end wrench over the pinion nut and pad the sides of the prop shaft bore to prevent distortion or damage from contact with the wrench.

10. Holding pinion nut with the wrench installed in Step 9, turn the drive shaft *counterclockwise* to loosen the pinion nut (**Figure 118**).

11. Remove the tools. Remove the pinion nut and pinion gear (**Figure 139**).

12. On DT 75 and DT 85 models, remove the drive shaft bearing housing and the shift rod.

13. On DT 115 and DT 140 models, remove the forward gear, thrust washer (**Figure 140**) and shim(s).

14. Remove the drive shaft, preload spring and thrust bearing as an assembly (**Figure 141**).

15. Remove the drive shaft spring collar from the gearcase bore (**Figure 142**).

16. Remove the 2 drive shaft thrust washers from the gearcase bore (**Figure 143**).

17. On DT 75 and DT 85 models, remove the forward gear and thrust washer (**Figure 144**).

18. On DT 115 and DT 140 models, remove the clutch rod and shift rod guide from the gearcase (**Figure 145**).

19. Slide all components except the clutch dog from the propeller shaft.

20. Disassemble the propeller shaft clutch assembly as described in this chapter.

21. Clean and inspect all parts as described in this chapter.

22. On DT 115 and DT 140 models, if inspection of the forward gear bearing indicates replacement is necessary:

   a. Install bearing housing remover (part No. 09930-39410) for 1985 models or (part No. 09930-39411) for 1986-on models using the 2 long bolts from tool (part No. 09930-39430).

   b. Remove forward gear bearing housing with a slide hammer (part No. 09930-30102) (**Figure 146**).

NOTE
Suzuki does not have or recommend a special tool for bearing installation.

   c. Install a new bearing with an appropriate installer.

23. Remove and discard the propeller shaft bearing housing oil seals and bearing (**Figure 126**, typical).

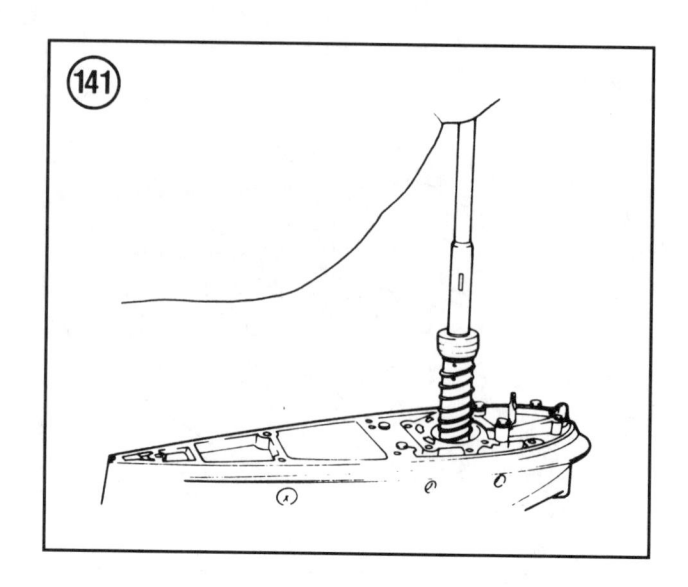

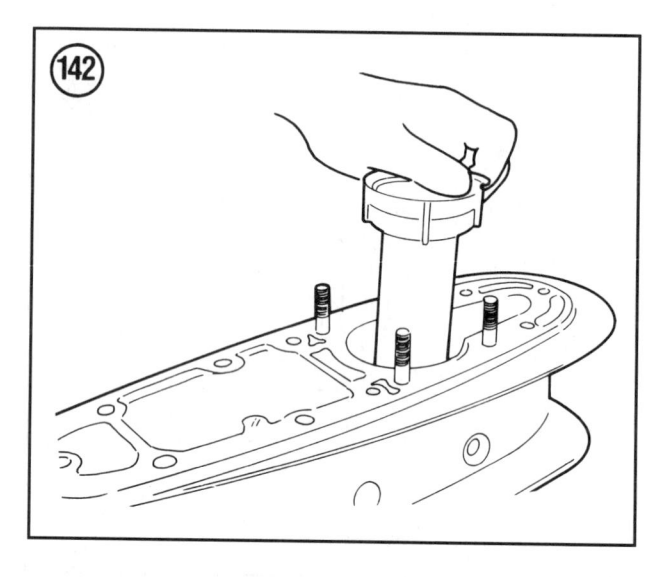

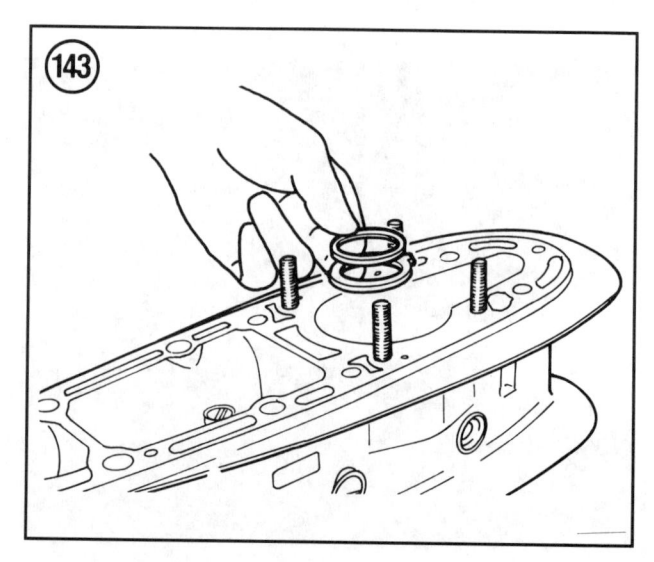

GEARCASE

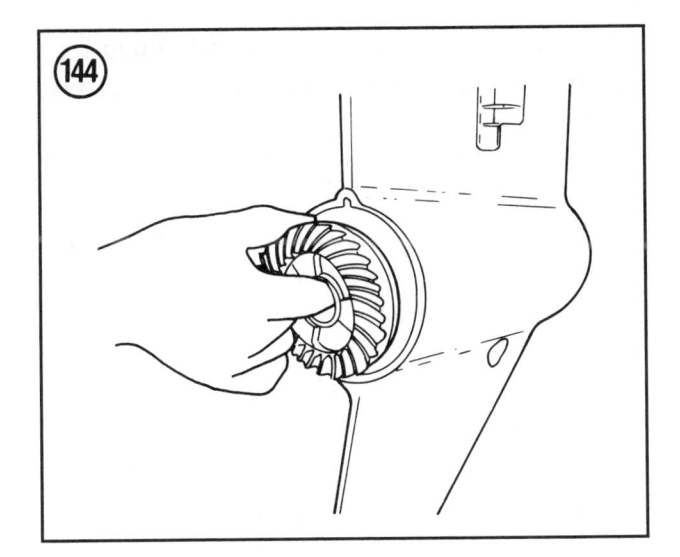

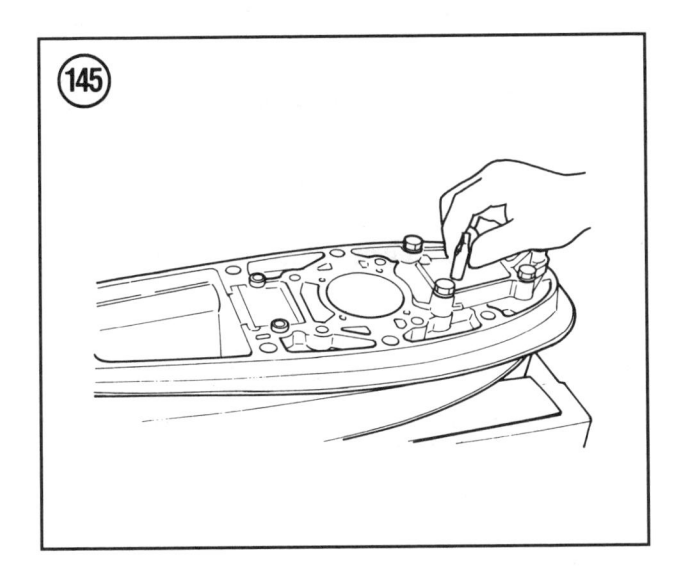

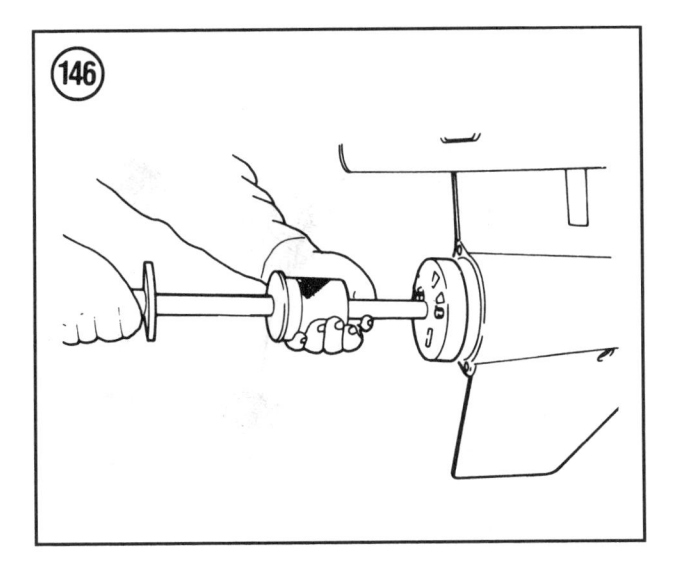

24. Install a new propeller shaft housing bearing with an appropriate installer.
25. Install new propeller shaft housing oil seals and a new housing O-ring. Coat seal lips and O-ring with water-resistant grease (part No. 99000-25160) or equivalent.
26. Remove and discard the drive shaft bearing housing oil seal (**Figure 127**, typical). Install a new seal with an appropriate installer. Coat seal lips with water-resistant grease.
27. Assemble the propeller shaft components according to **Figure 131** or **Figure 132**.
28. Lubricate all parts with Suzuki Outboard Motor Gear Oil.
29. On DT 115 and DT 140 models, install the clutch rod and shift rod guide into the gearcase.
30. Install the 2 drive shaft thrust washers into the gearcase bore.
31. Install the drive shaft spring collar into the gearcase bore.
32. Install the drive shaft, preload spring and thrust bearing as an assembly (**Figure 141**).
33. Install the shift rod assembly.
34. Install the shim, thrust washer and thrust bearing on the forward gear in that order. Install the forward gear assembly into the gearcase.
35. Install the pinion gear into the propeller shaft bore and mesh with the forward gear.
36. Holding the pinion gear in place, install the drive shaft assembly into the gearcase. Rotate shaft to align its splines with those of the pinion gear and seat gear on shaft.
37. Coat the pinion nut threads with Thread Lock 1342. Position nut over drive shaft threads in prop shaft bore. Start nut by hand.
38. Hold pinion gear nut with an appropriate size box-end wrench and pad side of prop shaft bore to prevent damage from contact with the wrench.
39A. On DT 75 and DT 85 models, attach drive shaft holder tool (part No. 09950-79510) to the top of the drive shaft.
39B. On DT 115 and DT 140 models, attach drive shaft holder tool (part No. 09921-29410) to the top of the drive shaft. Rotate drive shaft

and tighten the pinion gear nut to specification (**Table 1**).

40. Check forward gear backlash as described in this chapter.

41. Temporarily install the propeller shaft/bearing housing assembly. Check pinion gear depth, reverse gear backlash and gear tooth contact pattern as described in this chapter.

42. Coat gearcase and drive shaft bearing housing mating surfaces with Silicone Seal (part No. 99000-31120) or equivalent and install the housing. Tighten fasteners to specification (**Table 1**).

43. Remove the propeller shaft/bearing housing assembly. Check clutch pushrod dimension as described in this chapter.

44. Lubricate the propeller shaft bearing housing O-ring and the shift mechanism pushrod with water-resistant grease.

45. Install the propeller shaft/bearing housing assembly into the propeller shaft bore. Use shaft housing installer (part No. 09922-59410) and installer handle (part No. 09922-59420).

46. On DT 115 and DT 140 models, coat the bearing housing bolts with Thread Lock 1342 and tighten securely.

47. Install the water pump as described in this chapter.

48. Install the gearcase as described in this chapter. Fill with recommended type and quantity of lubricant. See Chapter Four.

49. Check gearcase lubricant level after engine has been run. Change the lubricant after 10 hours of operation (break-in period). See Chapter Four.

Gearcase Disassembly/Assembly (V4 and V6)

Refer to the following illustrations for this procedure:

a. **Figure 147**: V4.
b. **Figure 148**: V6.

1. Remove the gearcase as described in this chapter.

2. Secure the gearcase in a suitable holding fixture or a vise with protective jaws. If protective jaws are not available, position the gearcase upright with the skeg between wooden blocks in a vise.

3. Straighten the tabs on the lockwasher (**Figure 149**). Remove the key from the groove in the gearcase and lockwasher.

4. Install the propeller shaft stopper remover/installer (part No. 09951-18710) as shown in **Figure 150**. Remove the stopper and lockwasher.

5. Move the clutch rod and shift into neutral.

6. Remove the bolts holding the shift unit (**Figure 151**) and remove the shift unit.

7. Install the bearing housing remover (part No. 09930-39411) and housing remover arms (part No. 09950-58710) onto the housing as shown in **Figure 152**.

8. Remove the propeller shaft/bearing housing assembly.

9. Remove the water pump as described in this chapter.

10. Attach the drive shaft holder tool (part No. 09921-29410) to the top of the drive shaft.

11. Fit an appropriate size box-end wrench over the pinion nut and pad the sides of the prop shaft bore to prevent distortion or damage from contact with the wrench.

12. Holding pinion nut with the wrench installed in Step 11, loosen the pinion nut by turning the drive shaft.

13. Remove the tools. Remove the pinion nut and pinion gear (**Figure 139**).

14. Remove the forward gear, thrust washer (**Figure 140**) and shim(s).

15. Remove the bolts holding the drive shaft bearing housing (**Figure 153**).

16. Carefully remove the drive shaft bearing housing and the shift rod.

17. Remove the drive shaft thrust washers (A, **Figure 154**), protector (B, **Figure 154**) and the drive shaft spring collar (C, **Figure 154**) from the gearcase bore.

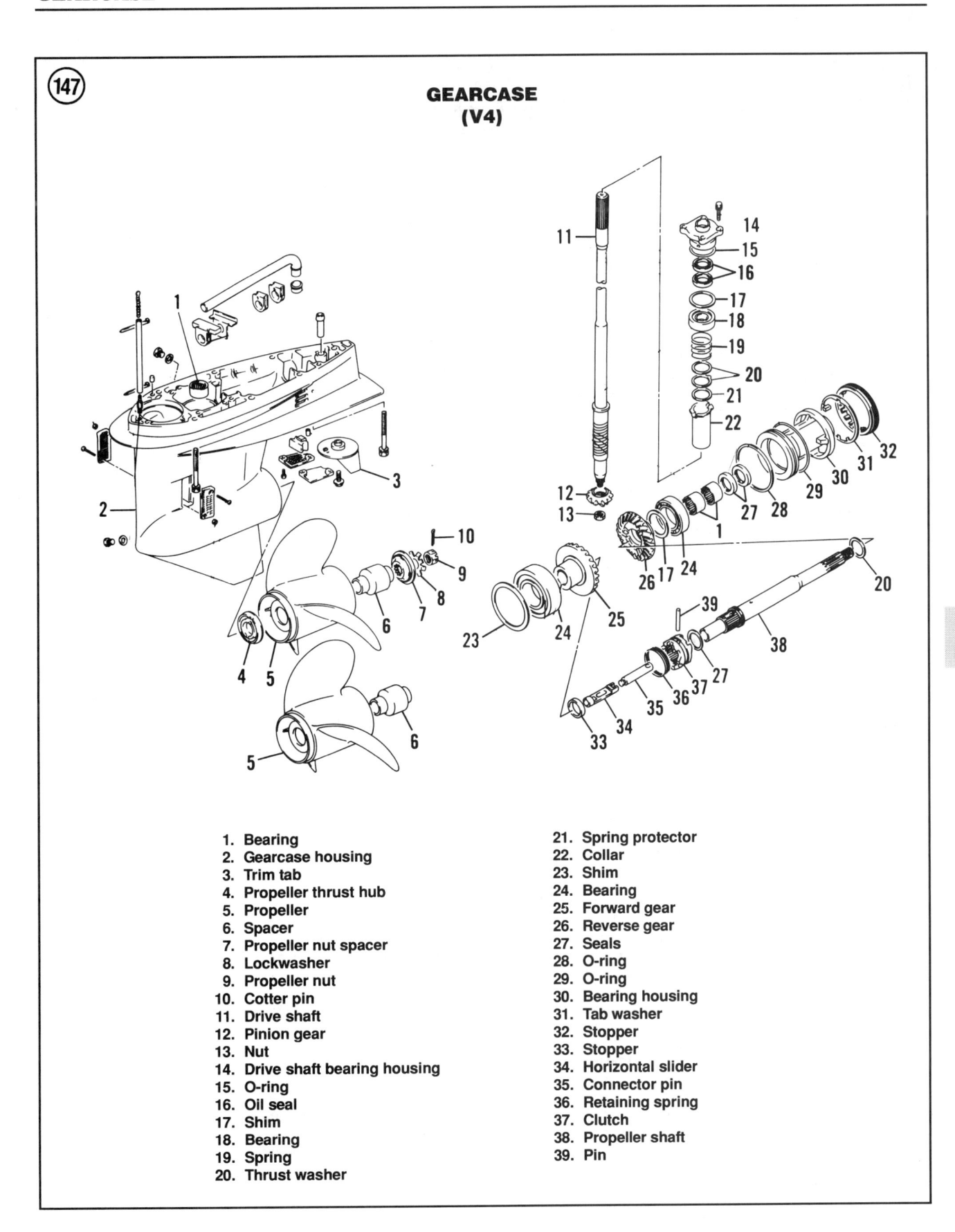

GEARCASE

GEARCASE (V4)

1. Bearing
2. Gearcase housing
3. Trim tab
4. Propeller thrust hub
5. Propeller
6. Spacer
7. Propeller nut spacer
8. Lockwasher
9. Propeller nut
10. Cotter pin
11. Drive shaft
12. Pinion gear
13. Nut
14. Drive shaft bearing housing
15. O-ring
16. Oil seal
17. Shim
18. Bearing
19. Spring
20. Thrust washer
21. Spring protector
22. Collar
23. Shim
24. Bearing
25. Forward gear
26. Reverse gear
27. Seals
28. O-ring
29. O-ring
30. Bearing housing
31. Tab washer
32. Stopper
33. Stopper
34. Horizontal slider
35. Connector pin
36. Retaining spring
37. Clutch
38. Propeller shaft
39. Pin

CHAPTER NINE

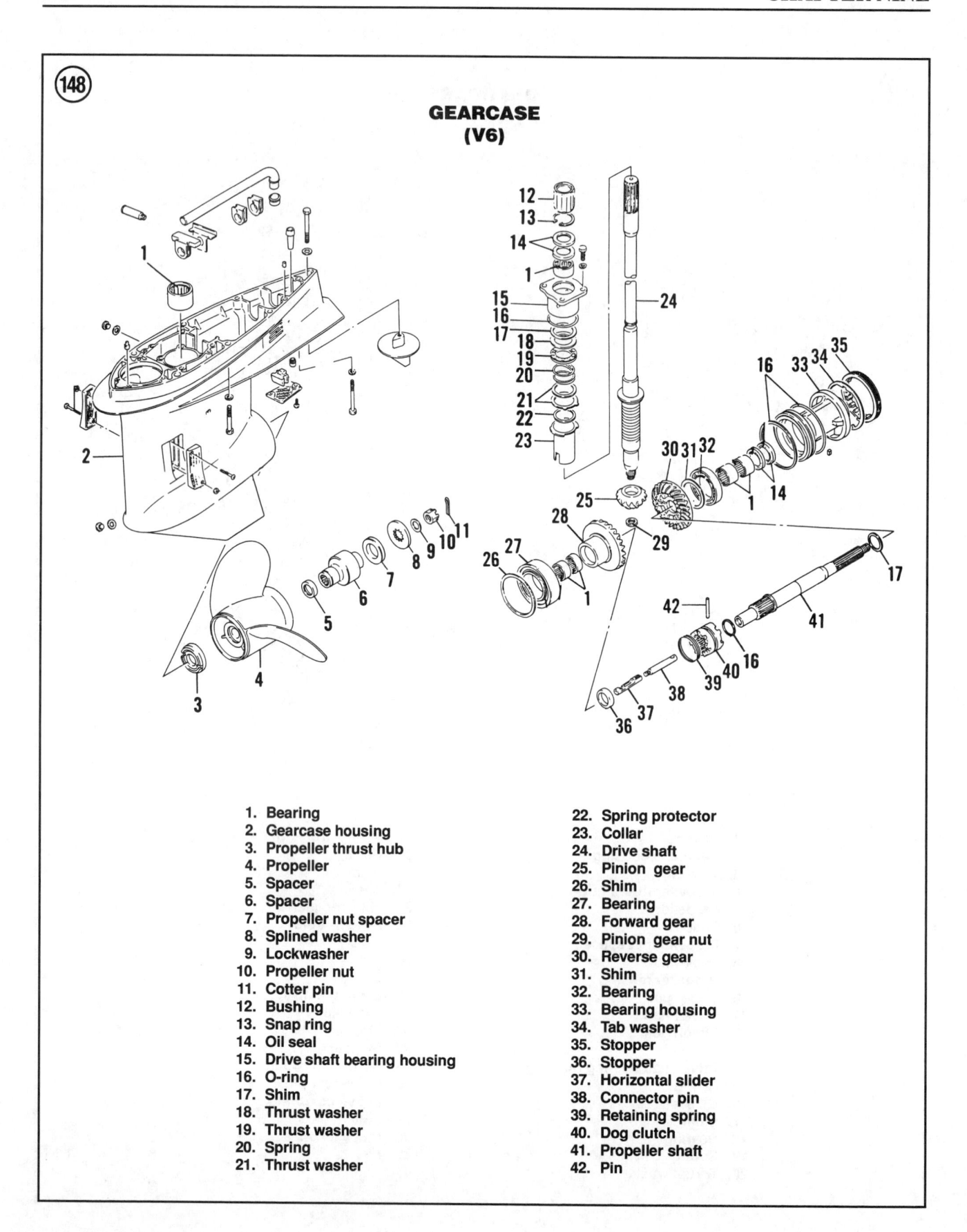

148 GEARCASE (V6)

1. Bearing
2. Gearcase housing
3. Propeller thrust hub
4. Propeller
5. Spacer
6. Spacer
7. Propeller nut spacer
8. Splined washer
9. Lockwasher
10. Propeller nut
11. Cotter pin
12. Bushing
13. Snap ring
14. Oil seal
15. Drive shaft bearing housing
16. O-ring
17. Shim
18. Thrust washer
19. Thrust washer
20. Spring
21. Thrust washer
22. Spring protector
23. Collar
24. Drive shaft
25. Pinion gear
26. Shim
27. Bearing
28. Forward gear
29. Pinion gear nut
30. Reverse gear
31. Shim
32. Bearing
33. Bearing housing
34. Tab washer
35. Stopper
36. Stopper
37. Horizontal slider
38. Connector pin
39. Retaining spring
40. Dog clutch
41. Propeller shaft
42. Pin

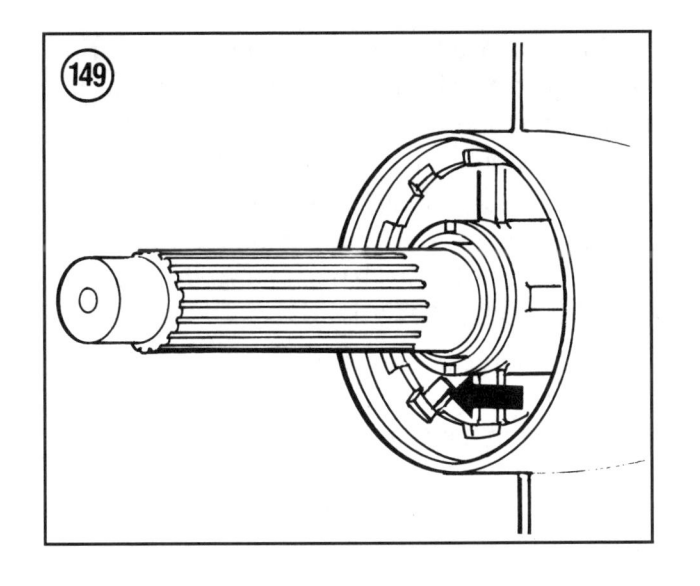

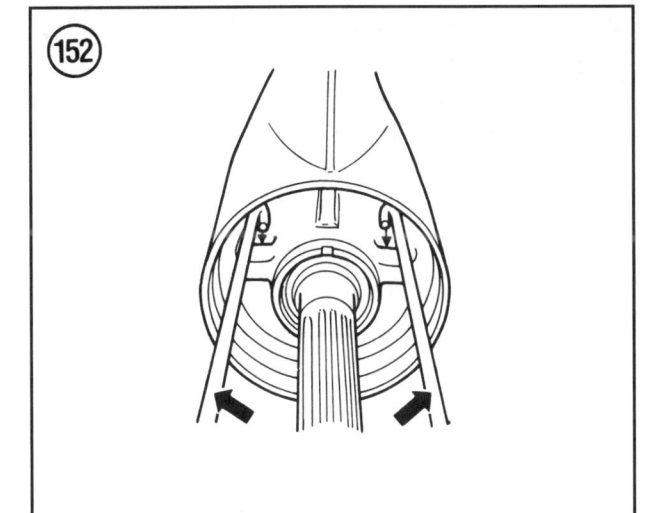

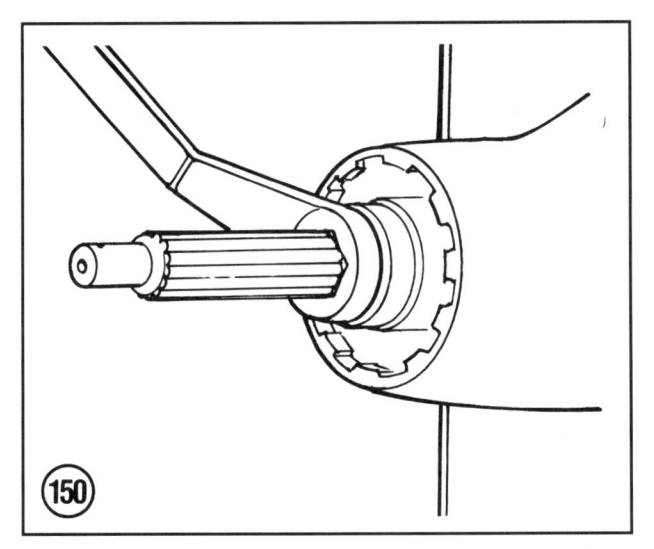

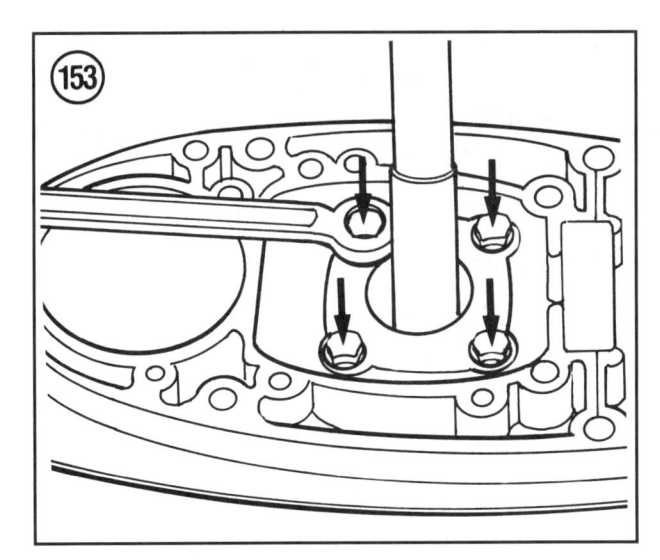

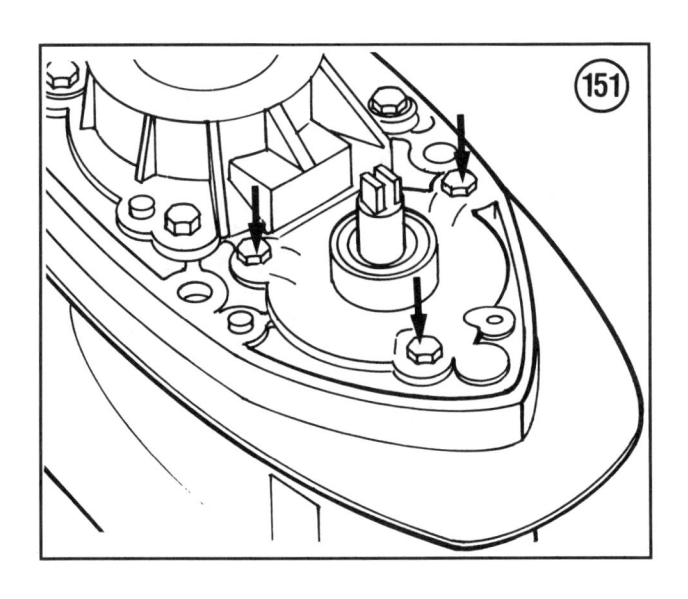

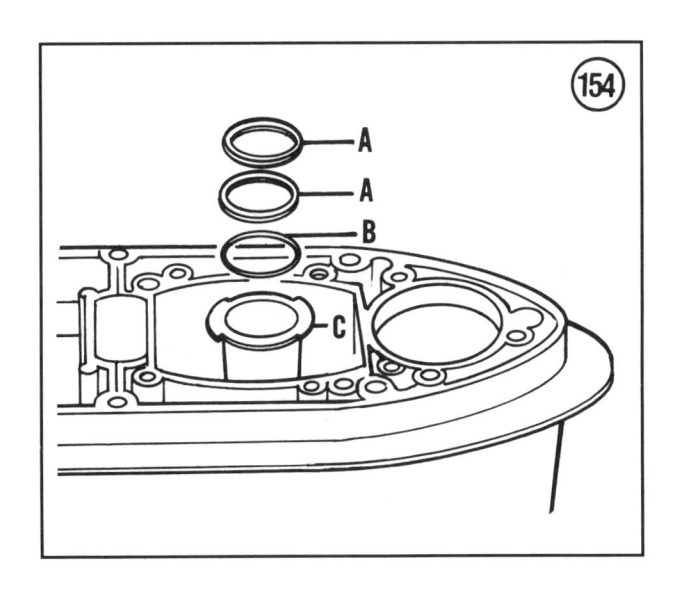

GEARCASE

435

18. Slide all components except the clutch dog from the propeller shaft.

19. Disassemble the propeller shaft clutch assembly as described in this chapter.

20. Clean and inspect all parts as described in this chapter.

21. Inspect the pinion gear adjustment of the drive shaft as described in this chapter.

22. Remove and discard the propeller shaft/bearing housing oil seal (A, **Figure 155**) and O-ring (B, **Figure 155**).

23. Inspect the propeller shaft/bearing housing bearings (**Figure 156**). Replace if necessary.

24. Install new propeller shaft/bearing housing oil seal and a new housing O-ring. Coat seal lips and O-ring with water-resistant grease (part No. 99000-25160) or equivalent.

25. Assemble the propeller shaft components according to **Figure 147** or **Figure 148**, noting the following.

26. Lubricate all parts with Suzuki Outboard Motor Gear Oil.

27. Align the groove in the drive shaft spring protector with the tongue in the gearcase and install drive shaft spring protector into the gearcase.

28. Install the protector and the drive shaft thrust washers (**Figure 154**).

29. Carefully install the drive shaft and the drive shaft bearing housing.

30. Install the bolts holding the drive shaft bearing housing (**Figure 153**). Tighten the bolts securely.

31. Install the forward gear, thrust washer (**Figure 140**) and shim(s).

32. Install the pinion gear into the propeller shaft bore and mesh with the forward gear.

33. Coat the pinion nut threads with Thread Lock 1342. Position nut over drive shaft threads in prop shaft bore. Start nut by hand.

34. Attach drive shaft holder tool (part No. 09921-29410) to the top of the drive shaft. Rotate drive shaft and tighten the pinion gear nut to specification (**Table 1**).

35. Prior to installing the propeller shaft/bearing housing assembly into the gearcase, rotate the propeller shaft so that the flat surface on the end of the horizontal slider is facing UP.

36. Install the propeller shaft/bearing housing into the gearcase. Use shaft housing installer (part No. 09922-59410) and installer handle (part No. 09922-59420).

37. Make sure the groove in the gearcase and the bearing housing are aligned. Install the key into the groove in the gearcase and bearing housing.

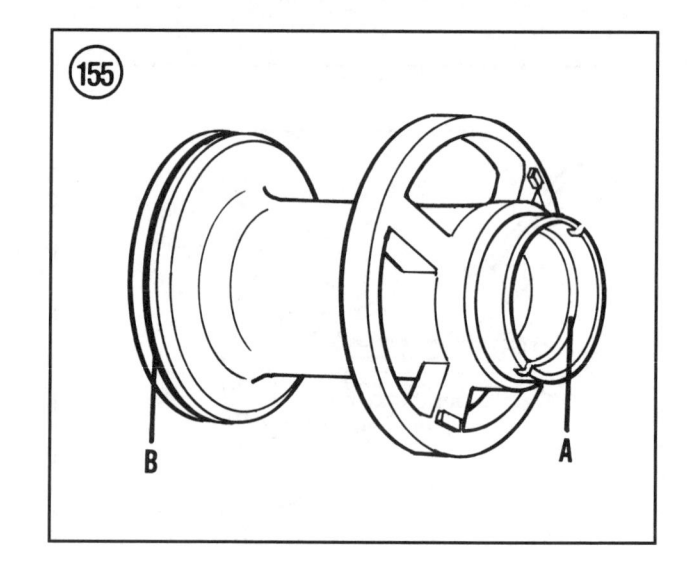

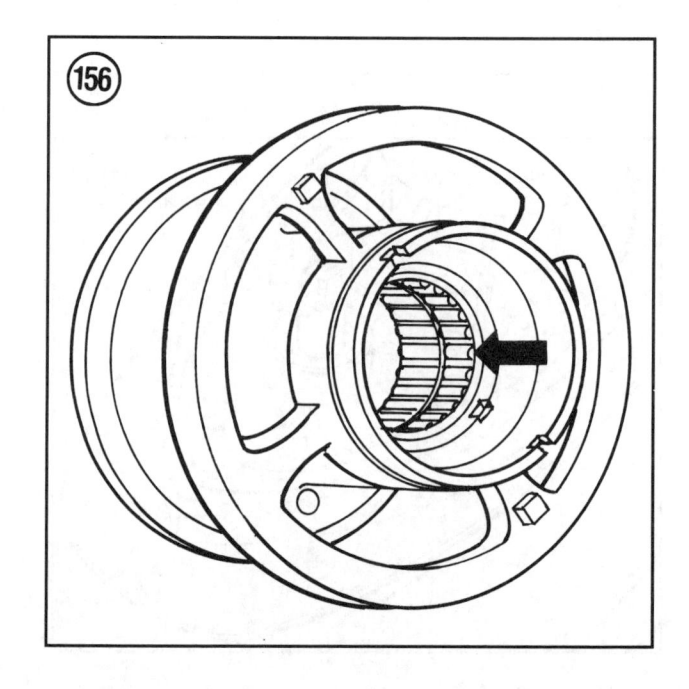

GEARCASE

38. Install the lockwasher and align the groove in the lockwasher with the tongue in the gearcase.

39. Check forward and reverse gear backlash as described in this chapter.

40. Apply Suzuki Bond 1207B, or equivalent, to the threads of the stopper.

41. Position the stopper with the "OFF" mark facing toward the outside and install the stopper.

42. Use the same special tool used to removed the stopper and tighten the stopper to specification (**Table 1**). If necessary, slightly tighten the stopper until it is aligned with the lockwasher as shown in **Figure 157**. Bend down the lockwasher tab to lock the stopper in place.

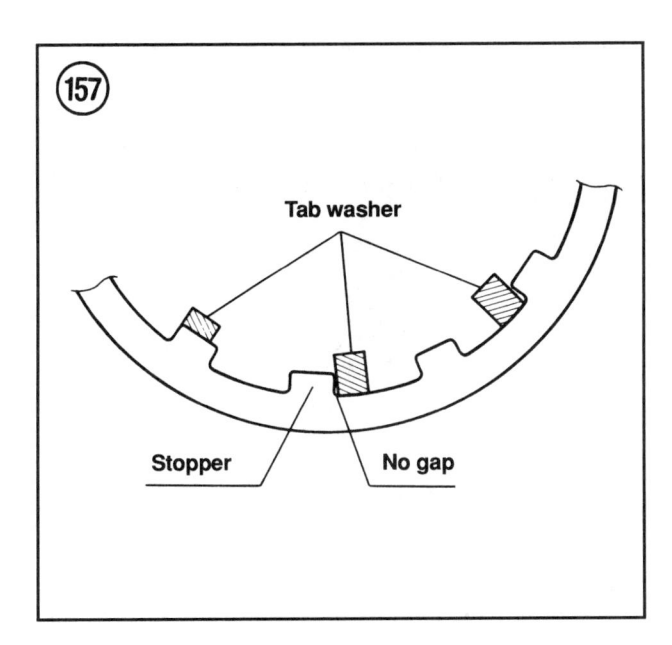

43. Apply water-resistant grease to the O-ring seal on the shift unit. Make sure the shift rod is in NEUTRAL.

44. Make sure the locating dowel is in place in the gearcase and install the shift unit and bolts. Tighten the bolts securely.

45. Install the water pump as described in this chapter.

46. Install the gearcase as described in this chapter. Fill with recommended type and quantity of lubricant. See Chapter Four.

47. Check gearcase lubricant level after engine has been run. Change the lubricant after 10 hours of operation (break-in period). See Chapter Four.

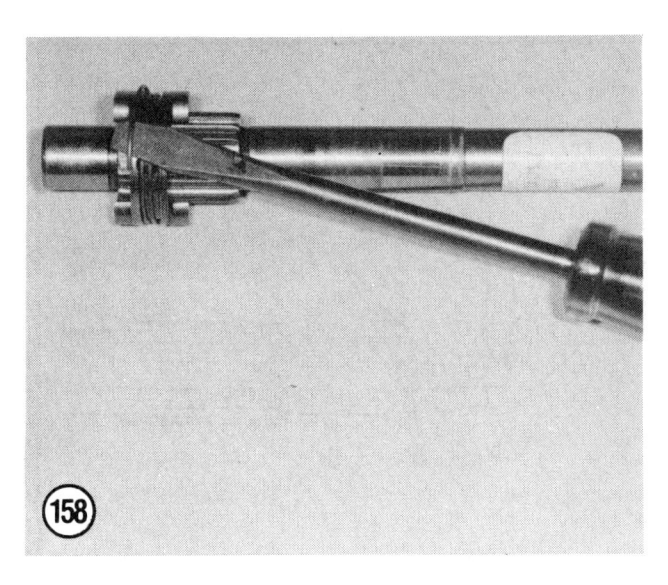

PROPELLER SHAFT CLUTCH

Disassembly/Assembly
(DT 4, DT 6; 1985-1987 DT 8)

NOTE
The propeller shaft on DT 2 models cannot be disassembled. If any component of the shaft is defective, replace the shaft assembly.

NOTE
While removing the clutch pushrod, note its direction within the propeller shaft. Some pushrods are directional (one end flat, other end pointed) while others are symmetrical (either flat or pointed on both ends). The push rod must be reinstalled in the propeller shaft in the same direction as removed.

1. On models so equipped, remove the clutch push rod from the front of the propeller shaft.

2. On DT 4 models, remove the retaining spring from the dog clutch (**Figure 158**). This will allow access to the dog clutch retaining pin.

3. Insert an appropriate size punch and drive the retaining pin from the dog clutch (**Figure 159**).

4. On DT 6 and DT 8 models, insert a pencil or similar tool into the end of the propeller shaft and compress the return spring, then remove the clutch dog pin with a punch or awl (**Figure 160**).
5. Slide the clutch dog off the propeller shaft and remove the push pin and return spring.
6. Clean and inspect all parts as described in this chapter.
7. Lubricate all parts with Suzuki Outboard Motor Gear Oil.
8. Insert the push pin (A, **Figure 161**) and the return spring (B, **Figure 161**) into the propeller shaft bore (C, **Figure 161**).
9. Install the clutch dog onto the propeller shaft with the stamped letter "F" facing the front of the shaft (**Figure 162**).
10. Align the push pin hole with the propeller shaft slot (**Figure 163**).
11. Temporarily install the clutch pushrod and compress the return spring. Align the clutch dog pin hole with the push pin hole, then insert the retaining pin through the clutch dog and push pin holes.
12. On DT 6 and DT 8 models, center the retaining pin in clutch dog and stake each side of clutch dog pin hole to retain pin.
13. On DT 4 models, install the pin retaining spring around the dog clutch.
14. Remove the clutch pushrod and apply a liberal coat of water-resistant grease to it.
15. Reinstall the pushrod into propeller shaft bore in the same direction as noted during removal in Step 1.

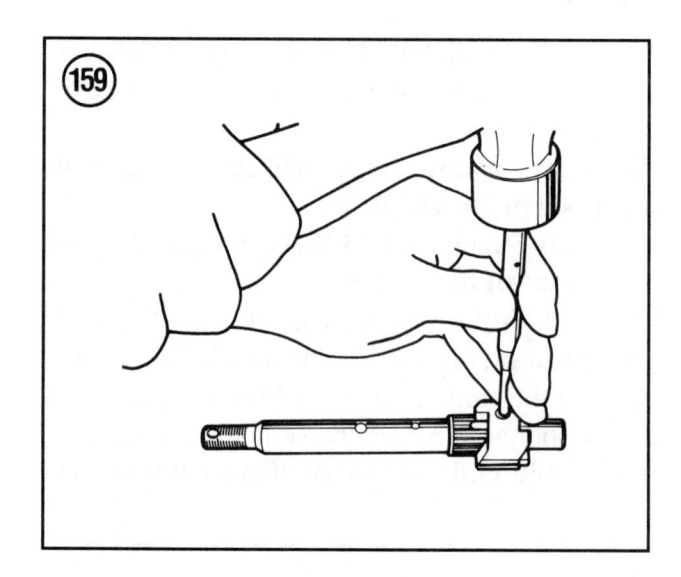

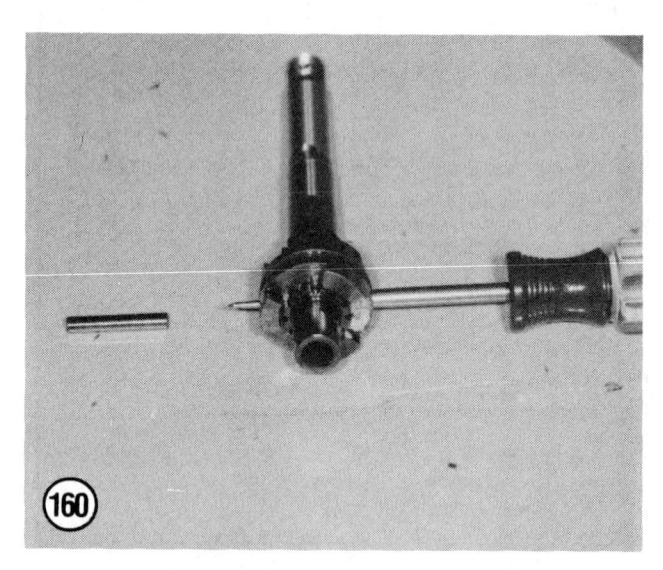

Disassembly/Assembly (1988-on DT 8; DT 9.9 through DT 85, DT 115, DT 140)

Refer to **Figure 164** for this procedure.

NOTE
While removing the clutch pushrod, note its direction within the propeller shaft. Some pushrods are directional (one end flat, other end pointed) while others are

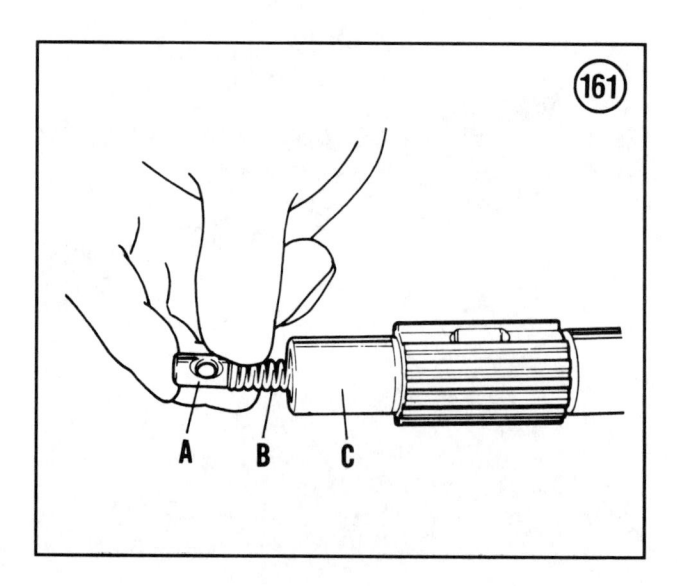

GEARCASE

439

⑯②

⑯③

Push pin hole

Propeller shaft slot

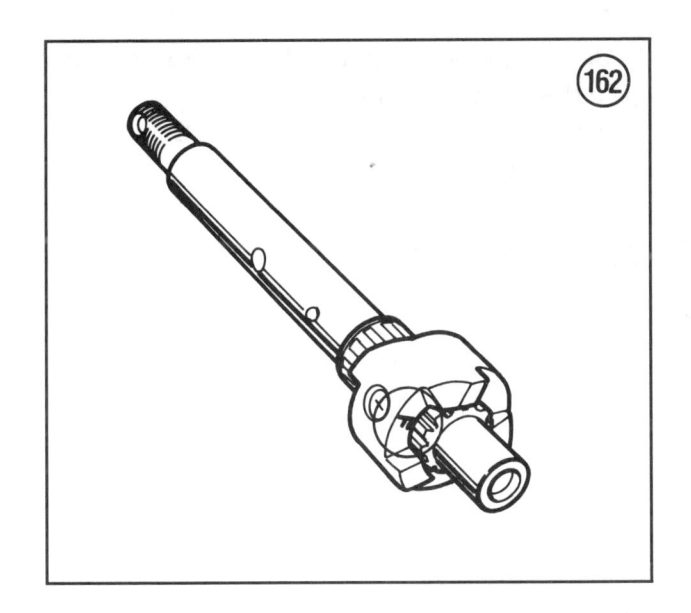

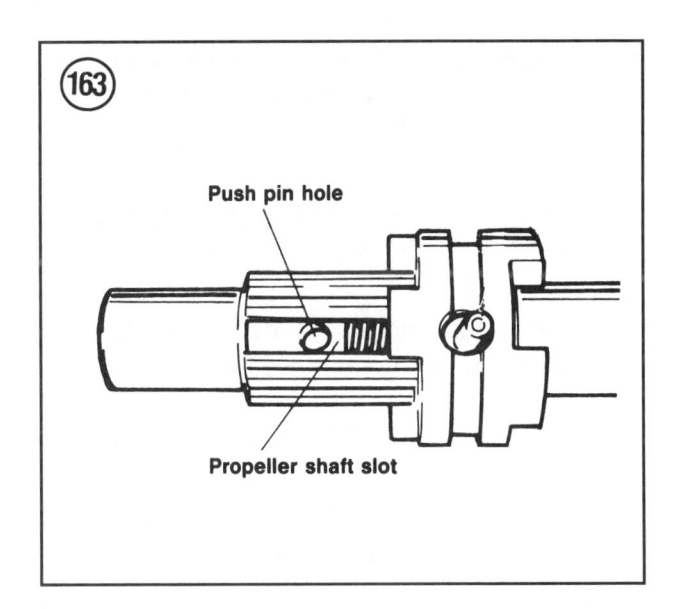

⑯④

**PROPELLER SHAFT
(1988-ON DT 8; DT 9.9 THROUGH DT 85; DT 115 AND DT 140)**

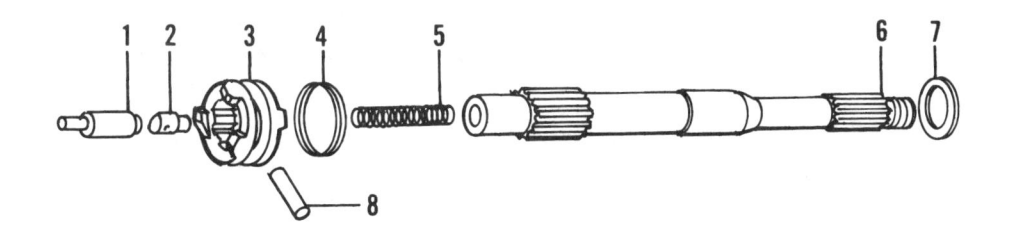

1. Clutch pushrod
2. Push pin
3. Dog clutch
4. Retaining spring
5. Return spring
6. Propeller shaft
7. Thrust washer
8. Retaining pin

9

symmetrical (either flat or pointed on both ends). The push rod must be reinstalled in the propeller shaft in the same direction as removed.

1. On models so equipped, remove the clutch push rod (A, **Figure 165**) from the front of the propeller shaft.
2. Remove the retaining spring from the dog clutch. This will allow access to the dog clutch retaining pin.
3. Insert an appropriate size punch and drive the retaining pin from the clutch dog (**Figure 166**).
4. Slide the clutch dog off the propeller shaft and remove the push pin and return spring.
5. Clean and inspect all parts as described in this chapter.
6. Lubricate all parts with Suzuki Outboard Motor Gear Oil.
7. Insert the push pin (A, **Figure 161**) and the return spring (B, **Figure 161**) into the propeller shaft bore (C, **Figure 161**).
8. Install the clutch dog onto the propeller shaft with the stamped letter "F" facing the front of the shaft (**Figure 167**).
9. Align the push pin hole with the propeller shaft slot (**Figure 163**).
10. Temporarily install the clutch pushrod and compress the return spring. Align the clutch dog pin hole with the push pin hole, then insert the retaining pin through the clutch dog and push pin holes.
11. Center the retaining pin in the dog clutch and install the retaining spring around the dog clutch.
12. Remove the clutch pushrod and apply a liberal coat of water-resistant grease to it.
13. Reinstall the pushrod into propeller shaft bore in the same direction as noted during removal in Step 1.

Disassembly/Assembly (V4 and V6)

Refer to **Figure 168** for this procedure.

1. Detach the horizontal slider (A, **Figure 169**) from the connector pin.
2. Slide the stopper (B, **Figure 169**) off the propeller shaft.

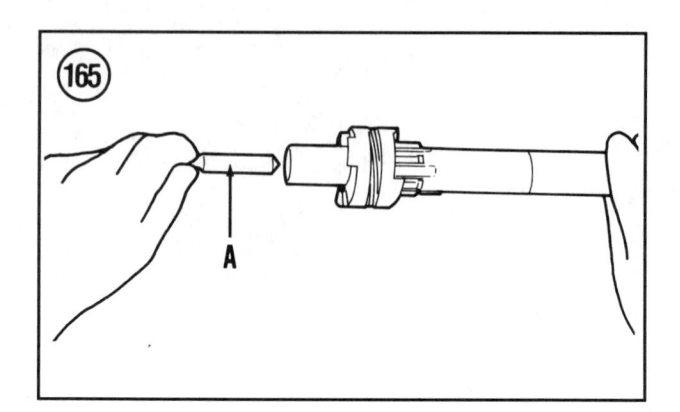

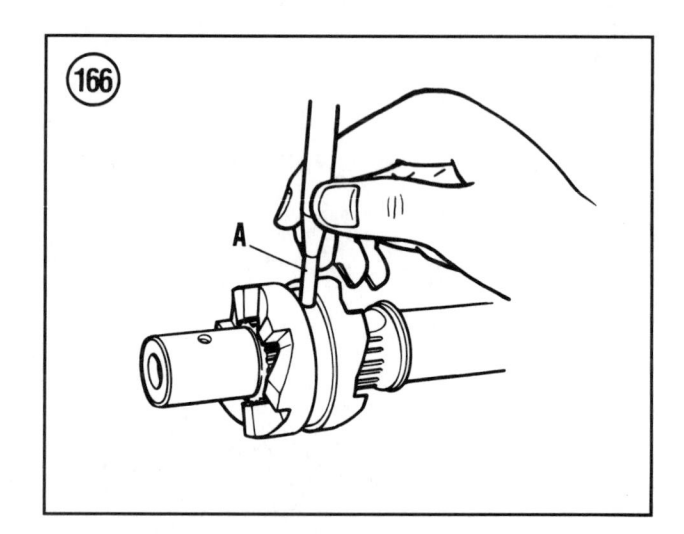

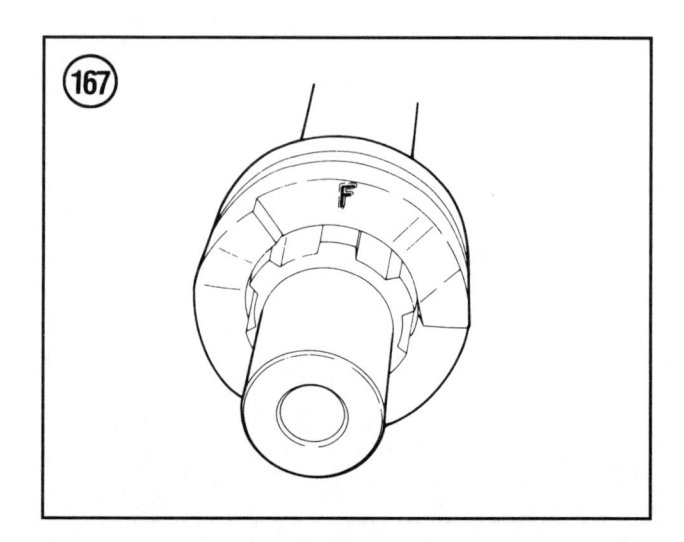

GEARCASE

441

⑯⑧

**DRIVE SHAFT
(V4 AND V6)**

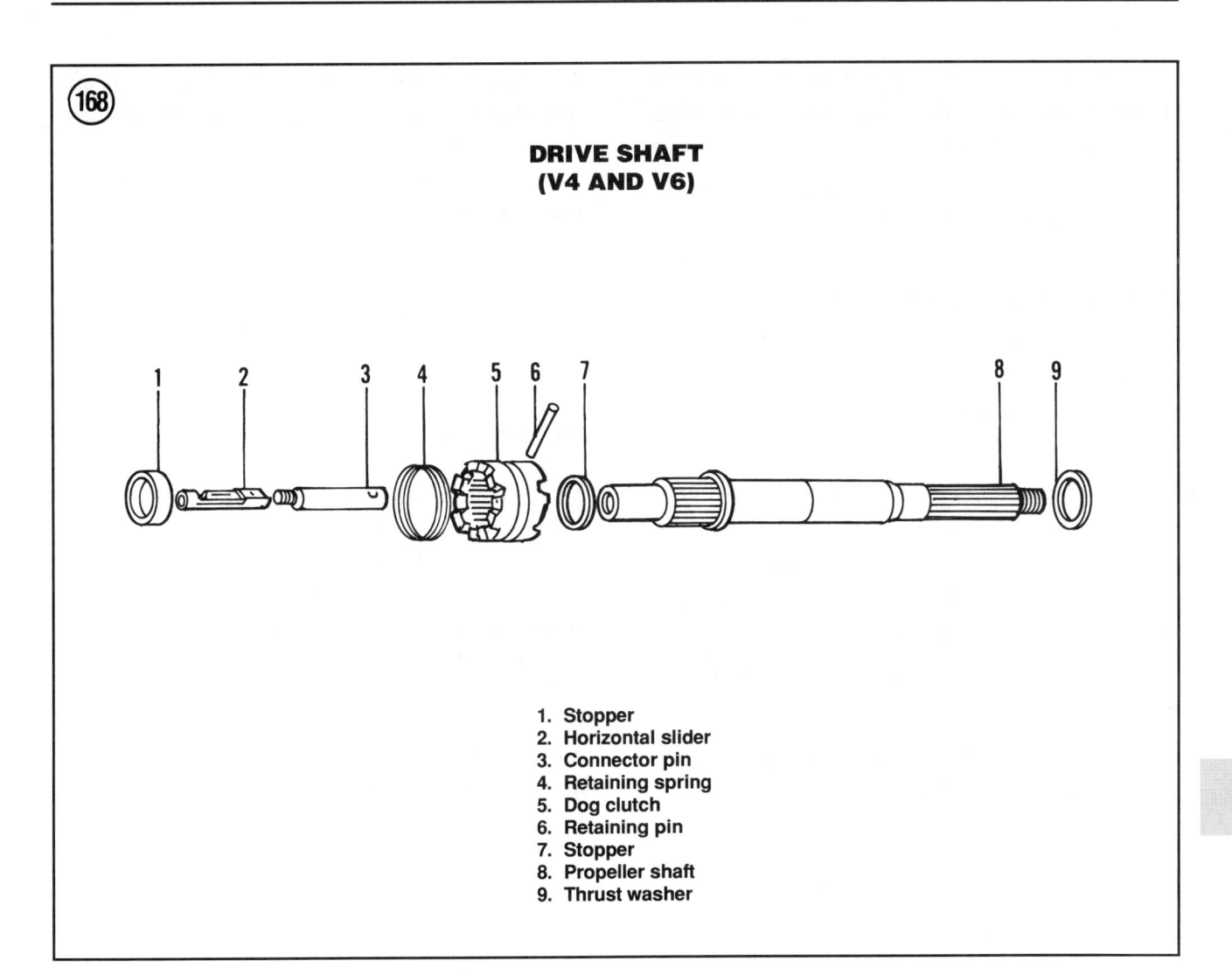

1. Stopper
2. Horizontal slider
3. Connector pin
4. Retaining spring
5. Dog clutch
6. Retaining pin
7. Stopper
8. Propeller shaft
9. Thrust washer

⑯⑨

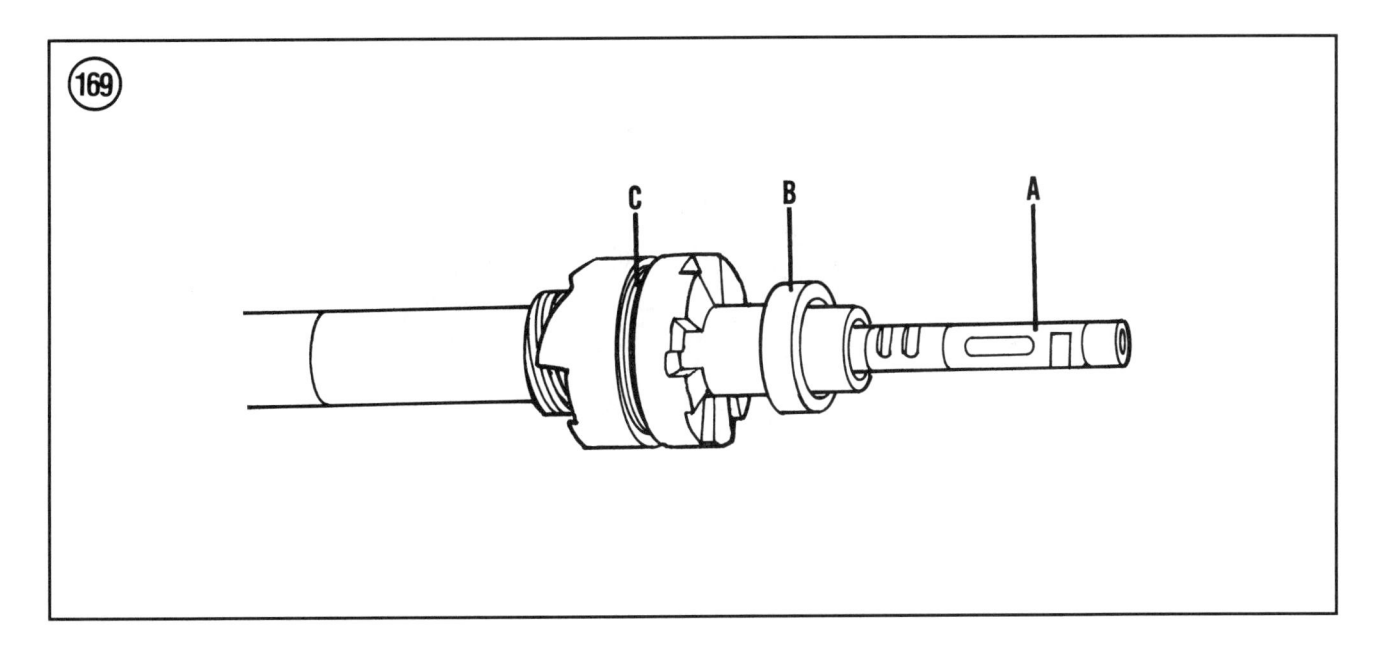

3. Remove the retaining spring (C, **Figure 169**) from the dog clutch. This will allow access to the dog clutch retaining pin.

4. Insert an appropriate size punch and push the pin (A, **Figure 170**) from the clutch dog.

5. Remove the connector pin (B, **Figure 170**) from the end of the propeller shaft.

6. Slide the clutch dog (C, **Figure 170**) off the propeller shaft.

7. Clean and inspect all parts as described in this chapter.

8. Lubricate all parts with Suzuki Outboard Motor Gear Oil.

9. Install the dog clutch (C, **Figure 170**) onto the propeller shaft with the stamped letter "F" facing the front of the shaft (**Figure 167**).

10. Install the connector pin (B, **Figure 170**) into the end of the propeller shaft.

11. Align dog clutch pin hole with connector pin hole, then insert the pin through the clutch dog and connector pin holes.

12. Center retaining pin in the dog clutch and install the retaining spring (C, **Figure 169**) onto the dog clutch.

13. Slide the stopper (B, **Figure 169**) onto the propeller shaft.

14. Connect the horizontal slider (A, **Figure 169**) onto the connector pin.

15. Apply a liberal coat of water-resistant grease to the horizontal slider and the connecting pin.

Cleaning and Inspection

1. Clean all parts in fresh solvent. Blow dry with compressed air, if available.

2. Clean all nut and screw threads thoroughly if Silicone Seal has been used. Soak nuts and screws in solvent and use a fine wire brush to remove residue.

3. Remove and discard all O-rings, gaskets and seals. Clean all Silicone Seal residue from mating surfaces.

4. Check drive shaft splines for wear or damage. If the gearcase has struck a submerged object,

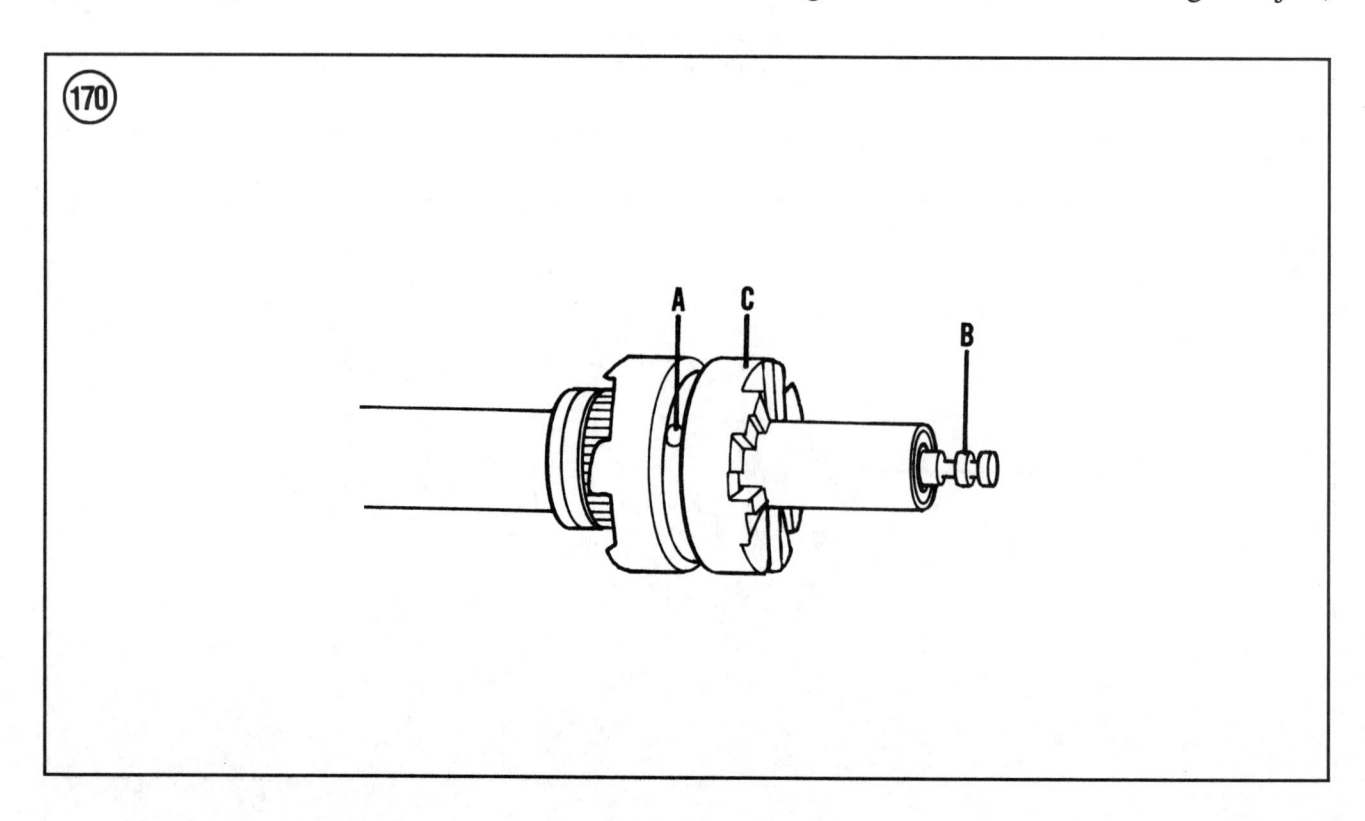

GEARCASE

the drive shaft and propeller shaft may suffer severe damage. Replace the drive shaft as required and check crankshaft splines for similar wear or damage.

5. Check the propeller shaft splines and threads for wear, rust or corrosion damage. Replace shaft as necessary.

6. Install V-blocks under the drive shaft bearing surfaces at each end of the drive shaft. Slowly rotate the drive shaft while watching the crankshaft end. Replace the shaft if any wobble is noted.

7. Repeat Step 6 with the propeller shaft. Also check the shaft surfaces where oil seal lips make contact. Replace the shaft as required.

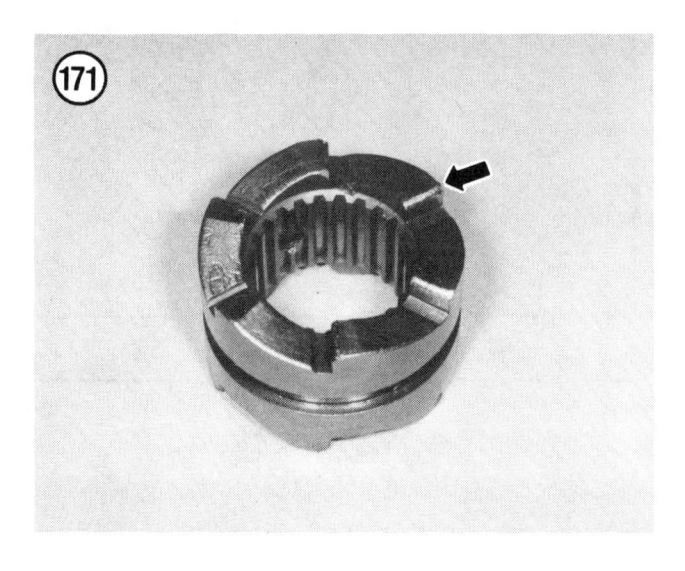

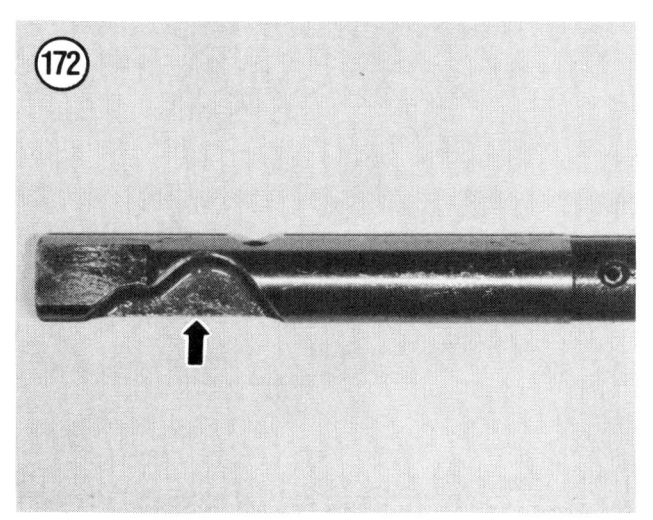

8. Check the propeller shaft bearing housing bearing for wear or damage. Replace bearing as required. If bearing wear is excessive, replace bearing housing.

9. Check the bearing housing contact points on the propeller shaft. If shaft shows signs of pitting, grooving, scoring, heat discoloration or embedded metallic particles, replace shaft and bearings.

10. Check the water pump as described in this chapter.

11. Check all shift components for wear or damage:

   a. Look for excessive wear on the clutch dog and forward/reverse gear engagement surfaces. If clutch dogs (**Figure 171**) are pitted, chipped, broken or excessively worn, replace the gear(s) and clutch dog.

   b. On models so equipped, roll the clutch return spring on a flat surface to check straightness. Replace the spring if it does not roll freely.

   c. Check the clutch dog for cracks at shear point and rounded areas that contact forward gear clutch dogs. Replace as required.

   d. Check the shift pin and cam for excessive wear (**Figure 172**, typical). Excessive wear on shift cam crests can allow the engine to drop out of gear. Replace as required.

12. Clean all roller bearings with solvent and lubricate with Suzuki Outboard Motor Gear Oil to prevent rusting. Check all bearings for rust, corrosion, flat spots or excessive wear. Replace as required.

13. Check the forward, reverse and pinion gears for wear or damage. If any teeth are pitted, chipped, broken or excessively worn, replace all gears as a set. Check pinion gear splines for excessive wear or damage. Replace as required.

14. Check the propeller for nicks, cracks or damaged blades. Minor nicks can be removed with a file, taking care to retain the shape of the propeller. Replace any propeller with bent, cracked or badly chipped blades.

PINION GEAR DEPTH AND FORWARD/REVERSE GEAR BACKLASH

Proper pinion gear engagement and forward and reverse gear backlash are important for smooth operation and long service life. Three shimming procedures may be used to set up the gearcase properly:

a. The pinion gear must be shimmed to the correct depth.
b. The forward gear must be shimmed to the pinion gear for proper backlash.
c. The reverse gear must be shimmed to the pinion gear for proper backlash.

Not all models will require all 3 shimming procedures.

The gearcase on large displacement outboards requires the use of measuring equipment. Smaller displacement models are checked by touch or sight and require some degree of experience to determine if the amount of backlash is within specifications.

Suzuki does not provide shimming instructions for the DT 4.

Pinion Gear Depth and Forward/Reverse Gear Backlash (DT 4, DT 6; 1985-1987 DT 8)

1. Depress and hold the drive shaft.
2. Reach into the prop shaft bore and lightly push the pinion gear upward.
3. Check pinion and forward gear tooth engagement by feel. The entire length of the gear teeth should be in contact as shown in **Figure 173**. There should be a minimum of up and down play in the drive shaft.
4. If the pinion gear is too high or too low, remove the drive shaft and increase or decrease the shim thickness to correct the gear height.
5. When the proper pinion gear depth has been established, depress and hold the drive shaft.
6. Move the forward gear back and forth with a finger. The free play felt is forward gear backlash (**Figure 174**) and should be 0.004-0.008 in. (0.1-0.2 mm).

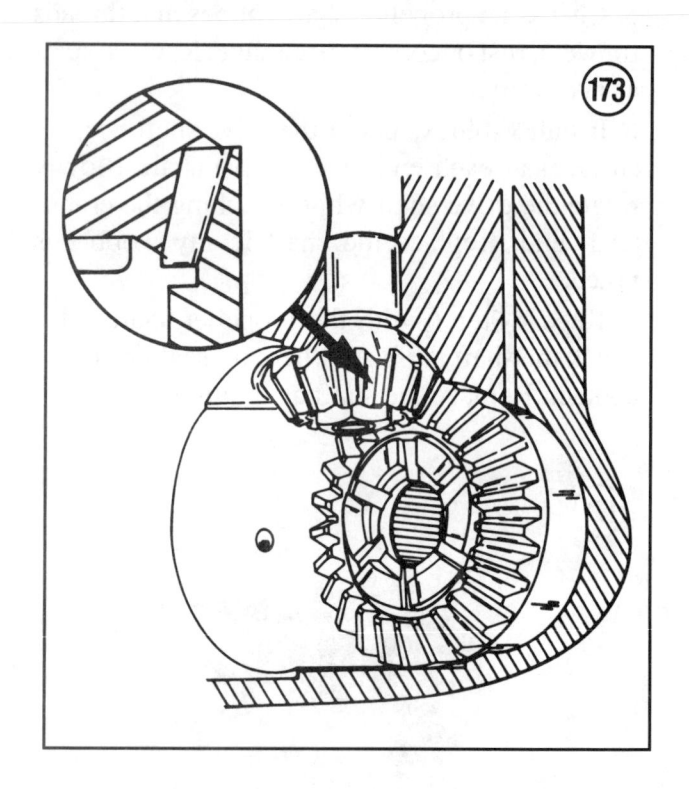

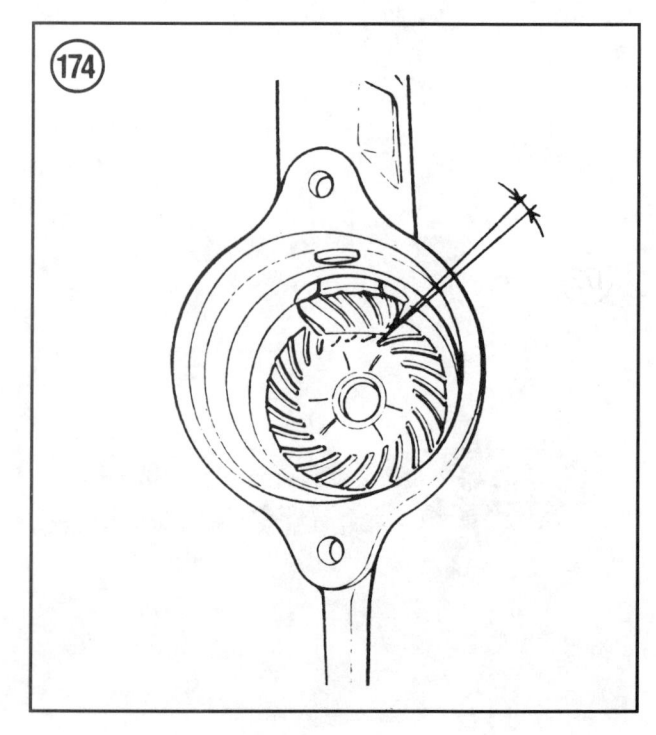

GEARCASE

7. If forward gear backlash is incorrect, remove the forward gear and increase or decrease the shim thickness to correct the backlash.

8. If forward gear backlash requires adjustment, recheck pinion gear depth after changing the forward gear shim.

9. To check reverse gear backlash:
   a. Check drive shaft thrust play by lifting up and depressing drive shaft.
   b. Temporarily install the propeller shaft/bearing housing assembly.
   c. Repeat Step 9a. The amount of thrust play should be the same in this step as in Step 9a. If it is less, remove the propeller shaft/bearing housing assembly. Decrease the shim thickness at the reverse gear and repeat this procedure until thrust play is identical before and after installing the propeller shaft/bearing housing assembly.

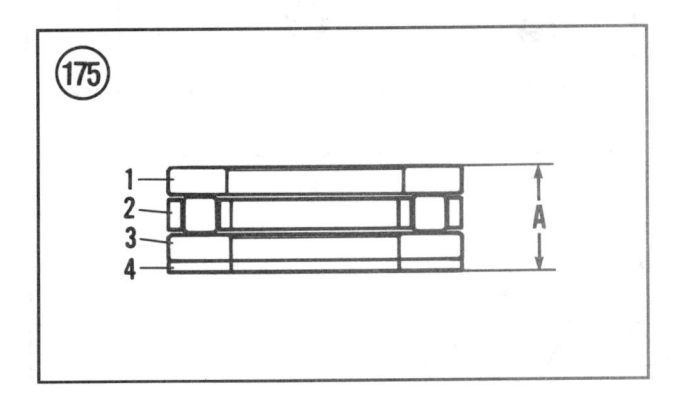

10. Check the propeller shaft thrust clearance as described in this chapter.

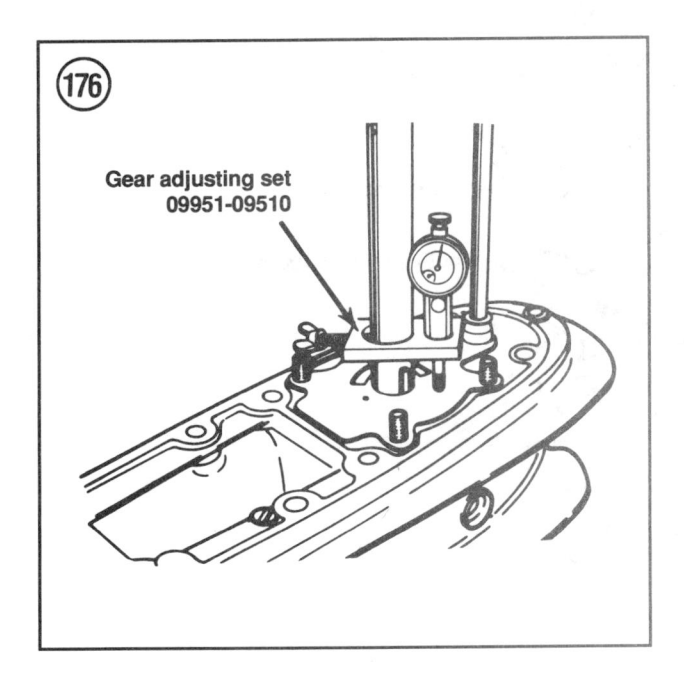

Pinion Gear Depth and Forward Gear Backlash (1988-on DT 8, DT 9.9)

1. Refer to **Figure 175** and select the pinion gear back-up shim as follows:
   a. Stack the upper washer (1), thrust washer (2), lower washer (3) and shims (4) and measure the total height.
   b. The specified total height (A) is 0.39 ±0.0020 in. (10 ±0.05 mm).
   c. If the total height is incorrect, select the shims (4) so that the height will be correct. More than one shim may be used to achieve the correct height
   d. Shims come in the following thicknesses: 0.5, 0.6, 0.7, 0.8, 0.9 and 1.0 mm.

2. Install the pinion gear, forward gear, drive shaft, drive shaft bearing housing, etc., into the gearcase as described in this chapter.

3. Install the gear adjusting set (part No. 09951-09510) as shown in **Figure 176**.

4. Depress and hold the drive shaft down and set the dial indicator gauge to zero with its rod pushed in approximately 1/8 in.

NOTE
The specified drive shaft end play is 0.004-0.012 in. (0.10-0.30 mm).

5. Slowly pull the drive shaft up as far as possible and read the indicator gauge.

NOTE
Shims come in the following thicknesses: 0.5, 0.6, 0.7, 0.8, 0.9 and 1.0 mm.

6A. If the end play exceeds 0.012 in. (0.30 mm), *increase* the thickness of the forward gear shim(s).

6B. If the end play is less than 0.004 in. (0.10 mm), *decrease* the thickness of the forward gear shim(s).

7. After the forward gear shim(s) have been changed, repeat Step 4 and Step 5 and write down the amount of backlash. This is dimension "C" and is referred to in the next procedure.

8. Perform *Reverse Gear Backlash (1988-on DT 8, DT 9.9)* in this chapter.

Reverse Gear Backlash (1988-on DT 8, DT 9.9)

1. Install gear adjusting set (part No. 09951-09510) as shown in **Figure 176**.

2. Depress and hold the drive shaft down and set the dial indicator gauge to zero with its rod pushed in approximately 2 mm.

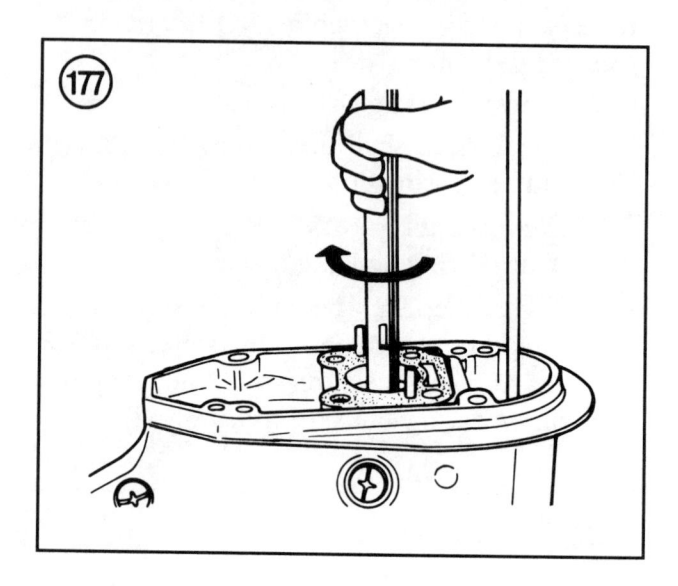

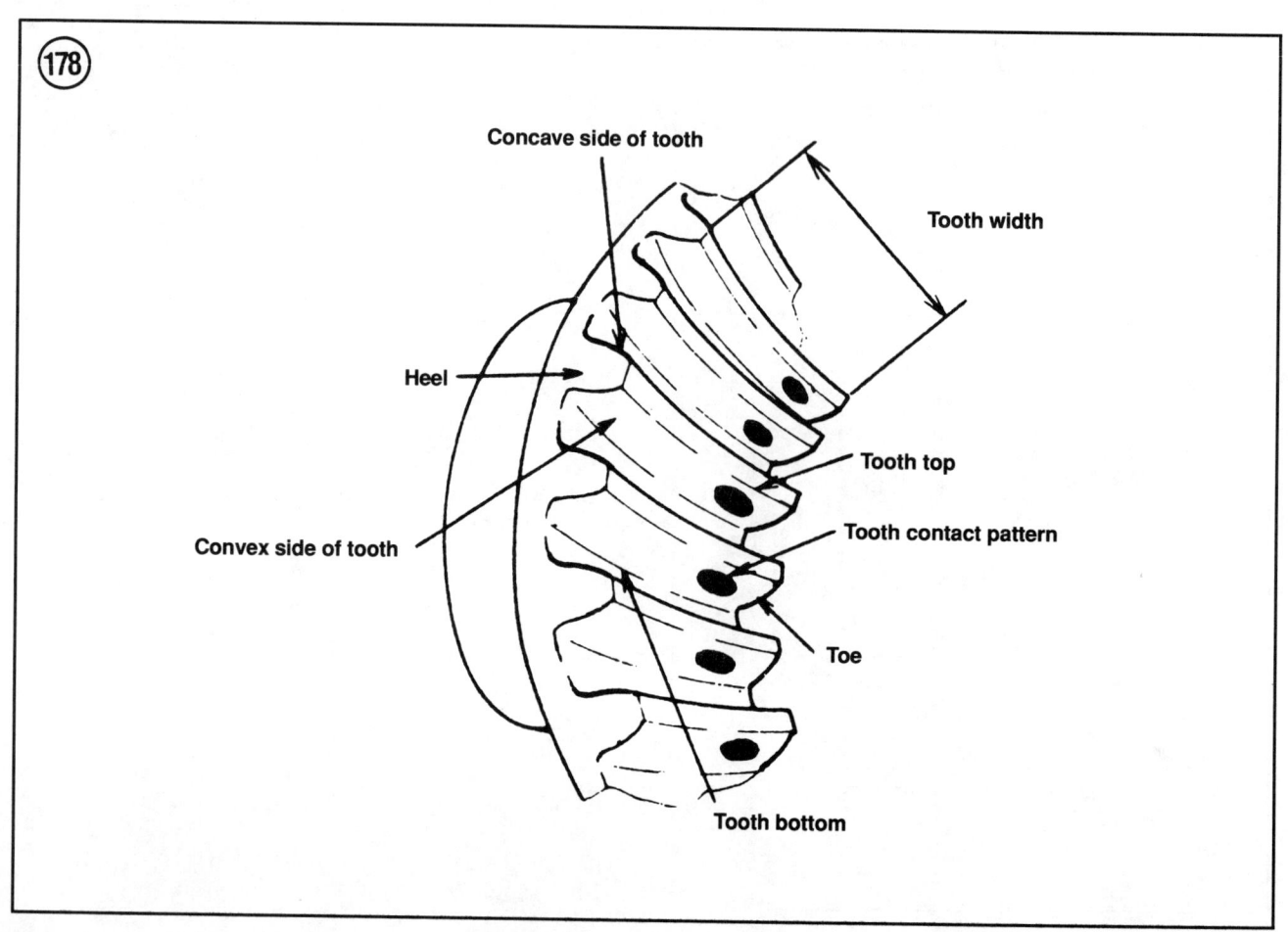

GEARCASE

3. Slowly pull the drive shaft up as far as possible and read the indicator gauge. This is the maximum play and is referred to as dimension "D."

4. If dimension "D" is equal to dimension "C," in the preceding procedure, the backlash is within specifications.

NOTE
Shims come in the following thicknesses: 0.5, 0.6, 0.7, 0.8, 0.9 and 1.0 mm.

5. If dimension "D" is smaller than dimension "C," *decrease* the reverse gear back-up shim(s).

6. Check the propeller shaft thrust clearance (end play) as described in this chapter.

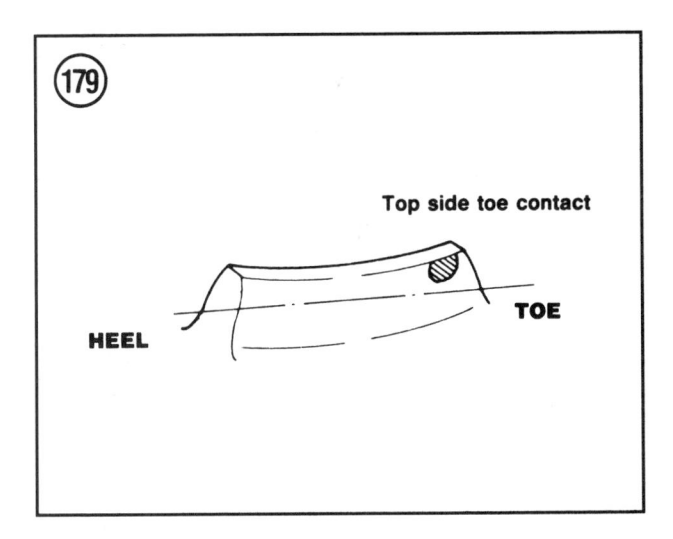

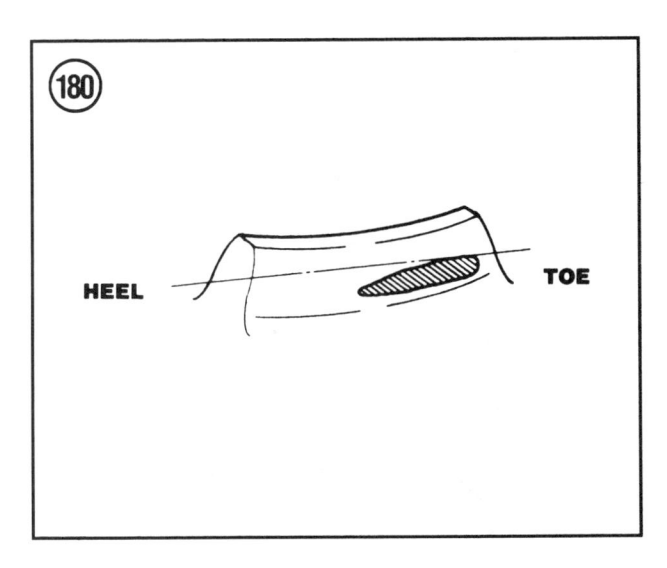

Pinion Gear Depth and Forward Gear Backlash (1985-1987 DT 9.9; 1985-on DT 15)

1. Install gear adjusting set (part No. 09951-09510) as shown in **Figure 176**.

2. Depress and hold the drive shaft while setting the dial indicator gauge to zero.

NOTE
The specified drive shaft end play is 0.002-0.012 in. (0.05-0.30 mm).

3. Slowly pull the drive shaft up as far as possible and read the indicator gauge.

4A. If the end play exceeds 0.012 in. (0.30 mm), *increase* the thickness of the pinion and forward gear shim(s).

4B. If the end play is less than 0.002 in. (0.05 mm), *decrease* the thickness of the pinion and forward gear shim(s).

5. Lightly coat the forward gear teeth with gear marking compound.

6. Temporarily install the propeller shaft/bearing housing assembly and engage the forward gear. Prevent the propeller shaft from moving and slowly rotate the drive shaft about 5 full turns (**Figure 177**) in a clockwise direction.

7. Remove the propeller shaft, drive shaft, pinion gear and forward gear.

8. Check the tooth contact pattern on the forward gear. It should resemble that shown in **Figure 178**:

   a. If it resembles that shown in **Figure 179**, decrease the forward gear shim thickness and increase the pinion gear shim thickness.

   b. If it resembles that shown in **Figure 180**, increase the forward gear shim thickness and decrease the pinion gear shim thickness.

   c. Clean the marking compound from the gear teeth and repeat Steps 5-7 to recheck pattern. Repeat this procedure as required until the tooth contact pattern is correct.

d. When pattern is correct, remove and clean all marking compound from gear teeth for final reassembly.

9. To check reverse gear backlash:
   a. Check drive shaft end play by lifting up and depressing the drive shaft (**Figure 181**).
   b. Temporarily install the propeller shaft/bearing housing assembly.
   c. Repeat Step 9a. The amount of end play should be the same in this step as in Step 9a. If it is less, remove the propeller shaft/bearing housing assembly. *Decrease* the shim thickness at the reverse gear.
   d. Repeat this procedure until end play is identical before and after installing the propeller shaft/bearing housing assembly.

10. Check the propeller shaft end play as described in this chapter.

Pinion Gear Depth, Forward and Reverse Gear Backlash (DT 20, DT 25, DT 30, DT 35, DT 40, DT 55, DT 65, DT 75, DT 85, 1985 DT 115 and DT 140)

Forward gear backlash

1. Drain the gearcase oil as described in Chapter Four.
2. Into the gearcase drain hole, install the Suzuki special tool, Gear Adjusting Gauge (part No. 09951-09510). Tighten securely.
3. Correctly position the gauge rod end on the convex side of the forward gear tooth heel end (**Figure 182**). Make sure the rod end is not touching the adjacent tooth.
4. Set the gauge to zero.

CAUTION
*In the following step, only **slightly** rotate the propeller shaft in each direction. If there is an excessive amount of rotation, the gauge rod end may be broken when the adjacent gear tooth comes in contact with it.*

5. Hold onto the end of the propeller shaft and slightly rotate the shaft in each direction and note the amount of gear backlash (**Figure 183**).
6. The specified backlash is as follows:
   a. DT 75, DT 85: 0.002-0.012 in. (0.05-0.3 mm).
   b. DT 115, DT 140: 0.006-0.012 in. (0.15-0.3 mm).
   c. All other models: 0.0039-0.0079 in. (0.1-0.2 mm).

7A. If backlash is *greater* than specified, install thicker forward and pinion gear backup shim(s).
7B. If backlash is *less* than specified, install thinner forward and pinion gear backup shim(s).

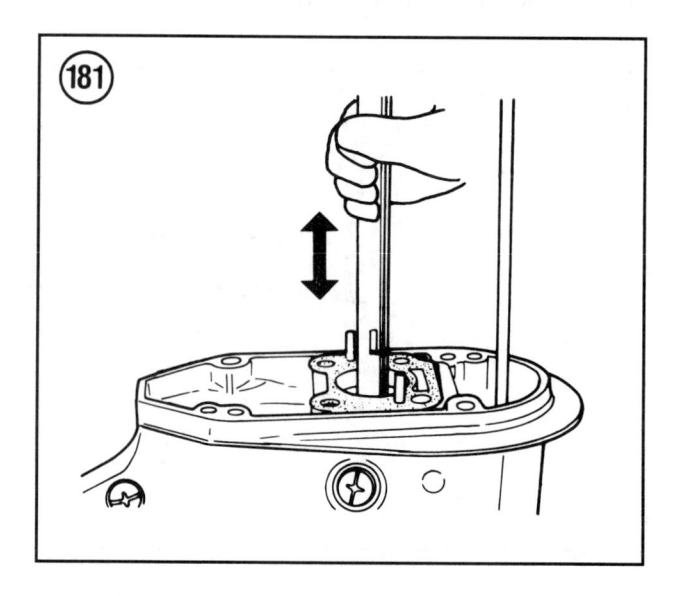

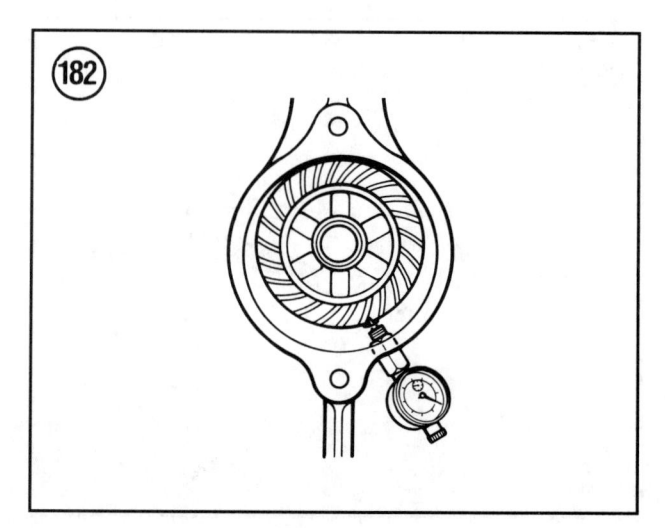

GEARCASE

8. Remove the Gear Adjusting Gauge and reinstall the drain plug.

9. Refill the gearcase oil as described in Chapter Four.

10. Perform *Propeller Shaft End Play* as described in this chapter.

Pinion gear depth

1. Lightly coat the forward gear teeth with gear marking compound.

2. Temporarily install the propeller shaft/bearing housing and engage the forward gear. Prevent the propeller shaft from moving and slowly rotate the drive shaft about 5 full turns (clockwise).

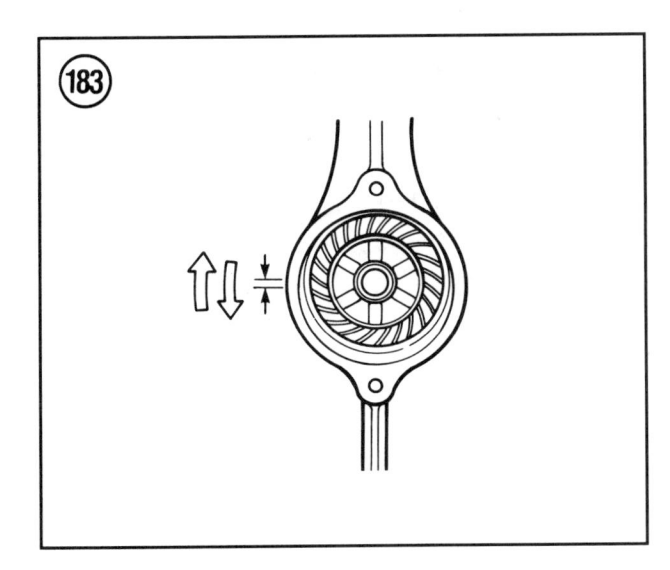

3. Remove the propeller shaft/bearing housing assembly, drive shaft, pinion gear and forward gear.

4. Check the tooth contact pattern on the forward gear. It should resemble that shown in **Figure 178**:

   a. It if resembles that shown in **Figure 179**, decrease the forward gear shim thickness and increase the pinion gear shim thickness.

   b. If it resembles that shown in **Figure 180**, increase the forward gear shim thickness and decrease the pinion gear shim thickness.

   c. Clean the marking compound from the gear teeth and repeat Steps 2-4 to recheck pattern. Repeat this procedure as required until the tooth contact pattern is correct.

   d. When pattern is correct, remove and clean all marking compound from gear teeth for final reassembly.

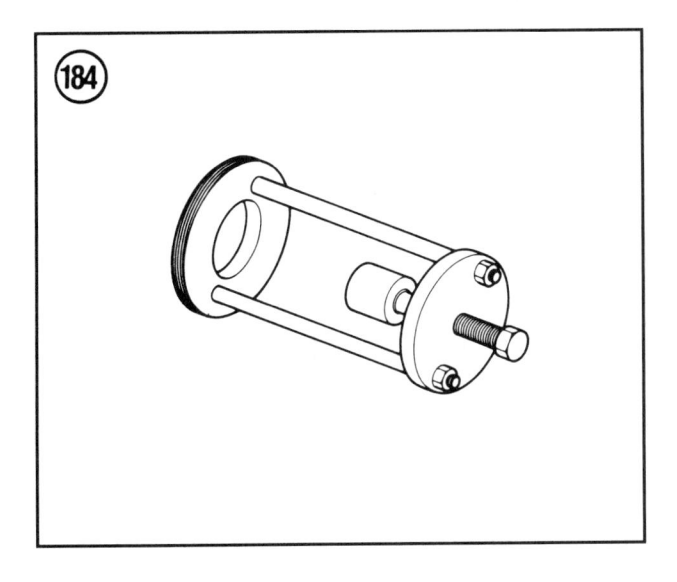

Reverse gear backlash

1. Check drive shaft end play by lifting up and depressing drive shaft.

2. Temporarily install the propeller shaft/bearing housing assembly.

3. Repeat Step 1. The amount of end play should be the same in this step as in Step 1. If it is less, remove the propeller shaft/bearing housing assembly. Adjust the thickness of the reverse gear-to-bearing shim and repeat this procedure until end play is identical before and after installing the propeller shaft/bearing housing assembly.

Forward and Reverse Gear Backlash (V4 and V6)

1. Install the Suzuki special tool (**Figure 184**), Gear Holder (part No. 09951-98710) into the gearcase (A, **Figure 185**).

2. Turn the special tool center bolt (B, **Figure 185**) *clockwise* and tighten securely.

3. Install the Suzuki special tool, Backlash Indicator Tool (part No. 09952-08710) onto the drive shaft (A, **Figure 186**).

4. Install the Suzuki special tool, Dial Gauge (part No. 09900-20606) into the gearcase (B, **Figure 186**).

5. Slightly rotate the drive shaft *clockwise* and then *counterclockwise* by hand.

6. The specified backlash is 0.020-0.026 in. (0.5-0.65 mm).

NOTE
*Only use 1 shim in this procedure. Do **not** use 2 shims to accomplish the correct amount of backlash.*

7A. If backlash is *greater* than specified, install a thicker forward gear backup shim.

7B. If backlash is *less* than specified, install a thinner forward gear backup shim.

NOTE
The backup shims are available in varying thicknesses in 0.05 mm increments. The thickness varies from 0.40 to 1.20 mm with the 1.00 mm backup shim being the standard thickness.

8. Turn the special tool center bolt (B, **Figure 185**) *counterclockwise* and tighten securely.

9. Slightly rotate the drive shaft clockwise and then counterclockwise by hand.

10. The specified backlash is 0.028-0.033 in. (0.7-0.85 mm).

NOTE
*Only use 1 shim in this procedure. Do **not** use 2 shims to accomplish the correct amount of backlash.*

11A. If backlash is *greater* than specified, install a thicker reverse gear backup shim.

11B. If backlash is *less* than specified, install a thinner reverse gear backup shim.

NOTE
The backup shims are available in varying thicknesses in 0.10 mm increments. The thickness varies from 1.0 to 1.9 mm for V4 models or 0.90 to 1.8 mm for V6 models.

12. Remove the special tool installed in Step 1.

13. Install the Suzuki special tool, Dial Gauge and Holder (part No. 09951-09510) onto the propeller shaft (A, **Figure 187**).

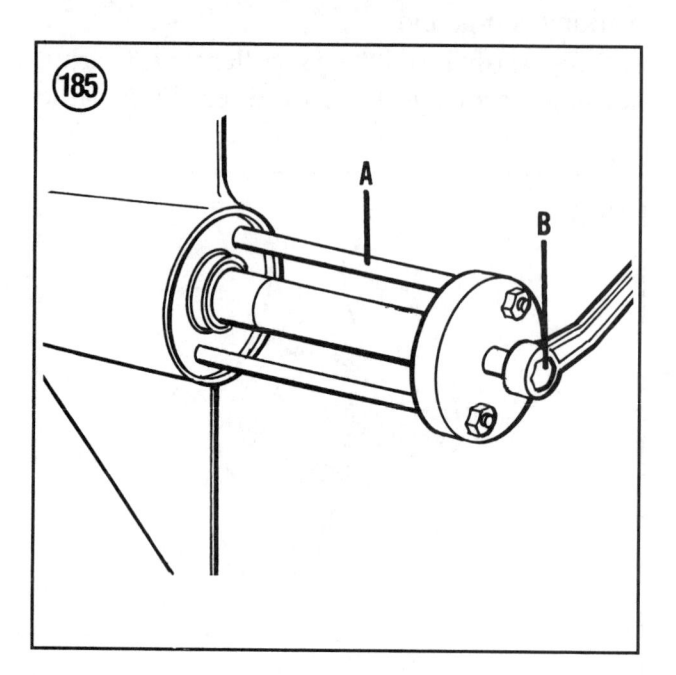

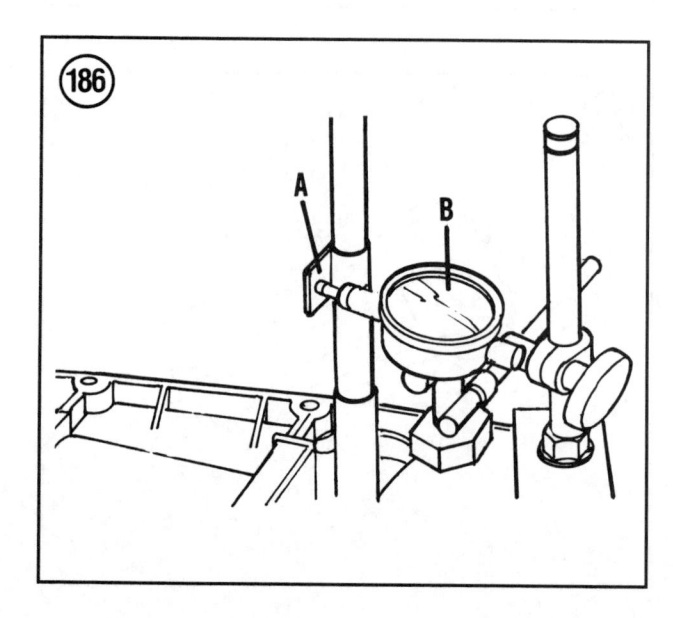

GEARCASE

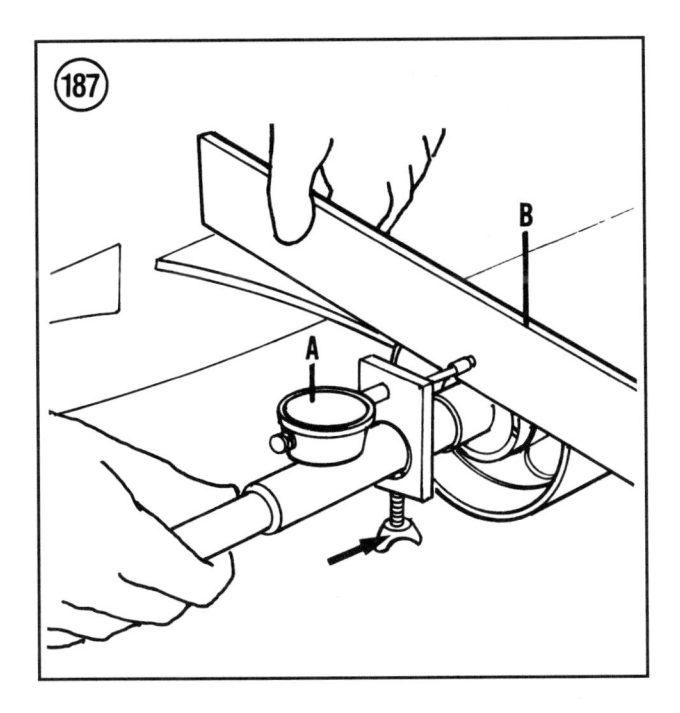

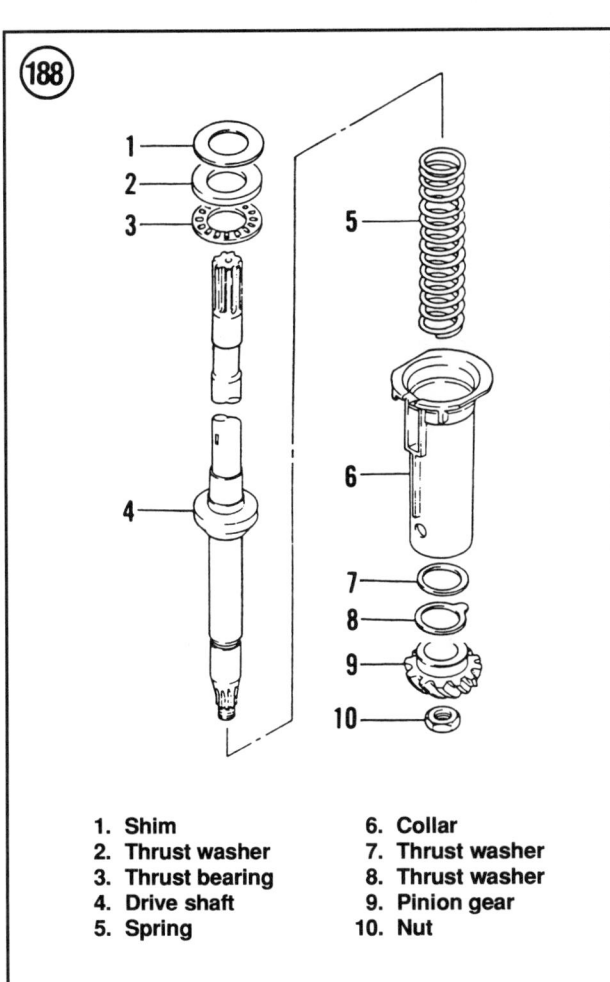

14. Place a machined straightedge (B, **Figure 187**) across the gearcase housing.

15. Grasp the propeller shaft and push it in as far as possible and hold it in this position.

16. With the propeller shaft in this position, move the dial indicator down on the shaft until at least 1/8 in. of the rod is pushed into the dial indicator. Set the indicator gauge to zero.

17. Now slowly pull the propeller shaft out as far as possible. Read the dial indicator gauge.

18. The specified end play should be 0.004-0.008 in. (0.10-0.20 mm).

19A. If end play is *greater* than specified, install a thicker reverse gear thrust washer.

19B. If end play is *less* than specified, install a thinner reverse gear thrust washer.

NOTE
The thrust washers are available in varying thicknesses in 0.20 mm increments. The thickness varies from 1.5 to 2.9 mm.

20. Remove the special tool installed in Step 13.

PINION GEAR ADJUSTMENT (1986-ON DT 115, DT 140; V4 AND V6)

Refer to **Figure 188** for DT 115 and DT 140 models or **Figure 189** for V4 and V6 models for this procedure.

1. Disassemble the drive shaft assembly.

NOTE
The standard thickness shim (1) is 1.00 mm thick. For the purposes of this procedure, install a shim of lesser thickness (e.g. 0.90 mm).

2. Refer to **Figure 190** and assemble the following parts onto the drive shaft: the bearing (3), thrust washer (2), shim (1) (see previous NOTE regarding thickness), the drive shaft bearing housing assembly (12-15, **Figure 189**).

---

1. Shim
2. Thrust washer
3. Thrust bearing
4. Drive shaft
5. Spring
6. Collar
7. Thrust washer
8. Thrust washer
9. Pinion gear
10. Nut

CHAPTER NINE

3. Place the special Suzuki Shimming Tool (part No. 09951-08710) horizontally in a vise and clamp it securely.

4. Insert the drive shaft assembly through the closed end of the special Suzuki tool.

5. Install the pinion gear (9) and nut (10). Tighten the nut securely.

6. Install the drive shaft bearing housing mounting bolts and washers. Tighten the bolts securely to fasten the drive shaft into the tool securely.

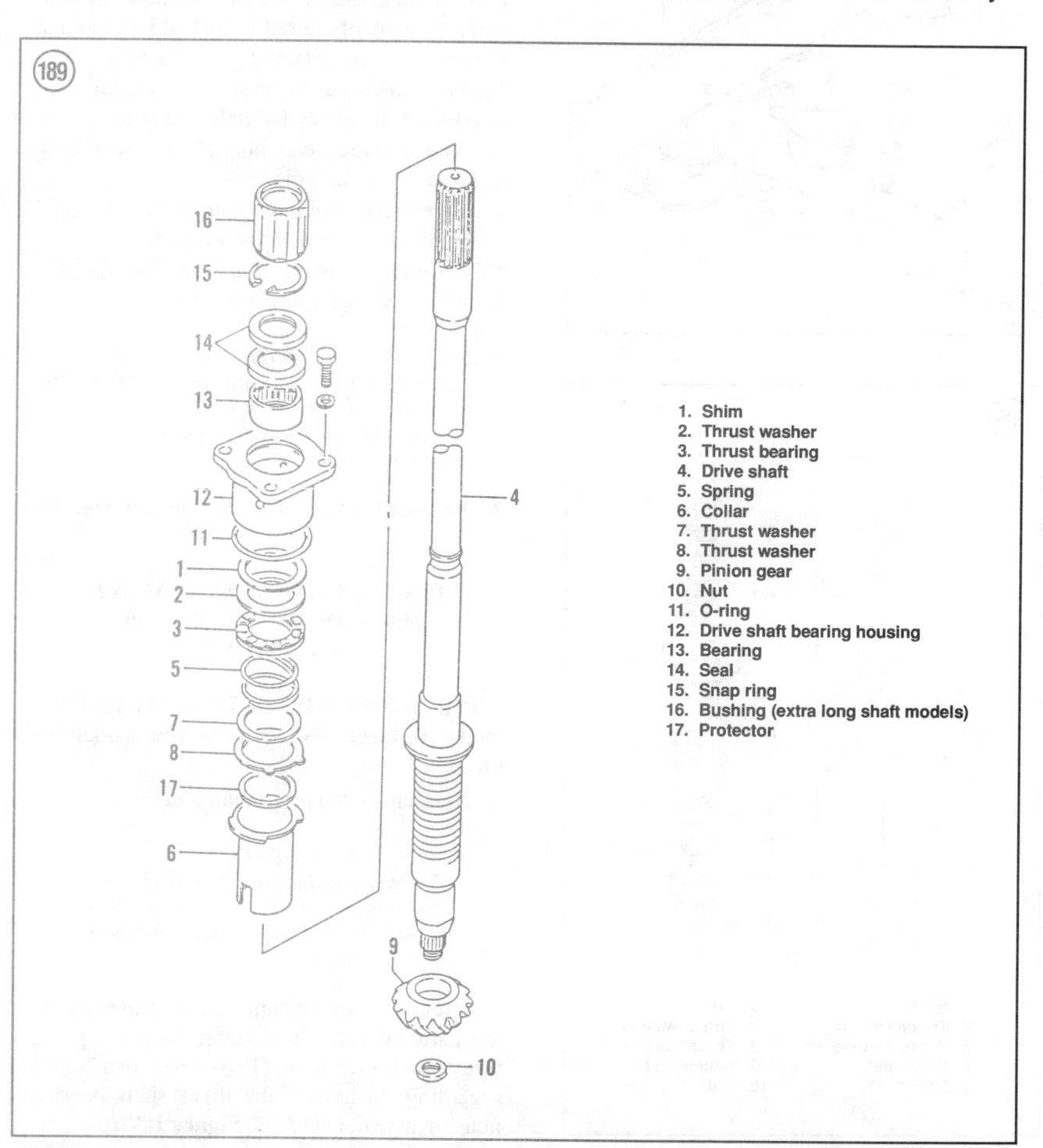

189

1. Shim
2. Thrust washer
3. Thrust bearing
4. Drive shaft
5. Spring
6. Collar
7. Thrust washer
8. Thrust washer
9. Pinion gear
10. Nut
11. O-ring
12. Drive shaft bearing housing
13. Bearing
14. Seal
15. Snap ring
16. Bushing (extra long shaft models)
17. Protector

GEARCASE

NOTE
The pinion gear depth is correct when there is no clearance between the machined end of the special tool and the flat surface on the pinion gear.

7. With the drive shaft secured in the special tool, use a flat metric feeler gauge and measure the clearance "B," from the flat surface of the pinion gear and the machined surface of the shimming tool.

8. If there is a clearance (distance "B") in Step 7, it must be eliminated by using a new thicker shim (9).

9. Add distance "B," taken in Step 7, to the thickness of the shim (9) installed in Step 2.

Example:
Distance "B" = 0.20 mm (Step 7)
+ Shim (9) = 0.90 mm (Step 2)
New shim thickness = 1.10 mm

10. Remove the drive shaft assembly from the special tool and disassemble it.

NOTE
*Only use 1 shim in this procedure. Do **not** use 2 shims to accomplish the correct shim thickness.*

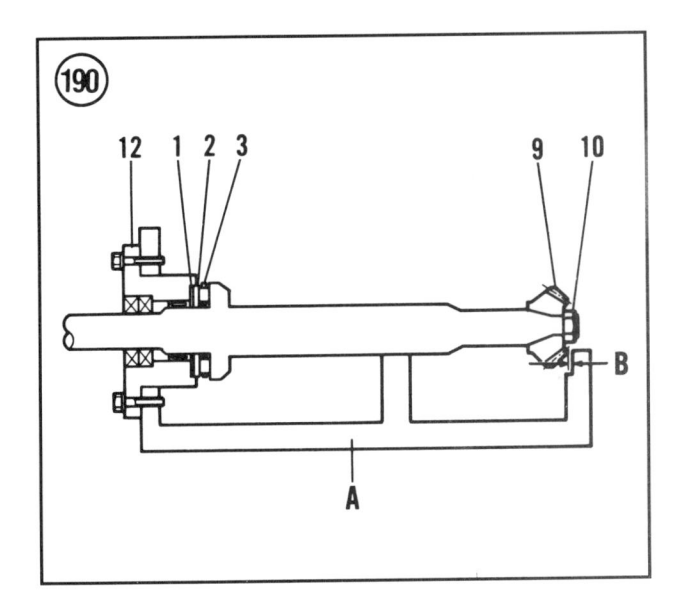

(190)

NOTE
The backup shims are available in varying thicknesses in 0.10 mm increments for DT 115, DT 140 models or in 0.05 mm increments for V4 and V6 models. The thickness varies from 0.6 to 1.3 on DT 115, DT 140 models, or 1.0 to 1.9 mm for V4 models or 0.90 to 1.8 mm for V6 models.

11. Install the new shim (1) of the correct thickness determined in Step 9 and reassemble the drive shaft.

12. Repeat Step 6 and Step 7. In Step 7, distance "B" should now be eliminated and there should be no clearance between the flat of the pinion gear and the machined surface of the special tool. If necessary, repeat this procedure until the clearance has been eliminated.

PROPELLER SHAFT END PLAY (DT 8 THROUGH DT 65 MODELS ONLY)

This measurement should be taken on these models after completing *Pinion Gear and Forward/Reverse Gear Backlash* as described in this chapter.

NOTE
These are the only models that require this type of adjustment.

1. Temporarily install the propeller shaft and bearing housing assembly.

2. Place a machined straightedge across the gearcase housing.

3. Install the Suzuki special tool, Dial Gauge and Holder (part No. 09951-09510) onto the propeller shaft.

4. Grasp the propeller shaft and push it in as far as possible and hold it in this position (**Figure 191**).

5. With the propeller shaft in this position, move the dial indicator down on the shaft until at least

1/8 in. of the rod is pushed into the dial indicator. Set the indicator gauge to zero.
6. Now slowly pull the propeller shaft out as far as possible (**Figure 192**). Read the dial indicator gauge.
7. The specified end play is 0.008-0.016 in. (0.20-0.40 mm).
8A. If end play is *greater* than specified, install a thicker reverse gear thrust washer.
8B. If end play is *less* than specified, install a thinner reverse gear thrust washer.
9. Remove the special tool installed in Step 3.

CLUTCH PUSHROD DIMENSION (1985-1989 DT 75, DT 85; 1985 DT 115, DT 140)

This check is made on these models after completing the tooth contact pattern check. Refer to **Figure 193** for DT 75 and DT 85 models or **Figure 194** for DT 115 and DT 140 models for this procedure.

NOTE
These are the only models that require this type of adjustment.

1. Place the shift mechanism in the neutral position.
2. With the propeller shaft/bearing housing assembly out of the gearcase, use a vernier caliper to measure and record the distance between the clutch engagement surface of the forward gear to the rear edge of the gearcase (**Figure 195**). This is dimension "A."
3. Temporarily install the propeller shaft (without bearing housing) into the gearcase until the front of the pushrod just touches the shift cam detent. See point D, **Figure 193** or **Figure 194**.
4. Use vernier caliper to measure and record the distance between the rear of the dog clutch and the rear edge of the gearcase (**Figure 196**). This is dimension "B."
5. Subtract dimension "B" from dimension "A" and record the result.

6. Subtract the thickness of the dog clutch from the figure obtained in Step 5 and record the result. This is dimension "C." The DT 75 and DT 85 dog clutch is 1.397 in. (35.5 mm) thick; the DT 115 and DT 140 dog clutch is 1.653 in. (42 mm) thick.

GEARCASE

193

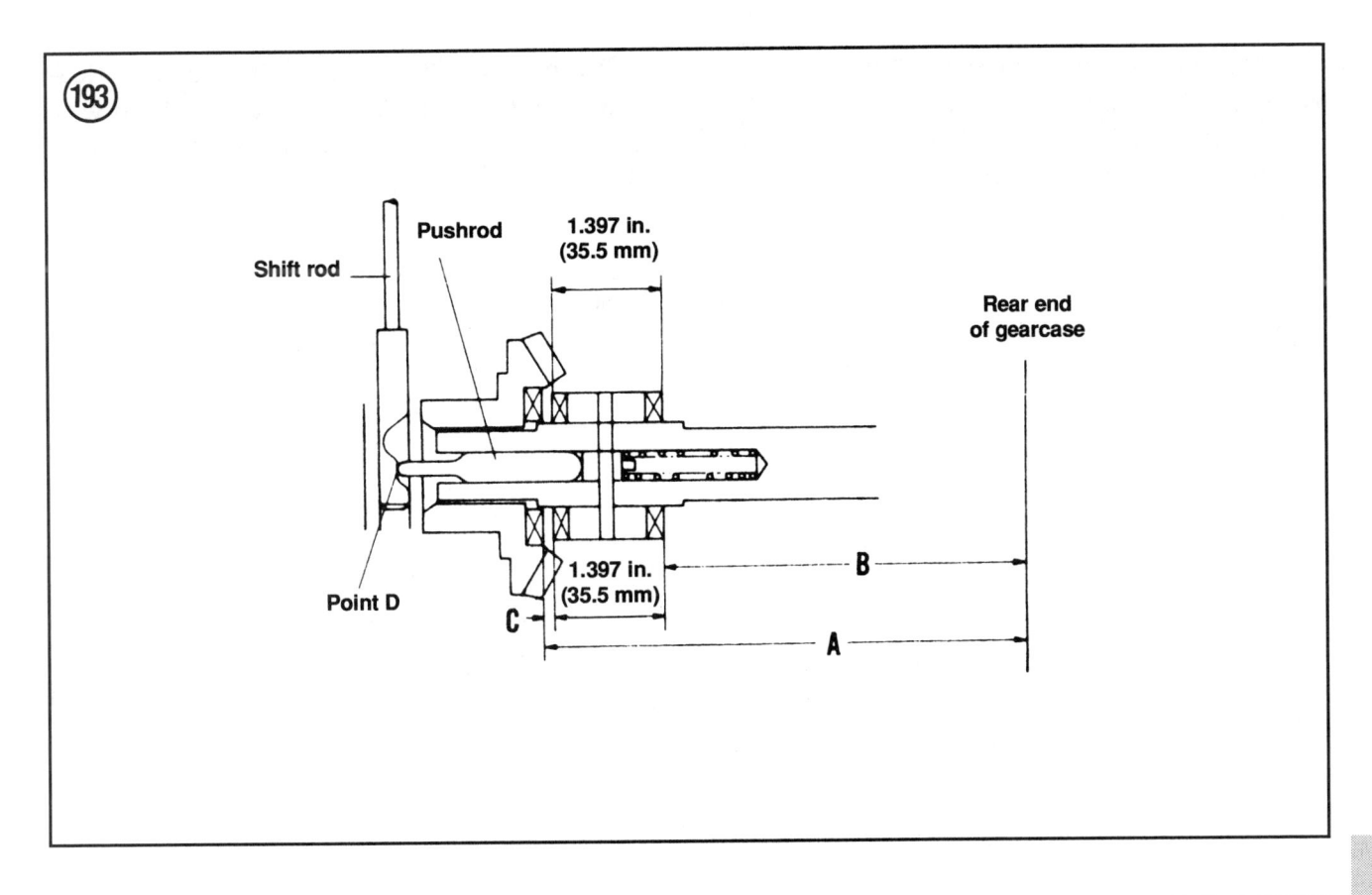

Shift rod — Pushrod — 1.397 in. (35.5 mm) — Rear end of gearcase — Point D — 1.397 in. (35.5 mm) — C — B — A

194

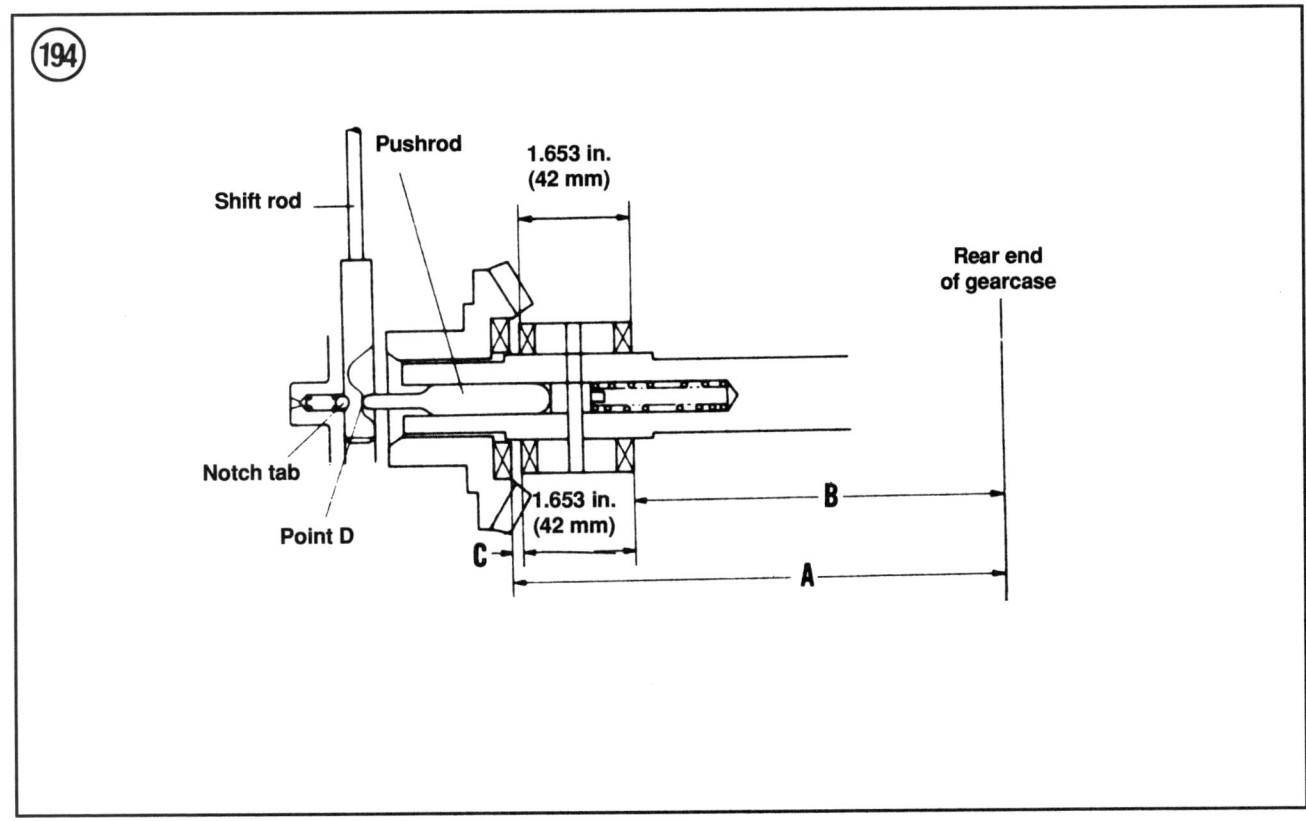

Shift rod — Pushrod — 1.653 in. (42 mm) — Rear end of gearcase — Notch tab — Point D — 1.653 in. (42 mm) — C — B — A

7. If the figure obtained in Step 6 (dimension C) is 0.035-0.055 in. (0.9-1.4 mm) (DT 75 and DT 85) or 0.051-0.067 in. (1.3-1.7 mm) (DT 115 and DT 140), the clutch pushrod length is satisfactory.

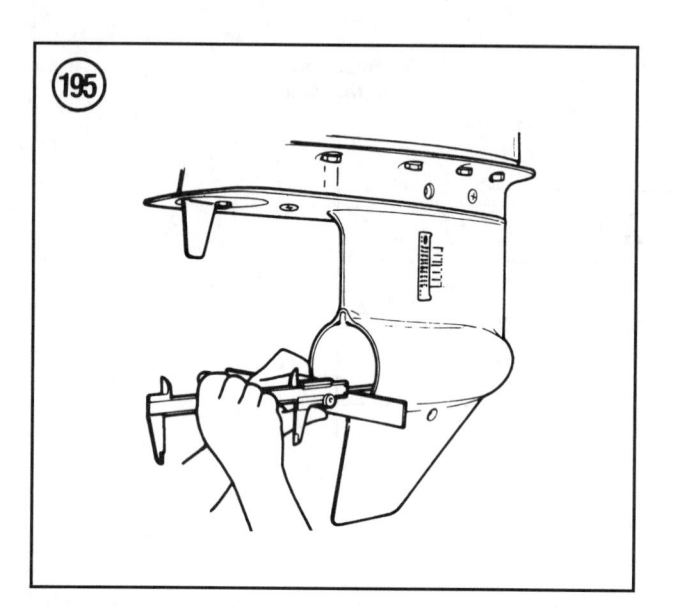

8. If dimension "C" is not within the specifications in Step 7, replace the clutch pushrod with a longer or shorter one as required. See your dealer for available pushrod lengths.

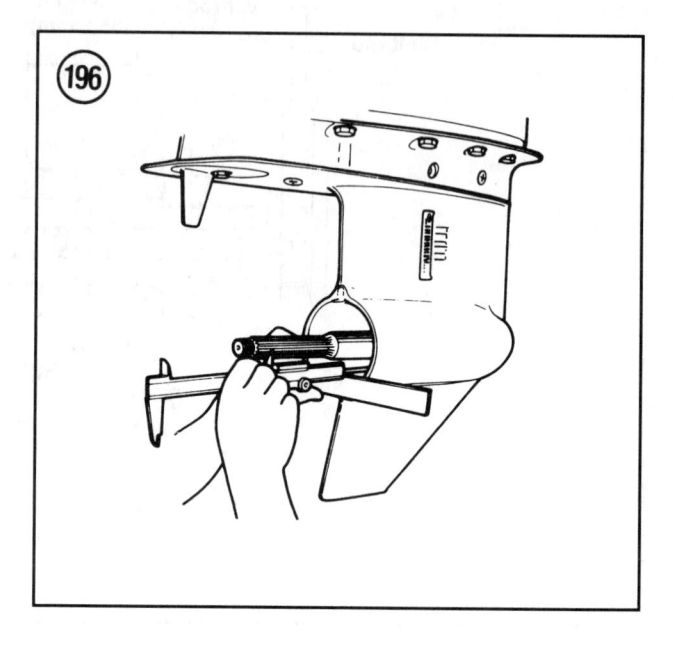

GEARCASE

Table 1 TIGHTENING TORQUES*

Fastener**	ft.-lb.	N·m
Bearing housing stopper		
V4	77-125	105-170
V6	133-147	180-200
Gearcase-to-drive shaft bolts		
DT 8–DT 40	11-14	15-20
DT 55, DT 65		
8 mm	11-14	15-20
10 mm	25-30	34-41
DT 115, DT 140, V4, V6	37-44	50-60
Pinion gear nut		
DT 20, DT 25 (1985-1988),		
DT 30 (1985-1987)	13-14	18-20
DT 25 (1989-on), DT 30 (1988-on)	20-22	27-30
DT 35, DT 40	22-29	30-40
DT 55, DT 65	11-14	15-20
DT 75, DT 85	44-51	60-70
DT 115, DT 140		
1985	59-66	80-90
1986-on	73-81	100-110
V4	59-73	80-100
V6	103-110	140-150
Propeller shaft housing bolt		
DT 9, DT 15	4.5-7	6-10
DT 20, DT 25 (1985-1988),		
DT 30 (1985-1987)	20-22	27-30
DT 25 (1989-on), DT 30 (1988-on)	21-22	29-31
DT 35, DT 40, DT 55, DT 65	37-44	50-60
Water pump cover bolts or nuts		
DT 9.9, DT 15	4.5-7	6-10
DT 20, DT 25, DT 30	13-14	18-20
DT 35, DT 40	4.5-7	6-10
DT 55, DT 65, DT 75, DT 85	11-14	15-20
DT 115, DT 140	10-13	13-18
Standard torque values		
Bolt head marked "4"		
5 mm	1.5-3	2-4
6 mm	3-5	4-7
8 mm	7-11	10-16
10 mm	16-25	22-35
Bolt head marked "7"		
5 mm	2-4	3-6
6 mm	6-8	8-12
8 mm	13-20	18-28
10 mm	29-44	40-60
Stainless steel bolt		
5 mm	1.5-3	2-4
6 mm	4.5-7	6-10
8 mm	11-14	15-20
10 mm	25-30	34-41

* Use standard torque values for any fastener not specifically listed.
** Suzuki provides tightening torque specifications for these models only.

Chapter Ten

Automatic Rewind Starters

All manual start (and some electric start) models are equipped with a rope-operated rewind starter. Some smaller displacement electric start models not equipped with a rewind starter have a flywheel drive cup and starter rope for emergency starts.

Two types of starters are used: a Bendix or spool starter bracket-mounted to the side of the power head (**Figure 1**) and an overhead starter mounted in a housing above the flywheel (**Figure 2**).

Pulling the rope handle causes the starter spindle or spool shaft to rotate against spring tension, moving the drive pawl or pinion gear to engage the flywheel and turn the engine over. When the rope handle is released, the spring inside the assembly reverses direction of the spindle or spool shaft and rewinds the rope around the pulley or spool for the next start up.

A neutral starter interlock (NSI) prevents starter operation unless the shift lever is in NEUTRAL. A circuit in the ignition system provides the interlock feature on most models. On models without the ignition system NSI circuitry, interlock is accomplished by a mechanical interlock device. Only the mechanical interlock device affects the servicing of the starters discussed in this chapter.

Automatic rewind starters are relatively troublefree; a broken or frayed rope is the most common malfunction.

BENDIX STARTER

This starter type is used on some DT 6 and some 1985-1987 DT 8 models. Refer also to *Overhead Starter* in this chapter.

Removal/Installation

1. Remove the engine cover.
2. Disconnect the spark plug wires to prevent the engine from accidentally starting.

AUTOMATIC REWIND STARTERS

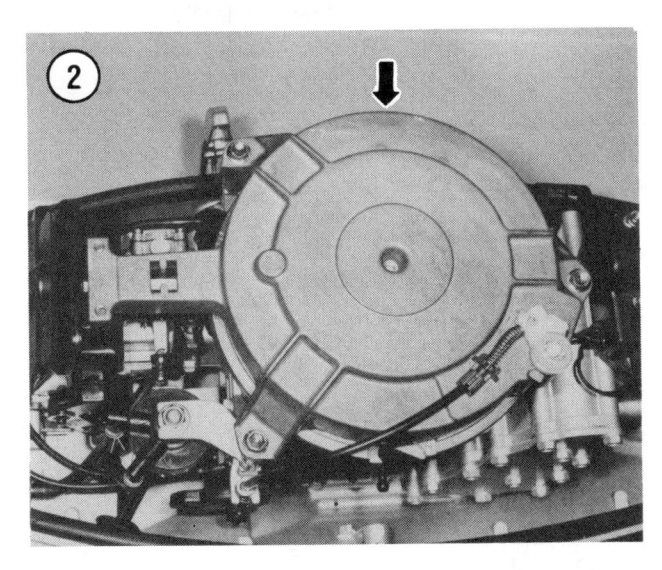

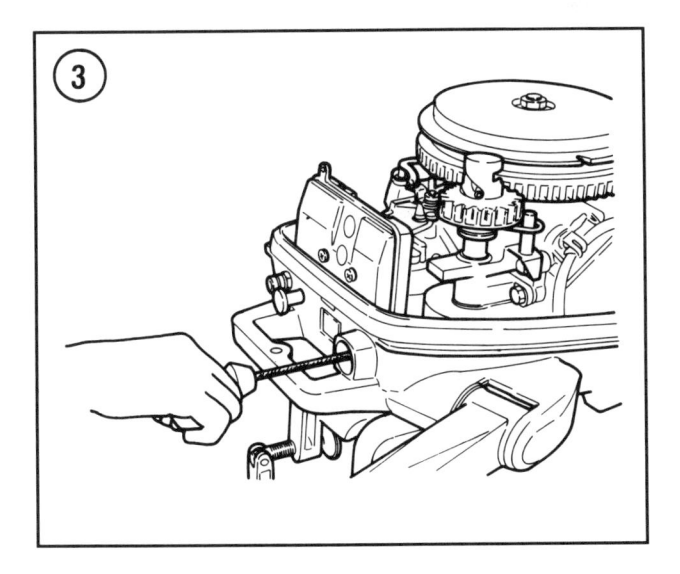

3. Pull the starter handle and rope from the engine lower cover (**Figure 3**).

4. Untie the rope knot and remove the handle. Allow the rope to wind itself onto the spool.

5. Remove the bolts holding the starter housing to the power head (**Figure 4**). Remove the starter housing.

6. Installation is the reverse of removal, noting the following.

7. Pull the starter rope to make sure that the pinion gear and flywheel teeth mesh properly.

Disassembly/Assembly

Refer to **Figure 5** for this procedure.

WARNING
Disassembling this starter mechanism without holding the spring in place can result in the spring unwinding violently and can cause serious personal injury. Wear safety glasses and gloves during this procedure.

1. Place the starter housing on a flat surface, support the pinion gear and drive out the roll pin with a pin punch (**Figure 6**).

2. Remove the pinion gear, bushing and spring assembly.

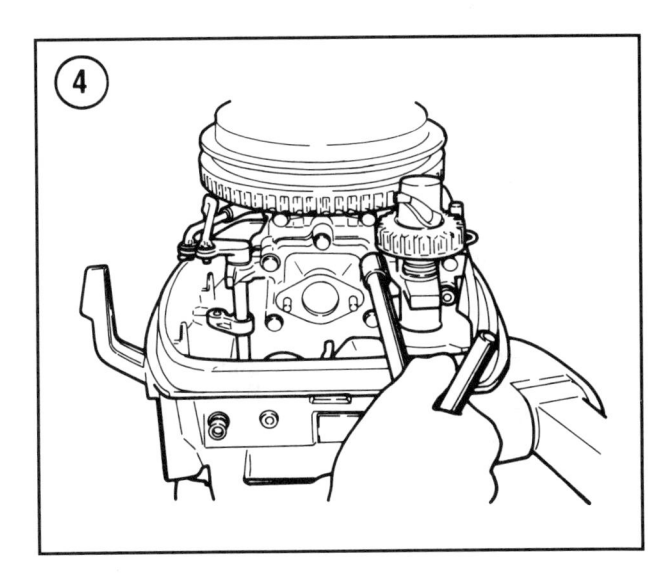

3. Remove the snap ring inside the housing bore with snap ring pliers (**Figure 7**).

4. Cover the starter housing with shop cloths and carefully withdraw the starter spool from the housing (**Figure 8**).

5. If the spring requires replacement, place starter housing on the floor (right side up) and gently tap on its top. The spring will drop down and unwind inside the housing. Remove the housing and discard the spring.

6. If the spring was removed, insert the looped end (A, **Figure 9**) of a new spring into the cutout in the housing (B, **Figure 9**). Carefully wind the spring into the housing bore clockwise, then lightly lubricate with water-resistant grease.

7. Make sure the bushings are properly installed on the starter housing.

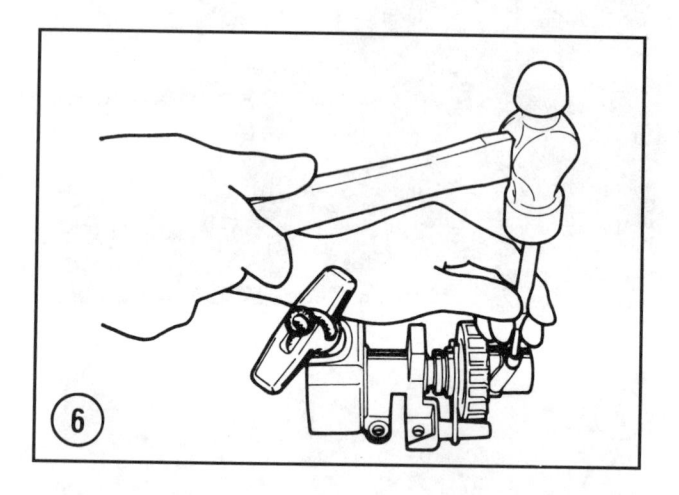

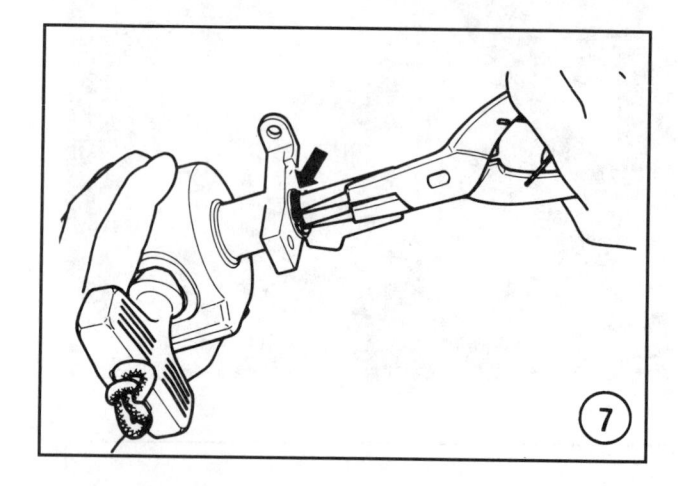

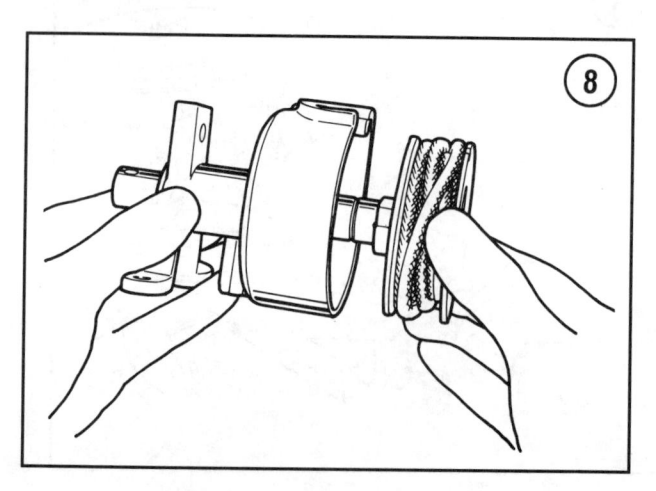

BENDIX STARTER (DT 5–DT 8)

1. Pinion gear
2. Friction spring
3. Snap ring
4. Bushing
5. Housing
6. Spring
7. Roll pin
8. Starter spool
9. Handle

AUTOMATIC REWIND STARTERS

8. Insert the end of the rope into the notch provided in the spool. Wind the rope tightly around the spool pulley in a clockwise direction.

9. Align the free spring loop (A, **Figure 10**) with the pawl cutout (B, **Figure 10**) on the spool, then carefully install spool into housing until spring loop engages spool pawl (**Figure 10**).

10. Install the starter spool snap ring in the housing bore.

11. Install the pinion gear, bushing and spring on the spool.

12. Align the pin holes in the spool, bushing and gear. Support the pinion gear and reinstall the roll pin with a suitable punch.

13. Pull the starter rope through the cutout in the spool (**Figure 11**) and tie a slip knot in the end. Wind the rope into the housing.

14. Install the starter as described in this chapter.

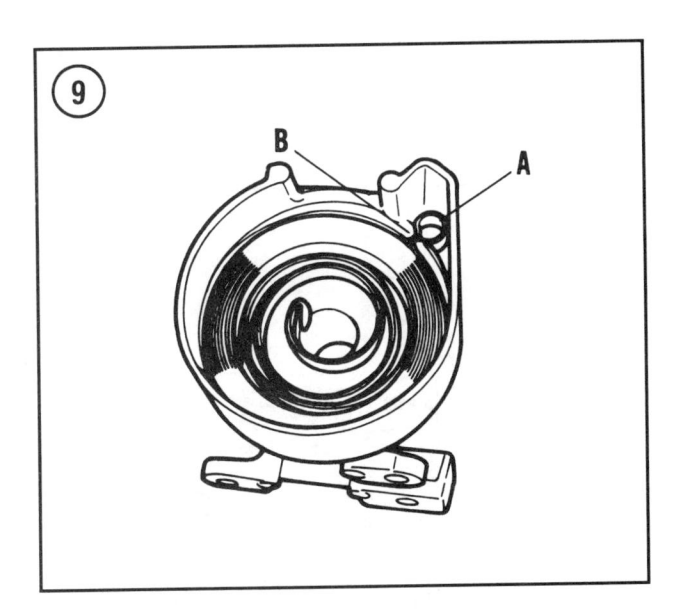

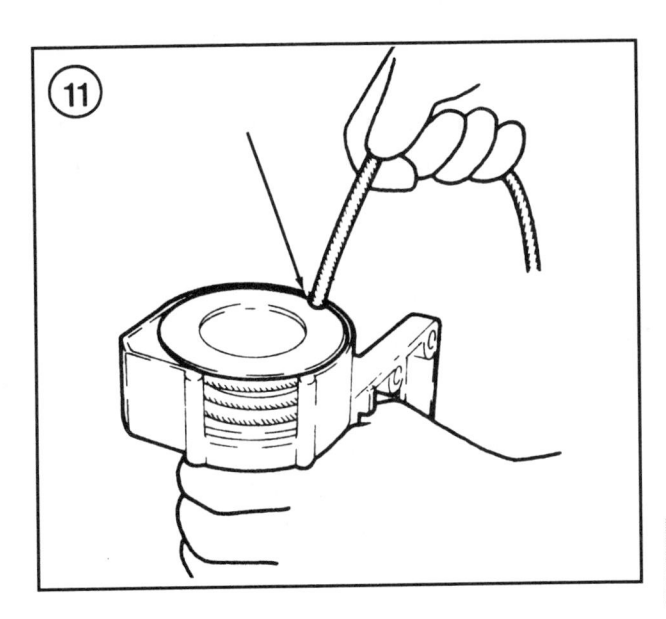

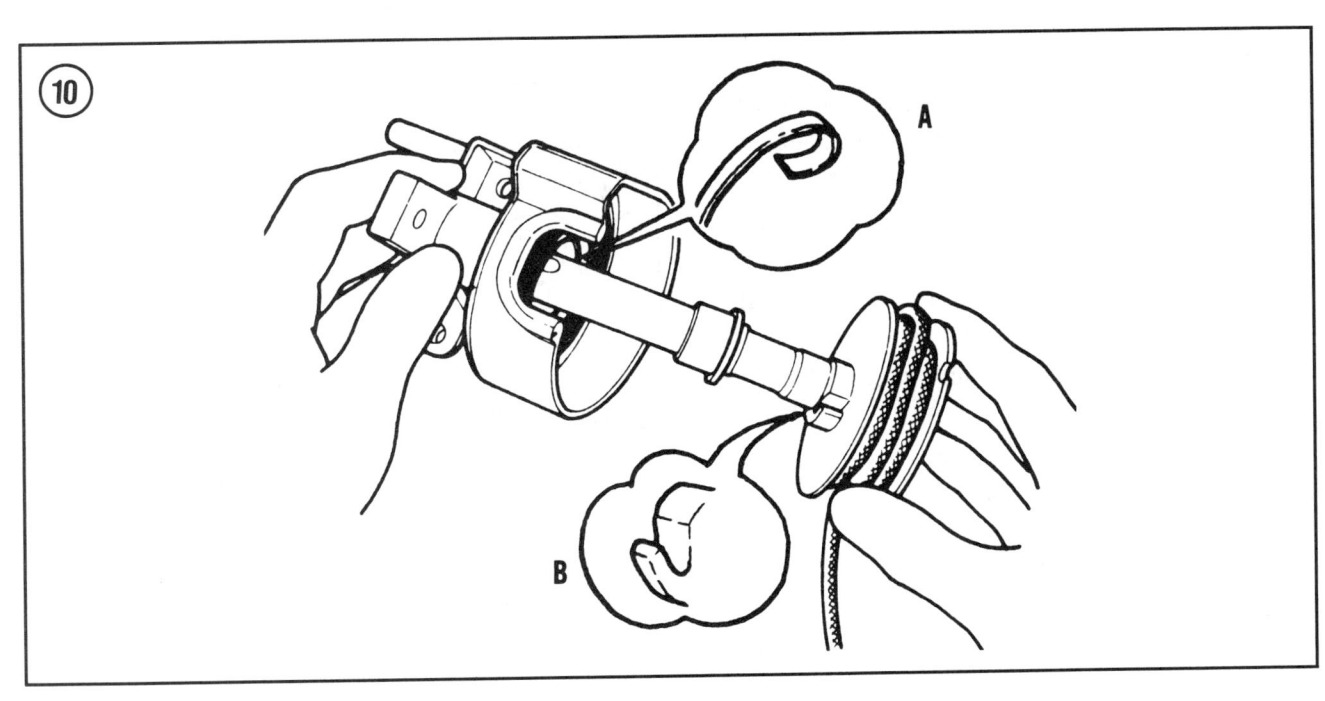

OVERHEAD STARTER
(DT 2, DT 4, DT 6, DT 8, DT 9.9, DT 15, DT 20, DT 25, DT 30, DT 35, DT 40)

This type of starter is mounted in a housing on top of the power head. The starter rope is wound around a spring-loaded pulley. A pawl plate attached to the pulley engages the flywheel drive cup when the rope is pulled. A mechanical neutral safety interlock device is used on some models to prevent the engine from being started in gear.

Several variations of the basic design have been used on Suzuki outboards. Disassembly and assembly of each variation is covered separately.

Removal/Installation
(All Models)

1. Remove the engine cover.
2. Disconnect the spark plug wire(s) to prevent accidental starting of the engine.
3A. On 1988-on DT 8, 1985-on DT 9.9 and DT 15 models, loosen the starter interlock cable locknut and disconnect the cable from the throttle (**Figure 12**).
3B. On DT 20, DT 25, DT 30 models, remove the starter interlock cable holder and disconnect the cable from the interlock arm (**Figure 13**).
3C. On DT 35 and DT 40 models, perform the following:
   a. Disconnect the starter interlock cable from the throttle limiter (**Figure 14**).
   b. Remove the oil tank as described in Chapter Twelve.
4. Remove the bolts holding the starter housing to the power head (**Figure 15**). Remove the starter housing.
5. Installation is the reverse of removal. Pull starter rope to engage the pawl assembly with the drive cup before tightening the bolts. Adjust the

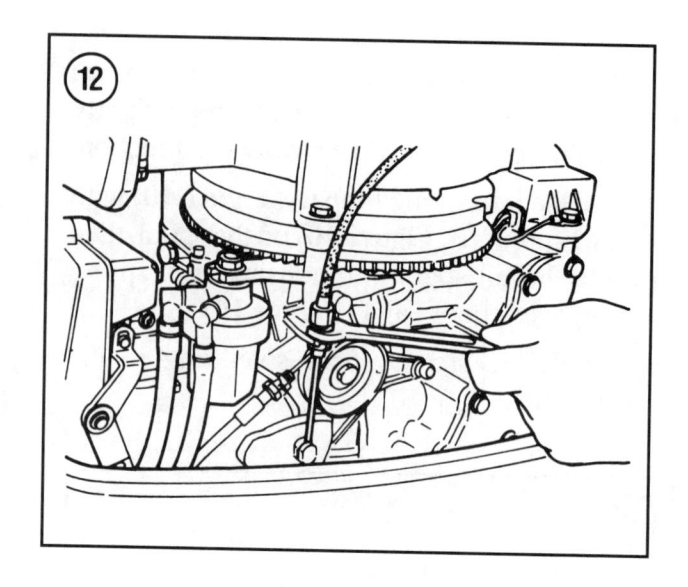

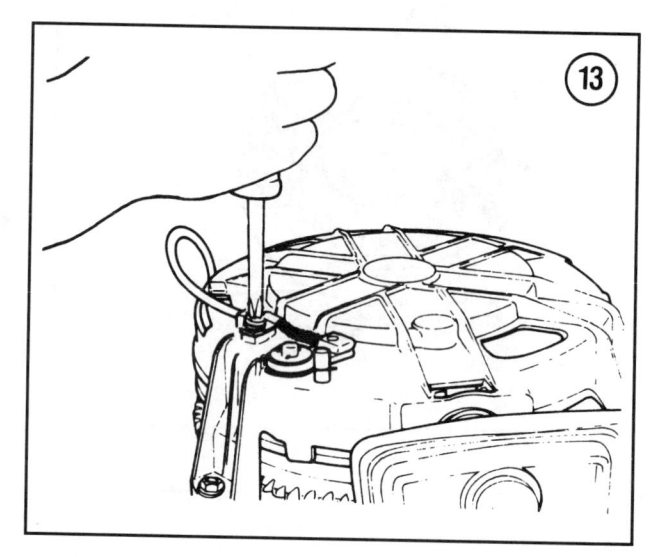

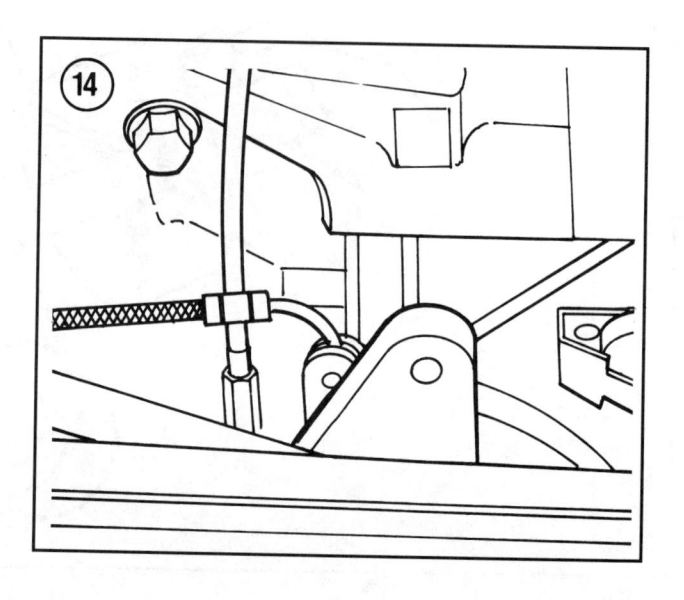

AUTOMATIC REWIND STARTERS

starter interlock (if so equipped) as described in this chapter.

Disassembly/Assembly
(DT 2, DT 4, DT 6; 1985-1987 DT 8)

WARNING
Disassembling this starter mechanism without holding the spring in place can result in the spring unwinding violently and can cause serious personal injury. Wear safety glasses and gloves during this procedure.

Refer to the following illustrations for this procedure:

a. **Figure 16**: DT 2.

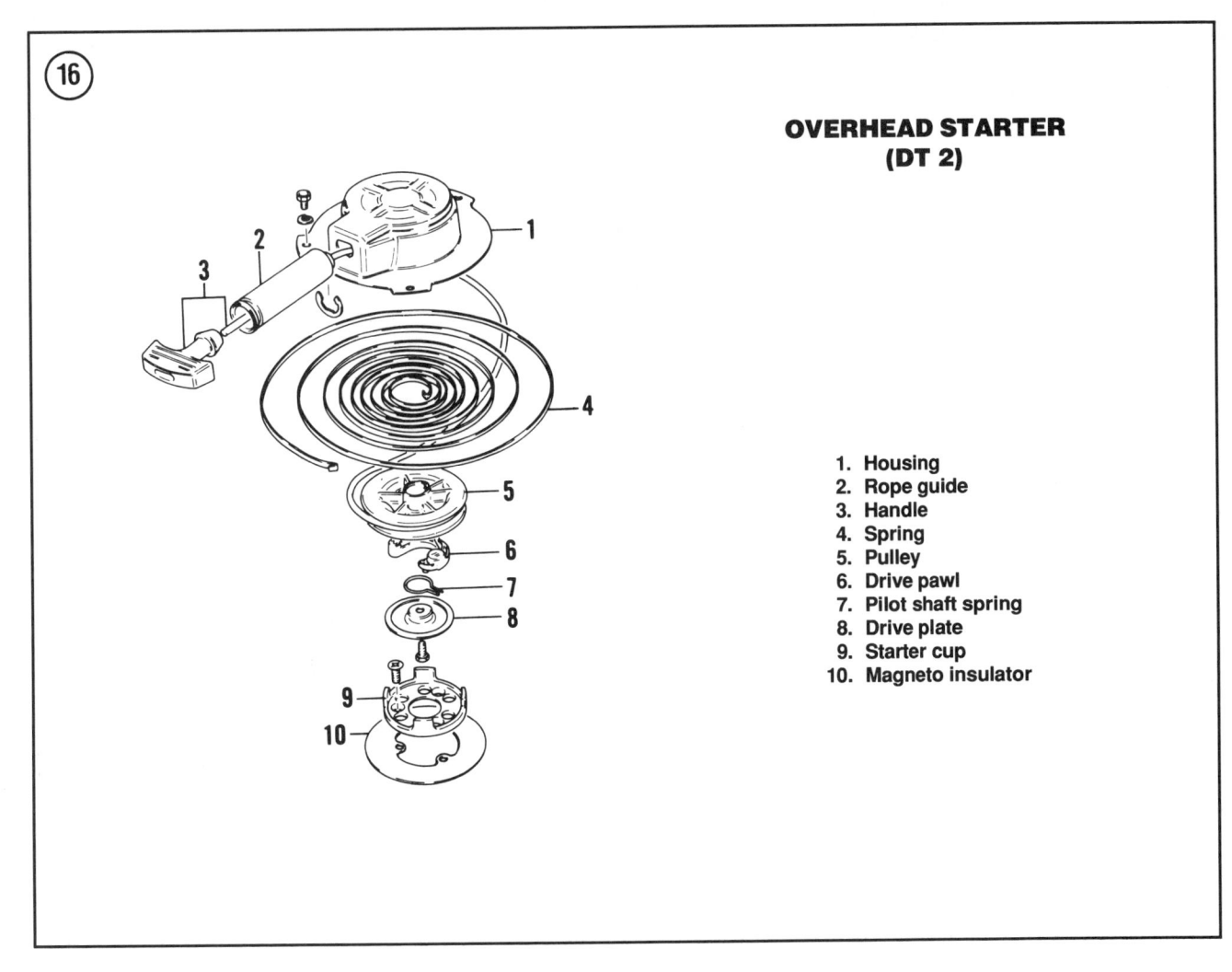

OVERHEAD STARTER (DT 2)

1. Housing
2. Rope guide
3. Handle
4. Spring
5. Pulley
6. Drive pawl
7. Pilot shaft spring
8. Drive plate
9. Starter cup
10. Magneto insulator

b. **Figure 17**: DT 4.

c. **Figure 18**: DT 6, 1985-1987 DT 8.

1. Invert the starter housing and pull the rope out as far as possible. Hold in this position and apply downward pressure on the starter pulley to prevent it from pulling the rope back in.

2. Hook the rope on the pulley notch. Let the pulley slowly rotate *clockwise* to release the spring tension (**Figure 19**).

3A. On DT 2 and DT 4 models, remove the bolt in the center of the pulley (**Figure 20**).

3B. On DT 6 and DT 8 models, pry the snap ring from the shaft in the center of the pulley (**Figure 21**).

4A. On DT 2 models, remove the drive plate (A, **Figure 22**) and pawl (B, **Figure 22**).

4B. On DT 4 models, remove the pawl, friction spring, friction plate and return spring.

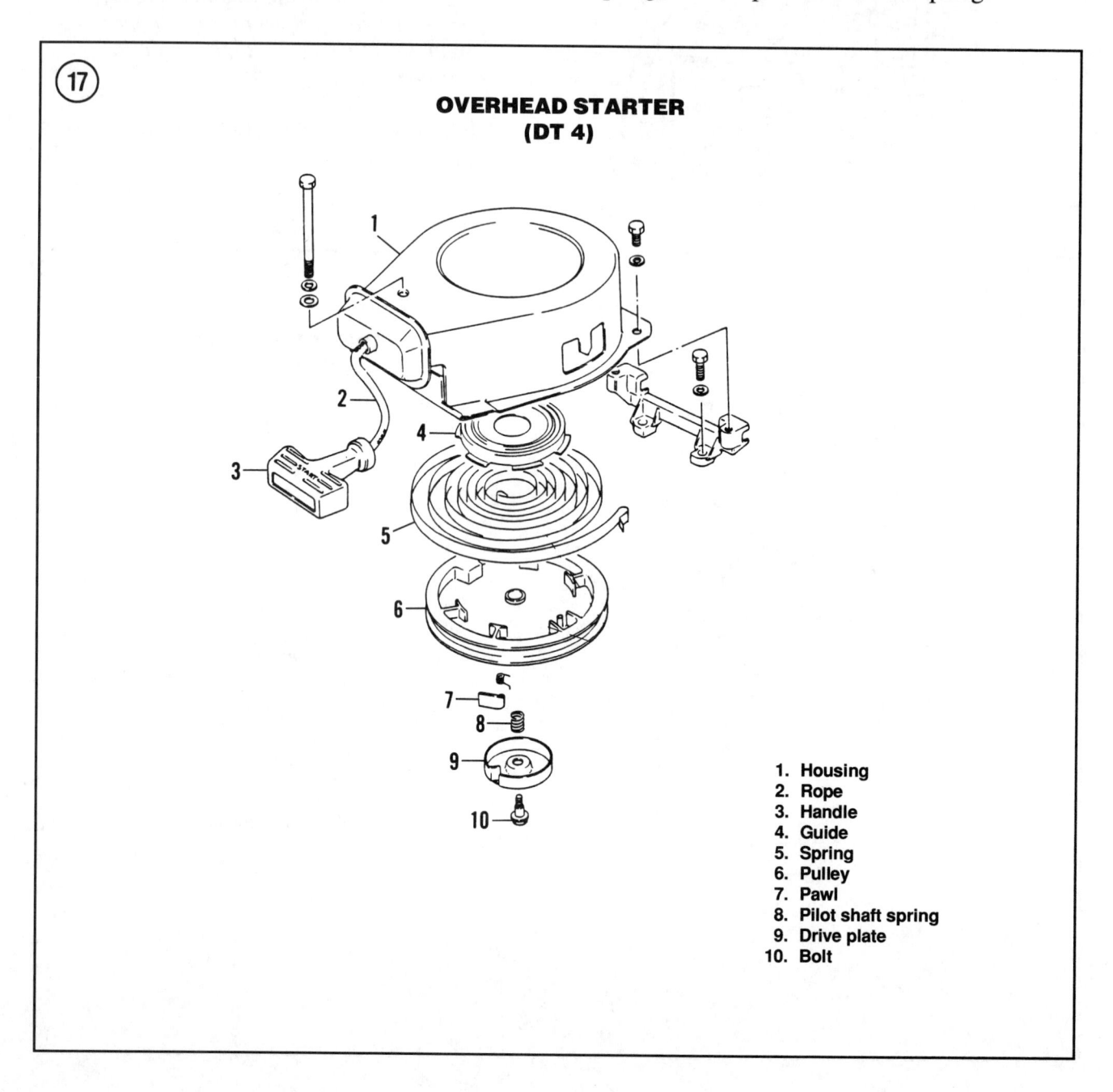

OVERHEAD STARTER (DT 4)

1. Housing
2. Rope
3. Handle
4. Guide
5. Spring
6. Pulley
7. Pawl
8. Pilot shaft spring
9. Drive plate
10. Bolt

AUTOMATIC REWIND STARTERS

465

⑱

**OVERHEAD STARTER
(DT 6; 1985-1987 DT 8)**

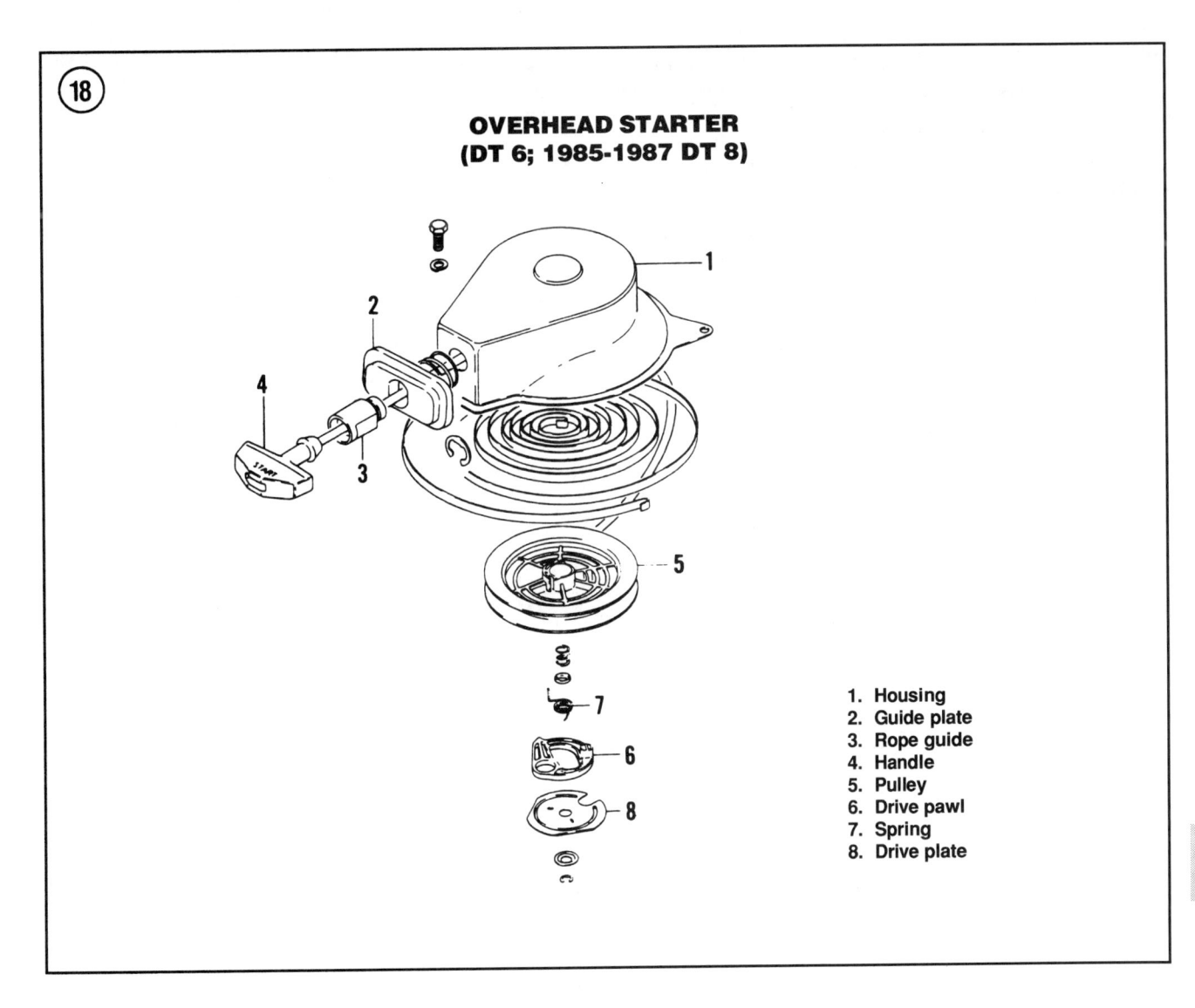

1. Housing
2. Guide plate
3. Rope guide
4. Handle
5. Pulley
6. Drive pawl
7. Spring
8. Drive plate

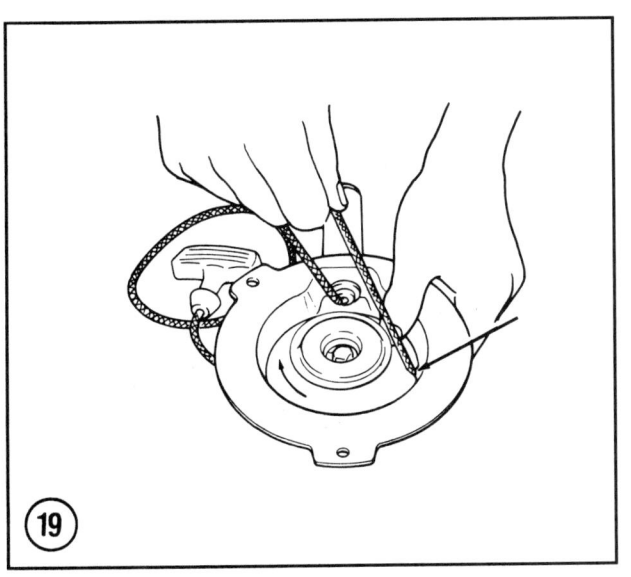

10

⑲

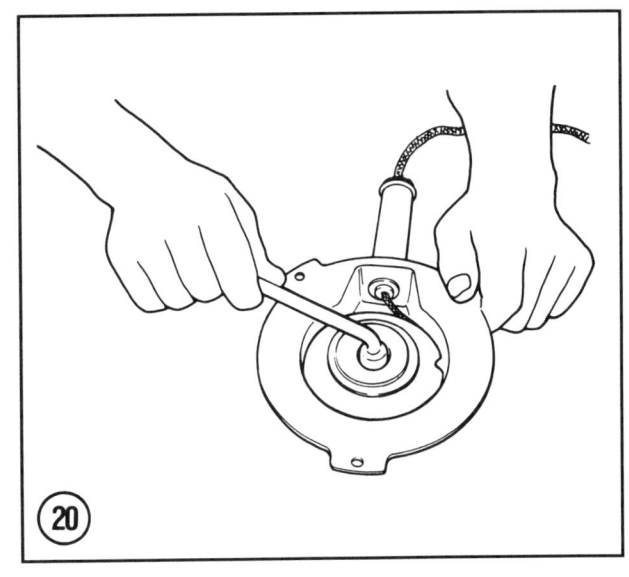

⑳

4C. On DT 6 and DT 8 models, remove drive plate, return spring and spacer (**Figure 23**). Remove the drive pawl (**Figure 24**)

5. Carefully lift the pulley from the housing. If pulley does not come out easily, insert a screwdriver in the pulley hole as shown in **Figure 25** and push the spring loop out of the pulley groove.

6. If the spring requires replacement, place starter housing on the floor (right side up) and gently tap on its top. The spring will drop down and unwind inside the housing. Remove the housing and discard the spring.

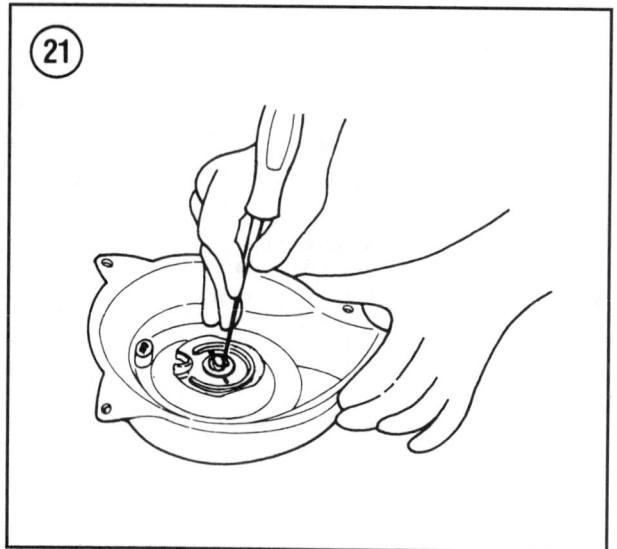

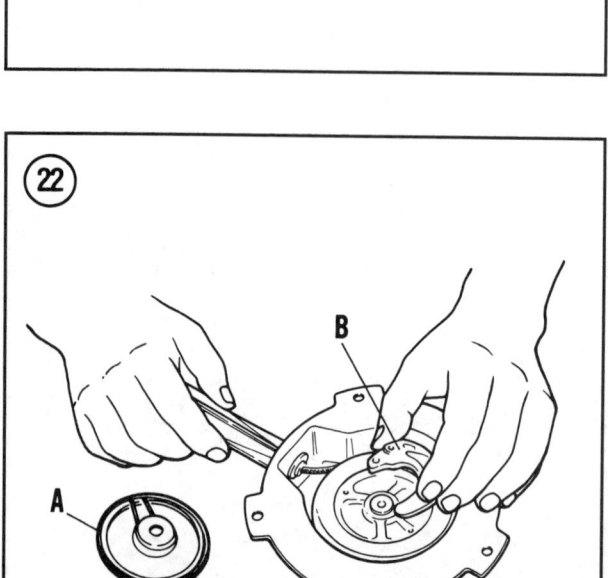

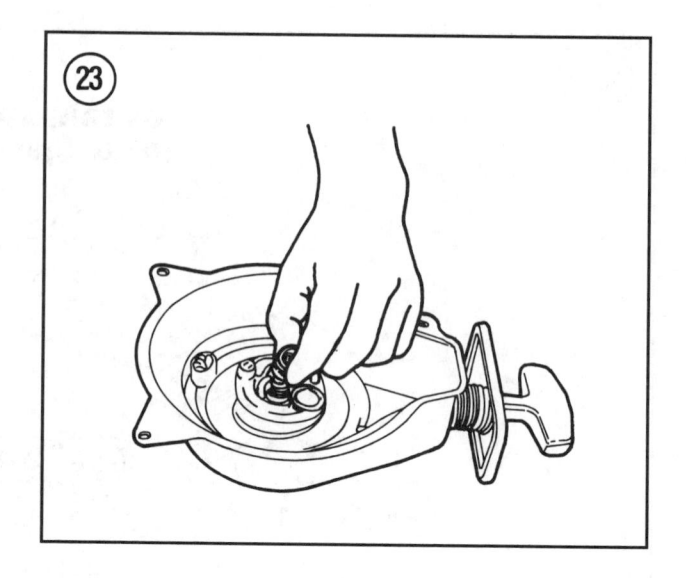

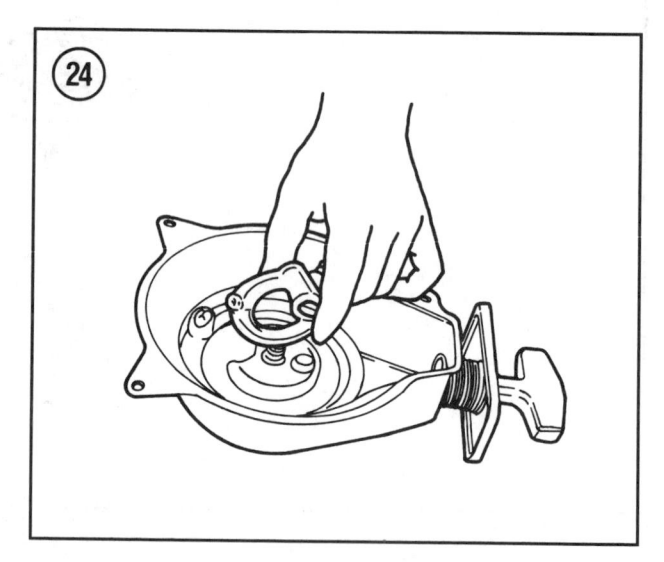

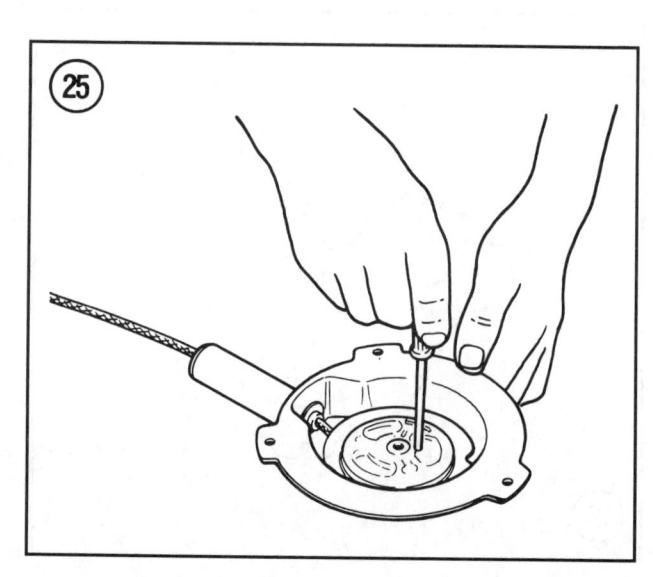

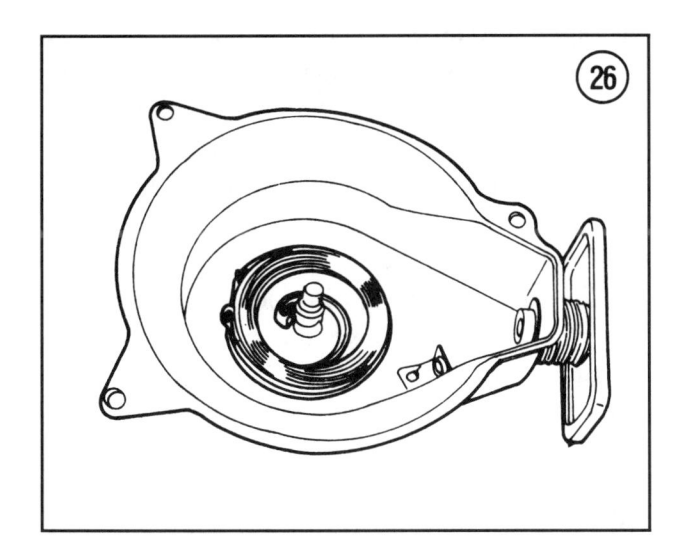

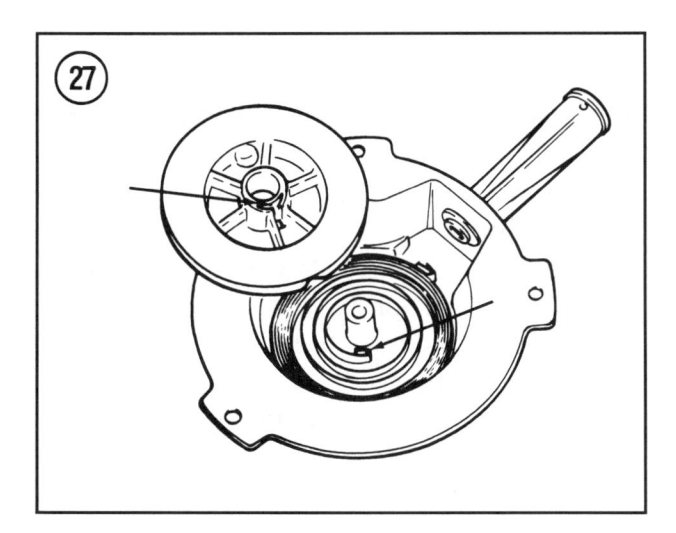

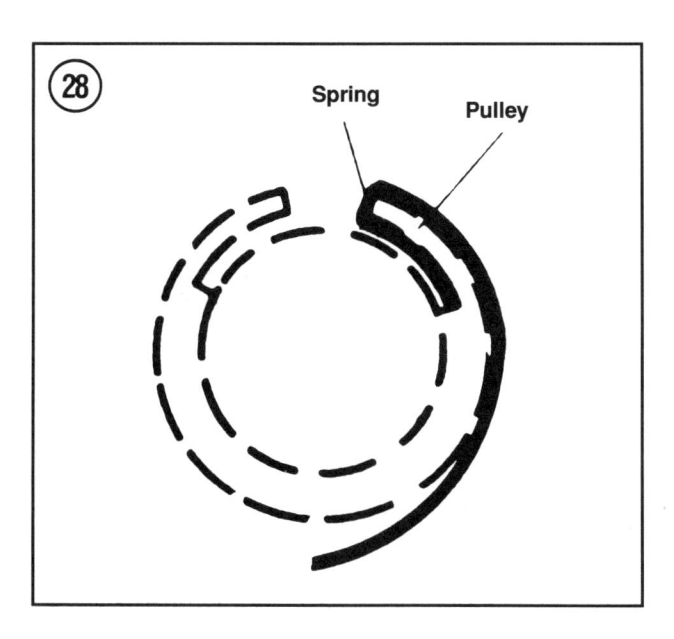

AUTOMATIC REWIND STARTERS

7. If the spring was removed, insert the looped end of a new spring into the housing bore and carefully wind the remainder of the spring into the housing bore in a clockwise direction. When properly installed, it will rest in the bore as shown in **Figure 26**.

8. If the rope requires replacement, attach the new rope to the starter pulley.

9. Install the pulley into the housing so the bent end of the spring will engage the groove in the pulley (**Figure 27**, typical).

10. Rotate the drum slightly *counterclockwise*. If the spring end and pulley groove have engaged properly, tension will be felt.

11. If no tension is felt in Step 10, insert a screwdriver in the pulley hole and rotate the pulley slightly, using the screwdriver to guide the spring over the pulley groove as shown in **Figure 28**.

12. Lubricate all parts with water-resistant grease.

13A. On DT 2 models, perform the following:
  a. Install the washer (A, **Figure 29**).
  b. Position the drive pawl (B, **Figure 29**) into housing and engage pilot shaft plate spring over the drive pawl pivot pin (C, **Figure 29**).
  c. Seat pilot shaft plate on drive pawl. Install and tighten the pulley bolt securely.

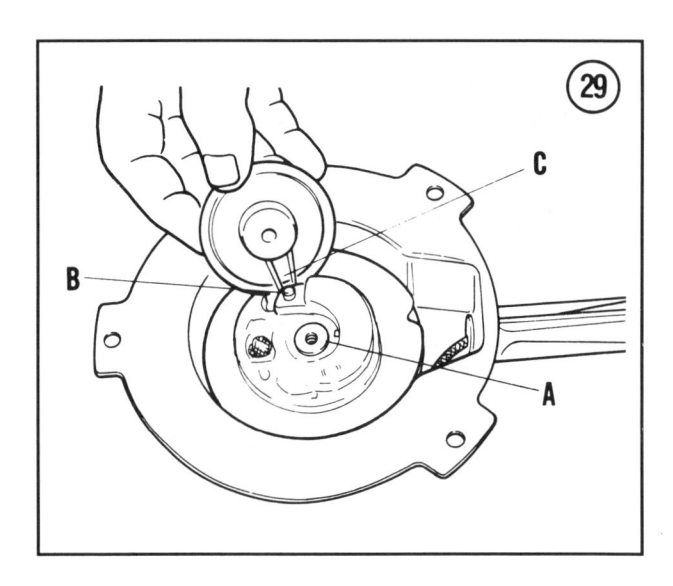

13B. On DT 4 models, perform the following:
   a. Install the return spring with the hooked end to the top side and insert it into the hole provided in the starter pulley.
   b. Install the friction plate and friction spring.
   c. Install the pawl making sure it fits into the notch.

13C. On DT 6 and DT 8 models, perform the following:
   a. Install the drive pawl and spacer. Engage one end of the return spring with the drive plate hole (A, **Figure 30**).
   b. Engage the other end of the spring with the drive pawl hole (B, **Figure 30**).
   c. Install the washer and snap ring.

14. Feed the rope through the housing and attach the handle.

15. Hook the rope on the pulley notch and rotate the pulley *counterclockwise* the following number of turns:
   a. DT 2: 6 turns.
   b. DT 4: 6 1/2 turns.
   c. DT 6 and DT 8: 6 turns.

16. Apply finger pressure to the pulley, then unhook the rope from the pulley notch and slowly allow the pulley to wind the rope under spring tension.

17. Install the starter housing as described in this chapter.

Disassembly/Assembly (1988-on DT 8; DT 9.9; DT 15)

> **WARNING**
> *Disassembling this starter mechanism without holding the spring in place can result in the spring unwinding violently and can cause serious personal injury. Wear safety glasses and gloves during this procedure.*

Refer to **Figure 31** for 1985-1987 DT 9.9 and 1985-on DT 15 models or **Figure 32** for 1988-on DT 8 and DT 9.9 models for this procedure.

1. Invert the starter housing and pull the rope out as far as possible. Hold in this position and apply downward pressure on the starter pulley to prevent it from pulling the rope back in.

2. Hook the rope on the pulley notch. Let the pulley slowly rotate clockwise to release the spring tension (**Figure 33**).

3. Remove the cotter pin and washer from the interlock arm. Remove the interlock arm (A, **Figure 34**) from the pivot (B, **Figure 34**). Remove the interlock spring and pivot.

4A. On 1985-1987 DT 9.9 and 1985-on DT 15 models, remove the stopper spring (A, **Figure 35**) and stopper arm (B, **Figure 35**).

4B. On 1988-on DT 8 and DT 9.9, remove the stopper spring, stopper arm and NSI cable.

5. Remove the drive plate bolt (**Figure 36**).

6. Remove the drive plate with friction spring (**Figure 37**).

7. Remove the drive pawl and spring from the pulley (**Figure 38**).

8. Carefully lift the pulley from the housing making sure the spring stays in the housing. If pulley does not come out easily, carefully insert the tip of a thin-blade screwdriver under the pulley and disengage the spring loop from the pulley.

9. On 1988-on models, remove the rope guide and holder guide.

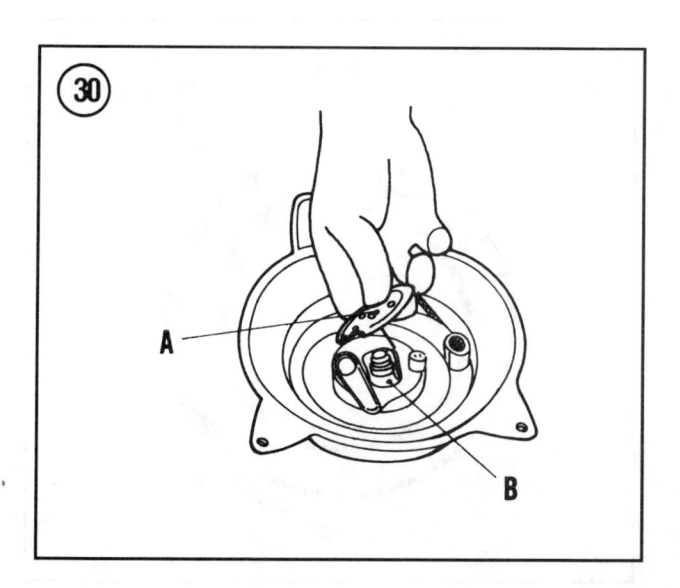

30

AUTOMATIC REWIND STARTERS

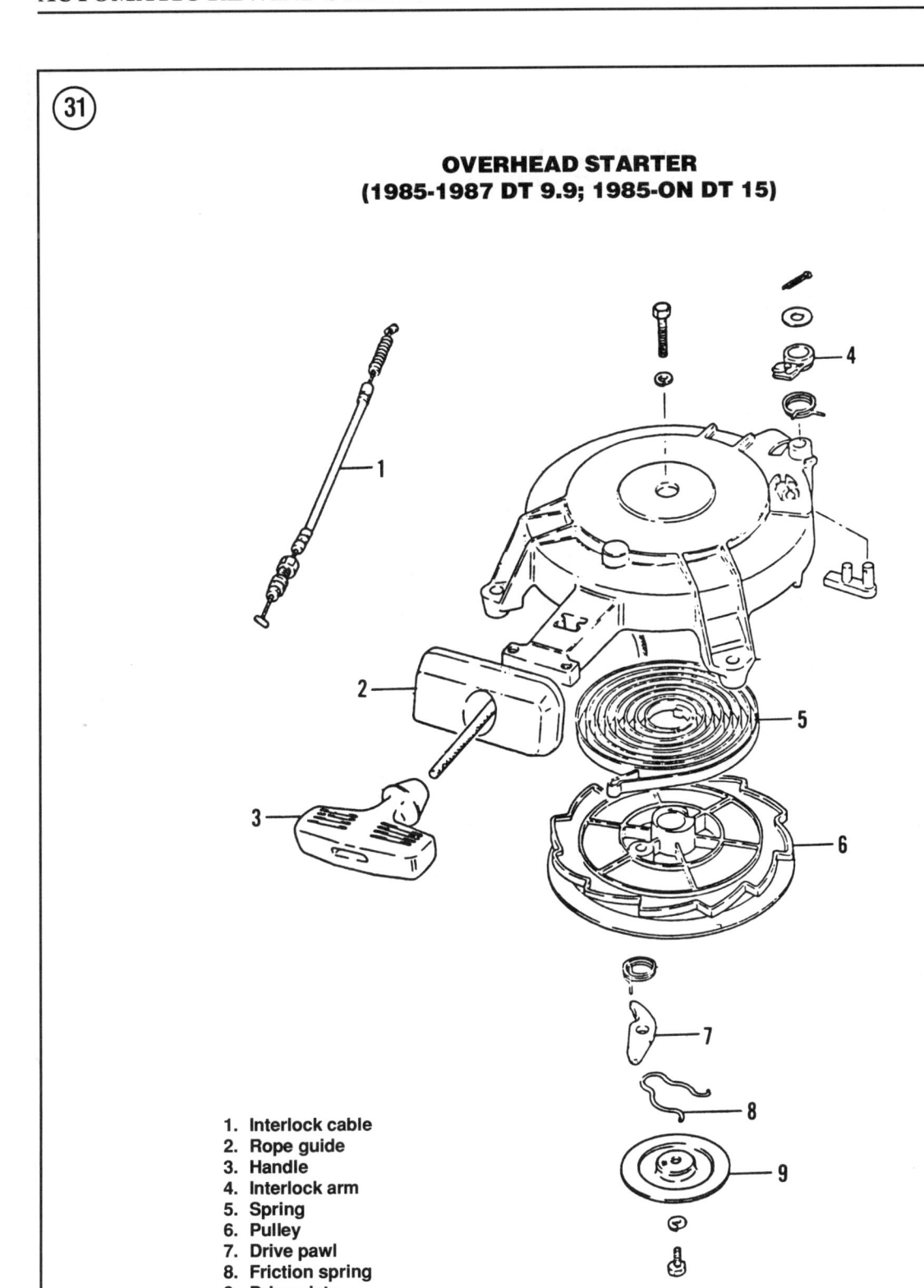

**OVERHEAD STARTER
(1985-1987 DT 9.9; 1985-ON DT 15)**

1. Interlock cable
2. Rope guide
3. Handle
4. Interlock arm
5. Spring
6. Pulley
7. Drive pawl
8. Friction spring
9. Drive plate

CHAPTER TEN

**OVERHEAD STARTER
(1988-ON DT 8 AND DT 9.9)**

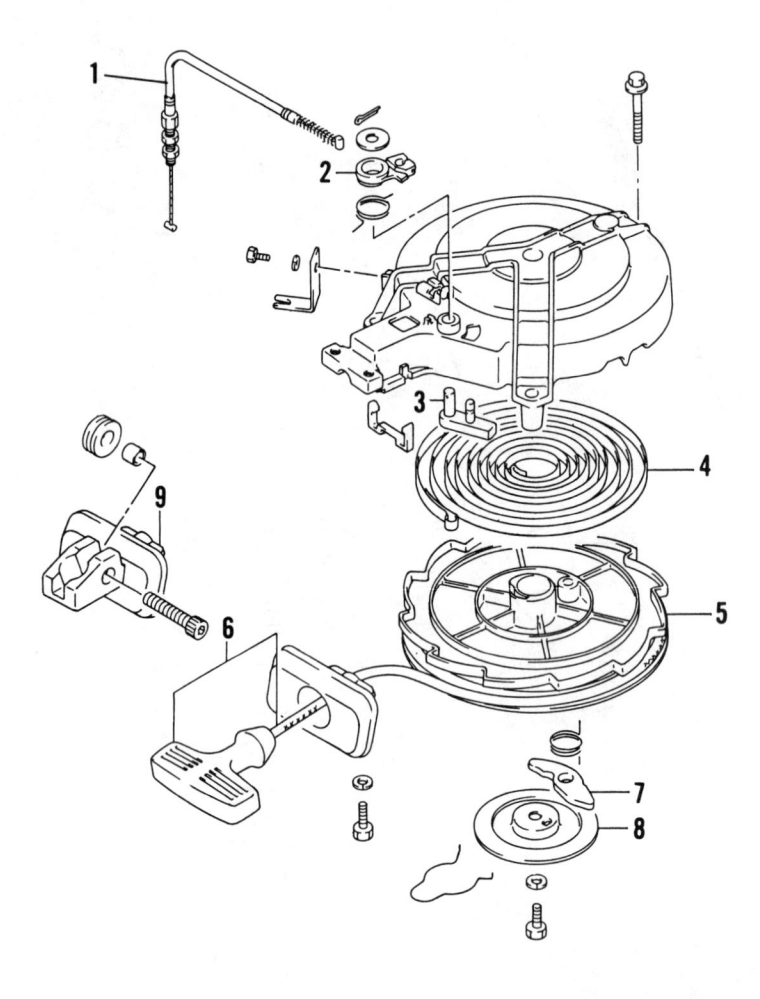

1. Neutral start interlock (NSI) cable
2. Stopper arm
3. Interlock arm
4. Spring
5. Pulley
6. Handle
7. Drive pawl
8. Plate
9. Rope guide

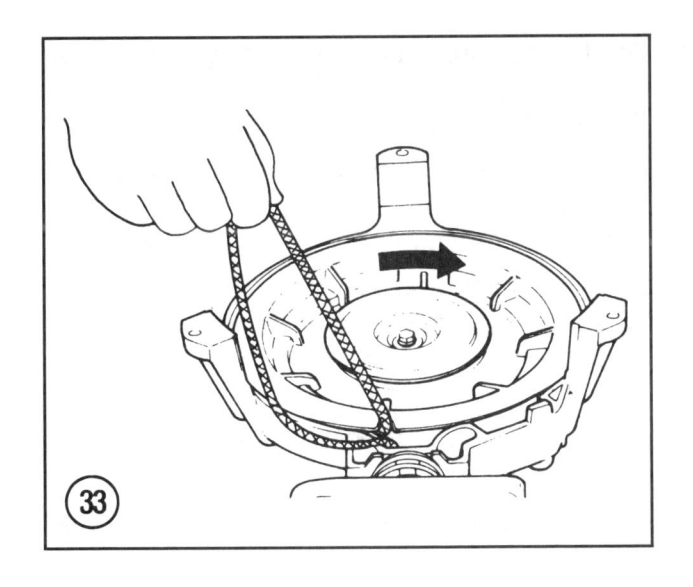

AUTOMATIC REWIND STARTERS

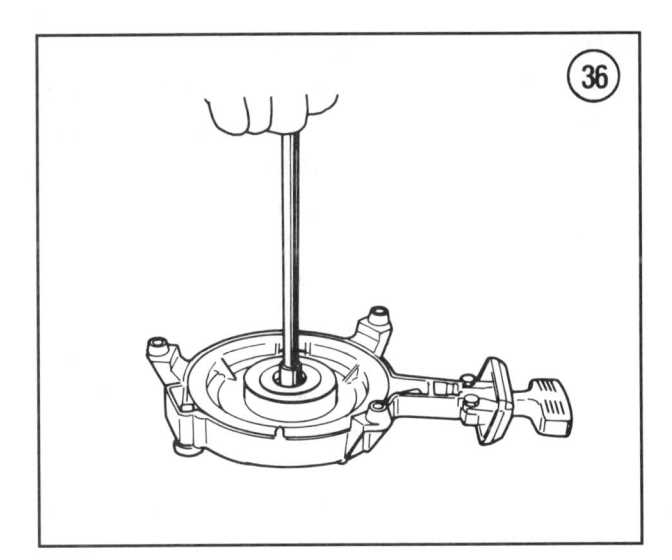

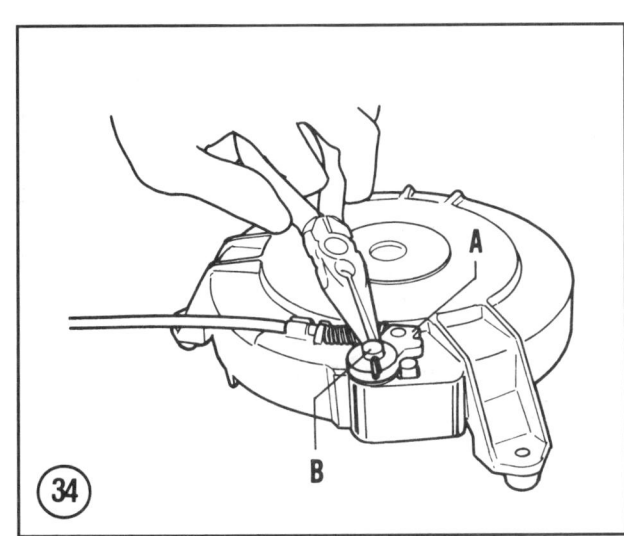

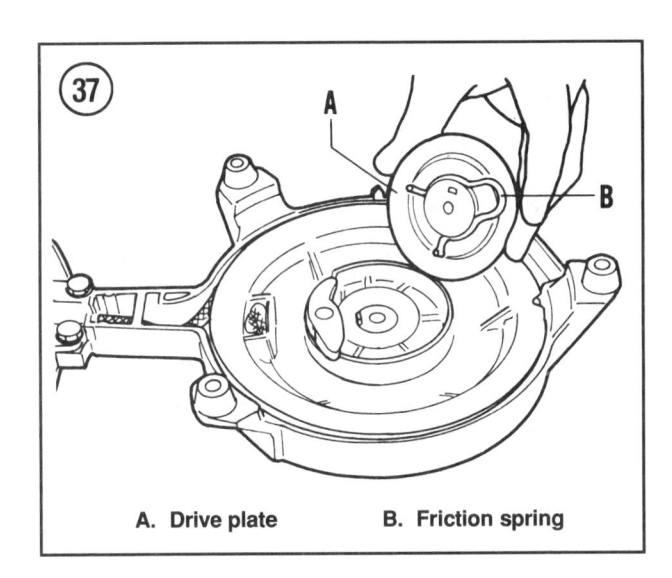

471

A. Drive plate   B. Friction spring

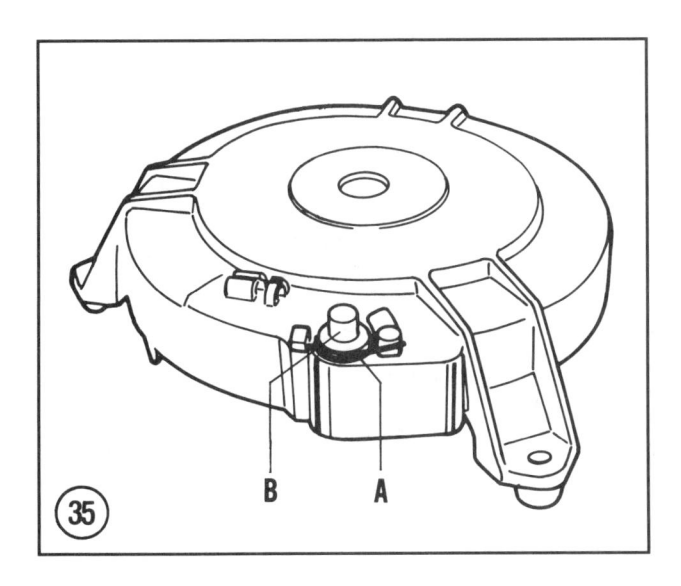

33

34

35

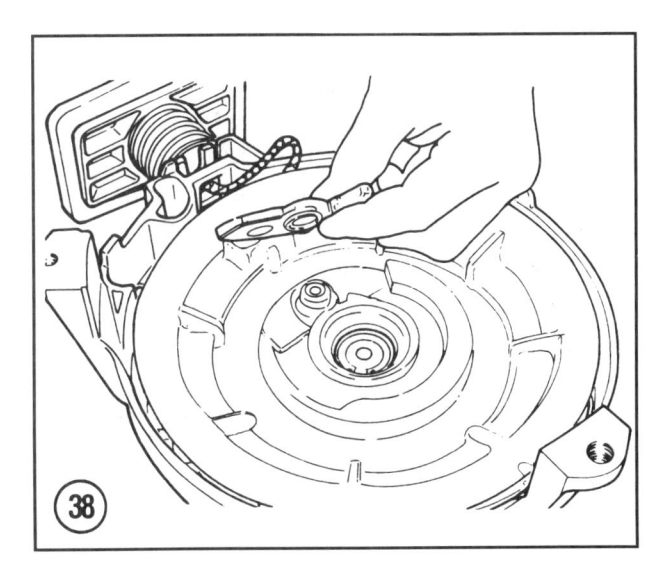

36

37

38

10

10. Remove the interlock arm.

11. If the spring requires replacement, place starter housing on the floor (spring side facing down) and gently tap on its top. The spring will drop down and unwind inside the housing. Remove the housing and discard the spring.

12. Lubricate all parts with water-resistant grease.

13. If the spring was removed, insert the looped end of a new spring into the housing bore (**Figure 39**) and carefully wind the remainder of the spring into the housing bore (**Figure 40**).

14. If the rope requires replacement, attach the new rope to the starter pulley.

15. On 1988-on models, install the holder guide and rope guide.

16. Install the interlock device (A, **Figure 41**).

17. Install the pulley into the housing so the bent end of the spring (B, **Figure 41**) will engage the groove in the pulley (C, **Figure 41**).

18. Rotate the pulley slightly counterclockwise. If the spring end and pulley groove have engaged properly, tension will be felt.

19. If no tension is felt in Step 18, carefully remove the pulley, realign the spring loop and pulley groove and repeat Step 18.

20. Insert the short bent end of the drive pawl spring in the pulley hole, then connect the long bent end to the drive pawl groove (A, **Figure 42**). Install drive pawl.

21. Install drive plate with friction spring. Align drive plate hole (A, **Figure 43**) with housing boss (B, **Figure 43**), and spring end with pulley groove. Seat drive plate in position.

22. Install pulley bolt and washer. Tighten securely.

23. Feed the rope through the housing and attach the handle.

24. Wind the rope counterclockwise around the pulley 2 1/2 turns, then hook the rope on the pulley notch. Rotate the pulley counterclockwise another 4 full turns.

25. Apply finger pressure to the pulley, then unhook the rope from the pulley notch and slowly allow the pulley to wind the rope under spring tension.

26A. On 1985-1987 DT 9.9 and all DT 15 models, install the reel stopper and the stopper spring.

26B. On 1988-on DT 8 and DT 9.9, install the NSI cable, reel stopper and the stopper spring.

27. Install the starter interlock pivot and spring. Connect the interlock arm to the pivot. Install the washer and a new cotter pin. Bend the ends over completely.

28. Install the starter housing as described in this chapter.

29. Adjust the starter interlock as described in this chapter.

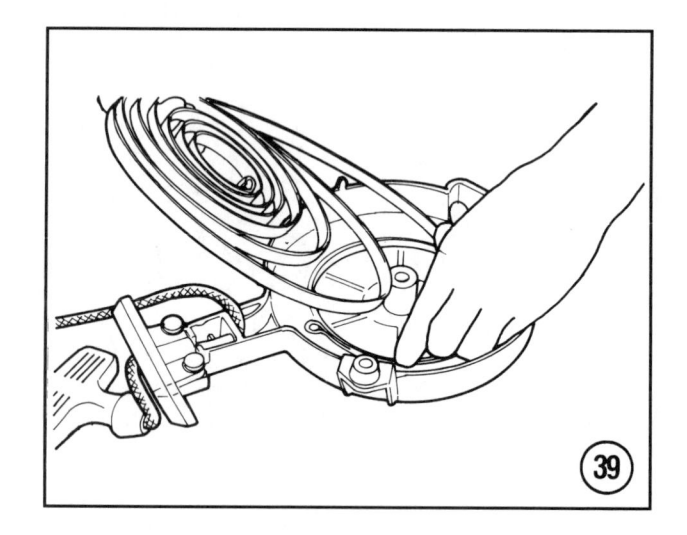

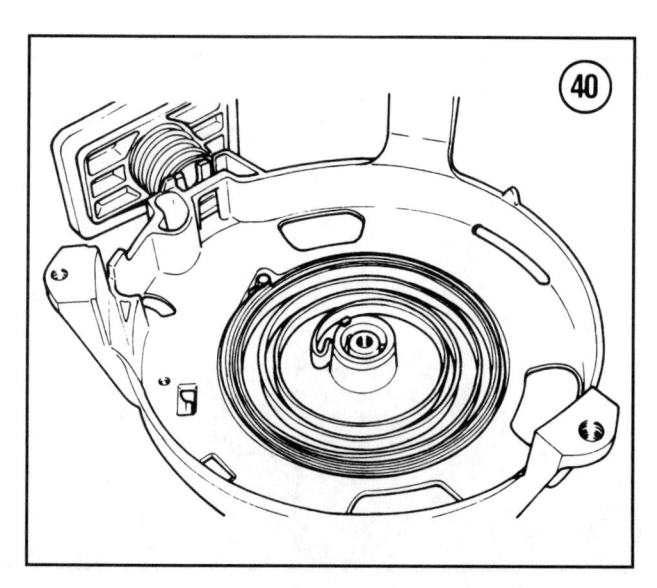

AUTOMATIC REWIND STARTERS

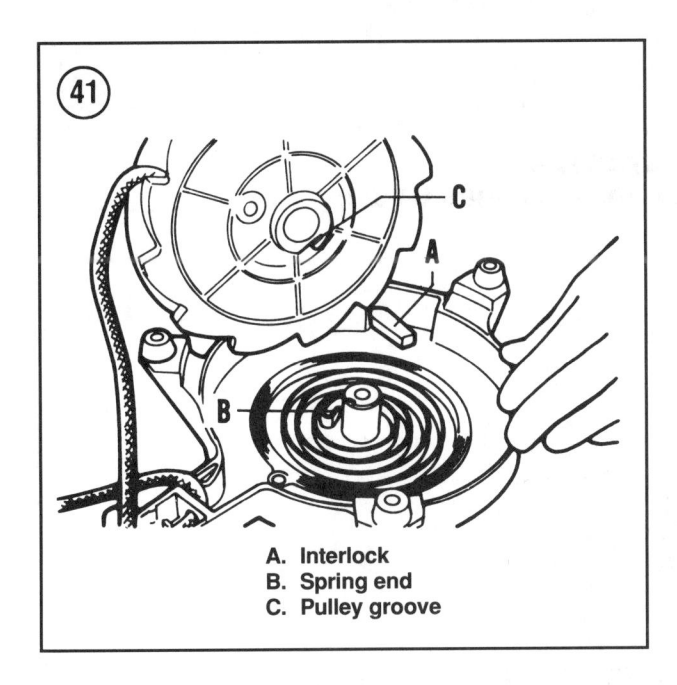

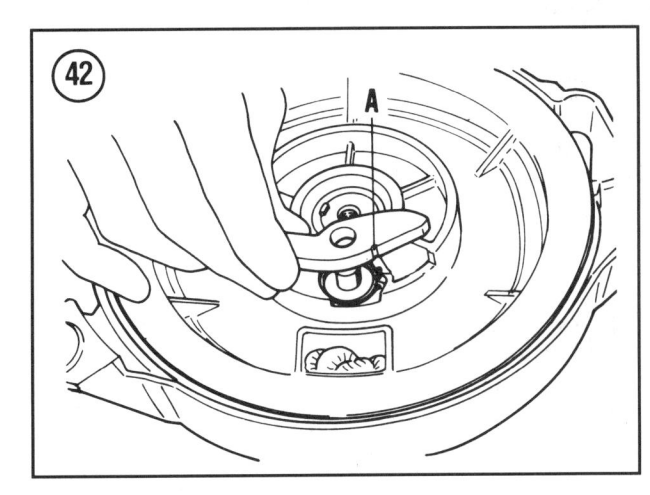

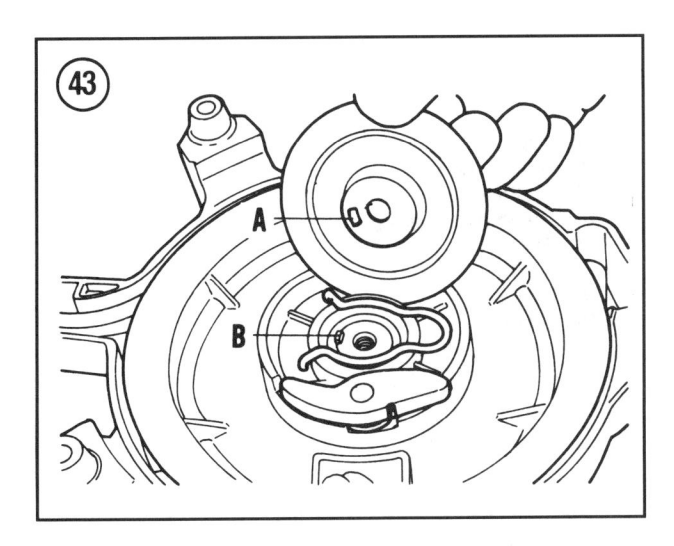

41

A. Interlock
B. Spring end
C. Pulley groove

42

43

Disassembly/Assembly (DT 20; DT 25, DT 30)

WARNING
Disassembling this starter mechanism without holding the spring in place can result in the spring unwinding violently and can cause serious personal injury. Wear safety glasses and gloves during this procedure.

Refer to **Figure 44** for DT 20, DT 25 and DT 30 (2-cylinder) models or **Figure 45** for DT 25 and DT 30 (3-cylinder) models for this procedure:

1A. On 2-cylinder models, perform the following:
   a. Remove the cotter pin and washer from the interlock arm.
   b. Remove the stopper arm (A, **Figure 46**) from the interlock arm (B, **Figure 46**).
   c. Remove the stopper spring (A, **Figure 47**) and the interlock arm (B, **Figure 47**).

1B. On 3-cylinder models, perform the following:
   a. Remove the cotter pin and washer from the interlock arm.
   b. Remove the stopper arm, spring and interlock arm.

2. Invert the starter housing and pull the rope out as far as possible. Hold in this position and apply downward pressure on the starter pulley to prevent it from pulling the rope back in.

3. Hook the rope on the pulley notch. Let the pulley slowly rotate clockwise to release the spring tension (**Figure 33**).

4. Remove the drive plate bolt (**Figure 48**).

5. Remove the drive plate with friction spring (**Figure 37**). On 2-cylinder models, also remove the spacer.

6. Remove the drive pawl and spring from the pulley (**Figure 38**).

7. On 2-cylinder models, remove the snap ring from pulley bore with snap ring pliers (**Figure 49**).

(44)

OVERHEAD STARTER
(1985-1987 DT 20 AND DT 30; 1985-1988 DT 25)

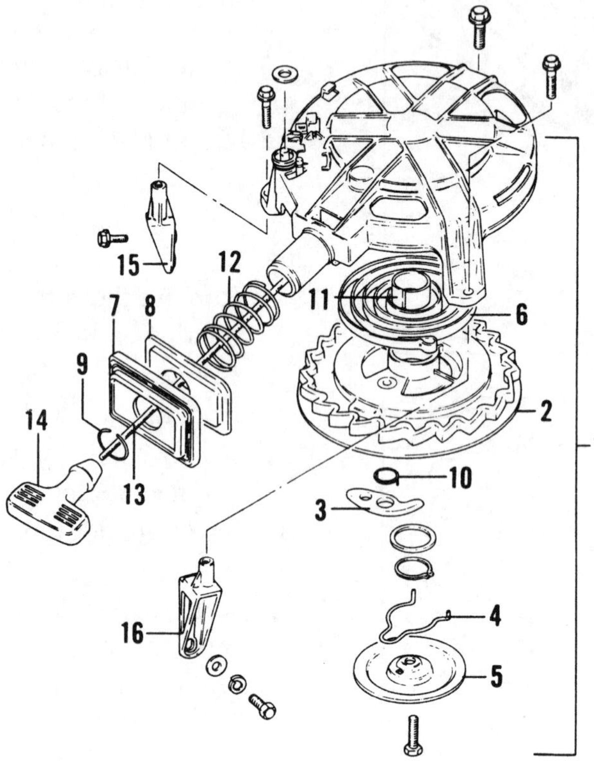

1. Starter assembly
2. Pulley
3. Drive pawl
4. Spring
5. Drive plate
6. Spring
7. Guide plate gasket
8. Guide plate
9. Clip
10. Return spring
11. Bushing
12. Guide plate spring
13. Rope
14. Handle
15. Brace
16. Brace

AUTOMATIC REWIND STARTERS

**OVERHEAD STARTER
(1989-ON DT 25, 1988-ON DT 30)**

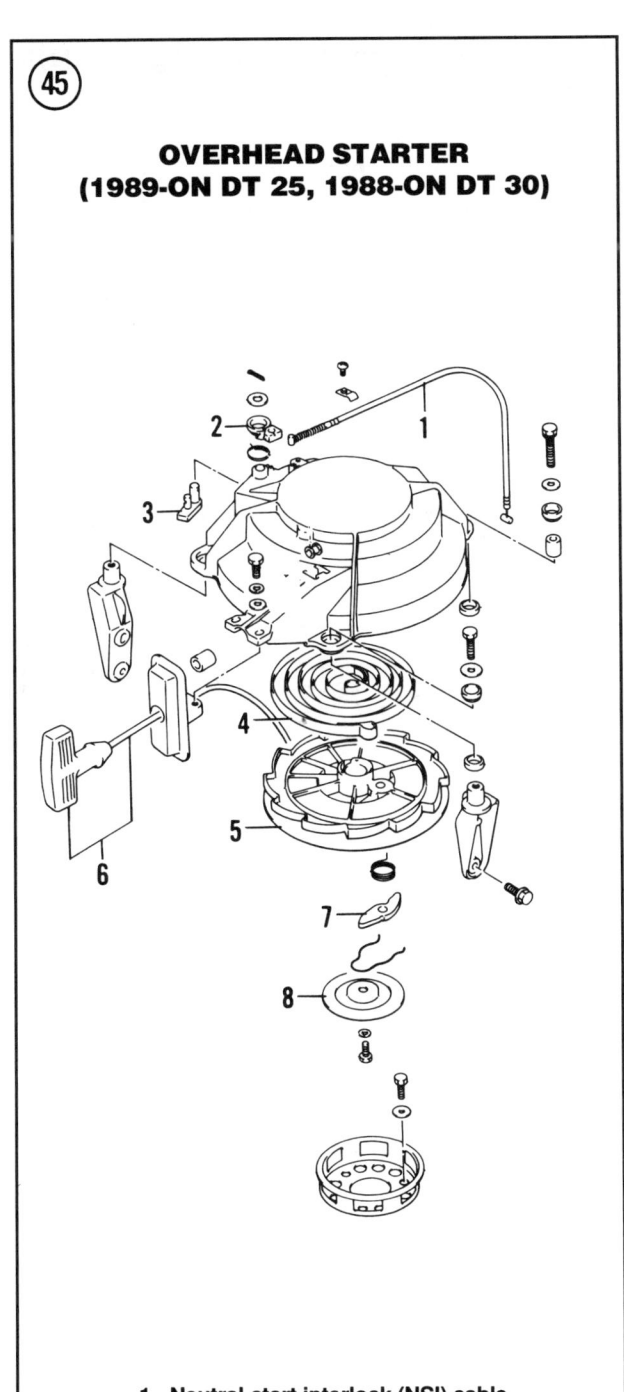

1. Neutral start interlock (NSI) cable
2. Stopper arm
3. Stopper lever
4. Spring
5. Pulley
6. Handle
7. Drive pawl
8. Plate

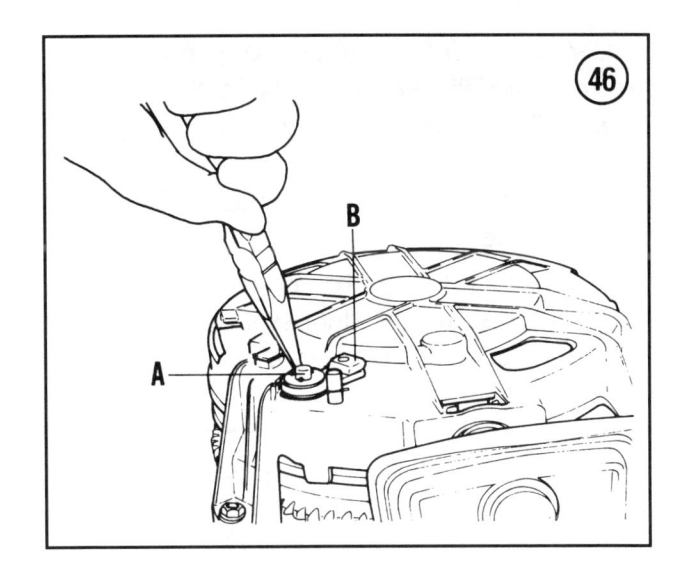

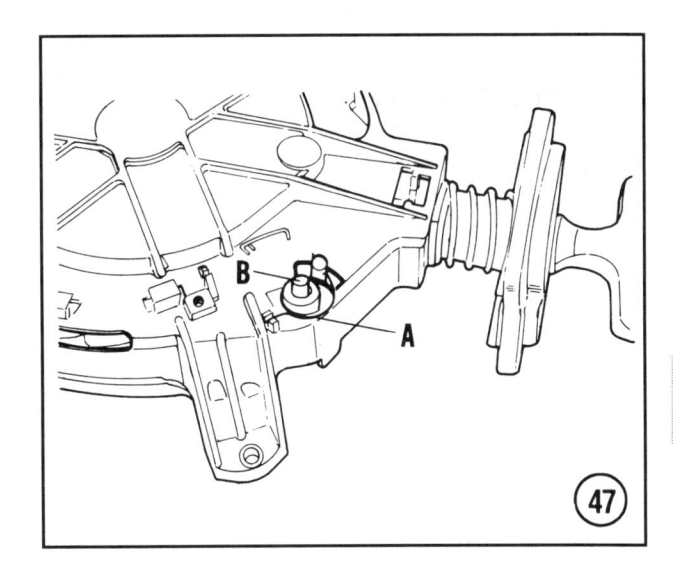

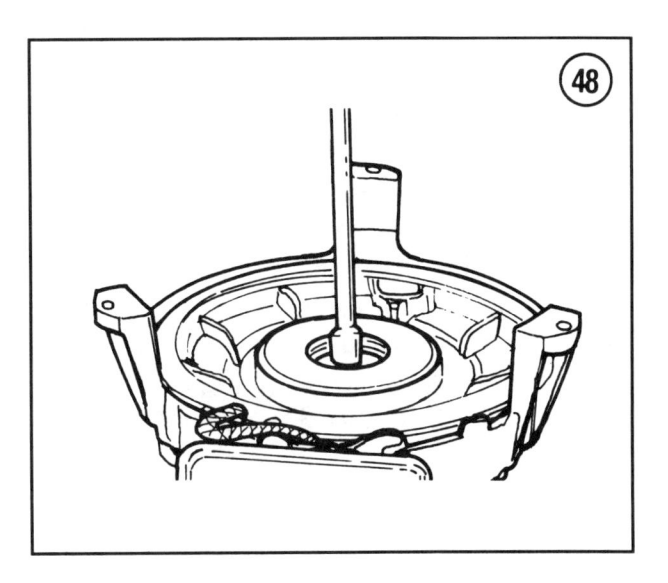

8. Carefully lift the pulley from the housing making sure the spring stays in the housing (**Figure 50**). If pulley does not come out easily, carefully insert the tip of a thin-blade screwdriver under the pulley and disengage the spring loop from the pulley.

9. If the spring requires replacement, place starter housing on the floor (spring side facing down) and gently tap on its top. The spring will drop down and unwind inside the housing. Remove the housing and discard the spring.

10. Lubricate all parts with water-resistant grease.

11. If the spring was removed, insert the looped end of a new spring into the housing bore and carefully wind the remainder of the spring into the housing bore (**Figure 40**).

12. If the rope requires replacement, attach the new rope to the starter pulley.

13. Install the pulley into the housing so the bent end of the spring (B, **Figure 51**) will engage the groove in the pulley (A, **Figure 51**). Rotate the pulley slightly counterclockwise. If the spring end and pulley groove have engaged properly, tension will be felt.

14. If no tension is felt in Step 13, carefully remove the pulley, realign the spring loop and pulley groove and repeat Step 13.

15. On 2-cylinder models, install the snap ring in the pulley bore and make sure it is correctly seated.

16. Place the short bent end of the drive pawl spring (A, **Figure 52**) onto the drive pawl then the other end (B, **Figure 52**) into the hole (C, **Figure 52**) in the pulley. Install drive pawl.

17. Install drive plate, spacer (2-cylinder models) with friction spring. Align drive plate hole (A, **Figure 53**) with housing boss (B, **Figure 53**), then align the spring end (C, **Figure 53**) with pulley groove (D, **Figure 53**). Seat drive plate in position.

18. Install pulley bolt and washer. Tighten securely.

19. Feed the rope through the housing and attach the handle.

20. Wind the rope counterclockwise around the pulley 2 1/2 turns, then hook the rope into the pulley notch. Rotate the pulley counterclockwise another 4 full turns.

21. Apply finger pressure to the pulley, then unhook the rope from the pulley notch and slowly allow the pulley to wind the rope under spring tension.

22A. On 2-cylinder models, perform the following:

   a. Install the interlock arm (B, **Figure 47**) and the stopper spring (A, **Figure 47**).

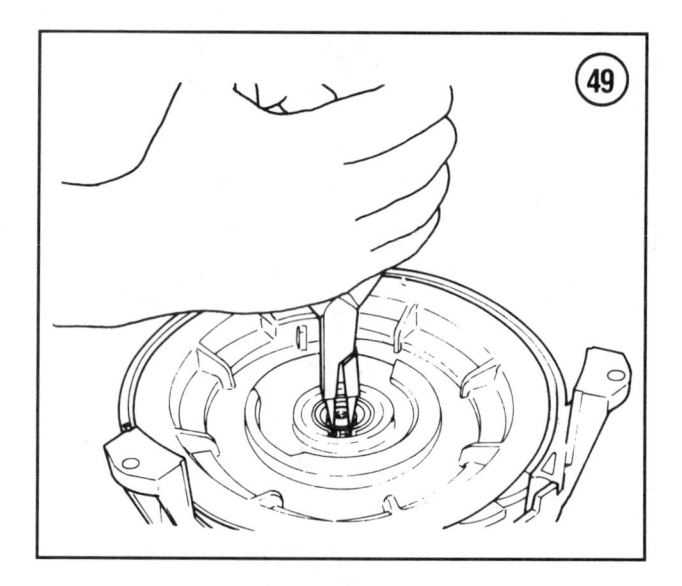

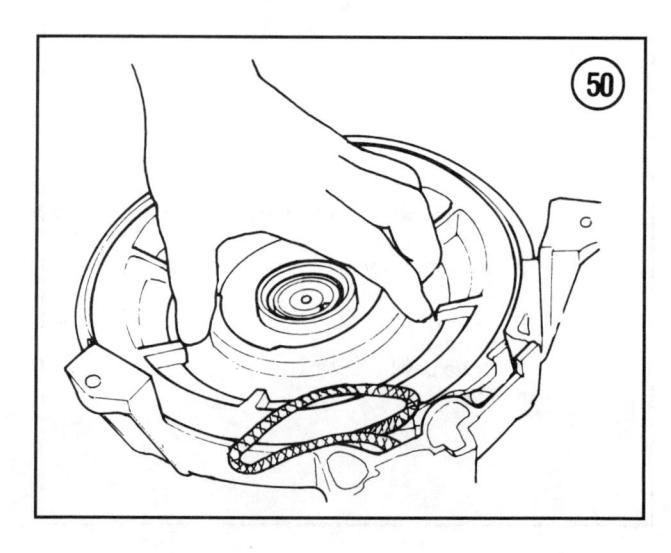

AUTOMATIC REWIND STARTERS

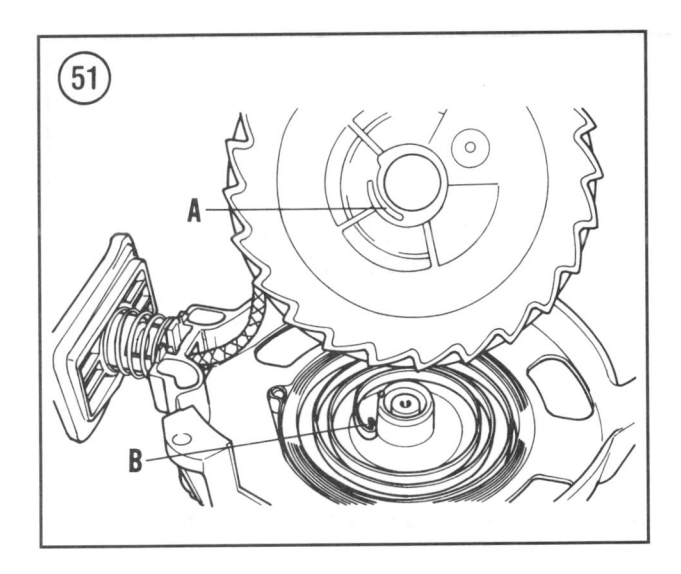

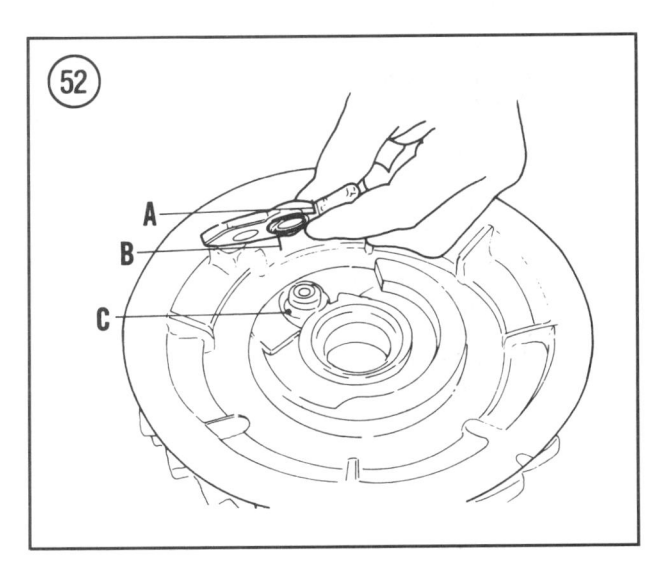

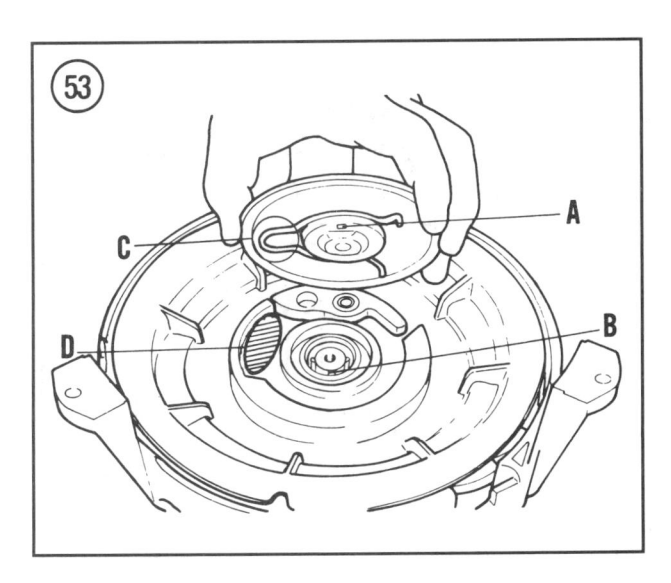

b. Install the stopper arm (A, **Figure 46**) into the interlock arm (B, **Figure 46**).

c. Install the washer and new cotter pin into the interlock arm. Bend the ends over completely.

22B. On 3-cylinder models, perform the following:

a. Install the interlock arm, spring and stopper arm.

b. Install the washer and new cotter pin into the interlock arm. Bend the ends over completely.

23. Install the starter housing as described in this chapter.

24. Adjust the starter interlock as described in this chapter.

Disassembly/Assembly (DT 35 and DT 40)

WARNING
Disassembling this starter mechanism without holding the spring in place can result in the spring unwinding violently and can cause serious personal injury. Wear safety glasses and gloves during this procedure.

Refer to **Figure 54** for this procedure.

1. Remove the screw securing the cable holder and remove the cable holder.

2. Remove the cotter pin and washer from the interlock arm.

3. Remove the stopper arm (A, **Figure 55**) from the interlock arm (B, **Figure 55**).

4. Remove the stopper spring (A, **Figure 56**) and interlock arm (B, **Figure 56**).

5. Invert the starter housing and pull the rope out as far as possible. Hold in this position and apply downward pressure on the starter pulley to prevent it from pulling the rope back in.

6. Hook the rope on the pulley notch. Let the pulley slowly rotate clockwise to release the spring tension (**Figure 57**).

OVERHEAD STARTER (DT 35, DT 40)

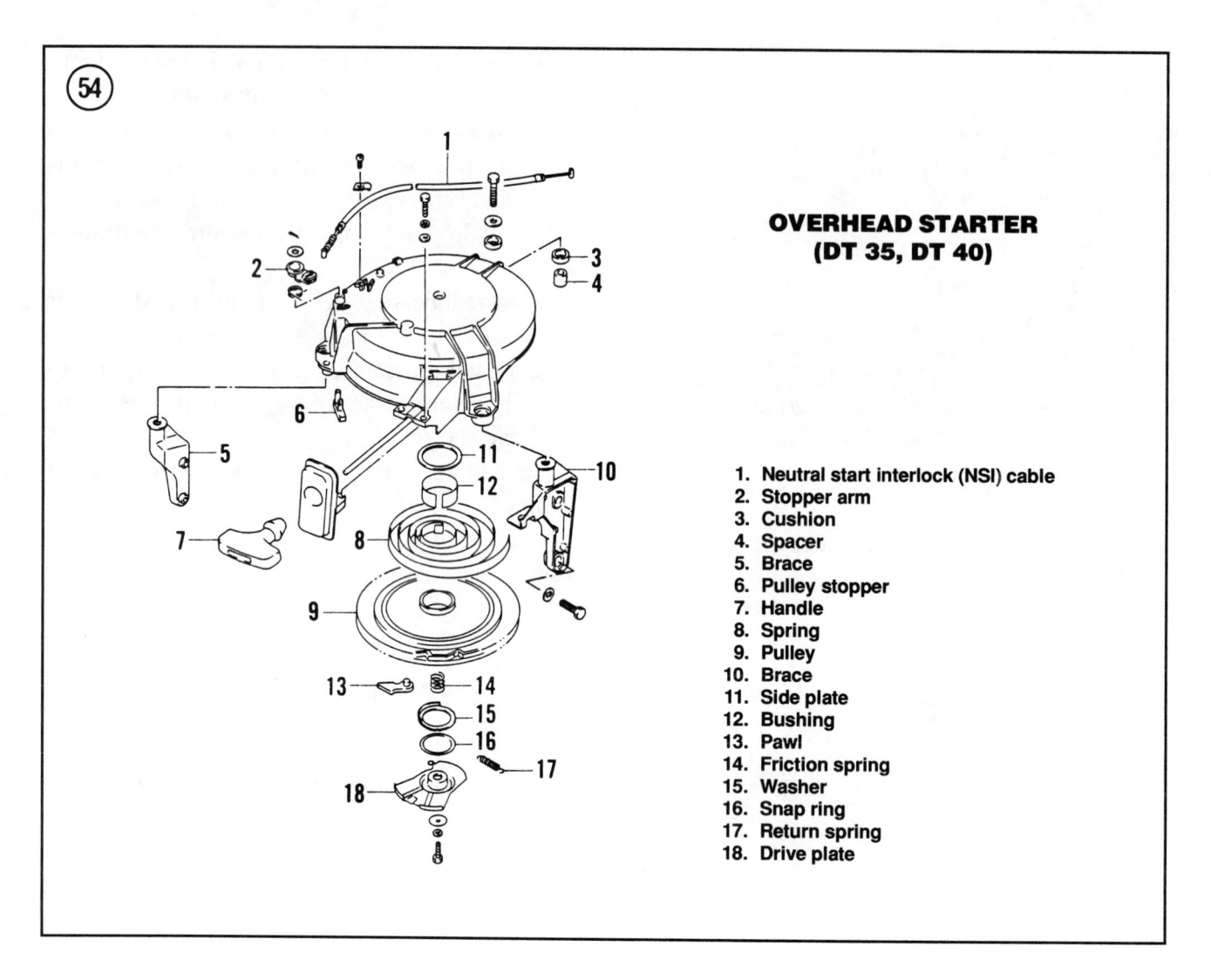

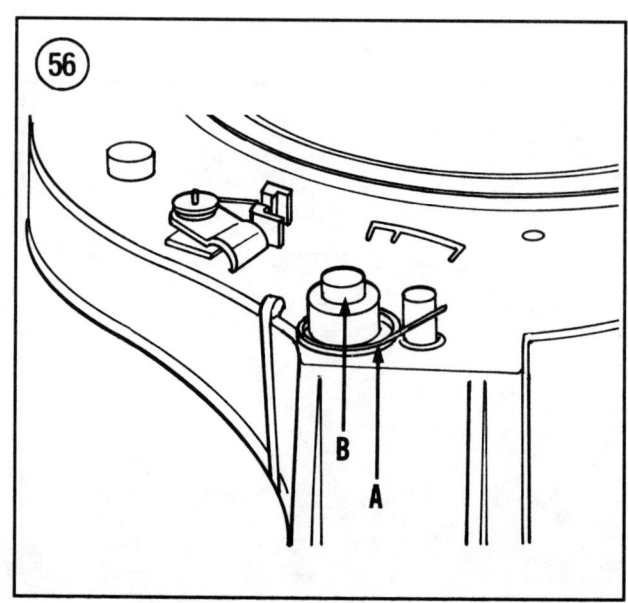

1. Neutral start interlock (NSI) cable
2. Stopper arm
3. Cushion
4. Spacer
5. Brace
6. Pulley stopper
7. Handle
8. Spring
9. Pulley
10. Brace
11. Side plate
12. Bushing
13. Pawl
14. Friction spring
15. Washer
16. Snap ring
17. Return spring
18. Drive plate

AUTOMATIC REWIND STARTERS

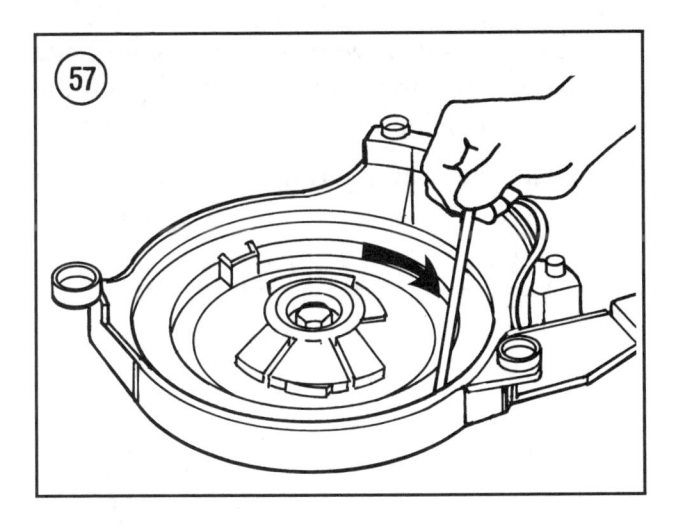

7. Remove the drive plate bolt and washers (A, **Figure 58**).

8. Remove the drive plate (B, **Figure 58**).

9. Remove the friction spring (A, **Figure 59**), the pawl (B, **Figure 59**) and return spring (C, **Figure 59**).

10. Remove the snap ring from pulley bore with a flat bladed screwdriver (**Figure 60**).

11. Carefully lift the pulley from the housing making sure the spring stays in the housing (**Figure 61**). If pulley does not come out easily, carefully insert the tip of a thin-blade screw-

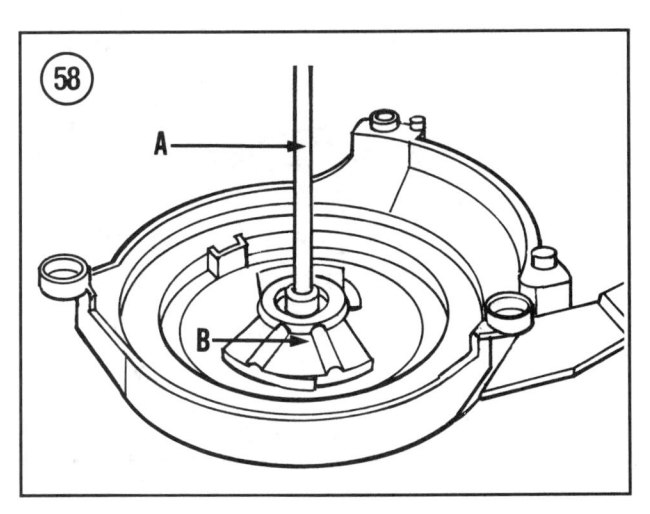

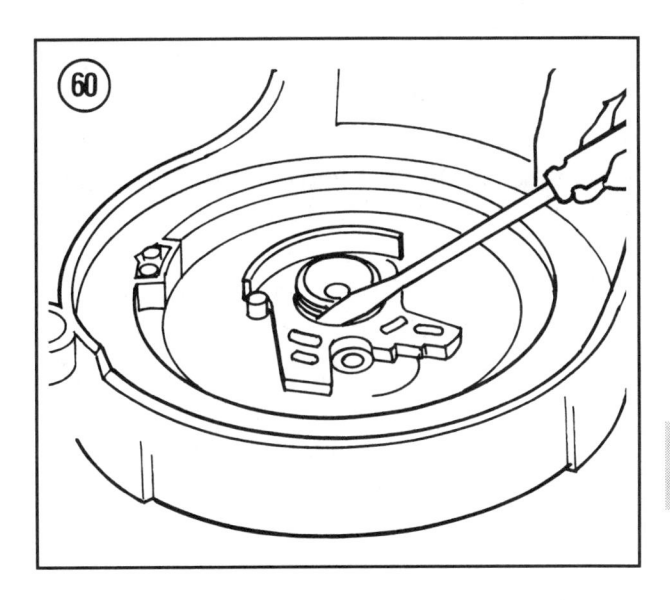

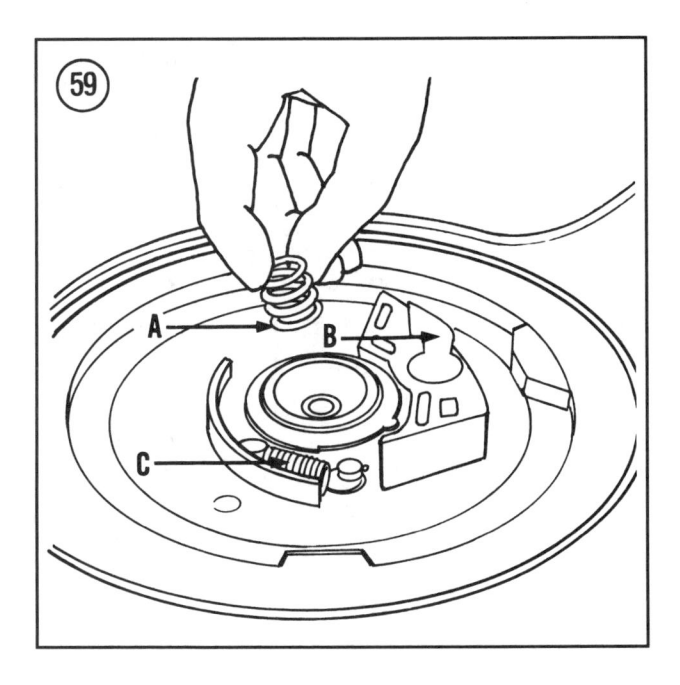

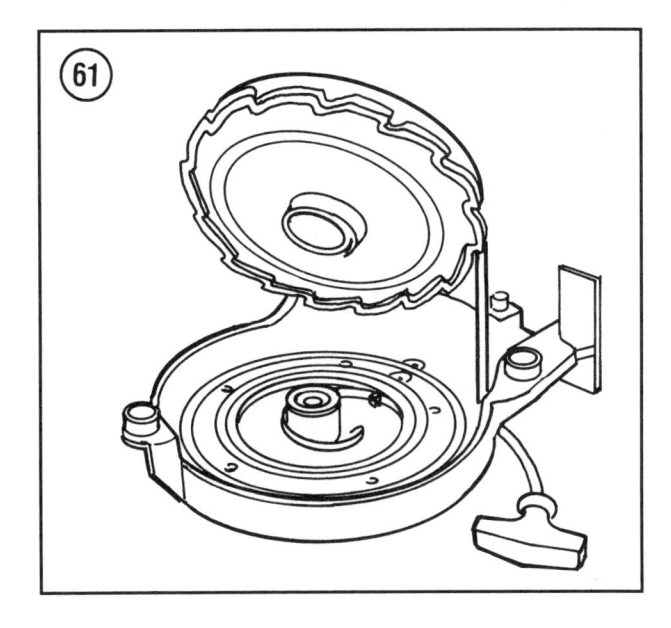

driver under the pulley and disengage the spring loop from the pulley.

12. If the spring requires replacement, place starter housing on the floor (spring side facing down) and gently tap on its top. The spring will drop down and unwind inside the housing. Remove the housing and discard the spring.

13. Lubricate all parts with water-resistant grease.

14. If the spring was removed, insert the looped end of a new spring into the housing bore and carefully wind the remainder of the spring into the housing bore (**Figure 62**).

15. If the rope requires replacement, attach the new rope to the starter pulley.

16. Install the pulley into the housing so the bent end of the spring (B, **Figure 63**) will engage the groove in the pulley (A, **Figure 63**). Rotate the pulley slightly counterclockwise. If the spring end and pulley groove have engaged properly, tension will be felt.

17. If no tension is felt in Step 16, carefully remove the pulley, realign the spring loop and pulley groove and repeat Step 16. Install the snap ring into the pulley bore. Make sure it is seated correctly.

18. Install the washer with the half-shoulder side (**Figure 64**) facing down and index it into the groove in the pulley.

19. Install the return spring, the pawl and the friction spring.

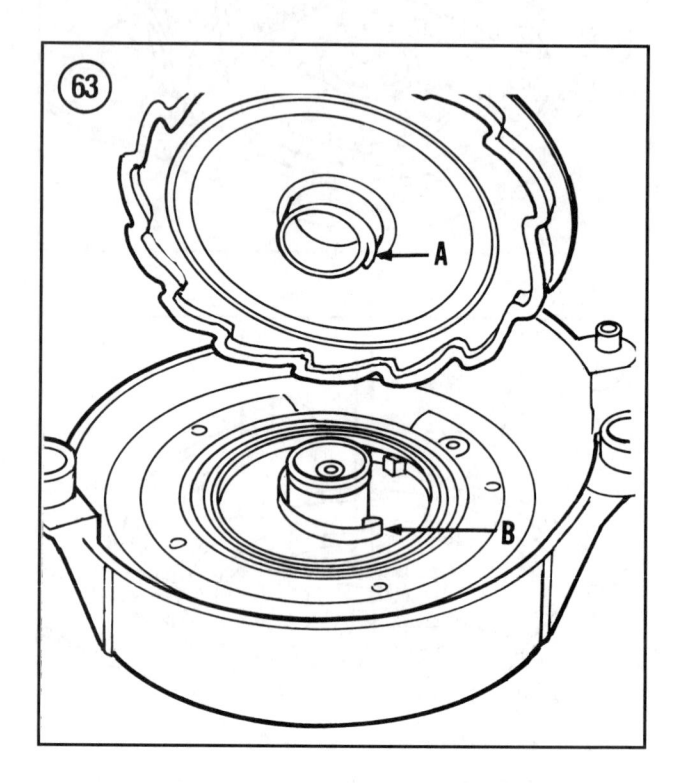

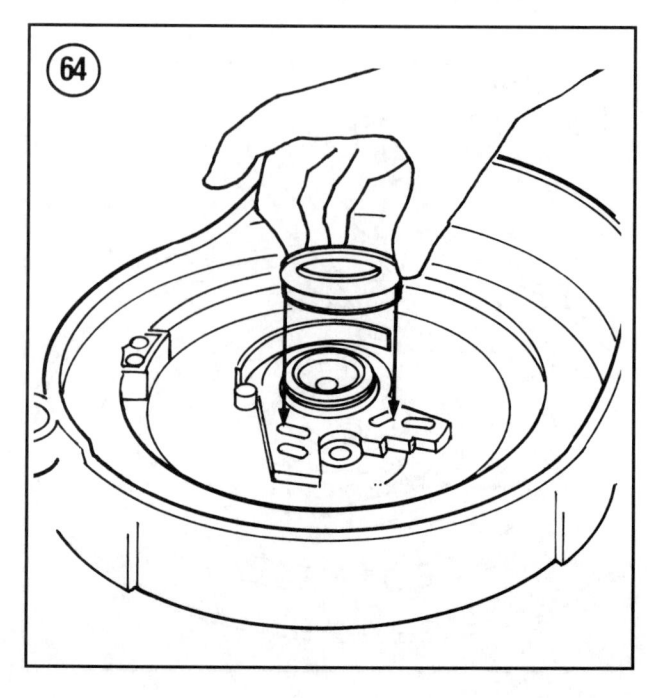

AUTOMATIC REWIND STARTERS

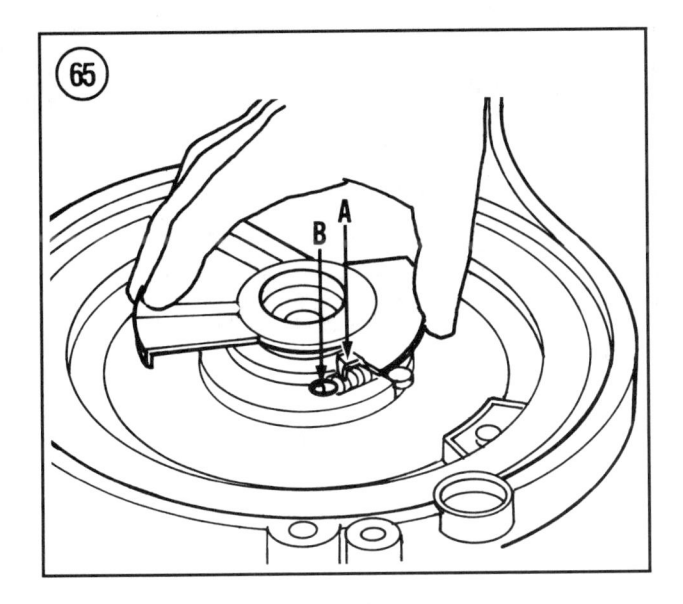

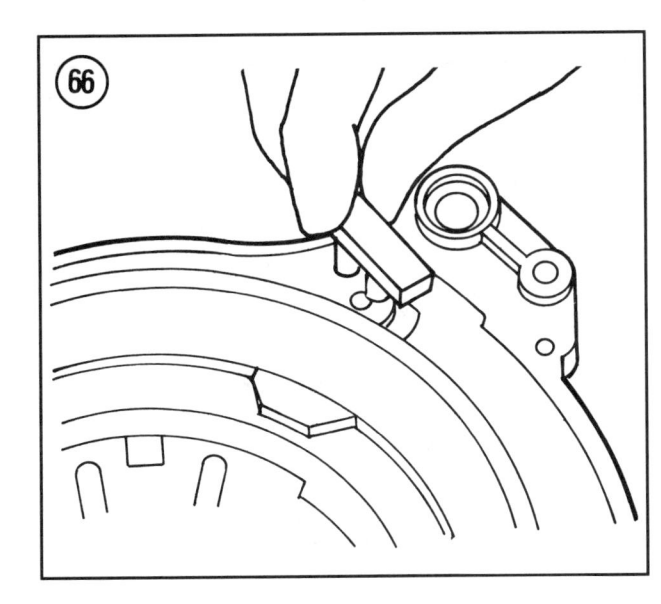

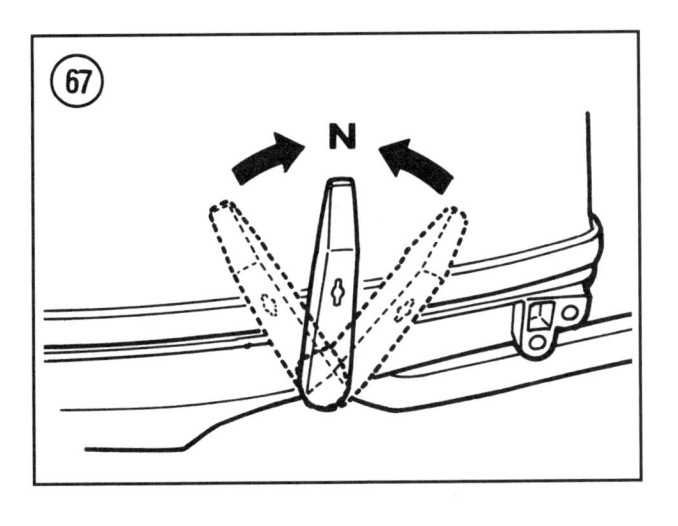

20. Install the drive plate and attach the hook portion of the drive plate (A, **Figure 65**) into the loop of the return spring (B, **Figure 65**).

21. Install the washers and the drive plate bolt. Tighten securely.

22. Install the pulley stopper (**Figure 66**) and install the pulley stopper spring (A, **Figure 56**).

23. Install the stopper arm (A, **Figure 55**) onto the pulley stopper (B, **Figure 55**).

24. Install the washer and new cotter pin onto the pulley stopper. Bend the ends over completely.

25. Feed the rope through the housing and attach the handle.

26. Wind the rope counterclockwise around the pulley 2 1/2 turns, then hook the rope on the pulley notch. Rotate the pulley counterclockwise another 4 full turns.

27. Apply finger pressure to the pulley, then unhook the rope from the pulley notch and slowly allow the pulley to wind the rope under spring tension.

28. Install the starter housing as described in this chapter.

29. Adjust the starter interlock as described in this chapter.

ADJUSTMENT

Starter Interlock Adjustment (1985-1987 DT 9.9; 1985-on DT 15)

1. Make sure the shift lever is in NEUTRAL (**Figure 67**).

2. Loosen the interlock cable locknut (**Figure 68**) and turn the adjusting nut until the slit on the starter housing stopper arm aligns with the forward mark on the housing (1, **Figure 69**).

3. Move the shift lever first to FORWARD, then to REVERSE. If the interlock arm slit does not align with the rear mark on the housing (2, **Figure 69**) in each position, repeat Step 2.

Starter Interlock Adjustment (1988-on DT 8 and DT 9.9)

1. Make sure the shift lever is in NEUTRAL.
2. Turn the adjusting nut until the length of wire from the base of the cable bracket on the power head to the lug on the end of the cable is 4.1 in. (105 mm).
3. Attach the cable to the NSI lever.

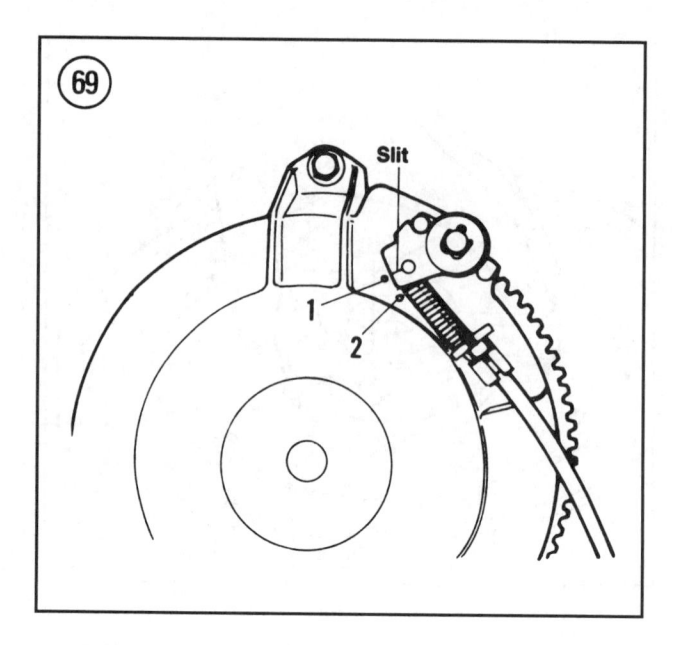

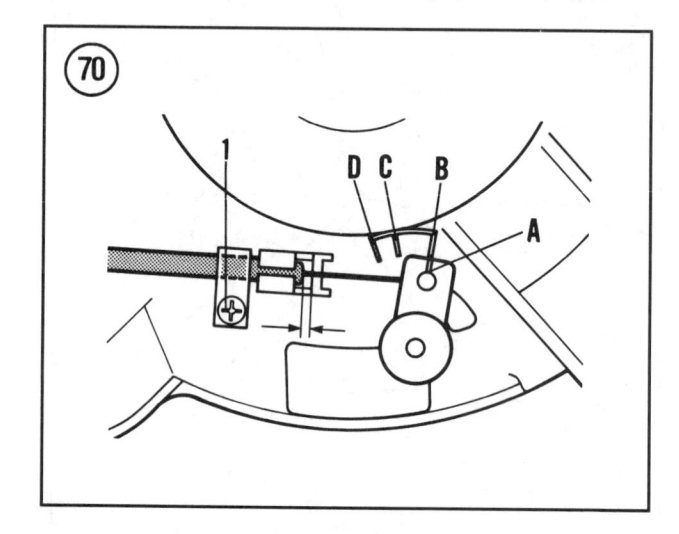

Starter Interlock Adjustment (DT 25 and DT 30 [3-cylinder])

1. Make sure the shift lever is in NEUTRAL.
2. Connect the end of the interlock cable to the interlock arm on the starter housing.
3. Loosen the clamp screw (1, **Figure 70**).
4. Align the slit (A, **Figure 70**) on the stopper arm with the forward mark (B, **Figure 70**) on the housing, then tighten the cable clamp screw.
5. Move the shift lever first to FORWARD, then to REVERSE. If the starter arm slit (A, **Figure 71**) falls between the marks (C and D, **Figure 71**) on the housing, the cable is adjusted correctly.
6. If it does not align correctly in each position, repeat Step 3 and Step 4.

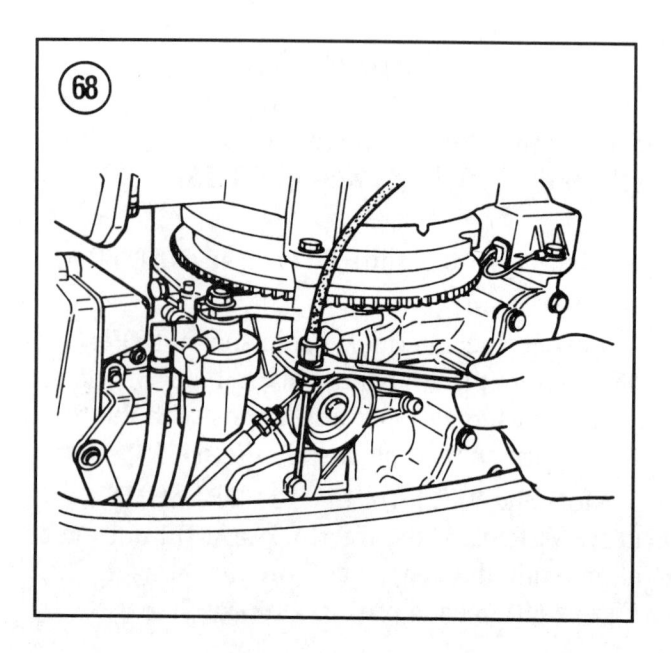

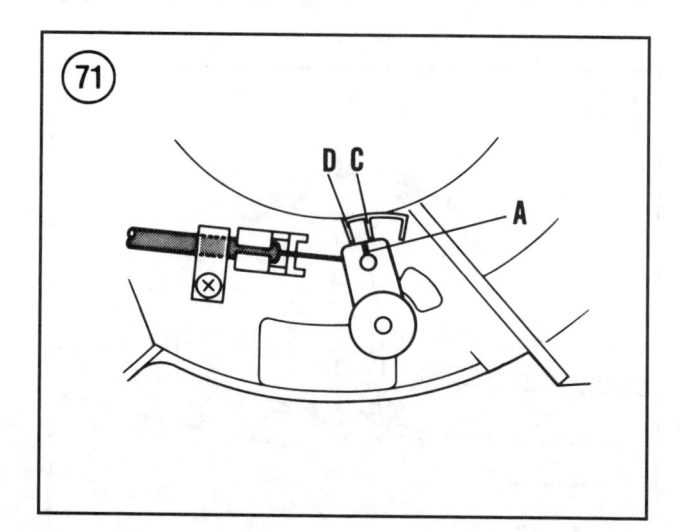

AUTOMATIC REWIND STARTERS

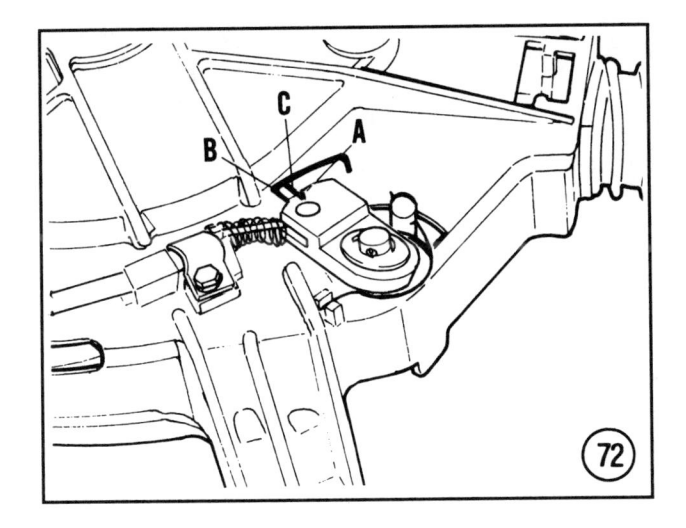

**Starter Interlock Adjustment
(DT 20, DT 25, DT 30; DT 35,
DT 40 [2-cylinder])**

1. Move the shift lever from FORWARD to REVERSE.
2. Check that the starter arm slit (A, **Figure 72**) falls between the marks (B and C, **Figure 72**) on the housing.
3. If adjustment is necessary, loosen the clamp screw and move the cable until alignment is correct.
4. Tighten the clamp screw securely.

Chapter Eleven

Power Trim and Tilt System

The usual method of raising and lowering the outboard motor is a mechanical one, consisting of a series of holes in the transom mounting bracket. To trim the outboard, an adjustment stud (**Figure 1**) is removed from the bracket, the outboard is repositioned and the stud reinserted in the proper set of holes to retain the unit in place.

A power trim and tilt system is standard on all DT 55 and DT 65 models as well as all DT 75 through DT 225 models. The 1988-on DT 35 and 1986-on DT 40 models are equipped with a tilt only feature.

Power trim provides low-effort control when the boat is underway or at rest.

On the 1988-on DT 35 and 1986-on DT 40 models the single trim cylinder is fitted inside the clamp brackets. The DT 55 and DT 65 and the 1985-1986 DT 75 and DT 85 models are fitted with twin trim/tilt cylinders mounted inside the clamp brackets (**Figure 2**). All other models use a dual trim and single tilt cylinder assembly mounted inside the clamp brackets (**Figure 3**).

This chapter includes maintenance, troubleshooting procedures and hydraulic pump and trim/tilt cylinder replacement for both power trim/tilt designs.

Components
(1988-on DT 35, 1986-on DT 40)

CAUTION
*The power tilt system is **not** designed to be used as a power trim system. If it is used for power trim, it will damage the system.*

This system consists of a hydraulic pump (containing an electric motor, oil reservoir, oil pump and valve body) and 1 hydraulic tilt cylinder. A tilt switch and the necessary hydraulic and electrical lines complete the system.

Operation
(1988-on DT 35, 1986-on DT 40)

Moving the tilt switch to the UP position closes the pump motor circuit. The motor drives

POWER TRIM AND TILT SYSTEM

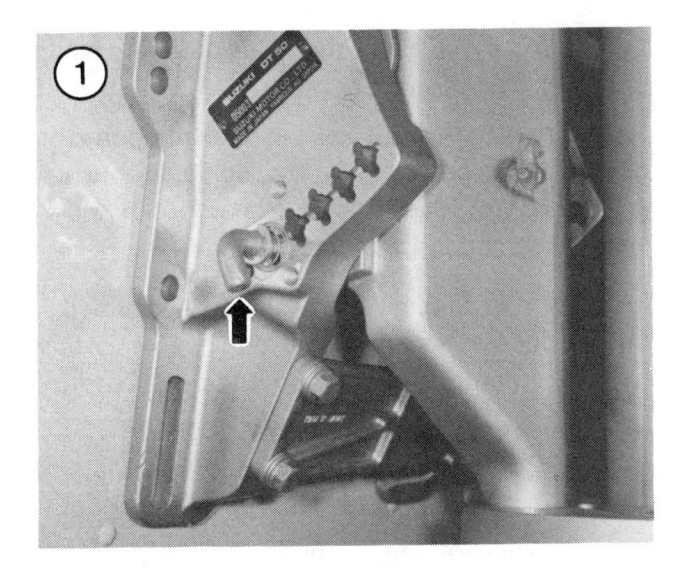

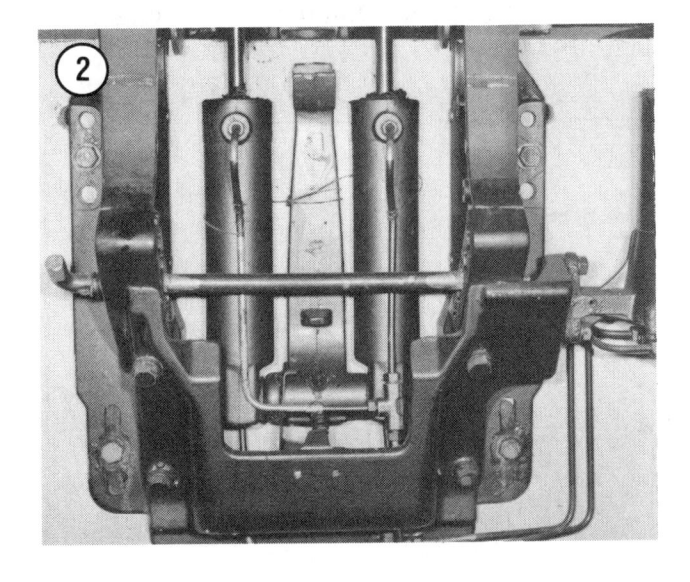

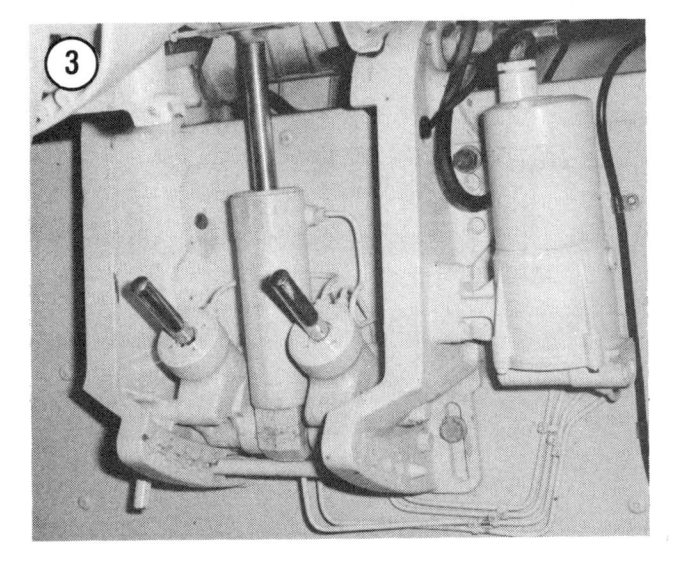

the oil pump, forcing oil into the up side of the tilt cylinder. The outboard will move upward until it reaches its maximum position.

Moving the tilt switch to the DOWN position also closes the pump motor circuit. The reversible motor runs in the opposite direction, driving the oil pump to force oil into the down side of the tilt cylinder bringing the outboard back down to the desired position.

The power tilt system will temporarily maintain the outboard at any angle within its range to allow shallow water operation at slow speed, launching, beaching or trailering.

To prevent damage to the outboard from striking an underwater object, a relief valve in the hydraulic pump opens to allow the outboard to pivot upward quickly and return slowly, absorbing the shock.

CAUTION
The manual release valve is not equipped with a stopper. If the valve is rotated more than 2 full turns counterclockwise, it is possible that the valve will work loose from the valve body.

The pump is equipped with a manual release valve. Opening this valve by turning its hex head up to 2 full turns *counterclockwise* permits manual raising and lowering of the outboard if the electrical system fails.

Components
(1985-on DT 55 and DT 65,
1985-1986 DT 75 and DT 85)

This system consists of a hydraulic pump (containing an electric motor, oil reservoir, oil pump and valve body) and 2 hydraulic trim/tilt cylinders (**Figure 4**). Up/down solenoids are mounted on the power head (**Figure 5**). An indicator sender and gauge, trim/tilt switch and the necessary hydraulic and electrical lines complete the system. **Figure 5** shows the electrical schematic for the 1985-on DT 55 and DT 65 and **Figure 6** is for the 1985-1986 DT 75 and DT 85.

Operation
(1985-on DT 55 and DT 65, 1985-1986 DT 75 and DT 85)

Moving the trim switch to the UP position closes the pump motor circuit. The motor drives the oil pump, forcing oil into the up side of the trim/tilt cylinders. The outboard will move upward until it reaches its maximum position.

Moving the trim switch to the DOWN position also closes the pump motor circuit. The reversible motor runs in the opposite direction, driving the oil pump to force oil into the down side of the trim/tilt cylinders bringing the outboard back to the desired position.

The power trim/tilt system will temporarily maintain the outboard at any angle within its range to allow shallow water operation at slow speed, launching, beaching or trailering.

To prevent damage to the outboard from striking an underwater object, a relief valve in the hydraulic pump opens to allow the outboard to pivot upward quickly and return slowly, absorbing the shock.

An optional trim gauge sending unit is located on the inside of the starboard clamp bracket. Access to the sending unit requires the engine to be fully tilted.

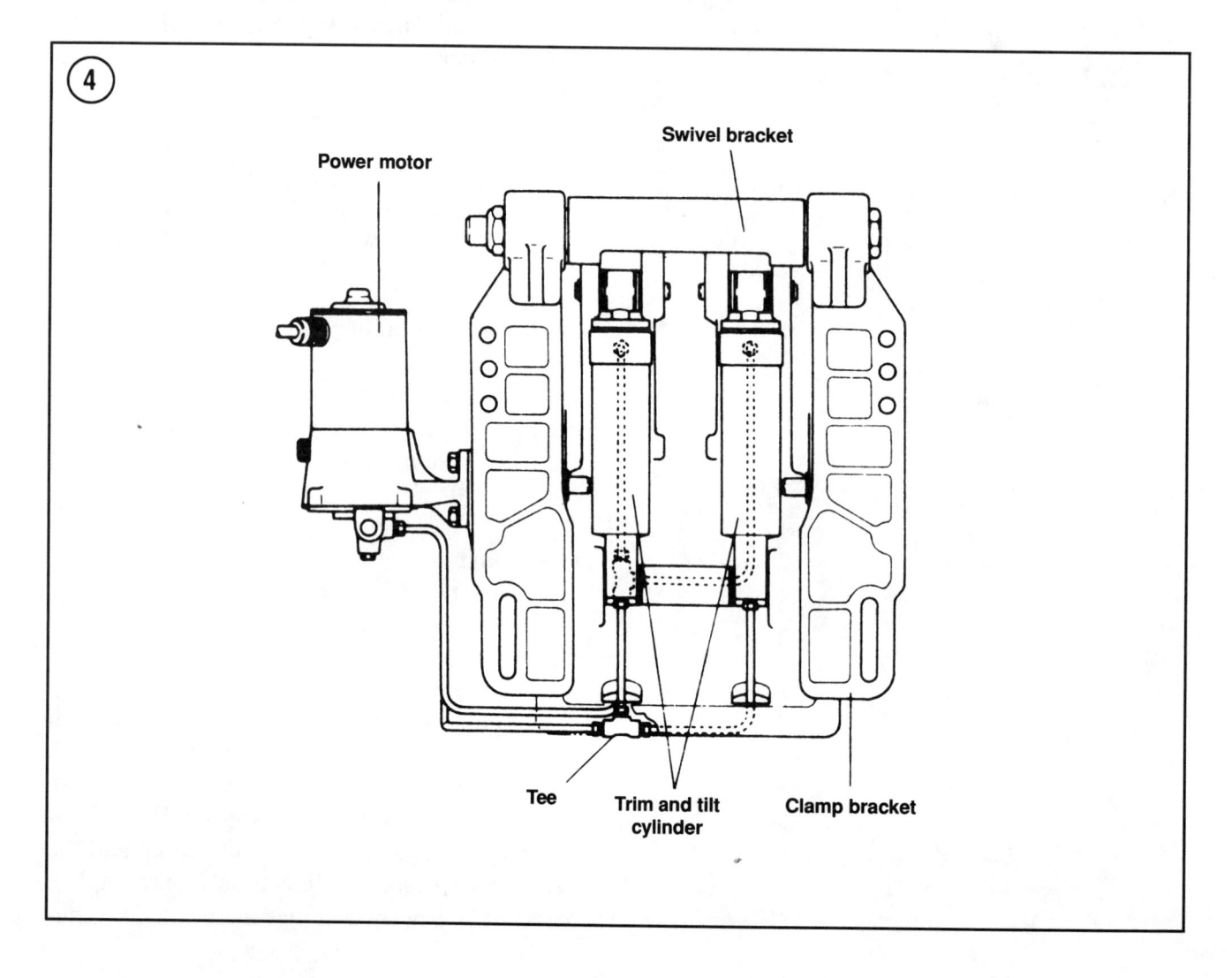

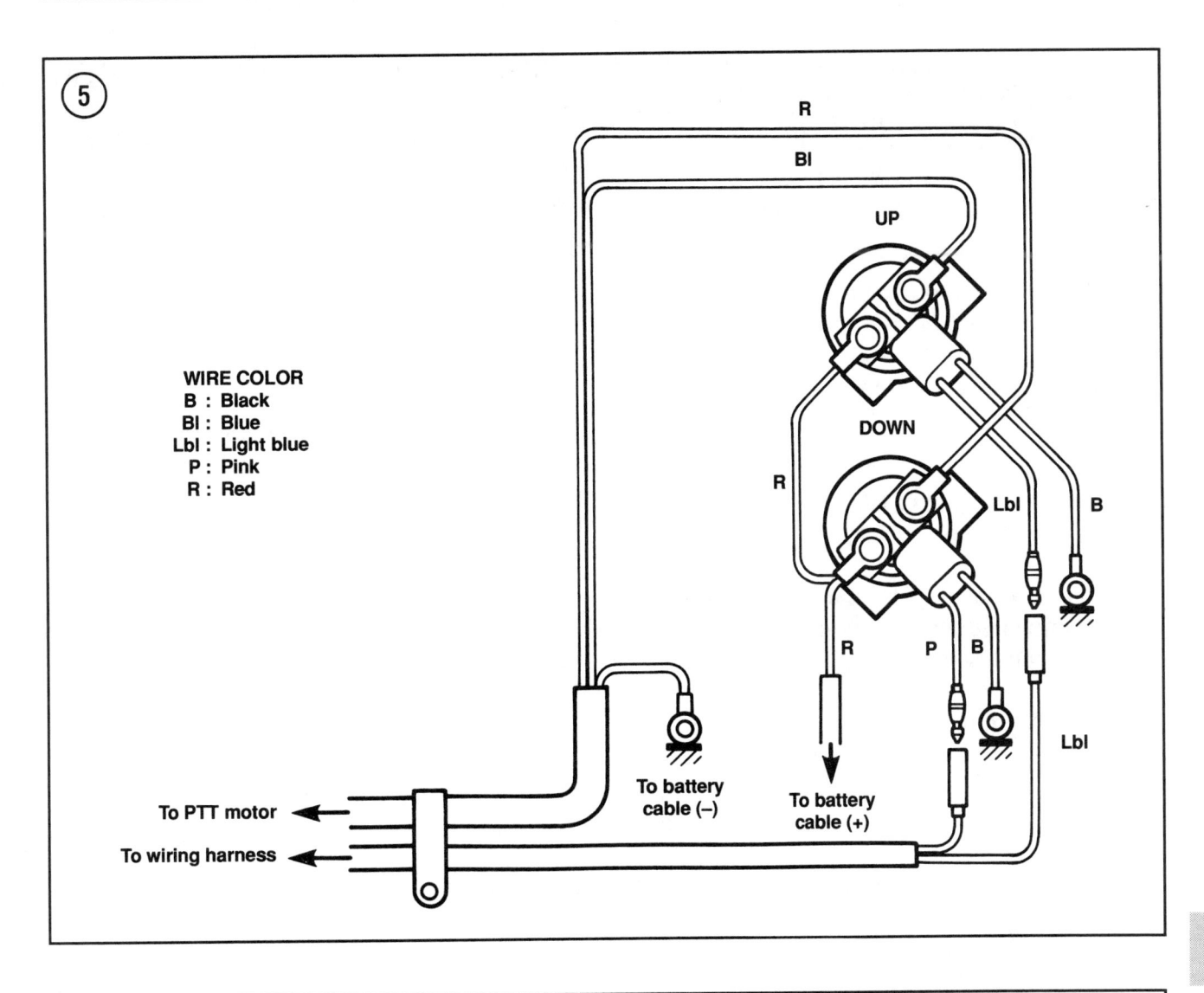

POWER TRIM AND TILT SYSTEM

⑤

WIRE COLOR
- B : Black
- Bl : Blue
- Lbl : Light blue
- P : Pink
- R : Red

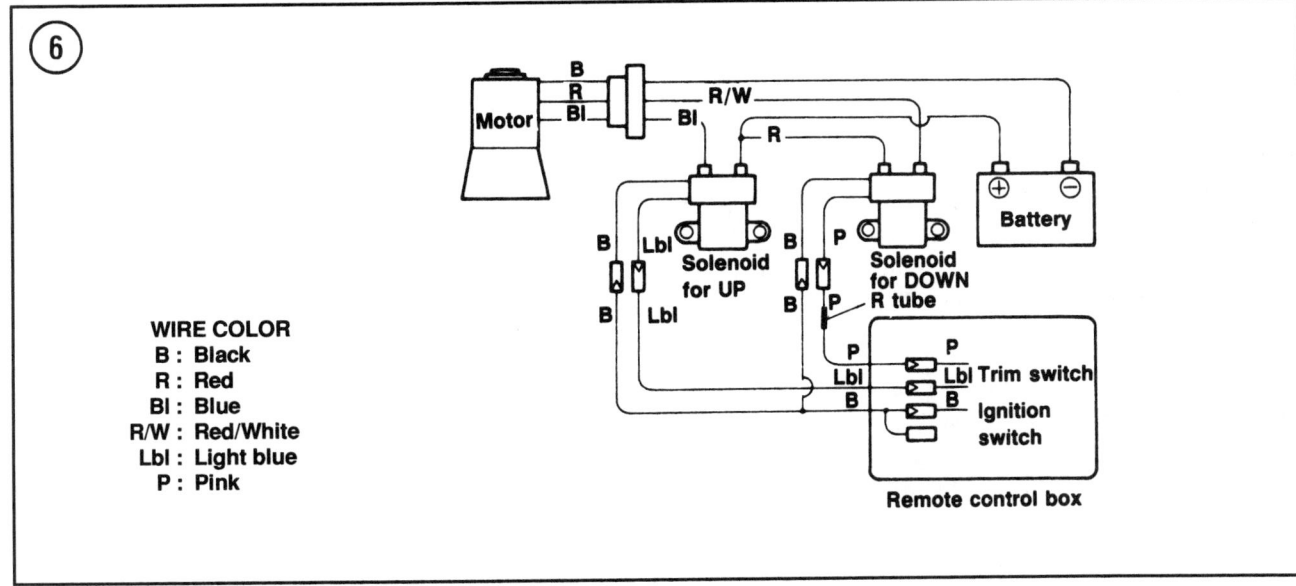

To PTT motor
To wiring harness
To battery cable (−)
To battery cable (+)

UP
DOWN

⑥

WIRE COLOR
- B : Black
- R : Red
- Bl : Blue
- R/W : Red/White
- Lbl : Light blue
- P : Pink

Motor
Solenoid for UP
Solenoid for DOWN
R tube
Battery
Trim switch
Ignition switch
Remote control box

CAUTION
The manual release valve is not equipped with a stopper. If the valve is rotated more than 2 full turns counterclockwise, it is possible that the valve will work loose from the valve body.

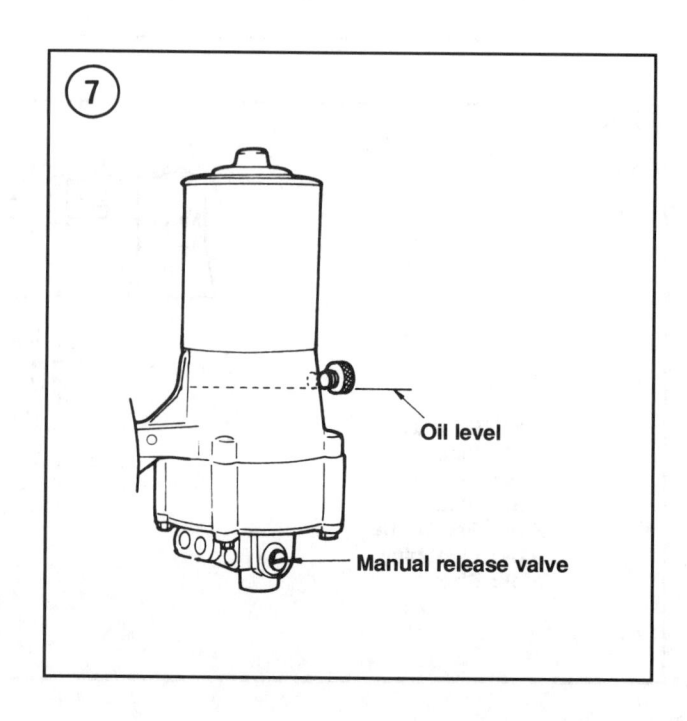

The pump is equipped with a manual release valve (**Figure 7**). Opening this valve by turning its hex head up to 2 full turns *counterclockwise* permits manual raising and lowering of the outboard if the electrical system fails.

Components
(1987-on DT 75 and DT 85, 1985-on DT 115 and DT 140)

This system consists of a manifold containing a combination hydraulic tilt cylinder, which also serves as a shock absorber, and 2 hydraulic trim cylinders. A hydraulic pump (containing an electric motor, oil pump, reservoir and all valving) is attached to the manifold (**Figure 8**). The up/down solenoids are mounted on the power head (**Figure 9**). A trim/tilt switch, indicator gauge, sending unit and the necessary hydraulic and electrical lines complete the system. **Figure 10** shows the electrical schematic for the 1987-on DT 75 and DT 85, **Figure 6** for 1985 DT 115 and DT 140 and **Figure 11** is for the 1986-on DT 115 and DT 140.

Operation
(1987-on DT 75 and DT 85, 1986-on DT 115 and DT 140)

Moving the trim switch to the UP position closes the pump motor circuit. The motor drives the oil pump, forcing oil into the up side of the trim cylinders. The trim cylinder pistons push on the swivel bracket thrust pads to trim the outboard upward. Once the trim cylinders are fully extended, the hydraulic fluid is diverted into the tilt cylinder, which now moves the outboard throughout the remaining range of travel.

Moving the trim switch to the DOWN position also closes the pump motor circuit. The reversible motor runs in the opposite direction, forcing oil into the tilt cylinder and bringing the outboard back to a position where the swivel brackets rest on the trim cylinder pistons. The pistons then lower the outboard the remainder of the way.

The power trim/tilt system will temporarily maintain the outboard at any angle within its range to allow shallow water operation at slow speed, launching, beaching or trailering.

To prevent damage to the outboard from striking an underwater object, a relief valve in the hydraulic pump opens to allow the outboard to pivot upward quickly and return slowly, absorbing the shock.

A trim gauge sending unit is located on the inside of the starboard clamp bracket (**Figure 12**). Access to the sending unit requires the outboard to be fully tilted.

CAUTION
The manual release valve is not equipped with a stopper. If the valve is rotated more than 2 full turns counterclockwise, it is possible that the valve will work loose from the valve body.

⑦

Oil level

Manual release valve

POWER TRIM AND TILT SYSTEM

The pump is equipped with a manual release valve. Opening this valve by turning its hex head up to 2 full turns *counterclockwise* permits manual raising and lowering of the outboard if the electrical system fails.

Components
(V4, V6)

This system consists of a manifold containing a combination hydraulic tilt cylinder, which also serves as a shock absorber, and 2 hydraulic trim cylinders. A hydraulic pump (containing an elec-

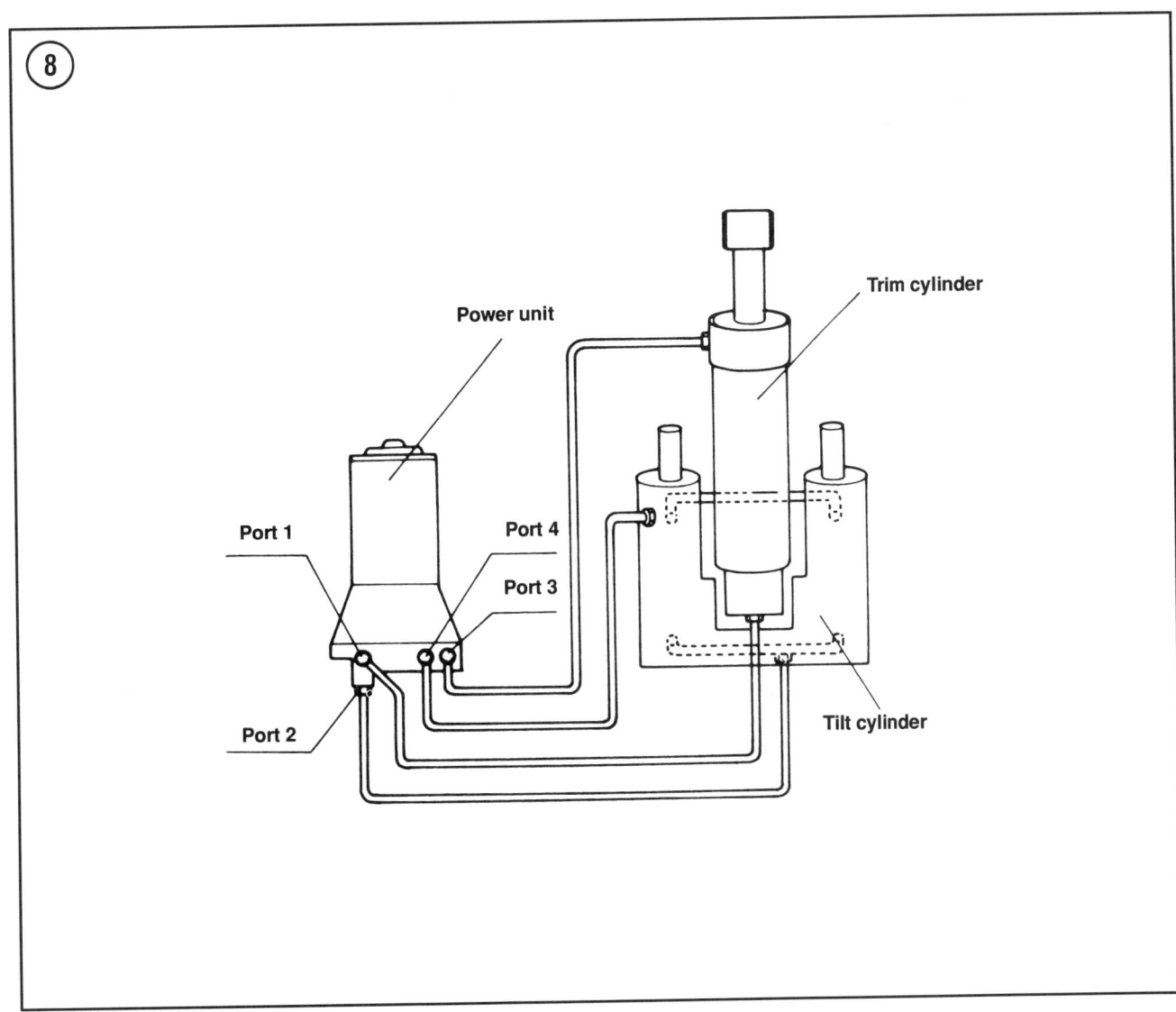

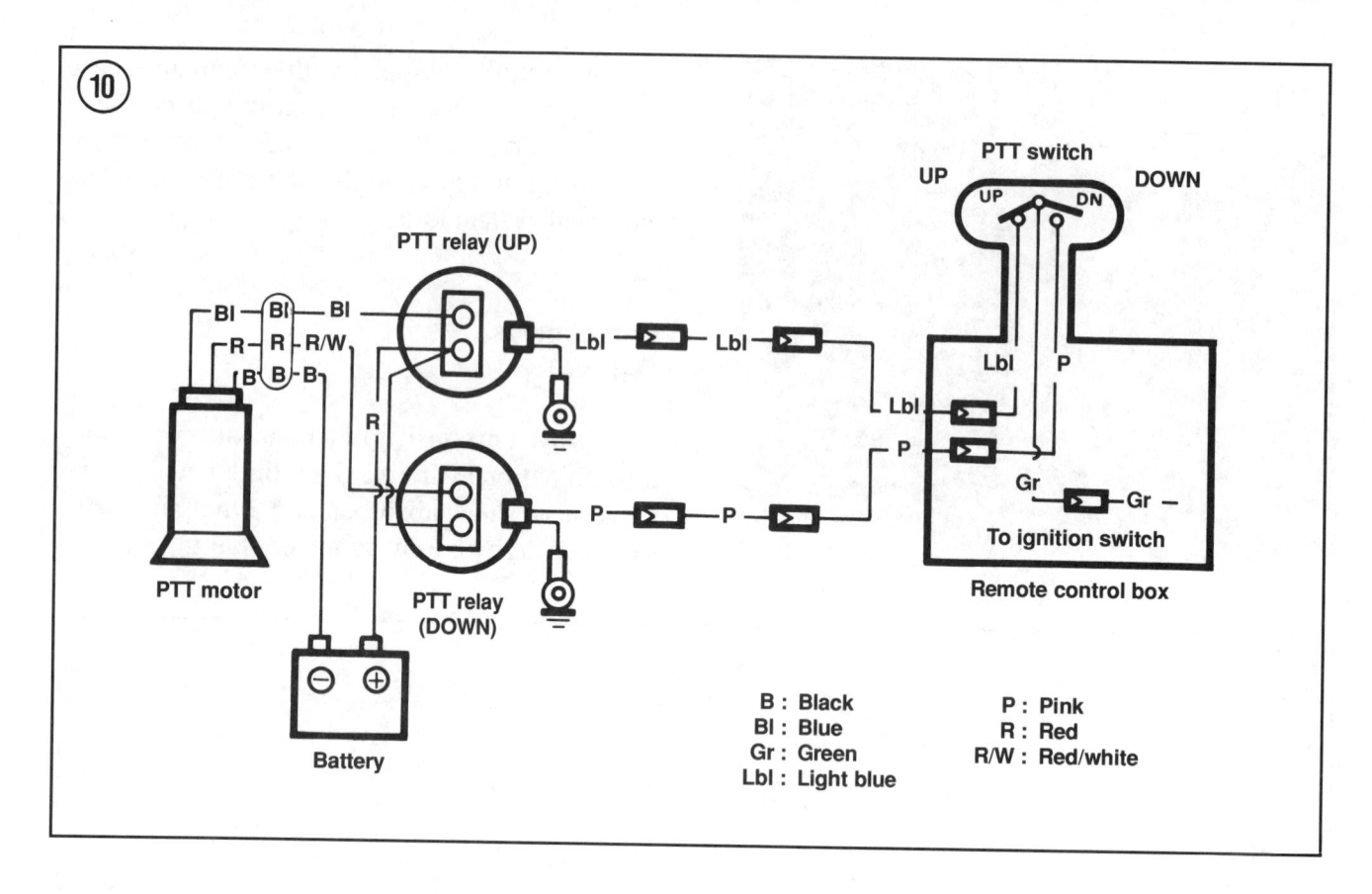

CHAPTER ELEVEN

⑩

PTT relay (UP) — **PTT switch** (UP / DOWN)

PTT motor — PTT relay (DOWN) — Battery — Remote control box — To ignition switch

B : Black
Bl : Blue
Gr : Green
Lbl : Light blue
P : Pink
R : Red
R/W : Red/white

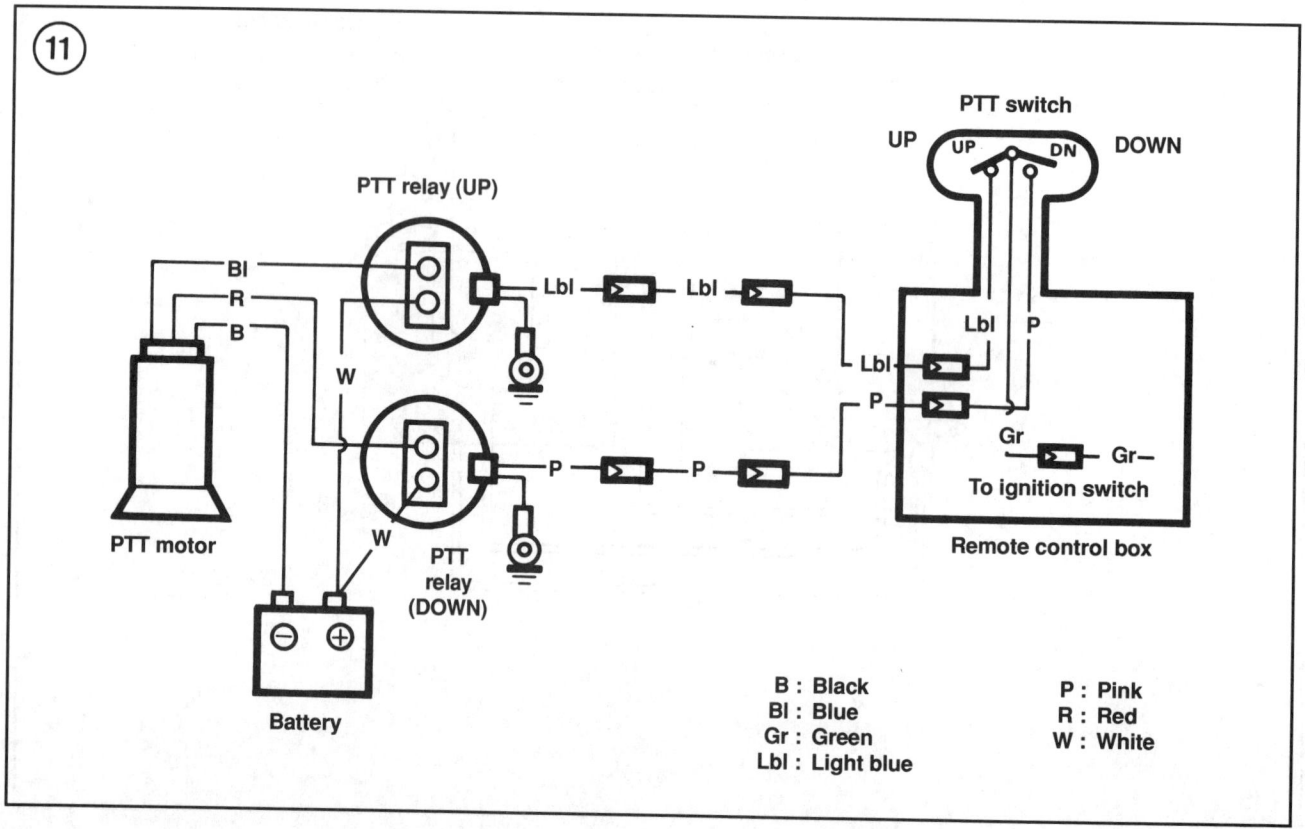

⑪

PTT relay (UP) — **PTT switch** (UP / DOWN)

PTT motor — PTT relay (DOWN) — Battery — Remote control box — To ignition switch

B : Black
Bl : Blue
Gr : Green
Lbl : Light blue
P : Pink
R : Red
W : White

POWER TRIM AND TILT SYSTEM

tric motor, oil pump, reservoir and all valving) is attached to the manifold (**Figure 13**). The up/down solenoids are mounted on the upper end of the crankcase. A trim/tilt switch, indicator gauge, sending unit and the necessary hydraulic and electrical lines complete the system. **Figure 14** shows the electrical schematic for the V4 models and **Figure 15** is for V6 models.

Operation
(V4, V6)

Moving the trim switch to the UP position closes the pump motor circuit. The motor drives the oil pump, forcing oil into the up side of the trim cylinders. The trim cylinder pistons push on the swivel bracket thrust pads to trim the outboard upward. Once the trim cylinders are fully extended, the hydraulic fluid is diverted into the tilt cylinder, which now moves the outboard throughout the remaining range of travel.

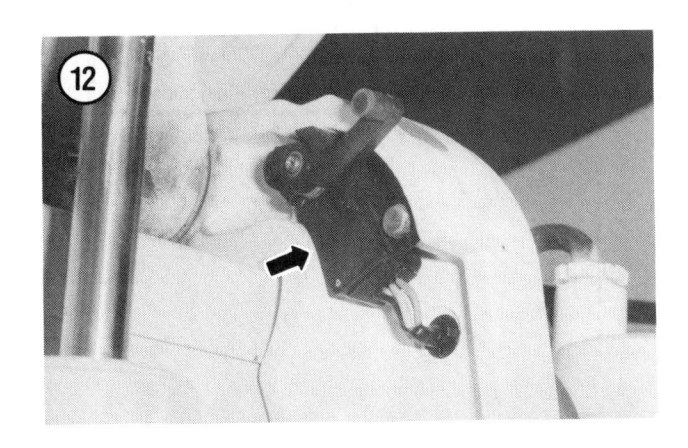

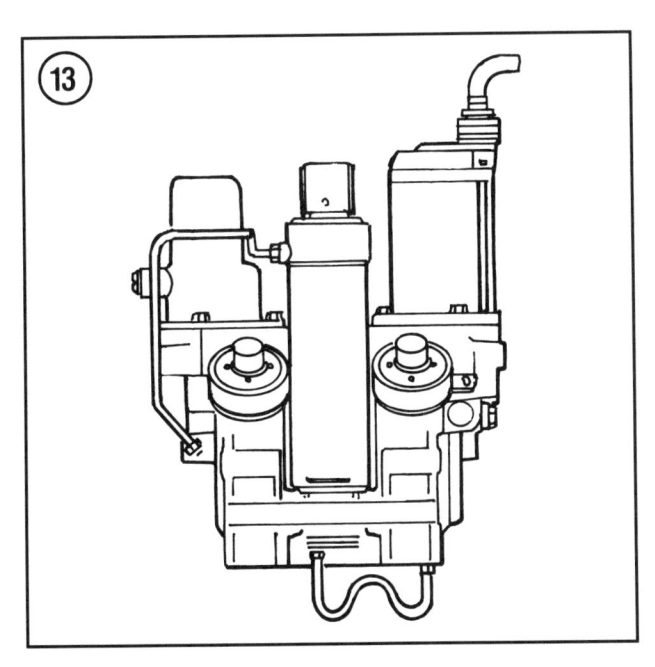

Moving the trim switch to the DOWN position also closes the pump motor circuit. The reversible motor runs in the opposite direction, forcing oil into the tilt cylinder bringing the outboard back to a position where the swivel brackets rest on the trim cylinder pistons. The pistons then lower the outboard the remainder of the way.

The power trim/tilt system will temporarily maintain the outboard at any angle within its range to allow shallow water operation at slow speed, launching, beaching or trailering.

To prevent damage to the outboard from striking an underwater object, a relief valve in the hydraulic pump opens to allow the outboard to pivot upward quickly and return slowly, absorbing the shock.

A trim gauge sending unit is located on the inside of the starboard clamp bracket (**Figure 12**). Access to the sending unit requires the outboard to be fully tilted.

The pump is equipped with a manual release valve. Opening this valve by turning it 2 full turns *clockwise* permits manual raising and lowering of the outboard if the electrical system fails.

HYDRAULIC PUMP

Fluid Check
(All Models Except V4 and V6)

The outboard must be in the correct position in order to check the fluid level correctly. The correct outboard position for each specific model is as follows.

Full DOWN position:
 a. DT 35 and DT 40 (up to serial No. 613244).

CHAPTER ELEVEN

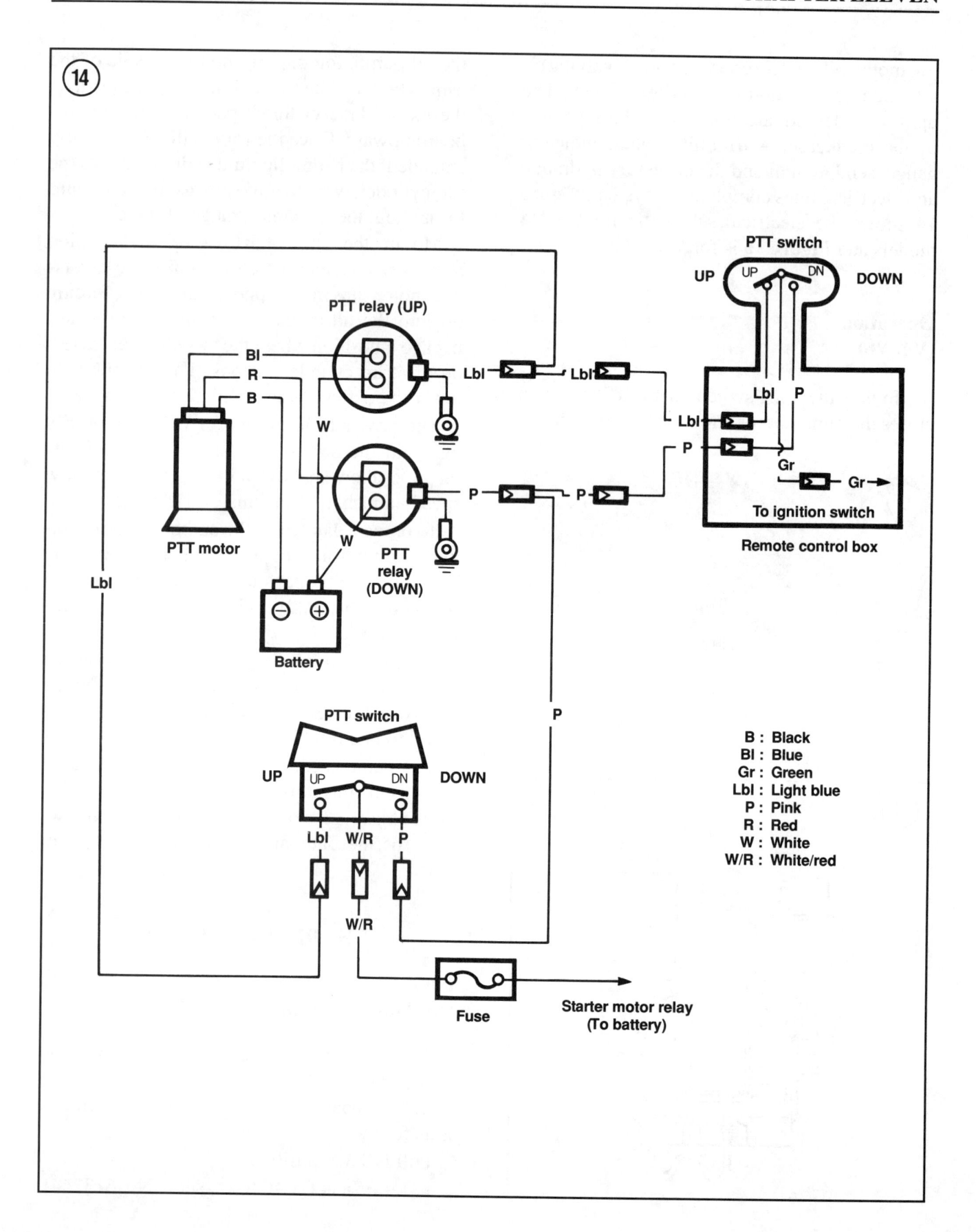

⑭

POWER TRIM AND TILT SYSTEM

b. 1985-1986 DT 55, DT 65, DT 75, DT 85, DT 115 and DT 140.

Full UP position:

a. DT 35 and DT 40 (from serial No. 613245 on).
b. 1987-on DT 55, DT 65, DT 75 and DT 85.
c. 1987-on DT 115 and DT 140.

CAUTION
*The pump reservoir must be filled with the outboard in the **recommended position only** or the valve body assembly may be damaged.*

1. Tilt the outboard to its recommended full UP or full DOWN position as previously noted.
2A. On all 1985-1986 models, clean area around pump fill plug and remove the single fill plug (**Figure 16**).
2B. On all 1987-on models, clean area around pump fill plugs and remove both fill plugs (**Figure 17**).
3. Visually check the fluid level in the pump reservoir. It should be at the bottom of the fill hole threads.
4. Top up if necessary with DEXRON automatic transmission fluid.
5. Install the fill plug(s) and tighten securely.

Fluid Check
(V4 and V6 Models)

The outboard must be in the full UP position in order to check the fluid level correctly.

CAUTION
*The pump reservoir must be filled with the outboard in the **full UP position only***

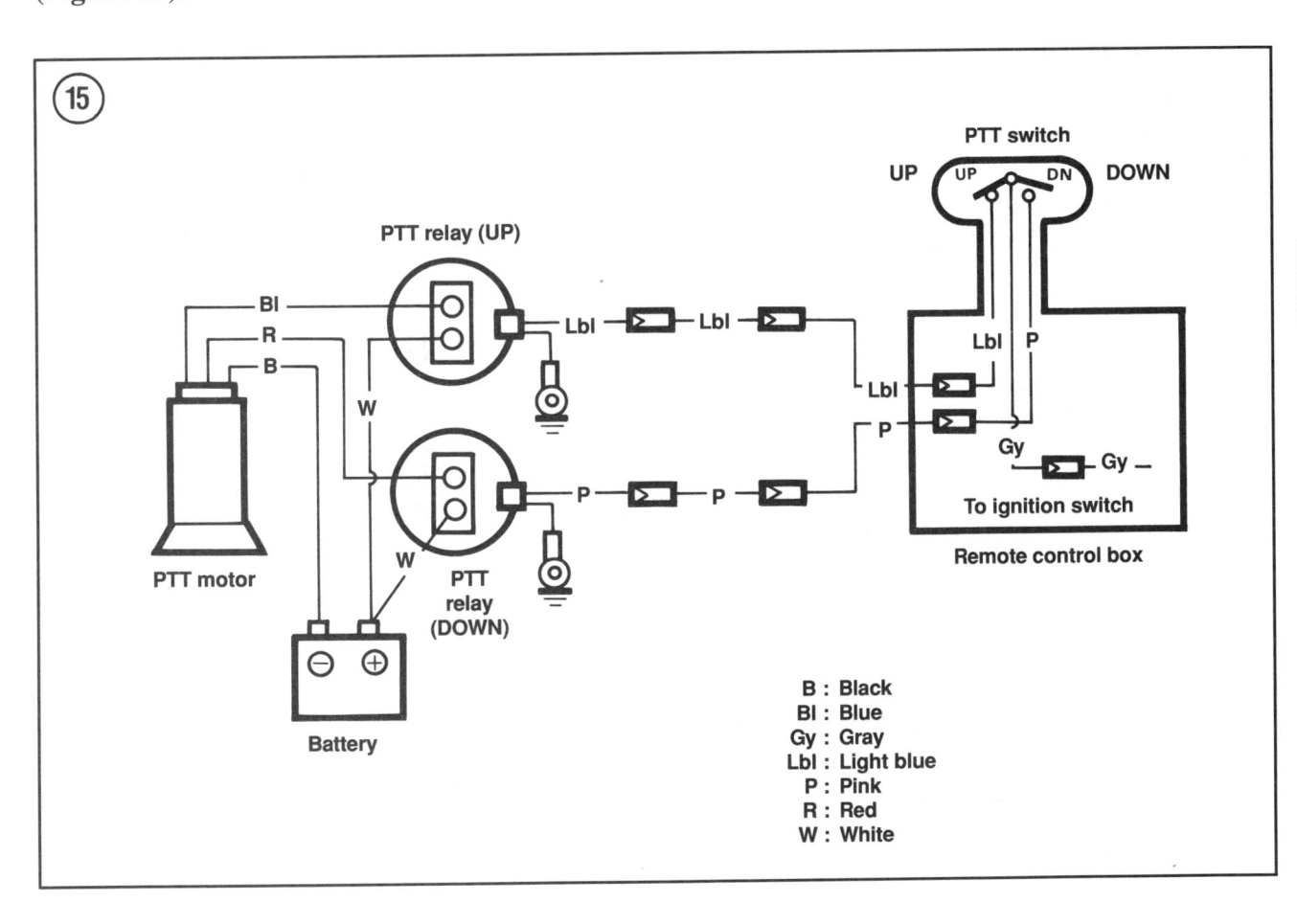

⑮

B: Black
Bl: Blue
Gy: Gray
Lbl: Light blue
P: Pink
R: Red
W: White

or the valve body assembly may be damaged.

1. Tilt the outboard to the full UP position.
2. Clean area around pump fill plug and remove the single fill plug (**Figure 18**) located on the port side clamp.
3. Visually check the fluid level in the pump reservoir. It should be at the bottom of the fill hole threads.
4. Top up if necessary with DEXRON automatic transmission fluid.
5. Install the fill plug and tighten securely.

Hydraulic Pump Fluid Refill and Bleeding

Follow this procedure when a large amount of fluid has been lost due to an overhaul of the system or leakage that has been corrected.

If any of the internal parts of the trim or tilt cylinders or the interconnecting hydraulic lines have been removed or replaced the system must be bled with the use of special tools. Refer this operation to a Suzuki dealer.

1. Remove the fill plug(s) and top off the reservoir with DEXRON automatic transmission fluid.
2. Install the fill plug(s) and operate the engine up and down several times.
3. Remove the fill plug(s) and add DEXRON to the reservoir as required to keep the fluid level at the bottom of the fill plug hole(s).
4. Repeat Step 2 and Step 3 until the fluid level stabilizes at the bottom of the fill plug hole(s). Install and tighten the fill plug(s) securely.

TROUBLESHOOTING

Whenever a problem develops in the power trim/tilt system, the initial step is to determine if the problem is in the electrical or hydraulic system. Electrical and hydraulic tests are given in this chapter. If the problem appears to be in the

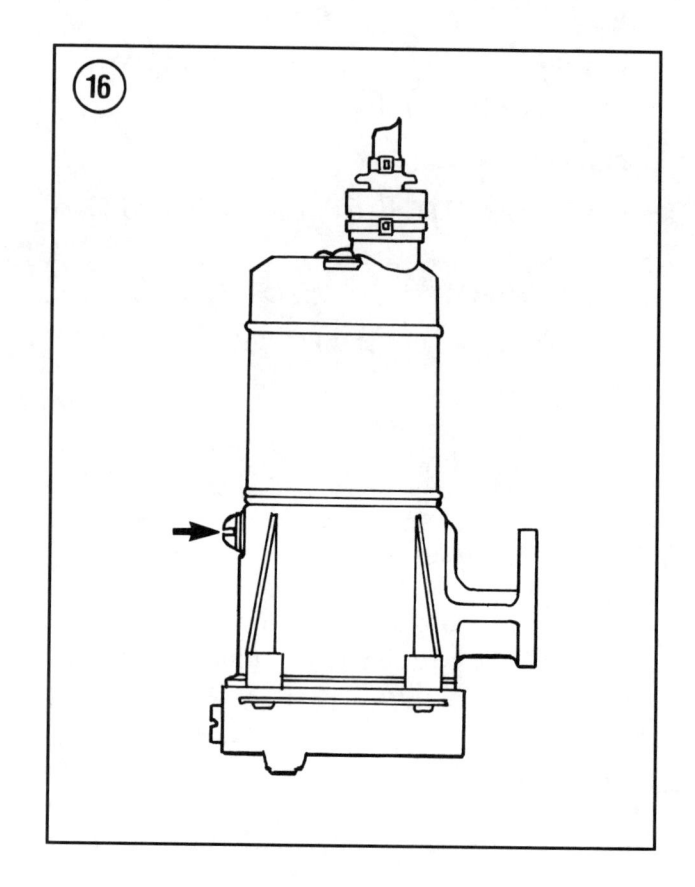

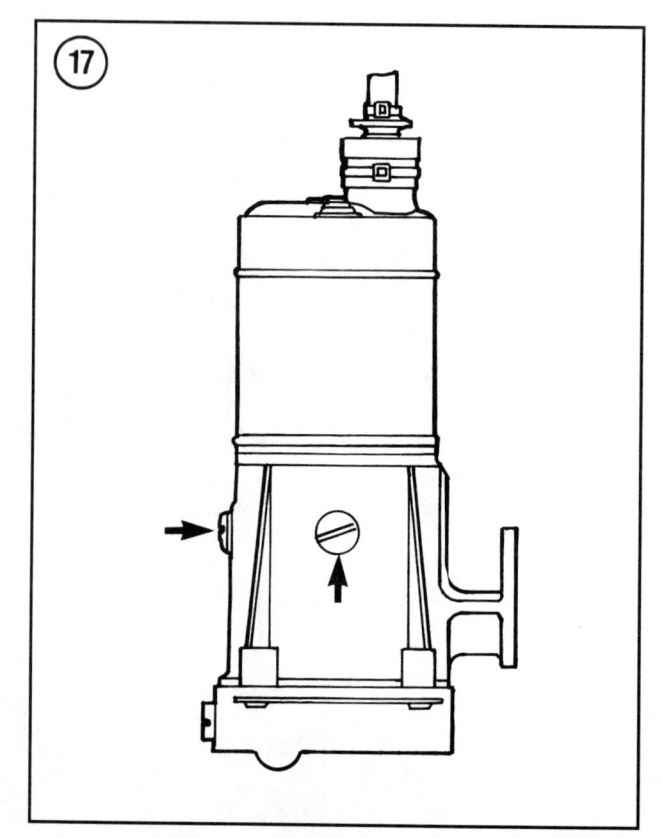

POWER TRIM AND TILT SYSTEM

hydraulic system, refer it to a dealer or qualified specialist for necessary service.

Basic Checks

1. Make sure the plug-in connectors are properly engaged and that all terminals and wires are free of corrosion. Tighten and clean as required.
2. Make sure the battery is fully charged. Charge or replace as required.
3. Check the system fuse, if so equipped, and replace if necessary.
4. Check the hydraulic fluid level as described in this chapter. Top off if necessary.
5. Make sure the manual release valve is fully closed.

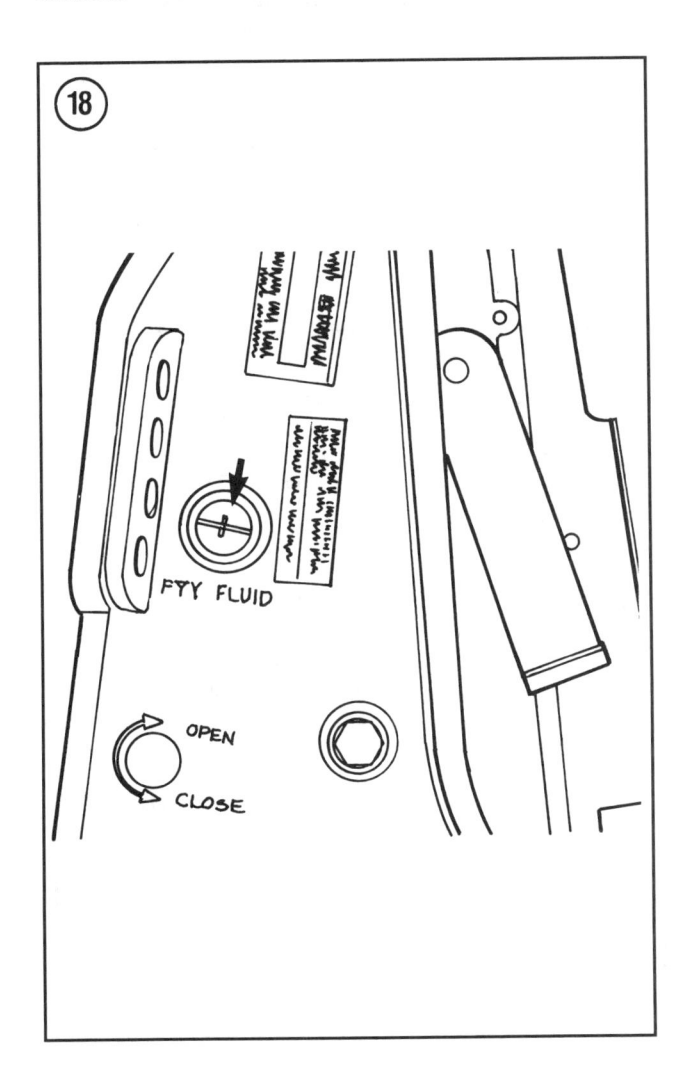

Motor Will Not Run

1986-on DT 35, DT 40

1. Disconnect the pump motor electrical connector from the wiring harness.
2. Connect the red voltmeter lead to the blue wiring harness lead. Connect the black voltmeter lead to the black wiring harness lead.
3. Move the tilt switch to the UP position. If the voltmeter does not read at least 12 volts, replace the up solenoid.
4. Move the red voltmeter lead to the pink wiring harness lead.
5. Move the tilt switch to the DOWN position. If the voltmeter does not read at least 12 volts, replace the down solenoid.
6. If the voltmeter readings are as specified in Step 3 and Step 5 and the motor still does not run, replace the motor.

DT 55, DT 65

Refer to **Figure 5** for this procedure.

1. Disconnect the pump motor electrical connector from the wiring harness.
2. Connect the red voltmeter lead to the blue wiring harness lead. Connect the black voltmeter lead to the black wiring harness lead.
3. Move the trim switch to the UP position. If the voltmeter does not read at least 12 volts, replace the up solenoid.
4. Move the red voltmeter lead to the red wiring harness lead.
5. Move the trim switch to the DOWN position. If the voltmeter does not read at least 12 volts, replace the down solenoid.
6. If the voltmeter readings are as specified in Step 3 and Step 5 and the motor still does not run, replace the motor.

1985 DT 115, DT 140 and 1985-1986 DT 75, DT 85

Refer to **Figure 6** for this procedure.

1. Disconnect the pump motor electrical connector from the wiring harness.
2. Connect the red voltmeter lead to the blue wiring harness lead. Connect the black voltmeter lead to the black wiring harness lead.
3. Move the trim switch to the UP position. If the voltmeter does not read at least 12 volts, replace the up solenoid.
4. Move the red voltmeter lead to the red/white wiring harness lead.
5. Move the trim switch to the DOWN position. If the voltmeter does not read at least 12 volts, replace the down solenoid.
6. If the voltmeter readings are as specified in Step 3 and Step 5 and the motor still does not run, replace the motor.

1987-on DT 75, DT 85; 1986-on DT 115, DT 140; V4, V6

Refer to the following illustrations for this procedure:

 a. **Figure 11**: 1987-on DT 75, DT 85; 1986-on DT 115, DT 140.
 b. **Figure 14**: V4.
 c. **Figure 15**: V6.

1. Disconnect the pump motor electrical connector from the wiring harness.
2. Connect the red voltmeter lead to the blue wiring harness lead. Connect the black voltmeter lead to the light blue wiring harness lead.
3. Move the trim switch to the UP position. If the voltmeter does not read at least 12 volts, replace the up solenoid.
4. Move the red voltmeter lead to the red wiring harness lead.
5. Move the trim switch to the DOWN position. If the voltmeter does not read at least 12 volts, replace the down solenoid.
6. If the voltmeter readings are as specified in Step 3 and Step 5 and the motor still does not run, replace the motor.

PUMP OIL PRESSURE TEST

The manual release valve must be completely closed while performing these tests or the results will be false.

1985-on DT 35, DT 40

Trim down pressure test

Refer to **Figure 19** for this procedure.
1. Place a suitable container under the fitting to be removed in the following steps to catch any hydraulic fluid that may leak out.
2. Disconnect the hydraulic line from ports A and B.
3. Install a test gauge between ports A and B.
4. Depress the power trim switch to the DOWN position until the trim rod reaches the bottom of its stroke.

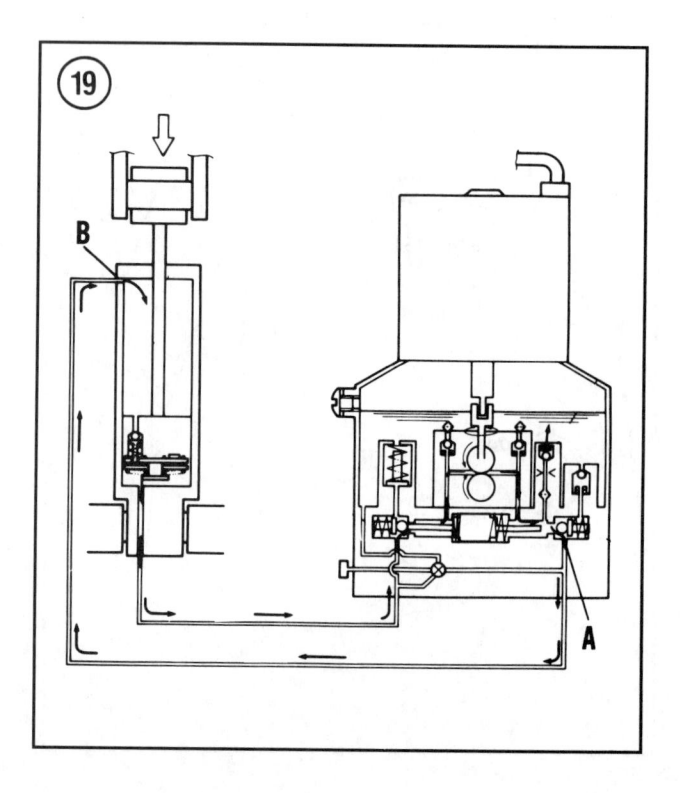

POWER TRIM AND TILT SYSTEM

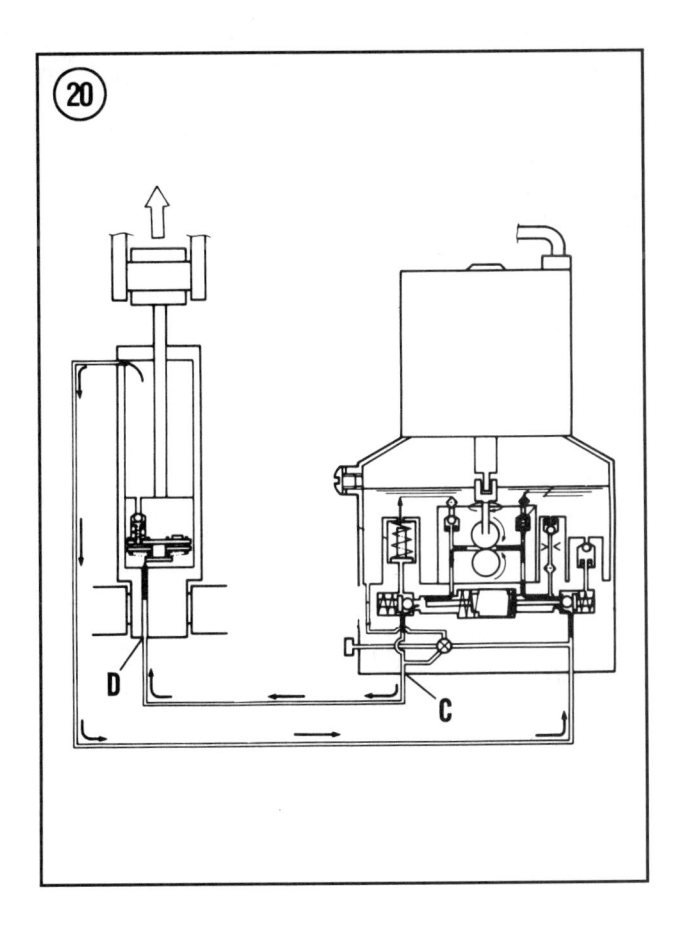

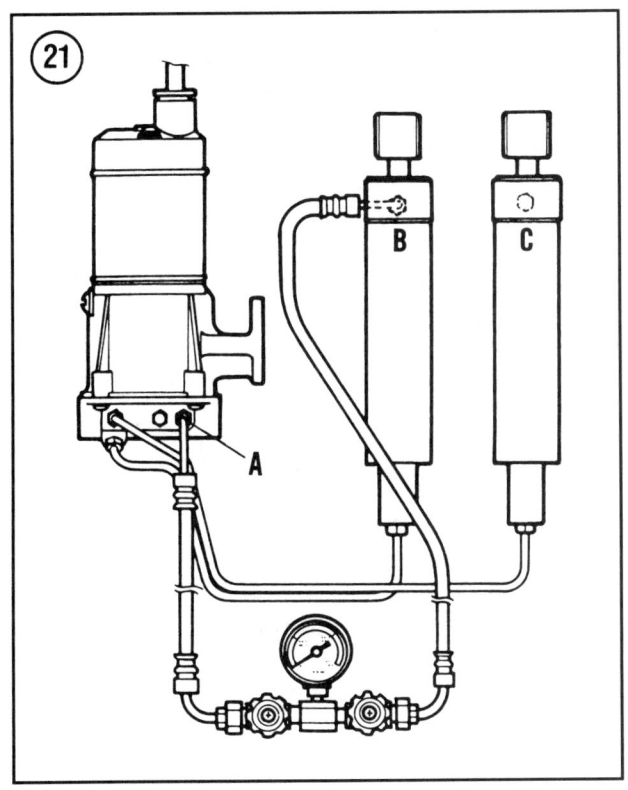

5. The specified oil pressure is 435-1,015 psi (3,000-7,000 kPa).
6. If the pressure reading does not remain steady or does not reach the correct pressure, the pump and valve body are defective and must be replaced.
7. Connect the hydraulic line onto ports A and B and tighten the fittings securely.
8. Check the oil level, refill and bleed the system as described in this chapter.

Trim up pressure test

Refer to **Figure 20** for this procedure.
1. Place a suitable container under the fitting to be removed in the following steps to catch any hydraulic fluid that may leak out.
2. Disconnect the hydraulic line from ports C and D.
3. Install a test gauge between ports C and D.
4. Depress the power trim switch to the UP position until the trim rod reaches the top of its stroke.
5. The specified oil pressure is 1,885-2,465 psi (13,000-17,000 kPa).
6. If the pressure reading does not remain steady or does not reach the correct pressure, the pump and valve body are defective and must be replaced.
7. Connect the hydraulic line onto ports C and D and tighten the fittings securely.
8. Check the oil level, refill and bleed the system as described in this chapter.

1986-on DT 55, DT 65

NOTE
Suzuki does not provide service information for 1985 models.

Tilt down pressure test

Refer to **Figure 21** for this procedure.

1. Place a suitable container under the fitting to be removed in the following steps to catch any hydraulic fluid that may leak out.
2. Disconnect the hydraulic line from ports A and B.
3. Install a test gauge between ports A and B.
4. Depress the power tilt switch to the DOWN position until the tilt rod reaches the bottom of its stroke.
5. The specified oil pressure is 406-812 psi (2,800-5,600 kPa).
6. If the pressure reading does not remain steady or does not reach the correct pressure, the pump and valve body are defective and must be replaced.
7. Connect the hydraulic line onto port B and tighten the fitting securely.
8. Disconnect the hydraulic line from port C.
9. Install a test gauge between ports A and C.
10. Depress the power tilt switch to the DOWN position until the tilt rod reaches the bottom of its stroke.
11. The specified oil pressure is 406-812 psi (2,800-5,600 kPa).
12. If the pressure reading does not remain steady or does not reach the correct pressure, the pump and valve body are defective and must be replaced.
13. Connect the hydraulic line onto ports A and C and tighten the fittings securely.
14. Check the oil level, refill and bleed the system as described in this chapter.

Tilt up pressure test

Refer to **Figure 22** for this procedure.
1. Place a suitable container under the fitting to be removed in the following steps to catch any hydraulic fluid that may leak out.
2. Disconnect the hydraulic line from ports F and G.
3. Install a test gauge between ports F and G.
4. Depress the power tilt switch to the UP position until the tilt rod reaches the top of its stroke.
5. The specified oil pressure is 1,885-2,465 psi (13,000-17,000 kPa).
6. If the pressure reading does not remain steady or does not reach the correct pressure, the pump and valve body are defective and must be replaced.
7. Connect the hydraulic line onto ports F and G and tighten the fittings securely.
8. Disconnect the hydraulic line from ports D and E.
9. Install a test gauge between ports D and E.
10. Depress the power tilt switch to the UP position until the tilt rod reaches the top of its stroke.
11. The specified oil pressure is 1,885-2,465 psi (13,000-17,000 kPa).

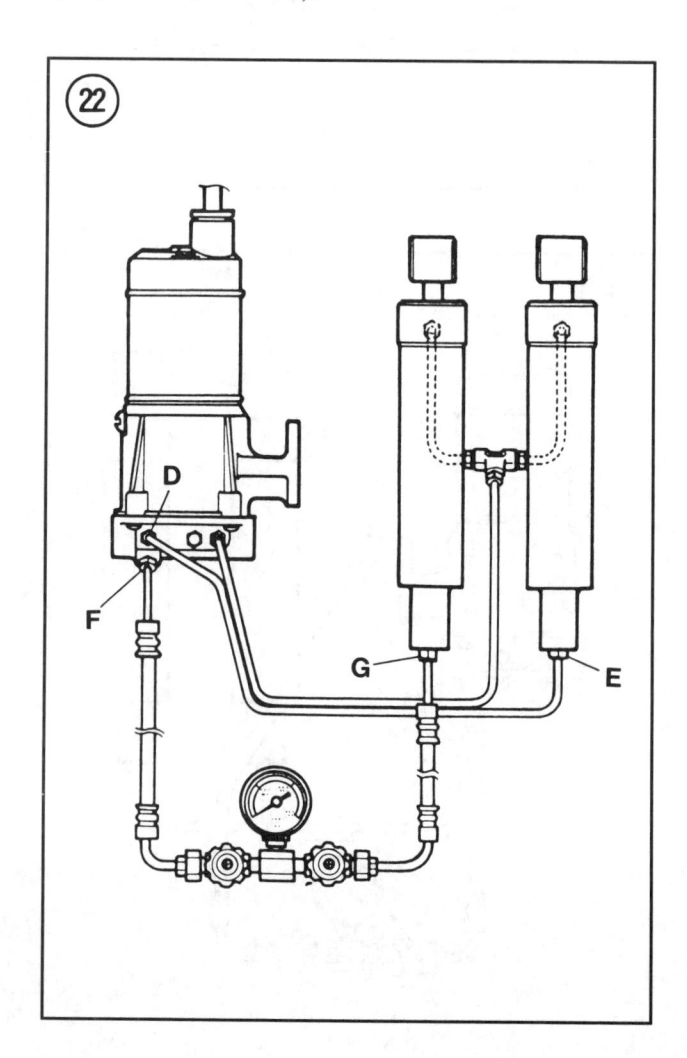

POWER TRIM AND TILT SYSTEM

12. If the pressure reading does not remain steady or does not reach the correct pressure, the pump and valve body are defective and must be replaced.
13. Connect the hydraulic line onto ports D and E and tighten the fittings securely.
14. Check the oil level, refill and bleed the system as described in this chapter.

1986-on DT 115, DT 140

Tilt down pressure test

Refer to **Figure 23** for this procedure.
1. Place a suitable container under the fitting to be removed in the following steps to catch any hydraulic fluid that may leak out.
2. Disconnect the hydraulic line from ports A and B.

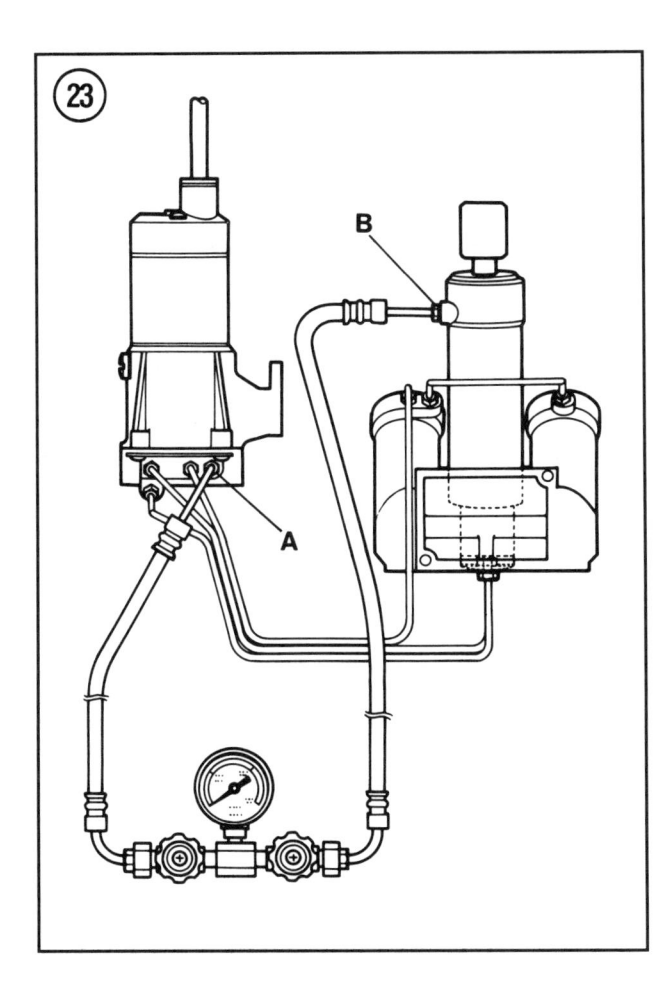

3. Install a test gauge between ports A and B.
4. Depress the power tilt switch to the DOWN position until the tilt rod reaches the bottom of its stroke.
5. The specified oil pressure is 406-812 psi (2,800-5,600 kPa).
6. If the pressure reading does not remain steady or does not reach the correct pressure, the pump and valve body are defective and must be replaced.
7. Connect the hydraulic line onto ports A and B and tighten the fittings securely.
8. Check the oil level, refill and bleed the system as described in this chapter.

Tilt up pressure test

Refer to **Figure 24** and **Figure 25** for this for this procedure.
1. Place a suitable container under the fitting to be removed in the following steps to catch any hydraulic fluid that may leak out.
2. Disconnect the hydraulic line from ports C and D.
3. Install a test gauge between ports C and D.
4. Depress the power tilt switch to the UP position until the tilt rod reaches the top of its stroke.
5. The specified oil pressure is 1,885-2,465 psi (13,000-17,000 kPa).
6. If the pressure reading does not remain steady or does not reach the correct pressure, the pump and valve body are defective and must be replaced.
7. Connect the hydraulic line onto ports C and D and tighten the fittings securely.
8. Disconnect the hydraulic line from ports E and F.
9. Install a test gauge between ports E and F.
10. Depress the power tilt switch to the UP position until the tilt rod reaches the top of its stroke.
11. The specified oil pressure is 1,885-2,465 psi (13,000-17,000 kPa).

12. If the pressure reading does not remain steady or does not reach the correct pressure, the pump and valve body are defective and must be replaced.

13. Connect the hydraulic line onto ports E and F and tighten the fittings securely.

14. Check the oil level, refill and bleed the system as described in this chapter.

V4 and V6

Pump pressure test

Refer to **Figure 26** for this procedure.

1. Raise the engine to the full tilt position and lower the tilt lock lever.
2. Place a suitable container under the fitting to be removed in the following steps to catch any hydraulic fluid that may leak out.
3. Disconnect the hydraulic inlet line from port A and the outlet line from port B.
4. Install a test gauge between ports A and B.
5. Close test gauge valve "B" and open test gauge valve "A."
6. Depress the power tilt switch to the UP position until the tilt rod reaches the top of its stroke.
7. The specified oil pressure is 1,280-1,700 psi (8,826-11,720 kPa).
8. Close test gauge valve "A" and open test gauge valve "B."
9. Depress the power tilt switch to the DOWN position until the tilt rod reaches the bottom of its stroke.
10. The specified oil pressure is 406-812 psi (2,800-5,600 kPa).
11. Connect the hydraulic line onto ports A and B and tighten the fittings securely.

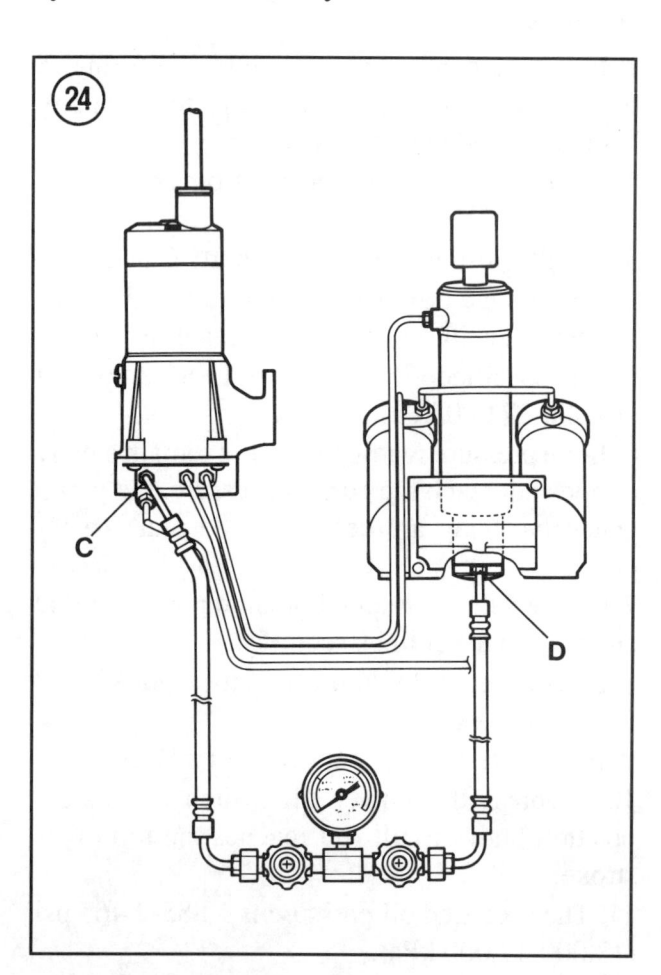

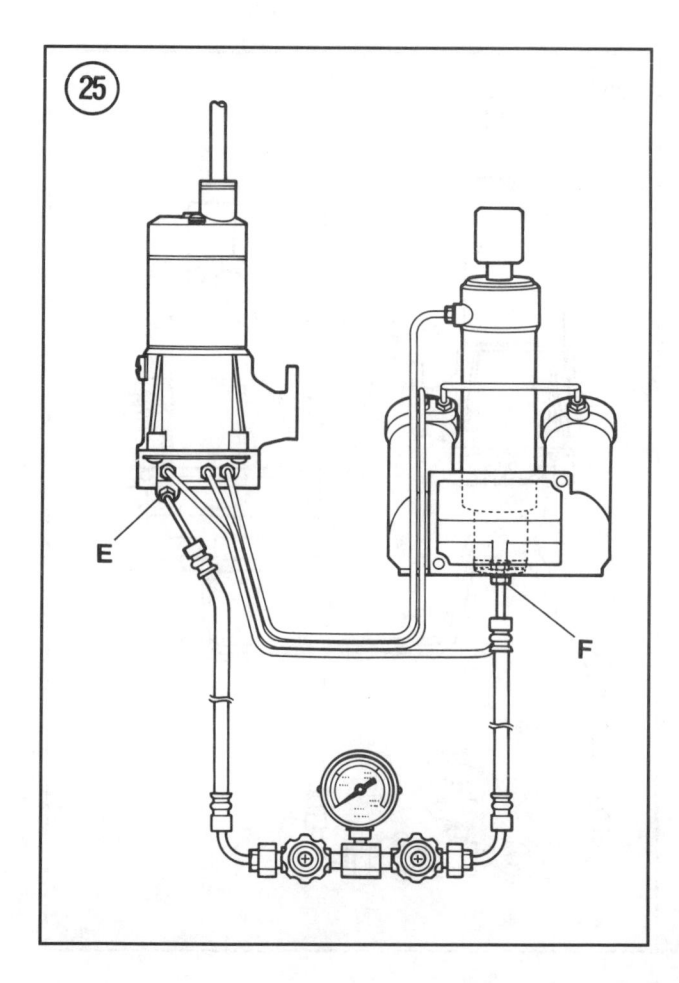

POWER TRIM AND TILT SYSTEM

12. If the pressure reading does not remain steady or does not reach the correct pressure, the pump and valve body are defective and must be replaced.

13. Check the oil level, refill and bleed the system as described in this chapter.

Tilt cylinder pressure test

Refer to **Figure 27** for this procedure.

1. Raise the engine to the full tilt position and lower the tilt lock lever.
2. Place a suitable container under the fitting to be removed in the following steps to catch any hydraulic fluid that may leak out.
3. Disconnect the hydraulic inlet line from ports A and C.
4. Install a test gauge between ports A and C.

5. Depress the power tilt switch to the UP position until the tilt rod reaches the top of its stroke. Don't release the switch until Step 6.
6. Close test gauge valve "B" prior to releasing the power tilt switch.
7. If the pressure reading does not remain steady, the tilt cylinder is defective and must be disassembled and inspected for a damaged seal ring or scored cylinder. Replace if necessary.
8. Remove the test gauge and install the hydraulic inlet line onto ports A and C. Tighten the fittings securely.
9. Disconnect the hydraulic outlet line from ports B and D.
10. Install a test gauge between ports B and D.
11. Depress the power tilt switch to the DOWN position until the tilt rod reaches the bottom of its stroke. Don't release the switch until Step 12.

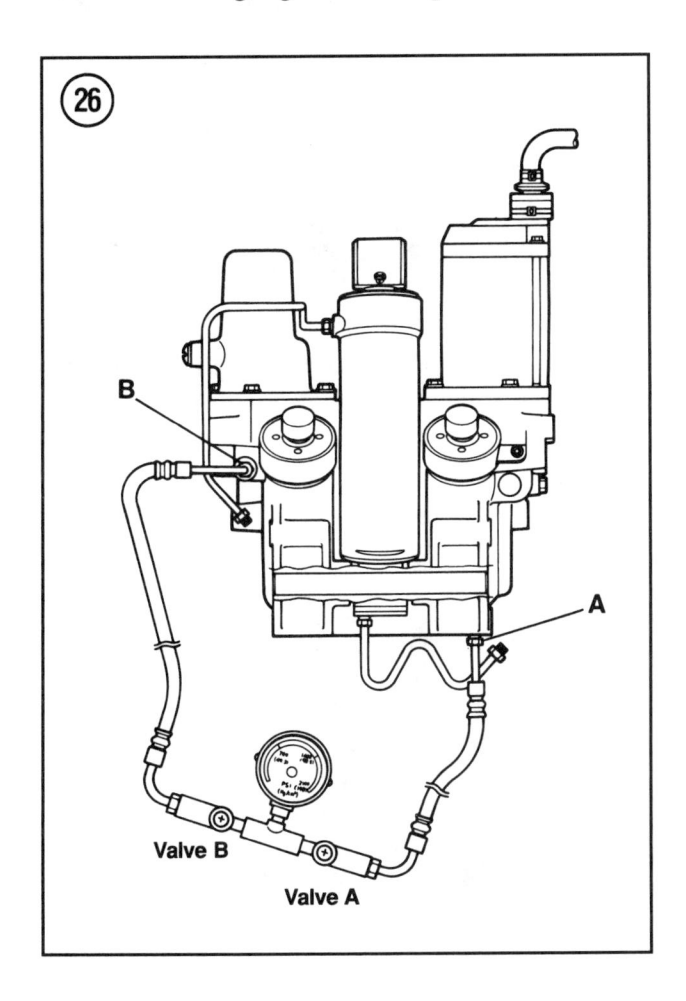

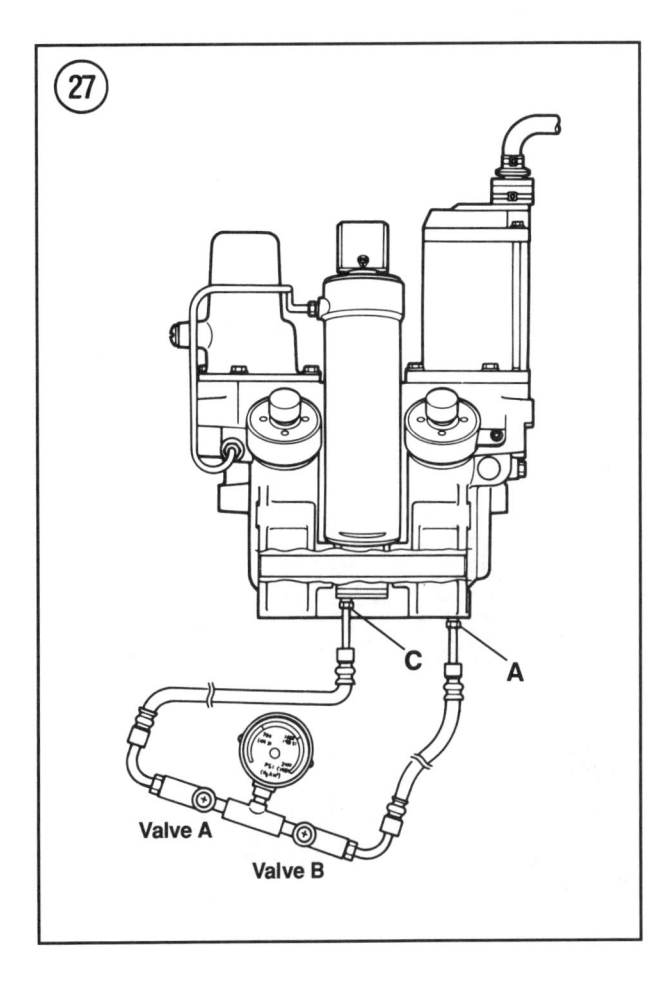

12. Close test gauge valve "A" prior to releasing the power tilt switch.

13. If the pressure reading does not remain steady, the tilt cylinder must be disassembled and inspected for a damaged seal ring, scored cylinder or shock valve problem. Replace if necessary.

14. Remove the test gauge and install the hydraulic inlet line onto ports B and D. Tighten the fittings securely.

COMPONENT REPLACEMENT

Service to the trim/tilt cylinders is much easier and quicker if the outboard motor has been removed from the boat. On some boat installations, outboard motor removal may be necessary to provide sufficient access to the hydraulic fittings to prevent damage to the lines.

Hydraulic Pump/Motor Removal/Installation (All Models Except V4 and V6)

NOTE
On V4 and V6 models, the motor is an integral part of the trim and tilt cylinder assembly and cannot be removed separately.

1. Disconnect the pump/motor electrical connector from the wiring harness.
2. Place a container under the pump and disconnect all hydraulic lines from the motor assembly. Cap the lines to prevent leakage.
3. Remove the bolts holding the pump to the clamp bracket (**Figure 28**). Remove the pump.
4. Installation is the reverse of removal.

Trim/Tilt Cylinder Removal/Installation (DT 55, DT 65; 1985-1989 DT 75, DT 85)

1. Securely support the motor in the full up position.

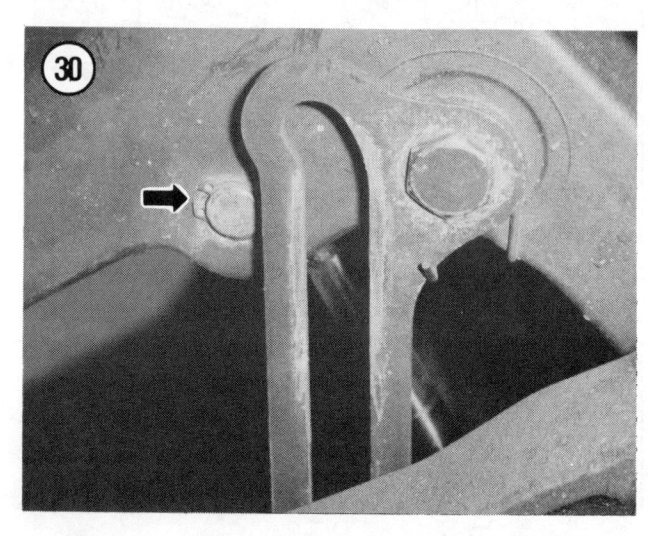

POWER TRIM AND TILT SYSTEM

2. Disconnect the hydraulic line at the cylinder. Cap the line to prevent leakage.

NOTE
Some installations use a single piston rod pin to hold both cylinder pistons in place; others use individual pins for each cylinder.

3. Remove the cotter pin on the inner side of each cylinder piston rod pin (**Figure 29**).

4. Remove the snap ring on the outer side of each rod pin (**Figure 30**).

5. Remove the piston rod pin(s).

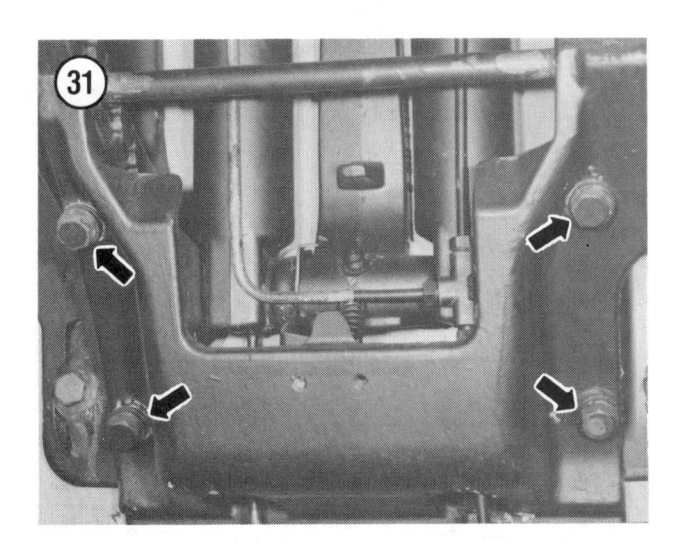

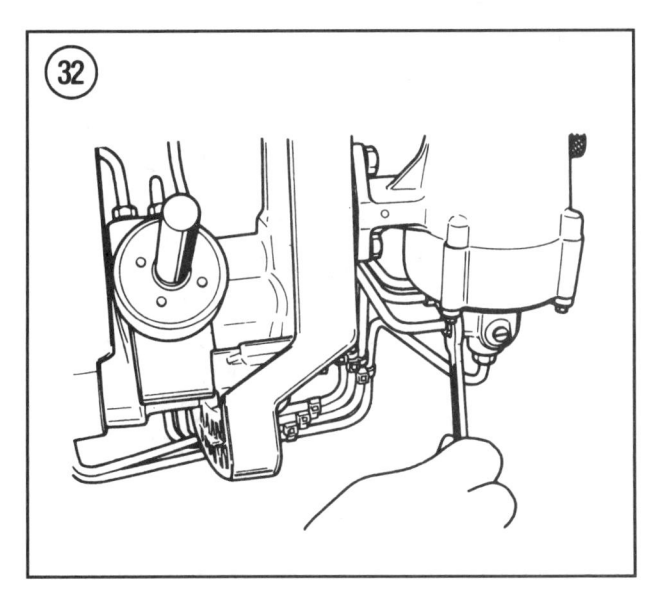

6. Remove the bolts and nuts holding the cylinder support bracket to the clamp bracket assembly (**Figure 31**).

7. Remove the cylinder support bracket with the cylinders attached.

8. Remove the lower cylinder shaft fasteners. Slide the shaft from the support bracket and cylinders. Remove the cylinders.

9. Installation is the reverse of removal. Note the following during installation.

10. Coat the piston rod pin(s) and lower cylinder shaft with water-resistant grease.

11. Refill the hydraulic pump and bleed the system as described in this chapter.

Trim/Tilt Cylinder Removal/Installation (1990-on DT 75, DT 85; DT 115; DT 140)

1. Securely support the motor in the full up position.

2. Disconnect the hydraulic lines at the pump (**Figure 32**). Cap the lines to prevent leakage.

3. Remove the pump as described in this chapter.

4. Remove the snap ring at each end of the tilt cylinder upper shaft (**Figure 33**). Carefully drive the upper shaft out.

5. Remove the tilt lock pin.

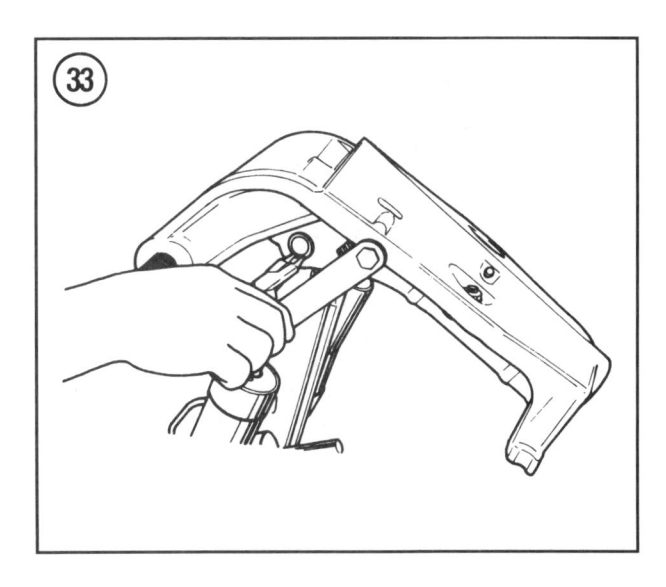

6. Remove the bolts on each clamp bracket holding the manifold in place (**Figure 34**). Remove the manifold.
7. Drive the tilt cylinder lower shaft from the manifold. Remove the tilt cylinder (**Figure 35**).
8. Installation is the reverse of removal. Note the following during installation.
9. Coat the upper and lower shafts with water-resistant grease.
10. Refill the hydraulic pump and bleed the system as described in this chapter.

Trim/Tilt Cylinder/Motor Assembly Removal/Installation (V4, V6)

On V4 and V6 models, the motor is an integral part of the trim and tilt cylinder assembly and is removed as a total assembly.
1. Disconnect the battery cables.
2. Remove the engine cover.
3. Remove the tilt/trim assembly electrical cable clamp on the starboard side of the engine.
4. Disconnect the tilt/trim assembly electrical cable connector from the wiring harness.
5. Remove the cover on the electrical component holder.
6. Remove the electrical cable clamp (A, **Figure 36**) from the starboard side of the engine lower cover.
7. Remove the trim sensor electrical cable harness and the rubber grommet (B, **Figure 36**) from the starboard side of the engine lower cover.
8. Turn the manual release valve 2 full turns clockwise (**Figure 37**).
9. Raise the motor to the full up position and lower the tilt lock lever. Secure the motor in this position.
10. Remove the snap ring (**Figure 38**) at each end of the tilt cylinder upper shaft. Carefully drive the upper shaft out.

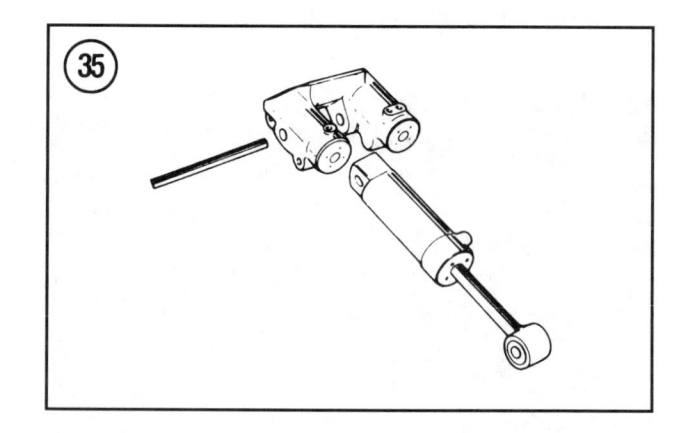

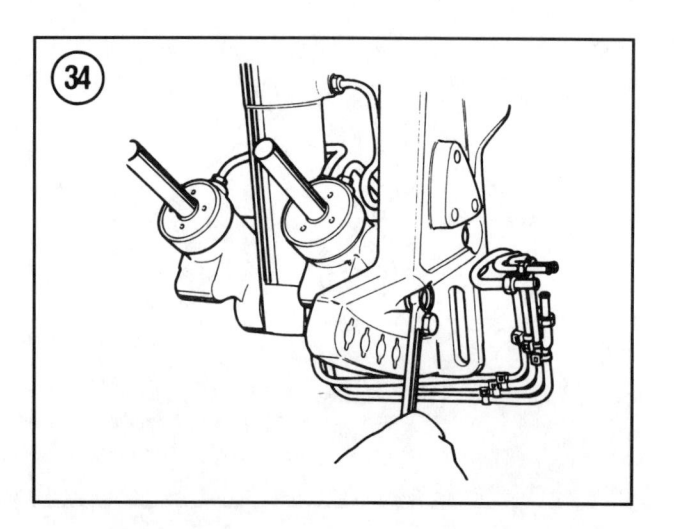

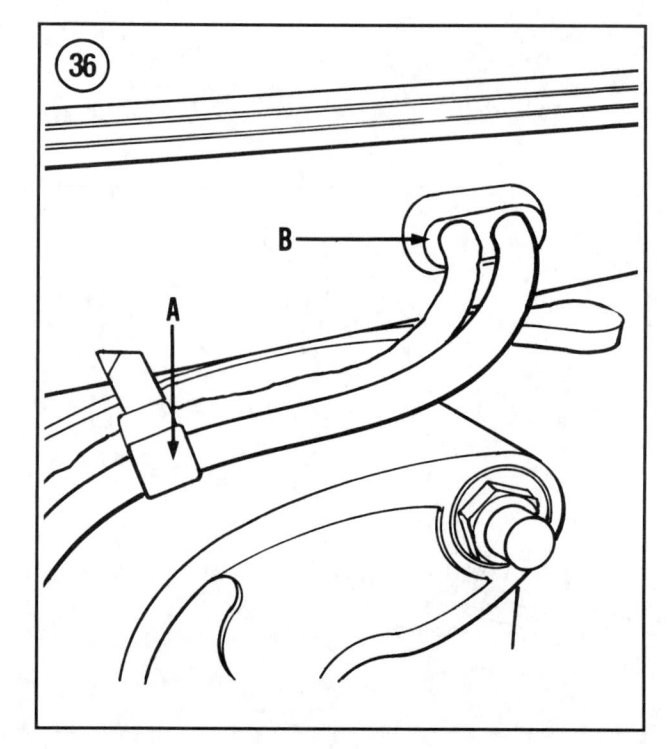

POWER TRIM AND TILT SYSTEM

11. Remove the clamp bracket shaft nut (A, **Figure 39**) and the tilt lock pin nut (B, **Figure 39**).

12. Remove the bolts (C, **Figure 39**) securing the starboard clamp bracket and remove the clamp bracket.

13. Remove the bolts (**Figure 40**) securing the port clamp bracket.

14. Remove the tilt/trim assembly.

15. Installation is the reverse of removal. Note the following during installation.

16. Coat the upper and lower shafts with water-resistant grease.

17. Refill the hydraulic pump and bleed the system as described in this chapter.

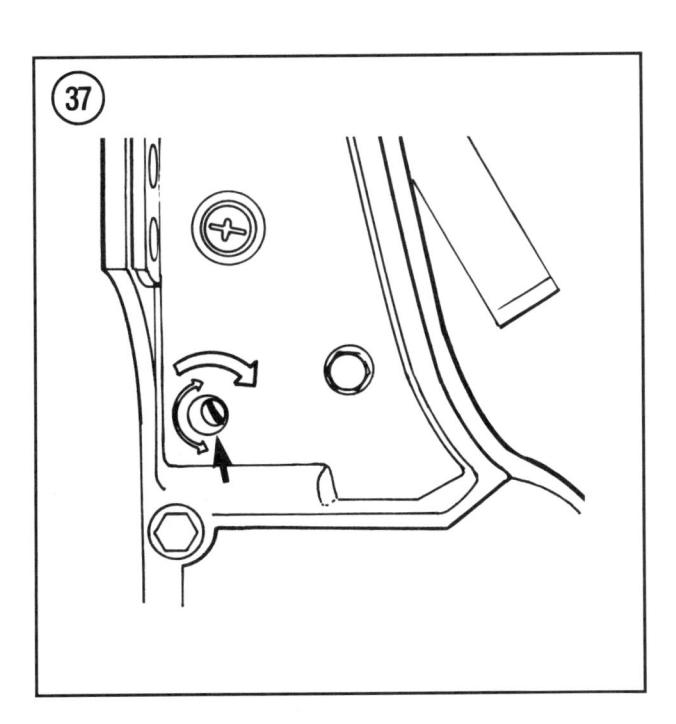

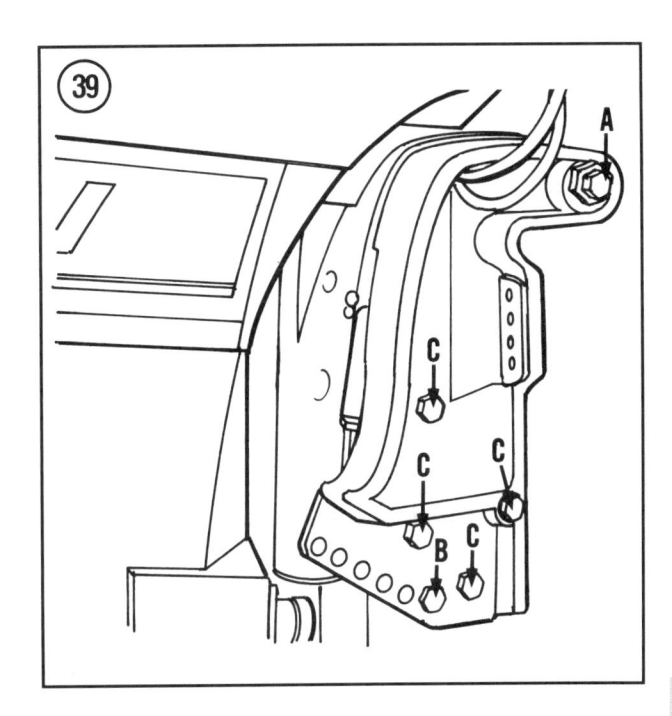

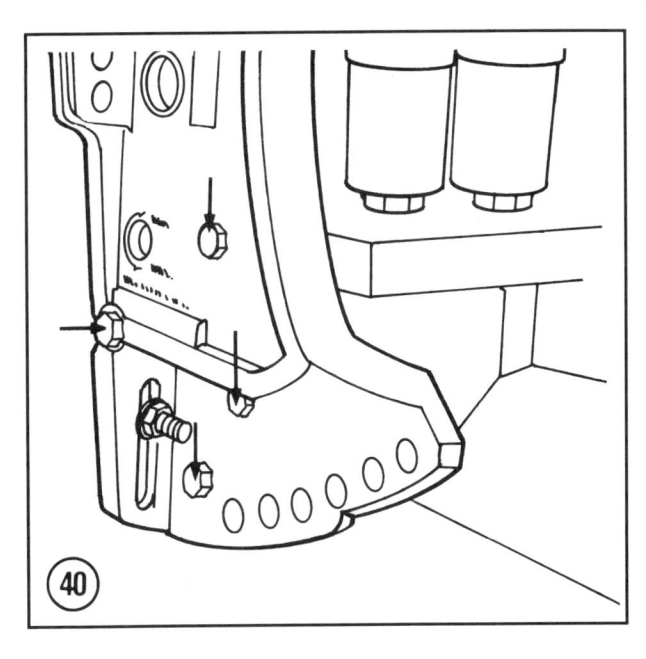

Chapter Twelve

Oil Injection System

The fuel:oil ratio required by outboard motors depends upon engine demand. Without oil injection, oil must be hand-mixed with gasoline at a 50:1 ratio to assure that sufficient lubrication is provided at all operating speeds and engine load conditions. This ratio is adequate for high-speed operation, but contains more oil than required to lubricate the engine properly during idle and low-speed operation.

With oil injection, the ratio of oil provided with the fuel sent to the engine cylinders can be varied instantly and accurately to provide the optimum ratio for proper lubrication at any operating speed or engine load condition.

DT 35 and DT 40, 1988-on DT 8 and DT 9.9, 1989-on DT 15, all 3- and 4-cylinder and all V4 and V6 models covered in this book are equipped with a mechanical oil injection system using a crankshaft-driven injection pump and integral oil tank. The pump draws oil from the oil tank and supplies it under pressure to intake manifold nozzles where it is sprayed into the air:fuel mixture. **Figure 1** shows a typical oil injection system. There are minor differences among the various models covered in this book. A control rod connects the pump to the throttle and varies the pump stroke according to throttle opening. The fuel:oil ratio ranges between 125:1 at idle to 50:1 at wide-open throttle.

NOTE
In order to comply with any applicable Suzuki warranty on 1990 and later models, you must use Suzuki Outboard motor oil or NMMA Certified TC-W II outboard motor oil. This type of motor oil is also highly recommended for use in all outboard motors covered in this book. This type of motor oil has shown no tendency to gel or cause filtration problems. The use of the normally recommended ashless oils can cause gelling which will lead to filter blockage in the oil injection system. This blockage can cause insufficient lubrication to the cylinders and can result in engine seizure.

This chapter covers the operation and service required by the oil injection system.

Table 1 is at the end of the chapter.

OIL INJECTION SYSTEM

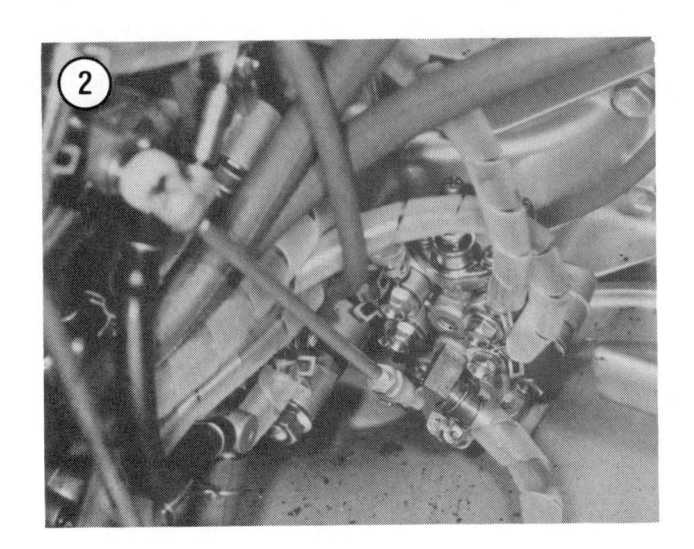

SYSTEM COMPONENTS

Models equipped with oil injection use a self-contained oil pump on the power head in addition to the fuel pump. **Figure 2** shows a typical oil pump installation. There are minor differences among the various models covered in this book.

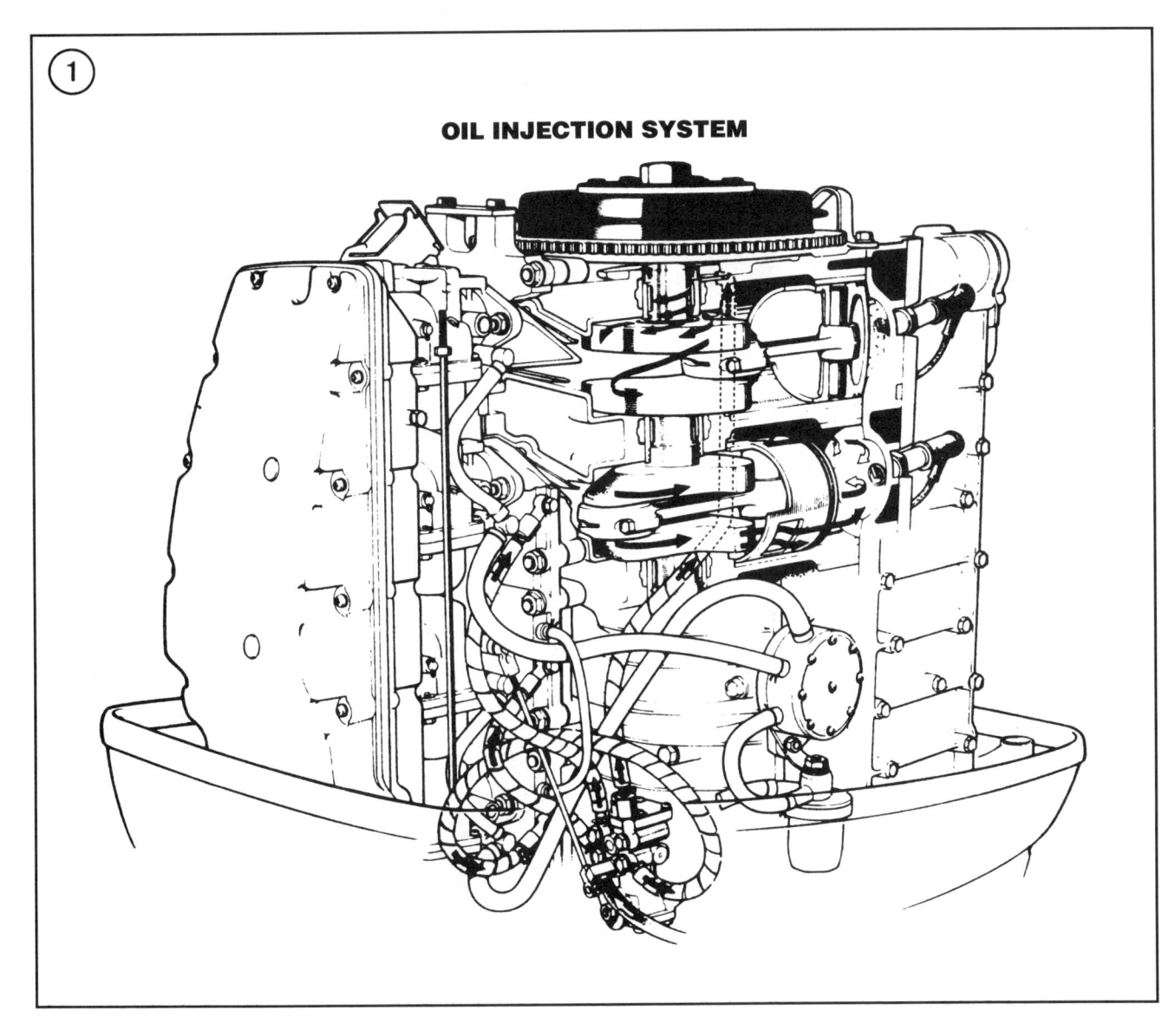

The pump is connected to the throttle by a control rod. An oil tank or reservoir is mounted on the power head (**Figure 3**, typical). When full, the tank contains sufficient oil for approximately 2 to 5 hours, depending on model, of continuous wide-open throttle operation. A warning buzzer located in the remote control box monitors oil level in the tank and sounds when the oil level becomes low.

OPERATION

Suzuki's injection system supplies oil to the engine separately from the fuel. A drive gear on the crankshaft engages a driven gear in the oil pump. This driven gear transmits crankshaft rotation through a series of reduction gears inside the pump, controlling the stroke of the pump plunger according to crankshaft speed. Since the pump is mechanically linked to the throttle, it supplies the proper amount of oil according to engine speed.

NOTE
The oil injection pump is sealed at the factory and cannot be serviced. If defective, it must be replaced. Any attempt to disassemble the pump will void the factory warranty.

Oil travels to and from the oil pump through transparent lines. This permits visual confirmation that the system is functioning properly.

Break-in Procedure

During the first 5 hours of operation (break-in period) of a new or rebuilt engine, the fuel in the fuel tank should be a 50:1 fuel-oil mixture (see Chapter Four) *in addition* to the lubricant supplied by the injection pump. Mark the oil level on the translucent oil tank mounted to the power head, then periodically check to make sure that the system is working (oil level diminishing) before switching over to plain gasoline at the end of the break-in period. This applies both to engines that have been overhauled and new engines out of the shipping box.

LOW OIL LEVEL WARNING LIGHT AND BUZZER

A low oil level warning light and buzzer are installed in the remote control box. The light and buzzer serve a dual function on oil-injected models. The sending unit in the power head oil tank is connected to the warning system through the key switch and grounded to the engine. When the oil level in the tank drops to a pre-determined level, the warning light on the monitor will turn from green to red, alerting the user of a low oil level. If the user continues to operate the engine, a buzzer will then sound in double "beeps" at 7 second intervals. After operating the engine in this condition for 10 seconds, the engine rpm will automatically be lowered to approximately 3,000 rpm and the rev limit light on the monitor will come on.

The oil injection system is equipped with a low oil warning reset system that will allow the operator to run the engine, in the low oil condition, at full throttle for about 30 minutes. To reset the warning system, remove the engine cover and press the *red reset button* on the electrical parts cover. This must be performed with the engine running.

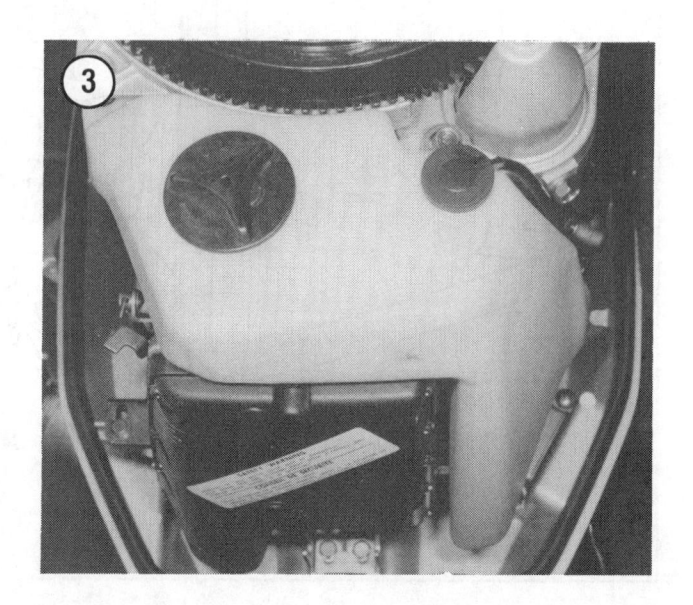

OIL INJECTION SYSTEM

OIL FLOW SENSOR (1987-ON DT 55, DT 65; 1988-ON DT 75, DT 85; V4 AND V6)

The inline oil filter is equipped with an oil flow sensor. If the oil flow is restricted or slows down due to an obstruction within the inline oil filter, the integral micro-switch will close. Refer to **Figure 4** (all models except V6) and **Figure 5** (V6). This will activate the low oil flow and rev limit lights on the monitor. The warning buzzer will sound and the engine speed will automatically decrease.

Sensor Cleaning

If the inline oil filter becomes clogged, it can be disassembled and cleaned.

1. Disassemble the oil filter and remove the strainer.
2. Clean the strainer in solvent (**Figure 6**).
3. Inspect the O-ring, diaphragm and strainer for damage and replace as necessary.
4. Assemble the oil filter and tighten the screws securely.

AIR/OIL MIXING VALVE (DT 25, DT 30 [3-CYLINDER]; 1987-ON DT 35, DT 40; V4 AND V6)

An air/oil mixing valve has been added to the oil injection system to accelerate the atomization of the injection oil. Refer to **Figure 7**. After the oil has left the oil pump and passed the check valve, it then goes through this valve where air from the crankcase is mixed into the stream of oil. This action atomizes the oil into very fine particles which are then injected into the intake channel in the crankcase.

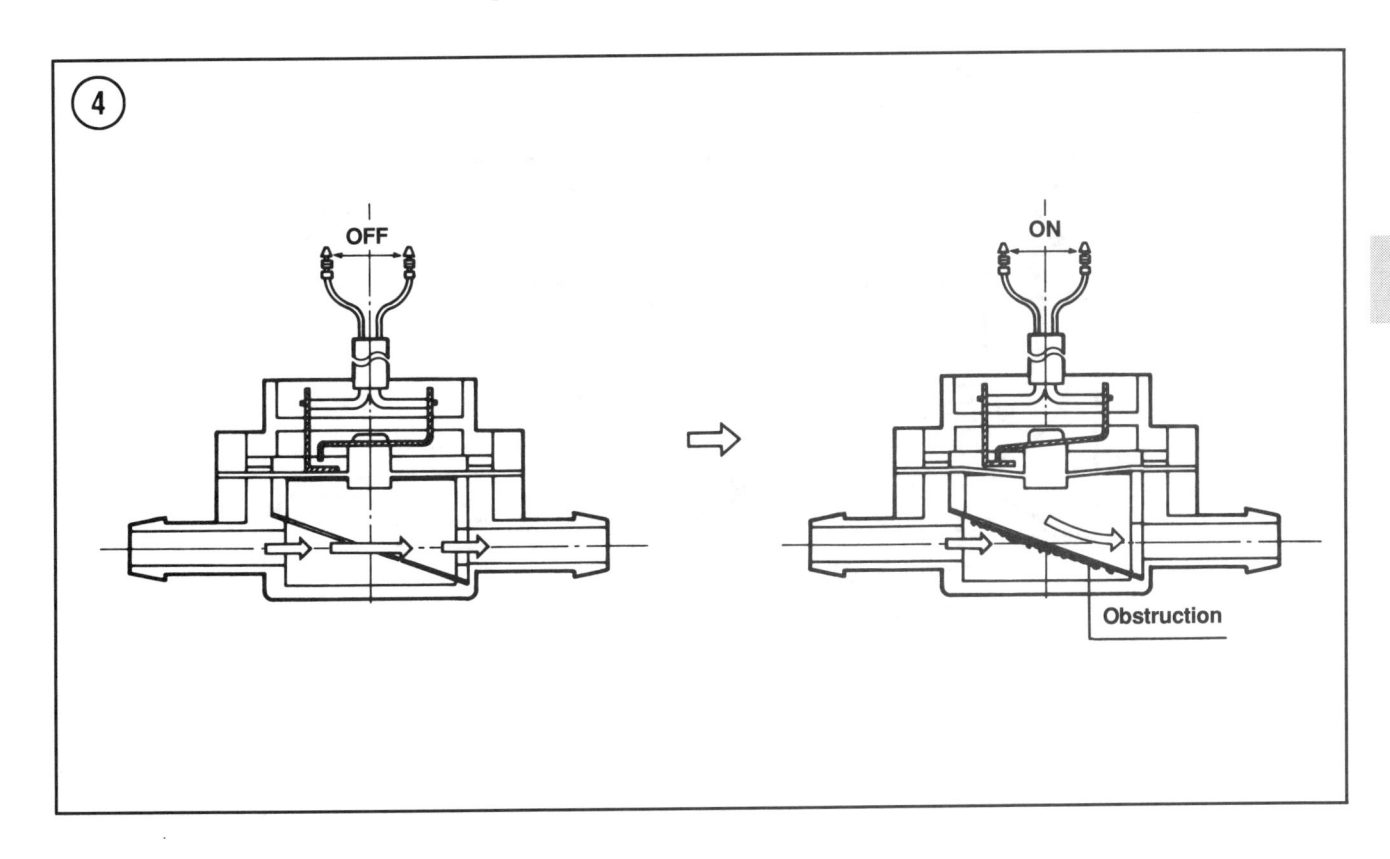

When the air/oil mixture leaves the mixing valve, there will be air present in the oil, visible in the transparent oil line leading to the crankcase—this is normal and is *not* an indication that the system must be bled.

OIL PUMP SERVICE

Oil Pump Bleeding

The oil injection system must be bled to remove any air whenever the system is serviced or when the outboard has not been used for a lengthy period of time.

DT 35, DT 40

1. Remove the engine cover.
2. Make sure that the engine is in an upright position. If tilted in a trailering position, place the engine upright.
3. If the oil pump has been removed, fill the oil lines with oil before reconnecting them to the pump fittings.
4. Place a suitable container under the bleed bolt to catch the oil that will flow out of the oil pump.

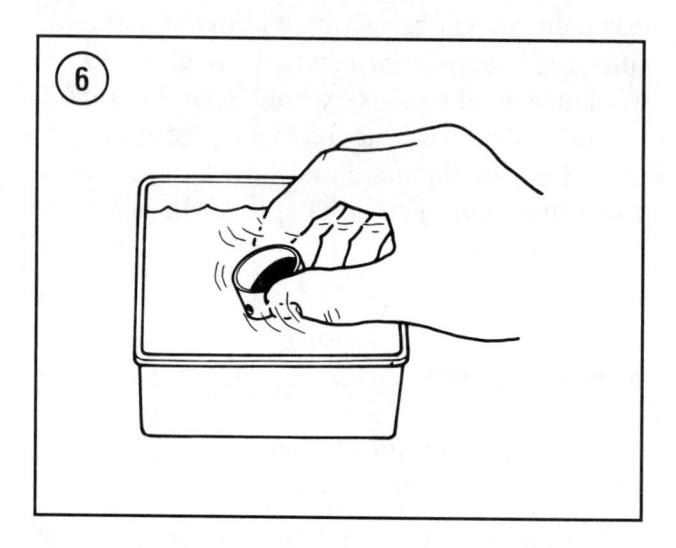

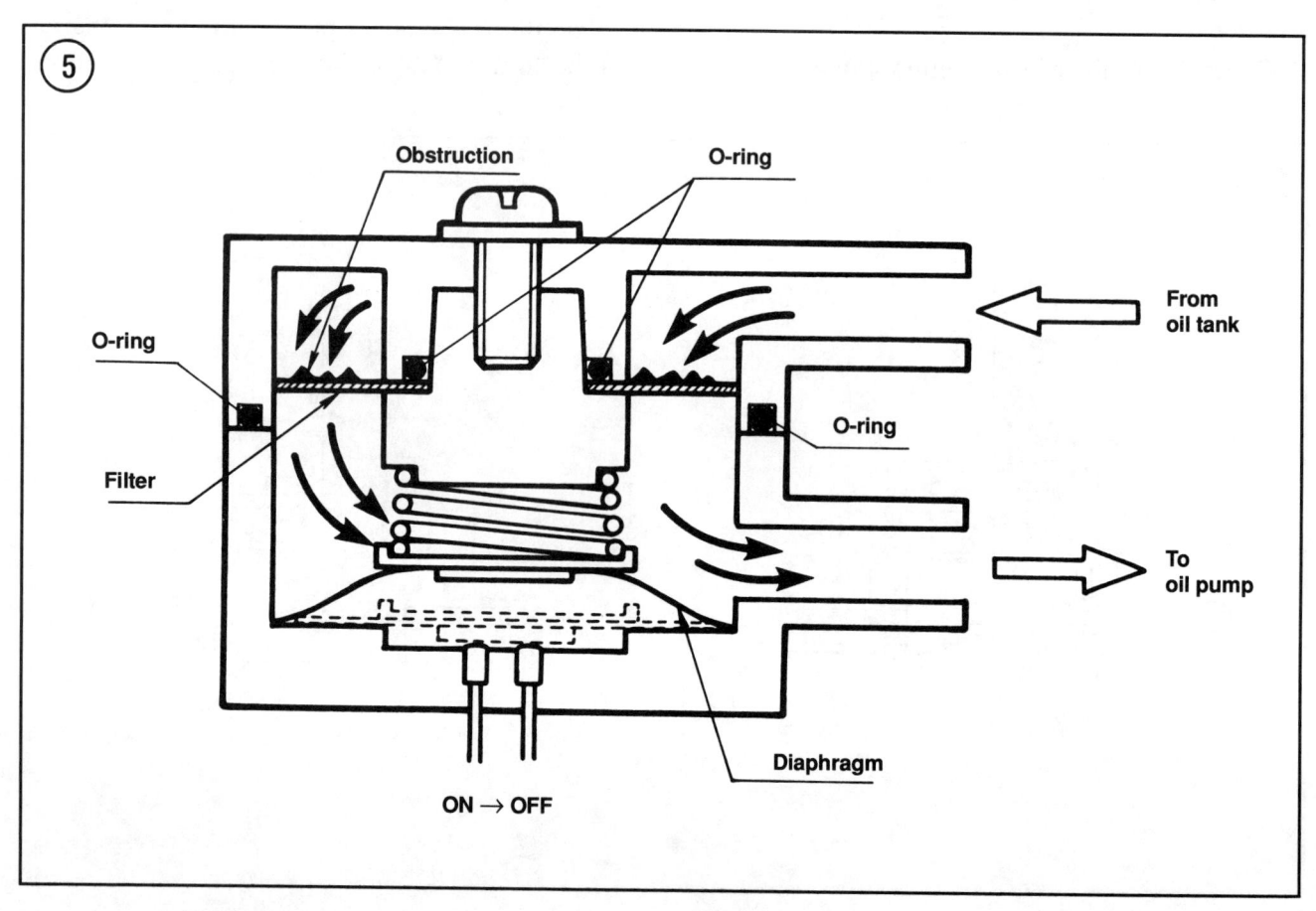

OIL INJECTION SYSTEM

5. Open the bleed bolt on the injection pump 3-4 turns *counterclockwise* (**Figure 8**).

6. Wait several seconds and turn the bleed bolt *clockwise* to close it. Tighten bolt snugly.

7. Install the engine cover.

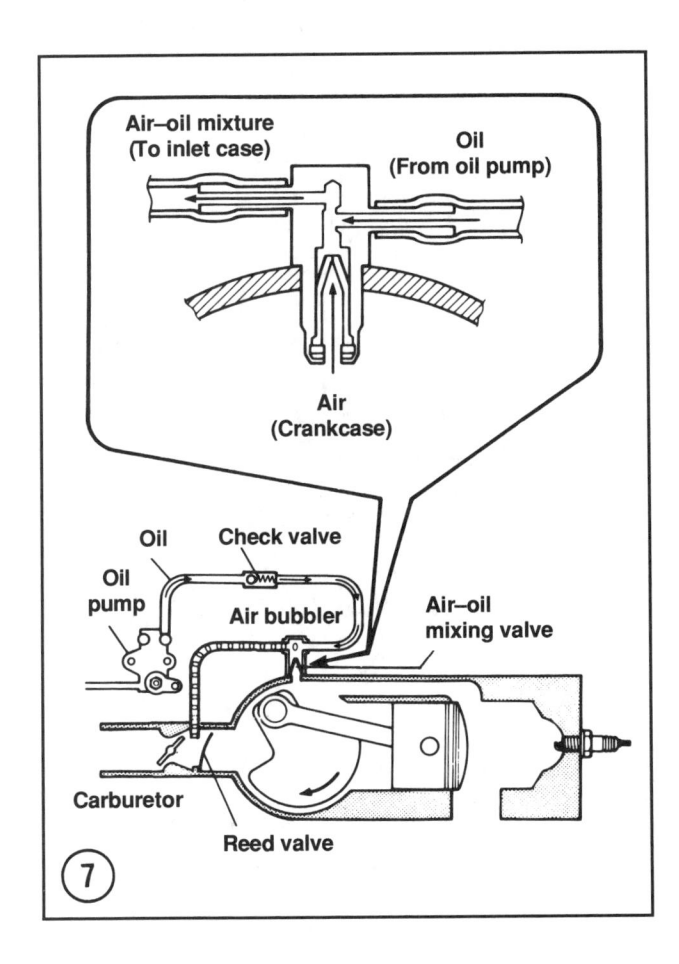

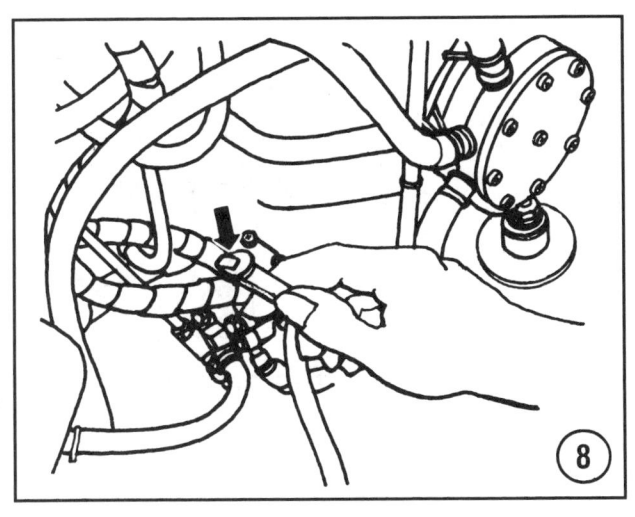

DT 55, DT 65, DT 75, DT 85; DT 115, DT 140

1. Remove the engine cover.
2. Make sure that the engine is in an upright position. If tilted in a trailering position, place the engine upright.
3. If the oil pump has been removed, fill the oil lines with oil before reconnecting them to the pump fittings.
4. Fill the oil tank to the recommended level.
5. Place a suitable container under the bleed bolt to catch the oil that will flow out of the oil pump.
6. Open the bleed bolt on the injection pump 3-4 turns counterclockwise.
7. Disconnect the oil line from the oil tank. Plug the oil line outlet in the oil tank to prevent loss of oil from the tank.
8. Hold the inline oil filter level in a horizontal position.
9. Pour oil into the line on the inline oil filter that was disconnected from the oil tank. Continue to pour oil into the line until the oil leaving the oil pump bleed bolt is free of air bubbles.
10. Turn the bleed bolt clockwise to close it. Tighten bolt securely.
11. Connect the oil line onto the oil tank.
12. Install the engine cover.

All other models

1. Fill the fuel tank with 50:1 gasoline/oil mixture.
2. Remove the engine cover.
3. Fill the oil tank to the recommended level.
4. Place a suitable container under the bleed bolt to catch the oil that will flow out of the oil pump.
5. Open the bleed bolt on the injection pump 3-4 turns counterclockwise.
6. Start the engine and let it idle at 650-700 rpm.
7. Let the engine idle until the oil leaving the oil pump bleed bolt is free of air bubbles.
8. Turn the bleed bolt clockwise to close it. Tighten bolt securely.
9. Install the engine cover.

Oil Pump Control Rod Adjustment

This adjustment should only be required if the oil pump is removed for service.

NOTE
Suzuki provides control rod adjustment procedures and specifications for these models only.

1989-on DT 25, 1988-on DT 30

Refer to **Figure 9** for this procedure.
1. Remove the engine cover.
2. Adjust the throttle linkage as described in Chapter Five.
3. Loosen the locknuts on the control rod.

NOTE
The connectors at each end of the control rod must be at 90° to each other after the correct dimension is achieved in the next step.

4. Adjust the control rod to an initial length of 3.05 in. (77.5 mm) measured between the center of each hole in the connectors. Refer to Dimension A.
5. Install the control rod onto the anchor pins on the oil pump lever and the bottom carburetor lever.
6. Move the throttle to the *fully closed* position.
7. With the throttle in the closed position, the oil pump lever stop must have 0-0.04 in. (0-1 mm) clearance, but it should not touch the boss on the oil pump. Readjust the control rod to achieve this clearance.
8. Hold the connectors and control in place and tighten the locknuts securely.
9. Install the engine cover.

1985-on DT 55, DT 65

Refer to **Figure 10** for this procedure.
1. Remove the engine cover.
2. Adjust the throttle linkage as described in Chapter Five.
3. Move the throttle to the *fully closed* position.
4. Rotate the control rod in either direction until there is a gap between the oil pump lever and the stopper of approximately 0.04 in. (1 mm).

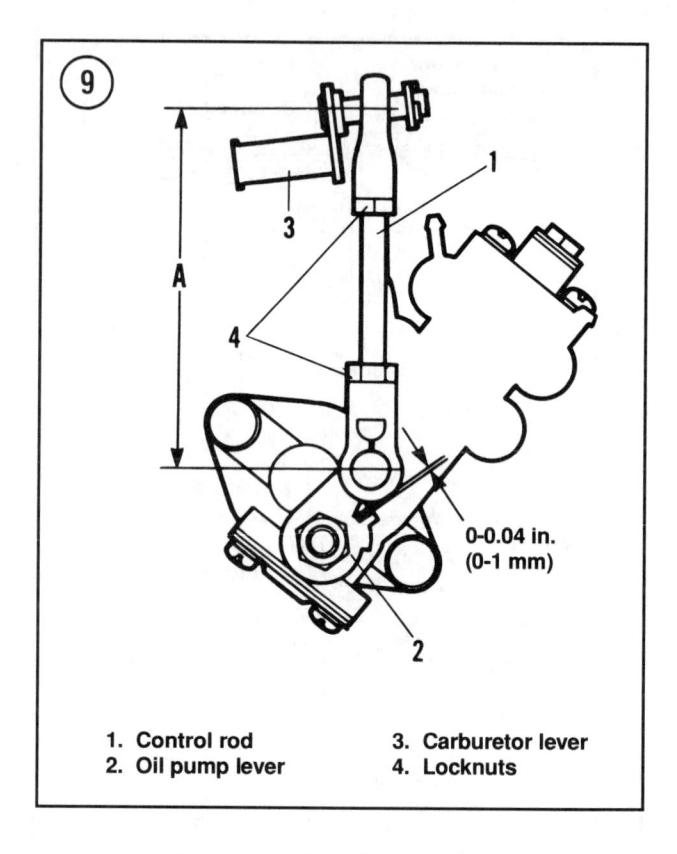

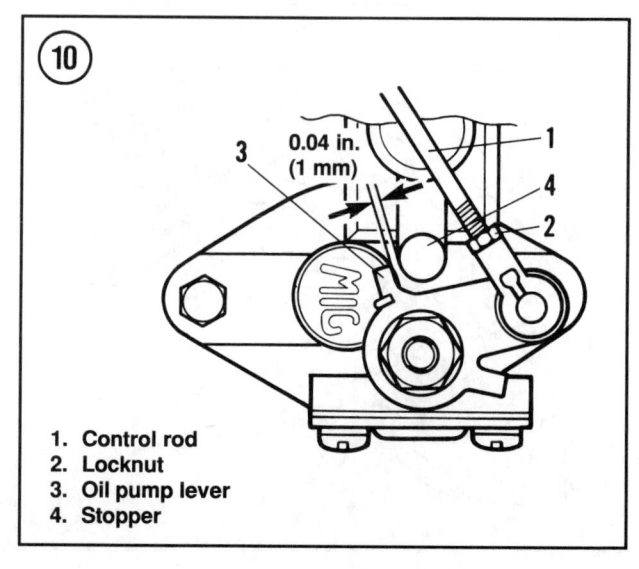

Figure 9
1. Control rod
2. Oil pump lever
3. Carburetor lever
4. Locknuts

Figure 10
1. Control rod
2. Locknut
3. Oil pump lever
4. Stopper

OIL INJECTION SYSTEM

Figure 11
1. Control rod
2. Oil pump lever
3. Carburetor lever
4. Locknuts

0.04 in. (1 mm)

Figure 12
1. Control rod
2. Oil pump lever
3. Carburetor lever
4. Locknuts

Less than 0.04 in. (1 mm)

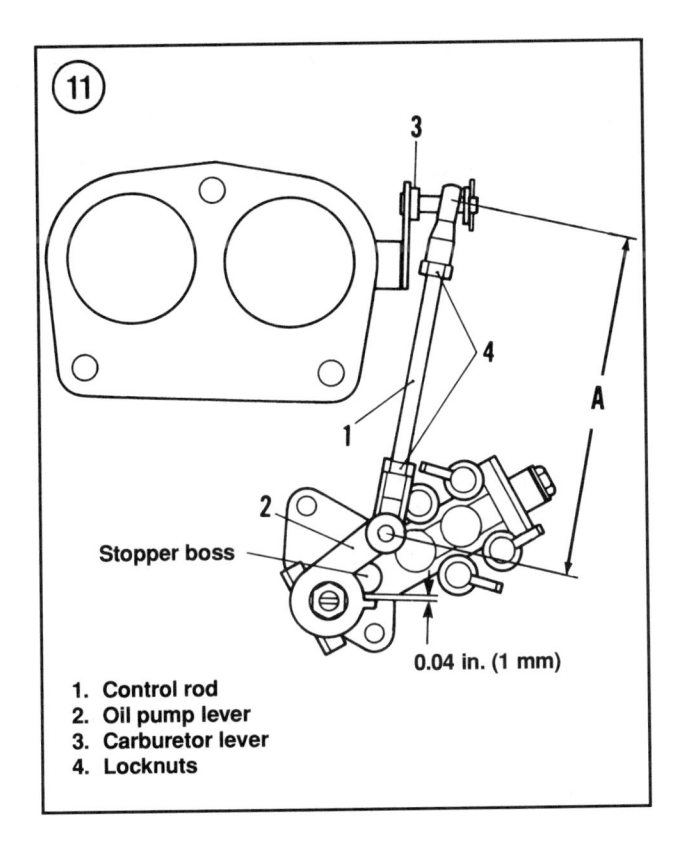

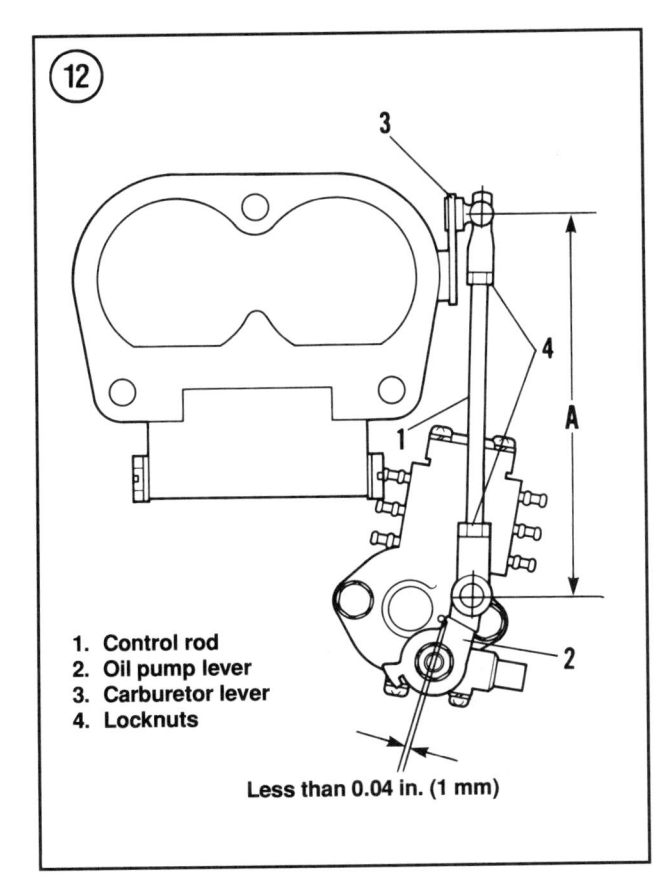

V4 and V6

Refer to **Figure 11** for V4 models or **Figure 12** for V6 models for this procedure.

1. Remove the engine cover.
2. Adjust the throttle linkage as described in Chapter Five.
3. Loosen the locknuts on the control rod.

NOTE
The connectors at each end of the control rod must be at 90° to each other after the correct dimension is achieved in the next step.

4A. On V4 models, adjust the control rod to an initial length of 4.9 in. (124 mm) measured between the center of each hole in the connectors. Refer to Dimension A.

4B. On V6 models, adjust the control rod to an initial length of 5.4 in. (137.5 mm) measured between the center of each hole in the connectors. Refer to Dimension A.

5. Install the control rod onto the anchor pins on the oil pump lever and the bottom carburetor lever.
6. Move the throttle to the *fully closed* position.
7. With the throttle in this position, the oil pump lever stop must have 0-0.04 in. (0-1 mm) clearance (less than 0.04 in. [1 mm] on V6 models), but it should not touch the boss on the oil pump. Readjust the control rod to achieve this clearance.
8. Hold the connectors and control in place and tighten the locknuts securely.
9. Install the engine cover.

Oil Pump Delivery Rate Test

To perform this test you will need a special Suzuki tool and a stop watch.

CAUTION
The engine should be operated with a 50:1 fuel-oil mixture in the fuel tank during this procedure.

1988-on DT 8, DT 9.9; 1989-on DT 115

1. Start the engine and run at idle for 5 minutes or until it reaches operating temperature. Shut the engine off.

2. Remove the engine cover.

3. Remove the oil tank as described in this chapter.

4. Install the engine oil measuring cylinder (Suzuki part No. 09900-21602) as shown in **Figure 13**. Fill the cylinder with oil.

5. Bleed the oil system as described in this chapter.

6. Install a tachometer according to manufacturer's instructions.

7. Start the engine and maintain an engine speed of 1,500 rpm.

8. Select a clean-cut value on the scale of the oil measuring cylinder as a reference point. Note this location and start a stop watch.

9. Run the engine at the specified speed for the next 5 minutes. Measure the amount of oil discharged from the reference point noted in Step 8 to the point at the end of 5 minute period. Refer to **Table 1** for the specified oil pump discharge rate.

NOTE
Oil pump output test results may vary depending upon testing error and ambient temperature. To ensure accurate results, repeat test 2-3 times, or until results are consistent.

10. If the oil discharged is not within specifications, check the injection lines for possible leakage or restrictions. If none are found, replace the oil pump.

11. Disconnect the portable tachometer.

12. Install the oil tank as described in this chapter.

13. Install the engine cover.

All other models

1. Start the engine and run at idle for 5 minutes or until it reaches operating temperature. Shut the engine off.

2. Remove the engine cover.

3. Remove the oil tank as described in this chapter.

4. Disconnect the oil pump control rod from the carburetor.

5. Install the engine oil measuring cylinder to the inlet line and fill with oil. Use Suzuki part No. 09941-8710 for 1986-on DT 115, DT 140, V4 and V6 models or part No. 09900-21602 for all other models.

6. Bleed the oil system as described in this chapter.

7. Install a tachometer according to manufacturer's instructions.

8. Manually move the oil pump control rod to the *fully closed* position.

9. Start the engine and increase and maintain an engine speed of 1,500 rpm.

10. Select a clean-cut value on the scale of the oil measuring cylinder as a reference point. Note this location and start a stop watch.

11. Run the engine at the specified speed and for the period of time listed in **Table 1**. Measure the amount of oil discharged at *fully closed* position from the reference point noted in Step 10 to the point at the end of period of time listed in **Table**

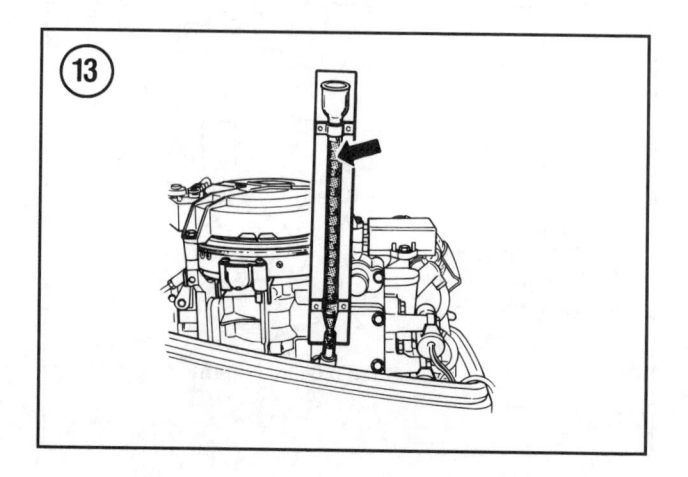

OIL INJECTION SYSTEM

1. Refer to **Table 1** for oil pump discharge specifications.

NOTE
Oil pump output test results may vary depending upon testing error and ambient temperature. To ensure accurate results, repeat test 2-3 times, or until results are consistent.

12. Shut the engine off and refill the engine oil measuring cylinder.
13. Restart the engine.
14. Manually move the oil pump control rod to the *fully open* position.
15. Increase and maintain an engine speed of 1,500 rpm.
16. Select a clean-cut value on the scale of the oil measuring cylinder as a reference point. Note this location and start a stop watch.
17. Run the engine at the specified speed and for the next period of time listed in **Table 1**. Measure the amount of oil discharged at *fully open* position from the reference point noted in Step 16 to the point at the end of period of time listed in **Table 1**. Refer to **Table 1** for oil pump discharge specifications.

NOTE
Oil pump output test results may vary depending upon testing error and ambient temperature. To ensure accurate results, repeat test 2-3 times, or until results are consistent.

18. If the oil discharged is not within specifications, check the injection lines for possible leakage or restrictions. If none are found, replace the oil pump.
19. Disconnect the portable tachometer.
20. Connect the oil pump control rod onto the carburetor.
21. Install the oil tank as described in this chapter.
22. Install the engine cover.

COMPONENT REPLACEMENT

Oil Tank Removal/Installation

1. Remove the engine cover.
2A. On 1985 DT 115, DT 140 models, remove the oil tank band and support rod nut, if so equipped (**Figure 14**).

2B. On all other models, remove the oil tank fasteners.

3. Move the tank away from the power head and disconnect the electrical sensor leads. See 1, **Figure 15**, typical.

4. Loosen the clamp holding the oil line to the tank (**Figure 15**). Pull the line from the tank fitting. Cap the tank fitting and plug the oil line to prevent leakage.

5. Installation is the reverse of removal.

**Oil Pump Removal/Installation
(DT 55, DT 65, DT 75, DT 85;
DT 115, DT 140)**

*CAUTION
Proper oil line routing and connections are essential for correct oil injection system operation. The line unions which connect to the power head and oil pump look the same but contain check valves of differing calibrations. Oil lines must be installed between the pump and power head correctly and connected to the proper cylinder fitting on the intake manifold.*

1. Disconnect the negative battery cable.
2. Remove the engine cover.
3. Remove the oil tank as described in this chapter.
4. Disconnect the oil pump control rod at the throttle.
5. Label the power head location of each oil line banjo fitting with masking tape and a felt-tipped

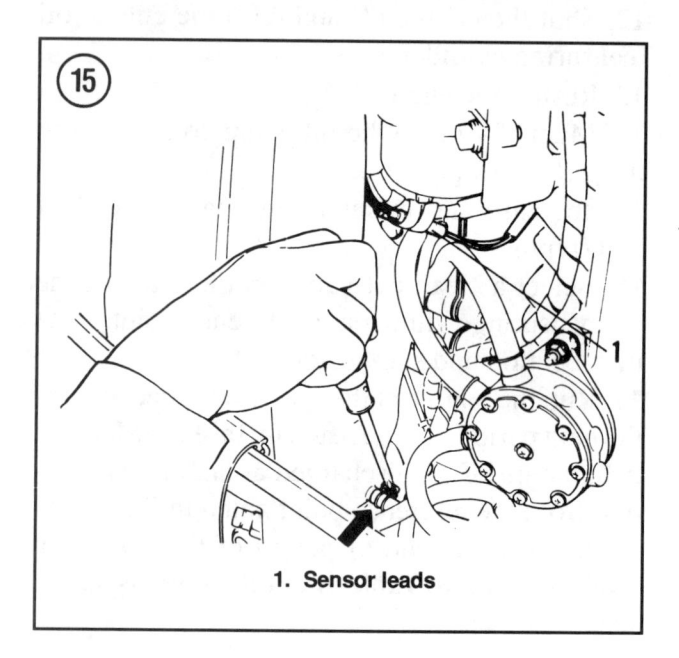

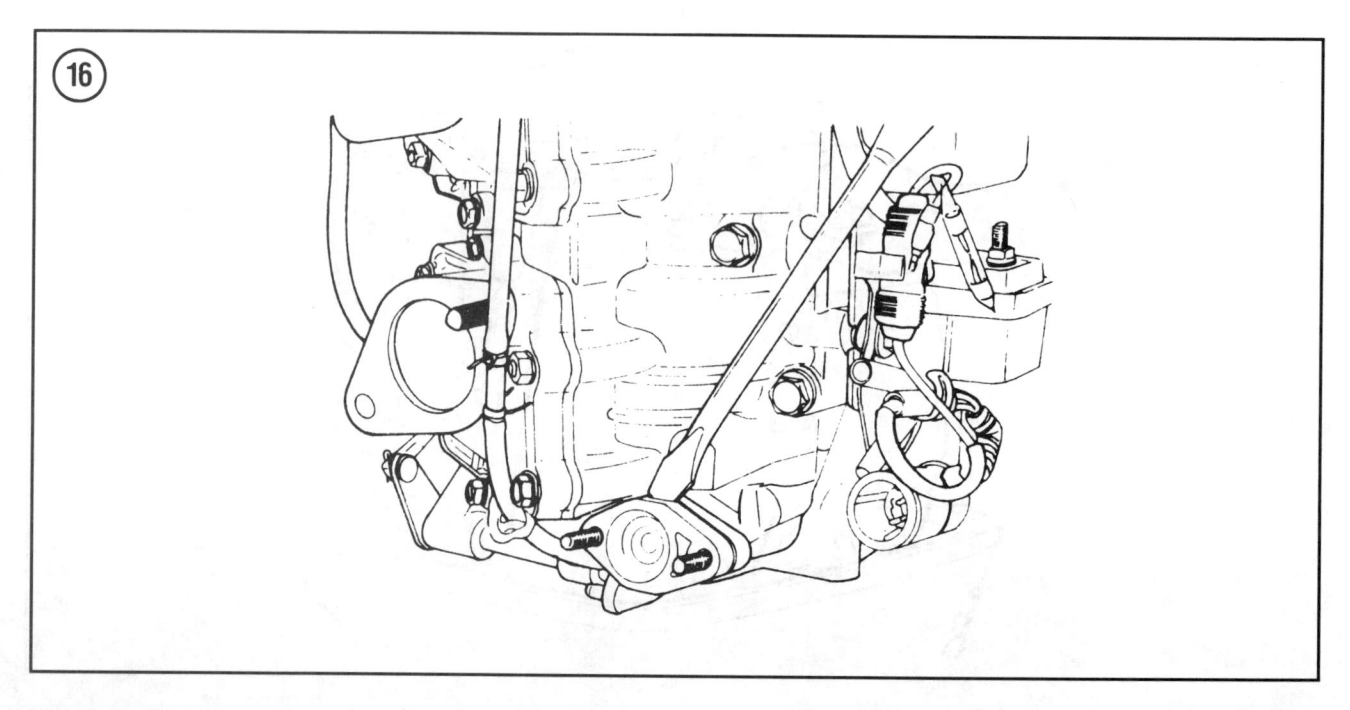

OIL INJECTION SYSTEM

pen for correct reinstallation reference, then disconnect each fitting.

6. Remove the oil pump mounting nuts. Remove the oil pump from the power head.

7. Carefully work the driven gear retainer and gasket loose using a putty knife or screwdriver blade (**Figure 16**), then remove the retainer, gasket and driven gear assembly from the power head.

8. Clean the power head and retainer mating surfaces of all gasket residue.

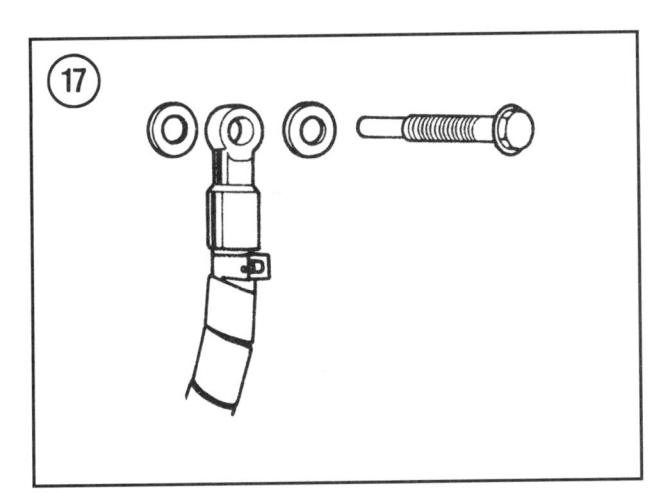

9. If the pump is to be replaced but the oil lines and banjo fittings reused:
   a. Unbolt one line at a time from the pump and discard the banjo fitting gaskets.
   b. Use new gaskets on each side of the banjo fitting (**Figure 17**) and install it to the same position on the new pump.
   c. Be sure to install the banjo fitting with its brass side facing the pump.

CAUTION
If the oil lines are disconnected from the banjo fitting to replace the check valve or oil line, they must be reinstalled with the same type of clamps as removed. The use of a screw type clamp will damage the vinyl line while a tie strap will not provide sufficient clamping pressure.

10. If the oil pump and banjo fittings are to be reused but the oil lines require replacement:
    a. Compress the oil line clamp with pliers, slide it away from the banjo fitting and pull the old line off the fitting (**Figure 18**).
    b. Compress a new oil line clamp with pliers and fit it over the new oil line.

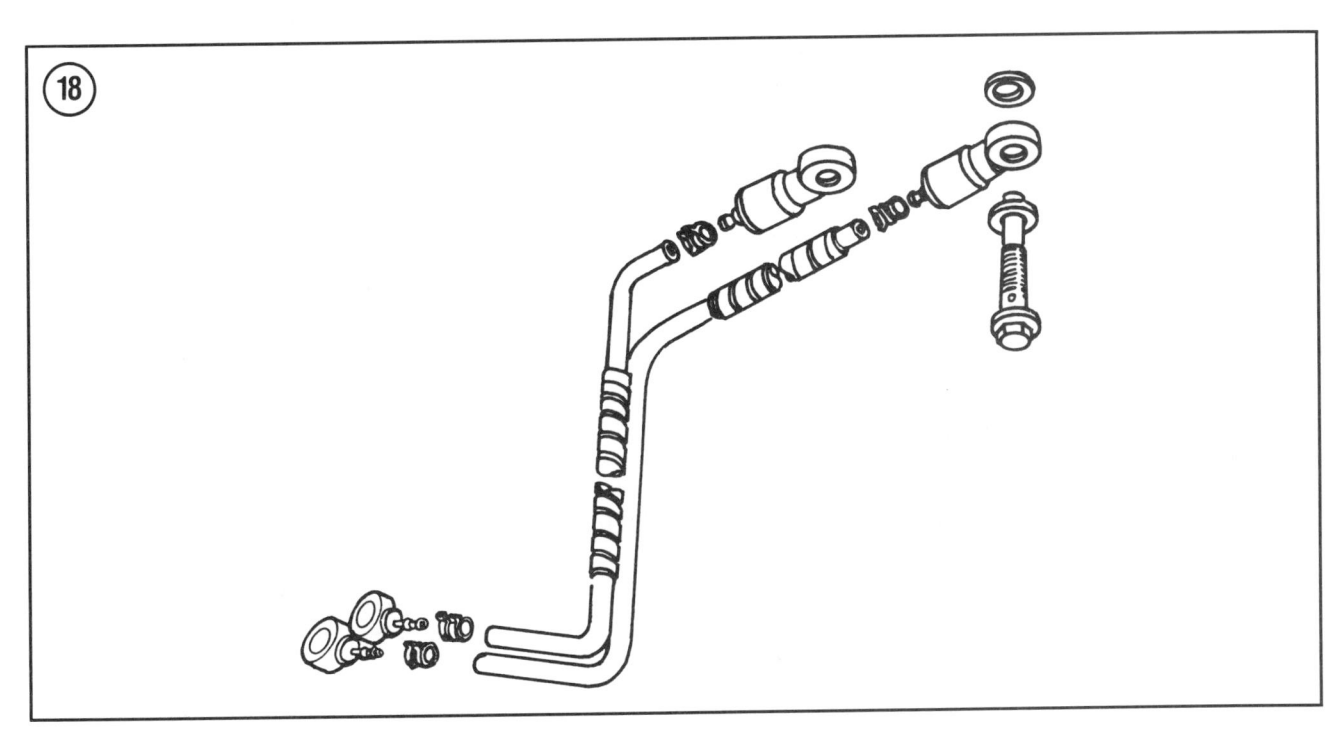

c. Compress the clamp and slide it into position over the banjo fitting nipple.

d. Repeat the procedure to install the line to the other banjo fitting.

11. If the banjo fitting check valve is to be replaced, disconnect the oil line as described in Step 10 and install the line to the new banjo fitting.

12. Align the oil pump shaft with the driven gear retainer groove and fit the pump to the retainer (**Figure 19**). Install the assembly to the power head with a new gasket and tighten the attaching nuts securely.

13. If oil lines were removed or replaced, loosen the banjo fitting bolts slightly and position the lines as shown in the following illustrations:

   a. **Figure 20**: DT 55, DT 65.
   b. **Figure 21**: DT 75, DT 85.
   c. **Figure 22**: 1985 DT 115, DT 140.
   d. **Figure 23**: 1986-on DT 115, DT 140.

   Tighten the bolts securely.

14A. On DT 75 and DT 85 models, install the oil lines to the intake manifold with new banjo fitting gaskets as shown in **Figure 21**.

14B. On 1985 DT 115 and DT 140 models, install the oil lines to the intake manifold with new banjo fitting gaskets as shown in **Figure 24**.

14C. On DT 55, DT 65 and 1986-on DT 115 and DT 140 models, compress the clamp and slide the oil lines into position over the nipples on the intake manifold and make sure the clamps are secure.

15. Connect the control rod and perform the *Injection Pump Discharge Adjustment* described in this chapter.

16. Compress the oil line clamps and disconnect the outlet line. Fill the oil line with Suzuki CCI 50:1 Outboard Oil, then reconnect the lines to the fittings. Repeat for each oil line.

17. Bleed the injection pump as described in this chapter.

18. Carefully check all lines and connections for oil leakage before starting the engine in Step 19.

19. Check the injection pump delivery rate as described in this chapter.

Oil Pump Removal/Installation (1988-on DT 8, DT 9.9; 1989-on DT 15; DT 25, DT 30 (3-cylinder); DT 35, DT 40; V4 and V6)

CAUTION
Proper oil line routing and connections are essential for correct injection system operation. The line connections to the power head and oil pump look the same but contain check valves of differing calibrations. Oil lines must be installed between the pump and power head cor-

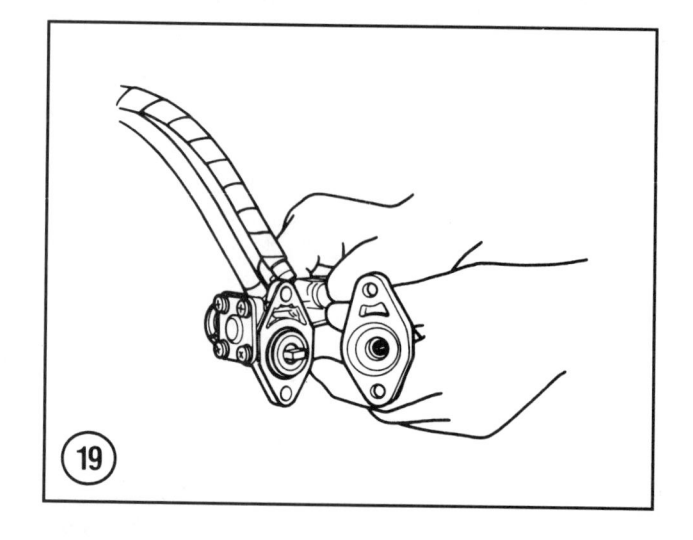

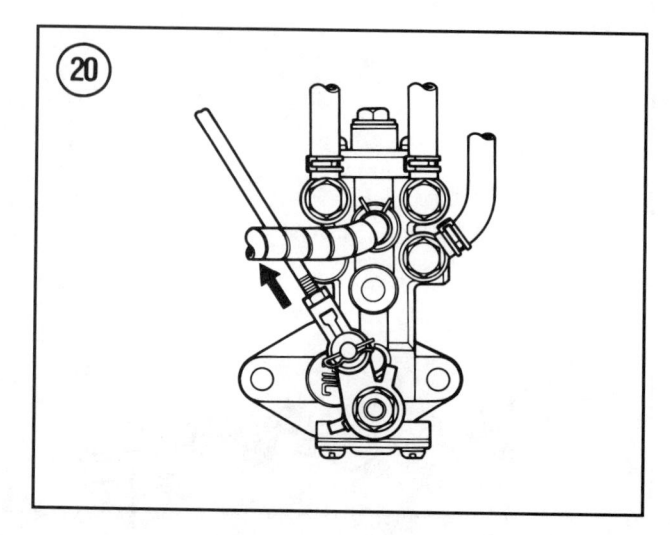

OIL INJECTION SYSTEM

rectly and connected to the proper cylinder fitting on the intake manifold.

NOTE
Suzuki has determined that there is a possible problem with the oil line clamps on 1986-1988 V6 models. This is noted in Suzuki Service Bulletin SB89-12, dated August 8, 1989. A new clamp with a slightly smaller inner diameter should be installed on all oil lines to guard against the accidental dislodging of an oil line from a fitting. The new clamp part No. is 09401-06405. It is suggested that these new oil line clamps be installed by the Suzuki service department. If you choose to install the new oil line clamps yourself, cut about 1/4 in. (6 mm) off of the end of the oil line to ensure a good tight fit.

1. Disconnect the negative battery cable.
2. Remove the engine cover.
3. Remove the oil tank as described in this chapter.
4. Disconnect the oil pump control rod at the throttle.
5. Label the power head location of each oil line with masking tape and a felt-tipped pen for correct reinstallation reference.
6. Compress the oil line clamp with pliers, slide it away from the fitting on the intake manifold or the air/oil mixing valve.

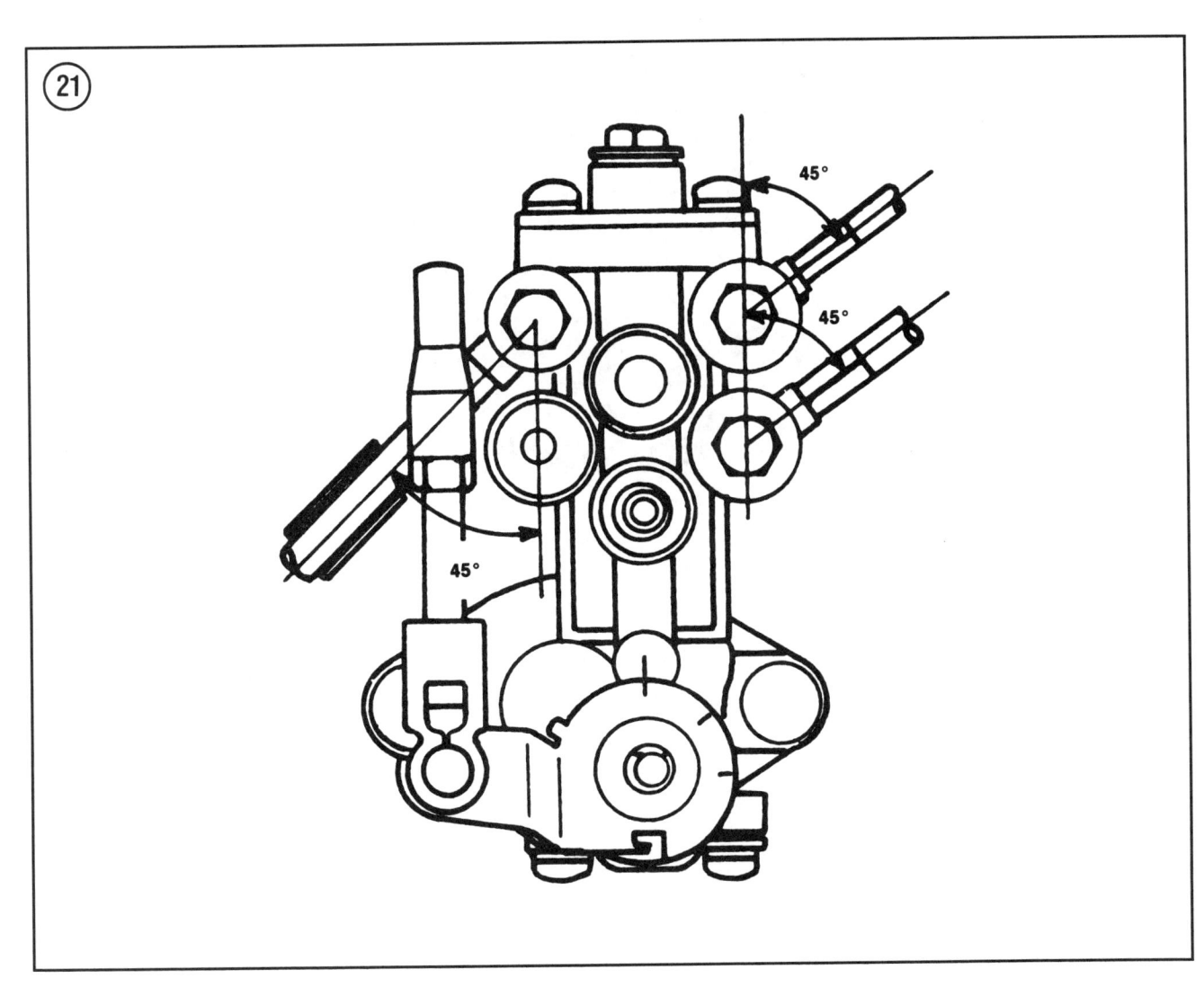

CHAPTER TWELVE

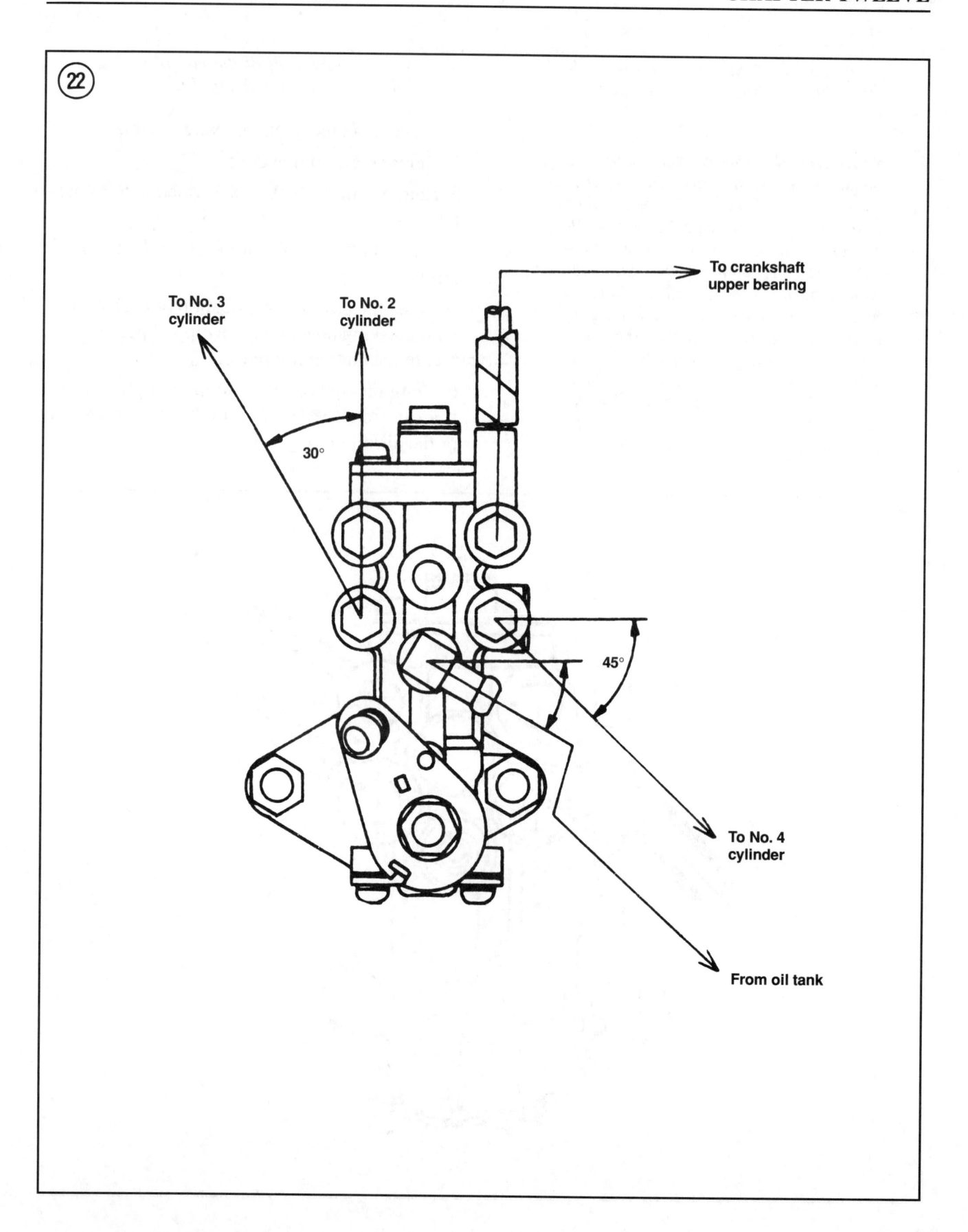

OIL INJECTION SYSTEM

521

(23)

Lubrication hose

Fuel filter

7. Remove the oil pump mounting nuts or bolts. Remove the oil pump from the power head.

8. Carefully work the driven gear retainer and gasket loose using a putty knife or screwdriver blade (**Figure 16**), then remove the retainer, gasket and driven gear assembly from the power head.

9. Clean the power head and retainer mating surfaces of all gasket residue.

10. If the pump is to be replaced but the oil lines reused:

   a. Label the oil pump location of each oil line with masking tape and a felt-tipped pen for correct reinstallation reference.
   b. Compress the oil line clamp with pliers, slide it away from the fitting on the oil pump.

CAUTION
If the oil lines are disconnected from the fitting to replace the check valve or oil line, they must be reinstalled with the same type of clamps as removed. The use of a screw type clamp will damage the vinyl line while a tie strap will not provide sufficient clamping pressure.

11. If the oil pump is to be reused but the oil lines require replacement:

   a. Compress the oil line clamp with pliers, slide it away from the fitting and pull the old line off the fitting.
   b. Compress a new oil line clamp with pliers and fit it over the new oil line.
   c. Compress the clamp and slide it into position over the fitting.

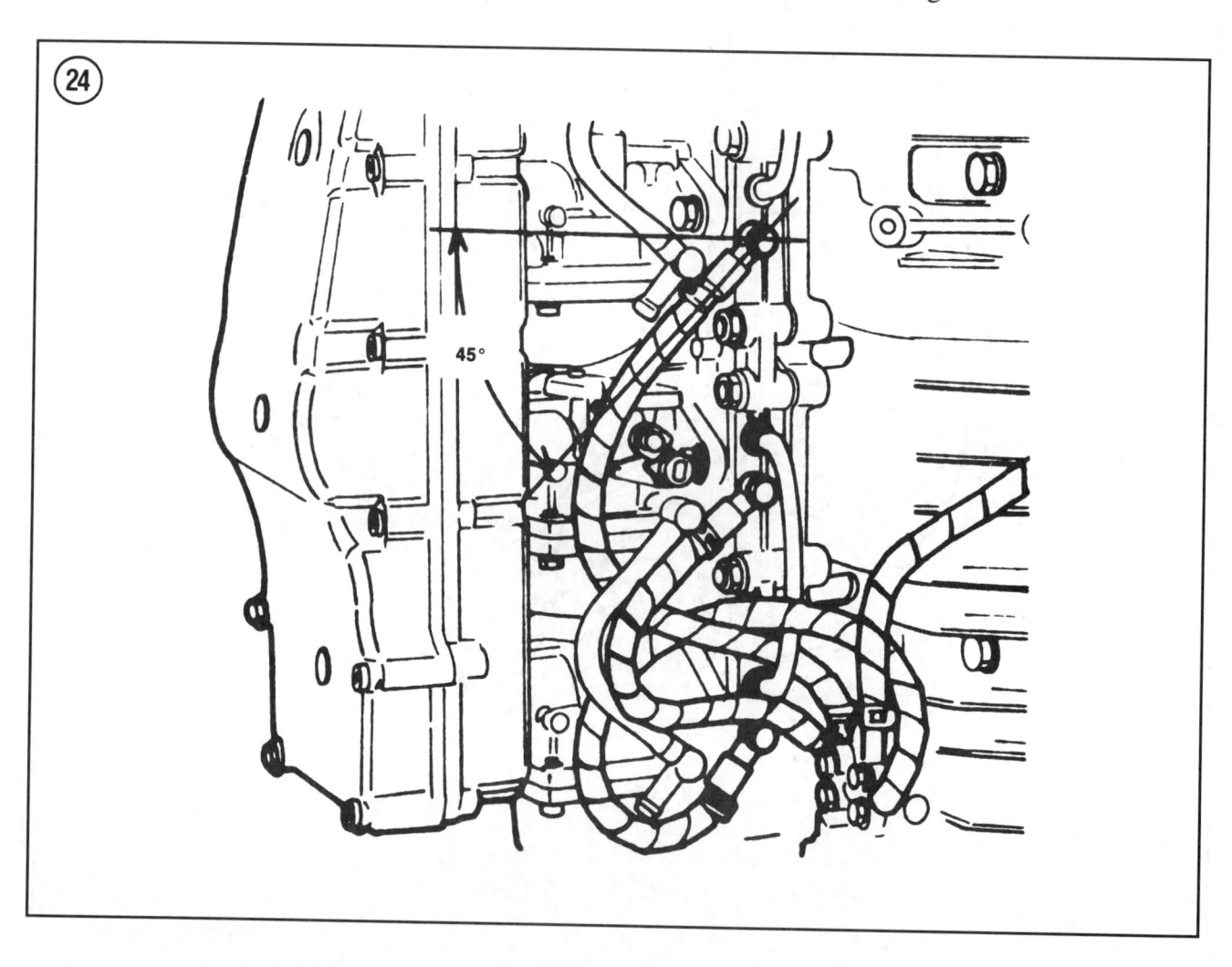

(24)

OIL INJECTION SYSTEM

d. Repeat the procedure for all other oil lines.

12. If a check valve is to be replaced, disconnect it from the oil line as described in Step 11 and install a new check valve noting the directional arrow on the valve.

13. Install the driven gear into the power head.

14. Install the retainer into the power head, then align the tab in the oil pump with the groove in the oil pump shaft and fit the pump to the retainer.

15. Install and tighten the attaching nuts, or bolts, securely.

16. Install the oil lines as follows:
   a. Fill the oil line with Suzuki CCI 50:1 Outboard Oil.
   b. Compress the oil line clamp with pliers, slide it onto the fitting on the intake manifold or the air/oil mixing valve. Repeat for all oil lines.

17. Connect the pump control rod.

18. Bleed the injection pump as described in this chapter.

19. Carefully check all lines and connections for oil leakage before starting the engine in Step 20.

20. Check the injection pump delivery rate as described in this chapter.

Oil Pump Discharge Adjustment

This adjustment should only be required if the oil pump is removed for service.

1. Remove the engine cover.

2. Move the throttle to the wide-open position.

3. Check the full-open mark on the oil pump control lever. It should align with the match mark on the pump housing. See **Figure 25**, typical.

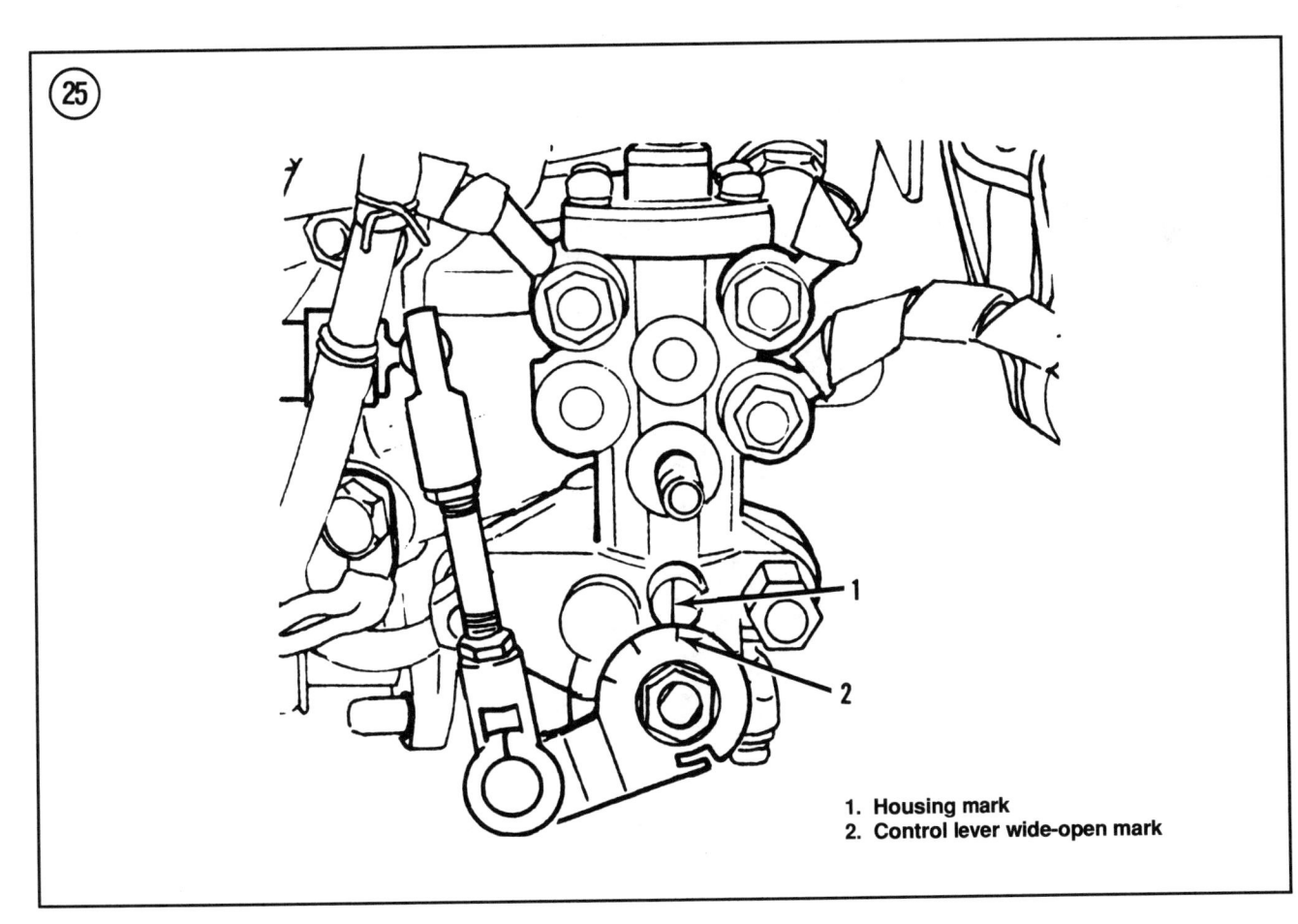

25

1. Housing mark
2. Control lever wide-open mark

CHAPTER TWELVE

4. Close the throttle. The full-closed mark on the control lever should align with the pump housing match mark. See **Figure 26**, typical.

5. If the marks do not align in Step 3 or Step 4, open or close the throttle as required and proceed as follows:

   a. DT 40-DT 65—Loosen the control cable locknut and turn the cable adjusting nut until the marks align, then retighten the locknut. See **Figure 27**.
   b. DT 75, DT 85, DT 115 and DT 40—Disconnect the control lever rod from the control lever. Loosen the rod connector locknut and rotate connector as required to align match marks when reinstalled. See **Figure 25** or **Figure 26**, typical, as required.

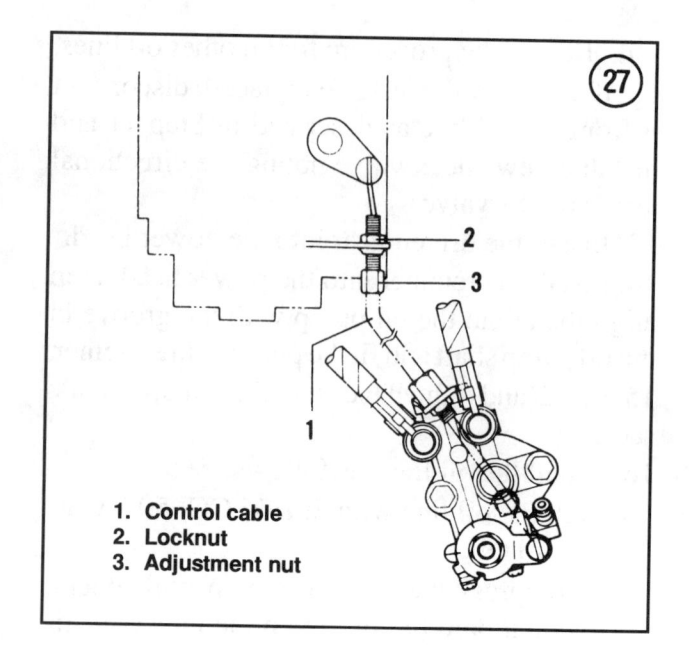

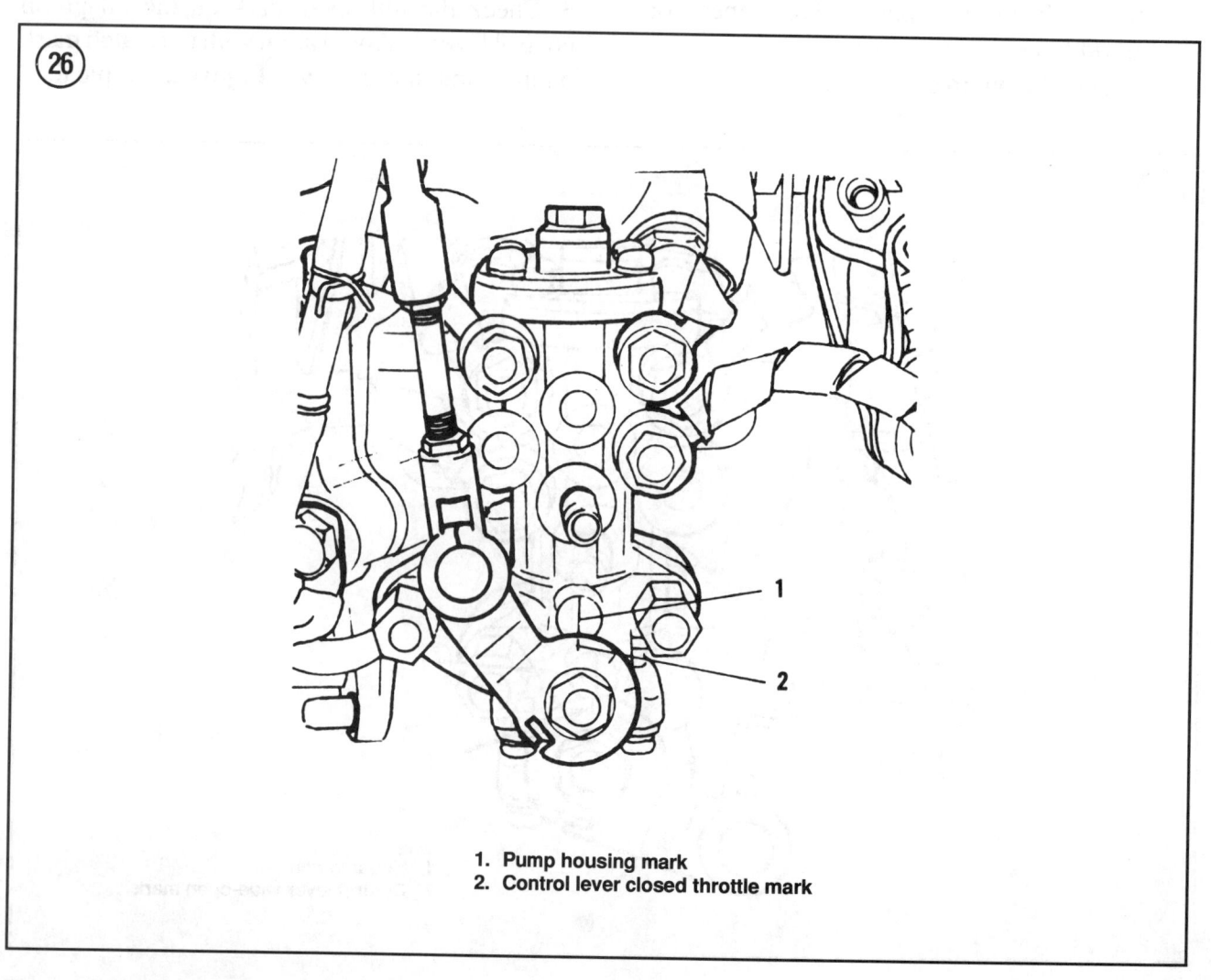

27
1. Control cable
2. Locknut
3. Adjustment nut

26
1. Pump housing mark
2. Control lever closed throttle mark

OIL INJECTION SYSTEM

Table 1 OIL PUMP DISCHARGE RATE AT 1,500 RPM

Model	Pump Position	Discharge Quantity (mL)	Test Duration
DT 8, DT 9.9 (1988-on)	closed	1.2-1.8	5 min.
DT 15 (1989-on)	closed	1.5-2.5	5 min.
DT 25, DT 30 (3-cylinder)	closed	1.0-1.9	5 min.
	open	1.2-1.7	2 min.
DT 40 (1985)	closed	2.2-2.8	5 min.
	open	3.2-4.0	2 min.
DT 35 (1987-on), DT 40 (1986-on)	closed	1.3-2.3	5 min.
	open	2.9-4.4	2 min.
DT 55 (1985)	closed	2.3-3.1	5 min.
	open	3.4-5.1	2 min.
DT 55 (1986-on)	closed	1.9-3.2	5 min.
	open	3.4-5.1	2 min.
DT 65 (1985)	closed	2.0-3.2	5 min.
	open	4.0-5.9	2 min.
DT 65 (1986-on)	closed	1.9-3.2	5 min.
	open	4.0-5.9	2 min.
DT 75, DT 85 (1985)	closed	3.7-5.6	5 min.
	open	5.9-8.9	2 min.
DT 75 (1986-on)	closed	2.2-3.7	5 min.
	open	5.0-7.5	2 min.
DT 85 (1986-on)	closed	2.2-3.7	5 min.
	open	5.9-8.9	2 min.
DT 115 (1985)	closed	6.0-7.3	5 min.
	open	10.0-12.0	2 min.
DT 115 (1986-on)	closed	3.8-6.8	5 min.
	open	6.5-9.7	2 min.
DT 140 (1985)	closed	8.0-9.8	5 min.
	open	13.4-16.0	2 min.
DT 140 (1986-on)	closed	3.8-6.8	5 min.
	open	8.0-12.0	2 min.
V4	closed	2.5-4.5	5 min.
	open	6.0-9.0	2 min.
DT 150, DT 150SS	closed	4.7-8.7	5 min.
	open	11.5-17.2	2 min.
DT 175, DT 200, DT 225	closed	4.7-8.7	5 min.
	open	13.3-19.8	2 min.

Chapter Thirteen

Jet Drives

Suzuki first introduced the jet drives on new models in 1989. There are four models and they are as follows: PU40, PU55, PU85 and PU140. The jet drive model numbers reflect the equivalent prop-driven engine's horsepower.

Service to the engine and its related components and power tilt and trim assembly is the same as on prop-driven models. Refer to the appropriate chapter and service section for the engine models or component you are servicing. Only service on the jet drive assembly is covered in this chapter.

The jet drive serial number is stamped on the starboard side of the pump housing above the thrust gate pivot and the jet drive model number is stamped into the port side of the pump housing above the thrust gate pivot. See **Figure 1**.

MAINTENANCE

Outboard Mounting Height

A jet drive outboard motor must be mounted higher on the transom plate than an equivalent stock prop-driven outboard motor. However, if the jet drive is mounted too high, air will be allowed to enter the jet drive resulting in cavitation and power loss. If the jet drive is mounted too low, excessive drag, water spray and loss in speed will result.

To set the initial height of the outboard motor, proceed as follows:

1. Place a straightedge against the boat bottom (*not* keel) and abutt end of straight edge with the jet drive intake.
2. The fore edge of the water intake housing should align with the top edge of the straightedge (**Figure 2**).
3. Secure the outboard motor at the setting, then test run the boat.
4. If cavitation occurs (overrevving; loss of thrust), the outboard motor must be lowered 1/4 in. (6.35 mm) at a time until uniform operation is noted.
5. If uniform operation is noted with initial setting, the outboard motor should be raised at 1/4 in. (6.35 mm) intervals until cavitation is noted. Then lower the motor to last uniform setting.

NOTE
The outboard motor should be in a vertical position when the boat is on plane. Adjust motor trim setting as needed. If

JET DRIVES

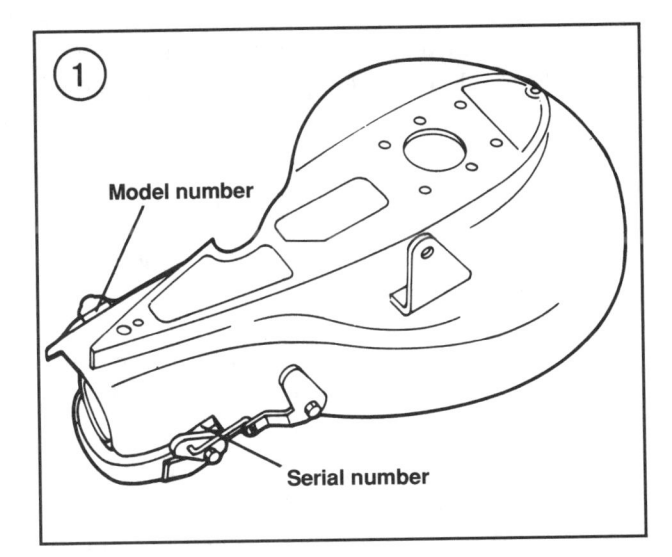

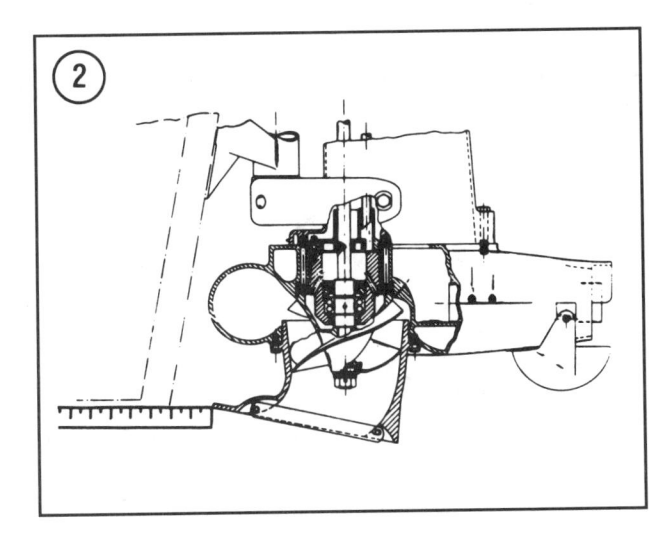

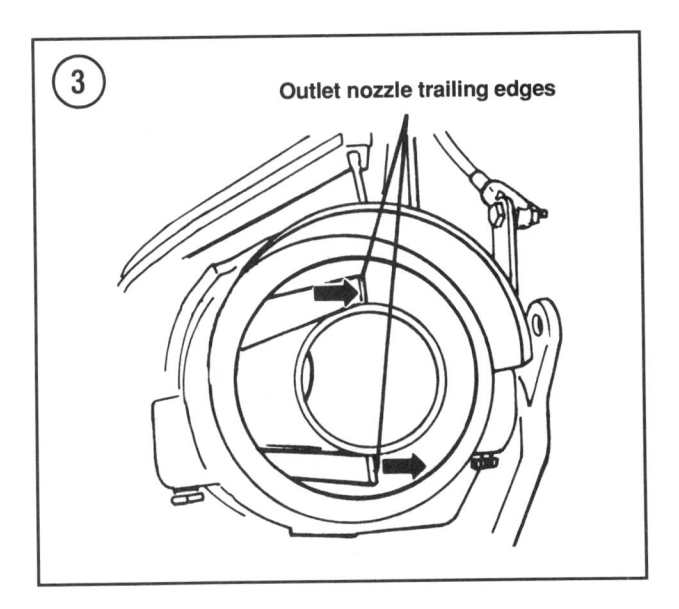

the outboard trim setting is altered, the outboard motor height must be checked and adjusted, if needed, as previously outlined.

Steering Torque

A minor adjustment to the trailing edge of the drive outlet nozzle may be made if the boat tends to pull in one direction when the boat and outboard motor are pointed in a straight-ahead direction. Should the boat tend to pull to the starboard side, bend the top and bottom trailing edge of the jet drive outlet nozzle 1/16 in. (1.6 mm) toward the starboard side of the jet drive. See **Figure 3**.

Bearing Lubrication

The jet drive bearing(s) should be lubricated after *each* operating period, after every 10 hours of operation and prior to storage. In addition, after every 50 hours of operation, additional grease should be pumped into the bearing(s) to purge any moisture. The bearing(s) is/are lubricated by first removing the cap on the end of the excess grease hose from the grease fitting on the side of the jet drive pump housing. See **Figure 4**. Use a grease gun and inject grease into the fitting until grease exits from the capped end of the excess grease hose. Use only Suzuki Jet Grease (part No. 99954-53885) or a suitable equivalent grease to lubricate the bearing(s).

Note the color of the grease being expelled from the excess grease hose. During the break-in period some discoloration of the grease is normal. After the break-in period, if the grease starts to turn a dark or dirty gray, then the jet drive assembly should be disassembled as outlined under *Jet Drive* and the seals and bearing(s) inspected and replaced as needed. If moisture is noted being expelled from the excess grease hose, then the jet drive should be disassembled and the seals replaced and the bearing(s) inspected and replaced as needed.

Directional Control

The boat's operational direction is controlled by a thrust gate via a cable and lever. When the directional control lever is placed in the full forward position, the thrust gate should completely uncover the jet drive housing's outlet nozzle opening and seat securely against the rubber pad on the jet drive pump housing. When the directional control lever is placed in full reverse position, the thrust gate should completely close-off the pump housing's outlet nozzle opening. Neutral position is located midway between complete forward and complete reverse position.

NOTE
On remote control models, the control box shift cable must be approximately 12 in. (30.5 mm) longer for a jet drive model than a propeller driven model as the control cable connects directly to the jet drive thrust gate linkage.

The directional control cable is properly adjusted if after placing the directional control lever in the full forward position, the link between the thrust gate and the lower arm of the control cable pivot bracket are in alignment (Figure 5). The thrust gate should seat securely against the rubber pad on the jet drive pump housing.

WARNING
Always use the lower hole of the thrust gate lever to attach the control linkage. See arrow, Figure 5.

On manual models, an adjustment neutral stop is provided. To adjust, first find true neutral with the engine at idle position and retain this setting. Then loosen the nut (**Figure 6**) securing the stop position and move the stop to contact the directional control lever. Securely tighten the nut to retain this adjustment.

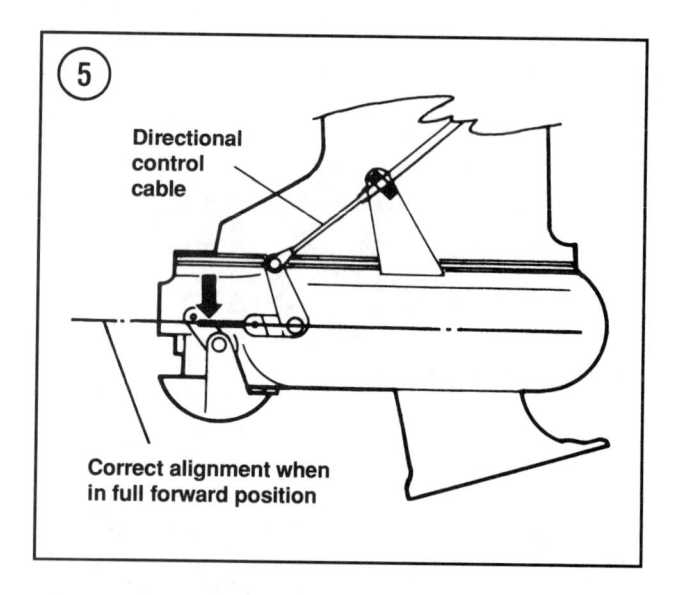

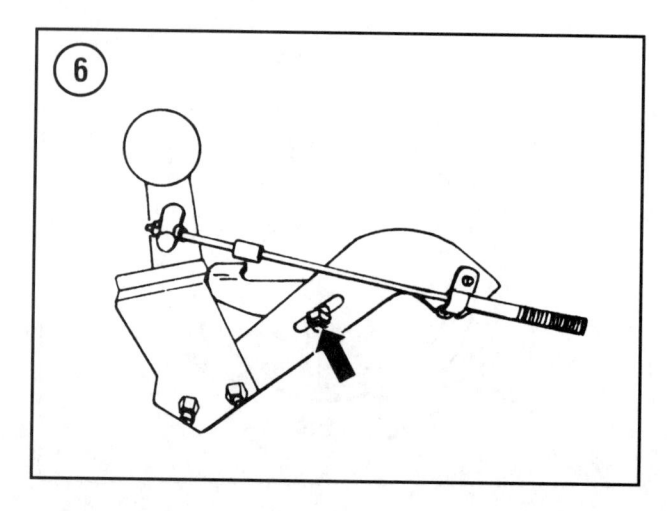

JET DRIVES

Impeller Clearance

1. Disconnect the spark leads as a safety precaution to prevent accidental starting of the engine.
2. Insert a selection of feeler gauge thicknesses through the clearance between the impeller blades and the intake housing. See **Figure 7**.
3. The impeller-to-intake housing clearance should be approximately 0.031 in. (0.8 mm).
4. If the clearance is incorrect, remove the six intake housing mounting bolts (14, **Figure 11** or 14, **Figure 12**). Remove the intake housing.
5. Bend the ears on the washer retaining the jet drive impeller nut to allow a suitable tool to be installed on the impeller nut. Remove the nut, eared washer, lower shims, impeller and upper shims. Note the number of lower and upper shims.

NOTE
Eight 1/32 in. (0.8 mm) shims are used to adjust the impeller-to-intake housing clearance on PU 55, PU 85 and PU 140 models. Nine 1/32 in. (0.8 mm) shims are used to adjust the impeller-to-intake housing clearance on PU 40 models.

6. If clearance is excessive, remove the shims as needed from below the impeller (lower shims) and position above the impeller.

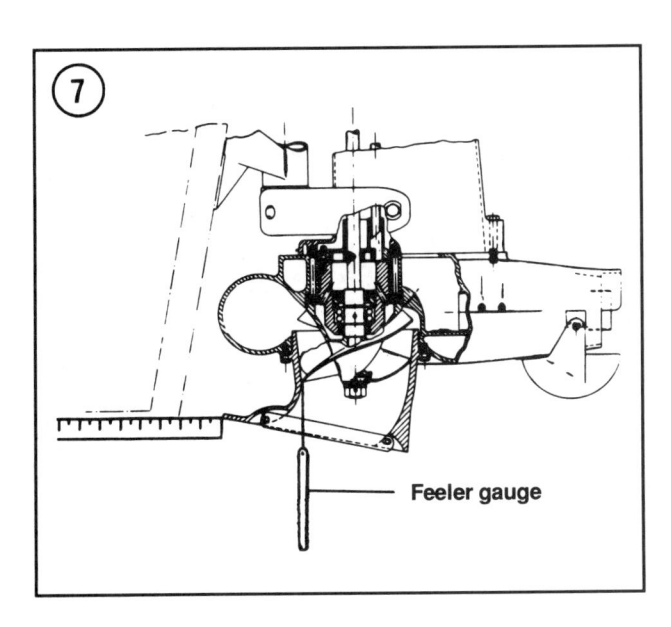

7. Install the impeller with the selected number of shims below the impeller.
8. Install the eared washer and impeller retaining nut on the drive shaft. Tighten the nut with hand pressure while ensuring that the eared washer does not lodge in the drive shaft thread and jam the nut. Tighten the nut to the torque specification shown in **Table 1**.
9. Grease the threads of the six intake housing mounting bolts with Suzuki Jet Grease (part No. 99954-53885) or equivalent.
10. Install the intake housing on the jet drive pump housing with the lowest part of the intake grill facing aft. Tighten the six intake housing bolts to the specification shown in **Table 1**.
11. Repeat Steps 2 and 3 to recheck impeller-to-intake housing clearance and Steps 4 through 10 if recommended clearance is not measured.
12. After correct clearance is measured, remove the six intake housing bolts and remove the intake housing.
13. Make sure the jet drive impeller retaining nut is correctly tightened to the specification shown in **Table 1**.

NOTE
*If the ears on the washer located behind the jet drive impeller retaining nut do not align with the flats on the nut, do not change position of the nut. Remove the nut and turn over the washer, reinstall the nut and then retighten to the specification shown in **Table 1**. **Make sure** the selected number of lower shims do not fall free.*

14. Bend the ears on the washer located behind the jet drive impeller retaining nut against the nut flats to retain the torque setting on the nut.
15. Grease the threads of the six intake housing bolts with Suzuki Jet Grease (part No. 99954-53885) or equivalent.
16. Install the intake housing on the jet drive pump housing with the lowest part of the intake grill facing aft. Tighten the six intake housing bolts to the specification shown in **Table 1**.

17. Install directional control cable in jet drive support bracket and reconnect cable end to lower arm of the control cable pivot bracket.
18. Refer to *Directional Control* in this chapter and adjust cable to provide correct operation of the thrust gate.
19. Complete reassembly in the reverse of disassembly.

Cooling System Cleaning

The cooling system can become clogged by sand and salt deposits if it is not flushed occasionally. Clean the cooling system after each use in salt water.

1. Remove the plug and gasket from the port side of the jet drive pump housing to gain access to flush passage. See **Figure 8**.

NOTE
An adapter for flushing the cooling system is available from Quicksilver (Mercury Marine) Parts and Accessories. The adapter part number is 24789-A1.

2. Install flushing adaptor into flush passage.
3. Connect a suitable freshwater supply to the adapter and turn on to full pressure (maximum output).

CAUTION
When the outboard motor is running, make sure a stream of water is noted being discharged from the motor's telltale outlet. If not, stop the motor immediately and diagnose the problem.

CAUTION
Do not operate the outboard motor at high rpm when connected to the flush adapter.

4. Start the motor and allow the freshwater to circulate for approximately 15 minutes.
5. Stop the motor, turn off the auxiliary water supply and disconnect the water supply from the adapter.
6. Remove the adapter.
7. Replace the gasket if damaged, then install on the plug.
8. Install the plug and gasket into jet drive pump housing flush passage and tighten securely.

NOTE
To flush the jet drive impeller and intake housing, direct a freshwater supply into the intake passage housing area.

WATER PUMP

Refer to **Figure 11** or **Figure 12** for this procedure.

The water pump is mounted on top of the aluminum spacer on all models. The impeller is driven by a key which engages a groove in the drive shaft and a cutout in the impeller hub. As the drive shaft rotates, the impeller rotates with it. Water between the impeller blades and pump housing is pumped up to the power head through the water tube.

The offset center of the pump housing causes the impeller vanes to flex during rotation. At low speeds, the pump acts as a positive displacement type; at high speeds, water resistance forces the vanes to flex inward and the pump becomes a centrifugal type. See **Figure 9**.

All gaskets should be replaced whenever the water pump is removed. Since proper water

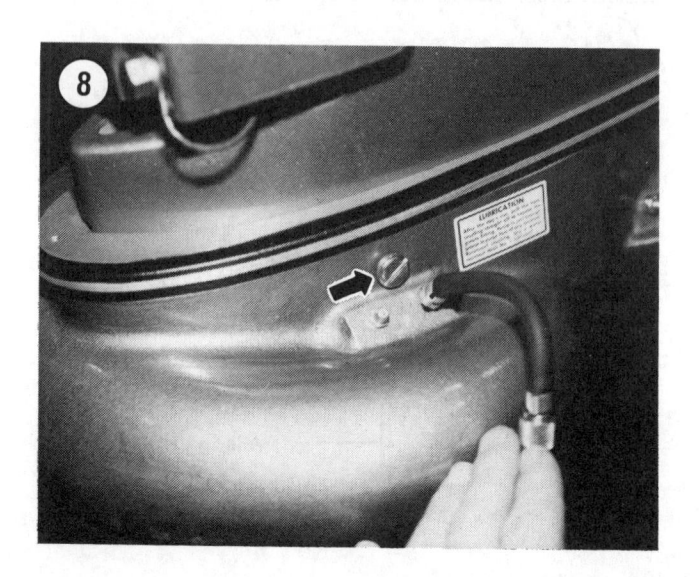

JET DRIVES

pump operation is critical to outboard operation, it is also a good idea to install a new impeller at the same time. Note that the impeller will only slide over the Woodruff key in one direction.

Removal and Disassembly

Refer to **Figure 11** or **Figure 12** for this procedure.

1. Remove the jet drive assembly as described in this chapter.
2. Remove the four bolts and washers retaining the pump housing.
3. Withdraw the pump cover off of the impeller. Rotate the pump cover *counterclockwise* if needed to assist in cover removal.

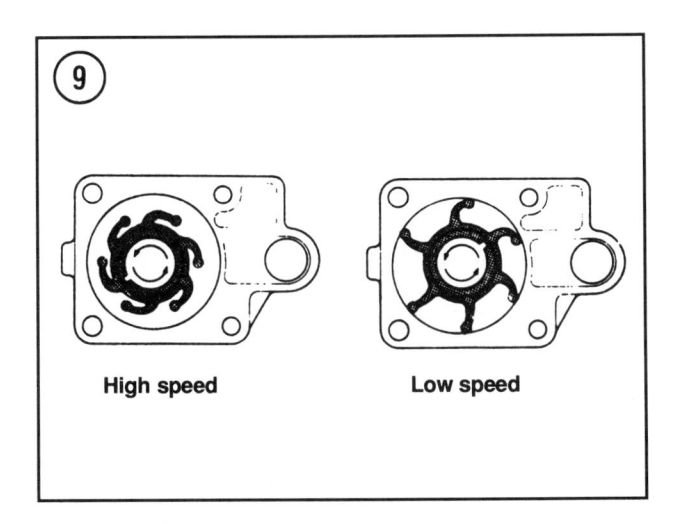

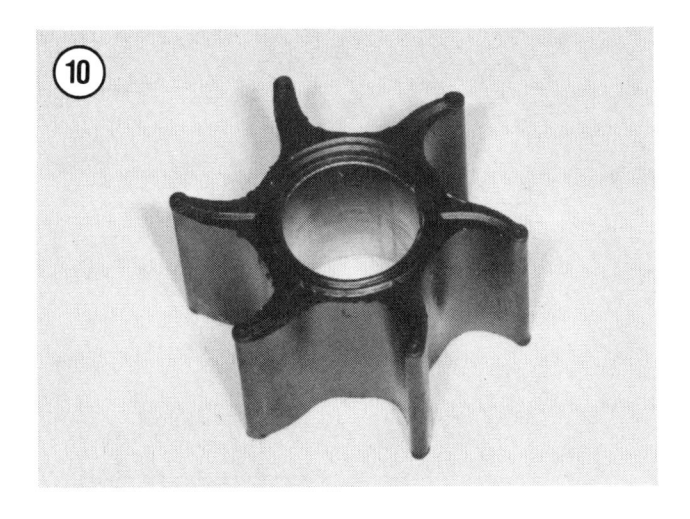

CAUTION
*Do not rotate the pump cover **clockwise** to assist removal or damage to the impeller will result and necessitate it being replaced.*

NOTE
In some cases, the impeller may come off with the pump cover. In extreme cases, the impeller hub may have to be split with a hammer and chisel to remove it in Step 4.

4. If impeller does not come off with the pump cover, slide it upward on the drive shaft. Remove the impeller.
5. Remove the impeller Woodruff drive key.
6. Remove the water pump face plate with top and bottom gaskets. Separate the face plate from the gaskets. Discard the gaskets.

Cleaning and Inspection

Refer to **Figure 11** or **Figure 12** for this procedure.

1. Remove the water pump cartridge insert from the pump cover.
2. Remove the water tube seal from the pump cover.
3. Clean all metal parts in solvent and blow dry with compressed air, if available.
4. Clean all gasket residue from all mating surfaces.
5. Check the pump cover for damage or distortion from overheating.
6. Check the face plate and cartridge insert for grooves or rough surfaces. Replace if any defects are found.
7. If original impeller is to be reused, check bonding to the hub. Check the side seal surfaces and blade ends for cracks, tears, wear and a glazed or melted appearance. See **Figure 10**, typical. If any of these defects are noted, do *not* reuse the original impeller.

CHAPTER THIRTEEN

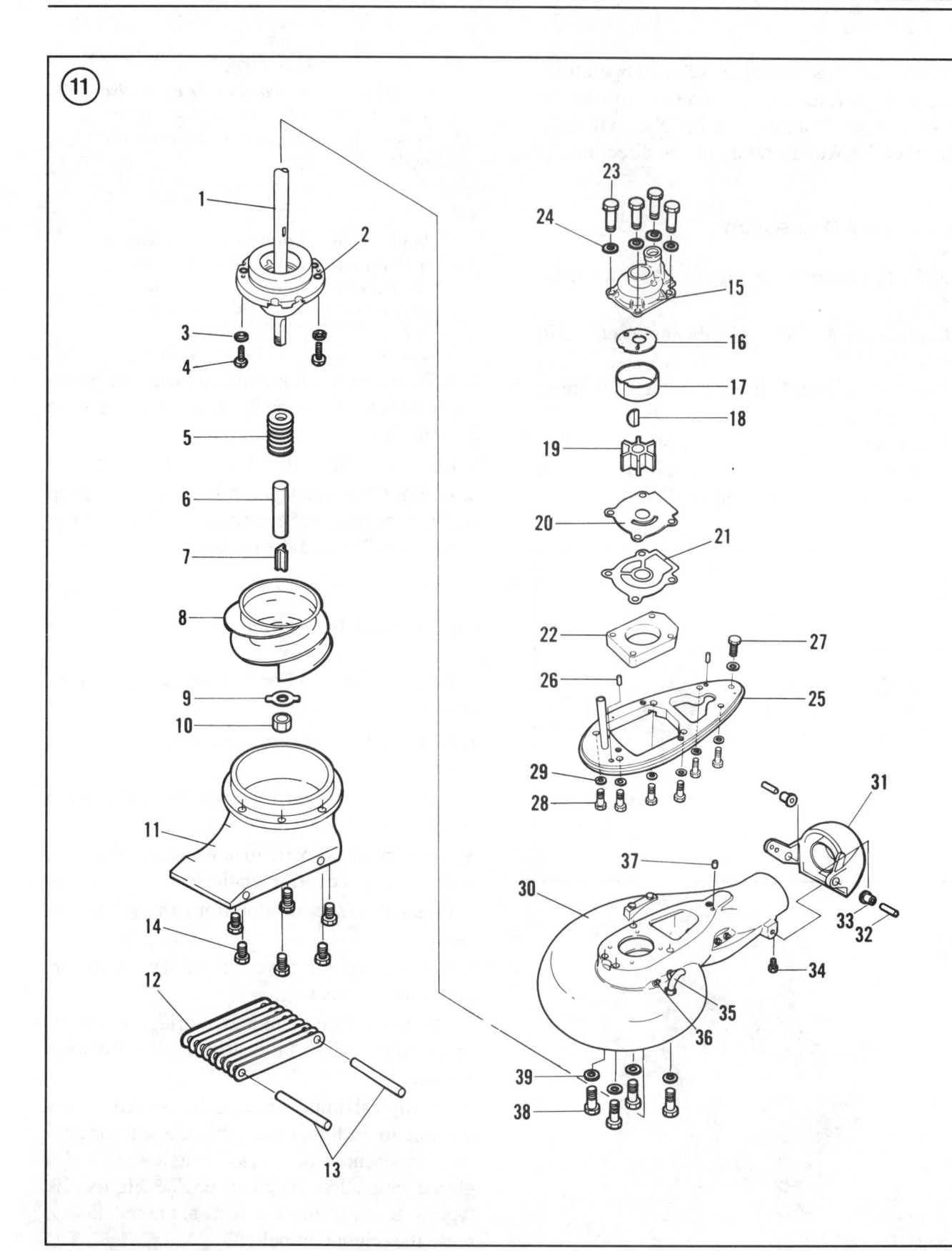

JET DRIVES

JET DRIVE ASSEMBLY
(MODEL PU 40)

1. Drive shaft and bearing housing assembly
2. O-ring
3. Lockwasher
4. Screw
5. Shims
6. Nylon sleeve
7. Impeller drive key
8. Impeller
9. Tab washer
10. Nut
11. Intake housing
12. Intake grill
13. Grill rods
14. Allen screws
15. Water pump housing
16. Plate
17. Pump insert
18. Impeller key
19. Impeller
20. Plate
21. Gasket
22. Water pump adapter
23. Bolts
24. Lockwashers
25. Adapter plate
26. Dowel pin
27. Bolts
28. Bolts
29. Lockwasher
30. Jet pump housing
31. Reverse gate
32. Pivot pin
33. Nylon sleeve
34. Bolt
35. Grease hose
36. Grease fitting
37. Dowel pin
38. Bolt
39. Lockwasher

534

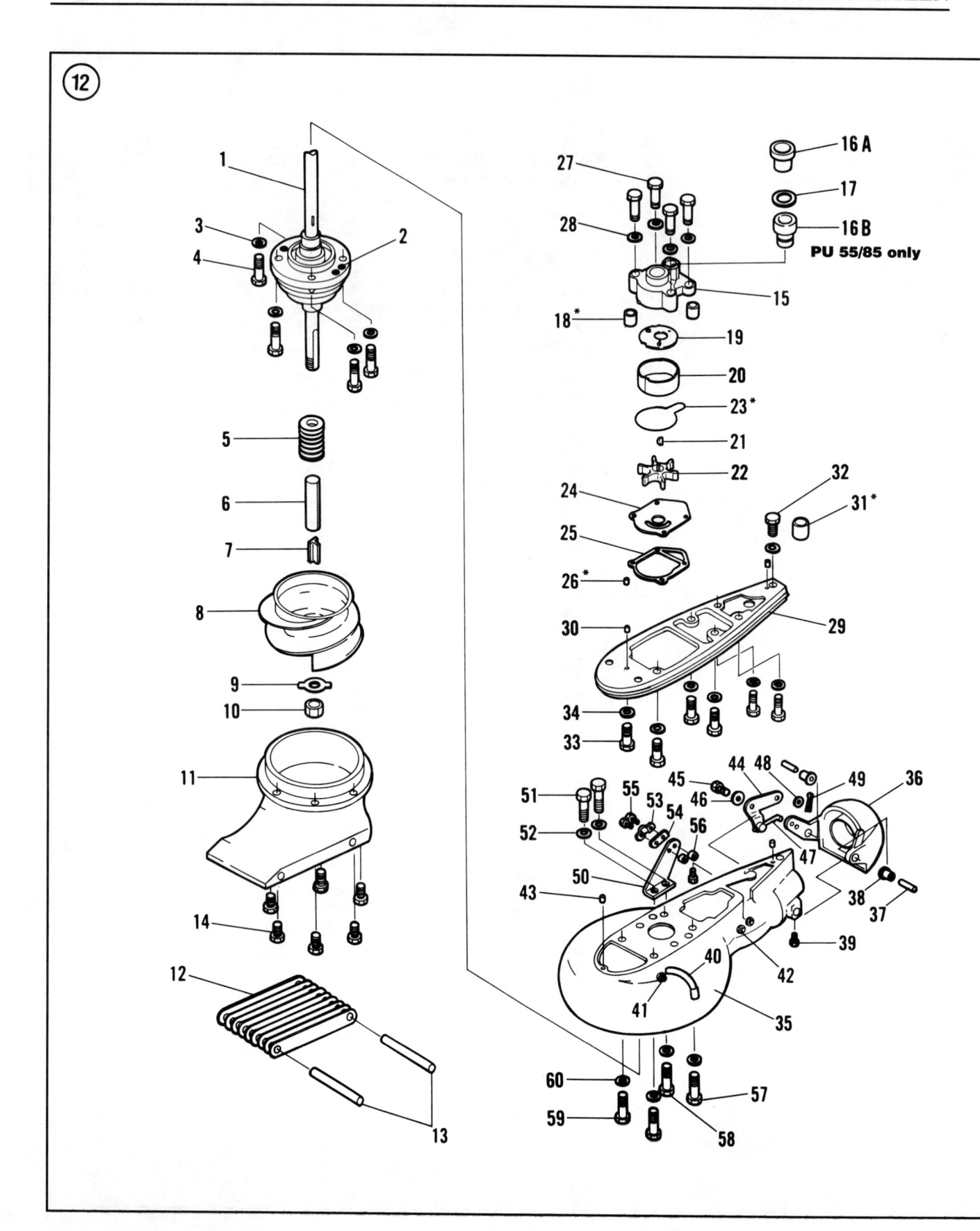

CHAPTER THIRTEEN

JET DRIVES

JET DRIVE ASSEMBLY
(MODELS PU 55, PU 85 AND PU 140)

1. Drive shaft and bearing housing assembly
2. O-ring
3. Lockwasher
4. Bolt
5. Shims
6. Nylon sleeve
7. Impeller drive key
8. Impeller
9. Tab washer
10. Nut
11. Intake housing
12. Intake grill
13. Grill rods
14. Bolts
15. Water pump housing
16A. Brass water tube extension (PU 140)
16B. Brass water tube extension (PU 55, PU 85)
17. O-ring
18. Water pump adapter (PU 140)
19. Plate
20. Pump insert
21. Impeller drive key
22. Impeller
23. Seal ring (PU 140)
24. Plate
25. Gasket
26. Dowel pin (PU 140)
27. Bolt
28. Lockwasher
29. Adapter plate
30. Dowel pin
31. Spacer (PU 140)
32. Bolt
33. Bolt
34. Lockwasher
35. Jet pump housing
36. Reverse gate
37. Pivot pin
38. Nylon sleeve
39. Bolt
40. Grease hose
41. Grease fitting
42. Clamp
43. Dowel pin
44. Shift arm
45. Bolt
46. Washer
47. Rod
48. Washer
49. Cotter pin
50. Bracket
51. Bolt
52. Washer
53. Clamp
54. Clamp plate
55. Bolt
56. Lockwasher
57. Bolt
58. Bolt
59. Bolt
60. Lockwasher

Assembly and Installation

Refer to **Figure 11** or **Figure 12** for this procedure.

1. If the pump cover cartridge insert was removed, reinstall cartridge with tab on top engaging the slot in pump cover.
2. If the water tube seal was removed, replace the seal or reinstall the original seal, if no damage was noted during inspection, into the pump cover.
3. Install the lower gasket, pump face and upper gasket.
4. Install the impeller Woodruff key.
5. Lightly lubricate the inner surfaces of the impeller with Suzuki Jet Grease (part No. 99954-53885) or equivalent. Slide the impeller onto the drive shaft. Align the slot in the impeller hub with the drive key and seat the impeller onto the water pump face plate.
6. Check the impeller installation by rotating the drive shaft *clockwise*. Impeller should rotate with the drive shaft. If not, remove and reposition the impeller to properly engage the drive key.
7. Slide the pump cover over the drive shaft. Push downward on the cover while rotating the drive shaft *clockwise* to assist the impeller in entering the cover without damage.
8. Grease the threads of the four bolts retaining the pump housing to the jet drive pump housing with Suzuki Jet Grease (part No. 99954-53885) or equivalent. Install the pump housing bolts and washers and tighten the bolts in small increments following a crisscross pattern until the torque specified in **Table 1** is obtained.
9. Install the jet drive assembly as described in this chapter.

JET DRIVE

When removing the jet drive mounting fasteners, it is not uncommon to find that they are corroded. Such fasteners should be discarded and new ones used when the jet drive is installed. The threads on all mounting bolts should be greased with Suzuki Jet Grease (part No. 99954-53885) or equivalent.

Removal

1. Disconnect the spark plug leads to prevent accidental starting of the engine during jet drive removal.
2. Tilt the outboard motor to full out position and engage the tilt lock lever.
3. Remove the directional control cable from the lower arm of the control cable pivot bracket. Remove the cable from the jet drive support bracket.
4. Remove the intake grate and the six intake housing mounting bolts. Remove the intake housing. Refer to **Figure 11** or **Figure 12**.
5. Bend the ears on the tab washer retaining the jet drive impeller nut to allow a suitable tool to be installed on the impeller nut. Remove the nut and the tab washer.

NOTE
Eight 1/32 in. (0.8 mm) shims are used to adjust the impeller-to-intake housing clearance on PU 55, PU 85 and PU 140 models. Nine 1/32 in. (0.8 mm) shims are used to adjust the impeller-to-intake housing clearance on PU 40 models.

6. Remove the shims located below the impeller and note the number. Remove the impeller and the shims located above the impeller and note the number.
7. Slide the impeller sleeve and drive key off the drive shaft.
8. Remove the four bolts located on the inside of the jet drive pump housing and adjacent to the bearing housing. Also remove the bolt at the external bottom aft end of the intermediate housing which secures the jet drive pump housing to the adapter plate. Withdraw the jet drive assembly.

JET DRIVES

Installation

Refer to **Figure 11** or **Figure 12** for this procedure.

1. Install the fore and aft dowel pins in the jet drive housing.
2. Lubricate the drive shaft splines and the water sub seal with Suzuki Jet Grease (part No. 99954-53885) or equivalent. Wipe any excess grease off the top of the shaft.
3. Grease all the mounting bolts with Suzuki Jet Grease (part No. 99954-53885) or equivalent.
4. Install the jet drive assembly and tighten the securing bolts to the specification shown in **Table 1**.
5. Grease the drive shaft threads, drive key and jet drive impeller bore with Suzuki Jet Grease (part No. 99954-53885) or equivalent.
6. If the original bearing(s), bearing housing, jet drive pump housing, impeller and intake housing are being reinstalled, install the shims located above and below the impeller in the same number and location as removed. If any of the previously listed components were replaced, then install all of the adjustment shims below the impeller (next to the retaining nut) and proceed to *Impeller Clearance* in this chapter after completion of installation.

NOTE
Eight 1/32 in. (0.8 mm) shims are used to adjust the impeller-to-intake housing clearance on PU 55, PU 85 and PU 140 models. Nine 1/32 in. (0.8 mm) shims are used to adjust the impeller-to-intake housing clearance on PU 40 models.

7. Install the impeller sleeve into the jet drive impeller bore.
8. Install shims above the impeller as noted in Step 6.
9. Slide the drive key and impeller onto the drive shaft.
10. Install the shims below the impeller as noted in Step 6.
11. Install the tab washer and impeller retaining nut on the drive shaft. Tighten the nut with hand pressure while ensuring that the tab washer does not lodge in the drive shaft threads and jam the nut. Tighten the nut to the specification listed in **Table 1**.
12. Grease the threads of the six intake housing mounting bolts with Suzuki Jet Grease (part No. 99954-53885) or equivalent.
13. Install the intake housing on the jet drive pump housing with the lowest part of the intake grill facing aft. Tighten the six intake housing bolts to the specification shown in **Table 1**. Install the intake grate.
14. Refer to *Impeller Clearance* in this chapter.
15. Remove the six intake housing bolts and remove the intake housing.
16. Make sure the jet drive impeller retaining nut is correctly tightened to specification shown in **Table 1**.

NOTE
*If the tabs on the washer located behind the jet drive impeller retaining nut do not align with the flats on the nut, do not change position of the nut. Remove the nut and turn over the washer, reinstall the nut and then retighten to the specification shown in **Table 1**. **Make sure** the selected number of lower shims do not fall free.*

17. Bend the tabs on the washer located behind the jet drive impeller retaining nut against the flats to retain the torque setting of the nut.
18. Reinstall the intake housing as previously outlined in Step 12 and Step 13.
19. Install the directional control cable in the jet drive support bracket and reconnect the cable end to the lower arm of the control cable pivot bracket.
20. Refer to *Directional Control* in this chapter and adjust the cable to provide correct operation of the thrust gate.
21. Complete reassembly in the reverse order of disassembly.

BEARING HOUSING

It is recommended that service to the bearing housing assembly be performed by an authorized Suzuki Outboard Motor dealer. You can save some labor cost by removing and installing the jet drive assembly as outlined in this chapter and then taking the assembly to a dealer for service. Should you elect to service the bearing housing yourself, read the following procedure completely before proceeding.

Refer to **Figures 11-14** for this procedure.

1. Remove the jet drive assembly as outlined under *Removal* in the *JET DRIVE* section in this chapter.
2. Remove the water pump assembly and Woodruff key as outlined under *WATER PUMP* in this chapter.
3. Withdraw the aluminum spacer.
4. Remove the bolts and washers securing the bearing housing to the jet drive housing. Withdraw the bearing housing assembly and drive shaft from the jet drive housing and place on a clean work bench.
5. Remove the snap ring from the bore in the top of the bearing housing.

CAUTION
Do not apply excessive heat to the bearing housing as the grease seals may be damaged.

WARNING
Components will be hot. Take necessary safety precautions to prevent personal injury.

6. Apply heat to the bearing housing in increments. After each application of heat, strike the impeller end of the drive shaft against a wooden block. If the bearing housing has been heated to a sufficient temperature, the housing should slide down off the bearing(s). If not, apply additional heat to the bearing housing and reattempt removal of the bearing housing.
7. Withdraw the upper seal carrier and spacer from the drive shaft.
8. Press the bearing off the drive shaft.
9. Slide the thrust washer off the drive shaft.
10. Remove the collar and the thrust ring from the drive shaft if needed.
11. Remove the grease seals and retaining rings from the bearing housing.
12. Remove the grease seals and retaining rings from the upper seal carrier.
13. Clean and inspect the bearing housing, upper seal carrier and the drive shaft.

NOTE
Whenever the bearing is pressed from the drive shaft, it is recommended that the bearing be replaced as it is damaged during removal.

14. Place the collar over the drive shaft with the grooved end of the collar facing the thrust ring on the shaft (**Figure 14**).
15A. Model PU 40—Install the thrust washer onto the drive shaft with the gray teflon coated side facing toward the jet drive impeller.
15B. All Models—Place the bearing on the shaft so the wide thrust shoulder on the bearing outer race is facing down as shown in **Figure 14**.

NOTE
Press only on the bearing inner race in Step 16 to prevent damage to the bearing.

16. Using a suitable tool, press the bearing onto the shaft until the thrust ring on the drive shaft is locked into place in the collar.
17. Install the inner seal retaining rings, one into each lower groove of the housing.
18. Lightly lubricate the seal bore in the housing and seal carrier and the outer diameter of the inner seals with Suzuki Jet Grease (part No. 99954-53885) or equivalent.
19. Place the inner seals into the housing with the seal lips facing outward. Press the seals into the housing until seated against the retaining rings.

JET DRIVES

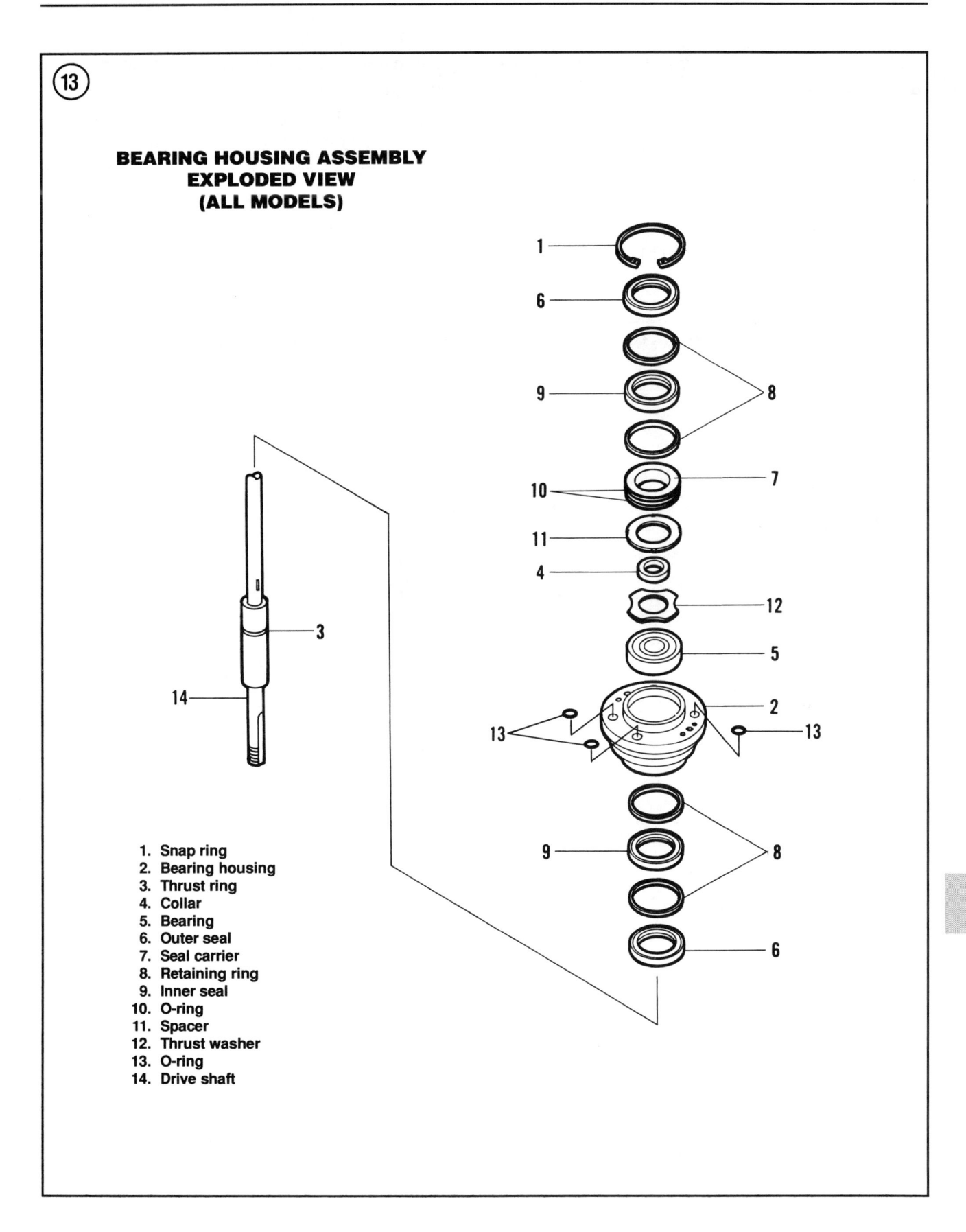

BEARING HOUSING ASSEMBLY EXPLODED VIEW (ALL MODELS)

1. Snap ring
2. Bearing housing
3. Thrust ring
4. Collar
5. Bearing
6. Outer seal
7. Seal carrier
8. Retaining ring
9. Inner seal
10. O-ring
11. Spacer
12. Thrust washer
13. O-ring
14. Drive shaft

CHAPTER THIRTEEN

20. Install the outer retaining rings into the outer grooves of the housing.

21. Lightly lubricate the seal bore in the housing and seal carrier and the outer diameter of the inner seals with Suzuki Jet Grease (part No. 99954-53885) or equivalent.

22. Place the outer seals into the housing with the seal lips facing outward. Press the seals into the housing until seated against the outer retaining rings.

23. Thoroughly pack the cavity between the seals with Suzuki Jet Grease (part No. 99954-53885) or equivalent.

24. Lubricate the 2 O-rings with Suzuki Jet Grease (part No. 99954-53885) or equivalent, and install them into the grooves on the upper

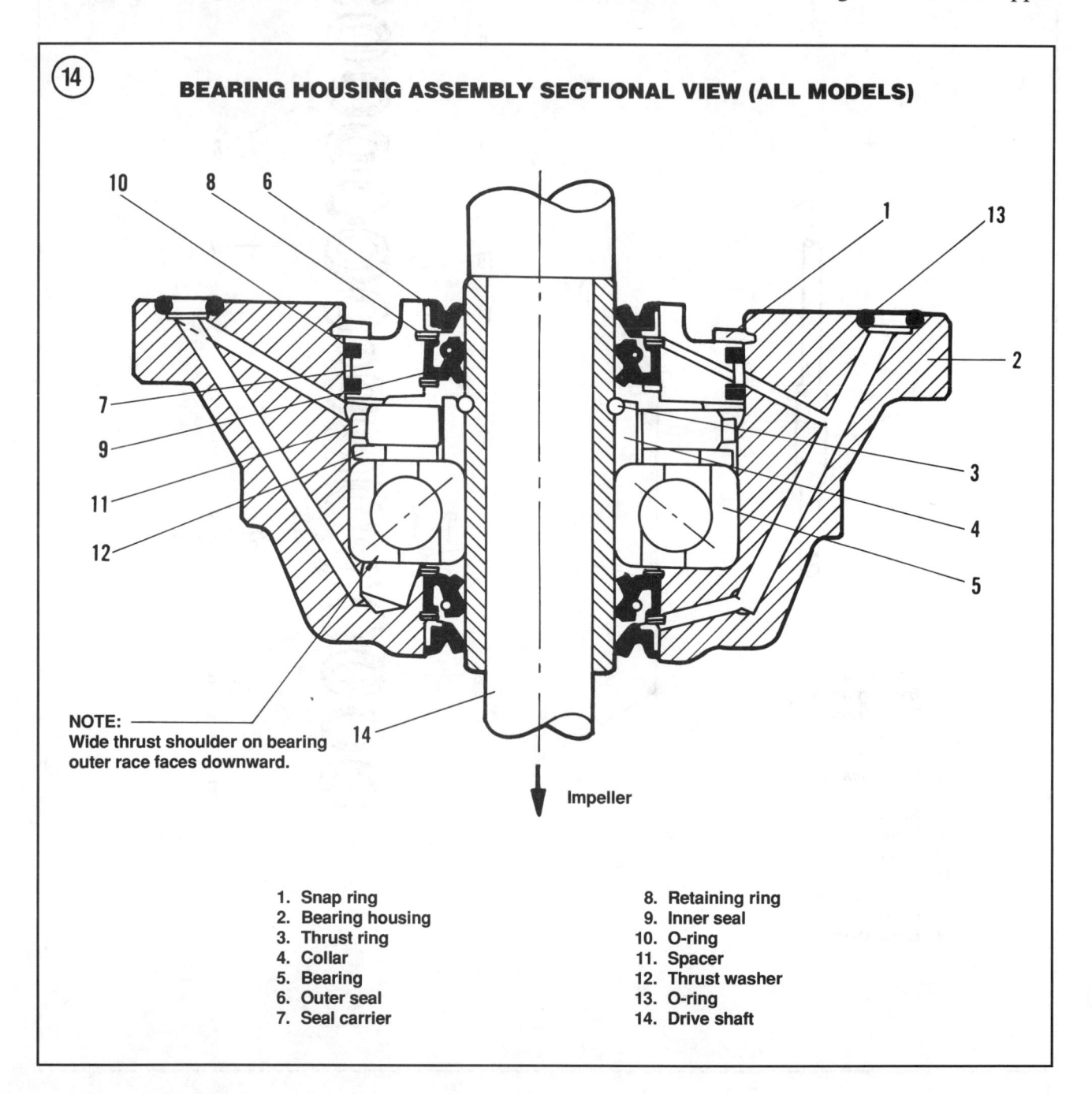

BEARING HOUSING ASSEMBLY SECTIONAL VIEW (ALL MODELS)

NOTE:
Wide thrust shoulder on bearing outer race faces downward.

1. Snap ring
2. Bearing housing
3. Thrust ring
4. Collar
5. Bearing
6. Outer seal
7. Seal carrier
8. Retaining ring
9. Inner seal
10. O-ring
11. Spacer
12. Thrust washer
13. O-ring
14. Drive shaft

JET DRIVES

seal carrier. Lubricate the bearing housing bore to ease seal carrier installation.

25. Install the bearing housing onto the drive shaft and bearing assembly. Make sure the housing is properly aligned with the bearing. If the housing and bearing are properly aligned, only a small amount of pressure will be required to seat the bearing in the housing.

NOTE
A small amount of heat applied to the bearing housing will ease bearing and shaft assembly installation. Use caution not to overheat the housing as seal damage will result. Do not use open flame to heat the housing.

26A. PU 40—Install the spacer over the collar and on top of the thrust washer.

26B. PU 55, PU 85 and PU 140—Install the thrust washer with the gray Teflon coating facing the bearing. Install the spacer over the collar and on top of the thrust washer.

27. Carefully press the upper seal carrier into the housing. Only light pressure should be required to seat the seal carrier if properly aligned with the housing.

28. Install the snap ring with the beveled side facing up.

29. Place new O-rings into position on the mating surface of the bearing housing and install the housing assembly into the jet drive pump housing.

30. Apply Suzuki Jet Grease (part No. 99954-53885) or equivalent, to the bearing housing retaining bolts.

31. Install the housing retaining bolts and tighten to specification in **Table 1**.

32. Install the aluminum spacer on top of the jet drive housing.

33. Install the water pump assembly as outlined under *Water Pump* in this chapter.

34. Install the jet drive assembly as outlined under *Installation* in this chapter.

35. Grease the bearings as outlined under *Bearing Lubrication* in this chapter.

INTAKE HOUSING LINER

Replacement

1. Remove the six intake housing mounting bolts. Remove the intake housing. See **Figure 11** or **Figure 12**.
2. Identify the liner bolts for reassembly in the same location, then remove the bolts.
3. Tap the liner loose by inserting a long drift punch through the intake housing grille. Place the punch on the edge of the liner and tap with a hammer.
4. Withdraw the liner from the intake housing.
5. Install the new liner into the intake housing.
6. Align the liner bolt holes with respective intake housing holes. Tap the liner into place with a hammer if necessary.
7. Apply Suzuki Jet Grease (part No. 99954-53885) or equivalent grease to liner retaining bolts prior to installation.
8. Install the liner retaining bolts and tighten to the specification shown in **Table 1**.
9. Remove any burrs from the liner retaining bolt area.
10. Grease the threads of the six intake housing mounting bolts with Suzuki Jet Grease (part No. 99954-53885) or equivalent grease.
11. Install the intake housing on the jet drive pump housing with the lowest part of the intake grill facing aft. Tighten the six intake housing bolts to the specification listed in **Table 1**.
12. Refer to *Impeller Clearance* in this chapter.

Table 1 is on the following page.

TABLE 1 JET DRIVE TIGHTENING TORQUES

Fastener	ft.-lbs.	N·m
Adapter plate-to-intermediate housing		
8 mm	11	15
10 mm	22	30
Bearing housing-to-jet drive housing		
PU 40	Hand tighten	Hand tighten
PU 55, PU 85 and PU 140	5	7
Impeller nut	17	23
Intake housing	5	7
Intake housing liner	11	15
Jet drive-to-adapter plate	11	15
Water pump housing	11	15

Grease all fasteners with Suzuki Jet Grease (part No. 99954-53885) or a suitable equivalent prior to installation.

INDEX

A

Air/oil mixing valve 509-510
Anti-corrosion maintenance 98-99
Automatic rewind starters 458-483
    Bendix starter 458-461
    overhead starter 462-481
    starter interlock adjustment ..481-483

B

Battery 195-203
    care and inspection 198-199
    charging 201-202
    installation in aluminum
        boats 197-198
    jump starting 202
    new battery installation 203
    storage 201
    testing 199-201
Battery and starter motor check 109
Battery charging coil 38
Battery charging system 205-206
    charging coil replacement 206
    rectifier, or voltage regulator/
        rectifier replacement 206
Bearing housing 538-541
Bendix starter 458-461
    disassembly/assembly 459-461
    removal/installation 458-459
Breakerless transistorized ignition
    (1990-on DT 2)
    igniter unit removal/installation .. 222
    stator base
        removal/installation 222-223
Breaker point ignition
    adjustment 108-109
    component testing
        (1985-1989 DT 2) 41-42
    replacement 108
    system service
        (1985-1989 DT 2) 107-108
Breaker point test 41

C

Carburetor
    centerbowl (DT 4, DT 6;
        1985-1987 DT 8) 154-156
    centerbowl dual-throat
        (V4 and V6) 166-171
    centerbowl single-throat 160-166
    cleaning and inspection 171-173
    integral fuel pump 157-159
    overhaul/adjustment 150
    removal/installation 148-150
    throttle plunger carburetor
        (DT 2) 151-154
Carburetor fuel pump 142
CDI unit (all models) 240-247
    individual CDI unit removal/
        installation 241-247
Centerbowl carburetor
    (DT 4, DT 6;
    1985-1987 DT 8) 154-156
    float adjustment 155-156
    idle speed adjustment 156
    overhaul 154-155
Centerbowl dual-throat carburetor
    (V4 and V6) 166-171
    float adjustment 169-170
    fuel level test 168-169
    idle speed adjustment 170
    overhaul 166-168
    throttle linkage synchronizing 170-171
Centerbowl single-throat carburetor
    (DT 20 through DT 85;
    DT 115, DT 140) 160-166
    carburetor balancing 165-166
    float adjustment 161
    idle speed adjustment 162-165
    overhaul 161
Charge coil resistance test
    Pointless Electronic Ignition 43-50
    Integrated circuit (IC) and Micro
        Link ignition systems 66-68
Charging system 37-39
Choke solenoid resistance check ... 110

Clutch pushrod dimension
    (1985-1989 DT 75, DT 85;
    1985 DT 115, DT 140) 454-456
Condenser test 41
Connecting rod bearings 337-343
Continuity test
    key ignition switch 36
    neutral start interlock switch 36
    push button start switch 36
    starter relay 35-36
    stop switch 37
Crankshaft 337-339, 343-345
Cylinder head, cylinder block
    and crankcase 334-337

E

Electric starting systems 206-207
Electrical connectors 195
Electrical systems 195-247
    battery charging system 205-206
    battery 195-203
    ignition systems 220-247
    lighting system 203-205
    starter motor 207-220
Engine
    detonation 77
    fasteners and torque 250-251
    flat spots 78
    flushing 100
    misfiring 77-78
    noises 79
    operation 2
    overheating and lack of
        lubrication 76-77
    piston seizure 78-79
    poor idling 77
    power loss ....
    preignition 77
    serial number 250
    temperature and overheating 76
    timing and synchronization ...116-117
    troubleshooting 76-79
    vibration, excessive 79
    water leakage in cylinder 78

14

INDEX

F

Fasteners	2-8
Fasteners and torque	250-251
Flywheel	251-261
inspection	261
removal/installation	
DT 2 and DT 4	251-252
DT 6; 1985-1987 DT 8	252-253
1988-on DT 8, DT 9.9	253-254
1985-1987 DT 9.9; 1985-on DT 15	254-255
1986-1988 DT 20; 1985-1988 DT 25; 1985-1987 DT 30; 1987-1989 DT 35; 1985-on DT 40, DT 55, DT 65	256-257
1989-on DT 25; 1988-on DT 30	257-258
DT 75, DT 85; 1985 DT 115	258-259
1986-on DT 115, DT 140; V4; V6	259-260
Fuel filter	148
Fuel injection system (DT 225 V6)	
air temperature sensor	184
atmospheric pressure sensor	184
control unit precautions	184-185
cylinder wall temperature sensor	184
depressurizing the fuel system	185
fuel injectors	183
fuel pressure regulator	183
fuel system precautions	185
high pressure fuel filter	183
removal/installation	185-186
high pressure fuel pump	182-183
system description	181
throttle body	181
throttle valve sensor	184
vapor separator	181-182
vapor separator removal/installation	185
Fuel line, connector and primer bulb	179
Fuel pump	107
Fuel pump and filter	141-148
Fuel system	140-186
carburetors	148-173
engine fuel filter service (carburetted models)	105-107
fuel line, connector and primer bulb	179
fuel lines	105
fuel mixing	91-92
fuel selection	88-89
fuel tank	176-179
recommended fuel mixture	89-91
reed valve assembly	173-175
sour fuel	89
troubleshooting	74-76
Fuel tank	176-179
portable fuel tank filter cleaning	177-179
portable fuel tank fuel gauge inspection	177-179
removal/installation	176

G

Galvanic corrosion	11-14
Galvanic protection and bonding wires	99
Gasket sealant	10-11
Gasohol	89
Gearcase	373-456
clutch pushrod dimension	454-456
disassembly/assembly	
DT 2	401-404
DT 4	404-408
DT 6; 1985-1987 DT 8	409-412
1988-on DT 8; DT 9.9, DT 15	412-416
DT 20, DT 25, DT 30 (2-cylinder); DT 25, DT 30 (3-cylinder); DT 35, DT 40, DT 55, DT 65	416-425
DT 75, DT 85, DT 115, DT 140	425-432
V4 and V6	432-437
lubrication	92-94
pinion gear adjustment	451-453
pinion gear depth and forward/reverse gear backlash	444-451
propeller	373-376
propeller shaft clutch	437-443
propeller shaft end play	453-454
removal/installation	
DT 2	393-394
DT 4, DT 6; 1985-1987 DT 8	394-396
1988-on DT 8; DT 9.9 through DT 40	396-398
DT 55, DT 65	398-399
DT 75, DT 85, DT 115, DT 140, V4, V6	399-401
water pump	376-393
Gearcase and water pump check	105

H

Hydraulic pump	491-494
fluid check	
all models except V4 and V6	491-493
V4 and V6 models	493-494
fluid refill and bleeding	494

I

Ignition coil (all models)	238-240
combined ignition coil/CDI unit removal/installation	240
individual ignition coil removal/installation	238-240
Ignition coil/igniter assembly resistance test	42-43
Ignition coil power and leakage tests	42
Ignition coil resistance test	41-42
Ignition systems	220-247
breakerless transistorized ignition (1990-on DT 2)	222-223
CDI unit (all models)	240-247
ignition coil (all models)	238-240
Integral fuel pump carburetor (1988-on DT 8; 1985-on DT 9.9, DT 15)	
float adjustment	158
idle speed adjustment	159
Integrated Circuit (I.C.) ignition system	229-234
magneto breaker point ignition (1985-1989 DT 2)	220-222
Micro Link ignition system (V4 and V6)	234-238
PEI ignition	223-228
troubleshooting	39-40
Intake housing liner	541
Integrated circuit (IC) and Micro Link ignition systems	53-74
charge coil resistance test	66-68
gear counter coil resistance test	71-72
I.C. power source coil resistance test	72
idle speed adjustment switch	72
ignition coil resistance tests	71
peak output tests	
1989-on DT 25; 1988-on DT 30	53-55
1986-on DT 55, DT 65	55-57
1988-on DT 75, DT 85	57-58
peak voltage output tests	
1986-on DT 115, DT 140	58-60
V4 models	60-63
V6 models	63-66
pulser coil resistance tests	68-71
resistance tests	66
throttle switch test	72
throttle valve sensor	72-74
Integrated Circuit (I.C.) ignition system (1989-on DT 25; 1988-on DT 30; 1986-on DT 55, DT 65; 1988-on DT 75, DT 85; 1986-on DT 115, DT 140; 1985-1988 V6)	
condenser charging coil, gear counter coil, battery charging coil or pulser coil replacement	231-234

INDEX

I.C. ignition system (continued)
- operation 229-230
- stator base/alternator stator .. 230-231

J

Jet drives 526-541
- bearing housing 538-541
- bearing lubrication 527-528
- cooling system cleaning 530
- directional control 528
- impeller clearance 529-530
- installation 537
- intake housing liner 541
- outboard mounting height ... 526-527
- removal 536
- steering torque 527
- water pump 530-536

L

Lighting system (1985-1986 DT 9.9, DT 15, DT 40) 203-205
- lighting coil replacement 205
- troubleshooting 37

Loss of high speed rpm (all DT 15) 140

Lubricants 8-10

Lubrication 88-96
- fuel mixing 91-92
- fuel selection 88-89
- gasohol 89
- gearcase lubrication 92-94
- other lubrication points 94-95
- recommended fuel mixture
  - models without oil injection .. 90-91
  - oil injected models 89-90
- saltwater corrosion of gearcase bearing cage or spool 95-96
- sour fuel 89

M

Magneto breaker point ignition (1985-1989 DT 2)
- operation 220-221
- primary/secondary ignition coil removal/installation 222
- stator base removal/installation 221-222

Maintenance, anti-corrosion 98-99

Micro Link ignition system (V4 and V6) 234-238
- condenser charging coil, gear counter coil, battery charging coil or pulser coil replacement 237-238
- operation 236-237
- stator base/alternator stator 237
- troubleshooting 53-74

N

Neutral start interlock switch continuity test 36

O

Off idle lean condition 140-141

Oil injection system 506-524
- air/oil mixing valve (DT 25, DT 30 (3-cylinder); 1987-on DT 35, DT 40; V4 and V6) 509-510
- break-in procedure 508
- component replacement 515-524
- low oil level warning light and buzzer 508
- oil flow sensor (1987-on DT 55, DT 65; 1988-on DT 75, DT 85; V4 and V6) 509
- oil pump discharge adjustment 523-524
- oil pump removal/installation 516-523
- oil pump service 510-515
- oil tank removal/installation 515-516
- operation 508
- system components 507-508

Oil pump removal/installation
- DT 55, DT 65, DT 75, DT 85; DT 115, DT 140 516-518
- 1988-on DT 8, DT 9.9; 1989-on DT 15; DT 25, DT 30 (3-cylinder); DT 35, DT 40; V4 and V6 518-523

Oil pump service 510-515
- oil pump bleeding 510-511
- oil pump control rod adjustment 512-513
- oil pump delivery rate test ... 513-515

Overhead starter 462-481
- disassembly/assembly
  - DT 2, DT 4, DT 6; 1985-1987 DT 8 463-468
  - 1988-on DT 8; DT 9.9, DT 15 468-473
  - DT 20; DT 25, DT 30 473-477
  - DT 35 and DT 40 477-481
- removal/installation 462-463

P

Performance test (on boat) 111

Pinion gear adjustment (1986-on DT 115, DT 140; V4 and V6) 451-453

Pinion gear depth and forward/reverse gear backlash
- DT 4, DT 6; 1985-1987 DT 8 444-445

pinion gear depth and forward gear backlash
- 1988-on DT 8, DT 9.9 ... 445-446
- 1985-1987 DT 9.9; 1985-on DT 15 447-448
- DT 20, DT 25, DT 30, DT 35, DT 40, DT 55, DT 65, DT 75, DT 85, 1985 DT 115 and DT 140
  - forward gear backlash 448-449
  - pinion gear depth 449
  - reverse gear backlash 449
- reverse gear backlash (1988-on DT 8, DT 9.9) 446-447
- V4 and V6 449-451

Pistons 338-343

Pointless electronic ignition (PEI) (DT 4; DT 6; DT 8; DT 9.9; DT 15; 1985-1987 DT 20, DT 30; 1985-1988 DT 25; 1987-1989 DT 35; 1985-on DT 40; 1985-1987 DT 75, DT 85; 1985 DT 115, DT 140)
- charge coil, lighting coil, battery charging coil or trigger coil replacement 228
- charge coil or lighting/charging coil replacement 228
- operation 226-227
- stator base/alternator stator 227
- timer base removal/installation ... 228
- trigger coil replacement (1985-1987 DT 75, DT 85 and 1985 DT 115, DT 140 228

Pointless electronic ignition (PEI)
- component testing 43-53
- charge coil resistance test 43-50
- ignition coil resistance tests ... 50-52
- resistance tests 43
- throttle switch test 52-53

Power head 249-368
- assembly
  - DT 2 345-346
  - DT 4 346
  - DT 6; 1985-1987 DT 8 ... 346-349
  - 1988-on DT 8; 1985-on DT 9.9, DT 15 349-352
  - DT 20, DT 25, DT 30 (2-cylinder); 1987-1989 DT 35; 1985-on DT 40 352-355
  - inline 4-cylinder engines .. 361-364
  - 3-cylinder engines 355-361
  - V4 models 364-366
  - V6 models 366-368
- connecting rod bearings 337-343
- crankshaft 337-339, 343-345
- cylinder head, cylinder block and crankcase 334-337
- disassembly
  - DT 2 300-302
  - DT 4 303

14

Power head (continued)
  disassembly (continued)
    inline 4-cylinder engines .. 322-327
    2-cylinder engines 303-316
    3-cylinder engines 316-322
    V4 models 327-330
    V6 models 330-334
  pistons.................. 338-343
  removal/installation
    DT 2 261-262
    DT 4 262-263
    DT 6; 1985-1987 DT 8.... 263-265
    1988-on DT 8, DT 9.9 265-267
    1985-1987 DT 9.9;
      DT 15 267-269
    DT 20, DT 25, DT 30
      (2-cylinder)........... 269-274
    DT 25, DT 30 (3-cylinder) . 274-275
    DT 35, DT 40 276-277
    DT 55, DT 65 277-280
    DT 75, DT 85 280-286
    1985 DT 115, DT 140 286-289
    1986-on DT 115, DT 140.. 289-291
    V4.................... 292-296
    V6.................... 296-300
Power trim and tilt system 484-505
  component replacement 502-505
  components
    1988-on DT 35, 1986-on DT 40 484
    1985-on DT 55 and DT 65; 1985-
      1986 DT 75 and
      DT 85 485-486
    1987-on DT 75 and DT 85, 1985-
      on DT 115 and DT 140 488
    V4, V6 489-491
  operation................ 484-491
    1988-on DT 35, 1986-on
      DT 40 484-485
    1985-on DT 55 and DT 65;
      1985-1986 DT 75 and
      DT 85 486-488
    1987-on DT 75 and DT 85,
      1986-on DT 115 and
      DT 140 488-489
    V4, V6 491
  hydraulic pump............ 491-494
  hydraulic pump/motor removal/
    installation 502
  motor will not run
    1986-on DT 35, DT 40....... 495
    DT 55, DT 65 495
    1985 DT 115, DT 140 and 1985-
      1986 DT 75, DT 85 496
    1987-on DT 75, DT 85; 1986-on
      DT 115, DT 140; V4, V6 ... 496
  pump oil pressure test....... 496-502
  trim/tilt cylinder/motor assembly
    removal/installation 504-505
  trim/tilt cylinder removal/installation

      DT 55, DT 65; 1985-1989
        DT 75, DT 85 502-503
      1990-on DT 75, DT 85;
        DT 115; DT 140 503-504
    troubleshooting............ 494-496
Propeller shaft clutch 437-443
  cleaning and inspection 442-443
  disassembly/assembly
    DT 4, DT 6; 1985-1987
      DT 8 437-438
    1988-on DT 8; DT 9.9 through DT
      85, DT 115, DT 140 438-440
    V4 and V6............. 440-442
Propeller shaft end play (DT 8 through
    DT 65 models only) 453-454
Propeller 14-20, 373-376
Pump oil pressure test......... 496-502
  1985-on DT 35, DT 40
    trim down pressure test.... 496-497
    trim up pressure test 497
  1986-on DT 55, DT 65
    tilt down pressure test..... 497-498
    tilt up pressure test 498-499
  1986-on DT 115, DT 140
    tilt down pressure test......... 499
    tilt up pressure test 499-500
  V4 and V6
    pump pressure test 500-501
    tilt cylinder pressure test... 501-502

R

Rectifier test 38-39
Reed valve assembly 173-175
  inspection 175
  reed and reed stop replacement .. 175
  removal/installation 174-175
Remote fuel pump
  disassembly/assembly....... 143-148
  removal/installation 141-142
RPM, loss of high speed........... 140

S

Safety 21
Saltwater corrosion of gearcase
  bearing cage or spool 95-96
Service hints 28-30
Spark plugs.................. 102-105
  correct heat range 103
  gapping................... 104-105
  installation.................... 105
  removal................... 103-104
Special tips 30-31
Starter interlock adjustment 481-483
  1985-1987 DT 9.9;
    1985-on DT 15 481
  1988-on DT 8 and DT 9.9 482
  DT 25 and DT 30 (3-cylinder) ... 482

  DT 20, DT 25, DT 30; DT 35,
    DT 40 (2-cylinder) 483
Starter motor 207-220
  brush replacement
    DT 8 through DT 40 with
      2-brush starter........ 208-212
    1985-on DT 75, DT 85 and 1985
      DT 115 with 3- and 4-brush
      starter................ 213-215
    1989-on DT 25; 1988-on DT 30;
      1985-on DT 35, DT 40; 1985-on
      DT 55, DT 65; 1986-on DT 115,
      DT 140, V4 and V6 with 2- and
      3-brush starter........ 215-219
  removal/installation
    attached starter relay 207-208
    remote starter relay 208
    starter relay replacement........ 220
    starter relay testing 220
Starter relay continuity test...... 35-36
Starter relay resistance check 109
Starter relay testing 109
Starting system, troubleshooting . 34-37
Storage 96-97
Submersion, complete.......... 97-98

T

Test equipment 26-28
Throttle adjustment.......... 132-136
  1989-on DT 25, DT 30 132
  1985-on DT 55, DT 65 132-135
  1986-on DT 115, DT 140 135
  V4 models.................... 136
  V6 models (carburetted) 136
Throttle plunger carburetor
  (DT 2) 151-154
  float adjustment................ 153
  idle speed adjustment 154
  mid-range jet needle
    adjustment.............. 153-154
  overhaul.................. 151-153
Throttle valve sensor adjustment
  DT 225 EFI.................. 139
  1989-on DT 25, DT 55, DT 65,
    DT 115, DT 140; 1988-on
    DT 30, DT 75, DT 85 137
  V4 and 1987-on V6......... 138-139
Timing adjustment 118-131
Timing and synchronization ... 116-131
  dynamic timing adjustment
    DT 4 119
    1985-1987 DT 20, DT 30;
      1985-1988 DT 25 125-126
    1985-1987 DT 75, DT 85 128
    1987-1989 DT 35;
      1985-on DT 40 126-128
    1985 DT 115, DT 140........ 131
  dynamic timing check
    DT 6; 1985-1987 DT 8 121

INDEX

Timing and synchronization (continued)
  dynamic timing check (continued)
    1988-on DT 8, DT 9.9 121-123
    1985-1987 DT 9.9;
      1985-on DT 15........... 124
  equipment required............ 117
  static timing adjustment
    DT 2................... 118-119
    DT 6; 1985-1987 DT 8.... 119-121
    1985-1987 DT 9.9;
      1985-on DT 15........... 123
    1985-1987 DT 20, DT 30;
      1985-1988 DT 25...... 124-125
    1985-1987 DT 75, DT 85..... 128
    1987-1989 DT 35;
      1985-on DT 40........... 126
    1985 DT 115, DT 140 129-130
Tools, hand 21-26
Torque specifications 2
Transistorized ignition system
  (1990-on DT 2)............... 42-43
Troubleshooting 33-79
  breaker-point ignition component
    testing (1985-1989 DT 2).... 41-42
  charging system 37-39
  engine.................... 76-79
  fuel system 74-76
  ignition system.............. 39-40
  integrated circuit (IC) and Micro
    Link ignition systems....... 53-74

lighting system 37
operating requirements........ 33-34
pointless electronic ignition
  (PEI) component testing..... 43-53
starting system 34-37
transistorized ignition system
  (1990-on DT 2)............ 42-43
Tune-up.................... 100-111
  battery and starter motor
    check.................... 109
  breaker point ignition 107-109
  choke solenoid resistance
    check.................... 110
  compression check 101, 103
  engine fuel filter service..... 105-107
  performance test (on boat) 111
  spark plugs 102-105
  starter relay testing 109
  wiring harness check 110-111

V

Voltage regulator test 39

W

Water pump (gearcase)........ 376-393
  assembly and installation
    DT 2.................... 389

    DT 4, DT 6; 1985-1987
      DT 8 389-390
    1988-on DT 8; DT 9.9
      and DT 15 390-391
    DT 20, DT 25, DT 30 (2-cylinder);
      DT 35, DT 40; DT 25, DT 30
      (3-cylinder); DT 55, DT 65. . 391
    DT 75, DT 85 391-392
    DT 115, DT 140 392
    V4, V6................ 392-393
  cleaning and inspection..... 388-389
  removal and disassembly
    DT 2 378
    DT 4, DT 6, 1985-1987 DT 8 .. 379
    1988-on DT 8; DT 9.9
      and DT 15 379-382
    DT 20, DT 25, DT 30 (2-cylinder);
      DT 35, DT 40; DT 25, DT 30
      (3-cylinder); DT 55,
      DT 65 382-383
    DT 75, DT 85 383-385
    DT 115, DT 140 385-387
    V4, V6................ 387-388
Water pump (jet drives)........ 530-536
  assembly and installation ... 532-536
  cleaning and inspection..... 531-535
  removal and disassembly ... 531-535
Wiring diagrams 548-625
Wiring harness check 110-111

14

DT 2 (1985–1989)

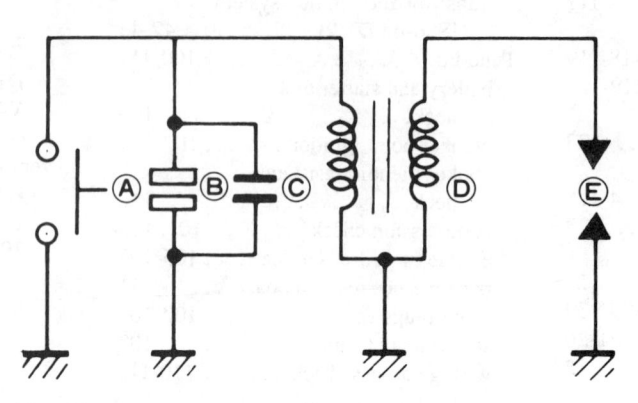

Ⓐ : Engine stop switch Ⓓ : Ignition coil
Ⓑ : Contact point Ⓔ : Spark plug
Ⓒ : Condenser

DT 2 (1990-1991)

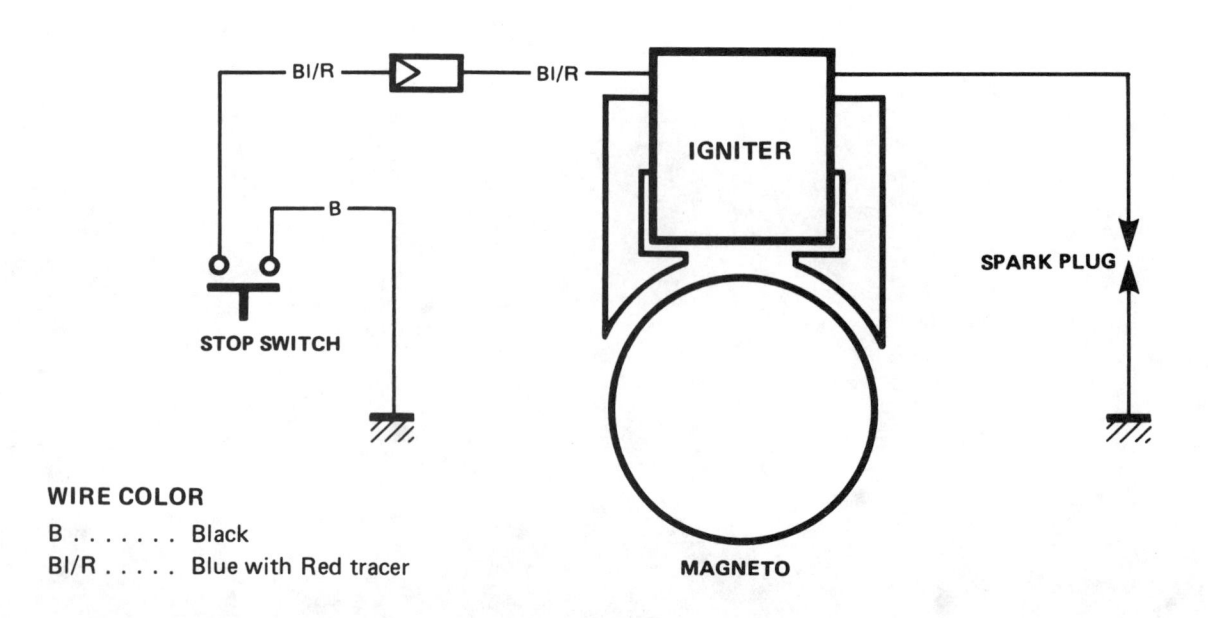

WIRE COLOR

B . . . . Black
Bl/R . . . . Blue with Red tracer

WIRING DIAGRAMS

DT 4 (1985–1989)

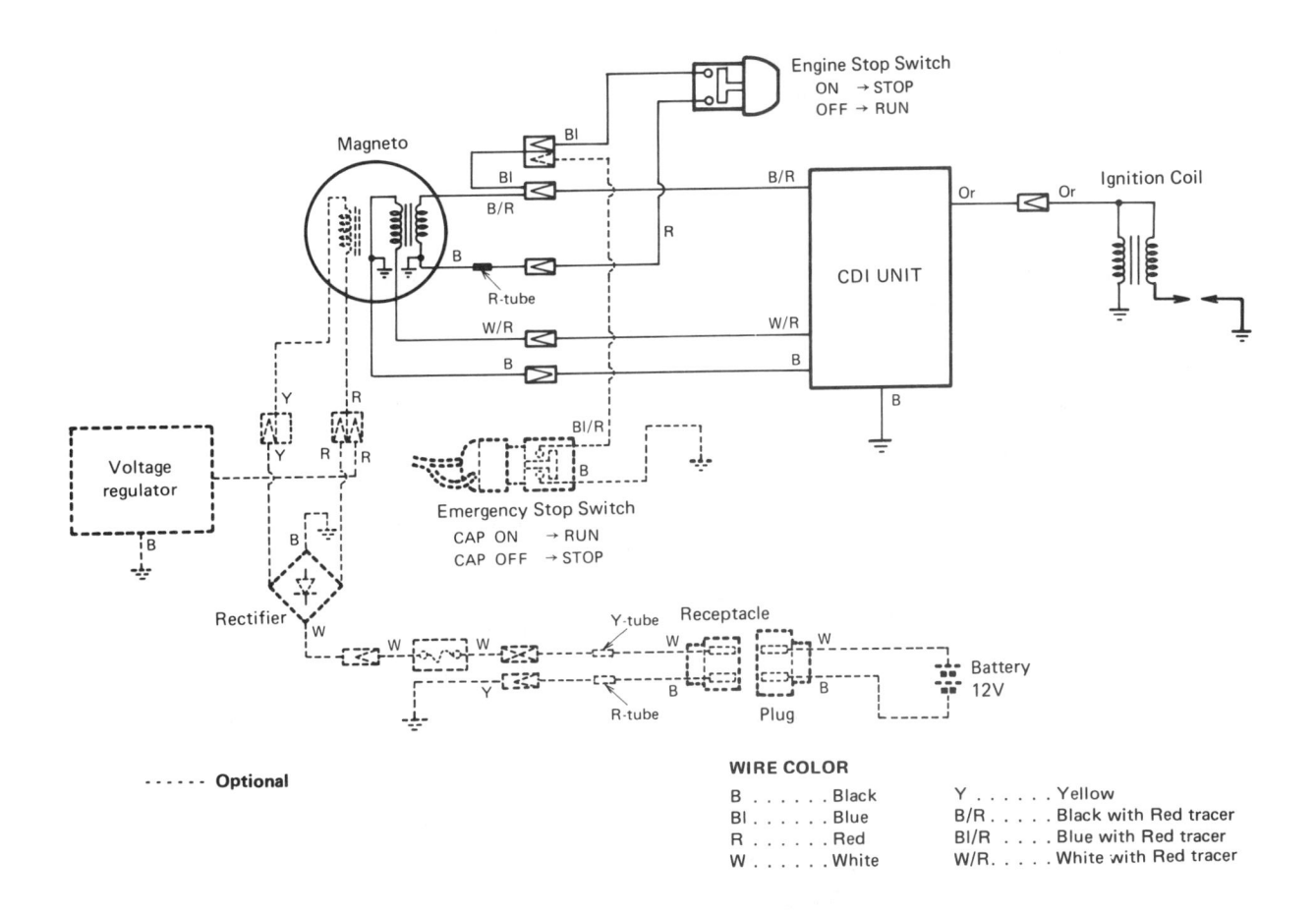

WIRE COLOR

B	Black	Y	Yellow
Bl	Blue	B/R	Black with Red tracer
R	Red	Bl/R	Blue with Red tracer
W	White	W/R	White with Red tracer

- - - - - - Optional

WIRING DIAGRAMS

DT 4 (1990-1991)

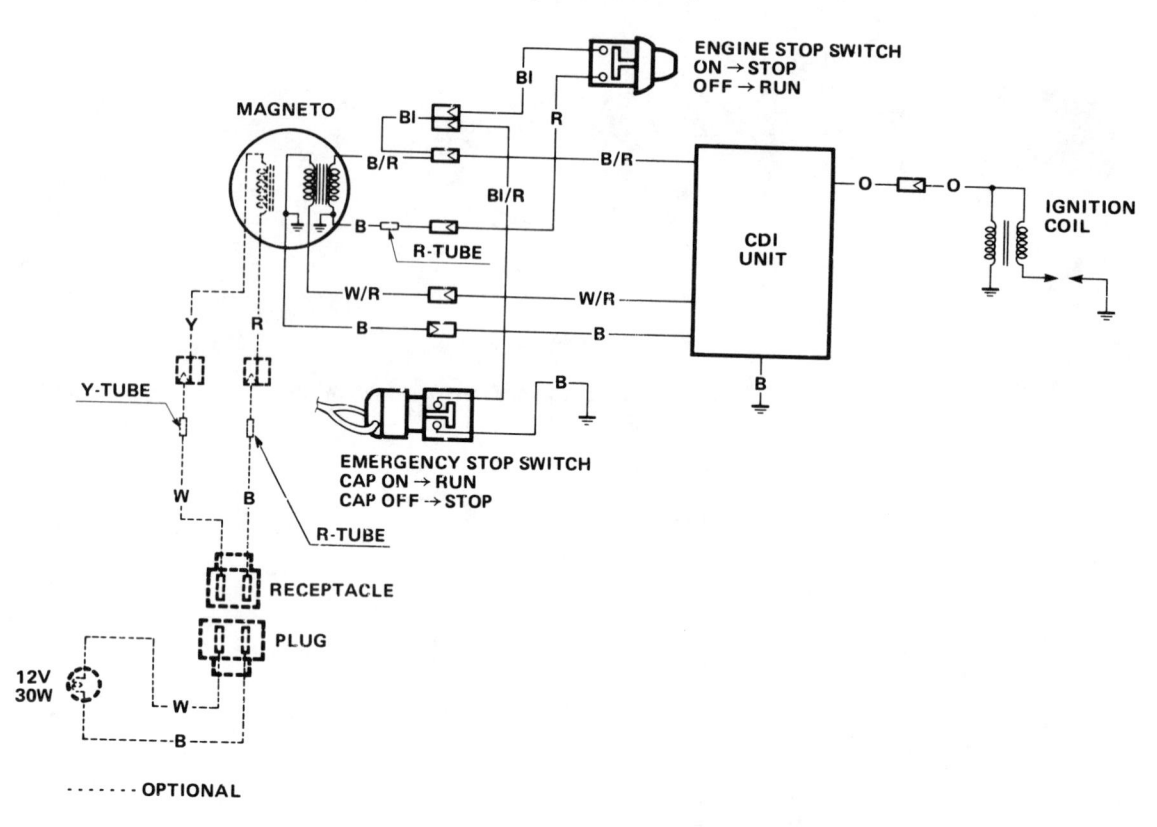

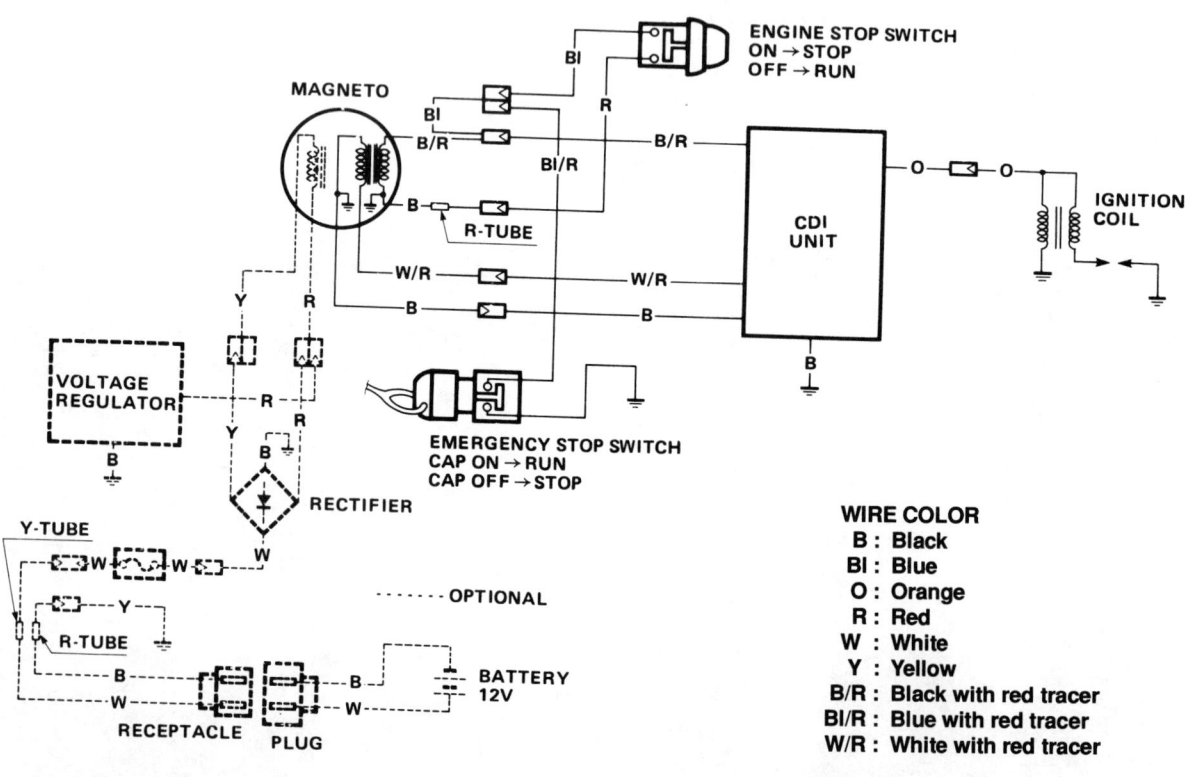

WIRE COLOR
- B : Black
- Bl : Blue
- O : Orange
- R : Red
- W : White
- Y : Yellow
- B/R : Black with red tracer
- Bl/R : Blue with red tracer
- W/R : White with red tracer

WIRING DIAGRAMS

DT 6 (1985-1986)

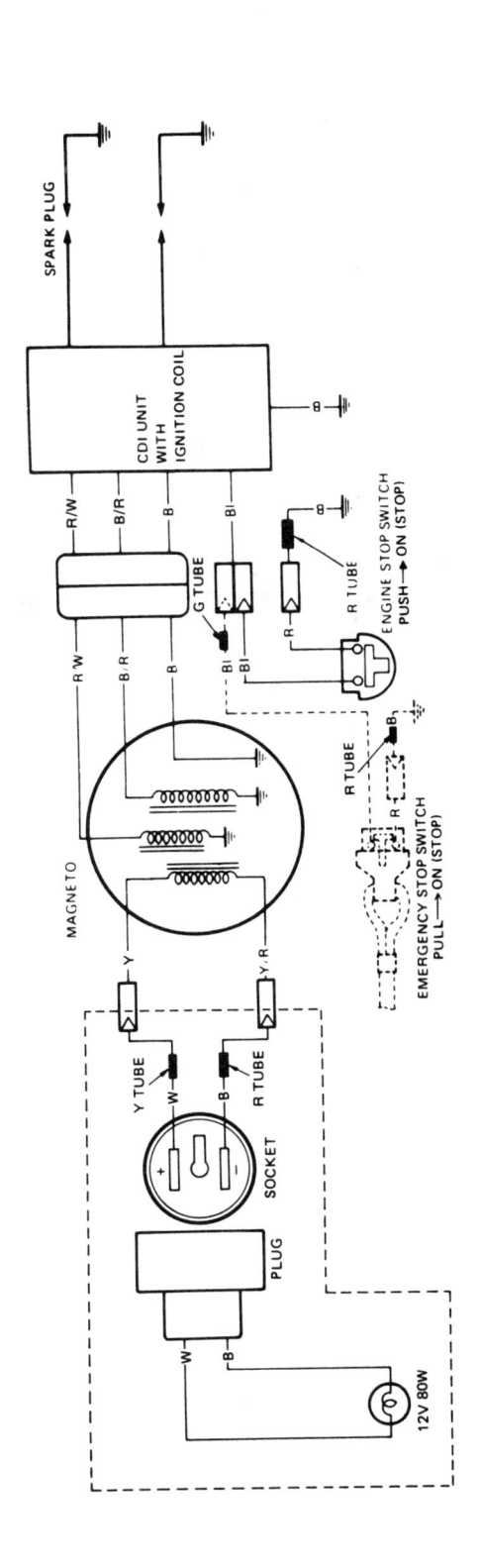

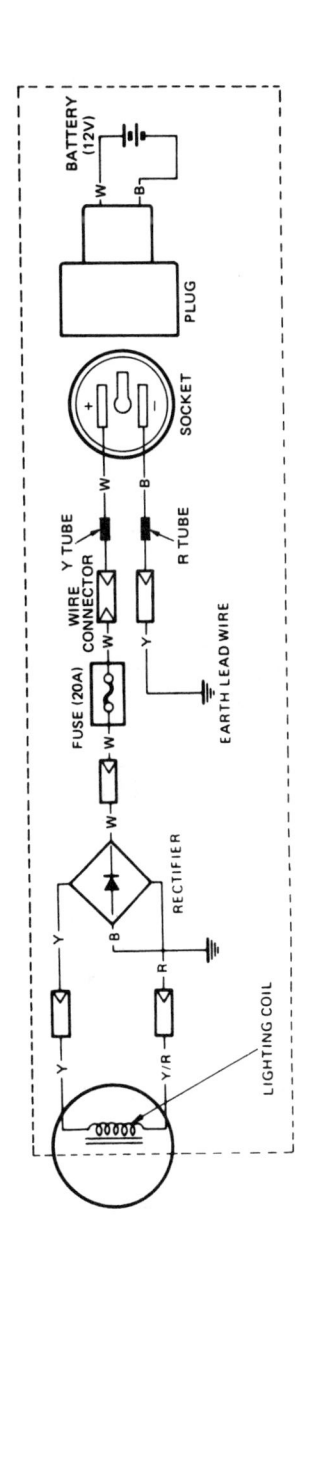

...... optional

WIRING DIAGRAMS

DT 6 (1987–ON), DT 8 (1985–1987)

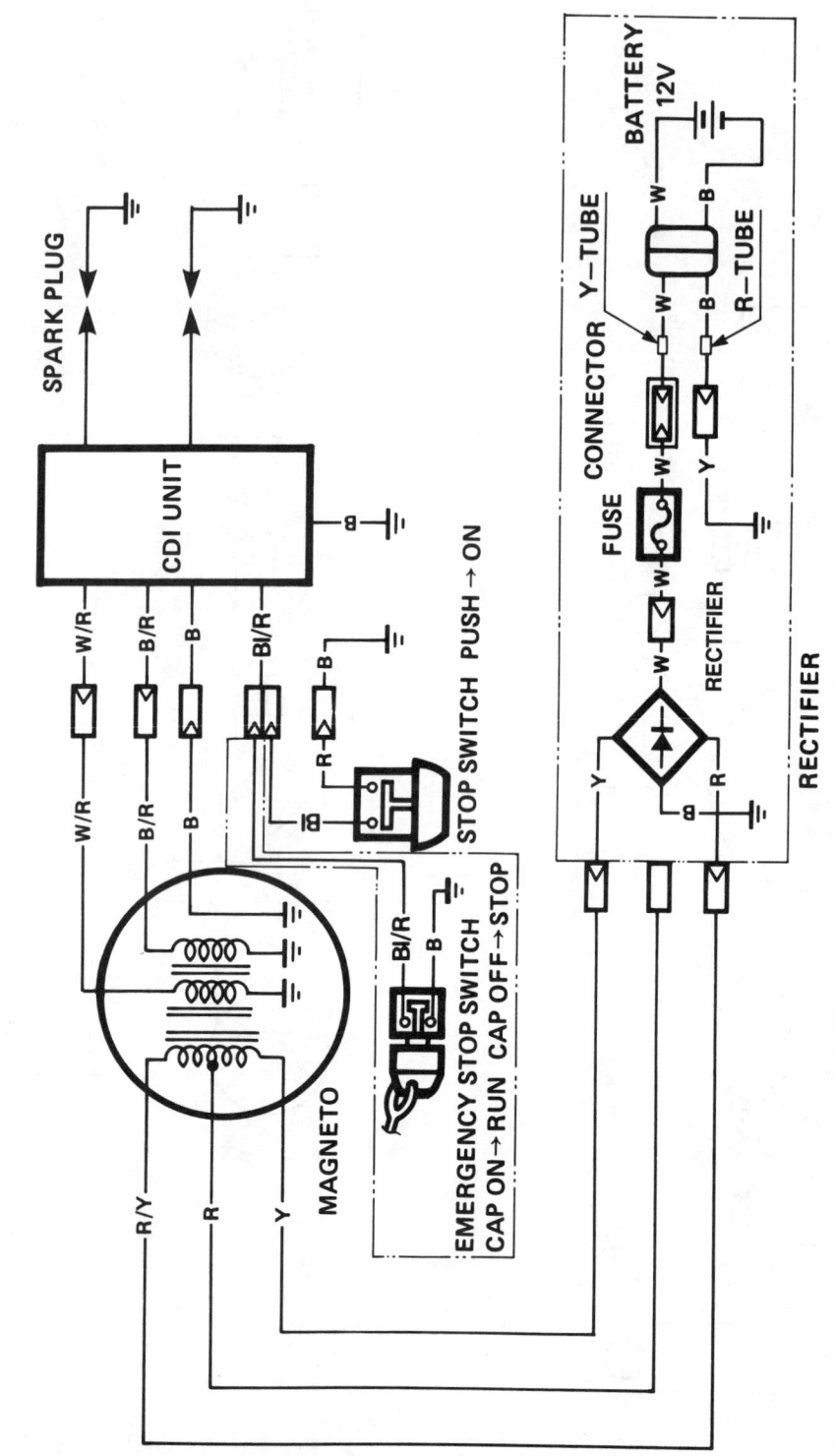

WIRE COLOR

- B . . . . . Black
- Bl . . . . . Blue
- R . . . . . Red
- W . . . . . White
- Y . . . . . Yellow

- B/R . . . . Black with Red tracer
- Bl/R . . . . Blue with Red tracer
- R/Y . . . . Red with Yellow tracer
- W/R . . . . White with Red tracer

WIRING DIAGRAMS

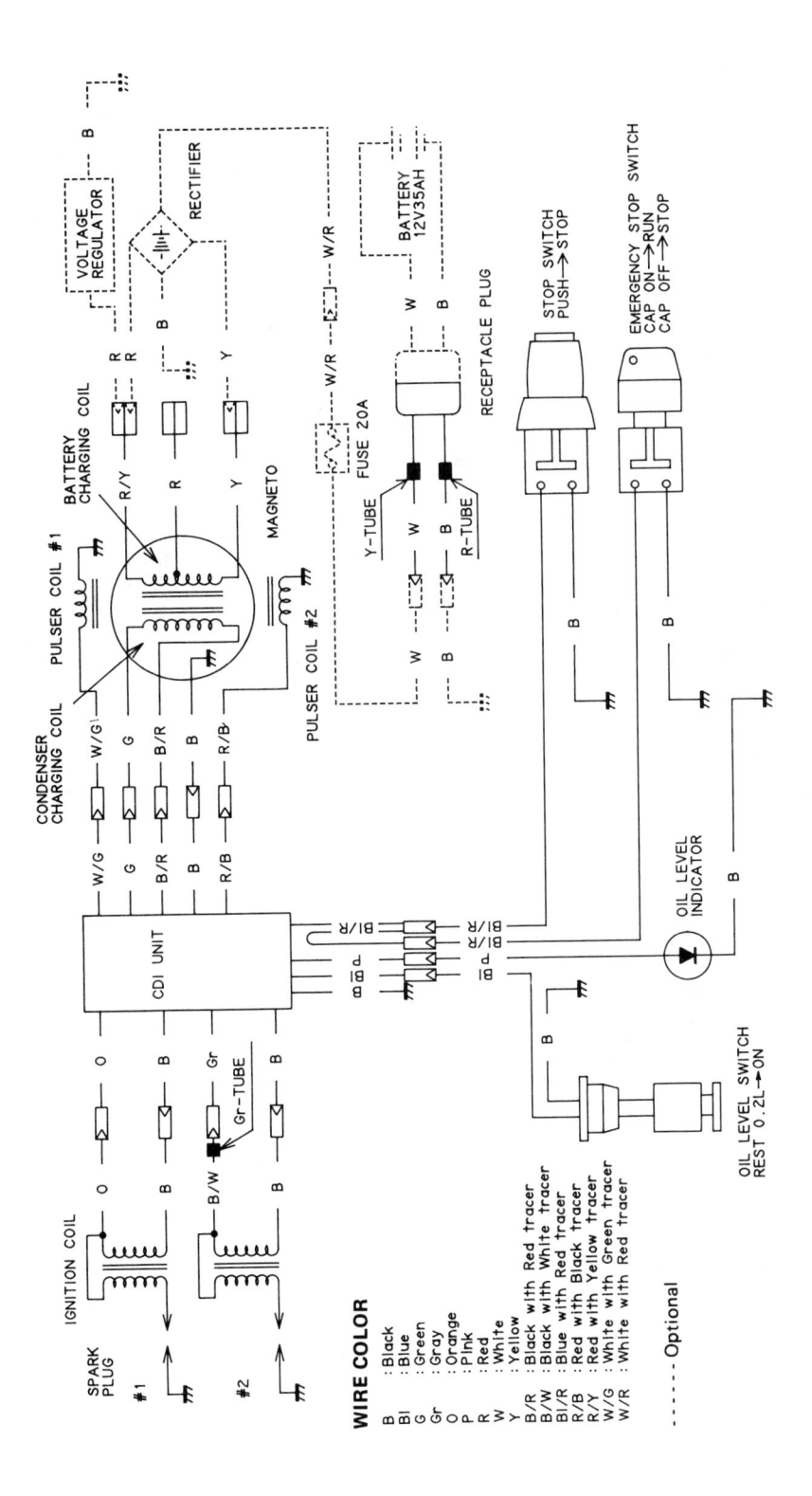

553

DT 8 C/DT 9.9MC (1988)

WIRE COLOR

- B : Black
- Bl : Blue
- G : Green
- Gr : Gray
- O : Orange
- P : Pink
- R : Red
- W : White
- Y : Yellow
- B/R : Black with Red tracer
- B/W : Black with White tracer
- Bl/R : Blue with Red tracer
- R/B : Red with Black tracer
- R/Y : Red with Yellow tracer
- W/G : White with Green tracer
- W/R : White with Red tracer

------ Optional

15

DT 8 C/DT 9.9MC (1989-1991)

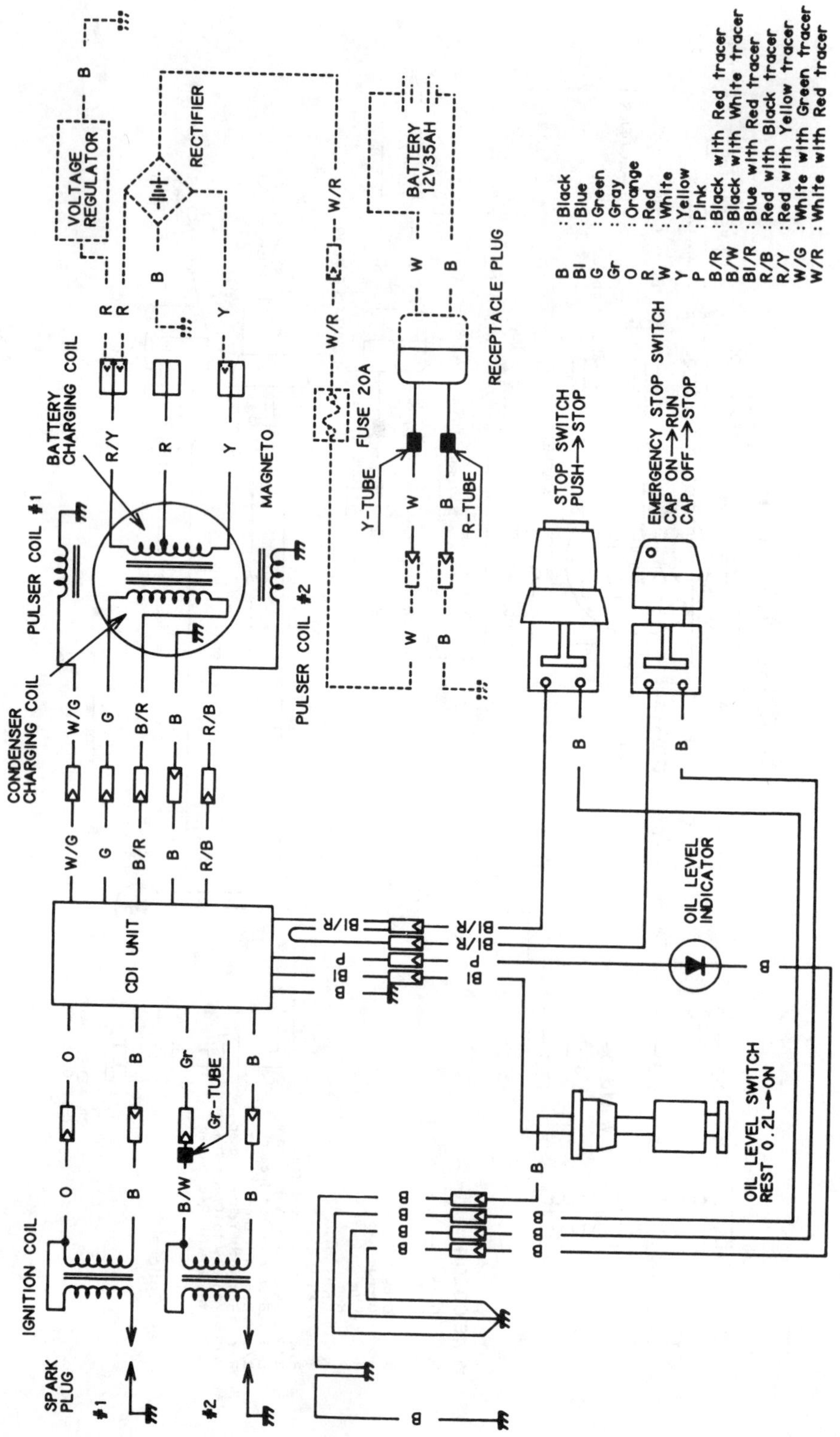

WIRING DIAGRAMS

DT 8 C/DT 9.9MC (1989–1991) WITH OPTIONAL BUZZER

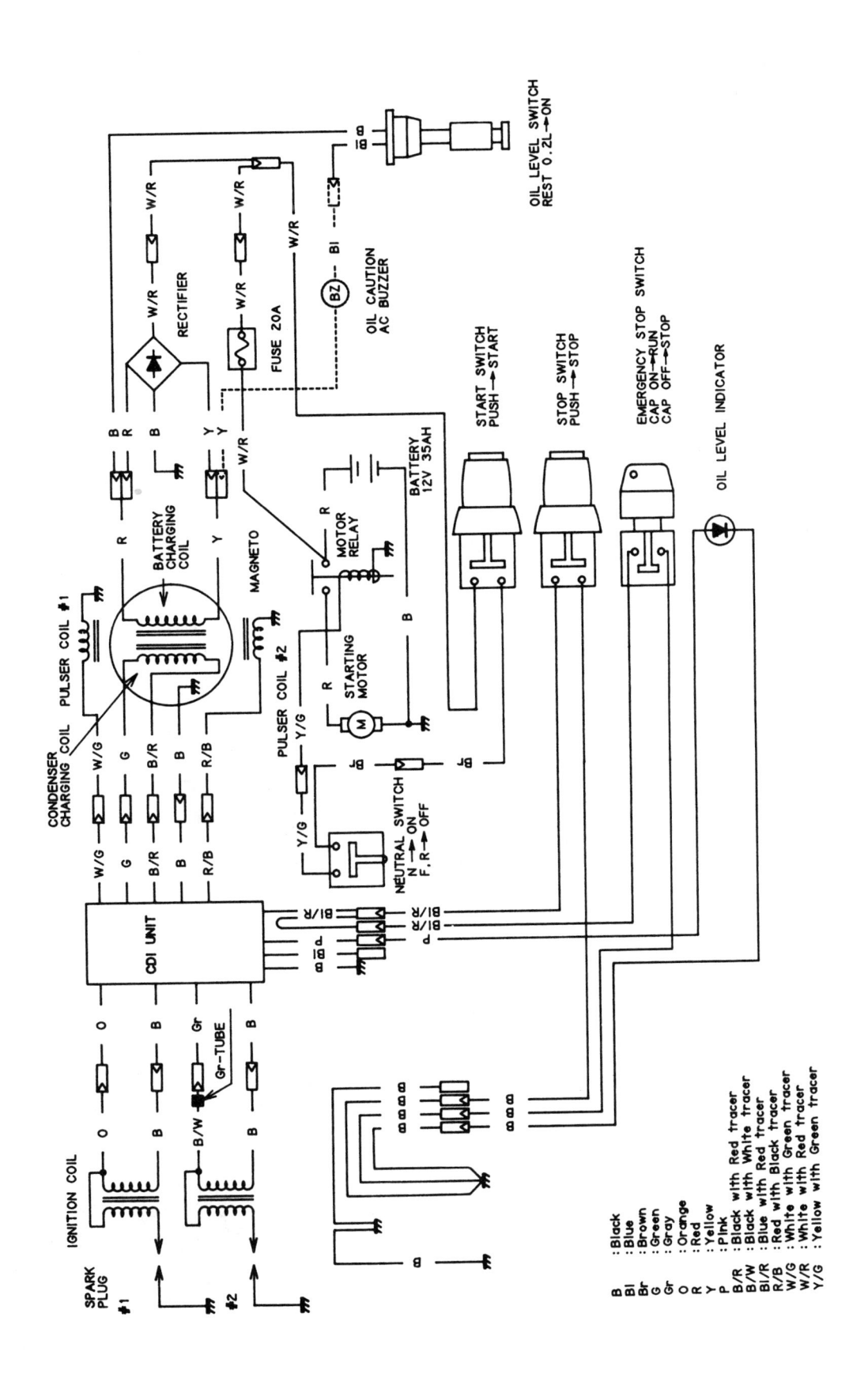

WIRING DIAGRAMS

DT 9.9CE/DT 9.9CEN (1989–1991) WITH OPTIONAL REMOTE CONTROL

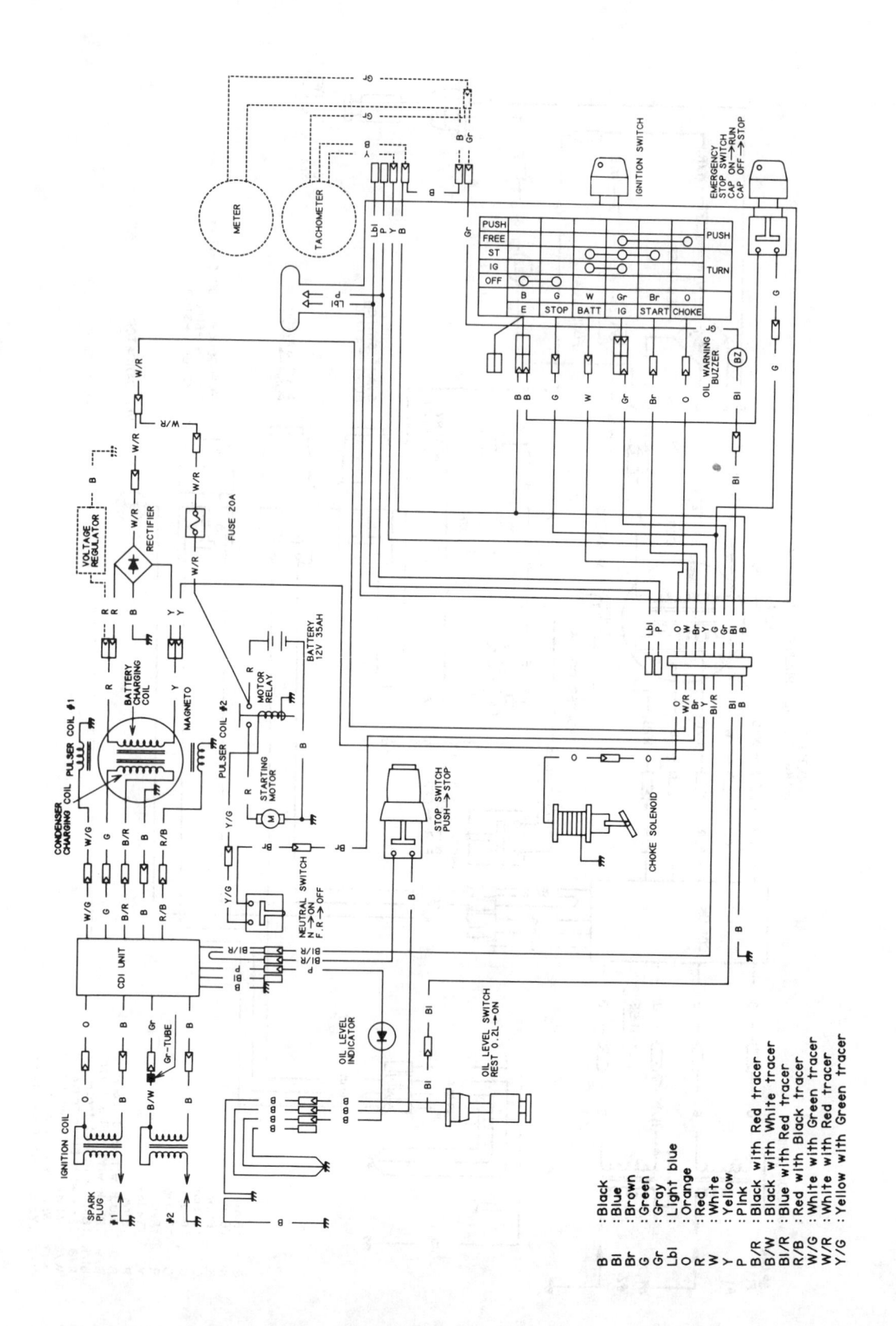

B	: Black
Bl	: Blue
Br	: Brown
G	: Green
Gr	: Gray
Lbl	: Light blue
O	: Orange
R	: Red
W	: White
Y	: Yellow
P	: Pink
B/R	: Black with Red tracer
B/W	: Black with White tracer
Bl/R	: Blue with Red tracer
R/B	: Red with Black tracer
W/G	: White with Green tracer
W/R	: White with Red tracer
Y/G	: Yellow with Green tracer

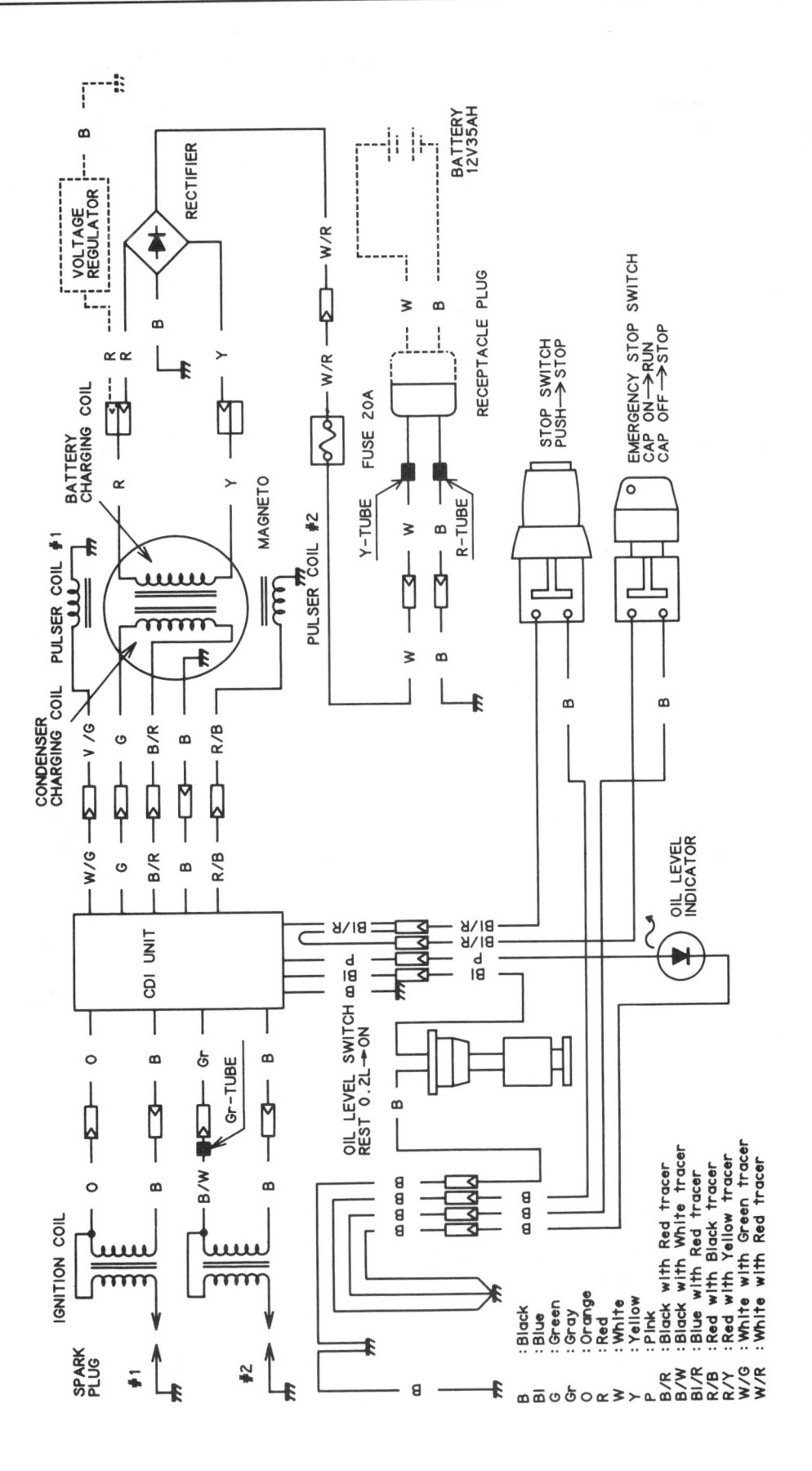

WIRING DIAGRAMS

DT 9.9CN (1989-1991)

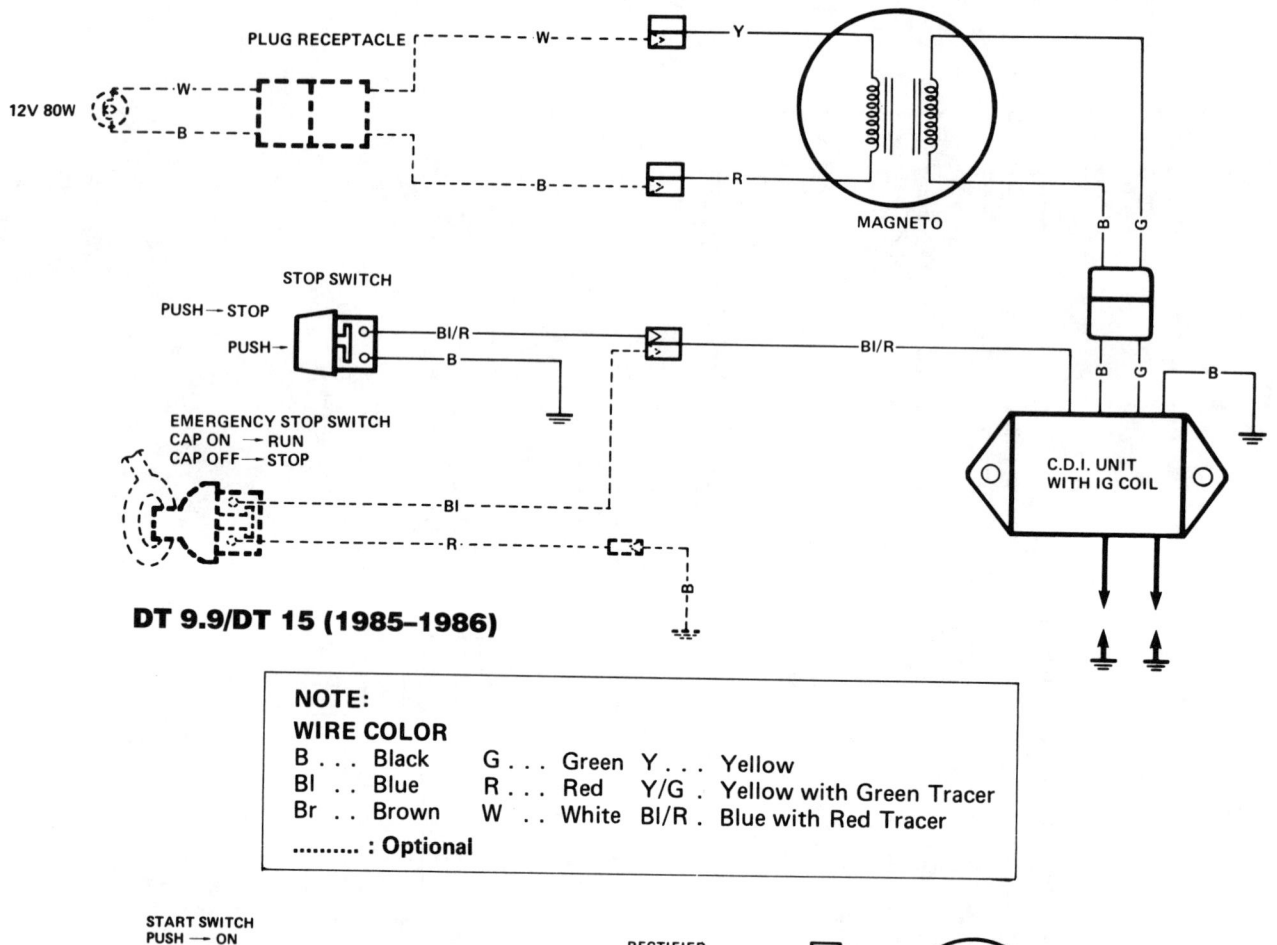

WIRING DIAGRAMS

DT 9.9/DT 15 (1985–1986)

NOTE:
WIRE COLOR
- B ... Black
- Bl .. Blue
- Br .. Brown
- G ... Green
- R ... Red
- W .. White
- Y ... Yellow
- Y/G . Yellow with Green Tracer
- Bl/R . Blue with Red Tracer
- .... : Optional

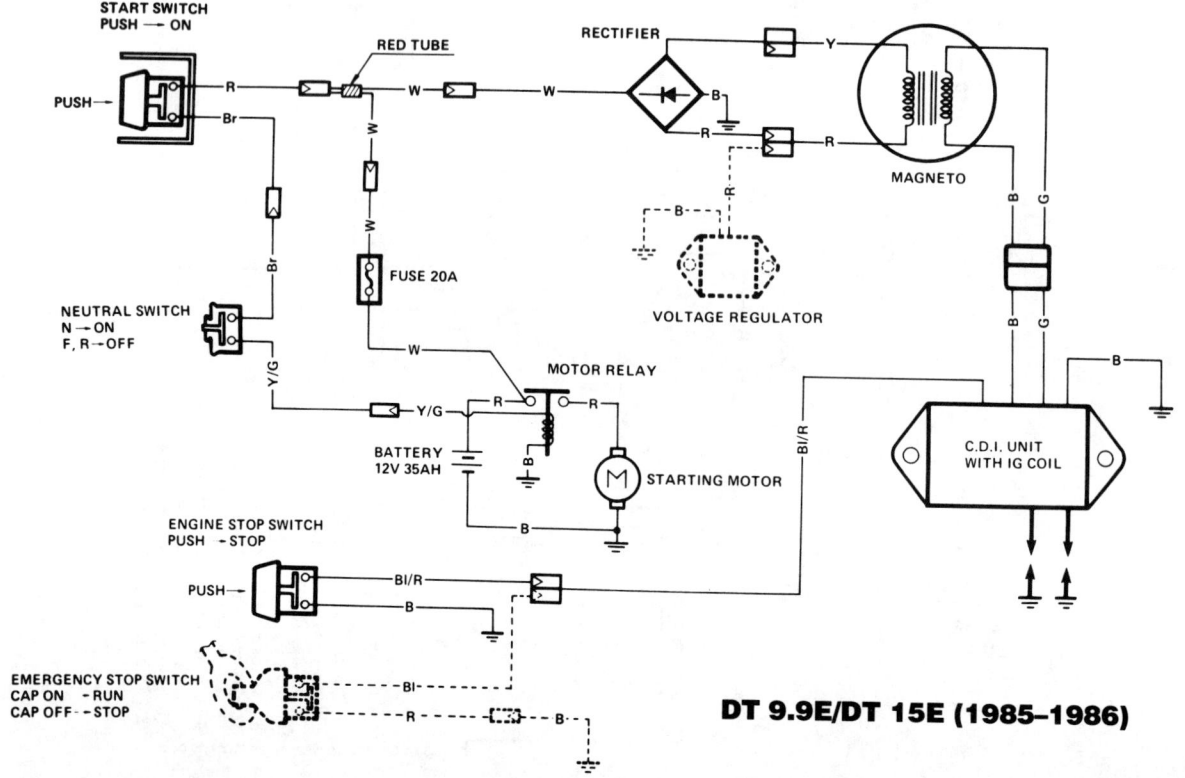

DT 9.9E/DT 15E (1985–1986)

WIRING DIAGRAMS

559

DT 9.9M/DT 15M (1987-1988)

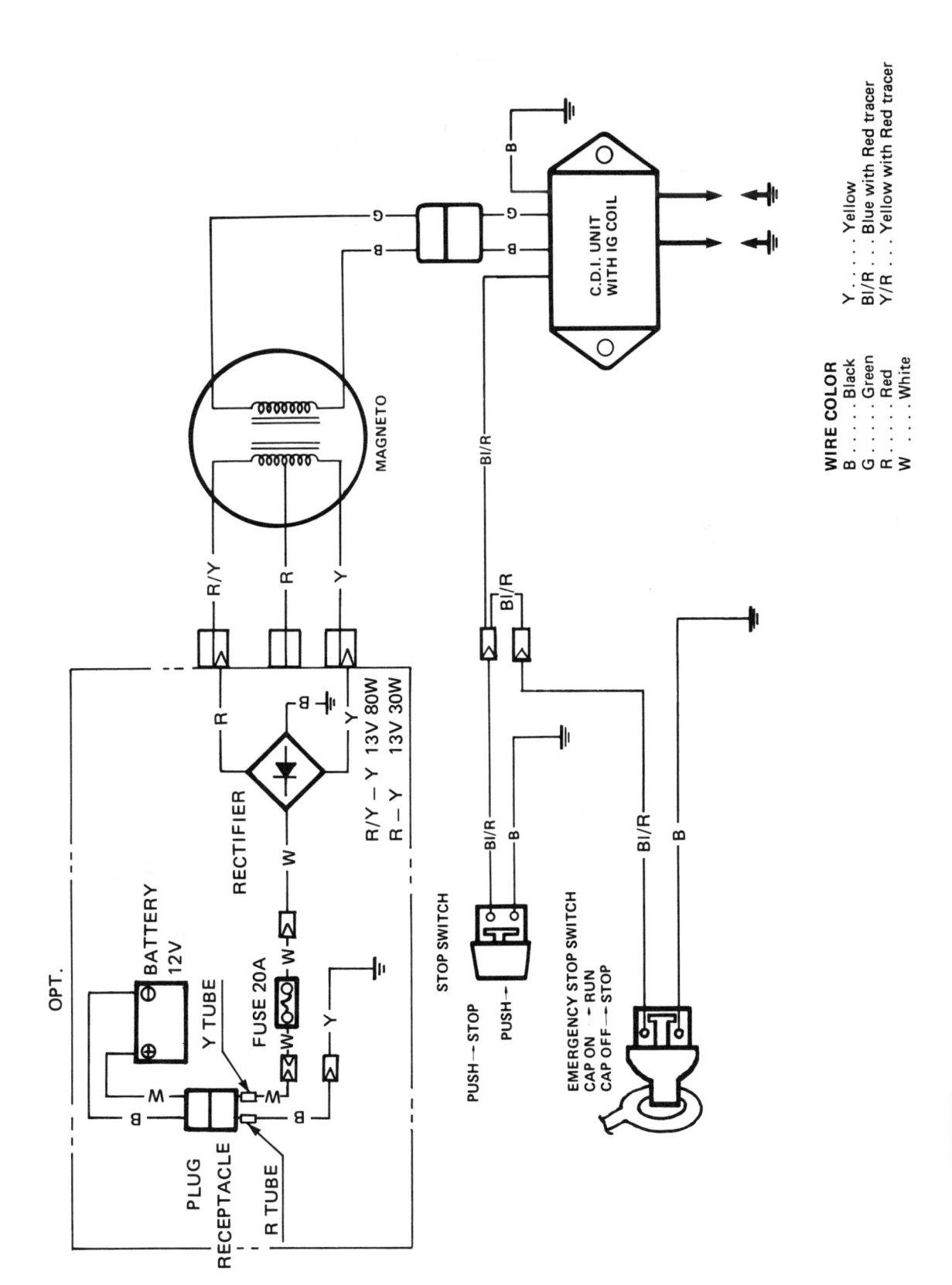

WIRE COLOR
B ..... Black
G ..... Green
R ..... Red
W ..... White

Y ..... Yellow
Bl/R ..... Blue with Red tracer
Y/R ..... Yellow with Red tracer

15

WIRING DIAGRAMS

DT 9.9E/DT 15E (1987-1988)

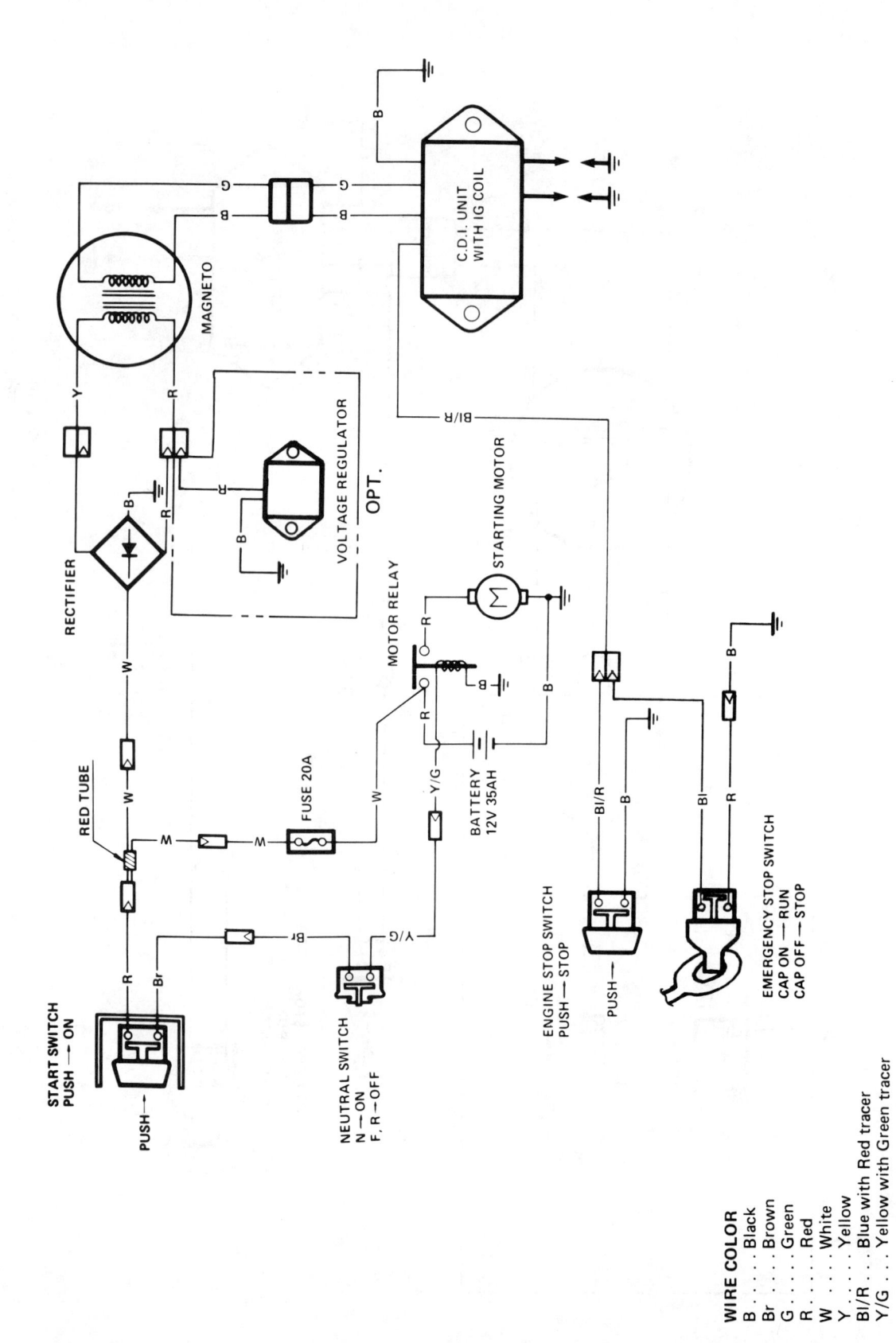

WIRE COLOR
- B .... Black
- Br ... Brown
- G .... Green
- R .... Red
- W .... White
- Y .... Yellow
- Bl/R .. Blue with Red tracer
- Y/G ... Yellow with Green tracer

WIRING DIAGRAMS

DT 9.9E/DT 15E (1987–1988) WITH REMOTE CONTROL

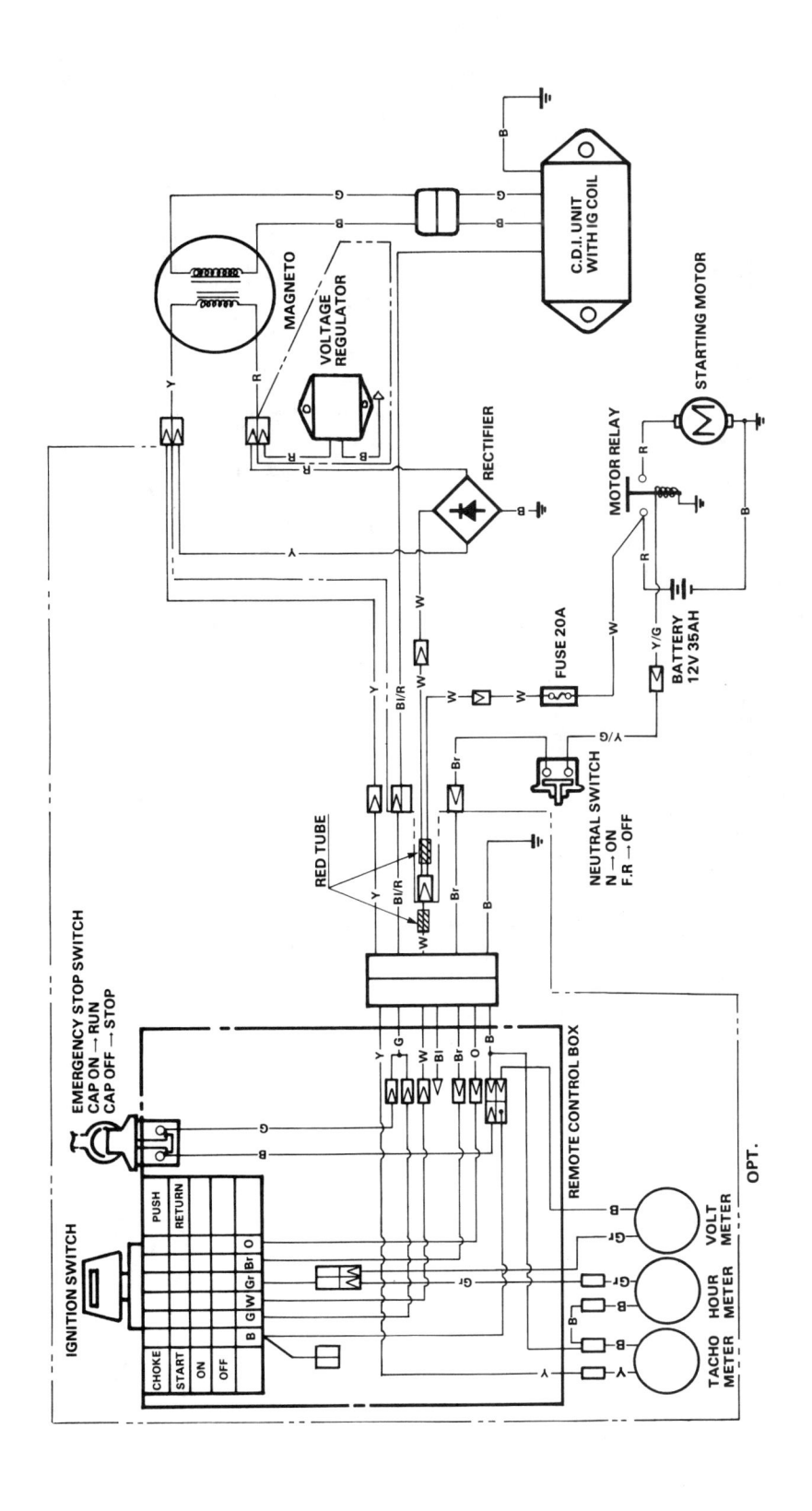

WIRE COLOR

- B .... Black
- Bl .... Blue
- Br .... Brown
- G .... Green
- Gr .... Gray
- O .... Orange
- R .... Red
- W .... White
- Y .... Yellow
- Bl/R .... Blue with Red tracer
- Y/G .... Yellow with Green tracer

WIRING DIAGRAMS

DT 15MC (1989–1991)

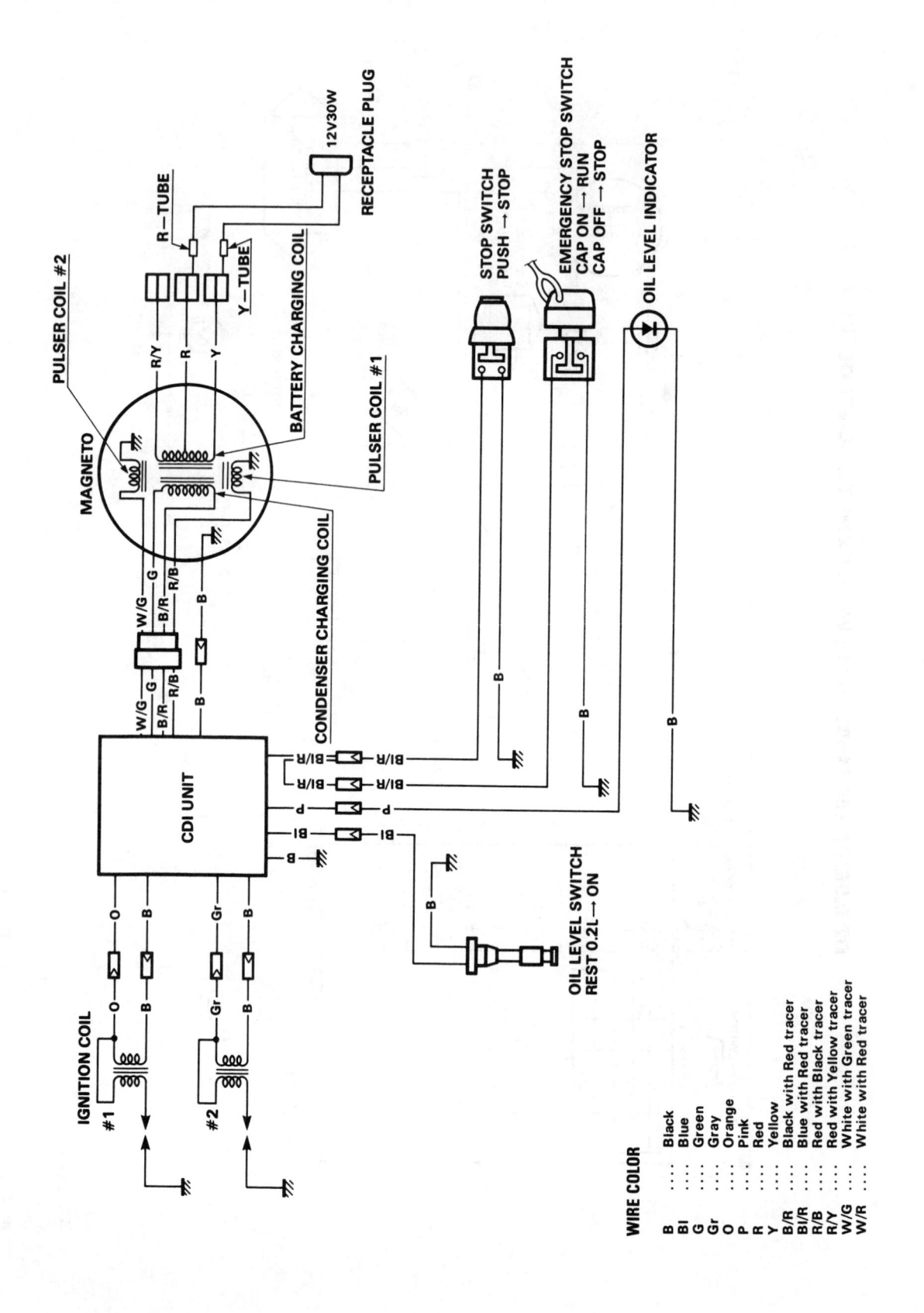

WIRE COLOR

B	:	Black
Bl	:	Blue
G	:	Green
Gr	:	Gray
O	:	Orange
P	:	Pink
R	:	Red
Y	:	Yellow
B/R	:	Black with Red tracer
Bl/R	:	Blue with Red tracer
R/B	:	Red with Black tracer
R/Y	:	Red with Yellow tracer
W/G	:	White with Green tracer
W/R	:	White with Red tracer

WIRING DIAGRAMS

DT 15MC (1989-1991) WITH RECTIFIER AND BATTERY

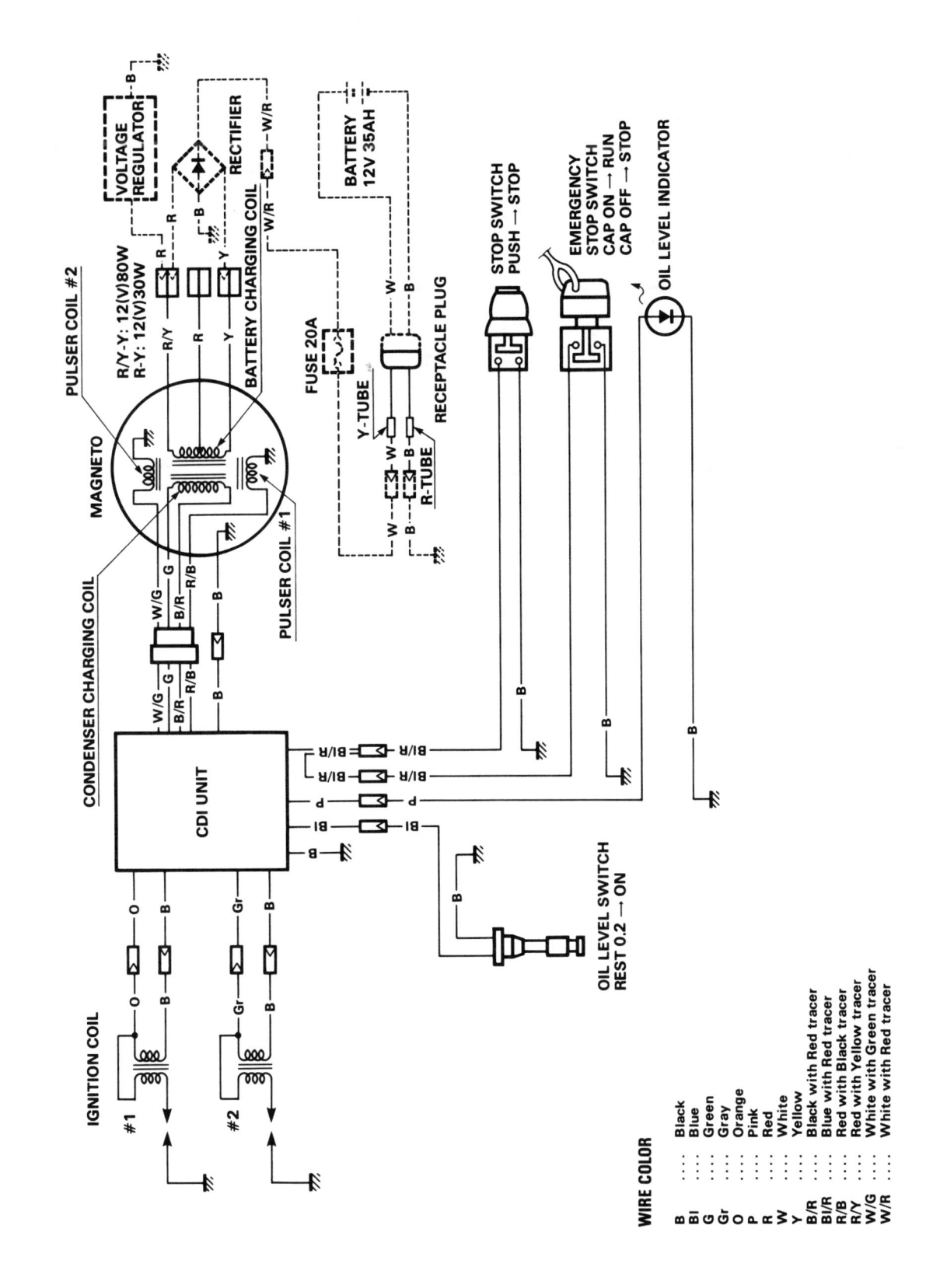

WIRE COLOR

B	.....	Black
Bl	.....	Blue
G	.....	Green
Gr	.....	Gray
O	.....	Orange
P	.....	Pink
R	.....	Red
W	.....	White
Y	.....	Yellow
B/R	.....	Black with Red tracer
Bl/R	.....	Blue with Red tracer
R/B	.....	Red with Black tracer
R/Y	.....	Red with Yellow tracer
W/G	.....	White with Green tracer
W/R	.....	White with Red tracer

15

WIRING DIAGRAMS

DT 15CE (1989–1990)

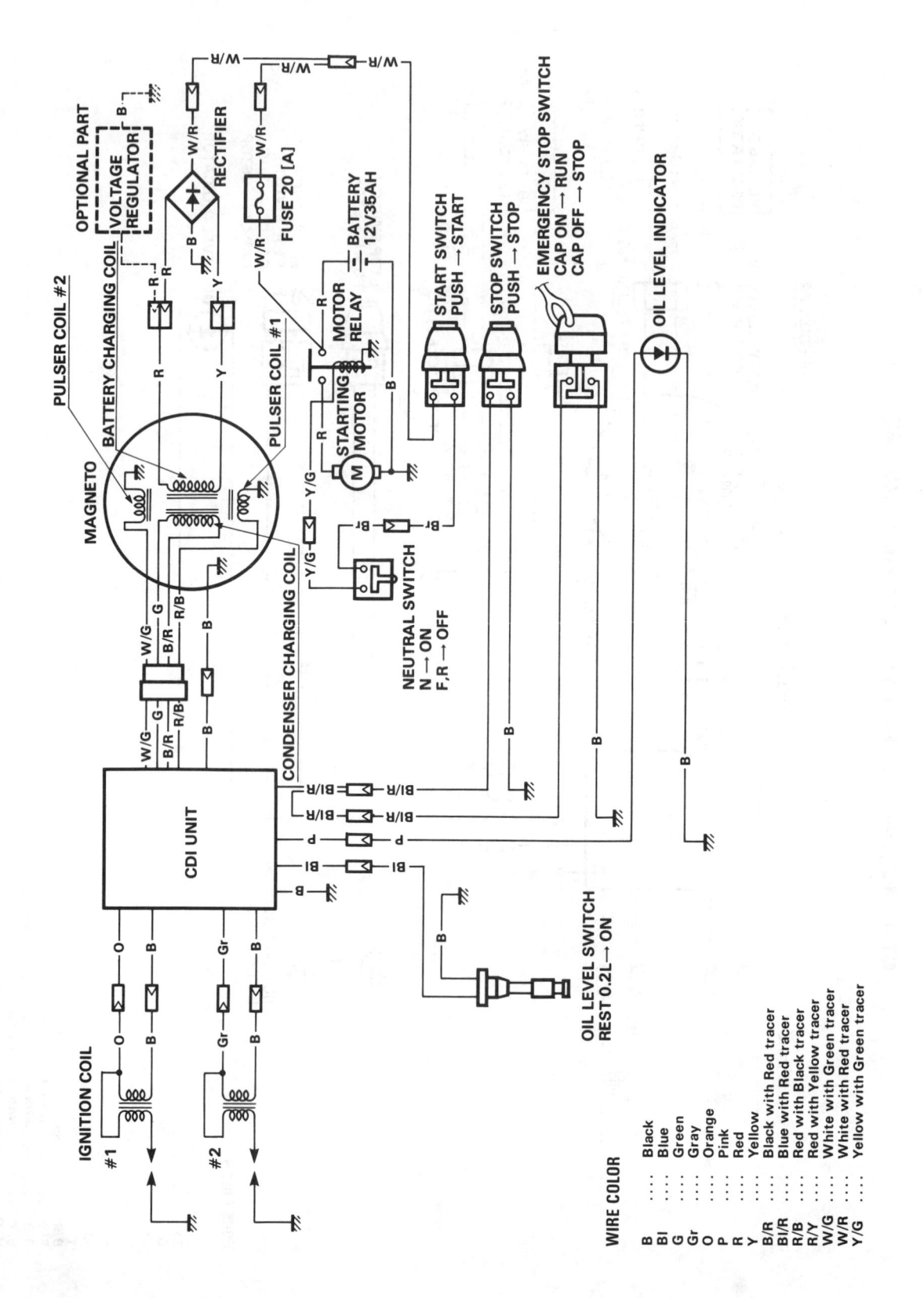

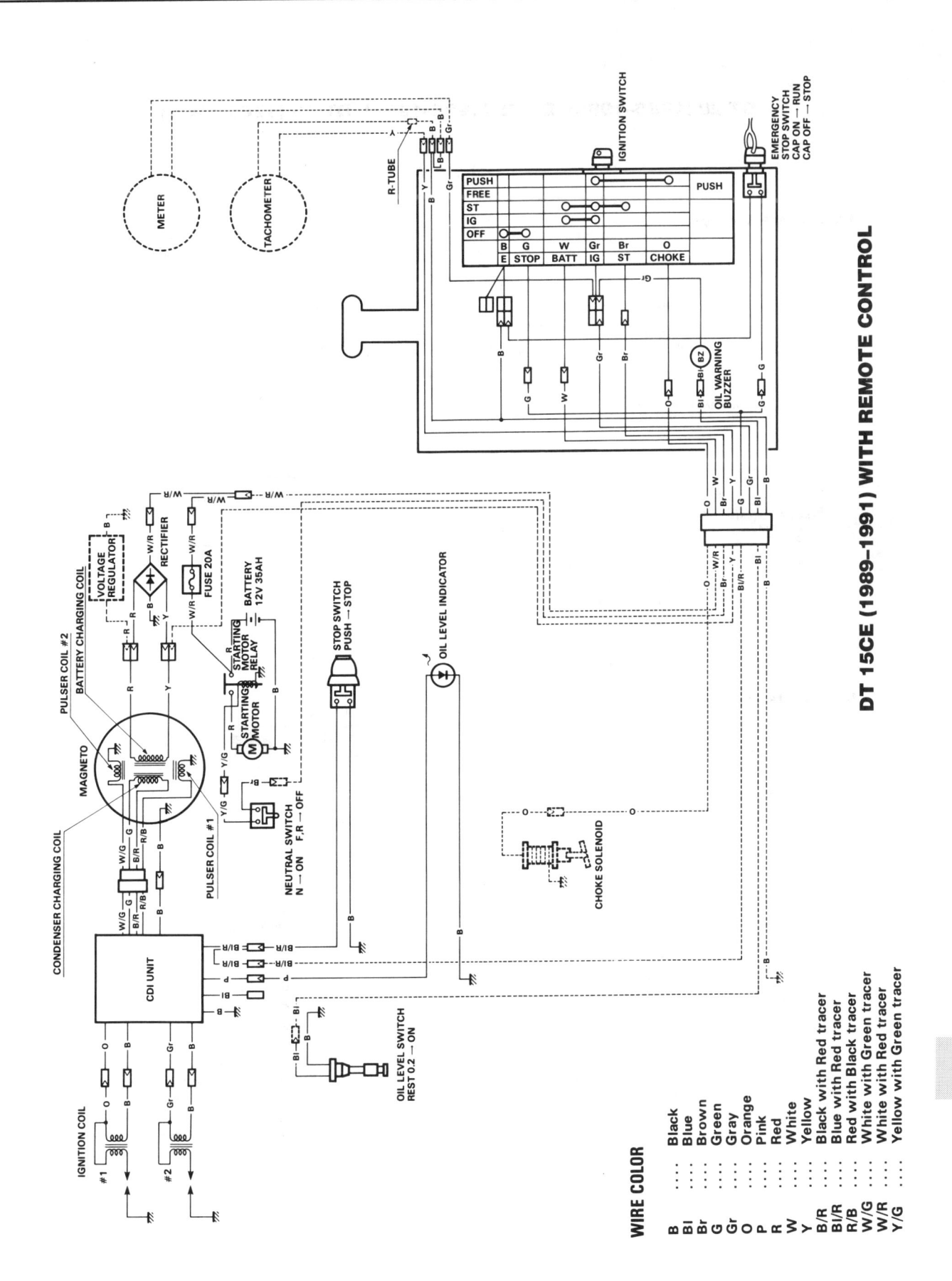

WIRING DIAGRAMS

DT 15CE (1989–1991) WITH REMOTE CONTROL

WIRE COLOR

B	......	Black
Bl	......	Blue
Br	......	Brown
G	......	Green
Gr	......	Gray
O	......	Orange
P	......	Pink
R	......	Red
W	......	White
Y	......	Yellow
B/R	......	Black with Red tracer
Bl/R	......	Blue with Red tracer
R/B	......	Red with Black tracer
W/G	......	White with Green tracer
W/R	......	White with Red tracer
Y/G	......	Yellow with Green tracer

WIRING DIAGRAMS

DT 20 (1986–1988), DT 25 (1985–1988), DT 30 (1985–1988)

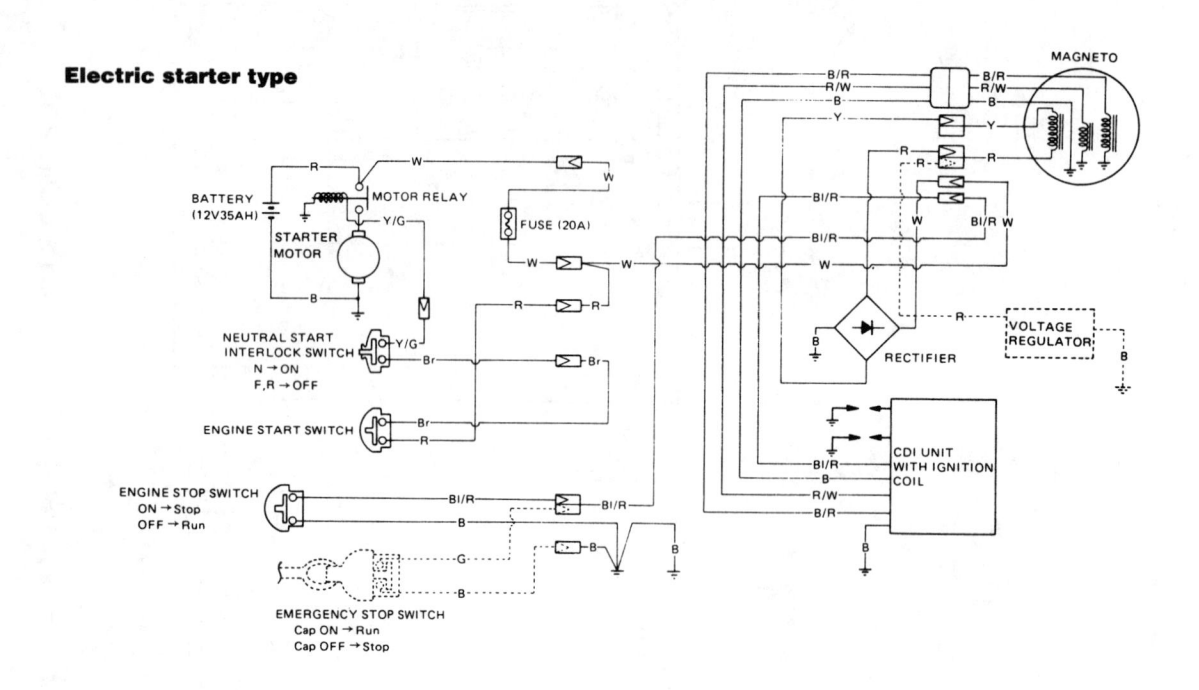

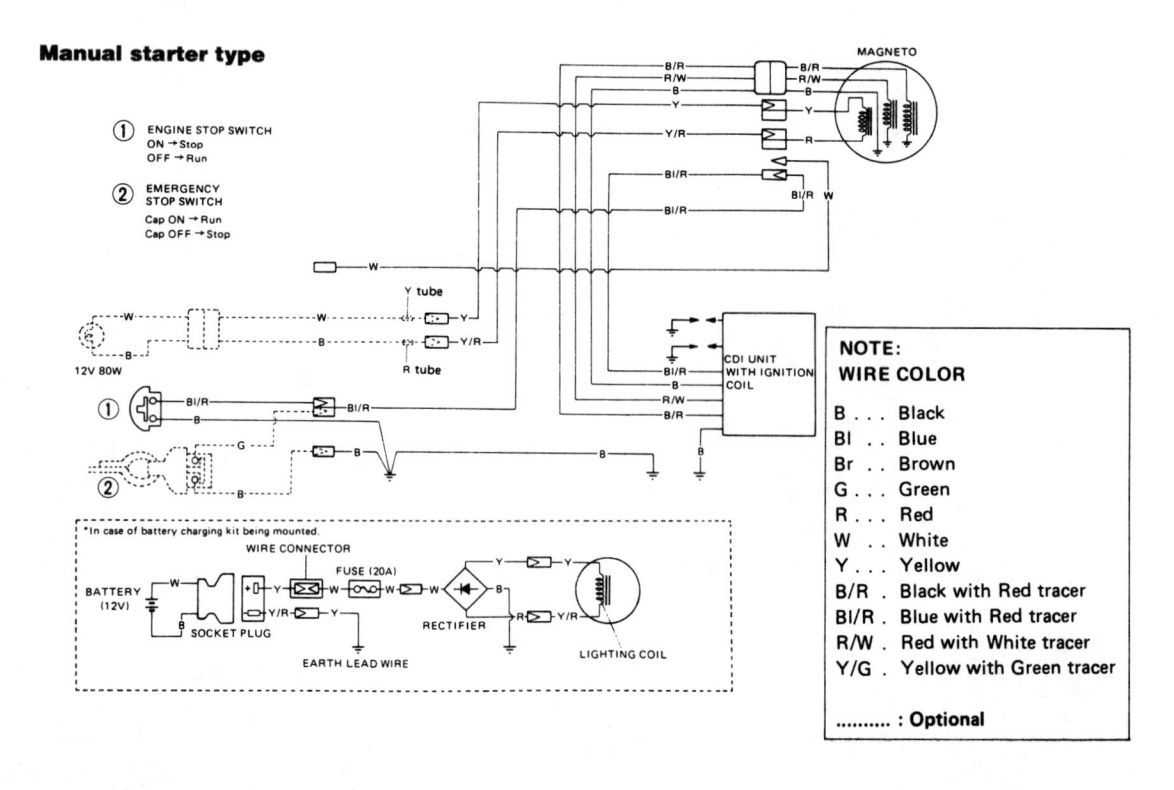

Electric starter type

Manual starter type

NOTE:
WIRE COLOR

- B ... Black
- Bl .. Blue
- Br .. Brown
- G ... Green
- R ... Red
- W .. White
- Y ... Yellow
- B/R . Black with Red tracer
- Bl/R . Blue with Red tracer
- R/W . Red with White tracer
- Y/G . Yellow with Green tracer

.... : Optional

WIRING DIAGRAMS

DT 20M, DT 25M (1986–1987)

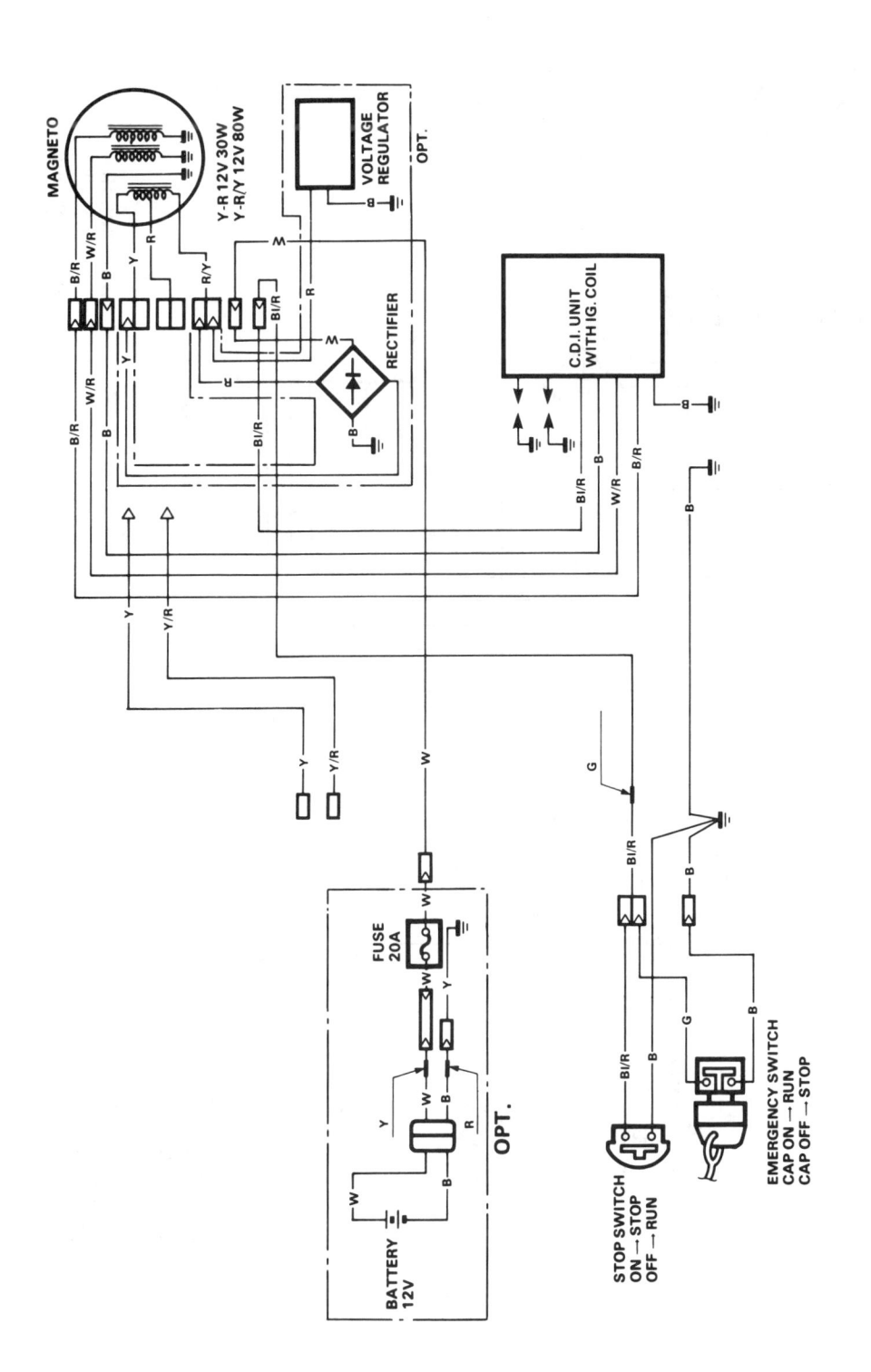

WIRE COLOR
- B ..... Black
- G ..... Green
- R ..... Red
- W ..... White
- Y ..... Yellow

- B/R ..... Black with Red tracer
- Bl/R ..... Blue with Red tracer
- R/Y ..... Red with Yellow tracer
- W/R ..... White with Red tracer
- Y/R ..... Yellow with Red tracer

15

WIRING DIAGRAMS

DT 20E, DT 25E (1986-1987)

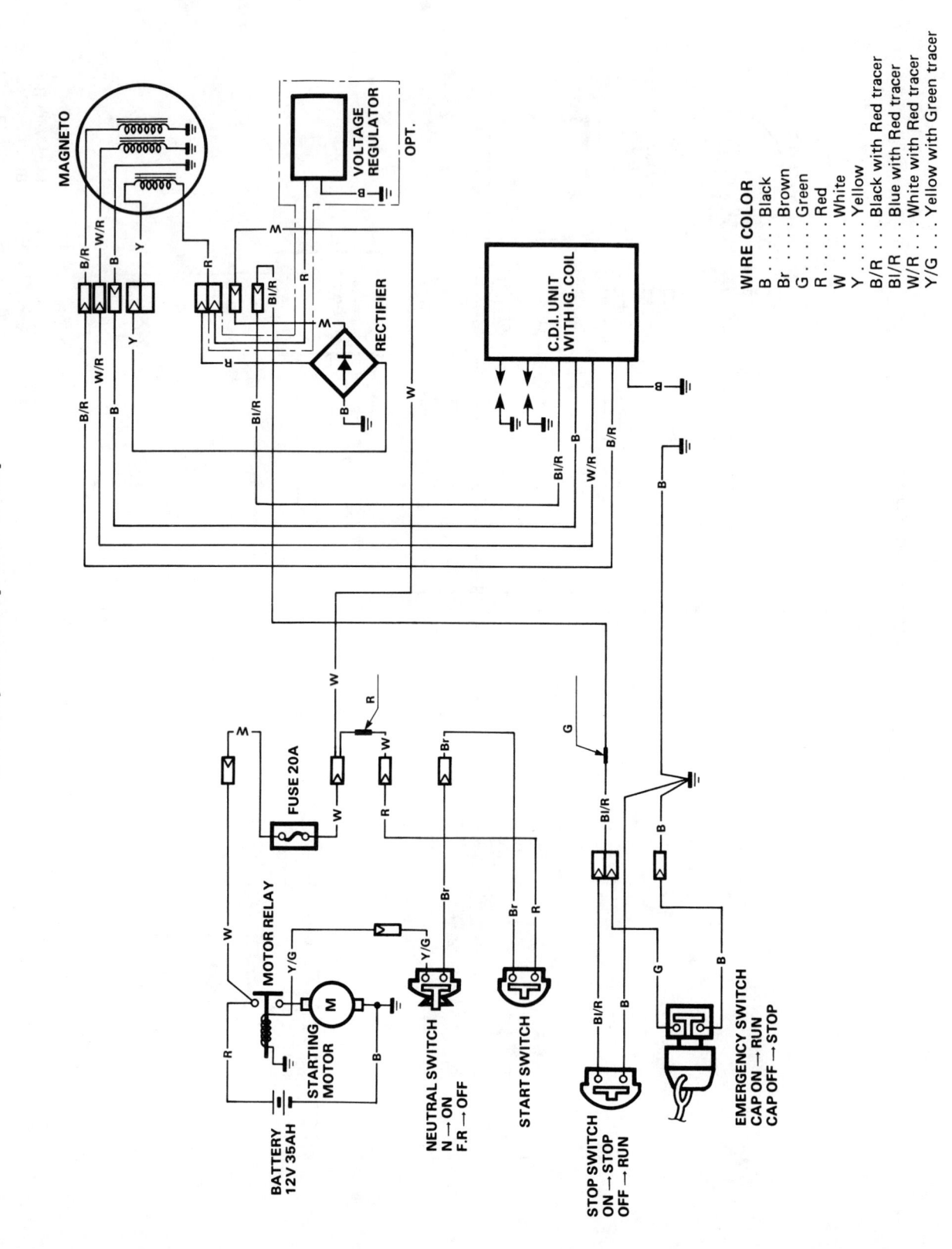

WIRE COLOR
- B . . . . . Black
- Br . . . . Brown
- G . . . . . Green
- R . . . . . Red
- W . . . . White
- Y . . . . . Yellow
- B/R . . . Black with Red tracer
- Bl/R . . . Blue with Red tracer
- W/R . . . White with Red tracer
- Y/G . . . Yellow with Green tracer

WIRING DIAGRAMS

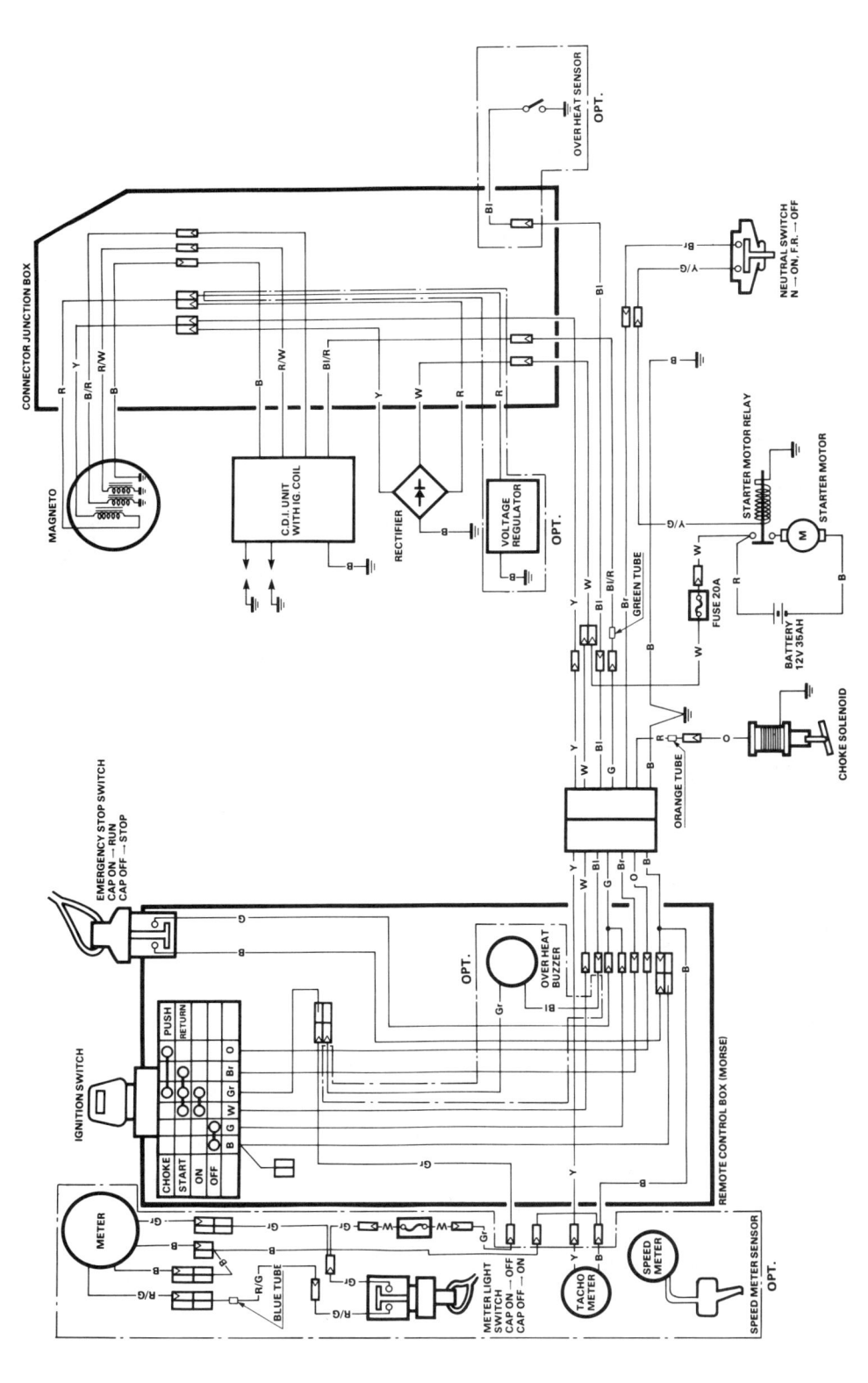

DT 30R (1986–1987)

WIRE COLOR

B	. . . . .	Black
Bl	. . . . .	Blue
Br	. . . . .	Brown
G	. . . . .	Green
Gr	. . . . .	Gray
O	. . . . .	Orange
R	. . . . .	Red
W	. . . . .	White
Y	. . . . .	Yellow
B/R	. . . . .	Black with Red tracer
Bl/R	. . . . .	Blue with Red tracer
R/G	. . . . .	Red with Green tracer
R/W	. . . . .	Red with White tracer
Y/G	. . . . .	Yellow with Green tracer

WIRING DIAGRAMS

DT 25MC, DT 30MC (1988–1989)

WIRE COLOR

B	: Black	Lg/R	: Light green with Red tracer
G	: Green	P/Bl	: Pink with Blue tracer
Gr	: Gray	R/B	: Red with Black tracer
Lbl	: Light blue	R/G	: Red with Green tracer
Lg	: Light green	R/Y	: Red with Yellow tracer
O	: Orange	R/W	: Red with White tracer
P	: Pink	O/G	: Orange with Green tracer
R	: Red	W/B	: White with Black tracer
Y	: Yellow	W/R	: White with Red tracer
W	: White	Y/B	: Yellow with Black tracer
B/W	: Black with White tracer	Y/G	: Yellow with Green tracer
B/R	: Black with Red tracer	V/W	: Violet with White tracer
Bl/R	: Blue with Red tracer		
G/R	: Green with Red tracer		

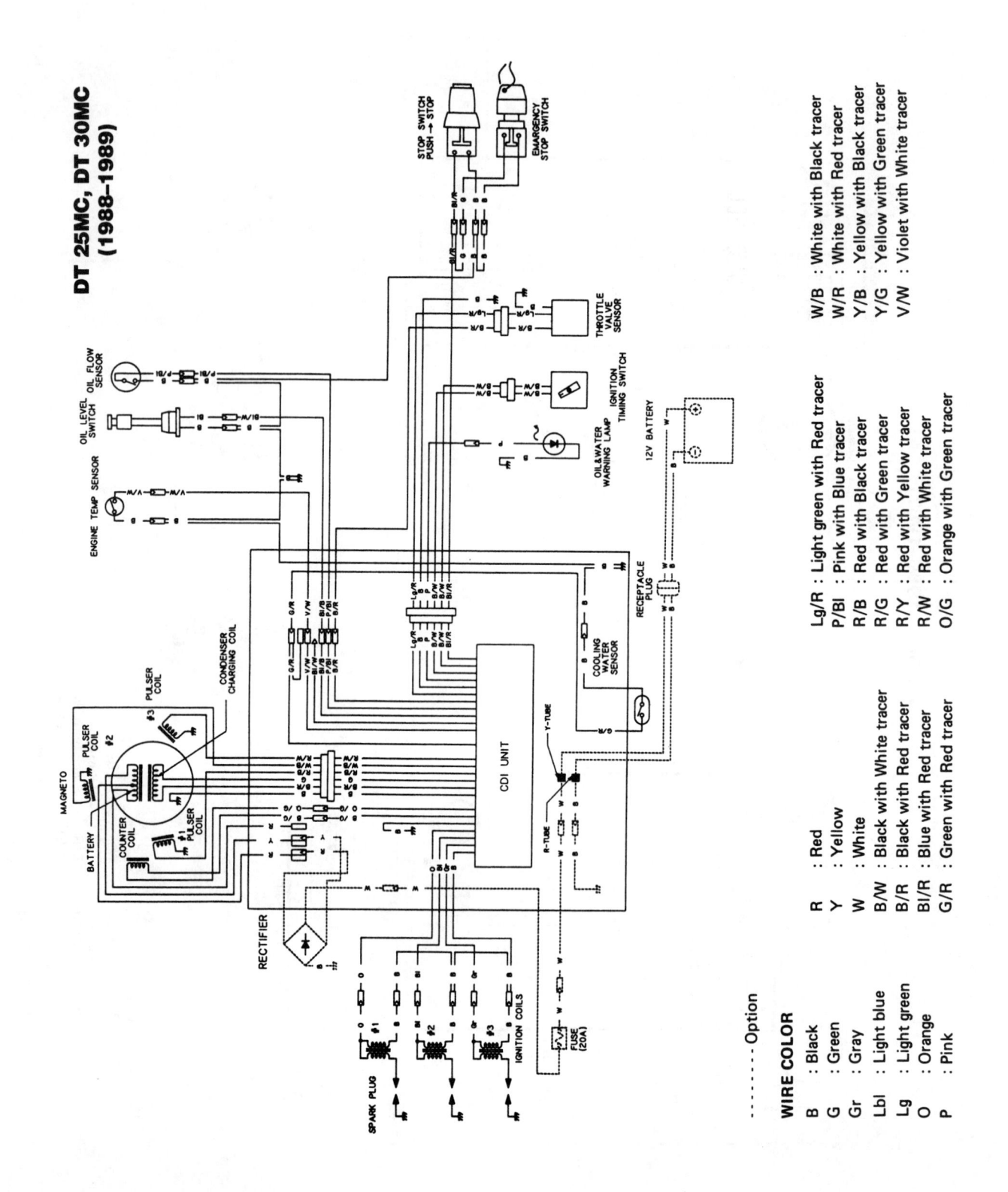

-------- Option

WIRING DIAGRAMS

DT 30R (1988–1989)

Lg/R	: Light green with Red tracer	W/B	: White with Black tracer
P/Bl	: Pink with Blue tracer	W/R	: White with Red tracer
R/B	: Red with Black tracer	Y/B	: Yellow with Black tracer
R/G	: Red with Green tracer	Y/G	: Yellow with Green tracer
R/Y	: Red with Yellow tracer	V/W	: Violet with White tracer
R/W	: Red with White tracer		
O/G	: Orange with Green tracer		

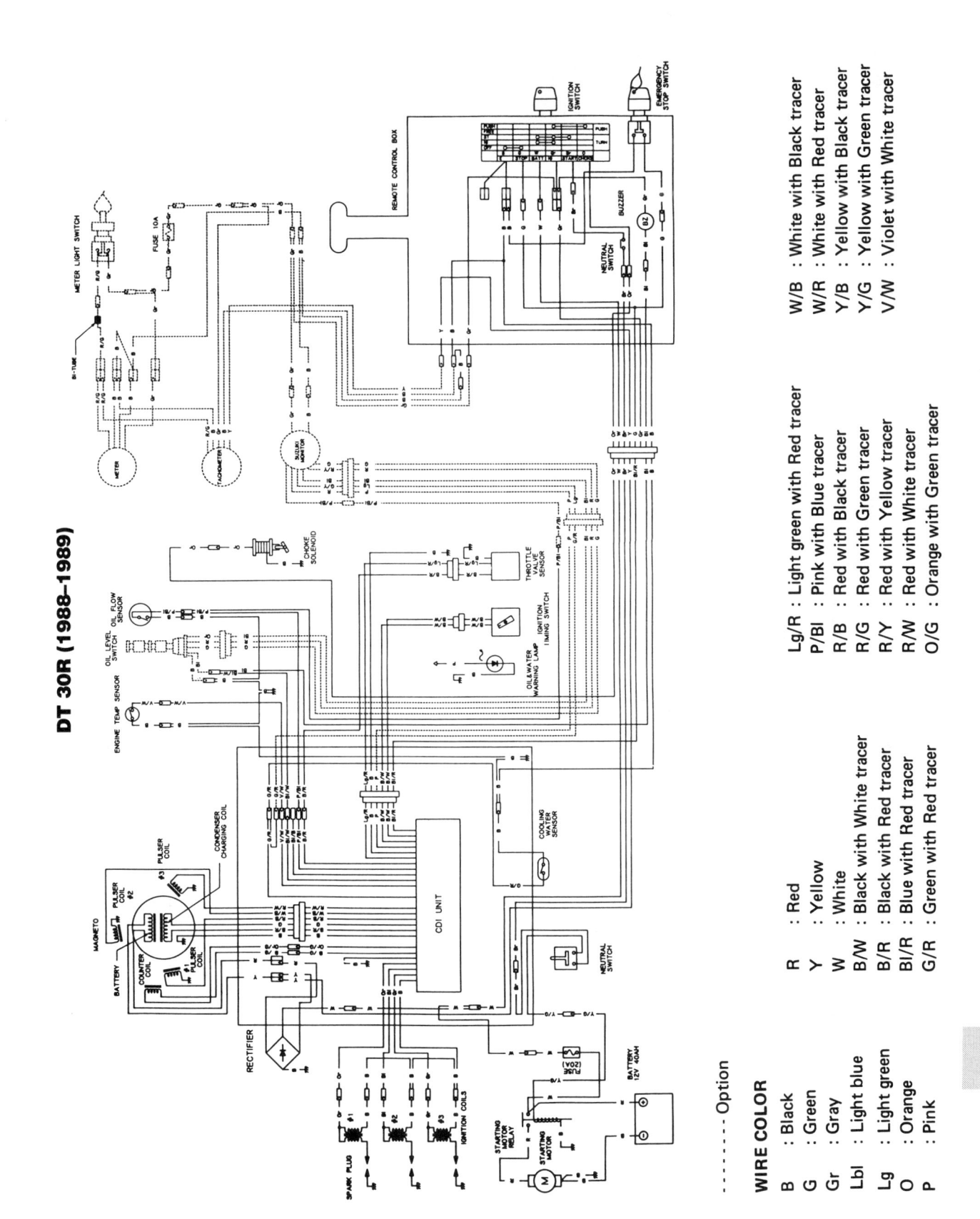

------- Option

WIRE COLOR

B	: Black	R	: Red
G	: Green	Y	: Yellow
Gr	: Gray	W	: White
Lbl	: Light blue	B/W	: Black with White tracer
Lg	: Light green	B/R	: Black with Red tracer
O	: Orange	Bl/R	: Blue with Red tracer
P	: Pink	G/R	: Green with Red tracer

15

WIRING DIAGRAMS

DT 25CE/ DT 30CE (1989)

Lg/R	: Light green with Red tracer	W/B	: White with Black tracer
P/Bl	: Pink with Blue tracer	W/R	: White with Red tracer
R/B	: Red with Black tracer	Y/B	: Yellow with Black tracer
R/G	: Red with Green tracer	Y/G	: Yellow with Green tracer
R/Y	: Red with Yellow tracer	V/W	: Violet with White tracer
R/W	: Red with White tracer		
O/G	: Orange with Green tracer		

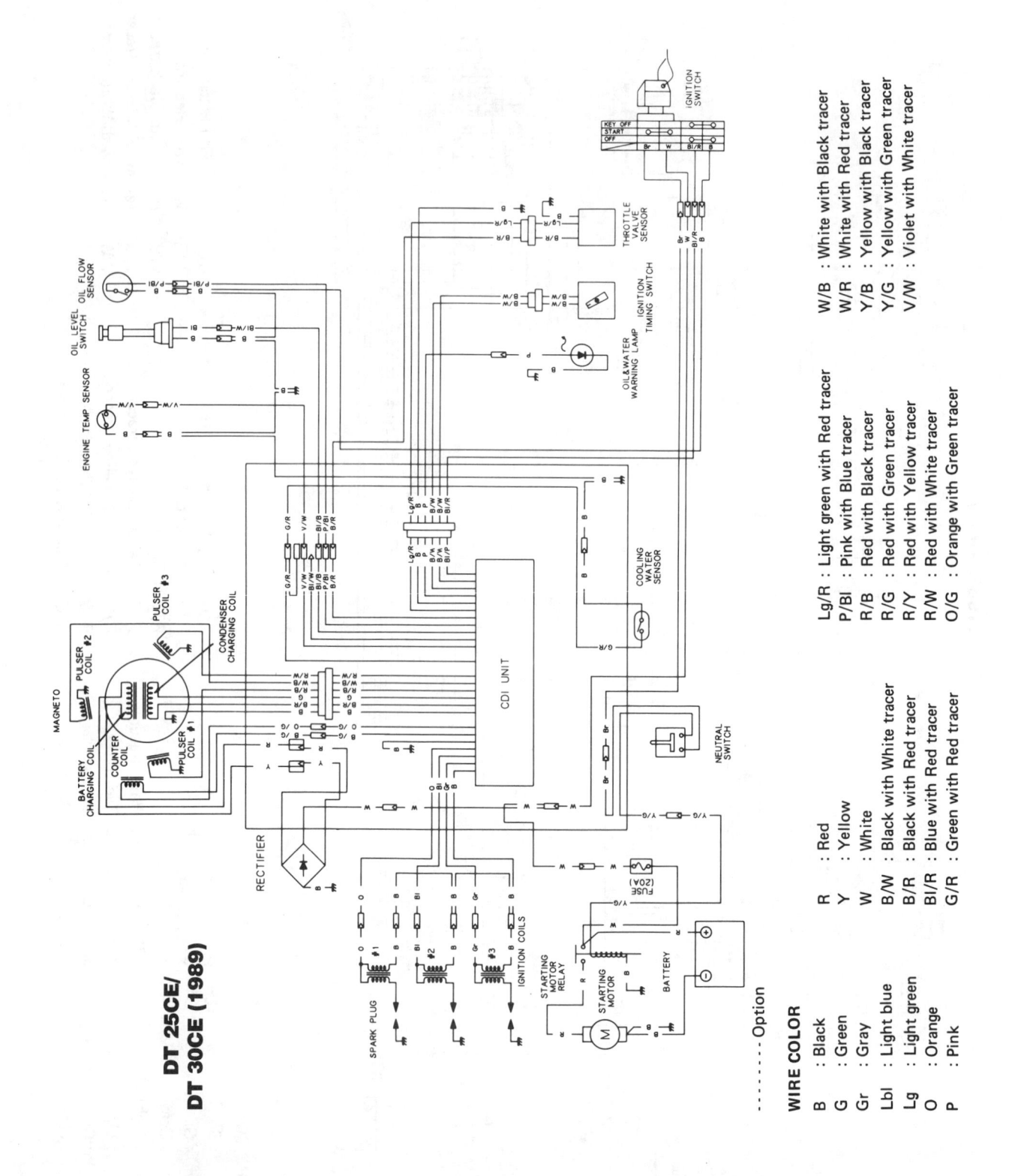

-------- Option

WIRE COLOR

B	: Black	R	: Red
G	: Green	Y	: Yellow
Gr	: Gray	W	: White
Lbl	: Light blue	B/W	: Black with White tracer
Lg	: Light green	B/R	: Black with Red tracer
O	: Orange	Bl/R	: Blue with Red tracer
P	: Pink	G/R	: Green with Red tracer

WIRING DIAGRAMS

DT 25CE / DT 30CE (1991)

Lg/R	: Light green with Red tracer	W/B	: White with Black tracer
P/Bl	: Pink with Blue tracer	W/R	: White with Red tracer
R/B	: Red with Black tracer	Y/B	: Yellow with Black tracer
R/G	: Red with Green tracer	Y/G	: Yellow with Green tracer
R/Y	: Red with Yellow tracer	V/W	: Violet with White tracer
R/W	: Red with White tracer		
O/G	: Orange with Green tracer		

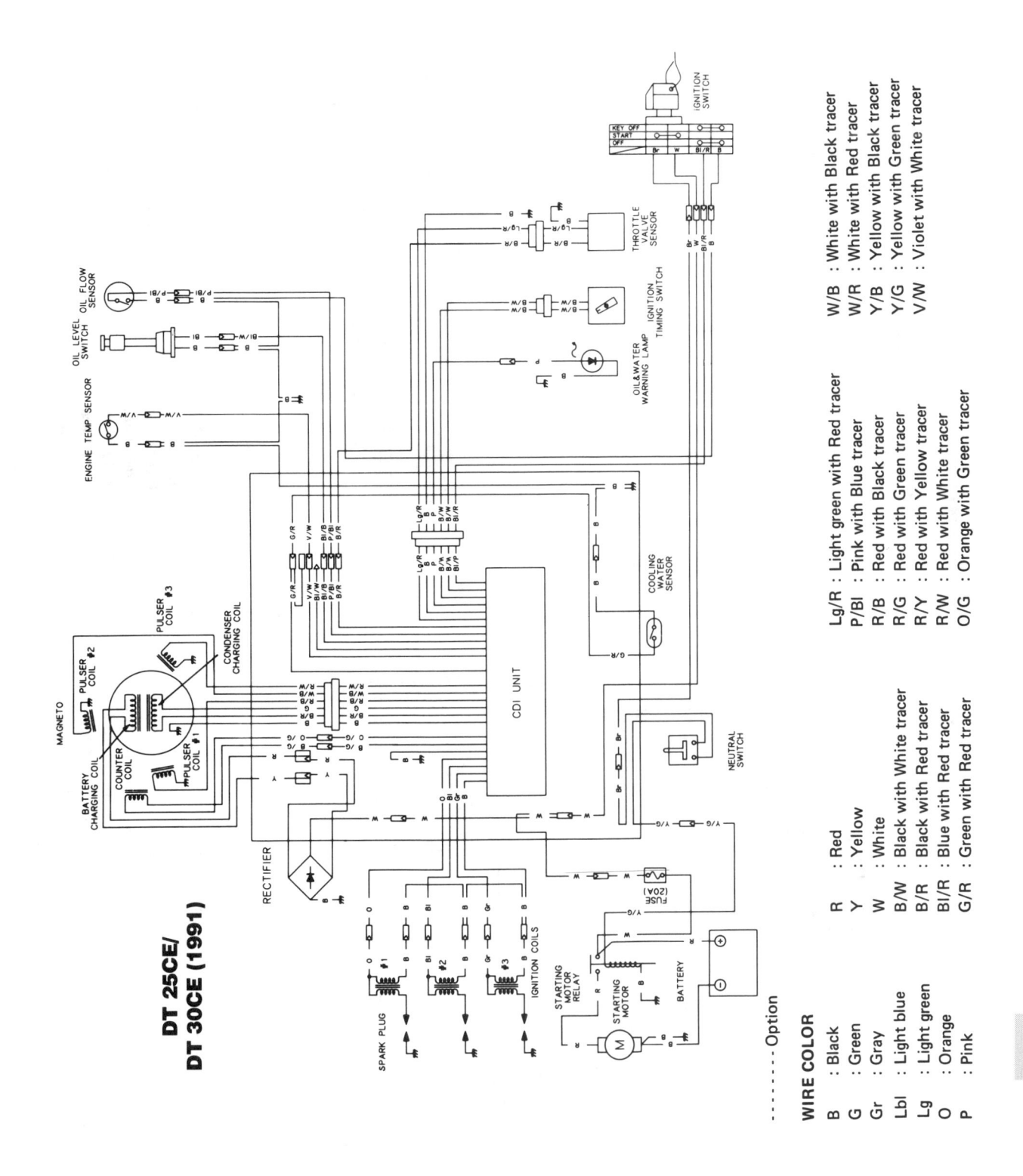

- - - - - - - - Option

WIRE COLOR

B	: Black	R	: Red	B/W	: Black with White tracer
G	: Green	Y	: Yellow	B/R	: Black with Red tracer
Gr	: Gray	W	: White	Bl/R	: Blue with Red tracer
Lbl	: Light blue			G/R	: Green with Red tracer
Lg	: Light green				
O	: Orange				
P	: Pink				

WIRING DIAGRAMS

DT 25CR/DT 30CR (1991)

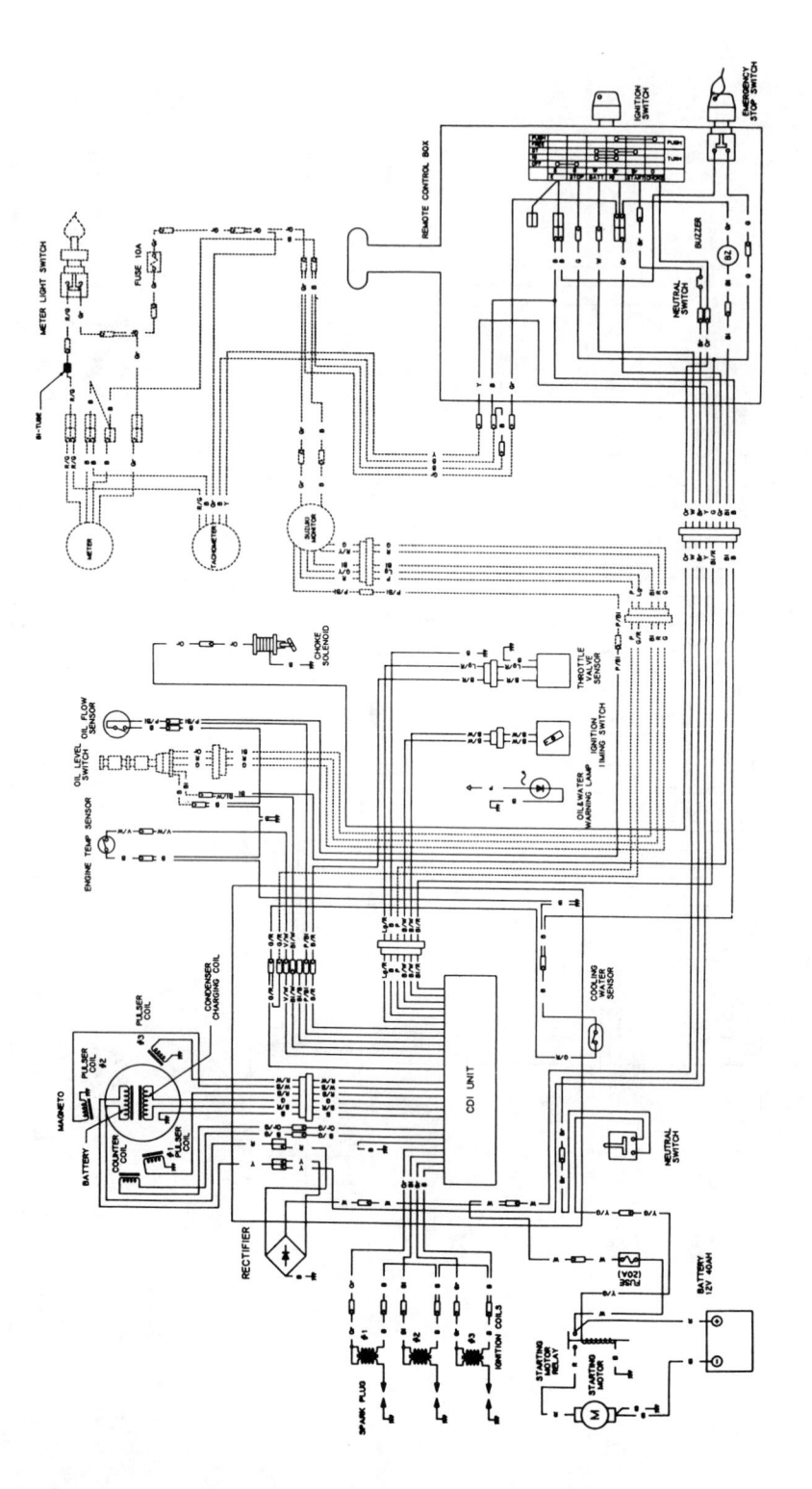

------- Option

WIRE COLOR

B	: Black	R	: Red
G	: Green	Y	: Yellow
Gr	: Gray	W	: White
Lbl	: Light blue	B/W	: Black with White tracer
Lg	: Light green	B/R	: Black with Red tracer
O	: Orange	Bl/R	: Blue with Red tracer
P	: Pink	G/R	: Green with Red tracer

Lg/R	: Light green with Red tracer
P/Bl	: Pink with Blue tracer
R/B	: Red with Black tracer
R/G	: Red with Green tracer
R/Y	: Red with Yellow tracer
R/W	: Red with White tracer
O/G	: Orange with Green tracer
W/B	: White with Black tracer
W/R	: White with Red tracer
Y/B	: Yellow with Black tracer
Y/G	: Yellow with Green tracer
V/W	: Violet with White tracer

Wiring Diagrams

DT 35TC (1987-1989)

WIRE COLOR

B	Black	Lbl	Light blue
Bl	Blue	O	Orange
Br	Brown	P	Pink
G	Gray	R	Red
Gr	Green	W	White
Y	Yellow		
Bl/R	Blue with Red tracer		
R/Y	Red with Yellow tracer		
R/G	Red with Green tracer		
W/R	White with Red tracer		
Y/G	Yellow with Green tracer		

- - - - : From the serial number 716025 does not have yellow (Y) lead wire.

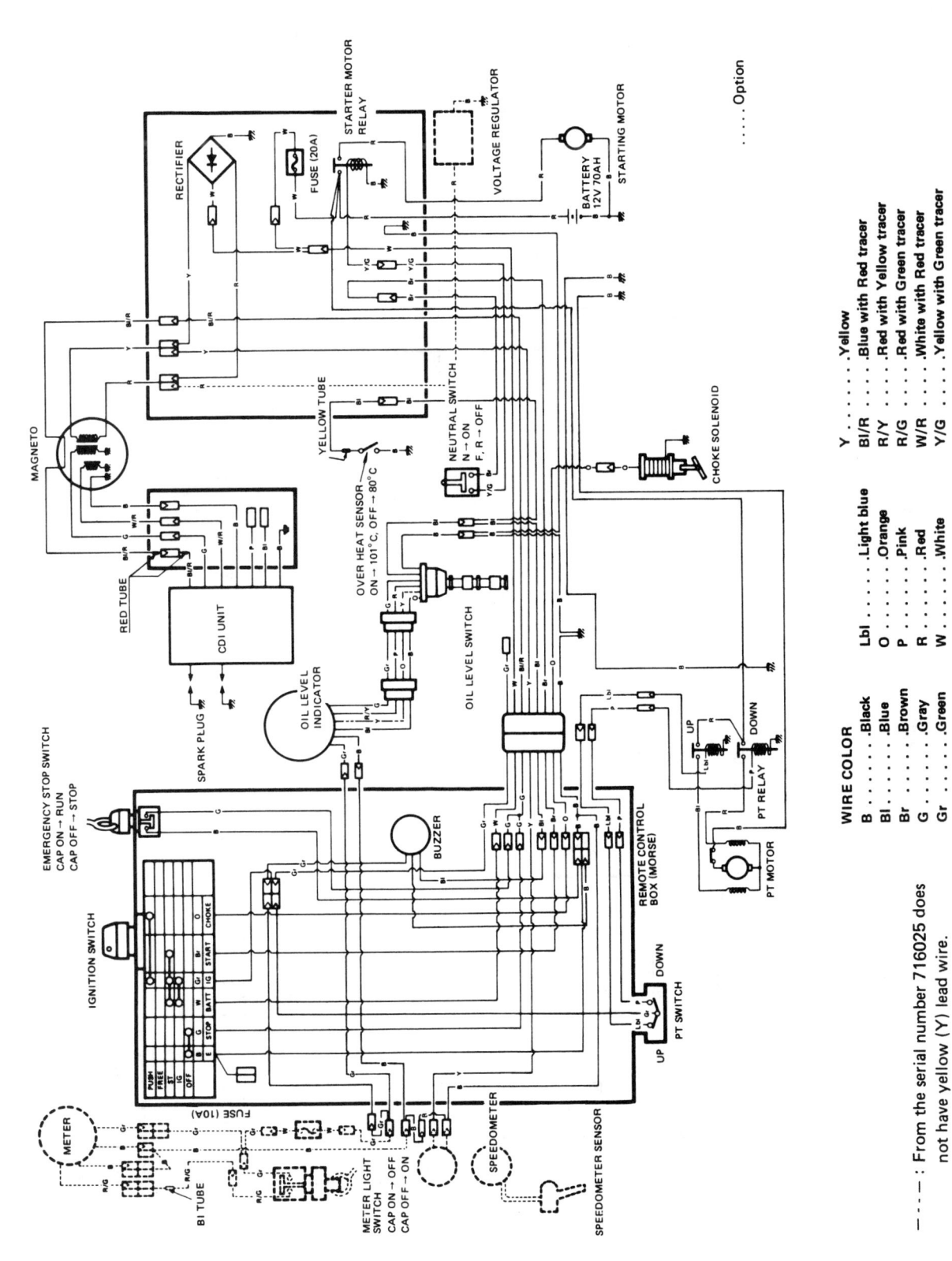

WIRING DIAGRAMS

DT 35CR (1987–1989)

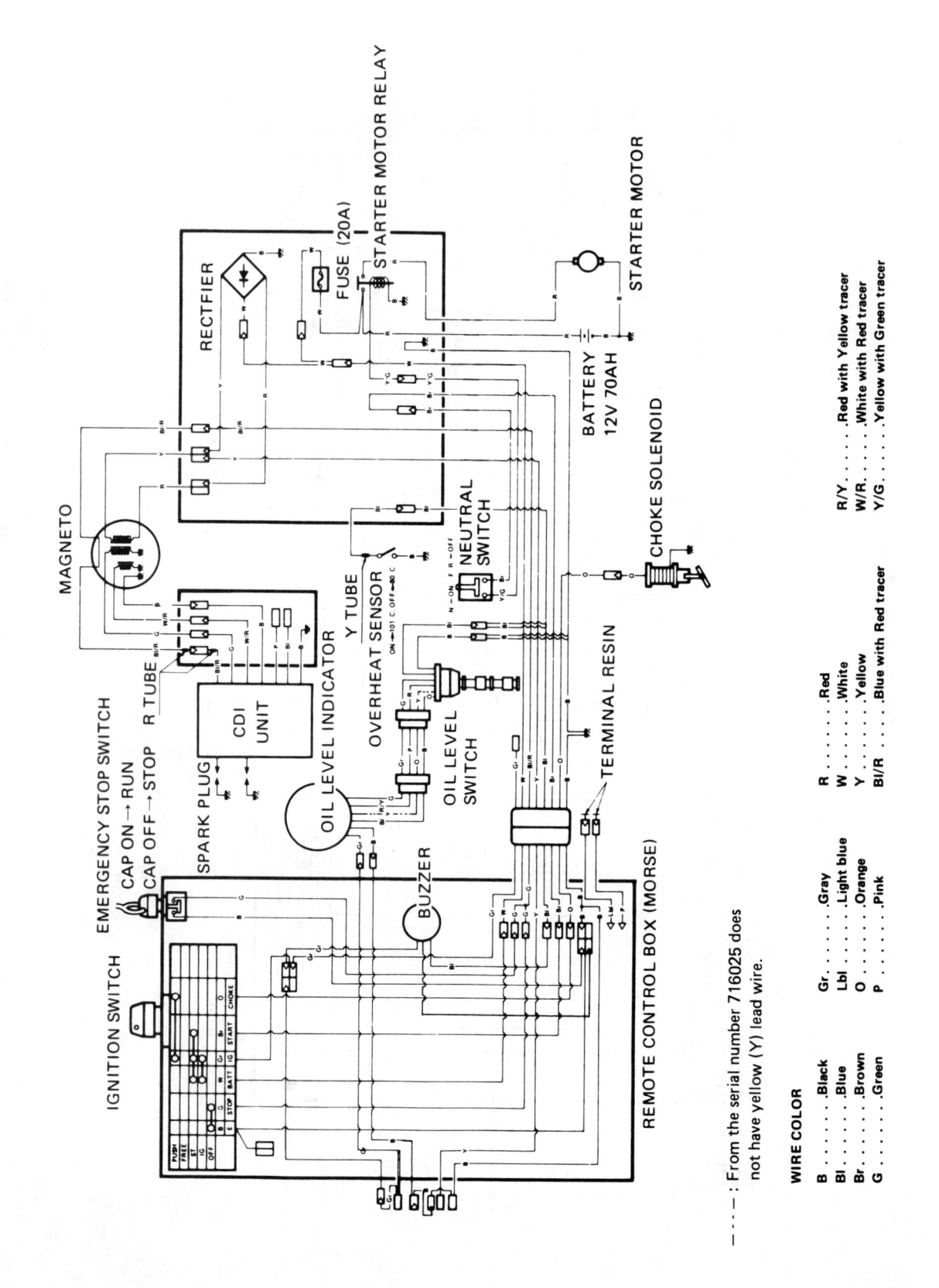

Full-page wiring diagram showing: Magneto, Rectifier, Fuse (20A), Starter Motor Relay, Starter Motor, Battery 12V 70AH, Choke Solenoid, Neutral Switch, Overheat Sensor, Y Tube, R Tube, CDI Unit, Spark Plug, Emergency Stop Switch (CAP ON → RUN, CAP OFF → STOP), Oil Level Indicator, Oil Level Switch, Buzzer, Terminal Resin, Ignition Switch, Remote Control Box (Morse).

- - - - : From the serial number 716025 does not have yellow (Y) lead wire.

WIRE COLOR

B	Black	Gr	Gray	R	Red	R/Y	Red with Yellow tracer
Bl	Blue	Lbl	Light blue	W	White	W/R	White with Red tracer
Br	Brown	O	Orange	Y	Yellow	Y/G	Yellow with Green tracer
G	Green	P	Pink	Bl/R	Blue with Red tracer		

WIRING DIAGRAMS

DT 35MC (1987-1989), DT 40MC (1991)

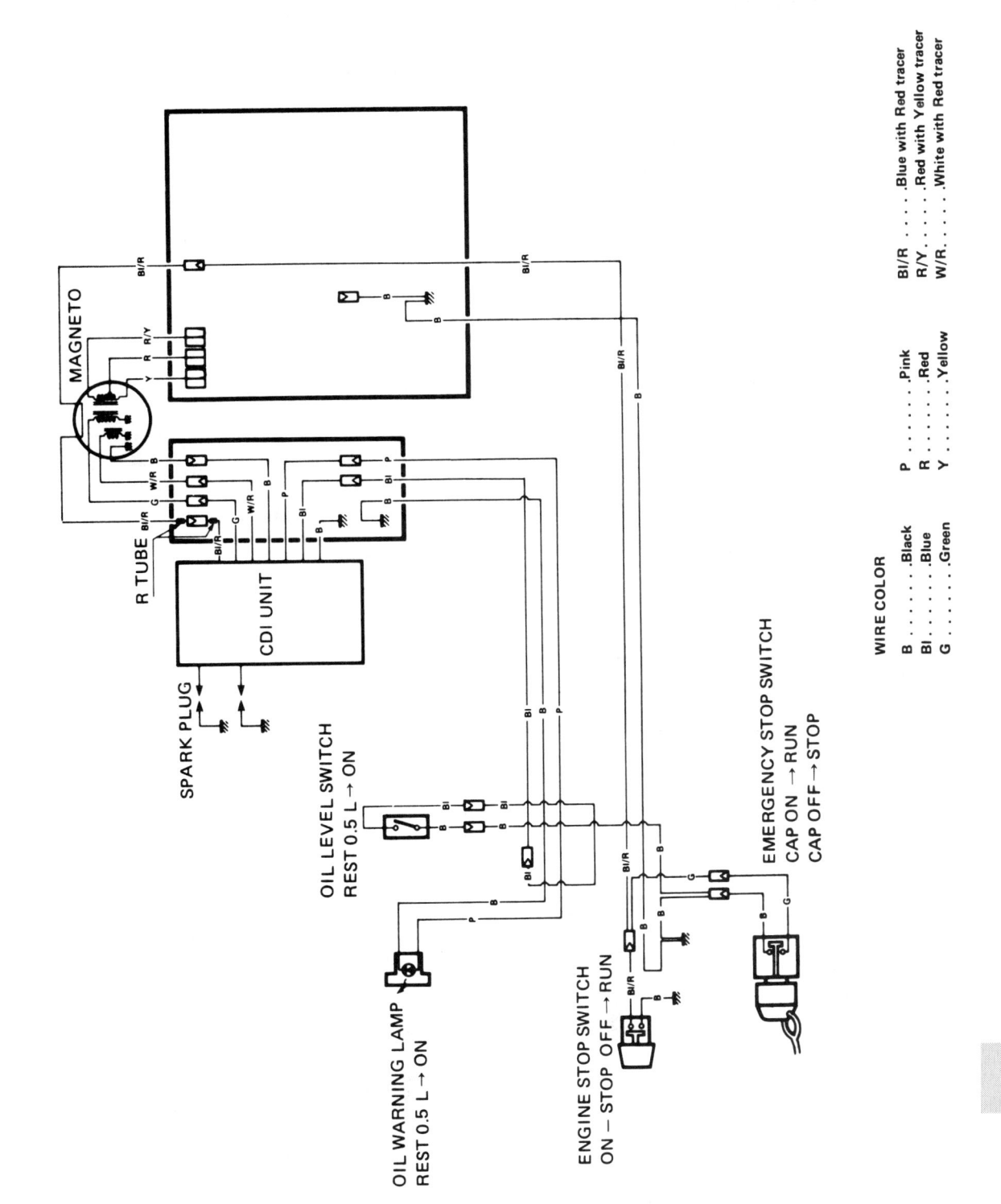

WIRE COLOR

B	. . . . Black	P	. . . . Pink	Bl/R . . . . . Blue with Red tracer
Bl	. . . . Blue	R	. . . . Red	R/Y . . . . . Red with Yellow tracer
G	. . . . Green	Y	. . . . Yellow	W/R . . . . . White with Red tracer

Wiring Diagrams

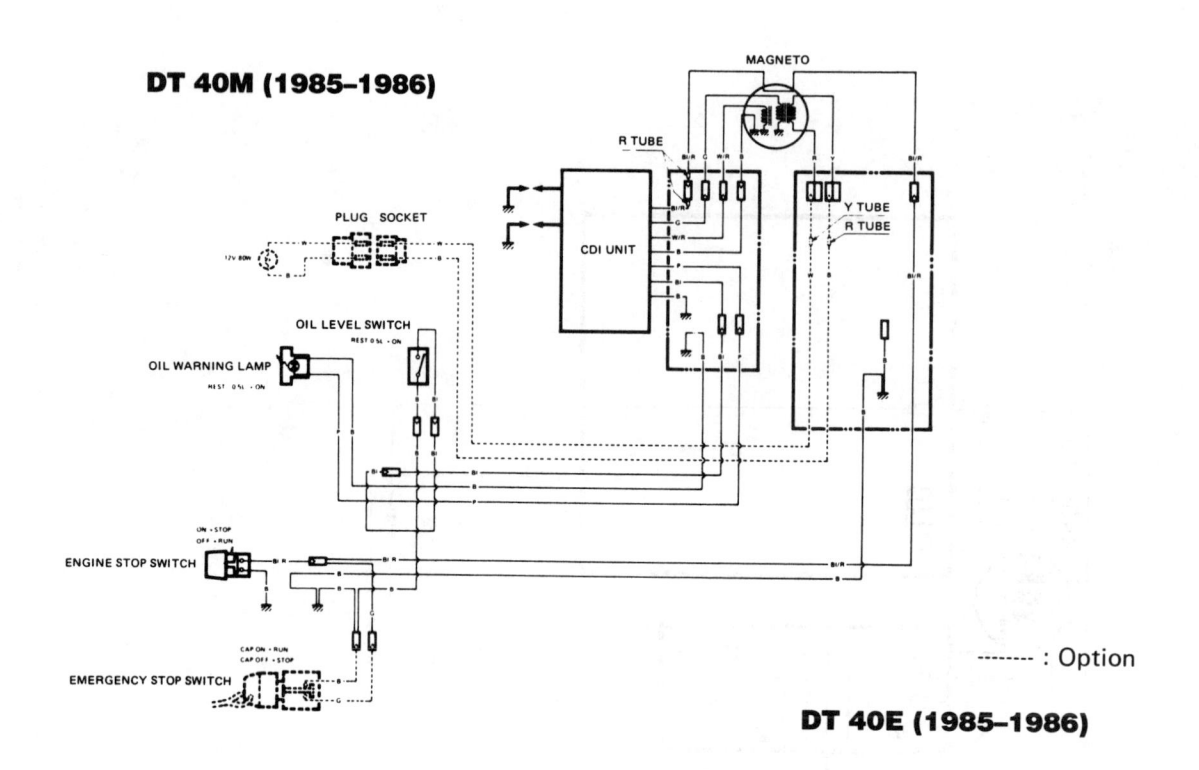

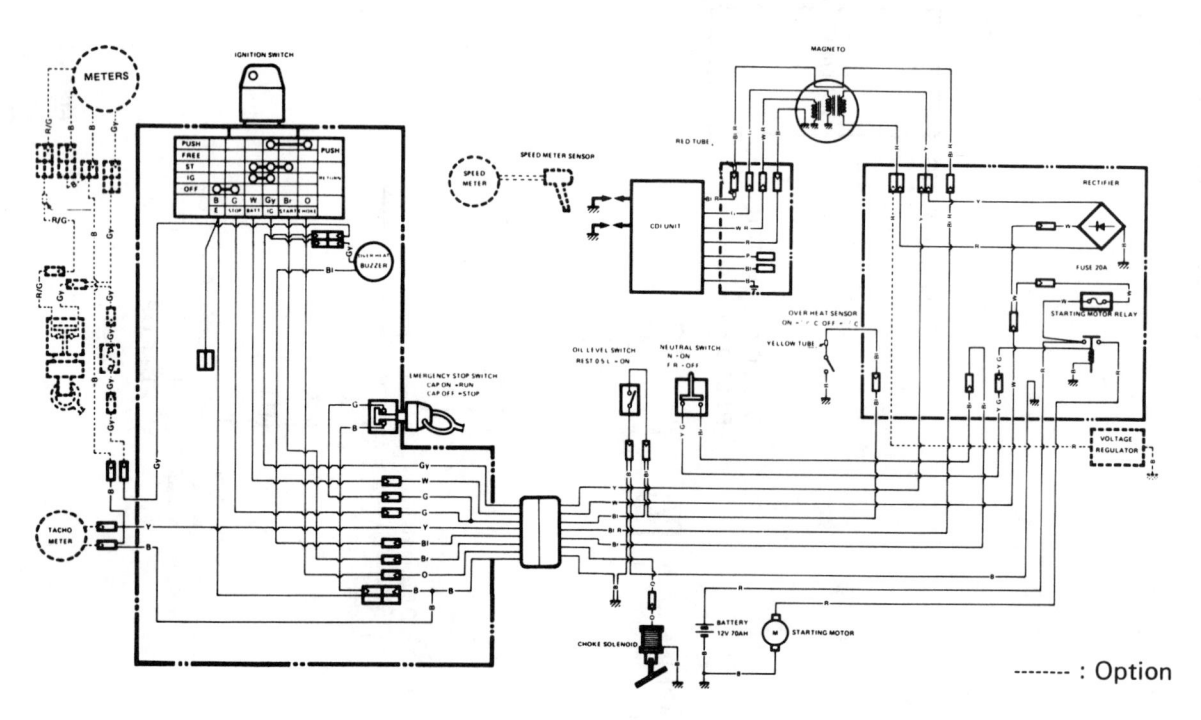

DT 40M (1985–1986)

------ : Option

DT 40E (1985–1986)

------ : Option

WIRE COLOR

B . . Black	O . . Orange	W . . White	Br . . Brown	W/R . . White with Red tracer
G . . Green	P . . Pink	Y . . Yellow	G/Y . . Green with Yellow tracer	Bl/R . . Blue with Red tracer
Gy . . Grey	R . . Red	Bl . . Blue	O/R . . Orange with Red tracer	Y/G . . Yellow with Green tracer

WIRING DIAGRAMS

DT 40CE (1990-1991)

---- OPTION

WIRE COLOR

- B ..... Black
- Bl ..... Blue
- Br ..... Brown
- G ..... Green
- P ..... Pink
- R ..... Red
- Y ..... Yellow

- W ..... White
- Bl/R ..... Blue with Red tracer
- P/Bl ..... Pink with Blue tracer
- W/R ..... White with Red tracer
- Y/G ..... Yellow with Green tracer

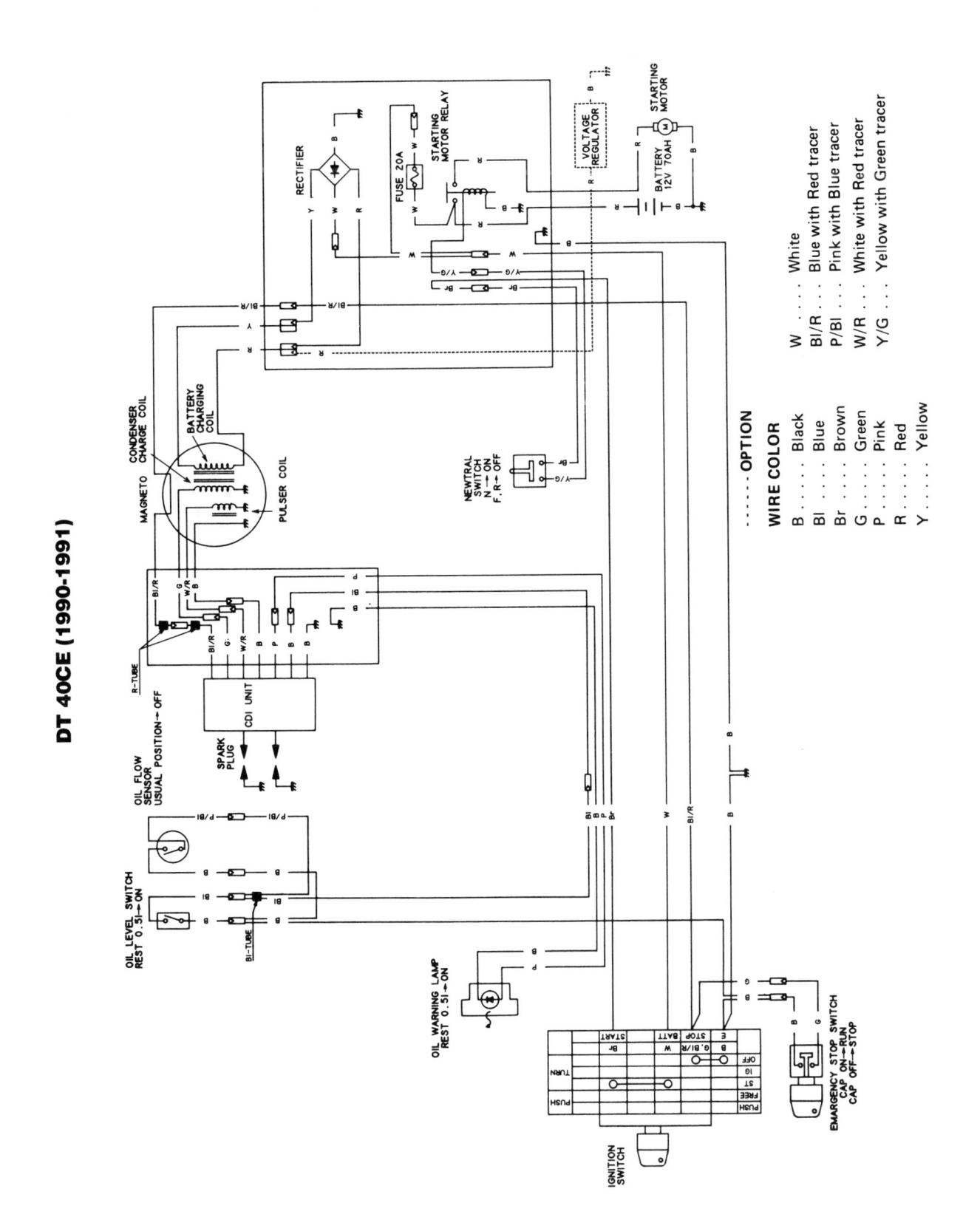

WIRING DIAGRAMS

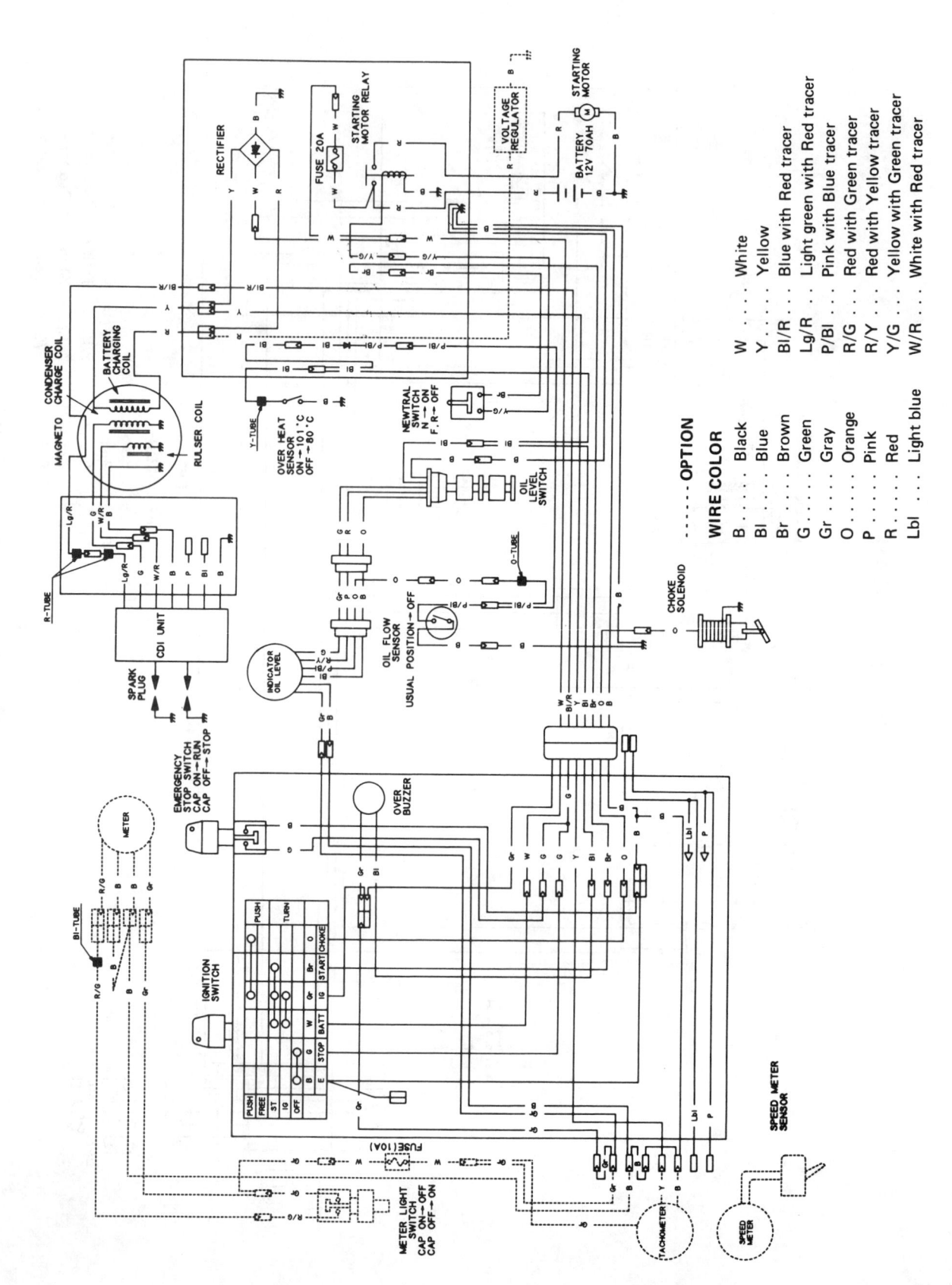

WIRING DIAGRAMS

DT 40TC (1990-1991)

WIRE COLOR

B	Black	O		Orange
Bl	Blue	P		Pink
Br	Brown	R		Red
G	Green	Lbl		Light blue
Gr	Gray	W		White
		Y		Yellow

Bl/R	. . . .	Blue with Red tracer
P/Bl	. . . .	Pink with Blue tracer
R/G	. . . .	Red with Green tracer
R/Y	. . . .	Red with Yellow tracer
Y/G	. . . .	Yellow with Green tracer
W/R	. . . .	White with Red tracer

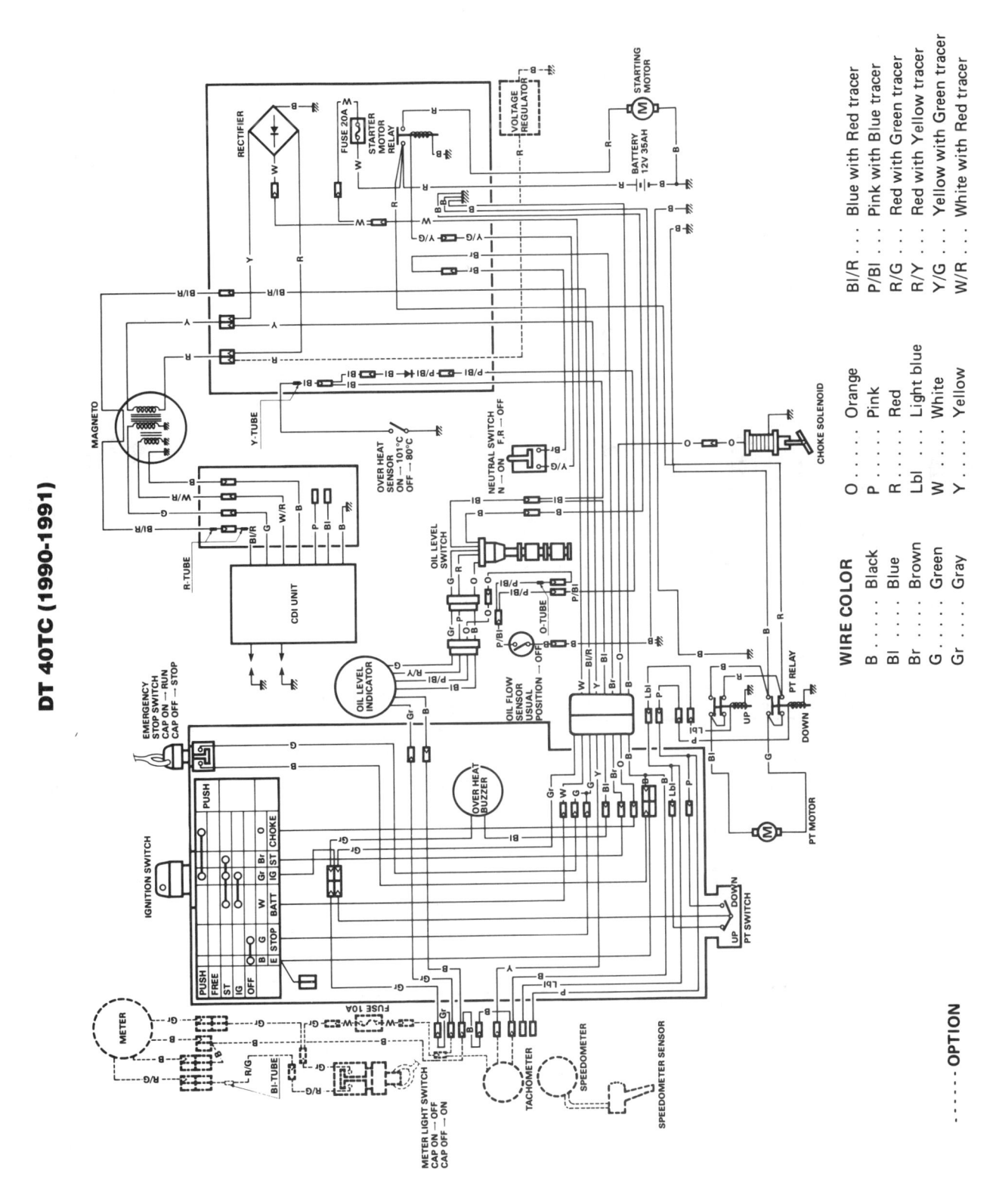

- - - - - - OPTION

WIRING DIAGRAMS

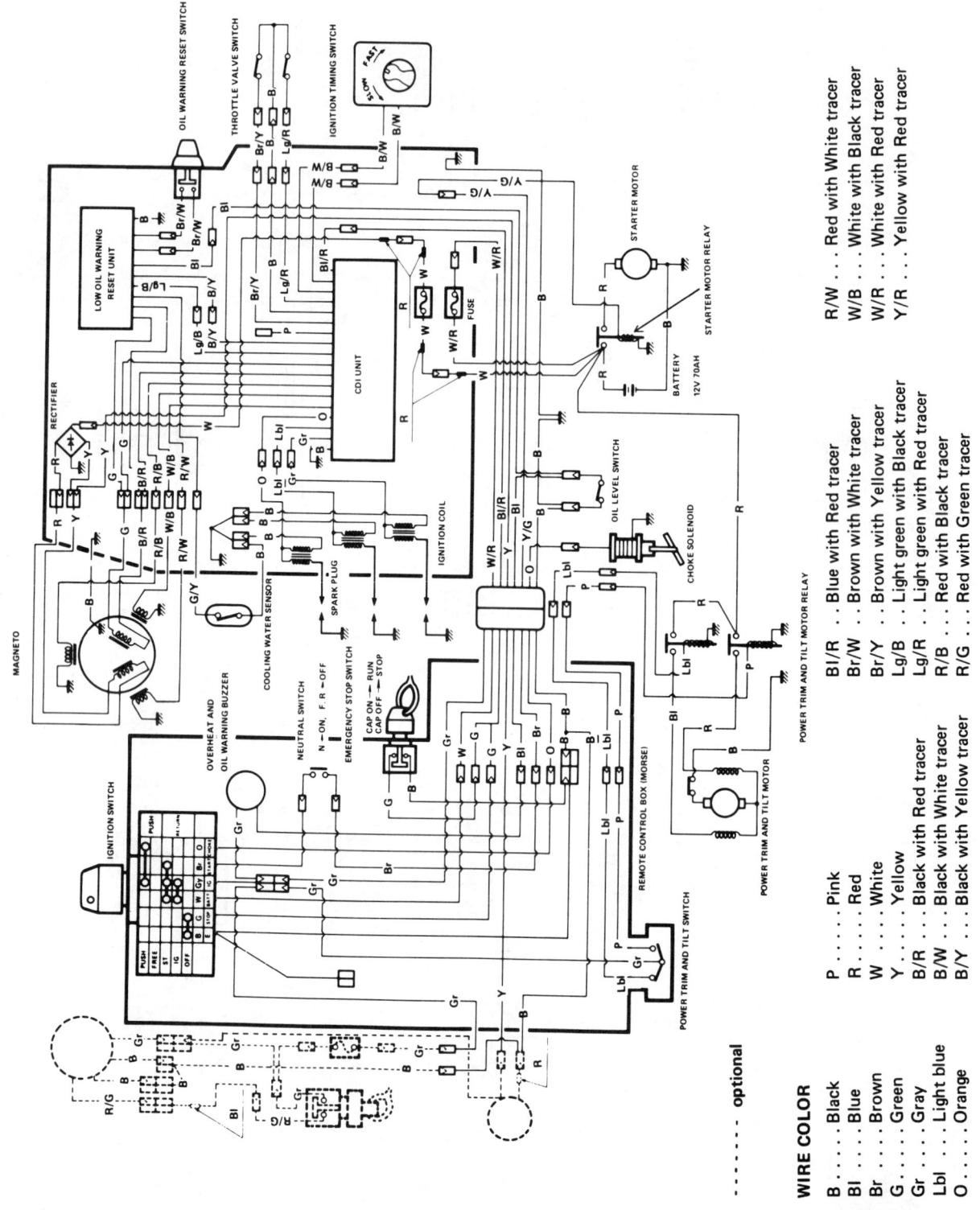

DT 55TC/DT 65TC (1985)

------ optional

WIRE COLOR

B	. . . .	Black	P	. . . . Pink
Bl	. . . .	Blue	R	. . . . Red
Br	. . . .	Brown	W	. . . . White
G	. . . .	Green	Y	. . . . Yellow
Gr	. . . .	Gray	B/R	. . . Black with Red tracer
Lbl	. . . .	Light blue	B/W	. . . Black with White tracer
O	. . . .	Orange	B/Y	. . . Black with Yellow tracer

Bl/R	. . Blue with Red tracer
Br/W	. . Brown with White tracer
Br/Y	. . Brown with Yellow tracer
Lg/B	. . Light green with Black tracer
Lg/R	. . Light green with Red tracer
R/B	. . . Red with Black tracer
R/G	. . . Red with Green tracer

R/W	. . Red with White tracer
W/B	. . White with Black tracer
W/R	. . White with Red tracer
Y/R	. . Yellow with Red tracer

WIRING DIAGRAMS

DT 55C/DT 65C (1985)

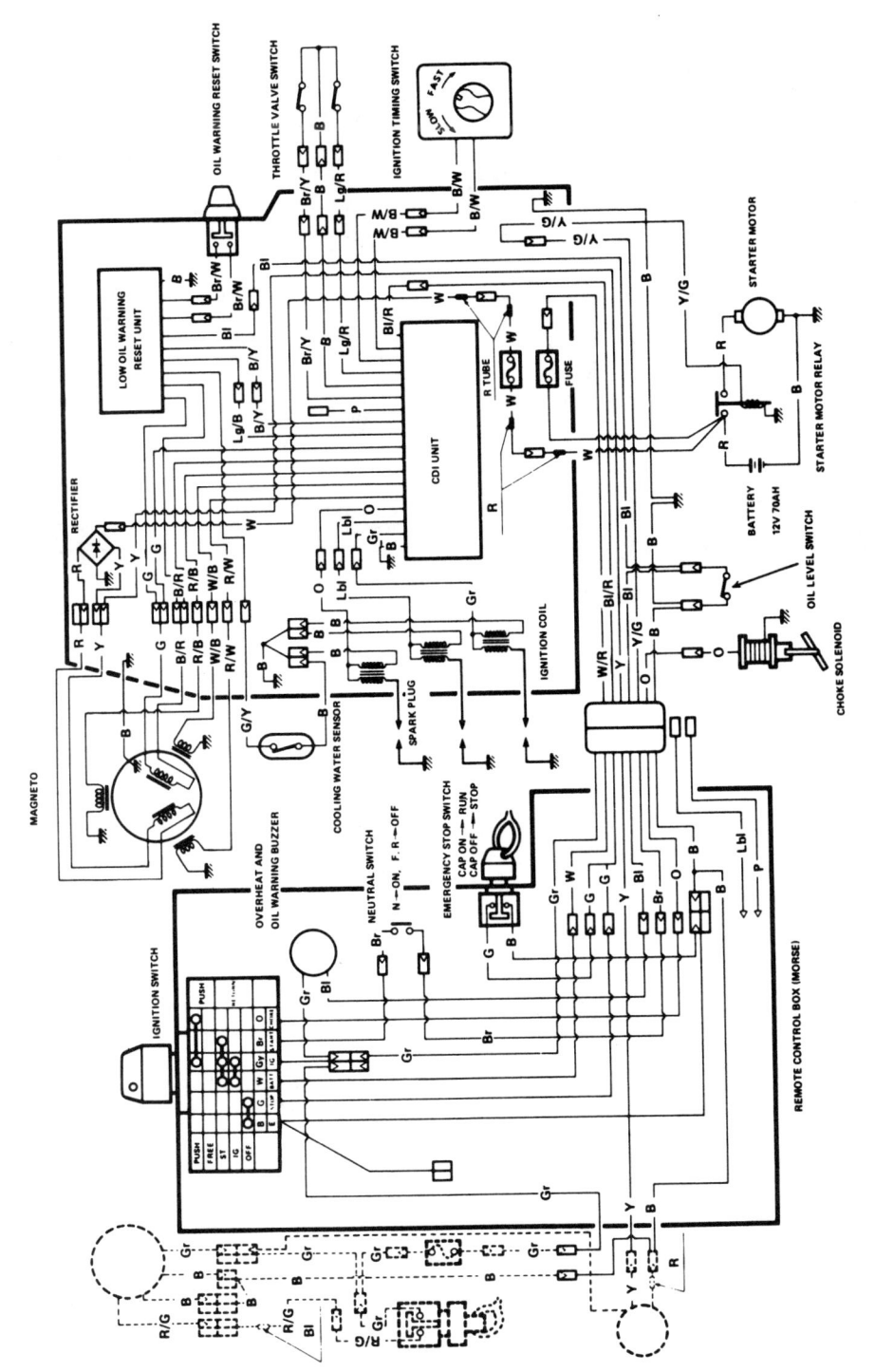

WIRE COLOR

- B . . . . . Black
- Bl . . . . Blue
- Br . . . . Brown
- G . . . . . Green
- Gr . . . . Gray
- Lbl . . . Light blue
- O . . . . . Orange
- P . . . . . Pink
- R . . . . . Red
- W . . . . White
- Y . . . . . Yellow
- B/R . . Black with Red tracer
- B/W . . Black with White tracer
- B/Y . . Black with Yellow tracer
- Bl/R . . Blue with Red tracer
- Br/W . . Brown with White tracer
- Br/Y . . Brown with Yellow tracer
- Lg/B . . Light green with Black tracer
- Lg/R . . Light green with Red tracer
- R/B . . Red with Black tracer
- R/G . . Red with Green tracer
- R/W . . Red with White tracer
- W/B . . White with Black tracer
- W/R . . White with Red tracer
- Y/R . . Yellow with Red tracer

----- optional

WIRING DIAGRAMS

DT 55C/DT 65C (1986)

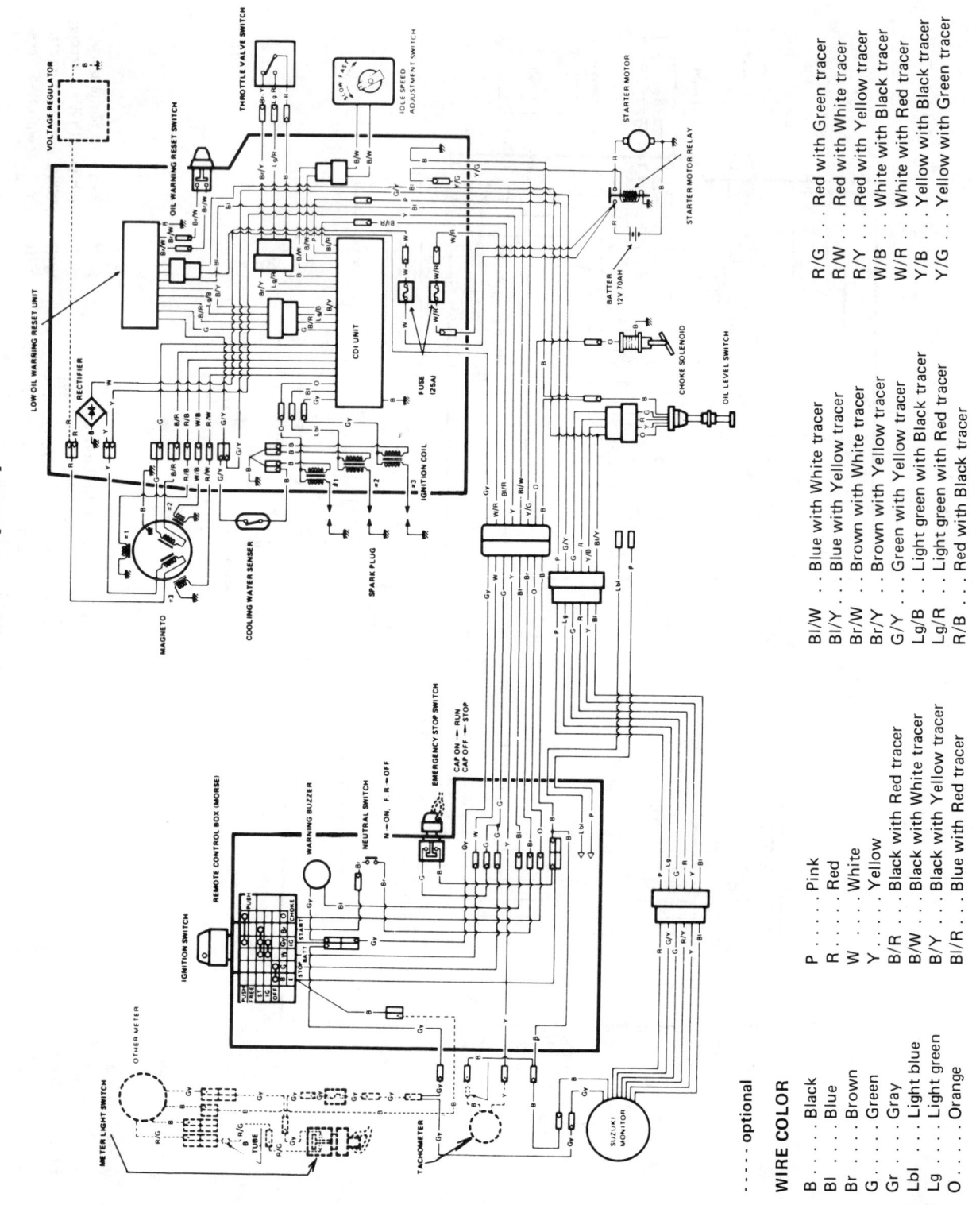

P	. . . . . Pink	Bl/W	. . Blue with White tracer	R/G . . . Red with Green tracer
R	. . . . . Red	Bl/Y	. . Blue with Yellow tracer	R/W . . . Red with White tracer
W	. . . . . White	Br/W	. . Brown with White tracer	R/Y . . . Red with Yellow tracer
Y	. . . . . Yellow	Br/Y	. . Brown with Yellow tracer	W/B . . . White with Black tracer
B/R	. . . Black with Red tracer	G/Y	. . Green with Yellow tracer	W/R . . . White with Red tracer
B/W	. . Black with White tracer	Lg/B	. . Light green with Black tracer	Y/B . . . Yellow with Black tracer
B/Y	. . Black with Yellow tracer	Lg/R	. . Light green with Red tracer	Y/G . . . Yellow with Green tracer
Bl/R	. . Blue with Red tracer	R/B	. . Red with Black tracer	

- - - - optional

WIRE COLOR

B	. . . . Black
Bl	. . . Blue
Br	. . . Brown
G	. . . . Green
Gr	. . . Gray
Lbl	. . Light blue
Lg	. . . Light green
O	. . . . Orange

WIRING DIAGRAMS

DT 55TC/DT 65TC (1986)

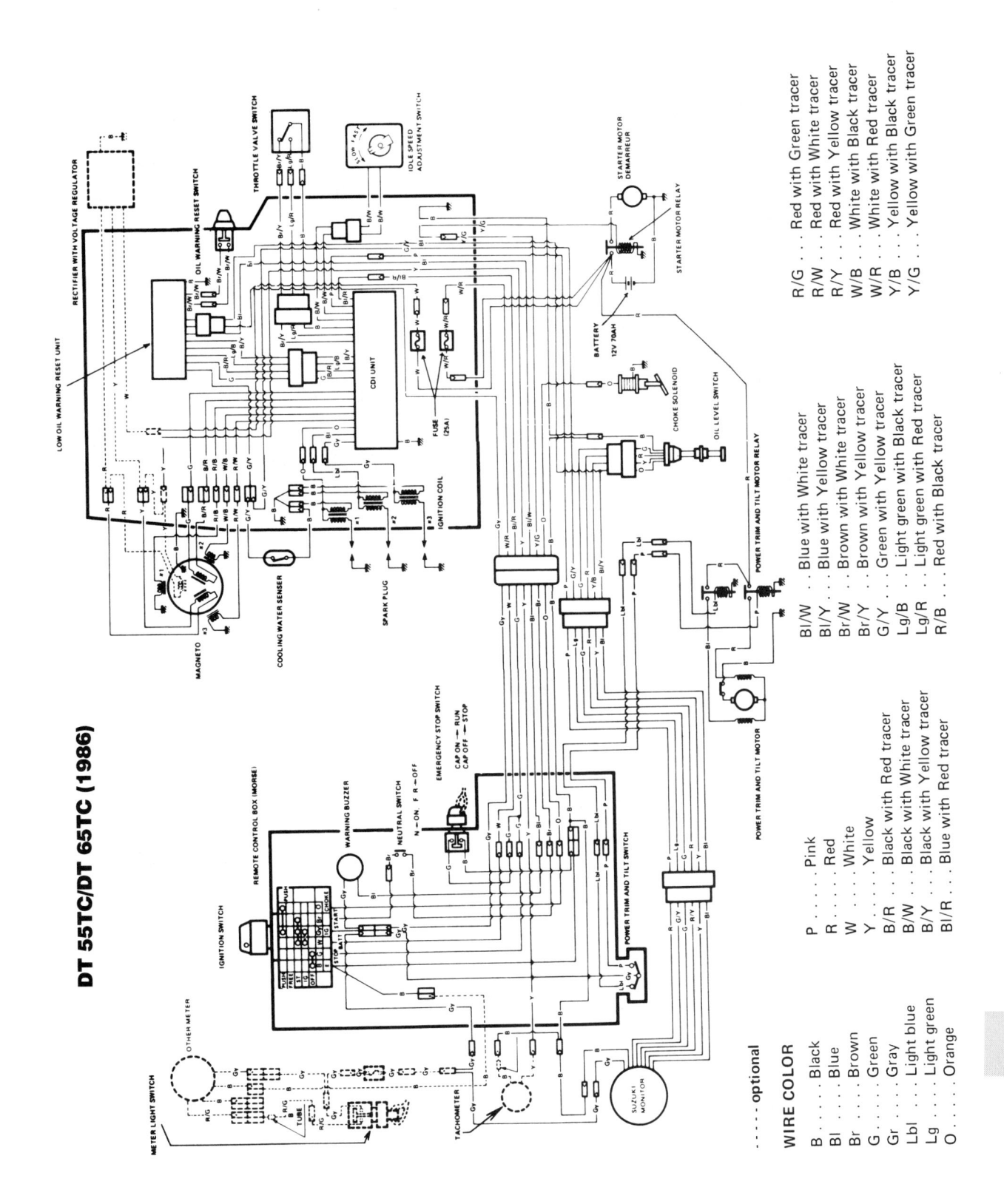

WIRE COLOR

B	. . . .	Black
Bl	. . . .	Blue
Br	. . . .	Brown
G	. . . .	Green
Gr	. . . .	Gray
Lbl	. . . .	Light blue
Lg	. . . .	Light green
O	. . . .	Orange

P	. . . .	Pink
R	. . . .	Red
W	. . . .	White
Y	. . . .	Yellow
B/R	. . . .	Black with Red tracer
B/W	. . . .	Black with White tracer
B/Y	. . . .	Black with Yellow tracer
Bl/R	. . . .	Blue with Red tracer

Bl/W	. . . .	Blue with White tracer
Bl/Y	. . . .	Blue with Yellow tracer
Br/W	. . . .	Brown with White tracer
Br/Y	. . . .	Brown with Yellow tracer
G/Y	. . . .	Green with Yellow tracer
Lg/B	. . . .	Light green with Black tracer
Lg/R	. . . .	Light green with Red tracer
R/B	. . . .	Red with Black tracer

R/G	. . . .	Red with Green tracer
R/W	. . . .	Red with White tracer
R/Y	. . . .	Red with Yellow tracer
W/B	. . . .	White with Black tracer
W/R	. . . .	White with Red tracer
Y/B	. . . .	Yellow with Black tracer
Y/G	. . . .	Yellow with Green tracer

- - - - optional

15

WIRING DIAGRAMS

DT 55C/DT 65C (1987-1988)

R/W	.. Red with White tracer			
R/Y	.. Red with Yellow tracer			
W/B	.. White with Black tracer			
W/R	.. White with Red tracer			
Y/B	.. Yellow with Black tracer			
Y/G	.. Yellow with Green tracer			

Bl/W	.. Blue with White tracer
Br/W	.. Brown with White tracer
Br/Y	.. Brown with Yellow tracer
G/Y	.. Green with Yellow tracer
Lg/B	.. Light green with Black tracer
Lg/R	.. Light green with Red tracer
P/Bl	.. Pink with Blue tracer
R/G	.. Red with Green tracer

P	.... Pink
R	.... Red
W	.... White
Y	.... Yellow
B/R	.. Black with Red tracer
B/W	.. Black with White tracer
B/Y	.. Black with Yellow tracer
Bl/R	.. Blue with Red tracer

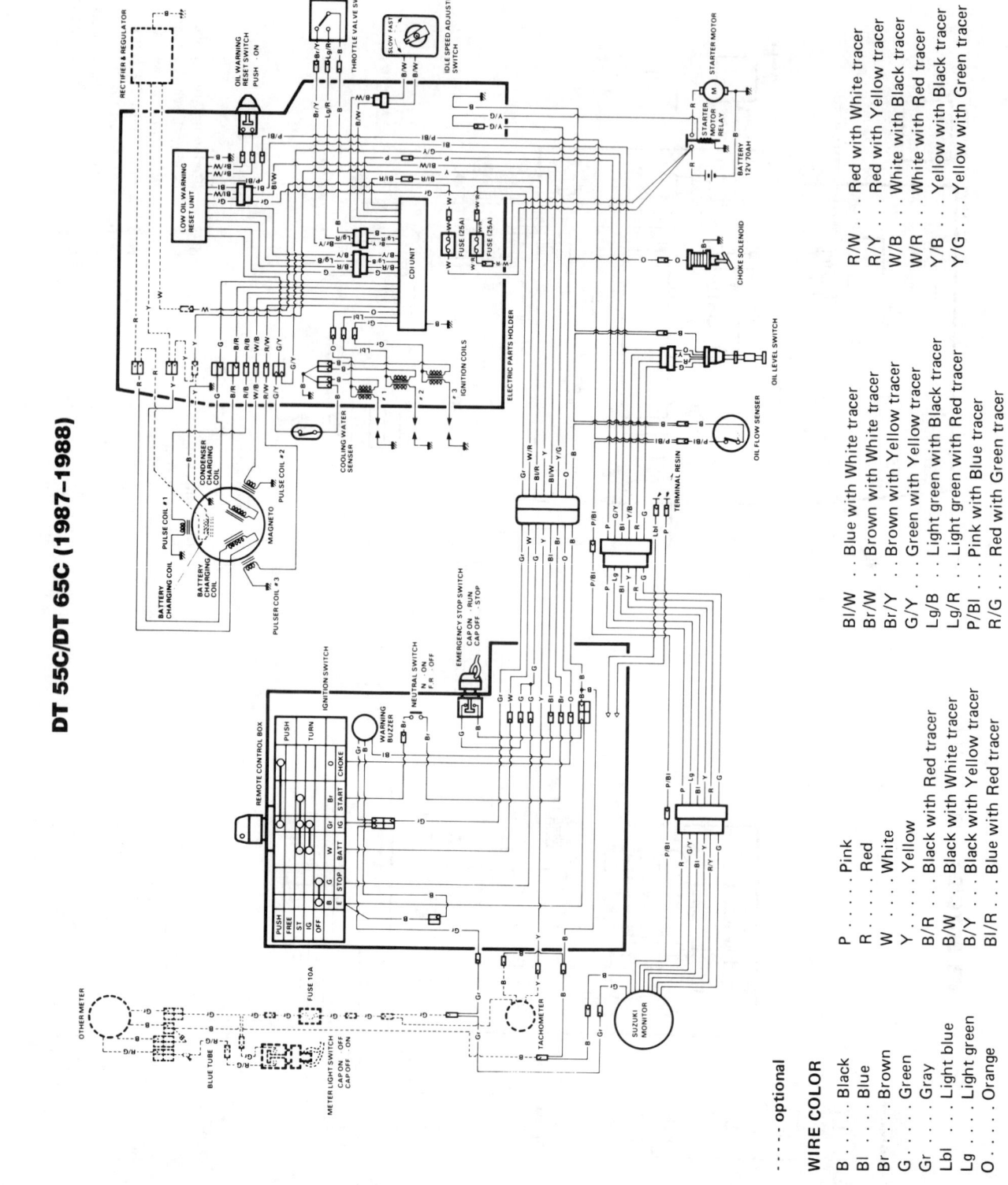

- - - - - optional

WIRE COLOR

B	.... Black
Bl	.... Blue
Br	.... Brown
G	.... Green
Gr	.... Gray
Lbl	.. Light blue
Lg	.... Light green
O	.... Orange

WIRING DIAGRAMS

DT 55TC/DT 65TC (1987-1988)

Bl/W	Blue with White tracer	R/W	Red with White tracer
Br/W	Brown with White tracer	R/Y	Red with Yellow tracer
Br/Y	Brown with Yellow tracer	W/B	White with Black tracer
G/Y	Green with Yellow tracer	W/R	White with Red tracer
Lg/B	Light green with Black tracer	Y/B	Yellow with Black tracer
Lg/R	Light green with Red tracer	Y/G	Yellow with Green tracer
P/Bl	Pink with Blue tracer		
R/G	Red with Green tracer		

- - - - optional

WIRE COLOR

- B ..... Black
- Bl .... Blue
- Br ... Brown
- G .... Green
- Gr ... Gray
- Lbl .. Light blue
- Lg ... Light green
- O .... Orange
- P ..... Pink
- R ..... Red
- W .... White
- Y ..... Yellow
- B/R .. Black with Red tracer
- B/W .. Black with White tracer
- B/Y .. Black with Yellow tracer
- Bl/R .. Blue with Red tracer

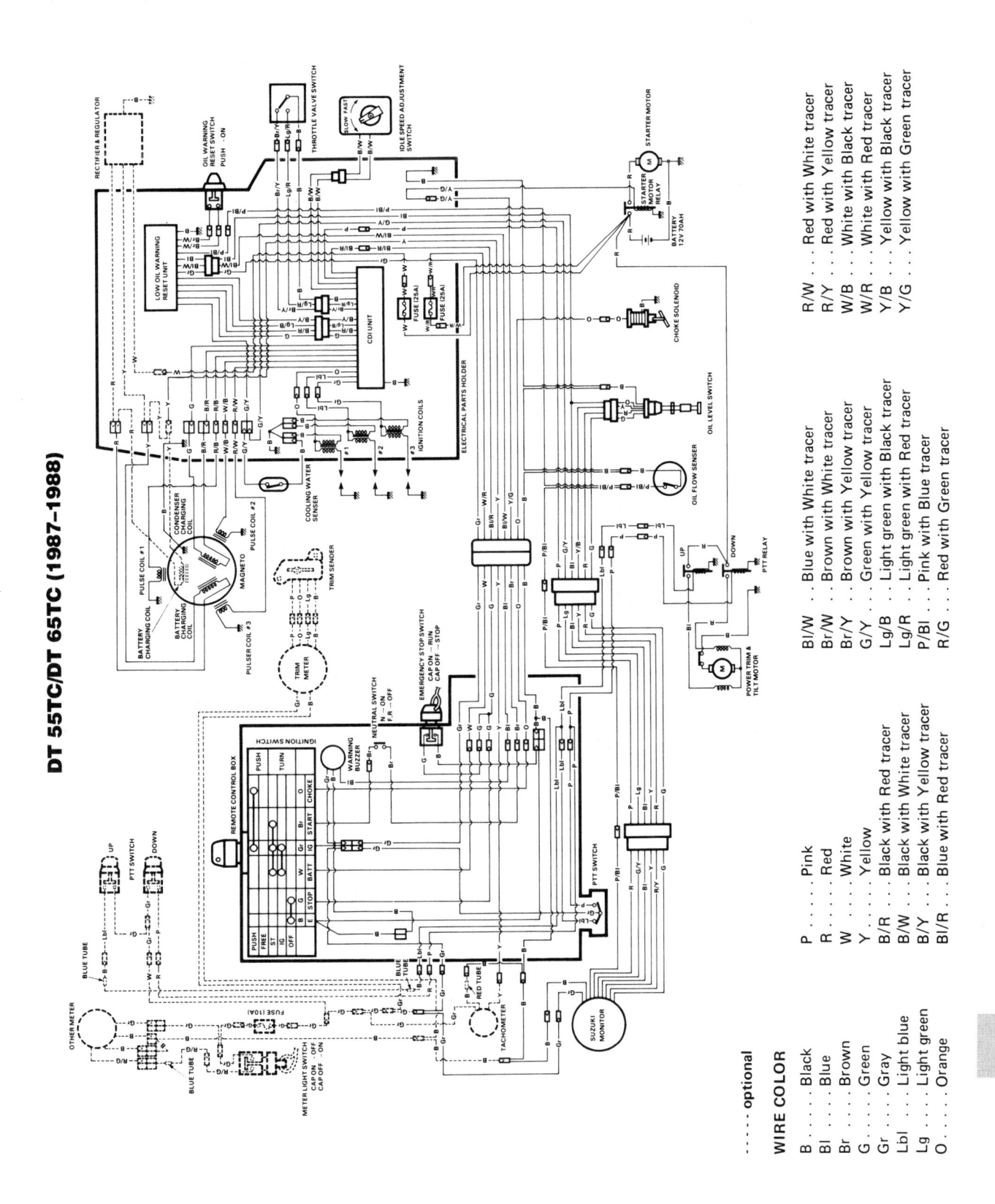

15

WIRING DIAGRAMS

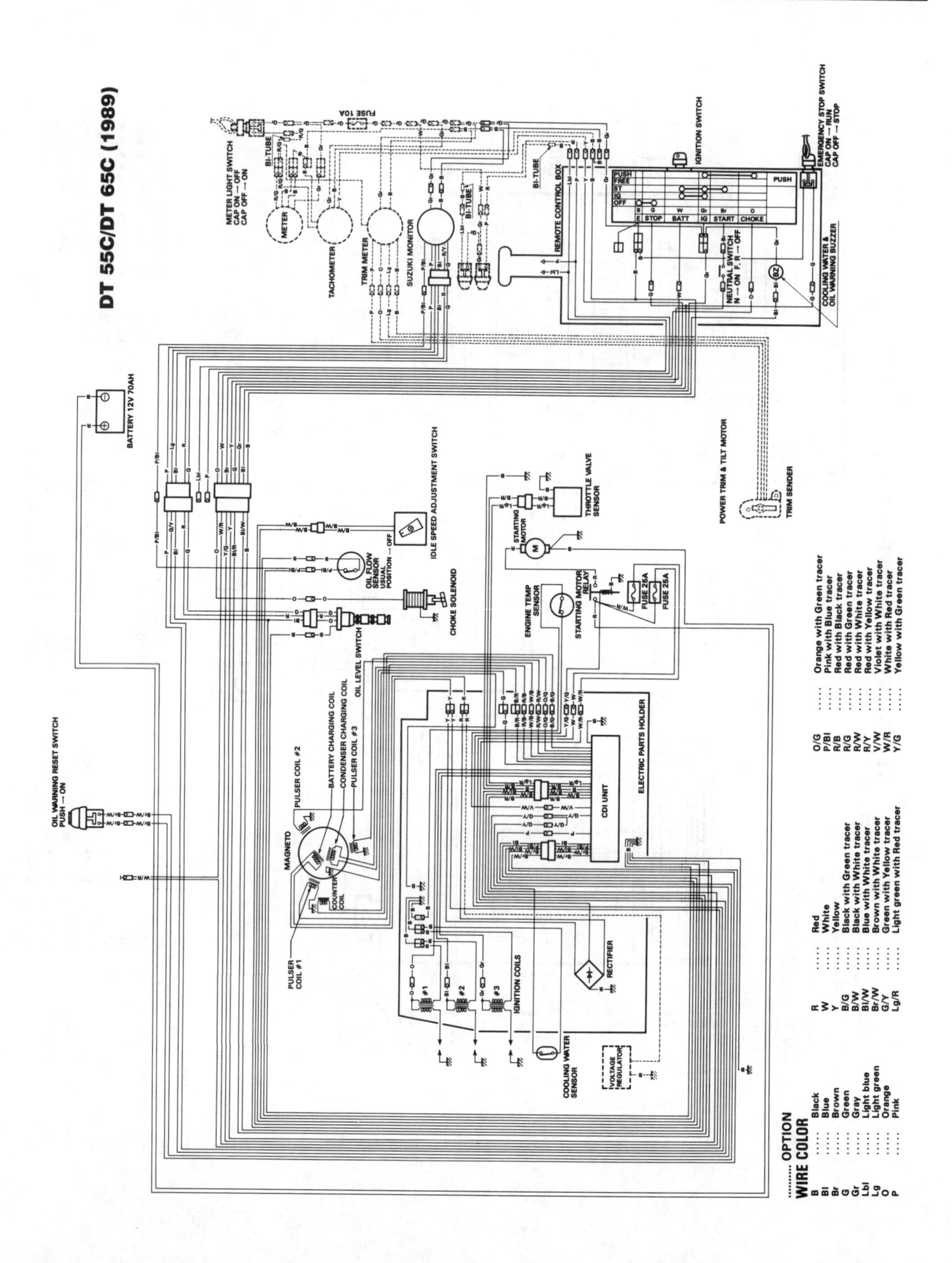

Wiring Diagrams

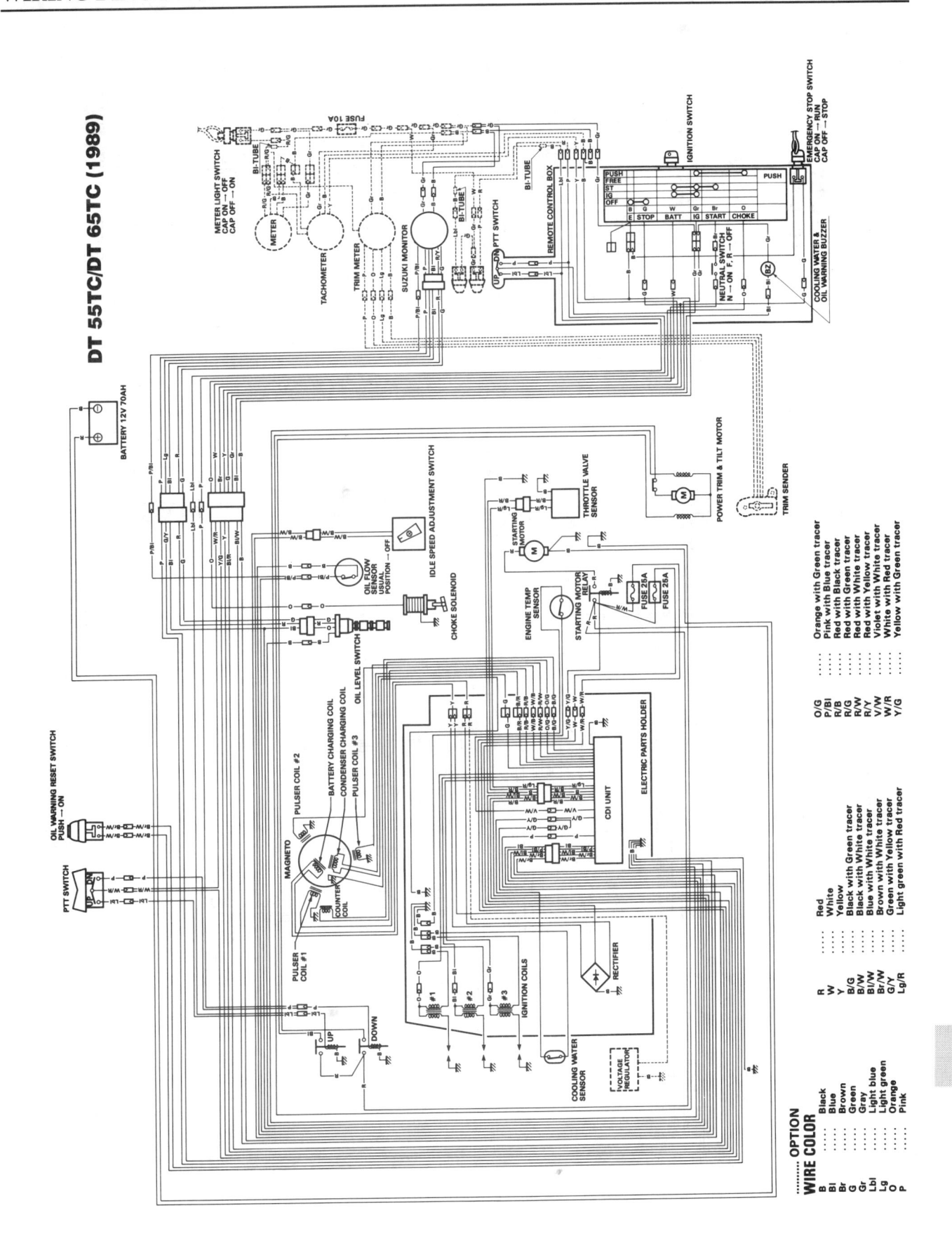

DT 55TC/DT 65TC (1989)

WIRING DIAGRAMS

DT 55C/DT 65C (1990-1991)

OPTION

WIRE COLOR

B	Black
Bl	Blue
Br	Brown
G	Green
Gr	Gray
Lbl	Light blue
Lg	Light green
O	Orange
P	Pink

R	Red
W	White
Y	Yellow
B/G	Black with Green tracer
B/W	Black with White tracer
Bl/W	Blue with White tracer
Br/W	Brown with White tracer
G/Y	Green with Yellow tracer
Lg/R	Light green with Red tracer

O/G	Orange with Green tracer
P/Bl	Pink with Blue tracer
R/B	Red with Black tracer
R/G	Red with Green tracer
R/W	Red with White tracer
R/Y	Red with Yellow tracer
V/W	Violet with White tracer
W/R	White with Red tracer
Y/G	Yellow with Green tracer

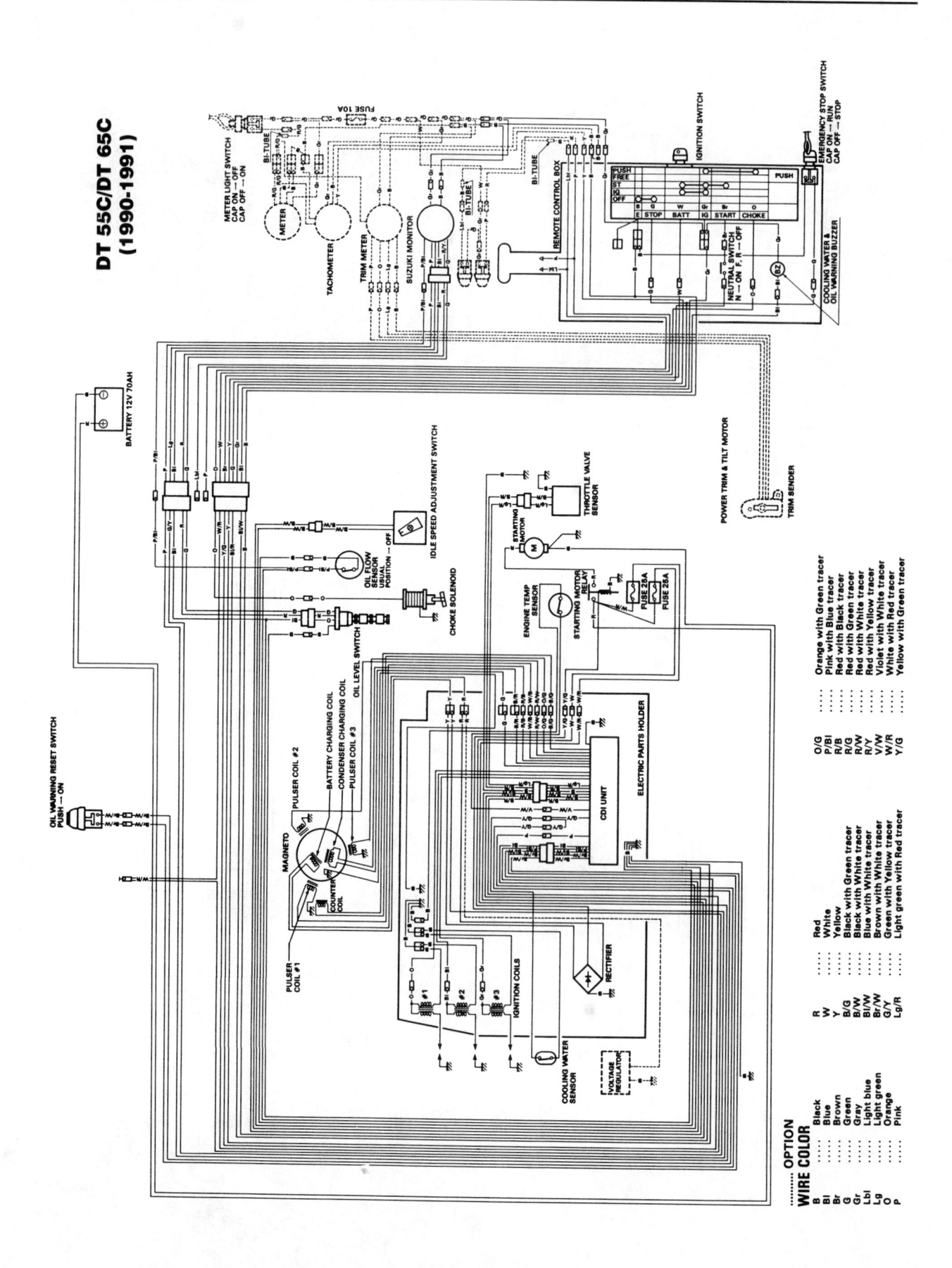

WIRING DIAGRAMS

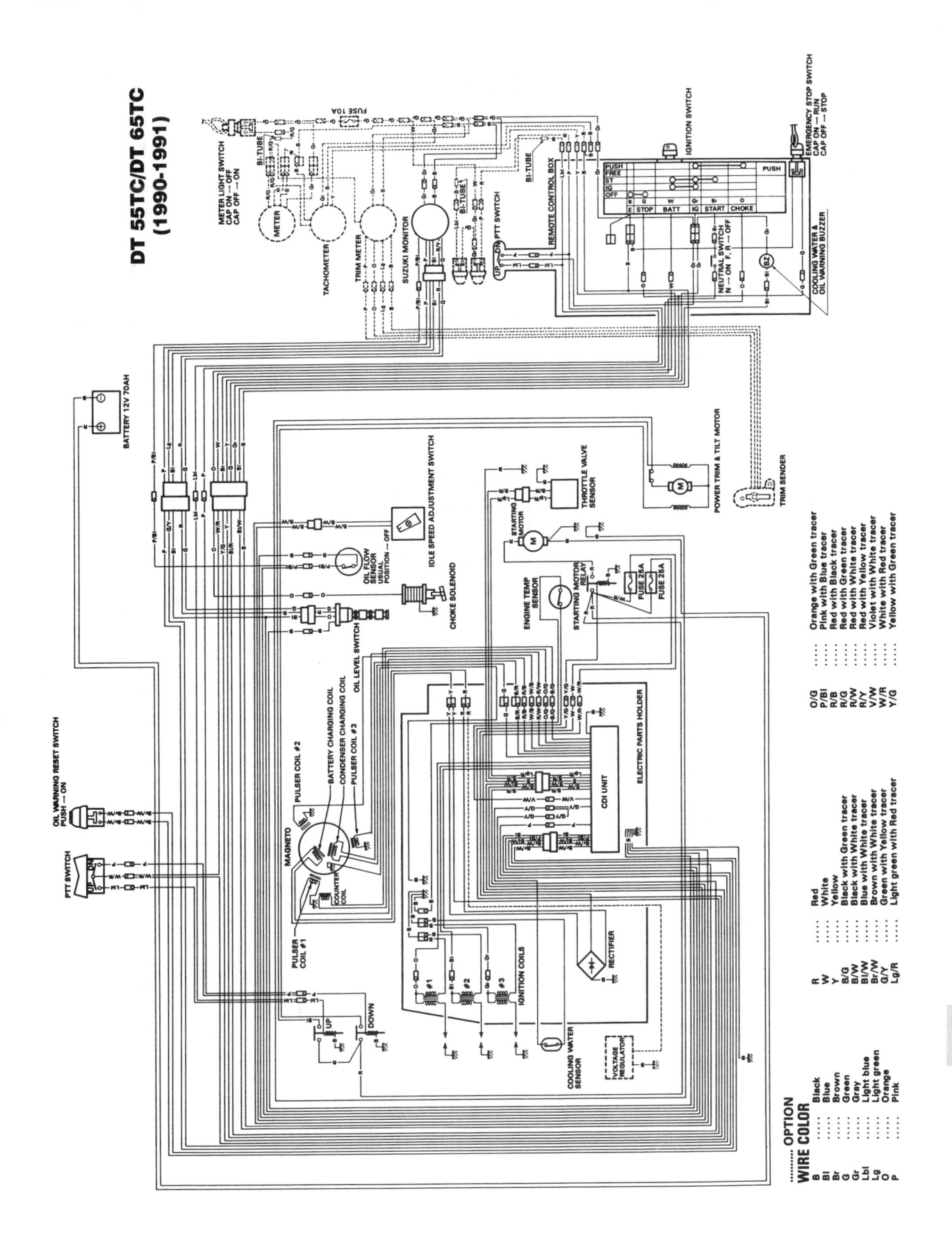

591

WIRING DIAGRAMS

DT 75 (1985–1986)

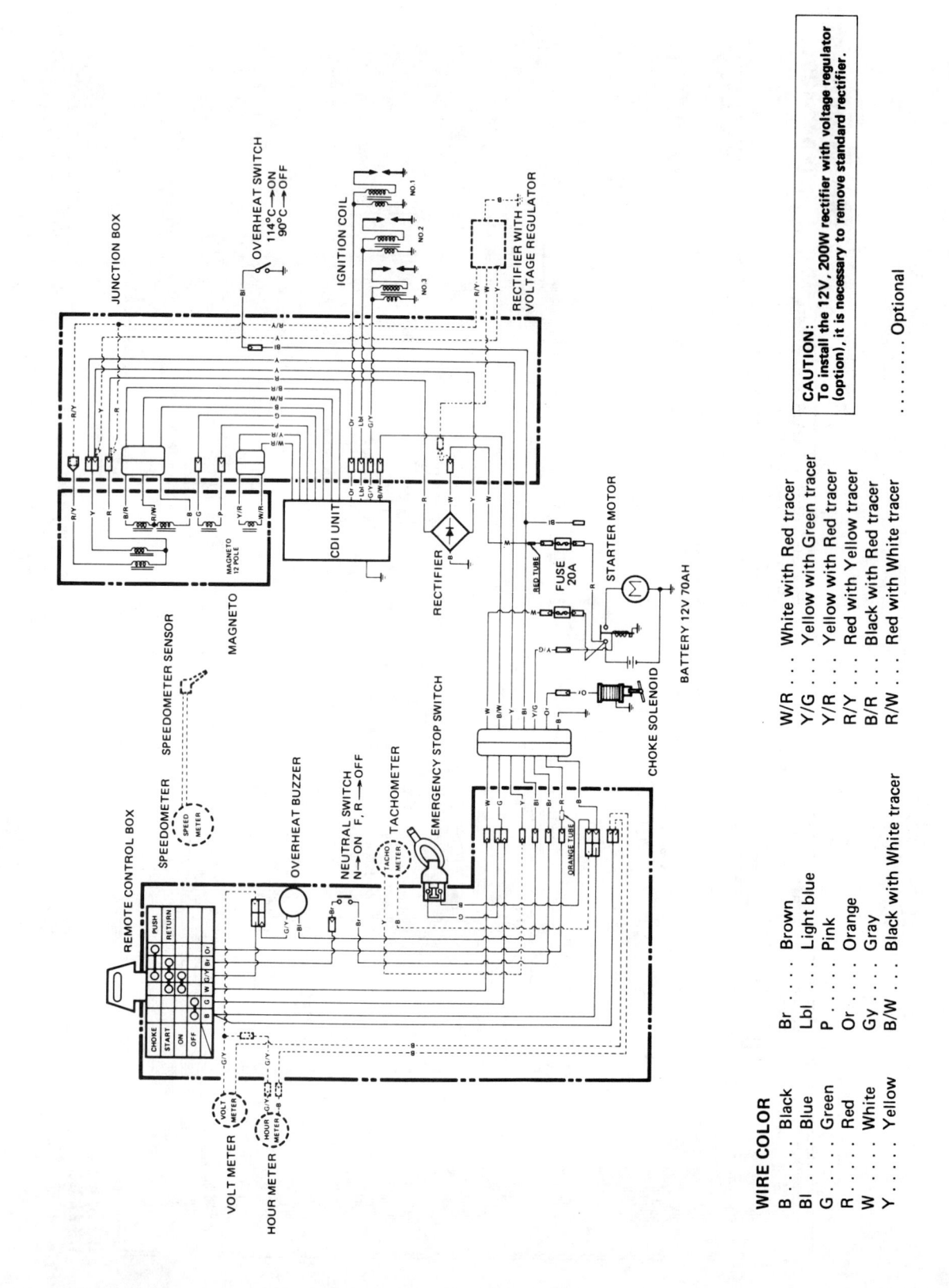

CAUTION:
To install the 12V, 200W rectifier with voltage regulator (option), it is necessary to remove standard rectifier.

· · · · · · · · Optional

WIRE COLOR

B	. . . .	Black	Br	. . . . . Brown
Bl	. . . .	Blue	Lbl	. . . . . Light blue
G	. . . .	Green	P	. . . . . Pink
R	. . . .	Red	Or	. . . . . Orange
W	. . . .	White	Gy	. . . . . Gray
Y	. . . .	Yellow	B/W	. . . Black with White tracer

W/R . . . White with Red tracer
Y/G . . . Yellow with Green tracer
Y/R . . . Yellow with Red tracer
R/Y . . . Red with Yellow tracer
B/R . . . Black with Red tracer
R/W . . . Red with White tracer

WIRING DIAGRAMS

DT 75C (1985–1986)

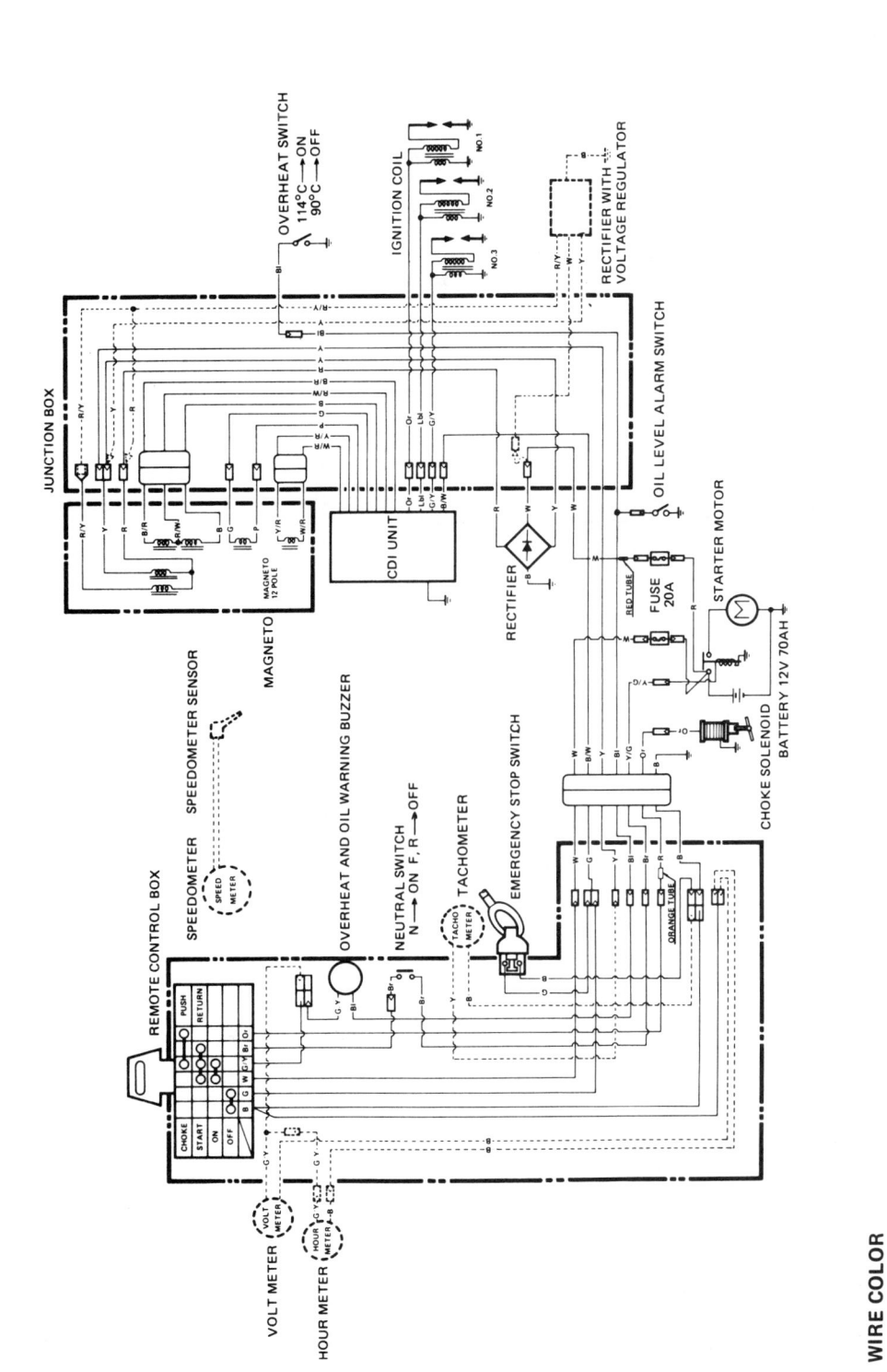

WIRE COLOR

B	. . . . .	Black
Bl	. . . . .	Blue
G	. . . . .	Green
R	. . . . .	Red
W	. . . . .	White
Y	. . . . .	Yellow
Br	. . . . .	Brown
Lbl	. . . . .	Light blue
P	. . . . .	Pink
Or	. . . . .	Orange
Gy	. . . . .	Gray
B/W	. . . . .	Black with White tracer
W/R	. . . . .	White with Red tracer
Y/G	. . . . .	Yellow with Green tracer
Y/R	. . . . .	Yellow with Red tracer
R/Y	. . . . .	Red with Yellow tracer
B/R	. . . . .	Black with Red tracer
R/W	. . . . .	Red with White tracer

CAUTION:
To install the 12V, 200W rectifier with voltage regulator (option), it is necessary to remove standard rectifier.

· · · · · · · · Optional

WIRING DIAGRAMS

DT 75TC (1985–1986)

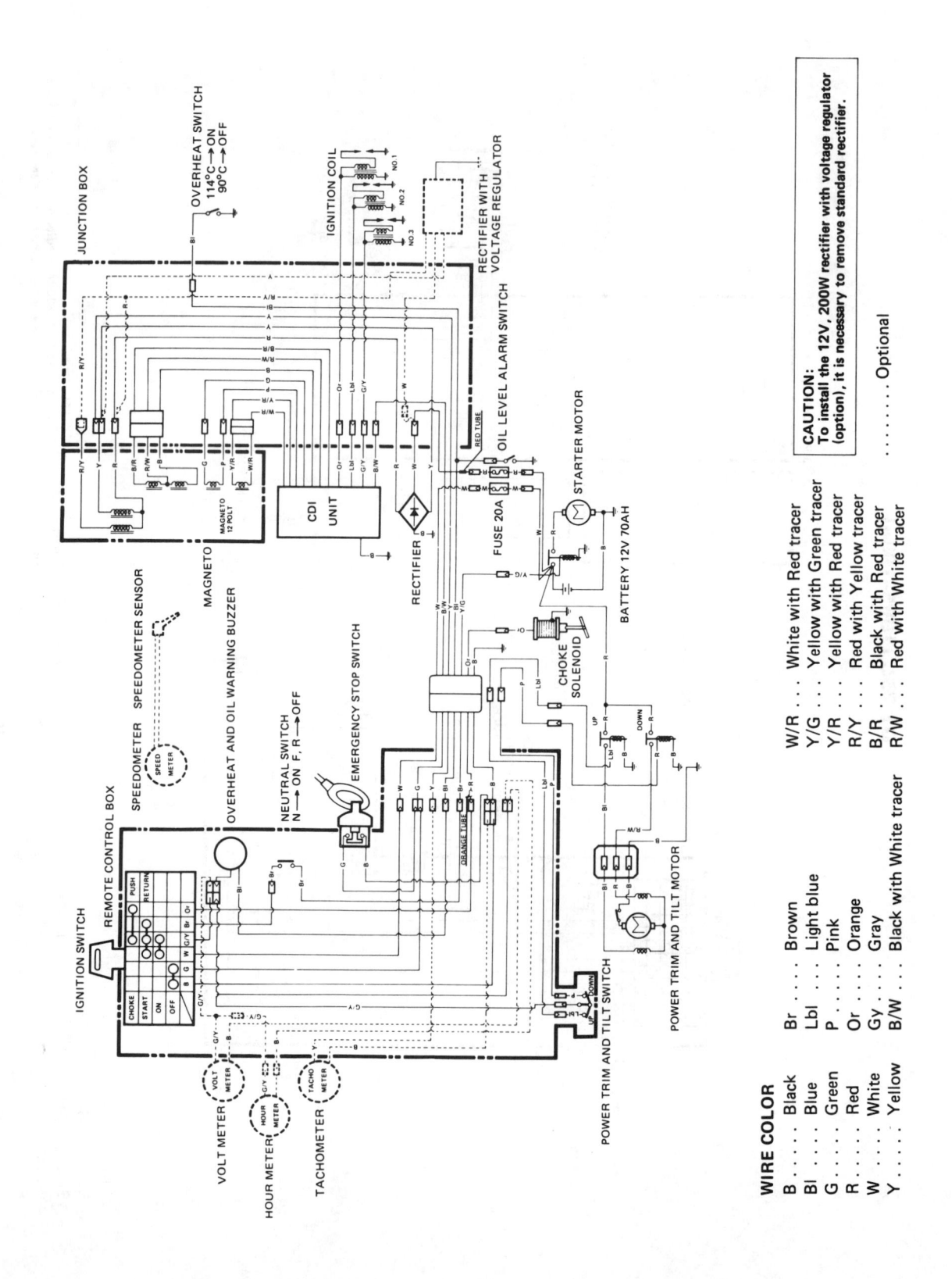

CAUTION:
To install the 12V, 200W rectifier with voltage regulator (option), it is necessary to remove standard rectifier.

· · · · · · · Optional

WIRE COLOR

B	· · · · ·	Black
Bl	· · · · ·	Blue
G	· · · · ·	Green
R	· · · · ·	Red
W	· · · · ·	White
Y	· · · · ·	Yellow
Br	· · · · ·	Brown
Lbl	· · · · ·	Light blue
P	· · · · ·	Pink
Or	· · · · ·	Orange
Gy	· · · · ·	Gray
B/W	· · · · ·	Black with White tracer
W/R	· · · · ·	White with Red tracer
Y/G	· · · · ·	Yellow with Green tracer
Y/R	· · · · ·	Yellow with Red tracer
R/Y	· · · · ·	Red with Yellow tracer
B/R	· · · · ·	Black with Red tracer
R/W	· · · · ·	Red with White tracer

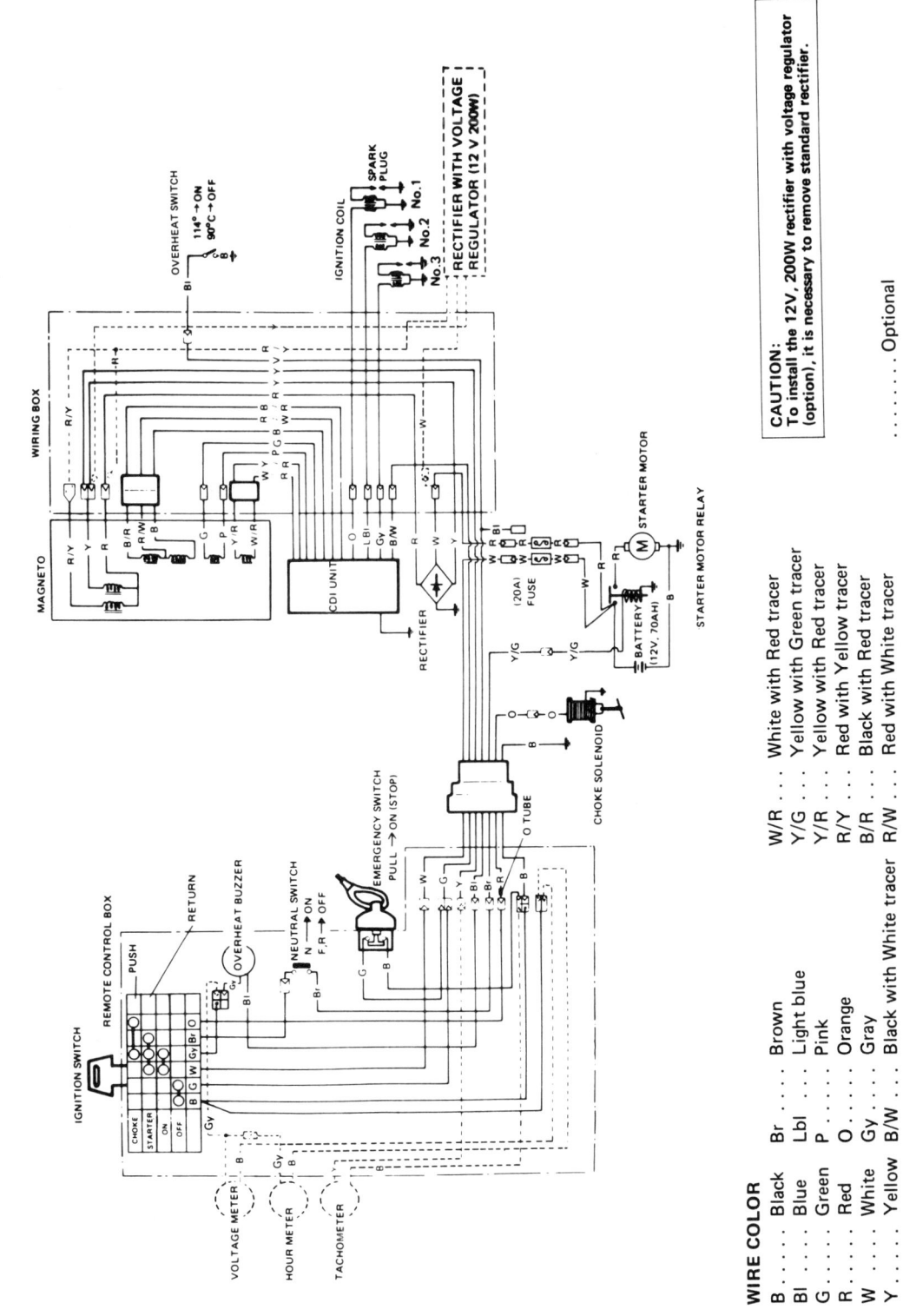

…

WIRING DIAGRAMS

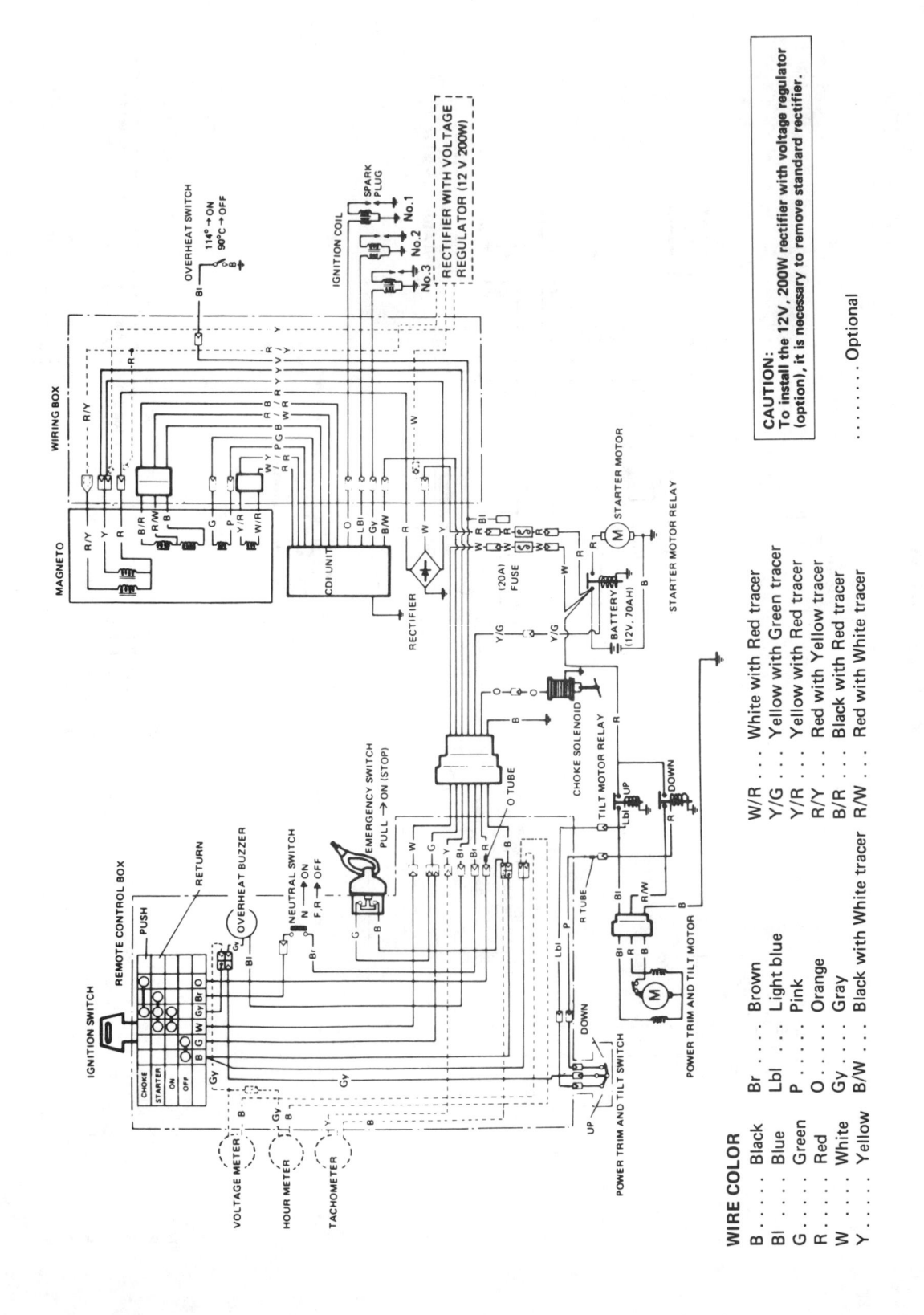

596

WIRING DIAGRAMS

DT 85TC (1985-1986)

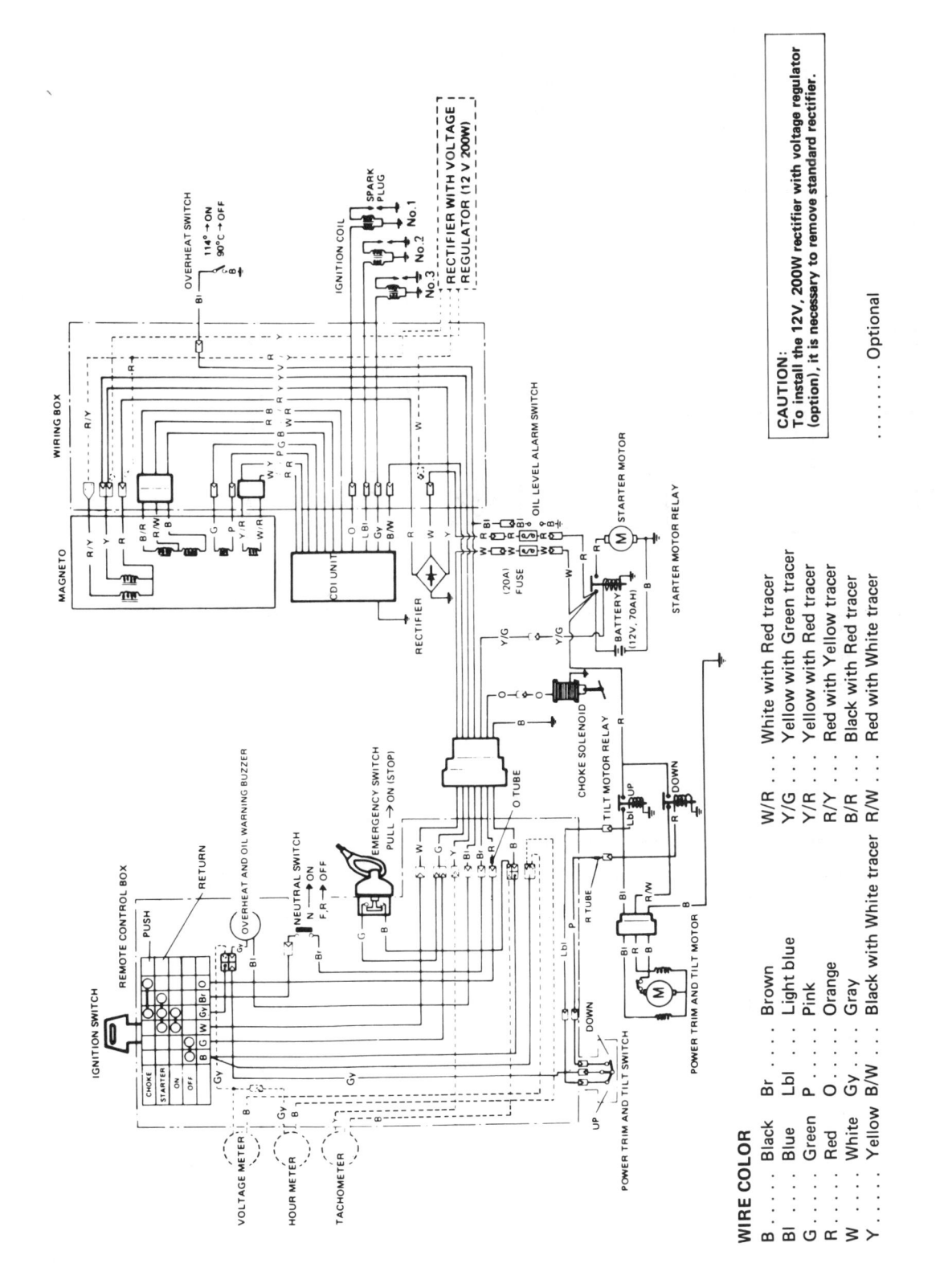

CAUTION:
To install the 12V, 200W rectifier with voltage regulator (option), it is necessary to remove standard rectifier.

- - - - - - - Optional

WIRE COLOR

B	Black	Br	Brown	W/R	White with Red tracer
Bl	Blue	Lbl	Light blue	Y/G	Yellow with Green tracer
G	Green	P	Pink	Y/R	Yellow with Red tracer
R	Red	O	Orange	R/Y	Red with Yellow tracer
W	White	Gy	Gray	B/R	Black with Red tracer
Y	Yellow	B/W	Black with White tracer	R/W	Red with White tracer

DT 75TC/ DT 85TC (1987)

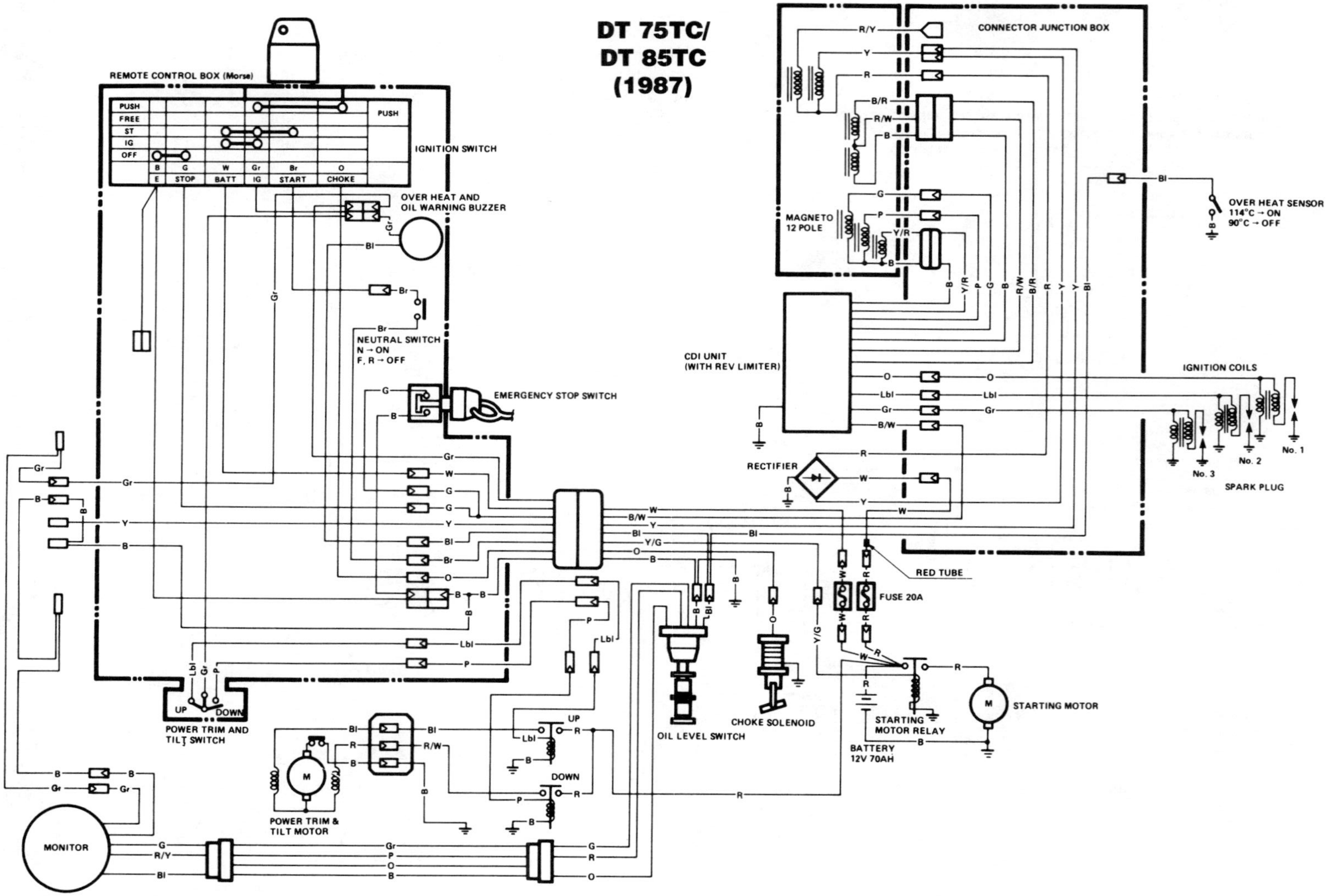

WIRE COLOR

B . . . . Black
Bl . . . . Blue
Br . . . . . . Brown
G . . . . Green
Gr . . . . . . Gray
Lbl . . . . . . Light blue
O . . . . Orange
P . . . . Pink
R . . . . Red
W . . . . White
Y . . . . Yellow
B/R . . . . . Black with Red tracer
B/W . . . . . Black with White tracer
R/W . . . . . Red with White tracer
R/Y . . . . . Red with Yellow tracer
Y/G . . . . . Yellow with Green tracer
Y/R . . . . . Yellow with Red tracer

WIRING DIAGRAMS

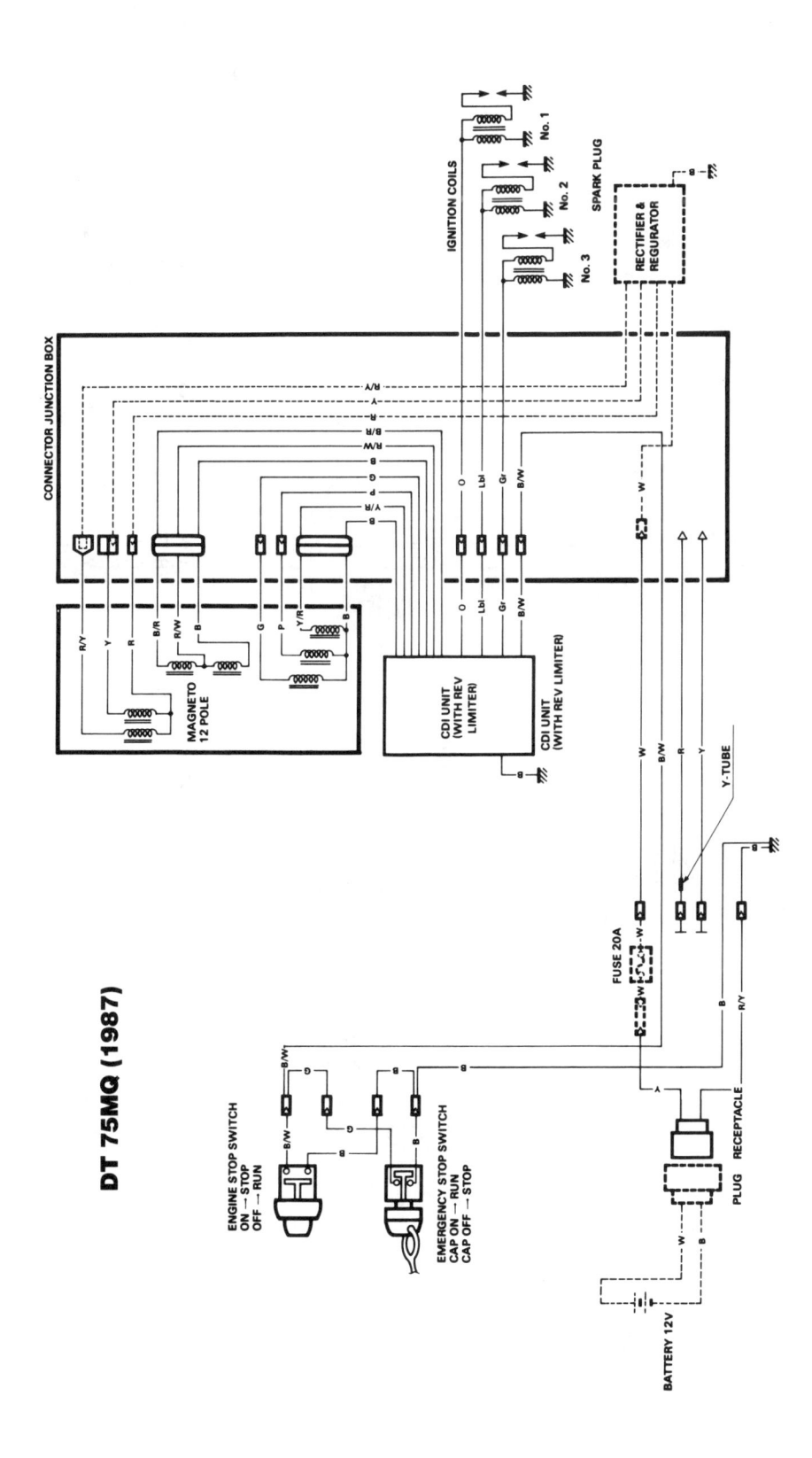

DT 75MQ (1987)

WIRE COLOR

B	Black
G	Green
Gr	Gray
Lbl	Light green
O	Orange
P	Pink
R	Red
W	White
Y	Yellow
B/R	Black with Red tracer
B/W	Black with White tracer
R/W	Red with White tracer
R/Y	Red with Yellow tracer
Y/R	Yellow with Red tracer

15

WIRING DIAGRAMS

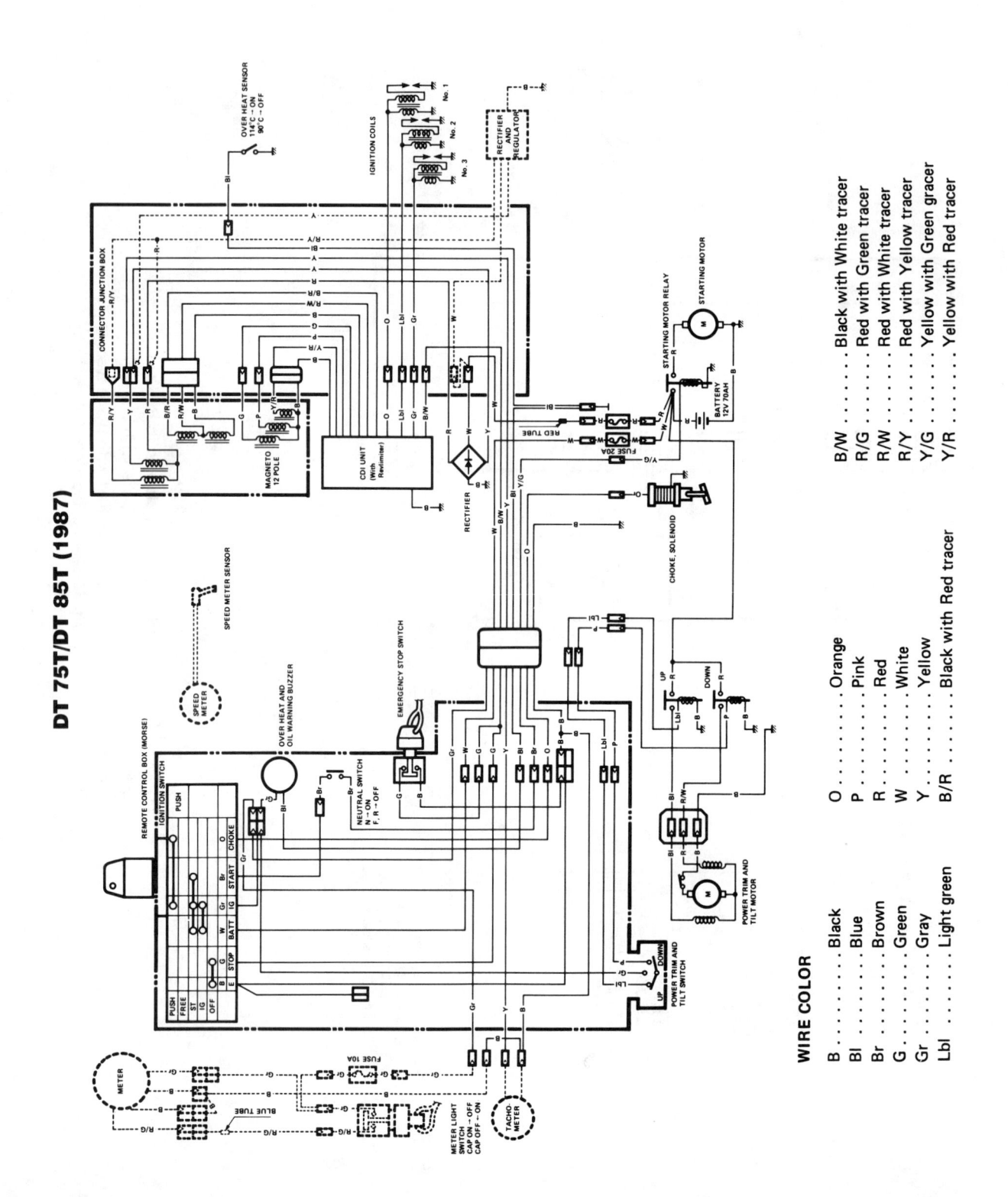

WIRING DIAGRAMS

DT 75HQ (1987)

Bl/R	Blue with Red tracer
R/W	Red with White tracer
R/Y	Red with Yellow tracer
Y/G	Yellow with Green tracer
Y/R	Yellow with Red tracer

P	Pink
R	Red
W	White
Y	Yellow
B/R	Black with Red tracer
B/W	Black with White tracer

WIRE COLOR

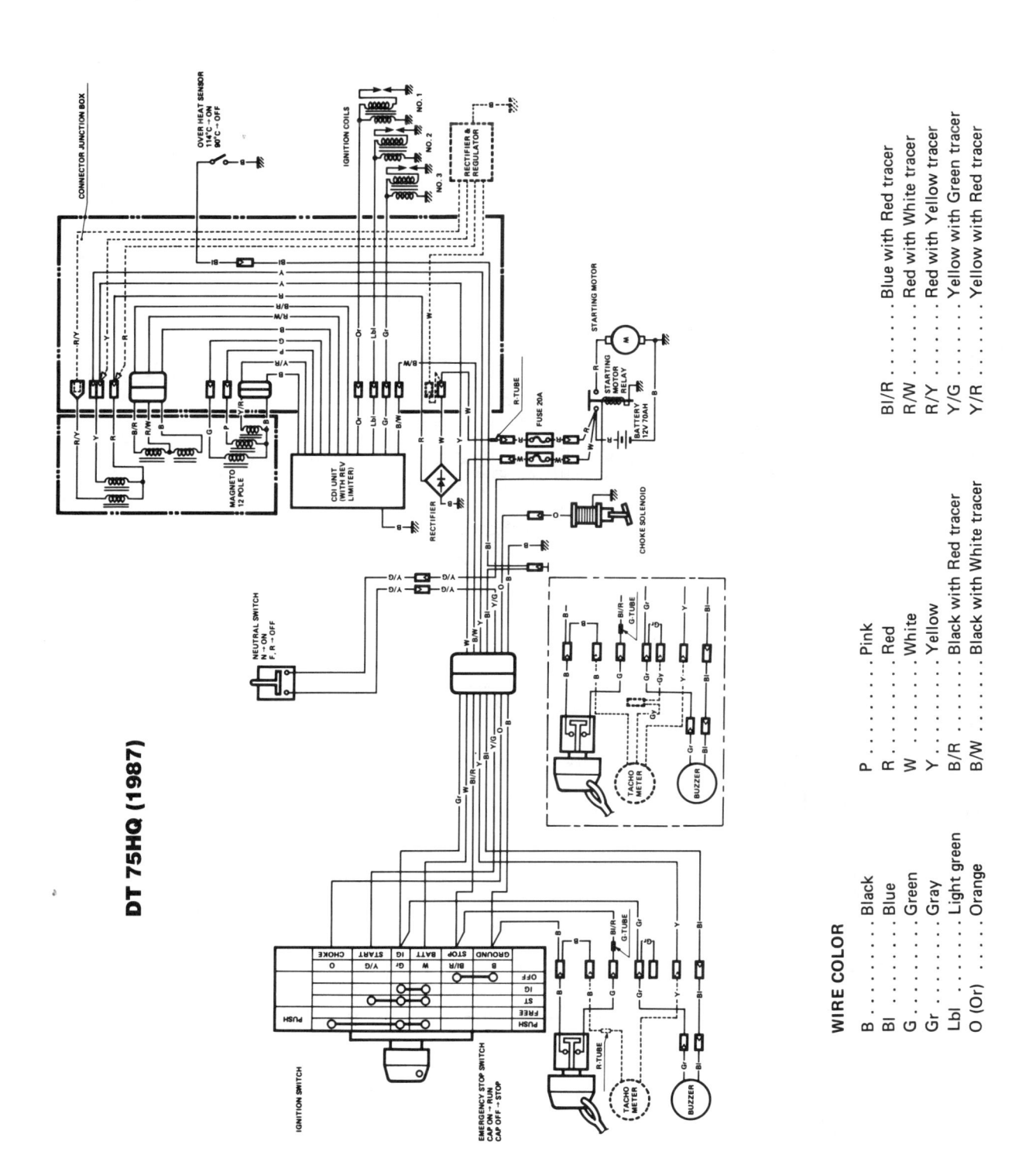

B	Black
Bl	Blue
G	Green
Gr	Gray
Lbl	Light green
O (Or)	Orange

WIRING DIAGRAMS

DT 75CQ/DT 85CQ (1987)

WIRE COLOR

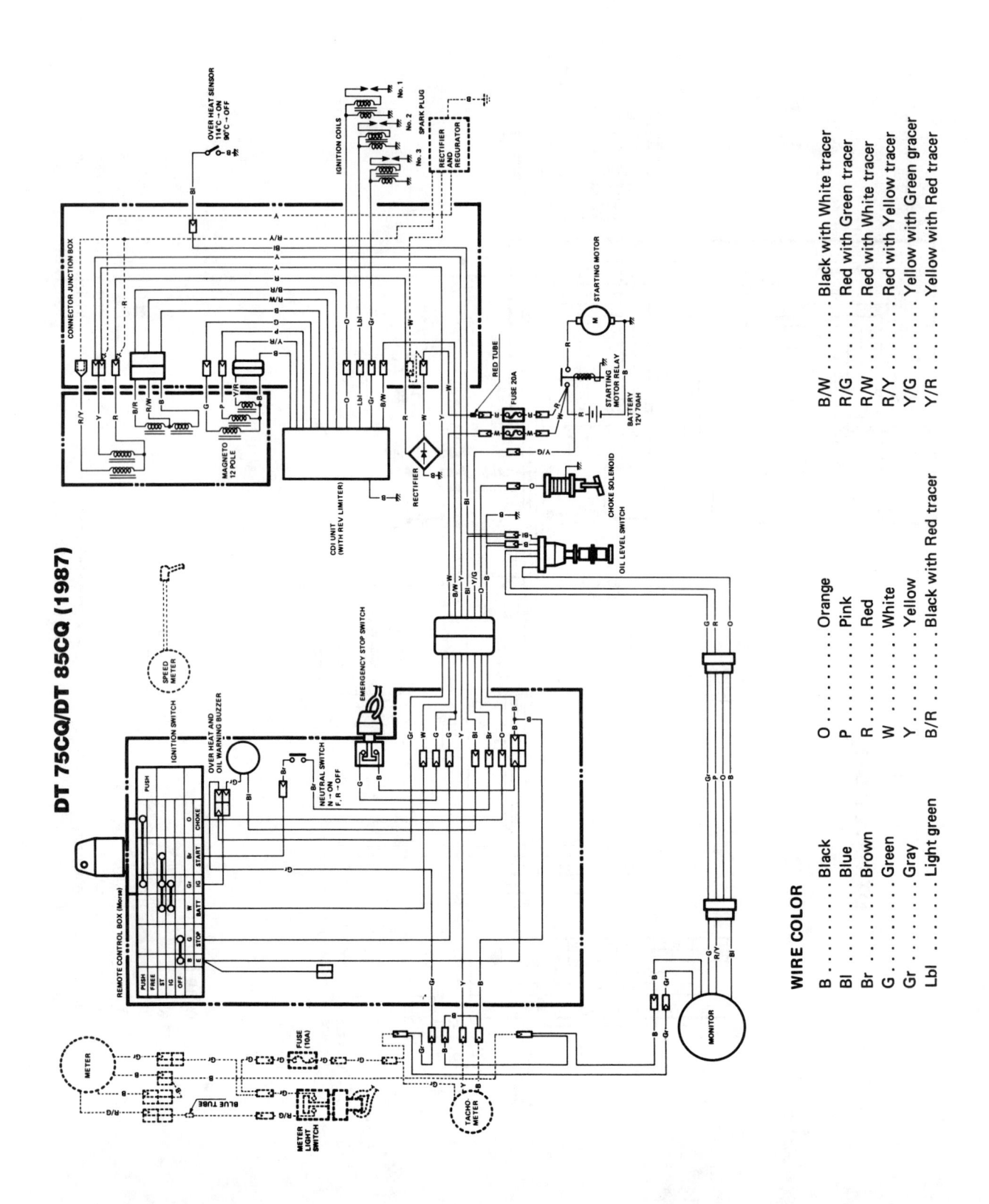

B	Black	O		Orange
Bl	Blue	P		Pink
Br	Brown	R		Red
G	Green	W		White
Gr	Gray	Y		Yellow
Lbl	Light green	B/R		Black with Red tracer

B/W		Black with White tracer
R/G		Red with Green tracer
R/W		Red with White tracer
R/Y		Red with Yellow tracer
Y/G		Yellow with Green tracer
Y/R		Yellow with Red tracer

Wiring Diagrams

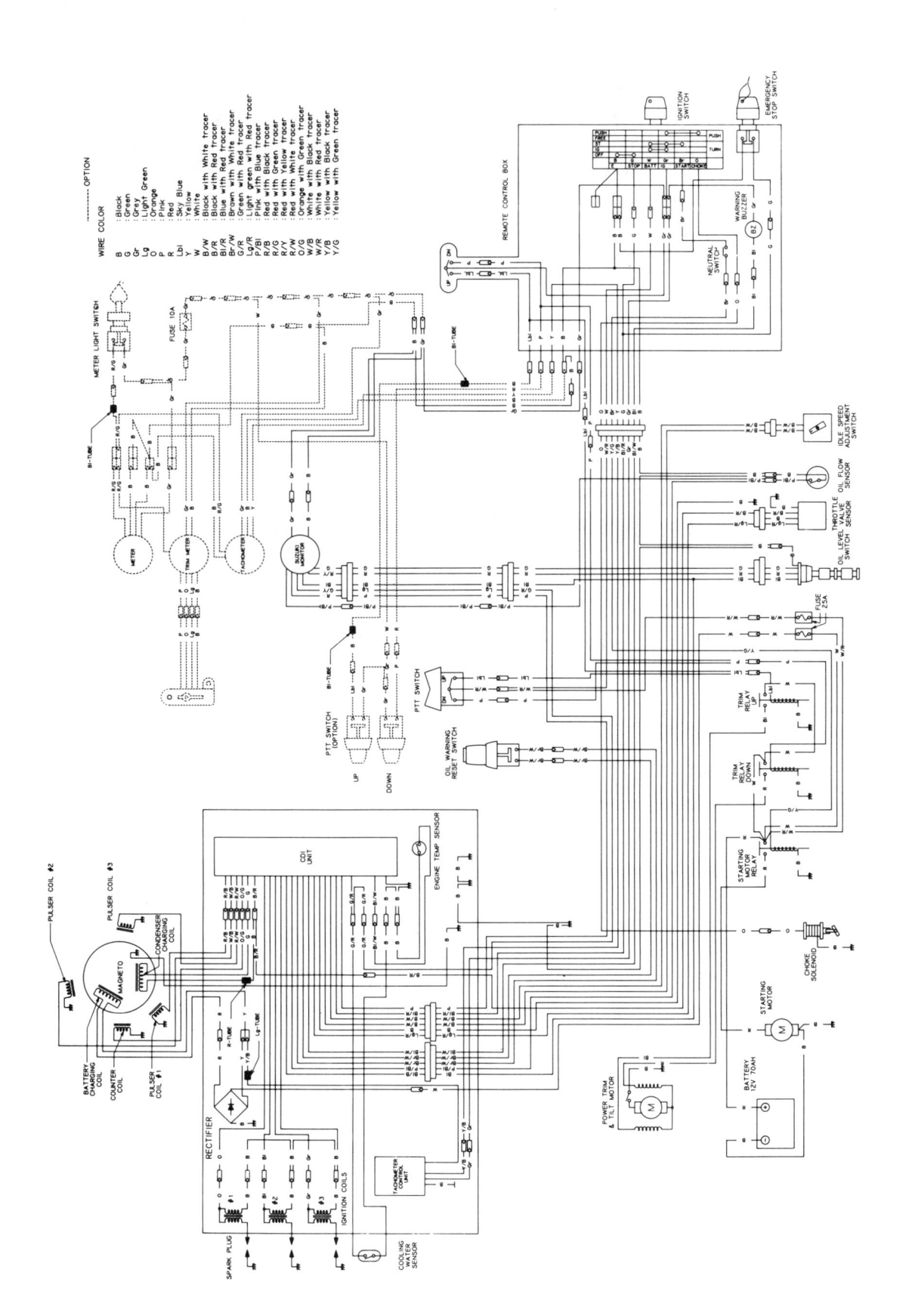

DT 75C/DT 85C (1988)

WIRING DIAGRAMS

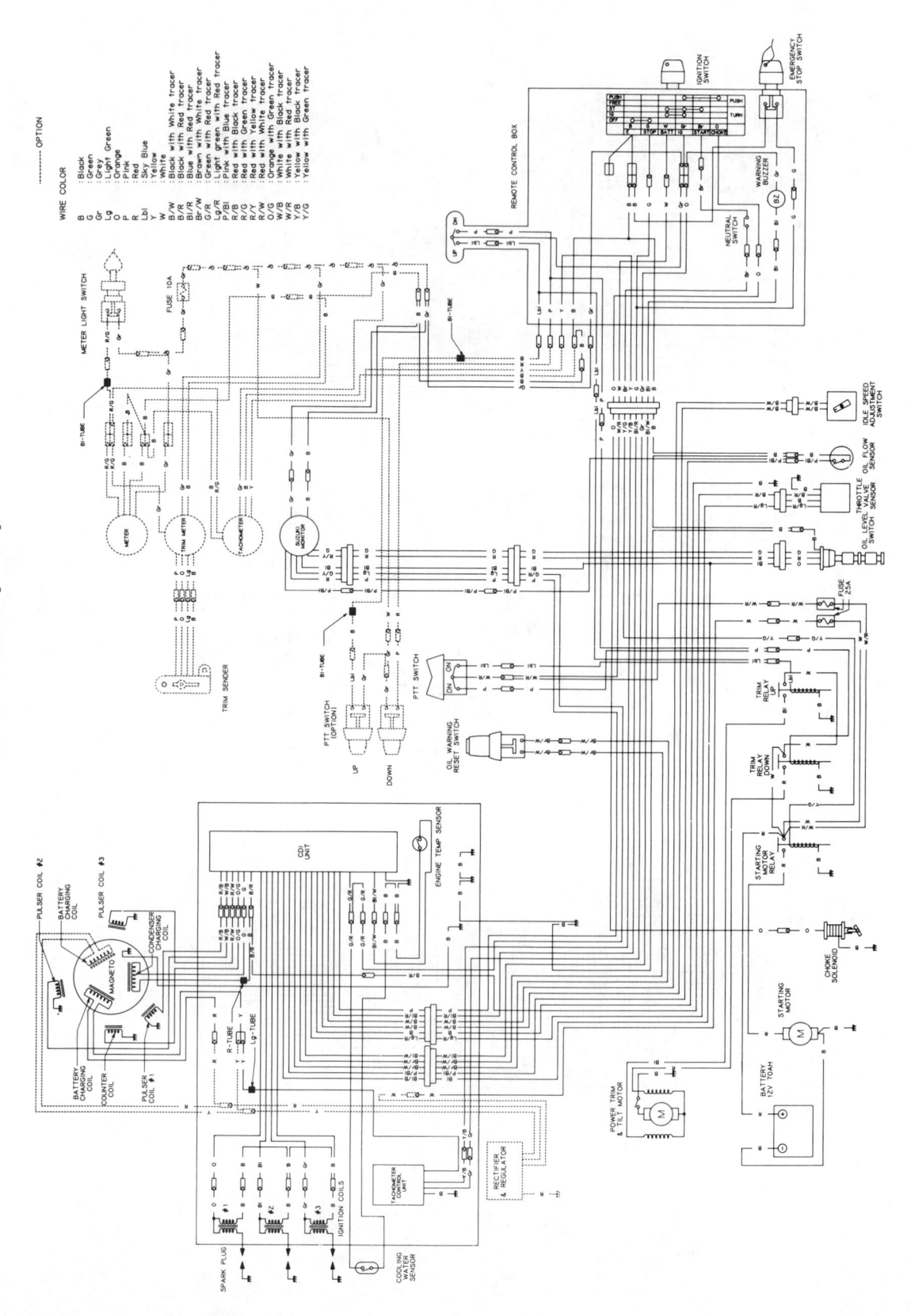

DT 75TC/DT 85TC (1988)

WIRING DIAGRAMS

DT 75TC/DT 85TC (1989)

------- : Option

NOTE: The tachometer and trim meter are standard equipments for Australian version.

WIRE COLOR

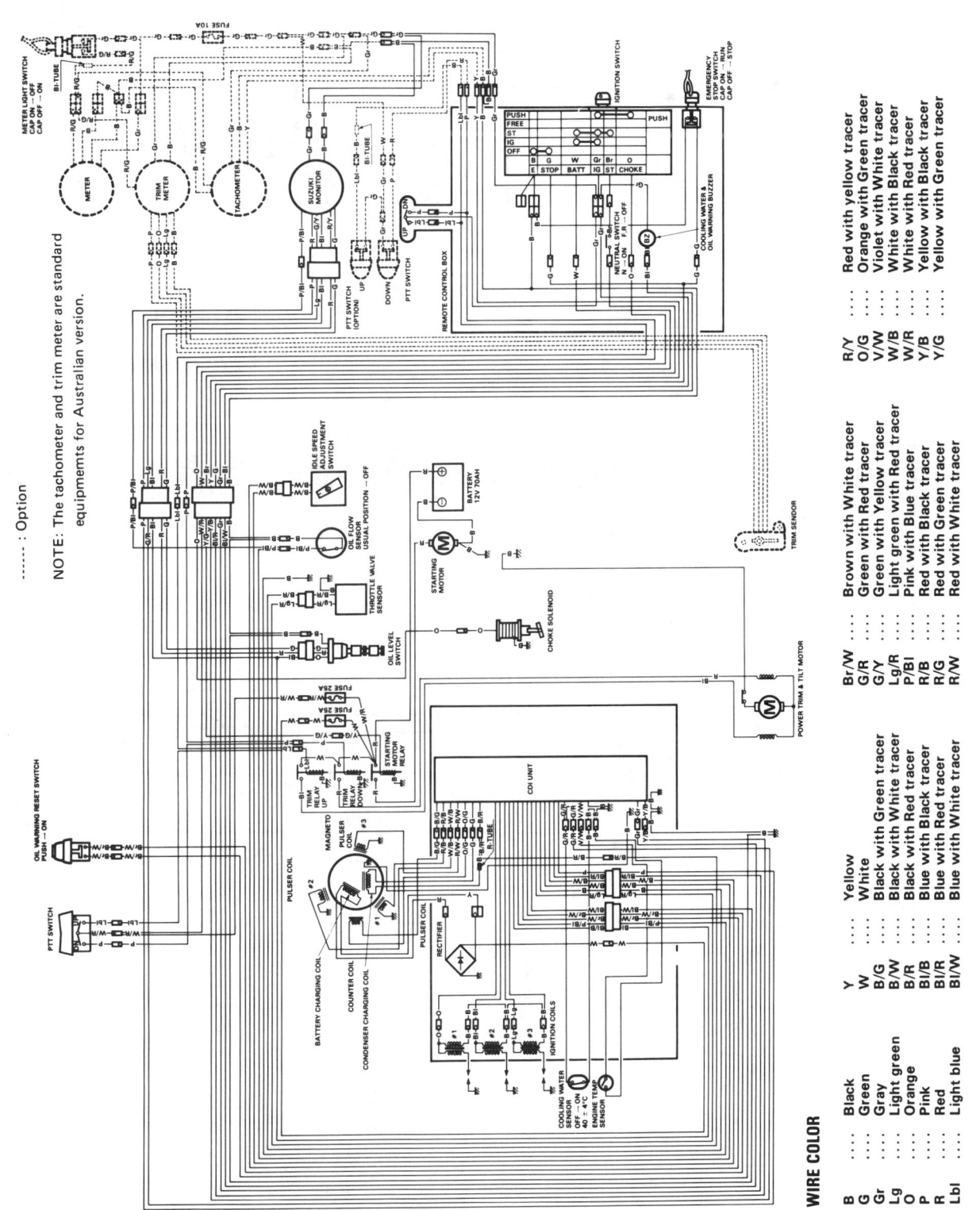

B	....	Black
G	....	Green
Gr	....	Gray
Lg	....	Light green
O	....	Orange
P	....	Pink
R	....	Red
Lbl	....	Light blue
Y	....	Yellow
W	....	White
B/G	....	Black with Green tracer
B/W	....	Black with White tracer
B/R	....	Black with Red tracer
Bl/B	....	Blue with Black tracer
Bl/R	....	Blue with Red tracer
Bl/W	....	Blue with White tracer
Br/W	....	Brown with White tracer
G/R	....	Green with Red tracer
G/Y	....	Green with Yellow tracer
Lg/R	....	Light green with Red tracer
P/Bl	....	Pink with Blue tracer
R/B	....	Red with Black tracer
R/G	....	Red with Green tracer
R/W	....	Red with White tracer
R/Y	....	Red with yellow tracer
O/G	....	Orange with Green tracer
V/W	....	Violet with White tracer
W/B	....	White with Black tracer
W/R	....	White with Red tracer
Y/B	....	Yellow with Black tracer
Y/G	....	Yellow with Green tracer

15

WIRING DIAGRAMS

DT 75TC/DT 85TC (1990-1991)

------ : Option

R/Y	Red with yellow tracer
O/G	Orange with Green tracer
V/W	Violet with White tracer
W/B	White with Black tracer
W/R	White with Red tracer
Y/B	Yellow with Black tracer
Y/G	Yellow with Green tracer
Br/W	Brown with White tracer
G/R	Green with Red tracer
G/Y	Green with Yellow tracer
Lg/R	Light green with Red tracer
P/Bl	Pink with Blue tracer
R/B	Red with Black tracer
R/G	Red with Green tracer
R/W	Red with White tracer

WIRE COLOR

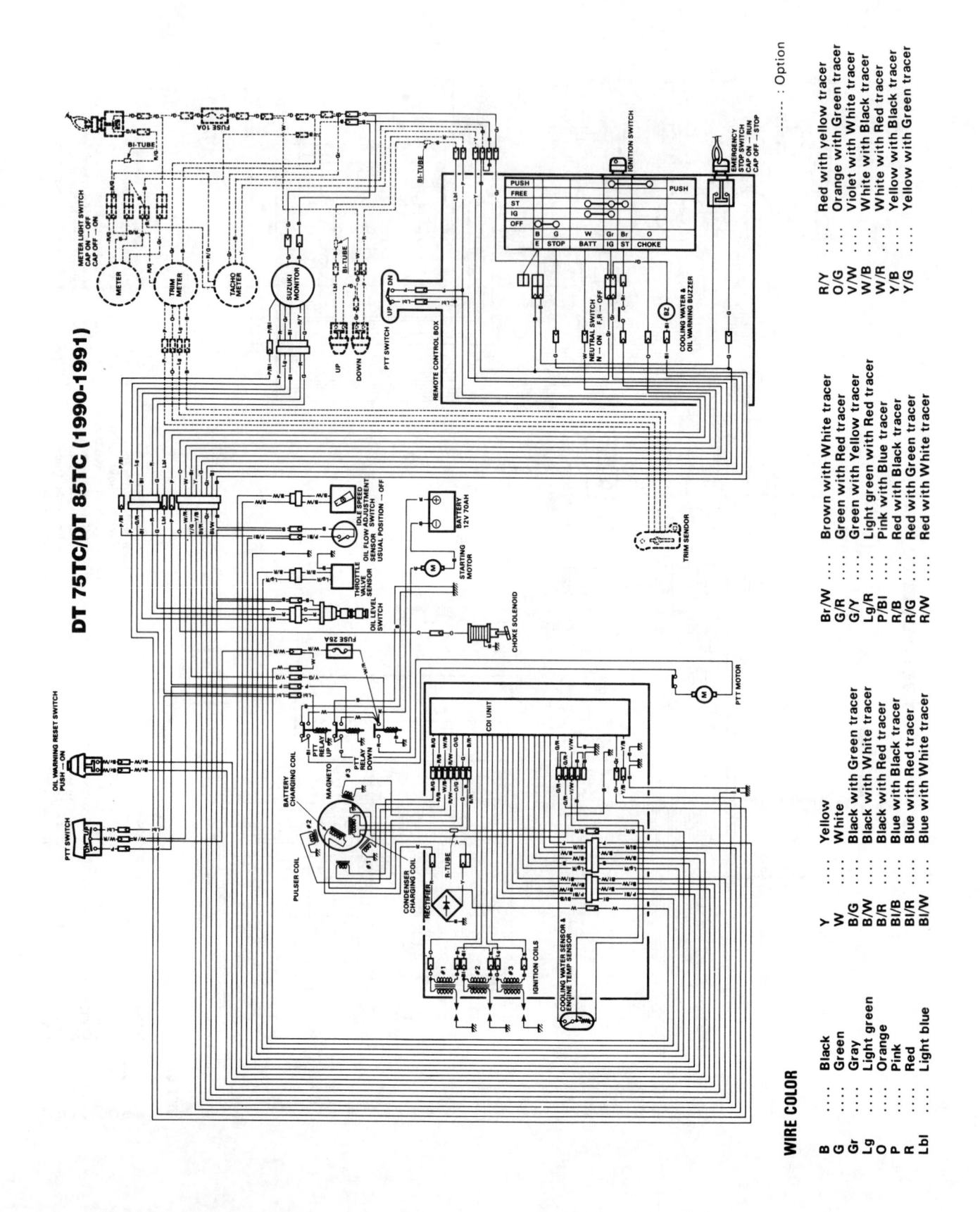

B	Black
G	Green
Gr	Gray
Lg	Light green
O	Orange
P	Pink
R	Red
Lbl	Light blue
Y	Yellow
W	White
B/G	Black with Green tracer
B/W	Black with White tracer
B/R	Black with Red tracer
Bl/B	Blue with Black tracer
Bl/R	Blue with Red tracer
Bl/W	Blue with White tracer

WIRING DIAGRAMS

DT 75TC/DT 85TC (1990-1991)

Br/W	:	Brown with White tracer
G/R	:	Green with Red tracer
G/Y	:	Green with Yellow tracer
Lg/R	:	Light green with Red tracer
P/Bl	:	Pink with Blue tracer
R/B	:	Red with Black tracer
R/G	:	Red with Green tracer
R/W	:	Red with White tracer

R/Y	:	Red with yellow tracer
O/G	:	Orange with Green tracer
V/W	:	Violet with White tracer
W/B	:	White with Black tracer
W/R	:	White with Red tracer
Y/B	:	Yellow with Black tracer
Y/G	:	Yellow with Green tracer

----- : Option

WIRE COLOR

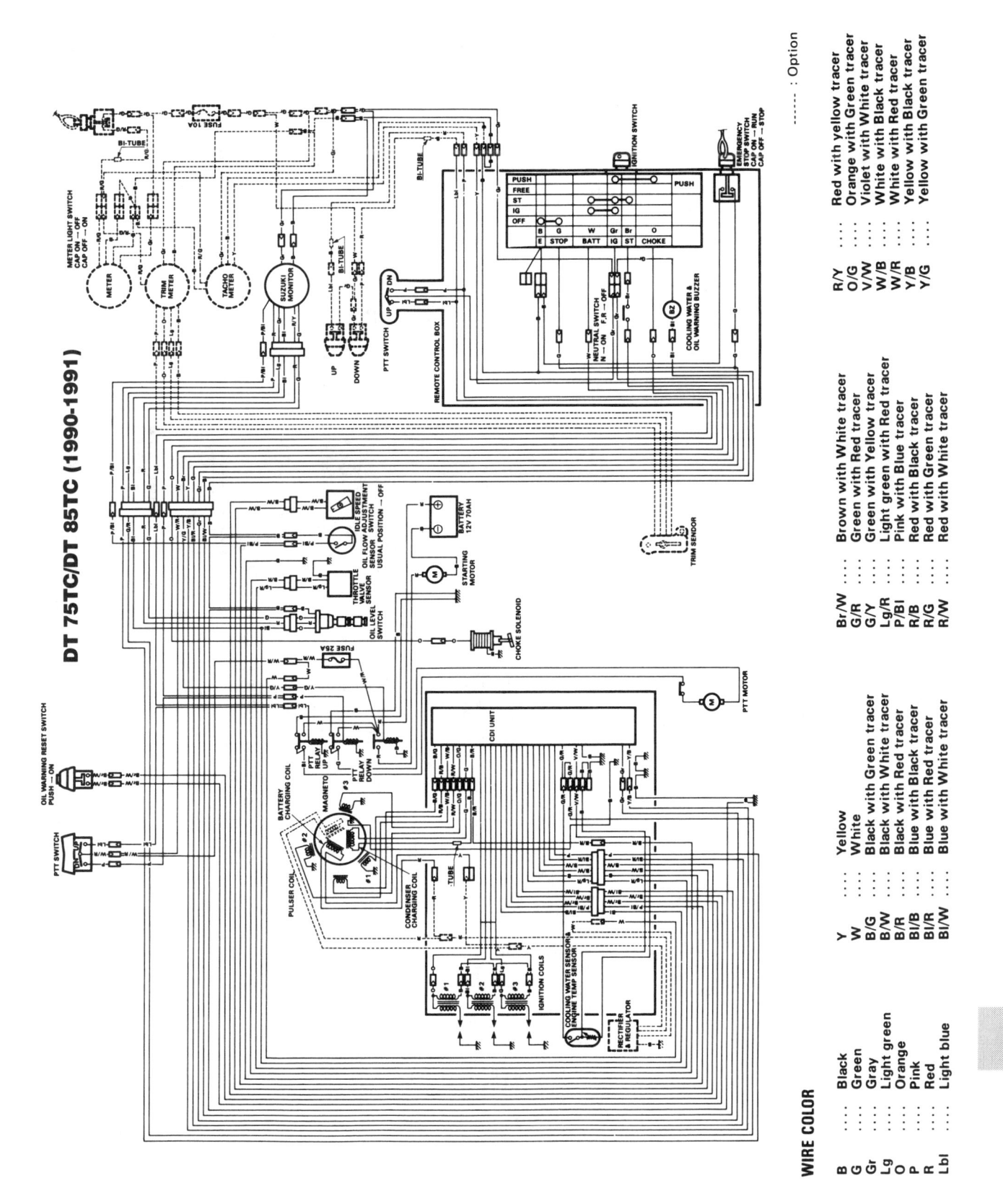

B	:	Black
G	:	Green
Gr	:	Gray
Lg	:	Light green
O	:	Orange
P	:	Pink
R	:	Red
Lbl	:	Light blue
Y	:	Yellow
W	:	White
B/G	:	Black with Green tracer
B/W	:	Black with White tracer
Bl/B	:	Blue with Black tracer
Bl/R	:	Blue with Red tracer
Bl/W	:	Blue with White tracer

WIRING DIAGRAMS

DT 115/DT 140 (1985)

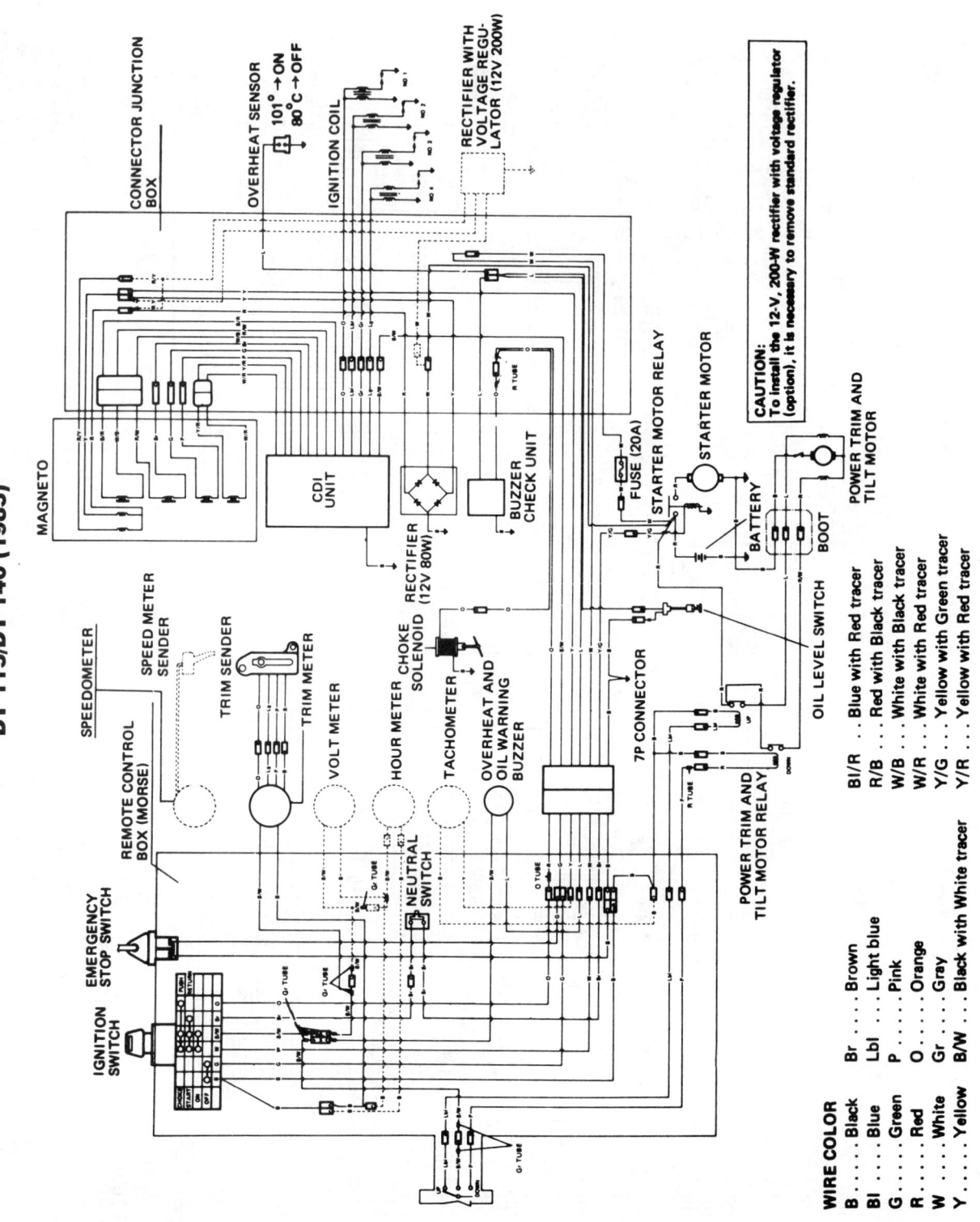

CAUTION: To install the 12-V, 200-W rectifier (option), it is necessary to remove standard rectifier.

WIRE COLOR

B	Black
Bl	Blue
G	Green
R	Red
W	White
Y	Yellow
Br	Brown
Lbl	Light blue
P	Pink
O	Orange
Gr	Gray
B/W	Black with White tracer
Bl/R	Blue with Red tracer
R/B	Red with Black tracer
W/B	White with Black tracer
W/R	White with Red tracer
Y/G	Yellow with Green tracer
Y/R	Yellow with Red tracer

WIRING DIAGRAMS

609

DT 115/DT 140 (1986)

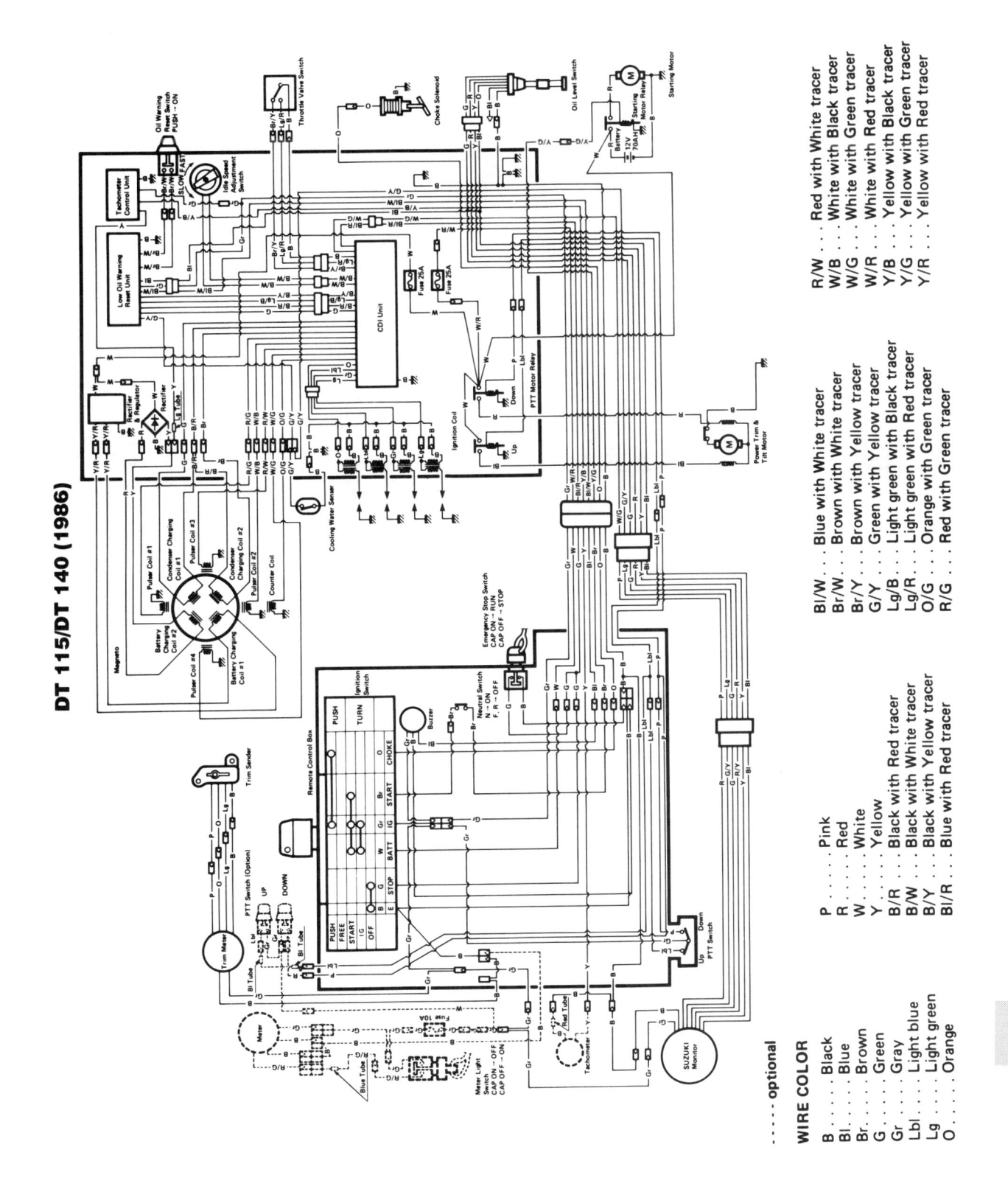

R/W	. . .	Red with White tracer		
W/B	. . .	White with Black tracer		
W/G	. . .	White with Green tracer		
W/R	. . .	White with Red tracer		
Y/B	. . .	Yellow with Black tracer		
Y/G	. . .	Yellow with Green tracer		
Y/R	. . .	Yellow with Red tracer		

Bl/W	. . .	Blue with White tracer
Br/W	. . .	Brown with White tracer
Br/Y	. . .	Brown with Yellow tracer
G/Y	. . .	Green with Yellow tracer
Lg/B	. . .	Light green with Black tracer
Lg/R	. . .	Light green with Red tracer
O/G	. . .	Orange with Green tracer
R/G	. . .	Red with Green tracer

P	. . . . .	Pink
R	. . . . .	Red
W	. . . . .	White
Y	. . . . .	Yellow
B/R	. . .	Black with Red tracer
B/W	. . .	Black with White tracer
B/Y	. . .	Black with Yellow tracer
Bl/R	. . .	Blue with Red tracer

- - - - optional

WIRE COLOR

B	. . . . .	Black
Bl	. . . . .	Blue
Br	. . . . .	Brown
G	. . . . .	Green
Gr	. . . . .	Gray
Lbl	. . . .	Light blue
Lg	. . . .	Light green
O	. . . . .	Orange

15

WIRING DIAGRAMS

DT 115/DT 140 (1987)

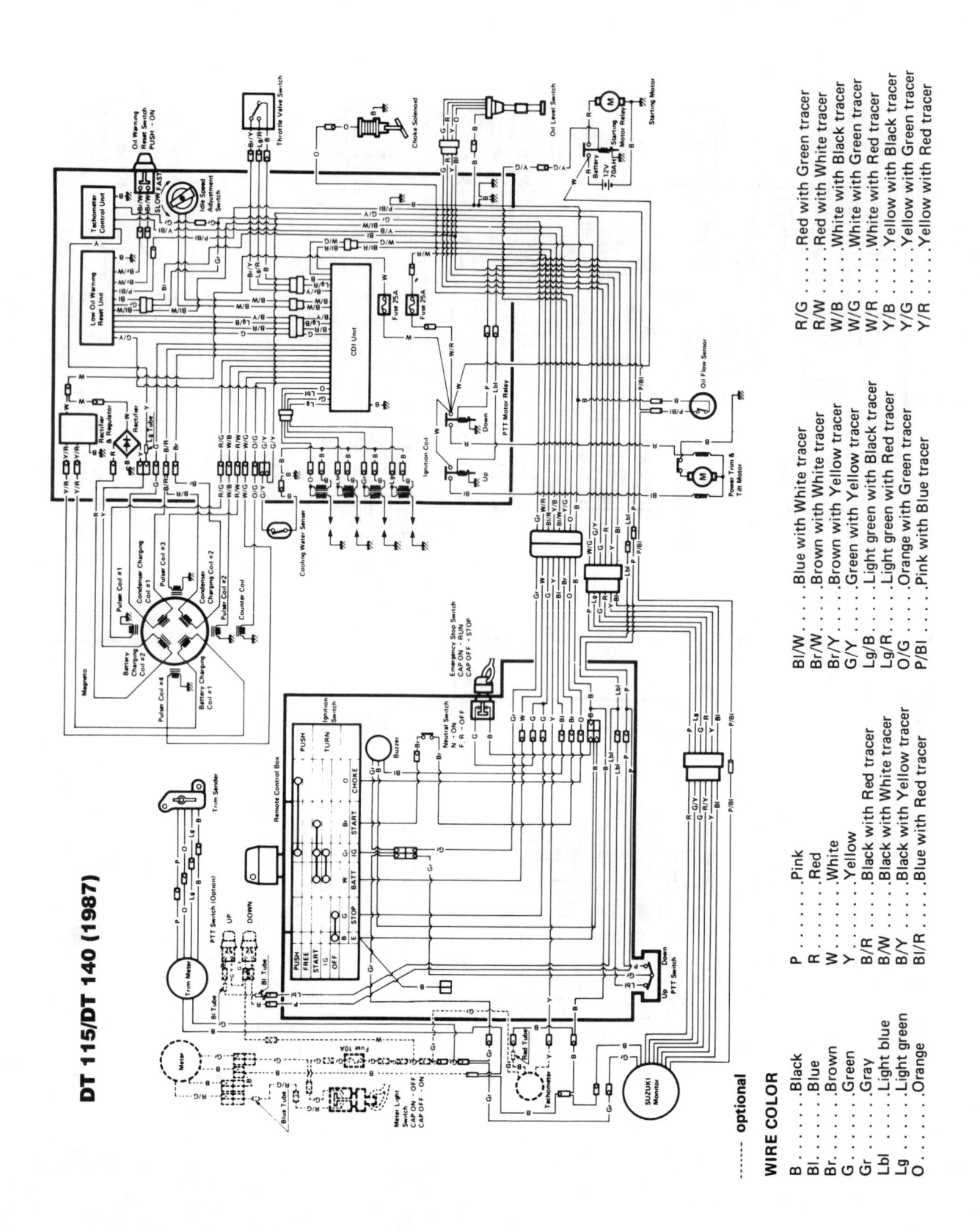

WIRE COLOR

B	......	Black
Bl	......	Blue
Br	......	Brown
G	......	Green
Gr	......	Gray
Lbl	......	Light blue
Lg	......	Light green
O	......	Orange
P	......	Pink
R	......	Red
W	......	White
Y	......	Yellow
B/R	......	Black with Red tracer
B/W	......	Black with White tracer
B/Y	......	Black with Yellow tracer
Bl/R	......	Blue with Red tracer
Bl/W	......	Blue with White tracer
Br/W	......	Brown with White tracer
Br/Y	......	Brown with Yellow tracer
G/Y	......	Green with Yellow tracer
Lg/B	......	Light green with Black tracer
Lg/R	......	Light green with Red tracer
O/G	......	Orange with Green tracer
P/Bl	......	Pink with Blue tracer
R/G	......	Red with Green tracer
R/W	......	Red with White tracer
W/B	......	White with Black tracer
W/G	......	White with Green tracer
W/R	......	White with Red tracer
Y/B	......	Yellow with Black tracer
Y/G	......	Yellow with Green tracer
Y/R	......	Yellow with Red tracer

----- optional

WIRING DIAGRAMS

611

DT 115/DT 140 (1988)

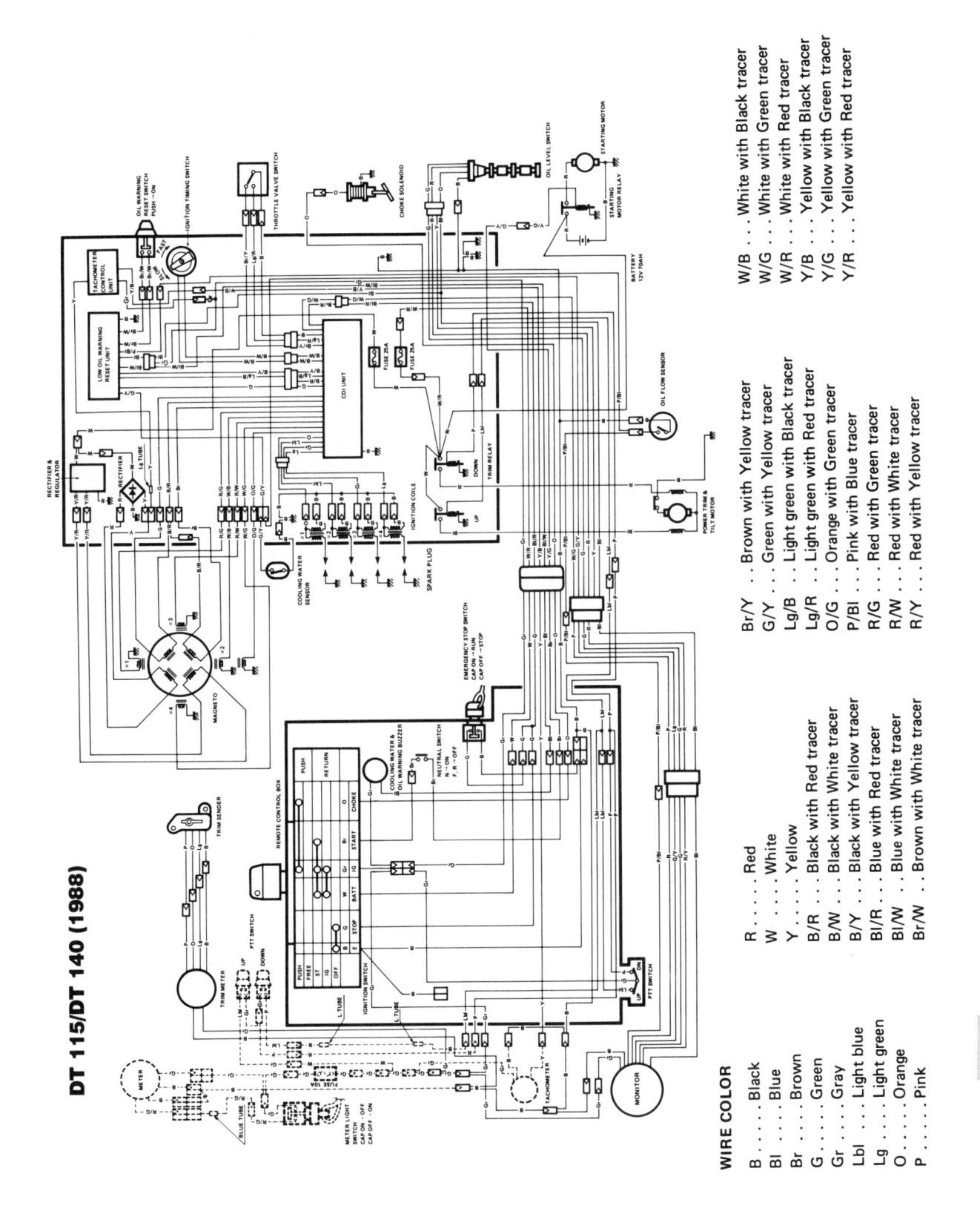

W/B	White with Black tracer		
W/G	White with Green tracer		
W/R	White with Red tracer		
Y/B	Yellow with Black tracer		
Y/G	Yellow with Green tracer		
Y/R	Yellow with Red tracer		

Br/Y	Brown with Yellow tracer
G/Y	Green with Yellow tracer
Lg/B	Light green with Black tracer
Lg/R	Light green with Red tracer
O/G	Orange with Green tracer
P/Bl	Pink with Blue tracer
R/G	Red with Green tracer
R/W	Red with White tracer
R/Y	Red with Yellow tracer

R	Red
W	White
Y	Yellow
B/R	Black with Red tracer
B/W	Black with White tracer
B/Y	Black with Yellow tracer
Bl/R	Blue with Red tracer
Bl/W	Blue with White tracer
Br/W	Brown with White tracer

WIRE COLOR

B	Black
Bl	Blue
Br	Brown
G	Green
Gr	Gray
Lbl	Light blue
Lg	Light green
O	Orange
P	Pink

15

WIRING DIAGRAMS

DT 115/DT 140 (1989)

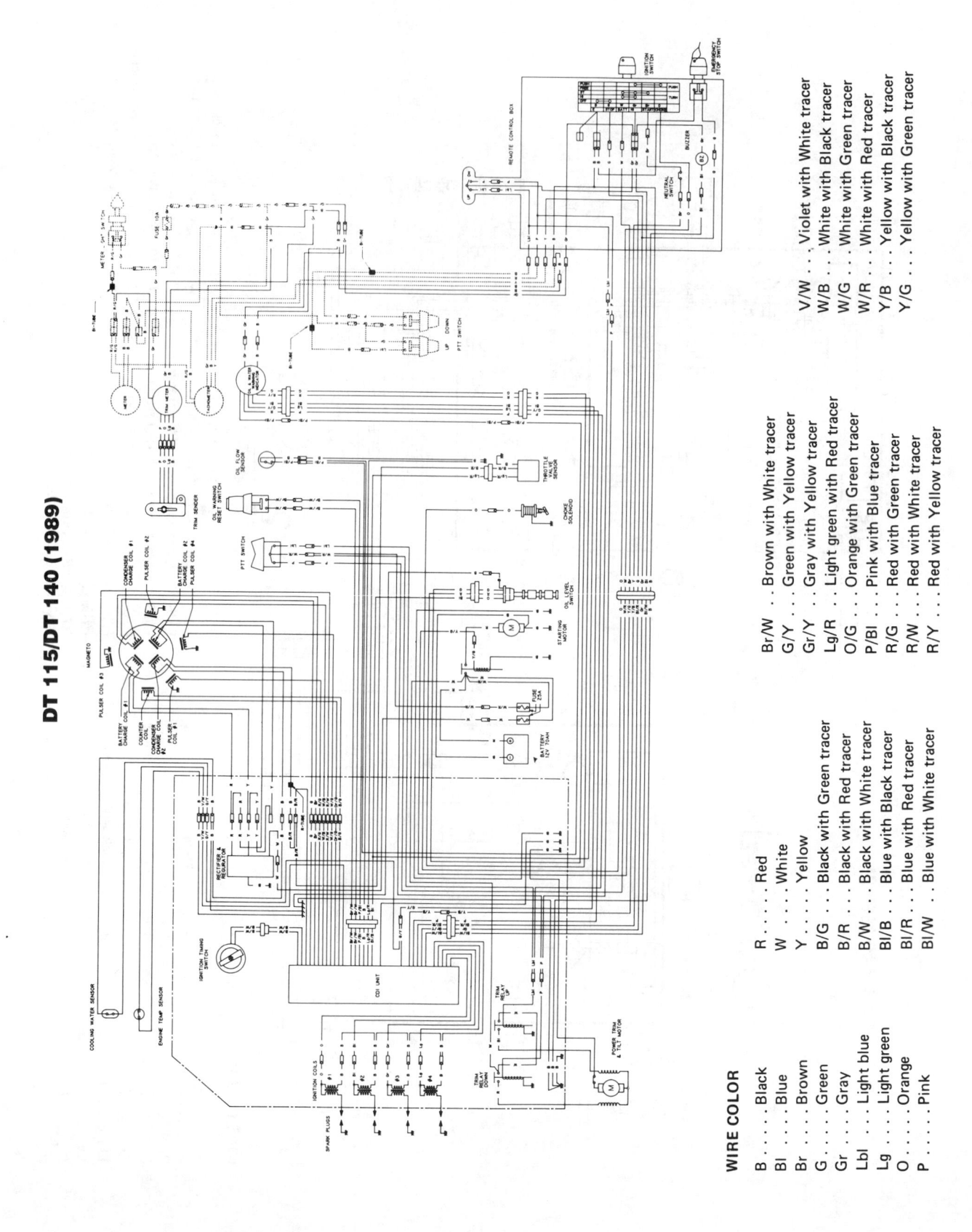

R	. . . .	Red
W	. . . .	White
Y	. . . .	Yellow
B/G	. .	Black with Green tracer
B/R	. .	Black with Red tracer
B/W	. .	Black with White tracer
Bl/B	. .	Blue with Black tracer
Bl/R	. .	Blue with Red tracer
Bl/W	. .	Blue with White tracer
Br/W	. .	Brown with White tracer
G/Y	. .	Green with Yellow tracer
Gr/Y	. .	Gray with Yellow tracer
Lg/R	. .	Light green with Red tracer
O/G	. .	Orange with Green tracer
P/Bl	. .	Pink with Blue tracer
R/G	. .	Red with Green tracer
R/W	. .	Red with White tracer
R/Y	. .	Red with Yellow tracer
V/W	. .	Violet with White tracer
W/B	. .	White with Black tracer
W/G	. .	White with Green tracer
W/R	. .	White with Red tracer
Y/B	. .	Yellow with Black tracer
Y/G	. .	Yellow with Green tracer

WIRE COLOR

B	. . . .	Black
Bl	. . . .	Blue
Br	. . . .	Brown
G	. . . .	Green
Gr	. . . .	Gray
Lbl	. . .	Light blue
Lg	. . .	Light green
O	. . . .	Orange
P	. . . .	Pink

Wiring Diagrams

613

DT 115/DT 140 (1990-1991)

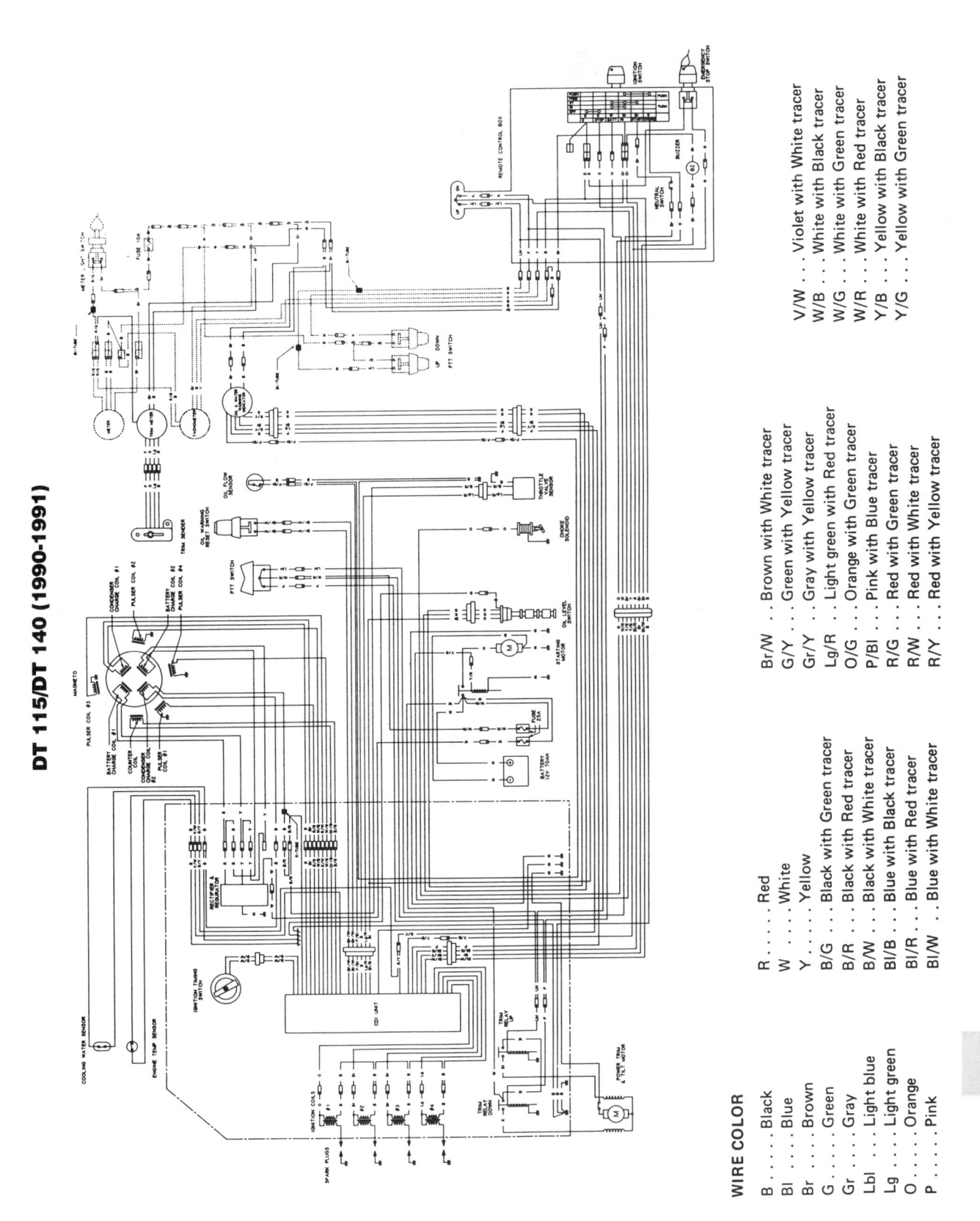

WIRE COLOR

B	Black
Bl	Blue
Br	Brown
G	Green
Gr	Gray
Lbl	Light blue
Lg	Light green
O	Orange
P	Pink

R	Red
W	White
Y	Yellow
B/G	Black with Green tracer
B/R	Black with Red tracer
B/W	Black with White tracer
Bl/B	Blue with Black tracer
Bl/R	Blue with Red tracer
Bl/W	Blue with White tracer

Br/W	Brown with White tracer
G/Y	Green with Yellow tracer
Gr/Y	Gray with Yellow tracer
Lg/R	Light green with Red tracer
O/G	Orange with Green tracer
P/Bl	Pink with Blue tracer
R/G	Red with Green tracer
R/W	Red with White tracer
R/Y	Red with Yellow tracer

V/W	Violet with White tracer
W/B	White with Black tracer
W/G	White with Green tracer
W/R	White with Red tracer
Y/B	Yellow with Black tracer
Y/G	Yellow with Green tracer

15

WIRING DIAGRAMS

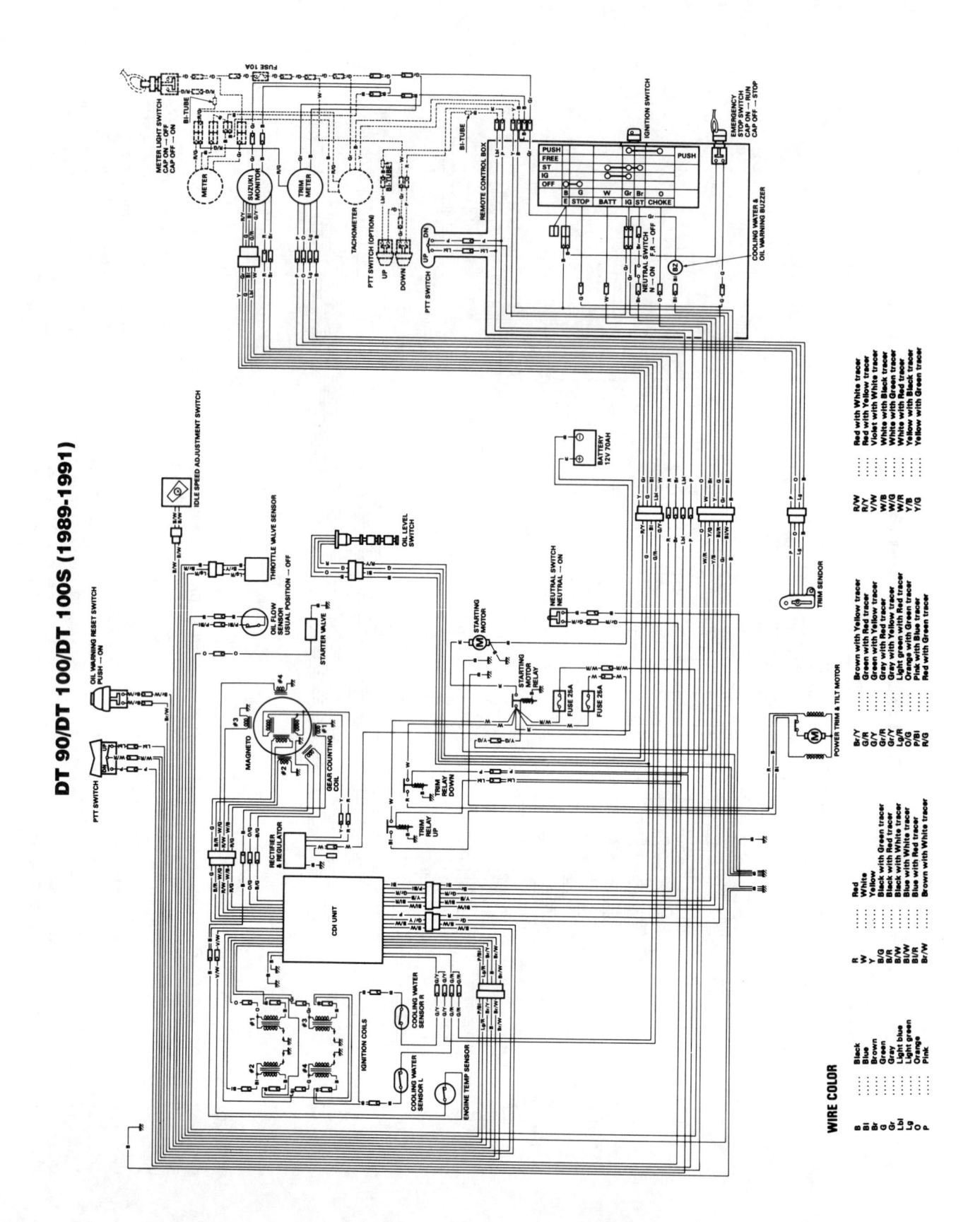

DT 90/DT 100/DT 100S (1989-1991)

WIRING DIAGRAMS

DT 150/DT 200 (1986)

WIRE COLOR

- B ..... Black
- Bl ..... Blue
- Br ..... Brown
- G ..... Green
- Gr ..... Gray
- Lbl ..... Light blue
- Lg ..... Light green
- O ..... Orange
- P ..... Pink
- R ..... Red
- W ..... White
- Y ..... Yellow
- B/R ..... Black with Red tracer
- B/W ..... Black with White tracer
- B/Y ..... Black with Yellow tracer
- Bl/R ..... Blue with Red tracer
- Bl/W ..... Blue with White tracer
- Bl/Y ..... Blue with Yellow tracer
- Br/W ..... Brown with White tracer
- Br/Y ..... Brown with Yellow tracer
- G/R ..... Green with Red tracer
- G/Y ..... Green with Yellow tracer
- Lg/B ..... Light green with Black tracer
- Lg/R ..... Light green with Red tracer
- O/G ..... Orange with Green tracer
- R/B ..... Red with Black tracer
- R/G ..... Red with Green tracer
- R/W ..... Red with White tracer
- R/Y ..... Red with Yellow tracer
- W/B ..... White with Black tracer
- W/G ..... White with Green tracer
- W/R ..... White with Red tracer
- W/Y ..... White with Yellow tracer
- Y/B ..... Yellow with Black tracer
- Y/G ..... Yellow with Green tracer

----- optional

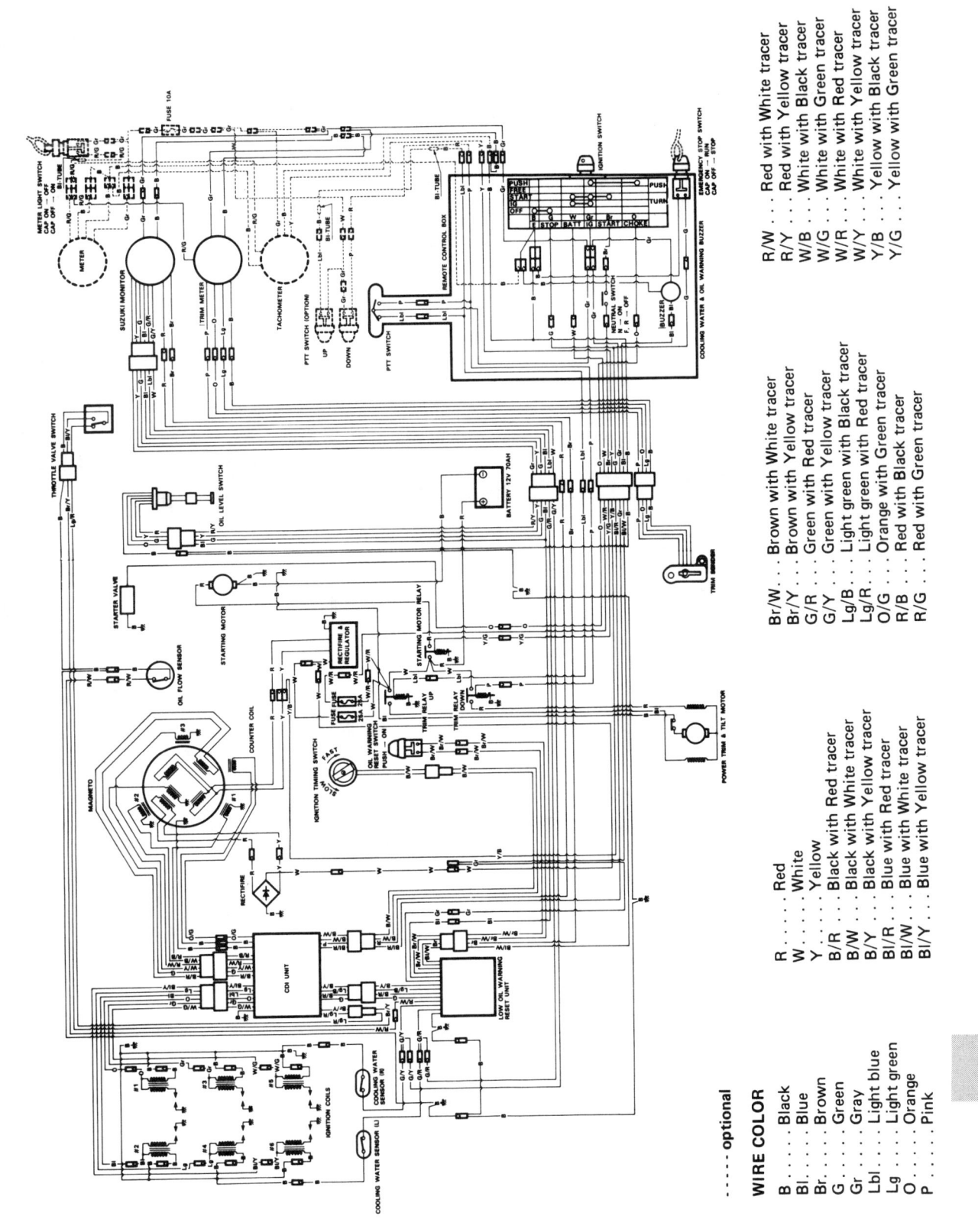

WIRING DIAGRAMS

DT 150SS (1986)

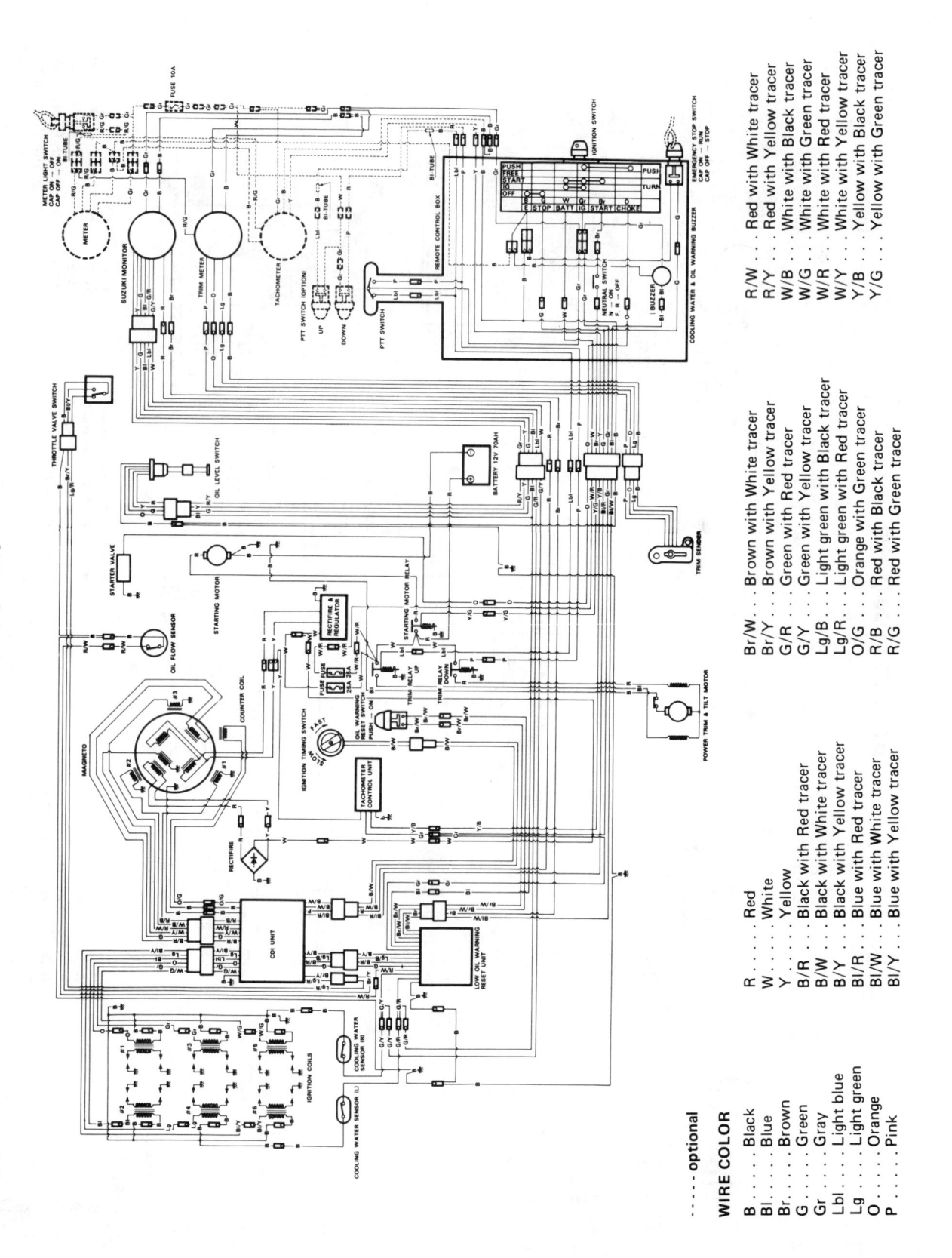

WIRE COLOR			
B	Black	R	Red
Bl	Blue	W	White
Br	Brown	Y	Yellow
G	Green	B/R	Black with Red tracer
Gr	Gray	B/W	Black with White tracer
Lbl	Light blue	B/Y	Black with Yellow tracer
Lg	Light green	Bl/R	Blue with Red tracer
O	Orange	Bl/W	Blue with White tracer
P	Pink	Bl/Y	Blue with Yellow tracer

Br/W	Brown with White tracer	R/W	Red with White tracer
Br/Y	Brown with Yellow tracer	R/Y	Red with Yellow tracer
G/R	Green with Red tracer	W/B	White with Black tracer
G/Y	Green with Yellow tracer	W/G	White with Green tracer
Lg/B	Light green with Black tracer	W/R	White with Red tracer
Lg/R	Light green with Red tracer	W/Y	White with Yellow tracer
O/G	Orange with Green tracer	Y/B	Yellow with Black tracer
R/B	Red with Black tracer	Y/G	Yellow with Green tracer
R/G	Red with Green tracer		

----- optional

WIRING DIAGRAMS

DT 150/DT 175/DT 200 (1987)

- - - - optional

WIRE COLOR

B	. . .	Black
Bl	. . .	Blue
Br	. . .	Brown
G	. . .	Green
Gr	. . .	Gray
Lbl	. . .	Light blue
Lg	. . .	Light green
O	. . .	Orange
P	. . .	Pink

R	. . .	Red
W	. . .	White
Y	. . .	Yellow
B/R	. . .	Black with Red tracer
B/W	. . .	Black with White tracer
B/Y	. . .	Black with Yellow tracer
Bl/R	. . .	Blue with Red tracer
Bl/W	. . .	Blue with White tracer
Bl/Y	. . .	Blue with Yellow tracer
Br/W	. . .	Brown with White tracer
Br/Y	. . .	Brown with Yellow tracer
G/R	. . .	Green with Red tracer
G/Y	. . .	Green with Yellow tracer
Lg/B	. . .	Light green with Black tracer
Lg/R	. . .	Light green with Red tracer
O/G	. . .	Orange with Green tracer
R/B	. . .	Red with Black tracer
R/G	. . .	Red with Green tracer
R/W	. . .	Red with White tracer
R/Y	. . .	Red with Yellow tracer
W/B	. . .	White with Black tracer
W/G	. . .	White with Green tracer
W/R	. . .	White with Red tracer
W/Y	. . .	White with Yellow tracer
Y/B	. . .	Yellow with Black tracer
Y/G	. . .	Yellow with Green tracer

15

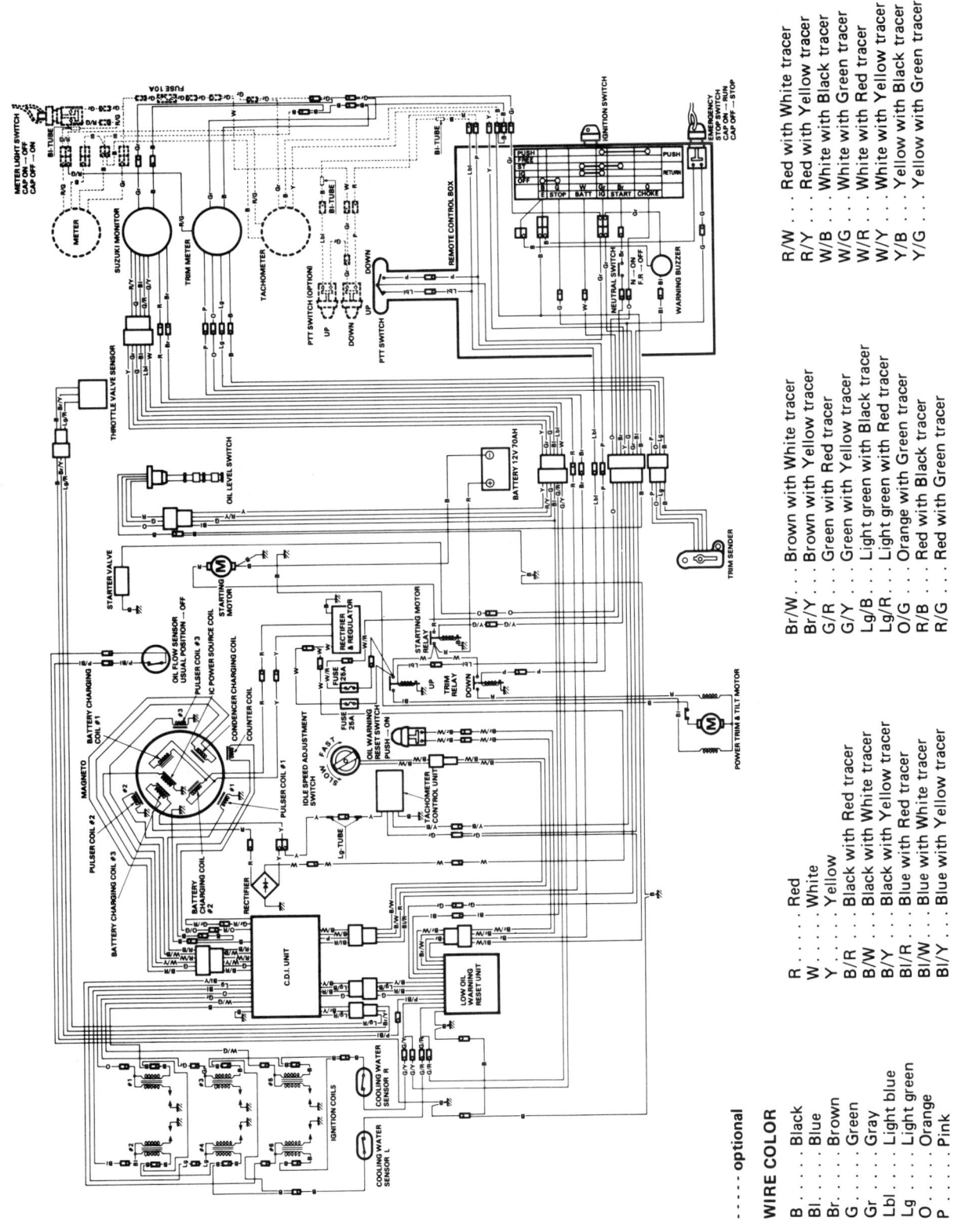

WIRING DIAGRAMS

DT 150SS/DT 200AE (1987)

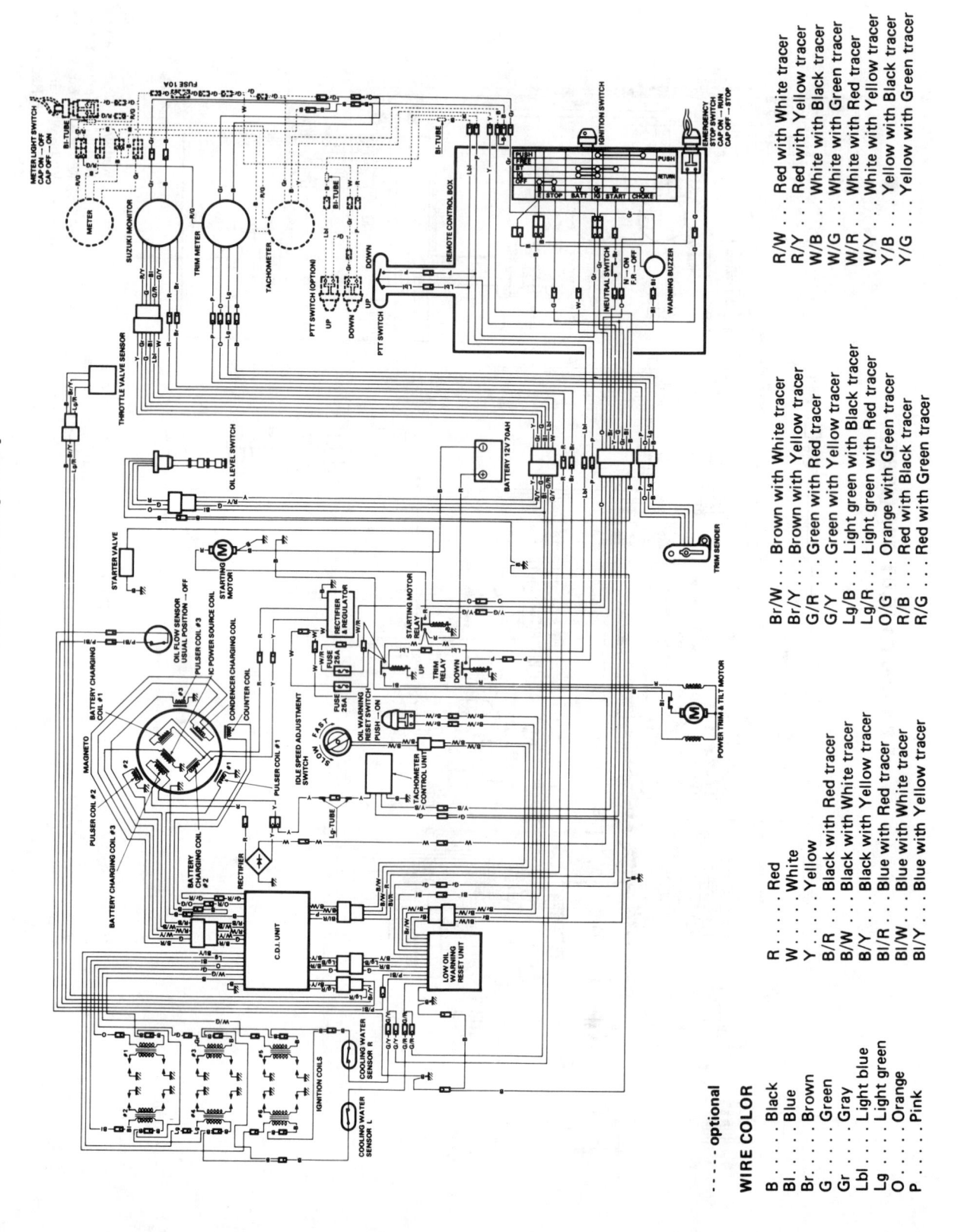

WIRE COLOR

- B . . . . . Black
- Bl . . . . Blue
- Br . . . . Brown
- G . . . . Green
- Gr . . . . Gray
- Lbl . . . Light blue
- Lg . . . . Light green
- O . . . . Orange
- P . . . . Pink

- R . . . . . Red
- W . . . . White
- Y . . . . . Yellow
- B/R . . . Black with Red tracer
- B/W . . . Black with White tracer
- B/Y . . . Black with Yellow tracer
- Bl/R . . . Blue with Red tracer
- Bl/W . . . Blue with White tracer
- Bl/Y . . . Blue with Yellow tracer

- Br/W . . . Brown with White tracer
- Br/Y . . . Brown with Yellow tracer
- G/R . . . Green with Red tracer
- G/Y . . . Green with Yellow tracer
- Lg/B . . . Light green with Black tracer
- Lg/R . . . Light green with Red tracer
- O/G . . . Orange with Green tracer
- R/B . . . Red with Black tracer
- R/G . . . Red with Green tracer

- R/W . . . Red with White tracer
- R/Y . . . Red with Yellow tracer
- W/B . . . White with Black tracer
- W/G . . . White with Green tracer
- W/R . . . White with Red tracer
- W/Y . . . White with Yellow tracer
- Y/B . . . Yellow with Black tracer
- Y/G . . . Yellow with Green tracer

- - - - - optional

WIRING DIAGRAMS

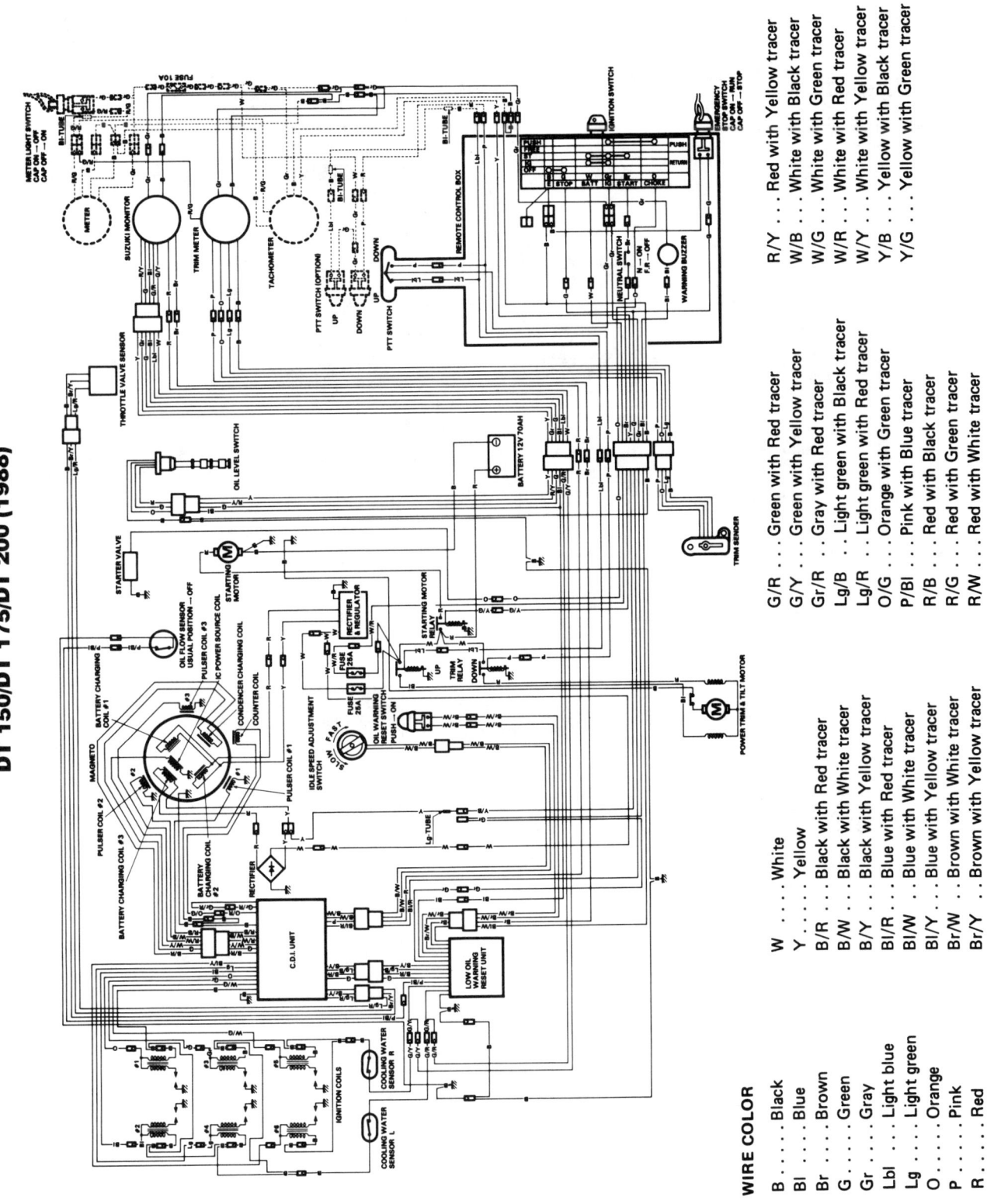

DT 150/DT 175/DT 200 (1988)

WIRE COLOR

B	Black	R/Y	Red with Yellow tracer
Bl	Blue	W/B	White with Black tracer
Br	Brown	W/G	White with Green tracer
G	Green	W/R	White with Red tracer
Gr	Gray	W/Y	White with Yellow tracer
Lbl	Light blue	Y/B	Yellow with Black tracer
Lg	Light green	Y/G	Yellow with Green tracer
O	Orange		
P	Pink		
R	Red		
W	White	G/R	Green with Red tracer
Y	Yellow	G/Y	Green with Yellow tracer
B/R	Black with Red tracer	Gr/R	Gray with Red tracer
B/W	Black with White tracer	Lg/B	Light green with Black tracer
B/Y	Black with Yellow tracer	Lg/R	Light green with Red tracer
Bl/R	Blue with Red tracer	O/G	Orange with Green tracer
Bl/W	Blue with White tracer	P/Bl	Pink with Blue tracer
Bl/Y	Blue with Yellow tracer	R/B	Red with Black tracer
Br/W	Brown with White tracer	R/G	Red with Green tracer
Br/Y	Brown with Yellow tracer	R/W	Red with White tracer

15

WIRING DIAGRAMS

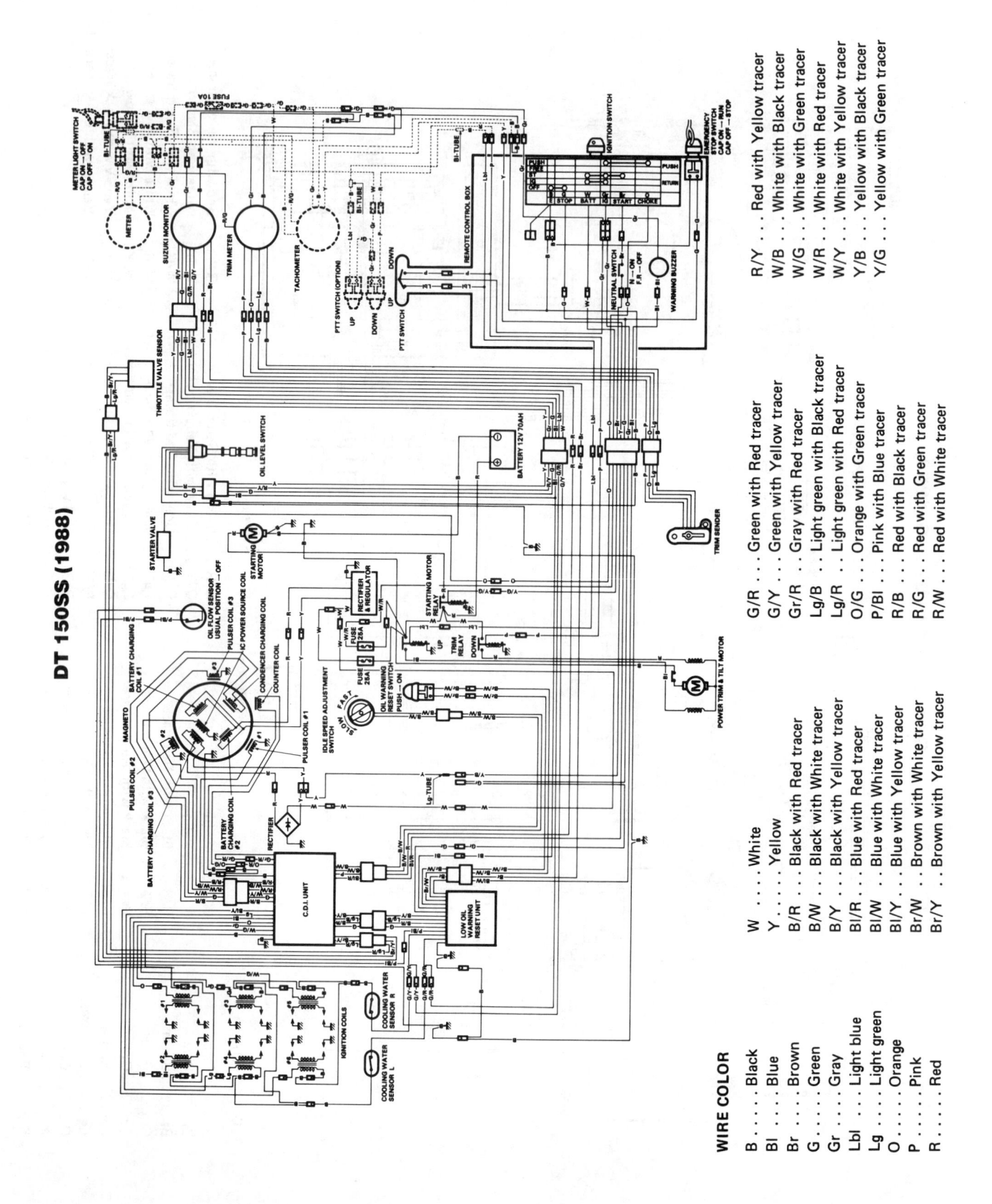

DT 150SS (1988)

WIRE COLOR			
B	Black	G/R	Green with Red tracer
Bl	Blue	G/Y	Green with Yellow tracer
Br	Brown	Gr/R	Gray with Red tracer
G	Green	Lg/B	Light green with Black tracer
Gr	Gray	Lg/R	Light green with Red tracer
Lbl	Light blue	O/G	Orange with Green tracer
Lg	Light green	P/Bl	Pink with Blue tracer
O	Orange	R/B	Red with Black tracer
P	Pink	R/G	Red with Green tracer
R	Red	R/W	Red with White tracer
W	White	R/Y	Red with Yellow tracer
Y	Yellow	W/B	White with Black tracer
B/R	Black with Red tracer	W/G	White with Green tracer
B/W	Black with White tracer	W/R	White with Red tracer
B/Y	Black with Yellow tracer	W/Y	White with Yellow tracer
Bl/R	Blue with Red tracer	Y/B	Yellow with Black tracer
Bl/W	Blue with White tracer	Y/G	Yellow with Green tracer
Bl/Y	Blue with Yellow tracer		
Br/W	Brown with White tracer		
Br/Y	Brown with Yellow tracer		

WIRING DIAGRAMS

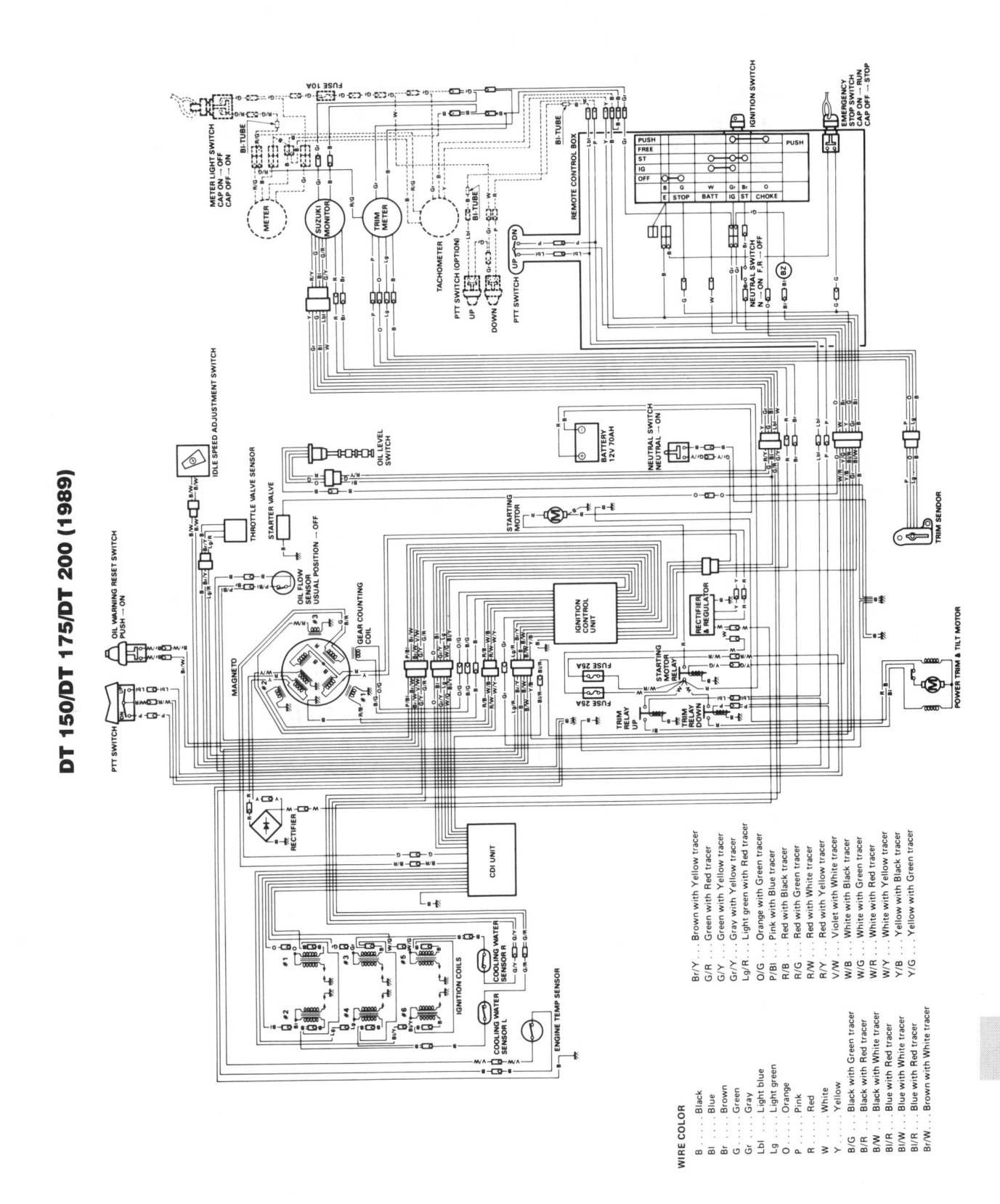

621

DT 150/DT 175/DT 200 (1989)

WIRE COLOR

B	Black	Br/Y	Brown with Yellow tracer
Bl	Blue	G/R	Green with Red tracer
Br	Brown	G/Y	Green with Yellow tracer
G	Green	Gr/Y	Gray with Yellow tracer
Gr	Gray	Lg/R	Light green with Red tracer
Lbl	Light blue	O/G	Orange with Green tracer
Lg	Light green	P/Bl	Pink with Blue tracer
O	Orange	R/B	Red with Black tracer
P	Pink	R/G	Red with Green tracer
R	Red	R/W	Red with White tracer
W	White	R/Y	Red with Yellow tracer
Y	Yellow	V/W	Violet with White tracer
B/G	Black with Green tracer	W/B	White with Black tracer
B/R	Black with Red tracer	W/G	White with Green tracer
B/W	Black with White tracer	W/R	White with Red tracer
Bl/R	Blue with Red tracer	W/Y	White with Yellow tracer
Bl/W	Blue with White tracer	Y/B	Yellow with Black tracer
Br/R	Brown with Red tracer	Y/G	Yellow with Green tracer
Br/W	Brown with White tracer		

15

WIRING DIAGRAMS

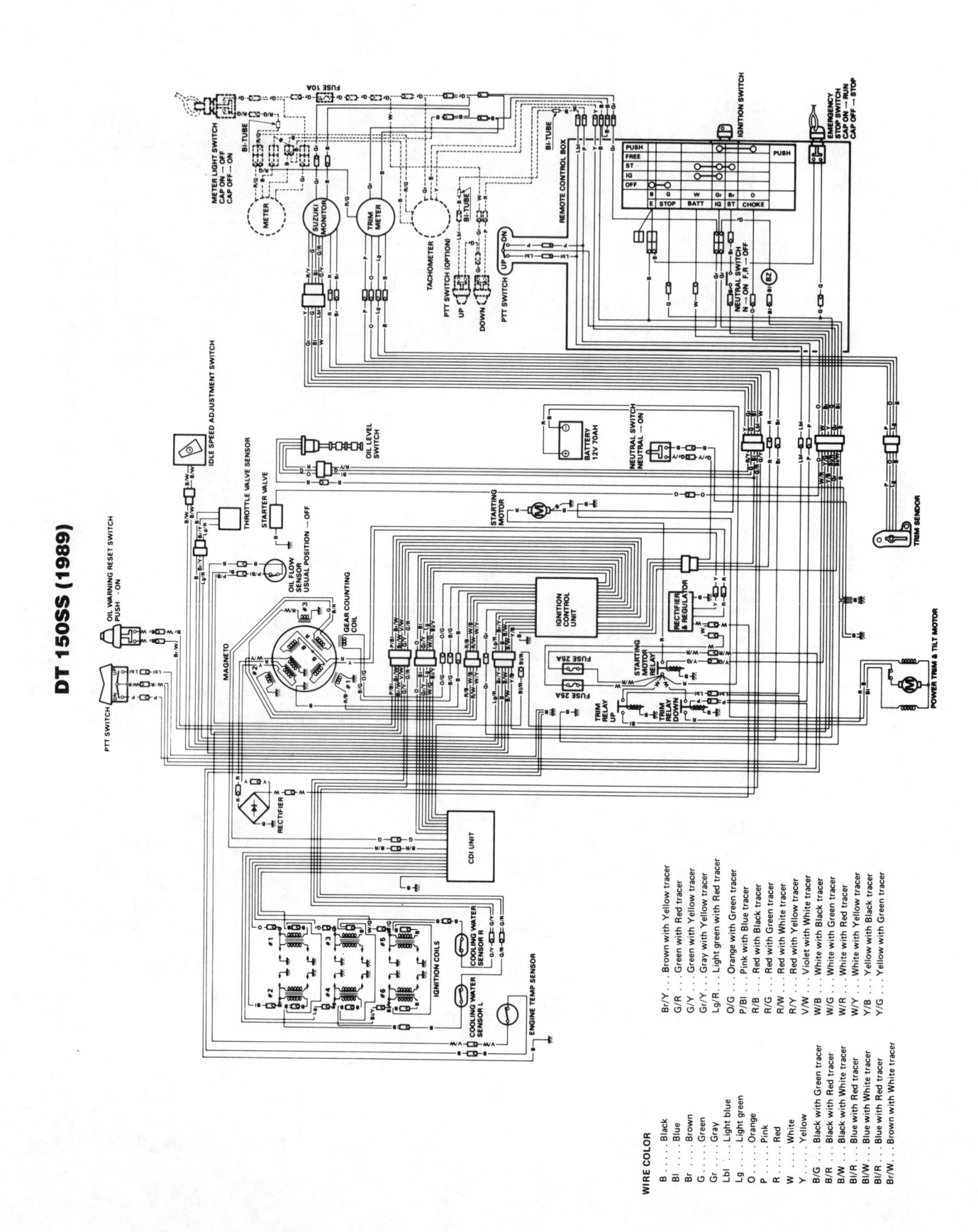

DT 150SS (1989)

WIRE COLOR

B	Black	Br/Y	Brown with Yellow tracer
Bl	Blue	G/R	Green with Red tracer
Br	Brown	G/Y	Green with Yellow tracer
G	Green	Gr/Y	Gray with Yellow tracer
Lbl	Light blue	Lg/R	Light green with Red tracer
Lg	Light green	O/G	Orange with Green tracer
O	Orange	P/Bl	Pink with Blue tracer
P	Pink	R/B	Red with Black tracer
R	Red	R/G	Red with Green tracer
W	White	R/W	Red with White tracer
Y	Yellow	R/Y	Red with Yellow tracer
B/G	Black with Green tracer	V/W	Violet with White tracer
B/R	Black with Red tracer	W/B	White with Black tracer
B/W	Black with White tracer	W/G	White with Green tracer
Bl/R	Blue with Red tracer	W/R	White with Red tracer
Bl/W	Blue with White tracer	W/Y	White with Yellow tracer
Br/R	Brown with Red tracer	Y/B	Yellow with Black tracer
Br/W	Brown with White tracer	Y/G	Yellow with Green tracer

Wiring Diagrams

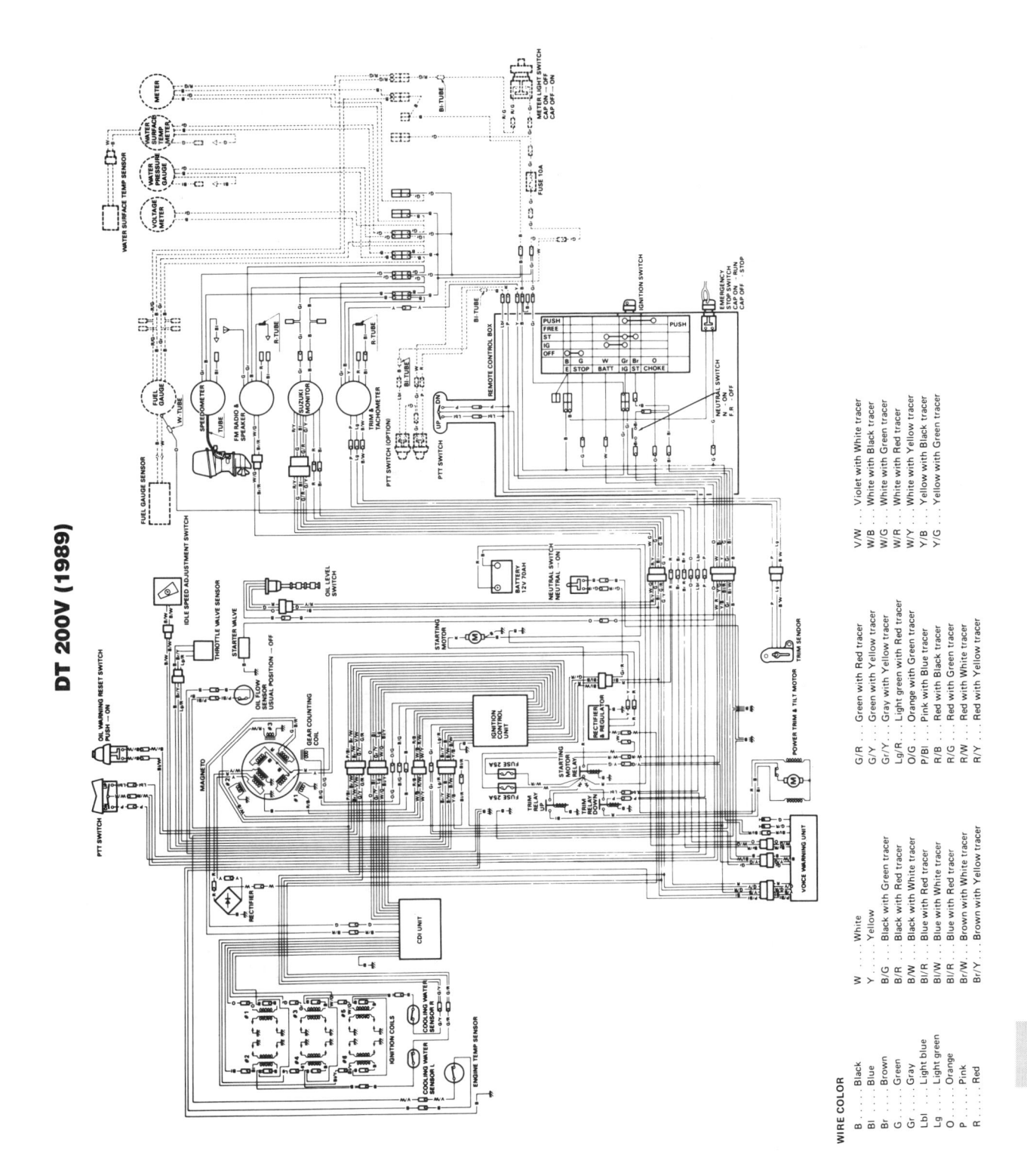

DT 200V (1989)

WIRING DIAGRAMS

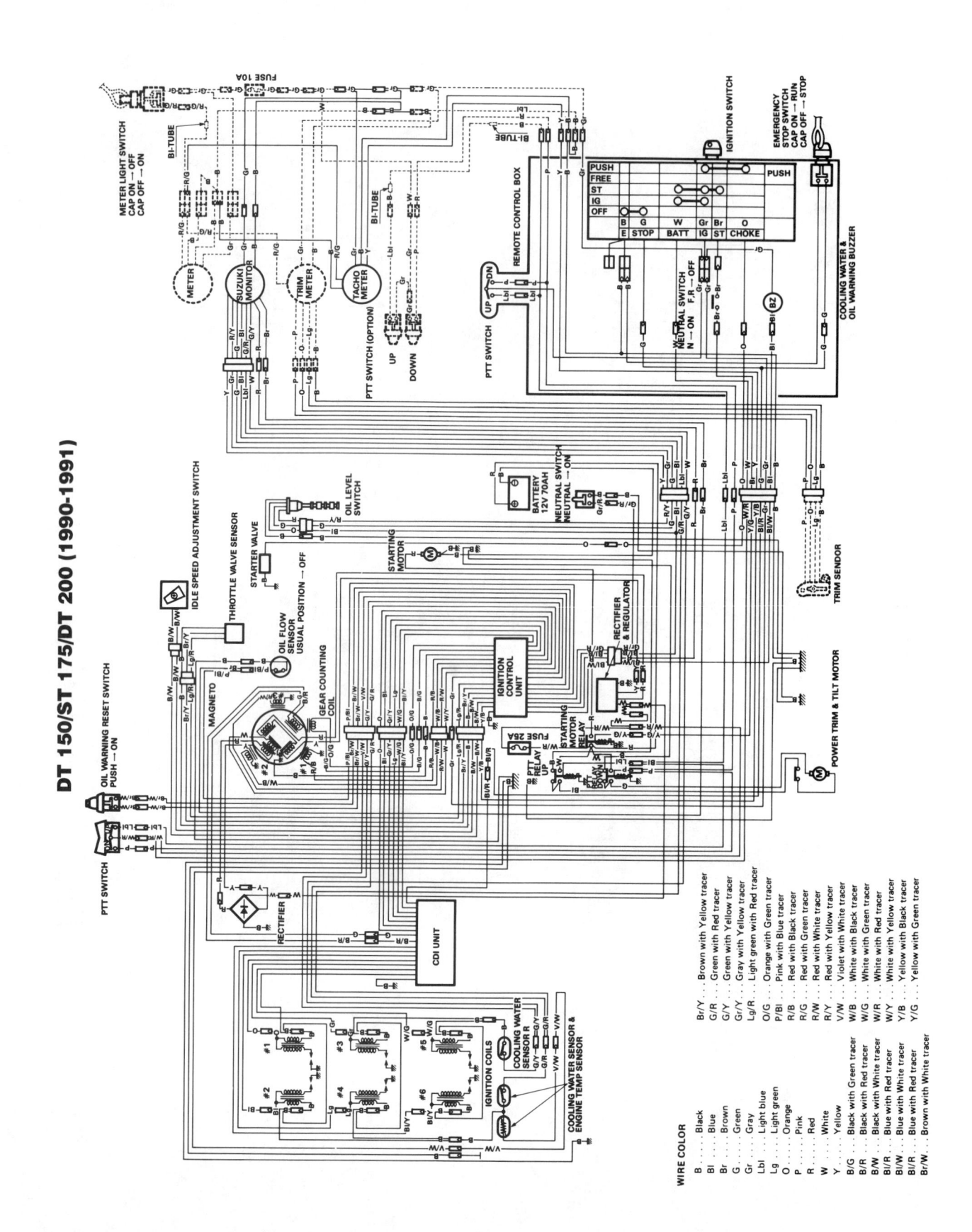

DT 150/ST 175/DT 200 (1990-1991)

WIRE COLOR

B	. . . .	Black
Bl	. . . .	Blue
Br	. . . .	Brown
G	. . . .	Green
Gr	. . . .	Gray
Lbl	. . . .	Light blue
Lg	. . . .	Light green
O	. . . .	Orange
P	. . . .	Pink
R	. . . .	Red
W	. . . .	White
Y	. . . .	Yellow
B/G	. . . .	Black with Green tracer
B/R	. . . .	Black with Red tracer
B/W	. . . .	Black with White tracer
Bl/R	. . . .	Blue with Red tracer
Bl/W	. . . .	Blue with White tracer
Bl/Y	. . . .	Blue with Yellow tracer
Bl/B	. . . .	Blue with Black tracer
Br/W	. . . .	Brown with White tracer
Br/Y	. . . .	Brown with Yellow tracer
G/R	. . . .	Green with Red tracer
G/Y	. . . .	Green with Yellow tracer
Gr/Y	. . . .	Gray with Yellow tracer
Lg/R	. . . .	Light green with Red tracer
O/G	. . . .	Orange with Green tracer
P/Bl	. . . .	Pink with Blue tracer
R/B	. . . .	Red with Black tracer
R/W	. . . .	Red with White tracer
R/Y	. . . .	Red with Yellow tracer
V/W	. . . .	Violet with White tracer
W/B	. . . .	White with Black tracer
W/G	. . . .	White with Green tracer
W/R	. . . .	White with Red tracer
W/Y	. . . .	White with Yellow tracer
Y/B	. . . .	Yellow with Black tracer
Y/R	. . . .	Yellow with Red tracer
Y/G	. . . .	Yellow with Green tracer

WIRING DIAGRAMS

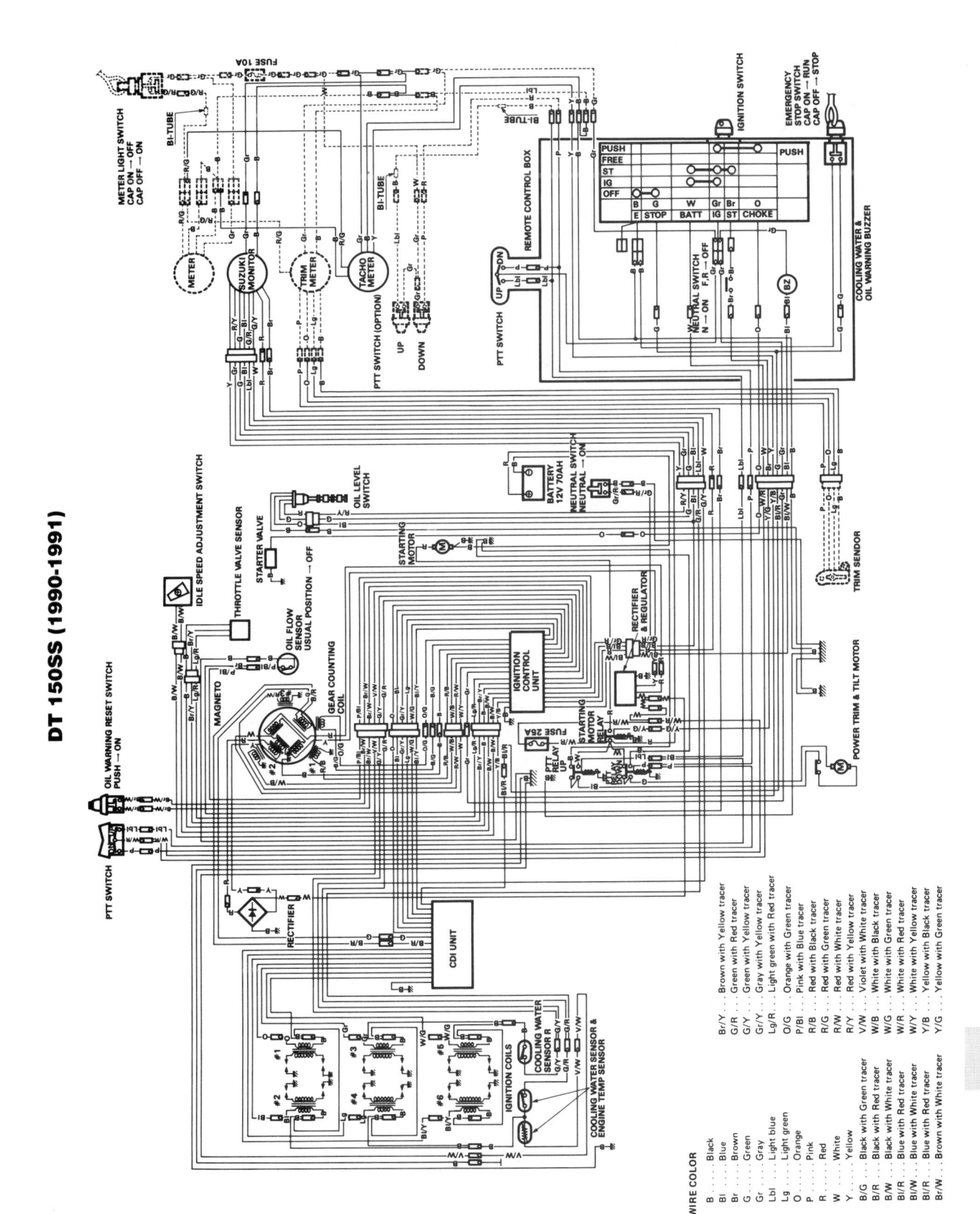

DT 150SS (1990-1991)

WIRE COLOR

B	Black	Br/Y	Brown with Yellow tracer
Bl	Blue	G/R	Green with Red tracer
Br	Brown	G/Y	Green with Yellow tracer
G	Green	Gr/Y	Gray with Yellow tracer
Gr	Gray	Lg/R	Light green with Red tracer
Lbl	Light blue	O/G	Orange with Green tracer
Lg	Light green	P/Bl	Pink with Blue tracer
O	Orange	R/B	Red with Black tracer
P	Pink	R/G	Red with Green tracer
R	Red	R/W	Red with White tracer
W	White	R/Y	Red with Yellow tracer
Y	Yellow	V/W	Violet with White tracer
B/G	Black with Green tracer	W/B	White with Black tracer
B/R	Black with Red tracer	W/G	White with Green tracer
B/W	Black with White tracer	W/R	White with Red tracer
Bl/R	Blue with Red tracer	W/Y	White with Yellow tracer
Bl/W	Blue with White tracer	Y/B	Yellow with Black tracer
Br/W	Brown with White tracer	Y/G	Yellow with Green tracer

15

626 WIRING DIAGRAMS

DT 200V (1990-1991)

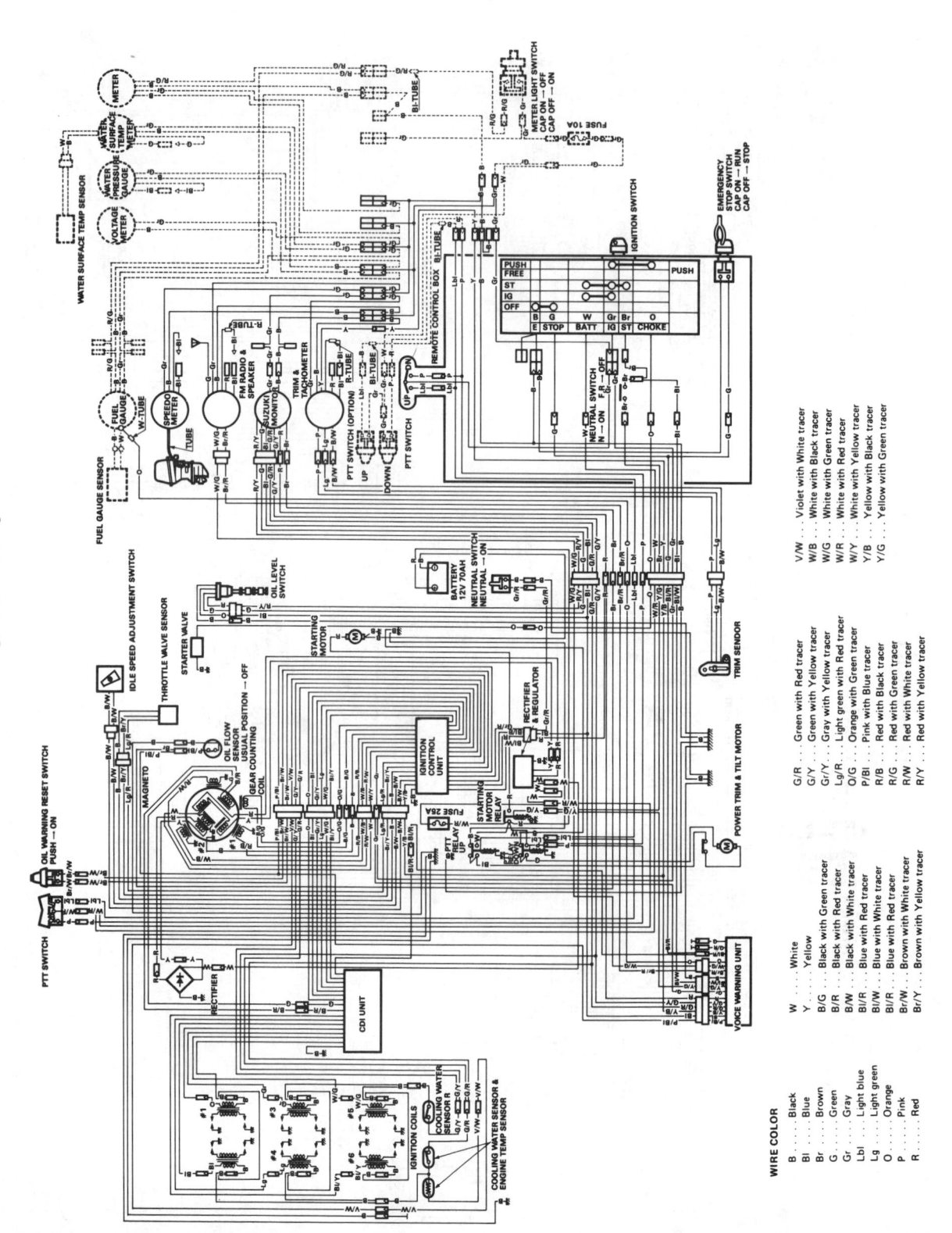

WIRE COLOR

B	Black	G/R	Green with Red tracer
Bl	Blue	G/Y	Green with Yellow tracer
Br	Brown	Gr/Y	Gray with Yellow tracer
G	Green	Lg/R	Light green with Red tracer
Gr	Gray	O/G	Orange with Green tracer
Lbl	Light blue	P/Bl	Pink with Blue tracer
Lg	Light green	R/B	Red with Black tracer
O	Orange	R/G	Red with Green tracer
P	Pink	R/W	Red with White tracer
R	Red	R/Y	Red with Yellow tracer
W	White	V/W	Violet with White tracer
Y	Yellow	W/B	White with Black tracer
B/G	Black with Green tracer	W/G	White with Green tracer
B/R	Black with Red tracer	W/R	White with Red tracer
B/W	Black with White tracer	W/Y	White with Yellow tracer
Bl/R	Blue with Red tracer	Y/B	Yellow with Black tracer
Bl/W	Blue with White tracer	Y/G	Yellow with Green tracer
Br/W	Brown with White tracer		
Br/Y	Brown with Yellow tracer		

WIRING DIAGRAMS

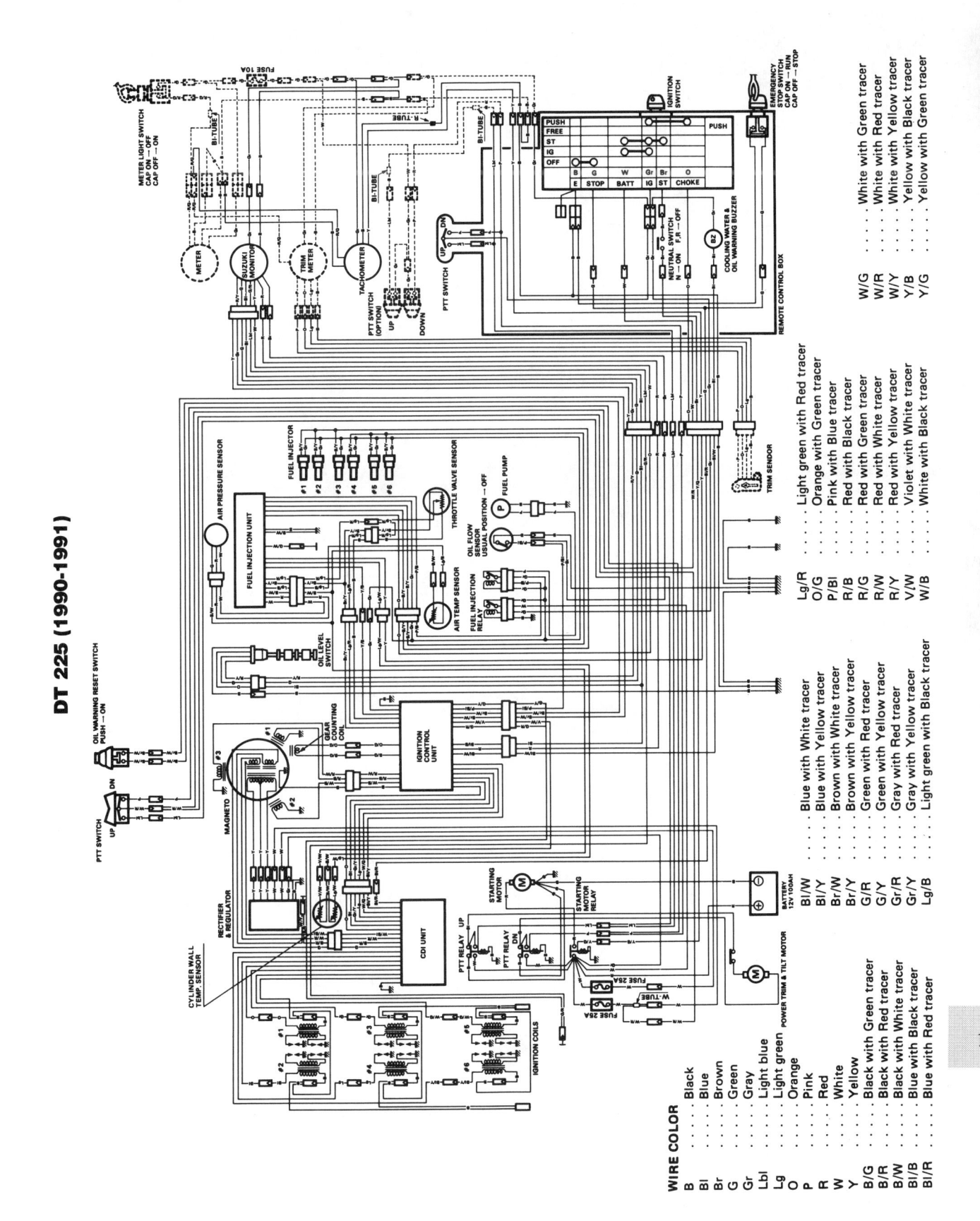

DT 225 (1990-1991)

WIRE COLOR		
B	. . . .	Black
Bl	. . . .	Blue
Br	. . . .	Brown
G	. . . .	Green
Gr	. . . .	Gray
Lbl	. . . .	Light blue
Lg	. . . .	Light green
O	. . . .	Orange
P	. . . .	Pink
R	. . . .	Red
W	. . . .	White
Y	. . . .	Yellow
B/G	. . . .	Black with Green tracer
B/R	. . . .	Black with Red tracer
B/W	. . . .	Black with White tracer
Bl/B	. . . .	Blue with Black tracer
Bl/R	. . . .	Blue with Red tracer
Bl/W	. . . .	Blue with White tracer
Bl/Y	. . . .	Blue with Yellow tracer
Br/W	. . . .	Brown with White tracer
Br/Y	. . . .	Brown with Yellow tracer
G/R	. . . .	Green with Red tracer
G/Y	. . . .	Green with Yellow tracer
Gr/R	. . . .	Gray with Red tracer
Gr/Y	. . . .	Gray with Yellow tracer
Lg/B	. . . .	Light green with Black tracer
Lg/R	. . . .	Light green with Red tracer
O/G	. . . .	Orange with Green tracer
P/Bl	. . . .	Pink with Blue tracer
R/B	. . . .	Red with Black tracer
R/G	. . . .	Red with Green tracer
R/W	. . . .	Red with White tracer
R/Y	. . . .	Red with Yellow tracer
V/W	. . . .	Violet with White tracer
W/B	. . . .	White with Black tracer
W/G	. . . .	White with Green tracer
W/R	. . . .	White with Red tracer
W/Y	. . . .	White with Yellow tracer
Y/B	. . . .	Yellow with Black tracer
Y/G	. . . .	Yellow with Green tracer

NOTES

NOTES

NOTES

NOTES

MAINTENANCE LOG

Service Performed — **Mileage Reading**

Oil change (example)	2,836	5,782	8,601		